CAMBRIDGE LIBRARY COLLECTION

Books of enduring scholarly value

Botany and Horticulture

Until the nineteenth century, the investigation of natural phenomena, plants and animals was considered either the preserve of elite scholars or a pastime for the leisured upper classes. As increasing academic rigour and systematisation was brought to the study of 'natural history', its subdisciplines were adopted into university curricula, and learned societies (such as the Royal Horticultural Society, founded in 1804) were established to support research in these areas. A related development was strong enthusiasm for exotic garden plants, which resulted in plant collecting expeditions to every corner of the globe, sometimes with tragic consequences. This series includes accounts of some of those expeditions, detailed reference works on the flora of different regions, and practical advice for amateur and professional gardeners.

A Dictionary of the Economic Products of India

A Scottish doctor and botanist, George Watt (1851–1930) had studied the flora of India for more than a decade before he took on the task of compiling this monumental work. Assisted by numerous contributors, he set about organising vast amounts of information on India's commercial plants and produce, including scientific and vernacular names, properties, domestic and medical uses, trade statistics, and published sources. Watt hoped that the dictionary, 'though not a strictly scientific publication', would be found 'sufficiently accurate in its scientific details for all practical and commercial purposes'. First published in six volumes between 1889 and 1893, with an index volume completed in 1896, the whole work is now reissued in nine separate parts. Volume 2 (1889) contains entries from *cabbage* (introduced to India by Europeans) to *Cyperus* (a genus of grass-like flowering plants).

Cambridge University Press has long been a pioneer in the reissuing of out-of-print titles from its own backlist, producing digital reprints of books that are still sought after by scholars and students but could not be reprinted economically using traditional technology. The Cambridge Library Collection extends this activity to a wider range of books which are still of importance to researchers and professionals, either for the source material they contain, or as landmarks in the history of their academic discipline.

Drawing from the world-renowned collections in the Cambridge University Library and other partner libraries, and guided by the advice of experts in each subject area, Cambridge University Press is using state-of-the-art scanning machines in its own Printing House to capture the content of each book selected for inclusion. The files are processed to give a consistently clear, crisp image, and the books finished to the high quality standard for which the Press is recognised around the world. The latest print-on-demand technology ensures that the books will remain available indefinitely, and that orders for single or multiple copies can quickly be supplied.

The Cambridge Library Collection brings back to life books of enduring scholarly value (including out-of-copyright works originally issued by other publishers) across a wide range of disciplines in the humanities and social sciences and in science and technology.

A Dictionary of the Economic Products of India

Volume 2: Cabbage to Cyperus

George Watt

CAMBRIDGE
UNIVERSITY PRESS

University Printing House, Cambridge, CB2 8BS, United Kingdom

Published in the United States of America by Cambridge University Press, New York

Cambridge University Press is part of the University of Cambridge.
It furthers the University's mission by disseminating knowledge in the pursuit of education, learning and research at the highest international levels of excellence.

www.cambridge.org
Information on this title: www.cambridge.org/9781108068741

This edition first published 1889
This digitally printed version 2014

ISBN 978-1-108-06874-1 Paperback

A

DICTIONARY

OF

THE ECONOMIC PRODUCTS OF INDIA.

BY

GEORGE WATT, M.B., C.M., C.I.E.,

REPORTER ON ECONOMIC PRODUCTS WITH THE GOVERNMENT OF INDIA.
OFFICIER D'ACADÉMIE; FELLOW OF THE LINNÆAN SOCIETY; CORRESPONDING MEMBER OF THE ROYAL HORTICULTURAL SOCIETY, &c., &c.

(ASSISTED BY NUMEROUS CONTRIBUTORS.)

IN SIX VOLUMES.

VOLUME II.,

Cabbage to Cyperus.

Published under the Authority of the Government of India, Department of Revenue and Agriculture.

CALCUTTA:

PRINTED BY THE SUPERINTENDENT OF GOVERNMENT PRINTING, INDIA.

1889.

CALCUTTA :
GOVERNMENT OF INDIA CENTRAL PRINTING OFFICE,
8, HASTINGS STREET.

DICTIONARY

OF

THE ECONOMIC PRODUCTS OF INDIA.

CARBAGE.

Cabbage, see under **Brassica (oleracea) capitata.** 1

The Cabbage was introduced into India by Europeans at an early date. It is now cultivated in the plains, during the cold weather, and in spring and summer on the hills. It is largely grown in the vicinity of towns and cantonments, and is as much eaten by Natives as by Europeans. Natives cook the cabbage into curry. The "drum head" form is that most generally cultivated by the people of India. (*Cameron, Mysore.*)

Cabbage-rose, see **Rosa alba,** *Linn.*; ROSACEÆ.

CABINET-WORK, FURNITURE, AND GENERAL CARPENTRY, 2

Timbers used for—

Abies dumosa.
A. Smithiana.
Acacia arabica.
A. melanoxylon.
Acer pictum.
Acrocarpus fraxinifolius.
Adenanthera pavonina.
Adina cordifolia.
Alangium Lamarckii.
Albizzia Julibrissin.
A. Lebbek.
A. odoratissima.
A. stipulata.
Alnus nitida.
Alseodaphne ? petiolaris.
Alstonia scholaris.
Amoora spectabilis.
Anogeissus latifolia.
Anthocephalus Cadamba.
Aporosa dioica (The Coco-wood of commerce).
Aquilaria Agallocha (Eagle-wood of commerce).
Areca Catechu.
Artocarpus Chaplasha.
A. hirsuta.
A. integrifolia.
A. Lakoocha.
A. nobilis.
Atalantia missionis.
Averrhoa Carambola.
Barringtonia acutangula.
Bassia latifolia.
B. longifolia.
Bruguiera gymnorhiza.
Buchanania latifolia.
Bursera serrata.
Calamus Rotang.
Calophyllum inophyllum.
Carallia integerrima.
Careya arborea.
Cassia timoriensis.
Cedrela Toona.
Cedrus Deodara.
Ceratonia Siliqua.
Chickrassia tabularis.

Cabinet-work.

Chloroxylon Swietenia.
Cinnamomum glanduliferum.
Cordia Macleodii.
Coriaria nepalensis.
Cupressus sempervirens.
Dalbergia latifolia.
D. Sissoo.
Dichopsis polyantha.
Diospyros cordifolia.
D. ebenum.
D. Kurzii.
D. montana.
Dipterocarpus turbinatus.
Dolichandrone stipulata.
Ehretia lævis.
Elæodendron glaucum.
Erythrina indica.
Excæcaria Agallocha.
E. sebifera.
Feronia elephantum.
Ficus bengalensis.
F. retusa.
Garcinia Cambogia.
G. Morella.
Gluta elegans.
G. travancorica.
Gmelina arborea.
Grevillea robusta.
Guazuma tomentosa.
Gyrocarpus Jacquini.
Hardwickia binata.
Heritiera littoralis.
Holarrhena antidysenterica.
Homalium tomentosum.
Hopea parviflora
Ixora parviflora.
Juglans regia.
Lagerstrœmia microcarpa.
Lophopetalum Wallichii.
Melia Azadirachta.
M. Azedarach.
Meliosma Wallichii.
Memecylon edule.
Mesua ferrea.
Michelia Champaca.
M. excelsa.
M. oblonga.
Mimusops Elengi.
Morus cuspidata.
M. serrata.
Murraya Kœnigii.
Myrsine semiserrata.
Nauclea rotundifolia.
Nephelium Longana.
Odina Wodier.
Ougeinia dalbergioides.
Parrotia Jacquemontiana.
Pentace burmanica.
Phyllanthus Emblica.
Pistacia integerrima.
Platanus orientalis.
Podocarpus bracteata.
P. latifolia.
Pœciloneuron indicum.
Poinciana elata.
Polyalthia cerasoides.
Premna tomentosa.
Prosopis glandulosa.
P. spicigera.
Prunus Puddum.
Pterocarpus indicus.
P. Marsupium.
P. santalinus.
Pyrus lanata.
Quercus semecarpifolia.
Rhododendron arboreum.
Rhus Cotinus.
Santalum album.
Shorea robusta.
Stephegyne parvifolia.
Stereospermum chelonoides.
S. xylocarpum.
Strychnos Nux-vomica.
Swietenia Mahagoni.
Talauma Rabaniana.
Tamarindus indica.
Taxus baccata.
Tecoma undulata.
Tectona grandis.
Terminalia Chebula.
Thespesia populnea.
Ulmus integrifolia.
Vitex leucoxylon.
Wrightia tomentosa.

Cacalia Kleinia, *Herb. Madr.*, see **Notonia grandiflora,** *DC.*; COMPOSITÆ.

C. Kleinia, as in *O'Shaughnessy*, see **Onosma bracteatum,** *Wall*; BORAGINACEÆ.

Cacao, see **Theobroma Cacao,** *Linn.*; STERCULIACEÆ.

Cactus indicus, *Roxb.*, see **Opuntia Dillenii,** *Haw.*; CACTEÆ.

Cactus tree of the lower Himalaya (referred to by some writers) is Euphorbia Royleana, *Boissier*, which see.

CADABA, *Forsk.; Gen. Pl., I., 108.*

Cadaba farinosa, *Forsk.; Fl. Br. Ind., I., 173;* CAPPARIDEÆ. 3

In Murray's *Plants and Drugs of Sind* this plant has been mentioned, but its medicinal properties have not been described. It is common in Sind and in the Panjáb.

Caden, see **Phœnix sylvestris,** *Roxb.;* PALMÆ.

CADMIUM.

Cadmium is imported into India as a drug. 4

CÆSALPINIA, *Linn.; Gen. Pl., I., 565.* 5

A genus of LEGUMINOSÆ and of the Sub-Order CÆSALPINIEÆ, containing some 40 species; inhabitants of the tropics of both hemispheres. There are in India some 9 or 10 species.

Robust erect trees, shrubs, or woody prickly climbers. *Leaves* large, abruptly bi-pinnate. *Flowers* showy, yellow, in copious axillary racemes. *Calyx* deeply cleft, with the disk confined to its base, the lobes imbricated, the lowest being largest and cucullate. *Petals* spreading, usually orbicular with a distinct claw, the uppermost smaller than the others. *Stamens* 10, free, declinate; *anthers* oblong, uniform, versatile. *Ovary* sessile or sub-sessile, few-ovuled; *style* filiform, sometimes clubbed at the tip, stigma terminal. *Pod* oblong or ligulate, thin and flat, or thick and sub-turgid, indehiscent or dehiscent, smooth or in the sub-gen. *Guilandina* armed all over with wiry spines.

The genus was named after Andreas Cæsalpinus, who was chief physician to Pope Clement VIII., in the latter part of the sixteenth century.

Cæsalpinia Bonducella, *Fleming; Fl. Br. Ind., II., 254.* 6

THE FEVER-NUT; PHYSIC-NUT; NICKAR.

Syn.—GUILANDINA BONDUCELLA, *Linn;* G. BONDUC, *W. & A.; Dals. & Gibs, Bomb. Fl., 79, in part.*

Vern.—*Katkaranj* (or *katkaraunj*), *also katkaranga, katkaleja, karanjú, karanjavá, katkalijá, katkuliji-ságar-ghóta,* HIND.; *Nátá, nátá karanja, nátá-koranza,* BENG.; *Bagni,* SANTAL; *Yang, kup,* LEPCHA; *Kath karunjá,* OUDH; *Karaunj,* KUMAON; *Kílgach, katakaránja,* AJMERE; *Kirbut,* SIND; *Kákáchiá, gájga, kachki,* GUJ.; *Sagaragota,* CUTCH; *Ságur ghota, gaja,* BOMB.; *Ságaragotá, gajagá, rohedo,* MAR.; *Gajá, gutchka, gudgega,* DUK.; *Gajkai, gajagakayi, gajega balli,* KAN.; *Kazhar-shikkáy, kalichi, kalishikkáy, gechchakkáy,* TAM.; *Gach-chakáya, gachcha,* TEL.; *Kazhanchik-kuru, kinanchik-kuru,* MALA.; *Kuberákshi, pútikaranja, latákaranja,* SANS.; *Kháyahe-i-iblis,* PERS.; *Akitmakit, kitmakit,* ARAB.; *Kumburu-atta* (or *Kumburu-wel*), SINGH.; *Kaliendza, kalein,* BURM.; *Búva gore,* MALAY. Dr. Dymock says this plant bears the vulgar name of *Kháya-i-Iblis* (Devil's testicles). The specific name of the plant, as given by botanists, is derived from the Arabic name *Bunduk*, a little ball or filbert.

References.—*Roxb., Fl. Ind., Ed. C.B.C., 356; Stewart's Pb. Pl., 69; Atkinson's Him. Dist., 730; Brandis, For. Fl., 156; Gamble, Man. Tim., 135; Pharm. Ind., 68, 445, 446; Modeen Sheriff, Supp. Pharm. Ind., 80; U. C. Dutt, Mat. Med. Hind., 314; Dymock, Mat. Med. W. Ind., 2nd Ed. 248; Bidie's Raw Prod., Paris Exb. Cat., 7; Drury, Us. Pl., 237; S. Arjun's Bomb. Drugs, 48; Spons' Encylop., 794; Guilandina Bonduc in Balfour's Cyclop., Ed. 1885.*

Habitat.—Found all over India, especially in Bengal, Burma, and South India; ascends to 2,500 feet in altitude in Kumaon.

Properties and Uses—

OIL. From seed. 7 **Oil.**—The seeds contain an oil, which is mentioned by **Ainslie** as useful in convulsions and palsy. "The oil expressed from the seeds is used as
From leaves. 8 a cosmetic; it is said to soften the skin and remove pimples." (*Surgeon-Major W. Dymock, Bombay.*) An oil is also said to be prepared from the leaves.

MEDICINE. Seeds. 9 **Medicine.**—THE SEEDS OR NUTS.—The seeds are viewed as possessing well-marked antiperiodic properties, and are largely used by the natives instead of quinine. For this purpose they are pounded with black pepper, from 5 to 30 grains being regarded as the proper dose. **Ainslie** seems first to have drawn the attention of Europeans to this powder, but even up to the present date it has not apparently taken the position which it deserves as a tonic and febrifuge. It was made officinal in the *Indian Pharmacopœia*, the dose of powder being 18 to 15 grains.

"*In Intermittent Fevers*, especially in those of the natives, this remedy has been found very useful. It is best given in the following form:
Powder. 10 Take of Bonduc seeds, deprived of their shells and powdered, one ounce; black pepper powdered, one ounce; mix thoroughly, and keep in a well-stoppered bottle. Of this the dose is from 15 to 30 grains three times a day for adults. In smaller doses it is a good tonic in *Debility after Fever and other diseases.*" (*Waring, Bazar Medicines, p. 27.*)

Thwaites says: "Every part of the plant is used medicinally in Ceylon." **O'Shaughnessy** remarks, that "the seeds afford a powerful tonic and very valuable febrifuge." "Nitric acid reddens the nut and subsequently gives it a yellow colour." **Dr. Irvine**, in his *Medical Topography of Ajmere* (p. 143), says: "The seed is very bitter; used very generally as a tonic, febrifuge, and deobstruent: common at Ajmere. Natives foolishly suppose the seed will cure a scorpion-sting."

In an official report, the Madras Committee for the proposed revision of the Indian Pharmacopœia remark that "the seeds are very useful and cheap (antiperiodic, antipyretic, and tonic), valuable in all ordinary cases of simple, continued, and intermittent fevers. They have also been found useful in some cases of asthma. They resemble **Aconitum heterophyllum** in their action, but are preferable to it for cheapness. The root-bark is inferior to the seeds, and the root quite useless." Fever-nut may be substituted for Pulv. Jacobi as a febrifuge, and for Valerian as an anti-spasmodic, and for Gentian and Culumbo as a tonic. "In Persia and India the seeds are considered to be hot and dry, useful for dispersing swellings, restraining hœmorrhage, and keeping off infectious diseases. They are also given internally in leprosy, and are thought to be anthelmintic." (*Surgeon-Major W. Dymock, Bombay.*) **Dr. Ch. Rice** writes to the author that the "seeds are used among the Malays as astringent tonics in bowel complaints. They have also been reported as facilitating child-birth."

In debility after fevers and other diseases, "the bark of the root of
Root-bark. 11 Bonduc shrub in 10-grain doses is reported to be even more effectual than the seeds themselves." (*Waring.*) It would thus appear that a difference of opinion prevails regarding the properties of the root, but all authors agree in extolling the virtues of the seeds.

An ointment is made from the powdered seeds, with castor oil, and
Ointment. 12 applied externally in hydrocele. **Dr. Dymock** says the seeds are in Bombay sold for R12 a cwt.

Leaves. 13 THE LEAVES.—"In disorders of the liver the tender leaves are considered very efficacious." (*Mr. T. N. Mukharji's Amst. Cat.*) **Drury** says that in Cochin China the leaves are reckoned as a deobstruent and emmenagogue, and that an oil expressed from them is useful in convulsions, palsy, and similar complaints. **Dr. Ch. Rice** informs the author that "the young leaves are used in intermittent fevers and for expelling intestinal worms."

At the late Colonial and Indian Exhibition a pale orange-coloured nut was shown by the West Indian Colonies as obtained from a special cultivated form of this plant. This is not known in India, the nuts being all of a slaty-olive green. Nearly every British Colony showed at the Exhibition the nuts of this plant, but in most cases as ornamental nuts only.

Chemical Composition.—"According to the medical reports alluded to in the *Pharmacopœia of India* (1868), Bonduc seeds, and still more the root of the plant, act as a powerful antiperiodic and tonic. CHEMISTRY. 14

"The active principle has not yet been adequately examined. It may perhaps occur in larger proportion in the bark of the root, which is said to be more efficacious than the seeds in the treatment of intermittent fever.

"In order to ascertain the chemical nature of the principle of the seeds, one ounce of the kernels was powdered and exhausted with slightly acidulated alcohol. The solution, after the evaporation of the alcohol, was made alkaline with caustic potash, which did not produce a precipitate. Ether now shaken with the liquid completely removed the bitter matter, and yielded it in the form of an amorphous white powder, devoid of alkaline properties. It is sparingly soluble in water, but readily in alcohol forming intensely bitter solutions; an aqueous solution is not precipitated by tannic acid. It produces a yellowish or brownish solution with concentrated sulphuric acid, which acquires subsequently a violet hue. Nitric acid is without manifest influence. From these experiments, we may infer that the active principle of the Bonduc seed is a bitter substance not possessing basic properties." (*Flück. and Hanb., Pharmacog., pp. 212-13.*)

SPECIAL OPINIONS.—§ "The kernel of the seeds is decidedly tonic and antiperiodic, but much inferior in this respect to the cinchona preparations. It is useful in dispensary practice where economy is a desideratum." (*Surgeon R. D. Murray, M.B., Burdwan.*) "*Nátá* is decidedly antiperiodic, but feeble in its action, requiring 3 to 3½ grs. of the powdered seed to check an ordinary intermittent fever." (*Surgeon R. L. Dutt, M.D., Pubna.*) "I have often used it; as an antiperiodic, it is certainly of value. The powdered seed smoked in a *húkka*, in lieu of tobacco, is said to be very efficacious in colic." (*Surgeon-Major C. W. Calthrop, M.D., Morar.*) "In doses of 5 to 20 grains, the powdered seeds constitute an efficient antiperiodic, little inferior to cinchona febrifuge." (*Surgeon S. H. Brown, M.D., Hoshangabad, Central Provinces.*) "Tonic, prescribed with warm butter-milk and asafœtida in dyspepsia." (*Surgeon W. Barren, Bhuj, Cutch.*) "Used in charitable hospitals as an antiperiodic, but it is decidedly inferior to cinchona febrifuge, and in large doses it produces nausea." (*Brigade Surgeon S. M. Shircore, Moorshedabad.*) "Much used as a tonic and mild antiperiodic, in 5-grain doses, in dispensary practice." (*Surgeon G. Price, Shahabad.*) "The kernel, in the form of a powder, has been extensively used by me in the treatment of ague, specially the tertian and quartian varieties, and found useful in about 50 per cent. of cases." (*Assistant Surgeon Bhagwan Dass, Rawal Pindi, Panjáb.*) "Its antiperiodic properties are not well marked. I have found it useful in convalescence, after fevers." (*Assistant Surgeon Shib Chunder Bhuttacharji, Chanda, Central Provinces.*) "The seeds are said to be useful in colic (dose one seed), and the ash as an external application to ulcers." (*Surgeon Joseph Parker, M.D., Poona.*) "The burnt seeds are used with alum and burnt areca-nut as a dentifrice, useful in spongy gums, gum-boils, &c., also MEDICINE. 15

MEDICINE. in intermittent fever and debility." (*Brigade Surgeon J. H. Thornton, B.A., M.B., Monghyr.*) "Sometimes used as a febrifuge in doses of about 30 grains, but has a nauseous taste and does not appear to be an efficacious remedy." (*Assistant Surgeon Jaswant Rai, Mooltan.*) "I have tried the Bonduc with black pepper for ague, and was satisfied that it had antiperiodic properties." (*Surgeon-General William Robert Cornish, F.R.C.S., C.I.E., Madras.*) "I tried these seeds in malarious fevers in the Upper Godavarí District, but found they had very slight antiperiodic properties." (*Surgeon-Major H. J. Haslitt, Ootacamund, Nilgiri Hills.*) "Root-bark is also antiperiodic and tonic." (*Apothecary Thomas Ward, Madanapalle, Cuddapah.*) "Decoction, made of terminal stems with leaves; is used as an antiperiodic." (*Assistant Surgeon Anund Chunder Mukerji, Noakhally.*) "The kernel of the seeds of **Cæsalpinia (Guilandina) Bonducella** is a very useful and cheap drug, and is antiperiodic, antipyretic, tonic, and antispasmodic. It has been used with good results in mild cases of intermittent and continued fevers, and also in asthma and general debility. Doses ʒi to ʒii as an antiperiodic and antispasmodic, from 40 to 90 grains as an antipyretic, and from 10 to 30 grains as a tonic." (*Honorary Surgeon Modeen Sheriff, Khan Bahadur, Triplicane, Madras.*) "A cake made of 30 grains of the powdered kernel, the contents of one egg, and fried in *ghí*, is taken twice a day in cases of acute orchitis, ovaritis, and scrofula. The kernel made into a paste is used locally for scorpion-stings." (*Surgeon W. A. Lee, Mangalore.*) "The leaves, boiled with castor oil or *ghí*, are thickly applied on painful and swollen testicles. The tender leaves are said to be most efficacious." (*Honorary Surgeon P. Kinsley, Ganjam, Madras.*) "The best vegetable antiperiodic used by natives. This drug might prove very useful if its active properties were brought into a concentrated form as an extract or otherwise." (*Surgeon W. G. King, M.B., Madras Medical Dept.*) "The nuts ground down and made into a paste are useful in dissolving glandular swellings, buboes, and swelling of the testicles in the acute stage." (*Honorary Surgeon Peter Anderson, Guntur, Madras Presidency.*) "Both the nut and the leaf are used internally and externally with good effect in recent orchitis. It is powerless against hydrocele." (*Native Surgeon Ruthnam T. Moodelliar, Chingleput, Madras Presidency.*)

DOMESTIC. Necklaces. 16 Amulets. 17 Rosaries. 18

Domestic Uses.—The nuts are used for making into bracelets, necklaces, rosaries, &c. (*Guide to the Kew Museum.*) "Necklaces of the seeds strung upon red silk are worn by pregnant women as a charm to prevent abortion." (*Surgeon-Major W. Dymock, Bombay.*) "The seeds are used by children in place of marbles." (*J. Cameron, Bangalore.*) "In Egypt the nuts are strung as necklaces and used by women as amulets against sorcery. In Scotland, where they are frequently thrown upon the seashore, they are known as Molucca Beans." (*Christy, New Commercial Plants, No. 4, p. 48.*)

Dr. Ch. Rice writes to the author that "in the Malay Archipelago they are used for counters and playthings, especially in the game known as *tjongka*.

19 **Cæsalpinia coriaria,** *Willd.*

THE AMERICAN SUMACH OR DIVI-DIVI.

Vern.—*Libi-dibi*, BOMB.; *Amrique-ka-sumáq*, DUK.; *Shúmak*, TAM.; *Sumáque-amriquah*, ARAB., PERS.; *Vilayati-aldekayi*, KAN.

References.—*Brandis, For. Fl., 157; Gamble's Man. Timb., 135; Government of India, Official Correspondence; Indian Forester, Vol. IX., 99-103, Vol. X., p. 77; Tropical Agricult., Vol. III., 137; Vol. II., 646, 730, 734, 955; Spons' Encycl., 1983; Simmonds' Commercial Products of Veg. King., 503; Wiesner, Rohstoffedes Pflanz., 754; Christy's New Com.*

Pl., Part II., 21; Drury's Us. Pl., 92; Müeller's Extra-Tropical Plants, 56.

Habitat.—A small tree, native of South America and the West Indies, found in marshy situations in New Grenada, Mexico, Venezuela, North Brazil, and Jamaica. Introduced into India and now almost acclimatised in South India, and cultivated in Dharwar, Belgaum, and Kanara in Bombay. (*Gaz., XV., Part I., 65.*) It is also sparingly cultivated in the North-Western Provinces and in Bengal and Burma.

Properties and Uses—

Tan.—The sinuous pods of this plant have, within recent years, begun to take an important place amongst tanning materials. The chief drawback seems to be in the fact that if the seeds are not removed, the oil which they contain, induces an injurious fermentation, which results in a discoloration of the skin being tanned. To obviate this danger two suggestions have been made with regard to the form in which the tan should be exported,—to reduce the dried pods to a powder, and in doing so to separate the seeds; or to prepare from the fresh pods a tanning resinous extract. Either of these suggestions would most probably minimise the danger, and would have the effect of lessening the charges on freight. Recently, a large experimental consignment of the pods has been sent to the Secretary of State for India by the Bombay Government. This consignment was accidentally seen by the writer, and it is likely that a highly favourable report will be furnished. The samples of Indian Divi-divi shown at London in connection with the Colonial and Indian Exhibition were pronounced far inferior to the usual supply to be had in the market. The tanners who visited the Exhibition would not look at them, while they professed themselves anxious to investigate some of the paler-coloured barks exhibited, such as **Acacia Catechu** and **A. leucophlœa,** and the pods of **A. arabica.** TAN. Pods. 20 Powder. 21

A considerable amount of interest has, within the past few years, been taken in the subject of the introduction or extended cultivation of the Divi-divi in India. The following extracts from a memorandum on this subject, published by the Government of India, Revenue and Agricultural Department, may be reproduced here:— Extracts. 22

"**Dr. Wallich** introduced the Divi-divi plant into India about the year 1830, and it has now been thoroughly acclimatised in South India, which, in soil and climate, resembles its original home. As the plantation near the Government Harness Factory at Cawnpore proves, however, it can be, with a little care, successfully cultivated in the drier climate of Upper India. But the hot winds in the summer and the frost in the cold weather are, unfortunately, very destructive to the young seedlings. The seeds therefore should, in the first instance, be sown in a nursery in May or June, before the commencement of the rains, and the seedlings should not be transplanted until they are at least 3 feet high, by which time, it is supposed, they will be sufficiently strong to endure all the variations of weather to which Northern India is subject. Irrigation will be necessary as long as the trees are not sufficiently grown up to dispense with it. The *ghará* system of irrigation can be advantageously employed. In South India the tree takes three years [? 10 years—Ed.] to arrive at maturity, but in a drier climate it will perhaps require a longer period. Indian botanists recommend the cultivation of Divi-divi as a profitable as well as an ornamental plant.

"The tree is cultivated for its seed-pods, which contain a large quantity of a most powerful and a quickly-acting tanning material, rather too strong to become a substitute for oak or *babúl* bark, but very valuable as a cleaning and brightening agent in the after-process of currying, when it takes the place of sumach or **Rhus coriaria.** Both in England and at the

Cawnpore Government Factory it is used as a substitute for sumach, which is a dearer article.

"The actual demand for Divi-divi pods is not known. England imports about 4,000 tons every year, in addition to about 12,000 tons of sumach. But as Divi-divi is gradually ousting the latter, its demand appears to be capable of great expansion. For the same reason France, which now annually imports more than five million kilogrammes of sumach, will probably become a large market for this article. It has been proved beyond doubt that the tree can be profitably cultivated in India, and the quality of a consignment sent two years from the Khandesh Farm (Bombay) was pronounced by experts in England to be very superior, and fetched a higher price than that imported from the West Indies. One great advantage in its cultivation is, that the tree requires no care after it has once grown up, and the proceeds are net gain, *minus* the trifling cost of picking the pods. The ground underneath can be utilised for raising fodder grass, and the falling leaves as fuel or manure, thus meeting the three great wants of the Upper Provinces. An acre of Divi-divi is supposed to yield not less than one ton of marketable produce, valued in India about R100, in England R150." (*Mr. T. N. Mukharji, Revenue and Agricultural Department.*)

23 CULTIVATION.—The information given in the above extract might be supplemented. The plant is, however, only being experimentally tried in India as yet, and considerable difference of opinion prevails as to the best methods for its cultivation, the nature of the soil most favourable to it, and as to the prospects of the plant becoming a commercial success. In **Mr. Duthie's** experiments at the Saharanpur Botanic Gardens the trees are planted 15 feet apart each way. The largest plantation of Divi-divi in India is that belonging to Mr. J. B. Carbozo, of Perambore, Madras. This gentleman has a plantation of about 600 trees; the trees being 22 feet apart. **Dr. King** in his report for 1879-80 says: "Many years ago the cultivation of the tree, yielding the well-known Divi-divi tan (**Cæsalpinia coriaria**) was undertaken in this garden. There are now many large Divi-divi trees here. They yield pods freely every year and the seeds are regularly distributed gratis to all applicants.

Dr. Bidie of Madras thinks the tree grows best at an elevation of 2,000 feet above the level of the sea; by others, a dry and light soil suits it best; and again, its favourite soil is a heavy clay associated wh **Acacia leucophlœa.** Some writers do not think it can ever be cultivated on a large scale in Bengal, because the climate is too moist and the soil not suitable, while **Baron von Mueller** recommends its introduction in the salt-marshes of Australia. From the brief notice already given of a consignment of the pods from Bombay, it may be added that certain tracts of the Western Presidency offer the most favourable prospects in India for Divi-divi cultivation.

A correspondent in *The Madras Times* says: "The Divi-divi pods are employed for tanning purposes, as they contain 50 per cent. of pure tannin. I have been told that in Bangalore there is a large plantation of this tree [**Mr. Cameron** of Bangalore informs the author that this is incorrect, only a few trees exist there], and that its pods are extensively used for giving the skins that superior smoothness remarkable in the Madras and Bangalore skins. I have also used Divi-divi in dyeing, and I have employed a solution of acetate of iron after the bath in Divi-divi tincture. The ink used in most of the Government offices in Fort St. George is made from this plant. The cultivation of this elegant shrub is very easy. The seeds should be sown in March, and the young plants can be removed from the nursery during the following rainy season; they require some watering till they have attained the height of 3 feet, after which no more care is necessary. The plant grows luxuriantly in a clayish calcareous soil, but very slowly in red soil, as I have observed at the Red Hills near Madras." **Dr. Ch. Rice** draws the author's attention to the fact that **Professor Flückiger**, in his *Pharmakugnosi* [*2nd*] *I., 245,* mentions that ink was reported to have been made from Divi-divi pods as early as the beginning of the 16th century, and in speaking of this fact **Fernandez de Oriedo** calls the plant *Arbol de la Sinta.*

The Madras Agri.-Horticultural Society made the following observations in their Proceedings of March 1884: "The pods undoubtedly contain large quantities of tannin, and as the tree grows freely in waste land, such as is very abundant in this Presidency, they might be worth collecting, freeing of the seeds, and shipping in the form of a clean powder closely packed in bags; but to be really remunerative and to show conclusive results, experimental shipments should be tried on a much larger scale than has yet been attempted, and means of continuing the supply must be available, as manufacturers will not try expensive experiments unless with some certainty of being able to get more of the substance tested, in the event of success."

Medicine.—According to **Dr. Bidie**, the pods are astringent. The powder prepared from them is of a light-yellow colour and astringent taste; it has been brought forward as an antiperiodic by **Dr. Cornish**, who administered it in ninety-four cases of intermittent fever, many of these severe, and with excellent results, the dose ranging from 40 to 60 grains. MEDICINE. Pods. 24

§ "Powerfully astringent." (*Brigade Surgeon G. A. Watson, Allahabad.*) "I have had no new experience of this drug. The tree is being largely planted about Madras for its pods, which are marketable. It makes a good black ink." (*Surgeon-General William Robert Cornish, F.R.C.S., C.I.E., Madras.*) "Powder of the pods astringent, antiperiodic, tonic. Dose one to two drachms as an antiperiodic." (*Apothecary Thomas Ward, Madanapalle, Cuddapah.*) "Used for tanning leather, and makes very good ink." (*Surgeon-Major J. Byers Thomas, Waltair, Vizagapatam.*)

Structure of the Wood.—Not in general use (*Bomb. Gaz., XV., Pt. I.* (*65*); weight 56 lbs. TIMBER. 25

Cæsalpinia digyna, *Rottl.; Fl. Br. Ind., II., 256.* 26

Syn.—C. OLEOSPERMA, *Roxb.; Ed. C.B.C., 356.*

Vern.—*Vákeri-múl,* HIND.; *Umúl-kúchi,* BENG.; *Núni gatcha,* TEL.; *Vákeri-chebháte, vákeri-múla,* BOMB.; *Sunletthé,* BURM.

Habitat.—A prickly tree of the Eastern Himálaya, Eastern and Western Peninsulas, and Ceylon.

Tan.—**Dr. H. McCann**, in his *Dyes and Tans of Bengal,* says that in Cuttack the pods of what appears to be this plant are sold as a tan under the name of *Kunti.* The word *Kunti-paras* would appear to be sometimes, in Orissa, applied to **Butea frondosa**, but incorrectly, as *kunti* means tanning pods, and *paras* **Butea frondosa**; in other words, the term *Kunti-paras* would mean the tanning pods of **Butea** and not of **Cæsalpinia**. —(*L. Liotard.*) TAN. 27

Oil.—**Roxburgh** says that an oil is expressed from the seeds, which is used for lamps. OIL. 28

Medicine.—The root is prescribed in native practice, and has marked astringent properties. (*S. Arjun, Bomb. Drugs.*) **Dr. Dymock** ((*Mat. Med. W. Ind., 2nd Ed. 251,*) says: "it is given in phthisis and scrofulous affections; when sores exist it is applied externally as well, a kind of tuberous swelling which is found on the root is preferred." MEDICINE. 29

30 **Cæsalpinia Nuga,** *Ait.; Fl. Br. Ind., II., 255.*

Syn.—C. PANICULATA, *Roxb.; Fl. Ind., Ed. C.B.C., 358; Wight, Ic., t. 36; Dalz. and Gibs., Bomb. Fl., 79; Brandis, For. Fl., 157.*

Vern.—*Kaku-múllu in Rheede's Hort. Mal.; Deya-wawúl-atteya,* SINGH.; *Sakauk,* BURM.; *Aroci-mata-hiang,* SUNDAW.

Habitat.—A scandent, armed shrub, common in Eastern Bengal (Sylhet to Chittagong), the Eastern and Western Peninsulas, and Ceylon.

MEDICINE. 31 **Medicine.**—The roots of this plant are said to be diuretic. **Dr. Rice** draws attention to the fact that the root of this plant has been reported as useful in gravel and stone in the bladder, and that the juice of the stem has been used externally and internally in eye diseases. For the same purposes are used also the roasted fruits, which have a bitter taste. The finely-powdered leaves have also been administered to women immediately after delivery as a tonic to the uterus.

32 **C. pulcherrima,** *Swartz.; Fl. Br. Ind., II., 255. Roxb., Fl. Ind., Ed. C.B.C., 356.*

THE BARBADOS PRIDE.

Syn.—POINCIANA PULCHERRIMA, *Linn.*

Vern.—*Krishnachúrá,* BENG., SANS.; *Ratnagandi,* KAN.; *Daungsop,* BURM.

Habitat.—An introduced plant, commonly occurring in nearly every garden throughout India; it forms a large elegant bush; remains in flower all the year; one variety is red and the other yellow.

MEDICINE. 33 **Medicine.**—The leaves, flowers, and seeds are largely used in native medicine." (*J. Cameron, Bangalore.*)

DOMESTIC. 34 **Domestic Uses.**—It is sacred to Siva. Ink is made from the charred wood. (*J. Cameron, Bangalore.*)

35 **C. Sappan,** *Linn.; Fl. Br. Ind., II., 255.*

THE SAPPAN WOOD; SAMPFEN WOOD.

Vern.—*Bakam, tairi, patang,* HIND., BENG.; *Teri,* SANTAL; *Bokmo,* URIYA; *Patang na lakaru, patang-nu-lákdo, bakam,* GUJ.; *Patang-ki-lakri,* DUK.; *Patang,* MAR.; *Patanga, vattángi, vattéku, vartangi,* TAM.; *Bakamu, bakapu, okánu-katta, patanga-katta, bakánu-chekka, bukkapu-chekka,* TEL.; *Patanga-chekke, sappanga,* KAN.; *Chappanum,* MALA.; *Pattánga,* SANS.; *Bakam, bakam-i-kirmyz,* or *baqam,* PERS.; *Baqqam,* ARAB.; *Teinn-nyet, tainniya* or *tainngiya,* BURM.; *Patangi,* SINGH.; *Davon-setjang,* MALAY.

References.—*Roxb., Fl. Ind., Ed. C.B.C., 356; Bedd., Fl. Sylv., 90; Brandis, For. Fl., 156; Kurz, Burm. For. Fl., 405; Gamble, Man. Timb., 135; Modeen Sheriff's Supp. Pharm. Ind., 81; Drury's Us. Pl., 93; Dymock's Mat. Med., W. Ind., 2nd Ed., 251; Pharm. Ind., 79; U. S. Dispens., Ed. 15th, p. 1590; Flück. and Hanb., Pharmacog. 216; Ainslie, Mat. Indica, II., 450; Buck, Dyes and Tans, N.-W. P., pp. 43, 57; McCann's Dyes and Tans of Beng., pp. 1, 4, 49; Crooke's Dyeing, 332; Spons' Encycl., 867; Smith's Econ. Dict.; Treasury of Botany, by Moore & Lindl.; Wiesner, Rohstoffe, 555.*

Habitat.—A small thorny tree of the Eastern and Western Peninsula and Pegu and Tenasserim; cultivated in Central India.

Properties and Uses—

DYE. 36 **Dye.**—The wood yields a valuable red dye, which is largely exported. This is also said to be prepared from the PODS (*tairi*), from the WOOD, from the BARK, or from all together, and the root is reported to afford a yellow dye. The bark of **Bauhinia racemosa** is said to be used as a mordant in Burma.

The pods are used in Monghyr, along with proto-sulphate of iron, to give a black colour. Sappan (or bakam) wood is largely used in calico-printing, its price being about R12 a cwt. Chips of the wood steeped in water yield a red colour. This is intensified by alkalies. Combined with turmeric and sulphate of iron, it gives the colour known as *Kalejai* (or liver-colour, "*lit de-vin*"). With indigo it gives (*sausni*) purple. Sappan colour, however, is not permanent, being formed through the presence of the soluble substance Brazilin. Tannin and alum are used as mordants in attaching the dye to cotton, and a mixture of alum and cream of tartar in the case of wool. "It is much used in Pegu for giving a red tint to silk, and in Madras for dyeing straw-plait for hat-making." (*Spons, Encycl.*)

Extraction of the Dye.—The PODS (*tairi*) are pounded and then put into cold water. After 2 or 3 hours the mixture is rubbed and afterwards mixed with a solution of proto-sulphate of iron (*hirakoshi*). The resulting colour is a blackish grey. One seer of *tairi* and 2 chittacks of *hirakosh* (sulphate of iron) are sufficient to dye 60 yards of cloth 1 yard wide. In the case of the WOOD, it is either cut into pieces or pounded and then boiled in water from 5 to 8 hours; 12 chittacks of *bakam* wood are boiled in 25 seers of water till 10 seers remain. The solution is put aside, and the same wood is again boiled in another 25 seers of water down to 10 seers. These two resulting solutions are then mixed up and allowed to cool. This is the process adopted in the Rajshahye District. To extract the dye from the BARK, it is boiled down till the solution attains the necessary consistency and tint. **Dye-tincture. 37**

Mr. Thomas Wardle, in his *Report on the Dyes and Tans of India, 1887*, says (page 21) that "the wood, used to a considerable extent in this country in wool-dyeing and calico-printing, is very rich in a beautiful red colouring matter, soluble in water which, by the application of the many processes which may be used, will produce beautiful red and chocolate colours on almost any fabric." Again (page 31), speaking of the pods, he calls them "a kind of Divi-divi," and remarks "that they are very astringent and much used for tanning and dyeing purposes in this country" (England), producing rich blacks with salts of iron. **Chip. 38**

Gulal.—§ "Sapan wood is used with alum to communicate to starch one of the red colours, the *Gulál*—or red powder used in the *Holi* festival." (*Surgeon-Major W Dymock, Bombay.*) **Gulal. 39**

Medicine.—**Ainslie** says a decoction of the wood has the property of a powerful emmenagogue. The wood, though chiefly used as a dye, is described as a useful astringent, containing much tannic and gallic acids, and has been recommended by **O'Shaughnessy,** and later by the *Indian Pharmacopœia*, as a good substitute for Logwood. "It is supposed to cure a swelling in the belly, and the disease known among native doctors as congealed blood." (*Bomb. Gaz., VI., 14.*) According to **Dr. Irvine**, it is used as an astringent in medicine. (*Mat. Med., Patna, p. 15.*) "The natives use it as a blood-purifier; as an astringent, it is prescribed in diarrhœa, dysentery, &c." (*Baden Powell, Pb. Prod., 344.*) **Professor Warden** of the Medical College, Calcutta, has favoured the author with the following note regarding sappan wood:—"The extract of Sappan contains a crystalline principle, which, if distilled, or fused with potash, yields Resorcin. This body is also obtained by fusing galbanum resin with potash. Sappan extract gives a larger yield than galbanum resin." (*Pharmacographia, &c.*) **MEDICINE. Wood. 40**

SPECIAL OPINIONS.—§ "Has been used as an astringent tonic in atonic diarrhœa." (*Assistant Surgeon Bhugwan Das, Rawal Pindi, Panjáb.*) "An excellent wool dye; it is also an astringent, being used instead of logwood. It is useful in some forms of skin disease—lichen especially,—and is given internally in decoction." (*Surgeon-Major George Cumberland Ross, Delhi.*) "Emmenagogue." (*Surgeon W. Barren, Bhuj, Cutch.*)

CAJANUS indicus. The Pigeon Pea.

TIMBER. 41 **Structure of the Wood.**—Sapwood white, heartwood red. The wood takes a fine polish and does not warp or crack.

Weight from 52 to 61 lbs. per cubic foot.

Mr. J. Cameron reports that the lac insect has recently taken to this plant in Bangalore.

42 **Cæsalpinia sepiaria,** *Roxb.; Fl. Br. Ind., II., 256.*

THE MYSORE THORN.

Vern.—*Urn, úri, arlu, relú, kando, aila,* HIND.; *Phulwái, úran* (JHELAM), *kando* (KASHMIR), *dodúr* (CHENAB), *relme, didrian, dhar-ki-karer,* (RAVI), *ándi, arleí, daghauri* (BIAS), *ongwá* (SUTLEJ), PB.; *Chillára* or *chillúr,* BOMB., MAR.; *Hotsigé,* KAN.; *Sukyanbo,* BURM.

References.—*Roxb., Fl. Ind., Ed. C.B.C., 357; Stewart, Pb. Pl., 60; Brandis, For. Fl., 156; Kurz, For. Fl., Burm., I., 406; Gamble, Man. Timb., 135.*

Habitat.—A large, climbing, prickly bush on the Himálaya, and extending to Ceylon and Ava; it ascends to 4,000 feet in altitude.

LAC. 43 **Gum.**—"Lac is gathered on this tree in Baroda." (*Bomb. Gaz., VII., 39.*)

TAN. Bark. 44 **Tan.**—The bark is much used for tanning in the Konkan.

OIL. Pods. 45 **Oil.**—"The young pod contains an essential oil." (*Bomb. Gaz., XV., pt. I., 65.*)

MEDICINE. 46 **Medicine.**—In Chumba the bruised leaves are applied to burns.—(*Stewart.*)

DOMESTIC. 47 **Domestic Uses.**—Makes an impenetrable fence; said to have been planted for this purpose by Hyder Ali round fortified places. (*Stewart.*)

The Chinese are said to use the seeds and pods of several species of **Cæsalpinia** as soap-nuts. This property does not appear to have been attributed to any of the species, wild or cultivated, in India.

48 **CAJANUS,** *DC.; Gen. Pl., I., 541.*

A genus of LEGUMINOSÆ in the Sub-Order PAPILIONACEÆ, containing only one species. Some doubt exists as to whether this species is originally a native of India or of tropical Africa. It was introduced into the West Indies from Africa; and outside Africa, India, and the Malaya its cultivation is quite historic.

The generic name *Cajanus* is derived from the Malayan name for the plant (*Katjang*).

49 **Cajanus indicus,** *Spreng.; Fl. Br. Ind., II., 217.*

PIGEON, NO-EYE (small form) or CONGO PEA (large form), DAL OR CADJAN PEA.

Syn.—CYTISUS CAJAN, *Linn.*; CAJANUS INDICUS, *Spr.*; C. FLAVUS, *DC.*; C. BICOLOR, *DC.*

Vern.—*Tuvar, arhar* or *arar dal,* HIND.; *Arhar, thor* or *tor, thur, dal, rahan, thohar,* N.-W. P. and OUDH; *Arhar, oror, orol,* BENG.; *Kohlú, kehú,* SIMLA; *Dángrí* (GUJARAT), *arhar, dinger, tohar,* PB.; *Dhíngra, kúndí,* KANGRA; *Tur,* C. P.; *Tuvéro, turdál,* GUJ.; *Túra, tuver,* BOMB.; *Túrí, túr,* MAR.; *Tuvvar, túr,* DUK.; *Thovaray, tuvarai,* TAM.; *Kandalu,* TEL.; *Togari, tovaray,* KAN.; *Tuvara,* MALA.; *Adhakí tubaríká,* SANS.; *Shakull,* PERS.; *Sház,* ARAB.; *Pay-in-chong, pai-si-gong, pésigón,* BURM.; *Rata-tora,* SINGH.

References.—*Roxb., Fl. Ind., Ed. C.B.C., 567; Stewart's Pb. Pl., 60; Baden Powell, Pb. Pr., I., 242; Kurz, For. Fl. Burm., I., 377; Gamble, Man. Timb., 123; Thwaites, En. Ceylon Pl., 90; Modeen Sheriff, Supp. Pharm. Ind., 81; U. C. Dutt, Mat. Med. Hind., 150; Drury's Us. Pl.,*

The Pigeon Pea.

CAJANUS indicus.

94; Bidie's Mad. Raw Prod., Paris Exh. Cat., 74; Duthie and Fuller's Field and Garden Crops of the N.-W. P. and Oudh, Part II., 20; Atkinson's Him. Dist., 696; Church's Food-grains of India, 169; Balfour, Cyclop., Ed., 1885; Smith's Dict., 320; Treasury of Botany.

Habitat.—Extensively cultivated throughout India even up to an altitude of 6,000 feet. The *Flora of British India* regards this bush as doubtfully wild in India, and **DeCandolle,** in his *Origin Cult. Pl.,* views it as more probably a native of tropical Africa, introduced perhaps 3,000 years ago into India.

Properties and Uses—

Medicine.—The pulse is said to be easily digested and therefore suitable for invalids. It is regarded as hot and dry; it, however, produces costiveness. The leaves are used in diseases of the mouth. MEDICINE. 50

SPECIAL OPINIONS.—§ "Said to be heating, disagrees with some individuals, causing acidity and heartburn." (*Assistant Surgeon Shib Chunder Bhuttacharji, Chanda, Central Provinces.*) "The pulse and leaves are mixed and made into a paste, which is warmed and then applied over the mammæ to check the secretion of milk." (*Surgeon W. A. Lee, Mangalore.*) "The tender leaves are chewed in cases of aphthæ and spongy gums." (*Brigade Surgeon J. H. Thornton, B.A., M.B., Monghyr.*) "A poultice made with its seeds will check swellings." (*Ummegudien Native Doctor, Mettapollium, Madras.*)

Food.—This pulse is highly "esteemed by the natives, who hold that it is third in rank among their leguminous seeds, though apt to produce costiveness." (*Drury.*) "This is the pea which most commonly enters into the formation of the vegetable curry of the Hindus." (*Balfour.*) It is sold either in the form of split peas or as a flour. Sometimes the green pods are cooked in curries. Sweet cakes are often made from the flour. FOOD. Seed. 51

There are two chief varieties: **C. flavus,** with the standard or upper petal plain yellow (known in the vernacular as *thúr*); and **C. bicolor,** with the standard veined with purple (the *arhar*). The latter is the most commonly cultivated in the North-West Provinces and Oudh, while in the Central Provinces and the Deccan *thúr* takes the place of *arhar*. **Mr. J. Cameron** of Bangalore writes that in the Mysore State there are three forms of **Cajanus indicus**:—a large form confined to garden cultivation, known as *turuk-togari,* and two smaller field forms known respectively as *walada-togari* and *sauna-togari*. *Thúr* is also extensively cultivated in the Nizam's Dominions and in the Madras Presidency.

In the North-West Provinces and Oudh, *arhar* is grown mostly as a subordinate crop along with *juar, bajra,* and cotton, but it is also, though to a comparatively much smaller extent, grown by itself. Hence, when it is cultivated as a mixed crop, the soil on which it is grown requires to be chosen and prepared in a way answering equally to the necessities of its associate. When sown with *juar* it requires the heaviest, and when with *bajra* the lightest, of soils; but a light moist soil is generally most favourable for the *arhar,* enabling it to strike its roots freely. About 6 seers of seed are required for an acre, if sown singly, and 2 seers when along with other crops. **Cajanus** is sown at the commencement of the rains, the pulse ripening in March or April. The average outturn is 7 maunds of grain and 16 maunds of *bhúsa* per acre, off land on which *arhar* is the only crop, and 1 to 5 maunds where grown along with other crops. **Professor Church,** in his *Food Grains of India,* states that "the yield of seed varies from 7 to 16 maunds per acre." The writer can find no record of a higher yield than 7 maunds. The outlay on cultivation is about the same as that for millets. N.-W. P. 52

In the North-West Provinces it has been calculated that there are $35\frac{1}{2}$

lakhs of acres on which this is cultivated as a joint crop, and perhaps 1¼ are under *arhar* solely. "It occupies the ground for a longer period than any other crop except sugarcane, being sown at the commencement of the rains, and not cut till the *rabi* harvest time in March and April." "It is cut with the *rabi* crops and allowed to be stacked on the threshing-floor until the threshing and cleaning of the former are completed. The leaves and pods are first of all stripped off the stems and then heaped together, and the grain threshed out either by bullocks treading or by being beaten with a stick." "Frost is the principal enemy with which *arhar* has to contend. A single cold night often utterly ruins the crops of a whole district, and in the following morning the cultivators may be seen sadly cutting down the withered plants as fodder for their cattle. Its liability to damage is, however, greatly dependent on the strength of the plants, and hence the crop grown on manured land near the village site will often remain green and flourishing after a frost which has withered up those on outlying fields." (*Duthie and Fuller, Field and Garden Crops.*)

C. P. 53 "A good deal of *túr* is grown (in Nagpore); it is often raised in the same field as cotton, generally five ridges of cotton to one of *túr.*" (*C. P. Gaz., 327.*) "In Ráipur two kinds of *arhar* or *túr* are known, the small and early *arhar* called *haroná*, and the larger and later kind called *mihí.* Both are sown at the same time, but the former ripens about two months before the latter."

BOMBAY. 54 In Thána it is grown as an early crop in uplands, often with **Eleusine corocana** and **Panicum miliaceum,** and also as a dry-weather crop in late or *rabi* soil, and in the better rice-fields. Both crops ripen in about four months, the early in November and the late in February. (*Bomb. Gaz., XIII., 289.*)

PANJAB. 55 According to **Stewart,** "The yellow and parti-coloured kinds are not uncommon, the one as a cold-weather and the other as a hot-weather crop in the eastern and central Panjáb, and extend sparingly to the Trans-Indus." **Baden Powell** says that in the Panjáb it is a less esteemed pulse than the others, and is liable to produce costiveness.

BENGAL. 56 "*Arhar* is sown with the *áus dhan* or early paddy, usually in the same field, and is cut in January. It will grow on almost any soil." (*Bengal Administration Report, 1882-83.*)

The vaiiety **C. flavus** or *thûr* is a much smaller plant, and it flowers at least three months earlier than the variety **C. bicolor** or *arhar.* According to **Roxburgh,** the former requires only three months to ripen its crop but yields only one hundred-fold, while the latter takes nine months, from sowing to ripening of seed, and yields about six hundred-fold. The former is sown in September and the latter in June. The small form is known in Jamaica as the No-eye pea, and the large as the Congo pea. In addition to these two varieties there are under each two or three other cultivated conditions which differ from each other chiefly in the colour and size of the pea. The fact of there being so few forms would seem to be a powerful argument in favour of the idea that **Cajanus** is but a comparatively modern cultivated crop and not so ancient as some authors view it, but on the other hand it is surprising that the wild stock should have disappeared, at least it does not occur in India.

57 **Chemical Composition**—The following analysis shows the percentage of nitrogen, starch, and oil contained in this pulse :—

Nitrogenous matter (albuminoids)	19·83 to 20·38
Starch or carbonaceous matter	61·90 to 64·32
Oil or fat	1·10 to 1·12

Calabar Bean. CALABAR bean.

Professor Church, in his *Food Grains of India*, publishes more recent analyses than the above, from which it would appear that the amount of albuminoids is slightly higher than has hitherto been supposed. A pound of the pea would, according to his analysis, contain 1 oz. and 361 grains of water, 3 oz. and 208 grains of albuminoids, and 9 oz. and 11 grains of starch. According to **Church** the nutrient ratio of *dal* would be about 1 : 3; the nutrient value 80.

The reader will be enabled to compare the relative quantities of these constituents in other species of pulse from the following table:—

Name.	Nitrogenous matter.	Starchy matter.	Fatty or oily matter.
Cicer arietinum	18·05 to 21·23	60·11 to 63·62	4·11 to 4·95
Cyamopsis psoralioides	29·80	52·89	1·40
Dolichos biflorus	23·03 to 23·47	61·02 to 61·83	0·70 to 0·87
Dolichos Lablab	22·45 to 24·55	60·52 to 60·81	0·81 to 2·15
Vigna Catiang	24·00	59·02	1·41
Ervum Lens	24·57 to 26·18	59·34 to 59·96	1·00 to 1·92
Glycine Soja	37·74 to 41·54	29·54 to 31·08	12·31 to 18·90
Lathyrus sativus	31·50	54·26	0·95
Phaseolus aconitifolius	23·80	60·78	0·64
Phaseolus Mungo	23·54 to 24·70	59·38 to 60·36	1·11 to 1·48
Phaseolus Mungo, var. radiatus	22·48	62·15	1·46
Pisum sativum	21·80 to 25·20	61·90 to 64·32	1·32 to 1·12

(*Baden Powell, Panjáb Products, I., 243.*)

SPECIAL OPINIONS.—§ "*Arhar* is a highly nutritious pulse, quite equal to peas in nutritive value; a good substitute for animal food. It does not, however, agree with weakly people of relaxed habits, but produces diarrhœa or dyspepsia." (*Surgeon-Major R. L. Dutt, M.D., Pubna.*)

"It is difficult of digestion and very unsuited to people who are subject to acidity and heartburn. I have always found it so." (*Surgeon K. D. Ghose, Bankura.*) **Professor Church** states that the irritant and laxative character is greatly reduced by the grain being properly freed from the husk. "It is also not unusual to find that the higher priced and finer qualities of this pea have been slightly oiled before sale to improve their appearance. This practice is not unknown in reference to the wheat in the south of Europe" (*Food-grains of India*). May not this fact account for the purgative property attributed to *dal*?

Fodder.—The leaves are considered excellent fodder for cattle. FODDER. 58

Domestic Uses.—"The wood is in Bengal sold as fuel." (*Bengal, Administration Report, 1882-83.*) The stalks are used for roofing, for basket-making, and for the tubular wicker-work fascines (*bira* or *ajar*) erected within wells so as to prevent the earth from falling in. (See also under **Adhatoda Vasica.**) **Roxburgh** says "the dry wood is excellent fuel; besides it is one of the best for producing fire by friction." **Birdwood** remarks that "the stalks are used in the preparation of gun-powder in the Government works at Mazagon." (*Bombay Products, 1862, page 17.*) Employed in the Bengal gun-powder works for charcoal. (*Balfour.*) DOMESTIC. 59

Cajuput oil, see **Melaleuca Leucadendron**, *Linn.*; MYRTACEÆ.

Calabar bean, see **Physostigma venenosum**, *Balf.*; LEGUMINOSÆ.

CALABAR SKINS.

60 **Calabar Skins** or SIBERIAN SQUIRREL SKINS.

PETITGRIS, *Fr.*; GRANWERK, *Germ.*; VAOR VAJO, *It.*; BJELKA, *Rus.* GRIS PEPUENO, *Sp.*

The Siberian squirrels' skins are imported into India in considerable numbers. They are of various shades, and in India are used for caps, and skin jackets, and in Europe for making muffs, tippets, &c. See **Squirrels**. Also under FURS.

CALAMANDER WOOD.

61 **Calamander Wood.**—A beautiful kind of rose-wood obtained from Ceylon, the timber of **Diospyros quæsita**, which see.

Calambac, see **Aquilaria Agallocha.**

62 ## CALAMUS, *Linn.*; *Gen. Pl., III., 931.*

A genus of palms, generally scandent, with long, thin, trailing stems, sometimes as much as 600 feet in length. There are 200 species known; nearly all are inhabitants of tropical and sub-tropical Asia, abounding in the Malayan Peninsula; a few are met with in tropical Australia and Africa, and according to Griffith 58 species are natives of India. A few are met with in Madras, but they chiefly occur in the hot moist forests of the eastern division of India, *e.g.*, in Chittagong, Sylhet, Assam, and Burma, extending along the Himálaya as far north as to the Dehra Dún.

They are characterised by having alternate, scattered, and distant, lateral pinnate leaves, the pinnæ plat linear and often armed along the edges and nerves; sheath large, with rigid dark-coloured prickles; mid-ribs, sheaths, or panicles often terminating in long whip-like thongs which are armed with prickles, either scattered or in oblique lines or rings. Flowers monœcious or diœcious, in long axillary or extra-axillary panicles with sheathing bracts. MALE FLOWERS—*Calyx* campanulate, 3-dentate. *Petals* 3, valvate in bud. *Stamens* 6, surrounding a rudimentary 3-fid ovary; *anthers* sagittate and adnate at the back. FEMALE FLOWERS often pedicellate, supported by 3 or more imbricate bracts; calyx and petals as in the male flowers. *Ovary* 3-celled, surrounded by 6 sterile stamens, closely embraced by a number of imbricate, reflexed scales; *style* 3, recurved; *ovules* solitary, erect. *Fruit*, 1-, rarely 2-celled, nearly dry, with a hard, shining wall of reflexed scales (the rind); albumen more or less ruminated.

The generic name *Calamus* is the Latin and the Greek Κάλαμος, a reed or cane.

For a more general and popular account of the genus, see under "**Canes.**"

63 **Calamus acanthospathus,** *Griff., Pl. cxc., fig. 1*; PALMÆ.

Reference.—*Gamble's Man. Timb., 423.*

Habitat.—Khásia Hills.

64 **C. andamanicus,** *Kurz, For. Fl. Burm., II., 519.*

Vern.—*Chowdah*, AND.

References.—*Gamble, Man. Timb., 424.*

Habitat.—Met with in the Andamans.

TIMBER. 65 **Structure of the Wood.**—**Dr. Kurz** describes it as "an evergreen lofty, scandent, rattan-palm, the sheathed stems being as thick as the arm and the canes up to an inch in diameter."

The Dragon's-Blood.

CALAMUS Draco.

CANES. 66

Calamus arborescens, *Griff., Pl. clxxxviii.*

Vern.—*Danoung, danôn* or *zanôn, theing, kyenbankyen,* BURM.

References.—*Gamble, Man. Timb., 423; Kurz, For. Fl., Burm., II., 516.*

Habitat.—An erect, elegant cane, often stoloniferous, met with in Pegu.

67

C. collinus, *Griff., Pl. clxxvi.; Gamble, Man. Timb., 423.*

Habitat.—An erect cane, met with in the Khásia Hills and in Upper Assam.

68

C. (Dæmonorhops, *Mart.*) Draco, *Willd.; Blume in Rumphia, II., t. 131-32.*

The DRAGON'S-BLOOD; CALAMUS.

Vern.—*Aprang, rangbharat, damlakwayþi, dam-ul-akhwain, jaida rumi, hiradukhi,* HIND.; *Hirá dakhan, hira-dukhi,* BOMB., MAR., GUJ., CUTCH; *Rotan-jurung,* MALA.; *Konda-murga-rattam,* TAM.; *Dam-ul-akhwain* (blood of two brothers), *dam-et-tinnin, dam-eth-thuaban* (Dragon's blood), ARAB.; *Dam-el-akwain, khún(i)-sidvesham,* PERS.; *Kyeing-ni,* BURM.; *Jarnang,* PENANG; *Rotan-jerenang,* MALAY.

The majority of these names would appear to refer mainly to the resin: the terms *rotan* or *rattam,* &c., referring to the stem or cane. The Persian name *khún-i-siaveshum* (blood of incestuous lovers) is most probably, as pointed out by **Dr. Dymock**, in allusion to the love of Siavesh (the father of Cyrus) for his step-mother.

References.—*Roxb., Fl. Ind., Ed. C.B.C., 719; Flück. & Hanb., Pharmacog., 672; Griffith, Pl. cci., A. and B.; Drury, Useful Plants of India, 97; Dymock, Mat. Med., W. Ind., 2nd Ed., 806; U. S. Dispens., 15th Ed., 1636; Spons' Encyclop., 1648; Balfour's Cyclop.; O'Shaughnessy's Beng. Disp., 642; Flückiger, Pharmakognosie (Berlin, 1883), p. 97—102; Dobbie and Henderson, Pharm. Journ., Nov. 1883.*

Habitat.—A native of the Indian Archipelago. **Dr. Dymock** says: "The drug comes to Bombay from Singapore, and is valued at 9 annas to R1 per lb, according to quality." It is said to grow in swampy forests near Palembang, Eastern Sumatra; also in Southern Borneo and in Penang, &c. The Dragon's-blood of modern commerce comes chiefly from Borneo. There are, however, two distinct forms of Dragon's-blood—the modern and the ancient.

Properties and Uses—

GUM. 69

Gum.—This gum is sold in dark-red friable masses, from which a blood-red powder is obtained; this is often met with in the bazar packed in the interior of canes.

The fruits of **C. Draco** are clustered, each covered with beautiful imbricating scales, which are coated with a red resinous substance. The fruits are collected, placed in long bags, and violently shaken; the resinous powder is thus separated and is thereafter baked into sticks or cakes—the forms in which the modern drug reaches Europe. An inferior quality is also said to be prepared by boiling the crushed fruit. **Griffith**, quoting **Rumphius**, says that in Palembang: "The natural secretion of the fruit constitutes the best *D'jarnang* or Dragon's-blood; a second and rather inferior kind is produced from the fruits, from which the natural secretion has been removed by heat and bruising. The third and most inferior appears to be the refuse of this last process. It is perhaps doubtful whether this article is procured from the plant by incisions."

Other species of **Calamus** also yield Dragon's-blood, and from incisions on the stem a resinous substance resembling Dragon's-blood is obtained from **Dracæna Draco,** a tree of the LILIACEÆ and a native of the Canary Islands. A famous specimen of this tree, one often referred to by writers

The Dragon's-Blood.

CANES.

on this subject, once existed at Oratava in Teneriffe, but it was unfortunately destroyed in the hurricane of 1867. The dragon's-blood afforded by this plant is met with as a secretion at the base of the leaves. A similar red gum is also said to be obtained from **Pterocarpus Draco,** a tree of the West Indies and South America, and also from **Croton Draco,** *Schlecht.*

Varnish. 70

The various forms of Dragon's-blood are used in varnishing and staining wood. The substance is chiefly judged by the dealers according to colour and the high percentage of resinous matter soluble in alcohol. It is of inferior quality when it gives a dull brick-red mark when rubbed on paper, or has an earthy look on fracture.

MEDICINE. 71

Medicine.—DRAGON'S-BLOOD.—In the first mention we have of this drug it is spoken of as exported to the East from Arabia and Socotra. **Ibn Batuta** makes no mention of it as found in 1325 and 1349 in Java and Sumatra. **Barbosa,** writing in 1514, speaks of it as a product of Socotra. The ancient drug referred to by **Dioscorides and Pliny,** under the name of *Κιννάβαρι*—the costly pigment and medicine of Africa—was in reality obtained from Socotra. This was the produce of **Dracæna schizantha,** *Baker,* and according to **Captain Hunter** of **D. Ombet,** *Kotschy,* and according to **Professor Balfour** of **D. Cinnabari,** also. It would thus appear that the Dragon's-blood of the ancients was a resinous extract from the stem of a Dracæna, and thus to have been a substance now treated as false Dragon's-blood, the true article being got from the fruits of a Calamus. A small quantity of the *Κιννάβαρ,* (Cinnabar) of **Dioscorides** still comes into commerce by way of Zanzibar and Bombay, and is bought and sold at the present day under the same name, *Dam-ul-akhwain,* as recorded by the Arabs in the tenth century.

Dragon's-blood is used medicinally in India as an astringent, and chiefly by the Muhammadans in the treatment of diarrhœa and dysentery. It is also used as an astringent application to the eyes. **Dr. Irvine,** in his *Medical Topography of Ajmere,* says: "A beautiful kind of Kino is brought from Bombay, and considered very astringent; it is given in intestinal hœmorrhages; and is also used in enamelling on gold: four tolahs for one rupee." This allusion is no doubt to dragon's-blood.

In European medical practice at the present day dragon's-blood is chiefly used as a colouring agent for plasters and tooth-powders.

SPECIAL OPINIONS.—§ "**Dracæna schizantha,** *Baker,* yields Zanzibar Dragon's-blood; and **D. Cinnabari,** Socotrian Dragon's-blood." (*Surgeon-Major W. Dymock, Bombay.*)

"The Burmese *Kyeing-ni* produces a red exudation like Dragon's-blood. **Dr. Mason** presumes this to be **C. Draco.**" (*J. C. Hardinge, Rangoon.*)

§ "Astringent, used as a dressing for ulcers." (*Surgeon W. Barren, Bhuj, Cutch.*)

72 **Chemical Composition.**—" Dragon's-blood is a peculiar resin, which, according to **Johnston,** answers to the formula $C_{20} H_{20} O_4$. By heating it and condensing the vapour an aqueous acid liquid is obtained, together with a heavy oily portion of a pungent burning taste and crystals of benzoic acid. The composition of these products has not yet been thoroughly ascertained, but the presence of acetone, *Toluol,* $C_6 H_5 (C H_3)$, *Dracyl* of Glénard and Boudault (1844), *Styrol,* $C_8 H_8$ (*Draconyl*), has been pointed out; the latter perhaps due to the existence in the drug of metastyrol as suggested by Kovalewsky. Both these hydrocarbons are *lighter* than water; yet we find that the above oily portion yielded by dry distillation sinks in water—a circumstance possibly occasioned by the presence of benzoic alcohol, $C_6 H_5 (CH_2 OH)$.

CANES.

"As benzoic acid is freely soluble in petroleum ether, it ought to be removed from the drug by that solvent: on making the experiment we got traces of an amorphous red matter, a little of an oily liquid, but nothing crystalline. Cinnamic acid, on the other hand, is always present, according to **Hirschsohn** (1877). As to the watery liquid, it assumes a blue colour on addition of perchloride of iron, whence it would appear to contain phenol or pyrogallol rather than pyrocatechin."

"By boiling Dragon's-blood with nitric acid, benzoic, nitro-benzoic, and oxalic acids are chiefly obtained, and only very little picric acid. **Hlasiwetz** and **Barth** melted the drug with caustic potash, and found among the products thus formed phloroglucin, para-oxybenzoic, protocatechuic, and oxalic acids, as well as several acids of the fatty series. Benzoin yields similar products." (*Flück. and Hanb., Pharmacog., 674.*)

The more recent investigations of **Messrs. Dobbie and Henderson** have shown that none of the forms of Dragon's-blood which they examined contained benzoic acid. They, however, found cinnamic acid in the resins of **Calamus Draco** and of **Dracæna Cinnabari.** They presume that the error of supposing the presence of benzoic acid arose through confounding it with cinnamic acid or possibly from working with a resin in which benzoic acid had been formed by partial oxidation. They established the chemical characters of four kinds of dragon's-blood, the origins of two of which were authentic, namely—

Dragon's-blood from **Calamus Draco.**—Is of a brick-red colour, melts at 80° C., giving off highly irritating fumes; is insoluble or nearly so in cold caustic soda, ammonia, lime-water, and sodium carbonate, but dissolves when boiled in these reagents. It may be represented by the formula $C_{18} H_{18} O_5$.

Dragon's-blood from **Dracæna Cinnabari.**—Is vermilion-coloured, melts at 80° C., giving off aromatic irritating fumes; is readily soluble in cold caustic soda, ammonia, lime-water, and sodium carbonate. It may be represented by the formula $C_{18} H_{18} O_4$. (*Pharm. Journ., 1883.*) This is probably the true *dam-ul-akhwain* of the Arabs; it occurs in tears covered with a dull-red powder.

Calamus erectus, *Roxb., Fl. Ind., Ed. C.B.C., 719.* 73

Vern.—*Sungotta*, SYLHET; *Theing, thaing*, BURM.

References.—*Kurz, For. Fl., Burm., II., 516; Gamble, Man. Timb., 423; Drury's Useful Plants of India, 97; Balfour, Cyclop.*

Habitat.—An erect cane found in Sylhet, Chittagong, and Pegu.

FOOD. 74

Food.—It is said that in Sylhet the poor classes use the seed of this cane as a substitute for betel-nut.

C. extensus, *Roxb., Fl. Ind., Ed. C.B.C., 720.* 75

Vern.—*Dengullar*, SYLHET; *Nelapoka*, TEL.

References.—*Gamble, Man. Timb., 424; Drury's U. P. of India, 96.*

Habitat.—Met with in Sylhet, and said to often attain a length of 600 feet. Extensively used in the northern Cachar and Mánipur Hills for suspension-bridges.

FOOD. 76

Food.—Seed eaten as a substitute for betel-nuts.

C. fasciculatus, *Roxb., Fl. Ind., Ed. C.B.C., 721.* 77

Vern.—*Bara bet*, BENG.; *Perambu*, MALA., TAM; *Amla, vetasawmu*, TEL.; Dutt gives *Ambuvetasa ?* (= a rattan growing in water) SANS.; but Dr. Ch. Rice informs the autho that this determination is incorrect,

CANES. and that the Sanskrit name of this species is more likely to be *Vetra; Kyeingkha, kyenka*, BURM.

References.—*Griffith, Pl. 195, A. & B.; Brandis, For. Fl., 559; Gamble, Man. Timb., 423; Kurz, For. Fl., Burm., 517; Balfour, Cyclop.; U. C. Dutt, Mat., Med., Hind., 290.*

Habitat.—Met with on the plains and hills of Bengal, Orissa, Chittagong, Burma, and the Andaman Islands.

DOMESTIC. 78 **Domestic Uses.**—Walking-canes are made of the stems of this species, which often exists as an erect, tufted plant, at least until it is able to obtain support, when it becomes a climber. **Griffith** says it is distinguished from all the other species of **Calamus** by the direction of its clustered pinnules (resembling **Zalacca**) with spinous margins and keels. The male spikes are also shorter and broader than those of other species.

79 **Calamus flagellum,** *Griff.; Gamble, Man. Timb., 423.*

Vern.—*Rabi bet*, NEPAL; *Reem*, LEPCHA; *Nagagola bet*, ASS.

Habitat.—Met with in Sikkim and Assam.

80 **C. floribundus,** *Griff., Pl. cxcvii.; Gamble, Man. Timb., 423.*

Habitat.—Met with in Upper Assam.

81 **C. gracilis,** *Roxb., Fl., Ed. C.B.C., 721.*

Vern.—*Mapúri bet*, BENG.; *Kraipang*, MAGH; *Hundi bet*, ASS.

References.—*Griffith, Pl. cxcvi.; Gamble, Man. Timb., 423; Drury, Useful Plants of India, 97; Kurz, For. Fl., Burm., 520; Thwaites, En., Ceylon Pl., 330; Balfour, Cyclop.*

Habitat.—Met with in Assam, Chittagong, and South Ceylon.

82 **C. grandis,** *Griff., Pl. ccx.; Gamble, Man. Timb., 424; Kurz, II., 523.*

Syn.—DŒMONOROPS GRANDIS, *Kurz (Enum., 30).*

Vern.—*Rotang sumambo, rotang chry*, MALACCA.

Habitat.—Met with in Malacca and the Andaman Islands; stem about 2 inches in diameter.

83 **C. Guruba,** *Mart.*

Vern.—*Kyeing-nee, kyeinni*, BURM.

References.—*Gamble, Man. Timb., 424; Kurz, For. Fl., Burm., 522.*

Habitat.—Met with in Chittagong and Burma.

84 **C. Helferianus,** *Kurz, ii., 521 (Enum., 39); Gamble, 424.*

Habitat.—Met with in Tenasserim or the Andamans.

85 **C. humilis,** *Roxb., Fl. Ind., Ed. C.B.C., 719.*

Reference.—*Gamble, Man. Timb., 423.*

Habitat.—An erect cane of Chittagong.

86 **C. hypoleucus,** *Kurz, For. Fl., Burm., II., 523.*

Syn.—DŒMONOROPS HYPOLEUCUS, *Kurz (Enum., 29).*

Reference.—*Gamble, Man. Timb., 424.*

Habitat.—Met with in Tenasserim.

CANES.

Calamus inermis, *T. And.; Gamble, Man. Timb., 424.* 87

Vern.—*Dangri bet,* NEPAL; *Brool,* LEPCHA.

Habitat.—Frequent in Sikkim and Bhután. Furnishes the finest alpen-stocks.

C. Jenkinsianus, *Griff., Pl. clxxxvi. A., fig. 3; Gamble, Man. Timb., 424, & xxx.* 88

Syn.—CYMBOSPATHES JENKINSIANUS, *Gamble.*
Vern.—*Gola bet,* ASS.; *Gallah,* CACHAR.

Habitat.—Met with in the Sikkim Terai, the Duars, and Assam.

C. latifolius, *Roxb., Fl. Ind., Ed. C.B.C., 719.* 89

Vern.—*Korak bet,* CHITTAGONG; *Sain,* MAGH.; *Ya-ma-ta,* BURM.
References.—*Griffith, Palms, Br. Ind., 68, Pl. cxcviii.; Brandis, For. Fl., 560; Gamble, Man. Timb., 423, 424; Kurz, For. Fl., Burm., 518.*

Habitat.—Met with in Chittagong, Burma, and the Andamans.

Structure of the Wood.—This cane is much used in Burma for tying timber in rafts, and making the cables which stretch across the river at the Salween rope station. An immense climber, with the stems about as thick as a walking-cane. 90

C. leptospadix, *Griff., Pl. lcxciv. A.; Gamble, Man. Timb., 423.* 91

Vern.—*Dangri bet,* NEPAL; *Lat,* LEPCHA.

Habitat.—Found in Sikkim and the Khásia Hills.

This species approaches **C. tenuis,** *Roxb.*; it is, however, distinguished from all the others by the long, filiform, distant branches of the spadix, and lanceolate rather large limbs of the primary and especially of the secondary spathes and the pennicillate bracts. (*Griffith.*)

C. longipes, *Griff., cciii. A. & B.; Gamble, Ma . Timb., 424.* 92

Vern.—*Gola bet,* SUNDERBUNDS.

Habitat.—Dr. King has identified this plant, proving the existence in India of a species hitherto supposed to be confined to Malacca.

C. longisetus, *Griff., Palms, Br. Ind., 44, Pl. clxxxix. A.; Thwaites, En. Ceylon, Pl. 330.* 93

Habitat.—An erect palm, very much resembling **C. arborescens**; met with in Pegu and Ceylon.

C. macracanthus, *T. And.; Gamble, Man. Timb., 424.* 94

Vern.—*Phekori bet,* NEPAL; *Ruebee, greem,* LEPCHA.

C. macrocarpus, *Griff., Pl. clxxx. VI. A., figs. 1 & 2; Gamble, Man. Timb., 423.* 95

Syn.—C. ERECTUS, *Roxb.*
Habitat.—An erect cane, met with in the Bhután Duars.

C. Mastersianus, *Griff., Pl. ccvi.; Gamble; Man. Timb., 424.* 96

Syn.—C. GURUBA, *Kurz.*
Vern.—*Sundi-bet, quabi-bet,* ASS.

Habitat.—Met with in Assam, and, according to Griffith, is the smallest cane in Assam, being less than half an inch in diameter.

CANES.

97 **Calamus mishmiensis,** *Griff.; Gamble, Man. Timb., 423.*

Habitat.—Met with in the Mishmi Hills.

98 **C. montanus,** *T. And.; Gamble, Man. Timb., 424.*

Vern.—*Gouri-bet,* NEPAL; *Rue,* LEPCHA.

Habitat.—Found in Sikkim and Bhután. Yields the best cane for suspension-bridges; used also in Sikkim for dragging logs.

99 **C. nutantiflorus,** *Griff., Pl. ccviii.; Gamble, Man. Timb., 424.*

Habitat.—Met with in Assam.

100 **C. palustris,** *Griff.*

Syn.—C. LATIFOLIUS, *Kurz, ii., 518 (Enum., 34).*

Habitat.—Met with in Mergui.

101 **C. paradoxus,** *Kurz, ii., 521 (Enum., 40).*

Reference.—*Gamble, Man. Timb., 424.*

Habitat.—Met with in Martaban.

102 **C. polygamus,** *Roxb., Fl. Ind., Ed. C.B.C., 721.*

Vern.—*Húdúm,* CHITTAGONG.

Reference.—*Gamble, Man. Timb., 423.*

Habitat.—Met with in Chittagong.

103 **C. quinquenervius,** *Roxb., Fl. Ind., Ed. C.B.C., 720.*

Vern.—*Hurnur-gullar,* SYLHET.

Reference.—*Gamble, Man. Timb., 424.*

Habitat.—Met with in Sylhet.

104 **C. Rotang,** *Linn. (in part); Roxb., Fl. Ind., Ed. C.B.C., 720.*

THE RATTAN CANE.

Syn.—C. ROXBURGHII, *Griff.* It seems probable that **C. Rotang,** *Linn.,* included originally more than one species: following **Martius** it is desirable, therefore, to retain the name as restricted to this species. **C. Rotang,** *Willd.,* as in *Roxb., Flora India,* is the plant here described. He presumed that the Indian form was the same as **Linnæus' Rotang,** found in Malacca, from the fact that the canes could not be distinguished.

Vern.—*Bet, chachi bet,* BENG., HIND.; *Pepa, prabba,* C. P.; *Bet,* BOMB.; *Veta, beta,* MAR.; *Natar,* GUJ.; *Perambu,* TAM.; *Beta mu, bettam, bettapu chettu, niru prabba, pemu, pepu, prabba chettu, prabhali,* TEL.; *Rotan,* MALA.; *Vetasa* (by some authors the following, although these names should be more applicable to a stronger species, such as **C. fasciculatus,** *vetra, vetus*), SANS.; *Bed,* PERS.

The generic name in Ceylon for Calamus is *waiwel,* SINGH.

References.—*Griffith, Pl. cxii.; Brandis, For. Fl., 559; Gamble, Man. Timb., 423; U. C. Dutt, Mat. Med. Hind., 323; S. Arjun, Cat. of Bom. Drugs, 146; Drury's Us. Pl., 96; Thwaites, En. Ceylon Pl., 330; U. S. Dispens., 15th Ed., 1636; Balfour, Cyclop.; Treasury of Botany; &c., &c.*

Habitat.—Met with in Bengal, Assam, South India, Burma, and in the hotter parts of Ceylon. It delights in rich, moist soil, where there are bushes and trees for it to climb on. (*Roxb.*) It flowers at the beginning of the rains and ripens during the cold season.

CANES. FIBRE. 105

Fibre.—This is the species which yields the best and stoutest rattan canes of commerce. Other species are, however, used as substitutes. It is split into strips and platted or woven into baskets, chairs, sofas, and carriages. It is made into ropes, or is stretched entire across rivers, as the main supports of cane suspension-bridges. For further information see CANES.

FOOD. 106

Food.—It flowers during the rains, and the fruit, which ripens in the cold season, consists of a fleshy substance surrounding the seed. This fleshy substance is eaten by the natives, who also eat the young tender shoots, regarding them as a delicacy.

Calamus Roxburghii, *Griff., Palms, Br. Ind., 55, Pl. cxii.* 107

Syn.—C. ROTANG, *Roxb. (non Linn), Fl. Ind., 720; Thwaites, En. Ceylon Pl., 330.*

See **C. Rotang**, *Linn.*, above.

C. Royleanus, *Griff., Pl. cxci.* 108

Syn.—C. ROTANG, *Linn., in part.*

References.—*Brandis, For. Fl., 559; Gamble, Man. Timb., 423; Drury, Us. Pl., 67.*

Habitat.—Met with in Dehra Dún and in Northern Bengal.

C. rudentum, *Lour.* 109

Vern.—*Ma-waiwel*, SINGH.

References.—*Roxb., Fl. Ind., Ed. C.B.C., 719.*

Habitat.—A native of the Malaya and of Ceylon.

Fibre.—Dr. Trimen writes that this species is used by the people of Ceylon for ropes. "It is split into strings and used for platting beds, chairs, baskets. Long rattans are also employed for bridges across streams and rivulets." FIBRE. 110

C. schizospathus, *Griff.; Gamble, Man. Timb., 423.* 111

Vern.—*Rong*, LEPCHA.

Habitat.—An erect cane, native of Sikkim and the Khásia Hills.

Structure of the Wood.—Stem about 2 inches in diameter, with hard wood and closely-packed fibro-vascular bundles. TIMBER. 112

C. Scipionum, *Lour.; Brandis, For. Fl., 560.* 113

THE MALACCA CANE. (See also under CANES.)

Habitat.—A native of Sumatra and Cochin China.

The canes are largely imported into India, after having been smoked, a process which gives them their beautiful brown colour.

Calamus, sweet, see **Andropogon Schœnanthus, A. 1117.**

C. tenuis, *Roxb., Fl. Ind., Ed. C.B.C., 721.* 114

Syn.—C. MONOICUS, *Roxb., Fl. Ind., Ed. C.B.C., 722.*

Vern.—*Bandhari bet*, CHITTAGONG; *Kring*, MAGH.; *Jalla bet*, ASS.; *Jali*, CACHAR.

References.—*Griffith, Pl. cxciii. A., B., & C.; Brandis, For. Fl., 559; Gamble, Man. Timb., 423, & xxx.; Kurz, For. Fl., Burm., 520; Thwaites, En. Ceylon Pl., 330.*

Habitat.—A monœcious climbing cane, met with in Assam, Sylhet, Chittagong, Pegu, and in the hotter parts of Ceylon.

115 **Calamus tigrinus,** *Kurz, For. Fl., Burm., 519.*

Vern.—*Leme,* BURM.; *Amdah,* AND.

Reference.—*Gamble, Man. Timb., 424.*

Habitat.—Found in Burma and the Andamans.

The Vernacular names given to Canes sent to the Paris Exhibition, the scientific names of which have not been determined.

Persons who have the opportunity of doing so may find it possible to supply fresh specimens of these plants along with leaves and fruits so as to admit of their identification.

"From Chittagong were received *Kerak jayat* and *golak;* the first is probably **C. latifolius**; from Assam, *Riphin, ringer, risan, raidana, bent, sowka bent, rangi, pakhori, howka, charainari, lijai;* from Cachar, *Jali, soondi,* and *gallah;* from Burma, *Theinkyeng, dyauthaukyeng, engkyeng, toungkyeng, kyengbot, yanoung, khaboung;* from South Kanara, *Betha* and *naga betha;* and from the Andamans, *Boledah, jobetahdah, jobetah,* and a palm called *chardah.*" (*Gamble, Man. Timb., 425.*)

CALAVANCE.

116 **Calavance.**—**Colonel Yule** tells us that this name was once in common use in English, and may, perhaps, to this day be used at sea for a kind of bean, perhaps the Indian **Vigna Catiang**, or a species of **Phaseolus.** The word comes from the Spanish *garbanzos,* which **DeCandolle** says is the Castilian name for **Cicer arietinum** (gram). See *DeCandolle's Origin Cult. Plants, p. 323.*

Calcium, see under **Lime**; also **Marble and Limestone.**

CALENDULA, *Linn.; Gen. Pl., II., 454.*

117 **Calendula officinalis,** *Linn.; Fl. Br. Ind., III., 357; Bot. Mag., t. 3204;* COMPOSITÆ.

MARIGOLD.

Vern.—*Aklel-ul-mulk, zergul, saldbargh,* PB.; *Htat-ta-ya,* BURM.

"*Aklel-ul-mulk* is **Astragalus hamosus**, a leguminous plant." (*Assistant Surgeon Sakhárám Arjun Ravát, L.M., Girgaum, Bombay.*)

References.—*Stewart, Panjáb Plants, 123; Balfour, Cyclop.*

Habitat.—Found in the fields of the Panjáb and Sind, scarcely indigenous; Pesháwar. (*Aitchison.*) **Stewart** says it is called *zergul* in the Trans-Indus tracts, where it is "common, wild in some parts."

DYE. 118 **Dye.**—An extract of the flowers is, by **Bellew,** said to be used to colour butter and cheese. It is probable that some of the properties assigned to this plant should more correctly be attributed to the *genda,* **Tagetes patula.** Both plants are used as dyes and are often mistaken the one for the other.

OIL. 119 **Oil.**—**Baden Powell,** in his *Panjáb Products,* mentions this as an oil-yielding plant. The oil is said to be used for medicinal purposes.

FODDER. 120 **Fodder.**—**Bellew** mentions the belief that when browsed on by cows, this plant is supposed to increase the flow of milk.

Calf-skins, see HIDES AND SKINS.

CALICO.

121

Calico. Cotton cloth originally made at Calicut.

Vern.—*Kapra,* HIND.; *Tuni,* TAM.; *Gudha,* TEL.; *Kapin-kapas,* MALAY.

The earlier writers speak of the cotton fabrics of India as "linens." When introduced to modern Europe they received the name of **Calicos,** after the town of Calicut, in the Madras Presidency, where they were then extensively made. At first the use of cotton fabrics was prohibited in England, the downfall of the trade in woollen goods being anticipated from the introduction of these cheaper textiles. Soon, however, this opposition was removed; but instead of the centres of woollen manufacture becoming the seats also of the cotton industry of England, the opposition of the woollen manufacturers drove the new industry to Lancashire. There could have been no happier accident, since it has been abundantly shewn that both in climate and soil no better situation could have been chosen. Indeed, had cotton manufacture been attempted in the more midland and eastern counties of England, it may be doubted how far the unprecedented success which rapidly ensued could have occurred. The time-honoured handlooms of India had then to give place to competition with the delicate and beautiful machinery of England, which, step by step and year by year, was made to do all and more than the patience and experience of the Indian weaver could accomplish. The calico and muslin trade of the world migrated from India to Lancashire. The exports from India, which once alarmed the British manufacturer, came to a sudden end. The tide turned, and wave by wave the imports from Great Britain increased until the cotton piece goods and yarns of Lancashire took complete possession of the Indian market. India now annually receives £25,000,000 worth of cotton goods and yarns, and exports about £14,000,000 worth of raw cotton. But indications are not wanting that India is recovering lost ground. It is feared over-competition has in Europe given birth in many cases to a depreciated article, and not in India only has the outcry gone forth against the weighted and starched piece goods which now leave the shores of Europe for the foreign markets. This want of confidence has recalled into new existence the hand-looms of India, and the weavers using the European yarns are now turning out an article which, it is admitted on all hands, may be less elegantly finished but is certainly not inferior in quality to the imported piece goods. This demand for yarns has enabled first one then another cotton mill to spring into life and activity. There are now cotton mills scattered all over India, keenly competing not in the yarn trade only, but in the piece goods as well, and last year £4,000,000 worth of Indian cotton goods and yarns were exported to Zanzibar and China to compete in these markets against the British goods. It may confidently be hoped that India will in a not very distant future greatly diminish her imports of foreign cotton goods if she does not at the same time make herself felt in the other cotton markets of the world. England has to fear English capital, English skill, and English energy established in India, where labour is at the same time cheap and the raw article produced at the very door of the factory. Economy of time and a saving of two freights may yet work the same revolution in the cotton trade of India as has become an established principle in jute.

For further information see **Cotton** and **Gossypium.**

CALICOPTERIS.

122

Calicopteris floribunda, *Lam.;* COMBRETACEÆ.

Syn.—GETONIA FLORIBUNDA, *Roxb., Fl. Ind., ii., 428.*

Vern.—*Kokoranj,* C. P.; *Bandi, murududu,* TEL.; *Marsada, boli,* MYSORE.

The Callicarpa.

A large climbing shrub of Central and Southern India.

Wood yellowish white, moderately hard, with numerous broad medullary patches of soft pith-like texture.

Calisaya Bark, see **Cinchona Calisaya**; RUBIACEÆ.

CALLICARPA, *Linn.*; *Gen. Pl., II., 1150.*

123 **Callicarpa arborea,** *Roxb., Fl. Br. Ind., IV., 567*; VERBENACEÆ.

Vern.—*Ghiwala, dera, shiwali,* KUMAON; *Ghivala,* CUTCH; *Bormala,* BENG.; *Búndún,* KOL.; *Dum kotokoi,* SANTAL; *Bogodi, gogdi,* KHARWAR; *Boropatri,* URIYA; *Sakrela,* MAL., S. P.; *Goehlo,* NEPAL; *Kodo, kozo,* MECHI; *Súnga,* LEPCHA; *Doika,* RAJBANSHI; *Khoja,* ASS.; *Makanchi,* GARO; *Turmong,* MAGH.; *Doung-sap-pya, daung-sat-pya,* BURM.

References.—*Roxb., Fl. Ind., Ed. C.B.C., 131; Brandis, For. Fl., 368; Kurz, For. Fl., Burm., II., 274; Gamble, Man. Timb., 282; Atkinson, Him. Dist., 730; Balfour, Cyclop., Ed. 1885.*

Habitat.—A moderate-sized tree, with brownish, rough, grey bark, met with in Kumaon, Oudh, Eastern Bengal, and Burma; chiefly in second-growth forests.

MEDICINE. 124 **Medicine.**—The bark is aromatic and bitter, and is applied in decoction to cutaneous diseases.

§ "Tonic, carminative." (*Surgeon W. Barren, Bhuj, Cutch.*)

TIMBER. 125 **Structure of the Wood.**—Grey, moderately hard, even-grained. Annual rings visible. Polishes beautifully, but is not used except for charcoal.

126 **C. cana,** *Linn.*; *Fl. Br. Ind., IV., 568.*

Vern.—*Arusha,* CHITTAGONG.

References.—*Roxb., Fl. Ind., Ed. C.B.C., 131; Royle, Fib. Pl., 310; Balfour, Cyclop.*

Habitat.—A shrub of Bengal; common in forests and along road-sides in the Terai and Duars, extending probably southwards to the Ganges.

FIBRE. 127 **Fibre.**—**Royle,** in his *Fibrous Plants of India,* says that a fibre is prepared from this plant, called *Arúsha* in Chittagong. **Captain Thomson,** reporting of this fibre, says: "It is much too weak for either sailcloth or cordage. It, however, possesses all the free and kindly nature of flax, and even smells like flax. It is easily worked, with little or no waste, &c." (*Royle, page 311.*)

TIMBER. 128 **Structure of the Wood.**—White, soft. Annual rings marked by a line of closer pores.

C. incana, *Roxb.,* Syn. for **C. macrophylla,** *Vahl.,* which see.

129 **C. lanata,** *Linn.*; *Fl. Br. Ind., IV., 567.*

Syn.—C. WALLICHIANA, *Walp.*; *DC. Prod., XI., 641; Wight, Ic., t. 1736, fig. 5, & Ic. t. 1480; Beddome, clxxii.*

Vern.—*Bastra,* HIND.; *Massandari,* BENG.; *Aisar,* BOMB.; *Coat comul,* TAM.; *Tondi, teregam,* MALA.

References.—*Roxb., Fl. Ind., Ed. C.B.C., 131; Brandis, For. Fl., 368; Bedd., For. Man., 173; Fl. Sylv., 123; Thwaites, En. Ceylon Pl., 243; Ainslie, Mat. Ind., II., 180; O'Shaughnessy, Beng. Dispens., 456; Drury, Us. Pl., 97; Grah., Cat. Pl., Bomb., 156;* **C. cana,** *Dalz. & Gibs., Bomb. Fl., 200, non Linn.; Dymock, Mat. Med., W. Ind.* **(C. cana),** *p. 716; Balfour, Cyclop., Ed. 1885.*

Habitat.—A shrub of Western and Southern India and the Circars.

MEDICINE. 130 **Medicine.**—**Ainslie** says that this plant is reckoned by the Javanese amongst their emollients. The bark, according to that author, possesses a peculiar sub-aromatic and slightly bitterish taste, and may probably be found to have other medicinal virtues. The Malays consider the plant as a diuretic. **Drury** mentions that in Upper Hindústan the root is employed in cutaneous affections. **Dr. Trimen** writes to the author that in Ceylon "the leaves, roots, and bark are used locally by the natives for skin diseases; they are very bitter."

Food.—"The bark, which is sub-aromatic and slightly bitter to taste, is chewed by the Singhalese instead of betel leaves." (*Drury.*) FOOD. 131

Callicarpa longifolia, *Lamk.; Fl. Br. Ind., IV., 570.* 132

References.—*Roxb., Fl. Ind., Ed. C.B.C., 132; Brandis, For. Fl., 369; Kurz, For Fl. Burm., II., 275;* **C. cana,** *Wall. Cat.*

Habitat.—A shrub of the Malaya Peninsula, Penang, and Nicobar Islands; and *var.* **lanceolaria** of Eastern Bengal, Khásia hills, Chittagong, and Burma.

C. macrophylla, *Vahl.; Fl. Br. Ind., IV., 568.* 133

Syn.—C. INCANA, *Roxb.;* C. ROXBURGHII, *Wall.;* C CANA, *Gamble's Darj. List and Man. Timbers, but non Linn.*

Vern.—*Pattharman, bá-pattra, bauna,* JHELAM; *Sumáli,* CHENAB; *Denthar, drúss,* RAVI; *Daya, shiwali,* KUMAON; *Mathara, mattranja,* BENG.

References.—*Roxb., Fl. Ind., Ed. C.B.C., 132; Brandis, For. Fl., 368; Kurz, For. Fl., Burm., 274; Gamble, Man. Timb., 282,* also *283; Stewart, Pb. Pl., 165; Baden Powell, Pb. Pr., 571; Balfour, Cyclop.*

Habitat.—A tall shrub of Northern and Eastern India, found as far north as Hazára, and ascending the Himalaya to 6,000 feet, and abundant in Bengal.

The *Flora of British India* establishes two varieties of this species—

var. **griffithii** from Bhután.

var. **sinensis** from Canara.

Medicine.—§ "In Hazára the heated leaves are applied to rheumatic joints (whence the name *bá-pattra,* from *bá,* rheumatism)." (*Surgeon-Major W. D. Stewart, Cuttack.*) MEDICINE. 134

C. rubella, *Lindl.; Fl. Br. Ind., IV., 569.* 135

Vern.—*Sugrúmúk,* LEPCHA.

Habitat.—A small tree of the North-East Himálaya to the hills of Martaban.

C. vestita, *Wall.; Fl. Br. Ind., IV., 567.* 136

Syn.—C. LANATA, *Gamble's Man. Timbers, and Darjeeling List, non Linn.*

CALLIGONUM, *Linn.; Gen. Pl., III., 95.*

Calligonum polygonoides, *Fl. Br. Ind., V., 22;* POLYGONACEÆ. 137

Vern.—*Balanja, berwaja, tatúke,* TRANS-INDUS and AFGHÁNISTAN; *Phok, phog, phogallí* (flowers), *tirní* (root), PB., SIND.

References.—*Brandis, For. Fl., 372; Gamble, Man. Timb., 303; Stewart, Pb. Pl., 183; Murray, Drugs & Pl., Sind, 99; Baden Powell, Pb. Pr., 264, 571; Balfour, Cyclop.*

Habitat.—A slow-growing shrub of the arid zone of Sind, the Panjáb, and Rájputana, distributed into Afghánistan and Western Asia. It has a pleasing appearance; its leafless branches and small pink flowers being in May succeeded by the small fruit.

MEDICINE. 138 **Medicine.**—The roots are bruised, and, boiled in combination with *Catechu* (*Kath*), are used as a gargle for sore gums. (*Murray.*)

FOOD. 139 **Food.**—The flowers, when fallen, are gathered and eaten as food by the natives. The abortive flowers are either made into bread or are cooked with *ghí* or cocoanut oil.

FODDER. 140 **Fodder.**—The shoots are relished by goats and camels.

TIMBER. 141 **Structure of the Wood.**—Most old stems are hollow. Heartwood reddish brown, very hard. The wood is chiefly used for fuel. The branches and twigs are used for the walls and roofs of huts.

CALLITRIS, *Vent.; Gen. Pl., III., 424.*

142 **Callitris quadrivalvis,** *Vent.; DC. Prod., XVI., 2, 452;* CONIFERÆ.

THE SANDARACH OR ARAR. (See **Sandarac.**)

Syn.—THUJA ARTICULATA, *Vahl.*

Vern.—*Sandaras, sandarach* (resin), SIND; *Chandrasa,* MAR.; *Sundaras,* GUJ.; *Sandarús,* PERS.

References.—*Brandis, For. Fl., 535; Gamble, Man. Timb., 394; U. S. Dispens., 1744; Flückiger, Pharmakognosie (Berlin, 1883), p. 94; Murray, Drugs & Pl., Sind, 27; S. Arjun, Bomb. Drugs, 133; Spons, Encyclop., 1681, 1684, 2012; Balfour, Cyclop., Ed. 188; Smith, Economic Dict., 367; Kew Cat., 130.*

Habitat.—A large tree of the forests of Algeria; might be introduced into India.

RESIN. 143 **Resin.**—"The resin exudes naturally from the bark of the stem, but the common practice is to make incisions in the stem, particularly near the base, by which the flow is much increased. The juice rapidly hardens on exposure, and is collected by the Moors, and carried by them to Mogador for export to Europe. It occurs in commerce mostly in cylindrical tears, which are occasionally agglutinated. It is not softened by boiling water, and is not soluble in caustic soda or acetic acid; it is partially soluble in benzol, rectified petroleum, chloroform and turpentine-oil, very slightly in carbon bisulphide and boiling linseed oil, but completely so in alcohol and ether. Formerly of wide renown in medicine, the resin is now valued, in Europe at least, principally as an ingredient of varnishes, to increase the hardness and glossiness; powdered, under the name of "pounce," it is used for preparing the surface of parchment and paper to receive writing. Its approximate price in the London drug market is 60*s.* to 115*s.* a cwt." (*Spons, Encycl.*) If rubbed over a piece of paper from which a word or letter has been erased, the space may be written over again without the ink spreading.

The resin is imported into India. An admirable sample was shown at the Colonial and Indian Exhibition, London, from Karachi. It consisted of large plates or masses of a clear straw colour.

MEDICINE. 144 **Medicine.**—The powder of *Sandarac* is used as a fumigatory after child-birth. (*Murray.*) It was formerly given internally and used in the preparation of various ointments and plasters. (*U. S. Dispens.*)

SPECIAL OPINIONS.—§ "Advantageously used in hæmoptysis, malena, and chronic diarrhœa, as an astringent." (*Surgeon E. W. Savinge, Rajabundry, Godavari District.*) "It makes a good varnish with which to polish furniture; this is prepared by mixing and boiling with oil (linseed)." (*Assistant Surgeon Bhugwan Das, Rawal Pindi, Panjáb.*)

TIMBER. 145 **Structure of the Wood.**—The tree "is remarkable on account of the compact, heavy, and very fragrant heartwood, which has a rich brown colour, and takes a most beautiful polish. The tree coppices readily, and the forest fires which are lighted by the Arab herdsmen (as they are by the pastoral population of India) frequently kill the stem to the ground, when abundant shoots spring up from the root, which attains a great age, and often a considerable size. These masses of roots have a beautifully mottled grain; exquisite ornaments and small articles of furniture are made of them, and veneers for the most elegant cabinet-work. This wood was one of those called Κέδρο,, *citrus*, was highly prized by the Romans, and fabulous prices were paid for tables made of it." (*Brandis.*) There are vases made of this wood in the Kew Museum presented by Prince Jerome Bonaparte. According to **Lindley**, the timber "is durable, very hard, fragrant, and of a mahogany colour, for which reason it is largely used in the construction of mosques and similar buildings in the north of Africa." (*Treasury of Botany.*)

Calomel, Mercurous Chloride ($Hg_2\,Cl_2$), see **Hydrargyrum** and **Mercury.**

Calonyction speciosum, *Chois.*, see **Ipomœa bona-nox,** *Linn.;* CONVOLVULACEÆ.

CALOPHYLLUM, *Linn.; Gen. Pl., I., 175.* 146

Calophyllum inophyllum, *Linn.; Fl. Br. Ind., I., 273; Wight, Ic., t. 77;* GUTTIFERÆ.

THE ALEXANDRIAN LAUREL.

Vern.—*Sultána champa, surpan, surpunka, undi,* HIND.; *Sultána champa, punnág,* BENG.; *Polang, punung,* URIYA; *Surangí, purreya, duggerfúl (purraya, dugurphort), undí,* SIND; *Udí, undí,* BOMB.; *Udi,* CUTCH; *Surfan, undí, surpanka,* DEC.; *Undí, undela, wúndí, surangí, nág chámpa, pumag, undag,* MAR.; *Bintangor, punna, ponna,* MALA.; *Pinnay, punnai, punagam, punnai-virai, punnágam,* TAM.; *Púna, púnás, ponna, pumágamu, ponna-chettu, ponna vittulu,* TEL.; *Wúma, surabunne* or *suragonne, vuma, wuma, pinekai, pinne, pinna bija,* KAN.; *Punnága,* SANS.; *Domba, domba-gahá, teldomba, domba gass,* SINGH.; *Pongnyet, pengnyet, phung-nyet, phounniya,* BURM.; *Betan,* MALAY.; *Dílo,* FIJI; *Tamanu,* TAHITI.

References.—*Roxb., Fl. Ind., Ed. C.B.C., 437; Kurz, For. Fl., Burm., I., 95; Gamble, Man. Timb., 25; Thwaites, En. Ceylon Pl., 51; Dalz. & Gibs., Bomb. Fl., 31; Grah., Cat. Pl., Bomb., 26; Voigt, Hort. Sub. Cal., 87; A note by J. E. O'Conor on, 1875; Pharm. Ind., 32; Modeen Sheriff, Supp. Pharm. Ind., 81; U. S. Dispens., 15th Ed., 1760; U. C. Dutt, Mat. Med. Hind., 232, 314; Dymock, Mat. Med., W. Ind., 2nd Ed., 86; Ainslie, Mat. Ind., II., 311; O'Shaughnessy, Beng. Dispens., 238; Murray, Drugs & Pl., Sind, 70; Bidie, Cat. Raw Pr., Paris Exb., 21, 61, 103; S. Arjun, Bomb. Drugs, 22; Drury, Us. Pl., 98; Cooke, Gum & Gum-resins, 108; Cooke, Oils and Oil-seeds, 32; Birdwood, Bomb. Pro., 13, 259, 278; Lisboa, Us. Pl. of Bomb., 12, 214; Spons, Encycl.; Balfour, Cyclop., Ed. 1885; Smith, Econ. Dic., 51; Treasury of Botany.*

Habitat.—Cultivated, especially near the sea-coast, throughout India as an ornamental tree; indigenous to the Western Peninsula, Orissa, South India, Ceylon, Burma, and the Andaman Islands. Distributed to Malay, Australia, Polynesia, and Eastern Africa. In flower and fruit most part of the year, and thrives on dry sandy beaches where little else will grow.

History of Tacamahaca.—The name Tacamahaca has been indiscriminately applied to the resins of several plants, some of which have 147
a doubtful existence, such as of **Icica Tacamahaca, Elaphrium tomentosum,**

Populus balsamifera, Calophyllum inophyllum, and **Calophyllum Calaba.** These are generally referred to two classes—the true *Tacamahaca* from Curacoa and Venezuela, and the East Indian or *Tacamahaca orientale.* The *United States Dispensatory* notices three kinds of resin under the name Tacamahaca, **C. inophyllum** being supposed to afford one of these. **Birdwood,** in his *Catalogue of Bombay Products,* remarks that it is stated that a resin exudes from the roots of this plant, and that this is the Tacamahaca of the Isle of Bourbon. He also quotes **Lindley,** who affirms that the true East Indian Tacamahaca is produced by **Calophyllum Calaba,** and that Myanas resin is referred to the same species. **Guibourt** describes several varieties of Tacamahaca, one being derived from **C. Tacamahac,** which grows in the Islands of Bourbon and Madagascar. **Ainslie,** on the authority of **Lamarck,** says that the resin called Tacamahaca is obtained from **C. inophyllum.** On the other hand, **Dr. Cooke,** in his *Report on Gums and Resins,* after reviewing the opinions on the subject of the supposed Indian source of this gum, says: "There is no evidence substantially in favour of **Calophyllum inophyllum,** or any other Indian tree, and it may safely be affirmed that the *Tacamahaca orientale* is not obtained from India, nor is it known in Indian commerce. The resin of **Calophyllum inophyllum** must therefore be held guiltless of Tacamahac."

Smith (*Dictionary, Economic Plants*) says: "A gum resin exudes from the bark of the tree, which is one of the kinds of Tacamahaca gums of commerce."

Properties and Uses according to Indian writers:—

GUM. 148 **Gum.**—Speaking of **Calophyllum inophyllum, Mr. Baden Powell** remarks that it yields a black resinous substance. Specimens of this were sent from Madras to the Panjáb Exhibition. **Dr. Dymock** informs the author, however, that he has prepared this gum-resin by incising a tree, and that it yields about an ounce per tree of yellowish green translucent gum. He further states that this gum is also obtained from the fruit in small quantities, chiefly in the form of very small tears. It is "soft and entirely soluble in rectified spirit; it has a parsley odour, and has been confounded with *Tacamahaca,* the exudation of **C. Calaba,** not a native of India." (*Compare with account given by Dymock, Mat. Med., W. Ind., 2nd Ed., 86.*) **Mr. Gamble,** citing **Sebert** in *Les Bois de la Nouvelle Calédonie,* remarks that it gives a yellowish green pleasantly-scented resin.

OIL. 149 **Pinnay, Pun,** or **Domba Oil.**—The fresh seeds, when shelled, yield a large quantity of fragrant dark-green oil, amounting, according to some reports, even to 60 per cent. by weight. The seeds are collected twice a year—in August and again in February. The oil varies in colour from greenish yellow to deep green; it possesses a disagreeable flavour, and an odour which is described as fragrant by some, but unpleasant by others. It is used for lamps and for caulking vessels (*Cooke*), but it is chiefly valued as a medicine, being employed as an external application in rheumatic affections. "It is prepared to some extent in Bombay, Tinnevelly, and other parts of India, and used as a lamp oil." "Formerly the seed and oil were exported from Madras to the Straits and Ceylon, but it has now ceased to be an article of export" (*O'Conor*). In Tanjore 437 acres are said by **Balfour** to be under this tree, the yield of oil being 2,671 maunds, which fetches R20-4 a maund, and is sold at 4 to 5 annas a seer. It is a curious fact in connection with this oil, that though it cannot compete with castor oil for industrial purposes in the Calcutta market, it fetches in Burma about four times the Calcutta price of castor oil. The low price in India generally is owing to its present crude condition, no method of refining having been attempted. **Simmonds** says that as much as 60 per cent. of oil may be expressed from the seed, and that this oil congeals when cooled below 50°. Pinnay oil is extensively used in Travancore, especially for lamps, and is largely manufactured in the southern district. **Babu Nand Kishor Das**, Deputy Collector, Puri, thus reports on the manufacture of oil as practised in Orissa :—"The seeds being gathered are beaten with a small wooden hammer, which causes the separation of the shell from the kernel. This is then sliced and dried in the sun and put into the mill. The mill used for the purpose is the common country oil-mill." "A given weight of seed will produce one third of its weight of oil. The ordinary price of the oil is nowadays R8 a maund (of 80 *tolas*). It would thus appear that it does not by any means yield the large amount of oil (60 per cent.) attributed to it by some writers, and as it is inferior to castor oil, it is not likely to ever take a very high place as an illuminating or saponaceous oil. **Mr. Gamble** remarks that in Orissa the tree is much cultivated. **Mr. F. W. Oliver**, Deputy Conservator of Forests, British Burma, is of opinion that there is some doubt as to the statement that this oil is used for caulking boats. He thinks that the oil referred to is more probably the *pruot hnyet* or dammar.

The oil deposits a quantity of solid fat after having been kept a short time. It is best known in Europe as Domba oil. (*Dr. Dymock, Bombay.*)

Medicine.—The oil obtained from the nut kernel is highly esteemed by the natives as an external application in rheumatism. "The bark of the tree when wounded exudes a small quantity of bright green gum, which is not collected, nor does it appear to be made use of in any way." (*Dymock, Mat. Med., W. Ind.*) **O'Shaughnessy**, however, mentions the use of the resin as a remedy for indolent ulcers. **Rheede** says that the tears which distil from the tree and its fruit are emetic and purgative. **Drury** remarks that: "the gum which flows from the wounded branches, being mixed with strips of the bark and leaves, is steeped in water, and the oil which rises to the surface used as an application to sore eyes." **Horsfield** adds that in Java the tree is supposed to possess diuretic properties. There seems, however, to be considerable doubt about the medicinal properties of the gum obtained from this tree by reason of its being incorrectly identified with true Tacamahaca. MEDICINE. 150

SPECIAL OPINIONS.—§ "The leaves soaked in water are employed as an application to inflamed eyes, in the Archipelago." (*Dr. Ch. Rice, New York.*) "The fixed oil obtained from the kernels of the seeds is said to cure scabies." (*Surgeon-Major Bankabehari Gupta, Pooree.*) "According to Hindú writers, the bark is astringent and useful in internal hæmorrhages." (*U. C. Dutt, Civil Medical Officer, Serampore.*) "Stimulant dressing for ulcers." (*Surgeon W. Barren, Bhuj, Cutch.*) "The juice of the bark is used as a purgative, and is said to be very powerful in its action." (*Surgeon J. C. H. Peacock, Nasik.*) "In rheumatism the natives use this oil as an external application, but the recovery is slow." (*Honorary Surgeon Easton Alfred Morris, Negapatam.*)

Structure of the Wood.—Reddish brown, moderately hard, close-grained. Used for masts, spars, railway sleepers, machinery (*Kurz*); also for cabinet-work. (*Sebert.*) Canoes are made of it. (*Bomb. Gaz., X., 38.*) It is used in ship-building. (*Lisboa.*) "In the Samaon islands the large *ava* bowl is made from the *tamanu*" wood. (*Balfour.*) TIMBER. 151

Calophyllum polyanthum, *Wall.; Fl. Br. Ind., I., 274.* 152

Vern.—*Kandeb*, BENG.; *Kironli*, NEPAL; *Sunglyer*, LEPCHA.

References.—*Kurz, For. Fl., Burm., I., 95; Gamble, Man. Timb., 25; Voigt, Hort. Sub. Cal., 87.*

Habitat.—An evergreen tree of Northern and Eastern Bengal, the Khásia Hills, Chittagong, and Burma, ascending to 5,000 feet.

TIMBER. 153 **Structure of the Wood.**—Similar to that of **C. spectabile.** **Mr. Chester** of the Forest Department says it is used largely in Chittagong for masts, spars, and rafters, and sometimes in small boat-building and canoes. Weight from 38 to 40 lbs. a cubic foot.

154 **Calophyllum spectabile,** *Willd.; Fl. Br. Ind., I., 271; Wight, Ic., t. 128 & III.*

Syn.—C. MOONII, *Wight;* C. AMŒNUM, *Wall.;* C. TETRAPETALUM, *Roxb.*

Vern.—*Pantα-ka, kyandoo,* BURM.; *Dakar táladá,* AND.; said to be known as *Lal chuni in* HIND.

References.—*Roxb., Fl. Ind., Ed. C.B.C., 438; Kurz, For. Fl. Burm., I., 94; Gamble, Man. Timb., 25; Thwaites, En. Ceylon Pl., 52; Bedd., Fl. Sylv., XXII.*

Habitat.—A tall evergreen tree of Tenasserim and the Andaman Islands.

TIMBER. 155 **Structure of the Wood.**—Light red, shining, cross-grained, moderately hard. Is used for masts and spars; also for planking, for which purpose it has lately been employed in building barracks in the Andamans.

156 **C. tomentosum,** *Wight, Ic., t. 110; Fl. Br. Ind., I., 274.*

THE POON SPAR, SIRPOON TREE.

Syn.—C. ELATUM, *Beddome, XXII. & t. 2.*

Vern.—*Pún, sirpon,* BOMB.; *Pún, púne, púngú,* MALA.; *Pongú,* TAM.; *Siri púne kuve, surponne bobbi,* KAN.; *Nágani,* MAR.; *Kina,* SINGH.

References.—*Gamble, Man. Timb., 26; Thwaites, En. Ceylon Pl., 51; Dymock, Mat. Med., W. Ind., 2nd Ed., 87; Drury, Us. Pl., 98; Cooke, Oils and Oil-seeds, 32; Lisboa, Us. Pl. of Bomb., 13; Spons, Encyclop., 1392; Balfour, Cyclop., Ed. 1885; Treasury of Botany.*

Habitat.—A large, tall, evergreen tree, often growing to a height of 150 feet, met with in the evergreen forests of the western coast from Kanára southward; and in Ceylon, ascending to altitude 5,000 feet.

Property and Uses—

GUM. 157 **Gum.**—**Dr. Dymock** informs the writer that this tree yields a black opaque gum, which, in the bazar, occurs much mixed with pieces of bark; it is feebly astringent, and very soluble in cold water. The solution is brownish yellow, exhibiting a strong blue fluorescence.

"If this gum is steeped in water for some time, the solution becomes very dark in colour. Alum, followed by carbonate of soda, throws down apparently some of the brown colouring matter without interfering with the fluorescence, since after precipitation the solution, although lighter in colour, is very strongly fluorescent. A solution purified by alum in this way has its fluorescence immediately destroyed by acids and restored again by alkalies. Examining its absorption spectrum it is found that while fluorescent the solution gives a broad absorption band at the violet end of the spectrum extending to about G; this band disappears on destroying the fluorescence by acids, but reappears on the addition of alkalies. The solution of the gum does not appear to rotate polarised light. The gum itself communicates only a very faint fluorescence to rectified spirit. (*Lyon.*) I am not aware of either of these gums having been applied to any industrial or medicinal uses, but as they are collected by the natives, it is probable that they are supposed by them to have some medicinal virtues." (*Dymock, Mat. Med., W. Ind., 2nd Ed., 87-88.*)

OIL. 158 **Oil.**—The seeds in Ceylon yield an abundance of oil known as *Keenatel.*" It is probably used as a lamp-oil.

TIMBER. 159 **Structure of the Wood.**—Same as that of **C. spectabile.** This tree affords the Poon Spars of commerce; these are much used for masts, and often fetch large prices. The timber is also used for building and bridge-work.

"A single tree has been known to realize more than £100 (R1,000)." (*Bomb. Gaz., XV., 64.*)

Calophyllum Walkeri, *Wight, Ic., t. 45; Fl. Br., Ind., I., 275.* 160

Syn.—C. DECIPIENS, *Wight, Ill., i., 128.*

References.—*Thwaites, En. Ceylon Pl., 51; Cooke, Oils and Oil-seeds, 32; Balfour, Cyclop.*

Habitat.—A large tree, found in South India and Ceylon.

Oil.—The seeds yield an oil, used for burning. OIL. 161

C. Wightianum, *Wall.; Fl. Br. Ind., I., 274; Beddome, Flora Sylvat., t. 90; Wight's Ill. I., 128. also Ic., t. 106.* 162

Syn.—C. SPURIUM, *Chois., and of Drury, Us. Pl.;* C. DECIPIENS, *Wight, Ic., t. 106 (not of Thwaites).*

Vern.—*Kalpún, kull-ponné, bobbi,* KAN.; *Cheru pinnay, pútengi,* TAM.; *Tsirou-panna,* MALA.; *Cherupinai* (as in **Lisboa**), *Sarapuna* (as in *Dymock's Glossary*), BOMB.; *Irai,* MAR.

References.—*Gamble, Man. Timb., 26; Dalz. & Gibs., Bom. Fl., 32; Dymock, Mat. Med., W. Ind., 2nd Ed., 87; Drury, Useful Plants; Cooke, Gums and Gum-resins, 109; Cooke, Oils and Oil-seeds, 33; Lisboa, Us. Pl. of Bomb., 12, 214; Spons, Encyclop., 1379, 1387, 1624, 1683, 2020, 2021; Balfour, Cyclop., Ed. 1885; Treasury of Botany.*

Habitat.—An evergreen tree of the Western Ghâts, from the Konkan to Travancore.

Gum.—"The gum occurs in large, translucent, irregular lumps of a yellowish colour; it is of horny texture, somewhat brittle, without odour; the taste is soapy. When placed in water it gradually softens, and finally disintegrates into a fine granular matter which floats in the form of flaky particles of a dirty-white colour, and numerous oil-globules which gradually collect upon the surface; the water dissolves a small portion and becomes slightly viscid." (*Dymock, Mat. Med., W. Ind.*) GUM. 163

Oil.—The seeds yield an oil not differing very much from that of **C. inophyllum.** It is used as a lamp-oil. OIL. 164

Medicine.—**Bonton,** in his *Medical Plants of Mauritius,* says that the resin obtained from this plant acts as a "vulnerary, resolutive, and anodyne." The oil obtained from the seeds is used as a medicine in leprosy and cutaneous affections, and in infusion mixed with honey in scabies and rheumatism. MEDICINE. 165

Food.—The fruit, when ripe, is red and sweet. It is eaten by the natives. (*Drury.*) FOOD. 166

Structure of the Wood.—Hard, red. **Beddome** and also **Lisboa** say the timber is in Kánara much esteemed, and is valuable for engineering purposes. TIMBER. 167

Calosanthes indica, *Blume,* see **Oroxylon indicum,** *Vent.;* BIGNONIACEÆ.

CALOTROPIS, *R. Br.; Gen. Pl., II., 754.* 168

THE SWALLOW-WORTS.

A genus of ASCLEPIADEÆ containing only three species; these are inhabitants of tropical Asia and Africa.

Erect shrubs, glabrous but with a hoary powder. ***Leaves*** **opposite, broad, subsessile.** ***Flowers*** **medium-sized in umbelliform or sub-racemose cymes.** ***Corolla*** **valvate, broadly campanulate, coronal-scales 5, fleshy, laterally com-**

pressed, radiating from the large staminal column dorsally spurred. *Stamens* 5, adhering around the staminal axis; anther-cells with a solitary pollen mass in each, pendulous, flattened; tips of the anthers membranous, inflexed. *Follicles* paired, thick, short; seeds comose.

169 **Calotropis Acia,** *Ham.*; ASCLEPIADEÆ.

Syn.—ASCLEPIAS HERBACEA, *Roxb., Fl. Ind., Ed. C.B.C.*, 258.

Habitat.—A form met with in Eastern Bengal and Sikkim, having petiolate leaves, the blade tapering into the petiole and with a globular corolla-tube.

This is much less known than either of the following species, and no particulars of its properties and uses are available.

170 **C. gigantea,** *R. Br.; Fl. Br. Ind., IV., 17; Wight, Ill., t. 155, 156 A.*

Syn.—ASCLEPIAS GIGANTEA, *Willd.*

Vern.—*Madár, ák, ág, árk, ákond, ákan, mudhár, safed-ák,* HIND.; *Ākanda, gurtákand, swet-ákond,* BENG.; *Ahauna,* SANTAL; *Auk,* NEPAL; *Ākra, rúi, akandá, mándárá,* BOMB.; *Ākanda, rui, akda cha jhada,* MAR.; *Akado, ákdámu jhadá, dhola akdo,* GUJ.; *Bij-elosha,* SIND; *Yercum, erukkam, erukku,* TAM.; *Jilledu, jilleru, nella-jilledu, mandaramu, jilleduchettu, yekka,* TEL.; *Yekka, ekkemále, arkagida, yekka-da-gidá, yekkada beru, yokada,* KAN.; *Erukku, yerica, erica, belerica,* MALA.; *Kadráti,* GOND.; *Āk, ákra,* DUK.; *Ushar, ashur, oshmor, osch-or,* ARAB.; *Khark,* PERS.; *Arka* (red-flowered form), *pratápasa, mandára, álarka* (white-flowered form), *svaytaurkum,* SANS.; *Mayo-beng, mayo-má-yo-pin, ma-yo-pin, mayo, maioh, mayobin,* BURM.; *Vara, vara-gaha, múdu wará,* SINGH.; *Waduri,* JAVA.

References.—*Roxb., Fl. Ind., Ed. C.B.C., 251; Wight, Contri. Botany, India, 53; Brandis, For. Fl., 331; Kurz, For. Fl., Burm., II., 200; Gamble, Man. Timb., 265, xxvi.; Stewart, Pb. Pl., 144; Thwaites, En. Ceylon Pl., 196; Dalz. & Gibs., Bomb. Fl., 149; Voigt, Hort. Sub. Cal., 540; Pharm. Ind., 141, 457, 458; Moodeen Sheriff, Supp. Pharm. Ind., 82, 364; Flück. & Hanb., Pharmacog., 424; U. S. Dispens., 15th Ed., 1595; Hooker, Him. Jour., I., 86; U. C. Dutt, Mat. Med., Hind., 196; Grah., Cat. Bomb. Pl., 120; Dymock, Mat. Med., W. Ind., 2nd Ed., 512, 519; Ainslie, Mat. Ind., I. 486, 488; O'Shaughnessy, Beng. Dispens., 453; Murray, Pl. and Drugs of Sind, 160; Waring, Bazar Med., 92; Bidie, Cat. Raw Pr., Paris Exbn., 12, 117; S. Arjun, Bomb. Drugs, 190; Drury, Us. Pl., 99; Royle, Fib. Pl., 306, 308, 310; Atkinson, Him. Dist., 788, 794; McCann, Dyes and Tans of Bengal, 56; Birdwood, Bomb. Prod., 52, 209, 249, 269; Lisboa, Us. Pl. of Bomb., 211, 232, 260, 279, 282, 290, 291; Kew Reports, 1877, p. 37; Kew Official Guide to the Museums, p. 97; Spons, Encyclop., 818, 933, 1627; Balfour, Cyclop., Ed. 1885; Smith, Dic., 278; Treasury of Botany.*

Habitat.—An erect, spreading, perennial shrub, chiefly frequenting waste lands. It ascends to 3,000 feet on the Himálaya and extends from the Panjáb to South India, Assam, Ceylon, and Singapore, and is distributed to the Malay Islands and South China. It is not very common in Burma, and as represented by the doubtfully distinct species, **C. procera,** it is distributed to Persia and tropical Africa.

History.—"The ancient name of the plant, which occurs already in the Vedic literature, was *Arka* (wedge), alluding to the form of the leaves, which were used in sacrificial rites. From one of the Sanskrit names of the plant, namely, *Mandára* (according to **Dr. Ch. Rice**), "*Madár* is a corruption; the latter is frequently mentioned in the writings of **Susruta.**" (*Flück. & Hanb. in Pharmacog.*) **Roxburgh** and also **U. C. Dutt** tell us that two varieties were known to the Sanskrit writers, *viz.*, the *white-flowered* or *alarka* (this is most probably **C. procera**), and the *purple-flowered* or *arka* (**C. gigantea**). **Rheede** calls the former *bel-ericu* and the latter *ericu.*

GUTTA-PERCHA.

Mir Muhammad Husain notices three kinds,—*1st*, a large form with white flowers, large leaves, and much milky juice, found near towns; *2nd*, a form with smaller leaves and flowers, white on the outside but lilac within; and *3rd*, a still smaller kind with pale greenish-yellow flowers. (*Dymock.*) The 1st and 2nd are most probably forms of **C. gigantea,** and the 3rd, **C. procera.**

C. gigantea was first figured and described by **Rheede** (*Hort. Malab., II., t. 31*) in 1679, and an accurate and detailed description was given by **Roxburgh** at the beginning of the present century under the name **Asclepias gigantea. Robert Brown** subsequently showed that it was incorrect to refer this plant to **Asclepias,** and he accordingly founded the genus **Calotropis,**—a genus which embraces, as far as at present known, two or probably three species. **C. procera** was first described from a specimen collected in Egypt by **Prosper Alpinus** (1580-84), and figured by him on his return to Italy (*De Plantis Ægypti, 1592*). It is also the **Apocynum syriacum** figured by **Olusius.** (*Flück. & Hanb., Pharmacog.*)

The drug prepared from one or other of these species was apparently well known to the Arabians. **Ibn Baytar** (*Sontheimer's translation in 1842*) describes the drug. Muhammadan writers at the present day refer to it under its Arabic name *Ushar;* in Persian it is known as *Khark.* The medicinal properties were first made known to Europe in 1826.

A tradition of Oomarcote narrates that the great **Emperor Akbar** was born under an *Ák* bush; hence his name. (*Birdwood.*) The word *bar* is applied to the liquor said to be prepared from *Ák* juice.

Properties and Uses—

The SAP yields a form of Gutta-percha; it is also used as a TAN and DYE: a MANNA is said to exude from the plant; the bast FIBRE and FLOSS from the seeds are well-known fibres; the ROOT bark and SAP are MEDICINAL; a LIQUOR is reported to be prepared from the juice; the WOOD is used for gunpowder charcoal, and various parts of the plant are employed for SACRED, DOMESTIC, and AGRICULTURAL purposes.

THE MILKY SAP—A SOURCE OF GUTTA-PERCHA.

MILKY SAP. Gutta-percha. 171

From the milky sap may be prepared a kind of gutta-percha. A specimen of this substance was sent from Madras to the Panjáb Exhibition. **Dr. Riddell** (*Journ. Agri.-Hort. Society of India, Vol., VIII.*) first drew attention to this gutta, and was followed by **Royle** in his *Fibrous Plants,* and still later by **Mr. Baden Powell,** in the *Panjáb Products.* It is probable that both this and the next species yield the same gutta substance, and as it is doubtful how far the published experiments refer to the one or the other, it has been thought advisable to give in one place a compilation of the entire literature. It is probable, however, that **Dr. Riddell's** experiments were entirely conducted with **C. procera** and not with **C. gigantea.**

The inspissated and sun-dried milky sap from the stem constitutes the gutta-percha. The *Madár* is in fact one of the most interesting and most hopeful plants, not belonging to the natural order **Sapotaceæ,** which can be said to yield a substance resembling gutta-percha. **Dr. Riddell,** the then Superintendent Surgeon to the Nizam's Army, was apparently the first to separate and experiment with this Sap; his results were published, in the first instance, by **Captain** (since Colonel) **Meadows Taylor** in a letter to the Secretary, Agri-Horticultural Society of India, Vol. VIII. Afterwards **Dr. Riddell** republished his discovery in *The Bombay Times* in 1852. As these letters may not be accessible to persons likely to be

CALOTROPIS gigantea.

The Swallow-worts.

GUTTA-PERCHA.

interested by this subject, the more important parts narrating the actual experiments are quoted below :—

"MY DEAR SIR,—I observe in the last number of the Society's *Transactions* that the *madár* (*Asclepias gigantea*) affords a very valuable kind of hemp or flax; and I have now the pleasure to communicate to you another valuable property it possesses, which has been lately discovered by a friend here, under whose permission I make the present communication to you.

"Dr. Riddell, the Officiating Superintendent Surgeon of the Nizam's Army, had for some time been employed in extracting or determining by chemical experiments the well-known medicinal properties of this plant; and during his investigation, having had occasion to collect the milky juice or sap, and expose it to the air, he found, as it gradually dried, that it became tough and hard, and not unlike gutta-percha. This induced him to treat the juice as that of the gutta-percha tree is done, and the result has been the obtaining of a substance apparently precisely analogous to gutta-percha, of which I have the pleasure to send you a specimen, bearing the impression of his seal, marked No. 1.

"The mode of preparing this substance is as follows :—

"The juice or sap to be collected by incision. An open slit may be made on the back of the plant and a pot tied to it, when the juice will flow into the pot; or it may be collected by cutting the bark and catching as much as flows out at once. Dr. Riddell calculates that ten average-sized plants or bushes will yield as much juice as will make a pound of gutta-percha substance, but it is not known yet how far the plant will bear tapping without injury, nor how often, or at what intervals, the extractions of juice might be made.

"The juice extracted may either be exposed to the sun in a shallow vessel, or left to dry in the shade: by the former process, the substance becomes a little darker than by the latter.

"When it has attained a tough consistency, it may be well worked up in very hot water with a wooden kneader, or boiled; either process serves to remove an acrid property of the juice, as also all other matter but the gutta-percha itself. It is believed that the more it is boiled and worked up, the harder it will eventually become when cool.

"Comparison with the true gutta-percha gives the following results :—

"Sulphuric acid—chars it.

"Nitric acid—converts it into a yellow resinous substance.

"Muriatic acid—has very little effect upon it.

"Acetic acid—has no effect.

"Alcohol—ditto.

"Spirit of turpentine—dissolves it into a viscid glue which, when taken up between the finger and thumb, pressed together, and then separated, shows numberless minute and separate threads.

"The above chemical tests correspond exactly with the established results of the real gutta-percha.

"The substance, however hard it may have become, becomes immediately flexible in hot water, and readily takes any form required, receiving and retaining impressions of seals, ornaments, &c. It has been made into small cups and other vessels which are not found to alter in form.

"A test I suggested myself was, would it unite with gutta-percha, and this was satisfactorily proved in my presence. A piece of the real gutta-percha, of similar size with a piece of the new substance, was softened in hot water, and united readily.

"The tests by acids on the mixed substance did not differ from those on either of the two original substances. . . .

"If the 'muddár' could be profitably grown for its hemp alone, it is evident, if this new substance proves in practice what it now appears to be, that an acre of cultivation of it would produce a large quantity of juice, and thus materially enhance its value. The poorest land suffices for its growth, but I have no doubt that if cultivated and plentifully irrigated, not only would the yield of juice be larger, but the growth of the plant, and the fineness of its fibre when made into hemp, materially increased." (*Meadows Taylor, Agri.-Horti. Society's Journal, Vol. VIII.*)

Dr. Riddell subsequently wrote :—

"As regards my experiments with the 'muddar' juice, they are as follows: Having collected about 18 fluid ounces, I had it strained through a cloth, and exposed 13½ ounces of it to solar evaporation on a flat dish. In three days it became firm, separating itself from the dish and easily removed. I then placed it in boiling water,

C. 171

GUTTA-PERCHA.

and worked it well about with a spatula, and when cool enough to handle, kneaded it with my fingers; when cool I found it to weigh a little more than six ounces. I then boiled it, and, as it cooled, worked it well again; and on weighing the substance, found it had lost one ounce. It was then pulled out into shreds and boiled a second time, kneading it whilst cooling, and four ounces, two drachms, apothecaries' weight, was obtained of what I call 'muddar' gutta-percha.

"The next experiment was with four ounces of the juice, which weighed four ounces, apothecaries' weight; and placing it in a basin, I poured about one quart of boiling water on it, stirring it up, and then leaving it to stand, when it broke into curds which fell to the bottom. I then partially poured off the fluid, and filtered the residue through paper, and on its being sufficiently dry to be removed, found it to weigh one ounce, six drachms. It was then worked well in hot water two or three times, and formed into a mass which gave six drachms, thus losing one ounce. On the whole it will be seen that the most economical method of preparing the juice is by solar evaporation, the residue being nearly double that of the second experiment."

Mr. Liotard publishes, in his "Memorandum on the materials in India suitable for the Manufacture of Paper," the opinion of **Professor Redwood** upon Madár-gutta. The Professor considers it possesses many properties in common with Gutta-percha of commerce. The specimen so reported on was collected by **Captain G. E. Hollings**, Deputy Commissioner, Shahpur (in the Panjáb) in the year 1853, little more than one year after the date of the original discovery of this gutta. We have learned nothing further for thirty years, and it is possible that a considerable wealth of fibre and gutta may have been lying unused along every roadside and over every rubbish-heap.

MADAR-ALBAN. 172

Chemistry of Madár-gutta.—§ "The chief obstacle to Calotropis gutta-percha lies in its being a good conductor of electricity, and therefore unfitted for cable purposes, otherwise it would at once assume commercial importance." (*Colonel D. G. Pitcher, Lucknow.*) **Dr. Duncan** in 1829 discovered in Madár-gutta a substance which he called *Mudarine.* This was said to have the property of coagulating by heat and becoming again fluid with cold. This statement has never been confirmed, but **Dr. Warden** published, in 1885, his discovery of a white crystalline mass closely resembling the substance named *Alban* by **Payen.** This **Dr. Warden** named *madár-alban.* A yellow resin associated with the *madár-alban* was found to agree with **Payen's** *Fluavil* as found in true gutta-percha. Speaking of these discoveries **Dr. Dymock** says: "The fact that the sap of the *madár* plant contains, in addition to caoutchouc, two principles, analogous to *alban* and *fluavil,* of gutta-percha, is a point of some interest, as *madár* gutta-percha has been recommended as a substitute for the commercial article." (*See further paragraph on Chemistry of the Drug.*)

VARNISH. 173

A Varnish-like Exudation.—Some time ago the writer observed the *ák* plants in Chutia Nagpur completely covered with multitudes of small green insects. The bushes did not look over-healthy, and (apparently as a result of the action of the insects) a gummy liquid exuded from them and trickled down to the ground below. The writer was travelling in company with **Sir Monier Williams** and one or two other gentlemen, so that this curious discovery was investigated by several persons, none of whom had ever observed the peculiarity alluded to before, although many years resident in the district. We were crossing the dry sandy basin of the Upper Barákar, and our attention was drawn to this curious fact by the ground under the bushes appearing wet. Stones were picked up but found to be quite dry, although completely varnished with the liquid falling from the bushes. The author is not aware of this varnish-like exudation having been recorded before, but unfortunately was unable to investigate its chemical nature. May it not, however, be in some way connected with the excretion of manna described by Arabian and Persian writers? (*See page 47.*)

THE DYE.

TAN. Dye. 174

Dye.—The milky sap is well known in tanning. It is made into a paste with the flour of the small millet (**Penicillaria spicata**), and is used previously to colouring the skin with lac dye. Alone it imparts a yellow colour to the skin, and destroys the offensive smell of fresh leather. **Dr. Dymock** adds that the tanners use the juice to remove the hair from the skin. For this purpose **Brandis** says it should be mixed with salt. **Irvine** refers, in passing, to the dye possessed by this plant. The natives are said to adulterate safflower with the powdered flour of the root.

THE MADÁR FIBRES.

FIBRE. Floss. 175

Fibre.—This plant, as also the next species, yields two distinct fibres—(1) a silk cotton from the seeds known commercially as "Madár floss;" and (2) a rich white bast fibre from the bark.

1. The Floss as a Textile Fibre.—The coma of hairs or floss from the seed constitutes one of the so-called "vegetable-silks" or "silk-cottons" (see under **Bombax malabaricum**). This particular silk-cotton is generally regarded as too short a staple to be spun, and is very difficult to deal with, owing to its extreme lightness; but it is soft, very white, and has a beautiful silky gloss. It is employed to some extent like the Dutch *kapok* for stuffing pillows, and is much used by the natives for the small pillows given to children, and to fever patients, having a reputation of being cool and soothing. **Brandis** says that in Borneo thread is spun from it.

Balfour remarks that "The silky down of the pods is used by natives on the Madras side in making soft cotton-like thread. It is susceptible of being spun into the finest yarn for cambric, and has been used for the manufacture of a light substitute for flannel by **Messrs. Thresher** and **Glennie** of London." "In 1856 **Major Hollings** exhibited carpets manufactured in the jail at Shahpur in the Panjáb from the follicles in the seed-pod." [This may be presumed to mean the floss of the seeds from the follicles.—*G. W.*] No efforts appear to have been made in India to improve the quality of the *madár* floss, although there would seem to be no reason why, under careful cultivation and selection, the length of the staple might not be greatly improved. In *Spons' Encyclopædia* occurs the following passage regarding this floss: "It is said to be sometimes woven into shawls and handkerchiefs, and to form a good paper-stock." The fibre, being short, was found by **Mr. Moncton** very difficult to spin, but when a mixture of one fifth of cotton was made, a good wearing cloth, capable of being washed and dyed, was produced. (*Royle.*) **Kurz,** in his *Forest Flora of British Burma,* says that strong ropes are made of this fibre. In **Mr. Liotard's** "Memorandum on Materials suitable for the Manufacture of Paper," the hope is held out that **Messrs. Thresher** and **Glennie** (a firm already alluded to as having experimented with this fibre) had at last overcome the difficulty of short staple and of lightness, and had been able to spin the floss mixed with cotton and wool. In a letter published in the *Kew Report* for 1881, they state, however, that owing to the shortness of the fibre and its extreme lightness they were forced to the conclusion that "it was practically useless." As opposed to this verdict **Mr. Hollins** recently informed the author that he had at last fairly overcome the difficulty of shortness of staple and lightness in weight. He had invented a machine which drew the floss mechanically into combination with cotton. The resulting yarn, **Mr. Hollins** states, has many advantages and peculiarities not possessed by cotton or wool alone, and he is thus now prepared to take steps to establish a large and important industry in this beautiful floss. (*See page 41.*)

The Floss as a Paper-Fibre.—Several authors refer to the possibility of using this silk-cotton as a paper-stock, but unless cultivated its collection would be far too expensive to admit of this. Although wild, no single province could supply sufficient to keep an ordinary paper-mill working for more than a few days, and indeed the yield per plant is so small as to preclude the possibility of its ever being produced even by cultivation at a price likely to find a market as a paper-fibre. It might, however, be quite otherwise with the textile industries, and were the staple improved a remunerative price might be obtained for it, as a silk substitute. There seems every reason to hope that, taking the many uses of this plant into account, its cultivation would prove profitable. The bast fibre from the stem is one of the finest, in point of quality, in India; the gutta-percha from the sap; the root-bark as a medicine; are all marketable even at the present day. FIBRE. Floss. 176

The Bark Fibre—its separation, &c.—The bast fibre, incidentally alluded to in the preceding pages, has attracted considerable attention, and is constantly spoken of as one of the best Indian fibres. The great difficulty appears to consist in the inability to rapidly and cheaply separate and clean the fibre, and in the shortness of the fibre when separated. **Royle** says: "The mode of separation of the fibre is tedious, and may for the present oppose some obstacle to the ready supply of this material. **Captain Hollings** states that the sticks of the Madár were cut about 12 or 18 inches in length; the outer bark was then carefully peeled off, and the fibre picked from the inner part of it. Several threads were then placed side by side, and twisted into a twine by rubbing them between the hands. No water is used (indeed, is injurious); everything is done by manipulation. In a subsequent paper **Captain** (afterwards in other reports **Major**) **Hollings** observes that the best plan is to select the straightest branches, which are always the largest; to let them dry for at least 24 hours before any attempt is made to separate the fibre; on the second or third day the sticks are slightly beaten, especially at the joints, which ensures the bark, with the fibre attached, being peeled off without breaking. The workmen then bite through the bark about the centre of its length; they then hold the tissue of threads in one hand and separate the bark with the other. He did not find that any of the ordinary methods of separating fibre were useful, but it is probable that some of the mechanical methods of separating flax would be effectual with this fibre when in a dry state." It is reported that the preparation of Madár fibre wears away the nails from the operator's fingers. If steeped in water, the fibre rots quickly, so that it cannot be separated by retting, but steaming may be practised and with good results. Bark. 177

Since the facts briefly epitomised above were first published (now nearly forty years ago), very little has been done to extend our knowledge of this fibre. **Mr. Strettell** has suggested that the machinery used in the separation of Agave fibre might be found to serve the purpose of preparing the Madár fibre, and it is probable that **Messrs. Death and Ellwood's** Universal Fibre Extractor or some other such modern machine might be found serviceable.

The writer is favoured by **Mr. L. Liotard** with the following note regarding his recent experiments in the separation and examination of Indian fibres, and more particularly the opinion he has now arrived at regarding *madár* bast fibre:—

"In the autumn of 1884, while testing different machines in their power of extracting the fibres of various fibre-yielding plants, I devoted attention to the *ákunda* or *madár* amongst other plants. I had already studied this shrub previously, to a certain extent, and had formed a hope-

FIBRE.

ful idea of it. But the trials just alluded to have induced me to alter considerably my previous opinion. I can now confidently state that the hopes expressed by previous writers, and by myself, that the *madár* would be one of the best fibre-producers of this country, will never be realised. Its fibre is certainly fine, strong, white, and silky, and could doubtless be extracted in a merchantable condition (though none of the machines tested by me produced any good results with it); but the obstacles to its profitable utilisation on a large scale outweigh its natural good qualities.

Without entering into many details, I may mention two of the chief obstacles:—

"(1) the very small proportion of the fibre to weight of the stems, the proportion being only 1·56 per cent; and

"(2) the shortness of the fibres, extending as they usually do from joint to joint, the joints being from 3 to 6 inches apart.

"These two chief obstacles are sufficient to justify a withdrawal of the *madár* from the list of hopeful fibre-bearing plants of India. I have been considering the fibre in connection with textiles and strings; and it follows that it would be still less suited as a material for making paper, for in the manufacture of paper a material is required which, besides possessing tenacity, fineness, and purity, has also the advantage of cheapness. *Madár*, owing to its very small proportion of fibre, and to the presence of a milk of a dangerous nature (both of which facts must necessarily raise the cost of extraction of the fibre), can never be utilised profitably as a paper material to any extent, and should, in my opinion, be considered as one of the last materials to which a paper manufacturer would have recourse."

A verdict so decisive and pronounced by a gentleman who has devoted much time to the study of Indian fibres should be gainsaid with caution, but opinions differ very considerably as to the prospects of *madár* bast fibre becoming of commercial importance. The attempts made by manufacturers hitherto would seem not to have been conducted on a sufficiently extended scale to justify the expression of strong expectations or to dispel such hopes.

The recent experiments conducted by the author in conjunction with **Mr. Cross** of Lincoln's Inn, London, have revealed the fact that by nitrating the fibre a substance, which can scarcely be distinguished from silk, may be produced. This, in the first stage of its preparation, is an admirable gun-cotton, but its explosive nature may be destroyed without injuring the beauty of the texture. Under chemical treatment the fibre behaves admirably, and with different reagents various results are obtained, but it may be concluded that the opinion we arrived at confirms the verdict already given that the mechanical difficulties are too great and the ultimate fibriles too short to justify high hopes being entertained of *madár* bast fibre becoming of any great commercial importance, although its great beauty makes one resign it with regret.

178 **Strength of Madár.**—The comparative strength of *madár* fibre has been repeatedly shown, and the following table contains the results of the experiments made by **Dr. Wight**:—

Name of the fibre.	Weight in ℔ the fibre can sustain.
The fibre of Cocos nucifera	224 ℔
„ Hibiscus cannabinus	290 „
„ Sansviera zeylanica	316 „
„ Gossypium herbaceum	346 „
„ Agave americana	362 „
„ Crotalaria juncea	407 „
„ Calotropis gigantea	552 „

Of the fibres experimented with by **Wight**, the *madár* was by far the strongest.

Madár bast fibre as a paper material.—Mr. G. W. Strettell, of the Forest Department, in his *New Source of Revenue for India*, states that the *madár* should "afford a material for paper as good as, and cheaper than, Esparto." In this opinion he is strongly supported by the Curator of the Victoria and Albert Museum, Bombay, who pronounces this as one of the finest of Indian fibres, its extended use being restricted only by the difficulty of extraction. In the *Kew Report for 1881*, however, an opinion is expressed by **Mr. Routledge** quite opposed to this; he believes that "neither it (*madár*) nor any other exogenous plant of similar character can ever compete with Esparto, nor be produced at a sufficiently low cost to admit of its being employed as paper-making material." With Esparto selling at £4 a ton, landed in London, it is hopeless to look to this (or indeed to any fibre which requires to be prepared) to ever become an article of export trade for the English paper market. It may, however, come to be of some use as an Indian paper fibre. Paper is reported to be in fact prepared from it in the following districts: Bellary in Madras, and Furruckabad and Meerut in the North-Western Provinces. [Colonel Pitcher throws doubt upon the accuracy of this last statement.] The plant is abundant in the Panjáb, and, together with the next species, is there, to a small extent, made into paper. PAPER. 179

Cultivation of the Madár Plant for its Bast Fibre and Floss.—"It thrives on soils where nothing else will grow, and needs neither culture nor water; hence it is admirably adapted for bringing waste land under tillage, and for protecting reclaimed desert from drifting sands. These reasons alone should suffice to encourage the cultivation of the plant, apart from its value as a fibre-producer. Its great abundance in a wild state may render cultivation unnecessary for a time. It is stated that an acre, stocked with plants 4 feet apart each way, will yield 10 tons of green stems, or 582 ℔ of fibre, as prepared by the present native process, which wastes 25 per cent. The cost of cultivation of the same area is placed at £2 9*s*. 8*d*., after which the only recurring expenditure would be for harvesting the plant. When raised from seed, it is said by some to require two years before being ready for cutting; but if cut close to the ground, it grows again rapidly, yielding a second crop within 12 months from the first." (*Spons' Encyclop.*) 180

Royle's account of this fibre is the most complete statement published. All subsequent authors appear to have simply reproduced **Royle's** account in other sentences, or in *their* investigations to have corroborated all that **Royle** had published years before; little or nothing has been added. Speaking of the advisability of cultivating the *madár* plant, **Royle**, quoting a communication from **Mr. Moncton**, says: "It is difficult to conceive anything less productive than dry sand, and yet the *madár* thrives on it. Should its cotton be found useful, the waste lands of India could be covered with it, as it requires no *culture* and no *water*. It comes to maturity in a year, but is perennial; when once planted or sown, it would require no further care; and where thickly planted, might be made the means of reclaiming poor soils, as the leaves and some of the upper branches rot, while the root and the stem remain. **Colonel Tremenheere**, of the Engineers, has suggested that the *madár* should be used as a hedge to protect desert land brought under cultivation from the encroachment of drift sand. This would give a healthful impetus to the cultivation of the plant itself." (*Royle, Fibrous Plants, 308.*) (*For further particulars see* SAND-BINDING PLANTS.)

Since the above was set up in proof the author has had many opportunities, in connection with the late Colonial and Indian Exhibition, held in

MEDICINE.

London, to discuss with manufacturers the prospects of *madár* floss. A Lancashire spinner stated that he had now completely overcome the difficulties offered by this floss, and was prepared to purchase any quantity. Being a wild plant, collected over a wide area, the supply is limited and irregular. The question now arises, will it pay to cultivate *madár* floss? The spinner referred to, acting upon a suggestion made to him, has placed a small sum of money in the hands of a Missionary in Chutia Nagpur, and a few acres of land are experimentally being thrown under this new crop. The report, so far, has been most encouraging, and the hope is even held out that, by careful selection of seed and a little extra attention in cultivation, the character of the floss may be changed and its length of staple improved. It is therefore confidently expected that very soon, and as a direct outcome of the Colonial and Indian Exhibition, India will commence to produce on a commercial scale this new textile fibre. **Mr. Cameron** of Mysore writes to the author that a demand has recently arisen for this floss, **Messrs. Collyer & Co.** of London offering 5*d*. a pound for it. This is nearly twice as much as was being paid during the Exhibition time for *Semul* cotton (see **Bombax**).

MEDICINAL PROPERTIES.

181 **Chemical Properties.**—Much difference of opinion still prevails regarding the relative medicinal values of **C. gigantea** and **C. procera.** **Dr. Wight** and with him the majority of authors have decided in favour of the latter, but all agree that the difference consists only in degree. The active principle seems to reside in a peculiar bitter principle, but no alkaloid occurs in the drug. The able authors of the *Pharmacographia* carefully re-performed **Dr. Duncan's** experiments (published in *The Edinburgh Med. and Surg. Journ., July 1st, 1829*), but failed to discover his *Mudarine.* From 200 grammes of the powdered bark they obtained 2·4 grammes of an acid resin, soluble in ether as well as alcohol. On separating the crude resin from the aqueous liquid and adding much absolute alcohol, an abundant precipitate of mucilage was obtained. The liquid was then found to contain the bitter principle alluded to above, which may be separated by means of tannic acid. This was further treated with carbonate of lead, dried and boiled with spirits of wine; thereafter on evaporation an amorphous bitter mass was obtained, not soluble in water, but nearly so in absolute alcohol. By purification with chloroform or ether, this substance was at last produced colourless. Since the date of these experiments, **Drs. Warden** and **Waddell**, of Calcutta, have carried our knowledge of the chemistry of this subject still further by the discovery (*as suggested by Flückiger and Hanbury*) of a principle which has the properties of **List's** *Asclepione.* (*Gmelin's Chemistry, XVII., 368.*) Further investigations, however, conducted by **Dr. Warden**, have revealed that this was not quite the case, since the substance he has now found would be more correctly represented by the formula $C_{17}H_{28}O$, whereas **List's** *Asclepione* would be represented by $C_{20}H_{34}O_3$. (*Pharm. Jour., 1885.*)

Bark. 182 **Therapeutic properties of the Bark** are described as "alterative tonic, diaphoretic, and in large doses emetic. The drug is prescribed in leprosy, constitutional syphilis, mercurial cachexia, syphilitic and idiopathic ulcerations, in dysentery, diarrhœa, and chronic rheumatism." (*Pharm. Ind.*)

Moodeen Sheriff says: "Almost every part of this plant is used in medicine, but the bark of the root and dry milky juice are by far the best." Considerable difference of opinion prevails as to the relative value of the bark and of the juice. The *Indian Pharmacopœia* accepts the bark; Hindú writers seem to prefer the root-bark, and Muhammadans the juice. The *Pharmacopœia* further directs that the roots should be collected in

Madár as a Drug.

April and May from plants grown in sandy soil; after carefully washing, to remove all earth and sand, they should be allowed to slowly dry in the shade until the sap no more flows from incisions made in the bark. The bark should then be carefully removed, dried, and reduced to a powder, and preserved in well-corked bottles. **Moodeen Sheriff** adds that the roots from old are superior to those from young plants.

MEDICINE. Root-bark. 183

"The root-bark is said to promote the secretions, and to be useful in skin diseases, enlargements of the abdominal viscera, intestinal worms, cough, ascites, anasarca, &c. The milky juice is regarded as a drastic purgative and caustic, and is generally used as such in combination with the milky juice of **Euphorbia nerlifolia.** The flowers are considered digestive, stomachic, tonic, and useful in cough, asthma, catarrh, and loss of appetite." "The leaves mixed with rock-salt are roasted in closed vessels, so that the fumes may not escape. The ashes thus produced are given with whey in ascites and enlargements of the abdominal viscera." "The root-bark, reduced to a paste, with sour *congí*, is applied to elephantiasis of the legs and scrotum. The milky juice of this plant and of **Euphorbia nerlifolia** are made into '*tents*' with the powdered wood of **Berberis asiatica,** and introduced into sinuses and fistula in ano." (*U. C. Dutt, Mat. Med. of the Hindús.*)

Milky-juice. 184

Flowers. 185

Leaves. 186

According to **Dr. Casanora,** *madár* stimulates the capillaries and acts powerfully on the skin, and is accordingly recommended as a remedy in the obstinate cutaneous diseases of tropical climates, such as elephantiasis and leprosy.

The *Pharmacopœia of India* gives an interesting *résumé* of the medical opinions held regarding this drug up to 1868. Its *emetic* properties are testified to by **Drs. Duncan, O'Shaughnessy, Bonavia, Hutchinson, Durand, Stewart, Newton,** and **Ross.** Its value in the treatment of *leprosy* by **Drs. Robinson, Playfair, Ross, Ainslie, Rogers,** and **Irvine.** Its efficacy in *syphilitic affections* by **Dr. Casanora,** and in *dysentery* by **Dr. Durand.** In another paragraph will be found a most interesting series of medical opinions which have been specially communicated for this work, and which bring it abreast of the most recent researches with the properties and uses of *madár*.

Properties of the Juice or Milky Sap.—Ainslie, Modeen Sheriff, and most other authors regard the juice as more powerful than the bark, but less valuable, owing to its being irregular in its action. **Dymock** says "the juice is described as a caustic, a purge for phlegm, depilatory, and the most acrid of all milky juices." (Compare this with the remarks further on, under the heading '*an alcoholic liquor said to be prepared from this sap.*') Medicinally it is recommended for skin diseases, ringworm of the scalp, and to destroy piles. Mixed with honey, it is viewed as useful in the treatment of aphtha of the mouth, and a piece of cotton-wool dipped in the juice and inserted into a hollow tooth is reported to cure toothache. "**Hakim Mir Abdul Hamed,** in his Commentary upon the *Tuhfat,* strongly recommends it in leprosy, hepatic and splenic enlargement, dropsy and worms. A peculiar method of administration is to steep different kinds of grain in the milk and then administer them. The milk itself is a favourite application to painful joints, swellings, &c.; the fresh leaves also, slightly roasted, are used for the same purpose. Oil in which the leaves have been boiled is applied to paralysed parts; a powder of the dried leaves is dusted upon wounds to destroy excessive granulation and promote healthy action." (*Dymock.*)

Roxburgh in his *Flora of India* gives the following account of the medicinal properties of this plant, from which it will be seen that nearly a century ago its properties were as well known to Europeans as they are

MEDICINE.

at the present day: "A large quantity of an acrid milky juice flows from wounds made in every part of these shrubs; the natives apply it to various medicinal purposes; besides which, they employ the plant itself and the preparations thereof to cure all kinds of fits: Epilepsy, Hysterics, Convulsions from Coitu immediately after bathing; also Spasmodic disorders, such as the lock-jaw, Convulsions in children, Paralytical complaints, cold sweat, poisonous bites, and venereal complaints."

SPECIAL OPINIONS.—The writer is unable to publish more than a very limited selection from the numerous opinions which he has been favoured with regarding this drug. The plant in some form is employed medicinally in every province, and is indeed one of the most extensively used drugs in India.

§ "The medicinal properties of **Calatropis gigantea** have been known to the natives of this country from the earliest period, and it is held in great esteem by the Hindú practitioners in the treatment of some venereal and skin diseases,—so much so, that it is called by some of them 'the vegetable mercury.' There are two varieties of this plant in Southern India,—one with blue or bluish-purple flowers, and the other with cream-white. Almost all the parts of **Calotropis gigantea** are used in medicine, but the dry milky juice, fresh flowers, and the root-bark are by far the best and most useful. In whatever way the milky juice is collected and dried, its smell and taste are the same, *viz.*, nauseous and unpleasant, but its colour and external appearance differ to a slight extent according to the method adopted for its collection. If it is collected in shallow earthen plates and dried under shade (which is the best way for the purpose), it is formed into thin layers, which, when quite dry, can be easily separated from the plates more or less entirely with a bolus knife, and are very brittle. The colour of these layers is grey or pale brown, but if the juice is collected and dried in a cup or deep vessel, it assumes the shape of the latter, and its colour is much deeper externally and paler internally. The dry juice is insoluble in water (hot or cold) and in alcohol, but soluble in ether, honey, and some fixed oils. Although the dry layers of the juice are brittle, they are converted into a soft mass when rubbed or bruised for the purpose of reducing them into powder; they also become soft when exposed to the heat of the sun; the dry juice, therefore, can only be administered in the form of pills.

"The drug is an irritant poison in large quantities, and produces severe vomiting or purging, and sometimes both, in more than four-grain doses; in medicinal doses (one to three grains, gradually increased), however, it is an efficient antispasmodic, alterative, and nervine tonic. It is a very useful therapeutic agent in some cases of asthma, epilepsy, and paralysis.

"Used in equal proportion with black pepper, the fresh flowers of **Calatropis gigantea** have lately been found to be a very useful and cheap remedy for asthma, hysteria, and epilepsy. They are antispasmodic, alterative, and nervine tonic, like the dry juice of the same plant, and their usefulness in the first-named disease deserves particular attention. They should be formed into pills, five grains each, well dried and kept in a corked bottle. One pill is to be used every fourth or sixth hour, according to the urgency of symptoms. For medicinal purposes the root-bark of **Calatropis gigantea** should be selected from as old plants as possible, and in the hot or dry weather. The bark should not be removed as soon as the root is dug out, but about 24 hours afterwards. The thick, rough, and corky epidermis with which the bark is covered is quite inert, and should be scraped off with a knife before the root-bark is reduced to powder. The powder thus prepared is white and bears a great resemblance to rice-flour. It has a nauseous and slightly acrid smell, and a bitterish taste. It should be preserved in a corked bottle. As an emetic and expectorant, the powder is

MEDICINE.

one of the best substitutes for ipecacuanha in this country, and has been found useful in many of the diseases for which the latter is indicated, including dysentery. As an alterative tonic, it has a beneficial influence over some forms of secondary syphilis, and is also temporarily useful in some recent cases of leprosy and a few other skin diseases. As a diaphoretic it often relieves pyrexia by producing perspiration in cases of simple and slight fevers." (*Honorary Surgeon Modeen Sheriff, Khan Bahadur, Triplicane, Madras.*)

The following abstract from a detailed account of the use of *madár* juice in the cure of snake-bite may be found interesting; this is the only instance, in a very extensive series of Medical opinions, in which *madár* is recommended for this purpose:—

"Collect the white tomentum from the leaves, and make a paste with this and the milky juice, from which prepare pills about the size of a common marble. This will be found to be a good remedy for snake-bite. Roll a pill in a betel-leaf and ask the bitten person to chew and swallow it. Repeat the dose every half hour till numbness begins, when the dose may be lessened and given every hour. In no case does it require more than nine doses to effect a cure. If the bitten person be unconscious and not able to swallow, the pill may be dissolved in water and given in that form. When convalescence has set in give a purgative, and continue with light food and no stimulants for a few days. If the patient vomits after taking the pill, recovery will be rapid." (*V. Ummegudien, Native Doctor, Mettapolium, Madras.*)

"It is a common sight in Oudh, of a morning, to see the people collecting the sap to be placed on a sore or skin disease." (*Colonel D. G. Pitcher, Lucknow.*) [This may be seen all over India, but it is a remarkable fact, at the same time, that if placed on an open cut on the skin it causes great burning and produces a bad sore.—*G. Watt.*

"The fresh juice is used with common salt in bruises and sprains, and the fresh leaves warmed are used as poultices in rheumatism, gout, and rheumatic anthritis, to relieve painful joints. The juice is an irritant, and in large quantities an irritant poison." (*Brigade Surgeon J. H. Thornton, Monghyr.*) "The dried bark mav be considered a substitute for ipecacuanha, and used as such, but it is very inferior to that invaluable drug." (*Brigade Surgeon S. M. Shircore, Moorshedabad.*) "The juice or milk of the plant is used as a rubefacient. In doses of from 5 to 10 grains with $\frac{1}{8}$ grain of opium given twice or three times a day, it proves as efficient as ipecacuanha in cases of dysentery. It produces great heat in the stomach, but is less liable than ipecacuanha to produce vomiting." (*Assistant Surgeon Jaswant Rai, Mooltan.*) "I have used powder of *madár* (or *Pulvis Calotropis*) in 10-grain doses three times a day, suspended in water with mucilage, in cases of chronic rheumatism, with considerable success." (*Surgeon-Major H. W. E. Catham, Ahmednagar.*) "Powdered root-bark has been much used in doses of from 5 to 10 grains in dysentery with success. Natives apply the milky juice of the stem and leaves to loosen foreign bodies on the skin, and also to make imputed bruises." (*Assistant Surgeon Bhugwan Dass, Rawal Pindi, Panjáb.*) "Powdered madár root 12 gr., black pepper 12 seeds, powdered together, are found useful in jaundice; they are given morning and evening. Also half a seer of whey of milk with half an ounce of *khari nimuk* (sodium carbonate). The patient will be cured in a week." (*Native Doctor Mir Comer Ali, Bhagnipur, Etawah.*) "A decoction of the root is used by the Santals in infantile convulsions, and also in wandering of the mind during fever. The leaves are also applied externally to relieve pain, being for that purpose kept warm by hot, dry applications." (*Rev. A. Campbell, Santal Mission, Chutia Nagpur.*)

C. 186

MEDICINE.

"A valuable remedy with similar effects to ipecacuanha, but not so good. The bark of the root should be gathered in April. The dried juice is also of value." (*Assistant Surgeon Nepal Singh, Saharanpore.*) "Fresh leaves and juice used in guinea-worm as local application. Given externally, produces dryness of the throat and running from the eyes, nose, &c." (*Dr. Darasha Hormarji Baria, L.M.S., Bombay.*) "Fifteen grains of the powdered root-bark, combined with a grain of opium, successfully used in acute dysentery. Milky juice from the flowering tops cures scabies rapidly." (*Assistant Surgeon Shib Chunder Bhuttacharji, Chanda, Central Provinces.*) "An excellent substitute for ipecacuanha in cases of dysentery. Much used in dispensary practice, diaphoretic in small doses (5 grains), emetic in large doses (30 grains)." (*Surgeon G. Price, Shahabad.*) "The juice of the ripe leaves is dropped into the ear for otorrhœa. The flowers are expectorant and are commonly used in the form of pill for bronchitis, phthisis, &c. The juice is also applied as a rubefacient for rheumatic pains. The leaves are used as a poultice for pain and constipation." (*Surgeon-Major J. Robb, Ahmedabad.*) "A valuable remedy in acute dysentery, and an efficient substitute for ipecacuanha. It has been known to effect a cure when ipecacuanha has failed." (*Surgeon R. D. Murray, Burdwan.*) "Alterative, emetic; dose 1 to 3 grains; used in leprosy, secondary syphilis, chronic diarrhœa and dysentery, rheumatism, intestinal worms, mercurial cachexia, bronchitis, elephantiasis." (*Hospital Assistant Choona Lall, Jubbulpore.*) "The dried and powdered pistils and stamens, in doses of 2 to 3 grains repeated hourly, useful in cholera. The vomiting is checked or moderated. The leaves are used as applications to rheumatic pains." (*Narain Misr, Hoshangabad, Central Provinces.*) "The powdered root-bark, smoked like tobacco, is used by native physicians in syphilis. The flower-buds, in doses of 5 grains, combined with black pepper and salt, are useful in dyspepsia with palpitation, and in cholera. In the latter disease they are used to check vomiting. The leaves are used as a local application in rheumatic affections." (*Hospital Assistant Lal Mahomed, Hoshangabad, Central Provinces.*) "The bark is said to be useful for chronic rheumatism, but I did not find it to be so." (*Surgeon-Major H. J. Hazlitt, Salem, Madras.*) "Mixed with pepper the leaves are used in Mysore for cleaning the teeth. The milky juice is also used with salt to allay toothache." (*J. Cameron, Mysore.*) "*Madár* leaves are very useful in relieving pain and swelling due to the presence of guinea-worm, and also in other inflammatory swellings. The leaves are smeared with sweet-oil and then heated by holding near a fire, and applied one over the other until a dozen or more have been placed on the affected part." (*Surgeon G. G. Ward, Mhow.*) "Is called '*Jilledo ochettú*' in Telugu. This is one of the articles used by natives to procure abortion. This is effected by brushing the mouth of the womb through the vagina with the milk or juice of the plant. Root-bark in powder or infusion or decoction is useful as an emmenagogue." (*Surgeon-Major E. W. Levinge, Rajamundry, Godavery District.*) "The powdered root-bark is much employed in the hospital in all obstinate forms of skin diseases and leprosy. It is a useful alterative; as an emetic also it acts well. In skin diseases it has been used in combination with **Hydrocotyle asiatica**, and, in cases attended with anæmia, iron preparations, such as the sulphate or peroxyde. Dose, &c., *vide* 'Indian Pharmacopœia,' page 141." (*Apothecary J. G. Ashworth, Kumbakonam, Madras.*) "The dried bark of the root is a good substitute for ipecacuanha. The dried flowers are used in Mysore, in from 1 to 2 grain doses, along with sugar, in leprosy, secondary syphilis, and in gonorrhœa, with milk diet." (*Surgeon-Major John North, Bangalore.*) "The leaves, smeared with castor oil and heated, are applied to the scrotum in epididymitis." (*Surgeon James McCloghry, Poona.*) "The green leaves, tied in bundles and cut into halves, are used as a fomentation by heating the cut ends in a pot in which castor oil has been warmed; useful in rheumatic affections, and largely used by the natives." (*Honorary Surgeon P. Kinsley, Chicacole, Ganjam, Madras.*) "The freshly-pounded root-bark is used by natives as an alterative, and the milky juice as a vesicant in rheumatism. In abscess of foot, the natives heat a brick and place half a dozen leaves over the heated brick and warm the foot on it." (*Honorary Surgeon Easton Alfred Morris, Negapatam, Tranquibar.*) "Externally used as a blister. The leaves made into a decoction are used for rheumatism." (*Surgeon-Major D. R. Thomson, Madras.*) "Decoction of the root or bark is useful as an antiperiodic. The flowers, made into poultice, relieve pain in the heels." (*Surgeon John Lancaster, Chittore.*) MEDICINE.

It is probable that the above special medical opinions refer to both this and the next species. As already stated, the majority of writers agree in regarding both as of equal medicinal merit. The reader is referred to **C. procera** under "Medicine" for one or two special opinions regarding that species.

MADÁR LIQUOR AND MANNA.

MANNA. 187 **Food and Liquor.**—The *Ak* is said by the Arabs and Persians to yield a sugar or manna: this fact is briefly alluded to by **Royle** (*Him. Bot., 275*) and by **Birdwood**, but definite information regarding this property does not appear to have been published. It may be doubted, if indeed produced from **Calotropis** in Persia, whether this excretion occurs in India at all. There are other instances of a plant producing a product in one country which it fails to do in another; witness **Cannabis sativa** for example. The manna said to be obtained from this plant is known in the bazars as *Sakkur-el-ushar*, and is said to be produced through the parasitic action of **Larinus ursus.**

§ "Most of the Arabian writers agree in describing a sugar or honey-dew which is produced upon the plant, probably by an Aphis as suggested by **Dr. Watt's** observation in Chutia Nagpur. The different kinds of Larinus build nests or cocoons (on various species of **Echinops**) which contain sugar, *e.g.*, the Persian *Shakar-i-tíghál*, for a description of which (with figures) see *Hanbury's Science Papers.*" (*Dr. W. Dymock, Bombay.*) (*Compare with the account at page 37 of the varnish-like juice alluded to by Dr. Dymock.*)

LIQUOR. 188 An intoxicating liquor is by some authors said to be prepared from the juice of this plant, and by others the juice would appear to be used as an auxiliary only in the fermentation of an alcoholic-producing substance. Speaking on this subject, **Sir G. Birdwood** (*Bomb. Prod., 208*) says: "The intoxicating liquor *Bar* is prepared by the tribes of the Western Ghâts. It is the last plant in the world from which an intoxicating spirit might be expected; and **Barth** also states of the tribes of the Tagamah, that they ferment their *Giya* with its milk-sap."

Mr. Lisboa (*Useful Plants of Bombay*), on the other hand, says: "**Barth** states that the pagan tribes of Central Africa also prepare from this plant their *giya.*" At present, therefore, it is impossible to gather whether the juice itself, by fermentation, directly affords an intoxicating liquor, or is only used as a ferment like yeast, or as a bitter adjunct similar to hops and Acacia bark. It is remarkable that, for whatever purpose it may be used, the practice of employing *ák* juice (or *bar*) in the preparation of intoxicating liquors should be known to the inhabitants of Western India in common with certain African tribes, but apparently be un-

LIQUOR.

known to the people on the eastern side of the peninsula. This would seem to point to the probability of the practice having reached India from Africa, and so, as far as India is concerned, and from a historic point of view, it would be of comparatively little interest. It should be remembered, however, that the sacred *Soma* of the ancient Sanskrit writers has by many botanists been associated with a species of **Sarcostemma**, a genus belonging to the same tribe of Swallow-worts, and not very far removed from **Calotropis.** We have abundant evidence of the antiquity of the practice of using auxiliaries in the indigenous processes of distillation (and *madár* juice may prove of this nature), but the subject of the careful investigation of the *bar* and *giya* liquors is commended to the attention of officers who may have the opportunity of removing the ambiguity indicated above.

TIMBER.

TIMBER. 189

Structure of the Wood.—The plant rarely produces wood of any size; it is, however, valued for making charcoal, and is employed as gunpowder charcoal in Kathiawar and in the Deccan. (*C. P. Gaz., 504.*) It is also made into gunpowder charcoal in the Godaveri District.

DOMESTIC AND SACRED USES.

DOMESTIC. 190

Domestic and Sacred Uses.—MANURE.—"The leaves and stalks serve for reclaiming *reh* (covered with saline efflorescence) lands. These leaves are strewn about the ground and covered with earth, and then crushed by being stamped upon. Water is then let on the land enough to flood it. When the water subsides, the crushing is repeated, and the land again inundated. The decomposition of the leaves somehow or other 'kills the salt' as the natives say. In fact, land that was thus treated for two successive years became so free from saline matter as to vield a very fair crop." (*Lisboa, Us. Pl., Bomb., 233.*) "In Mysore the branches are much sought after as a manure for paddy-fields. It is estimated that wet land thus manured will yield a much superior crop." (*J. Cameron, Mysore.*) The leaves and twigs are used in Madras to manure the fields. (*Indian Forester, IX., 35.*) **Col. Pitcher** writes to the author that he has chemically examined the *madár* leaves in order to discover whether or not they possess any substance which might have an effect on *reh* soil. He has arrived at what seems a correct opinion, that they can have no special virtue over straw or any other green manure. Indeed, land would appear to be, in the Lucknow District, reclaimed by rice-straw manure; but the opinion that *ák* leaves have a special merit for this purpose seems a general one.

"The flowers are used in the worship of *Mahadeo* and *Hanuman.*" (*Bomb. Gaz., VII., 42.*) In Bengal the sections of the bluish corona of the flower are carefully picked from the corolla and strung into garlands which are worn at certain religious ceremonies. The following extracts from **Mr. Lisboa's** *Useful Plants of Bombay* will be found instructive:—

"In *Chaturmás Máhatmá*, Chapter XX., in the narration of *Gallava Rushi*, taken from *Skand Purán*, this tree is mentioned to be the transformation of *Surya*, or the Sun. It is used in various ceremonies, both religious and those of time-hallowed custom. The leaves are used as *patri*, in the same way as those of *shemi*, in the worship of *Ganpatti*, *Haritáliká*, *Pithori*, &c. They are also employed in *shusti pujan* (a ceremony performed on the sixth day after confinement for propitiating *Jewti*, the goddess of Destiny) by females. When a Hindú is to marry a third time, it is believed that the third wife will soon die: in order to avoid such a calamity, the man is first married to this tree, which is then cut down. This ceremony is believed to ensure the longevity of the fourth, but really the third wife whom he now marries.

"It is ordered in the *Shrávan Máhátma* to worship *Máruti* (who is also known as *Hanuman*), or the Monkey-god, on every Saturday, with a garland of the flowers of this tree, which are then offered to him. The twigs are also ordered to be used as substitutes for tooth-brushes in the *Smritisar Granth.* They are also employed as *Samidhas* for the feeding of sacred fires, as mentioned before."

Mir Muhammad Husain gives a good description of this plant, and mentions the fact that the wandering Arabs and Tartars make their *Makhad* twist or *Yálish* tinder from the seed floss.

Calotropis procera, *R. Br.; Fl. Br. Ind., IV., 18; Wight, Ic., t. 1278.* 191

Syn.—C. HAMILTONII, *Wall.*

Vern.—*Safed-ák, ák, ág, madár, ákadá,* HIND.; *Ák, shakar-ul-úshar, shakar-al-lighal,* PB.; *Spulmei, spalmak, pashkand,* TRANS-INDUS; *Ák,* SIND; *Mándára,* MAR.; *Alarka,* SANS.; *Vellerku,* TAM.; *Ma-yo-pin, mehobin,* BURM.; *Spalwakka,* AFG.

Moodeen Sheriff, as well as **U. C. Dutt,** gives the same vernacular names for both the species of **Calotropis.**

References.—*Brandis, For. Fl., 331; Kurz, For. Fl., Burm., II., 200; Gamble, Man. Timb., 265; Dalz. & Gibs., Bomb. Fl., 149; Stewart, Pb. Pl., 144; Aitchison, Cat. Pb. Pl., 90; Voigt, Hort. Sub. Cal., 540; Pharm. Ind., 141, 457, 458; Moodeen Sheriff, Supp. Pharm. Ind., 82; Flück. & Hanb., Pharmacog., 424; Bentl. & Trim., Med. Pl., t. 176; U. C. Dutt, Mat. Med., Hind., 196; Dymock, Mat. Med., W. Ind., 2nd Ed., 512; O'Shaughnessy, Beng. Dispens., 454; Waring, Bazar Med., 92; S. Arjun, Bomb. Drugs, 84; Drury, Us. Pl., 102; Baden Powell, Pb. Pr., 361, 408, 472, 500, 501, 571; Royle, Fib. Pl., 306, 308; Atkinson, Him. Dist., 730, 788, 794; Lisboa, Us. Pl., Bomb., 233, 260; Balfour, Cyclop.; Smith, Dict., 278, 431; Treasury of Botany; Kew Official Guide to the Museum, p. 97.*

Habitat.—A shrub found in the drier parts of India, chiefly in the Sub-Himálayan tract, from the Indus to Jhelum; Oudh, Central India, and the Deccan; also in Burma and distributed to Persia and tropical Africa.

Gum.—As in preceding species. GUM. Gutta-percha. 192

Medicine.—As under **Calotropis gigantea.** Root of this species specially mentioned as used by the Pathans for tooth-brush, having the merit of curing toothache at the same time. The same opinion prevails in Mysor and other parts of India with regard to preceding species. **Dr. Hove,** in his account of Bombay, alludes to **Calotropis** as a native cure for guinea-worm, which he says is not so effectual as the juice of the aloe. MEDICINE. Root. 193 Milk. 194

Special Opinions.—§ "The fresh milk is employed in the Panjáb for the purposes of infanticide. [The mouth of the uterus is brushed with fresh twigs of the plant in other parts of India.—ED.] In a drachm dose the fresh juice will kill a large dog in 15 minutes; its action, though slower, resembles that of hydrocyanic acid, but commences with foaming at the mouth" (*Brigade Surgeon J. E. T. Aitchison, Simla*). "The juice is first rubbed on the skin, and subsequently ashes are put to darken the patch, and make it look like echymoris or bruise" (*Assistant Surgeon Bhugwan Dass, Rawal Pindi, Panjab*). "The flowers are used in cases of cholera." (*Surgeon-Major D. R. Thomson, Madras*). Flowers. 195

Fodder.—Used as a camel fodder (*Sind Gaz., 522*). According to **Dr. Stocks,** in his *Plants of Sind* (*Records of the Govt. Bombay, XVII., 606*), one of the four plants which the camel will not eat (*See* **Camel Fodder**). FODDER. 196

Domestic Uses.—In Oudh this species is regarded as an ill-favoured weed, notwithstanding its usefulness. DOMESTIC. Tooth-brushes. 197

CALTHA, *Linn.; Gen. Pl., I., 6.*

198 **Caltha palustris,** *Linn.; Fl. Br. Ind., I., 21;* RANUNCULACEÆ.

THE MARSH MARIGOLD.

Vern.—*Mamiri, baringú,* PB.

References.—*Stewart's Pb. Pl., 2; Smith's Dic., 268; Treasury of Botany.*

Habitat.—Marshes of the western temperate Himálaya, from Kashmír to Nepal; altitude 8,000 to 10,000 feet.

MEDICINE. 199 **Medicine.**—In Hazara the root is considered poisonous.

Caltrops terrestrial, see **Tribulus terrestris**; aquatic, see **Trapa bispinosa.**

Calumba Root, see **Jateorhiza palmata,** *Miers.;* MENISPERMACEÆ.

CALYCOPTERIS, *Lam.; Gen. Pl., I., 686.*

200 **Calycopteris floribunda,** *Lamk.; Fl. Br. Ind., II., 449; Roxb., Cor. Pl., t. 87;* COMBRETACEÆ.

Syn.—GETONIA FLORIBUNDA, *Roxb., Fl. Ind., Ed. C. B. C., 379.*

Vern.—*Kohoranj,* C. P.; *Ukshi,* MAR.; *Bandi murududu,* TEL.; *Marsada boli,* MYSOR.

References.—*Brandis, For. Fl., 220; Kurz, For. Fl., Burm., I., 468; Gamble, Man. Timb., 185; Dalz. & Gibs., Bomb. Fl., 91.*

Habitat.—A large, climbing shrub of Central and Southern India, and from Assam to Singapore. Found from plains up to 2,500 feet above sea.

MEDICINE. 201 **Medicine.**—Young twigs when cut give out watery fluid used medicinally.

TIMBER. 202 **Structure of the Wood.**—Yellowish white, moderately hard, tough, with numerous broad medullary patches of soft, pith-like texture. Used for making tool-handles.

Calysaccion longifolium, *Wight; Ill., I., 130, & Icon., t. 1999;* see **Ochrocarpus longifolius,** *Benth. & Hook. f.;* GUTTIFERÆ.

Calyptranthes, see **Eugenia.**

THE CAMEL.

203 **Camelus,** *Linn.*

This genus of Ruminant Mammals contains only two species—both Asiatic—the Dromedary and the Bactrian Camel. It is by zoologists referred to the CAMELIDÆ, a family represented in the New World by the LLAMA and ALPACA and two or three nearly allied animals, all belonging to the genus AUCHENIA. The CAMELIDÆ, in many respects, constitute an aberrant group of the RUMINANTIA, especially in their dentition and in the peculiarities of their feet. The two hinder supplementary toes, present in most ruminants, are altogether absent in the CAMELIDÆ. The soles of the feet are covered by a callous, horny, integument, connecting the two toes upon which the animal walks. They have no horns, and the nostril can be closed at will, a provision against the dust-storms of the regions in which camels are chiefly found.

TWO SPECIES. **Habitat.**—The two species—**Camelus dromedarius,** or the Arabian one-humped Camel, and **C. backtrianus,** or the Persian two-humped Camel—while very unlike each other are so nearly allied that the hybrid between them is fertile. But the progeny from the mule is said to be more unmanageable than the mule itself, and is accordingly very little bred. The Indian camel belongs to the former species, and, indeed, **C. dromedarius** is, by some writers, spoken of as the Indian, African, and Arabian camel. It is certainly found in all these countries, but in a state of domestication only, and seems to have accompanied the Muhammadans in their conquests, having been even taken to Spain by the Moors and to Pisa by one of the Dukes of Tuscany. More recently, the one-humped camel has been introduced to Australia and to New York, where it appears to have taken kindly to the sandy expanses of Nevada, a region in which thorny bushes abound, similar to those on which it browses in India and other warm countries. The Bactrian camel, on the other hand, requires a colder climate than the dromedary. The Russian Asiatic explorer, **Colonel Prejevalsky**, would appear to have discovered this animal east-south-east of Lob-nor, in what some authors deem a truly wild or indigenous condition, and others a naturalised state—an escape from domestication. Whichever view may be taken of this question, the Bactrian camel, in its domestication, is distributed from the point fixed by **Prejevalsky** as its indigenous habitat, throughout the region north and east of that inhabited by the dromedary, or from the Black Sea to China and northward to Lake Baikal. In Central Asia both species are found, as also the hybrid between them. There are numerous recognised breeds of both species, and there are even dromedaries so acclimatised to alpine rocky regions, that they are prized as beasts of burden by the inhabitants of such countries. The Bactrian camel is smaller than the dromedary, has longer, darker, and more plentiful hair, and the pads of its feet are much harder (an adaptation doubtless to the rocky region it inhabits) than those of the Arabian camel. **Palgrave**, however, informs us that dark-coloured or even black camels exist in Arabia, and that the term dromedary should be restricted to the pale-coloured, more elegantly-formed breed, which might be designated as the high-blooded race-horse of his species. According to some writers the camel is one of the oldest mammals now living, since fossil remains have been found in the Siwaliks of a species, which, but for its being a little larger than the Arabian camel, is scarcely distinguishable from it. How far this fact may be accepted as throwing light upon the original home of the animal is a matter of speculation. The Siwalik mountains, which skirt the foot of the Panjáb Himálaya, have now been satisfactorily established as belonging to the *pliocene* period of Geologists, although many earlier or *miocene* forms seem to have survived in the Siwalik pliocenes, just as many animal forms of the latter, including the camel, have continued to the present day. Thus wild camels may be accepted as having once upon a time existed in what is now Northern India, or in the region south of the present Himálayas; but at the present day the animal only occurs there in a state of domestication and need not by any means be the actual descendant of the Siwalik camel. It is remarkable, however, that no one has ever seen the one-humped camel in a wild state, and unless we are to accept the somewhat extreme view that they may after all be but varieties of one species (hence producing a fertile hybrid or cross-breed) **Prejevalsky's** home of the two-humped camel need have no bearing on the question of the nativity of the so-called Arabian camel.

A FOSSIL CAMEL. 204

Colonel Yule, in his most instructive "*Introductory Remarks*" to **Prejevalsky's** *Mongolia,* gives a valuable summary of the various references by authors to the wild camel. He says: "This is a somewhat interesting subject; for disbelief in the existence of the Wild Camel has been strongly expressed, and indeed not long since, by one of the greatest of scholars as well as geographical authorities on Central Asia. It is worth while, therefore, to observe that its existence by no means rests on the rumour heard by **Prejevalsky**. There is much other evidence; none of it, perhaps,

WILD CAMELS. 205

very strong taken alone, but altogether forming a body of testimony which I have long regarded, even without recent additions, as irresistible."

Since **Col. Yule** published the above opinion, **Prejevalsky** has himself shot the so-called wild camel near Lob-nor, so that it is now very generally accepted that a wild two-humped small and very woolly camel does exist in the region referred to.

Vernacular Names.—*Chameau*, FR.; *Kameel*, GER.; *Kamelos*, GR.; *Camello*, IT. and SP.; *Camelus*, LATIN; *Únt*, or *út*, HIND.; *Jamal* or *gamal*, ARAB.; *Ottagam*, TAM.; *Loti-pitta* or *wonte*, TEL.; *Unta*, MALAY.

The Names given to the Camel.

In most parts of India there are different names given to the camel according to its age and sex: thus it is *bota* (male) or *botí* (female) until it can carry a burden; then it is *únt* (male) and *úntní* (female). In Jhang after 8 years of age it is *armosh* or *út* (male) and *jharot* (female). From birth up to 8 years of age it receives the following names:—

	To 1 year.	To 2 years.	To 3 years.	To 4 years.	To 5 years.	To 6 years.	To 7 years.	To 8 years.
Male .	*Toda.*	*Mazat.*	{ *Trihun,* *Lihak.*	*Chhatr.*	*Doak.*	*Chhiga.*	*Nesh.*	*Nesh.*
Female .	*Todi.*	*Mazat.*	*Puraf.*	*Lihari.*	*Trokar.*		*Kuteli.*	

M. Kostenko tells us that in Turkistan the two-humped camel is called *tuya* and the one-humped *nar-tuya*.

References.—The following authors may be consulted:—Wellsted; Chesney; Stewart; Huc (Recollections of a Journey); Ferrier; Mignan; Pottinger (Beluchistan); Fontain (Egypt); Robinson; Postan; Kostenko; Prejevalsky; Palgrave, and the writings of many other travellers.

Breeds.

206

BREEDS AND RACES OF CAMELS.

This subject has already been alluded to while discussing the subject of the habitat of the camel. **Veterinary Surgeon Charles Steel**, in a paper read before the *United Service Institute of India*, on the camels employed in South Afghanistan during the expedition of 1878-79, states that of the "breeds of camels in India, the variety does not appear to be extensive; Rajputana supplies a great many, and from that district were derived those which were used during the siege of Delhi; our camels in South Afghanistan were almost all Sind, amongst which was a very small proportion of females, whereas, with the northern army, they are reported to have abounded; we had a small number of *Pahari* or hill camels, and a few specimens of the magnificent Persian.

"Some from Sind were very fine and powerful, distinguished by their height, length of leg, and paucity of hair, amounting in some instances to nudity, the disproportion in strength of fore and hind extremities being very remarkable, and their susceptibility to climatic changes very great; the *Pahari* is much more freely supplied with hair, is of lower stature as a rule, shorter in leg and of more proportionate development posteriorly; these certainly suffered much less from any cause, and I had no opportunity of making a *post mortem* examination.

"The Persian possesses a thick coat, splendid capillary appendages, especially about the neck, which has a deep and graceful curve; he has also a wide chest, and short legs, but I am sorry to observe that, as the climate increased in temperature, the ornamental hair began to fall off in patches, presenting a mangy appearance; this would probably be restored on the return of cold weather; there were only a few specimens, bought by officers above Kandahar as curiosities, so that there was little opportunity of judging as to their qualifications for transport." BREEDS.

MORTALITY AMONG THE CAMELS USED IN THE AFGHAN WAR.—The verdict passed by the various officers whose opinions were called for on the subject of the losses of camels during the Afghan campaign was most pronounced and uniform. The plains camels were preferable for the transport service on the hotter or Indian side, but were quite useless for the higher and colder regions. Of the plains camels those of Bikánír were superior to the Panjáb, and these again better than the camels from Sind. The majority of the camels that died at Thul during June seem to have succumbed to heat-apoplexy, while in the higher altitudes, death appears to have been caused through some affection of the lungs. The hill camels perished through the heat of the Bolan pass and the plains camels by the cold of the higher regions, but both had previously endured privation and excessive fatigue. It is reported that of one consignment of Panjáb camels nearly 39,000 died or were lost by desertion, but it is probable that if the losses among the Sind, Baluchistan, and other camels, from the commencement to the final termination of the campaign were to be added to that number, the total losses might be close upon 60,000. These facts are alluded to mainly with the object of showing how the various breeds of camels have been acclimatised to widely different conditions. Some are suitable for the caravan traffic over hot sandy regions, which has given to this beast of burden the appellation of the "ship of the desert," while others have been so far altered in their habits and character as to be useful on rocky and mountainous countries and be even capable of sleeping on ground from which the snow has been only removed for their accommodation. The principal breeds of camels have a great aversion to water, the animals sickening rapidly when brought into a damp atmosphere. To the majority of camels, a slippery path, and one crossing streams, is far more trying than rough rocky ground. But as contrasting this state of affairs the camels of Kathiawar are stated to graze in the mangrove swamps. It would thus seem desirable that greater attention should be paid to the selection of camels than appears hitherto to have been done, and it might be even commended as a desirable step to organise breeding stations on the hills for the rearing of camels designed for alpine work. It was apparently with some such idea as this that the late **Lord Mayo** attempted to introduce into India the South American Llama. Unlike the camel this animal is by nature accustomed to the hills, and if successfully acclimatised might be found serviceable in the trans-Himálayan trade, if not for certain military purposes. It is, however, a much smaller animal than the camel, and perhaps even greater results might be looked for in the direction of acclimatising and improving the Central Asiatic and Afghan breeds of camels. Mortality. 207 AVERSION TO WATER. THE LLAMA. 208

PANJÁB CAMELS.—The following extracts from the Gazetteers regarding Indian camels may be found useful. According to the *Panjáb Gazetteer* for Jhang there are in that district two breeds of camels. These are known as the *Thalwan* and the *Bars* or *Barí*. The Thal camel is a much lighter animal than the Bar, and cannot carry so heavy a load. The female of either breed comes into heat when it is three years old, from the middle of January to the middle of April, and it may breed from that date for 20 years, and during the same period the male may be worked but the female is rarely laden. A good male camel will carry a load of 8 maunds, and he will take double marches of from 20 to Panjab. 209

BREEDS. 30 miles a day comfortably. In Montgomery it is stated there are three breeds known as *Soháwa, Ganda,* and *Hazára*—terms which seem to apply to the colour of the animal. "The *Soháwa* camel has long lips, medium-sized head, thick skin, and is of a brown colour. The *Ganda* camel is grey, and has a large head, small mouth, and thin skin. The *Hazára* camel has a small tail and is of a red colour. This is the worst of the three kinds, as it has no endurance on a journey. The *Ganda* is the best." "The camels of this district are of no use for riding." "Large herds go down annually to Bhiwáni for employment." "If well treated a camel will live for 40 years." The coupling season is from December to March, and at 4 years of age the female brings forth her first young one, gestation having lasted for 12 months. She continues bearing 9 or 10 times, at intervals of 2 years. After a year the young one is weaned, but it begins to pick grass when it is only 22 days old. A camel will feed her young and yield 12 seers of milk a day besides. The owner milks the cow twice a day, leaving two teats for the young one. The milk yields curds and butter-milk, but not butter. It acts at first as a laxative to those unaccustomed to it, but is recommended as a simple medicine for those suffering from disease of the spleen.

Gestation.

The Dera Ismail Khan Settlement Report states that the Pawindah camels are superior to those reared in the Dera Ismail Khan district, being similar to those in the Panjáb generally. "No good riding camels are bred in the district, the few that there are being imported from Bhawalpur and Bikánír." By the age of 16 a she-camel will have had six foals, which is about the maximum number for Thal. The Chenab camels are said to give as many as eight or ten foals, owing to the superior quality of grazing. "In this district," the author of the report says, "it is not the custom to make *ghi* of camel's milk, though I believe that this is done in Marwat and elsewhere."

Sind. 210

SIND CAMELS.—The *Sind Gazetteer*, speaking of the Jerruck district, says: of the domestic animals the one-humped camel takes the first place as a beast of burden. "Close to the sea-coast they are scarce, but in the upper part of the delta droves of forty to fifty are frequently seen. The delta-bred camel is smaller and lighter in limb than his Arabian congener, and being better fed, is a much finer-looking animal. The Karmáti tribe breed a valuable description of camel in this division—one which in pace and hardiness is said to vie with that bred in the Thar and Párkar district."

Rajputana. 211

RAJPUTANA CAMELS.—The Bikánír camels are famous all over India for being the swiftest and best riding camels. The *Gazetteer* says: "it is a remarkable fact that the domestic animals of Bikánír are generally either finer or more serviceable than those of any other part of India. The horses, if not fine, are strong and wiry; and I have known a very ordinary-looking mare carry its rider eighty miles through sand one day and forty the next, and then without a rest continue moderate daily journeys. Well-fed riding camels will do even more than this. They do not, however, thrive well out of the sandy tract." The experience of the Afghan campaign of 1878-80 proved that the Bikánír camels were far superior to the Panjáb, and the latter superior again to the ordinary Sind animal, but of the Panjáb camels those obtained from Dera Ismail Khan were the worst, and those from Rawal Pindi the best. The camels of Rajputana, like those of Bikánír, have the reputation of being swift, and are thus good riding animals.

Bombay. 212

BOMBAY CAMELS.—Very little can be learned regarding the Bombay camels. In the Gujarat (*Gazetteer, Ahmedabad District*) it is stated that the Ahmedabad camels are less prized than those brought from Márwár

"Those, especially the very swift Thal camel, which can easily travel forty miles a day, are used for riding by Sindis. The largest number of Márwár camels is found in Dhandhuka and Wiramgám." In Kathiawar it is recorded that excellent camels are bred in Okha, Navángar, Mália, and the Machu Kántha, &c. "They abound where the mangrove grows freely, and graze in the swamps." "Camels' milk is used for feeding young horses, and in cases of diseased spleen." The idea that camel's milk strengthens and improves foals is very general all over the parts of India where the camel is met with, but the above statement that the camels of Kathiawar graze in the mangrove swamps is remarkable, as in all other districts of India the opinion prevails that the camel has a strong dislike to water and will not thrive in damp, swampy regions. So general is this belief that **Mr. Darwin**, in his "Animals and Plants under Domestication," was led to allude to the fact as pointing to the desert origin of the animal, since in domestication it has not been able to conquer its aversion to water. The Kathiawar breed may have overcome this feeling just as the one-humped camels reared in Afghanistan and other cold mountainous countries seem to have lost to a large extent their love for level, burning, hot, sandy plains and deserts. These mountain breeds have not only become acclimatised to colder regions, but they seem to have attained the greater muscular strength in the hind legs necessary to sustain continued marching on hilly ground. The weakness of the hind legs of the desert camel accounts for one of his most ungainly features—the great muscular development of the fore as compared with the hind legs. **BREEDS.** **Camels feeding in swamps.**

Arabian Camels.—**Palgrave** (*Central and Eastern Arabia, I., 324*) says: "The camel and the dromedary in Arabia are the same identical genus and creature, excepting that the dromedary is a high-bred camel, and the camel a low-bred dromedary, exactly the same distinction which exists between a race-horse and a hack; both are horses, but the one of blood, and the other not. The dromedary is the race-horse of his species—thin, elegant (or comparatively so), fine-haired, light of step, easy of pace, and much more enduring of thirst than the woolly, thick-built, heavy-footed, ungainly, and jolting camel. But both and each of them have only one hump, placed immediately behind their shoulders, where it serves as a fixing-point for the saddle or burden. Owing to this similarity, they are often confounded in the common appellations of '*Baa'reer*' or '*Nŏk*,' male or female camels, though yet more often the dromedary enjoys his special title of "*hejeen*" or "*dolool*." For the two-humped beast, it exists indeed, but it is neither an Arab dromedary nor camel; it belongs to the Persian breed, called by the Arabs '*Bakhtí*' or Bactrian." **Palgrave** further adds that to see a dromedary it is necessary to go to Arabia, "for these animals are not often to be met with elsewhere, not even in Syria; and whoever wishes to contemplate the species in all its beauty must prolong his journey to Omán, the most distant corner of the Peninsula, and which is for dromedaries what Nejed is for horses." According to this definition the riding camels of Rajputana are the dromedaries and Bikánír the Omán of India. **Arabian. 213** **Dromedary.**

Rutting and Breeding.

Breeding.

There seems to be much difference of opinion as to the duration of the period of gestation. According to some writers the female carries her young for 11 months only, by others for 12 or even 13 months. She comes into heat when she is three years old, and bears one foal every 2 years or so for 15 to 20 years. She suckles her young for 12 months, but about 20 days after birth the infant animal commences to nibble the **Gestation. 214** grass. The period of the year when rutting commences seems to have been so modified under domestication that the young are born in summer or at least during pleasantly hot weather. **Kostenko** tells us that in Turkistan the male gets *must* in the winter (from December to January), but in India this occurs from January to April. During this period the male refuses food and water and becomes unmanageable. The female is rarely worked, but is reserved for breeding purposes, and to supply the milk on which the camel breeders largely live. If well cared for a camel will live for 40 to 45 years.

Power of Endurance.

Privation. **215**

Privation from both food and drink.—Incidentally allusion has been made, in speaking of the anatomical peculiarities of the camel, to its power of endurance. It is perhaps only necessary to state here that the most conflicting experiences of travellers and observers prevail as to the power of endurance of the camel. It may be premised that an exaggerated acceptance of this notion must of necessity prove dangerous. If anything was demonstrated more clearly than another by the high mortality among the camels during the late Afghan campaign it was, that once the camel's endurance is overtaxed, it requires long rest and careful treatment to restore it to health. A remarkable feature of the camel is that it will go on toiling under its burden until it falls down dead. Much difference of opinion exists as to whether the camel is stupid enough to eat poisonous plants: if he is not normally so, privation seems to have that effect on him. The camels from the plains of India at all events were observed to eat plants which the hill camels would not touch, and which have the local reputation of being poisonous to the camel. In another paragraph will be found a list of the camel fodders and of the few plants which the camel will either not eat, or, if he does so, is poisoned thereby.

Eating poisonous plants.

Speaking of the power of doing without water, **Skinner** (*ii., pp. 112—113*) tells us that his camels had been 19 complete days without drinking. On the other hand, numerous writers affirm that three or at most five days without water will kill the camel unless the fodder given is green and moist. **Kostenko** tells us that camels eat only during the day; they eat quickly and are satisfied with 2-3 hours grazing. If subjected to privation the load must be gradually reduced, and when forage is obtainable the nature and amount should be carefully regulated for the first few days. If the camel is capable of enduring a very considerable amount of privation, it can also eat to excess during times of plenty. **Pottinger**, in preparing for an expedition, gave his camels 15lb of flour a day in addition to all the grass they could eat. So greedy is the camel of food, after a few days' desert marching, that **Sir Samuel Baker** says, when it arrives in good pasture, it often dies in a few hours from inflammation caused by repletion. Reference has already been made to the popular notion that the hump is a store of fat used by the animal when other food is not obtainable. It is a commonly reported opinion that no amount of grain will serve to sustain the camel if it is not provided with its usual amount of grazing. Its large stomach requires a bulky material to excite digestion. (*Compare with Food, &c., p. 58*).

Privation from Water. **216**

Death from Repletion. **217**

The Hump.

LOAD, &C.

Load, Distance, and Rate of Marching.

The carrying power of the camel will depend to a great measure on the stock it has come from, and the climate in which it is to be employed. The Central Asiatic camel is, as a rule, more vigorous and enduring than either the Indian or African. The load a camel will bear greatly depends on the nature of the work or which it is employed. For a short distance, and in its usual avocation, a healthy camel will carry about 1,100 to 1,200lb, but where produce or baggage has to be carried to a distance, 300 to 400lb will be found a heavy enough load. **Captain Lord** estimates seven camels to the ton, where marches of 20 miles a day (of 8 to 10 hours) have to be performed. Where more rapid movements are desired the burden should be proportionately lightened. In Algeria, Morocco, Tunis, Tripoli, 300 to 400lb is the load. In Egypt 350 to 550lb. In Syria, Asia Minor, Turkey in Asia, Persia, and Tartary 550 to 600lb; but large-sized bull-camels are usually employed. In Beluchistan, Kabul, Hindustan, Tibet, Burma, and Mongolia 300 to 400lb. In Crim-Tartary and the borders of Southern Asia 300 to 500lb; but in the latter case the Bactrian or two-humped camel is employed.

Average load, 400lbs. 218

Colonel J. I. Boswell says the Panjáb camels known as *Sangar* are capable of great endurance. They are in their prime from 4 to 12 years old, and should not be purchased beyond that age, although a good camel may be worked until it is 18 or 19 years of age. In agricultural districts they are often, however, worked until they are 20 or 25 years old, but may live to 40 or 45 years. They should not be worked during the hot weather, from May to October, and by the end of July it is customary to give them a good purgative, followed after the action of the drug by a dose of alum. It is also a common practice to anoint the body with oil at this period so as to prevent mange. **Kostenko** says that the Turkistan camel walks at the rate of $2\frac{2}{3}$ miles an hour with a full load, but if lightened he will go 3 to $3\frac{1}{2}$ miles. The trotting camel gets over $6\frac{2}{3}$ miles an hour. If these figures be correct it may be added that a good Bikánír camel trots much faster than the Turkistan animal. The trotting motion is said to be very easy, but the gallop extremely disagreeable. Swift camels are reported to get over 100 miles a day at a push, but the ordinary journey which they will keep up day after day is about 40 to 50 miles. **Fortune** mentions an instance of an Arab having accomplished a journey of 225 miles in 28 hours, thus keeping up 8 miles an hour continuously. **General Chesney** mentions that he crossed from Basrah to Damascus, a distance of $958\frac{1}{2}$ miles, in 19 days, a daily rate of 50 miles. It is worth adding in this connection that in 1791 **Mr. James Rennell** proposed, in the Transactions of the Royal Society, that, owing to the uniformity of pace kept up by the camel, that animal might be employed to measure distances during geographical exploration. He cited that the distance between Aleppo and Bussora had been accomplished by **Carmichael** in 322 hours, by **Cupper** in 310 hours, by **Hunter** in $299\frac{1}{2}$ hours, the variations being accounted for by the slightly different tracts followed.

Prime age 4-12 years. 219

Rate of marching $2\frac{2}{3}$ to 4 miles. 220

The Kirghiz often harness the camel to their carts, the shafts being fastened by a cord passing behind the foremost hump. When so yoked they will draw 730lb, but if harnessed to a properly-constructed cart they will draw 1,800 to 2,160lb. In Rajputana the camel is sometimes seen yoked to the plough.

Camels in harness. 221

Diseases.

The limited space at the writer's disposal has compelled the present article to assume the form of little more than an abstract of the literature on the subject. He is thus precluded from attempting to give even the commonest facts regarding the diseases of the camel or their modes of treatment. It is generally believed that the camel is liable to a number of diseases peculiar to itself, but is not subject to the attacks of infectious diseases which carry off other cattle. For military purposes by far the most

DISEASES. 222

serious disorder is the result of careless loading and a badly fitting saddle, namely, bruises on the back. With care these need not occur, and after seeing that the saddle fits well the next best preventive is to ascertain if the load be well balanced, for nothing is more annoying to the camel or a more fruitful source of sores than a load heavier on one side than the other. It has already been stated that many of the camels employed in the Afghan campaign succumbed to heat and others to cold, but it has been contended that the privation they endured for some time previously was the actual cause of death. This seems to be proved by the immunity enjoyed by the camels belonging to the officers, most of which returned in safety to India after passing through both the heat and the cold to which they were exposed while accomplishing for months heavy and forced marches. For an account of the diseases of the camel and their treatment, the reader is referred to a valuable memorandum written by **Dr. W. Gilchrist**, of the Madras Army, in 1842, and which to this day is perhaps the best treatise that has appeared.

Sores on the Back.

Kostenko says the disease known in Turkistan as *Sarpo* causes the soles of the animal's feet to fall off, and he adds, that as with all the other diseases to which the camel is subject, this is treated by the nomads by freedom from work and good food.

FOOD AND FODDER.

FOOD AND FODDER.
223

To keep a camel in health it should be allowed 6 hours' grazing and 2 seers of gram a day, or if grazing be scanty 3 seers of gram. **Kostenko** says the camel eats all the herbs and grasses unfit for other animals, and is indifferent as to the quality of the water it drinks—brackish, stagnant, or putrid. It can pass 2, 3, or 4 days without drinking and 3 days without food. While subjected to such privation, the load should be lightened and not re-imposed for at least a week after. The Bactrian camel has been referred only to by way of comparison, in the foregoing account, and it may here be disposed of by saying that in winter, when the country is covered with snow, it grazes chiefly on the willow. Of the plants which the Indian camel will graze on a few are more important than the others, because no other animal can subsist on them, and they are accordingly treated as more peculiarly camel fodders. It would be much easier, however, to enumerate the plants which the camel will not eat, or which are poisonous to it, than to mention those on which it may be fed. The latter would almost mean a list of the plants of India. The practical object will therefore be met by furnishing two lists, *viz.*, the plants mentioned by authors as more peculiarly camel fodders and the plants of which the camel will either not eat or on which at least it cannot subsist or which are poisonous to it.

CAMEL FODDERS.

224

1. **Acacia arabica,** *Willd.*; LEGUMINOSÆ.
2. **A. Farnesiana,** *Willd.*
3. **Ægiceras majus,** *Gærtn.*; MYRSINEÆ.
4. **Albizzia Lebbek,** *Benth.*; LEGUMINOSÆ.
5. **Alhagi maurorum,** *Desv.*; LEGUMINOSÆ.

THE CAMEL THORN OR SHÚTAR KHÁR.

Vern.—*Juwásá* or *junvásá* or *yavásá, javánsá,* HIND., PB., BOM.; *Tamiya, jawá, jawása, jawán,* PB.; *Dulallabhá,* BENG.; *Kandero,* SIND.; *Duralabha, girikarnika, yavása,* SANS.; *Shutar khár,* PERS.; *Alhaju,* ARAB.; *Zaz,* PUSHTU; *Zoz, zozán,* TRANS-INDUS.

A widely distributed shrub of the Ganges valley and the arid and northern zones; a native of south Africa, the deserts of Egypt, Arabia, Asia Minor, Beluchistan, and Central India. Abounds in many of the arid parts of the Panjáb plains; also common near Delhi.

FODDER. Camel Thorn.

In the hot season, when almost all the smaller plants are withered up, this puts forth its leaves and flowers. These are greedily eaten by the camel, and so much does that animal depend upon this plant that it has received the name of the camel-thorn. An officer, writing after the close of the Afghan campaign, says that the camels grazed upon this plant in the Pishin. The bush, he adds, is common all over the country. The Natives collect it in October or November and beat it up into *bhusha.* It is probable that about 50 to 60 maunds might be collected at Pishín and stored for winter use.

6. **Amarantus polygamus,** *Linn.;* AMARANTACEÆ.

7. **Anthrocnemum indicum,** *Moq.;* CHENOPODIACEÆ.

8. **Atriplex Stocksii,** *Boiss.;* CHENOPODIACEÆ.

9. **Avicennia officinalis,** *Linn.;* VERBENACEÆ.

10. **Bauhinia racemosa,** *Lam.;* LEGUMINOSÆ.

11. **Berberis,** various species; BERBERIDEÆ.

12. **Calligonum polygonoides,** *Linn.;* POLYGONACEÆ.

13. **Carduus nutans,** *Linn.;* COMPOSITÆ.

14. **Corchorus Antichorus,** *Ræusch.;* TILIACEÆ.

15. **Cressa cretica,** *Linn.;* CONVOLVULACEÆ.

16. **Crotalaria Burhia,** *Ham.;* LEGUMINOSÆ.

17. **Dalbergia Sissoo,** *Roxb.;* LEGUMINOSÆ.

18. **Dodonæa viscosa,** *Linn.;* SAPINDACEÆ.

19. **Eclipta alba,** *Hassk.;* COMPOSITÆ.

20. **Haloxylon multiflorum,** *Bunge.;* CHENOPODIACEÆ.

Syn.—ANABASIS MULTIFLORA, *Moq.*

Vern.—*Gora láni, láná* or *lánó,* SIND.; *Ghalme,* TRANS-INDUS.

The Lani

Common in the North-Western Panjáb and the Salt Range, and distributed to Afghánistan. Camels are fond of the plant.

21. **H. recurvum,** *Bunge.*

By mistake this plant was alluded to by **Stewart,** and following him by all subsequent authors, including the writer (see B. 162) as **Caroxylon Griffithii,** *Moq.*, an Afghán plant not found in India. **Haloxylon recurvum** is the plant from which *khár-sajjí* is chiefly made in India, and it is the salt plant most relished by the camel.

Khar-Sajji.

It is known in the Trans-Indus as *laghme,* and in Cis-Indus as *khár,* in Sind as *kárí láni.* A writer in the *Panjáb Gazetteer* says that camels thrive best if fed one day upon the *láná* and the next upon the *pílú* **(Salvadora oleoides).** The term *láná* appears to be almost generic for all the Chenopodiaceous plants alluded to in this list, but it is more especially applicable to this species.

FODDER.

22. Halocharis violaceæ, *Bunge.;* CHENOPODIACEÆ.

23. Indigofera pauciflora, *Delile;* LEGUMINOSÆ.

24. Kochia indica, *Wight.;* CHENOPODIACEÆ.

25. Lippia nodiflora, *Rich.;* VERBENACEÆ.

26. Leptadenia Spartium, *Wight.;* ASCLEPIADACEÆ.

27. Lycium europæum, *Linn.;* SOLANACEÆ.

28. Melia Azadirachta, *Linn.;* MELIACEÆ.

29. Mimosa rubicaulis, *Linn.;* LEGUMINOSÆ.

30. Mollugo hirta, *Thunb.;* FICOIDEÆ.

31. Phœnix dactylifera, *Linn.;* PALMÆ.

32. Pistacia integerrima, *J. L. Stewart;* ANACARDIACEÆ.

33. P. mutica, *Fisch. & They.*

34. Prosopis spicigera, *Linn.;* LEGUMINOSÆ.

35. Psoralea plicata, *Delile;* LEGUMINOSÆ.

36. Quercus Ilex, *Linn.;* CUPULIFERÆ.

THE HOLLY OAK.

The Oak.

Vern.—*Charrei, serei, balút, sháh balút,* AFG.; *Chúr, bán, kathún bán, írri, yirú khareo,* PB.; *Sperchereí, pargái, kharanja,* TRANS-INDUS.

Dr. Aitchison says that in Kuram the variety of this plant, devoid of spines, is given as a camel fodder. **Major Clifford** mentions that a bush like the holly found in Pishín was eaten by the camels, but only in June and July, after which it became too spiny. By another writer it is stated that a so-called **Ilex** eaten by the camels did not agree with them. It is difficult to discover what these writers allude to since the holly-oak does not occur in Pishín.

37. Rubia tinctorum, *Linn.;* RUBIACEÆ.

38. Salicornia brachiata, *Roxb.;* CHENOPODIACEÆ.

39. Salsola fœtida, *Del.;* CHENOPODIACEÆ.

Moti lani.

Vern.—*Moti láni,* PB.; *Mitho lani, samunaar láni,* SIND.

A camel fodder, but also used in the preparation of *khár sajjí,* especially near Jhelum.

40. S. Kali, *Linn.*

41. Salvadora oleoides, *Dcne.;* SALVADORACÆ.

Vern.—*Kabbar, jhar, diár, jál, váni jhál, ughai, koku, pílú, píl, plewane, mithi van,* HIND., PB., TAM.; *Pílú,* MAR.; *Sadní djar, mithi-diár,* SIND.

Pilu.

A large, evergreen shrub of the Panjáb and Sind, often forming the greater part of the vegetation of the desert, and ascending the Trans-Indus hills and Salt Range to 3,000 and 4,000 feet in altitude. Flowers in April, and its fruit ripens at the beginning of the hot weather. The fruit is sweetish and is largely eaten by the natives.

The leaves serve as fodder for camels.

42. **Salvadora persica,** *Linn.* FODDER.

Vern.—*Jit, kauri ván, kauri-jál, chhoti ván,* PB.; *Jál,* N.-W. P.; *Kabar* (*kuber* by Stocks), *khorí djhur, kharl djar,* SIND; *Pedda-warago-wenki,* TEL.; *Opa, ughai,* TAM.

A small, thick-stemmed, soft-wooded tree, wild in many of the drier parts of India, *e.g.,* Panjáb, Sind, Rajputana, North-West Provinces, Guzerat, Konkan, and the Circars. Produces flowers and very small black red juicy currant-like berries, having a strong aromatic smell, and pungent taste somewhat like mustard or garden cresses.

The shoots and leaves are pungent, and are occasionally eaten as salad; given as fodder to camels.

43. **Suæda fruticosa,** *Forsk.;* CHENOPODIACEÆ.

Vern.—*Chhoti láni, lúnak, phesak láné, baggi lána, dána,* CIS-INDUS; *Zamái,* TRANS-INDUS; *Aout láni, usak láni, lúnak,* SIND.

A sub-erect bush, common in North-West India from Delhi to the Indus, and distributed westward to Africa and America. Chhoti Lani.

Employed in the preparation of *khár sajjé,* but also extolled as a camel fodder. Major Clifford says, it is abundant at Chuckluk in Pishín.

44. **S. maritima,** *Damort.,* and **S. nudiflora,** *Moq.*

45. **Tamarix gallica,** *Linn.;* TAMARISCINEÆ.

Trianthema.—Four species belonging to this genus frequent the sandy tracts of the Panjáb and Sind, and, according to Stocks, one or all are known as *Fysur láni;* they are regularly eaten by camels.

The following are the better known species of this genus:—

46. **Trianthema cyrstallina,** *Vahl.;* FICOIDEÆ.

47. **T. monogyna,** *Linn.*

48. **T. pentandra,** *Linn.*

49. **Vitis carnosa,** *Lam.;* AMPELIDEÆ.

50. **Zizyphus nummularia,** *W. & A.;* RHAMNEÆ.

Vern.—*Mallá, bér, birár, jhari,* N.-W. P.; *Gangr, jangra,* SIND; *Mallá, kokni bér, mara bér, jand, jharberi sari, birotá,* PB.; *Kurkana,* TRANS-INDUS; *Karkanna,* AFG.

A densely-branched, small bush, met with in the drier parts of India. Rajputana Fodder.

Mr. F. Kinsman, of the Telegraph Department, informs the writer that this plant may be regarded as the most important camel fodder in a great part of Rajputana. The natives, to cut the plant, have invented a peculiar axe, with the cutting edge turned, so that it is parallel to the ground. The bushes are periodically lopped to the ground by means of this axe. The thorny twigs are then beaten to separate the leaves. These are given to the camels and the dried twigs used as firewood. The bush grows wild over the uncultivated sterile tracts of Rajputana, but it is left in the fields here and there or encouraged to grow as a hedge plant in order thus to afford both camel fodder and fuel.

51. **Zygophillum simplex,** *Linn.;* ZYGOPHYLLEÆ.

PLANTS POISONOUS OR AT LEAST NOT WHOLESOME TO CAMELS. 225

1. **Acorus Calamus,** *Linn.;* AROIDEÆ.

Vern.—*Bach,* HIND.; *Vekhanda,* BOMB.; *Vaj,* ARAB.; *Agri-turki,* PERS.; *Bari boj,* PB.

POISONOUS.

Bach Akri.

A semi-aquatic plant, met with in damp places in India, at altitudes from 3,000 to 6,000 feet.

It is reported that at Quetta and Pishín an Iris-like plant, eaten, during the Afghan campaign, by the camels from the plains, proved poisonous to them. The hill camels did not eat the plant. This seems to be the same plant which **Mr. Steel** speaks of under the name *akrí*, a word which may be taken as derived from the Persian name for this plant. **Mr. Steel** appears to think, however, that the camels in Quetta were not quite so stupid as to eat the *akrí*, but several officers report that they were poisoned by an Iris-like plant, which as here suggested may probably be **Acorus Calamus.** It is necessary to add that the name *akrí* is in the Panjáb applied to **Withania coagulans** (which see), a plant which bears no resemblance to an Iris whatsoever.

2. **Calotropis gigantea** and **C. procera,** *R. Br.;* ASCLEPIADACEÆ.

Vern.—*Ak, madár,* HIND., PB.. and SIND.; *Spalmei, spalmak,* AFG.; *Ushar,* ARAB.; *Khark,* PERS.

Stocks enumerates this among his four plants which the camel will *not* eat, but the *Sind Gazetteer* (page 522), under the account of the district Mehar, states that it is a camel fodder. It is probable **Dr. Stocks** is correct.

3. **Cannabis sativa,** *Linn.;* URTICACEÆ.

4. **Euphorbia neriifolia,** *Linn.;* EUPHORBIACEÆ.

5. **E. Royleana,** *Boiss.*

6. **E. Tirucalli,** *Linn.*

7. **Nerium odorum,** *Solander;* APOCYNACEÆ.

SWEET-SCENTED OLEANDER.

Vern.—*Kaner, kanira, ganhira,* HIND., PB.; *Karabi,* BENG.; *Kanhera, kanir,* BOMB.; *Diflí,* ARAB.; *Khar-zahrah* (the Asses-bane), PERS.

A common bush, with large pink or white flowers. **Dr. Stocks** says of this plant: "It is worthy of remark that the camel eats the **Nerium odorum** (*Zowr* or *jowr*), a remarkably poisonous plant, which in nearly every case proves fatal to him, as troops during the march have often found." Several officers have reported this same fact in connection with the high death-rate of camels during the Afghan campaign. Speaking of this subject, however, **Veterinary Surgeon Steel** discredits the idea of the camel eating poisonous plants, and says: "We observe that delicate appreciation of wholesome diet by means of the olfactory powers, which strongly argues against the probability of camels voluntarily eating poisonous herbs, as was suspected by some, when the mortality was so great at Quetta." (Compare with **Acorus**).

8. **Othonnopsis intermedia,** *Boiss.;* COMPOSITÆ.

Vern.—*Gúngú,* PUSHTU.

Mr. J. H. Lace, of the Forest Department, Quetta, reports that the Bíluchís regard this plant as poisonous to the camel.

9. **Peganum Harmala,** *Linn.;* RUTACEÆ.

Vern.—*Harmal,* ARAB.; *Isband,* PERS.; *Spelane, karmal,* PB.; *Isband,* HIND.

A small bush, much branched and densely clothed with dissected leaves. The whole plant strongly scented.

The camel will not eat this plant.

10. **Withania coagulans,** *Dunal;* SOLANACEÆ.

Vern.—*Akri, panir,* PB.; *Panir,* SIND; *Panir bad,* PERS.

While this species is not eaten by camels, the allied species, **W. somnifera,** is said to be browsed by goats, and it is possible it may therefore be also eaten by camels. Both species occur in Sind, the Panjáb, and are distributed to Afghánistan. (*Compare Nos. 1 and 7.*)

CAMEL-FLESH AND PRICES PAID FOR THE ANIMAL.

PRICES. 226

It is stated by writers on the subject that camel-flesh is very tough, but that the flesh of the sucking camel is passable. The camel owners are reported to kill and eat the animals that show signs of dying, and that only the rich during festive occasions can afford to kill a young camel. In India the price of a full-grown camel seems to average from R25 to R150. At Taskand a camel sells for about £6 to £10, and this price prevails over the greater part of Turkistan. **Palgrave,** speaking of the Nejdean camels, says, the "camel is somewhat slimmer and smaller than the northern, and the hair is finer. They are cheaper in proportion than sheep; twenty-five to thirty shillings is an average price."

CAMEL-HAIR.

HAIR. 227

The amount of hair or wool which the camel possesses seems to be inversely to the warmth of the country in which it is found. The two-humped camel has a longer and more abundant crop than the single-humped, and the wild camel most of all. It has already been stated that the natives near Lob-nor are said to hunt the wild camel on account of its hair, which is much valued for its softness. The single-humped camel, acclimatised to colder regions, loses its hair when brought into a warm country, but periodically all camels cast their hair, and the natives either wait for this or clip the hair shortly before the period at which it should be shed. This generally occurs in spring in Upper Asia, but not till May or June in India. The cold country camels yield as much as 12℔ of hair a year, but in India 2℔ is about the average. This is woven into the *boras* or sacks used by the camel-owners, but for this purpose it is usual to mix the camel-hair with goat-hair. It is also made into ropes. **Bellew** says it "is very highly prized for the manufacture of a superior kind of camlet" (*Kashmir and Kashgar, p. 348*). Both at the late Calcutta International and at the Colonial and Indian Exhibitions the Agra Jail exhibited carpets made entirely of camel-hair. **Baden Powell** says: "The soft under-wool is of a light-brown colour; it is made into *chogas* of a cheap kind, but they are soft, warm, and useful. The long hair is not made use of" in India, but "it is employed in Europe for making paint-brushes." In the manufacture of artists' hair-brushes or pencils, in addition to camel-hair the fine hairs of the sable, the miniver, the martin, the badger, and the polecat are also employed.

CAMEL-HIDE.

HIDE. 228

There seems to be little or no export trade from India in camel-hide. Locally it is employed for many minor purposes, such as the fastenings used by the camel drivers. With the hair on, it is also manufactured both in Europe and in India into trunks. The chief use to which it is put in India, however, is the manufacture of *kuppas,* or the huge skin jars employed in India for carrying oil or *ghí.* These are most probably made in the Lower Provinces (where the camel does not occur), of cow, buffalo, or

horse hide, but the writer can discover no account of the manufacture of the immense number of skin or leathern oil jars which form an almost
Kuppas. 229 characteristic feature of every bazár in Lower India. It would, however, appear that other skins are sometimes employed in addition to camel-hide, but as they are more expensive and more difficult to work, camel-hide is mainly used. The smaller ornamental jars employed for the household supply are, however, chiefly made of the intestinal integuments of the camel, cow, or horse. For this purpose the integuments are boiled until they are reduced to a glutinous skin or gluey mass. In both cases the vessel is formed in the same manner. Clay is taken and formed into the shape and size of the desired vessel. After this it is sun-dried to a desired extent. The fresh camel-hide freed from the hair, or the prepared intestinal membrane, is then drawn over the mould and beaten until it firmly adheres. The mouth is now formed by working the skin around a stick or bamboo and reflecting the lip in the characteristic shape. When quite dry the clay is broken up and carefully removed. The smaller jars or *kuppi* are also ornamented with patterns cut out in white parchment; after these have been stuck on, the vessels are varnished over the
Kuppi. 230 outside. **Mr. Baden Powell**, in his *Panjáb Manufactures*, says that at Rohtak the *kuppi* are often made in "fantastic shapes, some like jugs, others flattened and perforated apparently with large holes, which of course can open only longitudinally in the thin flat body of the vase." They are also largely made at Maghiana in the Jhang district and at Bikánír in Rajputáná, and at Cutch and Ahmedabad in Bombay. An interesting account of the *kuppa* manufacture of Lucknow will be found in **Hoey's** *Trade and Manufactures of North India, p. 138.* It would appear that the *kuppeaz* of that city use any hide available and construct the rim over a layer of mud permanently enclosed to give rigidity.

Kuppas may be of any desired size, the price varying accordingly. The larger sizes used by the merchants for holding oil, *ghí*, molasses, &c., are often "so large as easily to realize the familiar oil jars in the story of Ali Baba and the forty thieves" (*Baden Powell*). They may be made to hold one maund or six or eight maunds, but *kuppi* not more than sufficient to contain two ounces may also be procured.

Kuppas should not be mistaken for leather water bottles such as those made at Bikánír and used all over Northern India. Camel-skins sell for about R2 to R3 a piece.

MILK. 231

CAMEL'S MILK.

It is scarcely necessary to enlarge on this subject further than has already been done. It is regularly used by the camel-rearers; indeed, it forms an important item of their food. To those not accustomed to it, it is purgative, and is accordingly recommended as a medicine. It is supposed to give strength to horses, hence is commonly given to foals. According to some writers camel's milk will yield butter-milk but not butter, and by others it is said to afford butter also. The writer has at present no means of settling this point, but a matter of this nature might easily enough be disposed of and thus remove at least one of the numerous ambiguities that exist in the literature of the camel.

Halwa. 232 *Halwa*, a kind of sweet-meat made of camel's milk and honey, is brought from the Persian Gulf to India *viá* Bombay. Formerly this used to come in flat earthen plates, but it is now done up in tin boxes and is even sold by European provision store-keepers in India, so that a considerable trade seems to be done in the article, although it does not appear to be anywhere made in India. It is known in the bazars as *muscat-ka-halwa.*

TEA. 233

CAMELLIA, *Linn.; Gen. Pl., I., 187.*

A genus of trees or shrubs, containing in all some 14 species, and belonging to the Natural Order TERNSTRŒMIACEÆ. They are natives of Tropical Asia, India, and the Archipelago. Formerly, the Tea plant was retained as a separate genus (**Thea**), but it has been shown that this isolation serves no good purpose, since **Thea** only differs from **Camellia** proper in the absence of the tendency to produce inner free stamens in front of the petals, and in the flowers becoming pendulous and having persistent sepals with fewer bracts.

Leaves evergreen, serrate, coriaceous, or membranous *Flowers* axillary, solitary, or aggregated, sessile or only shortly stalked; petals often numerous, large, and forming elegantly-coloured flowers, with what might be almost called artificial regularity in their arrangement. *Sepals* 5-6, unequal, graduating from the bracts towards the petals. *Petals*, through their union with the stamens, slightly coherent at the base. *Stamens* numerous, the outermost in many rows slightly or distinctly monadelphous, adherent to the base of the petals: the innermost whorl of 5 to 12 stamens, which, in the **Camellia** proper, are often quite free; *anthers* at first extrorse, but as they mature becoming versatile. *Ovary* 3-5 celled; *styles* tubular, free to the base, or united below. *Capsule* woody, usually short and loculicidal, bursting along the flattened top to expose the seeds, which normally are three in number, but in some cultivated forms are often reduced to two or only one. *Seed* large with thick oleaginous cotyledons, but no albumen; radicle short, superior.

The genus **Camellia** is named in honour of **Camellus (Joseph Kamel)**, a Moravian Jesuit and Asiatic explorer. The cultivated or ornamental Camellias are mainly derived from **C. japonica,** a native of China and Japan; this was introduced into Europe in 1740. The Camellias are easy of cultivation in warm temperate climates, the best soil being a mixture of sandy-loam and peat. The pots should be well drained and the plants sparingly watered, except during the growing season. They are readily increased by cuttings or by inarching on the commoner kinds.

The Chinese tea-planters are said to propagate **C. Sasanqua** as a shelter for their tea plants. This small-leaved species has sweetly-scented red flowers, the odour of which is supposed to be communicated to the neighbouring tea leaves. Sometimes, however, the planters pluck the leaves and even the petals of this species, and mix these with the tea in order to produce a favourite-scented mixture. The black-scented teas, shipped from Canton, are said to be flavoured with the flowers of **Jasminum Sambac.** This is largely grown in the suburbs of Canton, and is there known as *Mok-lei*.

The seeds of **C. drupifera** (formerly known as **C. oleifera,** *Wall*) yield the largest amount of oil, but all the Camellia seeds contain a useful sweet oil. By far the most important of the Camellias, however, is that from which Tea is obtained.

Linnæus, in the middle of the eighteenth century, gave the Tea plant the name of **Thea sinensis (T. chinensis)**, but soon after, in the second edition of his *Species Plantarum*, he deemed it advisable to assume that the green and the black teas of commerce were obtained from different plants. He accordingly called the plant from which he supposed the green tea was obtained **Thea viridis,** and the black tea, **Thea bohea,** the latter specific name being derived from the "Wú-í or Bú-í Mountains in the north-west of Fuh-Kien, one of the districts most famous for its black tea" (*Yule*). These hypothetical Linnæan species were soon reduced to one, and that referred to the genus **Camellia,** under the name **Camellia theifera,** *Griff.* (**C. Thea,** *Link*). The so-called wild tea of Assam was next described as a separate species under the name of **Thea assamica,** *Masters*; but recent investigation has proved this to be but a large-leaved subtropical form of **C. theifera,** and it is open to doubt if it be even indigenous.

TEA. History of Assam Tea.

It is most probably only an escape from early cultivation, so far as Assam is concerned. The first scientific tea explorers of the forests around Sadiya, namely, **Drs. Wallich, McClelland,** and **Griffith**, describe it as undoubtedly a form of the true tea plant of China, which had degenerated by neglect of cultivation, and the wild conditions under which it was found in Assam. They concluded their report by recommending that fresh seed should be imported from China for the purpose of the contemplated Government cultivation of tea, since the stock found in Assam was of such inferior quality. In a correspondence with Assam tea planters, however, the writer has had this idea of the inferior quality, or rather degeneration, contested on the ground that the China plant, on its introduction into Assam, became smaller instead of larger leaved, whereas the Assam supposed escape from an early cultivation showed no such tendency and was on this account presumably a distinct plant from the China. This line of reasoning can scarcely be admitted, for, assuming that the Assam and the China both came originally from the same stock, they have each been altered by cultivation, under widely different climates, for centuries until they have become so different that what suits the one would not necessarily suit the other. Besides everything points towards the Assam being a much more modern stock than the China, and the wild plants of Manipur have larger leaves than the Assam, and even grow into large trees. **Captain** (afterwards General) **Jenkins**, prior to the deputation of the above Commissioners, reported that the Indo-Chinese tribes of Assam regularly cultivated the plant and made tea from its leaves. We thus need have no doubt as to its having been cultivated prior to the European effort, and with this knowledge, we may leave the disputed point as to whether or not it should be viewed as a truly indigenous or introduced plant. For convenience the writer prefers to adopt the popular expression "indigenous Assam," although personally he is of opinion that it has little claim to such a title. The wild plants of **C. theifera** in Assam have all the appearance of being escapes from cultivation, and it is a suspicious circumstance that they do not occur beyond the invaded and conquered territory now inhabited by the people who are reported to have cultivated tea at the time of the arrival of the first European visitors to Assam.

Manipur Tea. 234

In Manipur, however—a small Native State, 2° or 3° south of the region of the Assam and in the very latitude of the accepted Chinese home of the tea plant,—wild tea forms forests, the plants attaining to the size of trees of 30, 50, to even close upon 100 feet in height, and having leaves often approaching to a foot in length. It is probable that Manipur is thus the original home of the so-called Assam and Cachar indigenous teas if not also of those of China, centuries of cultivation in the temperate regions of China having given origin to the smaller leaves which the planters lay so much stress upon as a proof of independence of origin. (Compare with the remark under **Camellia theifera**, No. 244). The historic and even the accepted prehistoric colonization of Assam has many incidents in it that might be cited in support of this theory. The successive waves of Siamese and Shans which crossed Upper Burma and conquered Assam (surviving to this day in the Ahom and Kampti), and in still more recent times, the Burmese conquest of Manipur, Cachar, and Assam, might each have contributed something to the early introduction and cultivation of tea in Upper Assam and even in Cachar. The seed brought from China to India, in the early stage of Indian tea cultivation, was the seed of a stunted, small-leaved form which had been cultivated in China from very early times, and it naturally produced a plant which in many respects differed from the large spreading bushes to be seen in the damp glades of the hilly portions of Assam and Cachar.

TEA. Assam Hybrid. 235

The cross fertilization of these two forms gave origin to the popular race known as the "Assam hybrid," a term which scientifically must be viewed as incorrect, since it is not a cross between two species but between two forms of the same species. It is more accurately a cultivated form or race holding the same relation to the original species as do the races of wheat or of rice to the plants from which the multitudes of widely different kinds of these cereals have been derived. This is more than a mere technical distinction, since it accounts for many of the peculiarities of this widely cultivated "hybrid" stock (such as the ease by which it is propagated by seed), but it leaves absolutely the experimental production of a real hybrid between **C. theifera** and some of the other truly Indian wild (though hitherto non-tea-producing) species, a problem that would seem well worthy the attention of the practical planter. Whether any improvement in quality or healthiness of stock would result from the production of such a hybrid remains to be seen. Indeed, this may actually, although unknowingly, have been formed, but as far as the writer has been able to discover, no one has published an account of the production of hybrids or other special forms of tea such as have been produced with the garden Camellia. Effort would thus appear to have been diverted from the cultivation of the plant to the improvement and cheapening of the manufacture of tea, so that the past 50 years of Indian tea cultivation have seen no new forms produced, and perhaps little improvement in the methods of cultivation.

It is constantly protested by the planter that he can distinguish the Manipur stock from the Assam, the China, and the hybrid, the argument being that, therefore, they are quite distinct plants. The contention here urged does not for a moment discountenance the idea that these forms are recognisable. Local varieties exist of every widely distributed plant. Cultivation will modify almost any plant and even produce departures from the original type that are constant under certain conditions of climate and soil. The Manipur form may be as easily recognised from the so-called Assam indigenous (while both are in the plantation) as the cabbage may be instantly distinguished, even in the seedling state, from the cauliflower, but, in spite of all this, the forms of the tea plant need not possess so high a claim as these well-known vegetables, to be regarded as even varieties of a common species. The term variety is here used of course in its strictly scientific sense, and not in the loose popular manner in which it is but too frequently employed. A cultivated recognisable state of a plant is not necessarily a variety. A variety is a fixed natural departure from the specific type; in other words, it is what might be called a lower degree of species. According to this acceptation, all the forms of the mango, for example, must be thrown together as unworthy of the systematic position of constituting even one, still less many, varieties of the wild plant.

Recognisable cultivated forms. 236

This argument is of importance in its bearing upon the ambiguity which exists regarding the tea plant. Allow the "Assam indigenous," the "Assam hybrid," and the "Chinese" plants to become naturalised in the wild tea forests of Manipur, and it is extremely probable that they would in a comparatively short time be almost indistinguishable. The Chinese plant would doubtless retain its distinctive features longest, because it has been cultivated for a much greater period and acclimatised to a colder country than Manipur. Some of the forms of Chinese tea are accustomed to a climate with a short but severe snowy winter. There are in India, however, at least four perfectly distinct species of Camellia, which might be left for a comparatively indefinite period, growing side by side, without losing one particle of their distinctive features. One of these, with the true tea plant,

Assam Indigenous. 237

(conf. with 238).

TEA.

belongs to the section **Thea** of the genus **Camellia,** *viz.*, **C. caudata,** a species met in Bhután, the Mishmi hills, the Khásia hills, and even in Sylhet and Burma. Has any effort ever been made either to propagate this species, or to use it as a hardy stock for grafting, inarching, or hybridising with, or have its properties, as a possible source of tea, been tested? From a purely theoretical point of view it would seem desirable that this subject receive attention, for, should the suggested hopes of improvement even prove but visionary, an important scientific service would have been accomplished, while at the same time, in the event of a calamity like that which befel Ceylon coffee, overtaking the Indian tea industry, one of the problems which would then be certain to demand solution would have been disposed of. The Assam planters are nearly unanimous in saying that the indigenous Assam or even the Manipur is superior to the China-Assam hybrid. **Dr. J. Berry White,** for instance, in an instructive paper read before the *Society of Arts* (*May 27th, 1887*) remarks that—"It is a matter for profound regret that this garden (Chabwa) did not share the fate of its predecessor, for it proved the chief means of disseminating the pest of Assam—the miserable China variety—all over the province, not only by means of seed, but, owing to its prolific inflorescence,* the indigenous Assam plants in the vicinity were impregnated with its pollen, and thus produced the hybrid variety which now forms the great bulk of the plants found not only in India but also in Ceylon." **Dr. White** does not therefore show much favour either for the introduced China tea plant or for the so-called hybrid between it and the plant found in Assam. Other planters state that a first class hybrid is, however, at least as profitable a plant to grow as the pure Assam, since it will yield better at the beginning and the end of the season when the weather is too cold for the indigenous plant.

First Assam Tea Garden.

Comparing the results of the cultivation of China and indigenous Assam **Dr. Berry White** says of the Assam indigenous: "It is a tree growing from 25 feet to 35 feet high, with leaves six times larger than the China variety. Under like conditions, the yield of leaf from an acre of Assam tea will be not less than double that of the China plant, and the gain is not only in increased quantity, but the quality will, *cæteris paribus*, realise from 1*d.* to 2*d.* per ℔ more in Mincing-lane than the smaller quantity made from the China plant." It is, we may presume, a statement of fact and not of argument, that the Assam tea will fetch more in Mincing-lane, pound for pound, than Indian-grown China. But is the lesser yield, as **Dr. White** seems to think, due to inherent inferior quality or to insuitability to the Assam climate? Is the China plant, in other words, suited to Assam, and if not, is it possible by other means than hybridization to improve the Assam stock? The planter in Kangra or in the Nilghiris would most unquestionably say China stock was to him superior to Assam, because the Assam will not flush so well in cold as in hot climates. The quality of China tea grown in hot climates is, however, by most planters regarded as superior to that produced in cold regions, although, as compared with Assam, the yield is poor and unprofitable. The point aimed at by the writer is, however, strengthened by **Dr. White's** observation, namely, that we have been forcing the cultivation of a tea stock in unsuitable regions, and it is therefore high time that the whole subject of the indigenous and the long acclimatised forms of the Indian tea plant and its nearest allies should be carefully investigated by the planter, the botanist, and the chemist.

Having thus endeavoured to bring together one or two of the more important botanical ideas regarding the cultivated forms of the tea plant,

* This doubtless means prolific flowering: the flowers are axillary, solitary.

TEA.

and enumerated the various scientific and planter's names given to these, it may be as well to define very briefly each of the Indian species of **Camellia**, discussing their various economic and industrial properties before giving a brief history of the rise and present position of the Tea Industry. The reader is referred for further information to the article TEA.

Species of Camella. 238

Camellia caudata, *Wall., Pl. As. Rar., III., 36; Fl. Br. Ind., I., 293;* TERNSTRŒMIACEÆ.

References.—*Griff., Notul. IV., 559, t. 601; Trans. Agri.-Hort. Soc. Ind., V., 1838, t. A.; Kurz, Fl. Burm., I., 109; Gamble's Man. Timb., 30.*

Habitat.—A smallish bush, found in the Bhután, Mishmi, Khásia and Sylhet hills, and in Martaban; at altitudes from 3,000 to 5,000 feet above the sea.

Botanic Diagnosis.—*Leaves* with tapering points, hairy beneath and only 3 to 4 by ¾ to 1 inch in size. *Flowers* white solitary, nodding, with the stamens and styles hairy, as also the outer surfaces of the sepals and petals; sepals persistent.

This species is apparently not used for any industrial purpose, but it has been recommended in the preceding remarks as worthy of careful investigation as a possible source of improvement to the cultivated tea plant. It may indeed have had something to do with the production of the cultivated forms, in which the calyx is hairy, in addition to the normal ciliate margin of **C. theifera.**

C. drupifera, *Lour.; Fl. Br. Ind., I., 293.* 239

Syn.—C. KISSI, *Wall., As. Res., XIII., 429; Jour., As. Soc., Beng., IV., 48, t. 2; Pl. As. Rar., III., 36, t. 256;* C. KEINA, *Don, Prod., Nepál, 224;* C. MASTERSIA, *Griff., Notul. IV., 539;* C. SIMPLICIFOLIA, *Griff., Notul. IV., 560, t. 604;* C. CAUDATA, *Griff.* (non *Wall*); C. OLEIFERA, *Wall.*

Vern.—*Kissi, hingua,* NEP.; *Chashing,* BHUTIA and LEPCHA.

References.—*Kurz, For. Fl., Burm., I., 109; Gamble, Man. Timb., 30, also Darjeelig List, 9.*

Habitat.—A large evergreen shrub, with slender, much-divided branches, met with in Nepál and on the Eastern Himálaya generally in Bhután, the Khásia hills, Northern Cachar hills, Manipur, Tenasserim, and the Andaman Islands, at altitudes from 3,000 to 8,000 feet above the sea.

Botanic Diagnosis.—*Leaves* 3 to 4 by 1 to 1½ inches in size, tapering below and having also a long acuminate apex, margin serrulate, especially towards the apex, and often revolute. Twigs puberulent, with loose membranous scales embracing the buds. *Flowers* erect, white with the odour of the cherry-laurel. *Sepals* silky externally, deciduous (*i.e.*, not persistent). *Petals* emarginate. *Stamens* glabrous. *Styles* nearly free, woolly at the base.

This is closely allied to the sweetly-scented **C. Sasanqua** of China and Japan, to which allusion has been made as cultivated in China near the bushes in order to afford shade and to impart to the leaf the sweet scent of its flowers.

OIL. 240

Oil.—It is believed this species has never been cultivated in India; but apart from any possible service it might be found to render in the direction of the suggested improvement of tea through the production of a better hybrid, this plant would seem worthy of attention as an oil-seed-bearing species. At the Colonial and Indian Exhibition two or three samples of the oil from tea seed were shown and were much admired. Without any appreciable extra trouble this species might be reared as a hedge and yield a fairly remunerative oil crop at the same time. It is a

TEA. non-drying oil of a superior quality; it is used medicinally in Cochin China, and with the oil from **C. Sasanqua** is no doubt largely sold as tea-seed oil. The latter article is of considerable importance to the tea districts of China and is exported to Europe. It resembles olive oil, burns with a clear bright light, and is free from unpleasant odour.

Sasanqua Oil. 241 The oil of **Sasanqua** (*Sanskwa*=Japanese name) has an agreeable odour, and is used for many domestic purposes. It is obtained first by cold pressure, the pulp being boiled and again pressed.

The leaves are largely used by Japanese ladies for washing the hair. How far the art of perfuming teas in China is carried seems uncertain, but it is possible some of the special brands may owe more to the flowers of **C. Sasanqua** than is at present understood.

TIMBER. 242 **Structure of the Wood.**—Hard, close, and even-grained; weight 60lb per cubic foot.

243 **Camellia lutescens,** *Dyer; Fl. Br. Ind., I., 293.*

Habitat.—Mishmi Hills.

Botanic Diagnosis.—A shrub with much divided pale grey branches. *Leaves* caudate-acuminate, 2 to $3\frac{1}{2}$ by 1 to $1\frac{1}{4}$ inches, closely serrate. *Flowers* erect, crowded, white, becoming yellow, fragrant. *Sepals* caducous, pubescent internally. *Styles* short. *Stigmas* recurved.

Very little is known of this plant.

True Tea Plant. 244 **C. theifera,** *Griff., Notul. IV., 558, t. 601; Fl. Br. Ind., I., 292.*

TEA, *Eng.;* THÉ, *Fr.;* THI, *Germ.;* TE, *Dutch, It., Sp. & Scotch;* CHAI, *Rus. & Turk.*

Syn.—THEA SINENSIS (CHINENSIS), *Linn.;* THEA BOHEA (black tea) and T. VIRIDIS (green tea); CAMELLIA THEA, *Link.;* THEA ASSAMICA, *Masters, in Jour. Agri.-Hort. Soc., Ind., III. (1844), 63;* ASSAM TEA, *Wallich in Jour. A. Soc., Beng., IV., 48, t. 2;* CAMELLIA, Sp., *Griff., in Trans., Agri.-Hort. Soc., Ind., V. (1838), t. B.*

DeCandolle (*Orig. Cult. Pl., 117*) thinks with **Seemann** (*Trans., Linnæan Soc., XXII., 337, pl. 61*) that we are justified in retaining the genus **Thea,** while **Baillon** keeps the name **Thea** for the combined genus and sets aside **Camellia.** The *Genera Plantarum* and the *Flora of British India* adopt the course here followed, *viz.*, of referring the Tea plant and all its forms and allies to **Camellia. Thomson,** in a note published by **Ainslie** (*Mat. Ind., I., 440*), attempts to distinguish the so-called **Thea bohea** from **T. viridis.** The former, according to that author, has smaller leaves than the latter, with obscurely serrate margins, while the leaves of **T. viridis** are more or less oblique and denticulate. Characters so slight as these, even if constant, would not be sufficient to constitute species and they are most probably due to climate. It has, however, been shown by **Fortune, Ball,** and other writers on China, beyond all possible doubt, that the green and black teas are due to different modes of manufacture and not to different plants, although once upon a time this much-contested point carried about an equal number of advocates on either side. The modern experience of the Indian planter confirms this view, which is now all but universally accepted. The climate of one district of India (such as Kangra) or of the more northern latitudes of China, may favour the rapid process of manufacture more than the slow, and thus produce even as much or more green than black tea, but that these commercial products are derived from different species or even varieties of **Camellia (Thea)** is not now believed by any scientific writer, and the species **Thea bohea** and **T. viridis** have accordingly been entirely set aside.

Vern.—*Te, chhá,* CHINESE (**Crawford** regards the word *te* as of Malay origin, but **Yule** says it is Chinese, having, like many other words, reached the west through the Malaya); *Chhá,* ARAB., PERS., and HIND.; *Chedean,* COCHIN-CHINA (according to **Loureiro**); *Rata-tekola* (according

TEA.

to **Moon**) is the Ceylon name for **Thea bohea**. **Balfour** enumerates the following names said to be Chinese: *Ming-kutu, tu, ku-cha, kia, sheh,* and *chuen*; he further mentions the following Indian vernaculars, but these would appear to be tea garden names of a modern origin:—*Dullicham* (white wood), CACHAR; *Phlap* or *khlap, misa phlap* (in Muttack); *Hilkat*, ASSAM.

Bibliography of Indian Tea. 245

References.—In addition to the publications quoted above (under botanical synonyms) the following, among many other books and papers, &c., may be consulted. For convenience these have been arranged in the order of date of publication, commencing with those which appeared about the beginning of the present century.

Milburn's Oriental Commerce, &c., 2 Vols., 1813.

Moorcroft's Travels in the Himálaya, 1821.

New Camellia in Nepal; Wallich, in *As. Res., Vol. XIII., 428,* 1820.

Thea viridis, *Linn.*, by Ainslie, in *Mat. Ind., I., 430,* 1826.

Tea plant in Upper Assam—Report on, by Dr. W. Griffith, 1836.

Tea manufacture—Report on, by C. A. Bruce, Assam, 1839.

Thea, in *Royle's Ill., 125,* 1839.

Tea in Himálaya, Royle in *Prod. Res., Ind., 257 and 393,* 1840.

Tea in Java, Cultivation and Manufacture of; Translated from the Dutch, by J. Horsfield, 1841.

Tea in Robinson's Account of Assam, 1841.

Cultivation of Tea on the Himálaya, a lecture delivered by Dr. J. Royle, at the Royal Asiatic Society, 4th April 1844.

Tea, Report on the Cultivation and Manufacture of Tea in Kumaon and Garhwal by Dr. W. Jameson, 1843-45 (see also *Jour. Agri.-Horti. Soc. Ind., II. and IV.*)

Report of Committee on Com. Relations, China. (*Parl. Paper, 1847.*)

Tea Cultivation and Manufacture in China, E. S. Ball (London, 1846).

China Tea, Report on the progress of the culture of, on the Himálaya from 1835 to 1847, by Dr. J. Forbes Royle, London, 1849.

Tea, Kumaon, in Saunders' Monthly Mag., p. 389, 1851.

Tea Plantations of the N.-W. Provinces, by R. Fortune, Agra, 1851 (*Sel. Rec., N.-W. Prov.* (2nd series), *V., 401*).

Warrington, in *Chem. Gaz., p. 238,* 1852.

Tea-makers, Suggestions for the importation of, also Tea-seed from China to the N.-W. Provinces, by Dr. W. Jameson, Agra, 1852.

Tea, Countries of China, R. Fortune, Vol. II., 69, 233; London, 1852.

Tea Districts of India and China, by R. Fortune; London, 1853.

Tea in the Kangra District, by Dr. W. Jameson (*Sel. Rec., Pb. Govt., Vol. I.* (Nos. 1 to 6), *p. 287,* 1853).

Tea, Hooker's Him. Jour., Vol. I., 5, 144, 408; II., 341; London, 1854.

Tea cultivation in the N.-W. P. and Panjáb—Report on the present condition and future prospects (*Home Dept. Sel., XXIII., 1857*).

Tea discovery in Sylhet by Mahomed Warish, reported by J. P. Larkins, Magistrate (*Sel. Rec., Beng. Govt., XXV., 45, 1857*).

Dictionary of Commerce, &c., 1859.

Tea Plantat. in Panjáb. Corresp. (*Sel. Rec., Pb. Govt., IV., 1859*).

Tea and Tea Trade, Encyclop. Brit., XXI., 8th *Ed.*, p. 81, Edinb., 1860.

Tea Plantations of N.-W. Provinces; Report on, showing present condition and progress, Jameson, in *Supp. to Gaz., N.-W. P., 1861.*

Tea cultivation in Assam, Correspondence regarding (*Sel. Rec., Beng. Govt., XXXVII., pp. 1 to 73,* 1861).

Tea cultivation in India, by W. Nassau Lees; London, 1863.

Ship's Prize Essay on Cultiv. and Manuf. of Tea in Cachar, 1865.

Tea trade of Russia by Lambey, 1867.

Tea, in *Baden-Powell's Pb. Prod., 275.*

CAMELLIA theifera.

The Tea-plant.

TEA. Bibliography.

Tea in *Amer. Jl., Med. Sci.*, April, p. 525; Oct., p. 260; 1868.
Tea, State and Prospects of—Report by the Commissioners appointed to enquire in Assam, Cachar, and Sylhet, 1868 (*Beng. Govt.*)
Tea Cultivation in Kangra District (*Pb. Sel.* (*new series*), *V.*, *1-35, 1869*).
Tea Plantations of the N.-W. Provinces, by Dr. W. Jameson—Report for the years 1863, 1864, 1865, 1866, 1867, 1868, 1869 (*Sel. Rec., N.-W. Prov., V.* (*2nd series*), *422 to 433*).
Tea in the Nilghiris by Dr. G. Bidie, Report of Coffee Borer, p. 86, *Mad. Govt.*, 1869.
Tea Plantations in Kumaon, by Dr. G. King (*Sel. Rec., N.W.-P., II.*) No. 4 (2nd series), page 433, 1869.
Tea Cultivation, Evidence received from Gardens in Assam, Cachar, Sylhet, and Darjeeling, regarding, 1870.
Tea, in Hassal's Food, The Detection of Adulter. of, 1871 (also 1876).
Tea Pruning, Remarks on, by Dr. G. King, 1871.
Tea, Ure's Dict. of Arts, Man., &c., III., 870; London, 1872, also Supp. IV., p. 874.
Tea, Prize Essay on Cultiv. of, *Agri.-Horti. Soc. Jour., III., Part 2*, 1872.
Tea Operations in Assam, Report on, 1873-74.
Tea Cultivation in Bengal and Assam, Report on, by J. W. Edgar (*Bengal Govt., 1873; Parl. Paper, C. Pt. I., 982*, London, 1874).
Drury, *Useful Plants of India, pp. 422 and 477, Ed.; 1873.*
Allen, *Pharm. Journ.*, 1873 (also in *Chem. News, Vol. XXX.*, 1874).
Camellia Thea, *Link.*; Brandis, *For. Fl., 25;* 1874.
Tea, Coffee, and Cocoa, by Wanklyn, London, 1874 (also *Chem. News, XXVIII., 186*).
Thea, Ferminger, *Man. Gard.*, 416; Calcutta, 1874.
Brick Tea, *Sel. from Jour. As. Soc., Beng.*, Vols. I. to XXVIII, 825; Mad., 1875.
Dr. Campbell Brown on Tea in *Journ. Chem. Soc.*, p. 1217; 1875.
Thea, in Baillon's *Nat. Hist. Pl., IV.*, 235; 1875.
Tea Cultivation in Nilghiri Dist. by W. R. Robertson, p. 31, in Report on Agricultural Condition, &c., of Nilghiri Dist.; Madras, 1875.
Tea analysis by Wigner, 1876 (also *Pharm. Journ., 3rd series*), 1874.
Thea in *Treasury of Botany;* 1876.
Blyth, A. Wynter, Micro-Chemistry applied detection of foreign leaves in Tea (*Analyst*, 1877); (also Indian Tea in *Jour., Chem. Soc.*, 1875.)
Camellia, Kurz, *For. Fl., Burm., I., 108;* Calcutta, 1877.
Tea, by P. L. Simmonds, *Trop. Agri., p. 79;* London, 1877.
Tea in Assam by S. Baildon. Calcutta, 1877.
Tea Culture, Indian, by Burrell, *Jour. Soc. of Arts, XXV.*, 199; 1877.
E. Money, Cultivation and Manuf. of Tea, 3rd *Ed.;* London, 1878.
Watson, Dr. J. F., Prize Essay.
Tea, Cultivation of, in Kumaon, by J. H. Batten, in *Jour. Roy. As. Soc., X., 131*, also in *Jour. Agri.-Horti. Soc. Ind.*, V., Pt. IV.; 1878.
Tea Culture as a probable American Industry, by W. Saunders, 1879.
Camellia Thea, in Bent. and Trim., *Med. Pl., I.*, 34; London, 1880.
Chinese Tea Plant, in United States Agri. Report, 1877, p. 363; 1879, p. 27; 1881, p. 81 (as an insecticide.).
Camellia Thea, *Link.*, Müller, *Extra-Trop. Plants*, p. 59.
Tea, in Chemistry of Foods, by J. Bell, pp. 1-33; London, 1881.
Camellia Thea, *Link.*, Gamble's *Man. Timbs.*, 30; Calcutta, 1881.
Tea, Cyclop., pp. 1-355; Calcutta, 1881.
Tea, Smith, *Dic. Econ. Pl.*, p. 404; London, 1882.
Capabilities of New Zealand for Tea Culture, by W. Cochran, in *Jour. Soc. Arts*, XXX., 1882.

TEA. Bibliography.

Tea, Spons' Encyclop., p. 1994; London, 1882.
Tea, by J. F. Duthie, in Atkinson's *Him. Dists., Vol. X., N.-W. P. Gaz., pp. 887-908*: Allahabad, 1882.
The Tea Industry in India, by S. Baildon; 1882.
A Tea trade with Thibet, *Beng. Govt.*; 1883.
Tea Industry in N.-W. P. and Panjáb, by L. Liotard, 1882.
Tea trade with Thibet, by L. Liotard, 1883.
Tea in United States Dispens., Ed. 15th, 1762; 1883.
Tea, Folkard, *Plant-Lore*, p. 561; London, 1884.
Tea in De Candolle, *Origin Cult. Pl., p. 119*; London, 1884.
Tea-Mite and Tea-bug of Assam, by J. Wood-Mason; London, 1884.
Tea chests, Lead-lining, by A. Pedler; Calcutta, 1885.
Tea, Johnson's Chem. Com. Life, Ed. Church, p. 115; London, 1885.
Tea in Balfour's Cyclop., Vol. III., 829; London, 1885.
Tea, Yule-Burnell, Glossary of Anglo-Indian Terms, 688; Lon., 1886.
Indian Tea Association's Reports, 1883 to 1886.
Tea, by Stanton in the Vol. of Reports on Col. and Ind. Exh., 1885.
Tea Gazette, Indian, from 1880 to 1887; Tropical Agriculturist and other Indian Newspapers and Journals.
Indian Tea Industry, Dr. J. Berry White, *Soc. Arts Jour.*, p. 734; 1887.
Transactions, Agri.-Hort. Soc., India; Papers in Vols. II., pp. 181 and 195; III., p. 31; IV., pp. 1 and 29; V., pp. 94, 98, 102, 104, 105, 140, 155, and 160; VI., p. 10; VII., p. 1 (and Pro. pp. 45, 59); VIII., pp. 69 and 282.
Journal Agri.-Horti. Soc., India (old series). Vols. I., p. 288; II., p. 323; App. p. 408 (Chittagong); p. 337 (Assam); p. 5 (Paraguay); p. 161 (Ava); III., pp., 1 to 61 (Assam); App. p. 102 (Darjeeling); IV., p. 173; V., pp. 79 (Assam); 162, 204 (Java); App. pp. 132, (Assam), 146 (Kumaon); VI., p. 81 (Kumaon), 72 (Java), 123 (Darjeeling); App. p. 14 (Dehra Dun); VII., 292 (Java), 31 (Darjeeling); App. pp. 3—11 (Himálaya); VIII., p. 91 (Assam); App., p. 1 (Garhwal); IX., pp. 201 (Darjeeling); 207 to 342 (Sylhet); App., p. 64 (Scented Teas); X., pp. 107, 193, 227, and 229; XI., App., p. 28; XII. pp. 164, 229, 364 (Proc.) 37; XIII., pp. 31 and 181 (Proc.) 34; XIV., pp. 1, 37, 56, 119, and 282 (Proc.) 4, 22, 33, 64.
J. E. O'Conor's Reviews of the Trade of British India, 1874-75, pp. 47, 63; 1875-76, p. 20; 1876-77, pp. 11, 49; 1877-78, p. 32; 1878-79, pp. 9, 33; 1879-80, p. 42; 1880-81, pp. 39, 54; 1881-82, pp. 47, 76; 1882-83, pp. 61, 92; 1883-84, p. 39; 1884-85, pp. 21, 47; and 1885-86, pp. 28, 38, 39.

Distribution of Tea Plant. 246

Habitat.—As pointing to a common origin for the cultivated plant it is note-worthy that the name *Chhá* or some form of that word is given to tea in India, Persia, Russia, China, and Japan. But travellers in China do not appear to have observed the wild plant, and **DeCandolle** accordingly has come to the conclusion "that the tea plant must be wild in the mountainous region which separates the plains of India from those of China, but the use of the leaves was not formerly known in India." He further admits that "it is probable it exists also in the mountainous districts of south-eastern China, where naturalists have not yet penetrated." **Loureiro** (*Fl. Cochin, p. 414*) says that the tea plant is found in Cochin-China "cultivated and uncultivated," but he describes the leaves as lanceolate and acutely serrate, a description which would appear to agree better with **Camellia drupifera** than with the true tea plant. We now know that in Cochin-China that species is cultivated on account of its oil-bearing seeds. As in part supporting **DeCandolle's** conclusion that the "plant must be wild in the mountainous region which separates the plains of India

CAMELLIA theifera.

The Tea-plant.

TEA.

from those of China," it has been established beyond doubt that one if not two forms of the true tea plant occur in certain forest glades of Assam (Jaipur, Sudiya, &c.) and Cachar under such a condition as to lead to the supposition that they are either wild, that is, indigenous to these localities, or have become acclimatised as escapes from an early Indo-Chinese cultivation. In Manipur (a small Native State between Assam, Cachar, and Burma) the plant exists as a forest tree in such profusion as to leave no possible doubt that it is truly indigenous. It is note-worthy that Manipur occurs in the very latitude to which many authors fix the possible Chinese wild home of the plant. It is, perhaps, desirable, however, to prevent a too extended meaning being put upon **M. De Candolle's** "region which separates the plains of India from those of China." The genus **Camellia** has a distinct south-eastern range relative to India as a whole. The species which extends furthest to the west is **C. drupifera,** a plant met with in Nepál, and from there distributed east along the Sikkim and Bhután Himálaya, south-west to the Khásia and Cachar hills, and through Manipur to the mountains of Burma, and again south to the Andaman islands and south-east to Tenasserim and Cochin-China. **C. theifera** would appear to have its most westerly habitat in the mountainous tracts of Upper Assam, from which region it is distributed along the same range as the above species, and crossing the Khásia and Cachar hills and the Patkoi mountains, it reaches Manipur and Sylhet and extends even as far south-west as to the frontier of Hill Tipperah in South Sylhet. It is not cultivated by the present ruling race of Manipur, but the Shan tribes bordering on Manipur do cultivate the plant, and manufacture the leaves, in a crude way, into a form of wet tea. This is packed in bamboo tubes and sold all over Burma, in many parts of South-Western China, and even in India. Tea of this description is not made into a decoction, but is eaten as a preserve with other articles of food. The Western Tibetans boil tea with flour and butter and eat the mixture like a pudding, a habit somewhat similar to that followed by the Shans and Burmans of eating tea as a preserve instead of making a decoction from the leaves. The Shans have been known to manufacture this peculiar wet tea from almost time immemorial. One of the earliest Government records of this fact will be found in a report by **Colonel Hannay** on Bhamo and on the capacity of the Shan Countries (*dated January 1836, but reprinted in Sel. Rec., Beng. Govt., XXV., 1857*). Various early accounts also exist of a trade in tea between Assam and Burma with Yunnan, so that there seems little doubt the true tea plant is now, and has perhaps for centuries been, cultivated in that province of China. Numerous European travellers between India and China have referred to the tea industry of Yunnan. We have, however, little or no mention of the plant occurring in a wild state anywhere on the Chinese side of the line indicated as the known distribution of the tea plant except at the extreme south-eastern corner or in regions more or less adjacent to Manipur. It would thus appear that **DeCandolle's** opinion, as to the home of the tea plant being the mountainous region which separates the plains of India from those of China, is strictly speaking too extended. The plant in a truly wild condition occurs only in a small portion of the extreme easterly division of that mountainous tract, and further, as already remarked, as far as we have any direct evidence to bear on the question, it exists on the Indian and not on the Chinese slopes. Far away to the east, perhaps several hundred miles from the tea forests of Manipur, in South Eastern China, the great tea districts of China occur. We know very little indeed of tea in the intervening tract of rich mountainous and agricultural country. In the province of Si-Chuen several travellers have

(Margin notes: Shan Wet Tea. 247 (Conf. with 251). — Yunnan Tea. 248)

TEA.

reported tea as being found in an irregular state of cultivation. **Cooper** (*Trav. Pioneer of Commerce, page 171*), speaking of the flourishing city of Ya-tzow (or Ya-chau) says, "the greatest source of wealth to the city and surrounding district is the brick tea, which gives employment to thousands engaged in the manufacture and portage of tea from Ya-tzow to Ta-tsian-lu. The tree from which this peculiar kind of tea is manufactured grows chiefly along the banks of the Ya-ho, and unlike that which produces the tea exported to Europe, is a tall tree, often fifteen feet high with a large coarse leaf." This is very much like a description of the so-called indigenous Assam tea plant, but it recalls also in some respects the late unfortunate **Captain Gill's** description (*River of the Golden Sand*) of a curious tea plant (also grown in Western China) but which cannot possibly, from his description, be a species of **Camellia.** It would be worth knowing for certain if the brick tea of Western China, so largely exported to Lhasa and other parts of Tibet, be actually made from the leaves of a different plant from the ordinary tea of China. We thus know remarkably little of Western China and its teas.

A Tea Tree in Western China. 249

Brick Tea of Western China. 250

The tea-producing region of China lies between 23° and 25° North latitude and 115° and 122° East longitude. **Fountainier** (*Bulletin Soc. d' Acclim. 1870, 88*) says that the plant grows wild and abundantly in the mountains of Manchuria, and although not now believed to be indigenous, it occurs throughout the entire length of the islands of Japan. The Chinese and Japanese tea culivation thus extends from 20° to 24° North latitude. It is frequently found growing in regions subjected to a short but severe snowy winter, a fact which seems to have greatly influenced **Royle** and the other earlier advisers of the Government of India in selecting the Himálayan sites for experimental tea cultivation. Localities were actually selected where short snowy winters might be secured, and Assam and the Nilgiri Hills were viewed as second-rate regions.

Region of Chinese Tea Cultivation.

The Indian regions of wild (or so-called wild) distribution occur between 23° (in South Sylhet) and 28° North latitude (in Upper Assam). The Manipur tea forests are found on the mountains which separate the valley of Manipur from Burma and approximately between 24° and 25° North latitude. But the writer saw tea in the forests far to the north-east of Manipur, near the lofty mountain mass of Sarameti.

Region of Indian Tea Cultivation.

The region of Indian cultivated tea is much wider. It occurs in the Himálayan gardens of the Panjáb near 33° North latitude, and in South India it is grown between 10° and 13° North latitude. It has also been grown on the mountains of the Central Provinces and succeeds well on those of Central Bengal (Chutia Nagpur). In Ceylon it approaches to the 6° North latitude, but the principal region of Indian tea cultivation is between 23° and 32° North latitude.

The other species of Indian **Camellias** occur approximately along the same region as has been indicated for **C. drupifera**, only that they are much less abundant and are met with in isolated localities.

The writer feels disposed to regard Manipur and the Indo-Burman and Chinese adjacent regions as the true home of the tea-plant, and thus to view it as introduced into Assam and Cachar. He would even venture the suggestion that the crude mode of burying the tea leaf in the ground so as to produce the required fermentation, as practised to-day by the Shans in Upper Burma and on the borders of Manipur, may be the probable original method of preparing the article. This kind of tea has already been alluded to, and it is only necessary in this connection to add that the danger of decomposition when carried to great distances may have suggested the process of baking into brick tea. To be used, the brick tea requires to be softened by boiling, and hence might have originated

Probable History of the habit of Tea Drinking (Conf. with 247.) 251

TEA.

the Tibetan method of eating the tea leaves after they had been boiled in flour and butter. From this one might be pardoned drawing on imagination still further by supposing the enlightened Chinese to have improved the process of manufacture and to have refined the method of cooking by preparing an infusion from the leaves instead of eating them. As partly supporting this theory we have the astonishment expressed by several of the earlier writers that the Chinese only pour boiling water over their tea and do not cook the leaves. A large trade in Cardamoms exists between the capital of Kashmír and the neighbouring hill tribes who employ these to flavour their decoction of tea in place of the sugar used by the people of the west. **Major Ward** informs the writer that he has seen the shepherds of that region smoking tea instead of either eating it or drinking an infusion from it. An Assam tea planter writes to the author that he had once given him a coarse kind of tea to smoke in place of tobacco, and that although it seemed inferior stuff he was not able for some time to detect that it was tea and not tobacco that he had been actually smoking.

Smoking Tea.

The Spread of Tea Cultivation. 252

The stirring national migrations of the early inhabitants of Eastern Asia through the Burmo-Chinese regions, and the early trade-route which became established, with the more settled condition of the people, might easily be supposed to have carried the tea plant at an early date to China and to India more recently by the Siam invasions. As opposed to all this it may be urged that there are references to tea in Chinese botanical works (or to what appears to be tea) at a date prior to any known migrations from Burma to China or from China to Burma or Siam. But in none of the very early supposed references to tea is mention made of eating the leaves as pickle or after being cooked into pudding or of making a beverage from them by means of boiling hot water. May not the tea plant therefore or some allied Camellia, have been cultivated in ancient China for a perfectly distinct purpose to that for which it is now grown? This line of reasoning is only on a par with the fact that down to modern times we have no indications of the properties of the tea plant having been known to the enlightened ancient people of India.

Botanic Diagnosis.—A small bush while under cultivation or when found in high latitudes or high altitudes, but in warmer regions, such as in the damp forest glades of Assam and Cachar, and in the tea forest of Manipur, often becoming a tree from 30 to 50 or even 100 feet in height. *Leaves* variable, especially when cultivated, generally tapering at both extremities, elliptic-oblong, acute or cuspidate-acuminate, puberulous on the nerves below, 4 to 8 by $1\frac{1}{2}$ to $2\frac{1}{2}$ (in the wild plant often 12 to 15 by 4 to 6) inches in size. *Flowers* white, solitary, pendulous. *Sepals* persistent. *Styles* united for about $\frac{2}{3}$rds of their length.

Improvement of Tea stock. 253

In some of the cultivated states, the calyx is described as quite hairy, the leaves small, and even obtuse. This may possibly be due to a certain degree of hybridisation with **C. caudata**—a suggestion well worthy of investigation. Indeed, it may be repeated, with the greatest assurance, that the time has now come for the planter to devote a greater share of his time and attention to the study and improvement of his plant-stock than has hitherto been done.

CHINA TEA. 254

THE HISTORY OF THE CHINA TEA.

There is every reason to believe that, although the habitat of the tea plant may be somewhere on the Assam-Burman and Chinese frontier, the practice of preparing a beverage from its leaves existed for centuries in China before it was known in India. Apparently classical scholars have failed to find any allusion to the plant or to the beverage in the

TEA.

works of the early Sanskrit, Arabic, and Persian writers. Tradition would seem to point to the plant having come from India to China, but the legend upon which this idea mainly depends is told by the Japanese and seems unknown to the Chinese themselves. In his interesting little work (*On the Study and Value of Chinese Botanical Books*) **Dr. Bretschneider** says that the plant is alluded to by a writer as early as 2700 B.C., and that a commentator, alluding to this fact, adds (in the 4th century A.D.) that by means of hot water a beverage is obtained from the leaves of the plant.

The Beverage made in China in the 4th Century.

Thus the literature of China allows of little doubt as to the beverage having been known in that country at least since the 4th century, and very possibly from a much earlier date. According to most writers it began to be systematically cultivated in South-Eastern China about that period, and we have a definite reference to the industry in the annals of the T'ang Dynasty, 793 A.D., where allusion is also made to the article having been subjected to an imperial duty. **Macpherson** (*History of European Commerce with India*) remarks that **Soliman**, an Arabian merchant, who wrote an account of his travels in the East about A D. 850, describes the Chinese as using "tea (*sah*) as a beverage." Japanese writers admit that they got tea from China in the 9th century, and began to cultivate the plant for themselves in the year 1206 A.D. The Portuguese had dealings with the Chinese in the beginning of the 14th century, and it is probable they were the first to introduce tea to Europe. This is claimed, however, by some authorities for the Dutch, the article having been first shown in Amsterdam and thence sent to London. The earliest authentic European notice of tea occurs in **Ramusio's** introduction to **Marco Polo**, in the year 1545, where he mentions having learned of the beverage from the Persian merchant **Hajji Mahommed. Anderson** (in *History of Commerce*) quotes **Botero** as recording that in 1590 the Chinese "have an herb out of which they press a delicate juice which serves them as a drink instead of wine," and he infers, perhaps correctly, that this was tea. **Texeria**, a native of Portugal, is reported to have seen the dried leaves of tea in Malacca in the year 1600, and **Olearius** found tea being used in Persia in 1633.

Japan in the 9th Century.

Europe in the 16th Century.

Perhaps the most amusing and at the same time instructive incidents in the history of tea are recorded in the proceedings of the East India Company (*see Milburn's Oriental Commerce*). An officer of the Company wrote to his friend in Meaco in the year 1615, asking for "a pot of the best sort of chaw" (*Murray's China, II., 337*). Probably the earliest record of the importation of tea into England by the great East India Company is to be had in an entry in the Company's books in June 1664 of having presented the King with 2 ℔ and 2 oz. of "thea," which cost 40*s*. a ℔. Two years later the Company appears to have been more liberal, for a second present to His Majesty is recorded:—

Tea was in use in England in the 17th Century.

	£	s.	d.
"22¾ ℔ of thea at 50*s*. per ℔	56	17	6
For the two cheefe persons that attend His Majesty, thea .	6	15	6"

Not, however, until the year 1677 did the East India Company take steps to secure a regular and commercial supply of tea. The order the London Directors then issued was "for teas of the best kind to the amount of 100 dollars." This order seems to have been exceeded, and the market accordingly glutted, for we next read of complaints regarding the excessive consignment of 4,713℔ made in 1678 (*see Macpherson's Hist., European Com. with India, p. 131*). Tea sold in London about this period at from £5 to £10 sterling a pound. Shortly after (1657) cups of tea began to be sold in the public coffee-rooms of London, especially at

Commercial supply in 1677.

Imports 4,713 lbs.

TEA.

"Garraway's," and a duty was claimed from the vendor of 8*d.* a gallon. In *Pepys' Diary*, under date of 28th September 1660, there occurs the entry: "I did send for a cup of tea (a China drink) of which I had never drank before." **Yule-Burnell**, in their *Glossary of Anglo-Indian Terms*, give numerous other passages from early English writers in which mention is made of tea down to the year 1789.

A duty levied, 1689.

The first direct duty levied on the sale of tea was in the time of William and Mary (1689); it was then subjected to a tax of 5*s.* a pound and 5 per cent. on the value of the article *ad valorem.* This is perhaps the heaviest duty to which it has ever been subjected. As a result the consumption appears to have remained stationary, and what is noteworthy the East India Company drew its supplies at that time from Madras and Surat and not direct from China. Towards the close of the 17th century they began, however, to import direct, and the first consignment came from Amoy. The annual consumption in England, perhaps in Europe, seems to have assumed by this time the comparatively considerable proportions of 20,000℔. It is important to add that the East India Company had secured for themselves from the British Parliament the concession of being the only merchants allowed to import tea, and for nearly 180 years they enjoyed this monopoly, free trade in tea having only been allowed as late as 1833.

Tea Monopoly.

In 1703 the imports into Great Britain amounted to 105,000℔, and the article was sold at 16*s.* a ℔. In 1704, the Chinese, imitating the monopoly granted by the British Government to the East India Company, endeavoured to establish a Chinaman as the Emperor's merchant who alone would be permitted to sell tea to the Company. This audacity was characterised by the indignant Company of merchants as a "new monster in trade," but the monster was conquered by the payment of a bribe of £1,600 per ship. In 1728, we read that the Emperor of China was not, however, to be disposed of in this manner, and that an export duty of 4 per cent. was levied, which was raised, in 1736, by an additional 10 per cent. (*Aubor on China, p. 150.*)

Imports 1,000,000lbs.

In 1721 the imports into Great Britain of tea amounted to 1,000,000℔, and seven years later they had increased by another 100,000℔, the revenue therefrom having been £104,300. From 1722 to 1744 the duty was fixed at 4*s.* a ℔ excise, with, in addition, a customs due of 14 per cent. **Macpherson** has estimated that this amounted to 200 per cent. on the average price of the article. From 1784 to 1795 the duty was gradually remitted until it fell to only 12½ per cent. The imposition, at the beginning of the 18th century, of so heavy a tax naturally gave encouragement for excessive smuggling and developed a large trade in the art of adulteration. The evils urged by many writers of that period against the habit of tea drinking were doubtless justified from the poisonous or deleterious nature of the admixtures. In spite of all opposition, however, the habit of tea drinking steadily took hold of the English people, and at the present day there is perhaps no other article of food that is so little adulterated.

Adulteration.

During the 100 years from 1710 to 1810 the aggregate sales of tea by the East India Company amounted to 750,219,016℔, valued at £129,804,595 sterling, and of that amount 116,470,675℔ were re-exported to other countries. At the present day Great Britain consumes in three years as much tea as she thus required in a 100 years, little more than 70 years ago. But the effect of decreasing and again increasing the duty affords a very instructive lesson as to the influence of taxation on the luxuries of life. In 1745 it was reduced 50 per cent., with the result that whereas the sales for the five previous years had on an average been 768,520℔ and yielded a revenue of £175,222, for the five succeeding years (after the reduction)

TEA.

they were 2,360,000lb and gave an annual revenue of £318,080. This extremely favourable result, instead of suggesting the advisability of further reduction, seemed to excite only the cupidity of the rulers to obtain from the supposed educated taste of the people a greater revenue. From 1759 to 1784 the duty was steadily raised until it attained the alarming proportion of 119 per cent. on the value of the article. Smuggling and adulteration were of course renewed with greater energy than before. But in 1784 the duty was again reduced to 12½ per cent. For the three years previous to this reduction the average sales amounted to 5,721,655lb with a revenue of £700,000, and for the three succeeding years the sales were 16,044,603lb and realized, from a 1-10th duty, nearly 1½ the former revenue. Unfortunately, in 1795, the old course was resorted to, *viz.*, of raising required money by taxing tea; the article was again and again burdened, until in 1819 it was made to sustain 100 per cent. duty. The result was that during these 25 years the sales stood stationary at an average of 21,000,000lb and yielded an average revenue of 2½ million pounds sterling. The restriction in the sale of tea thus caused was greatly increased by the fact that the East India Company still retained its charter as the sole importers of tea, but in April 1834 a new state of affairs began to dawn. An Act of Parliament had abolished the East India Company's monopoly, and free trade considerably lowered the initial price of tea. At the same time the *ad valorem* duty was abolished and differential rates established, and all "bohea teas" were subjected to a customs duty of 1*s.* 6*d.* a lb, the better qualities of tea paying 2*s.* 6*d.* to 3*s.* a lb.

Revenue £2,500,000.

Removal of Tea Monopoly, 1834.

In 1836 the duty was again altered to a uniform charge of 2*s.* and 1*d.*, which rate, with the addition of 5 per cent. imposed in 1840, prevailed till 1851, and in 1853 it was so regulated that by 1856 it should become 1*s.* a lb. The Russian war temporarily disturbed this arrangement, and the English people paid cheerfully 1*s.* and 9*d.* a lb for their tea, until 1857, when it was again reduced to 1*s.* and 5*d.*, and in 1864 to 1*s.*, and was finally fixed in 1867 at 6*d.* a pound, at which rate it still remains. Coincidently with the reduction of duty occurred an equally important consideration—a fall in the price of the article. About the middle of the 17th century a pound of good tea cost in London as much as £10 sterling; at the present day a better article may be purchased for 2*s.* and 6*d.* a pound.

Present duty 6d. a lb.

Price of Tea.

The writer has purposely passed over, in their chronological places, the incidents connected with the history of the Indian tea industry, deeming it desirable to give, in the first place, a succinct account of tea as a whole, and then to treat of India by itself. By way of concluding this part of the history of tea, it may be repeated that, at the beginning of the 18th century, the imports of tea into Great Britain were only 20,000lb, but that in 1885 they amounted to 212,375,371lb, and were in 1883 even still higher. These facts forcibly illustrate the growth of the habit of tea-drinking during the past two centuries, and it is somewhat remarkable that this taste should have developed almost exclusively amongst the British people, for, with the exception of the Russians, the other nations of Europe drink very little tea. Indeed, were it not for the hold the luxury has long maintained in China and Japan, the growth of the British demand might be accepted as indicating a peculiar suitability of tea for the people of cold countries.

THE HISTORY OF THE INDIAN TEA INDUSTRY.

INDIAN TEA. 255

Difficulties with China early began to make the British Government realise the danger of having no other source of tea than China. Ultimately the whole energies of the Chinese section of the East India Com-

TEA.

Tea in America.

pany were concentrated in the tea trade. Friction with the Company soon gave vent to loud outcries in England which were re-echoed by the disaffection of America. Tea in fact became intimately connected with the severance of the American Colony from the Crown of England. Colonists, disguised as Indians, boarded British ships, laden with heavily-taxed tea and threw it over-board; this was one of their first acts of open rebellion. The taxation of tea thus became a serious problem, and in a half-hearted way the East India Company responded to the wish of the Government that efforts should be made to cultivate tea in India.

Tea seed sent to India in 1780.

Seed was accordingly sent from Canton in 1780 to Bengal, and this was handed over to **Colonel Kyd,** who germinated it and planted out a small nursery in his garden. As a historic fact of considerable interest, this garden ultimately became the Royal Botanic Gardens, Seebpore, near Calcutta. **Colonel Kyd,** one of the founders of horticulture in India, and one of the earliest botanists of whom we have mention, has a fitting memorial in the centre of the Seebpore Gardens. Reporting on his tea experiments he wrote to **Sir Joseph Banks** pointing out that the neighbourhood of Calcutta did not seem the most suited locality. In reply **Sir Joseph,** in 1788, addressed **Warren Hastings** as to the desirability of attempting China tea cultivation in Behar, Rungpur, and Kuch-Behar. Little more seems to have been heard of the subject, and even the discovery of tea in Assam and Manipur made by **Scott, Bruce** and others between 1819 and 1821, appears to have been allowed to pass without more than a cursory consideration.

Discovery of Tea in India, 1819—1821.

Dr. Buchanan Hamilton endeavoured to press upon the consideration of Government the importance of this discovery, but so little did the matter seem to affect the Directors of the East India Company, that the very correspondence regarding it is vague and indifferent. According to some writers, **Mr. David Scott,** the first Commissioner of Assam, discovered tea there; by others he is said to have received the plant through native agencies from Manipur. According to **Balfour,** he addressed **Mr. G. Swinton,** the then Chief Secretary to the Indian Government, on the subject and forwarded specimens to him. The writer has failed to find this correspondence in the published Records of the Government, but whether the discovery of tea was first recorded from Assam or from Manipur is almost immaterial. There seems no doubt whatever that **Mr. Scott** was the first European who drew attention to the existence of the tea plant on the eastern frontier of India. He is also said to have sent a plant of the Assam tea to **Mr. Kyd** (son of the **Colonel Kyd** already alluded to above), in order to compare it with **Kyd's** Chinese plants. That specimen found its way to **Dr. Wallich's** hands and is now, it would appear, in the Wallichian Herbarium in the Linnæan Society's Rooms, London, with **Mr. Scott's** letter attached to it.

Gold medal of the Society of Arts.

At this period so urgent were the home authorities to secure some other source of tea than China, that the Society of Arts, taking the matter up, offered their gold medal to any person who would produce the best sample of Indian or Colonial grown tea. Interest was thus awakened, but years passed before any one claimed the medal. In 1826 the brothers **Bruce,** inspired by **Scott** according to some authors, and acting independently according to others, rediscovered the tea plant in Assam; in consequence **Mr. C. A. Bruce** was awarded the Society of Arts' gold medal; he also obtained, from the Indian Government, a grant of land for tea cultivation. But another claimant for the medal appeared in the person of **Captain Charlton,** who asserted his claim as the real discoverer of the tea plant, and so far prevailed on the Agri.-Horticultural Society of India that they conferred on him *their* silver medal. It has been clearly proved by **Mr. Burrell** that **Scott** was prior to either of these

TEA.

pioneers, but there seems no doubt whatever that **Major** (and possibly also **Mr.**) **Bruce**, had prior claims to **Charlton** for being the re-discoverers of the indigenous tea of Assam.

Operations commenced.

About the time these discoveries were being made in the then (to Europe at least) *terra incognita* of Assam, animated discussions were taking place in England which ultimately culminated in the overthrow of the East India Company's monopoly. **Lord William Bentinck**, then Governor-General of India, took up warmly the matter of Indian tea cultivation. A committee was appointed, with **Dr. N. Wallich** as Secretary, to report on the situations best suited for the experimental cultivation of China tea in India. **Drs. Wallich and Royle** urged that the experiment should be first made at Kumáon, on the Himálaya, being guided by a consideration of the latitude, climate, soil, and vegetation of South Eastern China closely agreeing with certain portions of the Himálaya. One of the first acts of the committee was to despatch **Mr. G. J. Gordon** to China, in order to collect information regarding every feature of the Chinese cultivation and manufacture of tea, and to bring away plants and seed. That gentleman had scarcely commenced his enquiries when he was recalled by the announcement that the tea plant had been found in Assam. **Captain** (afterwards **General**) **Francis Jenkins** had become Chief Commissioner of Assam, and he went with energy into the **Bruces'** discovery of tea. Had **Mr. Scott's** still more early discovery received even a passing consideration, **Mr. Gordon** would, in all likelihood, never have been deputed to China, and several years would have been saved, and according to many planters the curse of Assam—China tea—would have never found its way there. As it was, **Dr. Wallich** at first refused to accept **General Jenkins'** plant, as being the true tea-yielding species, a fact which would point to **Dr. Wallich's** having in all probability paid little or no attention to **Mr. Scott's** discovery, although he appears to have deposited in his herbarium the specimen of the plant given him. In consequence of **Dr. Wallich's** doubting the accuracy of the identification of the Assam plant, a commission was appointed in 1836, consisting of **Drs. Wallich, McClelland,** and **Griffith** to visit Assam and report on the tea said to be found there. One of the most curious results of this commission was that the reiteration of the opinion that the Himálayan localities, formerly recommended, were preferable for experimental tea cultivation, and after those Upper Assam, and last of all the mountains of South India. They, however, concluded that it would be desirable to open out one garden in Assam, but recommended *that the China plant and not the degenerated Assam plant* should be tried. **Drs. Wallich, Royle,** and **Falconer** continued almost to the last to contend that the Himálayan localities would be preferable, but the claims of Assam were eventually recognised and urged by **Drs. McClelland and Griffith.**

Tea Commission appointed, 1836.

Himalayan Gardens recommended.

Experience has tended to show that the China plant grows better on the Himálaya than the Assam, and conversely, that better results are obtained in Assam and Cachar with the indigenous stock or some hybridized state from it than from the pure China. In a measure, therefore, the opinions given were correct, for **Drs. Wallich, Royle,** and **Falconer** were strong advocates for the pure China plant, and the localities selected by them for that plant were certainly preferable to the hotter and damper regions of Assam.

Seed sown in the Calcutta Botanic Gardens; plants sent to Kumaon, 1834.

By 1834 the plants raised in the Calcutta Botanic Gardens from the seed brought from China by **Mr. Gordon** were ready for issue to Kumaon, and were placed under the charge of **Dr. Falconer**, who had now succeeded **Dr. Royle** as Superintendent of the Saharanpur Botanic Gardens.

TEA.

First Assam Garden, 1835.

Indian Tea sent to England, 1838.

In 1835 the first experimental plantation in Assam was opened up by Government in Luckimpore, and in 1838 the first commercial sample of Indian-grown tea was forwarded to England; it amounted to 488lb. The Luckimpore experiment failed, and the plants were removed to Joypur in the Sheebsagar district. This was in 1840 sold to the Assam Company, the first tea concern, and to this day very much the largest Company in India. It was anything but prosperous during the first 15 years or its existence, and its shares fell so low that they could scarcely be sold. About 1852 its prospects began to improve, and with *its* success the tea industry appeared so promising and attractive, that speculators eagerly rushed into it. In 1851 the imports of Indian teas into England amounted to nearly ½ million pounds. In 1855 indigenous tea was discovered in Cachar, and in the following year it was found by **Mahomed Warish** in South Sylhet (*Beng. Govt. Sel., XXV., 45*). Previous to this (in 1853) attempts had been made to cultivate tea in Darjeeling, but the industry was not fully started there until 1856-57. Various attempts were made between 1835 and 1840 to introduce tea into Southern India, but little interest was taken in the experiments previous to 1865 (*Robertson's Rep., Nilgiri Dist., 1875, 31*).

In Chittagong and Chutia Nagpur tea cultivation was started about 1862-67. Ultimately tea cultivation spread over every district in India, where there was the least hope of success, but with a rapidity that was certain to culminate, as it did in the great disaster of 1865-67. It is needless to dwell on the causes of that disaster, but the reader is referred to

Tea Disaster, 1865-67.

Mr. Ware-Edgar's excellent and full report (*Reprinted as a Parliamentary paper, C. 982, 1874*). It may briefly be characterised to have been the result of reckless impetuosity, ignorant supervision, and positive dishonesty. Fortunes were made by the few who realised that the tide would turn. The better-situated gardens were purchased for fewer rupees than they had cost pounds sterling to construct. New companies were formed to work these gardens, and with the avowed purpose of growing tea for its own merits as a commercial article and not for the purpose of selling their gardens at a profit whenever popular favour returned to tea investment. Out of these trying times the industry rose on a firmer foundation, and the prosperity that has attended the labours of the planter has been recently and fittingly told by **Dr. J. Berry White** in the Journal of the Society of Arts. **Dr. White** has shown that the heavy expenditure on cultivation and manufacture has been so effectively reduced (and that it may be even still further lowered) that all fear of competition with China may be said to have been removed. But while this is so many planters hold the opinion that a danger exists in the outcry for reduction, since the point may be thereby reached of defective cultivation. China, once supposed an insurmountable obstacle to the Indian planter, has, however, been practically vanquished, for within the past few months India combined with Ceylon has been leading the market. Thus in little more than half a century India has come to supply half the world's demand for tea, and there is no reason to suppose that she has by any means reached her highest level. The latest returns show the shipments from China for this year as 30 million pounds below those of the preceding year. Hitherto the attention of the Indian planter has been directed to compete with China in the London market, while all the time the imports into India of cheap China teas have been steadily increasing. The time has now come when the Indian planter, to extend his trade, must consider the requirements of new markets.

Growth of Indian Tea Trade.

By way of strikingly illustrating the growth of the Indian tea industry the following table has been compiled from various trustworthy sources.

The Tea-plant. **CAMELLIA theifera.**

The British Government commenced to record separately Indian teas in 1852, but the table has been drawn up from 1864-65 to 1885-86. Briefly, it may be repeated the exports from India were in 1838 declared to be 488℔, while in 1886 they had attained the proportion of 68,784,249℔. TEA.

YEAR.	1 Quantity exported to all countries from India in ℔.	2 Value of the same in Rs.	3 Imports into Great Britain of Indian tea (from 1873 including Ceylon) in ℔.	4 Per centage of Indian to China teas consumed in Great Britain.
1864-65 . . .	3,457,430	28,02,840	2,510,000	3 to 97
1865-66 . . .	2,758,187	27,50,550	5,133,000	4 to 96
1866-67 . . .	6,387,088	36,03,268	7,084,400	6 to 94
1867-68 . . .	7,811,429	68,69,280	8,132,400	7 to 93
1868-69 . . .	11,480,210	95,13,704	10,448,320	10 to 90
1869-70 . . .	12,754,022	1,03,78,830	13,148,900	11 to 89
1870-71 . . .	13,232,232	1,12,05,167	15,351,600	11 to 89
1871-72 . . .	17,187,328	1,45,49,846	16,942,000	13 to 87
1872-73 . . .	17,789,911	1,57,76,907	18,404,000	15 to 85
1873-74 . . .	19,324,235	1,74,29,256	17,377,900	13 to 87
1874-75 . . .	21,137,087	1,93,74,292	25,605,100	16 to 84
1875-76 . . .	24,361,599	2,16,64,168	25,605,100	17 to 83
1876-77 . . .	27,784,124	2,60,74,251	29,383,700	19 to 81
1877-78 . . .	33,459,075	3,04,45,713	31,883,300	23 to 77
1878-79 . . .	34,432,573	3,13,84,235	36,007,100	22 to 78
1879-80 . . .	38,174,521	3,05,10,200	38,483,700	28 to 72
1880-81 . . .	46,413,510	3,05,42,400	45,764,900	30 to 70
1881-82 . . .	48,691,725	3,60,91,363	54,080,300	31 to 69
1882-83 . . .	57,766,225	3,69,94,965	61,666,500	34 to 66
1883-84 . . .	59,911,703	4,08,38,805	65,731,600	37 to 63
1884-85 . . .	64,162,055	4,04,47,592	68,159,600	39 to 61
1885-86 . . .	68,784,249	4,30,61,335	76,585,000	41 to 59

[NOTE.—The slight disparity in the figures in the first and third columns is due to various causes; India exports to other countries besides England; the official years did not correspond for the first 15 or 16 years, and from 1873 the returns of Ceylon teas are in column 3 included with those of India.]

It will suffice to conclude this brief history by stating that in the year 1884-85 there were in all 3,432 tea plantations in India, occupying a total area of 728,409 acres, of which 267,711 acres were actually under tea and 217,441 yielding crop, the average yield being 303℔ an acre.

Oil.—Tea-seed oil has already been alluded under **Camellia drupifera**, and it is only necessary to add that as this substance figures largely in Chinese and Japanese commerce it is commended to the attention of tea planters as a biproduct that might be worthy of their attention. (*See Spons' Encycl., p. 1411.*) An essential oil is also distilled from the leaves, quite distinct from the fatty oil. OIL. 256

The reader is referred to another volume under TEA for an account of the Methods of Cultivation and Manufacture of Tea and for other information regarding the Commercial Article, its Chemistry, Adulteration, and Trade Statistics.

Camphire, the sweet-smelling Camphire of Solomon, is, according to some authors, the *Henna* of Indian writers; see **Lawsonia alba**, *Lamk.*, LYTHRACEÆ. **Camphire** is by other writers a synonym for **Camphor.**

CAMPHOR.

257 **Camphor.**

CAMPHOR, *Eng.;* CAMPHRE, *Fr.;* KAMPHER, KAMPFER, *Germ.;* CANFORA, *It.;* ALCANFOR, *Sp.*

Vern.—*Káfúr, kapur, ghausar,* HIND.; *Karpúr, káppúr,* BENG.; *Kárpúra, kapur,* MAR.; *Kapúr, karpúr,* GUJ.; *Kápúr,* DUK.; *Karuppúram, karppúram, shúdan,* TAM.; *Karpúram,* TEL.; *Kappúram, kaporbarus, kapur, kafur,* MALA.; *Karpúra,* KAN.; *Karpúra, chandráhbo,* SANS.; *Káfúr,* ARAB., PERS.; *Pa-yók, payo, piyo, parouk,* BURM.; *Kapuru,* SING.

References.—*Roxb. (Shorea), Fl. Ind., Ed. C.B.C., 440; Pharm. Ind., 190; Camphor of Sumatra, by J. Macdonald, Esq., in As. Res., Vol. IV., pp. 19—33; Mason's Burma, 483; Flück. & Hanb., Pharmacog., 510-518; U. S. Dispens., 15th Ed., 330; U. C. Dutt, Mat. Med. Hind., 222; Dymock, Mat. Med. W. Ind., 2nd Ed., 93 & 665; Ainslie, Mat. Ind., I., 588; Waring, Bazar Med., 32; Year-Book of Pharm., 1873, p. 97; Spons, Encyclop., 571-778, 796, 1624; Balfour, Cyclop., Ed. 1885; Treasury of Botany; Ure, Dic. of Arts and Manuf.; Kew Official Guide, Bot. Gardens and Arboreum, 120, 125.*

Camphor.—The name 'Camphor' is applied to various concrete, white, odorous, and volatile products, all of vegetable origin and possessing similar properties. They would appear chemically to be secondary formations from the volatile oil of the particular plant from which they are derived. A number of plants belonging to widely different families are accordingly found to yield this substance. Of these, however, three may be regarded as important, but only one of these commercial at the present day.

FORMS OF CAMPHOR.

FORMOSA. 258 *1st.*—The FORMOSA or CHINESE CAMPHOR, and JAPANESE CAMPHOR. This is the most important—the commercial form of Camphor. It is prepared as a crystalline substance, deposited on cooling, from a decoction made from chips of the wood boiled by a process very similar to that adopted in the manufacture of catechu. The tree which affords this substance is known as the Camphor laurel, **Cinnamomum Camphora,** *F. Nees.,* of the Natural Order LAURINEÆ, a plentiful tree in the interior of the Island of Formosa, in Japan, and throughout Central China. The bulk of the Camphor from these countries reaches Europe from Canton, and is accordingly known by the collective name of Chinese Camphor; but a considerable proportion fully deserves that name, from the fact of its being extracted at Chinchew in the province of Fokien on the mainland of China. Camphor, however, is not extracted in the Chinese possessions of Formosa, but is prepared in the country still held by the aboriginal tribes, or in the belt of debateable territory which separates the Chinese possessions from the interior. Recently, through the action of the Chinese authorities, the Formosan trade in Camphor has been almost entirely ruined, and the reports of the London drug marts rarely, if ever, now mention this once valued Camphor. In Japan, the plant flourishes throughout the three principal islands, but the extract is chiefly prepared in the province of Tost in Sikok, the mild damp sea-air of that island being apparently favourable to the growth of the tree. In the districts of Satsuma and Bungo a considerable amount of Camphor is also manufactured.

BARUS. 259 *2nd.*—The BARUS CAMPHOR (from Barus, a town in Sumatra), also known as KAPUR BARUS, BORNEO CAMPHOR, and MALAY CAMPHOR, and, in the Indian Trade Returns, as BHIMSAINI or BARAS. It is obtained as coarse crystals, formed naturally in the stems of **Dryobalanops Camphora,** *Colebr.* (**D. aromatica,** *Gærtn.*), a tree closely allied to the

FORMS OF.

Indian *sál* and a member accordingly of the Natural Order DIPTEROCARPEÆ. This is a large and handsome tree, met with in the north-western coast of Dutch Sumatra (the Batta country) from Ayer Bagnis to Barus and Singkel, also in the northern part of Borneo and in the Island of Labuan. To obtain this substance the trees are felled and completely destroyed, being cut up into small splinters in the search for the camphor crystals. It is stated that only about one-tenth part of the trees thus ruthlessly destroyed are remunerative. The formation of this crystalline substance within the tissue of the wood is thought to be due to some cause similar to that which gives origin to the crystalline substance *Khersal* in the interior of the stems of **Acacia Catechu**, or to the formation of the highly-prized masses of the *Agar* resin in **Aquilaria Agallocha.** The crystals of Camphor are chiefly found in the interior of the stem, often existing in concrete masses, which occupy longitudinal cavities or fissures in the heart of the tree, from a foot to a foot and a half long. More frequently they fill the hollows and interstices within the timber, especially in the knots and swellings formed where branches issue from the stem. The old trees are generally the most productive; an average tree is said to yield 1 lb. In addition to occurring within the wood, the Camphor is also found in a concrete form underneath the bark. In searching for trees likely to yield Camphor, the natives pierce the stem to the heartwood, thus injuring the tree materially; but it is said that a tree left for seven or eight years will be then found to contain Camphor.

BLUMEA. 260

3rd.—The BLUMEA OR NGAI CAMPHOR. This is scarcely known out of China, but a consignment according to the **Rev. Mr. Mason** was sent to Europe from Tenasserim. In China this is prepared chiefly at Canton and in the Island of Hainan, the plant being a large, herbaceous, or bushy member of the COMPOSITÆ in the genus **Blumea.** It is probable that two if not three species are used in Burma for this purpose, the most abundant being the plant employed in China, *viz.*, **B. balsamifera,** *DC.* This species is common throughout the Eastern Himálaya, ranging from 1,000 to 4,000 feet in altitude. It occurs also in the Khásia Hills, in Chittagong, Pegu, and Burma, being distributed throughout the Eastern Peninsula to China. In some of the reports which have appeared of the Burmese Camphor, it is stated that **B. densiflora,** *DC.* (**B. grandis,** *Wall*), was the species used; but this seems improbable, since **B. balsamifera** is abundant in Burma, and **B. densiflora** apparently absent, or at least absent from the district where this Camphor is described as having been prepared. **Mason** says "the plant is so abundant, that Burma might supply half the world with camphor. Wherever trees are cut down this weed springs up, and often to the exclusion of almost everything else." **Dr. Dymock** has recently drawn attention to a camphoraceous Blumea common near Bombay, and used by the country-people to drive away fleas. (*See* **Blumea,** Vol. I, B. 539.)

PERFUMERY CAMPHORS. 261

4th.—CAMPHORS USED CHIEFLY IN PERFUMERY.—As already indicated, in addition to the three sources discussed above, the chemical substance camphor can be, and is to a limited extent, actually prepared from a number of other plants, among which the following may be enumerated: TOBACCO CAMPHOR, produced by distilling tobacco leaves with water; CAMPHOR OF THYME, a crystalline product of the fractional distillation of the essential oil of thyme—**Thymus serpillum,**—one of the commonest west temperate Himálayan plants; PATCHOULI CAMPHOR (a substance known in perfumery and homologous with BORNEO CAMPHOR), prepared from **Plectranthus** or **Pogostemon Patchouli,** two herbaceous plants both members of the LABIATÆ, which are met with in Sylhet, Burma, and the Malayan Peninsula and cultivated in many parts of India. There are, in addition, a number of other camphors, less intimately related to India, such as NEROLI CAMPHOR, prepared from the flowers of the bitter orange; BERGAMOT CAMPHOR, BARASA CAMPHOR, SASSAFRAS CAMPHOR, and ORRIS CAMPHOR.

In India, in addition to the species of **Blumea** above enumerated as yielding Ngai Camphor, there are many plants which smell strongly of camphor, some of which would most probably be found to yield that substance. Among these may be mentioned the common aquatic weed of the plains of Bengal, **Limnophila gratioloides**, *Br.*, the *Karpúr* of the Bengalis; and also the numerous species of aromatic Blumeas, some of which have already been alluded to.

HISTORY.

262 **History of Camphor.**—Having now very briefly discussed the sources of the various kinds of Camphor, it may not be out of place to say something here of the history of that substance. The authors of the *Pharmacographia* inform us that there is no evidence that Camphor was known to Europe during the classical period of Greece and Rome. The first mention of the substance "occurs in one of the most ancient monuments of the Arabic language, the poems of **Imru-i-Kais**, a prince of the Kindah dynasty, who lived in Hadramaut in the beginning of the sixth century." About this period no mention occurs in Chinese writings of Camphor, although the tree was well known and the timber described. In the thirteenth century **Marco Polo** saw forests in Fokien, South-Eastern China, of the trees which give camphor (*Yule, Book of Ser. Marco Polo, II.* (*1871*), *185*). It was not, however, until **Garcia de Orta** in 1563 pointed out that the Camphor of Europe came from China, that the existence of the two forms of Camphor became known. The earlier Arabian writers all clearly refer to the expensive Camphor of the Malaya, which, even at the present day, is a hundred times more expensive than that of China. In the sixth century, Borneo Camphor was regarded as the rarest and most expensive of perfumes. "**Ishak ibn Aman**, an Arabian physician living towards the end of the ninth century, and **Ibn Khurdadbah**, a geographer of the same period, were among the first to point out that camphor is an export of the Malayan Archipelago; and their statements are repeated by the Arabian writers of the Middle Ages, who all assert that the best camphor is produced in Fansúr. This place, also called Kunsúr or Kaisur, was visited in the thirteenth century by **Marco Polo**, who speaks of its camphor as selling for its weight in gold (*Flück. & Hanb., Pharmacog.*).

Yule and Burnell, in their *Glossary of Anglo-Indian Words*, inform us that the *Kaisúr* and *Káfúr-i-Kaisúrí* of some authors is the result of the perpetuation of a blunder, "originating in the misreading of loose Arabic writing. The name is unquestionably *fansúrí*. The Camphor *al-fansúrí* is mentioned as early as by **Avicenna** and by **Marco Polo**, and came from a place called *Pansúr* in Sumatra, perhaps the same as Barus, which has long given its name to the costly Sumatran drug."

The uniformity of the name Camphor, or some transparent derivative from a common root, shows that the substance was procured originally from one place, and it seems abundantly demonstrated that the Camphor first known to the world was that obtained from **Dryobalanops Camphora**, and not the Camphor of modern commerce, which is prepared from the wood of the Camphor laurel tree. **U. C. Dutt** mentions the fact that two sorts of Camphor are referred to by Sanskrit writers, "namely, *pakva* and *apakva*, that is, prepared with the aid of heat and without it. The latter is considered superior to the former. It would seem from the above description that by the term *apakva karpúra*, was probably meant the

Camphor obtained from Borneo from the trunk of **Dryobalanops aromatica**; and by the term *pakva karpúra,* the China Camphor obtained by sublimation from the wood of **Cinnamomum Camphora**" (*Hindú Mat. Med.*, 222). **Dr. Dymock**, in his *Materia Medica of Western India,* also accepts this opinion regarding the two kinds of Camphor mentioned by the Sanskrit writers. The fact that the earliest mention we have of the modern Camphor is in the thirteenth century would seem, however, to be opposed to this being the *pakva karpúra* of the Sanskrit writers, and the suggestion may be offered that the boiled Camphor referred to may have been Blumea or Ngai Camphor, a substance which at the period indicated may either have been manufactured in India or imported from China. The history of Ngai Camphor does not appear to have been sufficiently investigated, but it is quite possible that the strongly camphoraceous bush of China and India may have been the first plant resorted to as a substitute or adulterant for the prized Camphor of Sumatra. As a matter of fact, this Camphor is much more nearly related to the Malayan than to the China Camphor, and even at the present day it is ten times the price of the Formosa Camphor, and is extensively consumed in China, partly as a medicine and partly in perfuming the finer qualities of Chinese ink. **Moodeen Sheriff** mentions four kinds of Camphor as met with in the bazárs of South India, *viz.*, (*a*) *Káfúre-qaisúri,* (*b*) *Súratí káfúr,* (*c*) *Chíní-káfúr,* and (*d*) *Batái-káfúr.* HISTORY.

TRADE RETURNS AND COMMERCIAL HISTORY.

TRADE. 263

Commerce.—While some of the less important camphors do, to a limited extent, reach Europe and India, the commercial or Chinese form is that which has been called "Common Camphor." This arrives at the English and Indian markets chiefly in a crude state, and is in both countries resublimed. The Japan Camphor is preferred to the Chinese, as it is generally purer. These two kinds enter commerce in different conditions. Formosa or Chinese Camphor is met with in square chests lined with lead or tinned iron, and always in a semi-liquid state, water having been added (according to some writers, before shipment), from an idea that it prevents evaporation. Japan Camphor, on the other hand, is lighter in colour, even sometimes pinkish. It occurs in larger grains than the Chinese, and is quite dry. It arrives in double tubs (one within the other), without any metal lining. Hence it is sometimes called "Tub Camphor." It fetches a higher price than the Formosa Camphor.

Bombay and Calcutta import, in addition to crude Camphor, a small quantity of Japan sublimed Camphor "which comes in tin-lined cases, which hold about 90℔" (*Dymock*).

Indian Trade in Camphor.—Mr. J. E. O'Conor, in his Trade Review for 1875-76, gives the following note regarding the relative value of the Barus and China Camphors:—

"Camphor is of two kinds, Bhimsaini or Barus, and the ordinary sort. The first is the produce of the **Dryobalanops Camphora**, and is imported from Borneo and Sumatra, where only the tree is found, *viâ* the Straits. It is valued in the tariff at R80 per ℔, while the ordinary kind, imported chiefly from China, is worth not more than R40 to R65 per cwt. This enormous difference is accounted for by the reputation (scarcely merited) which the Bhimsaini kind enjoys of peculiar excellence." (*Para. 16, pages 9 and 10.*)

Of Borneo and Sumatra Camphor probably not more than 2 or 3 cwt. are annually imported into India.

CAMPHOR.

Trade Returns and Commercial History.

INDIAN TRADE IN CAMPHOR.

The Import and Re-export trade in Camphor between India and foreign countries for the past seven years was as follows :—

Year.	Value of Camphor			
	Imported into India.		Re-exported from India.	
	Bhimsaini or Barus.	Other kinds.	Bhimsaini or Barus.	Other kinds.
	R	R	R	R
1879-80	20,909	5,34,001	2,316	23,174
1880-81	22,924	5,53,732	140	26,559
1881-82	38,574	5,52,335	1,640	21,138
1882-83	43,618	8,68,794	529	25,231
1883-84	38,579	6,27,278	790	28,730
1884-85	35,501	6,83,333	270	13,432
1885-86	25,944	6,53,545	*Nil.*	16,779

In addition to the above, a small amount of Camphor is annually imported into the French possessions in India; during 1882-83 these imports were valued at R43,600, and in 1883-84, R39,103. It is noteworthy that a certain amount of the Camphor imported into India comes from Great Britain. This is the European refined Camphor found in India—an article far superior to the water-impregnated Indian refined Camphor.

Mr. O'Conor publishes, under the quotations of exports of articles of "Indian Produce and Manufacture," the following figures for Camphor (other than Bhimsaini or Barus) :—

Year.	Value.	Analysis of Exports for 1885-86.			
		Country to which exported.		Province from which exported.	
	R		R		R
1879-80 . .	7,514				
1880-81 . .	7,142				
1881-82 . .	6,510				
1882-83 . .	9,475	Ceylon . . .	4,905	Bombay . . .	1,607
1883-84 . .	6,682	Other Countries .	1,150	Madras . . .	4,448
1884-85 . .	6,135				
1885-86 . .	6,055	Total .	6,055	Total .	6,055

It is presumed that the so-called Indian Camphor referred to in the last table is crude China Camphor, refined (or manufactured) in India, and in that state re-exported. With the exception of a small amount of Blumea Camphor manufactured in Burma, it is doubtful if there be any other Camphor which, strictly speaking, could be called Indian.

Indian Refined. 264

Purification of Camphor.—**Dr. Dymock** gives the following account of the process as practised in Bombay: "The process of resublimation is a peculiar one, the object being to get as much interstitial water as possible into the camphor cake. The vessel used is a tinned cylindrical copper drum, one end of which is removable; into this is put 14 parts of crude camphor and 2½ parts of water; the cover is then luted with clay, and the drum, being placed upon a small furnace made of clay, is also luted to the top of the furnace. In Bombay four of these furnaces are built together, so that the tops form a square platform. The sublimation is completed in about three hours; during the process the drums are constantly irrigated with cold water. Upon opening them a thin cake of camphor is found lining the sides and top; it is at once removed and thrown into cold water. Camphor sublimed in this way is not stored, but distributed at once to the shopkeepers before it has had time to lose weight by drying. It is sold at the same price as the crude article, the refiner's profit being derived from the introduction of, water" (*Mat. Med., W. Ind., 1st Ed., 549*). This same practice seems to be followed at Delhi and at a few other cities in India, but the method is crude and unsatisfactory, when the purified article is compared with that imported into India from Europe. The European process of refining camphor was long kept a secret, and towards the end of the seventeenth century the entire camphor of Europe had to be sent to Holland to be sublimed. A monopoly was also held for some time in Venice, but at the present day camphor-refining is largely accomplished in England, Holland, Hamburg, Paris, New York, and Philadelphia.

European Refined. 265

In England the impure camphor is broken up and mixed with 3 to 5 per cent. of slaked lime and 1 to 2 per cent. of iron filings. After being well sifted, this mixture is introduced through a funnel into a series of glass flasks, almost completely buried in a sand-bath. Instead of treating these by means of a fire, where flame might ignite the gas given off during the process of sublimation, dishes of fusible metal, kept warm by a furnace below the room, are used. The heat is suddenly raised from 120° to 190° C., and kept at that point for half an hour, so as to expel the water from the camphor. The temperature is then raised to 204° C., and maintained at that point for 24 hours. When the crude camphor has melted, the sand is removed from the upper half of each of the flasks and a paper cork placed in the neck. This allows of a lower temperature in the exposed part, and the vapour of camphor not being permitted to escape, condenses on the inside of the exposed part of the flask as a pure cake, leaving all impurities in the bottom. Air if freely admitted would render the camphor opaque, but this is prevented by placing a glass bell-jar over the neck of each flask just as the vapour of camphor begins to be given off. The whole process lasts for about 48 hours, and when completed the flasks are removed from the sand-bath and cold water sprinkled on them. They are thus broken, and a large cake of refined camphor, 10 to 12 inches in diameter and 3 inches thick, and weighing 9 to 12℔, is removed from each *bombolo* or flask.

The *rationale* of the process consists in preserving the temperature uniformly at the point of volatilization; the quicklime retains resin or empyreumatic oil, the iron fixes on any sulphur that may be present, while a little charcoal is often added to remove colouring matter, and sand is sometimes mixed with the crude camphor, to allow of a more uniform escape, and thus save bumping or the sudden evolution of confined volumes of vapour. The process requires great care, for, in addition to the very inflammable nature of the vapour, a too rapid evolution might result in the refined camphor forming loose crystals instead of compact cakes.

Camphor Plants. 266

Cultivation of Camphor-yielding plants in India.—Definite information cannot be found regarding the experiments which have been made to introduce the camphor-yielding trees into India. In the report of the Lucknow Horticultural Gardens for 1882-83 it is mentioned that a tree there being cultivated has so far done well. It seems likely that, instead of importing yearly over six lakhs of rupees worth of China Camphor, India might become a source of supply, since there is every reason to suppose that, if extended experiments were made, the tree could be successfully introduced. The amount of Barus Camphor consumed in

India is not sufficiently great to tempt experiments being undertaken with **Dryobalanops Camphora,** but the extended cultivation and manufacture of Blumea and China Camphors would seem highly desirable.

CAMPHOR OIL.

OIL. 267 **Oil of Camphor.**—There are two very distinct substances known by that name in commerce. The first and most important is the oleo-resin or camphor-oil of Borneo. This is obtained by tapping the trees. Sometimes this accumulates to such an extent that (as with the South American copaiba tree) the trunk, no more able to resist the pressure of the fluid, spontaneously bursts open or has its tissue broken into large internal chambers, producing while this occurs a loud noise, "as if the tree were rent in twain." The *Pharmacographia* states that **Motley,** in cutting down a tree in Labuan in May 1851, pierced one of these reservoirs in the trunk of the tree, from which over 5 gallons of camphor-oil were obtained. This oil is termed *Borneene,* and it is isomeric with oil of turpentine, being represented by the formula $C_{10}H_{16}$, but in its crude state it holds in solution a certain amount of Borneol and resin.

The other so-called Camphor-oil is quite distinct and should not be confused with the above. It is known as Camphor-oil of Formosa. This is a brown liquid, holding in solution an abundance of common camphor, and is found to drain from the cases containing crude camphor. It has an odour of sassafras. From this so-called oil, or rather solution, camphor is precipitated on the temperature of the liquid falling.

CHEMICAL AND MEDICAL PROPERTIES OF CAMPHOR.

268 **Chemistry.**—It is not necessary to enter into this subject in great detail. For a full account of the chemistry of Camphor the reader is referred to works on chemistry, but more particularly to the *Pharmacographia* and the *United States Dispensatory,* as these are more likely to be accessible than the numerous and scattered papers in which this subject has been treated of from a purely chemical point of view.

1st.—ORDINARY CAMPHOR.—A white, translucent substance, of a crystalline structure, readily pulverised in the presence of a little alcohol or of sugar, ether, or chloroform, otherwise tough and difficult to pulverise. It has the sp. gr. 0·990 and melts at 175° C., boiling at 205° C. It possesses a penetrating odour and pungent taste, and burns with a luminous smoky flame. It is very volatile, the vapour condensing on the bottle in which it is kept. A small fragment, if thrown upon water, displays a peculiar circulatory movement, which is at once stopped by the addition of a little oil. This fact has been taken advantage of, in detecting the presence of oily substances in water. Camphor is only slightly soluble in water, but the amount may be increased by the addition of sugar. Carbonic acid also increases its solvent power. Ordinary alcohol will take 75 per cent. of camphor. When mixed with resins or concrete oils, camphor often partially or completely loses its odour. The formula given for this form of camphor is $C_{10}H_{16}O$; by treatment with various reagents it yields a number of interesting products. Prolonged boiling with nitric acid oxidises the camphor into *Camphoric acid,* $C_{10}H_{14}O_4$ and *Camphoronic acid,* $C_9H_{12}O_5$, water and carbonic acid being eliminated. When repeatedly distilled with chloride of zinc, it is converted into *Cymene* or *Cymol,* $C_{10}H_{14}$, a substance present in many essential oils.

2nd.—BARUS CAMPHOR.—This has the formula $C_{10}H_{18}O$. It is somewhat harder than the preceding, is less volatile, and does not consequently crystallise on the inside of the bottle containing it. It is also heavier, having the sp. gr. 1·009. It is easily pulverised without the aid of alcohol; it is, in fact, a more compact and brittle substance than ordinary

CHEMISTRY.

camphor. It requires for fusion 198° C. In optical properties an alcoholic solution is found to be 12½° *dextrogyre*. By the action of nitric acid it may be converted into ordinary camphor, and by continued oxidation, into *Camphoric acid*. Its medicinal properties are regarded as stronger than those of ordinary camphor.

3rd.—NGAI or BLUMEA CAMPHOR is chemically more nearly related to Barus Camphor than to Chinese Camphor. It has the same composition, and differs mainly in the fact that its alcoholic solution is *levogyre*, and remains so even after being oxidised by nitric acid, whereas Barus Camphor is converted into ordinary camphor.

MEDICINE. 269

Medicine.—Camphor possesses stimulant, carminative, and aphrodisiac properties, and is widely used in medicine, both externally and internally. Its primary action is that of a diffusible stimulant and diaphoretic; its secondary, that of a sedative, anodyne, and antispasmodic. In large doses it is an acro-narcotic poison. Camphor has been extensively used in the advanced stages of fevers and inflammation, insanity, asthma, angina pectoris, hooping-cough, and palpitations connected with hypertrophy of the heart; affections of the genito-urinary system, comprising dysmenorrhœa, nymphomania, spermatorrhœa, cancer, and irritable states of the uterus; chordee, incontinence of urine, hysteria, rheumatism, gangrene, and gout. It has also been employed as an antidote to strychnia, but with doubtful results. It is regarded as a medicine in impotence (*Pharm. of India; U. C. Dutt's Hind. Med.*). The Hindús consider camphor to be aphrodisiacal, but the Muhammadans hold a contrary opinion; both regard it as a valuable application to the eyelids in inflammatory conditions of the eye (*Dymock*).

The medicinal properties of camphor are too well known to require to be discussed here at great detail. The reader is therefore referred to the *Pharmacopœia of India*, pp. 190, 192, and other standard works on materia medica. As having a special bearing on India, however, the following extract may be republished from **Waring's** most useful little book, *Bazar Medicines*:—

"In chronic rheumatism, in addition to its use externally, it may be given internally in a dose of 5 grains with one grain of opium at bedtime; it affords relief by causing copious perspiration, which should be promoted by a draught of infusion of ginger and by additional bedclothes. An excellent vapour-bath for these cases may be made by substituting half an ounce of camphor placed on a heated plate for the chattie of hot water. Thus employed, it causes speedy and copious perspiration. Care, however, is necessary to prevent the patient inhaling the vapour, which is of comparatively little consequence when simple water is used.

"In asthma, camphor in 4-grain doses, with an equal quantity of asafœtida, in the form of pill, repeated every second or third hour during a paroxysm, affords in some instances great relief. Turpentine stupes to the chest should be used at the same time. Many cases of difficulty of breathing are relieved by the same means. These pills also sometimes relieve violent palpitation of the heart. In the coughs of childhood, camphor liniment, previously warmed, well rubbed in over the chest at nights, often exercises a beneficial effect. For young children, the strength of the liniment should be reduced one-half or more by the addition of some bland oil.

"In rheumatic and nervous headaches, a very useful application is one ounce of camphor dissolved in a pint of vinegar, and then diluted with one or two parts of water. Cloths saturated with it should be kept constantly to the part.

"In spermatorrhœa, and in all involuntary seminal discharges, no

MEDICINE.

medicine is more generally useful than camphor in doses of 4 grains with half a grain of opium taken each night at bed-time. In gonorrhœa, to relieve that painful symptom, chordeé, the same prescription is generally very effectual; but it may be necessary to increase the quantity of opium to one grain, and it is advisable to apply the camphor liniment along the under-surface of the penis as far as the anus. To relieve that distressing irritation of the generative organs which some women suffer from so severely, it will be found that 5 or 6 grains of camphor, taken in the form of pill twice or three times a day, according to the severity of the symptoms, will sometimes afford great relief. In each of these cases it is important to keep the bowels freely open.

"In painful affections of the uterus, camphor in 6 or 8-grain doses often affords much relief. The liniment should at the same time be well rubbed into the loins. In the convulsions attendant on child-birth, the following pills may be tried: Camphor and calomel, of each 5 grains. Beat into a mass with a little honey, and divide into two pills; to be followed an hour subsequently by a full dose of castor oil or other purgative.

"In the advanced stages of fever, small-pox, and measles, when the patient is low, weak, and exhausted, and when there are at the same time delirium, muttering, and sleeplessness, 3 grains of camphor, with an equal quantity of asafœtida, may be given even every third hour; turpentine stupes or mustard poultices being applied at the same time to the feet or over the region of the heart. It should be discontinued if it causes headache or increased heat of the scalp. Its use requires much discrimination and caution.

"To prevent bed-sores, it is advisable to make a strong solution of camphor in arrack or brandy, and with this night and morning to bathe, for a few minutes, the parts which, from continued pressure, are likely to become affected" (*Waring, Bazar Medicines*).

The Lancet (May 31st, 1884) gives an account of a simple process of curing coryza by the inhalation of camphor vapour through a paper tube, the whole face and head being covered so as to secure the full action.

Special Opinions.—§ "Daily employed in dispensary practice in the form of camphor-water as a vehicle for other medicines. When quinine is rejected by the stomach, the following formula may be used: Quinine gr. iii., camphor gr. $\frac{1}{3}$, opium $\frac{1}{8}$. To be made into a pill and given three or four times daily. A drachm of camphor dissolved in chloroform mixed with an ounce of simple ointment forms a soothing application for piles" (*Assistant Surgeon Jaswant Rai, Múltan*). "It is an irritant and rubefacient, good for a cold in the head with coryza, summer diarrhœa" (*Brigade Surgeon W. R. Rice, Jubbulpore*). "Largely used as a liniment for muscular pains. Is a good expectorant" (*Surgeon R. Gray, Lahore*). "Used in 3 or 4-grain doses and mixed with about $\frac{1}{4}$ grain of extract of belladonna. I have found this to be of very great value in neuralgic pains" (*Assistant Surgeon Doyal Chunder Shome, Campbell Medical School, Calcutta*). "Stimulant, expectorant, anodyne, antispasmodic, anaphrodisiac, and diaphoretic, doses 1 to 10 grains. I have used this in the following cases: (1) In acute bronchitis, with other ingredients. (2) In pneumonia, with amm. carb. and quinine. (3) In toothache and carious tooth, useful to relieve the pain if stuffed in the cavity. (4) In bilious headache, externally applied with vinegar and cold water. (5) In chronic rheumatism, either muscular or articular, if embrocated, mixed with mustard oil and opium. (6) In a few cases of cholera (cold stage) the use of the spirit of camphor with rum has proved successful. (7) In irritation and chordeé of gonorrhœa, if given with belladonna in the form of pill" (*Hospital Assistant Abdulla, Civil Dispensary, Jubbulpore*). "Stimulant and diaphoretic, useful in

MEDICINE.

neuralgic headache in doses of 5 grains; gives healthy action to foul ulcers when applied; 2 to 5 drops of a saturated solution in rectified spirit on sugar is very useful in indigestion and colic. I have known several instances in which it checked vomiting and purging in the first stage of cholera. It preserves clothing and other articles against insects and worms" (*Surgeon Shib Chunder Bhuttacharji, Chanda, Central Provinces*). "Useful in cholera" (*Surgeon H. D. Masani, Karáchi*). "In the form of spirit, camphor is very efficient in the early stages of catarrh of the inspiratory passages. In the collapse stage of cholera I have seen good effects from its use, and think it worthy of more extended trial in this disease" (*Surgeon S. H. Browne, Hoshangabad, Central Provinces*). "I have found that when given in 10-grain doses every fourth hour in cholera, good results follow. It is often administered with the fruit of the plantain to produce abortion; about 20 grains are said to be enough for this purpose" (*Surgeon W. F. Thomas, Mangalore*). "Stimulant, antispasmodic, diaphoretic, hypnotic, anaphrodisiac, doses 10 to 20 grains, used in toothache, chronic rheumatism, dysmenorrhœa, delirium, typhus fever, debility after fever, hooping cough, gangrene of the lungs, cholera, chorea, hysteria, epilepsy, puerperal convulsion, palpitation of the heart" (*Hospital Assistant Chuna Lal, Jubbulpore*). "Is taken in large doses to procure abortion" (*Surgeon-Major D. R. Thompson, Madras*). "Camphor is daily used as a stimulant, antispasmodic, sedative to the genito-urinary system, and parasiticide. The spirit of camphor is a useful remedy in cholera, in 1 to 5-drop doses" (*Assistant Surgeon Nundo Lal Ghose, Bankipur*). "*Camphor.* Used in 3 or 4-grain doses and mixed with about ¼ grain of extract of belladonna. I have found this to be of very great value in neuralgic pains" (*Assistant Surgeon Doyal Chunder Shome, Campbell Medical School, Sealdah, Calcutta*).

DOMESTI 270

Domestic Uses.—Mr. T. W. Lee, writing in the *Journal of Agriculture*, says that most seeds are greatly hastened in their germination by being soaked, previous to sowing, in soft water, to a pint of which a lump of camphor, about the size of a large nut, has been added. **Mr. Lee** tried this experiment on many vegetable seeds, such as peas, beans, &c., as well as palms, castor-oil seeds, and various other tropical seeds which have very hard seed-coats, many of which would require soaking in water for a long time before they would otherwise show signs of germination, but which, with the addition of camphor, sprout easily and rapidly. This same fact may be taken advantage of in stimulating cuttings of roses or other plants sent from one country to another. Rose-cuttings, for example, posted in England, carry safely to India, and the stimulation caused by dipping their freshly-cut ends in camphor-water helps greatly to enable them to take root when placed in the soil.

Camphora glandulifera, *Nees*, see **Cinnamomum glanduliferum**, *Meissn.*; LAURINEÆ.

Canada Balsam, see **Abies balsamea**, *Aiton.*; CONIFERÆ.

CANANGA, *Rumph.*; *Gen. Pl., I., 24.*

271

Cananga odorata, *H. f. & T. T.*; *Fl. Br. Ind., I., 56*; ANONACEÆ.

THE ILANG-ILANG of European perfumers.

Syn.—UVARIA ODORATA, *Lamb.*

Vern.—*Kadat-ngan, kadapgnam*, BURM.; *Ilang-ilang*, MALA.

References.—*Roxb., Fl. Ind., Ed. C.B.C., 454; Kurz, For. Fl., Burm., I., 33; Gamble, Man. Timb., 8; U. S. Dispens., 15th Ed., 1782; Spons' Encyclop., 1422; Smith, Dic., Econ. Pl., 218.*

ILANG-ILANG. **Habitat.**—A large evergreen tree of Burma (Ava and Tenasserim), distributed to Java and the Philippines. Cultivated in many parts of India on account of its sweet-smelling flowers.

OIL. 272 **Oil.**—An otto is prepared from the flowers known as Otto of Ilang. It is highly esteemed, as may be seen from the fact that it fetches in Europe about 18*s.* to 22*s.* per oz. It is frequently blended with pimento, orris, rose, tuberose, and jasmine in the preparation of handkerchief perfumes. The volatile oil contains a benzoic ether, phenol, and an aldehyd or ketone. The yield of oil is about 0·5 per cent. The so-called Macassar Hair Oil is said to be a solution of Ilang in cocoanut oil.

For further information see **Michelia.**

CANARIUM, *Linn.; Gen. Pl., I., 324.*

273 **Canarium bengalense,** *Roxb.; Fl. Br. Ind., I., 534;* BURSERACEÆ.

Vern.—*Gogul dhúp,* NEPAL; *Narockpa,* LEPCHA; *Tekreng,* GÁRO; *Bisjang, dhúna,* ASS.

References.—*Roxb., Fl. Ind., Ed. C.B.C., 504; Kurz, For. Fl. Burm., I., 209; Gamble, Man. Timb., 68, xi.; Voigt, Hort. Sub. Cal., 149; O'Shaughnessy, Beng. Dispens., 288; Royle, Him. Bot., 177; Cooke, Gums and Gum-resins, 7; Balfour, Cyclop.*

Habitat.—A tall tree, with a straight cylindrical stem; it is met with in the eastern moist zone, eastern Himálaya, Bengal, and Burma.

GUM. 274 **Gum.**—Yields a brittle, amber-coloured resin, resembling copal, which is used as incense. The natives set little value on it. In Calcutta bazárs it sells at two to three rupees per maund.

TIMBER. 275 **Structure of the Wood.**—Shining, white, when fresh cut, turning grey on exposure, soft, even-grained, does not warp, but decays readily. Weight 28℔ per cubic foot. It is much esteemed in Bengal for tea-boxes, and also for shingles. It is also valuable for building.

MEDICINE. 276 **Medicine.**—§ "The leaves and bark are used externally for rheumatic swellings.

FOOD. 277 **Food.**—" Fruit edible.

TIMBER. 278 **Structure of the Wood.**—" Strong and durable, used for common house building" (*Trimen*).

279 **C. commune,** *Linn.; Fl. Br. Ind., I., 531.*

JAVA ALMOND TREE.

Vern.—*Jangali bádám,* HIND.; *Jangali bédáná,* CUTCH; *Kagli mara, kagga libija, java badamiyanne,* KAN.; *Canari,* MALA.; *Rata-kækana,* SING.

References.—*Roxb., Fl. Ind., Ed. C.B.C., 504; Voigt, Hort. Sub. Cal., 148; Pharm. Ind., 51; Moodeen Sheriff, Supp. Pharm., 85; U. S. Dispens., 15th Ed., 536; Ainslie, Mat. Ind., II., 60; O'Shaughnessy, Beng. Dispens., 288; Drury, Us. Pl., 103; Cooke, Gums and Gum-resins, 109; Cooke, Oils and Oil-seeds, 10; Spons, Encyclop., 1392, 1649; Balfour, Cyclop., Ed. 1885; Smith, Dic., 12, 113, 163; Treasury of Botany; Ure, Dic. of Arts and Man.; Kew Cat., 28; Baillon, Nat. Hist., Pls. V., 298.*

Habitat.—A plant of the Malay Archipelago cultivated in India; introduced into Bengal, where it does not thrive well in winter.

GUM. 280 **Gum.**—The resin, imported under the name of Manilla Elemi, has long been supposed to be a product of this plant. The authors of the *Pharmacographia*, however, affirm that "The resin known in pharmacy as Elemi is derived from a tree growing in the Philippines, which

Blanco, a botanist of Manilla, described in 1845 under the name *Icica Abilo,* but which is completely unknown to the botanists of Europe. **Blanco's** description is such that, if correct, the plant cannot be placed in either of the old genera **Icica** or **Elaphrium,** comprehended by **Bentham** and **Hooker** in that of **Bursera,** nor yet in the allied genus **Canarium;** in fact, even the order to which it belongs is somewhat doubtful."

" Manilla Elemi is a soft, resinous substance, of granular consistence, not unlike old honey, and when recent and quite pure is colourless; more often it is found contaminated with carbonaceous matter which renders it grey or blackish, and it is besides mixed with chips and similar impurities. By exposure to the air it becomes harder and acquires a yellow tint. It has a strong and pleasant odour suggestive of fennel and lemon, yet withal somewhat terebinthinous. When moistened with spirit of wine, it disintegrates, and examined under the microscope is seen to consist partly of acicular crystals. At the heat of boiling water the hardened drug softens, and at a somewhat higher temperature fuses into a clear resin" (*Pharmacographia, p. 147*). **Manilla Elem. 281**

The *United States Dispensatory* (15th Ed.), page 536, says: "The Manilla Elemi is conjecturally referred to **Canarium commune.**" In their *Medicinal Plants* **Bentley** and **Trimen** give a detailed description of this plant. They say: "It is also cultivated in Java, and has been grown in the gardens at Calcutta, where, however, it did not thrive. We cannot certainly identify it as the source of Elemi, but it is probably the 'Terebinthus Luzonis prima' of Camelli, in 'Ray's History of Plants,' which he says is called *Laguaan, Lanvan,* and *Pagsaingan* by the natives, and *Arbol de la Brea* by the Spaniards." Elemi is said to be derived from the hypothetical plant **Icica Abilo** of **Blanco,** a botanist of Manilla, who published a description of the tree from which the resin was obtained in 1845 under that name. Its description cannot be identified, but although, as stated above, it has been supposed to be allied to **Canarium,** there is no actual evidence of this, and it is doubtful if **Icica,** as described by **Blanco,** should be even referred to the BURSERACEÆ.

The gum is used principally in the manufacture of varnishes, also in felting and in medicine.

Oil.—The nut yields a semi-solid oil on expression, similar in appearance to cocoanut oil. It is used for culinary purposes, and is regarded palatable. It is also burnt in lamps. **OIL. 282**

§ "The bark yields an abundance of limpid oil with a pungent turpentine smell, congealing to a buttery camphoraceous mass. It is stated to possess the same properties as copaiba (*O'Shaughnessy*)." (*Surgeon C. J. H. Warden, Professor of Chemistry, Calcutta*).

Medicine.—Ainslie remarks that the gum has the same properties as Balsam of Copaiva. It is applied in the form of an ointment to indolent ulcers. The oil expressed from the kernels might be substituted for almond oil. **Dr. Waitz,** in his *Diseases of Children in hot climates,* speaks favourably of the kernels in emulsion as a substitute for the European preparation, *Mistura Amygdalæ.* **MEDICINE. 283**

Special Opinion.—§ " A demulcent" (*Surgeon W. Barren, Bhuj, Cutch, Bombay*).

Food.—Cultivated in the Moluccas for its fruit, which is a three-sided drupe, containing, as a rule, only one perfect seed; this tastes somewhat like an almond. The oil expressed from the nuts, when fresh, is mixed with food in Java. Bread is also made from the nuts in the Island of Celebes. If eaten fresh or too frequently, the nuts often produce diarrhœa (*Drury*). **FOOD. 284**

285 **Canarium strictum,** *Roxb., Fl. Br. Ind., I., 534; Beddome, t. 128.*

THE BLACK DAMMAR TREE.

Vern.—*Kálá dammár,* HIND., BENG., GUJ.; *Dhúp, gúgul,* BOM.; *Dhúp ráldhup,* MAR.; *Karapu kongiliam, karapu dammar, congilium-marum, karuppu dámar,* TAM.; *Nalla-rójan,* TEL.; *Manda-dhup, raldhupada,* KAN.; *Thelli,* MALA.

References.—*Roxb., Fl. Ind., Ed. C.B.C., 504; Beddome, Fl. Sylv., I., p. 128; Gamble, Man. Timb., 68; Dalz. & Gibs., Bomb. Fl., 52; Voigt, Hort Sub. Cal., 149; Pharm. Ind., 53; Moodeen Sheriff, Supp. Pharm. Ind., 85; Dymock, Mat. Med., W. Ind., 135, also 2nd Ed., 167; Bidie, List of Raw Prod., Paris Exb., 24; Drury, Us. Pl., 104; Cooke, Gums and Gum-resins, 93; Birdwood, Bomb. Prod., 264; Lisboa, Us. Pl. of Bomb., 40; Balfour, Cyclop.; Smith's Dic., 150; Treasury of Botany.*

Habitat.—A tall tree of South India. Common about Courtallum in the Tinnevelly district and in Kanara.

GUM. 286 **Gum.**—It yields a brilliant resin called the Black Dammar of South India. This is obtained by making vertical cuts in the bark and setting fire to the bottom of the stem. This result is effected by lighting firewood piled to the height of a yard round the base of the trunk. The dammar exudes from the stem as high as the flames reached commencing about two years after the above operation. The flow is said to continue for ten years, between the months of April and November, and the resin is collected in January.

"This substance occurs in stalactitic masses of a bright shining colour when viewed *en masse,* but translucent and of a deep reddish-brown colour when held between the eye and the light; homogeneous, with a vitreous fracture; partially soluble in boiling alcohol, and completely so in oil of turpentine" (*Pharm. Ind.*).

The following is **Mr. Broughton's** report on Black Dammar: "This well-known substance offers little chance of usefulness, in Europe at least, when the many resins are considered that are found in the market at a far less price. It is used in this country for many small purposes, as in the manufacture of bottling-wax, varnishes, &c. Its colour when in solu-

BLACK DAMMAR. 287 tion is pale, if compared with its dark tint when in mass. Thus, though insoluble in spirit, its solution in turpentine forms a tolerable varnish. When submitted to destructive distillation, it yields about 78 per cent. of oil, resembling that obtained from common colophony; but I fear, in the majority of its possible applications, it possesses few advantages over ordinary resin at 7*s.* 6*d.* per cwt. **Major** (now Col.) **Beddome** estimates the price of Black Dammar on the coast of Kanara at R8 per 25℔ (or nearly ten times the price of resin in England). The number of substances suitable for varnishes have lately become very numerous in Europe. Common resin is now purified by a patent process, consisting of distillation with superheated steam, by which it is obtained nearly as transparent and colourless as glass, in such amount that a single firm turns out 60 tons per week."

MEDICINE. Burgundy Pitch. 288 **Medicine.**—The resin is used medicinally, according to **Dr. Bidie,** as a substitute for Burgundy Pitch in making plasters.

Special Opinions.—§ "Bathing in a tub painted inside with dammar is supposed to relieve the irritation of prickly heat" (*Surgeon-Major A. S. G. Jayakar, Muskat, Arabia*). "Employed as a liniment with gingelly oil, in rheumatic pains" (*Surgeon-Major J. J. L. Ratton, Salem*).

CANAVALIA, *Adans. (? DC.); Gen. Pl., I., 537.*

Canavalia ensiformis, *DC.; Fl. Br. Ind., II., 195; Wight, Ic., t. 753;* LEGUMINOSÆ. 289

SWORD BEAN. Sometimes called PATAGONIAN BEAN.

Syn.—C. GLADIATA, *DC.;* DOLICHOS GLADIATUS, *Willd.*, as in *Roxb., Fl. Ind., Ed. C.B.C., 559;* D. ENSIFORMIS, *Linn.*

Vern.—*Makham shim, mekhun,* BENG.; *Tihon,* SANTAL; *Sufêd* or *lâl kud-sumbal,* HIND.; *Sem,* PB. & N.-W. P.; *Gavari,* MAR.; *Gaivara,* BOM.; *Gavria,* GUJ.; *Burra shim, kudsumber abye,* DUK.; *Cambe-gida, tumbay kayi,* KAN.; *Segapu, vellay thun betten, thambatin, segapu thumbatin,* TAM.; *Tellay tumbetten kasa, chamma, tamma, vela* or *yerra tambalin,* TEL.; *Shimbi,* SANS.; *Poi noung ni, pai ka lag,* BURM.

References.—*Thwaites, En. Ceylon Pl., 68; Dalz. & Gibs., Bomb. Fl., Suppl., 23; Stewart, Pb. Pl., 61; Aitchison, Cat. Pb. & Sind Pl., 48; Voigt, Hort. Sub. Cal., 234; Murray, Drugs & Pl., Sind, 124; Drury, Us. Pl., 104; Atkinson, Him. Dist., 702; Birdwood, Bomb. Prod., 118; Lisboa, Us. Pl. of Bombay, 152; Balfour, Cyclop.; Smith, Dic., 304; Treasury of Botany; Kew Cat., 44; Church, Food-Grains of India, p. 144, Fig. 27.*

Habitat.—This climber is found along the eastern part of India from the Himálaya to Ceylon and Siam, wild or cultivated. The name "Overlook" is given to it by the West Indian negroes, and it is generally planted by them to mark the boundary of their plantations, from the superstitious belief that it will protect their property from plunder (*Smith*).

There are several forms of this plant met with in India, the seeds and flowers being of different colours (*Drury*). These, according to the *Flora of British India*, are referred to three distinct varieties:—

Var. 1st, **virosa,** *W. & A., Prod., 253; Dalz. & Gibs., Bomb. Fl., 69;* **Dolichos virosus,** *Roxb., Fl. Ind., Ed. C.B.C., 559.* Pods often 2-4 inches long, 4-6-seeded. Speaking of this form, **Roxburgh** says: "I do not find that any part of this species is in any shape useful to the natives or others; indeed, the natives of Coromandel, where the plant is common, reckon it poisonous, which is corroborated by **Van Rheede.**" This is known in Bengal as *Kath-shim,* or *Kala-shim* and *Gaivara* (*Gowara*) in Bombay. 290

Var. 2nd, **turgida,** *Grah. in Wall. Cat.;* **C. Stocksii,** *Dalz. & Gibs., Bomb. Fl., 69.* Pods large and turgid, 3 to 5 inches by $1\frac{1}{2}$ to 2 inches. 291

Var. 3rd, **mollis,** *Wall.* Found in Southern and Western India. The pods are smaller than in either of the above; when cultivated they are tender and eaten like French beans. 292

Food.—The young, tender, half-grown pods, apparently of only var. 3, are actually eaten, but these constitute the so-called French beans at the tables of Europeans. Natives also eat them in curry. The form with large white seeds is considered the most wholesome. Some five varieties are reported to be cultivated in Lucknow, of which the form known as *hilwa,* a white narrow-podded variety, is considered the best. **Mr. Cameron** informs the writer that the seeds of this pulse are highly relished in Mysor. **Atkinson** writes of the North-West Provinces that the *sem* is "consumed by all classes." FOOD. 293

Professor Church gives the analysis of this pulse (p. 144), and adds that its nutrient ratio is 1 : 2·2 and the nutrient value 80.

294 **Canavalia obtusifolia,** *DC.; Fl. Br. Ind., II., 196.*

References.—*Thwaites, En., Ceylon Pl., 88; Voigt, Hort. Sub. Cal., 235; Drury, Us. Pl., 105; Balfour, Cyclop.; Kew Cat., 44.*

Habitat.—Met with on the coasts of the Western Peninsula, Ceylon, and the Malaya Peninsula.

"Is a useful binder of loose sand" (*Balfour*).

295 ## CANELLA, *Sw.; Gen. Pl., I., 121, 970.*

Canella alba, *Murray; DC. Prod., I., 563;* CANELLACEÆ.

WHITE CINNAMON, *Eng.;* CANELLE BLANCH, *Fr.;* WEISSER ZIMMET, *Germ.;* CANELLA BIANCA, *It.;* CANELLA ALBA, *Sp.;* CANELLA BLANCA, *Sp.*

References.—*Voigt, Hort. Sub. Cal., 88; Pharm. Ind., 25; Flück. & Hanb., Pharmacog., 73; U. S. Dispens., 15th Ed., 337; Year-Book of Pharmacy, 1873, p. 43; Spons, Encyclop., 1419; Smith, Dic., 84; Treasury of Botany; Hanbury, Sc. Papers, 353; Kew Cat., 14.*

Habitat.—A West Indian aromatic plant, the bark of which is imported into India, and is sold by druggists; the tree might be cultivated in India.

OIL. 296 **Oil.**—"An essential oil, erroneously called 'white cinnamon,' is obtained by the aqueous distillation of the bark; it is a mixture of caryophyllic (engenic) acid, an oil resembling cajuput, and an oxygenised oil." (*Spons, Encyclop.*) It is a rare article, not known to commerce.

MEDICINE. Bark. 297 **Medicine.**—The bark is met with in rolls or quills two or three feet in length, having a bitterish acrid peppery taste. The odour is something like a mixture of cloves and cinnamon. The bark is an aromatic stimulant used to a limited extent in combination with other articles in constitutional debility, dyspepsia, scurvy, &c. (*Pharm. Ind.*) In the West Indies it is used as a condiment and has some reputation as an antiscorbutic.

CANES.

CANES. 298 **Canes.**

CANNE, *Fr.;* ROHR, *Germ.;* *Bhate*, HIND.; *Nathur*, GUZ.

The species of the genus Calamus—a genus of climbing palms—yields the canes of commerce. Few plants are more useful to the hill tribes of India and the Malay than are the various forms of cane, yet very little of a definite nature is known as to the peculiar properties and uses of the individual species. They afford "Dragon's-blood," and the "Malacca" and "Rattan Canes" of commerce, but it is probable that each of these articles is obtained from more than one species of Calamus. Reeds and small bamboos are sometimes, but incorrectly, spoken of as canes.

The species of Calamus are formidable but graceful objects, giving a delicate green effect to the tropical vegetation. Sometimes they occur as stunted erect bushes, constituting large impenetrable clumps; at other times, by means of their prickly tendrils, clasping firmly the bushes and trees of the forest, they ascend as gigantic climbers, often attaining to as much as 600 feet in length. The stems, leaves, and tendrils are covered with spines and prickles. The fruit hangs in great clusters, the inner

Canes often 600 feet long.

succulent layer of which is edible, constituting a favourite fruit with the hill tribes. It is a refreshing bitter sweet pulp, agreeable on a hot march, but otherwise not eaten by Europeans. The stems when freshly cut contain a large quantity of liquid, which may be collected by blowing through short pieces. The roots and young sprouts are eaten as vegetables and somewhat resemble asparagus. Canes owe their value to their great strength, and more particularly to the strength of the outer layer of woody structure. As substitutes for ropes they are invaluable, and in some respects even **Substitutes for Ropes.**
superior to ordinary ropes. For walking-sticks and canes, and for spear 299
and lance shafts, they are in great demand and are justly popular; light- **Shafts.**
ness, strength, and uniform structure and size, are properties of the greatest 300
importance.

The Asiatic Uses of Canes are varied and extensive. One of the most interesting, as illustrating the length and strength of these natural ropes,
is the construction of cane suspension-bridges. Good examples of these **Cane-bridges.**
may be seen in the Khásia and Northern Cachar hills. On the march 301
from Silchar to Manipur, for example, three have to be crossed, namely, over the Muku, the Barak, and the Irang rivers. Within the past few years, owing to heavy traffic, these have been strengthened by one or two wire-ropes, but cane bridges are by no means unfrequent in the mountainous tracts of the eastern side of India, and cane ladders are not uncommon in the South on the Animalis. Carefully selected canes, 300 or 400 feet long, constitute the chains, and bridges of that length are often thrown across rocky valleys 50 feet above the water. This height is necessary in order to be above the water-level in the sudden rising of the rivers which takes place during the rains. Each bridge generally consists of three parallel canes forming the pathway, the canes being knit together with bamboo or bark, so as to constitute a band not more than 18 inches in breadth, through which the rushing water may be seen below. The railing affords additional support; it consists of two canes carried about three or four feet above the pathway, one on either side. These are here and there connected by perpendicular canes passing under the pathway, and the whole structure is bound together by a network of bark-ropes or smaller canes. **Bridges.** With the weight of the traveller the bridge bends until it is often alarmingly near the water, and to prevent the railing closing on the person crossing the bridge, barriers are thrown across here and there, about 18 inches above the pathway; similar stays are also carried overhead. These barriers constitute the chief difficulty in crossing a cane bridge, for on raising the foot, the swaying structure and the rushing water produce the giddy sensation of walking up the river sideways.

In addition to bridges, long canes are used as ropes in towing heavy **Ropes.**
objects—stones, logs of timber, &c.—up the hill-sides. "In Java the cane is cut into fine slips, which are platted into excellent mats, or made into strong ropes." "It is stated that in China, as also in Java and Sumatra, and indeed throughout the Eastern Islands, vessels are furnished with cables formed of cane twisted or platted. This sort of cable was formerly extensively manufactured at Malacca" (*Royle, Fibrous Plants*). Dampier says: "Here we made two new cables of rattans, each of them four inches about. Our captain bought the rattans, and hired a Chinese to work them, who was very expert in making such wooden cables. These cables I found serviceable enough after, in mooring the vessel with either of them; for when I carried out the anchor, the cable being thrown out after me, swam like cork in the sea, so that I could see when it was tight, which we cannot so well discern in our hemp cables, whose weight sinks them down; nor can we carry them out but by placing two or three boats at some distance asunder, to buoy up the cable while the long-boat rows

out the anchor." Ropes are regularly made in China by splitting the
rattan and twisting the long fibres thus prepared into a rope of any desired
thickness. This is rarely if ever done in India, entire canes being always
used. The smaller canes are extensively employed in basket-work, both
Baskets. entire and cut. Useful chairs, sofas, and couches are made all over India
302 from cane, and cane *punkha* ropes are almost in universal use. In Bengal
Chairs. baskets (*dháma*) are made of entire canes by twisting the canes round
303 and round and fastening the one to the other by thin strips. The prac-
Mats. tice of cutting the cane into narrow strips for caning chairs may be regard-
304 ed as a European industry, but it is now practised all over India, the
Cane-work. chairs made in this way being light and cool. A strong and durable floor
305 mat for office purposes is constructed of small entire rattans, bound together, by means of cane-strings, the canes being arranged so as to be flat
Walking Sticks. and parallel.

306 The European Uses of Canes are even more varied than the Asiatic.
Umbrella handles. They are valued on account of their lightness, flexibility, and strength.
They are extensively used as walking-sticks, umbrella handles, and even
307 as a substitute for whalebone for umbrella and parasol ribs, each set of
Umbrella ribs. such ribs costing only from 1*d.* to 2½*d.* instead of 2*s.* 6*d.* to 3*s.* for whalebone.
308 Cane is also extensively employed in saddlery and harness, and a wicker-
Saddlery. work of rattan is now used in the construction of the German military
309 helmet, which is said to make it sword proof. But the chief purpose to
Harness. which cane is put in Europe is in furniture and basket making. In India,
310 canes are cut up by hand, the outer strips being separated at the expense
Furniture. of the central core. In Europe this central portion is saved, a patented
311 machine being used to split the rattans which cuts off the outer layer in
Central axis. bands of any required size or thickness, while leaving the central core in
312 the form of a perfectly round and even rod. This rod is utilised in the
Window blinds. construction of fancy baskets, chairs, and window-blinds, and has one pro-
313 perty not possessed by the strong outer bands, namely, that it takes with
Dyed cane. ease any desired colour. European authorities do not appear to be aware,
314 however, of the fact that the Nagás and other hill tribes of Assam dye
human and goats' hair a beautiful scarlet, as also tint with the same
colour the outer silicious layer of the rattan cane. Bands of stained
rattan they use for decorating ear-rings, bracelets, and leggings.

Prepared strips of rattan are extensively used in Europe as in India for
caning furniture, but a comparatively new and increasing trade in rattan
is the construction of baskets, which are rapidly displacing willow baskets;
Fibre from cane. these are used in cotton-mills, sugar refineries, and other factories, as
well as employed extensively by Railway Companies and by gardeners,
315 &c. Rattan baskets are peculiarly adapted for carrying carboys contain-
Cane-mattresses. ing acids, since the silica of the cane is not acted on by acids. (*Spons,*
316 *Encyclop.*) The waste product, after stripping the cane, is, by certain
manufactures, reduced to a fibre, and in this form is largely used for stuffing
mattresses. Cane mattresses are in great favour on the Continent, taking
the place of the coir of India.

Trade Returns of Canes.

Very little can be learned regarding the internal trade in rattan canes; but, from the fact of the imports (which come chiefly from the Straits Settlements) into Calcutta, Madras, Burma, and Bombay, far exceeding the exports, it seems that with improved facilities of communication a trade might easily be opened up with Eastern Bengal, Assam, and Burma which would to a large extent check the importation, from foreign countries, of a product of which India has herself an unlimited amount. The following

CANES

TRADE.

Trade Returns.

summary of the foreign trade in "Canes and Rattans" will be found instructive :—

Foreign Trade in Canes and Rattans.

Year.	Imports.		Exports and Re-exports.	
	Quantity.	Value.	Quantity.	Value.
	Cwt.	R	Cwt.	R
1879-80	20,617	1,93,035	7,483	73,582
1880-81	21,164	1,99,557	16,346	1,62,363
1881-82	29,559	2,92,754	23,801	2,06,544
1882-83	24,603	2,46,476	14,244	1,33,061
1883-84	28,183	2,51,203	20,836	1,34,884
1884-85	33,408	3,10,675	14,133	1,33,734
1885-86	21,213	1,77,536	6,485	56,844

Detail of Imports, 1885-86.

Province into which imported.	Quantity.	Value.	Country whence imported.	Quantity.	Value.
	Cwt.	R		Cwt.	R
Bengal . .	7,194	66,198	Siam	413	3,158
Bombay and Sind	9,871	79,095	Straits Settlements .	20,350	1,72,880
Madras . .	1,162	8,713	Other Countries .	450	1,498
British Burma .	2,986	23,530			
Total .	21,213	1,77,536	Total .	21,213	1,77,536

Detail of Exports, 1885-86.

Province from which exported.	Quantity.	Value.	Country to which exported.	Quantity.	Value.
	Cwt.	R		Cwt.	R
Bengal . .	1,525	20,770	United Kingdom .	3,827	35,030
Bombay . .	623	2,466	United States. .	427	8,435
Madras . .	637	1,254	Italy . . .	63	1,160
British Burma .	3,700	32,354	Cape Colony . .	469	6,128
			Mauritius . .	187	1,080
			Other Countries .	1,512	5,011
Total .	20,836	1,34,884	Total .	6,485	56,844

The reader is referred for further particulars to the information given under the species of Calamus. In concluding this account of Canes, it is necessary to briefly mention a few of the more common articles sometimes sold, though incorrectly, under the name of cane. The most important is the "male bamboo." Walking-sticks and alpine-stocks of bamboo are becoming very common, and these are very probably included in the above returns for "Canes and Rattans." Reeds or the culms of several species of grasses are also now used for this purpose; the Whangee cane of China

Substitutes for canes. 317

Whangee canes. 318

Palm walking sticks. is one of the greatest favourites of this class. These are the beautifully jointed stems, with a portion of the root, of **Phyllostachys nigra.** Specially
319 prepared palm walking-sticks may also be included under the heading
Male bamboo. of canes. These are chiefly prepared from the betel-nut palm, the palmyra
320 palm, and from the cocoa-nut palm, and are now-a-days largely used for umbrella handles. The "Malacca cane" is obtained from **Calamus Scipionum,** and the rattan from **C. Ratong** and one or two allied species; the former obtains its beautiful colour by being smoked.

321 **CANNA,** *Linn.; Gen. Pl., III., 654.*

Canna indica, *Linn.; Roxb., Fl. Ind., Ed. C.B.C., 1;* SCITAMINEÆ.

INDIAN SHOT.

Vern.—*Sabba jaya,* HIND.; *Kiwára,* N.-W. P.; *Sarba-jayá, lál sarbo jayá,* BENG.; *Hakik,* PB.; *Deva-keli,* MAR.; *Soogúndaraju gida, kélahú-húdingana,* KAN.; *Kullvalei-mani, kundamani cheddi,* TAM.; *Krishna-tamarah, guri genza chettu,* TEL.; *Katú-bala,* MALA.; *Ukilbar-ki-munker,* DUK.; *Sarvajayá, silarumba,* SANS.; *Budda-tha-ra-na,* BURM.; *Bútsarana,* SING.

References.—*Thwaites, En. Ceylon Pl., 320; Dalz. & Gibs., Bom. Fl. Suppl., 88; Voigt, Hort. Sub. Cal., 576; Flück. & Hanb., Pharmacog., 634; U. C. Dutt, Mat. Med. Hind., 317; Drury, Us. Pl., 105; Baden Powell, Pb. Prod., 382; Atkinson, Him. Dist., 730; Balfour, Cyclop.; Smith, Dic., 220; Treasury of Botany; Morton, Cyclop., Agri.*

Habitat.—Several varieties are common all over India and Ceylon, chiefly in gardens, where they are grown as ornamental and flowering plants; they are in flower all the year.

DYE. Seed. 322 **Dye.**—"The SEED is black, and round like a pea and yields a beautiful but evanescent purple dye." (*Dalz. & Gibs., Bomb. Fl.*)

MEDICINE. Root. 323 **Medicine.**—The ROOT is used as a diaphoretic and diuretic in fevers and dropsy (*Atkinson*), and also given as a demulcent. (*Irvine.*) It is considered acrid and stimulant (*Fleming*). When cattle have eaten any poisonous grass, which is generally discovered by the swelling of the abdomen, the natives administer to them the root of this plant, which they
Seed. 324 break up in small pieces, boil in rice-water with pepper, and give the cattle to drink (*Drury*). The SEED is cordial and vulnerary (*Baden Powell*).

FOOD. Root. 325 **Food.—Drury** says: "Nearly all the species contain starch in the root-stock, which renders them fit to be used as food after being cooked. From the root of one kind, **C. edulis,** a nutritious aliment (*Tous les mois*) is pre-
Starch. 326 pared. This is peculiarly fitted for invalids, not being liable to turn acid.
Aliment or arrow-root. 327 To prepare it, the starch is first separated by cutting the tubers in pieces and putting them in water; the water is poured off after a time, when the starch subsides."

§ "In the West Indies arrow-root has been obtained from **C. glauca,** called '*Tous les mois*' (*O'Shaughnessy*)." (*Surgeon C. J. H. Warden, Professor of Chemistry, Calcutta*).

DOMESTIC. Leaves. 328 **Domestic Uses.**—"The leaves are large and tough, and are sometimes used for wrapping up goods. The seeds are black, hard, and shining, re-
Seeds. 329 sembling shot, for which they are sometimes used. The natives make necklaces and other ornaments of them. In the West Indies the leaves
Necklaces. 330 are used to thatch houses" (*Drury*). [See also under **Beads,** Vol. I.—*Ed.*] "In Bangalore, the leaves are used by the natives in lieu of plates, to serve *rági* pudding and other dishes." (*J. Cameron, Esq.*)

CANNABIS, *Linn.; Gen. Pl., III., 357.* 331

Cannabis sativa, *Linn.; DC. Prodr., XVI., I., 30;* URTICACEÆ.

HEMP; INDIAN HEMP; CHANVRE, *Fr.;* HANF, *Germ.;* CANAPE, *It.;* KONAPLI, *Rus.;* CANAMO, *Sp.;* HAMP, *Dan.;* KANAS, *Keltic.;* CANNABIS, *Latin* and *Greek.*

Syn.—C. INDICA, *Lamk.*

Vern.—*Gánjé-ká-pér, kinnab, bháng, gánjá, charas, gúr, siddhí, sabzi, phúlgánjá,* HIND.; *Gánjá, bháng, sidhí,* BENG.; *Gúr-bhanga,* female and *phúl-bhanga* male fibre plant in N.-W. P.; *Sabji,* or *bhang, gánjá* and *charas,* N.-W. P.; *Bhangí, bháng, bengi, charas, kas, sabzi (gúlú,* the seeds, and *chel,* fibre), PB.; *Bangi,* KASHMIR; *Gánjá,* C. P.; *Gánja,* GUJ.; *Ganja,* CUTCH; *Bhángácha-jháda,* MAR.; *Sidhi, gánjé-ka-jhar, ganja,* DUK.; *Gánja-chedi, korkkar-muli, gánja-ilai, bangi-ilai* (leaves), *ganja-phal* (resin), *ganja-rasham, gánja, kalpam,* TAM.; *Gánjari-chettu, bangi-aku, kalpam-chettu, ganja-chettu, ganjai, gangah,* TEL.; *Tajeru cans-juva, kanchava-chetti, gingi-lacki-lacki, ginjil-achi-lachi,* MALA.; *Bhangi, bhangi-gida,* KAN.; *Ganjika, vajradru-vrikshaha, bhánga, vijáyá, gánjá, indrásana, ganjika, hangá, ujáyá, jáyá, vrij-patta, chapola, ununda, hursíní,* SANS.; *Kandir,* KASHGAR; *Kinnab, hináb, nabátul-qunnah, kanab,* ARAB.; *Darakhte-kinnab, darakte-bang, bang, nabátul-qunnab,* PERS.; *Bhénbin, ben, bin, séjáv-bin, sechaub,* BURM.; *Matkansha, ganjá-gahá, kansá-gahá,* SING.

The above vernacular names are either given to the plant or to the forms of the narcotic. It has been found impossible to separate them for certain, and they have accordingly been left for the present in what must be admitted an unsatisfactory form. Much apparent confusion exists in the various provincial forms of the same word. In the Godavery District the fibre is said to be cultivated under the name *Zanumu.*

References.—*DC. Prod., XVI., p. I., 30, published in 1869; Roxb., Fl. Ind., Ed. C.B.C., 718; Kurz, For. Fl., Burm., II., 420; Dalz. & Gibs., Bomb. Fl., Suppl., 79; Stewart, Pb. Pl., 215; Aitchison, Cat. Pb. and Sind Pl., 139; DC. Origin of Cult. Pl., 148; Voigt, Hort. Sub. Cal., 282; Pharm. Ind., 216, 463; Moodeen Sheriff, Suppl., Pharm. Ind., 85; Flück. & Hanb., Pharmacog., 546; Bentl. & Trim., Med. Pl., 231; Per. Mat. Med., by B. & R., p. 504; Cooke, Seven Sisters of Sleep, pp. 212-217; T. & H. Smith, in Pharm. Journ., VI., Series I., 171; Amer. Journ. Pharm., XLIX., 371; Lindl., Fl. Medica, 299; Macnamara, Report on Gánjá, No. 71, dated 11th June 1872; U. S. Dispens., 15th Ed., 338, 977; U. C. Dutt, Mat. Med. Hind., 235, 293; Dymock, Mat. Med., W. Ind., 603, and 2nd, p. 732; Ainslie, Mat. Ind., II., 108; O'Shaughnessy, Beng. Dispens., 579; Murray, Pl. and Drugs, Sind, 30; Bidie, List of Raw Prod., Paris Exh., 43, 83, 106; Johnson, Chemistry of Common Life, Ed. by Church (1880), p. 333; Hem Ch. Kerr, Report on Bengal Gánjá; Year-Book of Pharmacy, 1873, p. 100; 1874, pp. 72, 105; S. Arjun, Bomb. Drugs, 126; Drury, U. Pl., 106; Baden Powell, Pb. Prod., 292, 293, 504; Royle, Fib. Pl., 252, 314, 316; Jameson, in Journ. Agri.-Hort. Soc. of India, VIII., 167; Cooke, Oils and Oilseeds, 33; Liotard, Paper-making Mat., 5, 7, 10, 15, 23, 24, 28, 43; Atkinson, Him. Dist., 760, 799; Crooke, Dyeing, 63; F. Watson, Gums and Resins, 28; Birdwood, Bomb. Prod., 79, 212, 271, 318; Lisboa, U. Pl. of Bombay, 211, 233; Duthie & Fuller, Field and Garden Crops of the N.-W. P. & Oudh, Part I., 80; Spons' Encyclop., 934-8, 1305, 1391; Balfour, Cyclop., Ed. 1885, I., p. 569; Smith, Dic., 210; Treasury of Botany; Morton, Cyclop. Agri.; Ure, Dic. of Arts and Manufactures; Hanbury, Sc. Papers, 187; Kew, Official Guide, Mus., No. 1, p. 120; Morris, Godavery Dist., p. 69. All Government Excise and other Reports down to 1884-85.*

Habitat.—**Cannabis indica** has been reduced to **C. sativa**—the Indian plant being viewed as but an Asiatic condition of that species. This extends the region of the hemp-plant very considerably. It has been found

The History of the Indian Hemp.

wild to the south of the Caspian Sea, in Siberia, in the desert of Kirghiz. It is also referred to as wild in Central and Southern Russia and to the south of the Caucasus. The plant has been known since the sixth century B.C. in China, and is possibly indigenous on the lower mountain tracts. **Bossier** mentions it as almost wild in Persia, and it appears to be quite wild on the Western Himálaya and Kashmír, and it is acclimatised on the plains of India generally. Indeed, the intimate relation of its various Asiatic names to the Sanskrit *bhánga* would seem to fix the ancestral home of the plant somewhere in Central Asia. On the other hand the Latin and Greek *Cannabis* is apparently derived from the Arabic *Kinnab.* **De Candolle** says that "the species has been found wild, beyond a doubt, to the south of the Caspian Sea, in Siberia, near the 'Irtysch,' in the desert of the Kirghiz, beyond Lake Baikal, and in Dahuria." He is doubtful of its being a native of Southern and Central Russia, but suspects that its area may have extended into China, and is not sure about the plant being indigenous to Persia.

Hemp Acclimatised and Cultivated in India.

It has gone wild as a cold-season annual on rubbish-heaps in Bengal and in many other parts of the plains of India. It is specially reported as springing up spontaneously on the *churs* of the Subarnarekhá river and to be wild in the territory of the Mohurbhunge State on the frontier of Midnapur and also in Singbhum. It is cultivated more or less throughout India, either on account of the NARCOTIC derived from (*a*) the resin, *charas*; (*b*) the young tops and unfertilised female flowers—*gánjá* (or *gánja*); (*c*) the older leaves and fruit-vessels—*bháng*; or on account of the fibre, HEMP; or the ripe seed from which an OIL is prepared. *Gánjá* is derived from the cultivated plant, reared in Eastern Bengal, the Central Provinces, and Bombay; *Charas,* from the cultivated plant on the mountain tracts, such as in Nepal, Kashmír, Ladakh, Afghánistan; *Bháng* from the wild plant on the lower hills, especially in the North-West Provinces, the Panjáb, and Madras. In Europe, especially in Central and Southern Europe, the plant is cultivated on account of the fibre, and the seeds are eaten or made into oil. For some time the European form of the plant was supposed to be distinct from the Asiatic, the chief value of the latter consisting in its narcotic properties; but this distinction has now disappeared from the literature of the subject, since it could not be supported by botanical characters. The reduction became the more necessary when it was fully understood that, according to climate and soil, the Indian plant varied in as marked a degree as it differed from the European. On the mountains of upper India, for example, it yields a good fibre which the natives separate and weave into garments or twist into ropes, but its chief value in Kashmír and Ladákh consists in the fact that, just before maturing its flowers, the bark spontaneously ruptures and a resinous substance exudes. This is also found upon the young leaves, flowers, and fruits, and when rubbed off constitutes the narcotic *charas.* The same plant cultivated in the plains is found not to secrete its resin in this way, but instead it charges the young female flowers and twigs with the narcotic principle; this constitutes the *gánjá.* It has been observed that if even one or two male plants are left in a field, the whole crop of *gánjá* will be destroyed, since, with the fertilisation of the flowers, the *gánjá* almost entirely disappears. In other parts of India the narcotic property is not developed until the fruits are mature; leaves at this stage, and sometimes the fruits also, afford *bháng*. With **Cannabis indica** differing in so marked a degree according to the climate, soil, and mode of cultivation, it was rightly concluded that its separation from the hemp plant of Europe could not be maintained. We have here, in fact, one of the most notable illustrations of the effect of climate in changing the

FORMS OF HEMP.

chemical processes which take place in the structure and physiological peculiarities of a plant. In most instances, a plant taken by man from one climatic condition to another, either dies quickly, or if it survives, it exists in a sickly condition. A few plants however, such as the potato, the tobacco, the poppy, and the hemp, seem to have the power of growing with equal luxuriance under almost any climatic condition, changing or modifying some important function as if to adapt themselves to the altered circumstances. As remarked, hemp is perhaps the most notable example of this; hence it produces a valuable fibre in Europe, while showing little or no tendency to produce the narcotic principle which in Asia constitutes its chief value.

The plant for one or other of these purposes is now extensively cultivated throughout Persia; in India, from the level of the sea in Bengal to the inner Himálaya at an altitude of 10,000 feet; in China; in Arabia; and in Africa, from the extreme south to the north, and on the mountains as well as on the plains; in the north-eastern portions of America and on the table-land of Brazil. It is also to be met with in Northern Russia even as far as Archangel. In England it not unfrequently occurs as a weed, springing up most probably from rejected birdseed.

The modes of cultivation and the nature of the soil required, depend on the purpose for which the plant is cultivated. This subject will accordingly be discussed later on.

HISTORY OF HEMP.

332

The Narcotic.

Indian Literature.—"The earliest synonym appears to be *bhánga*, which occurs in the Atharva Veda—the last of the four scriptures of the Hindús. It is derived from a root which means 'to break,' and is supposed to imply the process of debarkation by which the fibres of the plant were separated from the stem. This would indicate that even at the remote period when the *Veda* in question was written, probably about 3,000 years ago, the use of hemp as a fibre-yielding plant was well known and the knowledge fully utilised The *Veda*, however, reckons it, along with the *Soma*, as one of the five plants 'which were liberators of sin,' and this would imply that its narcotic property was also well known. The word is used in the masculine form with a short final vowel, and not, as in later literature, with a long one. Both the masculine and feminine forms of the name are still in use by the people of the Himálayan regions, who call the male plant *Phúl bháng*, and the female *Gúl* (or *gúr*) *bhánga*. In the *Kausitaki Brahminia* of the *Rig Veda*, which is next in antiquity to the *Atharva*, the attributive form of the word *Bhánga*, derivable both from the feminine and the masculine forms, occurs both singly and in combination: one, *Bhángájala*, meaning a hempen net; and another, *Bhánga-sayana*, a 'bed-stead woven with hempen cords.' The masculine form occurs in the celebrated medical treatise of **Susruta**, in which the plant is described as a medicine for the accumulation of phlegm in the larynx and for some other diseases." "In the Institutes of **Manu** the feminine form is used, and the plant is noticed for its fibres. In later works the feminine form prevails" (*Mr. Hem Chunder Kerr*). The curious fact of the popular opinion, that the *gánjá*-yielding plants are males and the staminate or non-narcotic-yielding females, prevailing to the present day, would seem to corroborate the inference that in Asiatic countries during ancient times the plant was cultivated chiefly for its narcotic properties. Hence, in all probability, the habit of speaking of the narcotic in the masculine form of the name, and of the fibre in the feminine. As a matter of fact, the nar-

The History of the Indian Hemp.

HISTORY.

cotic-yielding is the reverse to the popular belief: the male or staminate plants are rooted up as being non-narcotic, and the pistilate or female are allowed to mature, so that, botanically speaking, it is the female which yields the narcotic, not the male. Not being fecundated, the female plants do not mature fruits and seeds, and are accordingly viewed as male. **Bretschneider** informs us that the male and female plants were recognised by the Chinese as early as 500 B.C. It may be here incidentally remarked that this distinction would seem to point to the idea that the ancient Chinese and Sanskrit writers were aware of the existence of male and female flowers centuries before the sexes of plants were realised in Europe.

The intoxicating property of the drug is implied in the names *ánandá*, "the joyous;" *harshini*, "the delight-giver;" *madiní*, "the intoxicator," and *gánjá* and *ganjákíní*, "the noisy." The probable importation of the narcotic in ancient times into India in a prepared form, as it comes at the present day from Yarkand, is indicated in the name *Káshmírí* often applied to it in early literature. It is thus probable that the knowledge of the narcotic, or at least of *charas*, was brought to India across the Himálaya.

The Narcotic. 333

Classical Literature of Europe.—The ancient SCYTHIANS seem to have been acquainted with the narcotic properties of the plant as well as with its fibre. HERODOTUS tells us that they excited themselves by "inhaling its vapour." "HOMER makes HELEN administer to TELEMACHUS, in the house of MENELAUS, a potion prepared from nepenthes, which made him forget his sorrows. This plant had been given to her by a woman of Egyptian Thebes; and DIODORUS SICULUS states that the Egyptians laid much stress on this circumstance, arguing that HOMER must have lived among them, since the women of Thebes were actually noted for possessing a secret by which they could dissipate anger or melancholy. This secret is supposed to have been a knowledge of the qualities of hemp" (*Johnston, Chemistry of Common Life, 337*).

Mythology. 334

Mythological History of the Narcotic.—"The notices of hemp in Arabic and Persian works are much more numerous. The oldest work in which it is noticed is a treatise by **Hassan**, who states that in the year 658 A.H., **Sheik Jafer Shirazi**, a monk of the order of HAIDER, learned from his master the history of the discovery of hemp. **Haider** lived in rigid privation on a mountain between Nishabar and Rama, where he established a monastery. After having lived ten years in this retreat, he one day returned from a stroll in the neighbourhood with an air of joy and gaiety; on being questioned, he stated that, struck by the appearance of a plant, he had gathered and eaten its leaves. He then led his companions to the spot, who all ate and were similarly excited. A tincture of the hemp leaf in wine or spirit seems to have been the favourite formula in which **Sheik Haider** indulged himself" (*Dymock, Mat. Med., W. Ind., 604*).

A curious story is told in the Hindú mythology about the origin of this plant. "It is said to have been produced in the shape of nectar while the gods were churning the ocean with the mountain called Mandára. It is the favourite drink of Indra, the king of gods, and is called *vijayá*, because it gives success to its votaries. The gods, through compassion on the human race, sent it to this earth, so that mankind by using it habitually may attain delight, lose all fear, and have their sexual desires excited. On the last day of the Durga Pooja, after the idols are thrown into water, it is customary for the Hindús to see their friends and relatives and embrace them. After the ceremony is over it is incumbent on the owner of the house to offer his visitors a cup of *bháng* and sweetmeats for tiffin (lunch)" (*U. C. Dutt's Mat. Med. Hind., 236*).

HISTORY.

More Recent Historic Facts regarding the Narcotic.—The use of hemp (*bháng*) in India was particularly noticed by **Garcia de Orta** (1563), and the plant was subsequently figured by **Rheede,** who described the drug as largely used on the Malabar coast. It would seem about this time to have been imported into Europe, at least occasionally, for **Berlu,** in his *Treasury of Drugs,* 1690, describes it as coming from Bantam in the East Indies, and "*of an infatuating quality and pernicious use.*"

"It was NAPOLEON's expedition to Egypt that was the means of again calling attention to the peculiar properties of hemp, by the accounts of **DeLacy** (1809) and **Rouger** (1810). But the introduction of the Indian drug into European medicine is of still more recent date, and is chiefly due to the experiments made in Calcutta by **O'Shaughnessy** in 1838-39. Although the astonishing effects produced in India by the administration of preparations of hemp are seldom witnessed in the cooler climate of Britain, the powers of the drug are sufficiently manifest to give it an established place in the Pharmacopœia" (*Flück. & Hanb., Pharmacog., 547-48*).

HISTORY OF THE HEMP FIBRE.

The Fibre. 335

The following extract may be here published as giving the most trustworthy facts which can be adduced regarding the history of the fibre: "According to **Herodotus** (born 484 B.C.), the Scythians used hemp, but in his time the Greeks were scarcely acquainted with it. **Hiero II.,** King of Syracuse, bought the hemp used for the cordage of his vessels in Gaul, and **Lucilius** is the earliest Roman writer who speaks of the plant (100 B.C.). Hebrew books do not mention hemp. It was not used in the fabrics which enveloped the mummies of ancient Egypt. Even at the end of the eighteenth century it was only cultivated in Egypt for the sake of an intoxicating liquid extracted from the plant. The compilation of Jewish laws known as the Talmud, made under the Roman dominion, speaks of its textile properties as of a little known fact. It seems probable that the Scythians transported this plant from Central Asia and from Russia when they migrated westward about 1500 B.C., a little before the Trojan War. It may also have been introduced by the earlier incursions of the Aryans into Thrace and Western Europe; yet in that case it would have been earlier known in Italy. Hemp has not been found in the lake-dwellings of Switzerland and Northern Italy" (*DC., Orig. Cult. Pl., 148*).

The Arabic name for the plant, *benj,* and the Persian *beng,* seem but corruptions from the Sanskrit *bhánga.* The common name for hemp in Arabic is, however, *kinnab* or *konnab,* admittedly the origin of the Greek *Kannabis* and of the Latin *Cannabis,* and from this again the English word canvas. So, in a like manner, the Arabic *hashish* would seem to be given in allusion to the green intoxicating liquor, and allied to this would appear to be the Hebrew *shesh,* a word supposed by some to be a name for flax, but may have rather been applied to hemp, since it literally means "to be joyous." Persons who drink the draught are called *hashâshins* (or *bhángis, e.g.,* drinkers of green liquid). As a curious illustration of the origin of names or of words, it may be here remarked that **Mr. Sylvester de Lacy,** the well-known Oriental scholar, derives the modern English word "assassin" from *hashâshín* (or *haschischin*), persons who fortify themselves with *hasísh* before performing certain ceremonies or perpetrating inhuman deeds. The word according to some would appear to have been originally used in Syria to designate the followers of "the old man of the mountains;" by others it came into European use during the wars of the Crusaders. Certain of the Saracen army having intoxicated themselves with the *hasish,* rushed fearless of death into the Christian camp, committing

Canvas.

Assassin.

great havoc. It seems probable that the English form of the word was adopted at the latter date, but that the more Arabic form was known in Europe for some time previous. Hemp is alluded to in the "Arabian Nights" under its more ancient Arabic name, *beng.*

CULTIVATION.

CULTIVATION. 336

It has already been incidentally remarked that the cultivation of **Cannabis sativa** in India is naturally referable to two sections: (*a*) Cultivation with a view to preparing some of the forms of the narcotic, and (*b*) cultivation on account of the fibre. It has also been stated that the hemp plant has, to a large extent, changed its character under Indian or rather Asiatic cultivation. It is very generally admitted, for example, that in the plains, while the narcotic principle is readily developed, the hemp fibre is but very imperfectly formed. Let it, however, be distinctly understood that by hemp is here exclusively meant the fibre of **Cannabis sativa.** This remark is all the more necessary when it is added that in the Government returns of the Trade and Navigation of British India, the fibre of **Cannabis sativa,** as well as that of **Crotalaria juncea, Musa textilis,** and perhaps the fibres also of one or two other plants, are commercially returned as hemp, and the manufactures therefrom as hemp manufactures. To obtain the true hemp fibre, a rich soil and a high state of cultivation is required in a temperate climate. The plant will grow anywhere in India, and may be said to be naturalised in every province. This fact seems to have influenced the minds of the earlier writers on this subject, who uniformly urge that since it grows so freely in a wild state, it might be cultivated to any desired extent as a source of fibre. **Dr. Stocks** (one of the most careful observers India has ever had) wrote in 1848:—"The plant grows well in Sind, and if it ever should be found advantageous (politically or financially) to grow hemp for its fibre, then Sind would be a very proper climate." The writer does not think that the question of its possible cultivation as a cold-season fibre-crop on the plains of India has been fully tested. There may be some localities where this might be found possible* and even remunerative; but so far as the published experiments go, like flax the hemp plant may be grown freely enough, but not as a source of fibre. The flax plant of the plains of India yields a superior oil-seed, and the hemp plant a valued narcotic, but neither would seem to justify the expectation of becoming a profitable fibre crop. This fact does not appear to have been fully realised by writers in Europe, and on the one hand the existing cultivation of hemp as a source of narcotic has been confused with a supposed fibre production, while on the other, the reports of the limited Himálayan cultivation as a source of fibre have been mistaken as the total Indian cultivation of the plant. The authors of the *Pharmacographia* say: "It is found in Kashmír and in the Himálaya, growing 10 to 12 feet high and thriving vigorously at an elevation of 6,000 to 10,000 feet." **Balfour,** in his new edition of the *Cyclopædia of India,* while stating incidentally that the "plant is grown in Persia, Syria, Arabia, and throughout India," enters into an account of its cultivation in Garhwál, with the apparent object of proving that it is more extensively grown there than in the Panjáb, but he makes no mention of the fact that the principal seats of hemp cultivation, as a commercial article, are in Eastern Bengal, the Central Provinces, and Bombay. The *Encyclopædia Britannica* has also fallen into the same mistake, and, indeed, illustrations might be multiplied to show that undue prominence has been given to the fact that the plant is grown in Garhwál, the

Expectations regarding Hemp Fibre.

* See a further page regarding Godavery District.

CULTIVATION.

Panjáb, and Kashmír, the more so since by most writers the true regions of Indian cultivation have been, to a large extent, overlooked.

Unfortunately, the available material is too meagre to allow of the subject being dealt with province by province, although there are doubtless different methods pursued in each. This difficulty, fortunately, does not exist with the Lower Provinces, since **Mr. Hem Chunder Kerr** in his *Report on the Cultivation of and Trade in Gánjá in Bengal (1877)*, has placed in the hands of the public a valuable treatise which deals both with the cultivation of the plant and the preparation of the narcotic. **Dr. Forbes Royle** in 1855 issued his *Fibrous Plants of India*, a work which treats of hemp at some detail, discussing mainly the cultivation pursued in Upper India. A more recent publication, **Messrs. Duthie** and **Fuller's** *Field and Garden Crops*, gives a brief account of the cultivation in the North-West Provinces. From these works, and the writer's own personal observations, supplemented by several less important publications, and Government reports, the following abstract regarding Indian hemp cultivation has been prepared.

(a) Cultivation for the Narcotic.

For the Narcotic. 337

Bengal Cultivation.—The method pursued in Eastern Bengal, according to **Mr. Hem Chunder Kerr**, is briefly as follows: After selecting the land, for hemp cultivation, the preparation of the soil commences in March-April, but where this can be afforded operations are started even earlier. The sites selected are those which are moist, but not shaded, and the soil a rich friable loam. The land is then ploughed from four to ten times, the object being to free it as far as possible of all weeds. Fresh earth from the surrounding ditches or from any neighbouring low-lying land is thrown over the field, and it is freely manured with cowdung. After a week this is ploughed into the soil, and the ploughing repeated as often as the means of the cultivator will admit of. The belief is that for hemp the land cannot be too often ploughed. The rains now set in, and the field is thoroughly flooded, the surrounding banks preventing the escape of the water. While inundated it is several times ploughed and harrowed. After the rains subside the ploughing is again repeated, and the soil thrown into ridges a foot high, the furrows being a foot in breadth.

Nursery.—It is customary for the cultivators to combine in the rearing of seedlings. This is done in a nursery selected on high land of a light sandy loam. The preparation of the nursery generally commences in May after the first shower of rain. The plot is repeatedly ploughed, and if need be harrowed, until, by August, the soil has been completely pulverised. On a sunny day the seed is sown broadcast, and by the latter end of September, the seedlings are about 6 to 12 inches high, and are then ready for transplantation. About 4 to 5 seers of seed are deemed necessary for every *bigha* of land to be cultivated with hemp.

Transplantation.—The seedlings are planted out 6 to 8 inches apart on the ridges prepared for their reception, after having been simply pulled up by the root; they are planted the same day in the field. (**Dr. Royle** says that for *bháng* the seedlings should be planted 9 or 10 feet apart.) Towards the end of October the ridges are opened out, manured with oil-cake and cowdung, and thereafter re-formed, so that freshly-manured soil is thrown up around the plants.

Treatment of the Plants.—Trimming of the plants commences by November. This consists in lopping off the lower branches so as to favour the upward growth of the shoots. The ridges are again re-dressed and manured, the furrows ploughed, and all weeds removed. At this stage the plants begin to form their flowers, when the services of an expert, known

The Cultivation of Hemp in India.

CULTIVATION.

as the *gánjá*-doctor (*poddár* or *parakdár*) are called in. This person passes through the field, furrow by furrow, cutting down all the male or staminate plants, or what are colloquially known as *mádí* (female) plants. Speaking of the importance of this operation **Mr. Hem Chunder Kerr** remarks: "The presence of a few *mádí* plants in the field suffices to injure the entire crop, inasmuch as all the plants run into seed, and the *gánjá* yielded by them is very inferior and scarcely saleable." The destruction of the *mádí* plants is, however, never so complete but that a few escape detection, the result being that a certain number of the female plants are fecundated, fruits and seeds being produced. These are thrashed out as far as possible in the manufacture of the drug, the quality of which may be judged of by the freedom from such impurities.

Fruits injure Ganja.

The female plants come to maturity about the beginning of January, but the *gánjá* is not fully developed till a month later. The crop is sold in the field to the *gánjá* dealers, who bring their own men to manufacture it. The crop intended to be made into what is technically known as flat *gánjá* is reaped a few days before that intended for the round form.

In another page will be found an account of the processes of manufacture of the various forms of *gánjá*, together with considerable details as to the extent of cultivation as a source of the various forms of the narcotic.

For the Fibre. 338

(*b*) Cultivation for the Fibre Hemp.

Indian Methods.—Dr. Royle very appropriately remarks: "There is every reason for believing that the plant is of Eastern origin, while there is no sufficient reason for thinking that the climate of Europe is so peculiarly suited to the production of its fibre as to exclude those of its native climes, especially where attention is paid to those where the plant is grown on account of its fibre, and those distinguished from the others where it is cultivated for its resinous and intoxicating secretion. The latter requires exposure to light and air. These are obtained by thin sowing, while the growth of the fibre is promoted by shade and moisture, which are procured by thick sowing." It has already been pointed out that the regions suited for *gánjá* cultivation are perfectly distinct from those where it might be possible to develope an industry in the fibre. However much it may be regretted it seems impossible to combine the two industries, and it is an accepted fact that, unless utilisable as a paper stock, the immense amount of stems annually destroyed by the *gánjá* cultivators must continue to be so.

Godavery Hemp. 339

At the same time **Mr. Morris**, in his account of the Godavery District, gives some interesting facts regarding the cultivation of hemp fibre. It is planted in November and cut by the end of March. It is grown in drills and never watered. Clay soils and those beyond the reach of inundation are those best suited. "About 2,200 bundles can be produced in one *putti* of land, each bundle yielding 1½ *viss* of fibre, or a total of 3,300 *viss* or 412½ maunds, and is valued at one rupee a maund. The expenses of cultivation are estimated at R8-8, and those of the preparation of fibre at R100 a *putti* of land. The bundles are buried in mud and left to rot for about a week when they are taken out and beaten in the water; and after all impurities are removed, the fibre is collected." The exports from the district are said to have been, in 1854-55, 4,269 cwt.

Unless there be some mistake, *Sunn* hemp having been called "**Cannabis sativa,**" for **Mr. Morris** gives that scientific name as well as the vernacular name *zanumu* for the fibre he is describing, this information is of the greatest interest, as it would show, what the writer was not aware of until recently, that hemp fibre was actually produced on the plains of India.

Cultivation of Hemp in India.

CULTIVATION. For the Fibre.

Early Experiments in Hemp cultivation.—In 1802 the Government of India made various experiments on an extended scale to establish hemp fibre cultivation. European seed was imported, and farms and factories established, but finally abandoned. Recourse was had to improving the cultivation of the Indian stock. The cultivation and manufacture was carried on at Rishra, Cassimpore, Maldah, Gorackpore, Mhow, Rohilkand, and Azimgarh, under the experienced supervision of European hemp-dressers. The results were everywhere unsatisfactory and the experiments abandoned.

Inquiry was re-instituted in 1871 as to whether the rejected stems from *gánjá* cultivation might not be utilised for fibre, but the enquiry in this case met again with an unfavourable result.

Possible Prospects.

The Possibility of more favourable Results.—In spite of these disheartening results, it cannot be definitely stated that it is impossible that hemp fibre can be produced in India. The efforts alluded to were mainly directed to combining the two industries of producing resin and fibre, and the regions selected were accordingly those where it is now believed the plant does not develope its fibre. There may, however, be localities where further investigation might prove that it was possible to organise a commercial fibre industry. It is well known, for example, that the people of Kumáon and Garhwál grow the plant on account of its fibre, and with the results of the experiments conducted at the beginning of the century before him, **Dr. Royle** still entertained the highest hopes of ultimate success. From a paper which appeared in 1839, in the *Transactions of the Agri-Horti. Society of India, Vol. VIII., p. 15*, the following passage may be reprinted, as it expresses pretty clearly **Dr. Royle's** view:—"This (hemp) might be cultivated in suitable situations in India, in a manner similar to that adopted in Europe, or like that practised with its substitutes in India. The effect would undoubtedly be to produce a sufficiently long fibre, which would also be softer and more pliable at the same time that it retained a great portion of its original strength, and probably in as large a quantity as is yielded by the *sunn* plant. Thus, an article might be produced which, judging from the Italian samples, might enter into competition with the Russian product, and at all events afford much more valuable cordage than the several (usually considered) inefficient substitutes which are so extensive in India, and which, imported into England, sell only for 15 to 20 shillings per cwt., at the same time that the Russian, Polish, and Italian hemps are sellng for 42 to 50 shillings per cwt." In his *Fibrous Plants* **Dr. Royle** alludes to successful experiments of hemp cultivation in the plains, especially at Chittagong. But in most cases, as was proved with the plant reared at Saharanpur, it is admitted that the plains crop is far inferior to that reared on the hills. The opinion is therefore arrived at that if the cost of transmission from the hills to the sea-ports can be reduced, it might be possible to greatly extend the cultivation of hemp on the lower Himálaya. **Mr. Atkinson** (*Himaláyan Districts, p. 800*) gives a long and instructive account of the methods of culture, compiled chiefly from **Huddleston** and **Batten's** notes. There are said to be two varieties common in Garhwál—the wild and the cultivated. The former is practically useless either for the fibre or the drug. The cultivated form is grown in high land having a northern exposure and in well prepared and abundantly manured land close to the village sites. Occasionally freshly cleared forest land gives a crop for a year or two without the aid of manure. The plant does not flourish below 3,000 feet, as the heat of the valleys is prejudicial to its growth, and it seems to thrive best at from 4,000 to 7,000 feet in altitude above the sea. After being well prepared and freed from weeds, the ground is sown in May or June. During the growth of the

The Cultivation of Hemp in India.

CULTIVATION. For the Fibre.

plants the ground is once or twice dressed, and, where necessary, the plants thinned so as to leave a few inches between each. The plants ultimately attain a height of 12—14 feet, and from September to November the crop is regarded as mature. In warm situations the crop is sown later than May to June, so as to save the seedlings from running into useless stalks during the heat and rain of the early period of its growth. In Nepal the sowing takes place from March to April; the plants flower in July, and reach their full growth by August: if allowed to grow longer the fibre becomes hard and not suitable for manufactures. **Messrs. Duthie and Fuller**, in discussing the North-West Himálayan hemp fibre cultivation, allude to the fact that a form of *charas* is at the same time prepared, and **Mr. Atkinson** adds that "the farm of *charas* in Kumáon alone, during 1880-81, was sold for R3,357." It is commonly reported that the cultivation of the hemp-narcotics is prohibited in the North-West Provinces. In an early paragraph (No. 339), it has been shown that hemp fibre would appear to be cultivated in the Godavery District.

Season of Sowing and Reaping.—**Messrs. Duthie and Fuller** remark:—"The seed is sown in May at the rate of 30 seers to the acre, and the plants are thinned out if they come up too closely, and are kept carefully weeded. By September they will have attained a height of 12 or 14 feet. In the hemp the male and female organs are contained in separate flowers and borne on separate plants. The male plants (called *phúl-bháng*) yield the best fibre, and they are cut a month or six weeks before the female plants (*gul-bháng*), which are allowed to stand until their seed ripens. The next process is the collection of the *charas*, which is done by rubbing the seed pods and leaves between the hands."

European Cultivation for the Fibre.—**Dr. Royle** and several other authors give accounts of the methods pursued in Europe in hemp cultivation. The limited space at the writer's disposal precludes this being entered into in detail, but the reader who may desire further information is referred to *Spons' Encyclopædia of Manufactures and Raw Materials* (1882), *Vol. I., 934*. It is stated by most writers that while the plant will grow almost anywhere in the temperate and sub-tropic regions of the globe, it can be made to yield a profitable fibre only when reared on rich soil, freely manured and repeatedly ploughed. "Alluvial lands, where sand and clay are intimately mixed, or friable loams containing much vegetable matter, are well suited. Stiff cold clays are to be avoided. Over-rich soils produce coarse but strong fibre; light poor soils, when well manured, will bear the crop for several years in succession."

Italian Hemp. 340

The best fibre is that raised in Italy, but it is generally stated that seed from Holland is most esteemed: well-grown English seed is perhaps equal to it. "Seed from the plains of India, though of good outward appearance, yields poor fibre for the first crop or two; but Himálaya seed is inferior to none." Constant change of seed is recommended and good seed is described as plump and of a bright-grey colour.

Male Fibre. 341

"The fibre afforded by the male plants is tougher and better than that yielded by the females; it is usual to divide the harvest. The males are gathered as soon as they have shed their pollen, about 13 weeks after sowing; each is uprooted singly, care being taken not to injure the stem.

"The fibre is separated either by retting or by breaking and scutching" (*Spons' Encycl.*).

Properties and Uses of Cannabis sativa.

ECONOMIC PROPERTIES.

From the STEMS, LEAVES, OR FLOWERS, and even the FRUITS, a RESINOUS EXTRACT, of a powerful narcotic character, may be prepared. The INNER BARK affords the valuable FIBRE HEMP. The SEEDS are occasionally eaten; they are much valued for feeding birds. An OIL is expressed from them which is of some importance, but can scarcely be called commercial.

RESIN OR NARCOTIC.

There are primarily three forms of this substance, but under each there exist also local modifications, special preparations from these, and adulterants or imitations. The three forms are known as *Gánjá*, *Charas*, and *Bháng*. *Gánjá* is the female flowering tops with the resinous exudation on these: *Charas* the resinous substance found on the leaves, young twigs, bark of the stem, and even on the young fruits: *Bháng*, the mature leaves and in some parts of India the fruits also, but not the twigs.

BENGAL MANUFACTURE.

GANJA. 342

(*1st*) GÁNJÁ.—This is known in the trade as consisting mainly of two forms: *Flat Gánjá* and *Round Gánjá*. Speaking of the manufacture of *gánjá* in Bengal Mr. Hem Chunder Kerr says:—"In February and March, when *gánjá* attains its maturity, the cultivator proceeds to make arrangements for reaping the crop and preparing the drug. His first step is to present himself to the supervisor, show him the license under which he has grown the crop, and obtain his permission to remove the crop from the field." For flat *gánjá*, cutting of the plants commences in the morning; for round *gánjá*, in the afternoon, and by the Hindus Thursday, and by the Muhammadans Friday, is considered the best day for commencing operations.

Flat. 343

Flat-Gánjá.—The stems are cut with a sickle about 6 inches above ground, and are tied together by their ends and placed across a bamboo, and in this manner carried to the place of manufacture. In the preparation of *flat-gánjá* (*cháptá gánjá*) they are exposed to the sun till about 1 or 2 o'clock in the afternoon. The twigs are then cut into lengths of about a foot, the non-flower-bearing portions being thus cut off and rejected. If by any chance male plants (*khásiá*) have been brought from the field, these, together with any fruit-bearing female twigs, are carefully picked out and thrown away, the seeds, if ripe, being first retained for next year's crop. At dusk the selected flowering heads are laid on the grass and left overnight to the influence of the dew. At 2 o'clock of the following afternoon they are assorted in bundles of from three to ten, according to their size. These are arranged on a mat in a circular form, with their points directed towards the centre and overlapping each other. The circle thus formed is about 14 feet in circumference. Upon this a number of persons, holding each other by resting their arms across each other's shoulders, commence to tread. They stamp and press down the flowering heads with their right foot, while holding them tight with their left. This operation is continued for three to four minutes, when the narcotic resin is pressed firmly among the flowers in the desired form. Fresh twigs are then arranged over those which have been pressed and the treading repeated. This process continues until the ring rises to about a foot in height. Care, however, is taken not to place the resinous flowering-tops exactly on the top of each other, since this would result in the formation of an inextricable mass. Before the trampling commences the persons to be so employed salute the *gánjá* before placing their feet on it. When the process is complete the treaders get down. Mats are placed over the stack, upon which one or two men sit down and remain for about half an hour. After this fresh mats are spread and the flowering twigs beaten two and two together so as to shake off the leaves or any fruits that may still remain and are re-arranged in a new circle, so that what was on the top before now forms the bottom

GANJA.

layer of the new circle. The treading is repeated stage by stage until the stack is again covered by the mats, and men take up their inexplicable seat on the top. After this each twig is trodden upon separately, being placed for that purpose on a canvas cloth; by sunset the process is completed for the day's manufacture. Next day the treading is repeated with slight modifications of little importance in the peculiar method followed. The ultimate result of all this labour is that the resin and flowers are firmly consolidated into flat patches near the apex of the twigs, and the leaves and fruit vessels (if such exist) carefully removed.

The twigs are then carried to the homestead and stacked, with the tips pointing inwards, and the stems thus exposed to be dried; when completed, the top of the stack is carefully covered in with mats.

Round. 344 *Round Gánjá.*—In the manufacture of round *gánjá* greater care is bestowed. A larger amount of the twigs and leaves are rejected. Instead of being arranged in a circle, they are placed on the ground in a straight line and just below a bamboo bar, on which the men rest their arms and thus support themselves while treading. Instead, however, of tramping, they now roll twig by twig so as to force the resinous matter into the form of a thin sausage shape near the apex of the twig. This rolling is repeated several times, and the twigs even taken up in the hands and individually trimmed, superfluous leaves, &c., being picked out, and when loose the resin pressed into the desired form by the fingers.

Chur or rora. 345 *Gánjá powder or chúr.*—When perfectly dry both the flat and round *gánjá* are next bailed in a prescribed manner, and during this operation a certain amount of loose particles of the resinous matter falls off: this is known locally as *chúr.* Under the excise rules a separate rate is fixed for *chúr.* It is held to be more powerful than round *gánjá*, and therefore the duty on it is R4 as compared with R3 a seer on round *gánjá.* The fragments which constitute *chúr* cannot be made to adhere, and although prepared at one and the same time with the pressed or rolled *gánjá* and from the self-same plants, it is probable that these fragments exist in a slightly different chemical state, and probably more nearly resemble *charas* than *gánjá. Chúr* is also known under the name of *rorá.*

Mr. E. T. Atkinson (in his *Himálayan Districts, p. 761*) says of the *gánjá* of the N.-W. Provinces: "The *gánjá* produced in Kumáon and Garhwál is considered of little value, and is not, so far as I am aware, exported. The *gánjá* consumed locally is imported from the lower districts. Two sorts of *gánjá* are sold in these Provinces—the *pattar* and the *bilúchar.* The *pattar* is imported chiefly from Holkar's territories and is of quality inferior to the Bengal *gánjá.* It is purchased at from R5 to 6 a maund in Indúr in the rough state," and "pays a duty of about 4 annas per maund on exportation to British territory." It is sold retail at from R3 to 4 a seer. The *bilúchar* variety is imported from Lower Bengal, and is sold at R10 to 12 a seer.

BOMBAY AND THE CENTRAL PROVINCES.

IMITATIONS OF GANJA.

Although definite information cannot at present be obtained as to the details of the process of manufacture of *gánjá* as followed in the Central Provinces and Bombay, it is probable that it differs but slightly from that narrated above as pursued in Bengal. **Dr. Irving**, in his *Materia Medica of Patna*, however, informs us that there are two imitations of *gánjá*, or perhaps more correctly, of *charas.* The one is obtained by evaporating

Expressed Juice. 346 the expressed juice of the plant, and the other an extract obtained by boiling the whole plant. To what extent these adulterants are sold separ-

Decoction. 347 ately or mixed with the pure drug it is difficult to learn; but as far as Bengal is concerned, it may confidently be stated that adulteration can alone take place when the intoxicant reaches the hands of the dealer. In the *golás* it is quite pure.

The mention of *chúr*, and of the extracts referred to by **Dr. Irving**, naturally lead to the consideration of—

(*2nd*) CHARAS.—This may be defined as the resinous substance which naturally exudes from the leaves, stems, and fruits of the hemp plant (see No. 355) in more northern or higher regions where the plant is accordingly grown in a colder climate than that of the *gánjá*-producing districts of the plains. In another page (No. 377) it will be seen that **Dr. Aitchison** says that the resin collected from the leaves and flowers is in Turkistan called *nasha*—the *charas* of trade. Before being exported it is, however, very much adulterated, so that *nasha* and *charas* are practically two very different substances. **Mr. Dalgleish** informs the writer that at Yarkand-Kashgar the *charas* is collected as a fine greyish white powder by beating the flowering twigs over a coarse cotton cloth spread on the ground. The crop is reaped about November and the powder stored in small 24lb bags. About May these are sold to the traders, who cut the bags open and spread out the now partially agglutinated powder on cloths under the sun. It softens and deepens in colour and is hard pressed into bags or bales 1⅓ maunds in weight (a half pony-load ready for exportation). The quality is judged of by the amount of oil seen through the degree of transparency in a fragment flattened on the hand until it is of the thickness of paper, or by rolling a small piece into a cord and exposing it to the sun for a few minutes. The oil is sucked on to the surface of the cord; the *charas* deepens in colour, but if pure, on being broken, is seen to be composed of minute granules of the appearance of pure steel. With age the oiliness is sucked out of the *charas* or by being exposed, it is then valueless. *Charas* is in Yarkand adulterated with linseed oil and a powder of the hemp leaves. CHARAS. 348

From the above description it would appear as if Yarkand *charas* was not the resinous exudation from the leaves and stems, as in Sind, Kashmír, &c., but a powder formed on the flowers similar to the *chúr* of Bengal (see No. 345) and the *garda* (No. 358) of the Panjáb. In Kumáon and Garhwál the cultivators who grow the plant as a source of fibre are said to rub the fruits between their hands so as to remove the resin. This is regarded as a form of *charas*, and the outturn of an acre is said to be four maunds fibre, worth R8, and three seers *charas*, R6. Of Nepál it is commonly reported that a very fine quality of *charas* known as *momea* is similarly prepared. (*See Church's Ed., Johnston's Chemistry of Common Life; Spons' Encyclop.; O'Shaughnessy, &c.*) **Dr. Gimlette**, Residency Surgeon, Katmandu, Nepál, writes to the author, however, that the word *momea*, as applied to *charas*, is unknown at the present day in the vicinity of the capital. Speaking of the modes of collecting *charas* as practised in Nepál, **Dr. Gimlette** adds: "I have been unable to verify the accounts of the collection of *charas* by means of leather coats worn by men who run about among the plants; this may possibly occur on the northern frontier of Nepál. Within a radius of 50 miles of Katmandu, the extract is collected by rubbing the young flowering tops between the hands *in situ*; one man is able thus to collect in a day 2 *tolas* weight. *Charas* is sold in the Katmandu bazar at R2, in the Tarai markets at R3 per *tola*" MOMEA. 349

It is a remarkable coincidence, however, that a drug *is* sold in the bazars of Katmandu under the name of *momea* or *mimea*. This, **Dr. Gimlette** says, is not a preparation of hemp. He thus describes it:—"*Momea* is a mixture of *dhup*, oil, and lymph, the latter specially induced in the human subject by a certain drug. The preparation is also brought from Tibet, where human fat is said to be used as an ingredient. This curious drug is Momea of Nepal. 350

The Narcotic—Indian Hemp.

MOMEA.

given internally in cases of wounds and ulcers along with *ghí;* dose one *masha.*" It is noteworthy, in connection with **Dr. Gimlette's** discovery regarding human fat used in the manufacture of Nepál *momea,* that amongst the ignorant classes of Northern India a superstition prevails that they may be captured and carried off to some distant land to be made into *momea.* This fact has been alluded to by various officers in their reports on the objections raised by the poorer classes against emigration. Speaking of this subject, **Colonel D. G. Pitcher,** in a report dated June 1882, writes: "The feeling of the native community on the subject of emigration is, for the most part, *nil,* or a ludicrous distorted image, in which the coolie hangs head downwards like a flying fox or is ground in mills for oil." "The natives have in some parts of the country a strong belief that coolies are suspended head downwards to facilitate the extraction from them of oil." In a recent expedition into Kulu this same superstition was brought out in an alarming manner (*See Pb. Gazette*).

Mumiai 351

It seems probable that the word *momea* has been erroneously applied to hemp, and in partial support of this suggestion, it may be added that the word *mumiai* is, in Afghánistan, applied to a substance which apparently has no relation to the narcotic of hemp. **Ball,** in his *Economic Geology,* says of this substance: "In Afghánistan it is believed that there are several localities where bituminous products occur, as they are commonly sold as drugs in the bazars of that country. According to **Captain Hutton** (*Cal. Jour., Nat. Hist., Vol. VI., 601*), a mineral pitch called *mumiai* by the natives, which is used for external application, is found in the Shah Makhsud range. A substance supposed to be this same *mumiai,* otherwise called Rock Chetny, which was obtained by **Lieutenant Conolly** as an exudation from a fissure in a rock in Ghazni, was analysed by **Mr. Peddington,** who concluded, in spite of its savoury name, that it was composed of the excreta of birds, more probably of bats, mixed with salts of lime. There was no trace of bitumen or sulphur. In fact, this substance was no doubt similar, as regards its origin, to the reputed discovery of petroleum in the Madras Presidency" (*Ball, page 126*). **Baden Powell** (*Panjáb Products, pages 22, 104*) alludes to *momyai* as a black substance, obtained from Kashmír, consisting "principally of clay, which, however, burns feebly and softens slightly to the flame of a lamp, giving

Momyai. 352

out a peculiar empyreumatic odour." In the chapter devoted to Panjáb medicinal substances, he again alludes to *momyai* while speaking of a substance exhibited under that name: "This is in reality a dry mass of tar. Real *momyai* is said to be rarely met with; it is supposed to be of

353

great efficacy in healing bones, and is in fact an "Osteocolla." It is said to come from Persia, where it exudes and floats on the surface of a certain spring where it is collected and monopolised by the Government, who sell it at a high price." **Mr. Ribbentrop,** Inspector-General of Forests,

354

showed the writer a substance which, in all probability, was *mumea,* or allied to *mumea.* This was procured in Chumba State, where it was found to exude from a crack on the face of a high rock.

There are thus numerous allusions to a substance or substances known in the bazars of India under the name *momea,* but in none of the published accounts of this drug is there the slightest reference to its being a product of Indian hemp, although, in the early literature of that narcotic, it is repeatedly stated that a pure waxy form of *charas* obtained from Nepál is sold under the name of *momea.*

Charas from Sind. 355

Charas is collected in Sind and in Central India by causing men to run through the hemp fields. They are said to be generally clad in

Central India. 356

leathern aprons to which the resin adheres, but in some cases are reported to have their bodies first oiled and then to run naked through the fields.

In either case the *charas* thus collected is scraped off and made into the cakes in which it is sold. The chief supply of *charas* to India comes, however, from the trans-Himálayan regions. In concluding the present account of this form of the narcotic, the following extract may be here given from Mr. Baden Powell's *Panjáb Products*, in which, it will be seen, the substance there referred to is most probably more nearly related to *chúr* than to *charas*. It is prepared from the *bháng*-yielding plant:— CHARAS. Trans-Himalaya. 357

"There is a kind of charas called '*garda*' which is much in use, and of this again there are three sorts:—'*Surkhai*,' '*Bhángra*,' and '*Khaki*.' When the *bháng* has been gathered and placed in a store-house as soon as it is dry, persons go in with their faces covered with a thin cloth, which enables them to breathe without inhaling the dust which results from the process they perform. Next the heaps of dry *bháng* are covered over with a fine cloth, and the operators, putting their hands under the cloth, begin stirring about the *bháng* and making hay of it. Soon a fine dust flies out and, filling the room, settles down on the surface of the cloth spread over the heaps. When all the dust has been shaken out and settled on the cloth, the cloth is itself taken out and shaken: a dust falls down which is of the best quality and of a reddish colour. This is collected and kneaded with a little water into a cake, and forms the best *charas*, which is called *Surkhai*: more frequently the dirt that is shaken off is of a greenish tint, like the *bháng* itself, and this collected forms '*bhángí charas*.' Lastly, the powder which adheres to the cloths, and is scraped and shaken off, forms the worst kind, called *khaki*. Garda or Panjab Charas. 358 Surkhai, Bhangra, and Khaki. 359

(*3rd*) Bháng or Siddhí, Sabjí, and Sabzí.—Apparently the wild plant is the chief source of this form of the drug, which consists of the mature leaves and in some parts of India of the fruits also. The resin is apparently not extracted from these and sold or used in that form; the leaves are directly employed in the manufacture of the preparations in which *bháng* constitutes the form of the narcotic. According to some writers the wild plant does not yield *bháng* (see Mr. Atkinson's *Himálayan Districts*). The comparative uniformity, throughout India, of the retail price of *bháng*, namely, 8 to 12 annas a seer in the provinces where duty is realised over and above the licenses to sell and 4 annas (except in the N.-W. Provinces) where this is not the case, is of considerable importance when the disproportion in the revenue credited to Government from this article is taken into consideration. BHANG. 360

Indian Preparations from Hemp.

Forms of Indian Hemp.—As already explained there are three forms of this poisonous drug: (*a*) *gánjá*, the agglutinated female flowering tops and resinous exudation on these; (*b*) *charas*, a resinous substance found on the leaves, young twigs, and bark; and (*c*) *bháng* or *siddhi*, the mature leaves, and in some parts of India the fruits also, and even the very young twigs, but not the stems. *Gánjá* and *charas* are smoked, and *bháng* is either used in the preparation of the green intoxicating beverage *hashísh*, or in the manufacture of the sweetmeat known as *májun* (vulgarly *májum*). *Bháng* is much weaker than either *gánjá* or *charas*, and by many is supposed to be much less injurious. It is sold in Bengal at from 4 to 8 or 12 annas a seer, whereas *gánjá* and *charas* cost from R4 to R25 a seer. *Bháng*, being collected largely from the wild plant, is extensively used all over India, the bulk of the consumption entirely escaping duty. This is mainly due to the fact that it would be impracticable to hold a man responsible for the existence of a wild plant growing within a certain radius of his hut, and it would be impossible to prohibit him gathering, from such a plant, the daily quantity used by himself and family. This is precisely the state Smoking mixtures. 361 Hashish. 362 Majun. 363 PRICES.

of affairs which prevails over a great part of India, and, indeed, on the lower slopes of the Himálaya and up to an altitude of 8,000 feet, the plant is often so plentiful as to be extensively used as bedding for cattle. The greatest difficulty exists, therefore, in regulating the consumption of *bháng*, but practically no such difficulty exists with regard to *gánjá* and *charas*. The last-mentioned narcotics can be produced only from the cultivated plant, and the consumption can therefore be regulated by law. The Excise Act provides that licensed persons may cultivate the plant, prepare the narcotics, and retail these to the consumer. The right to vend is sold by public auction, a person purchasing thereby the sole right, for one year, to all or so many of the shops in a district. Any person, other than a licensed dealer, having in his possession more than a very small quantity at one time is liable to prosecution and fine. This system of farming the wholesale and retail shops exists all over India,—Madras being an exception to the rule, since in South India, no revenue whatever is credited to Government from these drugs.

Bedding for Cattle.

Excise Arrangements.

The administrative arrangements which are made in each Province for levying excise uses on hemp will be found under the heading **Narcotics.**

THE FIBRE-HEMP.

FIBRE. 364

The reader is referred to the account given of the cultivation of the hemp plant in a preceding page. It will there be found that a considerable amount of information has been given as to the early experiments made to extend the cultivation in India of **Cannabis sativa** as a source of fibre; a possible still further development has also, to a certain extent, been dealt with. It has been urged that the regions where the plant is grown for its narcotic, *gánjá*, should be carefully distinguished from those where the plant may be found to form fibre. But an equally important fact remains to be investigated and thereafter clearly kept in view, namely, the age of the plant and season of the year when the fibre is at its best, in both the temperate and tropical regions of India. It cannot be disguised that the defects complained of in many of the reports on Indian hemp cultivation, against the quality of the fibre produced, are traceable to ignorance as to the period when lignification is reached by the Indian plant. The season of sowing, period of repeating, and modes of culture, practised in Europe have, apparently, been forced on the plant in India, and the suggestion is accordingly offered that the brittle character complained of, against the resulting fibre, may have been due to the fact of the plant reaching in India the mature state of the fibre at an earlier stage of its growth than in European countries. Thus, for example, it is reported that the plants experimented with on the plains of India, at Saharanpur, grew vigorously, attained a height of 12 feet, and gave every promise of proving successful. When reaped, **Dr. Falconer**, however, reported that "the hemp-fibre did not retain the strength or flexibility which characterise it in the Himálayas." Similar results were obtained at Agra and in various parts of Bengal. The chemistry of fibre and of the process of fibre-forming within the plant has, during recent years, reached a high development. To arrive at a definite understanding as to whether the plains of India can or cannot produce good hemp, it would be necessary to carry out a series of systematic experiments in certain selected districts in each province. The seed would have to be sown and the plant cultivated according to a uniform and pre-arranged plan. From a certain stage, say after the plants had attained a height of two feet, a certain number of the plants from each field would have to be microscopically and chemically examined once a fortnight, right through

When Mature.

Lignification.

Experiments to be performed in India.

FIBRE.

their subsequent growth, or until in each locality the period when lignification was reached by the plants had been determined. It would also be desirable, in such experiments, to make several sowings so as to determine the best period for sowing, since much would depend upon the cultivation being so regulated that the fibre would reach maturity during the period of the year best suited to the development of fibre. Without some such experiments being performed we must remain in ignorance, with hemp as with flax, whether or not the peculiar climate and soil exists in India on which either of these fibres could be produced on a commercial scale. Up to the present day, the experiments which have been made have either failed to discover such regions or were imperfectly conducted, for, with the exception of certain limited tracts of the Himálayas, no part of the plains of India can be said to have been discovered in which there is the least hope of hemp or flax cultivation becoming of much importance. (See remarks as to hemp in Godavery District No. 339).

In portions of the North-West Himálaya the hemp plant has been cultivated for its fibre for a very long time. **Mr. Atkinson** gives a brief but practical account of this industry in his *Himálayan Districts* (*p. 799*). "The possibility," says that author, "of attaining success in the cultivation of hemp in these provinces was pointed out by **Dr. Roxburgh** as early as 1800, and on the cession of these provinces, skilled Europeans were sent to carry on experiments in the Murádabad and Gorakhpur districts. In Garhwál and Kumáon its cultivation was encouraged, and for many years the East India Company procured a portion of its 'annual investment from the Kumáon hills in the shape of hemp.' With the abolition of the Company's trade the cultivation languished and is now entirely dependent on the local demand, which, however, is by no means small." "The male plant yields only fibre from which the *bhángela* cloth of Garhwál is manufactured: also called *kothla*, *bora*, and *gáji*, and the ropes (*sel*) for bridges."

Separation of Fibre.

Messrs. Duthie and Fuller, speaking of the hemp fibre of these provinces, say that after harvest "the stalks are then laid in water to promote a fermentation which will allow the bark to strip easily; on being taken out they are beaten with mallets to loosen the bark, which is then detached by hand in strips, and after a second beating breaks up into a fibre which is made up into hanks for sale. In some places the fibre is boiled in potash and bleached before spinning. The principal things manufactured from it are hemp cloth (*bhángra* or *bhángela*), and the ropes which are used for the swing-bridges over hill-streams. The cloth makes an admirable material for sacks, and is largely used in the grain trade on the Nepál frontier; and latterly, in the export of potatoes from Kumáon. It also furnishes a large portion of the hill population with a characteristic article of clothing—a hemp blanket worn like a plaid across the shoulders and fastened in front with a wooden skewer."

Defects of the Indian Fibre.

It is commonly stated by European writers that the fibre from the male plant is superior to that obtained from the female. It is urged that particular care should be taken to strip the plant in dry weather; should the fibre get wet, it is certain to heat and get almost totally spoiled. The method of platting the fibre into long tails as pursued by the hill-tribes of India lessens the value of the fibre very much, since it increases the labour in cleaning it, each hank requiring to be opened out by the hand.

HEMP SUBSTITUTES. 365

HEMP SUBSTITUTES.—Hemp is mainly, if not entirely, employed in the manufacture of ropes, twine, and nets. As has already been urged, care must be taken, however, to distinguish the true hemp fibre from the numerous substitutes for it which are often commercially grouped with the true article. Thus, for example, we have in India Sunn-hemp (**Crotalaria**

The Hemp Fibre of India.

FIBRE.

juncea), Jabbalpur-hemp (also **Crotalaria juncea**), Bowstring-hemp (**Sanseviera zeylanica**), and Manilla-hemp (**Musa textilis**). In European commerce there are, besides these, several other fibres known as hemps, such as Canada-hemp (**Apocynum cannabinum**), Kentucky-hemp (**Urtica canadensis,** and **U. cannabina,** &c.)

EUROPEAN HEMP. 366 EUROPEAN TRUE HEMP.—Russia, even to the present day, holds the first place among hemp-producing countries, although, since Russian hemp first became an article of extensive demand, nearly every country in the world has attempted the cultivation of the plant and the manufacture of its own ropes and twine. Where this competition proved comparatively hopeless, substitutes were brought forward, and at the present day the most extensively-used fibres in the rope trade may be said to be hemp, coir (or the fibre from the outer layer of the cocoanut), Manilla-hemp, cotton, and sunn-hemp. Italy produces the finest hemp; France is perhaps next in importance, then Great Britain, Servia, Germany, and of Asiatic countries China is reputed to produce good hemp.

INDIAN FOREIGN TRADE IN "HEMP."

The following figures as to the value of the Indian trade in "hemp" have been extracted from the *Statement of the Trade and Navigation of British India* regarding hemp, but the bulk of the imports of raw-hemp fibre are in all probability Manilla hemp, and of the exports Sunn-hemp :—

			Foreign Hemp imported.	Foreign Hemp exported.	Indian Hemp exported.
			R	R	R
Raw Hemp. 367	Raw Hemp	1881-82	1,10,875	...	5,59,112
		1882-83	1,82,993	...	4,30,325
		1883-84	1,76,765	...	6,85,316
		1884-85	2,14,118	...	5,82,679
		1885-86	1,96,052	...	9,88,825
Manufactures. 368	Manufactured Hemp (excluding cordage).	1881-82	10,179	4,182	1,409
		1882-83	27,090	8,857	3,176
		1883-84	32,570	4,548	6,510
		1884-85	41,356	150	3,129
		1885-86	42,810	323	3,205
Cordage. 369	Cordage and rope, excluding jute, but otherwise the bulk probably Manilla Hemp and true Hemp	1881-82	3,22,485	24,886	3,25,178
		1882-83	4,31,693	15,586	2,84,106
		1883-84	3,90,584	11,198	4,92,068
		1884-85	3,52,413	13,076	3,53,389
		1885-86	3,24,519	7,437	3,28,320

Foreign Trade in Manufactured and Unmanufactured Hemp, excluding Cordage.

Year.	Imports.	Exports and re-exports.
	Value.	Value.
	R	R
1881-82	1,21,054	5,64,703
1882-83	2,10,083	4,42,358
1883-84	2,09,335	6,96,374
1884-85	2,55,474	5,85,958
1885-86	2,38,862	9,92,353

The Indian Hemp.

CANNABIS sativa.

Detail of Imports, 1885-86.

FIBRE. Imports. 370

Province into which imported.	Value.	Country whence imported.	Value.
	R		R
Bengal	1,33,235	United Kingdom	83,431
Bombay	1,01,600	China	1,23,474
Madras	1,183	Phillipines . . .	2,609
Sind	2,844	Straits Settlements .	17,827
		Other Countries . .	11,521
TOTAL .	2,38,862	TOTAL .	2,38,862

Detail of Exports, 1885-86.

Exports. 371

Province from which exported.	Value.	Country to which exported.	Value.
	R		R
Bengal	3,11,551	United Kingdom . .	6,78,607
Bombay	6,31,444	Belgium	2,56,566
Madras	49,358	Persia	11,438
		Arabia	15,698
		Other Countries .	30,044
TOTAL .	9,92,353	TOTAL .	9,92,353

It has been found impossible to give the quantities, since the raw fibre is expressed in weight, cloth in pieces, and rope in balls of various lengths and weights.

OIL.

HEMP SEED OIL. 372

Oil.—The seeds, when expressed, yield a pale, limpid oil. They contain of this oil from 25 to 34 per cent. This oil is at first greenish or brownish-yellow, but the colour deepens when it is exposed to the air. The flavour is disagreeable, but the odour is mild. It is, however, said to make a very bad-smelling and deep-coloured boiled oil, and on this account it is never used in England, although extensively so in East and North Europe as a paint and varnish oil. In Russia it serves, in a great measure, the purpose of lamp-oil, but it is chiefly employed in the manufacture of soft soaps, inferior to that from linseed oil. It has a specific gravity of 0·9252 at 15°C.; it thickens at — 15°C., and solidifies at — 25° to — 27·7C. It dissolves in boiling hot water and in 30 parts of cold alcohol.

MEDICINE.

MEDICINE. 373 Indian Hemp. Extract Indian Hemp. 374 Tincture. 375

Medicine.—The various forms of the drug have already been described, namely, *bháng*, *gánjá*, and *charas*. *Gánjá*, or the dried flowering female tops, constitute the officinal parts in European pharmacy. From these an alcoholic extract is prepared known as *Extractum Cannabis Indicæ*, and from this a *Tincture* is also made. The use of Indian hemp in European prac-

MEDICINE.

tice has greatly decreased of late years owing to a feeling of insecurity as to the quality of the article. It is commonly recorded that no reliance can be put upon the uniformity in strength. The writer, at a meeting of the Royal Pharmaceutical Society of Great Britain, recently expressed the opinion that the heavy fiscal restrictions now imposed on Bengal *gánjá* had, in all probability, diverted the export trade from Bengal to Bombay, so that, instead of the carefully-cultivated Bengal article finding its way to Europe, the much inferior but infinitely cheaper *gánjá* of Bombay and the Central Provinces was, in all probability, that which was now used in European pharmacy. The *Chemist and Druggist*, commenting on this subject shortly after, recommended the suggestion as worthy of attention, and added: "The price of Bengal *gánjá* may be prohibitive, but the whole subject should be considered by authorities." There would seem little doubt that the high reputation the drug once enjoyed might be recovered by greater care in selecting the article, but there is perhaps no other commodity in India that is produced in a larger number of forms and qualities, or which in the hands of the retail dealer is subjected to a greater degree of adulteration. The only guarantee an exporter can have is to purchase his *gánjá* direct from the Government *golás* of Bengal, not even allowing the article to pass through the hands of a wholesale *gánjá*-dealer or "middle-man" of any kind. If the article be shipped under a permit direct from the *golá* it is believed little complaint would be raised as to the uniformity in strength, but none but that which is registered as of the first quality should be purchased for medicinal purposes. From what the writer has been able to learn it would be even preferable to use for European pharmacy the *chúr* or the dust obtained on packing and handling round *gánjá* rather than round *gánjá* itself; flat *gánjá* should be resorted to with caution, and *charas*, or *momea*, should never be employed, nor round *gánjá* in which ripe fruits are found with the flower heads.

Chur or Round Ganja best suited for Pharmacy. Flat Ganja and Charas should be avoided.

Medicinal Properties and Uses of Indian Hemp.—The *Pharmacopœia of India* describes the drug as primarily stimulant, and secondarily anodyne, sedative, and antispasmodic. It is also said to be narcotic, diuretic, and parturifacient. It has been used with advantage in tetanus, hydrophobia, delirium tremens, ebrietas, infantile convulsions, various forms of neuralgia, and other nervous affections. It has also been employed in cholera, menorrhagia and uterine hœmorrhage, rheumatism, hay fever, asthma, cardiac functional derangement, and skin diseases attended with much pain, and pruritus. In lingering and protracted labours depending upon atony of the uterus, it has been employed with the view of inducing uterine contractions.

It is admitted by most Indian physicians to be of special merit in the treatment of tetanus and cholera and has not the injurious after-effects (constipation and loss of appetite) which but too frequently result from the use of opium. Its action is, however, very similar to that of opium, and it is accordingly stated that a habitual opium-eater may take large quantities of hemp without injurious consequences.

Sir William O'Shaughnessy was the first European writer to draw prominent attention to the peculiar properties and actions of the hemp-narcotics. He experimented with these in Calcutta and published his results. The reader is referred to his *Bengal Dispensatory* and to a "Memoir on the preparations of Indian Hemp" in the *Transactions of Medical and Physical Society of Calcutta* for 1839, and to two papers in the *Journal of the Asiatic Society*, Vol. VIII., of the same year. Shortly after the appearance of these most exhaustive accounts, the drug began to be experimented with in Europe.

MEDICINE.

Ainslie, in his *Materia Indica*, 2nd Vol., gives an interesting account of the uses of the drug, and according to him the *májún* of South India, in addition to the ingredients of the Bengal article, contains datura and nux-vomica. **Dr. Stocks** says that the best form of *bháng* met with in Sind is that known as *bubakai*, from the town of Bubak near lake Manchhar. He further adds that the *májúm* of Sind is made up of some 20 to 30 different ingredients, of which datura and opium are frequent. In some parts of India a beer is brewed with *bháng*, and this, together with *bháng* itself, *májúm* and other preparations, are often employed in Native pharmacy.

Professor Christison published the following valuable remarks, derived from his personal experience with this drug: "I have for some years," he observes, "used a very good alcoholic extract, sent to me from Calcutta twenty years ago, and still as powerful as ever to subdue pain, obtain sleep, and put an end to spasm in circumstances under which morphia either did not suit or was objected to by the patient, and after wide experience with it I am quite satisfied that it is an excellent substitute for it, if given in sufficient doses. The difficulty is, to be always sure of the quality of the extract, or rather of the *gánjá* from which the extract is obtained. I have known two grains of my alcoholic extract, given in the form of tincture, put an end, promptly and permanently, to the agonising pain caused by biliary calculus impacted in ducts; and there can be no more unequivocal test than this of the potency of an anodyne. I have long been convinced, and new experience confirms the conviction, that for energy, certainty, and convenience, Indian Hemp is the next anodyne hypnotic and anti-spasmodic to opium and its derivatives, and often equal to it." **Dr. Christison** considers that a well-prepared alcoholic extract is the best of all forms for use, but it requires to be prepared from *gánja*, not too old, collected in the right district and at the right season. The ordinary resin (*charas*) is generally very impure and untrustworthy (*Pharm. India*).

Uniformity in quality.

The earlier writers express almost the same opinions as were arrived at by the modern European physicians. According to the author of *Makhzan*, "the leaves make a good snuff for deterging the brain; their juice applied to the head removes dandriff and vermin; dropped into the ear it allays pain and destroys worms; it checks the discharge in diarrhœa and gonorrhœa, and is diuretic." The powder of the leaves is applied externally to fresh wounds and sores, and is used to promote granulation; a poultice of the whole plant is applied to local inflammations, erysipelas, neuralgia, &c. The dose of the leaves is 48 grains when administered internally" (*Makhzan, Article Kinnab, see also in Dr. W. Dymock, Mat. Med. West India*).

The medicinal properties of hemp, in various forms, are the subject of some interesting notes by **Mirza Abdul Russac.** "It produces a ravenous appetite and constipation, arrests the secretions, except that of the liver, excites wild imagining, a sensation of ascending, with forgetfulness of all that happens during its use, and such mental exaltation, that the beholders attribute it to supernatural inspiration.

"Of its utility as an external application in the form of a poultice with milk, in relieving hœmorrhoids, and internally in gonorrhœa, to the extent of a quarter drachm of *bháng*." He further states that the habitual smokers of *gánjá* generally die of diseases of the lungs, dropsy, and anasarca, so do the eaters of *májún* and smokers of *siddhi*, but at a later period. The inexperienced, on first taking it, are often senseless for a day, some go mad, others have been known to die.

Dr. U. C. Dutt says that, according to the Sanskrit writers, "the leaves of **Cannabis sativa** are said to be purified by being boiled in milk

Leaves. 376

CANNABIS sativa.

The Indian Hemp as a Drug.

MEDICINE.

before use. They are regarded as heating, digestive, astringent, and narcotic." "In sleeplessness, the powder of the fried leaves is given in suitable doses for inducing sleep and removing pain."

Special Opinions.—§ "Used as anodyne, antispasmodic, diuretic; leaves may be employed in dose of 20 grains" (*Assistant Surgeon Nehal Sing, Shaharanpur*). "During the last twelve months I have used **Cannabis sativa** with great success in the treatment of acute dysentery; similar results were obtained by other medical officers of this station, who tried it at my request. The dose was, of the tincture 15 or 20 m. three times a day (*Surgeon S. J. Rennie, Cawnpore*). "I have found a poultice of the bruised fresh leaves, slightly warmed, very useful in affections of the eye, attended with photophobea. Natives also use the poultice in piles" (*Assistant Surgeon Bhagwan Dass* (*2nd*), *Punjab*). "The RESIN collected from the leaves and flowers in Turkestan is called *Nasha*, and this ought to be the
377 *Charas* of the trade, but it is terribly adulterated. The plant is called in Turkestan *kandir*, and the oil, *kandir yak*. The OIL extracted from the seed is in Kashmír considered as a valuable remedy, applied by rubbing in rheumatism" (*Surgeon-Major J. E. T. Aitchison, Simla*). "Used in the form of '*sidhí*,' in small quantities, it is a very good stomachic tonic, useful in atonic dyspepsia and diarrhœa. In large quantities exhilarant at first, depressent subsequently. Long continued use of *gánjá* is a prominent cause of insanity" (*Civil Surgeon D. Basu, Faridpur*). "Used to produce sleep in certain cases in which opium is contra-indicated. It does not induce nausea, constipation, or headache as opium does. Valuable as a remedy for sick headache, and especially in preventing such attacks. It removes the nervous effects of a malady. Useful in malarial, periodical neuralgias. Valuable in the treatment of the sleeplessness and restlessness of acute mania, in whooping-cough, and in asthma, in dysuria, and in relieving pain in dysmenorrhœa" (*Dr. E. G. Russell, Superintendent, Asylums, at Presidency General Hospital, Calcutta*). "Commonly used as a narcotic; a few grains of the leaves called *siddhí* rubbed in with cardamom and other spices to allay pain; taken as a drink habitually by many for intoxicating purposes; may be used as an anodyne; it increases appetite; is an aphrodisiac; and increases the activity of the brain, producing better flow of thoughts, and deep meditation, but often wild reveries and causeless laughter. A small quantity of the leaves, mixed with other drugs and spices, forms an useful compound in diarrhœa and indigestion of children" (*Assistant Surgeon Shib Chunder Bhattacharji, Chanda, Central Provinces*). "The leaves, which are known as *bháng*, are used to check diarrhœa; an extract is prepared from them; it is made into confection and used for narcotic purposes" (*Surgeon-Major Robb, Civil Surgeon, Ahmadabad*). "Used for asthma and in tetanus, dose $\frac{1}{4}$ to 2 grains, with sugar well fried in *ghí* and mixed with black pepper, given, in cases of chronic diarrhœa, with poppy seeds in dysentery, with asafœtida in hysteria" (*Surgeon W. Barren, Bhuj, Cutch*). "Very often used by natives in some parts of this Presidency as an aphrodisiac and I believe in some cases successfully, in the form of '*Májún*,' *i.e.*, a kind of pill-mass containing various drugs" (*Surgeon D. N. Parakh, Bombay*). "The leaves made into a poultice used in orchitis, also dried leaves warmed and used for fomentations" (*Civil Surgeon S. M. Shircore, Murshedabad*). "Used frequently by all hospital assistants, particularly for asthma and other paroxymal affections. In cases of chronic colic I have found the extract in one-grain doses with $\frac{1}{4}$ grain of Ipecac. to produce wonderful effects" (*Doyal Chunder Shome, Campbell Medical School, Sealdah, Calcutta*). More commonly used in this country to produce intoxicating effects than for its medicinal properties in smaller doses"

Dysentery. **Affections of the eye.** **Piles.** **NASHA.** **Oil used in Rheumatism.** **Acute Mania.** **Hysteria.** **Orchitis.** **Asthma.** **Chronic Colic.**

MEDICINE.

(*Dr. G. Price, Civil Surgeon, Shahabad*). "It is also used in the form of tincture for checking ague fits, and is employed by native physicians, in the form of electuary, as an aphrodisiac in cases of impotence, spermatorrhœa, and incontinence of urine. Taken in the form of *bháng* it acts on the kidneys as a powerful diuretic, and on the liver as a mild cholagogue" (*Civil Surgeon J. H. Thornton, B.A., M.B., Monghir*). "Dried tender leaves and flowering tops with sugar, black pepper powder, and with or without opium, proves highly beneficial in dysentery" (*Civil Surgeon E. W. Savinge, Rajamundry, Godaveri District*).

Ague Fits.
Impotence.

CHEMICAL COMPOSITION.

Chemical Composition.—"The most interesting constituents of hemp, from a medical point of view, are the *Resin* and *Volatile Oil*.

"The former was first obtained in a state of comparative purity by **T.** and **H. Smith** in 1846. It is a brown amorphous solid, burning with a bright white flame and leaving no ash. It has a very potent action when taken internally, two-thirds of a grain acting as a powerful narcotic, and one grain producing complete intoxication. From the experiments of **Messrs. Smith** it seems impossible to doubt that to this resin the energetic effects of **Cannabis** are mainly due.

"When water is repeatedly distilled from considerable quantities of hemp, fresh lots of the latter being used for each operation, a volatile oil lighter than water is obtained together with ammonia. This oil, according to the observations of **Personne** (1857) is amber-coloured, and has an oppressive hemp-like smell. It sometimes deposits an abundance of small crystals. With due precautions it may be separated into two bodies, the one of which named by **Personne** *Cannabene* is liquid and colourless, with the formula $C_{18}H_{20}$; the other, which is called *Hydride of Cannabene*, is a solid, separating from alcohol in platy crystals to which **Personne** assigns formula $C_{18}H_{22}$. He asserts that *Cannabene* has indubitably a physiological action, and even claims it as the sole active principle of hemp. Its vapour he states to produce when breathed a singular sensation of shuddering, a desire of locomotion, followed by prostration and sometimes syncope. **Bohling** in 1840 observed similar effects from the oil which he obtained from the fresh herb, just after flowering, to the extent of 0·3 per cent.

Cannabene 378

"It remains to be proved whether an *alkaloid* is present in hemp, as suggested by **Preobraschensky**.

"The other constituents of hemp are those commonly occurring in other plants. The leaves yield nearly 20 per cent. of ash.

"As to the resin of Indian hemp **Bolas and Francis**, in treating with nitric acid, converted it into *Oxycannabin*, $C_{20}H_{20}N_2O_7$. This interesting substance may, they say, be obtained in large prisms from a solution in methylic alcohol. It melts at 176°C., and then evaporates without decomposition; it is neutral. One of us (**F.**) has endeavoured to obtain it from purified resin of *charas*, but without success" (*Flück. and Hanb., Pharmacog., page 549*).

Dr. Dymock (in his *2nd Ed. of the Materia Medica of Western India*) goes into considerable detail on the chemistry of this drug. **Preobraschensky** discovered in China *haschisch*, a volatile alkaloid which he believed to be identical with nicotine. **Dragendroff** and **Marquiss** suggested that this may have been derived from tobacco mixed with the beverage. **Siebold** and **Bradbury**, while concurring with this explanation, announce their discovery of an alkaloid perfectly distinct from nicotine; they give it the name of *Cannabine*. **Dr. Hay**, some time before, published his conviction that hemp contained several alkaloids, the principal one being a substance he named *Tetano-cannabine*. More recently to all these published results of the chemical investigation of the narcotic resin of **Cannabis sativa, Drs. Warden and Waddell** of Calcutta have failed to discover the principle alluded to by **Dr. Hay.** But they obtained, by the destructive distillation of the alcoholic extract, an amber-coloured oil. "This oil had a mildly empyreumatic odour which was distinctly tobacco-like. Its taste was warm, aromatic, and somewhat terebinthinate. The oil contained phenol, ammonia, and several other of the usual products of destructive distillation.

"The nicotine-like principle contained in this oil appeared to be an alkaloid. It formed salts which evolved a strong nicotine-like odour when acted on by alkalies. But physiologically it was found to be inert, and therefore was evidently not identical with nicotine" (*Ind. Med. Gaz., Dec. 1884*).

FOOD.

FOOD. 379 **Food.—Messrs. Duthie and Fuller,** writing about the Himálayan tracts within the North-Western Provinces, say that the seed is not uncommonly roasted and eaten by the hill-men, and that after the oil is expressed the oil-cake is given to their cattle. **Dr. Stewart** writes that on the Sutlej the seeds are roasted and eaten in small quantities with wheat.

DOMESTIC AND INDUSTRIAL USES.

DOMESTIC. 380 **Cannabic Composition.**—"This material for architectural decoration is described by **Mr. B. Albans** to have a basis of *hemp* amalgamated with resinous substances, carefully prepared and worked into sheets of large dimensions. Ornaments in high relief and with great sharpness of detail are obtained by pressure of metal discs, and they are of less than half the weight of *papier-maché* ornaments, sufficiently thin and elastic to be adapted to wall surfaces, bearing blows of the hammer and resisting all ordinary actions of heat and cold without change of form. Its weather qualities have been severely tried in Europe, as for coverings of roofs, &c., remaining exposed without injury.

This composition is of Italian origin, and in Italy it has been employed for panels, frames, and centres. It is well fitted to receive bronze, paint, or varnish; the material is so hard as to allow gold to be burnished after gilding the ornaments made of it" (*Ure, I., 611*).

CANOES.

See **Boats,** Vol. I., B. 548.

381 **TIMBERS USED FOR CANOES, DUG-OUTS, TROUGHS, WATER-PIPES, DRINKING CUPS, &c.**

1. **Acer cæsium,** *Wall.* (drinking cups made in Tibet).
2. **A. oblongum,** *Wall.* (drinking cups).
3. **A. pictum,** *Thunb.* (drinking cups made of knotty excrescences).
4. **Adina cordifolia,** *Hook.* (dug-out canoes, combs, &c.).
5. **Æsculus indica,** *Colebr.* (water-troughs, platters, drinking-cups).
6. **Ailanthus excelsa,** *Roxb.* (catamarans and for floats, &c.).
7. **Amoora Rohituka,** *W. & A.* (Chittagong canoes).
8. **Artocarpus Chaplasha,** *Roxb.* (much used for canoes).
9. **A. Lakoocha,** *Roxb.* (canoes).
10. **A. nobilis,** *Thw.* (Ceylon canoes).
11. **Bœhmeria rugulosa,** *Wedd.* (Lepchas make cups, bowls, and tobacco-boxes).

12. **Bombax malabaricum**, *DC.* (Bengal and Burma canoes).
13. **Borassus flabelliformis**, *Linn.* (dug-outs, water-pipes, gutters).
14. **Calophyllum polyanthum**, *Wall* (Chittagong canoes).
15. **Caryota urens**, *Linn.* (water conduits and buckets).
16. **Cedrela Toona**, *Roxb.* (Bengal and Assam canoes).
17. **Celtis tetrandra**, *Roxb.* (Assam canoes).
18. **Cinnamomum granduliferum**, *Meissn.* (Assam boats and canoes).
19. **Cordia Myxa**, *Linn.* (Bengal canoes).
20. **Dipterocarpus alatus**, *Roxb.* (canoes).
21. **D. tuberculatus**, *Roxb.* (Burma canoes).
22. **D. turbinatus**, *Roxb.* (Burma canoes).
23. **Drimycarpus racemosus**, *Hook.* (mostly used in Chittagong for boats and canoes).
24. **Duabanga sonneratioides**, *Buch.* (canoes, cattle-troughs cut out of green wood).
25. **Dysoxylum Hamiltonii**, *Hiern.* (canoes).
26. **D. procerum**, *Hiern.* (Assam canoes).
27. **Givotia rottleriformis**, *Griff.* (catamarans).
28. **Gmelina arborea**, *Roxb.* (clogs, canoes, &c.).
29. **Gyrocarpus Jacquini**, *Roxb.* (preferred above all other woods for catamarans).
30. **Hopea odorata**, *Roxb.* (Burma canoes).
31. **Juniperus excelsa**, *M. Bieb.* (drinking cups).
32. **Lagerstrœmia Flos-Reginæ**, *Retz.* (boats and canoes).
33. **L. tomentosa**, *Presl.* (canoes).
34. **Mangifera indica**, *Linn.* (canoes and masula boats).
35. **Michelia Champaca**, *Linn.* (Assam canoes).
36. **Michelia oblonga**, *Wall.* (Assam canoes).
37. **Morus serrata**, *Roxb.* (Troughs).
38. **Odina Wodier**, *Roxb.* (Rice-pounders).
39. **Pajanelia multijuga**, *DC.* (Andaman Island canoes).
40. **Phœnix sylvestris**, *Roxb.* (water-tubes).
41. **Pinus excelsa**, *Wall.* (water-channels).
42. **P. Gerardiana**, *Wall.* (hollowed out for water-courses).
43. **Platanus orientalis**, *Linn.* (trays).
44. **Populus ciliata**, *Wall.* (water troughs).
45. **Sarcosperma arborea**, *Hook.* (Sikkim canoes).
46. **Schima Wallichii**, *Choisy.* (Assam canoes).
47. **Shorea obtusa**, *Wall.* (canoes).
48. **S. robusta**, *Gærtn.* (Hills of Northern Bengal, canoes).
49. **S. stellata**, *Dyer.* (canoes).
50. **Stereospermum chelonoides**, *DC.* (Assam canoes).
51. **Terminalia belerica**, *Roxb.* (canoes in South India for catamarans).
52. **Vateria indica**, *Linn.* (occasionally used for canoes).

CANSCORA, *Lam.; Gen. Pl., II., 811.*

Canscora decussata, *R. & Seb.; Fl. Br. Ind., IV., 104; Bot. Mag., t. 3066*; GENTIANACEÆ. 382

Syn. PLADERA DECUSSATA, *Roxb., Fl. Ind., Ed. C. B. C., 135.*

Vern.—*Sankháhuli*, HIND.; *Dánkuni*, BENG.; *Shun-khapushappi*, CUTCH; *Sankhapushpi, dandotpala*, SANS.

References.—*Thwaites, En. Ceylon Pl., 204; Voigt, Hort. Sub. Cal., 520; U. C. Dutt, Mat. Med. Hind., 201, 296, 316; Dymock, Mat. Med., W. Ind., 451*; also *2nd Ed., 542.*

Habitat.—Common throughout India from the Himálaya to Burma, ascending to 4,000 feet; is abundant in the plains of Bengal and not uncommon in Ceylon.

MEDICINE. 383 **Medicine.**—This plant is regarded as laxative, alterative, and tonic, and is much praised as a nervine tonic. Used in insansity, epilepsy, and nervous debility. The fresh juice of the plant, according to **Chakravatta,** in doses of about an ounce, is given in all cases of insanity. A paste made of the entire plant, including the roots and flowers, is recommended to be taken with milk as a nervine and alterative tonic (*U. C. Dutt's Mat. Med. Hind., 201*).

Special Opinions.—§ "This deserves a trial" (*Surgeon-Major C. J. McKenna*). "Laxative, tonic, expectorant" (*Dr. W. Barren, Bhuj, Cutch*).

384 **Canscora diffusa,** *Br.; Fl. Br. Ind., IV., 103; Wight, Ic., t. 1327 (not of Clarke).*

Syn.—PLADERA VIRGATA, *Roxb., Fl. Ind., Ed. C. B. C., 134.*

Vern.—*Kyouk pan,* BURM.

References.—*Thwaites, En. Ceylon Pl., 204; Dalz. and Gibs., Bomb. Fl., 158; Voigt, Hort. Sub. Cal., 520.*

Habitat.—Common throughout India, ascending to 4,000 feet, from Kumáon and Bhutan to Ceylon and Tenasserim.

MEDICINE. 385 **Medicine.**—Used as a substitute for **C. decussata.**

386 **C. sessiliflora,** *Roem. and Sch.; Fl. Br. Ind., IV., 104.*

387

CANTHARIS, *Latreille.*

Cantharis vesicatoria, *Latreille;* COLEOPTERA.

CANTHARIDES, BLISTERING BEETLE, SPANISH FLIES, *Eng.;* MOUCHES D'ESPAGNE, *Fr.;* SPANISCHE FLIEGEN, *Germ.;* CANTERELLE, *It.;* HISCHPANSKIE MUCHI, *Rus.;* CANTHARIDES, *Sp.*

Blistering Insect. 388 **References.**—*Pharm. Ind., 274; U. S. Dispens., 15th Ed., 342; Spons, Encyclop., 796; Balfour, Cyclop.; Ure's Dic. of Arts and Manufactures.*

Habitat.—A dried insect imported into India and sold by chemists. For indigenous insects used as substitutes, *see* **Mylabris cichorii,** *Fabr.*

389

CANTHIUM, *Lam.; Fl. Br. Ind., III., 131.*

The *Genera Plantarum* reduces the above genus to PLECTRONIA, *Linn.;* but CANTHIUM has been retained in the *Flora of British India,* which puts PLECTRONIA (in part) under CANTHIUM.

390 **Canthium didymum,** *Roxb.; Fl. Br. Ind., III., 132;* RUBIACEÆ.

Vern.—*Garbha gojha,* SANTAL; *Yerkoli,* TAM.; *Yellal, porawa-márá, Gal-karandu,* SING., KAN.

References.—*Roxb., Fl. Ind., Ed. C.B.C., 180; Kurz, Fl. Burm., II., 359; Thwaites, En. Ceyl. Pl., 152; Bom. Gaz., XV., 65.*

Habitat.—A shrub or small tree found in the Sikkim Himálaya at an altitude of 1,500 feet and distributed east to the Khasia and Jyntea mountains. It also is met with in Chutia Nagpur and in the Western Peninsula from the Concan southwards to the Malayan Peninsula and Ceylon.

Medicine.—Bark used by the Santals in fever (*Rev. A. Campell*). MEDICINE. 391

Structure of the Wood.—Hard, heavy, and close-grained; yellowish, with central masses of black. (*Bomb. Gaz.*) This is very much like the description of the wood, as given by **Brandis** and by **Lisboa** for **C. umbellatum.** TIMBER. 392

Canthium parviflorum, *Lamk.; Fl. Br. Ind., III., 136.* 393

Syn.—WEBERA TETRANDRA, *Willd.;* KANDEN KARA *in Rheede, Hort. Mal., V., t. 36.*

Vern.—*Kirni,* BOMB.; *Karai-cheddi,* TAM.; *Tsjéron kárá,* MAL.; *Bálusu, chettú, balsú,* TEL. (AINSLIE); *Kára,* SING.

References.—*Roxb., Fl. Ind., Ed. C. B. C., 179; Gamble, Man. Timb., 230; Ainslie, Mat. Med., II., 63; Dymock, Mat. Med., W. Ind., 713,* and *2nd Ed., 409; Lisboa, U. Pl., Bomb., 162; Thwaites, En. Cey. Pl., 152; Trimen's Cat., Ceyl. Pl., 44.*

Habitat.—A shrubby plant met with at altitudes of 4,000 feet, in the Western Peninsula from the Concan southwards to Ceylon.

Medicine.—Ainslie says: "A decoction of the edible leaves, as well as the root of this plant, is prescribed in certain stages of flux, and the last is supposed to have anthelmintic qualities, though neither have much sensible taste or smell" (*Mat. Med., II., 63*). MEDICINE. 394

Food.—The people of Western India are said to eat the fruits and employ the leaves as an adjunct to their curries. FOOD. 395

Structure of the Wood.—Hard and used for turning (*Gamble*). TIMBER. 396

C. umbellatum, *Wight, Ic., t. 1034; Fl. Br. Ind., III., 132.* 397

Syn.—PLECTRONA DIDYMA, *Benth. & Hook.; Brandis, For. Fl.*

Vern.—*Arsúl,* BOMB.; *Neckanie, nalla, balsú,* TAM. & TEL.; *Abalu,* KAN.; *Tolan,* URIYA.

References.—*Brandis, For. Fl., 276; Bedd., Flor. Sylv., 221; Dalz. & Gibs, Bomb. Fl., 113; Gamble, Man. Timb., 230 (under* **Plectonia didyma,** *Benth. & Hook.); Lisboa, U. Pl., Bomb., 87.*

Habitat.—An evergreen tree met with in the Western Peninsula (on the Ghâts at altitudes of 4,000 to 8,000 feet) and distributed south to Tenasserim and Ava.

Structure of the Wood.—Hard, close-grained, and heavy; yellowish white or chocolate-coloured with irregular masses of black wood in the centre (*Brandis*). According to Gamble, the wood is grey, hard, with very small, numerous and uniformly distributed pores; medullary rays fine and numerous. **Gamble** makes no mention of the irregular masses of black wood. (Compare with **C. didymum**). Weight 57℔ a cubic foot. Timber is used for agricultural purposes. TIMBER. 398

CANVAS.

Canvas. 399

SAILCLOTH, *Eng.;* KANEVAS and SEGELTUCH, *Germ.;* CANEVAS and TOILE-A-VOILE, *Fr.;* ZEILDOCK, *Dut.;* LONA, *It., Port., Sp.;* CANEVÁZZA, *It., Port.;* PARUSSINA, PARUSSNOE POLOTNO, *Rus.;* KITTAN, *Tam., Tel.*

A coarse, strong cloth, manufactured originally from hemp and subsequently from flax: hence it would appear the word canvas is derived from **Cannabis**—hemp. It is used for a variety of purposes, such as sail-cloths, tent-cloth, floor-cloths, hand-bags, shoes, &c.; a finer kind specially prepared is employed by artists for painting on.

Sails are usually made with the salvages and seams of the canvas running down parallel to the edges, though, when so constructed, they are very apt to give way during storms. This inconvenience may be obviated in a great measure by running the seams diagonally to the edges. **Messrs. Ramsay and Orr** effect the same purpose by an improvement in the manufacture of the cloth intended for sail-making, which consists in weaving the canvas with diagonal threads (*Ure, I., 613*).

400 In India the principal seats of canvas manufacture are Pondicherry, Cuddalore, and Travancore, where it is sold in bolts of 40 yards at from R20 to R25 the bolt; the coarser kinds selling from R8 to R15. A still coarser description of hard brown canvas is also produced in Bengal. In the Madras Presidency, excellent cotton canvas is manufactured by combining two or more threads together in the loom (*Balfour, I., 573*). Although originally, as stated, the term 'canvas' appears to have been restricted to a hemp or flax textile, it has been found possible to meet certain purposes of canvas by the manufacture of a fabric of jute or other pure or mixed fibres; this modern commercial textile is also designated as canvas. (*See* **Jute** and **Cannabis sativa**).

401

CAOUTCHOUC.

Caoutchouc is in England generally restricted to mean the pure hydrocarbon isolated from the other materials with which it forms the impure rubber of commerce. See **India-rubber.**

Capillare. See **Adiantum Capillus-Veneris,** *Linn.;* FILICES, *Vol. I.*

402

CAPPARIS, *Linn.; Gen. Pl., I., 108.*

Capparis aphylla, *Roth.; Fl. Br. Ind., I., 174;* CAPPARIDEÆ.

Vern.—*Karél, karér, kurrél, lete, karu,* HIND.; *Kari,* BEHAR, BOMB.; *Kirra kerin, karíl, karia, karis, tenti, delha, pinju,* PB.; *Kiral, kirrur, dorá kiram, kirab,* SIND; *Ker,* GUJ.; *Kerá, karíl,* MAR.; *Karyal,* DUK.; *Karíra,* SANS.; *Sodáda,* ARAB. (fruit); *Doro* (unripe fruit), *pukko* (ripe), SIND; *Tentí delpha, pinjú,* PB.; *Pussí* (flower), SIND; *Tít* (bud), SIND and PB.

References.—*Brandis, For. Fl., 14, 571; Gamble, Man. Timb., 15; Dalz. & Gibs., Bomb. Fl., 9; Stewart, Pb. Pl., 15; Aitchison, Cat., Pb. Pl., 10; Voigt, Hort. Sub. Cal., 75; Moodeen Sheriff, Supp. Pharm. Ind., 87; Dymock, Mat. Med., W. Ind., 2nd Ed., 64; O'Shaughnessy, Bengal Dispens., 206; Murray, Drugs & Pl., Sind, 53; S. Arjun, Bomb. Drugs, 190; Drury, Us. Pl., 110; Baden Powell, Pb. Prod., 330; Royle, Ill. Him. Bot., I., 72; Lisboa, Us. Pl. of Bomb., 4, 145; Balfour, Cyclop.; Kew. Cat., 13; Ind. For., IX., 174; Irvine, Mat. Med., Patna, 47.*

Habitat.—A dense, branching shrub of the Panjáb, of the North-West Provinces, and of the Dekkan,—chiefly in arid desert tracts. Distribution westward as far as Arabia, Egypt, and Nubia. It flowers in early spring, the fruit ripening in April.

MEDICINE. 403 **Medicine.**—The tender shoots and young leaves are made into a powder which is used as a blister in native medicine, and is also said to relieve toothache when chewed. "The plant is reckoned as heating and aperient; useful in boils, eruptions, and swellings, and as an antidote to poison; also in affections of the joints." **Dr. Dymock** says that the plant possesses somewhat similar properties to **C. spinosa.**

Special Opinions.—§ "The fruit when eaten causes obstinate constipation. It is used largely in the Harriana and Karnal districts as an astringent" (*Surgeon-Major C. W. Calthrop, Morar*). "The bark is described as bitter and laxative, and is said to be useful in inflammatory swellings" (*U. C. Dutt, Serampore*).

Food.—Dr. Stewart remarks that the buds are cooked when fresh as a pot-herb, and that the fruit is very largely consumed by the natives, "great numbers of whom go out for the purpose of collecting it both when green and after it is ripe. In the former state it is generally steeped for 15 days in salt and water, being put in the sun to ferment till it becomes acid, pepper and oil are then added. It is said that it will keep thus for a year, and is eaten to an ounce or two at a time, usually with bread. The ripe fruit is generally made into pickle with mustard or other oil (Hindús are not allowed to use vinegar), to be eaten with bread." The young flower-buds are preserved as pickle. FOOD. 404 Buds. 405 Fruit. 406 Pickle. 407 Flower-buds. 408

Special Opinions.—§ "The fruit is eaten" (*G. A. Watson, Allahabad*). "The flower-buds are made into pickle as a condiment" (*Surgeon-Major J. E. T. Aitchison, Simla*).

Structure of the Wood.—Light yellow, turning brown on exposure, shining, very hard and close-grained; weight 53℔ per cubic foot. Used for small beams and rafters in roofs, for the knees of boats, for oil-mills and agricultural implements, and is valuable owing to its not being attacked by white-ants—a fact due to its bitterness; it makes a good firewood. "The branches are commonly used for fuel, burning with a strong gaseous flame even when green, and are also used for brick-burning" (*Drury*). TIMBER. 409 Not eaten by White-ants.

Capparis grandis, *Linn. f.; Fl. Br. Ind., I., 176.* 410

Syn.—C. BISPERMA, *Roxb., Fl. Ind., Ed. C.B. C., 425.*

Vern.—*Puchownda, ragota,* BOMB.; *Kauntel,* MAR.; *Vellai toaratt, maram,* TAM.; *Guli, regguti, ragota, gullem chettu, regutti,* TEL.; *Tarate,* KAN.; *Waghutty,* MALA.; *Hkaw-kwa,* BURM.

References.—*Kurz, For. Fl., Burm., I., 64; Gamble, Man. Timb., 15; Thwaites, Enum. Ceylon Pl., 16; Dalz. & Gibs., Bomb. Fl., 10; Lisboa, U. Pl., Bomb., 5; Balfour, Cyclop.*

Habitat.—A small tree of the Chanda district and of the eastern part of the Dekkan, the Eastern Gháts and Carnatic, the Prome district in Burma, and the north-east of Ceylon.

Oil.—"Yields an oil which is used in medicine and for burning" (*Bomb. Gaz., XV., 65*). OIL. 411

Structure of the Wood.—White, moderately hard, durable; weight 46℔ per cubic foot. Much used by the natives in the Madras Presidency for plough-shares and rafters. **Roxburgh** says it is "heavy, hard, and durable; the natives employ it for various purposes." **Kurz** remarks that in Burma it is regarded as good for turning. TIMBER. 412

C. Heyneana, *Wall.; Fl. Br. Ind., I., 174.* 413

Vern.—*Chayruka,* HIND.

References.—*Dalz. & Gibs., Bomb. Fl., 9; Balfour, Cyclop.*

Habitat.—An erect shrub, distributed from the South Konkan and Kanara to Travancore; also met with in Ceylon.

Medicine.—The leaves are used for rheumatic pains in the joints, and the flowers are made into a laxative drink. MEDICINE. 414 Leaves. 415 Flowers. 416

C. horrida, *Linn. f.; Fl. Br. Ind., I., 178; Wight, Ic., t. 173.*

Syn.—C. ZEYLANICA, *Roxb., Fl. Ind., Ed. C.B.C., 425.*

Vern.—*Ardanda,* HIND., SIND, DUK.; *Ulta-kánta, bipuwa-kánta,* KUMAON; *His, karvila, hún garna,* PB.; *Karralura,* OUDH; *Katerni,*

GOND.; *Gitoran*, AJMIR; *Buru asaria*, SANTAL; *Bagnai*, MONGHYR; *Oserwa*, URIYA; *Wagatti, wag, anti*, BOMB.; *Gowindi*, MAR.; *Atanday, attandax, katalli kai*, TAM.; *Adonda, arudonda*, TEL.; *Hunkarú*, SANS.; *Welangiriya*, SING.; *Nah-maní-tan-yet, nah-mani-than-lyet, nwa-mani-than-lyet*, BURM.

References.—*Brandis, For. Fl., 15; Kurz, For. Fl., Burm., I., 62; Gamble, Man. Timb., 15, ii.; Thwaites, Enum. Ceylon Pl., 15; Dalz. & Gibs., Bomb. Fl., 10; Stewart, Pb. Pl., 16; Aitchison, Cat. Pb. Pl., 10; Voigt, Hort. Sub. Cal., 74; Murray, Drugs and Pl., Sind, 54; Royle, Ill. Him. Bot., I., 72; Atkinson, Him. Dist., 730; Lisboa, U. Pl., Bomb., 277; Balfour, Cyclop.*

Habitat.—A climbing, thorny shrub, growing in most parts of India and Burma and in the hot dry tracts of Ceylon.

MEDICINE. Leaves. 417 Bark. 418 Fruit. 419

Medicine.—The leaves are applied medicinally as a counter-irritant. A cataplasm made from them is useful in boils, swellings, and piles (*Atkinson*). "The bark along with native spirit is given in cholera. The leaves and the fruit are also used medicinally" (*Rev. A. Campbell, Report on the Economic Products of Chutia Nagpur*).

Special Opinion.—§ "A decoction of the leaves is used in syphilis" (*Surgeon-Major D. R. Thompson, 1st District, Madras*).

FOOD. 420 FODDER. 421

Food.—In the Southern Panjáb and Sind the fruit is made into pickle (*Stewart*). The twigs, shoots, and leaves are greedily eaten by goats and elephants.

TIMBER. 422

Structure of the Wood.—Yellowish white, moderately hard; weight about 47lb per cubic foot. Used as fuel.

423 **Capparis multiflora,** *Hook. f. & Th.; Fl. Br. Ind., I., 178.*

Vern.—*Suntri*, NEPAL.

References.—*Kurz, For. Fl., Burm., I., 61; Gamble, Man. Timb., ii.*

Habitat.—A climbing, thorny shrub of the Eastern Himálaya and Upper Burma.

TIMBER. 424

Structure of the Wood.—White, moderately hard.

425 **C. olacifolia,** *Hook. f. & Th.; Fl. Br. Ind., I., 178.*

Vern.—*Naski, hais*, NEPAL; *Jhenok*, LEPCHA.

References.—*Gamble, Man. Timb., 15, ii.*

Habitat.—A thorny shrub of the Sub-Himálayan tract from Nepal to Assam, chiefly in the undergrowth of *Sissu* forests along river banks.

TIMBER. 426

Structure of the Wood.—White, hard; weight about 44lb per cub. ft.

427 **C. sepiaria,** *Linn.; Fl. Br. Ind., I., 177.*

Vern.—*Hiún garna, hius*, PB.; *Kantá-gúr-kámai, káliakará*, BENG.; *Kanti kapali*, URIYA; *Kanthár*, GUJ.; *Nella-uppi*, TEL.; *Ahinsra, kákádani*, SANS.

References.—*Roxb., Fl. Ind., Ed. C.B.C., 425; Brandis, For. Fl., 15; Kurz, For. Fl., Burm., I., 66; Gamble, Man. Timb., iii.; Thwaites, Enum. Ceylon Pl., 16; Dalz. & Gibs., Bomb. Pl., 10; Aitchison, Cat., Pb. Pl., 10; Voigt, Hort. Sub. Cal., 75; Murray, Drugs and Pl., Sind, 54; Royle, Ill. Him. Bot., I., 72; Balfour, Cyclop.*

Habitat.—A shrub, growing in dry places in India and Burma.

MEDICINE. 428

Medicine.—The plant possesses febrifugal properties.

Special Opinions.—§ "Is said by Sanskrit writers to be useful in fevers caused by deranged bile and wind. It is also considered alterative and tonic and useful in skin diseases" (*U. C. Dutt, Civil Medical Officer, Serampore*).

TIMBER. 429

Structure of the Wood.—White, hard; pores moderate sized.

DOMESTIC. 430

Domestic Uses.—The branches make excellent hedges.

Capparis spinosa, *Linn.; Fl. Br. Ind., I., 173.* 431

THE EDIBLE CAPER.

Syn.—C. MURRAYANA, *Graham; Wight, Ic., t. 379.*

Vern.—*Kabra, ber,* HIND.; *Kábra,* LADAK, TIBET; *Ulta kanta,* KUMAON; *Kaur, kiári, baurí, ber, bandar, bassar, kakrí, kander, taker, barar, kerí, kábra, kabarra, barárí, bauri,* PB.; *Kalvári,* SIND; *Kabar,* BOMB.; *Kabarra, kabawa,* AFG.; *Kabar, kabur,* ARAB.; *Kebír,* PERS. (In Persia it is known as *Kabar, kúrak.*) *Kabár,* SYRIAN; *Kabarish,* TURKISH.

References.—*Brandis, For. Fl., 14; Kurz, For. Fl., Burm., I., 58; Gamble, Man. Timb., 14; Dalz. & Gibs., Bomb. Fl., 9; Stewart, Pb. Pl., 17; Aitchison, Cat. Pb. Pl., 10; U. S. Dispens., 15th Ed., 1597; Dymock, Mat. Med. W. Ind., 2nd Ed., 64; Ainslie, Mat. Ind., II., 150; O'Shaughnessy, Beng. Dispens., 206; Murray, Drugs and Pl., Sind, 54; Baden Powell, Pb. Pr., 572; Royle, Ill. Him. Bot., I., 72; Lisboa, U. Pl., Bomb., 145, 277; Balfour, Cyclop.; Treasury of Botany; Smith, Dict., 90; Irvine, Mat. Med. Patna, 87; Kew Official Guide; Mus. of Ec. Bot., 13.*

Habitat.—This is the plant which affords the Caper-berry of Europe. It occurs in India in the central and northern parts of the Panjáb and in Sind; is less frequent in Rájputana than **C. aphylla.**

Medicine.—Dr. **Stewart** remarks that in Kangra the roots are said to be applied to sores. The author of the *Makhzan-ul-Adwíya* considers the root-bark "to be hot and dry and to act as a detergent and astringent, expelling cold humors; it is therefore recommended in palsy, dropsy, and gouty and rheumatic affections; the juice of the fresh plant is directed to be dropped into the ear to kill worms, just as **Cleome** juice is used in India; all parts of the plant are said to have a stimulating and astringent effect when applied externally" (*Dymock*). **Ainslie** notices its use as an external application to malignant ulcers. The *United States Dispensatory* says: "The buds or unexpanded flowers, treated with salt and vinegar, form a highly esteemed pickle, which has an acid, burning taste, and is considered useful in scurvy. The dried bark of the root was formerly officinal. It is in pieces partially or wholly quilled, about one-third of an inch in mean diameter, transversely wrinkled, grayish externally, whitish within, inodorous, and of a bitterish, somewhat acid, and aromatic taste. It is considered diuretic, and was formerly employed in obstructions of the liver and spleen, amenorrhœa, and chronic rheumatism."

MEDICINE. Roots. 432 Root-bark. 433 Juice. 434 Buds. 435

Chemical Composition.—"The root-bark is said to contain a neutral bitter principle of sharp irritating taste, and resembling senegin. The flower-buds, distilled with water, yield a distillate having an alliaceous odour. After they have been washed with cold water, hot water extracts from them Capric acid ($C_{10}H_{20}O_2$), and a gelatinous substance of the Pectin group; Capric acid is sometimes found deposited on the calices of the buds in white specks having the appearance of wax (*Rochleder and Blas*)" (*Watts' Dict., Chemistry*).

CHEMISTRY. 436

Food.—In Europe this furnishes the Caper. **Mr. Edgeworth** found the buds (prepared in the style of "Capers") to answer very well as a substitute for the European congener. In India the ripe fruit is either eaten raw or made into pickle. In Sind and in some parts of the Panjáb, a compound of oil, mustard, fœnu-greek, &c., is used in pickling capers. In Ladak the leaves are eaten as greens.

FOOD. 437 Berries. Pickle. 438 Leaves. 439

Fodder.—The leaves and ripe fruits constitute a favourite food of goats and sheep.

FODDER. 440

441 **Capparis zeylanica,** *Linn.; Fl. Br. Ind., I., 174.*

Syn.—C. ACUMINATA, *Roxb.*; C. BREVISPINA, *DC.*

Vern.—*Kalo-kera,* BENG.; *Authoondy kai,* TAM.

References.—*Voigt, Hort. Sub. Cal., 74; Dalz. & Gibs., Bomb. Fl., 9; Balfour, Cyclop.*

Habitat.—Common in the Carnatic and Malabar, occasional in the Western Dekkan and in the drier parts of Ceylon.

FOOD. Pickle. 442 **Food.**—The green fruit is pickled.

CAPSELLA, *Mœnch.; Gen. Pl., I., 86.*

443 **Capsella Bursa-pastoris,** *Mœnch.; Fl. Br. Ind., I., 159;* CRUCIFERÆ.

SHEPHERD'S PURSE; PICKPOCKET, *Eng.*; BOURSE DE PASTURE, *Fr.*; HIRTENASCHE, *Germ.*

Habitat.—A weed in the vicinity of cultivation throughout the temperate regions of India; particularly abundant on the N. W. Himálaya.

MEDICINE. 444 **Medicine.**—"This very common weed is bitter and pungent, yields a volatile oil on distillation, identical with oil of mustard, and has been used
Oil. 445 as an antiscorbutic, also in hæmaturia and other hæmorrhages, as well as in dropsy" (*U. S. Dispensatory, 15th Ed., p. 1597*).

FOOD. 446 **Food.**—Apparently the natives of India are ignorant of the uses of this plant. No Indian writer, on economic subjects at least, alludes to it. **Balfour** says it "grows in Europe, Persia, Asia, and Japan; used by the natives as a pot-herb."

447 CAPSICUM, *Linn.; Gen. Pl., II., 892.*

The greatest confusion exists in Indian literature as to the cultivated species of Capsicum. Popularly the larger fruits are usually designated Capsicums, and the smaller Chillies. According to **Firminger**, the powdered seeds of the latter constitutes Cayenne pepper. That author, in his *Manual of Gardening for India,* states that there are a great many varieties of Capsicum grown in India, some of which are very ornamental when grouped together.

The writer can at most hope that he has thrown the various vernacular names approximately under their corresponding botanical species. Much remains still to be done in order to clear up the ambiguities which exist in the literature of the Indian Capsicums. Many of the vernacular names appear to be given to all the species alike.

448 **Capsicum annuum,** *Linn.; DC. Prodr., XIII., Pt. 1., 412;* SOLANACEÆ.

RED PEPPER.

Vern.—*Mattisa, wángrú, lál mirch, marcha, mirch, gáchmirch,* HIND. and PB.; *Lál-marich, lanká-marich, gách-marich,* BENG.; *Súrú-phamsah,* BHOTE; *Matitsa-wangrú,* KUMAON; *Mirtz-a-vangun, mirch-wángum,* KASHMIR; *Lál-mirich, marchu,* GUJ.; *Mirchu,* CUTCH; *Mirsingá,* MAR.; *Milagáy, mulagáy, mollaghai, mollagu,* TAM.; *Mirapakáya, merapu-kai,* TEL.; *Kappal-melaka, kapú-mologú,* MALA.; *Ménasiná-káyı,* KAN.; *Marichi-phalam,* SANS.; *Filfile, ahmur,* ARAB.; *Filfile-surkh, pilpile-surkh,* PERS.; *Miris, rata miris,* SING.; *Náyu-si, na yop,* BURM.

References.—*Roxb., Fl. Ind., Ed. C. B. C., 193; Stewart, Pb. Pl., 156; DC. Orig. of Cult. Pl., 289; Voigt, Hort. Sub. Cal., 510; Pharm. Ind., 180; Modeen Sheriff, Supp. Pharm. Ind., 87; Flück. & Hanb., Pharmacog., 452; U. S. Dispens., 15th Ed., 349; Dymock, Mat. Med. W. Ind., 531; O'Shaughnessy, Beng. Dispens., 467; Murray, Drugs and Pl, Sind, 157; Bidie, Cat. Raw Prod., Paris Exh., 14, 87; S. Arjun, Bomb. Drugs, 96; Drury, U. Pl., 111; Birdwood, Bomb. Prod., 222; Spons' Encyclop., 1803; Balfour, Cyclop.; Treasury of Botany; Smith', Dict., 91; Simmonds, Trop. Agri., 479; Kew Official Guide, Mus. of Ec. Bot., 100.*

Capsicum or Red Pepper. **CAPSICUM annuum.**

Habitat.—A native of equinoctial America, most probably of Brazil. Commonly cultivated for its fruit throughout the plains of India, and on the lower hills such as in Kashmír, and in the Chenab valley up to altitude 6,500 feet. When grown on the hills it is said to be very pungent. There are seven varieties, differing chiefly in the length, shape, and colour of the fruit, some being round, others oblong, obtuse, pointed or bifid, smooth or rugose; and red, white, yellow, or variegated. It is probable that most Indian authors have confused this species with **C. minimum**, which see.

History.—"This species has a number of different names in European languages, which all indicate a foreign origin, and the resemblance of the taste to that of pepper. In French it is often called *poivre de Guinée* (Guinea pepper), but also *poivre du Brésil, d'Inde* (Indian, Brazilian pepper), &c., denominations to which no importance can be attributed. Its cultivation was introduced into Europe in the sixteenth century. It was one of the peppers that **Piso** and **Maxgraf** saw grown in Brazil under the name *quija* or *quiya*. They say nothing as to its origin." (*DC. Orig. of Cult. Pl.*) "Chillies are not mentioned by any Sanskrit writer, consequently their introduction into India must have taken place at a comparatively recent date. It is probable that the Portuguese brought the fruit from the West Indies. Up to the present time the cultivation of the plant is carried on more extensively at Goa than at any other place on the western coast, and capsicums are well known in Bombay by the name of *Gowaí mirchi* (Goa pepper)" (*Dr. Dymock, Mat. Med. W. Ind.*). Hove alludes to **Capsicum** as grown in Bombay in 1787 and expresses no astonishment at its existence in India. 449

Cultivation of Capsicums.—"A light well-manured soil is the best for all kinds, in which the plants should be picked out at about four inches apart when they attain a growth of three inches; and afterwards put out into a bed of rich light earth when they attain six inches in height, giving them a good supply of water and keeping them clear from weeds" (*The Gardener*). 450

Medicine.—**Dr. Stewart** says that the fruit is used externally in the form of plasters and taken internally in cholera; it is eaten from a conviction that it counteracts the effects of bad climates. **MEDICINE. Plaster. 451**

As a drug, red pepper is considered by the natives as stomachic and stimulant, and is used externally as a rubefacient (*Dymock*). "It has been employed with success as a topical application to elongated uvula and relaxation of the pendulous veil of the palate. Made into a lozenge, with sugar and tragacanth, it is a favourite remedy for hoarseness with professional singers and public speakers. In putrid sore-throat whether symptomatic or strictly local, gargles of an infusion of red pepper are often very usefully resorted to" (*O'Shaughnessy, Beng. Dispens., 468*). "It is employed in medicine in combination with cinchona in intermittent and lethargic affections and also in atonic gout. It is a valuable adjunct to bitters, tonics, and other stimulants in weak states of the stomach; in cold leucophlegmatic habits, dyspepsia and flatulence, and as a gargle in relaxed states of the throat it is highly extolled and has also been used with success in the advanced stages of rheumatism. In native practice it is given, in conjunction with asafœtida and sweet-flag root, in cholera. By German physicians it is supposed to be particularly injurious in gonorrhœa" (*Murray's Pl. and Drugs of Sind*). **Lozenge. 452**

Dr. Sakharam Arjun says that the fruit is used as a stimulant in snake-bite.

Chemical Composition.—"**Bucholz**, in 1816, and about the same time **Braconnot**, traced the acridity of capsicum to a substance called *capsicin.* **CHEMISTRY. 453**

Capsicum or Red Pepper.

CHEMISTRY.

It is obtained by treating the alcoholic extract of ether, and is a thick yellowish red liquid, but slightly soluble in water. When gently heated it becomes very fluid, and at a higher temperature is dissipated in fumes which are extremely irritating to respiration. It is evidently a mixed substance consisting of resinous and fatty matters.

"**Felletar**, in 1869, exhausted capsicum fruits with dilute sulphuric acid and distilled the decoction with potash. The distillate, which was strongly alkaline and smelt like *conine*, was saturated with sulphuric acid, evaporated to dryness and exhausted with absolute alcohol. The solution, after evaporation of the alcohol, was treated with potash, and yielded by distillation a volatile alkaloid having the odour of conine.

"From experiments made by one of us (**F.**) we can fully confirm the observations of **Felletar**. We have obtained the volatile base in question, and find it to have the smell of conine. It occurs both in the pericarp and in the seeds, but in so small a proportion that we were unsuccessful in isolating it in sufficient quantity to allow of accurate examination.

"**Dragendorff** states (1871) that petroleum ether is the best solvent for the alkaloid of capsicum; he obtained crystals of its hydrochlorate, the aqueous solution of which was precipitated by most of the usual tests, but not by tannic acid.

"The colouring matter of capsicum fruits is sparingly soluble in alcohol, but readily in chloroform. After evaporation an intensely red soft mass is obtained, which is not much altered by potash; it turns first blue, then black, with concentrated sulphuric acid, like many other yellow colouring substances. By alcohol chiefly *palmatic* acid is extracted from the fruit, as shown by **Thresh** in 1877.

"The fruits of **Capsicum fastigiatum** have a somewhat strong odour; on distilling consecutively two quantities, each of 50℔, we obtained a scanty amount of flocculent fatty matter which possesses an odour suggestive of parsley. Both this matter, as well as the distilled water, were neutral to litmus-paper and the water tasteless. We separated the latter and exposed the remaining greasy mass to a temperature of about 50°C., when it for the most part melted. The clear liquid on cooling solidified and now consisted of tufted crystals, which we further purified by recrystallization from alcohol. Thus about two centigrammes were obtained of a neutral white stearoptene having a decidedly aromatic, not very persistent taste, and by *no means acrid*, but rather like that of essential oil of parsley. The crystals melted at 38°C. On keeping them for some days at the temperature of the water-bath, covered with a watch-glass, some drops of essential oil were volatilized which had the same taste and did not solidify; the crystals were consequently accompanied by a liquid oil. When kept for some days more in that condition, the crystals themselves began to be volatilized, and the part remaining behind acquired a brownish hue. This, no doubt, points out another impurity, as we ascertained by the following experiment. With boiling solution of potash, the stearoptene produces a kind of soap which on cooling yields a transparent jelly. If this is dissolved and diluted, it becomes turbid by addition of an acid. This probably depends upon the presence of a little fatty matter, a suggestion which is confirmed by the somewhat offensive smell given off by our stearoptene if it is heated in a glass tube.

"BUCHHEIM'S 'CAPSICOL' is in our opinion a doubtful substance.

"**Thresh** (1876-77) succeeded in isolating a well-defined, highly active principle, the *Capsaicin*, from the extract which he obtained by exhausting Cayenne pepper with petroleum. From the red liquor dilute caustic lye removes *capsaicin*, which is to be precipitated in minute crystals by passing carbonic acid through the alkaline solution. They may

CHEMISTRY.

be purified by recrystallizing them from either alcohol, ether, benzine, glacial acetic acid, or hot bisulphide of carbon; in petroleum *capsaicin* is but very sparingly soluble, yet dissolves abundantly on adilition of fatty oil. The latter being present in the pericarp is the cause why *capsaicin* can be extracted by the above process.

"The crystals of *capsaicin* are colourless and answer to the formula $C_9H_{14}O_2$; they melt at 59°C., and begin to volatilize at 115°C.; but decomposition can only be avoided by great care. The vapours of *capsaicin* are of the most *dreadful acridity,* and even the ordinary manipulation of that substance requires much precaution. *Capsaicin* is not a glucoside: it is a powerful rubifacient, and taken internally produces very violent burning in the stomach" (*Pharmacographia*).

Special Opinions.—§"Stimulant and rubefacient, useful in dyspepsia; recommended in infusion as an external application to the eye" (*Assistant Surgeon Nehal Sing, Shaharanpur*). "Chiefly used as a condiment and considered to be stomachic" (*Assistant Surgeon Anund Chunder Mookerji, Noakhally*). "Anti-malarious to a certain extent" (*H. D. Masani, Surgeon, H. M.'s 30th N. I., Bombay, Karachi*). "Carminative, cooling medicine. The decoction with opium and fried asafœtida seeds is used in cholera. In the form of gargle it is useful in stomatitis and sore-throat. It is an ingredient in what is called *masala* in the Deccan, Guzerat, and Cutch" (*W. Barren, Surgeon, H. M.'s 25th N. L. I., Bombay, Bhuj, Cutch*). "The capsule is innocuous; the seeds, as well known, are powerfully irritant" (*R. T. H., Morar*). "Chillies are applied by natives to dog-bites. An infusion made with 4 drams of chillies and a bottle of boiling water has been found useful in severe sore-throat" (*Assistant Surgeon Bhagwan Dass, Rawal Pindi*). "In delirium tremens in 20-grain doses" (*Surgeon-Major George Cumberland Ross, Delhi*). "Is used in liniments as a rubefacient; in cholera pills with camphor and asafœtida; as an application to elongated uvula and relaxed throat it is very useful" (*A Surgeon*). "Active principle, an acrid oil-capsaicin. In dyspepsia, a good pill is made with equal parts of capsicum, rhubarb, and ginger" (*C M. Russell, Civil Surgeon, Sarun, Bengal*). "Internally it has a stimulant action on the bowels and helps to relieve constipation" (*Surgeon-Major A. S. G. Jayakar, Muskat*).

FOOD. 454

Food.—The fruit when green is used for pickling and when ripe is mixed with tomatos, &c., to make sauces. It is also dried and ground for use like Cayenne pepper (*Treasury of Botany*).

The consumption of chillies is very great, and both rich and poor daily use them; they form the principal ingredient in all chutnies and curries; ground into a paste, between two stones, with a little mustard oil, ginger, and salt, they form the only seasoning which the millions of poor can obtain to eat with their rice (*Balfour's Cyclop.*). **Dr. Dymock** gives the value of Ghâti chillies at R$3\frac{1}{2}$ per maund, and Goway, R$2\frac{1}{2}$ to 4 per maund of 28℔ in Bombay.

Capsicum, fastigiatum, *Blume.* See **C. minimum,** *Roxb.*

455

C. frutescens, *Linn.; Fl. Br. Ind., IV., 239.*

SPUR PEPPER, CAYENNE PEPPER, GOAT PEPPER, AND CHILLIES. THE SHRUBBY CAPSICUM.

Vern.—*Lál* or *gách-marich, lál lanká murich, lanká,* BENG.; *Lál* or *gách-mirich, lál-mircha, lanká-mirchi,* HIND.; *Kursáni,* HIMÁLAYA; *Lal mirchi,* BOMB.; *Mirchi,* GUJ.; *Tambhuda mirchingay, mirchi,* MAR.; *Mullá-ghái,* TAM.; *Mirápa káia, golakonda, mirapah, sima, sudi-mirapa-kaia,* TEL.; *Ladu miro, chabai, kappal-melaka, chabe-lombok, ladamera,*

ladamera china, MAL.; *Menashiná káyi*, KAN.; *marichi-phalam, brahu* or *bran maricha*, ? SANS.; *Filfile-ahmar*, ARAB.; *Fúlfil-i-súrkh*, PERS.; *Gas miris*, SING.

References.—*Roxb., Fl. Ind., Ed. C.B.C., 193; Aitchison, Cat., Pb. Pl., 102; DC. Origin of Cult. Pl., 290; Voigt, Hort. Sub. Cal., 510; U. S. Dispens., 15th Ed., 349; U. C. Dutt, Mat. Med., Hind., 212; Ainslie, Mat. Ind., I., 306; O'Shaughnessy, Beng. Dispens., 468; Baden Powell, Pb. Prod., 301; Royle, Ill. Him. Bot., I., 280; Atkinson, Him. Dist., 705, 730; Balfour, Cyclop.; Treasury of Botany; Smith, Dic., 91; Simmonds, Trop. Agri., 479.*

Habitat.—An annual, cultivated throughout India. Supposed to have been recently, comparatively speaking, introduced from South America. According to the best authorities, this and the other species of Capsicum, now cultivated in India, have no Sanskrit names. Of the Indian cultivated species this is perhaps the commonest, as it is also the largest, being sometimes cultivated in the hedges around fields. It is grown during the cold weather on light sandy soil in most parts of the country, and especially so in Bengal, Orissa, and Madras. The fruit, when ripe, is generally of a bright red colour; it is then picked and laid out on mats to dry in the sun.

Cayenne Pepper. 456 Chillies. 457

Opinions differ slightly as to the plants which afford Cayenne pepper. Speaking of this species, **DeCandolle** says: "The great part of the so-called Cayenne pepper is made from it, but this name is given also to the product of other peppers. **Roxburgh**, the author who is most attentive to the origin of Indian plants, does not consider it to be wild in India" (*Orig. Cult. Pl.*) **Simmonds** writes that "the Cayenne pepper of commerce is obtained chiefly from the pulverised chillies or fruit pods of one or two species of Capsicum (**C. annuum**, *Linn.*, and **C. fastigiatum**, *Blume*). So also in the *Kew Official Guide* (p. 100) the dried and pulverised rind of the pods of **C. annuum** and its allies is said to make the best Cayenne pepper.

MEDICINE. 458

Medicine.—Chillies are used as medicine in typhus and intermittent fevers and in dropsy; they are regarded as stomachic and rubefacient. In native practice they are prescribed in gout, dyspepsia, cholera, and ague (*Atkinson*).

Special Opinions.—§ "When taken in curry in unusual quantities, chillies cause, in many instances, great irritation and burning in the rectum, especially after defœcation, attended also with scalding and frequent desire to urinate; mixed with ginger and mustard, they form a powerful rubefacient paste" (*Assistant Surgeon Shib Chunder Bhattacharji, Chanda, Central Provinces*). "A dose of ten grains of finely powdered capsicum

Seed. 459

seed, given with an ounce of hot water, two or three times a day, sometimes shows wonderful effects in cases of delirium tremens" (*Surgeon R. Gray, Lahore*). "Stimulant, aromatic, and stomachic. I use the tincture

Cholera mixture. 460

and powder largely in the preparation of cholera mixture and pills, also in gargles for sore-throat" (*Brigade-Surgeon S. M. Shircore, Murshedabad*). "A powerful stimulant used as a gargle in sore-throat, also in dyspepsia and loss of appetite" (*Brig.-Surgeon J. H. Thornton, Monghir*).

Food.—In every Indian bazar chillies may be purchased; although not natives of India, the cultivated forms, at the present date, are everywhere met with and constitute an indispensable ingredient in native curry.

Chilli Vinegar. 461

They are "much used for flavouring pickles. By pouring hot vinegar upon the fruits all the essential qualities are procured, which cannot be

Chilli Extract. 462

effected by drying them, owing to their oleaginous properties; hence chilli vinegar is in repute as a flavouring substance. In Bengal the natives make an extract from the chillies which is about the consistence and

Powder. 463

colour of treacle. A form of soluble Cayenne was sent from British Guiana in 1867 in the collection forwarded to the Paris Exhibition" (*Simmonds, Trop. Agri.*, 480).

The pods are dried on a hot plate or in a slow oven and then pounded in a mortar. This powder is then passed through a handmill until it is brought to the finest possible state; thereafter it is well sifted and preserved in corked glass bottles for use (*Treasury of Botany*).

Capsicum grossum, *Willd.; Fl. Br. Ind., IV.*, 239. 464

BELL PEPPER.

Vern.—*Kafri-murich*, BENG., HIND.

References.—*Roxb., Fl. Ind., Ed. C.B.C., 193; Flück. & Hanb., Pharmacog., 452; Dymock, Mat. Med. W. Ind., 2nd Ed. 640; Birdwood, Bomb. Prods., 222; DC. Orig. Cult. Pl., 290; Balfour, Cyclop.; Smith, Dic., 91; Simmonds, Trop. Agri., 479.*

Habitat.—Not much cultivated in India; native place uncertain.

Food.—Cultivated to a limited extent in gardens, but chiefly for Europeans, who either cut this capsicum in stews or have it opened, stuffed with certain spices, and pickled in vinegar. The thick fleshy skin is not so hot as that of the other species. FOOD. 465

C. minimum, *Roxb.; Fl. Br. Ind., IV., 239; Wight, Ic., t. 1617.* 466

BIRD'S-EYE CHILLI.

Syn.—C. FASTIGIATUM, *Blume;* C. BACCATUM, *Wall.*

Vern.—*Gách marich*, HIND.; *Dhan-lung ka-murich, lanká-morich, lál-morich*, BENG.; *Lál-mirich marchá*, GUJ.; *Mirchi, lál mirch*, DUK.; *Usi-mulaghai*, TAM.; *Sudmirapa kaia*, TEL.; *Chalie, loda-china.* MAL.; *Kappal-melaka*, MALABAR; *Fifil-i-surkh*, PERS.; *Filfil-i-ahmar*, (red-pepper), ARAB.; *Miris*, SING.; *Náyú-si, gna yoke, gna yoke-nopmyan, nayop*, BURM.

References.—*Roxb., Fl. Ind., Ed. C.B.C., 193; Voigt., Hort. Sub. Cal., 510; Pharm. Ind., 180; Flück. & Hanb., Pharmacog., 452, 453; U. S. Dispens., 15th Ed., 349; Bentl. & Trim., Med. Pl., t. 188; U. C. Dutt, Mat. Med., Hind., 221; Dymock, Mat. Med. W. Ind., 1st Ed., 531; Waring, Bazar Med., 35; Baden Powell, Pb. Prod., 363; Spons' Encyclo., 1803; Balfour, Cyclop.; Smith, Dic., 91; Simmonds, Trop. Agri., 479o.*

Habitat.—Cultivated throughout India, but not extensively; closely resembles **C. annuum**, but is distinguished by the more acute corolla lobes, the smaller seeds, and by the pod being erect, nearly cylindrical and yellow when ripe. It is generally known as Bird's-eye Chilli. This "is found in many parts of India, principally in the southern districts, growing in waste places, gardens, &c., in an apparently wild state. It is also found abundantly in Java and other parts of the Eastern Archipelago under similar conditions. There is, however, good reason to believe that, in common with the rest of the genus, it was originally brought from some part of the American Continent. It is now cultivated to a large extent in the tropics of both the old and new worlds" (*Bentley and Trimen*).

Medicine.—The *Pharmacopœia of India* describes the fruit as an acrid stimulant. "In atonic dyspepsia, and in diarrhœa arising from putrid or crude ingesta in the intestines, and in the vomiting of bilious remittent fever, it acts beneficially. In scarlatina it has been used with great repute in the West Indies. In various forms of cynanche, and in hoarseness or aphonia, depending upon a relaxed condition of the *chordæ vocales*, it has been found a useful adjunct to gargles. As a rubefacient and counter-irritant, the bruised fruit, in the form of poultice, acts energetically; added to sinapisms it greatly increases their activity." "Acts as an acrid stimulant, and externally as a rubefacient used in MEDICINE. 467 Gargles. 468

MEDICINE. putrid sore-throat and scarlatina; also in ordinary sore-throat, hoarseness, dyspepsia, and yellow fever, and in diarrhœa occasionally; also in piles" (*Baden Powell*).

Mixture. 469 "In *Scarlatina,* the following mixture has attained much repute in the West Indies: Take two table-spoonsful of bruised Capsicum and two tea-spoonsful of Salt; beat them into a paste and add half a pint of boiling Water; when cold, strain and add half a pint of Vinegar. Dose for an adult, one table-spoonful every four hours; to be diminished for children according to age or the severity of the attack. The same formula forms an excellent gargle in the *sore-throat which accompanies this disease* as well as in ordinary *relaxed sore-throat, hoarseness,* &c." (*Waring, Bazar Medicines*).

FOOD. 470 **Food.**—This small "chilli" is rarely used by natives, but by Europeans is steeped in vinegar and mixed with salt; in this form it is employed as a seasoning in stews, chops, &c.

CARAGANA, *Lam.; Gen. Pl., I., 505.*

471 **Caragana pygmæa,** *DC.; Fl. Br. Ind., II., 116; Royle, Ill., t. 34, fig. 2;* LEGUMINOSÆ.

Vern.—*Táma, dáma, tráma,* LADAK; *Shmalak,* SIND.

References.—*Brandis, For. Fl., 134; Stewart, Pb. Pl., 61; Balfour, Cyclop.*

Habitat.—A low shrub very much resembling furze. It inhabits the dry highlands of the Western Himálaya; altitude 8,000 to 17,000 feet.

FOOD. Roots. 472 FODDER. 473 **Fodder.**—It is browsed by goats and is much valued for fuel in the treeless regions where it is met with. **Balfour** states that in China the roots of **Caragana flava** are eaten in times of scarcity.

CARALLIA, *Roxb.; Gen. Pl., I., 680.*

474 **Carallia integerrima,** *DC.; Fl. Br. Ind., II., 439; Wight, Ic., t. 605; Beddome, Fl. Sylv., t. CXCIII.;* RHIZOPHOREÆ.

Syn.—C. LUCIDA, *Roxb., Fl. Ind., Ed. C. B. C., 396; Kurz, i., 451.*

Vern.—*Kierpa,* BENG.; *Júr,* KOL.; *Palamkat,* NEPAL; *Kujitekra,* ASS.; *Punschi,* BOMB.; *Pansi, phansi,* MAR.; *Karalli,* TEL.; *Andipunar, phansi,* KAN.; *Dawata, davette,* SING.; *Bya,* ARRACAN; *Maneioga, maní-aw-ga,* BURM.

References.—*Brandis, For. Fl., 219; Gamble, Man. Timb., 177, XX.; Thwaites, En. Ceylon Pl., 120; Dalz. & Gibs., Bomb. Fl., 96; Voigt, Hort. Sub. Cal, 42; Royle, Ill. Him. Bot., I., 210; Lisboa, U. Pl., Bomb., 73; Balfour, Cyclop.*

Habitat.—An evergreen tree with thin, dark-grey bark, found in the Eastern and Western moist zones; particularly in the Eastern Himálaya, Bengal, Burma, South India, the Andaman Islands, and Ceylon.

TIMBER. 475 **Structure of the Wood.**—Sapwood perishable; heartwood red, very hard, durable, works and polishes well; weight from 42 to 51℔ per cubic foot. In Calcutta used for house-building. In South Kanara employed for furniture and in cabinet-making, and in Burma for planking, furniture, and rice-pounders. It is tough and not easily worked, brittle and not durable, but has a pretty wavy appearance and is peculiar in structure (*Beddome*).

CARALLUMA, *R. Br.; Gen. Pl., II., 782.*

Fleshy, erect, nearly leafless herbs, with very thick subterete or angular stems. The generic **Carallum** is said to be derived from a South Indian vernacular name.

Caralluma adscendens, *Br.; Fl. Br. Ind., IV., 76;* ASCLEPIADEÆ. 476

Vern.—*Culli mulayan,* TAM.

References.—*Murray, Pl. and Drugs, Sind, 162; Balfour, Cyclop.*

Habitat.—Met with in arid places in the Dekkan Peninsula.

Food.—This fleshy plant is often eaten by the Natives in the form of pickles, or is made into chutney. FOOD. 477

C. edulis, *Benth.; Fl. Br. Ind., IV., 76.* 478

Syn.—BOUCEROSIA EDULIS, *Edge.*

Vern.—*Chúng, chunga pippá, pippá, pípa, silún, sittú, súhi-gandhal,* PB.

References.—*Stewart, Pb. Pl., 144; Aitchison, Cat., Pb. Pl., 90; Murray, Pl. and Drugs, Sind, 162; Baden Powell, Pb. Pr., 264; Balfour, Cyclop.*

Habitat.—Found in the arid tracts of the Panjáb and Sind.

Food.—The stems have a sub-acid or bitterish taste, and are eaten by the poorer class of natives as a relish to farinaceous food. They are sometimes sold in the bazárs of the South Panjáb (*Coldstream*). FOOD 479

§ "Eaten as a vegetable in the Shahpore District and Salt Range, where it is called *pípa*" (*Brigade-Surgeon G. A. Watson, Allahabad*).

C. fimbriata, *Wall.; Fl. Br. Ind., IV., 77.* 480

MONKEY'S HORN.

Vern.—*Makar-sing,* BOMB.

References.—*Dalz. & Gibs., Bomb. Fl., 155; Voigt, Hort. Sub. Cal., 535; Lisboa, U. Pl., Bomb., 165.*

Habitat.—Met with in arid rocky places of the Dekkan Peninsula, from the Konkan southwards, and also in the Ava district of Burma.

Food.—In the Bombay Presidency the plant is eaten as a vegetable. FOOD. 481

Carambola. See **Averrhoa Carambola,** *Linn.,* GERANIACEÆ.

CARAPA, *Aubl.; Gen. Pl., 338.*

Carapa moluccensis, *Lam.; Fl. Br. Ind., I., 567; Bedd., Fl. Sylv., t. 136;* MELIACEÆ. 482

Syn.—C. OBOVATA, *Bl.* (*Kurz, i., 226*); XYLOCARPUS GRANATUM, *Kœn.*

Vern.—*Poshúr, pussur,* BENG.; *Kandalanga,* TAM.; *Pinlayoung, pinlón, peng-lay-oang,* BURM.; *Kadol,* SING.

References.—*Roxb., Fl. Ind., Ed. C.B.C., 319; Gamble, Man. Timb., 74; Kurz, For. Fl., Burm., 226; Thwaites, En. Ceylon Pl., 61; Pharm. Ind., 56; Moodeen Sheriff, Supp. Pharm. Ind., 260; Cooke, Oils and Oilseeds, 10.*

Habitat.—A moderate-sized evergreen tree of the coasts of Bengal, Malabar, Burma, and Ceylon.

Gum.—It yields a clear, brown, brittle resin. GUM. 483

Oil.—The seeds yield, on expression, a whitish semi-solid fat. This remains fluid only at high temperatures. It is used as a hair-oil, and also for burning purposes. OIL. 484

MEDICINE. Bark. 485 **Medicine.**—"The bark, in common with other parts of the tree, possesses extreme bitterness, conjoined with astringency; it may probably prove a good astringent tonic. It is much employed by the Malays in cholera, colic, diarrhœa, and other abdominal affections" (*Pharm. Ind.*)

TIMBER. 486 **Structure of the Wood.**—White, turning red on exposure, hard. Weight about 45 to 50℔ per cubic foot.

Used in Burma for house-posts, handles of tools, and wheel-spokes. **Captain Baker**, in May 1829, in *Gleanings in Science*, spoke of *Pussuf* or *Pussúah* as being a jungle wood of a deep purple colour, extremely brittle and liable to warp. He said that native boats made of the best species last about three years, and that the wood, if of good quality, stands brackish water better than *sál*.

Caraway. See **Carum Carui**, *Linn.*; UMBELLIFERÆ.

487

CARBON.

Carbon.

Vern.—*Kóyelah*, HIND.; *Kóyalá*, BENG.; *Tsúing, tsuna*, KASHMIR; *Sallah*, BHOTE; *Kólasé*, MAR.; *Kóelo, kólso*, GUJ.; *Kólsá*, DUK.; *Kari*, TAM.; *Boggu*, TEL.; *Kari*, MAL.; *Iddallu*, KAN.; *Angáraha*, SANS.; *Zughál*, PERS.; *Fahm, or Faham*, ARAB.; *Anguru*, SING.; *Misu-e, midu-ye*, BURM.

References.—*Pharm. Ind.*, 289; *Moodeen Sheriff, Supp. Pharm. Ind.*, 87; *U. S. Dispens.*, 15th Ed., 351; *Baden Powell, Pb. Prod.*, 608-9; *Ure, Dict. of Arts and Manufactures*, 720.

MEDICINE. 488 **Medicine.**—Wood charcoal is antiseptic, deodorizing, and disinfectant. It has been employed successfully in dyspepsia, diarrhœa, dysentery, and intermittent fevers. It is also used as a dentifrice. Animal charcoal is deodorizing and antiseptic. It has been employed as an antidote in poisoning cases and as a poultice to foul swellings and ulcers.

Special Opinions.—§ "In place of animal charcoal, wood charcoal has been largely used in hospitals as a disinfectant. It purifies water and may be used in filters for that purpose" (*Assistant Surgeon Shib Chunder Bhattacharji, Chanda, Central Provinces*). "The charcoal of Areca nut is a good tooth-powder" (*V. Ummegudien, Mettapollium, Madras*). "Fine powder, with syrup or treacle, useful in sloughing dysentery" (*Surgeon-Major C. J. McKenna, Cawnpore*). "Animal charcoal is a blood-purifier, and as such is of great value in boils." (*Surgeon-Major A. S. G. Jayakar, Muskat, Arabia*). "Wood charcoal mixed with oil is used by carpenters as an external application for wounds" (*Assistant Surgeon Bhagwan Dass, Civil Hospital, Rawal Pindi, Panjáb*). "Used to stop bleeding from wounds" (*Honorary Surgeon P. Kinsley, Chicacole, Ganjam District, Madras Presidency*).

For further information see **Charcoal.**

489

CARBONATE OF LIME.

Carbonate of Lime.

CARBONATE OF LIME, MARBLE, LIMESTONE, CHALK, and LIME.

Vern.—LIME.—*Chúná, chúnah, chunnah*, HIND.; *Chún, chúná*, BENG.; *Chúnah, áhak*, (quicklime) *kalai* (slaked) PB.; *Chúno*, GUJ.; *Chúnná, káli chúna*, MAR.; *Chúnak, chúnnah*, DUK.; *Chúnámbú, shúnnámbu*, TAM.; *Súnnam, súnna*, TEL.; *Capúr, núra*, MALYAL; *Sunná*, KAN.;

Sudhá, chúrna, sankha-bhasm, kapardaka-bhasma, sukti-bhasma, sambuka-bhasma, SANS.; *Kils, ahú,* ARAB.; *Núrah, áhak,* PERS.; *Húnnú, hunu,* SING.; *Thón-phiyu,* BURM.; *Kapor,* MALAY.

LIMESTONE and MARBLE.—*Kalai-ka-pattar, safaid pattar, sang-i-marmar* (white marble), *shah maksadi, kankar, bájri, kothár, kálásar, chunaka-pattar, sarma safaid* (Iceland spar), *karya matti, sang-yahúdi* (Jew's stone), *sangi-khurús* (a fossil encrinite), *sang-i-irmali,* (a fossil nummulite), *hajr-ul-ya húdi, sang-i-sarmai, san-i-shadauj, sangcha, tabá-khir,* PB.

CHALK.—*Khari-mitti,* HIND., PB.; *Khari-máti,* BENG.; *Viláyati-chuna,* MAR.; *Chák, viláti-chunó,* GUJ.; *Viláyati-chunná,* DUK.; *Shimaa, shannámbu,* TAM.; *Shíma-sunnum,* TEL.; *Shimanúra,* MALAY; *Shima-sunná,* KAN.; *Ratauhunu,* SING.; *Mie-phéaú* or *me-biyu, thombiyu,* BURM.

UNSLAKED LIME.—*Kali ká-chúna,* HIND.; *Kar-shunnambu,* TAM.; *Ralla sunnamu,* TEL.

References.—*Page, Hand-book of Geology, &c.; Dana, Manual of Mineralogy, 112, 363; Ball, Geology of India, 455-472; Miller, Chemistry, pt. 2, 460-466; Pharm. Ind., 335, 336; Moodeen Sheriff, Supp. Pharm. Ind., 83, 117, Ainslie, Mat. Ind., I., 195; U. C. Dutt, Mat. Med. Hind., 82; U. S. Dispens., 15th Ed., 322, 853, 865; S. Arjun, Bom. Drugs, 163; Waring, Bazar Med., 85-89; Baden Powell, Pb. Prod., 36, 37, 39-40, 41, 99; Atkinson, Him. Dist., 294; Watt, The Art of Soap-making, 116; Spons' Encyclop., 232, 360, 459-60, 1460, 1529, 1848-49, 1864-65, 1930-3, 1969; Balfour, Cyclop.; Morton, Cyclop. Agri.; Ure, Dic. Indust., Arts and Manuf., II., 871-74; Johnson, How Crops Grow, 126, 172; Mallet, in the Man. of Geol. of India, Pt. IV., 149, (1887.)*

The further Bibliography of Lime, Limestone, Marble, and Kankar will be found in Ball's Economic Geology, pp. 625, 627.

The Minerals of India having been treated in considerable detail in Mr. Ball's "Economic Geology" and in the other voluminous publications of the Geological Survey, it is not intended to do more in this work than to indicate briefly the minerals of commercial value. Limestone, Lime, and Marble are, however, of such importance as to justify an account being given, the more so since the literature of these substances is scattered and not readily obtainable. Lime is also intimately associated with many industries, and plays a distinct part in the manufactures which fall fairly within the scope of the present work. It has therefore been thought desirable to give a brief abstract of the available information regarding Lime, Limestone, and Marble. See MARBLE.

I. The term MARBLE should be restricted to limestone capable of receiving a polish. The two unicoloured conditions are white marbles and black marbles, but streaked and parti-coloured marbles are frequent. The colouring is derived from accidental minerals, or from metallic oxides producing the colouring and veining, and from the presence of imbedded shells, corals, or other organisms (See **Marble**). Marble.

II. The quality or richness of a LIMESTONE is generally perceptible to the eye, but when this is not the case, it may be detected by the violence of the effervescence produced on the application of a little sulphuric or muriatic acid, or by heating a fragment before the blow-pipe so as to convert it into quicklime. Limestone.

III. CHALK is a soft and earthy-looking limestone, or a "soft amorphous carbonate of lime, which can be converted into quicklime by calcination and used for all the purposes of ordinary limestone" (*Page*). It is composed of 44 parts of carbonic acid and 56 parts of lime. Pure chalk dissolves readily in dilute muriatic acid, and gives no precipitate with the addition of ammonia water. Chalk.

Lime. IV. LIME is an oxide of the metal Calcium. It is known as quicklime before being slaked with water; the expression "quicklime" is in allusion to its corrosive property. It is literally Calcic Oxide (Ca O) or CARBONATE OF LIME deprived of its carbonic acid. On being slaked it is converted into CALCIC HYDRATE (CaH_2O_2), which on being mixed with sand forms mortar or cement. "As an earth, lime is properly disseminated in nature; as a rock, it enters largely into the composition of the earth's crust; it is less or more diffused in all its waters; it forms the principal ingredient (earth of bone) in the skeletons of the larger animals; and is secreted by many classes of the invertebratæ to form their shells, crusts, shields, corals, and other means of protection. Economically it is also of vast importance, being used in the manufacture of mortars and cements, in tanning, bleaching, deodorising, and the like, and also in agriculture as a fertiliser or promoter of vegetable decays" (*Page*).

FORMS OF LIME USED IN INDIA.

There are three kinds of lime used in India: (*a*) lime prepared from limestone, (*b*) lime found on the surface of the ground and known as ***kankar***, and (*c*) lime prepared from fresh-water or marine shells.

LIMESTONE. 490 (*a*) LIME FROM LIMESTONE.

Speaking of the distribution of limestone and marble, **Mr. Ball** in his "Economic Geology" says: "Limestones can hardly be said to be absent from any of the formations in India, though in some they are either rare or so impure as hardly to deserve the title. In the metamorphic series, bands of crystalline limestones occur locally in some abundance, but they are capriciously distributed, being often absent over large areas. In some of the groups of the next succeeding or transition series, namely, in the Kadapah, Bijáwar, and Arvali, the limestones attain a considerable development, and some of the varieties have yielded the marbles which have played such an important part in Indian architecture. In the lower Vindhyan series the limestones are more notable for their abundance, and the wide areas over which they spread, than for producing any marbles of particular beauty. In the upper Vindhyans, limestones are principally found in the Bhanrer group, where they sometimes attain as great a thickness as 260 feet, and are used both as a building stone and for lime.

"In the Gondwána series, limestones are rarely met with, and then chiefly in the Talchir and Ràniganj groups, where they occur as lenticular or concretionary masses.

"In the rocks of cretaceous age, within the peninsula, limestones of both sedimentary and coral reef origin occur. The other sources of lime are principally sub-recent and recent tufaceous deposits of ***kankar***, travertin, &c.

"In the extra-peninsular regions the principal formations containing limestones are of carboniferous, jurrassic, cretaceous, and nummulitic ages. Another source of lime . . . . is recent coral. On the whole it may be said that, although lime is a dear commodity at most of the centres of consumption owing to the cost of carriage, possible sources of lime occur in the greatest variety throughout the country, while, on the other hand, some of the marbles are probably unsurpassed for beauty by any to be obtained in any other part of the world."

Mr. Ball further gives, in the succeeding 16 pages of his work, a detailed account of the limestones and marbles, arranged according to provinces. The following abstract may be found useful:—

491 *In Madras*, good limestones and marbles occur at Trichinopoly, Coimbatore, Kadapah, Karnul, and Guntur. These, since the opening

of the railways, have largely replaced the *kankar* formerly employed for building purposes in the Presidency.

492 *In Bengal,* although not so valuable as the limestones of other parts of India, workable stones are found in the Mánbhum district, where the supplies are practically inexhaustible; also in Singhbhum, Hazáribágh, and Lohardaga. In the last-mentioned districts the limestones have a peculiar interest because of their proximity to iron ore.

493 *In the Central Provinces,* limestones occur at Sambalpur, Raipur, and Jabalpur, the latter consisting of the famous marble rocks of that name. Limestones also occur throughout the Vindhya range, the most accessible being in the neighbourhood of Warora. At Raipur a stone suitable for lithography has been found.

494 *In Kutch,* limestones of different ages are met with, but those most esteemed belong to the lower Jurassic group.

495 *In Southern Afghánistán,* limestones of cretaceous age abound, and in *Baluchistán* nummulitic limestones are found in the eastern frontier as well as in Northern Afghánistán. In the latter the Safed Sang takes its name from a beautiful Statuary marble.

496 *In the Panjáb,* marbles and limestones in considerable variety and from different geological formations are met with.

497 *In the North-West Provinces* and along the Tarái to Darjíling, limestones are not infrequent. An account of these may be found in **Atkinson's** *Economic Geology of the Hill Districts,* and in a paper by **Mr. Mallet** on the Geology of the Darjíling District and the Western Dúars. Speaking of the lime prepared from limestone, **Mr. Atkinson,** in his *Himálayan Districts, p. 295,* says: "Lime is manufactured at Naini Tal, at Jyúli in the Kharáhi range, half way between Bágeswar and Almora, at Chiteli, north of Dwárahát, at Simalkha, Baitálghát, and Dhikuli for Ránikhet, and on the new cart road to Rámnagar. Lime is also made in Borarau, Sor, Síra, Dhyánirao, and Charál. Two kinds of limestone are used in the Tarái district, the one being obtained from the quarries at the foot of the Kumáon hills, which give by far the best kind of lime; the other is the tufa deposit obtained in the small *nálas* of the tract itself: this latter kind, however, is of a very inferior quality. First-class limestone costs at the quarries R5 to R8 per 100 maunds; the tax levied by the Forest Department is R8 on that amount, and cartage may be averaged at half a rupee per mile for a 100 maunds. Thus the stone is landed at most points in the district for R30 per 100 maunds, and including the expense of burning, a maund of lime costs 10 to 12 annas. This lime will bear two or three portions of pounded brick or *súrkí*. Second-class lime ready for use now costs R25, and delivered in Naini Tal R50 to R100 per 100 maunds; it will, however, only bear a proportion of one part of pounded brick to two parts of lime."

498 *In Central India,* at Gwalior an abundant supply of flaggy limestones occurs.

499 *In Rájputana,* the Arvali group of transition rocks includes many varieties of marble, some of them being of great beauty. The Jhirri quarries of Alwar afford hard white marble. Black marble is met with at Mandla, near Rámghur; white as well as pink and grey marbles at Raialo in Taipur. But the most extensive marble quarries of Rájputana are at Makrana in Jodhpur. This marble has been celebrated for ages, the Táj of Agra being built of it.

500 *In Bombay,* there are numerous localities where limestone occurs, but no marble. In the Panch Mehála, good building limestones are obtained, but not hydraulic, and in Guzerat more or less calcareous rocks are met with.

LIMESTONE.

501 *In Assam,* in the Brahmaputra Valley, nummulitic limestones occur at several localities, the southern face of the Khásia and Jaintya Hills affording an inexhaustible source of supply, known in trade as Sylhet lime.

502 *In Burma,* nummulitic limestones occur in Arracan and Pegu, and in Tenasserim true carboniferous limestones are met with. In Upper Burma a beautiful white semi-transparent marble, extensively used for carving figures of *Gondama,* is said to be obtained from the hills in the Madeya district.

503 *In the Andaman Islands,* an important supply of lime, for Calcutta, is afforded by the coral reefs.

The writer has been favoured, by **Mr. H. B. Medlicott,** with the following brief account of the important commercial limestones of India :—

Lime is a scarce article in many parts of India. Much of the lime used in Calcutta is carried many hundred miles by river and railway. The want of a pure limestone flux at moderate cost has been the chief difficulty in working the iron furnaces in the Raniganj coal-field. The most general source of building lime in India is *kankar* or *kunkur* (meaning gravel), a granular or nodular stone found on the surface and in the sub-soil. It is purely of secondary origin, being formed on the spot by the evaporation of the ground-water, containing in solution more or less of carbonate of lime produced in the slow process of soil formation by the gradual decomposition of rock-particles. The production of it is very much a matter of climate, by alternating periods of soaking moisture and extreme dryness. Where this latter condition is most pronounced, as in North-Western India, the lumps of kankar often coalesce into a continuous mass, fit for use as building stone. A stone so formed must of course be impure and variable in quality, as to the quantity and nature of foreign matter according to the texture and composition of the bed in which the concentration of the lime is effected, but when these are favourable an excellent hydraulic lime is the result.

" LIMESTONE is, however, of widespread occurrence throughout British India, but as a rule the available deposits are at a distance from the centres of demand, and consequently the price of lime rules high. The most important sources, commercially, are :—

504 " *1st, Katni,* in the Jabalpur district; supplies a lime of excellent quality, which is carried as far as Calcutta (737 miles distant), and forms a large proportion of the lime used in that city.

505 " *2nd, Sylhet.*—Along the southern foot of the Sylhet hills there is an inexhaustible supply of lime in the limestones of the nummulitic series, which formerly supplied the whole of the demand of Calcutta and lower Bengal, and still does so to a large extent.

506 " *3rd, Rhotasgarh.*—The lower Vindhyan limestone near Rhotasgarh is quarried to a small extent and exported down the Són in boats; it was largely used in the works on the Són Canal.

507 " *4th, Himálayas.*—Along the foot of the Himálaya, boulders of limestone are collected and burnt in large quantities every year; the slaked lime is exported on camels, and supplies a large portion of the Panjáb.

508 " *5th, Andamans.*—There is a band of cream-coloured marble near Port Blair which may prove of economic importance, as it is at about the same distance from Calcutta as Katni, and the lime is of equally good quality.

" Other localities where limestone is known are numerous, but at present of merely local importance, or in most cases of no value whatever. A full list of them, as far as they are known, will be found in the *Manual of the Geology of India,* Vol. III., p. 449, *et seq.*"

KANKAR. 509

(*b*) KANKAR OR CONCRETIONARY LIME.

KANKAR (KUNKUR).—"Throughout the plains of Upper India the principal source of lime is the *kankar* which is found in nodules and layers of various sizes in the clays of the Gangetic alluvium. It yields an excellent but somewhat hydraulic lime" (*H. B. Medlicott. See also the remarks under Limestone.*)

"By Anglo-Indians the term '*kankar*' (which really means any kind of gravel) has been specially used for concretionary carbonate of lime, usually occurring in nodules, in the alluvial deposits of the country, and especially in the older of these formations. The commonest form consists of small nodules of irregular shape, from half an inch to 3 or 4 inches in diameter, and composed within of tolerably compact carbonate of lime, and externally of a mixture of carbonate of lime and clay. The more massive forms are a variety of calcareous tufa, which sometimes forms thick beds in the alluvium, and frequently fills cracks in the alluvial deposits or in older rocks.

"In the beds of streams immense masses of calcareous tufa are often found, forming the matrix of a conglomerate, of which the pebbles are derived from the rocks brought down by the stream. There can be no doubt that the *kankar* nodules, calcareous beds and veins, are all deposited from water containing in solution carbonate of lime derived either from the older rocks of various kinds, or else from fragments of limestone and calcareous formations contained in the alluvium" (*Blanford*).

510

"As a flux for iron, *kankar* has been tried on several occasions, and opinions are somewhat divided as to its applicability to the purpose; but owing to the uncertainty of its composition, it is distinctly less well adapted than rock-limestones which have a well-defined average composition, even though in the latter the proportion of carbonate of lime may average something less.

"Block *kankar* has been largely employed as a building-stone, more particularly in connection with the Ganges Canal Works" (*Ball*).

Most of the roads in Northern India, and indeed in India generally, are metalled with *kankar*.

(*c*) SHELL-LIME.

SHELL-LIME. 511

SHELLS.—**Ainslie**, in his *Materia Indica*, mentions lime produced by burning the sea-shells, called in Tamil *kullingie chunambu*. **Dr. U. C. Dutt**, speaking of the lime used in Hindú medicine, says: "We have lime from limestone called *churna*. Then we have lime from calcined cowries, conch-shells, bivalve-shells, and snail-shells, called respectively, *Kapardaka bhasma, Sankha bhasma, Sukti bhasma*, and *Sambuka bhasma*." A large quantity of this lime is annually produced by burning shells found in the marshes of the interior of Bengal, and also on the coast, especially near the Sunderbans. These shells are conveyed by country boats in enormous quantities. Two kinds of lime are produced: the *Sámbuka* from thin land-shells is regarded as inferior in quality, and is accordingly used by the poor for building purposes; the *Gora* or *Jamrál* (or thick shells, chiefly of marine origin) afford a superior quality. The latter "is considered more valuable for building purposes than that obtained from limestone, and fetches a higher price" (*T. N. Mukharji, Amsterd. Cat.*) Formerly, when stone-lime was not much in use, the former was employed by all classes of people as a building material.

The following account of shell lime was communicated to the Agri.-Horticultural Society of India by **Dr. A. Grant**, of the Bengal Medical Service: "Lime is procured at all the ports of the east coast of China

SHELL-LIME.

that I have visited by burning the shells of the genus OSTREA, which abound on the neighbouring shores. It is too valuable to be much used in agriculture, but the Chinese are not ignorant of its uses."

Dr. Anderson mentions (1883) the following shells as used in Murshidabad for making lime: **Unio marginalis, U. flavidens,** and **Ampularia globosa.**

LIME ESSENTIAL TO VEGETATION.

AGRICULTURAL USES. 512

Lime is invariably present in the ash of all agricultural plants. It is, however, difficult to decide from this fact alone, whether it is indispensable to vegetable life, since the substances found in ash are universally distributed over the earth's surface and are invariably present in all soils. Several experiments have been made by scientific men under various circumstances to establish fully the above facts, with results to a certain extent satisfactory. For further information on this subject the reader is referred to Johnson's *How Crops Grow*, pp. 166-172.

INDUSTRIAL PURPOSES.

INDUSTRIAL USES.

Dye.—Lime is universally used by the Mánipuris to assist in the transformation of green into blue indigo and to deepen the blue colour of indigo; and a small piece placed in the mouth of a vessel containing indigo is also supposed to preserve the dye. (See **Strobilanthes.**) Lime is employed in the Rajshahye district for dyeing thread dark blue; of this **Dr. McCann** gives the following account: "The thread is first washed

Dye adjunct. 513

with *sajjí matí* and dried. Then 4 *chittacks* of *koli chuna* (shell-lime), ½ seer of *patta sajjí matí*, 4 *chittacks* of *aoosh* wood **(Morinda tinctoria)**, 2 seers of cold water, are mixed together and allowed to stand for two or three hours. The solution is then decanted so as to separate it from the deposit, and 2 *chittacks* of indigo are rubbed down in it, and the whole is then put in a jar in which a little old indigo dye remains: 4 *chittacks* lime and 2 *chittacks* of *aoosh* wood are again added to this solution. The thread is then twice dipped in this solution and dried. It is again twice dipped and dried on the following day, when it acquires a permanent dark-blue colour. A small portion of the dye is always kept as a soft leaven or permanent to mix with new dyes." **Mr. (now Sir Edward) Buck**, in his *Dyes and Tans of the North-West Provinces*, gives a preparation of blue printing ink of permanent colour. A mixture of 4℔ of shell-lime, 10℔ of stone-lime, and 15℔ of impure carbonate of soda (*reh*), with 3 gallons of water, is

Calico-printing. 514

strained through grass; to this is added 1℔ of sulphurate of arsenic and 1℔ of indigo; the mixture is then boiled "till it assumes the metallic greenish-blue lustre of the peacock's tail. It is then thickened with *babul* gum and is then ready for printing." **Sir Edward** further remarks: "Lime is used in calico-printing, in combination with gum, as a 'resist-paste' It is also employed with sugar to excite fermentation in indigo and convert it into 'indigo-white,' in the presence of hydrogen."

A paint. 515

Carbonate of lime is used as an oil paint for in-door work, and as a water colour when mixed with gelatine. It is sometimes employed to dilute other pigments. Prepared chalk is used in chemical manufactories for the purpose of saturating acid liquids (*Crooke's Handbook of Dying*, *150*).

Tanning. 516

Tan.—As a preliminary in the process of tanning, lime is applied to hides for the removal of the hair. In England it is universally used for this purpose. It has at the same time a solvent action on the hide "The hardened cells of the epidermis swell up and soften the *rete malpighi*, and the hair-sheaths are loosened and dissolved, so that, on scraping with a blunt knife, both come away more or less completely with hair" (*Spons' Encycl., II., 1221*).

MEDICINAL USES.

Medicine.—According to **Dutt**, in the *Hindú Materia Medica* (*p. 82*) lime is used internally in dyspepsia, enlarged spleen, and other enlargements in the abdomen, and externally as a caustic. A mixture of lime, carbonate of soda, sulphate of copper and borax, is applied as a caustic to tumours and warts. It enters into the composition of several prescriptions for different forms of dyspepsia, such as *Amrita vati* and *Agnikumára rasa.* MEDICINE. 517

Ainslie says the Vytians prescribe lime water mixed with gingelly oil and sugar in obstinate cases of gonorrhœa. "Mixed with gamboge, quicklime is applied externally to painful and gouty limbs. It is also used as a caustic in the bites of rabid dogs" (*S. Arjun, Bomb. Drugs*). The exhaustive account of the medicinal properties of lime given by **Dr. Waring** in his *Bazár Medicines* (*p. 85*) may be here quoted, since by doing so it will practically be unnecessary to refer to other authors:— 518

"Lime in a medical point of view is of great importance as the basis of lime water, a mild and useful antacid; it is prepared by adding two ounces of slaked lime to one gallon of water in a stoppered bottle, shaking well for two or three minutes, and then allowing it to stand till the lime is deposited at the bottom. In cases of emergency, as burns, &c., half an hour is sufficient for this purpose; otherwise it should be allowed to stand for twelve hours at least before being used. It is only the clear water which holds a portion of lime in solution, which is employed in medicine. It is advisable always to keep a supply ready prepared, as it is useful in many ways, and it will remain good for a long time, if kept in well-stoppered bottles, so that the air cannot have access to it. The dose for adults is 1 to 3 ounces twice or thrice daily; it is best administered in milk.

"Another form, called saccharated solution of lime, thought to be better adapted for internal use in the diseases of childhood and infancy, is prepared by carefully mixing together in a mortar one ounce of slaked lime and two ounces of powdered white sugar, and adding to this a pint of water, as described above. It should be kept in a well-stoppered bottle. The dose of the clear water is from 15 to 20 drops or minims in milk, twice or thrice daily. 519

"*In acidity of the stomach, in heart-burn, and in those forms of indigestion arising from or connected with acidity of the stomach,* lime water in doses of $1\frac{1}{2}$ to 2 ounces, is often speedily and permanently effectual. It is particularly useful in indigestion when the urine is scanty and high coloured, and when vomiting and acid eructations are prominent symptoms. It is best given in milk.

"*In diarrhœa arising from acidity,* lime water frequently proves useful; it is best given in a solution of gum arabic or other mucilage, and in obstinate cases 10 drops of laudanum with each dose increase its efficacy; it may also be advantageously combined with Omun water. *In chronic dysentery* the same treatment sometimes proves useful. Enemas of lime-water diluted with an equal part of tepid milk or mucilage have also been used with benefit. It is especially adapted for the *diarrhœa and vomiting of infants and young children which result from artificial feeding*; in these cases a sixth or a fourth part of lime water may be added to each pint of milk. The saccharated solution of lime has also been found of great service in this class of cases.

"*Obstinate vomiting* sometimes yields to a few doses of lime water in milk, when other more powerful remedies have failed. It is worthy of a trial in the *vomiting attendant on the advanced stages of fever*; it has

MEDICINE. been thought to arrest even the black vomit of yellow fever. It is also a remedy of much value in *pyrosis or water-brash.*

"*To relieve the distressing irritation of the genital organs* (*Pruritus Pudendi*), bathing the parts well with tepid lime water three or four times a day sometimes affords much relief. *Leucorrhœa and other vaginal discharges* have in some instances been mitigated and even cured by the use of vaginal injections of a mixture of 1 part of lime water and 2 or 3 of water.

520 "*In scrofula,* lime water in doses of ½ ounce in milk, three or four times a day, proves beneficial in some cases; it is thought to be especially adapted for those cases in which abscesses and ulcers are continually forming. To be of service, it requires to be persevered in for some time. *Scrofulous and other ulcers attended by much discharge* have been found to improve under the use of lime water as a local application. For *syphilitic ulcers* or *chancres,* one of the best applications is a mixture of lime water ½ pint and calomel 30 grains; this, commonly known as black wash, should be kept constantly applied to the part by means of a piece of lint or rag moistened with it. Many forms of *skin disease,* attended with much secretion and with great irritation or burning, are benefited by lime water either pure or conjoined with oil. *To sore or cracked nipples* it proves very serviceable. Diluted with an equal part of water or milk, it forms a useful injection in *discharges from the nose and ears* occurring in scrofulous and other children.

"*In Consumption,* lime water and milk has been strongly recommended as an ordinary beverage. The same diet-drink has been advised in *Diabetes,* but little dependence is to be placed upon it as a *cure;* it may produce temporary benefit.

"*In Thread-worm,* enemas of 3 or 4 ounces of lime water, repeated two or three times, have sometimes been found sufficient to effect a cure.

"*In Poisoning by any of the Mineral Acids,* lime water given plentifully in milk is an antidote of no mean value, though inferior to some of the other alkalies. It may also be given in *Poisoning by Arsenic.*

521 "*To Burns and Scalds,* few applications are superior to Lime Liniment, composed of equal parts of lime water and a bland oil. Olive oil is generally ordered for this purpose, but linseed oil answers just as well, and where this is not at hand Sesamum oil forms a perfect substitute. When thoroughly shaken together, so as to form a uniform mixture, it should be applied freely over the whole of the burnt surface, and the parts kept covered with rags constantly wetted with it, for some days if necessary. This liniment on cotton-wool, applied to the pustules, is said to be effectual in preventing *Pitting in Small-pox.*"

LIME AS A CONDIMENT.

FOOD. In pan. 522 **Food.**—Lime forms one of the essential ingredients of the preparation known as *pán* which is universally chewed by the natives of India. Either the lime prepared from limestone or from shells may be used for this purpose. The latter, however, being an animal product, is not used by persons who are strict in their religious observances. It is also mixed with the pulp of the fruit of **Borassus flabelliformis,** in preparing the cake called *talpatali* (*see the remarks under* **B. 901**). The *Pharmacopæia of India,*

523 alluding to the use of lime in *pán,* says, "when used for any lengthened period, it considerably modifies the natural condition of the mucous covering of the mouth, and alters the appearance of the tongue so as to render it useless or fallacious as a means of diagnosis in disease. Its use in moderate quantities does not appear to act prejudicially on the system, but when largely indulged in, it lays the foundation of much visceral disease."

DOMESTIC AND OTHER USES.

DOMESTIC. Manure. 524

Manure.—As a manure, lime plays an important part. It is largely employed for this purpose, and is "particularly valuable upon very rich vegetable soils, such as those formed over peat bogs; its effects in these cases are partially due to the decomposition of the organic matter, which it renders soluble and capable of assimilation, while the lime itself is converted into carbonate" (*Miller's Chemistry, Part II., 466*). The black cotton soils are usually rich in most of the elements of plant food except lime. Lime therefore "acts beneficially on the soil itself. Owing to the general absence of lime in these black soils, the crops produced on them are not so diversified as is desirable. A dressing from 1,000 to 5,000℔ of lime may be applied per acre, according to the price at which the lime can be obtained" (*W. R. Robertson, Agriculture, 13*).

Lime is often employed as a deodorising agent. "It is mixed with decaying vegetable matter, and with animal bodies, with the view of hastening their destruction and preventing the escape of offensive and noxious effluvia. This effect lime produces by its tendency, in common with the other caustic alkalies, to carry the decomposition through the intermediate stages of putrefaction at once to the ultimate products" (*Morton, Cyclop., Agriculture, Vol. II., 266*).

Soap. 525

Soap.—Lime is used in preparing soap according to **Lunge's** method, which is described thus: "A flat-bottomed pan is preferred for making this soap, into which is introduced any given quantity of water and slaked lime equal to 12 per cent. of the weight of fatty matter. The whole is to be boiled and stirred when an insoluble hard lime soap and a solution of glycerine are produced, when the latter may be drawn off from the bottom of the pan. A certain quantity of water and commercial carbonate of soda (the latter being slightly in excess of the quantity of lime used) are next added, and the boiling and stirring continued, when the hard insoluble lime soap will be decomposed, and a 'granulated' carbonate of lime will deposit, leaving a soluble soda soap floating in flakes on the surface of the liquid. If the soda employed does not contain sufficient salt, a suitable quantity of sea-salt is to be added to promote the separation" (*Watt, The Art of Soap-making, 116*). Lime is also of great importance in CANDLE-MAKING. See the author's pamphlet on *Candles*.

Cement. 526

Mortar and Cement.—The use of lime in the preparation of mortars and cements is too well known to require any special description. The following paragraph from *Miller's Chemistry, Part II., 462*, is, however, quoted here, as it will be found instructive: "The great consumption of lime in the arts is for the purpose of making mortars and cements. Pure lime, when made into a paste with water, forms a somewhat plastic mass which sets into a solid as it dries, but gradually cracks and falls to pieces. It does not possess sufficient cohesion to be used alone as a mortar; to remedy this defect and to prevent the shrinking of the mass, the addition of sand is found to be necessary. Ordinary mortar is prepared by mixing one part of lime into a thin paste with water, and adding 3 or 4 parts of sharp sand of tolerable fineness; the materials are then thoroughly incorporated, and passed through a sieve to separate lumps of imperfectly burnt lime; a suitable quantity of water is afterwards worked into it, and it is then applied in a thin layer to the surfaces of the stones and bricks which are to be united. The bricks or stones are moistened with water before applying the mortar, in order that they may not absorb the water from the mortar too rapidly. The completeness of the subsequent hardening of the mortar depends mainly upon the thorough intermixture of the lime and sand."

In India instead of sand pulverised bricks are employed, a substance found more adapted than sand to the peculiarities of Indian life. This is known as *surkhi*. The industry of *surkhi*-making has passed into the hands of the natives who, instead of old fashioned pounders worked by the feet, now employ for *surkhi* grinding steam power to drive heavy rollers which work in a strong iron basin. For further information see **Cement**.

527 Carbonate of Potash.

POTASHES, PEARL-ASH; CARBONATE DE POTASSE, *Fr.*; KOHLENSÄURES KALI, *Germ.*

Vern.—*Sarjika*, BENG.; *Jon-khár, ivak-chhár* or *ouk-chhár*, HIND.; *Jhár-ká-namak, rák-ká-namak*, DUK.; *Jhádicha-mítha*, MAR.; *Mara-vuppu, shámbal-vuppu*, TAM.; *Mánu-vuppu, búdide-vuppu*, TEL.; *Káram, pappatak-káram, mara-uppa*, MALA.; *Marada-uppu*, KAN.; *Dáru-lavanam, yavakshára*, SANS.

References.—*Pharm. Ind.*, 311; *Modeen Sheriff, Supp. Pharm. Ind.*, 205; *Treatise on Chemistry by Roscoe and Schorlemmer, Vol. II.*, 91; *U. S. Dispens.*, 751, 1156; *Ure, Dict. of Arts and Manufactures*, 449; *Spons' Encyclop.*, p. 253; *Balfour's Cyclop.*

Potashes. 528 Pearl-ash. 529 (Conf. with A. 759.) The mon-oxide of the metal Potassium is known commercially as *Potash* (K_2O); theoretically this, by combining with a molecule of carbon dioxide (CO_2), forms the carbonate of potash ($K_2 CO_3$). The term *potash* is, however, loosely applied to the oxide and to the carbonate, the latter being more correctly *potashes*, and when calcined *pearl-ash*. The carbonate is soluble in about its own weight of water, but its solubility increases with heat. It is a hard, white solid, with a strong alkaline reaction in taste. It rapidly absorbs moisture if exposed to the atmosphere, forming thereby a thick oily liquid known as *Oleum tartari per deliquium*. If subjected to dry heat it melts at 800°, but loses a portion of its carbonic acid: at still higher temperatures it volatilizes. Acids decompose it with brisk effervescence of Carbonic acid, leaving behind salts of the acid employed with potassium. A concentrated solution of the salt on cooling yields crystals of the carbonate in which three proportions of water have combined with two of the salt. At 130° the whole of this water of crystallization may be expelled and the anhydrous carbonate obtained.

Sources.—For many years the entire source of carbonate of potash was the ashes of plants, land and marine. Although new sources of supply have to a large extent diverted the industry, about one-half of the total European supply is still drawn from plants; but this is steadily diminishing, the extermination of forests operating greatly in favour of the modern sources. Of plants it may in general terms be said that herbaceous annuals contain more pearlash than woody arborescent plants, but even of the same plant the succulent young parts are more highly charged than mature tissues. Of different plants, pines contain on an average only 0·45 per cent., oaks 0·75 to 1·5 per cent., vine shoots 5·50, ordinary straw 5·8, ferns from 4·25 to 6·26, Indian-corn stalks 17·5, nettles 25·03, wheat straw before earing 47·0, wormwood 73·0, and beet about the same amount.

These facts naturally suggest the plants best suited for the preparation of pearlash, and the immense development within recent years of the beet sugar industry naturally awakened an interest in carbonate of potash as a bi-product that might supplement the returns from beet cultivation. This has been actually turned to account. "When the juice containing the sugar has been extracted from the roots, we have to deal with a solution which contains something besides sugar and water. After it has been

SOURCES OF

clarified and the crystallizable sugar extracted, the remaining liquor is permitted to ferment, that the uncrystallizable sugar may be turned into alcohol and so utilized; but in the stills there will yet remain a waste liquor, and it is in this that abundance of potash salts occur. By evaporating this liquor in a long trough divided across into an evaporating and a calcining section, a salt is finally obtained, consisting of a mixture of potassium chloride, sulphate, and carbonate (together 50 or 60 per cent.) with insoluble matter and a good deal of sodium carbonate. The potassium carbonate forms about one-third of the weight of the calcined mass, and arises in a great measure from the destruction, during the calcining process, of the potassium oxalate, tartrate, and nitrate which occur naturally in the beetroot, and, consequently, in the liquor from the still" (*Prof. Church in British Manuf. Ind.*). This instructive account of the extraction of carbonate of potash from the waste of beet-root has been reproduced here because of its direct bearing on many of the native contrivances employed in India for the preparation of pearlash. It would be almost impossible to over-estimate the extent to which a crude carbonate of potash is employed by the people of India. In another volume under **Alkaline Ashes** (**A. 759**, also **A. 1626**) will be found an enumeration of the principal plants used by the natives of India for that purpose, and these should be compared with the plants given under **Berilla** (**B. 163**) as employed in the manufacture of carbonate of soda. Although in India immense tracts of mountainous land are injuriously covered with various species of wormwood (see **Artemesia**), except as a manure, the ashes of these plants are not apparently utilized. From the high percentage of carbonate of potash which the wormwoods contain, the preparation of pearlash might be confidently recommended to the poorer inhabitants of these regions as a useful new industry. A large export trade might reasonably be anticipated from the Himálayas to the plains of India, if not to foreign countries.

Wormwood Ash. 530

While this is possible, an equally profitable industry might also be organised in preparing the carbonate from the injurious amount of saltpetre that impregnates the soil of many parts of India. One of the methods recommended for obtaining pure carbonate of soda for the laboratory is to heat pure saltpetre in a porcelain or earthen crucible, adding small pieces of charcoal till deflagration ceases. This is the *rationale* of a process that might readily be employed in converting crude saltpetre into carbonate of potash. As a commercial fact, large quantities of carbonate are now manufactured from the sulphate; indeed, after the ashes of plants, this is the next most important source of the carbonate. A curious and recent source is the Suint or perspiration on the wool of sheep.

The Carbonate from Saltpetre. 531

from the Sulphate 532

From Suint. 533

Uses of Carbonate of Potash.—It is largely employed in the manufacture of soft soap: it is practically the chief source from which the hydrate is obtained; it is employed in the manufacture of flint glass, the commoner kinds of glass being prepared with carbonate of soda or a mixture of both these alkalies. "It is also largely used for cleansing purposes, in turkey-red dyeing, for the emulsion of oil, and, in printing, as a solvent of annatto, &c." (*Spons'* p. 260). For "the rectification of spirit, bleaching, and in medicine and for other purposes" (*Balfour*).

Soft Soap. 534

Hydrate. 535

Flint-glass. 536

Used in Turkey-red Dyeing. 537

Rectification of Spirit. 538

Bleaching. 539

Manufacture in India.—Although, as already stated, the ashes of plants are universally used, both in dyeing and in medicine, throughout India, every district or almost each artisan holding special merits as possessed by the ashes of this and that plant, still there are no large recognised centres where the carbonate (which alone must be held as the active principle in these ashes) is prepared for transport, still less export. The suggestion made above as to a possible Indian manufacture from worm-

CARBONATE of POTASH.

wood on the hills and from saltpetre on the plains seems, therefore, worthy of consideration.

Yearly Production.—The world's annual production is about one million hundredweights.

MEDICINE.
540 **Medicine.**—Carbonate of potash is antacid, then alterative and diuretic, and in over-doses poisonous. It is described in Hindú works on medicine "as stomachic, laxative, diuretic. It is used in urinary diseases, dyspepsia, enlarged spleen, and other enlargements of the abdominal viscera. A decoction of chebulic myrobalan and *rohitaka* bark is given, with the addition of carbonate of potash and long pepper, in enlarged spleen and liver, and in tumours in the abdomen called *gulma*. In strangury or painful micturition, carbonate of potash with sugar is considered a very efficacious remedy" (*U. C. Dutt, Mat. Med. Hind.*, 87).

Special Opinions.—§ "An impure carbonate of potash (*pápáda khara*) is also sold in the Bombay bazárs, and is used in the preparation of *pápáda* (*pápun*), or little cakes made with the meal of the different sorts of *dhall* and a little quantity of asafœtida; these are given as a digestive, but more as an article of food than medicine; the cakes are roasted over the fire and taken with rice" (*C. T. Peters, M.B., Zandra, South Afghánistán*).

For further information see **ALKALINE EARTHS, BARILLA, POTASH, REH** and **SALTPETRE.**

541 ## Carbonate of Soda.

Vern.—*Sajji, sajji-mitti, sajji-khár*, HIND.; *Sájji*, BENG.; *Chour-ki-matti, chour-ká-namak*, DUK.; *Sajjekhára*, MAR.; *Shach-chi-káram*, TAM.; *Lotá-sach-chi*, TEL.; *Qili, milhul-qili*, ARAB.; *Shikhár, tine-gásur*, PERS.; *Sarjikákshára*, SANS.

References.—*Pharm. Ind., 322; S. Arjun, Bomb. Drugs, 160, 161; U. S. Dispens., 1321; Ure, Dict. of Arts and Manufactures, 854.*

MEDICINE.
542 **Medicine.**—A substance too well known to require any special description. (See remarks under the preceding and under **BARILLA, SAJJI,** and **REH.**) It is antacid and then alterative. "A paste made of equal parts of *yavakshára* and *sajji-kakshára* with water is applied to abscesses for the purpose of opening them" (*U. C. Dutt*).

Special Opinions.—§ "Carbonate of soda (impure), *bángada khára*, being the residue left during the manufacture of glass bangles. A second form, which appears to be a purer carbonate of soda, is called *Suráti khára*; both are used in the treatment of dyspepsia" (*C. T. Peters, M.B., Zandra, South Afghánistán*).

CARBUNCLE.

543 ## Carbuncle.

"The Carbuncle of the ancients is garnet cut, as it is called, *en cabuchon*. The art is still practised in India, and the stones, when of good quality and well cut, are very beautiful and would meet with more esteem were it not that they happen to be cheap, which has put them within the reach of so large a circle that they are made but little account of. It is be-

Calcutta.
544 lieved, however, that there is still a small trade in them from Calcutta" (*Ball, Econ. Geo.*, 522). **Dr. Balfour** says they are common in South

South India.
545 India, where they are known as *Maníkiam* (*Tam. & Tel.*).

Bombay.
546 The garnet when cut as a Carbuncle is convex above and hollowed out below, so as to leave but a thin layer of the stone through which the

Burma.
547 light passes, revealing the bright colour. The finest carbuncles are said to come from Pegu and Ceylon. Conf. with **Carnelian.**

CARCHARIAS, *Müller and Henle.; Day, Fishes of India, 710.* 548

Carcharias.—Several species of sharks are employed by the natives of India in the preparation of a medicinal oil. It seems probable that the sharks specially selected for that purpose belong to the genus **Carcharias.** Of these **C. gangeticus** is the most ferocious: it ascends the rivers to about the limits of the tidal influence. **C. hemiodon** also goes up the rivers, specimens having been caught near Calcutta. Several other species are frequent in the Red Sea and Indian Ocean, particularly on the coast of Sind. (See SHARKS AND SHARK FINS.)

CARDAMINE, *Linn.; Gen. Pl., I., 70.*

Cardamine hirsuta, *Linn.; Fl. Br. Ind., I., 138;* CRUCIFERÆ. 549

References.—*Thwaites, En. Ceylon Pl., 14; Dalz. & Gibs., Bomb. Fl., 7; Stewart, Pb. Pl., 13; Treasury of Botany.*

Habitat.—A herb found in all the temperate regions of India; very abundant in Bengal during the cold weather.

Food.—The leaves and flowers constitute an agreeable salad, resembling water-cress. FOOD. 550

Cardamom, see **Amomum subulatum,** *Roxb.*,—the Greater Cardamom; and **Elettaria Cardamomum,** *Maton*—the Lesser Cardamom.

Cardamom seed oil, see **Amomum subulatum,** *Roxb.*

CARDIOSPERMUM, *Linn.; Gen. Pl., I., 393.*

Cardiospermum Halicacabum, *Linn., Fl. Br. Ind., I., 670; Wight, Ic., t. 508;* SAPINDACEÆ. 551

BALLOON-VINE, HEART PEA OR WINTER CHERRY.

Vern. *Latáphatkari, nayáphatki, noaphutki, sibjhúl,* BENG.; *Hab-ul-kalkal* (seed), PB.; *Karolio,* GUJ.; *Kánphúti, bodha, shib-jal,* BOMB.; *Múda-cottan,* TAM.; *Nalla gúlisienda, kanakaia búdha-kakara,* TEL.; *Jyautishmati, káraví,* SANS.; *Habb-ul-kalkal, taftaf,* ARAB.; *Ma-la-mai,* BURM.; *Painaira-wel,* SING.

References.—*Roxb., Fl. Ind., Ed C.B.C., 335; Ainslie, Mat. Ind., II., 204; Thwaites, En. Ceylon Pl., 54; Stewart, Pb. Pl., 31; U. C. Dutt, Mat. Med. Hind., 139; Murray, Pl. and Drugs of Sind, 68; Dymock, Mat. Med. W. Ind., 2nd Ed., 187; Lisboa, U. P. Bomb., 197; S. Arjun, Bomb. Drugs, 24; Baden Powell, Pb. Pr., 330; Balfour, Cyclop.; Treasury of Botany; Rheede, VIII., t. 28; Rumph., VI., t. 24, f. 2; Mason's Bur., 502, 752.*

Habitat.—A climbing herbaceous plant plentiful in the plains of India; chiefly in Bengal and the North-West Provinces; is distributed to Ceylon and Malacca. Tendrils are modifications of portions of the flower bud: fruit triquetrous inflated.

Medicine.—The ROOT is used in medicine as an emetic, laxative, stomachic, and rubefacient. It also possesses diaphoretic, diuretic, and tonic properties. In combination with other remedies it is prescribed by Hindú physicians in rheumatism, nervous diseases, piles, &c. The decoction of the root is considered aperient by native practitioners, who prescribe it in doses of half a tea-cupful twice daily. It is mucilaginous and slightly nauseous to the taste. The SEED is said to be officinal in the Panjáb (*Hab-ul-kalkal*). Mr. Baden Powell remarks: "it is used as a

MEDICINE. Root. 552

Seeds. 553

MEDICINE. Leaves. 554 tonic in fever, and a diaphoretic in rheumatism." The fried LEAVES are said to bring on the secretion of the menses. The following prescription is given by **Dr. Dutt** as a Hindu cure for amenorrhœa: Equal parts of *Jyautishmati* leaves, *sarjiká* (impure carbonate of potash), **Acorus Calamus** root (*vachá*), and the root-bark of **Terminalia tomentosa** (*asana*) reduced to a paste with milk; taken in doses of about a drachm for three days (*Mat. Med. Hindus*). "On the Malabar coast the leaves are administered in pulmonic complaints, and mixed with castor oil, are internally employed in rheumatism and lumbago." Mixed with jaggery and
Plant. 555 boiled in oil, they are a good specific in sore eyes. The whole PLANT, boiled in oil, is sometimes employed to anoint the body in bilious affections. **Rheede** says that rubbed up with water, it is applied to rheumatism and stiffness of the limbs. The plant, steeped in milk, has been used externally to reduce swellings and hardened tumours, and this
Juice. 556 treatment is reputed to have proved successful. "The JUICE of the plant taken in a dose of a table-spoonful daily, promotes the catamenial flow during the menstrual period." **Mr. Baden-Powell** adds that it is also a demulcent in gonorrhœa and in pulmonary affections. (*Ainslie; U. C. Dutt; Drury; S. Arjun.*)

FOOD. Leaves. 557 Seeds. 558 **Food.**—"In the Moluccas the LEAVES are cooked as a vegetable." (*Drury, U. Pl.*) **Lisboa** states that in the Bombay Presidency the leaves and shoots are "eaten as green." **Balfour** remarks that "popular superstition asserts that by eating the SEEDS, the understanding is enlightened and the memory rendered miraculously retentive."

CARDUUS, *Linn.; Gen. Pl., II., 467.*

559 **Carduus nutans,** *Linn.; Fl. Br. Ind., III., 361;* COMPOSITÆ.

THE THISTLE.

Vern.—*Kanchári, tíso, bádáward,* PB.; *Guli-bádáwurd,* KASHMIR.

References.—*Stewart, Pb. Pl., 123; Baden Powell, Pb. Pr., 356; Dymock, Mat. Med. W. Ind., 386; also 2nd Ed., 466.*

Habitat.—A tall stout thistle, found in the Western Himálaya, from Kashmir to Simla, at an altitude of 6,000 to 12,000 feet; also at Hazára in the Panjáb; and in Western Tibet, at an altitude of 13,000 feet.

MEDICINE. Flowers. 560 **Medicine.**—The flowers are considered febrifugal in Lahore; according to **Mr. Baden Powell**, in Kashmír, they are also used to purify the blood.

FODDER. 561 **Fodder.**—Eaten by camels greedily. When bruised, to destroy the spines, is given to cattle, and in dry seasons, when other food is scarce, is regularly used in this way. (Conf. vernacular names with those given for **Cratægus.**)

DOMESTIC. 562 **Domestic.**—**Murray** remarks that the leaves are employed to curdle milk.

CAREYA, *Roxb.; Gen. Pl., I., 721.*

Leaves alternate, not gland-dotted. *Flowers* large, 4-merous. *Stamens* numerous, in several series, slightly connate at the base; filaments filiform, innermost and outermost without anthers. *Ovary* 4-5-celled, crowned by an annular disc. *Fruit* large, globose, fibrous; dissepiments absorbed; seeds numerous.

A genus, containing only 3 species, and these confined to India; named in honour of the **Rev. Dr. Carey**—one of the distinguished Serampore Missionaries—a distinguished botanist and a contemporary of **Dr. Roxburgh's.**

Careya arborea, *Roxb., Fl. Br. Ind., II., 511; Bedd., Fl. Sylv., t. 205; Wight Ill., 99, 100;* MYRTACEÆ. 563

Vern.—*Kumbi, vákambu, kamba, kumbh, kúmbhi, khumbi* (fruit *gugaira*), HIND.; *Kúmbhi* (flowers), *vakhúmba*, PB.; *Kumbhi, vákamba, pilu*, BANDA; *Gumar*, MANDLA and BALAGHÁT; *Kumri*, CHHINDWARA; *Kumbha, kumbhásála, kumbya, vákumbhá*, MAR.; *Gummar*, GOND; *Boktok*, LEPCHA; *Dambel*, GÁRO; *Kumbha, kumbia, kumbi, kómbi wakumba*, (flowers) BOMB.; *Ayma, pailae, púta-tammi* (or *púta-tanni*), *posta-tammi* (*ari-maru, kasaddai, panichai*, in CEYLON); *pailæpúta tammi*, TAM.; *Kumbir*, SANTAL; *Asunda*, KOL.; *Kúm*, BHUMIJ; *Budá-durmi, buda darini, dudippi*, TEL.; *Kaval*, KAN.; *Gavuldu*, MYSOR; *Banbwe, bambway, ban-bway, bhan-bhwai, ban-bwai*, BURM.; *Kabúay, kumbi*, TALEING; *Tagúyi*, KAREN; *Kahata, ahatte*, SING.; *Kumbhi*, SANS.

References.—*Roxb., Fl. Ind., Ed. C.B.C., 447; Dalz. and Gibs., Bom., Fl., 95; Brandis, For. Fl., 236; Kurz, For. Fl., Burm., I., 499; Gamble, Man. Timb., 197; Thwaites, En. Ceylon Pl., 119; Stewart, Pb. Pl., 95; Dymock, Mat. Med. W. Ind., 327; S. Arjun, Bomb. Drugs, 55; Baden Powell, Pb. Pr., 273, 329, 572; Rev. A. Campbell, Report on Econ. Prod., Chutia Nagpur; Balfour, Cyclop.; Treasury of Botany; Dalz. & Gibs., Bomb. Fl., 95; Lisboa, U. Pl. of Bomb., 232; Cooke, Gums and Resins, 11.*

Habitat.—A large deciduous tree, with the leaves turning red in the cold season and the new foliage appearing in March and April just after the flowers have faded. Frequent in the Sub-Himálayan tract from the Jumna eastward, and in Bengal, Burma, Central, Western, and South India, ascending to 5,000 feet in altitude. Growth rapid, often giving as much as four rings per inch of radius.

Gum.—Yields a brown or greenish-brown gum, regarding which but
little is known (*Atkinson*). This forms with water a tolerably thick muci- GUM. 564
lage of a dark-brown colour (*Dymock*).

Dye and Tan.—Bark used for tanning. (*Kurz.*) The **Rev. A. Camp-**
bell says that in Manbhum the bark is used as a dye.

Fibre.—The bark yields a good fibre for coarse cordage. (*Gamble,* TAN. Bark. 565
Campbell, &c.) **Lisboa** remarks that the bark affords a "stuff suitable
for brown paper of good quality." Tasar silkworms feed on the leaves. DYE. Bark. 566
(*C. P. Gaz., 1870, 504.*)

Medicine.—The BARK is used as an astringent medicine; when moist- FIBRE. Bark. 567
ened it gives out much mucilage, and is on this account used in the preparation of emollient embrocations. "The bark is applied to the wound
in snake-bite and an INFUSION of the same is given internally." (*Rev. A.* Paper making. 568
Campbell, Manbhum.) The FLOWERS are given as a tonic in *sherbet* after
child-birth. They are "administered in infusion by native midwives to
heal ruptures caused by child-birth" (*Murray, Pl. and Drugs of Sind*). MEDICINE. Bark, 569
"The CALICES of the FLOWERS are sold in the shops under the name *Wa-*
kúmbha; they are clove-shaped, 4-partite, fleshy, of a greenish-brown Infusion. 570
colour, and about an inch long; when placed in water they become coated Flowers. 571
with mucilage and emit a sickly odour. The natives use them as well as
the JUICE of the fresh bark with honey as a demulcent in coughs and Juice. 572
colds" (*Dymock*). "The FRUIT is also astringent and generally aromatic,
and is used in the form of a decoction to promote digestion" (*S. Arjun,* Fruit. 573
Bomb. Drugs, 55).

Food.—The tree blossoms during the hot season, the seed ripening FOOD. Seed. 574
about three or four months after (*Roxb.*). The **Rev. A. Campbell** says
the fruit is eaten by the Santals, and is also used medicinally, as are the Fruit. 575
flowers. The fruit, known as *khuni*, is eaten in the Panjáb; it is also
given to cattle. The seeds are said to be more or less poisonous. Seeds. 576

TIMBER. 577 **Structure of the Wood.**—Sapwood whitish, large; heartwood dull red, sometimes claret-coloured, very dark in old trees, even-grained, beautifully mottled, seasons well, very durable, moderately hard. Weight from 43 to 60lb a cubic foot. **Mr. Gamble** says that the specimens brought by **Dr. Wallich** from Tavoy in 1828, and those brought from the Mishmi Hills by **Dr. Griffith** in 1836, were found to be quite sound on being cut up, though they had been stored in Calcutta for 50 years.

The wood is little used except for agricultural implements. **Drury** says "the cabinet-makers of Monghír use the wood for boxes. It takes a polish, is of a mahogany colour, well veined." It is being tried for railway sleepers on the Eastern Bengal and Northern Bengal State Railways, but the results of the experiment are not yet known. **Kurz** remarks that it is used in Burma for gun-stocks, house-posts, planking, carts, furniture, and cabinet-work but is too heavy for such purposes. It stands well under water and is much admired for axles. "It is frequently employed for wooden hoops, being very flexible" (*Drury, U. Pl.*). **Beddome** says it is a favourite wood in some parts of the country for charcoal.

DOMESTIC. Slow-match. 578 **Domestic Uses.**—The fibrous bark is used in Mysor as a slow-match to ignite gunpowder (*Cameron*). In many parts of India it is also used in the preparation of fusees for matchlocks. **Brandis** says these are prepared by pounding, cleaning, drying, and twisting the fibre into a thin cord. This fact might be taken advantage of in the manufacture of tinder. The Indian-made fusees burn at the rate of 12 inches an hour. In Ganjam the fibrous bark forms the scanty clothing of Byragis and other Hindús affecting sanctity (*Burm. Gaz., I., 129*). "The timber was formerly used for making the drums of sepoy corps." (*Drury, U. Pl.*)

Tinder 579

580 **Careya herbacea,** *Roxb.; Fl. Br. Ind., II., 510; Wight, Ic., t. 557.*

Vern.—*Bhui dalim,* BENG.; *Chuwa,* NEPAL; *Bhumi darimba,* SANS.

References.—*Brandis, For. Fl., 237; Kurz, For. Fl., I., 499; Gamble, Man. Timb., 197.*

Habitat.—A small undershrub with pink flowers which appear from February to March. Common in the Taraí from Kumaon to the Khasia Hills and Chittagong. Also plentiful throughout the plains of Bengal, Oudh, and the Central Provinces.

CARICA, *Linn.; Gen. Pl., I., 815.*

581 **Carica Papaya,** *L.; Fl. Br. Ind., II., 599;* PASSIFLOREÆ.

THE PAPAW OR PAPAYA TREE.

Vern.—*Pappaiyá, pepiyá, papeya,* BENG.; *Papaya* or *papiya amba, pepiya, popaiyá* or *popaiyah,* HIND.; *Arandkharbúza* or simply *kharbúza* (or CASTOR-OIL-MELON), PB.; *Popái,* DUK.; *Papaya,* MAR. and CUTCH; *Papai,* BOMB.; *Paputa, katha chibhado,* SIND; *Papia, papáyi, kath, chibda, eranda kakdi,* GUJ.; *Pappáyi, pappáli,* TAM.; *Bappayi* or *boppáyi, madana-anapakáya,* TEL.; *Perangi, perinji,* KAN.; *Pappáya,* MAL.; *Thinbaw, thimbawthi, thimbaw, simbo-si, timbo-si, pimbo-si,* BURM.; *Aanabahe-hindi, amba-hindi,* ARAB., PERS.; *Pæpol, papaw,* SING; *Kai-du-du,* COCHIN CHINA.

References.—*Roxb., Fl. Ind., Ed. C.B.C., 736; Brandis, For. Fl., 244; Kurz, For. Fl., Burm., I., 533; Gamble, Man. Timb., 207; Dalz. & Gibs., Bom. Fl. Supp., 37; Stewart, Pb. Pl., 99; DeCandolle, Orig. Cult. Pl., 293; Pharm. Ind., 97; U. S. Dispens., 15th Ed., 1598, 1720; Moodeen Sheriff, Supp. Pharm. Ind., 89; Waring, Bazar Med., 114; Dymock, Mat. Med., W. Ind., 294, and 2nd Ed., 356; Kew Official Guide to Museums, p. 70; Christy, New Commercial Plants, Parts II. and VII.; Fleming, List* *of Drugs in Asiatic Researches; Ainslie, Mat. Ind., II., 343; O'Shaughnessy, Beng. Dispens., 352; Irvine, Med. Topog. Ajmir, 149; S. Arjun, Bomb. Drugs, 60; Atkinson's Econ. Prod. Pt. V, 17; Lisboa, U. Pl. of Bomb., 157; Drury, U. Pl., 113; Tropical Agriculturist, 1882-83, pp. 715, 967; Balfour, Cyclop.; Smith, Dict., 311; Treasury of Botany, &c., &c.*

Habitat.—A sub-herbaceous, almost branchless tree, commonly cultivated in gardens throughout India; from Delhi to Ceylon. Fruits all the year round, but the fruit is most luscious during the summer and when cultivated in a hot moist climate: does not succeed well in the dryer parts of India. **DeCandolle** believes it to be a native of the shores of the Gulf of Mexico and of the West Indies and doubtfully of Brazil. All the other species of the genus are unquestionably American. The non-Asiatic origin of the Papaw is conclusively proved by its not having been known before the discovery of America; by its having no Sanskrit name, and by the modern Indian names being evidently derived from the American word *papaya*, itself a corruption of the Carib *ababai*. **Ainslie** says it is a native of both Indies, an opinion held by many propular writers, but not supported by modern botanists. **Atkinson** regards it as introduced into India by the Portuguese. **Brandis** tells us that its Burmese name, *thimbawthi*, means fruit brought by sea-going vessels. In 1626, seeds were sent from India to Naples, so that the tree must have been introduced into India at an early date or shortly after the discovery of America. It is generally diœcious, the female flowers sessile, and the male on long peduncles. Sometimes, however it is monœcious or the flowers even hermaphrodite.

Resin.—Exudes a white resin. (*Kurz.*) RESIN. 582

Fibre.—**Dr. Dymock** recommends the fibre from the stem to be examined: he is of opinion that it is used in America and Africa. FIBRE. 583

Medicine.—The JUICE of the fruit (in all countries where the tree is found) is regarded as medicinal. **Ainslie** writing of it in 1826 says: "The milky juice of the unripe fruit is supposed by the natives of the Isle of France to possess powerful anthelmintic properties, but I perceive by the *Hortus Jamaicensis* (*Vol. II., 37*) that in Jamaica it is reckoned as most injurious to the intestines: the same fruit when ripe is excellent and wholesome." "The anthelmintic properties of the milky juice of the UNRIPE FRUIT were first noticed in the seventeenth century by **Hernandez.** The attention of the medical profession was in India called to it in 1810 by **Dr. Fleming** (*Asiatic Researches*, Volume XI), who cites an interesting passage from the writings of **M. Charpentier Cossigni** in support of its alleged virtues. Further confirmatory evidence has more recently been added by **M. Bouton** (*Med. Plants of Mauritius*, 1857, p. 65), and it may justly be concluded that the statements as to its efficacy as an anthelmintic are founded on fact. The following mode of administration was employed by **Dr. Lemarchand**, of Mauritius (quoted by **Bouton**), and it would be desirable to adopt it in all future trials with this remedy. Take of the fresh Papaw milk and honey, of each a table spoonful; mix thoroughly; gradually add three or four table-spoonfuls of boiling water, and when sufficiently cool take the whole at a draught, following its administration two hours subsequently by a dose of castor oil, to which a portion of lime-juice or vinegar may be added. This may be repeated two days successively if required. The above is a dose for an adult; half the quantity may be given to children between seven and ten years of age, and a third, or a teaspoonful, to children under three years. If it cause griping, as it occasionally does, enemas containing sugar have been found effectual in relieving it. Taking the dose above named as correct, the statement of **Sir W. O'Shaughnessy** (*Bengal Disp.*, p. 352) that he had prescribed the

MEDICINE. Juice. 584

Unripe fruit. 585

Mode of administration.

CARICA Papaya.

The Papaya or Papaw.

MEDICINE.

Juice useful in Lumbrici. milky juice as an anthelmintic, in doses from 20 to 60 drops, without obvious effect, is fully explained. It is principally effectual in the expulsion of lumbrici. On tænia it is reported to have little effect. Anthel-
Seeds. 586 mintic virtues have also been assigned to the SEEDS, which have a pungent taste, not unlike that of mustard or cress, but the evidence of their efficacy
Useful as an Emmenagogue. is very inconclusive. A belief in their powerfully emmenagogue properties prevails amongst all classes of women in Southern India; so much so, that they assert that if a pregnant woman partake of them, even in moderate quantities, abortion will be the probable result. This popular belief is noticed in many of the reports received from India. In them it is also stated that the milky juice of the plant is applied locally to the os uteri with the view of inducing abortion" (*Pharm. Ind., pp. 97, 98*).

The opinions so liberally contributed for this publication, by the Indian medical officers (see below), give so much of personal experience regarding the properties of this drug that it is scarcely necessary to abstract an account of it from the publications usually consulted. The following passages may, however, be found useful.

A writer in the *Ceylon Observer* (30th July 1884) says: "Papain," papainum, or vegetable papsin, may be prepared from the juice of the green fruit of **Carica Papaya** by adding alcohol, which precipitates papain. This precipitate is dried and powdered, and is then quite ready for use. **Brunton** considers that, in its peptonising powers, it is superior to the ordinary animal pepsin, and it has the additional advantage of neither requiring the addition of an acid nor an alkali to convert the contents of the stomach into peptone. It is extensively used both in France and Germany, and has been given with good results even to children. It is an invaluable remedy in the dyspepsia of those recovering from attacks of cholera. No planter up country should have any difficulty in preparing it. It should be thoroughly dried in the sun on a hot plate and preserved in well-stoppered bottles. In France it is mixed with starch to preserve it, and is often made up in the form of papain lozenges, '*Drgées de Papaine*,' of which two form a dose. The '*Syrop de Papaine*' is also a favourite form, as well as papain wine and papain elixir. The alcoholic preparations of it are the most popular. This vegetable pepsin is said to digest the tapeworm in the intestines, and the juice of the **Carica Papaya** is a favourite domestic remedy for the expulsion of that worm in Brazil."

The author of *Makhzhan* mentions it as a remedy for hæmoptysis, bleeding piles, and ulcers of the urinary passages; it is also useful in dyspepsia; rubbing the milk in, two or three times, cures ringworm, or psoriasis (قوبا), causing a copious serous exudation attended with itching. **Evers** has employed the milk in splenic and hepatic enlargements with good results; a tea spoonful with an equal quantity of sugar divided into three doses was administered daily. (*Ind. Med. Gazette, Feb. 1875; Dymock, Mat. Med.*)

"**Dr. Boucher**, in conjunction with **Dr. Wurtz**, who has tried papaw so successfully, has obtained good results with this drug in diphtheria. A solution of 10 to 30 drops, applied as a paint, rapidly dissolved the false membranes. A number of cases in the hospital for infants were cured
Leaves. 587 by this treatment." (*Journal de Therapeutique.*) "Physicians in Germany, in cases of diphtheria, give papaine in doses of 4 to 5 grammes, and paint the throat with a solution (1·10) with great success" (*Christy's New Com. Pl.*). The **Rev. George Henslow** reported to the Royal Horticultural Society that he had carefully tested the property of the leaves to soften meat and found that it unquestionably did so, but that it imparted a peculiar flavour to the meat. In Africa the odour emitted by the flowers is

believed to be the cause of disease. **Firminger** (*Man. Gard. Ind.*) says that the tree comes into flower during the rains and emits at times a fine fragrance.

CHEMISTRY. 588

CHEMICAL COMPOSITION AND PROPERTIES.—The following abstract of the chemistry of this substance may be here given from **Dr. W. Dymock's** *Materia Medica of Western India* (*p. 358*): "**Herr Wittmack** has recently (1878) examined the properties of the milky juice with the following results: He obtained after repeated incision of a half ripe fruit 1·195 grammes of white milky juice of the consistence of cream. This dried in a watch glass to a hard vitreous white mass, having what appeared to be greasy spots on the surface, but what really were flocks of a gelatinous substance that always adheres to the more hardened material. The odour and flavour of the fresh juice recalled that of petroleum or of vulcanised India-rubber. The microscope showed it to be a fine grumous mass containing some larger particles and isolated starch grains. Iodine coloured the juice yellowish-brown. A portion of the juice was dissolved in three times its weight of water, and this was placed with 10 grammes of quite fresh lean beef in one piece in distilled water, and boiled for five minutes. Below the boiling point the meat fell into several pieces, and at the close of the experiment it had separated into coarse shreds. In the control experiments made without the juice the boiled meat was visibly harder. Hard boiled albumen, digested with a little juice at a temperature of 20° C., could after twenty-four hours be easily broken up with a glass rod: 50 grammes of beef in one piece, enveloped in a leaf of **C. Papaya** during 24 hours at 15°C. after a short boiling became perfectly tender; a similar piece wrapped in paper and heated in the same manner remained quite hard. Some comparative experiments were also made with pepsin, and the following are the conclusions arrived at by the author:—

(1) The milky juice of the **Carica Papaya** is (or contains) a ferment which has an extraordinarily energetic action upon nitrogenous substances, and like pepsin curdles milk; (2) this juice differs from pepsin in being active without the addition of free acid; probably it contains a small quantity, and further it operates at a higher temperature (about 60° to 65°C.) and in a shorter time (5 minutes at most); (3) the filtered juice differs chemically from pepsin in that it gives no precipitate on boiling, and further that it is precipitated by mercuric chloride, iodine, and all the mineral acids; (4) it resembles pepsin in being precipitated by neutral acetate of lead, and not giving a precipitate with sulphate of copper and perchloride of iron (*Pharm. Jour. (III.), 30th November 1878*). The active principle has since been separated and given the name of *Papaine*; it is now an article of commerce in Europe for medicinal purposes, and is said to be capable of digesting 200 times its weight of fibrine; it has been used as a solvent of diphtheritic false membrane, and also as a local application in old standing cases of chronic eczema, more especially of the palms of the hands, and where other remedies failed great benefit has attended its application in the following way:—12 grains of *Papaine* and 5 grains of powdered borax, in 2 drachms of distilled water, to be painted on the parts twice daily."

MEDICAL OPINIONS.

Special Opinions.—§ "The *papaya* is an important medicine. **Dr. Roy** performed certain experiments under **Dr. Parker** of Netley, and showed that the juice of the raw fruit is more efficacious in dissolving albumen than pepsin. Since I have used it in dyspepsia, with great benefit, I had a plantation of nearly two hundred trees in the grounds of Bankura jail. The raw fruit was scraped longitudinally and the milky juice collected. This was then put on a sand-bath. After 24 hours or so, a dull white powder was left. This I consider the best preparation for internal use;

CARICA Papaya.

The Papaya or Papaw.

MEDICAL OPINIONS.

one or two grains with sugar or milk after meals should be given to adults. A few drops of juice added to tough meat render it quite tender and fit for immediate cooking. This is very desirable in the case of invalids. Tincture of the juice does not keep well and is disagreeable to taste. A syrup of the powder may be made if required for children and delicate women" (*Surgeon R. L. Dutt, M.D., Pubna*). "The milk-like juice of the green or unripe fruit is a good digestive, and most efficacious in dyspepsia. I have frequently prescribed it with marked success. The ripe fruit is alterative, and if eaten regularly every morning, corrects that habitual constipation so common in India. The dry fruit is said to reduce enlarged spleen, but I administered it in several cases without any apparent benefit. The leaves are reputed to promote the secretion of milk. I tried this, and the result was not unfavourable, but I think the good effect was chiefly owing to the maintenance of a uniform heat. However, more experiments are necessary to decide the question. The leaves should be gently bruised and heated in a pan and applied warm to the breast. The dose of the milk-like juice is 30 drops, mixed with water, two or three times a day. The juice must be fresh, as it decomposes quickly, but it may be obtained by picking the green fruit on the tree and collecting the white fluid in a glass" (*R. A. Barker, M.D., Civil Surgeon, Dumka, Santal Parganas*). "The ripe fruit is very pleasant eating indeed. The leaves of this tree have the peculiar property of making tough meat tender. If a fowl, recently killed, be wrapped up in *papya* leaves for a couple of hours, and then cooked, it will be as tender as if it had been hung for 24 hours. I have seen spleen grow smaller in young persons who have been treated with the dried and salted fruit. The juice called *papaine* has digestive ferment properties and will remove thickened skin, as in eczema and corns. It is also said to be a certain remedy in cases of scorpion sting" (*Surgeon E. Borill, Motihari, Champaran*). "Fruit much eaten, and has an agreeable flavour; used to render meat tender; has the power of digesting and dissolving albumen" (*Brigade Surgeon G. A. Watson, Allahabad*). "The fruit is good in chronic diarrhœa" (*Assistant Surgeon Nehal Sing, Sahâranpur*). "Papya juice is used in dyspepsia as a vegetable substitute for pepsine" (*Surgeon R. Gray, Lahore*). "It has the property of rendering meat tender and of facilitating the process of cooking. It contains a vegetable peptine and can be used as pepsine" (*Brigade Surgeon J. H. Thornton, B.A., M.B., Monghir*). "The juice has great solvent properties. If dropped on raw meat, it dissolves it in a few minutes. The green fruit when boiled with meat renders it tender. The green fruit is used as a vegetable. Is a mild laxative and diuretic. The ripe fruit is cooling at first, but has a laxative effect afterwards. The fresh juice is rubefacient" (*Surgeon D. Basu, Faridpur*). "The juice of the green fruit is a reliable stomachic and is slightly laxative. It should *not* be given to pregnant women, as it is emmenagogue, and is occasionally used to produce criminal abortion. It digests raw meat in a short time" (*Assistant Surgeon Devendro Nath Roy, Campbell Medical School, Sealdah, Calcutta*). "Ripe fruit is laxative and useful in piles. Milky juice of the unripe fruit is a medicine for ringworm; it has rubefacient properties. Hospital Assistant Gopal Chunder Ganguli reports that meat softens when boiled with the unripe fruit cut into pieces; it is also used in the form of curry by the natives" (*Surgeon Anund Chunder Mukerji, Noakhally*). "A fowl or piece of meat rolled up in some *papya* leaves soon becomes tender. The ripe fruit is pleasant and useful in dyspepsia" (*Surgeon G. Price, Shâhabad*). "I have used the unripe fruit in enlarged spleen but with little or no effect" (*Surgeon D. Picachy, Purnia*). "The milky juice of the unripe fruit is said to possess digestive

The Papaya or Papaw.

CARICA Papaya.

MEDICAL OPINIONS.

properties" (*P. W. B., Dacca*). "The juice has the power of dissolving coagulated albumen" (*Surgeon A. Crombie, Dacca*). "Anthelmintic. A leg of mutton or a fowl left under a *papaya* tree for a night is said to become quite tender" (*Surgeon C. M. Russell, M.D., Sáran*). "The juice is applied in psoriasis and other skin affections of a similar character" (*Surgeon-Major W. Dymock, Medical Store-keeper, Bombay*). "The fresh juice of the fruit of this plant has the power of digesting meat if it is kept at about the temperature of the body. It is owing to this property that it has been proposed as a drug in cases of dyspepsia. The fruit is largely used by the natives when ripe in cases of piles. I have used it myself, but did not find it beneficial" (*Surgeon Roderick Macleod, Gaya*). "Introduced by me in the treatment of enlarged spleen.—*Vide Indian Medical Gazette* for February 1875, page 38" (*Surgeon B. Evers, M.D., Wardha*). "The juice is very effective in rendering meat soft, and an extract is very good for digestion; also considered useful in the treatment of round-worm" (*Surgeon-Major J. M. Zorab, Balasor*). "The milky juice of the unripe fruit obtained by superficial scarification contains an active principle which is analogous to pepsine in its physiological property, and has the virtue of dissolving all azotised matter. Its action on muscular fibre is peculiarly strong; hence it can be used in atonic dyspepsia. The unripe fruit is cooked as a vegetable and the ripe fruit eaten as dessert has the same effect, and acts as a mild chologogue and purgative; hence its use amongst the natives for piles, enlarged liver, and spleen. The active principle has been since separated by **Mr. Wurtz**, which he has termed papyaine (*vide* **Braithwaite's** *Retrospect of Medicine*, 1873,—my article on Papaya juice" (*Surgeon G. C. Roy, Bírbhúm*). "The ripe fruit has laxative properties, and is said to be useful in bleeding piles" (*Assistant Surgeon Shib Chunder Bhattacharji, Chanda, Central Provinces*). "I have tried the juice of the unripe fruit in many cases of enlarged spleen, but have not found it an effective remedy. It was given in the form of a bolus with sugar in doses of a drachm three times a day" (*Surgeon S. H. Browne, Hoshangabad, Central Provinces*). "The unripe fruit possesses ecbolic properties, and is often resorted to by natives to induce criminal abortion. The milky juice is irritant and is applied for the same purpose to the os uteri" (*Surgeon-Major J. McD. Houston, Travancore; and John Gomes, Esq., Medical Store-keeper, Trivandrum*). "When taken internally it produces abortion" (*Surgeon W. A. Lee, Mangalur*). "Said to produce abortion. Fruit eaten" (*Surgeon-Major J. J. L. Ratton, M.D., Salem*). "The unripe fruit, made into a curry, is eaten by women to excite secretion of milk. It also has the property of making meat of any kind tender when cooked with it" (*Honorary Surgeon P. Kinsley, Chicacol, Ganjám, Madras*). "Acts on the spleen" (*Surgeon W. A. Barren, Bhuj, Cutch*). "Very useful in cases of dyspepsia due to defective gastric secretion, well mixed with raw minced-meat" (*Surgeon H. D. Masani, Karáchi*). "The green fruit when eaten brings on an immediate flow of milk. Ripe fruit is said to bring on abortion, and if applied to tough meat makes it tender" (*Surgeon-Major Lionel Beech, Coconada*). "The fresh fruit of this small tree in South India is eaten to procure abortion" (*Surgeon-Major John North, Bangalor*). "Mature green fruit, sliced, dried, and powdered; dose 5 to 20 grains for dyspepsia" (*Apothecary Thomas Ward, Madanapalli, Cuddapah*). "The peculiarities of this fruit, and its effects as a solvent of meat, require to be scientifically investigated" (*Surgeon-General William Robert Cornish, F.R.C.S., C.I.E., Madras*). "The juice is used externally to prevent suppuration" (*Surgeon-Major D. R. Thomson, M.D., C.I.E., Madras*). "The bruised leaves applied as a poultice have an excellent influence in reducing elephantoid growth. The inspissated juice of the

CARICA Papaya.

The Papaya or Papaw.

MEDICAL OPINIONS.

fruit, in doses of 1 grain injected hypodermically, will remove the morbid tissue within the area of its contact. Fever is occasionally excited as well as local irritation, and hence this mode must be pursued carefully. I have used the inspissated juice also in the form of pills in 2-grain to 4-grain doses for the same disease. The result seemed favourable, but as other methods were used the matter is open to doubt" (*Surgeon W. G. King, M.B., Madras*). "The leaves are used externally for nervous pains. The leaf may be either dipped in hot water or warmed over a fire and applied to the painful part" (*Surgeon-Major W. Nolan, M.D., Bombay*). "The seeds are considered to be anthelmintic" (*Surgeon-Major J. Robb, Ahmedabad*).

The above opinions show how widely and uniformly the properties of the *papaya* are believed in by Native and even by European Medical Officers.

FOOD. Ripe fruit. 589 Green fruit. Curries and pickles. 590 Other modes of preparation. 591

Food.—When ripe the fruit attains the size of a small melon; the interior is soft, yellow, and sweetish; eaten by all classes and esteemed innocent and wholesome. When green it is cooked by the natives in their curries and also pickled. The ripe fruit has a flavour peculiar to itself; the better qualities are eaten without sugar, and by many persons are ranked among the first of eastern fruits. By others the *papaya* is eaten with pepper and salt. The seeds have a pleasantly pungent taste, not unlike mustard; hence in all probability the idea occasionally alluded to that this is the mustard tree of the scriptures. **Lisboa** says the fruit has a sweetish taste and makes an excellent tart. When boiled in slices it is eaten as a vegetable. **Don** says that in South America the fruit after being boiled and mixed with lime juice and sugar is used in place of apple sauce. **Sloane** remarks that the unripe fruit is cut into slices and soaked in water till the milky juice is removed. It is then boiled and eaten as turnips or baked as apples. A few drops of the milky sap of the papaw is said to render meat tender. The author of the *Makhzan* recommends that for this purpose the juice should be mixed with fresh ginger. In Barbadoes the flesh of animals is reported to be hung on the tree over night in order to soften it. This idea prevails all over India and is doubtless often resorted to by domestic servants. **Drury** confirms this and states that he has personally tested the accuracy of the popular notion. **Dr. John Davy** (*Edin. Ph. I., 1855*) declares that this is due to accidental causes. According to some writers the best plan to soften

Juice. 592

meat is to wrap it overnight in the *papaw* leaves, or to drop a little of the fresh juice into the vessel in which the meat is being cooked. **Brandis** mentions another process, namely, to wash meat with water impregnated with the milky juice. It is even stated that meat is rendered tender by causing the animals to eat the seeds before they are killed. The best qualities of *papaw* are said to be obtained from Singapore and Moulmain stock. "The green fruit, when peeled, boiled, cut into small pieces, and served with sweet oil, vinegar, salt and pepper, serves as a very palatable vegetable, and is very similar to squash in taste" (*Mr. L. Liotard*).

TIMBER. 593

Structure of the Wood.—The stem of this fast-growing tree is too spongy and fibrous to be regarded as affording timber. **Gamble** describes it as soft wooded.

DOMESTIC. 594

Domestic.—The juice is used by native ladies as a cosmetic to remove freckles. It is also exceedingly acrid, causing blisters and itching if applied to the skin (*Treasury of Botany*). "The leaves are employed by the Negroes in washing linen as a substitute for soap" (*O'Shaughnessy*).

Carica spinosa.

A branching tree met with in Guiana and Brazil, has a much more acrid juice than the other species. If dropped on the skin it causes disagreeable blisters. The fruit is not eaten, and its flowers have a carrion-like odour. The juice enjoys a considerable reputation as a useful drug in the treatment of enlarged spleen and as an anthelmintic. MEDICINE. Juice. 595

The tree might with advantage be introduced into India.

CARISSA, *Linn.; Gen. Pl., II., 695.*

A genus of densely-branched, spinous, erect shrubs, belonging to the APOCYNACEÆ. There are some twenty species, African, Asiatic, and Australian. **Sir J. D. Hooker** remarks of the five Indian species that they are probably mere forms of one or two very variable plants.

Leaves opposite, small, coriaceous. *Flowers* interminal or axillary, peduncled, 3-chotomous cymes. *Corolla-tube* cylindrical, naked. *Anthers* included, free from the stigma; cells rounded at the base. *Disc* none. *Ovary* 2-celled, 2 wholly combined carpels. *Style* filiform, stigma minutely 2-fid. *Ovules* 1-4 in each cell, berry ellipsoid or globose. *Seeds* usually 2, peltately attached to the septum without a wing or pencil of hairs.

Carissa Carandas, *Linn.; Fl. Br. Ind., III., 630; Wight, Ic., t. 426;* APOCYNACEÆ. 596

Syn.—C. CONGESTA, *Wight, Ic., t. 1289; Bedd., Fl. Sylv., Man., 156, Anal., t. 19, fig. 6.*

Vern.—*Karaundá, karúnda,* or *karonda, garinga, karroná, timukhia, gotho,* HIND.; *Kurumia, karamchá, bainchi, karenja, tair,* BENG.; *Timukhia,* N.-W. P.; *Gotho,* C. P.; *Karinda, karándá, karwand,* BOM.; *Karavanda, boronda,* MAR.; *Karamarda* (the tree), *timbarran,* GUJ.; *Kendakeri, kerendo kuli,* URIYA; *Karamarda* (the tree), *karamardaka, avinga* (fruit), *krishna-pakphula,* SANS.; *Kalaka, kalapa,* TAM.; *Kalivi kaya, waaka,* TEL.; *Karekai, heggarjige,* KAN.

References.—*Roxb., Fl. Ind., Ed. C.B.C., 231; As. Res. IV., 263; Brandis, For. Fl., 320; Kurz, For. Fl. Burm., II., 169; Gamble, Man. Timb., 261; Dalz. & Gibs., Bom. Pl., 143; Thwaites, En. Ceylon Pl., 191; Stewart, Pb. Pl., 141; U. C. Dutt, Mat. Med. Hind., 303; Dymock, Mat. Med. W. Ind., Ed. II. (507); Birdwood's Bom. Prod.; Atkinson's Econ. Prod., N.-W. P., Part V., 80; McCann, Dyes and Tans of Beng., 142; Balfour, Cyclop.; Smith, Dic., 92; Treasury of Botany; Firminger, Man. Gard., 256.*

Habitat.—A dichotomously-branched bush, cultivated for its fruit in most parts of India; said to be wild in Oudh, Bengal, and South India. In the Panjáb and Gujarat it frequents hedges, and forms spiny, low, dense bushes; is also found in Burma, Ceylon, and Malacca.

It flowers from February to April, and produces a small fruit which is grape-green when young, white and pink when approaching maturity, and nearly black when ripe. The fruit is ripe in July to August.

Dye.—Dr. McCann states that in Bhagalpur the fruit is used as an auxiliary in dyeing and tanning. The milky fluid which exudes from the wounded part of the fruit when gathered is very adhesive. DYE. Fruit. 597

Medicine.—The unripe FRUIT is astringent, and the ripe fruit cooling, acid, and useful in bilious complaints. The ROOT has the reputation of being a bitter stomachic. "It is used as a plaster in the Concan to keep off flies, and pounded with horsepiss, limejuice, and camphor as a remedy for itch." (*Dymock.*) MEDICINE. Fruit. 598 Root. 599

Special Opinions.—§ "It is considered to be antiscorbutic and much used in the form of curry and chutney by the natives" (*Assistant Surgeon Anund Chunder Mukerji, Noakhally*). "Antiscorbutic, expector-

MEDICINE. ant" (*Surgeon W. Barren, Bhuj, Cutch*). "The juice is irritant and capable of producing ulcers. The ripe fruit is a pleasant acid, goes well with food, and has, I believe, antiscorbutic properties" (*Surgeon-Major J. M. Zorab, Balasore, Orissa*). "The decoction of the leaves is very much used at the commencement of remittent fever" (*Surgeon-Major P. N. Mukerji, Cuttack, Orissa*).

FOOD. Pickle. 600 Preserves. 601 **Food.**—The fruit is made into pickle just before it is ripe, and is also employed in tarts and puddings; for these purposes it is superior to any other Indian fruit (*Firminger*). When ripe it makes a very good jelly (equal to red currant), for which it is cultivated in the gardens owned by Europeans. The natives universally eat the fruit when ripe, and excepting pickling they do not cook it.

TIMBER. 602 **Structure of the Wood.**—White, hard, smooth, close-grained.

DOMESTIC, Fences. 603 **Domestic Uses.**—Makes exceedingly strong fences. Its number of strong, sharp thorns, renders such hedges almost impassable. (*Roxb.*)

Carissa diffusa, *Roxb., Fl. Ind., Ed. C.B.C., 231;* Syn. for **C. spinarum,** *A. DC.*, which see.

604 **C. macrophylla,** *Wall.; Fl. Br. Ind., III., 631.*

Syn.—CARISSA LANCEOLATA, *Dalz.*; C. DALZELLII, *Bedd., Fl. Sylv., Man., 157.*

References.—*Dalz. & Gibs., Bom. Fl., 143; Lisboa, U. Pl. of Bom., 166.*

Habitat.—A large shrub with very strong, curved thorns, common on the Deccan peninsula; Coorg (*Heyne*); Konkan at Ramghat (*Dalzell*); Courtallum (*Wight*). The flowers are much larger than those of the other species.

FOOD. Fruit. 605 **Food.**—The fruit is eaten; it is about the size of a plum and ripens in May. **Beddome** says it is superior to that of **C. Carandas.**

606 **C. spinarum,** *A. DC., Fl. Br. Ind., III., 631; Wight, Ic., t. 427.*

Syn.—C. DIFFUSA, *Roxb.*

The *Flora of British India* regards this species as probably only a state of **C. Carandas,** concurring in this opinion with **Dr. Brandis.** It is mainly distinguished by its being a smaller plant, with shorter and more slender spines, more acute leaves, and a smaller berry.

Var. **hirsuta** is more pubescent than the type condition. It is **C. villosa,** *Roxb., Ed., Carey and Wall.*, and also of *Wight, Ic., t. 437*—a form which **Roxburgh** regarded as quite distinct from the others described by him and of little economic value.

Vern.—*Karaunda*, HIND.; *Gán, garinda, garna*, PB.; *San karunda, anka koli*, URIYA; *Karamadika*, SANS.; *Wakoilu*, TEL.; *Kanuwán*, ORAON.

References.—*Roxb., Fl. Ind., Ed. C.B.C., 231; Brandis, For. Fl., 321; Kurz, For. Fl. Burm., II., 169; Gamble, Man. Timb., 261; Thwaites, En. Ceylon Pl., 191; Stewart, Pb. Pl., 142; Drury, U. Pl., 116; Baden Powell, Pb. Pl., 361, 376, 572; Birdwood's Bom. Prod.; Balfour, Cyclop.; Dalz. & Gibs., Bom. Fl. Suppl., 53; Lisboa, U. Pl. of Bom., 166.*

Habitat.—A small, thorny, evergreen shrub, wild in most parts of India, especially in the drier zones and in the plains of the Panjáb, the Sub-Himálayan tract up to 4,000 feet, and in the Trans-Indus territory; also on the coast of South Andaman (*Kurz*), and in the hotter parts of Ceylon. **Birdwood** views it as introduced into Bombay and not indigenous. **Roxburgh** says it is a native of the Ganjam district and from thence northward to the mouth of the Hugli (**C. diffusa**).

Medicine.—This plant is mentioned by **Baden Powell** amongst his drugs of the Panjáb, but its supposed properties are not stated. MEDICINE.

"A Kangra authority remarks that the very old wood gets quite black and fragrant, and is sold at a high price as *aggar* or *úd-hindí*, an officinal wood generally referred to **Alœxylon Agallochum** (=**Aquilaria Agallocha**) which is given as a tonic and cholagogue" (*Dr. Stewart*). Wood. 607

Food.—The fruit is eaten in tarts. The leaves are greedily devoured by goats and sheep. FOOD. Fruit. 608 FODDER. 609

Structure of the Wood.—Hard, smooth, close-grained, said when very old (in Kangra) to be black and fragrant (*Brandis*). It is generally gregarious, often forming undergrowth in the forests of **Pinus longifolia**, of bamboo, and occasionally of teak. It is used for turning and combs. TIMBER. 610

Domestic Uses.—Largely used for dry fences, but spreads so rapidly where clearances have been made that it may impede the reproduction and growth of the forest. It coppices freely and makes excellent fuel. DOMESTIC. Fences. 611 Fuel. 612

CARMINE.

Carmine and Carminic Acid. 613

CARMIN, *Fr.*; KARMIN, *Germ.*; CARMINIO, *It.*

References.—*Balfour's Cyclopæd.*; *Ure's Dictionary of Arts, Manuf., and Mines.*

A pigment of a bright red colour, made from cochineal and alumina or bichloride of tin. This is prepared by throwing into a decoction of cochineal a certain proportion of the base employed. A salt is produced which is allowed to precipitate in shallow basins. The colourless liquid is decanted and the powder carmine dried and preserved. By the old German process carmine is prepared with alum.

The uses of Carmine have recently been greatly extended. It is employed for making fine red inks and for silk-dyeing. It is the finest red the water-painter, and more especially the miniature painter, possesses. The French carmine and rouge is preferred to the English. See **Cochineal**.

Carnation. See **Clove.**

CARNELIAN. 614

The Carnelian and the Agate are, perhaps, the best known members of what for convenience may be generically designated the CHALCEDONY group of quartzose Minerals. Quartz is the natural silicic anhydride, but hydrated silica also exists. This has led to the classification of the quartzose minerals into—

1st—Transparent Crystallised Quartz or Anhydrous Quartz, as represented by the ROCK CRYSTALS. These, when violet, are known as the Amethyst, and when yellow or sherry-coloured as the Cairngorm, but numerous intermediate shades also exist from red to black.

2nd—Uncrystallised or Crypto-Crystalline Anhydrous Quartz.—This corresponds to the Chalcedony series, but by most writers this is also made to include JASPER, an opaque rock of undefined nature rather than a definite mineral. The term AGATE is sometimes given generically to denominate this series, or Agate and Chalcedony are used as synonymous terms.

3rd—Uncrystalline Semi-transparent to Opaque Hydrated Quartz.—The OPAL may be given as the type of this group.

CARNELIAN. **The Carnelian.**

QUARTZ.

The quartzose stones referrible to the above sections are extensively used in India for ornamental purposes, in the lapidaries' art, in decorative architecture, and in the manufacture of cheap jewellery. They are popularly assigned a position with the "inferior gems"—the diamond, ruby, sapphire, emerald, pearl, &c., being classed as the "gems" or "precious stones and gems." Some of the better qualities of opal have assigned to them, however, a position with the gems, and indeed a bright colour-flashing opal is one of the prettiest of all stones. The quartzose minerals were apparently not known to the ancients, and when first brought to their attention obtained fabulous prices. **Pliny** mentions that fragments of a small Cambay cup were exhibited in the theatre of Nero, "as if," adds **Pliny**, "they had been the ashes of no less than Alexander the Great himself." **Balfour** remarks with much truth that "amongst the people of India the inferior gems are held in but little esteem; they value a gem for its intrinsic price, not for the workman's skill expended in shaping it, in which the chief value of all the inferior gems consists." While this is so the trade in these inferior gems, both internal and foreign, is far more extensive than it is possible, with our present means of determining, to definitely express. Indeed, the utmost that can be done in this direction, is to remind the reader of the elaborate decorations of the Taj Mahal of Agra and of the other similar memorials of the Moghul Empire, in order to convey an idea of the extent the art of lapidary decoration prevailed during that period, and to add that there is little to justify the conclusion that it has materially diminished during the present day. It has become more diffused, with the result that the art of the dealer and worker in stones is now plied ir every town and village of India. His efforts are directed to humbler objects, but his skill has in no wise diminished. Cambay and Broach hold their own as great emporiums of carefully-prepared carbuncles, carnelians, and agate cups. Vellum is still famed for its manufactures in rock crystals, and Jabalpur for its cheap lapidary work, while Rajputana and the Panjáb have an increased rather than a diminished industry in ornamental stones. The following facts regarding the extent of the foreign trade in certain of the inferior gems will convey some idea of the Indian lapidary industry. Chalcedonic stones are in Burma known under the generic name of *ma-hu-ya*.

EXPORTS.
615

Exports from India of Inferior Gems.—Under the heading Jade-stone Burma is said to have exported, since the beginning of the present decade, the following quantities and values :—

Years.	Quantity.	Value.
	cwt.	R
1880-81	3,371	8,03,890
1881-82	7,788	23,01,800
1882-83	4,159	9,00,900
1883-84	3,849	8,12,960
1884-85	3,738	5,60,050
1885-86	3,842	5,00,050
1886-87	2,890	5,61,000
Total	29,637	64,40,650

Thus, during the past seven years, British Burma has exported over half a million of pounds sterling worth of jade, an amount which has gone

EXPORTS.

exclusively to China and the Straits Settlements. This does not of course include the exports from the mines (in Mogoung district) by land to Yunan and China. **Dr. Anderson**, for example, in his Journey to Yunan, describes the very important industry in jade at Momein, where the stone is worked into ornaments. The Administration Reports of British Burma, which deal exclusively with that portion of the trade in jade which comes down the Irrawaddy to Rangoon, allude to jade as one of the standard articles of export trade from that province. It is shown, for example, that of the total trade in 1881-82, rice represented 82·08 per cent. of the exports, teak wood 7·64 per cent., cutch 2·56 per cent., and jade-stone 3·51 per cent. From the table given above it will be seen that the exports of jade during that year were exceptionally high, but it may safely be added that jade still holds a position as the fourth or fifth most important article of export from Burma, and that, with the extension of the British frontier, the trade in jade, in rock crystals, and in the nobler gems may in the future be considerably extended. The exceptional development of the trade in 1881-82 was due to the discovery of a new mine and the decrease that followed accounted for by the jade thus sent into the market having proved much inferior to the stone usually exported.

An inferior quality of jade-stone is also found at Mirzapur, and a very considerable trans-frontier trade is done in the Panjáb in Kara-kash jade from Turkistan, and in jade and imitations of jade or false jade from Kashmír. (See on a further page, under AGATE, variety *plasma*.)

We have alluded to jade in the present connection, not from an established belief that it belongs to the quartzose group of minerals, with which we are at present dealing, but because it is one of the so-called inferior gems. The chalcedony and rock crystal gems, however, are even as extensively employed in India as jade-stone, yet it has been found difficult to furnish definite facts regarding the extent of the internal and foreign trade in these. Perhaps the most interesting of the early accounts of the Cambay trade and industry in "Cambay stones" and Rájpípla Carnelians was written in 1787 by an explorer—**Dr. T. Hove**—who has not obtained from the writers of the past hundred years the high position which his botanical, zoological, and geological researches in Bombay merit. **Dr. Hove** states that while he was in Cambay a very considerable trade existed with Europe and Arabia in seal-shaped stones, and with China in pearl-shaped stone, as large as a pistol ball.

From *Milburn's Oriental Commerce* we learn that the sales, during the Honourable East India Company's time, fluctuated as much as they do at the present day. The average is now, however, much higher than during the first few years of the present century.

The following figures give some idea of the trade :—

The exports were valued in—

	R
1804 at	49,140
1808 at	54,240
Passing over 70 years they were in	
1874 valued at	84,370
1878 at	50,970
but the returns for the five years ending 1878 show an average of	70,000

We must now describe, as briefly as possible, the principal quartzose inferior gems :—

ROCK CRYSTAL. 616

1st.—ROCK CRYSTAL; *Mallet, Mineralogy, 62.*

Vern.—*Bilaur*, HIND.; *Phatak*, GUJRATI; *Tansala* (smoky Cairngorm), PB. The Burmese name for an Amethyst signifies "egg-plant, Sapphire."

References.—*Ball's Econ. Geol., 502; Balfour, Cycl. of India; Bomb. Gaz., VI., 201; Mason's Burma (1860), p. 579; Calcutta Jour. Nat. Hist., II.; Madras Jour., Lt. and Sci., XII., 172; Mysore Gaz., I., 20; Central Prov. Gaz., 506; Oldham, Jour. As. Soc., Beng., XXIII., 271.*

CHARACTER OF.—When pure this mineral consists chiefly of silicic acid; it is an oxide of the carbon-silicon group. The differently-coloured forms of rock-crystal owe their tints to the presence of small quantities of foreign minerals. These coloured crystals are known by various names such as the Amethyst, Cairngorm, Rose-quartz, Pellucid-quartz, False-topaz or Citrine, Smoky-quartz, Milky-quartz, Prase, Aventurine-quartz, &c.

COLOURING OF.—Artificially, all these and many other shades are, however, produced, so that a tinted crystal may be passed off on the ignorant as a ruby, topaz, amethyst, emerald, or sapphire. The following account of the processes adopted to colour rock crystals is reproduced from Dr. Balfour's *Cyclopædia of India* :—"If made red-hot, and plunged repeatedly into a tincture of cochineal, it becomes a ruby; if into a tincture of red sandal, it takes a deeper red tint; into tincture of saffron, a yellow, like the topaz; into a tincture of turnesol, a yellow like the topaz; into a mixture of tincture of turnesol and saffron it becomes an imitation of the emerald." Crystals coloured red are known in France as *rubaces* or false rubies.

PROVINCES WHERE MET WITH.—Rock Crystals are very abundantly met with in South India, as, for example, at Vellum in Tanjore, in the Godavery basin, and at Hyderabad. In the Bombay Presidency they are found at Tankára in Morvi. Blocks from one to twenty pounds are found as clear as glass and capable of taking a high polish (rock crystals are also imported into Cambay from Ceylon and China). They are by no means uncommon at Sambalpur in the Central Provinces. Agates and quartz in great beauty and variety have been reported from the Rajmahál hills in Bengal. In the Panjáb they are met with in the Gurgaon, Bannu, Sháhpúr, and Jhilam districts, and in Kashmír crystals of a large size have been found. In Burma the Shans are known to pass off coloured crystals as rubies: large crystals are found in their country. Milky-quartz occurs in Mergui.

ECONOMIC USES AND MANUFACTURES OF.—The lapidaries of Vellum have the reputation of being skilled as workers in the different varieties of rock crystal—both in the crystals found in the district and the cairngorms brought from Coimbatore. They manufacture brooch-stones cut in the brilliant, rose and other patterns, also watch-glasses and lenses. In Hyderabad the amethysts found in the Nizam's Territory are cut into beads and ring stones, the value being about the same as garnets. The crystals of Sambalpur are not worked and they have accordingly no local value. At the *loot* of the Delhi palace a number of vases, pitchers, drinking cups, &c., cut in transparent quartz, were found. These are supposed to have been cut out of large crystals found at the Arvali quartzites in the neighbourhood. The Shans of Upper Burma are said to be experts at making imitation gems from rock crystals.

2nd.—AGATE; *Mallet, Mineralogy, 70.* **AGATE. 617**

The name Agate is supposed to be derived from the *achates* (ἀχάτης) river in Sicily, or from *akik*, a river, in Arabic. AGATE *Fr.*; *Achat*, GERM.; *Akík*, ARAB; *Yamni*, HIND. (agate); *Chakmak* (a flint), HIND.; *Manka*, HIND. (cut agates and beads brought from Kandahár); *Asshar*, HIND. (Silica); *Pathanni*, HIND. (blood-stone).

They are commonly known to Europeans as Cambay stones or Godavery pebbles.

References.—*Hamilton, Capt. (1681), New account of the East Indies, I., 143; Hove, Dr. (1787), Explorations in Bombay; Sel. Rec., Bomb. Govt., XVI., pp. 49 to 51; Kennedy, Dr. (1826), Trans. Med. & Phys. Soc., Calc., III., 425; Wallace, Major (1854), Sel. Rec., Govt., Bomb., XXIII., 269; Tod's Travels; Campbell, J. M., Bomb. Gaz., III., 13; IV., 22; VI., pp. 199 to 208, & XXIII., 61; Mason, Rev. T. (1850), Prod. Burma, p. 18; Wynne, A. B. (1878), Mem. Geol., S. I., XIV., 268; Baden Powell, Pb. Prod., 97; Campbell, Capt. (1842), Cal. Jour., Nat. Hist., II., 282; Newbold, Jour. Roy. As. Soc., IX., 38; Congreve, Mad. Jour., Lit. and Sci., XXII., 237; Asiatic Res., XVIII., pt. I., 106; Blanford, H. F., Mem. Geol. S. I., IV., 217; Foote & King, Mem. Geol. S. I., 370; Ball, Econ. Geol., III., 503; Balfour, Cycl. Ind., 40, 285; Ure, Dict. Arts, Man., &c., I., 36; Encyclop. Brit., I., 277; Dana's Mineralogy.*

SOURCES.—Indian Agates are mainly obtained from the mines of Rewa Kantha in the Bombay Presidency, but they exist also in Bengal in the Rajmahál and Singbhum districts, in Hyderabad, and in the Central Provinces at Jabulpur.

Mr. Campbell thus writes of the Bombay Agates:—"Four Agates—the common, the moss, the Kapadvanj, and the veined—rank next to the Rájpipla Carnelian. The common Agate is of two kinds—a white half clear stone called *dola* or *cheshamdár*, and a cloudy or streaked stone called *jámo*. The colour varies, but is generally a greyish white. Both kinds come from north-east Káthiáwár, near Mahedpur in Morvi, three miles from Tankára. Of the stones which lie in massive blocks near the surface, the most perfect do not exceed five pounds in weight, while those of inferior quality, in many cases cracked, weigh as much as sixty pounds. These stones are brought to the Cambay dealers by merchants, who, paying a royalty to the Morvi Chief, hire labourers, generally Kolis, to gather them. When worked up, the common agate is a greyish white, and being hard, brittle, and massive, it takes a high polish."

"Like the common agate the moss agate, *sua bháji*, comes from Bud Kotra, three miles from Tankára in Morvi. Found in the plain about two feet under the surface, in massive layers, often cracked, and from half a pound to forty pounds in weight; they are gathered in the same way as the common agate. When worked up they take a fine polish, showing, on a base of crystals, sometimes clear, sometimes clouded, tracings as of dark-green or red-brown moss."

"Besides, from the town of Kapadvanj in Kaira, where, as its name shows, the Kapadvanj agate is chiefly found, this stone is brought from the bed of the river Májam, between the villages of Amliyára and Mándva, about 15 miles from Kapadvanj. It is found on the banks and in the beds of rivers, in round, kidney and almond shaped balls, from half a pound to ten pounds in weight." "The trade names for the chief varieties are *khániyu*, *ágiyu*, and *rátadin*."

"The most valued Cambay agate, the veined agate, *dorádár*, comes from Ránpur in Ahmedabad. Found near the surface, in pebbles of various shades not more than half a pound in weight; they are gathered in the same way as moss agates, and when worked up take a high polish,

AGATE.

showing either a dark ground with white streaks, or dark veins on a light black ground."

CHARACTER OF.—Agates are concretionary masses or nodules, which occur usually in hollows or veins in volcanic rocks. When cut across the sections show layers. "The colour markings are often in concentric rings of varying forms and intensity, or in straight parallel layers or bands. The colours are chiefly grey, white, yellow or brownish red." The composition of most of the forms of agate and carnelian is from 70 to 96 per cent. of silica, with varying proportions of alumina, coloured by oxide of iron or manganese.

COLOURING OF.—When the colours are indistinct or not deep enough they are readily intensified by artificial means. **Ure** says: "By boiling the colourless stone in oil, and afterwards in sulphuric acid, the oil is absorbed by the more porous layers of the stone; it subsequently becomes carbonised, and thus the contrast of the various colours is heightened. The red varieties, also, are artificially produced by boiling them in a solution of proto-sulphate of iron; after which, upon exposing the stones to heat, peroxide of iron is formed, and thus red bands or rings of varying intensities are produced. Carnelians are thus very commonly formed, the colouring matter of the true stone being a peroxide of iron."

In the concluding account of the carnelian a few extracts from Indian writers will be found as to the methods adopted in this country to colour agates and carnelians.

FORMS OF.—Some difference of opinion prevails amongst writers as to the stones which should be treated as forms of agate. The following are those most frequently described as such (separating the Carnelian by itself):—

1. "*Mocha stones*, originally brought from the East, are clear greyish chalcedonies, with clouds and dashes of rich brown of various shades. They probably owe their colour chiefly to art." *Mocha* stones are found in Dekkan traps. **Irving** (*Med. Top. of Ajmere*) mentions them as found in the bed of the Chambal.

2. "*Moss agates* are such as contain arborisations or dendrites of oxide of iron, some of which seem to be putrefactions of real vegetable forms."

Summers (*Sel. Rec., Bomb. Govt., IV., 28*) specially mentions these stones as met with at the village of Khijaria, and **Newbold** records moss agates as found in the beds of the Godavery, Kistna, and Bhima rivers.

3. "*Bloodstone* is a dark green agate containing bright red spots like blood drops.

4. "*Plasma*, a grass-green stone, found engraved in ruins at Rome, on the Schwartzwald, and on Mount Olympus, appears to be chalcedony coloured with chloride." According to some writers the *false jade* sold in Upper India has its nearest affinity to *plasma*. It is found at Kandahár and is brought down the Indus on rafts floated with inflated skins to Attock. It is then conveyed to Bhera, where it is extensively employed by the lapidary cutlers. *Plasma* has been reported as found in the Nizam's territory south to the Bhima river, and **Dr. Voysey** mentions a form of *plasma* as seen in the Dekkan trap of the Sáwilgarh hills.

5. "*Chrysoprase*, found in Silesia, is an agate coloured apple-green by oxide of nickel."

6. *The Scotch Pebble or Fortification agate.*—This is a form known chiefly by its zigzag pattern.

USES OF.—Agates are much used for ornamental and decorative purposes. These are made into brooches, rings, seals, and cups, knive handles, sword hilts, beads, smelling-bottles, snuff-boxes, paper-weights, buttons, paper-cutters, &c., &c. They were in India formerly much used for inlay.

ing in marble and to a certain extent are so employed at Agra and other places, where marble plates, boxes, &c., are made. Agates are also used for burnishing gold and silver and by the book-binders; they are made into the finer mortars used by the chemist, as well as employed for the pivots of chemical balances, &c. AGATE.

Some doubt seems still to exist as to the material of which the *murrhine* cup which Nero paid £56,000 for was made. Professor Muller seems to be of opinion that it was flourspar, but Ball very properly comments upon this opinion: 'if it was obtained at Ujein or Ouzein, or any other locality within the trappean area, it was almost certain to have been one of the chalcedonic minerals, *viz.*, carnelian or agate. Flour spar is not known to occur in the trap.'"

3rd—CARNELIAN (from *Caro-nis*, flesh, in allusion to the colour); *Mallet, Mineralogy, 72.* CARNELIAN. 618

CORNALINE, *Fr.*; KARNEOL, *Germ.*; CORNALINA, *It.*

Vern. *Sung-i-akik*, HIND. & PB.; *Ghori* (white and red Carnelians from Kandahár), PB.; *Ghár*, in its uncut state, and *Akik* when worked up, in Gujaráti. One of its Burmese names, according to Mason, is *Kyatthwe*, or fowl's blood. *Sard* is the name generally given to deepred carnelians.

References.—*Ball, Econ. Geol., 506; Balfour, Cycl., I., 555 & 583; Encycl. Brit., I., 277; Ure's Dict., Arts, &c., I., 656; Baden Powell, Pb. Prod., 97; Copeland, Bomb. Researches; Thomson, Mad. Jour., Lit. and Sci., V., 161.*

Mr. J. M. Campbell, in his *Gazetteer* of the Cambay States, gives an instructive account of the history and present position of the industry in agates and carnelians. Space cannot be afforded to do more than to single out, in the following remarks, the prominent features of that trade; the reader is referred for further information to Volume VI. of the *Bombay Gazetteer*. The works and Journals referred to under Agate may also be consulted.

CHARACTERS OF.—Dana defines the carnelian as a reddish variety of chalcedony, generally of a clear bright tint, but it is sometimes of a yellow or brown colour, passing into common chalcedony through greyish red. White carnelians also occur and are prized, but they are rare.

SOURCE.—The principal source of Carnelians are the mines of Rátanpur in the Rájpipla State, about 14 miles of Broach. Agates come mainly from Rewakantha, Kapadvanj, and Sukultirth on the Nerbadda, and from Rájkot in Káthiáwár. Carnelians are also found in Burma, Mergui, and abundantly so in Japan.

ARTIFICIAL COLOURING OF AGATES INTO CARNELIANS.—While collecting the pebbles the miners divide them into two primary classes—those that are not improved in colour by burning, and those that are. Of the former there are three chief varieties: (1) the Onyx, known as *mora* or *báwa ghori*; (2) the Cat's-eye, *cheshamdár* or *dola*; and (3) a yellow half clear pebble called *rori* or *lasania*. All other stones are baked to bring out their colour. "During the hot season, generally in March and April, the stones are spread in the sun in an open field. Then in May, a trench, two feet deep by three wide, is dug round the field. The pebbles are gathered into earthen pots, which, with their mouths down and a hole broken in their bottoms, are set in a row in the trench. Round the pots goat or cowdung cakes are piled, and the whole is kept burning from sunset to sunrise. Then the pots are taken out, the stones examined, and the good ones stowed in bags. About the end of May, the bags are

CARNELIAN. carried to the Nerbadda and floated to Broach. Here they are shipped in large vessels for Cambay, and are offered for sale to the Carnelian dealers.

"By exposure to the sun and fire, among browns the light shades brighten into white, and the darker deepen into chestnut. Of yellows, maize gains a rosy tint, orange is intensified into red, and an intermediate shade of yellow becomes pinkish purple. Pebbles in which cloudy browns and yellows were at first mixed are now marked by clear bands of white and red. The hue of the red carnelian varies from the palest flesh to the deepest blood red. The best are a deep clear and even red, free from cracks, flaws, or veins. The larger and thicker the stone, the more it is esteemed. White carnelians are scarce. When large, thick, even coloured, and free from flaws, they are valuable; yellow and variegated stones are worth little."

USES OF.—Carnelians are extensively used for seals. Many of the antique gems are engraved on carnelian.

ONYX. 619

4th—ONYX; *Mallet, Mineralogy, 73.*

ONYX, ONICE, *Fr.;* ONYX, *Germ.;* ONIQUE, *Sp.*

References.—*Ball's Econ. Geol., 503; Mason's Burma, 581; B. Heyne, Indian Tracts, p. 265; Newbold, Jour. Royal Asiatic Soc., IX., 37.*

The Onyx resembles the agate very closely, differing only in the fact that the colours are arranged in flat horizontal planes. This stone was once highly esteemed. A form of it, known as the *Onicolo,* is still used, although not so extensively as formerly, for Cameos. The surface colour of the *Onicolo* is bluish white and the underlying layer of a deep brown. This brown layer in the Cameo is made to form the ground, the figure being cut in the white. *Sard-Onyx* is Onyx with stripes of sard or deep red carnelian. **Irvine**, in his *Topography of Ajmere,* alludes to onyx as found in Rajputana. **Mason** says: "The Onyx is often seen in Burma, but the localities whence it comes are not known." This mineral has, however, been recorded as found in the Dekkan traps of the Godavery, Kistna, and Bhima, and also near Hyderabad.

USES OF.—Onyx is employed for brooches, signets, rings, seals, studs, and such like articles.

JASPER. 620

5th—JASPER; *Mallet, Mineralogy, 76.*

JASPE, *Fr.;* JASPISS, *Germ. & Dutch;* DIASPRO, *It.;* JASCHMA, *Russ.*

References.—*Mason's Burma, 581; Ball, Econ. Geol., 503.*

As already stated this stone has been referred to the present position more as a matter of convenience than of scientific classification. It is a quartzose mineral of a red or yellow colour. The former occurs among the Cambay stones from the Dekkan, and the latter is found in Tenasserim. A soft green jasper and also a striped jasper are found in Burma, and known as *na-ga-thway* (? *naga* the dragon, and *thwe* blood). **Mason** says: "Jasper is regarded as a variety of quartz, and is not uncommon. I have met with yellow jasper on the Tenasserim, and red jasper on the Toungoo Mountains." Jasper is abundant in the transition rocks of Kadapah; ribbon jasper is said by **Mr. Foote** to be largely produced in the Sandur hills in Bellary. Bright red jasper is also reported to be abundant in the transition rocks of the Narbada and Sone Valleys. Nodules of jasper are also common in conglomerate rocks.

JASPER.

USES OF.—Sometimes employed for seals.

HELIETROPE.

The HELIETROPE is by most writers treated as a form of jasper, but by some it is regarded as a form of bloodstone (see under AGATE No. 617). It may almost be said in general appearance to differ from green jasper merely in being spotted or streaked.

OPAL. 621

6th—OPAL; *Mallet, Mineralogy, 80.*

OPALE, *Fr.*; OPAL, *Germ.*; OPALO, *It.*; *Dhúdiá pathar*, HIND. Chalcedony and Opal are sometimes known as *Gomed sannibh*, HIND.

This is a compact uncrystalline semi-transparent to opaque hydrated silica. When of milky-white colour, opalescent, and exhibiting a rich play of colours, it is the *Noble Opal.* When not opalescent it is the *Common Opal.* The former are obtained chiefly from Hungary and Mexico, and the latter are fairly abundant in the Dekkan trap rocks. While this is so the first mention of the Noble Opal describes it as having come from the Indus. **Colonel Sikes** mentions opals as found at Poona, and **Captain Newbold** refers to those found on the plains of Bejapore and Sitabaldi.

On being first dug out of the earth opal is said to be soft, and to harden and diminish in bulk on being exposed to the atmosphere.

CAT'S-EYES. 622

7th—CAT'S-EYES; *Mallet, Mineralogy, 69.*

This stone is perhaps closely allied to Onyx; but by some writers it is placed nearer rock crystal. It is a translucent quartz, presenting a peculiar opalescent reflection, said to be due to the presence of asbestos. It is called cat's-eye from the resemblance it bears to the eye of a cat, an idea which the Burmans have also accepted, their name for the stone, *kyoung*, meaning cat. **Mason** says: "The stones are common and cheap, but the Burmese localities where they are found are not known." A stone collected at various localities on the Malabar Coast is generally accepted as a form of cat's-eyes. They are sent from Cambay to Bombay, thence to Persia and Arabia, and are best known in India under the name of *Chush-Maidar* or *Lasniyán*. *Lahasaniá* is the name given to peculiar green coloured cat's-eyes that are said to "have reflecting powers like those of the looking-glass" (*Raja Sourindro Mohun Tagore*). *Rorí* and *Lussuniá* are names given to a much valued pebble, found scantily with cat's-eyes in the Rájpipla mines of Bombay (*Select. Records, Bomb., New Series, No. IV., 31*).

LAPIDARIES' ART.

It is not proposed to deal with this subject in the present article, it having been deemed desirable to give in one place under "LAPIDARY" an abstract of all that is known regarding this industry, not merely as practised with the inferior gems but with all gems and ornamental stones. For convenience the reader may, however, be referred to the following works which deal more immediately with the cutting, &c., of the inferior gems :—

Bom. Gaz., VI., 201; Hoey, Trade and Manuf. of Northern India, pp. 54 and 119; Baden Powell, Pb. Manuf., 192; Kipling, Cat. Cal. Intern. Exh., Pb. Section, 28; Burma Admin. Rep., 1882-83, p. 64; Hendley, Indian Art Journ., Part 2, 28.

The above account of the inferior gems was in type before the writer received **Mr. Mallet's** Vol. IV. of the "Manual of Geology of India."

CAT'S EYES. He has therefore been unable to do more than give references to **Mr. Mallet's** account of these minerals, but the reader is referred to that work for fuller particulars.

See "CARBUNCLE," "DIAMOND," "JADE," "GARNET," "LAPIDARY," "PRECIOUS STONES," and RUBY."

Carob tree. See **Ceratonia Siliqua,** *Linn ;* LEGUMINOSÆ.

CAROXYLON, *Thumb.; Gen. Pl., III., 71.*

623 **Caroxylon fœtidum,** *Moq.; Fl. Br. Ind., V., 18;* CHENOPODIACEÆ.

Syn. SALSOLA FŒTIDA, *Del.*, which see; also under CAMEL FODDER, 39.

C. Griffithii, *Moq.; DC. Prodr., XIII., 2, 175.*

An Afghanistán plant, supposed by **Stewart** and several other writers to be the botanical name for the Sind and Panjáb *láná*, from which *Khár-sajji* is made. This is **Haloxylon recurvum,** *Bunge*, or the **Salsola lana,** *Stocks: Fl. Br. Ind., V., 15.* See also under **Camel Fodder 21,** and **Haloxylon recurvum.** Correct the mistake of **Caroxylon Griffithii** into **Haloxylon recurvum** in **BARILLA, B. 163.**

CARPESIUM, *Linn.; Gen. Pl., II., 336.*

624 **Carpesium abrotanoides,** *Linn., Fl. Br. Ind., III., 301;* COMPOSITÆ.

Syn.—CARPESIUM RACEMOSUM, *Wall.*

Vern.—*Wotiangil*, KASHMÍR; *Hukmandás*, PB.

Reference.—*Baden Powell, Pb. Pr., 357.*

Habitat.—A stout herb, met with abundantly in Kashmír, extending along the Himálaya to Sikkim; altitude 5,000 to 10,000 feet. Some of the specimens so named by **Wallich** belong to **Rhynchospermum verticillatum,** *Reinw.*, a plant which extends to the Khásia Hills and Burma, descending to lower altitudes than **Carpesium.**

DYE. 625 **Dye.**—Largely used in Kashmír as a yellow dye for silk. This dye-stuff is quite unknown outside Kashmír. It is described by **Vigne** and **Stewart** and by the *Flora of British India*, iii., 301. It is unknown to the dyers in the plains of India, and has never been experimented with in Europe. Specimens of the plant and of the dye-stuff are very much required, as well as definite information as to the method of extracting the dye.

Medicine.—Mr. Honigberger says that the plant is also used medicinally in the Panjáb. The plant is common at Simla, but its properties are quite unknown to the hill people.

CARPETS.

626 **Carpets and Rugs.**

TAPIS, *Fr.;* TEPPICHE, *Germ.;* TAPYTEN, *Dutch;* TAPPETI, *It.;* TAPETES, ALFOMBRAS, ALCITIFAS, *Sp.;* KOWRU, KILIMI, *Rus.*

The term Carpet is probably connected with the Latin *tapetes* from whence tapestry.

Vern.—*Dari* (small rug), *satranji* (large carpet), cotton; *Kálín* (large carpet), *gálícha* or *kálícha* (small rug), woollen, HIND.; *Ghalichah*, PERS.; *Janíkalam*, TAM.; *Jamcana*, TEL.; *Jemkhani* (in Belgaum), BOMB.; *Parmadani*, MALAY.

References.—*Birdwood, Memo., 29th Sept. 1879; Indian Arts, 284; Vincent J. Robinson, Eastern Carpets, also Journ. Soc. Art (1886), p. 447; Baden Powell, Manuf. and Arts, Panjáb, pp. 10 & 26; Dr. Forbes Watson's Rep.; Col. Davidson in Rep., Hyderabad Committee; Balfour, Cyclop. of Ind., I., 584; Ure, Dict. Art and Manuf., I., 657; Spons' Encycl., II., 2095, 2106; Encycl. Brit., V., 127; Calc. Intern. Exhib. Cats., Kipling, Punj. Sec., 42; Bidie, Mad., 34; Griffith, Bom., and Pitcher, N.-W. P., 61; Hyderabad, p. 18; Gazs., Bom. Vols. XII., 232; XVI., 169; XXI., 342; XXIII., 372; N.-W. P., Vol. I., 116; Punj. Vols. Ambala, 53; Amritsar, 44; Dera Ghazi Khan, 92; Dera Ismail Khan, 142; Lahore, 101; Multan, 107-8; Peshawar, 148.*

It is not contemplated in the present article to do more than draw attention to the main facts regarding the Indian Carpet Industry, the object being more to indicate the nature of the carpets made, the materials of which they are woven and the dyes employed in their coloration, than to treat of the historic and artistic features of the manufactured articles. Indian carpets may be classified either according to the nature of the materials of which they are made or the manner in which they are woven. There are cotton, woollen, silk, goat's-hair, yak's hair, and pashm carpets, or mixed carpets of any two or more of these materials. Then again, there are carpets woven by the warp horizontal, and others in which it is vertical. The former are chiefly cotton carpets and the latter nearly always woollen, although it is frequent in both classes to use cotton or hemp for the warp and wool or hair for the woof. The warp, with the single exception of the so-called Jabbalpur *darí*, is not coloured, but the woof is so manipulated that in both these classes of carpets it covers the warp. The Jabbalpur *darís* are almost precisely of the same character as the Kidderminister or Scotch carpets—a certain proportion of the pattern being developed by the coloured warp which may be either in bands of different shades or of one uniform colour. In such carpets longitudinal or checked patterns are produced, whereas in the ordinary *darí* or cotton carpet the patterns run across the warp.

Popularly the terms *darí* and *satranjí* are applied synonymously to cotton carpets, but in more precise language, the former is a rug or small cotton carpet and the latter a large one. *Darís* (=*dar*, a door, *darís* being literally door-mats) are used by the natives to sleep or to pray upon, and *satranjís* to cover the floor of a room or tent. Woollen carpets are known generically as *Kálíns* (carpets) or *Kálíchas* or *gálíchas* (rugs). The Anglo-Indians speak of "*darís*" for all cotton carpets and *carpets* for woollen carpets, but more particularly pile carpets or those woven on a vertical warp.

The following extracts from the *Bombay Gazetteer* (*Vol. XIII.*, *401*) express clearly the difference between the *darí* and the *kálín*, while at the same time they describe the processes which are pursued, with only slight variations, throughout India in the manufacture of these carpets:—

1st, DARIS.—"The cotton carpet loom which lies horizontally along the floor passes round stout poles at either end which are secured by ropes tied to strong wooden pegs driven into the ground. The weavers crouch on a broad wooden plank placed across the warp. This plank rests on stones at the side of the loom, and as the work goes on is moved forward. The design is formed in the same way as in weaving Persian carpets—by passing the different coloured threads through the strands of the warp, as called out by the overseer in charge. Instead of being cut off, these threads are left slack and driven home by a fork-like instrument called the heckle, the white warp threads being entirely hidden by the weft, which forms the colouring of the carpet. The loom has only two heddles. The striped cotton carpet loom differs from the coarse cloth-loom only by DARIS. 627

DARIS.

being broader and having a stronger reed or *phani*. The chief aim of the carpet-weaver is to hide completely the white warp-yarn, leaving unbroken belts of the coloured weft. For this purpose, each time the shuttle passes, the weaver inserts his index finger about the middle of the warp and pushes the weft-yarn forward to the middle of the reed or *phani*, making an angular arch with the fabric already woven. He then drives the weft-yarn home, thus using a greater length of weft-yarn than the breadth of the carpet.

"A cotton carpet costs from $3\frac{3}{4}d.$ to $7\frac{1}{2}d.$ ($2\frac{1}{2}$ annas to 5 annas) a square foot. There are (1882) twenty cotton carpet weavers."

Mr. Baden Powell describes the process by which the shuttle and with it the woof is carried across, and by which after each passage of the woof the warp is crossed and re-crossed. "This is effected by placing a long pole, supported at either end by two legs, trestle fashion, across the whole width of the warp. This pole is called a '*gori*' (which literally means 'mare,' and so called from its rude resemblance to a quadruped); from the 'gori' are hung two bamboos, each of which carry a number of threads, which are attached to the under and upper threads of the web respectively. When it is desired to cross the threads of the warp, it is simply necessary to pull up one of the bamboos and lower the other: as the bamboos are merely hung to the 'gori' by ropes at each end, the raising and lowering is easily done by tightening or loosening the suspending string by means of a stick attached. No regular shuttle is used. A number of workmen sit in a row, on that part of the *durree* (*darí*) which has already been completed, and pass the thread along between the lines of the warp, from hand to hand. The thread is wound in a long egg-shape on an iron skewer or needle.

"If the pattern is elaborate there will be a considerable number of these thread shuttles at work: each workman has charge of his own, and passes it along according to the pattern, taking the thread out and allowing the next workman to insert and withdraw his shuttle in the same manner, and so on; the threads as they are passed through the threads of the warp, are kept close together and the work is rendered compact and even by striking between the lines of the warp with a kind of fork, having a wooden handle and iron teeth and called a *kangí*."

Sir George Birdwood says: "*Darís* and *satrangis* are made of cotton, and in pattern are usually striped blue and red, or blue and white, or chocolate and blue; and often squares and diamond shapes are introduced, with sometimes gold and silver, producing wild picturesque designs like those seen on the bodice and apron worn by Italian peasant women. They are chiefly made in Bengal and Northern India, and, like the loom-made *dhotis* and *saris*, illustrate the most ancient ornamental designs in India, perhaps earlier even than the immigration of the Aryas. In Bombay cotton carpets of some note are made at Nasik, Bijapur, Khandesh, and Belgaum." In the North-West Provinces they are made at Lohamandi, Nayobosti, and Fatehpur Sikri, about 2,000 persons being so employed: small praying rugs are largely made at Agra.

Woollen *darís* are, however, also made in many parts of India, as in the Panjáb and Bombay. Those woven by the aboriginal races are small in size, thick in texture, and even painfully uneven in quality but they are elaborately and quaintly ornamented.

PILE CARPETS. 628

2nd, PILE CARPETS OR KÁLÍNS, the smaller sizes or rugs being known as GÁLÍCHAS.—In Europe it is customary to speak of oriental pile carpets as belonging to either of three classes—*Persian*, *Turkey*, and *Indian*. It seems likely, however, that, as pointed out by **Sir George Birdwood**, this industry was introduced into India by the Saracens. Hence in all probability

PILE CARPETS.

the fact that in India they are often spoken of as Persian carpets, the *daris* described above being viewed as Indian carpets. "Carpet-making, which from the Hindustani names for all the parts of the loom, seems to have been brought from North India, is almost entirely in the hands of the Musalmans (*Bom. Gaz., XIII., 401*)."

"Persian carpet-looms differ from plain carpet-looms in having the warp fastened vertically, instead of horizontally, in the absence of heddles and treddles, and in the absence of the reed *phani*. The loom consists of two uprights, from fifteen to twenty feet high and from ten to fifteen feet apart, supporting two beams, one fixed to the lower ends of the uprights and the other moveable. The warp-yarn is passed round these beams forming a huge embroidery-like frame. On one side of this frame from three to six workmen sit, while on the other side the overseer stands with a sketch or sample of the design before him. When all is ready, he calls out to the workmen the number of loops of each variety of coloured wool that have to be taken up for the first row. The workmen repeat in chorus what the overseer says, and fix up the loops, tie a knot, and cut the pieces off. As soon as the first row is ready, a weft-yarn is passed between the two sets of the warp, and is fixed tightly in its place by the aid of a fork-like instrument called the heckle. In this manner row after row is laid up, till the whole of the carpet is woven, when it is taken down from the loom, spread on the floor, and sheared.

"Persian carpets vary in price, according to texture and design, from 14*s.* to £1-8*s.* (R7—R14) the superficial square yard. There are (1882) seventy-five Persian carpet weavers" (*Bomb. Gaz., XIII., District 401*).

Present Position and Future Prospects of the Indian Pile Carpet Industry.

Mr. Kipling writes as follows:—"There is not the faintest doubt that a great trade is possible in Indian carpets, if they are good in design. One great secret of the demand for them is believed to be their durability as compared with carpets of English make. The warp is of strong elastic cotton threads, which are soft in texture and not made hard and tight by over-twisting and sizing. On these wool thread is tied, and the allowance of wool is very liberal. The looms are large enough to make any size of carpet, and there are, therefore, no seams. For ordinary English carpets the warp is of hard fine cords, and there is very frequently an underlay of jute, which does not appear either on the back or front of the carpet, but which gives substance and firmness to the fabric. Into this sub-structure the woollen threads are tightly woven, a long needle holding the loop, which, as it is cut by the withdrawal of the knife with which the needle is terminated, forms the pile. The demand for cheapness makes economy of wool a great point in the manufacture, and many English carpets are in reality a firm fabric of flax or cotton and jute with a slight covering of wool. The jute is exceedingly hard and sharp, and as the wool is pressed against it by use, the softer material wears and cuts away. In an Indian carpet, the whole fabric sinks together under the foot.

"Moreover, very few of the English jacquered power looms are more than three-quarters of a yard wide. Hence the necessity for seams, which are the first places to wear thread-bare.

"So it may be said that it is more economical, when buying a carpet, to give three or four times the English price for an Indian hand-woven fabric. It is not, of course, contended that bad Indian carpets are

PILE CARPETS.

impossible. There are several practices, such as *jhutha bharni*,—literally, a false weft, a way of taking up two threads instead of one, which are common even in some good jail factories, and which detract considerably from their value. But the general conditions of Indian carpet-weaving are distinctly more favourable to the production of a serviceable fabric than those which obtain in England. **Mr. Morris**, of the well-known Oxford Street firm of designers and decorators, has indeed started looms in England which are similar to those in use in this country, and young Englishwomen produce Hammersmith carpets of great beauty, but at a high rate. This instance, however, is scarcely necessary to prove a well-known fact—the demand for a good hand-made carpet. One of the difficulties that industrial schools like those of Hoshiarpur and Kasur have to contend with is the absence of continuous direction by any one who is in touch with the requirements of the largest consumers. It may be worth while to indicate briefly the sorts of carpets for which there is likely to be a regular demand For the very best there can only be a limited sale. Carpets at and above fifteen rupees a square yard must be not only of good quality and a fine count of stitch, but they must also be of a choice design. Where facilities exist for the production of these costly fabrics, and pains are taken to secure good designs, they can be profitably made; but a greater variety of patterns than such schools have contented themselves with is absolutely necessary. A cheaper carpet, with no more than nine stitches to the inch, and costing about seven rupees a square yard, is now wanted; and for such goods, if the colours are good and the designs are characteristic, there will be an almost unlimited sale. The jails have set a pattern which is followed too faithfully by industrial schools. This type is the design known as 'old shawl;' an equal and formless sprinkling of somewhat hot colour all over the field. And modern native designers are too apt to imitate mere minuteness. In the best Persian carpets and those of Warangal, which, though made in Southern India, are really of Persian origin, precisely as the cotton prints of Masulipatam are identical in tone and pattern with the 'persiennes' of Teheran, the designs are bold and full of variety, each carpet possessing a distinctive character and key-note. The slavish and spiritless copying of both jails and industrial schools does not seem to promise much for the future; but if models of a larger and more artistic quality of design are followed it may be that in time the natural aptitude for design which still exists will again be developed" (*J. Kipling, Esq., C.I.E., in Pb. Gaz., Hoshiarpur Dist., p. 111*).

Much difference of opinion seems to prevail amongst writers on Indian pile carpets as to the position this industry occupied 30 or 40 years ago. **Mr. Vincent Robinson**, adopting the views advocated by **Sir George Birdwood** in his *Indian Arts*, lays a large share of the acknowledged degeneration to the charge of the Indian jails. In his paper read before the Society of Arts (*March 19th 1886*) he says: "Twenty years ago—the reputation of India for its carpets having been established in Europe at the first Exhibition of 1851, and subsequently well developed by private enterprise—the Government of India, casting about in the midst of difficulties with taxation, blundered on the scene" and introduced carpet manufacture into the jails in the hope of thereby making those at least self-supporting instead of a burden to the country. He continues: "I have already shown that the reputation of these carpets was not a fresh creation; it was an art upon the practice of which thousands of our fellow-subjects in India depended for their livelihood. It had its traditions, its methods, its caste." The Government, through the hope of gain, rushed into the resuscitated industry. "Buildings were adapted, plant on so-called improved English or European models obtained and fixed, and the arma-

C. 628

PILE CARPETS.

ments of chemical laboratories with their processes introduced; and such a system of organised work set up as completely transformed not only the trade but actually the carpets themselves which were the foundation of it." But may it not fairly be asked, since pile carpet-weaving is admittedly a Muhammadan introduction to India, is it likely to have ever been bound by Hindu "caste influences" and well established "traditions" and "methods?" The Warangal weavers of Hyderabad, who are held by many to produce the best Indian carpets, are Muhammadans of the Sunni sect, who are said to be descendants of Persian settlers. So in Bombay, and indeed in most parts of India, the weavers are to this day Muhammadans and therefore never had any caste. Is it, however, a fact that in the jails European machinery has been extensively applied to the pile carpet industry? **Mr. Kipling** holds a different opinion to **Mr. Robinson**, and remarks: "it has been said that the Panjáb jails have injured the indigenous industry of carpet-weaving. It would be more like the truth to assert that they have created such as exists. It was not until the Exhibition of 1862, that the Panjáb was known beyond its border for the production of carpets, and then only by the productions of the Lahore jail executed for a London firm. There exist no specimens to show that the Multan industry, the only indigenous one of the province, was of either artistic or commercial importance. The success of the Lahore jail led to the introduction of the manufacture in other jails, and it is now taken up by independent persons." **Sir George Birdwood** would appear to have modified his opinion as to the injurious influence of the Indian jails, for, in his remarks on **Mr. Vincent Robinson's** address to the Society of Arts, he is reported to have said—"At one time I attributed this degeneration almost exclusively to the influence of the Government Schools of Art and the jails: but at present I feel that it is chiefly due to the influence of English commerce on the historical handicrafts of India." This seems a much more likely explanation, and that a considerable trade was done in western and southern India, in Indian pile carpets, previously to the Exhibition of 1851, is undeniable. Reference is repeatedly made to this trade in the records of the Hon'ble East India Company's proceedings. This, for example, is alluded to as follows in the *Gazetteer* for Cambay:—

"Cambay carpets had once a great name. Among the articles mentioned in the proclamation of 1630 'for restraining the excess of private trade to the East Indies,' are rich carpets of Cambay. Later on a chief part of the Senior Factor's duty at Cambay was to buy carpets 'valuable in Europe;' and in another place, Cambay carpets are spoken of as equal to any of Turkey and Persia. Though this trade has greatly fallen off, there are still four carpet factories, each paying the Nawab a yearly tax of £1-10-0 (R15)."

That the extent and character of the Indian pile carpet trade has declined is all but universally admitted.

Pile carpets are made of cotton at Hyderabad and at many other places, tufts of cotton yarn being used in place of wool. In the same way expensive pile carpets are made of silk, but more frequently silk is used with certain colours to bring out the pattern: especially so with camel-hair and pashima carpets. In Hyderabad cotton pile carpets are said to be preferred because "prettier and more durable, and they therefore fetch higher prices than the others."

Pile Carpets ARE MADE at a limited number of jails in each Presidency and Province and by a few private manufacturers scattered here and there over the country. The references given to the Gazetteers convey some idea of the distribution of the industry, but it may be concluded that

C. 628

PILE CARPETS. the carpets of Kashmír, Afghanistán, Lahore, Multan, Delhi, Jeypore, Agra, Mirzapore, Jhansi, Lucknow, Fatehgarh, Allahabad, Jabbalpur, Sind, Baluchistán, Bombay, Hyderabad, Warrangal, Malabar, Coconada, Masulipatam, Tanjore, Ellore, Salem, Midnapur, Calcutta, Murshedabad and Benares are best known.

For farther information the reader is referred to the articles "COTTON," "HAIR," PASHM," "SILK," and "WOOL." For the dyes used in carpet making to the article "DYES and DYEING."

Complete information as to the places at which various kinds of carpets, cotton and woollen, are made can be obtained from the authorities of the Indian Museum in Calcutta.

629 CARPINUS, *Linn.; Gen. Pl., III., 405.*

Carpinus faginea, *Lindl.; DC. Prodr., XVI., 2, 127;* CUPULIFERÆ.

Vern.—*Shirásh, ímar, bíjawwi,* PB.; *Gísh,* N.-W. P.

References.—*Brandis, For. Fl., 492; Gamble, Man. Timb., 390.*

Habitat.—A moderate-sized tree of the Himálaya, from Kumaon (and Nepal?) eastward, altitude 4,000 to 7,000 feet.

TIMBER. 630 **Structure of the Wood.**—Similar to the next species.

631 C. viminea, *Wall.; DC. Prodr., XVI., 2, 127.*

INDIAN HORNBEAM.

Vern.—*Charkhri, kái,* PB.; *Pumne, goria, chamkharak,* N.-W. P.; *Chukissi, konikath,* NEPAL.

References.—*Brandis, For. Fl., 492; Kurz., For. Fl. Burm., 477; Gamble, Man. Timb., 390; Stewart, Pb. Pl., 200; Baden Powell, Pb. Pr., 572; Balfour, Cyclop.*

Habitat.—A moderate-sized tree of the Himálaya, from the Ravi eastward, from 5,000 to 7,000 feet, frequent near water. Also met with in the Martaban Hills, altitude 5,000 to 6,000 feet, and, according to **Brandis,** on the Khásia Hills.

Structure of the Wood.—White, shining; no heartwood; warps in seasoning. Weight 50℔ per cubic foot; growth moderately slow. The stem is irregular in section, like that of the European *Hornbeam,* which it much resembles both in bark, wood, and general appearance. **Cleghorn** states that it is much esteemed by carpenters.

Carrot. See **Daucus Carota,** *Linn.;* UMBELLIFERÆ.

632 CART AND CARRIAGE BUILDING—Woods used for—

During the Colonial and Indian Exhibition two conferences were held to examine the timbers shown in the Imperial Indian Section. **Mr. Hooper,** the well-known London Coach Builder, remarked: "That a wood was much wanted in the carriage trade for making wheels intended for foreign countries. The export trade in English carriages to India, for example, had almost ceased, owing to the fact that the wood used in these carriages would not stand the climate. He thought if the woods employed in India for wheel-making could be cut up into working sizes and well seasoned before being exported to Europe, the carriage trade could be revived, as English-made carriages were preferred to the Indian." "**Sir E. C. Buck** added that during the Egyptian war it was noticed that the wheels which stood the climate best were those made in India, where the hot dry weather of the north seasoned the wood in a way very much superior to the artificial methods employed in Europe." The following are the timbers used in India for these purposes, more especially those marked*:— WOOD USED FOR CART AND CARRIAGE BUILDING.

Acacia ferruginea (carts).
A. melanoxylon (coaches, railway carriages).
Albizzia amara (carts).
Barringtonia acutangula (carts).
B. racemosa (carts.)
Bassia longifolia (carts.)
Berrya Ammonilla (carts).
Briedelia montana (carts).
B. retusa (carts).
Calamus Rotang (carriages).
Careya arborea (carts).
Cassia Fistula (carts).
Chloroxylon Swietenia (carts).
Cynometra ramiflora (carts).
***Dalbergia latifolia** (wheels; gun-carriages).
***D. Sissoo** (felloes, naves; carts).
Diospyros melanoxylon (carriage shafts).
Eugenia Jambolana (carts).
Ficus bengalensis (cart yokes).
Gmelina arborea (carriages, palanquins).
***Heritiera littoralis** (buggy shafts).
Hymenodictyon excelsum (palanquins).
***Lagerstrœmia Flos-Reginæ** (carts, gun-carriages).
***Lagerstrœmia parviflora** (buggy shafts).
Melia Azadirachta (carts.)
Michelia Champaca (carriages).
Miliusa velutina (carts).
Mimusops Elengi (carts).
Prosopis spicigera (carts).
***Pterocarpus indicus** (carts, gun-carriages).
P. Marsupium (carts).
Pterospermum suberifolium (carts).
Sandoricum indicum (carts).
Sapindus emarginatus (carts).
Schleichera trijuga.
Shorea robusta.
Strychnos Nux-vomica.
S. potatorum.
Tectona grandis (railway carriages).
Terminalia Arjuna.
T. belerica.
T. Chebula.
T. tomentosa.
Thespesia populnea (carts and carriages).
Ulmus integrifolia (carts).
Vitex altissima (carts).
Xylia dolabriformis (carts).
Zizyphus zylopyra (carts).

CARTHAMUS, *Linn.; Gen. Pl., II., 483.*

Carthamus oxyacantha, *Bieb.; Fl. Br. Ind., III., 386;* COMPOSITÆ. 633

Vern.—*Kantiári, kandiára, poli, kháreza, karar, poliyán,* PB.

References.—*Stewart, Pb. Pl., 123; Aitchison, Cat., Pb. Pl., 80; Baden Powell, Pb. Pr., 356; Cooke, Oils and Oilseeds, 34; Balfour, Cyclop.*

Habitat.—Wild in the North-West Provinces and the Panjáb, most common in the more arid tracts. **Mr. C. B. Clarke** thinks this may be the wild form of Safflower.

Oil.—Dr. **Stewart** says that near Peshávar and elsewhere in the Panjáb, an oil is extracted from the seeds which is used for illuminating purposes, as well as for food. Dr. **Stocks** probably alludes to this when he says, under the oil from the seed of **C. tinctorius**: "There is a wild seed" in Sind, "which is also called *Powári,* but it is of no use." OIL. 634

Medicine.—Dr. **Bellew** remarks that the oil is used medicinally. MEDICINE 635

Food.—The seeds are sometimes eaten by the natives parched, alone or with wheat, or are ground and mixed with wheaten flour. FOOD. 636

C. tinctorius, *Linn.; Fl. Br. Ind., III., 386.* 637

THE SAFFLOWER, WILD OR BASTARD SAFFRON, AFRICAN SAFFRON, AMERICAN SAFFRON, CARTHAMINE DYE, *Eng.;* CARTAME, SAFRAN BATARD, *Fr.;* DER SAFFLOR, FARBERDISTEL, FALSCHE

The Safflower.

SAFRAN, *Germ.;* ZAFFRONE, CARTAMO, *It. & Sp.;* POLERROI, *Russ.*

Vern.—*Kusum, kásumba, kar* (the seed), *barre,* HIND.; *Kusum, kusamphul, kajírah,* BENG.; *Galáp machú,* MANIPUR; *Kúsam, kúrtam, kushumbha, ma, sufir, karar,* PB.; *Barre, kar,* N.-W. P.; *Bundi,* RAJ.; *Kusumba, kardai,* BOMB.; *Kusúmbo,* GUJ.; *Kurdi, kararhi, kasdi, sadhi* (oil plant), *kardai,* MAR.; *Kusumba,* CUTCH; *Powári-jo-bij* (seed), *khoinbo* (the plant), SIND.; *Sendurgam, kushumbá, kushumba-virai, sendurkun,* TAM.; *Agnisikha, kúshumbá-vittulu,* TEL.; *Kusanbe* (or *kusambi*), *kusumba,* KAN.; *Heboo, su, hshú, supán, subán,* BURM.; *Qurtum, qirtum, usfar,* ARAB.; *Kazhirah, muasfir, kasakdánah,* PERS.; *Kusumbha, kamalottara, kúshumbha,* SANS.; *Kurtim,* EGYPT. The κνῆκος, κνίκος of the Greeks.

In Sind the seeds are called *Kardai* (*kurtum*), and in Panjáb *Khar, polian.*

References.—*Roxb., Fl. Ind., Ed. C.B.C., 595; Stewart, Pb. Pl., 124; Aitchison, Cat., Pb. Pl., 80; DC., Origin of Cult. Pl., 164; Moodeen Sheriff, Supp. Pharm. Ind., 89; U. S. Dispens., 15th Ed., 503, 1599; Stocks, Account of Sind; Murray, Drugs and Pl., Sind, 185; Bidie, Cat. Raw Prod., Paris Exh., 55, 113; S. Arjun, Bomb. Drugs, 76; Baden Powell, Pb. Pr., 355; Cooke, Oils and Oilseeds, 34; L. Liotard, Memo., Dyes, pp. 26-32; McCann, Dyes and Tans of Beng., 4-18; Wardle, Report on Indian Dyes; Report on the Dyes of Ajmír; Admin. Rep., Beng., 1882-83, p. 24; Crooke's Dyeing, 384; Birdwood, Bomb. Prod., 126, 163, 300; Duthie & Fuller, Field and Garden Crops, N.-W. P. and Oudh, 51; E. James, Ind. Industries, p. 128; Lisboa, U. Pl. of Bomb., pp. 163, 218, 246; Drury, Us. Pl. Ind., 116; Spons' Encycl., 864; Balfour, Cycl.; Simmonds, Trop. Agricult., 374; Smith, Dic., 361; Treasury of Botany; Morton, Cycl. Agri.; Ure, Dict. of Arts, Man. and Mines; Gazetteers:—Rajputana, 227; C. P. 326; Bomb. III., 52; VII, 97; X, 153, 164; Pb. (Hoshiarpur), 91; (Karnal), 182; (Lahore), 89; (Kangra), 154.*

Habitat.—An annual, herbaceous plant, with large orange-coloured flower-heads, cultivated as a dye-crop all over India, also in Spain, Southern Germany, Italy, Hungary, Persia, China, Egypt, the Sunda Islands, South America, and Southern Russia. Some doubt seems to exist as to the origin of this plant. It has never been found in a wild state, but botanists assign to it an origin in India, Africa, or Abyssinia. **De Candolle** (*Origin, Cult. Pl.*) says that the grave-cloths found on Egyptian mummies are dyed with carthamine. The Chinese received the plant only in the second century B.C., when **Chang-kien** brought it back from Bactriana. The Greeks and Latins were probably not acquainted with it, although **Birdwood** and other writers give κνῆκος as its Greek name. As the plant has not been found wild either in India or Africa, although probably cultivated in both these countries for thousands of years, **De Candolle** suggests that it may possibly be found indigenous to Arabia. The knowledge of the red dye which the flowers possess appears in India to be modern—a fact opposed to any idea of the plant having been first cultivated in India.

CULTIVATION 638

CULTIVATION.

A few years ago Safflower was an exceedingly important substance, but recently the aniline colours have driven it almost entirely out of the European market. "It still, however, holds its place with the natives as a brilliant though evanescent dye, and as they employ it largely for home use, it must still rank among the industries of the country, as many people are employed in its cultivation and preparation." (*E. James.*) Safflower is cultivated more or less all over India on account of the FLORETS, which are used as a dye-stuff, and the SEEDS as a source of oil. Although occasionally sown broadcast as a primary crop, safflower is

CULTIVATION.

chiefly grown as subsidiary to some other crop, participating, therefore, in the treatment given to its associate. On this account it is extremely difficult to obtain trustworthy details as to the area under safflower, the method and cost of cultivation, nature of soil necessary, or value of the crop.

BENGAL. 639

Sown Oct. to Dec.

(*a*) *In Bengal* it is chiefly grown in the Eastern division, where even still it constitutes a crop of some considerable value, although greatly decreased through the introduction of aniline dyes. In fact, the Indian safflower industry may be regarded as ruined, at least for the present, but similar fluctuations have occurred with other dye-stuffs, and it is quite possible the safflower trade may be resuscitated. Of Indian safflower, that from Dacca bears the highest reputation. It is there sown from the middle of October, and later sowings not till the beginning of December. The period of sowing varies slightly in different parts of Bengal: in Chittagong, for example, it is reported to be sown as late as January. Low *churs* are, as a rule, preferred, and especially where these are either new or have been left fallow for a few years. It appears to be an exhausting crop, and is reported to give a very poor yield if cultivated for three successive seasons on the same soil. In Hazáribágh it is grown on high lands, and in Jessor the high land crop is said to be earlier than that from the low. It requires, however, a light, well-ploughed, sandy soil, with a fair amount of moisture, and when grown on high lands it will not succeed unless it receives three or four supplies of water in its early state. It is accordingly chiefly cultivated on irrigated land as a subsidiary crop, and is carefully weeded for the first month. Rain is of course good for the plant until it attains a height of one foot, but very injurious afterwards, as it extracts the colour from the flower-heads. It is a common practice to nip off the central bud of each plant, when the flowers begin to appear, so as to cause the plants to become bushy, and thus produce a larger amount of flower-heads. The flowering commences about the end of January and is fairly established in February, and the plants are carefully picked every two or three days, the flowers being gathered just as they begin to get brightly coloured. Delay in plucking is said to considerably weaken the dye. The picking of the florets continues till the end of March, or in favourable seasons even till May. In removing the florets, the flower-heads are not much injured, and as they are fecundated before the time of removal, the seeds continue to mature within their small, white, angular, one-seeded fruits, and are ripe in April to May. They are then collected for the oil crop (*Agri.-Hort. Soc. Journ., VII., 191*).

Gathered March to May.

Area.

As already stated, safflower is grown but to a limited extent as a primary crop; it chiefly occurs in lines or along the margins of fields, especially of gram, wheat, barley, tobacco, chillies, or opium. When grown by itself, the seed is sown broadcast after the soil has been carefully prepared. It is, however, almost impossible to arrive at any definite idea of the area under this crop in Bengal, but the following figures are quoted from **Dr. McCann's** work (which is taken from the official returns sent to the Economic Museum): Dacca, 11,500 acres; Gya, 2,260 acres; Monghir, 2,000 acres; Midnapur, 15,000 acres; all other districts about 2,000 acres.

N.-W. P. AND OUDH. 640

(*b*) *In the North-West Provinces and Oudh*, safflower is not so extensively cultivated, and the dye produced is considered decidedly inferior to that from Eastern Bengal. It is also exceedingly local, being confined, apparently through the necessities of trade, to the Meerut Division and to Rohilkhand. The Bulandshahr district of the Meerut Division (which is nearest to Delhi) contains above 91 per cent. of the entire area which in the North-West Provinces is annually under safflower, and it has been computed that the total area under this crop is about 18,000 acres, of which

CULTIVATION 38 per cent. is irrigated land. The mode of cultivation is very similar to what has already been described for Bengal. Light soils are preferred; the plant is rarely grown alone, but is generally sown in the gram fields and disposed like rape in lines. It is extensively grown along with carrots near wells, participating in the rich cultivation bestowed on the latter.
Sown Oct. to Nov. It is also associated with cotton, wheat, or barley. In the North-West Provinces the sowings generally take place in October to November, so that the crop is obtained a little earlier than in Bengal.

"Lightning is popularly supposed to do great injury, if it occurs while the heads are in flower, and the plants are reported to suffer occasionally from the attacks of an insect known as the *ál,* the scientific name and affinities of which have not been ascertained" (*Duthie and Fuller*). In a report on the dyes and processes of dyeing in Ajmír it is stated that
Price. about 20,000 maunds of safflower are annually received from Delhi, the best quality being valued at R30 a maund and the inferior sort at R24.

BOMBAY.
641 (*c*) *In Bombay* it is reported to be cultivated in Ahmedabad, Kaira, Surat, Násik, Khándesh, Sholapur, and Broach. **Lisboa** says the cultivation "is very expensive and unremunerative if carried out by itself; it is, therefore, almost always grown as a subordinate crop along with barley, gram, &c., to which last the cultivator looks for his profits." Probably not more than 5,000 acres are annually under this crop in the whole of the
Area. Bombay Presidency. A considerable trade is done in Ahmednagar, where the plant is sown in strips along with millets, wheat, and other crops, the
Sown Oct., gathered March. seed being put into the ground in October and the crop of florets collected in March. **Mr. Liotard** states that the town of Nagar imports from the Nizam's dominions the prepared dye-stuff to the value of R12,000 annually, nearly two-thirds of which is forwarded to Bombay; and he adds that
Production. the neighbourhood produces about R8,000 worth of the dye. In Kaira it is stated that 41,134 maunds are annually produced, of which 25,600 maunds are used up locally. The official reports from Bombay state, however, that
Varieties. the crop is grown more for oil than for dye. In the Deccan two forms
Sadhi.
642 of the plant are grown—*sádhi,* a strong plant with thorny leaves grown chiefly for its oil-seeds; *kusumbyáchi,* a slenderer plant grown for its dye-
Kusambyachi.
643 yielding flowers (*Bomb. Gaz., XII., 164*). In Gujarát the "*kabri* or *kasumba* is grown both in *gorádu* and black soil. The land is ploughed for ten to twenty times before the sowing. The seed is thrown broadcast at the rate of 10℔ to the *bigha* and is reaped in February. The average yield is in seed 400℔ and in flowers 80℔" (*Bomb. Gaz., VII., 97*). Bombay safflower is commercially much inferior to that from Bengal.

PANJAB.
644 (*d*) *In the Panjáb,* safflower appears to be grown to a very limited extent, and entirely as a local article, there being no export. It is sown in September and reaped in April. In the Delhi district there were, during the settlement, 288 acres under the crop, and in Hoshiárpur 6,722 acres, especially in the northern part of the Garhshankar *Tahsíl.* It is generally grown as a mixed crop in lines with gram and requires a sandy soil. It is sown in September.

CENTRAL PROVINCES.
645 (*e*) *In the Central Provinces,* a little over 6,000 acres are annually under this *rabi* crop, and Raipur is stated to export the dye-stuff to about R10,000 a year.

The brief notices given above regarding the safflower of Bengal, the North-West Provinces, Bombay, the Panjáb, and the Central Provinces, may be accepted as pretty nearly correct; but the official reports for the remaining provinces and Native States are either incomplete or quite
Area. incorrect, and it seems probable that not more than 10,000 acres are under this crop in the remaining provinces of India.

CULTIVATION. BERAR. 646

(f) In Berar, safflower, however, appears to be cultivated to a very considerable extent; **Mr. Liotard** informs us that the area under it is over 40,000 acres. This statement is compiled from official returns, but is obviously incorrect, since cultivation on so extensive a scale would indicate a very important trade, whereas we are informed that the dye-stuff is not exported. In the reports from the Nizam's territories, safflower seems to be an imported article, but this is at variance with the statement of the imports from His Highness's dominions into Ahmednagar.

MYSORE. 647

(*g*) *In Mysore and also in Madras* it is cultivated very generally, but only in small patches, and there is no export trade.

BURMA. 648

(*h*) *In the Prome district* alone of Lower Burma there are said to be 260,000 acres annually under safflower. It is unnecessary to say this statement must be incorrect, since Burma has only a little over four million acres of arable land, of which three million acres are annually under rice. This remarkable agricultural peculiarity almost precludes an extensive cultivation of safflower, since rice-lands are not suitable for this crop, and besides, Burma, instead of exporting safflower, receives annually a small amount from the Straits Settlements.

VARIETIES.

CULTIVATED VARIETIES.—It has already been stated that, according to **Mr. C. B. Clarke, Carthamus oxyacantha**—a wild plant in the Panjáb—may possibly be the source from which by cultivation **C. tinctorius** has been derived. It is frequently observed that plants, which in a wild state are very spiny, show a tendency to lose the spines under cultivation. This might account for some of the peculiarities of the cultivated plant (**C. tinctorius**), and there exists the curious fact in further support of this, that there are two distinct cultivated varieties met with in India:—

Spiny Form. 649

(*a*) Very spiny form. This may be regarded as the typical condition. It is known as *kutela* in Patna and *kati* in Berar, and is supposed to give an inferior quality of dye. This is the *sadhi* or oil-yielding form of the Deccan alluded to above.

Spineless Form. 650

(*b*) Almost spineless form. This is known as *bhuili* in Patna, *bod-ki* in Berar, *murilia* (or shaved) in Azamghar and the *kusumbyáchi* in the Deccan. A superior quality of dye is derived from this form.

OUTTURN.

AVERAGE OUTTURN AND PROFIT OF CULTIVATION.—The average outturn of safflower sown thickly amongst carrots has been estimated at R15 along with R5 for seed, and allowing the other crop to pay its share of rent of land and expense of cultivation, as much as one-third of the earnings may be regarded as profit; but it is difficult to obtain trustworthy information regarding the profits from safflower cultivation, and it cannot pay now-a-days to cultivate it alone. **Dr. McCann** gives the profits in Bengal as from R3 to R15 a *bigha*.

PRESENT POSITION OF THE SAFFLOWER INDUSTRY.

TRADE. 651

Simmonds in his *Tropical Agriculture* says: "The cultivation of safflower, known as *Coosumban* in Bengal, is receiving attention at the hands of the local Government. The prosperity of Bengal, though it mainly depends upon the jute trade, is in some measure attributable to the demand for safflower." The writer proceeds to state that the value of the exports from Dacca alone "would be from nine to ten lakhs of rupees—£90,000 to £100,000. The cultivation is said to be largely extending." Then follows: "Safflower is grown, but to a limited extent, in Bengal, and does not grow promiscuously all over the district." **Mr. Simmonds'** work was published in 1877, and in that year the total exports from all India were only R1,48,806, or a little over £10,000, and in 1874-75 they were R6,50,827, so that the decline of the safflower trade was fully established at the time **Mr. Simmonds** described it as "largely extend-

TRADE.

ing." The total exports for 1886-87 were only R83,819. The following table gives the exports from India for the past fourteen years:—

Exports.

Year.	Safflower.	
	Quantity.	Value.
	Mds.	R
1873-74 . . .	13,206	7,58,906
1874-75 . . .	14,222	6,50,827
1875-76 . . .	4,080	1,63,528
1876-77 . . .	7,662	3,04,672
1877-78 . . .	3,698	1,48,806
1878-79 . . .	4,977	1,86,711
1879-80 . . .	2,411	1,81,456
1880-81 . . .	6,675	3,51,157
1881-82 . . .	2,293	94,754
1882-83 . . .	3,008	92,038
1883-84 . . .	2,333	64,492
1884-85 . . .	2,167	83,083
1885-86 . . .	1,898	68,991
1886-87 . . .	2,149	83,819

Instead of extending, the safflower industry in India has been steadily declining for the past 15 years, and (like lac-dye) may now be pronounced as almost ruined, being rapidly displaced by aniline dyes both as an article of internal and foreign trade. **Messrs. Duncan Bros. & Co.** report in June 1883 that "there is no land under safflower cultivation in Tipperah and Faridpur this year, as this crop does not now pay the growers. In the Dacca district the outturn of this season's crop will not exceed 600 maunds of flowers." Chinese safflower is considered superior to Indian, and of Indian the Bombay and Panjáb are viewed as very inferior.

THE DYE.

DYE. 652 Preparation.

Preparation of Dye-stuff.—The florets are picked as fast as they appear, since they lose their colour if left exposed to the sun. They are carefully dried in the shade and sometimes by being pressed; in this condition they are generally sold for the home market, or when perfectly dry they are powdered and sifted. The first and last harvests are always inferior to those gathered in the middle of the season. In the former case many undeveloped florets are collected, and in the latter the plant is becoming exhausted and does not produce such brilliant colours. Care in the preparation and preservation of the dye-stuff exercises a most important influence over the quality; but the produce of one district is often much superior to another—a fact accountable for either by the more favourable nature of the soil or the care bestowed in cultivation. If intended for export, after having been dried as above, the florets are either placed in a bag or on a basket or other contrivance, permitting of the easy escape of a supply of water which is kept poured on them while beaten or trodden on. This process is continued until the water passes through quite clear. At first it is bright yellow, because of the soluble yellow colour
Yellow. 653
which the florets contain in addition to their valuable red dye. The
Red. 654
yellow colour is useless, and as its presence greatly injures the quality of the red dye, especially in dyeing silk and wool (fabrics for which the yellow colours possess a strong affinity), the florets are repeatedly washed until

DYE.

they are quite freed from the yellow colour. River water (if clean) is regarded as preferable to tank water, but the presence of mud or other impurities in the water is most detrimental. The red colouring matter is completely soluble in dilute alkaline solutions, and care must be taken that the water used does not contain soluble alkaline salts; in fact, to be safe it should be slightly acidulated, otherwise a large proportion of the red colour may be removed during the process of washing away the yellow pigment. The tramping or kneading is continued at intervals for three or four days, the mass being allowed to get dry between the washings. To ascertain if all the yellow colour has been removed, a small quantity is thrown into a basin of clean water; if it does not impart yellow colour, the dye-stuff has been sufficiently washed. The pulpy mass is now squeezed between the hands into small, flat, round cakes, like biscuits; or it is sometimes, though less frequently, made into balls. These are known in the trade as "Stripped Safflower." When the cakes or balls have been carefully dried, they are ready for the market.

"Stripped Safflower." 655

The *Gazetteer* for the district of Karnal in the Panjáb describes the process generally followed in that province, in which apparently the florets are baked into cakes without removing the yellow colour. "When the florets open, the women pick out the petals; three days later they repeat the operation; and again a third time after the same interval. If hired they take a quarter of the picking as their wages. The petals are bruised the same day in a mortar, rolled between the hands, and pressed slightly into a cake. Next day they are rolled again, and then spread in the sun for two days to dry, or still better one day in the sun and two days in the shade. One seer of petals will give a quarter of a seer of dry dye. Any delay in the preparation injures the dye." This process is so very defective from that pursued in Bengal and other parts of India that it may be accepted as accounting for the lower price obtained for Panjáb safflower.

Reason of lower price paid for Punjab Safflower.

Mr. J. G. French, writing of Dacca district in Bengal in the *Agri.-Horticultural Society's Journal* for 1850, remarks: "Safflower is said to have been formerly grown solely for the evanescent yellow the petals so readily yield, while their valuable properties being unknown, they were then thrown away as so much rubbish." This is a remarkable fact, and would seem to indicate that safflower in India is, comparatively speaking, a modern dye-stuff, and this may even be viewed as of some importance in relation to the question of the origin of the plant. The supposition is almost justifiable that if the cultivation of this plant originated in India, its red dye would have been known for centuries.

Originally grown for yellow dye.

Adulteration of Safflower Cakes and Test for Purity.—The above process of preparing the dye-stuff may be accepted as applicable for all India, and indeed it differs but slightly from that pursued in Europe, China, and Egypt. The brighter the colour of the safflower cakes, the more valuable they are. Dullness may be due either to the inferior quality of the plant, defective modes of cultivation, carelessness in the washing process (the whole of the yellow colouring matter not having been removed) or to mud being allowed to mix with the florets through using dirty water. Direct adulteration is also practised, and this has proved most injurious to the continuance of even the present greatly reduced trade. The cultivators, apparently trying to compensate themselves for the fall in the price of safflower (due to the introduction of aniline dyes to many of the purposes for which safflower was formerly used), have, within the past few years, taken to adulterate the cakes with cow-dung, rice-flour, jute cuttings, and turmeric. "When safflower is damaged, as it easily is by salt water, it is employed by fraudulent dealers in the adulteration of shag tobacco." (*Morton's Cycl., Agri.*)

Adulteration. 656

Cowdung. 657

Rice flour. 658

Turmeric. 659

Tobacco, 660

CARTHAMUS tinctorius.

The Safflower.

DYE.

Estimation of Quality.
661 The quality of safflower cake is estimated by dyeing a known weight of cotton; about 4 ounces of safflower will dye 1℔ of cotton cloth light pink; 8 ounces will dye it full rose-pink; and from 12 ounces to 1℔ will dye it a full crimson. In order to take up this quantity, the cotton must be several times dyed in fresh solutions of the colouring matter.

Chemical History.—It is scarcely necessary to go into great detail regarding this, now almost unimportant, product. It has already been stated that the florets contain two colouring principles, or, to be more correct, three, namely, two yellows and one red, the latter being Carthamin or Carthamic acid, $C_{14}H_{16}O_7$. By treating safflower with cold water or oil, the first yellow colouring substance is removed (known as *pujan* of the North-West Provinces). This is soluble in water, and constitutes about 26 to 36 per cent. of the florets, while from 0·3 to 0·6 per cent. is the usual amount of Carthamin. The proportion of Carthamin present varies, however, in the inverse ratio to the amount of the soluble yellow principle. The second yellow colour is soluble only in an alkaline liquor.

Two yellows and one red.

Carthamin.

If the dye-stuff, after the removal of the soluble yellow principle, be acidulated with acetic acid, filtered, and first acetate of lead and next ammonia added, the second yellow colour will be precipitated along with lead salt. (*Crookes.*) Carthamin, the valuable red colour, may be extracted in a pure form by making use of its solubility in alkaline solutions and insolubility in pure or acidulated water. The alkali in most frequent use is carbonate of soda (or ordinary washing soda, 15 per cent. to the weight of the florets). In India pearl-ash is most frequently used, especially that prepared by incinerating *bajra* **(Penicillaria spicata)** or of *chir chira* **(Achyranthes aspera)**, (impure potassium carbonates), but the natural earth carbonate of soda or *sajjí-mátí* is also frequently employed for this purpose.

EUROPEAN DYE SOLUTIONS.

European Dye Solutions.

663 **Preparation of Dye Solution and European Methods of Dyeing with Safflower.**—In principle the method of preparation of the red dye as practised in India is the same as in Europe, and the following passages may therefore be extracted from **Crookes'** most valuable *Hand-book of Dyeing and Calico-printing*, and **Ure's** *Dictionary of Arts, Manufactures, and Mines*: "The mixture is well squeezed, and to the clear bright yellow liquid thus obtained an acid is added, whereby the carthamic acid, along with many impurities, is precipitated. It would be almost impossible to free the carthamic acid from these foreign substances, but this result is very readily brought about by immersing into the alkaline solutions, previous to the addition of an acid, a quantity of cotton-wool. This material attracts, by special action, the carthamic acid at the moment it is set free by the addition of an acid; and cotton-wool may first be washed in a weak acid, and next in water, and, lastly, again with a weak alkaline liquid, which re-dissolves the carthamin. After removal of the cotton-wool, plenty of which should be used, it is re-precipitated by an acid, very dilute citric or tartaric being the best. It falls down in the state of a beautifully rose-red flocculent matter, which may be collected on a filter, washed, and dried. In order to obtain a still purer material, the flocculent dry substance should be dissolved in alcohol, and this solution, after having been strongly concentrated, is poured into a large bulk of water, and the precipitate thus formed is again collected on a filter and dried. Care should be taken to avoid the use of too strong alkaline solutions, nor should they be exposed to air for any length of time, since this completely alters the colouring matter.

DYE.

"Carthamin in a pasty state, as obtained by the process just described, is met with in commerce suspended in water for direct use. The paste is dried upon suitable vessels—porcelain saucers, plates, or even upon polished cardboard.

"Carthamin is insoluble in ether, almost so in water, but soluble in alcohol, to which it imparts a cherry red. This alcoholic tincture dyes silk immediately, no mordant being required."

Ure gives the more general practice of extracting the carthamic acid; this, while of course not so accurate as that given above, is nevertheless the mode pursued where absolute purity is not necessary. The following passage may prove useful to Indian dyers or persons interested in the safflower industry: "Carthamus from which the yellow matter has been extracted, and whose lumps have been broken down, is put into a trough. It is repeatedly sprinkled with crude pearl-ash or soda, well powdered and sifted, at the rate of 6℔ for 120℔ of carthamus; but soda is preferred, mixing carefully as the alkali is introduced. This operation is called *amestror*. The *amestred* carthamus is put into a small trough with a grated bottom, first lining this trough with a closely-woven web. When it is about half filled, it is placed over a large trough, and cold water is poured into the upper one till the lower one becomes full. The carthamus is then set over another trough till the water comes from it almost colourless. A little more alkali is now mixed with it, and fresh water is passed through it. These operations are repeated till the carthamus be exhausted, when it turns yellow.

"After distributing the silk in hanks upon the rods, lemon-juice, brought in casks from Provence, is poured into the bath till it becomes of a fine cherry-colour; this is called 'turning the bath.' It is well stirred, and the silk is immersed and turned round the skein-sticks in the bath, as long as it is perceived to take up the colour. For *ponceau* (poppy-colour) it is withdrawn, the liquor is run out of it upon the peg, and it is turned through a new bath, where it is treated as in the first. After this it is dried and passed through fresh baths, continuing to wash and dry it between each operation, till it has acquired the depth of colour that is desired. When it has reached the proper point, a brightening is given it by turning round the sticks seven or eight times in a bath of hot water, to which about half a pint of lemon-juice for each pailful of water has been added.

"When silk is to be dyed *ponceau* or poppy-colour, it must be previously boiled as for white; it must then receive a slight foundation of arnatto. . . . The silk should not be alumed. The *nacarats* and the deep cherry-colour are given precisely like the *ponceaux*, only they receive no arnatto ground; and baths may be employed which have served for the *ponceau*, so as to complete their exhaustion. Fresh baths are not made for the latter colours, unless there be no occasion for the poppy.

"With regard to the lighter cherry-reds, rose-colour of all shades, and flesh-colours, they are made with the second and last runnings of the carthamus, which are weaker. The deepest shades are passed through first.

"The lightest of all these shades, which is an extremely delicate flesh-colour, requires a little soap to be put into the bath. This soap lightens the colour, and prevents it from taking too speedily and becoming uneven. The silk is then washed, and a little brightening is given it in a bath which has served for the deeper colours.

"All these baths are employed the moment they are made, or as speedily as possible, because they lose much of their colour upon keeping, by which they are even entirely destroyed at the end of a certain time. They are, moreover, used cold, to prevent the colour from being injured. It

DEY.

must have been remarked, in the experiments just described, that caustic alkalis attack the extremely delicate colour of carthamus, making it pass to yellow. This is the reason why crystals of soda are preferred to other alkaline matters.

"In order to diminish the expense of carthamus, it is the practice in preparing the deeper shades to mingle with the first and the second bath about one-fifth of the bath of archil" (*Ure's Dict. of Arts, Man., and Mines, Vol. I., 661*).

INDIAN DYE SOLUTIONS. 664

INDIAN DYE SOLUTIONS.

Indian Method of dyeing with Safflower.—As already stated, the method adopted in India is in theory identical with the European, but as practised it is crude, giving much inferior results when compared with the delicate shades prepared in Europe from this dye. The separation of the carthamic acid from mechanical impurities by precipitating it on cotton-wool and again dissolving off this pure dye by means of an alkali, does not appear to be known to the natives of India. The dye-stuff, after the removal of the yellow colour, is rubbed up, by the hand, with the pearl-ash, and thereafter strained through a cloth. The first straining is regarded as the best, and is reserved for giving the final shade in dyeing, but the process of rubbing up with an alkaline solution and straining is repeated three or four times, until no more colour can be extracted. No mordant is required when dyeing with safflower, but it is a common practice in India to dye the fabric first with the yellow liquid, then with the last straining of carthamin, and so on until, when depth of colour is required, the first straining is used to give the final immersion. Before the fabric is dipped in the carthamic liquid, however, a dilute acid is added in order to precipitate the red carthamic acid. This fine powder remains for a considerable time minutely diffused through the liquid, instead of subsiding to the bottom. It has no actual chemical affinity for fibres, but when a fabric is dipped in the red liquid, the fine powder is rapidly precipitated within the fabric, producing the well-known and brilliant shades of orange, pink, and even dark red. The acid used is generally lime-juice in the proportion of about 1℔ of lime-juice to 2℔ of dye solution. Sometimes the juice of the tamarind is employed in place of lime-juice. In Mánipur the fruits of **Garcinia pedunculata** are viewed as superior to lime-juice, and have the reputation of rendering the colour less fleeting.

Combinations. 665

Indian Dye Combinations.—Different depths of red colour are generally obtained either by longer immersion in the dye solution or by frequent repetitions in fresh solutions. Shades of orange are generally produced either by dyeing the fabric first with the yellow soluble colour (in some parts of Bengal known as *peworree* water, according to **McCann,** a name which, if actually applied, must be carefully distinguished from the yellow urine dye or *peori* or *peri rung*). Instead of the safflower yellow, a ground colour may be given with turmeric or any other yellow dye, when different shades of orange or *nárangí* will be obtained; so also combinations with arnatto, *kamala*, and *harsinghar*, the shades of orange passing into pink. Red is produced by three immersions in safflower dye, the 3rd straining, 2nd straining, and last of all the 1st straining, the cloth being allowed to dry between each, and finally washed out with turmeric. This is known in Farukhabad as *gulanár;* if instead of turmeric indigo be used, a magenta colour is produced, the *gulabbasi* of Agra. The *sappai* pink of Cawnpore is produced with *harsinghar* and safflower, the latter being weak if concentrated, orange or the *nárangí* of Etawah is the result; and a more yellow orange, *zafráni*, is produced, when the cloth is first dyed with *harsinghar* and afterwards with safflower. (*Buck's Dyes and Tans of*

DYE.

N.-W. P.) With **Terminalia Chebula** or **T. citrina** and protosulphate of iron, safflower gives a dark neutral tint; with safflower, sappanwood, and alum a purplish brown; and with indigo and safflower, greens and purples. (*McCann, Dyes and Tans of Beng.*)

Use of acids and alkalis. 666

An almost indefinite series of colours are obtained in India by various combinations with safflower. It should be carefully observed, however, that the red dye-stuff, carthamin, is only formed from the alkaline solution when an acid has been added. In many official reports it would appear that the writers have in some cases overlooked this important point, and considerable confusion has accordingly crept into the published accounts of this dye-stuff. In some instances writers expressly state that an acid is used to produce one colour, while in their accounts of other colours they make no mention of an acid, leading thus to the conclusion that in the latter case acid is not used. It is probable that the associated dye-stuffs employed along with the alkaline dye solution may have the power of precipitating the carthamin, but from the chemical facts which have already been given it will be clear that carthamin having no affinity of itself for fabrics, alkaline preparations in which the carthamin exists in a soluble condition cannot be used as dye solutions. It is necessary that this peculiarity be fully appreciated, otherwise the observer cannot give an accurate account of the indigenous modes of dyeing with safflower.

Alkaline solutions cannot act as Dyes. 667

FIXING. 668

Fixing Safflower Dye.—It is much to be regretted that no one has as yet discovered a mode of preventing the decoloration of safflower dye; its fleeting property appears to depend on the oxidation of the particles of carthamin held mechanically in the fabric. The inhabitants of different parts of India boast of possessing a secret of effecting this purpose, and careful observation on the part of local officers may help to throw some light on the subject. All that is necessary to re-establish the carthamine dye as an important industry is the discovery of some mode of preventing this oxidisation of carthamin. The fruit of **Garcinia pedunculata**, a common tree in Assam, has already been alluded to. Although there would not appear to be much hope of finding the property attributed to this fruit confirmed on careful examination, its extensive use as a dye auxiliary by most of the hill tribes of Assam certainly justifies this matter receiving careful attention. **Dr. McCann** informs us that the dyers of Chittagong district claim to be able to produce a "semi-permanent" safflower dye. This is done by adding safflower to water in which tamarinds and the ashes of burnt plantain rinds have been well soaked. The principle here employed is the mixing of the acid and alkali together, instead of first extracting the dye with an alkali and precipitating the carthamin with an acid upon the fabric. In some parts of India the pearl-ash and lime-juice are mixed together, and the liquid thus prepared is used to extract the carthamic acid direct. It is difficult to understand this action, but it is nevertheless frequently practised.

"The colours safflower yields, although the most beautiful and delicate shades which the art of the dyer can produce, have the disadvantage of being very unstable; nothing is easier than to detect safflower colours upon fabrics; the rose, pink, or crimson hue is at once turned yellow by a single drop of alkali, and is destroyed by any further washing: this property is actually made use of now and then as a discharge, so as to produce a yellow pattern upon a pink ground; weak acids do not affect the colours, but chlorine and sulphurous acid destroy the colour at once." (*Crookes.*) Safflower-dyed fabrics should not be washed with soap, as the colour is removed by the alkali of the soap.

ROUGE. 669

Rouge.—It is necessary to refer here very briefly to an important purpose for which safflower is employed, *viz.*, the manufacture of *rouge*

DYE.

végétale. This trade is unaffected by the aniline imitations of safflower, and constitutes an article of considerable importance. The dry carthamine precipitate is sometimes called India or China lake, and this mixed with finely-pulverised talc constitutes *rouge végétale.* (See **Carmine**; also **Carnelian**—the coloration of inferior gems.)

OIL. 670

THE OIL.

There are two kinds of seeds, or, to be more accurate, of fruits—one the cultivated, a white and glossy form, the other (*Karar*) is a smaller but similarly-shaped seed, mottled or dusted, brown-grey and white; both yield oil. As already explained, an oil-bearing variety is often specially cultivated, the flowers of which yield little or no dye. In Bombay in fact the plant is mainly grown for its oil. The oil is very clear and yellow; it is esculent, and would be peculiarly suitable for burning in lamps, on account of the little heat which it gives out (*Baden Powell*). It is used locally for culinary purposes, and is said to form an ingredient of the "Macassar Hair-oil." In the *Gazetteer* for Karnál it is stated that the oil is bitter, but that 40 seers of seed will give $3\frac{1}{2}$ seers of oil. **Lisboa** remarks that in Bombay the "seeds yield 28 per cent. of a light yellow oil, possessed of drying properties, and useful for ordinary culinary purposes and for lamps." The average yield of oil is said to be about 25 per cent.

Prices.

"In Bulandshahr the safflower yields about 7 maunds of seed per local *bigha.* The oil-cake is supposed to be the perquisite of the oil-presser in lieu of wages. A maund of seed yields 7 seers of oil, 14 seers of oil-cake, and 19 seers of husk or *bhusá*, and the oil sells at from 4 to 5 seers for the rupee, the cake at 36 seers, and the *bhusá* at 4 maunds" (*E. T. Atkinson*).

"The pure oil is seldom offered for sale. Though it lowers the quality of the oil, the outturn is generally increased by mixing its seeds with gingelly seed" (*Bomb. Gaz., 153*). Although the oil is apparently not exported from India a considerable trade is done with Liverpool and London in the seeds.

EXPRESSION. Dry cold. 671

Expression of Oil.—"The oil is expressed in the same manner as the other oil-seeds. After the husks have been removed, the husks of the seed are thick and would weigh about one-third of the weight of seed. When they are removed, 25 per cent. of the remainder will represent the extractable oil, which is of a light colour and burns well. I am only surprised that it has not been brought more into use for English lamps. I use scarcely any other oil."

Dry Hot. 672

Dry Hot extraction of Oil.—"There is also another way of extracting the oil, which is, I think, so peculiar that I will attempt to describe it. It renders the oil useless for burning purposes,—charring it, in fact; but this is the oil used by the native agriculturist for greasing his well-ropes, leathern well-buckets, &c., and in fact all leather-work used for exposure to water. It is a rude still, with process inverted. A hole is dug in the ground deep enough to receive an earthen jar or *gurrah* of any capacity, on the mouth of which is placed an earthen plate with a hole of about a quarter of an inch diameter bored in its centre. Above this is placed another similar jar nearly filled with the *bhurra* or *kussum* seed inverted upon the plate. The juncture of the three is luted with clay, and earth then filled in up to some inches above the juncture of the vessels,—in fact, up to the swell above the neck of the upper inverted vessel. Dried cow-dung is then heaped above the upper vessel and set on fire. The fire is kept in ignition for about half an hour, when it is removed. The

Process of extracting the oil after the Dry Hot method.

upper inverted vessel is found to be about half full of the charred seed, and the lower one, which was imbedded in the ground, about one-third full of a black sticky oil. By this process the oil as well as the seed is charred, but the natives assert that it is all the more valuable for the preservation of leathern vessels exposed to the action of water. It might be worth the while of chemists to enquire why this should be, and whether this kind of oil would be of any commercial value at home. The yield of oil by this process is more than a fourth larger than by the press" (*R. W. Bingham, Jour. Agri.-Hort. Soc., XII., 340*). OIL.

THE MEDICINE.

MEDICINE.

"This plant is the ***kusumbhu*** of Sanskrit writers, who describe the seeds as purgative, and mention a medicated OIL which is prepared from the plant for external application in rheumatism and paralysis. Muhammadan writers enumerate a great many diseases in which the seeds may be used as a laxative; they consider them to have the power of removing phlegmatic and adust humours from the system." (*Dymock's Mat. Med. W. Ind.*) **Ainslie** gives the following account of the plant: "A fixed oil is prepared from it which the *Vytians* used as an external application in rheumatic pains and paralytic affections, also for bad ulcers; the small SEEDS are reckoned amongst their laxative medicines, for which purpose I see they are also used in Jamaica (the kernels beat into an emulsion with honeyed water)." **Burham** tells us that a drachm of the dried FLOWERS taken internally cures jaundice (*Hort. Jamaica, I., 72*). **Loureiro** says that the SEEDS are considered as purgative, or *eccoprotic*, resolvent, and emmenagogue. In South America, as well as in Jamaica, the flowers are much used for colouring broths and ragouts. Oil. 673 Flowers. 674 Seeds. 675

"In Bombay the seeds, under the name of ***kardayi***, are pressed along with ground-nuts and sesamum to form the sweet oil of the bazars; they are also universally kept by the druggists for medicinal use" (*Dymock*). The "flowers are sometimes fraudulently mixed with saffron, which they resemble in colour, but from which they may be distinguished by their tubular form, and the yellowish style and filaments which they enclose. In large doses carthamus is said to be laxative; and administered in warm infusion, diaphoretic. It is used in domestic practice as a substitute for saffron in measles, scarlatina, and other exanthematous diseases to promote the eruption. An infusion, made in the proportion of two drachms to a pint of boiling water, is usually employed, and given without restriction as to quantity." (*U. S. Dispens.*) Sweetmeats. 676 Used to adulterate Saffron. 677

Special Opinions.—§ "The powdered seed made into a poultice are used to allay inflammation of the womb after child-birth. The oil is used as a liniment in rheumatism" (*Surgeon-Major C. W. Calthrop, M.D., Morar*). "The seeds are said to have properties like linseed, and to be useful in unhealthy ulcers" (*U. C. Dutt, Civil Medical Officer, Serampore*). "Decoction used as a diuretic. The seeds are laxative. The oil is used as a dressing for ulcers" (*Surgeon W. Barren, Bhuj, Cutch*). Decoction. 678

Food.—Poultry fatten on the seeds. **Mr. Atkinson** says the leaves are eaten as a spinach in some parts of the North-West Provinces. "The tender leaves of the cultivated plant, which are not spinous, are used as an article of food. The roasted seeds are eaten; they were much procured by well-to-do people during the late famine at Sholápur. The cake is excellent for fattening poultry" (*Lisboa, Us. Pl., Bomb., 163*). Safflower is sometimes used to dye cakes, biscuits, and toys, but as it is purgative it should not be too freely employed for this purpose. FOOD. Seeds. 679 Leaves. 680

CARUM Carui.

The Caraway.

CARUM, *Linn.; Gen. Pl., I., 890.*

681 **Carum Carui,** *Linn.; Fl. Br. Ind., II., 680;* UMBELLIFERÆ.

CARAWAY; FRUITS OU SEMENCES DE CARVE, *Fr.;* KÜMMEL, *Germ.*

Vern.—*Shiá-jirá* (U. C. Dutt), *zira,* HIND.; *Jira,* BENG.; *Zira siyah,* PB.; *Gúnyún,* KASHMIR and CHENAB; *Úmbú,* LADAK; *Zira,* N.-W. P.; *Wiláyati-zirah,* BOMB.; *Shimai-shombu, kékku-virai,* TAM.; *Shimai-sapu,* TEL.; *Karoyá, karawya,* ARAB.; *Karóya,* PERS.; *Sushavi* (U. C. Dutt), SANS. The *Κάρος* of **Dioscorides,** and the *Careum* of **Pliny.**

References.—*Stewart, Pb. Pl., 104; DC. Prodr., IV., 115; Pharm. Ind., 98; U. S. Dispens., 15th Ed., 362; Flück. & Hanb., Pharmacog., 304; Bentley & Trim., Med. Pl., 120; U. C. Dutt, Mat. Med. Hind., 174, 319; Dymock, Mat. Med. W. Ind., 304; O'Shaughnessy, Beng. Dispens., 358; Murray, Pl. and Drugs of Sind, 201; S. Arjun, Bomb. Drugs, 63; Irvine, Mat. Med. Patna, 94; Kew. Cat., 74; Atkinson, Him. Dist., 705, 730; Birdwood, Bomb. Prod., 39; Lisboa, U. Pl. of Bomb., 161; Spons' Encyclop., 1418; Smith, Dic., 92; Treasury of Botany; Morton, Cyclop. Agri.*

Habitat.—A herbaceous plant cultivated, for its seeds, as a cold-season crop on the plains of India and frequently on the hills, as a summer crop, as in Baltístán, Kashmír, and Garwhál, &c., at an altitude of between 9,000 to 12,000 feet. Distributed to Western and Northern Asia and Europe. The Greek and Latin names of the plant are said by some writers to be "derived from Caria, the native country of the plant" (*Birdwood*); by others this is doubted, since the plant does not occur in Caria.

CONDIMENT. 682

CONDIMENT.—The spice seems to have come into use in Europe about the 13th century, and it was known in England at the close of the 14th century, since it is mentioned "with coriander, pepper, and garlick in the *form of curry,* a roll of ancient English cookery compiled by the master cooks of Richard II., about A.D. 1390." "The oriental names of Caraway show that as a spice it is not a production of the East: thus we find it termed Roman (*i. e.,* European), Armenian, mountain or foreign Cummin; Persian or Andalusian Caraway; or foreign Anise. And though it is now sold in the Indian bazars, its name does not occur in the earlier lists of Indian spices" (*Pharmacog., 305-6*).

Many recent writers agree in viewing the true Caraway plant as wild in certain parts of the Northern Himálayas, and it is certainly cultivated both on the hills and in the plains, so that the expression that it is sold in the bazars is slightly misleading. Then again it is scarcely correct to say that the oriental names for Caraway show that it is not a production of the East. Such names as *Gúnyún* and *Úmbú,* and others given above, leave little doubt of the indigenous nature of the plant, so far as vernacular names alone can be accepted as of significance. The spice was well known to the Persians, and in the *Makhzan* the name *karawya* or *kuruya* is said to be derived from the Syrian *kárui.* It is quite customary to find in oriental names of plants that two species, or even two widely different things, are spoken of as "white" or "black" forms of the same. Accordingly, the name *zíra* (*zíraka* or *jíraka,* SANS.) may be viewed in the terminology of spices as generic. By some writers and in some parts of India that word is restricted to mean **Cuminum Cyminum,** while by others that spice is spoken of as the (white) *saféd zíra* (in the Punjab, &c.) and **Nigella sativa** as the (black) *kále jíra.* But in the same way there are many forms of black *jíra* or substitutes for black *jíra,* as, for example, **Vernonia anthelmintica,** and the form of **Carum** which will be found discussed under **C. nigrum.** The existence of the name *Wilayati-zíra,* that is, European *zíra,* should not by itself be viewed as excluding the true Caraway from an

The Caraway.

CARUM Carui.

CONDIMENT.

oriental origin since such a name might simply mean that in that part of the country it was first brought to the attention of the natives by the Europeans. Indeed, the facilities of trade offered by the Persian Gulf can easily be understood to have made the people of Bombay more familiar with an imported article than with a wild or event cultivated plant of the Panjáb Himálaya. Authors are about equally divided in the restriction of the word *síra* to **Carum Carui** on the one hand, and to **Cuminum Cyminum** on the other. (Conf. with **C. nigrum.**)

Dr. Dymock says that Caraways are brought from the Red Sea Ports to Bombay where they are sold at R1 per pound. **Dr. Stewart** alludes to a considerable trade from Afghánistan, Kashmír, and other parts of the Panjáb Himálaya to the plains of India. The imports of Caraway into Great Britain are about 20,000 cwts. a year and chiefly from Holland. It is also largely grown in Kent and Essex. TRADE. 683

Oil.—A valuable essential oil is obtained from the seeds, called Caraway Oil. This oil is colourless or pale yellow, thin, with a strong odour and flavour of the fruit. It is used in medicine and more extensively as a perfume for soaps. (*Spons'.*) OIL. 684

Perfumery.—**Piesse**, in his book on perfumery, remarks that the odoriferous principle obtained from the seeds by distillation, when dissolved in spirit, may be combined with lavender and bergamot for the manufacture of cheap essences in a similar way to cloves. PERFUMERY. 685

Medicine.—As a medicine the dried fruit possesses stimulant and carminative properties. It has been found useful in flatulent colic, atonic dyspepsia, and spasmodic affections of the bowels. Two preparations are given in the *Pharmacopœia af India, viz.*, Oil of Caraway and Caraway Water. The oil is used as an aromatic stimulant, and also as an adjunct and solvent in the preparation of the resinous cathartic pills. An infusion is also prepared by adding two drachms of the seeds to a pint of boiling water. MEDICINE. 686

"Muhammadan writers describe the fruits as aromatic, carminative, and astringent; from them they prepare an eye-ash which is supposed to strengthen the sight; they are also used as a pectoral, and considered diuretic and anthelmintic. A caraway bath is recommended for painful swelling of the womb, and a poultice for painful and protruding piles." (*Dymock's Mat. Med. W. Ind.*, *304*). Fruit. 687

Chemical Composition.—"Caraways contain a volatile oil, which the Dutch drug affords to the extent of 5·5 per cent., that grown in Germany to the amount of 7 per cent.; in Norway 5·8 per cent. have also been obtained from the indigenous caraways. The position and size of the vitæ account for the fact that comminution of the fruits previous to distillation does not increase the yield of oil. CHEMISTRY. 688

"**Volckel** (1840) showed that the oil is a mixture of a hydrocarbon $C_{10}H_{16}$ and an oxygenated oil, $C_{10}H_{14}O$. **Berzelius** subsequently termed the former *carvene* and the latter *carvol*.

"Carvene, constituting about one-third of the crude oil, boils at 173° C., and forms, with dry hydrochloric gas, crystals of $C_{10}H_{16} + 2\,HCl$. It has been ascertained by us that carvene, as well as carvol, has a dextrogyrate power, that of carvene being considerably the stronger; there are probably not many liquids exhibiting a stronger dextrogyrate rotation. Carvene is of a weaker odour than carvol, from which it has not yet been absolutely deprived; perfectly pure carvene would no doubt prove no longer to possess the specific odour of the drug. By distilling it over sodium, it acquires a rather pleasant odour; its specific gravity at 15° C. is equal to 0·861.

CHEMISTRY.

"Carvol at 20° C. has a specific gravity of 0·953; it boils at 224° C.; the same oil appears to occur in dill (**Fructus Anethi**), and an oil of the same percental constitution is yielded by the spearmint. The latter, however, deviates the plane of polarization to the left. If four parts of Carvol, either from caraways, dill, or spearmint, are mixed with one part of alcohol, specific gravity 0·830, and saturated with sulphuretted hydrogen, crystals of $(C_{10}H_{14}C)_2SH_2$ are at once formed as soon as a little ammonia is added." (*Pharmacog.*)

Special Opinions.—§ "Stimulant and laxative. The white variety is lactagogue" (*Assistant Surgeon Nehal Singh, Saharanpore*). "Have used it to increase the flow of milk with no decided effect" (*Surgeon D. Picachy, Purneah.*)

FOOD. Seed. 689 Roots. 690

Food.—The seed is used parched and powdered, or raw and entire. In the former case it is employed to flavour curries; in the latter it is put in cakes. It is used in confectionery and in flavouring drinks. It also "produces a spirit cordial" (*Morton*). The roots of the caraway plant are very agreeable and are much eaten in the north of Europe (*O'Shaughnessy*).

Special Opinions.—§ "As a condiment with curries" (*Surgeon C. M. Russell, M.D., Sarun*). "Carminative; largely used in curry-powder" (*Assistant Surgeon Shib Chunder Bhattacharji, Chanda, Central Provinces*).

691 **Carum copticum,** *Benth.; Fl. Br. Ind., II., 682; Wight, Ic., t. 566.*

THE BISHOP'S WEED; LOVAGE; AJAVA SEEDS; AMYZAD, *Dutch*; SISON, *Fr.*; AMEOS, *Port.*

Syn.—AMMI COPTICUM, *Boiss.*; LIGUSTICUM AJAWAIN, *Fleming*; L. AJOUAN, *Roxb.*; PTYCHOTIS COPTICA, *DC.*; P. AJOWAN, *DC.*; SISON AMMI, *Jacq.*; BUNIUM AROMATICUM, *Linn.*

Vern.—*Ajowan, ajwain*, HIND.; *Jowan? juvaní*, BENG.; *Ajamo*, GUJ.; *Chohara*, CUTCH; *Owa*, MAR.; *Jawind*, KASHMIR; *Aman, oman*, TAM.; *Omamí, omamu*, TEL.; *Omu, oma*, KAN.; *Ajwán, owa*, BOMB.; *Ova*, MAR.; *Yamani* (and according to Ainslie *Ajmódum, brahmadarbha*), SANS.; *Kamue mulúki, tálib-el-khubz*, ARAB.; *Zinián, nánkhwah* (appetiser) PERS.; *Assamodum* (in Ainslie), SING.; *Chœlle*, EGYPT.

References.—*Roxb., Fl. Ind., Ed. C.B.C., 271; Flem., Cat. Ind. Med., 70; Dalz. & Gibs., Bomb. Fl. Supp., 41; Stewart, Pb. Pl., 107; Aitchison, Cat. Pb. Pl., 67; DC. Prodr., IV., 108; Voigt, Hort. Sub. Cal., 21; Pharm. Ind., 99; Moodeen Sheriff, Supp. Pharm. Ind., 51, 52, 90; Flück. & Hanb., Pharmacog., 302; U. S. Dispens., 1433; U. C. Dutt, Mat. Med. Hind., 172, 173, 324; Dymock, Mat. Med. W. Ind., 2nd Ed., 365; Ainslie, Mat. Ind., I., 38; O'Shaughnessy, Beng. Dispens., I., 357; Murray, Drugs and Pl. of Sind, 196; Waring, Bazar Med., 123; S. Arjun, Bomb. Drugs, 65; Drury, U. Pl., 359; Baden Powell, Pb. Prod., 301; Atkinson, Him. Dist., 705, 730; Lisboa, U. Pl. of Bomb., 161, 223; Spons' Encycl., 791; Smith, Dict., 7; Irvine, Mat. Med. Patna, 5; Top. Ajmir, 124; Kew Cat., 74.*

Habitat.—Cultivated extensively in India on account of its seeds; from the Panjáb and Bengal to the South Deccan. This seems to be the αμμι of the Greeks. It is first mentioned in Europe as brought from Egypt about 1549 and had come into medical use in London about 1693, since it is mentioned by Dale.

OIL. 692

Oil.—The seeds yield an oil on distillation with water, which is used medicinally in cholera, colic, and indigestion.

MEDICINE. 693

Medicine.—In native practice the seeds are much valued for their antispasmodic, stimulant, tonic, and carminative properties. "They are considered to combine the stimulant quality of capsicum or mustard with the bitter property of chiretta, and the antispasmodic virtues of asafœtida."

The Bishop's Weed.

MEDICINE.

(*Waring's Bazar Med.*) They are administered in flatulence, flatulent colic, atonic dyspepsia, and diarrhœa, and are often recommended for cholera. They are used most frequently in conjunction with asafœtida, myrabolans, and rock-salt. A decoction is supposed to check discharges, and it is therefore sometimes prescribed as a lotion, and often constitutes an ingredient in cough mixture. **Ainslie** remarks that the small, warm, aromatic seed resembling anise seed in its virtues, is much used by the native doctors as a stomachic, cardiac, and stimulant, and also by the veterinary practitioners in India in the diseases of horses and cows. **Dr. Bidie** is strongly in favour of the extended use of this medicine. "As a topical remedy it may be used with advantage, along with astringents, in cases of relaxed sore-throats. For disguising the taste of disagreeable drugs and obviating their tendency to cause nausea and griping, I know of no remedy of equal power." The seeds have come into special notice in England and Germany for the manufacture of *Thymol*, enormous quantities of which are now made and used as an antiseptic (*Smith*).

Thymol. 694

OMUM WATER—or distilled water from the seeds—is sold in the bazars of India, also a crystalline essential oil (*Ajwan-ke-phul*). This is chiefly prepared at Oojein and elsewhere in Central India. (*Pharm. Ind.*) It is identical with thymol ($C_{10}H_{14}O$), and is used as a powerful carminative in flatulence and as an antispasmodic in hysterical pains. Of late, it has been extolled as a powerful antiseptic superior to carbolic acid (*Home Dept. Cor., Nov. 1880*). "In cholera, much reliance is placed by the natives and Anglo-Indians on omum water, and although it appears to have no claim to the character of a specific in that disease which popular opinion assigns to it, there can be little doubt that it exercises considerable power, especially in the early stage, of checking the diarrhœa and vomiting, and at the same time of stimulating the system. It is not to be trusted to alone, but forms an admirable adjunct to other remedies. In habitual drunkenness and dipsomania, omum seems worthy of trial." (*Waring's Bazar Med.*) **Dr. Stocks** was the first to draw attention to a crystalline substance sold in the bazars of the Deccan and Sind, known as *Ajwain-ka-phul*. This is prepared from the fruits of **Carum copticum** or forms spontaneously on the surface of the distilled water. (*Pharm. Ind.*)

OMUM. 695

CHEMISTRY.

Chemical Composition.—The authors of the *Pharmacographia* say: "The fruits on an average afford from 4 to 4·5 per cent. of an agreeable, aromatic, volatile oil; at the same time there often collects on the surface of the distilled water a crystalline substance, which is prepared at Oojein and elsewhere in Central India, by exposing the oil to spontaneous evaporation at a low temperature. This steoroptene, sold in the shops of Poona and other places of the Deccan under the name of *Ajwain-ka-phul, i.e.*, flowers of ajwain, was shown by **Stenhouse** (1855) and by **Hains** (1856) to be identical with thymol, $C_6H_3\left\{\begin{matrix}OH\\CH_3\\C_3H_7\end{matrix}\right\}$, as contained in **Thymus vulgaris**.

"We obtained it by exposing oil of our own distillation, first rectified from chloride of calcium, to a temperature of 0° C., when the oil deposited 36 per cent. of thymol in superb tabular crystals, an inch or more in length. The liquid portion, even after long exposure to a cold some degrees below the freezing point, yielded no further crop. We found the thymol thus obtained began to melt at 44° C., yet using somewhat larger quantities, it appeared to require fully 51° C. for complete fusion. On cooling, it continues fluid for a long time, and only recrystallizes when a crystal of thymol is projected into it.

CHEMISTRY.

"Thymol is more conveniently and completely extracted from the oil by shaking it repeatedly with caustic lye, and neutralizing the latter.

"The oil of ajwain, from which the thymol has been removed, boils at about 172°, and contains cymene (or cymol) $C_{10}H_{14}$, which, with concentrated sulphuric acid, affords cymen-sulphonic acid, $C_{10}H_{13}SO_2OH$. The latter is not very readily crystallizable, but forms crystallized salts with baryum, calcium, zinc, and lead, which are abundantly soluble in water. In the oil of ajwain no constituent of the formula $C_{10}H_{16}$ appears to be present; mixed with alcohol and nitric acid, it at least produces no crystals of terpin.

"The residual portions of the oil, from which the cymene has been distilled, contains another substance of the phenol class different from thymol."

Special Opinions.—§ Sometimes used by the natives for colds; useless as far as my experience goes (*Surgeon-Major C. J. McKenna, Cawnpore*). "Much used in flatulence, diarrhœa, and with other drugs in dyspepsia. Very useful in flatulence and with dyspepsia, especially administered in powder mixed with other antispasmodics" (*Surgeon G. Price, Shahabad*). "Exported to England for the sake of the thymol it contains, and which is used in surgery as an antiseptic. Native doctors in Madras famine relief camp used to give 'omum water' for dysentery. I don't think it was of any use, nor, for the matter of that, was any other drug, but they and their patients had great faith in it" (*G. B. Madras*). "Aromatic, stimulant, antispasmodic, tonic, sialagogue; used in dyspepsia, vomiting, griping, diarrhœa, flatulence, faintness" (*Hospital Assistant Choonna Lall, Jubbulpore*). "Carminative and stomachic; mixed with black pepper and salt, and taken in empty stomach, relieves flatulence and colic and promotes digestion" (*Assistant Surgeon Shib Chunder Bhattacharji, Chanda, Central Provinces*). "The water distilled from the seeds is very useful as a carminative, and is largely used by the natives, being administered to newly-born infants as a carminative and stimulant. This plant is commonly cultivated in this district, being largely used as a condiment" (*Surgeon S. H. Browne, M.D., Hoshangabad, Central Provinces*). "The seeds form a constant ingredient in all native mixtures for rheumatism. In combination with cardamoms and nutmegs in powder, and mixed with the mother's milk, they are commonly given to newly-born children" (*Narain Misser, Hoshangabad, Central Provinces*). "A very good carminative" (*Honorary Surgeon E. A. Morris, Negapatam*). "Stimulant, anti-scorbutic, heating medicine" (*Surgeon W. A. Barren, Belgaum, Bombay*).

FOOD. 696 **Food.**—The seeds are aromatic, and form an ingredient of the preparation known as *pán*.

697 **Carum nigrum**, ? *Royle; Him. Bot., 229.*

BLACK CARAWAY.

Syn.—Stewart, Baden Powell, &c., refer the name CARUM GRACILE, *Bth.*, to this species, or rather place both under C. CARUI.

Vern.—*Sháh-zirah, shah-zirāh*, HIND., DUK.; *Shimai-shiragam, pilappu shiragam*, TAM.; *Sima-jilakara*, TEL.; *Shima-jirakam*, MALAY; *Shime-jerage*, KAN.; *Kaméne-kirmání*, ARAB.; *Zirahe-siyáh, zirahe kirmáni, siyáh-zirah*, PERS.

References.—*Pharm. Ind., 99; Baden Powell, Pb. Prod., 351; Moodeen Sheriff, Supp. Pharm. Ind., 90; Dymock, Mat. Med. W. Ind., 305; S. Arjun, Bomb. Drugs, 63; Birdwood, Bomb. Drugs, 39.*

Habitat.—**Royle** mentions that seeds under the name of *Zeera seeah* are imported from Kunawar, and that these are "a kind of caraway." To these seeds he gave the name **Carum nigrum**, without apparently having either seen the plant or ascertained anything more about them. **Stewart** seems to have gone into the subject for he reduces **Royle's C. nigrum** to **C. Carui**. In this view he appears to be supported by **Mr. C. B. Clarke** in the *Flora of British India*, since **Royle** is by that author quoted as having found the true caraway in Kashmír and Garwhal. In what has been already said under **C. Carui** this opinion has been supported; but at the same time it must be added that **Dr. Dymock** and many other writers continue to allude to a black form of caraway. **Dr. Dymock** says: "*Sajíra* or *Siáh-zírah* (*Bomb.*) has more slender and darker-coloured fruits than the true caraway; a transverse section shows a similar structure. The flavour approaches that of Cummin, and the Persian name which it bears signifies black cummin. It is probably the article described in Persian works on Materia Medica as *Kirmání* or black cummin." 698

"*Siáh-zirah* is imported in large quantities from the Persian Gulf ports, average value R8 per Surat maund of 37½lb." "In 1881-82 the imports into Bombay from Persia amounted to 2,683 cwts. valued at R71,886. The exports were 5 cwts. to Mauritius and 4 cwts. to Aden."

Under **C. Carui** it has already been stated that a considerable trade is done between the North-Himálayan and trans-Himálayan regions with the plains of India in what has been accepted as the **true** caraway. These two seeds are distributed all over India, the Europeans using the latter and the Natives, especially the Muhammadans, preferring the former. Whether or not the *Siáh-zirah* of Bombay may prove a variety, or perhaps even a species, of **Carum** remains to be determined when the seeds are cultivated, but there would appear no reason to doubt that that seed is not what **Royle** had in view when he published the name **C. nigrum**. The information here given refers exclusively to the Persian Gulf seed, and it has been placed under the botanical name published by **Royle** in order to forcibly draw attention to the fact that recent writers have, as it would appear, been confusing two very distinct seeds under one botanical name.

It is thus probable that the vernacular names given under **C. Carui** and **C. nigrum** (above) may have to undergo a rearrangement when the two species, or two varieties of one species, to which they refer, have been finally determined. The name *Siáh-zirah* is applied to both as matters stand at present, but the other Panjáb names refer to the cultivated and wild forms of the Panjáb plant only. It is of course possible that the Persian Gulf *Siáh-zírah* may prove a form of the Panjáb plant, and that both may be even referable to **Carum Carui**.

Medicine.—The seeds are used as a carminative. MEDICINE. 699

Food.—In Western India the seeds are used mostly by Muhammadans as a spice, and also to a certain extent by the Marathas (*Dymock*). FOOD. 700

Carum Roxburghianum, *Benth.; Fl. Br. Ind., II., 682; Wight, Ic., t. 567.* 701

Syn.—Apium involucratum, *Roxb.*; Ptychotis Roxburghiana, *DC.*

Vern.—*Bazrul-karafs*, Arab.; *Tukhme-karafs*, Pers.; *Ájmúd, ájmúdá, ajmot, ajmod*, Hind.; *Ájmudáh, ájmúdah-ajván*, Dec.; *Randhuni*, C. P.; *Asham tágam, ashamtá-óman*, Tam.; *Ajumóda-vomam, ashumadága-vóman, ajumoda vomaru*, Tel.; *Ajmúd, rándhoni*, or *radhuni, chanu*, Beng.; *Ajmódá-vóvá, koranza*, Mar.; *Ájmodá-vómá*, Kan.; *Ajmod, bodiajamo*, Guj.

References.—*Roxb., Fl. Ind., Ed. C.B.C., 273; Dalz. & Gibs, Bomb., Fl., 106; DC. Prodr., IV., 109; Pharm. Ind., 108; Moodeen Sheriff, Supp. Pharm. Ind., 91; U. C. Dutt, Mat. Med. Hind., 173; Atkinson, Him. Dist., 705; Birdwood, Bomb. Prod., 39; Lisboa, U. Pl. of Bomb., 161; Irvine, Mat. Med. Patna, 6; Ajm. Med. Top., 124; Fleming, Cat., Ind. Med.*

Habitat.—A herbaceous plant extensively cultivated throughout India, from Hindustan and Bengal to Singapore and Ceylon.

MEDICINE. 702 **Medicine.**—The seeds of this species are useful in hiccup, vomiting, and pain in the bladder. They form an ingredient of carminative and stimulant preparations, and are useful in dyspepsia.

Special Opinions.—§ "Carminative. It is an essential ingredient of native cookery, and is generally called *Randhuni*" (*Assistant Surgeon Shib Chunder Bhattacharji, Chanda, Central Provinces*).

FOOD. Seeds. 703 **Food.**—Often raised in gardens during the cold season for the seed which is used in flavouring curry; also used by the Europeans as a substitute for parsley (*Royle*). Extensively cultivated in Gajarat (*Lisboa*).
Leaves. 704 Leaves though of an unpleasant smell are now and then used by Europeans as a substitute for parsley (*Voigt*).

705 **Carving, Fancy work, Images, &c.**—

Timbers used for :—

Berberis nepalensis, *Spreng.* (useful for inlaying).
Buxus sempervirens, *Linn.* (carving).
Cedrela Toona, *Roxb.* (carving).
Celastrus spinosus, *Royle* (carving and engraving).
Chickrassia tabularis, *Adr. Juss.* (carving).
Cocos nucifera, *Linn.* (fancy work).
Cratæva religiosa, *Forst.* (models).
Cupressus torulosa, *Don.* (images).
Dalbergia cultrata, *Grah.* (carving).
D. latifolia, *Roxb.* (carving and fancy work).
D. Sissoo, *Roxb.* (carved work).
Diospyros Ebenum, *Konig.* (used for inlaying).
D. melanoxylon, *Roxb.* (fancy work and carving).
Euonymus grandiflorus, *Wall.* (carving).
E. Hamiltonianus, *Wall.* (carving into spoons).
Givotia rottleriformis, *Griff.* (carving figures).
Gmelina arborea, *Roxb.* (carving images).
Hardwickia binata, *Roxb.* (ornamental work).
Holarrhena antidysenterica, *Wall.* (carvings).
Kydia calycina, *Roxb.* (carving).
Melia Azadirachta, *Linn.* (idols).
Pistacia integerrima, *J. L. Stewart* (carving, ornamental work).
Premna tomentosa, *Willd.* (fancy work).
Santalum album, *Linn.* (carving).
Stephegyne parvifolia, *Korth.* (carved articles).
Symplocos cratægoides, *Ham.* (carving).
Tecoma undulata, *G. Don* (carving).
Ulmus integrifolia, *Roxb.* (carving).
Viburnum erubescens, *Wall.* (carving).
Wightia gigantea, *Wall.* (Buddhist idols).
Wrightia tinctoria, *R. Br.* (carving).
W. tomentosa, *Rom. & Sch.* (carved work).

CARYOPHYLLUS, *Linn.; Gen. Pl., I., 719.*

706 **Caryophyllus aromaticus,** *Linn.; DC. Prodr., III., 262;*
CLOVES. [MYRTACEÆ.

Syn.—EUGENIA CARYOPHYLLATA, *Thunberg.*

Vern.—*Mekhak*, PERS.; *Lavanga, langa*, BENG.; *Lóng, laung*, HIND. *Laung, karanfal*, PB.; *Raung*, KASHMIR; *Lawanga, lavinga*, MAR., GUJ.; *Lavang*, BOMB.; *Kiramber, kirámbu, ilavangap-pú, karuváp-pui crambú*, TAM.; *Lavangalu*, TEL.; *Chanki*, MAL.; *Labang*, DEC.; *Lavanga*, SANS.; *Varrala*, SING.

References.—*Roxb., Fl. Ind., Ed. C.B.C., 401; Kurz, For. Fl. Burm., I., 471; Gamble, Man. Timb., 188; DC. Origin of Cult. Pl., 128; Pharm. Ind., 91; Fluck. & Hanb., Pharmacog., 280; U. S. Dispens., 15th Ed., 363; Bentley & Trim., Med. Pl., 112; Ainslie, Mat. Ind., I.,*

593; U. C. Dutt, Mat. Med. Hind., 164, 307; Dymock, Mat. Med. W. Ind., 2nd Ed., 328; O'Shaughnessy, Beng. Dispens., 334; Murray, Pl. and Drugs of Sind, 192; Baden Powell, Pb. Pr., 349; Waring, Bazar Med., 44; S. Arjun, Bomb. Drugs, 56; Birdwood, Bomb. Pr., 35; Lisboa, U. Pl. of Bomb., 156; Dalz. & Gibs, Bomb. Fl. Suppl., 34; Spons, Encyclop., 1807; Balfour, Cyclop.; Smith, Dic., 120; Treasury of Botany; Ajmír Med. Top., 144, 145; Royle, Him. Bot., 217.

Habitat.—A native of the Moluccas. Cultivated in Southern India. The Dutch tried to restrict its cultivation to the Island of Amboyna, but in the course of time it got introduced into India and other tropical countries. The flower-buds of this plant yield the cloves of commerce.

CULTIVATION.

Cultivation and yield.—"In cultivating cloves, the mother-cloves (fruits) are planted in rich mould about 12 inches apart, screened from the sun, and duly watered. They germinate within 5 weeks, and, when 4 feet high, are transplanted at distances of 30 feet. There should be a certain amount of sand in the soil to reduce its tenacity, and less manure is required than for nutmegs. The tree naturally selects a volcanic soil, and a sloping position. The yield commences at about the 6th year, and is at its maximum in the 12th year, when the average annual produce may be estimated at 6-7lb of marketable fruit from each tree. There is usually a crop every year, but in Sumatra the trees often bear only twice in 3 years. When past its prime, the tree has a ragged appearance. Its existence in Sumatra is supposed to be limited to a duration of about 20 years, except in very superior soil, when it may perhaps last 24 years; yet in Amboyna, it does not bear till the 12th-15th year, and continues prolific to the age of 75-150 years. Hence, it is necessary to plant a succession of seedlings when the old trees have attained their 8th year, this octennial system being adhered to throughout. The slight hold which the trees have upon the soil, renders it very desirable that they should be provided with shelter from strong winds. . . . . The harvesting of the flower-buds (cloves) commences immediately they assume a bright red colour. The best and most usual plan is to pluck them singly by hand, moveable stages facilitating the operation in the case of the upper branches. Sometimes, however, they are beaten off by long bamboos, and caught in cloths spread below. The plucked cloves undergo a process of drying, which confers a brown hue, and prepares them for packing. In Sumatra, simple exposure to the sun for several days on mats is the common method; but elsewhere they are occasionally also smoked on hurdles covered with matting near a slow wood-fire; and very rarely they are scalded in hot water before smoking. They are ready for packing when they break easily betwen the fingers." (*Spons' Encycl.*)

OIL. 707

Oil.—Every part of the plant abounds with aromatic oil. The flower-buds and flower-stalks of cloves yield, when distilled with water, an essential oil. The process of distillation is largely carried on in England. It is a colourless or a yellowish oil, having a powerful odour and flavour of cloves. It easily combines with grease, soap, and spirit, and is extensively made use of in the manufacture of perfumery. In Germany it is often adulterated with carbolic acid. The essence of cloves is obtained by dissolving oil of cloves in the proportion of four ounces of oil to one gallon of spirit.

Description of the Drug.—"The varieties of cloves occurring in commerce do not exhibit any structural differences. Inferior kinds are distinguished by being less plump, less bright in tint, and less rich in essential oil. In London price-currents, cloves are enumerated in the order of value thus: Penang, Bencoolen, Amboyna, Zanzibar" (*Pharmacog., 284*). The cloves met with in the Indian bazars are generally old and worthless. Those suited for medical use should have a strong, fragrant odour, a bitter,

CARYOPHYLLUS aromaticus. **Cloves.**

DESCRIPTION OF EHT DRUG.

spicy, pungent taste, and should emit a trace of oil when pressed with the nail (*Waring's Bazar Medicines*). "The Americans have introduced into commerce an imitation: deal-wood cut into appropriate pieces are soaked in a solution of true cloves" (*Lisboa*). "Clove-stalks, the *vikunia* of the natives, are largely shipped from Zanzibar, and used in the manufacture of mixed spice and for adulterating ground cloves. . . . . Mother-cloves or fruits are also exported, probably for a similar purpose" (*Spons' Encycl., 1808*).

MEDICINE. Buds. 708

Medicine.—The dried flower-buds which constitute the cloves of commerce are, aromatic, stimulant, and carminative; they are used in atonic dyspepsia and in gastric irritability, especially in the vomiting of pregnancy. **Ainslie** recommends the use of the infusion made with one drachm of bruised cloves and half a pint of boiling water, to be taken in the dose of from ℥i to ℥iis, thrice daily in languors and in dyspepsia. A five-grain pill made of equal parts of jalap and powdered cloves generally opens the bowels. "Cloves are much used in Hindu medicine, as an aromatic adjunct. They are regarded as light, cooling, stomachic and digestive, and useful in thirst, vomiting, flatulence, colic, &c. An infusion of cloves is given to appease thirst" (*U. C. Dutt, Mat. Med. Hind, 164*). A mixture of equal parts of the infusion of cloves and chiretta has an excellent effect in debility, loss of appetite, and in convalescence after fevers. "The oil, *Lavanga-tela,* is used externally in rheumatic pains, headache, and toothache" (*S. Arjun*), and "is a frequent ingredient of pill-masses." "Muhammadan writers describe cloves as hot and dry, and consider them to be alexipharmic and cephalic, whether taken internally or applied externally; they also recommend them for strengthening the gums and perfuming the breath, and on account of their pectoral, cardiacal, tonic, and digestive qualities. They have a curious superstition to the effect that one male clove eaten daily will prevent conception" (*Dymock's Mat. Med. W. Ind., 329*).

Chemical Composition.—"Few plants possess any organ so rich in essential oil as the drug under consideration. The oil known in pharmacy as *Oleum Caryophylli,* which is the important constituent of cloves, is obtainable to the extent of 16 to 20 per cent. But to extract the whole, the distillation must be long continued, the water being returned to the same material.

"The oil is a colourless or yellowish liquid with a powerful odour and taste of cloves, sp. gr. 1·046 to 1·058. It is a mixture of a hydrocarbon and an oxygenated oil called *Eugenol,* in variable proportions. The former, which is termed *light oil of cloves,* and comes over in the first period of the distillation, has the composition $C_{15}H_{24}$, a sp. gr. of 0·918, and boils at 251° C. It deviates the plane of polarization slightly to the left, and is not coloured on addition of ferric chloride; it is of a rather terebinthinaceous odour.

"Eugenol, sometimes called *Eugenic Acid,* has a sp. gr. of 1·087 at 0° C., and possesses the full taste and smell of cloves. Its boiling point is 247·5°. With alkalis, especially ammonia and baryta, it yields crystallizable salts. Eugenol may therefore be prepared by submitting the crude oil of cloves to distillation with caustic soda; the 'light oil' then distils the eugenol, being now combined with sodium, remains in the still. It will be obtained on addition of an acid and again distilling. Eugenol is devoid of rotatory power, whence the crude oil of cloves, of which eugenol is by far the prevailing constituent, is optically almost inactive. The constitution of eugenol is given by the formula $C_6H_3\left\{\begin{matrix} OCH_3 \\ OH \\ CH.CH.CH_3 \end{matrix}\right\}$. It belongs

MEDICINE.

to the phenol class, and has also been met with in the fruits of **Pimenta officinalis,** in the Bay leaves, in Canella bark, in the leaves and flower-buds of **Cinnamomum zeylanicum,** and in Brazilian clove bark **(Dicypellium caryophyllatum,** *Nees*).

"Eugenol can be converted into *Vanillin.*

"The water distilled from cloves is stated to contain, in addition to the essential oil, another body, *Eugenin,* which sometimes separates after a while in the form of tasteless, crystalline laminæ, having the same composition as eugenol. We have never met with it.

"According to **Scheuch** (1863), oil of cloves also (sometimes) contains a little Salicylic acid, $C_6H_4 \left\{ \begin{matrix} OH \\ COOH \end{matrix} \right\}$, which may be removed by shaking the oil with a solution of carbonate of ammonium.

Caryophyllin, $C_{20}H_{32}O$, is a neutral, tasteless, inodorous substance, crystallizing in needle-shaped prisms. We have obtained it in small quantity, by treating with boiling ether cloves, which we had previously deprived of most of their essential oil by small quantities of alcohol. **E. Mylius** (1873) obtained from it, by nitric acid, crystals of *Caryophyllinic Acid,* $C_{20}H_{32}O_6$.

"*Carmufellic Acid,* obtained in colourless crystals, $C_{12}H_{20}O_{16}$, in 1851, by **Muspratt** and **Dansan** after digesting an aqueous extract of cloves with nitric acid, is a product of this treatment and not a natural constituent of cloves.

"Cloves contain a considerable proportion of gum; also a tannic acid not yet particularly examined" (*Pharmacog.*, *285*).

Special Opinions.—§ "Cloves relieve tickling cough when kept in the mouth. They are stimulant" (*Surgeon-Major W. Moir, Meerut*). "Mixture formed by rubbing cloves with honey on a copper plate, is applied by means of a feather to the conjunctiva of the lower eyelid in cases of conjunctivitis. Oil extracted from cloves is useful in toothache" (*Surgeon Anund Chunder Mukerji, Noakhally*). "Clove-stalks (*vikuria*) are also imported for shipment to Europe, where they are distilled. India is largely supplied from Zanzibar with cloves and mother-cloves (*nar laung*)" (*Surgeon-Major W. Dymock, Bombay*). "Said to be stimulant and carminative" (*Surgeon-Major C. J. McKenna, Cawnpore*). "Useful in heartburn and colic. Burnt in the flame of a lamp, &c., then taken they relieve irritation of the throat and hacking cough" (*Brigade Surgeon G. H. Thornton, B.A., M.B., Monghir*). "Stimulant and carminative" (*Assistant Surgeon Shib Chunder Bhattacharji, Chanda, Central Provinces*). "The oil of cloves is a well-known application to relieve toothache used in the same way as creosote" (*Brigade Surgeon W. H. Morgan, Cochin*). "Useful in carious tooth" (*Surgeon H. D. Masani, Karachi*).

FOOD. 709

Food.—The dried flowers (cloves) are used to a limited extent as a hot spice throughout India. They are also chewed in *pán.*

TRADE. 710

Foreign Trade in Cloves.

Year.	IMPORTS.		EXPORTS AND RE-EXPORTS.	
	Quantity.	Value.	Quantity.	Value.
	℔	R	℔	R
1880-81	2,583,852	14,40,739	1,064,115	6,20,331
1881-82	2,653,836	12,64,254	735,892	3,49,879
1882-83	3,878,232	13,09,518	1,230,104	3,74,857
1883-84	3,893,159	10,61,206	1,068,906	2,75,564
1884-85	4,791,006	11,09,841	1,649,040	3,67,249

CARYOTA urens.

Sago Palm.

TRADE.

Imports for 1884-85.

Presidency to which imported.	Quantity.	Value.	Country from which imported.	Quantity.	Value.
	lb	R		lb	R
Bombay . .	4,598,419	10,50,680	Zanzibar . .	4,776,842	11,05,877
Bengal . .	190,526	58,283	Aden . .	11,767	2,908
Britsh Burma .	1,288	425	Other Countries .	2,397	1,056
Madras . .	773	453			
TOTAL .	4,791,006	11,09,841	TOTAL .	4,791,006	11,09,841

Exports for 1884-85.

Presidency from which exported.	Quantity.	Value.	Country to which exported.	Quantity.	Value.
	lb	R		lb	R
Bombay . .	1,618,465	3,55,692	United Kingdom .	1,112,224	2,32,739
Bengal . .	29,165	10,090	China—Hongkong .	349,698	84,966
Madras . .	1,390	1,462	Straits . . .	124,101	33,543
Sind . . .	20	5	Turkey in Asia .	15,137	3,887
			Aden . . .	7,000	1,790
			France . . .	7,000	1,750
			Other Countries .	33,880	8,574
TOTAL .	1,649,040	3,67,249	TOTAL .	1,649,040	3,67,249

Very little can be said regarding the present position of the new industry of cultivating cloves in South India. Good samples were, however, shown at the Colonial and Indian Exhibition.

CARYOPTERIS, *Bunge; Gen. Pl., II., 1157.*

Caryopteris Wallichiana, *Schauer; DC. Prodr., XI., 625;* [VERBENACEÆ.

Vern.—*Moni, mohâni,* KUMAON; *Shechin,* NEPAL; *Malet,* LEPCHA.

References.—*Brandis, For. Fl., 370; Gamble, Man. Timb., 299.*

Habitat.—A large shrub with thin, grey, papery bark, peeling off in vertical strips; met with on the outer Himálaya, from the Indus to Bhután, ascending to 3,000 feet.

Structure of the Wood.—Dark grey, moderately hard, with the scent of cherry-wood.

CARYOTA, *Linn.; Gen. Pl., III., 918.*

CARYOTA URENS. 711

Caryota urens, *Linn.; Gamble, Man. Timb., 420;* PALMÆ.

KNOWN IN BOMBAY AS THE HILL PALM; also "SAGO PALM."

Vern.—*Mari,* HIND.; *Rungbong, simong,* LEPCHA; *Bara flawar,* ASS.; *Salopa,* URIYA; *Mari-ká-jhár,* DEC.; *Bherawa, berli, bhirli mahad, berli* *mád, bherlá máda, berli mhár, ardhi supári,* MAR.; *Birli-mhad, birli mhar,* BOM.; *Shiwajatá, shaukar jata,* GUJ.; *Birli mád,* KONKAN; *Mhár mardi, mari, jirágú, goragú, gorregu,* TEL.; *Conda-panna, erim-panna, utalipanha, kúndal-panai,* TAM.; *Bhyni, beina, bagni, baini,* KAN.; *Shunda pana,* MAL.; *Kittúl, nepora,* SING.; *Hlyamban,* MAGH; *Minbo, minbaw, kimbo,* BURM.

References.—*Roxb., Fl. Ind., Ed. C.B.C., 668; Brandis, For. Fl., 550; Kurz, For. Fl. Burm., II., 530; Voigt, Hort. Sub. Cal., 637; Thwaites, En. Ceylon Pl., 329; Dalz. & Gibs., Bomb. Fl., 278; Pharm. Ind., 248; Moodeen Sheriff, Supp. Pharm. Ind., 92; Ainslie, Mat. Ind., I., 452; S. Arjun, Bomb. Drugs, 147; Lisboa, U. Pl. of Bomb., 135, 212, 237; Spons, Encyclop., 938, 1827, 1904; Balfour, Cyclop.; Smith, Dic., 307, 862; Royle, Fib. Pl., 98; Treasury of Botany; Journ. Soc. of Arts, XXXII., 734; Mason's Burma, 811; Bomb. Gaz., XI., 30; XIII., 21; XIV., 252. Settlement Reports, C. P., Upper Godavery, 39, Chanda, 108.*

Habitat.—A beautiful palm, with smooth, annulated stem, met with in the forests of the western and eastern moist zones. On the Western Ghâts, it extends to near Mahableshwar. In the Settlement Reports of the Chanda district it is stated that this palm abounds in the south-eastern corner of Aheree, and might with advantage be extended to all parts of the district, for it thrives well wherever it is planted. It is common in Burma, Bengal, and Orissa, ascending in Sikkim to 5,000 feet.

Fibre.—"The leaves give the *Kittul Fibre,* which is very strong and is made into ropes, brushes, brooms, baskets, and other articles; the fibre from the sheathing petiole is made into ropes and fishing-lines" (*Gamble*), and is said to be suitable for paper manufacture. **FIBRE. 712**

At the Colonial and Indian Exhibition (1886-87) much interest was taken in *salopa* fibre sent from Orissa, Burma, and Kolaba in Bombay. A corset manufacturer applied at the office of the Indian section for a fibre which might take the place of whalebone in corset-making. He was shown the *salopa* (*kittul*) fibre and also the similar cord-like fibres from the interior of the stems of the cocoanut and palmyra palms. It was suggested that if either of these were to be sewn in bands into the fabric of the corset the desired object would be obtained. The idea met with approval, and within a few days the manufacturer exhibited a sample and expressed the utmost confidence that if he could procure a continuous supply of the fibre a large trade might be done. He was referred to the Commissioner for the Ceylon Court, since a considerable trade was being done with that Colony in its *kittul* fibre. Shortly after, however, he returned with the report that while the *kittul* fibre was perhaps preferable for the brush-maker, the softer nature of the *salopa* fibre of India made it preferable for his purpose. These facts are alluded to in the hope of awakening interest in an Indian fibre that has been much neglected. For a good few years past Ceylon has done a by no means inconsiderable trade in *kittul* fibre, but no person seems to have thought of India as a possible source of supply. Since its introduction into commerce in 1860, the uses of *kittul* have greatly extended, and there would, therefore, seem everything to justify the expectation that India might with advantage enter into competition with Ceylon. The *kittul,* or as it is called in Orissa the *salopa* fibre, is the cord-like fibro-vascular bundles which surround the base of the leaf sheath. **Mr. A. Robottom** was the first to introduce *kittul* fibre to European commerce: that gentleman is reported in *Spons' Encyclopædia* to have stated that Indian *kittul* is inferior to Ceylon. At the Colonial and Indian Exhibition he pointed out to the writer a sample of the much inferior *kittul*-like fibre from **Arenga saccharifera** (see **A. 1336**) as the *kittul* he had formerly seen as sent from India. He admitted that the sample of *salopa* shown him at the Exhibition was as good as any he had ever seen from Ceylon, and seemed confident a large trade could be done in the Indian fibre.

It is commonly reported that in Ceylon the black fibre from the leaf-stalks is manufactured into ropes which are of great strength and durability, being used for tying wild elephants. A woolly material found

Tomentum stem fibres. at the base of the leaves is sometimes used for caulking ships in Burma. In some parts of India the cord-like fibre from the stem of this and other palms is employed as a bow-string or as a fishing line (see **B. 667**). (*Royle, Fib. Pl.*)

MEDICINE. 713 **Medicine.**—"An excellent spirit is obtained by the fermentation and distillation of the toddy obtained from this elegant palm, which is not uncommon on the west coast of the Madras peninsula. It is well adapted for pharmaceutical purposes." "A glass of the freshly-drawn toddy, taken early in the morning, acts as a laxative" (*Pharm. of India*). "The nut is used as an application to the head in cases of hemicrania, from an idea of the supposed efficiency of the half nut in curing the affected half of the head" (*S. Arjun, Bombay Drugs*).

FOOD. 714 **Food.—Roxburgh** writes: "This tree is highly valuable to the natives of the countries where it grows in plenty. It yields them, during the hot season, an immense quantity of toddy or palm wine. I have been informed that the best trees will yield at the rate of 100 pints in the 24 hours. The sap in some cases continues to flow for about a month. When fresh, the toddy is a pleasant drink, but it soon ferments, and when distilled becomes arrack, the gin of India. The sugar called jaggery is obtained by boiling the toddy. The pith or farinaceous part of the trunk of old trees is said to be equal to the best sago; the natives make it into bread, and boil it into thick gruel; these form a great part of the diet of those people, and during the late famine (1830?), they suffered little while those trees lasted. I have reason to believe this substance to be highly nutritious. I have eaten the gruel, and think it fully as palatable as that made of the sago we get from the Malay countries." (*Fl. Ind.*)

"The trees are tapped when they are from fifteen to twenty-five years old. Besides bruising and binding it, the spathe, which is called *kote*, is heated to make the juice flow. Every three or four days a white cottony substance called *kaph*, which forms in the centre of the spathe, is removed. The stem of the tree is so soft that notches cannot be cut, and the tapper climbs by the help of branches tied to the trunk. Tapping goes on for eight months in the year. It is stopped during the rainy season (June to October), because the tree becomes slippery and the spathe cannot be heated. The trees are not allowed to rest, but are tapped until they are exhausted. In good ground they last for ten years, and in poor soil for four or five. After this they are useless. In yield, or in the value of the juice, the big-trunked palm differs little from the palmyra. Since 1879, when the tree-tax was raised from 1*s*. 6*d*. to 6*s*. (*annas 12 to R3*), the number of trees tapped has greatly fallen" (*Bomb. Gaz.* (Kolaba), *XI., p. 30*).

TIMBER. 715 **Structure of the Wood.**—The outer part of the stem is hard and durable, and the vascular bundles crowded, black, very large. The wood is strong and durable; it is used for agricultural purposes, water conduits, and buckets. It is "useful for building purposes" (*Thwaites*). "Is in general use for field-tools" (*Bomb. Gaz., XV., I, 65*).

716 **Cascarilla bark,** the bark of **Croton Eluteria**, EUPHORBIACEÆ.

A native of the Bahamas. The bark is imported into India.

CASEARIA, *Jacq.; Gen. Pl., I., 796.*

Casearia esculenta, *Roxb.; Fl. Br. Ind, II., 592;* SAMYDACEÆ. 717

Syn.—C. LÆVIGATA, *Dalz.*, in *Hooker's Jour. Bot., IV., 107;* C. CHAMPIONII and C. ZEYLANICA, *Thwaites.*

Vern.—*Kunda-jungura,* TEL.; *Wal-wareka,* SING.

References.—*Roxb., Fl. Ind., Ed. C.B.C., 377; Drury, U. Pl., 119; Dalz. & Gibs., Bomb. Fl., 11; Thwaites, En. Ceylon Pl., 19.*

Habitat.—A shrub or small tree, frequent in Malabar from Bombay to Coorg; common in Ceylon. It is also met with in Burma from Moulmein to Singapore. **Roxburgh** mentions it as met with on the Circar mountains.

Medicine.—"The roots are purgative, and as such used by the hill people." (*Roxb.*) MEDICINE. 718

Food.—"The leaves are eaten in stews by the natives." (*Roxb.*) FOOD. 719

C. glomerata, *Roxb.; Fl. Br. Ind., II., 591.* 720

Vern.—*Lúrjúr,* SYLHET; *Burgonli,* NEPAL; *Sugvat,* LEPCHA.

References.—*Roxb., Fl. Ind., Ed. C.B.C., 370; Kurz, I., 530; Gamble, Man. Timb., 206.*

Habitat.—A shrub or (in the interior of Sikkim) a tree 20 to 30 feet in height. Frequent in Bhután and on the Khásia Hills at an altitude of 3,000 feet.

Structure of the Wood.—Yellowish white, moderately hard, rough, weighing between 45 and 48℔ per cubic foot. Used for building, charcoal, and occasionally for tea-boxes. TIMBER. 721

C. graveolens, *Dalz.; Fl. Br. Ind., II., 592.* 722

Vern.—*Chilla, náro, aloál, kathera, pimpri,* HIND.; *Rari,* KOL.; *Beri,* KHARWAR; *Newri,* SANTAL; *Girchi, túndri,* GOND.; *Rewat,* KURKU; *Moda,* MAR.

References.—*Brandis, For. Fl., 243; Gamble, Man. Timb., 206; Dalz. & Gibs., Bomb. Fl., 11; Lisboa, U. Pl. of Bomb., 81 and 265.*

Habitat.—A shrub or small tree, 20 feet in height, found in Garhwál and Kumaon; Sikkim at an altitude of 1,500 feet; Deccan Peninsula and in Burma.

Structure of the Wood.—Light yellow, moderately hard, rough; weight between 40 and 50℔ per cubic foot. Not put to any economic purpose. TIMBER. 723

Domestic Uses.—The fruit is used to poison fish. An infusion of the leaves is also said to have a poisonous effect on human beings. DOMESTIC. 724

C. tomentosa, *Roxb.; Fl. Br. Ind., II., 593; Wight, Ic., t. 1849.* 725

Syn.—C. ANAVINGA, *Dalz. & Gibs., Bomb. Fl., 11;* C. CANZIALA, *Ham.;* C. OVATA, *Roxb.;* C. ELLIPTICA, *Willd.*

Vern.—*Chilla, chilara, bairi, bhari,* HIND.; *Maun,* MANBHUM; *Roré,* KOL.; *Beri,* KHARWAR; *Chorcho,* SANTAL; *Munkuro-kuri,* MAL; *Girari,* URIYA; *Thundri,* GOND; *Khesa,* KURKU; *Men, wasa, gamgudu,* TEL.; *Lainja, massei, karei,* MAR.

References.—*Roxb., Fl. Ind., Ed. C.B.C., 377; Brandis, For. Fl., 243; Kurz, I., 529; Gamble, Man. Timb., 206; Stewart, Pb. Pl., 44; Lisboa, U. Pl. of Bomb., 81 and 272; Drury, U. Pl., 118; Thwaites, En. Ceylon Pl., 19.*

Habitat.—A shrub or small tree, attaining a height of 25 feet, common throughout India and Ceylon.

Medicine.—The bark is bitter and used as an adulterant for the (**Mallotus philippinesis** or) *Kámela* powder. "The pounded fruit yields a MEDICINE. 726

MEDICINE. milky, acrid juice, employed to poison fish" (*Brandis*). The leaves are used in medicated baths, and the pulp of the fruit is a very useful diuretic (*Lindley*).

Special Opinion.—§ "Bark applied externally in dropsy" (*Rev. A. Campbell, Santal Mission, Bengal*).

TIMBER. 727 **Structure of the Wood.**—Yellowish white, moderately hard, rough, close-grained; weight 41℔ per cubic foot; used to make combs.

Cashew-nut. See **Anacardium occidentale,** *Linn.*; ANACARDIACEÆ.

Cassareep, and

Cassava Bread, and **Tapioca,** see **Manihot utilitissima,** *Pohl.*; EUPHORBIACEÆ.

CASSIA, *Linn.*; *Gen. Pl., I., 571.*

The word Cassia is taken from the Latin and the Greek **Κασσία**, and from this has been derived CASSIA the Italian, and CASSE, the French. In the Scriptures two or three different things appear all to be rendered as Cassia. The genus is of considerable importance from a medical point of view.

728 **Cassia Absus,** *Linn.*; *Fl. Br. Ind., II., 265.*

Vern.—*Tashmízaj, chashmízaj, hab-es-soudán,* ARAB.; *Chashmízak, chashúm, cheshmak,* PERS.; *Chákút, chákśú, banár,* HIND., DEC.; *Mulaippál-virai, karunká-nam, káttukkol, edikkol,* TAM.; *Chanupála-vittulu,* TEL.; *Karin-kolla,* MALA.; *Cháksie,* BOMB.; *Kán-kuti,* MAR.; *Chimar* or *chimr, chínól,* GUJ.; *Chowun,* SIND; *Kalu-kollu, bú-tóra,* SING.

References.—*Roxb., Fl. Ind., Ed. C.B.C., 351; Gamble, Man. Timb., 136; Thwaites, En. Ceylon Pl., 96; Stewart, Pb. Pl., 61; Aitchison, Cat. Pb. Pl., 51; Pharm. Ind., 78; Moodeen Sheriff, Supp. Pharm. Ind., 92; Dymock, Mat. Med. W. Ind., 2nd Ed., 267; O'Shaughnessy, Beng. Dispens., 309; Murray, Drugs and Pl., Sind, 132; Stocks' account of Sind; Bidie, Raw Prod., Paris Exh., 27; S. Arjun, Bomb. Drugs, 45; Drury, U. Pl., 119; Baden Powell, Pb. Pl., 343; Atkinson, Him. Dist., 731; Birdwood, Bomb. Prod., 27; Balfour, Cyclop.; Treasury of Botany, 232.*

Habitat.—An erect annual, 1-2 feet high, having grey, bristly, viscose hairs. Found growing at the foot of the Western Himálaya, and from thence distributed to Ceylon.

History.—The seeds of this plant were used by the ancient Egyptians in the treatment of ophthalmia, and through them the Roman and the Greek, and from the latter the Muhammadan writers, became aware of their properties. **Dioscorides** speaks of them under the name of *Akákális.* Their Arabic names, *Hab-us-Soudán* and *Tashmízai,* are corruptions of the Persian *Chashmízak.* According to **Ibn Baitar,** the Soudan seeds are the best and the largest. (*Dr. Dymock, Mat. Med. W. Ind.*)

MEDICINE. Seeds. 729 **Medicine.**—A preparation from the SEEDS is applied beneath the eyelids in the treatment of ophthalmia. **Dr. Stocks** says that in Sind "the kernels are put into the inflamed eye made up with water." "For this purpose the seeds are reduced to a fine powder, and a small portion, a grain or more, is introduced beneath the eyelids. In an epidemic of purulent ophthalmia, which visited Brussels in 1882, **Dr. Harbauer** gave a fair trial to this treatment, and the results were on the whole confirmatory of its alleged efficacy. **Dr. G. Smith,** Superintendent of the Eye Infirmary at Madras, in his report, characterises it as a dangerous

MEDICINE.

application in catarrhal ophthalmia and granular lids, adding that its application causes great pain. As met with in the bazárs, these seeds are of a black, shining colour, somewhat flat, of an oval or oblong form, pointed at one extremity, about one-sixth of an inch long, having a bitter taste." (*Pharm. Ind.*) They are very bitter, somewhat aromatic and mucilaginous, and, as such, have been found very useful in mucous disorders. An extract is prepared from them and used to purify the blood. **Dr. Irvine,** in his *Materia Medica of Patna,* says that the receptacle of the seed possesses stimulant and diuretic properties (dose 5 grains to 1 scruple). According to some authors, a plaster made from the seeds is a useful application to wounds and sores, especially of the penis. Extract. 730

Special Opinions.—§ "Seeds are found efficacious in ring-worm" (*Surgeon C. J. W. Meadows, Burrisal*). "Cathartic, dose ½ to 3 drachms, used in habitual constipation, or in constipation caused by pregnancy, with confection of rose and liquorice, have proved effective. In dyspepsia, flatulent colic, and bilious headache, it is given as a compound powder, containing ginger, black rock-salt, *amla* and *buru*, and *chotty hus*" (*Hospital Assistant Abdulla, Civil Dispensary, Jubbulpore*).

According to **Dr. Dymock**, the Bombay supply comes from Sind and Cutch; value, R4 a Surat maund of 37½lb.

Cassia acutifolia, *Delile.*

European Senna. 731

THE ALEXANDRIAN SENNA of Commerce.

Syn.—C. SENNA, *β. Linn.*; C. LANCEOLATA, *Nectoux, non Forsk. nec W. & A.*; C. LENITIVA, *Bisch.*; SENNA ACUTIFOLIA, *Batka.* See also the remarks under C. LANCEOLATA, *Forskhal.*

Habitat.—A native of Nubia (at Sukkot, Mahas, Dongola, Berber), of Kordofan and Sennaar, and other parts of Africa.

For Indian Senna see **C. angustifolia, C. Burmannii,** and **C. obovata.**

C. alata, *Linn.*; *Fl. Br. Ind., II., 264.*

732

Vern.—*Dádmardan, dádmári,* BENG.; *Dádmurdan, dát-ká-pát,* HIND.; *Dádamardana,* BOMB., MAR.; *Dát-ká-pattá, viláyati-agati,* DUK.; *Dadrughna,* SANS.; *Shimai-agati, vandukolli,* TAM.; *Sima avisl,* TEL.; *Shima-akatti,* MALA.; *Shime-agase,* KAN.; *Attora,* SING.; *Timbó-mezali* or *simbo-maizali, maizali-gi,* BURM.

References.—*Roxb., Fl. Ind., Ed. C.B.C., 354; Gamble, Man. Timb., 136; Dalz. & Gibs. (Supp.), Bomb. Fl., 29; Thwaites, En. Ceylon Pl., 97; Pharm. Ind., 77; Moodeen Sheriff, Supp. Pharm. Ind., 77; U. C. Dutt, Mat. Med. Hind., 155; Ainslie, Mat. Ind., II., 361; O'Shaughnessy, Beng. Dispens., 308; Bidie, Raw Prod., Paris Exh., 26; S. Arjun, Bomb. Drugs, 190; Drury, U. Pl., 119; McCann, Dyes and Tans of Beng., 1158; Lisboa, U. Pl. of Bomb., 254; Balfour, Cyclop.*

Habitat.—A small shrub, with very thick, finely downy branches, cosmopolitan in the tropics; met with in Lower Bengal, Western Peninsula, Burma, and Malacca. Very probably introduced into India from the West Indies, as it does not appear to occur far away from human dwellings.

Tan.—"Specimens of *Sunari* bark used in tanning in Cuttack sent as **Cassia Fistula** proved on examination to be **Cassia alata**" (*McCann's Dyes and Tans*). The numerous samples of this bark, shown at the late Colonial and Indian Exhibition, were highly commended by the tanners who attended the conference on tanning materials. TAN. Bark. 733

Medicine.—The *Pharmacopœia of India* gives the following account of the medicinal properties and uses of this plant: The tree "is often cultivated by the natives for the sake of its LEAVES, which are held in high esteem as a local application in skin diseases. A belief in their powers in this cha- MEDICINE. Leaves. 734

racter prevails also in the West Indies, Brazil, Mauritius, Java, and other tropical countries. Their efficiency, especially in *Herpes circinatus*, is confirmed by **Dr. McKenna** (*Madras Med. Jour., Vol. I., p. 431*), **Dr. Arthur** (*Indian Ann. of Med. Science, 1856, Vol. III., p. 632*), and others. Favourable statements as to their efficiency in this class of cases are contained in the reports of **Dr. G. Bidie**, **Dr. W. J. Van Someren**, **Dr. L. Stewart**, and **Dr. Rean.** As a general rule, they appear to be more effectual in recent cases than in those of long standing. The Bengal Pharmacopœia contains the following formula for an ointment of the leaves, which is described as being almost a specific in ring-worm: 'Take of the fresh leaves of **Cassia alata** a sufficiency, bruise into a paste, and incorporate with an equal weight of simple ointment.' A more effectual mode of application, however, is thoroughly to rub in, over the affected part, the bruised leaves worked into a paste with a small portion of lime-juice. In

Tincture. 735 many cases it is productive of excellent effects. The leaves taken internally act as an aperient. **Mr. J. Wood** reports that a tincture of the dried leaves has been found to operate in the same manner as senna; and **Dr. Pulney Andey** states that an extract prepared from the fresh leaves is a good substitute for extract of Colocynth. It is desirable that further trials should be made with them."

Roxburgh remarks that, according to the Telinga and Tamil physicians, the leaves cure all poisonous bites as well as venereal affections, and strengthen the body. The fresh leaves are often employed to cure ring-worm. They are well rubbed into the parts affected, once or twice a day, and generally with great success. In Jamaica, a poultice made of the flowers is used by the natives in cases of ring-worm (*Dr. Wright*).

Roots. 736 **Special Opinions.**—§ "The ROOTS with *hur* and borax made into paste are used as a specific in ring-worm" (*Assistant Surgeon T. N. Ghose, Meerut*). "The fresh leaves bruised form an excellent application for ring-worm" (*Brigade-Surgeon J. H. Thornton, B.A., M.B., Monghyr*). "I have used it with good effect in ring-worm" (*Surgeon R. D. Murray, Burdwan*). "I have pretty largely used the fresh leaves bruised on patches of ring-worm met with in this district, with great success. I did not intend to blister the part, but let the patients rub the leaves on the part for a few minutes every day. In most cases the part became natural in about ten days. There is a tendency to relapse, but if the leaves are applied for a few days after the apparent cure, the disease does not reappear" (*Surgeon D. Basu, Faridpur*). "The efficacy of the leaves is increased by the addition of common salt" (*Surgeon-Major J. M. Zorab, Balasore*). "Expectorant, tonic, and astringent, used as a mouth-wash in stomatitis" (*Surgeon-Major J. M. Houston, Travancore; and John Gomes, Esq., Medical Store-keeper, Travandrum*). "Used in ring-worm, but its efficacy is uncertain" (*Brigade-Surgeon S. M. Shircore, Moorshedabad*). "Efficacious in ring-worm" (*Assistant Surgeon Shib Chunder Bhuttacharji, Chanda, Central Provinces*). "Leaves fresh rubbed on parts affected with ring-worm with great benefit" (*Surgeon-Major J. J. L. Ratton, Salem*).

737 **Cassia angustifolia,** *Vahl.; Fl. Br. Ind., II., 264.*

INDIAN OR TINNEVELLY SENNA.

Syn.—C. LANCEOLATA, *Roxb., W. & A.*, and (?) *Wall.*, but not C. LANCEOLATA, *Forskhal*, as in *Brandis, For. Fl., 166;* C. ELONGATA, *Lem.-Lis.;* SENNA OFFICINALIS, *Roxb.;* S. ANGUSTIFOLIA, *Batka.*

Vern.—*Saná-e-hindi*, ARAB. and PERS.; *Hindi-sana, hindi-saná-ká-pát*, HIND.; *Sanna makki, shón-pát, són-pát*, BENG.; *Sená makhi, middiáwal*, GUJ.; *Nát-kí-sana, nat-kí-sana-ká-pattá*, DUK.; *Bhúitarvada, mulkácha,* *shóná-makhi*, MAR.; *Náttu-nilá-virai, nilá virai, nila-vákai*, TAM.; *Néla-tangédu*, TEL.; *Níla váká*, MALA.; *Nelávarike*, KAN.; *Sana-kola, nilá-vari, nelá-vari*, SING.; *Puve-kain-yoe*, BURM.

References.—*Royle, Him. Ill., pp. 186, 187, also Pl., 37; Wight & Arn., Prod., Vol. I., 288; Aitchison, Cat. Pb. Pl., 51; Pharm. Ind., 65; Moodeen Sheriff, Supp. Pharm. Ind., 94; Flück. & Hanb., Pharmacog., 216; U. S. Dispens., 15th Ed., 1296; Bentl. & Trim., Med. Pl., 90; Royle, Mat. Med., 2nd Ed., p. 402; Ainslie, Mat. Ind., II., 249; O'Shaughnessy, Beng. Dispens., 306; Bidie, Raw Prod., Paris Exh., 7; S. Arjun, Bomb. Drugs, 46; Drury, U. Pl., 121; Birdwood, Bomb. Prod., 27; Spons, Encyclop., 825; Balfour, Cyclop.; Smith, Dic., 375; Treasury of Bot., 232; Kew Official Guide, Museums, p. 50; Dymock, Mat. Med. W. Ind., 268.*

Habitat.—The plant abounds in the Yemen and Hadramant in Southern Arabia; it is also found on the Somali coast. According to **Brandis** (who gives incorrectly **C. angustifolia**, *Vahl.*, as a syn. for **C. lanceolata**, *Forsk.*), this in addition is a native of Sind and of the Panjáb, and is cultivated in many parts of India. The *Flora of British India* says **C. angustifolia** "has no claim to be considered indigenous to India." **C. lanceolata**, *Forsk.*, is a native of Arabia. It seems probable that the mistake made by **Dr. Brandis** gave origin to the statement (see *Pharmacographia*, also *Bentley and Trimen, Med. Pl.*) that **C. angustifolia** is indigenous to Sind and the Panjáb.

The cultivated plant, as met with in India, is the Tinnevelly Senna of commerce, and the uncultivated, the Bombay Senna or *Senna Mekki* or *Sana-maki, Sona-maki* of the East. The last mentioned is imported into India from Arabia.. In Bombay it is cultivated at Poona to supply the requirements of Government Hospitals and not as an article of commerce. **Stocks** say it is grown in Sind.

Botanic Diagnosis.—This species is closely related to the preceding, but the leaflets are usually 5 8-jugate, are narrower, being oval, lanceolate, tapering from the middle towards the apex; they are longer, often nearly 2 inches long, and are either quite glabrous or furnished with a very scanty pubescence. The legume is narrower (7-8 lines broad), with the base of the style distinctly prominent on its upper edge.

Description of the Drug.—This plant thus affords two of the commercial forms of senna:—

1st. TINNEVELLY SENNA.—This is the leaf obtained from the plant carefully cultivated in South India and (at Poona) in Bombay. Owing to greater care in its collection, Tinnevelly senna is of better quality than the Arabian article. The leaves are also larger, being 1-2 inches long, of a yellowish-green colour on the upper surface and a duller tint on the under, glabrous or thinly pubescent on the under-side, with short, adpressed hairs. The leaflets are less rigid in texture than the Alexandrian senna, and have a tea-like odour. Tinnevelly senna has of late fallen off in quality, the leaves being so small as to resemble Alexandrian senna. **Tinnevelly. 738**

Dr. Dymock says that large quantities of Tinnevelly senna are now sent to Bombay, and that so successfully does this Indian article compete in the market, that the importation of Arabian senna is rapidly declining, Tinnevelly senna being exported to Europe in its place.

2nd. ARABIAN, MOKA, BOMBAY, or EAST INDIAN SENNA.—As already stated, this drug is derived from the wild plant as met with in Southern Arabia, and is imported from Moka, Aden, and the other Red Sea ports to Bombay, and thence re-exported to Europe. From being collected and dried without care, this is mostly an inferior commodity, fetching in London as low a price as ⅛*d.* or ½*d.* a ℔. It is now, however, never adulterated. **Arabian. 739**

Arabian Senna.

MEDICINE. Leaves. 740

Medicine.—Senna was first made known by the Arabs in the ninth century: it is extensively employed as a simple and active purgative. The Alexandrian is generally regarded as more powerful than Tinnevelly and the Arabian or Moka much inferior to either of these. The objections urged against the drug are its taste and the tendency to gripe which it manifests, combined with a somewhat irritant action. These dangers are, however, greatly lessened by administering the drug in the form of an alcoholic preparation, thus very considerably removing the taste. The griping is greatly checked by combination with salines such as bitartrate of potash, tartrate of potash, or sulphate of magnesium, along with an aromatic, as in the preparation commonly known as "black draught." Dr. Sakharam Arjun says that the leaves are sometimes chewed in *pán*, "and thus a combination of a laxative and an aromatic corrective is at the same time obtained."

Dr. Waring (*Bazár Medicines*) says: "The imported senna met with in the bazárs is usually of very inferior quality, consisting of broken pieces of old leaves, pieces of stem, and other rubbish. That grown in India, especially in Tinnevelly, is preferable to that imported from Arabia, which is called *Sana-mukhí* or *Mecca-senna.* The leaves should be unbroken, clean, brittle, pale green or yellow, with a heavy smell. It is a good, safe aperient, and may be given as follows: Take of senna leaves one ounce; of bruised ginger and cloves, each half a drachm; boiling water, ten ounces. Let it stand for one hour and strain. This is a good aperient in all cases of *constipation*, in doses of one and a half to two ounces: half this quantity, or less, is required for children, according to age."

In a list of Economic Plants sent to the Calcutta International Exhibition a sample of this plant from Cuddapah was described as given in decoction for fevers and also to cattle.

CHEMISTRY.

Chemical Composition.—The purgative property is considerably increased by combination with bitters. This fact has been confirmed by many observers. The purgative properties are due essentially to a glucoside acid named *Cathartic Acid.* This, which is almost insoluble in water or strong alcohol, is readily soluble in ether or chloroform. In senna it is, however, combined with calcium and magnesium, and in this form it is very soluble in water, although still insoluble in alcohol. The objectionable taste is removed, therefore, by alcoholic decoction, although the cathartic acid is only slightly altered. Senna yields rapidly one or more of its properties to urine, and 20 or 30 minutes after partaking the drug the urine will indicate these properties by being reddened on the addition of ammonia. Senna taken by wet-nurses with equal rapidity influences the milk, purging the sucking infant. If injected into the blood, senna acts as a cathartic.

For further particulars see "Alexandrian Senna" under **C. acutifolia,** and for Senna substitutes see **C. obovata.**

Special Opinions.—§ "Bombay senna, prepared from the same plant as the senna imported from Arabia, has been for many years the only senna obtainable in this market. It now seems likely to be driven out of the market by the lower qualities of Tinnevelly senna, which are cleaner and can be purchased at one anna a ℔" (*Surgeon-Major W. Dymock, Bombay*). "Powdered leaves are used in secondary syphilis" (*Surgeon-Major J. J. L. Ratton, M.D., Salem*). "Senna leaves are always purchased in the bazárs and esteemed for their cathartic properties" (*A Surgeon*). "An efficient purgative, commonly taken by the natives as a cold infusion, causes griping and abundant flow of mucus" (*Assistant Surgeon Shib Chunder Bhuttacharji, Chanda, Central Provinces*). "Not much used in these days" (*Brigade-Surgeon S. M. Shircore, Moorshedabad*).

Cassia auriculata, *Linn.; Fl. Br. Ind., II., 263.* 741

THE TANNER'S CASSIA.

Syn.—SENNA AURICULATA, *Roxb.*

Vern.—*Tarwar, tarvar,* HIND., DUK.; *Tarota,* BERAR; *Taravada,* MAR.; *Awal, aval,* GUJ.; *Awala,* CUTCH; *Avári, ammera verai, ávirai,* TAM.; *Tangédu, thâgedu tangar,* TEL.; *Avareke, tengedu, tangádi-gida, ávara-gidá, taravadagida,* KAN.; *Avára, ponnáviram,* MALA.; *Ranavará,* SING.

References.—*Roxb., Fl. Ind., Ed. C.B.C., 354; Brandis, For. Fl., 165.; Kurz, For. Fl. Burm., I., 393; Gamble, Man. Timb., 136; Thwaites, En Ceylon Pl., 96; Dalz. & Gibs., Bomb. Fl., 81; Pharm. Ind., 78; Moodeen Sheriff, Supp. Pharm. Ind., 93; Dymock, Mat. Med. W. Ind., 2nd Ed., 264; Ainslie, Mat. Ind., II., 931; O'Shaughnessy, Beng. Dispens., 309; Murray, Drugs and Pl., Sind, 132; Bidie, Raw Prod., Paris Exh., 27, 10; S. Arjun, Bomb. Drugs, 46; Drury, U. Pl., 120; Baden Powell, Pb. Pl., 343; Birdwood, Bomb. Prod., 27; Lisboa, U. Pl. of Bomb., 198, 243; Spons, Encycl., 1693; Balfour, Cyclop.; Treasury of Botany, 232; Kew Official Guide, Museums, p. 49; Liotard, Dyes and Tans of India, App. II., VI.; Wardle, Dyes and Tans of Ind., 33; Reports on the Col and Ind. Exhib., published by the Soc. of Arts, 405; Report of the Commercial Conferences held in the Indian section of the Col. and Ind. Exhib.*

Habitat.—A tall shrub, with the virgate branches and under-side of the leaves finely grey-downy. Wild in the Central Provinces, the Western Peninsula, South India, and Ceylon; often planted elsewhere.

Gum.—It is said in *Spons' Encyclopædia* to yield a medicinal resin, very scarce; but **Dr. Dymock** informs the writer he has never seen this supposed resin, although he has frequently handled the bark. In Bengal a brownish sap hardens on the surface of wounds on the bark; this may be the so-called resin. GUM. 742

Tan and Dye.—The bark is one of the most valuable of Indian tans, and is also, like myrabolans, used to modify dyes. It is said to give a buff colour to leather. **Bidie** remarks that "when the Government Tannery existed at Húnsúr, this bark was used almost exclusively for tanning purposes." This bark was highly commended by the Tanners who attended the conference on tanning materials held at the Colonial and Indian Exhibition in London. It was regarded as a little too dark-coloured, but the leather shown as tanned by it was admired. It was recommended that an effort should be made to have an extract prepared from this bark for export to Europe similar to Cutch. **Mr. Wardle** in his recent report says: "The bark does not produce much dye, only light browns and drabs peculiar to Indian dye-stuffs, the one with no mordant on unbleached tussur producing more colour than the rest with mordants, except perhaps the tussur mordanted with alum, which has a similar shade." The bark is said to sell at 2 annas a pound, retail, in Madras. At Bangalore it is said to be sold at R60 a ton but that the price is rising owing to an increasing demand. The flowers yield a yellow colouring matter, apparently not used economically. DYE & TAN. Bark. 743 Flowers. 744

§ "Skins of animals are tanned by soaking them in water in which the bark of this shrub has been infused for several days" (*Honorary Surgeon P. Kinsley, Chicacole, Ganjam*).

Fibre.—Specimens of the bark were sent to the Calcutta Exhibition from Cuddapah, Madras, as a tanning material, but an excellent fibre was prepared from a surplus of this bark and made into rope. The fibrous property of the plant does not appear to have been investigated. FIBRE. 745

The caterpillar of a large species of silkworm feeds on the leaves of this plant. (*Roxb.*)

The Tanner's Cassia.

MEDICINE. Seeds. 746 **Medicine.**—"The SEEDS of this common Indian plant, like those of **C. Absus**, are a valued local application in that form of purulent ophthalmia known in India by the name of 'country sore-eye.' **Dr. Kirkpatrick** (*Cat. of Mysore Drugs*, No. 258) expresses his opinion that they constitute an undoubtedly useful application in these cases, when not severe. They are smooth, flattish seeds, of an oval, oblong, or obscurely triangular form, obtusely pointed at one extremity, and varying in colour from brown to dull olive-green: they are tasteless and inodorous. The BARK is highly
Bark. 747 astringent; and **Dr. Kirkpatrick** states (*op. cit.*, No. 475) that he has employed it in the place of oak bark for gargles, enemas, &c., and found it a perfect substitute for the imported article. Both the seeds and bark appear worthy of further trials. A spirituous liquor is prepared in some parts of India by adding the bruised bark to a solution of molasses, and allowing the mixture to ferment" (*Waring, Pharm. Ind., pp. 78, 79*).

Leaves. 748 A decoction or infusion of the LEAVES of this plant is much esteemed as a cooling medicine by the Singhalese, and also as a substitute for tea (*Thwaites; Murray*). **Ainslie** says that the Vytians reckon the seeds amongst their refrigerants and attenuants, and prescribe them in electuary, in cases in which the habit is preternaturally heated or depressed, in doses of a small teaspoonful twice daily. **Dr. Ainslie** also records his opinion in favour of the use of the seeds in the treatment of ophthalmia, and he adds that for this purpose the powdered seeds are generally blown into the eyes.

Plant. 749 **Special Opinions.**—§ "Bark substituted for oak-bark. Seeds powdered a good local application for ophthalmia" (*Apothecary Thomas Ward, Madanapalle, Cuddapah*). "Antiscorbutic, antibilious; *trifala*, which is
Flower-buds. 750 made up of dry *awala, gall*, and *hirada*, is used as a diuretic and also as an expectorant" (*Surgeon W. Barren, Bhuj, Cutch*). "The whole plant, or any part of it, is used in diuresis and diabetes with fair results. The decoction of the flower-buds is an agreeable form in which it is taken in the morning as a drink in the diseases mentioned. It acts as an astringent tonic. The flower-bud is also rubbed into a paste with some unboiled rice and rubbed on the skin before bathing. It is a cooling detergent" (*Native Surgeon T. Ruthnam Moodelliar, Chingleput, Madras*). "The seed testa removed and kernels thereafter powdered finely; this powder if blown into the eye, is useful in conjunctivitis." (*Surgeon-Major A. F. Dobson, Bangalore*).

FOOD. Leaves. 751 **Food.**—The leaves are eaten as a green vegetable in times of famine (*Lisboa*).

DOMESTIC. Tooth-brushes 752 **Domestic Uses.**—The branches are largely used by natives as tooth-brushes, and are esteemed as preferable to those of any other plant for this
Root. 753 purpose. The root is of great use to workers in iron for tempering the metal (*Ainslie*).

754 **Cassia Burmannii**, *Wight* (in *Madras Jour., VI., t. 5*).

Vern.—The same as those of **C. angustifolia**, *Vahl*.

References.—*Brandis, For. Fl., 165; Gamble, Man. Timb., 136; Dalz. & Gibs., Bomb. Fl., 81; Aitchison, Cat. Pb. Pl., 52; Pharm. Ind., 65; Moodeen Sheriff, Supp. Pharm. Ind., 94; Ainslie, I., 389; O'Shaughnessy, Beng. Dispens., 307; S. Arjun, Bomb. Drugs, 47.*

Habitat.—A glabrous, shrubby plant, 1-4 feet in height, often procumbent; pod much curved into a kidney-shape, with a crest in the middle of the valve opposite each seed; leaflets 4-8 pairs. Frequent in the Panjáb (Salt Range, ascending to 2,500 feet, where it is known as *sanna*) and Trans-Indus (where it is called *jíjan*), according to **Brandis**; it

The Purging Cassia.

extends to Sind and the Western Peninsula. Distributed to Arabia, Egypt, Nubia, and Abyssinia.

MEDICINE. Plant. 755

Medicine.—The whole plant is sold in the bazárs as a substitute for the true senna under the name of country senna. Its action is of course similar, though much inferior, to Tinnevelly or Mecca senna.

It seems probable that many Indian authors have confused this with **C. angustifolia** in the published descriptions of that drug. (Conf. with **C. obovata**, *Colladon.*)

Cassia Buds. See **Cinnamomum Tamala,** *Nees;* LAURINEÆ.

C. Fistula, *Linn., Fl. Br. Ind., II., 261; Wight, Ic., t. 269.* 756

THE INDIAN LABURNUM, THE CASSIA FISTULA or PURGING CASSIA, *Eng.;* CASSE OFFICINALE, CASSE MONDEE, CASSE, *Fr.;* ROHRENKASSIE PURGIERCASSIE, FISTELKASSIE, *Germ.;* CASSIA, *It.;* CANA FISTULA, *Sp.*

Syn.—CATHARTOCARPUS FISTULA, *Pers.;* CASSIA FISTULA, *Willd.*, as in *Roxb., Fl. Ind.*

Vern.—*Amaltás, girmálah,* HIND., DUK.; *Alash, ali, karangal, kiar, kaniár,* PB.; *Raj-briksh, kitola,* KUMAON; *Raj briksha,* NEPAL; *Chimkani,* SIND; *Sundáli, sonali, ámultás, bandarláti,* BENG.; *Nuruic',* SANTAL; *Sonawir,* MAL. (S. P.); *Hari,* KÓL.; *Dunrás,* KHARWAR; *Rajbirij,* NEPAL; *Sonalu,* GARO; *Bonurlati, bonurlauri,* PALAMOW; *Sunaru,* ASS.; *Bandolat,* CACHAR; *Sandari* or *súnari,* URIYA; *Kitwáli, kitoli, itola, shimarra, sím,* N.-W. P.; *Warga,* OUDH; *Jaggarwah, raila, hirojah, karkacha,* C. P.; *Raella,* BAIGAS; *Jaggra, jugarúa, kambar, rera,* GOND; *Banag, bangru,* KURKU; *Báhavá, bháwá baya, bawa,* MAR.; *Garmal* or *garmálá,* GUJ.; *Konraih-káy, skarak-konraik-káy, kone,* TAM.; *Reylu, réla-rálá, réla-káyalu, suvarnam,* TEL.; *Konnak-káya,* MALA.; *Kakee,* KAN.; *Khiyár-shanbur, katha-ul-Hind,* ARAB.; *Khiyár-chanbar,* PERS.; *Suvarnaka, áragbadha, rájataru,* SANS.; *Ahalla* or *ahilla,* SING.; *Gnooshway, gnoo-kyee,* BURM.

References.—*Roxb., Fl. Ind., Ed. C.B.C., 348; Brandis, For. Fl., 164; Kurz, For. Fl. Burm., I., 391; Bedd., Fl. Sylv., 91; Gamble, Man. Timb., 136; Thwaites, En. Ceylon Pl., 95; Dalz. & Gibs., Bomb. Fl., 80; Aitchison, Cat. Pb. Pl., 51; Pharm. Ind., 65; Mooaeen Sheriff, Supp. Pharm. Ind., 93; Fluck. & Hanb., Pharmacog., 221; U. S. Dispens., 15th Ed., 368; Bentl. & Trim., Med. Pl., 87; U. C. Dutt, Mat. Med. Hind., 155; Dymock, Mat. Med. W. Ind., 209; Kew Official Guide to the Museums, p. 49; Ainslie, Mat. Ind., I., 61; Murray, Drugs and Pl., Sind, 130; Bidie, Raw Prod., Paris Exh., 6; Baden Powell, Pb. Pl. 343, 572; Atkinson, Him. Dist., 722, 731, 779, 782, also Econ. Prod., N.-W. P., Pt. V., 61; Buck, Dyes and Tans. of N.-W. P., 81; McCann, Dyes and Tans of Bengal, 66, 145, 158, 167; Birdwood, Bomb. Prod., 28; Lisboa, U. Pl. of Bomb., 254; Spons' Encycl., 798; Balfour, Cyclop.; Smith, Dict., 339; Treasury of Botany, 233; Ure, Dict. of Arts and Manuf., 747; Wardle's Dyes and Tans of India, 50 and 54; Liotard's Dyes and Tans of Ind., App. VII; Gimlette's Report on Nepal Exhibits to Col. and Ind. Exh.; Campbell's Report on Econ. Prod., Chutia Nagpur.*

Habitat.—A moderate-sized, deciduous tree of the Sub-Himálayan tracts, and common throughout India and Burma, ascending to 3,000 feet in altitude. Very common on the lower mountainous tracts skirting the Himálaya (from the Khásia Hills to Pesháwar), and extending through Chutia Nagpur and Central India to Bombay. It chiefly occurs as a small spreading tree, not more than 20 feet in height, leafless in March, the long pendulous racemes of bright yellow flowers and fresh green leaves appearing together in April, but sometimes a second flowering occurs in autumn. The long, brown, pendulous, sausage-like pods, 1-1½ feet in

CASSIA Fistula.

The Purging Cassia.

length, ripen in the cold season. **U. C. Dutt** thinks this must be *Rajataru* of the Sanskrit writers, the king of trees.

GUM. 757 **Gum.**—From the stem exudes a red juice which hardens into a gummy substance. This is generally known as *kamarkas.* Its economic uses, if any, are at present unknown to authors on Indian economic science, but it is stated to be astringent. A specimen was contributed to the Paris Exhibition from Travancore.

DYE AND TAN. Bark. 758 **Dye and Tan.**—The bark is used in tanning, chiefly along with **Terminalia. Dr. McCann** reports that in the district of Lohárdagá, in Bengal, a light-red dye is obtained from the bark, with alum as a mordant; 2 chittacks of bark with 2 tolas of alum being boiled together. The colour is deepened by the use of pomegranate rind. **Mr. Wardle** reports that the bark contains only a very small quantity of colouring matter. It yielded yellowish drab with tusser silk, light fawn with corah and eri silks, and light yellow-brown with wool. The wood ash is used as a mordant in dyeing. In Dacca and in Cuttack the bark is used as a tan. **McCann** describes the process of tanning as follows: "Skins, after being treated with lime and cleaned, are soaked in the astringent solution prepared by pounding the bark of *sunari* **(Cassia Fistula)**, bark of *Asan* **(Terminalia tomentosa)**, and pods of *kunti* **(Cæsalpinia digyna)**, and soaking in water for 24 hours. The process of soaking is repeated three times." **Mr. (now Sir E.) Buck** says it is used to a small extent in Cawnpore and at Bijnor. Experiments were tried at the Government factory, the result being that *amaltás* bark was pronounced a very valuable tanning material. The North-Western Provinces do a small trade in exporting the *amaltás* bark.

MEDICINE. Pulp. 759 Root-bark. 760 **Medicine.**—The PULP of the fruit and also the ROOT-BARK are used medicinally. They constitute, especially the former, one of the commonest and most useful of domestic medicines—a simple purgative. This drug is also used as a mild cathartic. The *Makhzan-ul-Adwiya* recommends that the pods be warmed to extract the pulp, which should then be rubbed up with almond oil for use. It is a safe purgative for children and pregnant women. In small doses (3·9 to 7·8 gr.) it may be prescribed as a laxative, and larger doses (31·1 to 62·2 gr.) as a purgative. (*U. S. Dispens.*) It is described as lenitive and useful in relieving thoracic obstructions. It is often combined with tamarinds, and in this preparation is regarded as a good purge for adust bile. Externally it is useful in gout and rheumatism (*Dymock*).

Flowers. 761 Bark. 762 Leaves. 763 Root. 764 It is also employed in the essence of coffee. The FLOWERS are made into a confection, known as *gul-kand*, and viewed as a febrifuge. "The BARK and LEAVES rubbed up and mixed with oil are applied to pustules" (*Drury*). As in most other species of this genus, they are valued as an external applicant in skin diseases, especially in ring-worm. **Mr. Campbell** says that the Santals use an infusion of the leaves as a laxative. **Dr. Irvine** (*Med. Top. of Ajmír*) states that he found the ROOT act as a strong purgative. **Thwaites** says that every part of the plant is used as a purgative by the Singalese. According to **Bellew**, the root is given as a tonic and febrifuge in the Panjáb (*Dr. Stewart, Pb. Pl., 62*).

The name *Cassia Fistula* (Latin) and *Κασίας σῦριγξ* (Greek) was first applied to a form of cinnamon very similar to the Cassia Lignea of the present day, the name Fistula having been given because of the bark being rolled up. The tree which now goes by that name was described by **Abul Abbas Annabati** of Sevilla in the thirteenth century, and the fruit is mentioned as a medicine by **Joannes Actuarius,** who flourished in Constantinople towards the close of that century. The drug was a familiar remedy in England in the time of Turner, 1568 (*Flück. and Hanb., Pharmacog., 222*). It is never prescribed at the present day in England, except in the form of the well-known Lenetive Electuary (*Confectio Sennæ*), of which it is an ingredient. MEDICINE.

Special Opinions.—§ "A very useful and safe purgative when procurable. The pulp does not keep fresh more than a few weeks, even within the unbroken pod" (*Brigade Surgeon S. M. Shircore, Moorshedabad*). "The fruit imported into Yarkand is there called *Foluse*" (*Surgeon-Major J. E. T. Aitchison, Simla*). "A poultice made of the leaves is said to relieve the chilblains which are common in Upper Sind. It has been beneficially used in facial paralysis and rheumatism when rubbed into the affected part. *Internally* it is given as a derivative in paralysis and brain affections. The leaves are principally used" (*Surgeon H. P. Dimmock, Shikarpur*). "The tender leaves rubbed into a paste and taken internally act as a laxative" (*Native Surgeon T. Ruthnam Moodeliar, Chingleput, Madras*). "The confection is given in cases of diabetes mellitus" (*Surgeon W. F. Thomas, Mangalore*). "*Gulkand* is a cooling laxative which I frequently use in constipation, especially in delicate women. Half an ounce with warm milk at bed-time is enough for a dose" (*Surgeon-Major R. L. Dutt, Pubna*). "The pulp of the ripe pod is commonly used as a purgative mixed with tamarind pulp; taken as a drink at night, this acts on the bowels mildly the following morning" (*Assistant Surgeon Shib Chunder Bhuttacharji, Chanda, Central Provinces*). "In the flatulent colic of children, it is commonly applied round the navel to produce motions. The new leaves worked down to a paste are applied in ringworm" (*Assistant Surgeon T. N. Ghose, Meerut*). "A good purgative, extensively used by natives" (*Honorary Surgeon Easton Alfred Morris, Negapatam*). "A favourite laxative and purgative amongst natives" (*Assistant Surgeon Nehal Sing, Saharunpore*).

Food.—The leaves, parched, are said to be eaten as a mild laxative with food. "The flowers are largely used by the Santals as an article of food" (*Campbell*). The pulp of the pods is largely used in Bengal to flavour native tobacco. FOOD. Leaves. 765 Flowers. 766 Pulp mixed with tobacco 767

Structure of the Wood.—Sapwood large, heartwood varying in colour from grey or yellowish red to brick-red, extremely hard. The difference between the wood of this tree and that of **Ougeinia dalbergiodes** consists in the fact that in the former the patches of white soft tissue form continuous belts, whereas in the latter they are rhomboidal, pointed at the ends, and form interrupted belts. TIMBER. 768

The wood is very durable, but rarely of sufficiently large size for timber. It makes excellent posts, and is good for carts, agricultural implements, and rice-pounders.

Cassia glauca, *Lam.; Fl. Br. Ind., II., 265.* 769

Vern.—*Konda tantepu chettu*, TEL.; *Wal-ahalla*, SING.

References.—*Roxb., Fl. Ind., Ed. C.B.C., 352; Kurz, For. Fl. Burm., I., 394; Gamble, Man. Timb., 136; Thwaites, En. Ceylon Pl., 96; Balfour, Cyclop.*

Habitat.—A small tree of the eastern part of South India and of Burma to Ceylon and Malacca.

Medicine.—The bark mixed with sugar and water is given in diabetes, and a preparation of the bark and leaves, mixed with cummin seed, sugar and milk, is given in virulent gonorrhœa (*Balfour*). MEDICINE. Bark. 770 Leaves. 771

C. lanceolata, *Roxb.; Wall.; W. & A.* (but not of *Forskhal*); also [C. angustifolia, *Vahl.*]

C. lanceolata, *Nectoux*, see C. acutifolia, *Delile*.

Country or Italian and Jamaica Senna.

772 **Cassia lanceolata,** *Forskhal.*

This species is, by the majority of authors, viewed as quite distinct from either **C. acutifolia** or **C. angustifolia**. It is a native of Arabia, and doubtless to a certain extent is used as a substitute or adulterant for the Mecca senna. It differs chiefly from **C. acutifolia** in having glandular petiolets; the plants are, however, very nearly allied, and, as **Forskhal's** description is anterior to **Delile's** account of **C. acutifolia,** both might be reduced to one, which in that case would have to receive the name **C. lanceolata,** *Forskhal.* Most Indian authors give **C. lanceolata,** *Forskhal,* but in the writer's opinion incorrectly, as a synonym for **C. angustifolia,** *Vahl.*

C. Lignea. See **Cinnamomum Tamala,** *Nees;* LAURINEÆ.

773 **C. marginata,** *Roxb.; Fl. Br. Ind., II, 262; Wight, Ill., t. 83.*

Syn.—C. ROXBURGHII, *DC.*

Vern.—*Urimidi, uskiamen,* TEL.; *Ngoomee,* BURM., *Ratoo-waa,* SING.

References.—*Roxb., Fl. Ind., Ed. C.B.C., 350; DC. Prod., II., 489; W. & A., Prod., 286; Gamble, Man. Timb., 137; Thwaites, En. Ceylon Pl., 95; Bedd., Fl. Sylv., t. 180.*

Habitat.—A small deciduous tree, with deeply cracked, brown bark, found in the Western Peninsula, and in Madras, Ceylon, and Burma (Thoungyeen forests).

TIMBER. 774 **Structure of the Wood.**—Heartwood light brown, very hard. The wood is well adapted for turning, naves of wheels, and handles of tools.

775 **C. mimosoides,** *Linn.; Fl. Br. Ind., II., 266.*

Vern.—*Patwa-ghás,* SANTAL.

Habitat.—Grows on the Himálaya, ascending 5,000 to 6,000 feet in Kumaón, and on the hills of Bengal and of the Khásia, to Ceylon and Malacca.

MEDICINE. Root. 776 **Medicine.**—§ " Root given for spasms in the stomach (*Rev. A. Campbell, Santal Mission, Pachamba*).

777 **C. nodosa,** *Ham.; Fl. Br. Ind., II., 261.*

Vern.—*Gnu-theing,* BURM.

References.—*Mason's Burm., 404, 770.*

Habitat.—A common species in the Eastern Himálaya, Manipur, and Burma.

It has the properties assigned to most of the wild species.

778 **C. obovata,** *Colladon; Fl. Br. Ind., II., 264; Wight, Ic., t. 575.*

Syn.—CASSIA SENNA, *Linn.*; ?SENNA OBTUSA, *Roxb.*

Known, in India, as COUNTRY SENNA; and as ITALIAN, TRIPOLI, and JAMAICA SENNA; from its being one of the first species made known to Europe; it was cultivated in Italy during the 16th century.

Vern.—*Bhúi Tarwar,* BOMB.

References.—*Roxb., Fl. Ind. (Ed. C.B.C.), 352; W. and A. Prod., 288; Moodeen Sheriff, Supp. Pharm. Ind., 94, in part; Fluck aud Hanb., Pharmacog., 218; Bentley and Trim., Med. Pl. ,89; U.S. Dispens., 1296; Ainslie, Mat. Med., II., 249; Treasury of Botany; Dymock, Mat. Med. W. Ind., 263.*

Habitat.—The Western Peninsula, Mysore, and South India, e special-ly the Coromandel coast. A small shrub, with the leaves smaller (leaflets 3-6 pairs) than in **C. Burmannii,** and the pods not near so prominently tubercled over the seeds as in that species.

The writer is by no means certain that he is correct in regarding the plant known in Europe as **C. obovata** as distinct from the Indian corresponding species, still less, in viewing **Roxburgh's Senna obtusa** as more nearly allied to **C. obovata,** *Colladon* (which, indeed, may prove to be only an acclimatised plant) than the more northern form which the writer has retained under **Wight's** name of **C. Burmannii.** The *Flora of British India* may probably be correct, however, in throwing these plants into one species; but it is deemed desirable to stimulate more careful investigation by forcibly placing a doubtful point like this before the reader.

Medicine.—The LEAVES are collected and sold as country Senna. Formerly cultivated in Italy, the West Indies, &c., 3 parts in 10 of Alexandrian Senna being used as an adulteration in the commercial article. This habit has now for some time been discontinued, as also the cultivation of the plant (Conf. with **C. Burmannii.**) MEDICINE. Leaves. 779

Cassia occidentalis, *Linn.; Fl. Br. Ind., II., 262.* 780

THE NEGRO COFFEE.

Vern.—*Kasóndi, bari-kasóndi* or *kásundá,* HIND. and DUK.; *Hikal,* BOMB.; *Kásamarú,* SANS.; *Kálkáshundá,* BENG.; *Náttam-takarai, peyá-veri,* TAM.; *Kasindhá,* TEL.; *Natram-takara,* MALA.; *Kalan, mezali, maizali,* BURM.; *Peni-tóra,* SING. The same vernacular names are generally given to this species as to **C. Sophera.**

References.—*W. & A., Prod., 290; Bot. Reg., t. 83; Roxb., Fl. Ind., Ed. C.B.C., 352; Thwaites, En. Ceylon Pl., 95; Dalz. & Gibs., Bomb. Fl., 81; Aitchison, Cat. Pb. Pl., 52; Pharm. Ind., 78; Moodeen Sheriff, Supp. Pharm. Ind., 94; Dymock, Mat. Med. W. Ind., 2nd Ed., 262; O'Shaughnessy, Beng. Dispens., 309; S. Arjun, Bomb. Drugs, 46; Drury, U. Pl., 121; Lisboa, U. Pl. of Bomb., 198; Spons', Encycl., 707, 798; Balfour, Cyclop.; Treasury of Botany; Kew Official Guide, Museum, p. 50; Kew Reports, 1877, p. 39; and 1881, pp. 34-35.*

Habitat.—A diffuse, sub-glabrous under-shrub, scattered from the Himálaya to the Western Peninsula, Bengal, South India, and Burma to Ceylon. Probably introduced. Distribution cosmopolitan in the tropics.

Medicine.—The LEAVES, ROOTS, and SEEDS are used medicinally; and by Hindú and Muhammadan writers they are supposed to have the same properties as **C. Sophera.** They are "alexipharmic, useful in the expulsion of corrupt humours and to relieve cough, especially whooping-cough. In the French-African colonies the seeds are called 'Negro coffee;' they are employed in France and in the West Indies as a febrifuge, chiefly in the form of vinous tincture (℥ii to Cij of Malaga wine). An infusion of the root is considered by the American Indians to be an antidote against various poisons. Torrefaction is said to destroy the purgative principle in the seeds, and make them taste like coffee. The whole plant is purgative. Dose of the leaves about 90 grains" (*Dr. Dymock, Mat. Med. W. Ind.*). MEDICINE. Leaves. 781 Root. 782 Seed. 783

"In the West Indies the ROOT is considered diuretic, and the leaves, taken internally and applied externally, are given in cases of itch and other cutaneous diseases, both to men and animals. The negroes apply the leaves, smeared with grease, to slight sores, as a plaster. The root is said by **Martius** to be beneficial in obstructions of the stomach, and in incipient dropsy." (*Drury, U. Pl.*)

Chemical Composition.—**Professor Clonet** has analysed the seeds. The following abstract of his views and results taken from the *Year-Book of Pharmacy for 1876, p. 179,* will be found instructive:— CHEMISTRY.

"Fatty matters (olein and margarin), 4·9; tannic acid, 0·9; sugar, 2·1; gum, 28·8; starch, 2·0; cellulose, 34·0; water, 7·0; calcium sulphate and

MEDICINE.

phosphate, crysophanic acid, 0·9; malic acid, sodium chloride, magnesium sulphate, iron, silica, together, 5·4; and achrosine, 13·58 parts in 100. The latter substance was obtained by exhausting the powder of seeds, previously treated with ether, by means of alcohol of 60 per cent; the alcohol is distilled off, the syrupy residue treated with absolute alcohol, which dissolves out various constituents, leaving a solid brown-red mass, having when dry a resinous fracture, and being soluble in water, to which it communicates a garnet colour. It contains C, H, O, N, and S, but its exact composition has not been determined. (It is most likely a mixture of various bodies.) It is soluble also in weak alcohol, and in acids and alkalies. The colour cannot be fixed upon tissues by any known mordant. This circumstance induced the author to term it *achrosine*, or 'not-colouring,' although being coloured itself."

Special Opinions.—§ "Leaves pounded and made into a paste are applied to fresh wounds to bring on their healing by first intention" (*Assistant Surgeon Anund Chunder Mukarji, Noakhali*). "The mature seeds are used as an external application in ring-worm" (*Surgeon J. H. Thornton, B.A., M.B., Monghir*). "The seeds are used in the treatment of scabies" (*Surgeon-Major C. W. Calthrop, M.D., Morar*).

FOOD. Seeds. 784 NEGRO COFFEE.

Food.—In the Kew Reports interesting information is given regarding the use of the seeds of this plant as a substitute for coffee. The following passages may be republished here:—

"NEGRO COFFEE.—The Commissioners of Customs forwarded to me in the early part of the year a sample of an article imported at the port of Liverpool from Bathurst, River Gambia, under the above name. They were identified at Kew as the seeds of **Cassia occidentalis.** According to Livingstone, these are used under the name of '*Fedegozo* seeds' on the Zambesi as a substitute for coffee. **Monteiro,** however, states in his 'Angola and the River Congo' (*Vol. II., p. 249*) that *Fedegoza* seeds are used only medicinally as a substitute for quinine. The seeds are roasted and ground, and their infusion taken either alone or generally mixed with coffee" (*1877, p. 39*).

"These seeds occasionally find their way into the European market. The following extract from a letter from **Dr. Nicholls** of Dominica, dated September 27, 1881, shows that their use is well known amongst the negro inhabitants of that island:—

"**Cassia occidentalis** is, I find, an excellent coffee substitute. It is called in Dominica by the following names: '*l'herbe puante*,' '*café marron*,' and 'wild coffee.' I have often heard of the negroes using the seeds of a native plant as coffee, but it is only lately that I have enquired into the subject, with results that will, I believe, be of interest to you.

"I collected some seeds and directed my cook to roast and grind them, so that I might taste the 'coffee.' Other matters engaging my attention, I forgot the circumstance until several days afterwards, when one evening my wife enquired how I liked my after-dinner cup of coffee. I turned to her enquiringly, when she laughingly said, 'That is your wild coffee.' I was indeed surprised, for the coffee was indistinguishable from that made of the best Arabian beans, and we in Dominica are celebrated for our good coffee. Afterwards some of the seeds roasted and ground were brought to me, and the aroma was equal to that of the coffee ordinarily used in the island.

"I intend to send you a good quantity of the '*café marron*' in its stages of preparation, in order that you may have an opportunity of undergoing my experience, and afterwards, you will, I think, be willing to raise **Cassia occidentalis** above the rank of a weed. I may inform you that the plant itself is used by the native 'doctors' medicinally in the form of a decoction, and it has the reputation of being a good diaphoretic. I will enquire into the matter, experiment physiologically on myself, and report the result to you. The weed is very common, indeed troublesome to the sugar-planters, so if it turns out to be valuable it can be obtained in large quantities." (*1881, pp. 34-35.*) FOOD.

Cassia Oil. See **Cinnamomum zeylanicum.**

C. siamea, *Lamk.; Fl. Br. Ind., II., 264.* 785

Syn.—C. FLORIDA, *Vahl.;* SENNA SUMATRANA, *Roxb.*

Vern.—*Kassod,* BOMB.; *Beati, manje konne,* TAM.; *Sime tangadi,* KAN.; *Waa,* SING.; *Maizalee,* BURM.

References.—*Roxb., Fl. Ind., Ed. C.B.C., 353; W. & A. Prod., 288; Kurz, For. Fl. Burm., I., 392; Gamble, Man. Timb., 138; Thwaites, En. Ceylon Pl., 96; Bedd., Fl. Sylv., t. 179; Kew Official Guide, Museum, p. 49; Mason's Burma, 404.*

Habitat.—A moderate-sized tree, with smooth bark, found in South India, Burma, and Ceylon. Distributed to the Malayan Peninsula and Siam.

Structure of the Wood.—Sapwood whitish, rather large. Heartwood dark brown, nearly black; very hard and very durable. Used in Burma for mallets, helves, and walking-sticks. In South India it is little known, but it is considered one of the best kinds of fuel for locomotives in Ceylon. (*Beddome.*) TIMBER. 786

C. Sophora, *Linn.; Fl. Br. Ind., II., 262.* 787

Syn.—SENNA SOPHERA and S. ESCULENTA, *Roxb.;* C. CHINENSIS, *Jacq.;* SENNA PURPUREA, *Roxb.*

Vern.—*Banár, kásundá, bás-ki-kasóndi,* HIND.; *Kál-káshundá,* BENG.; *Sarí-kasóndi, jangli-takla,* DUK.; *Kuwádice,* GUJ.; *Ran-tánkala,* MAR.; *Ponná-virai, periya-takarai, perá-virai,* TAM.; *Paidi-tangédu, núti-kashindha, kásá-mardhakamu, tagara-chettu,* TEL.; *Ponnám-takara,* MALA.; *Kásamarda,* SANS.; *Uru-tora,* SINGH.

References.—*Roxb., Fl. Ind., Ed. C.B.C., 352; Thwaites, En. Ceylon Pl., 95; Dalz. & Gibs., Bomb. Fl., 81; Pharm. Ind., 78; Moodeen Sheriff, Supp. Pharm. Ind., 95; U. C. Dutt, Mat. Med., Hind., 155; Dymock, Mat. Med. W. Ind., 212; Murray, Drugs and Pl., Sind, 133; Bidie, Raw Prod., Paris Exh., 26; Drury, U. Pl., 122; Baden Powell, Pb. Pl., 343; Atkinson, Him. Dist., 731; Lisboa, U. Pl. of Bomb., 153, 198; Balfour, Cyclop.*

Habitat.—A closely allied species to **C. occidentalis,** from which it differs by its more shrubby habit, more numerous, smaller and narrower leaflets, and shorter, broader, and more turgid pods. (*Fl. Br. Ind.*) Cosmopolitan in the tropics; common throughout India, from the lower Himálaya to Ceylon and Penang.

Medicine.—The BARK, LEAVES, and SEEDS are used as a cathartic, and the JUICE of the leaves is viewed as a specific in ring-worm, specially when made into a plaster in combination with sandal-wood. A paste made from the root is sometimes used instead of the juice of the leaves. The powdered seed is used for the same purpose and also for itch. The Sanskrit name means "destroyer of cough;" it is supposed by Hindús to have expectorant properties. It is noticed in Muhammadan works as a remedy for snake-bite, the root being given along with black pepper. The bark, in the form of infusion and the powdered seeds, mixed with honey, are given in diabetes (*Drury*). "An ointment made of the bruised seeds and leaves and of sulphur is used in itch and ring-worm" (*Taylor's Top. of Dacca*). MEDICINE. Bark. 788 Leaves. 789 Seeds. 790 Juice. 791

CASSIA Tora.

The Fœtid Cassia.

CHEMISTRY. **Chemical Composition.**—" This plant, like several others of the same genus, owes its medicinal activity to the presence of chrysophanic acid, sometimes called Rhein, form $C_{14}H_6O_2''$ (O.H.$_2$) This substance belongs to the anthracene group of carbon compounds, and, like alizarin, is regarded as dioxyan thraquinone, $C_{14}H_6O_2''\left\{\begin{matrix}OH\\OH\end{matrix}\right\}$. It crystallizes in six-sided prisms, is tasteless, and may be sublimed without decomposition; it is contained in Goa powder (50 per cent.), rhubarb, most varieties of dock, *Lichen orcella, Permelia parietina, Cassia alata, C. occidentalis, C. Tora, &c.* As met with in commerce, it is in the form of a light-yellow powder, soluble in benzol, chloroform, turpentine, and in the fixed and volatile oils to a large extent, sparingly soluble in ether and alcohol, and insoluble in water, glycerine, and in solid paraffin. It is dissolved by sulphuric and nitric acids (in the latter to a less extent), by caustic potash and by ammonia; fuses at 123°·3 C., and boils at 232°·2 C. At the latter temperature it is decomposed into a dark-green resinous substance, which is largely soluble in ether. Oil Jecoris dissolves twice its weight of the acid, yielding a mixture containing 70 per cent. Oil olivæ, Oil Pini sylvest., Creasotum, Oil Terebinth., Oil Lavand., and Vaseline, dissolve readily their own weight of acid, yielding mixtures containing 52 per cent.

"Taking advantage of its solubility in the fixed oils, a considerable saving may be effected by preparing ointments direct from Araroba. Oil olivæ thoroughly exhausts that substance, yielding the acid after removal of the oil by ether in a state of purity. The Singhalese doctors take advantage of this fact, and fry the leaves of **Cassia alata, C. Tora, C. occidentalis,** and **C. Sophera** in gingelly or castor oil. The strained product is used as an ointment for ring-worm and other skin diseases" (*J. Laker Macmillan, Phar. Journ., 15th March 1879; Dymock, 261*).

Special Opinions.—§ "The sap is used locally for ring-worm and other skin diseases; a good specific for 'dhobies-itch.' The flowers are eaten by the natives" (*Surgeon W. Barren, Bhuj, Cutch*). "The leaves of this shrub are made into an infusion which is taken internally for gonorrhœa in its sub-acute stage" (*Surgeon W. F. Thomas, Mangalore*). "Anthelmintic, used externally for syphilis" (*Surgeon-Major D. R. Thompson, M.D., C.I.E., Madras*).

FOOD. Leaves. 792 **Food.**—"The leaves are eaten by men and animals" (*Atkinson*). The disagreeable smell is removed by boiling.

793 **Cassia, sp (?)**

TIMBER. 794 **Major Ford** sent from the Andaman Islands, in 1866, a sample of a hard, durable wood, olive-brown, with a structure very similar to that of **Ougeinia dalbergioides.** Evidently some common Andaman wood, and known by the name of *Gnúgyí.* (*Gamble, Man. Timb.*)

795 **C. timoriensis,** *DC.; Fl. Br. Ind., II., 265.*

Vern.—*Arremene,* SING.; *Toung maisalee,* BURM.

References.—*Kurz, For. Fl. Burm., 393; Gamble, Man. Timb., 138; Thwaites, En. Ceylon Pl., 96.*

Habitat.—A handsome, small, evergreen tree, met with in Burma and Ceylon.

TIMBER. 796 **Structure of the Wood.**—Dark brown, nearly black, resembling that of **C. siamea;** used in Ceylon for building and furniture.

797 **C. Tora,** *Linn.; Fl. Br. Ind., II., 263.*

THE FŒTID CASSIA.

Vern.—*Chakundá, panevár,* HIND. and BENG.; *Chakaoda arak',* SANTAL; *Pawár, panwár, pawás, chakunda,* PB.; *Panwar,* N.-W. P.; *Takálá,*

The Fœtid Cassia.

tarotá, tákla, tánklí, MAR.; *Kawário, kovariya*, GUJ.; *Tánkalá, kowaria, kovariya*, BOMB.; *Tarota*, DEC.; *Prabúnátha, dádamari, dadamardan*, SANS.; *Sanjsabóyah*, ARAB.; *Sang sabóyah*, PERS.; *Ushit-tagarai, tarotah*, TAM.; *Tagarishu-chettu*, TEL.; *Dan-gywe, dang-we, kujne*, BURM.; *Peti-tora*, SING.; *Mwango, swahili*, S. AFRICA.

References.—*Roxb., Fl. Ind., Ed. C.B.C., 351; Thwaites, En. Ceylon Pl., 96; Dalz. & Gibs., Bomb. Fl., 81; Aitchison, Cat. Pb. Pl., 52; Pharm. Ind., 78; Moodeen Sheriff, Supp. Pharm. Ind., 95; U. C. Dutt, Mat. Med. Hind., 153; Dymock, Mat. Med. W. Ind., 216; Ainslie, Mat. Ind., II., 405; O'Shaughnessy, Beng. Dispens., 309; Bidie, Raw Prod., Paris Exh., 26, 110; S. Arjun, Bomb. Drugs, 47, 190; Drury, U. Pl., 122; Baden Powell, Pb. Pl., 343; Atkinson, Him. Dist., 731; McCann, Dyes and Tans of Beng., 124, 142; Lisboa, U. Pl. of Bomb., 153, 198, 243, 291; Balfour, Cyclop.; Wardle, Report on Dyes & Tans of India.*

Habitat.—A gregarious annual under-shrub, from 1 to 2 feet in height, found everywhere in Bengal, and widely spread and abundant throughout the tropical parts of India.

DYE. Seeds. 798

Dye.—**Baden Powell, Atkinson,** and other writers say that the seeds of this shrub are used as "a blue dye." There are annually from 15 to 20 tons of these seeds collected in South India, which are sold for 2 to 3 annas a pound. The statement of their yielding a blue dye is apparently derived originally from **Ainslie,** who remarks that "in Coimbatore the seeds are had recourse to, in combination with the *pala* (**Wrightia tinctoria**, *Br.*), in preparing a blue dye." **Mr. Hutchins,** Assistant Conservator of Forests, Mysore, reports that the average collection of seeds of this plant is about 12 tons in Nundidroog, and imagines that they act the part of starch in the indigo solution. It is a little difficult to understand what **Mr. Hutchins** means by "indigo solution." **Wrightia tinctoria** yields of course the chemical substance indigo; but from the use of the popular word "indigo," one would infer that either the blue dye was extracted from the **Wrightia** in such quantities in Mysore as to justify the word "indigo," or that **Mr. Hutchins** was alluding to the use of **Cassia Tora** seeds along with the true commercial indigo. The latter conclusion, if correct, is exceedingly curious and apparently practically unknown to the indigo dyers of Bengal. **Dr. McCann** says, however, that the seeds are used in Maldah along with indigo, so that their property, whatever it may be, is much more widely known in India than might at first sight have been supposed. The peculiar action of starch upon the dye is difficult to understand. The natives of Assam and Manipur use lime along with their indigo (the produce of **Strobilanthes flaccidifolius**), and it seems likely that the reactions with the indigos of different plants may be peculiar and specific. In **Mr. C. T. Gardner's** report of Ichang (China) for the year 1883 occurs the following passage: "Two green dyes are made from the barks of the **Rhamnus** and **Cassia.**" In China apparently **Cassia** bark is used, but as the barks of several species of that genus afford yellow dyes, it is probable that the green referred to is the result of the combination of a **Cassia** yellow with a blue (or indigo) dye extracted from the **Rhamnus.** The much-talked-of green indigo of China is, however, often said to be prepared from a species of **Rhamnus,** and to be a pure colour and not a combination. The use of Cassia seeds or bark with indigo dyes seems worthy of careful chemical examination. Meantime **Mr. T. Wardle,** while examining the dyes of India, had occasion to try the seeds of this plant, and found that they afforded a most useful yellow dye suitable for tasar silk. **Mr. Wardle** does not appear to have investigated the question of their special property, if any, of being used along with indigo, but from his results it is natural to infer that they would produce a green shade with indigo instead of assisting the blue.

MEDICINE. Leaves. 799 Seeds. 800 **Medicine.**—The LEAVES are used as an aperient; both LEAVES and SEEDS constitute a valuable remedy in skin diseases, chiefly for ringworm and itch. This is known in Sanskrit as *Chakramarda*. **Dr. Dymock** says: "**Chakradatta** directs the seeds to be steeped in the juice of **Euphorbia neriifolia**, and afterwards to be made into a paste with cow's urine as an application to cheloid tumours. He also recommends the seeds, together with those of **Pongamia glabra**, as a cure for ringworm." Muhammadan writers "consider the seeds and leaves to have solvent properties in those forms of skin disease accompanied by induration, such as leprosy, cheloid, psoriasis, &c; and mention their having been used with advantage in the plague (*waba*)." **O'Shaughnessy** remarks that the leaves "are much used for adulterating senna." There is no evidence that this is done at the present day in India. **O'Shaughnessy** also records his testimony in favour of the properties of this plant in curing itchy eruptions: "The ROOT rubbed to pulp with lime-juice has almost specific powers in the cure of ringworm." He adds: "Like all the allied species of Cassia, this seems to owe its virtues to its astringency." **Ainslie**, one of the earliest European writers who alludes to this drug, says: The mucilaginous and fœtid-smelling leaves of the **Cassia Tora** are gently aperient, and are prescribed in the form of a decoction, and in doses of about two ounces, for such children as suffer from feverish attacks while teething; fried in castor oil, they are considered as a good application to foul ulcers; the seeds, ground with sour-butter milk, are used to ease the irritation of itchy eruptions; and the root, rubbed on a stone with lime-juice, the *Vytians* suppose to be one of the best remedies for ringworm."

Root. 801

§ "I have used the powdered leaves of a **Cassia** shrub common in waste places in Madras with success in dhobie's itch" (*Deputy Surgeon-General G. Bidie, C.I.E., Madras*). "The seeds are an article of commerce in Bombay, under the name of *Kovariya-bij*" (*Surgeon-Major W. Dymock, Bombay*). "Decoction of the leaves used as a wash in skin diseases (Zanzibar)" (*Surgeon-Major John Robb, Surat, Bombay Presidency*).

FOOD. Seeds. 802 **Food.**—The small SEEDS of this plant are eaten in times of scarcity. Recently they have been brought to notice, in British Burma, as worthy of use as a substitute for coffee when roasted and ground, very similar to the way in which "Negro coffee" seeds are used in tropical Africa, Central America, and the West Indies (see **C. occidentalis**).

Coffee substitute. 803

Leaves. 804 The tender leaves are boiled and eaten as a pot-herb. They are largely used during times of famine (*Lisboa*). The Santals regularly use this pot-herb, both leaves and fruit (*Campbell*).

§ "The seeds are said to yield a decoction which is reported to be in every respect as good as coffee" (*Mr. C. D. Hardinge, Rangoon*). "A kind of coffee is made from this in Arracan" (*Prof. Romanis, Rangoon*).

Cassis, see **Ribes nigrum.**

CASSYTHA, *Linn.; Gen. Pl., III., 164.*

805 **Cassytha filiformis**, *Linn.; Fl. Br. Ind., V., 188; Wight, Ic., t. 1847;* LAURINEÆ.

Vern.—*Amarbeli*, HIND.; *Akásbel*, BENG.; *Alagjari*, SANTAL; *Akáswel*, BOMB.; *Amarvéla*, MAR.; *Kotan*, DUK.; *Cottan*, TAM.; *Paunch figa*, TEL.; *Acatsjabulli*, MALA.; *Akásvalli*, SANS.; *Shway-nway-pin*, BURM.

References.—*Roxb., Fl. Ind., Ed. C.B.C., 342; Dalz. & Gibs., Bomb. Fl., 223; Thwaites, En. Ceylon Pl., 258; U. C. Dutt, Mat. Med. Hind., 290; Dymock, Mat. Med. W. Ind., 555; Birdwood, Bomb. Prod., 227; Murray, Pl. and Drugs of Sind, III.; Drury, U. Pl., 123; S. Arjun, Bomb. Drugs, 115; Treasury of Botany; Balfour, Cyclop.*

Habitat.—A small parasitic plant, much resembling a Cuscuta, for which it is often mistaken; met with in almost every part of the coast of India and very general from Banda to Bengal. It is common in the hotter parts of Ceylon, especially near the sea (*Thwaites*). Distributed to Arabia, Africa and America, and through the Polynesian islands to Australia.

Medicine.—"*Akáswel* is used in native practice as an alterative in bilious affections and for piles" (*Dymock*). "It is put as a seasoning into butter-milk, and much used for this purpose by the Brahmins in South India" (*Ainslie*). "The whole plant pulverised and mixed with dry ginger and butter is used in the cleaning of inveterate ulcers. Mixed with gingelly oil, it is employed in strengthening the roots of the hair. The juice of the plant mixed with sugar is occasionally applied to inflamed eyes." (*Drury, U. Pl.*) A decoction of the plant is occasionally prescribed in retention of the placenta (*S. Arjun, Bomb. Drugs*). MEDICINE. Plant. 806

Special Opinions.—§ "It is used by the natives in a vapour bath for dropsical affections, the boiling decoction being placed under the bed" (*Assistant Surgeon Bhugwan Das, Rawal Pindi, Panjáb*). "Sanskrit writers describe it as a tonic and alterative, and regard it as possessing the property of increasing the secretion of semen" (*U. C. Dutt, Civil Medical Officer, Serampore*).

Domestic.—"A portion of the plant is by the Santal tied round the neck, arm, and ancles, as a cure for rickets" (*Rev. A. Campbell, Report, Chutia Nagpur*). DOMESTIC. Charm. 807

CASTANEA, *Gærtn.; Gen. Pl., III., 409.* 808

Castanea vulgaris, *Lam.; DC. Prodr., xvi., 2, 114, 683;* CUPULIFERÆ.

THE SWEET CHESTNUT OR SPANISH CHESTNUT; CHÂTAIGNIER, *Fr.;* EDELKASTANIE, *Germ.*

Syn.—C. VESCA, *Gærtn.*

References.—*Brandis, For. Fl., 491; Gamble, Man. Timb., 379; DC., Origin of Cult. Pl., 353; Smith, Dic., 110.*

Habitat.—"A large, long-lived, deciduous tree, of rapid growth, more rapid than the oak; introduced in the Himálaya, and grown in various localities, and especially in a large number of places in the Panjáb and the hills of the North-West Provinces, in Darjíling, and the Khásia Hills" (*Gamble*).

Cultivation.—"It has been sown or planted in several parts of the south and west of Europe, and it is now difficult to know if it is wild or cultivated. However, cultivation consists chiefly in the operation of grafting good varieties on the trees which yield indifferent fruit. For this purpose the variety which produces but one large kernel is preferred to those which bear two or three, separated by a membrane, which is the natural state of the species." (*DeCandolle, Orig. Cult. Pl.*) CULTIVATION 809

Food.—The nuts are eaten. When ground into meal they form an important article of food for the poor. **Mr. Atkinson** says the tree was introduced by **Sir John Strachey** in Kumáon, and in Dehra by **Dr. Jameson**, where the fruits are now brought into the market. FOOD. 810

Structure of the Wood.—Sapwood white, heartwood dark brown. Weight from 32 to 54 lb per cubic foot. "The timber is not so durable as that of oak; in the south of Europe it is used for building, furniture, and cask-staves; but the legends of the roofs of old churches and other buildings made of chestnut timber, in France and England, are mythical; wherever examined, such timber has been found to be oak. It coppices TIMBER. 811

vigorously; along the Vosges it is grown for vineyard poles, in Kent and Sussex for hop-poles" (*Brandis*).

CASTANOPSIS, *Spach.; Gen. Pl., III., 409.*

Several species of this genus are met with on the mountains of Eastern India, but none are reported to be used for tanning. This is probably an oversight, since the European members possess this property to a considerable extent, **Castanea vesca** containing 14 to 20 per cent. of tannic acid.

812 **Castanopsis indica,** *Alph. DC., Prodr., XVI., 2, 109;* CUPULIFERÆ.

Syn.—CASTANEA INDICA, *Roxb., Fl. Ind., Ed. C.B.C., 674; Kurz, ii., 478;* QUERCUS SERRATA, *Roxb. (l. c., 641, probably).*

Vern.—*Banj katús,* NEPAL; *Kashiorón,* LEPCHA; *Serang,* ASS.; *Charang,* GARO; *Tailo,* CACHAR; *Nikari, gol-shingra,* SYLHET; *Theet-khya,* BURM.

References.—*Brandis, For. Fl., 490; Gamble, Man. Timb., 388; Kurz, For. Fl., Burm., 478; Balfour, Cyclop.*

Habitat.—A moderate-sized, evergreen tree, met with in Nepal, Eastern Bengal, Assam, and Chittagong, ascending to 5,000 feet.

FOOD. 813 **Food.**—"The fruit is eaten; it much resembles the filbert both in shape and in flavour, but has a thinner shell" (*Gamble*).

TIMBER. 814 **Structure of the Wood.**—Grey, hard. It splits well, and is very largely used for shingles in Darjeeling. It coppices freely, and is often pollarded and the branches burnt for manure.

815 **C. rufescens,** *Hook. f. & Th.; Gamble, Man. Timb., 389.*

Vern.—*Dalné katús,* NEPAL; *Sirikishu,* LEPCHA; *Hingori,* ASS.

Habitat.—A very large evergreen tree of Sikkim Himálaya, from 6,000 to 9,000 feet.

FOOD. 816 **Food.**—The fruit is small, but edible and of good flavour.

TIMBER. 817 **Structure of the Wood.**—Grey, hard. Annual rings marked by narrow belts of firmer texture. It is used in Darjíling for house-building, agricultural implements, and other purposes, exactly as that of **Quercus pachyphylla,** which it very closely resembles. It makes excellent shingles, and is more valuable as planking and posts wherever exposed to wet than other species of this genus.

818 **C. tribuloides,** *Alph. DC., Prodr., XVI., 2, III.; Wight, Ic., t. 770.*

Syn.—CASTANEA TRIBULOIDES, *Kurz (ii., 480);* QUERCUS FEROX, and Q. ARMATA, *Roxb., Fl. Ind., Ed. C.B.C., 673.*

Vern.—*Túmari, katonj,* KUMAON; *Musré katús, kotur, chisi, maku, shingali,* NEPAL; *Bar hingori, kanta singar,* ASS.; *Dingsaot,* KHASIA; *Singhara,* TIPPERAH; *Kanta lal batana,* CHITTAGONG; *Kyantsa,* BURM.

References.—*Gamble, Man. Timb., 389; Brandis, For. Fl., 490; Balfour, Cyclop.*

Habitat.—An evergreen tree, met with in South-East Kumáon, Nepal, Eastern Bengal, ascending from the plains to 6,000 feet, in Chittagong and hills in Burma, above 3,000 feet.

FOOD. 819 **Food.**—The fruit is eaten.

TIMBER. 820 **Structure of the Wood.**—Grey, moderately hard. Annual rings marked by darker lines. Used for planking and shingles, being good and durable.

The tree coppices admirably, and with **Castanopsis indica, Quercus spicata,** and **Engelhardtia** might be grown on the hills wherever firewood and charcoal forests are required.

CASTANOSPERMUM, *A. Cunn.; Gen. Pl., I., 556.*

"A genus of plants so named in consequence of the supposed resemblance of the seeds to the sweet chestnuts of Europe."

Castanospermum australe, *A. Cunn.;* LEGUMINOSÆ. 821

THE MORETON BAY CHESTNUT.

References.—*Drury, U. Pl., 124; Balfour, Cyclop.; Smith, Dic., 110; Treasury of Botany.*

Habitat.—A tree of the sub-tropical regions of Australia, occasionally planted for ornament; introduced into India about thirty years ago.

Food.—The seeds are eaten by the natives of Australia, but are unpalatable to Europeans (*Smith*). FOOD. 822

Structure of the Wood.—White, with a yellowish tinge; hard. TIMBER. 823

CASTILLOA, *Cerv.; Gen. Pl., III., 372.*

Castilloa elastica, *Cerv.;* URTICACEÆ. 824

THE ULE TREE.

References.—*Brandis, For. Fl., 427; Kurz, For. Fl., Burm., II., 419; Smith, Dic., 87, 89; Spons' Encyclop., 1659-61; Reports of Bot. Gardens, Nilgiri Hills, for 1881-82, 1882-83, and 1885-86.*

Habitat.—A lofty forest tree of the Bread-fruit family, native of America; lately introduced into Ceylon and some parts of India. In *Kew Report for 1877*, p. 15, is given an account of the attempts made to introduce this plant into India. Burma, Assam, Ceylon, and the lower slopes of the Nilgiris have now been pronounced as suitable for its cultivation.

Mr. Lawson reports of the Nilgiri plants: "In these days of uncertain coffee crops and low prices, planters are anxious to cultivate any plant that will return a small interest on their outlay." "I have no doubt many localities in the Wynaad and on the slopes of the hills will be found to suit the **Castilloa**, and where it will yield a profitable return to the cultivator." **Colonel Campbell Walker** writes of **Castilloa** cultivation in Calicut: "It has been found easy to raise these trees from cuttings. I hope they will in the future form no unimportant item in the forestry of this place. The other rubber-producing plants have so far been a failure, either from their not yielding as much rubber as they do in America, or because we have not yet learnt how to tap the trees properly."

Gum.—The tree exudes, on tapping, a milky juice which, when thickened, forms what is called the Central American rubber. In some countries the trees are cut down, and thereafter rings are made by cutting out a few inches of the bark. The trunk is then raised to a certain angle, and vessels are placed under each ring into which the milk flows; it thickens on exposure to the air, but it is said to do so more quickly by the addition of the juice of **Ipomœa bona-nox.** GUM. 825

For further particulars of this gum see under "**India-rubber.**"

Castor Oil, see **Ricinus communis,** *Linn.;* EUPHORBIACEÆ.

CASUARINA, *Forst.; Gen. Pl., III., 402.*

826 **Casuarina equisetifolia,** *Forst.; DC. Prodr., XVI., 2, 338;* CASUARINACEÆ.

THE BEEFWOOD OF AUSTRALIA.

Syn.—C. MURICATA, *Roxb., Fl. Ind., Ed. C. B. C., 623.*

Vern.—*Jangli sarv,* HIND.; *Jáu,* BENG.; *Viláyatisaro, wiláyati-sarú, saroka jhar,* BOMB.; *Júrijur, mujjun,* SIND.; *Sarpúhala, sarova, suru,* MAR.; *Janglijháú, jangli-saru, jangli-saru-chal,* DUK.; *Chouk, shavuku-maram, shavuku-pattay,* TAM.; *Serva, chavuku-mánú, chavuku-patta,* TEL.; *Kásrike,* MYSOR; *Sura,* KAN.; *Aru, chavaka-maram,* MALA.; *Tin-yu,* BURM. Many of the Indian names are modern adaptations; Conf. with **Tamarix.**

References.—*Gamble, Man. Timb., 346; Brandis, For. Fl., 435; Kurz, For. Fl., Burm., II., 494; Dalz. & Gibs., Bomb. Fl., Suppl., 82; Pharm. Ind., 217; Moodeen Sheriff, Supp. Pharm. Ind., 96; Dymock, Mat. Med. W. Ind., 2nd Ed., 750; Ainslie, Mat. Med. Ind., II., 443; Murray, Drugs and Pl., Sind, 27; Liotard, Dyes of India, 10; Wardle, Dyes of India, 1, 45; Bidie, Cat. Raw Prod., Paris Exh., 44; S. Arjun, Bomb. Drugs, 131; Drury, U. Pl., 124; Baden Powell, Pb. Pr., 573; Lisboa, U. Pl. of Bomb., 132; Kew Cat., 121; Hutchins, Report on, in Madras, 1883; Report, Agri. Dept., Madras, 1878-79, pp. 38-39; Balfour, Cyclop.; Smith, Dic., 294; Treasury of Botany.*

Habitat.—A large, evergreen tree, with leafless, drooping branches and branchlets, which are deciduous, and perform the functions of leaves. Found on the coast of Chittagong, Burma, the Malay Archipelago, North Australia, and Queensland; cultivated all over India, except in the north-western portion of the Panjáb. Thrives best in the sandy tracts near the sea-shore. Introduced into the plains of India as a road-side tree (valuable on account of the rapidity of its growth) about the beginning of the present century, and from its resemblance to the **Tamarix** received the vernacular names of that plant.

CULTIVATION 827 **Cultivation.**—"It has been largely planted in North Arcot, South Arcot, Madras, and other districts of the Madras Presidency, for fuel, for which it is excellent, but it requires to be near the sea-coast and to have water at the roots, at least 10 feet from the surface of the ground. Trees planted in sandy soil often suffer much from drought the first two or three years, the tap-root then finds its way down to about 10 feet, and reaching water the tree begins to thrive. It is of course best near the sea, but fine trees may be seen in places in Northern India, especially at Saharanpur and Amballa" (*Gamble*).

The Madras Agricultural Report for 1878-79 gives particulars of the cost of cultivation of an acre containing 1,200 trees. The initial cost is put down at R85: with interest at 10 per cent. for four years this raises the gross capital to R119. At this time half the trees (600) should be removed. Valuing these at 8 annas each the capital is returned and a balance left of R181. Two years later another 200 trees are removed, worth R1 each, and in the eighth or ninth year the land may be cleared; the remaining trees, at the lowest estimate, after paying all expenses on the same, would realize R600.

GUM. 828 **Gum.**—Reported to yield a good resin.

DYE. 829 **Dye.**—The bark is used in tanning (*Birdwood, Bomb. Prod.,* and *Bidie, Mad. Exh. List for 1855*). A brown dye is extracted from it according to **Balfour. Mr. Wardle** remarks: "The bark contains a small quantity of colouring matter, and produces in dyeing light-reddish drab colours on each of the fabrics on which I have experimented." He further adds: "The shades produced by this dye-stuff are very good though faint, but the dye-stuff contains too small an amount of colouring matter to be of any great value in the dye-house." **Lisboa** says that it is used in Bombay as a mordant. DYE.

Medicine.—The bark is slightly astringent, and is employed in infusion as a tonic; according to **Dr. Gibson** it is an excellent and at the same time a readily available astringent, useful in the treatment of chronic diarrhœa and dysentery (*Murray*). MEDICINE. 830

Structure of the Wood.—White, brown near the centre, very hard; it cracks and splits. It is hard and heavy, and difficult to cut; weighs from 55 to 62 lb per cubic foot. "**Casuarina** seems to coppice well, and undoubtedly is, in suitable localities, and considering its extremely quick growth and the qualities of its wood, one of the most important trees we have for fuel and other plantations" (*Gamble*). "The wood is used for fires, as it burns readily, and the ashes retain the heat for a long time. It is much valued for steam-engines, ovens, &c." (*Treasury of Botany.*) Clubs made of the hard wood are used in Fiji for beating the bark of the PAPER MULBERRY (**Broussonetia papyrifera,** *Vent.*) for the manufacture of Tapa cloth (*Kew Official Guide to Museums, 121*). The natives of Australia make their war-clubs from this wood (*Smith*). TIMBER. 831

Domestic Uses.—"The burnt ash is made into soap" (*Smith*). DOMESTIC. Ash. 832

Catechu, see—

[A. 139.] (*a*) **Acacia Catechu,** *Willd.*, LEGUMINOSÆ (black catechu).
(*b*) **Uncaria Gambier,** *Roxb.*, RUBIACEÆ (pale catechu).
[A. 1298.] (*c*) **Areca Catechu,** *Linn.*, PALMÆ (palm catechu).

Cattle and Buffaloes, see **Oxen.**

Cat, Civet, see **Tigers and Panthers.**

Catha. Several species exist in India, but by the *Flora of British India* they have been all reduced to **Celastrus,** which see.

Catha edulis yields the *Kat or Kafter* of the Arabs, the leaves of which if chewed are said to prevent sleep. Sometimes imported into India, largely so to Aden, where they are used as a substitute for Tea. 833

Cat's-eyes, see **Chalcedony.**

Cat's-skins, see **Skins.**

Cauliflower and Brocoli, see **Brassica (oleracea) botrytis, B 851.**

Caustic Potash, see **Potassium,** also **Carbonate of Potash, C. 527.**

Caustic Soda, see **Sodium,** also **Carbonate of Soda.**

CEDRELA, *Linn.; Gen. Pl., I., 339* 834

The *Flora of British India* has reduced at least three if not four easily-recognisable trees to one species, not even retaining the old specific names to denote varieties. If dried specimens in the Herbarium do not exhibit the characters of the **Cedrelas,** there is no mistaking the living plants. **C. serrata,** *Royle*, is so dissimilar from **C. Toona,** *Roxb.*, that were they to be found growing side by side, through the aid of a glass, they could be distinguished miles off. The former is a sparsely-branched tall tree, with palm-like clusters of pale green leaves, at the ends of its ascending branches, from which when in flower a panicle three or four feet long is suspended. This is the characteristic form of the North-Western Himálaya at altitudes from 4,000 to 8,000 feet. It frequents damp shady streamlets, growing so gregariously as to exclude all other trees.

CEDRELA serrata.

The Toon woods.

In the Monograph of the **Meliaceæ** published in 1878 by **Casimir de Candolle,** the species of **Cedrela** formerly grouped under the one head of **Cedrela Toona,** *Roxb.*, have been separately described.

They are thus distinguished :—

Ovary glabrous—

Leaflets petioled . . . . . . **C. serrata,** *Royle.*
Leaflets subsessile . . . . . **C. glabra,** *C. de Cand.*

Ovary hairy—

Leaflets acute at the base . . . . **C. Toona,** *Roxb.*
Leaflets round at the base . . . **C. microcarpa,** *C. de Cand.*

Mr. Gamble, in his *Manual of Timbers, XII.*, remarks that in his "*Trees, Shrubs, and Climbers of the Darjíling District,* three varieties were spoken of and separated as follows :—

No. 1. Deciduous; flowering March; fruiting June; bark grey-brown, smooth exfoliating; found in the plains on low land.

No. 2. Evergreen; flowering October-November; fruiting February and March; bark dark-brown, rough, not exfoliating; found in the lower hills up to 4,000 feet.

No. 3. Evergreen; flowering June; fruiting November-December; bark light-reddish brown, exfoliating in long flakes; found in the upper hills from 5,000 to 7,000 feet and of great size.

"No. 1 is **C. Toona,** *Roxb.;* No. 2 probably **C. microcarpa,** *C. de Cand.;* No. 3 probably **C. glabra,** *C. de Cand.* It would, however, have probably been better to describe No. 1 as 'deciduous in the cold season,' and Nos. 2 and 3 as 'deciduous in the rains.' There is perhaps a fifth species.

"They may also be distinguished as follows by the capsule :—

Capsule smooth . { capsule round . . **C. Toona.**
,, long, pointed . **C. microcarpa.** }
Capsule covered with corky tubercles . . **C. glabra.**

"Of the Northern Bengal specimens which we have examined, E 360 and E 2333 will be **C. glabra,** while E 655, E 2332, E 3599, E 3619, and E 3623 will be **C. microcarpa.** Some of the Assam, Chittagong, and Burma specimens are probably **C. microcarpa.**

"No. B 3378 from the Salween, 2,000 feet, is probably **C. multijuga,** *Kurz, i., 229.*—Vern. *Toungdama,* BURM.; *Nee,* KAREN (Trade name, like the other Toon-woods, *Thtitktado*). It has a light, soft, pink wood, with the usual characteristic scent strongly perceptible, and structure resembling that of the other species of Toon, the pores being perhaps more scantily distributed. Weight 35·5℔ per cubic foot."

The preceding remarks may for the present be accepted as indicating the Nepal plant, **C. glabra,** *DC.*, and the Sikkim **C. microcarpa,** *DC*, as distinct from the following :—

835 **Cedrela serrata,** *Royle; Ill., p. 144, t. 25; Monog., DC., I., 742;* [MELIACEÆ.

Syn.—C. TOONA; *Roxb.* (*Hook., Fl. Ind., i, 568, in part*).

Vern.—*Drawi, dalli, dál, dauri, khishing, khinam,* N.-W. H.

Habitat.—A tree of the N. W. Himálaya up to 8,000 feet; particularly abundant in the valleys a little below Simla. Found by the writer in the forest of the Naga hills and Manipur, and especially abundant on the moraine-like walls which extend from Japvo at an altitude of 6,000 feet.

TIMBER. 836 **Structure of the Wood.**—Heartwood light red, even, but open-grained, fragrant. Annual rings distinctly marked by broad belts of numerous large pores,

The Toon-woods.

Domestic Uses.—Used about Simla, for the hoops for sieves, for bridges, and for many such purposes. The shoots and leaves are lopped for cattle fodder. DOMESTIC. 837 FODDER. 838

Cedrela Toona, *Roxb.; Fl. Br. Ind., I., 568; Wight, Ic., t. 161.*

THE TOON OF INDIAN MAHOGANY TREE; MOULMEIN CEDAR.

Vern.—*Tún, túni, lím, mahá-ním, mahá-limbo, túnká-jhár, túna, lúd,* HIND.; *Túni, tún, lúd, túnna,* BENG.; *Kujya,* TIPPERA; *Somso,* BHUTIA; *Katangai,* KÓL.; *Maha limbu,* URIYA; *Máhlun,* SATPURAS; *Drawí, chiti-sirin, tún, gúl-tún* (flowers), *drab, deri, chútí-sirin, der, dorí, bísrú, guldár, darab, khúshing, khanam,* PB.; *Túni, babich, labshi,* NEPAL; *Simal,* LEPCHA; *Poma, henduri poma, tún, jía, tungd,* ASS.; *Deodari, kúruk,* MAR.; *Deodari, kúruk, túndu, tún, tunna, mahá-ním, limb, tuni, túpa, kudaka,* BOMB.; *Túnu-maram, tún-maram, malí, wunjúli, thúna-maram, wunjúli-maram,* TAM.; *Nandi-chettu, nandi,* TEL.; *Arana-maram,* MALA.; *Súli, máli,* SALEM; *Kal kilingi,* NILGIRIS; *Sandani vembu,* TINNEVELLY; *Tundú, kempú gandagheri, tunda, sauola-mara, kanda gariga mara, devdari,* KAN.; *Nogó, belandi,* COORG; *Tunna, kuberaka, kachla, nandí vriksha, tunna-kuberaka,* SANS.; *Chikado, tsithado,* MAGH; *Shurúsbed,* CHAKMA; *Thit-kadoe,* BURM.

References.—*Roxb., Fl. Ind., Ed. C.B.C., 213, 633; Brandis, For. Fl., 72; Kurz., For. Fl. Burm., I., 228; Bedd. Fl. Sylv., t. 10; Gumble, Man. Timb., 77, 79, xii.; Dalz. & Gibs., Bomb. Fl., 38; Stewart, Pb. Pl., 34; Aitchison, Cat. Pb. Pl., 30; Voigt, Hort. Sub. Cal., 137; Pharm. Ind., 55; Moodeen Sheriff, Supp. Pharm. Ind., 96; U. C. Dutt, Mat. Med. Hind., 311, 321; Dymock, Mat. Med. W. Ind., 2nd Ed., 177; Ainslie, Mat. Ind., II., 429; Murray, Drugs and Pl., Sind., 83; Baden Powell, Pb. Pr., 334, 573; Cooke, Gums and Gum-resins, 12; Liotard, Dyes, &c., Ind., 82; Wardle, Dyes of Ind., 9, 49; McCann, Dyes and Tans of Beng., 74; Buck, Dyes and Tans, N.-W. P., 28, 23; Birdwood, Bomb. Prod., 325; Lisboa, U. Pl. Bomb., 46, 241, 258; Balfour, Cyclop.; Treasury of Bot.; Kew, Cat., 29; Fleming's Med. Pl. and Drugs in As. Socy. Res., Vol. XI., 163; Med. Top., IX., 93.*

Habitat.—A large tree, about 50 to 60 feet in height, growing in the tropical Himálaya from the Indus eastward, and throughout the hilly districts of Central and South India to Burma; ascending to 3,000 feet in the N.-W. Himálaya and in Sikkim (?) to 7,000 feet. Distributed to Java and Australia.

Gum.—It yields a resinous gum, of which little is known at present. **M. Nees von Essenbeck** has published an account of some experiments with the bark, which indicate the presence in it of a resinous astringent matter, a brown astringent gum, and a gummy brown extractive matter, resembling *Ulmine.* (*Balfour.*) GUM. 839

Dye.—The flowers yield a red and a yellow dye (in Bengal generally known as *Gulnari*) said to be used in Mysore for dyeing cotton. This must be to a small extent only, since **Dr. Bidie** omits it from his list of Madras dyes sent to Paris. The flowers are boiled to extract the colour which is known as *basanti* in the North-West Provinces. It is fleeting and apparently only used by the poorer classes. In Burma it is used in conjunction with safflower. **Sir E. Buck,** in his *Report on the Dye-stuffs, N-W. Provinces,* says that a red dye is obtained from the seeds; and **Dr. McCann,** in his *Report on the Dyes of Bengal,* remarks that the seeds are used as a dye-stuff at Palamau. Apparently *Tún* is not used with mordants, and is rarely combined with other dyes. The sulphur yellow (*basanti*) of Cawnpore is produced from tún, turmeric, lime, and acidulated water. "It was a commoner practice under native rulers than it appears to be now to wear *basanti*-coloured clothes in the spring, whence its name *basant*" or spring time. Safflower and tún are combined in Tirwa. **Dr. McCann** DYE. Flowers. 840 Seeds. 841

The Toon-woods: Moulmein Cedar.

DYE. says the cloth previously dyed yellow is changed into red by the *pán* eaten by Hindús.

Mr. Wardle reports: "These flowers contain a large quantity of yellow colouring matter;" they "appear well adapted" for tasar silk.

MEDICINE. Bark. 842 **Medicine.**—The bark has astringent properties, and is a mild febrifuge, useful in diarrhœa and dysentery, especially of children. The following is an extract from the *Pharmacopœia of India* on the medicinal properties of the bark: "The bark of this tree is a powerful astringent, and may be resorted to when other remedies of the same class are not available. **Dr. Waitz** (*Diseases of Children in Hot Climates*, p. 225) used with success an extract of the bark in chronic infantile dysentery. **Blume** attributes valuable antiperiodic virtues to it, and in this character it is favourably noticed by **Dr. J. Kennedy** (*Ann. of Med.*, 1796, Vol. I., p. 387). **Dr. Æ. Ross** speaks of it as a reliable antiperiodic, and **Dr. J. Newton** as a good substitute for cinchona. The dose of the dried bark is about an ounce daily in the form of infusion. The powder of the bark was found by **Dr. Kennedy** to be of great service as a local astringent application in various forms of ulceration." According to **Dr. Dymock**, the native physicians use the bark in combination with bonduc nuts as a tonic and antiperiodic, a fact also mentioned by **Ainslie** in his *Materia Indica*. The

Flowers. 843 FLOWERS are called *Gul-tún* in Bombay and considered emmenagogue. "The bark was used in Java by **Blume** in epidemic fevers, diarrhœa, and other complaints. **Horsefield** gave it in dysentery, but only in the last state, when inflammatory symptoms had disappeared" (*Balfour*).

FOOD. 844 **Food.**—The seeds are used to feed cattle. The young shoots and leaves are lopped as cattle-fodder.

TIMBER. 845 **Structure of the Wood.**—Brick-red, soft, shining, even but open grained, fragrant, seasons readily, does not split nor warp. Annual rings distinctly marked by a belt of large and numerous pores.

It is durable and is not eaten by white-ants; is highly valued and universally used for furniture of all kinds, and is also employed for door-panels and carving. From Burma it is exported under the name of *Moulmein Cedar*, and as such is known in the English market. It

Price. there fetches about R65 per ton; according to **Major Seaton** the cost of cutting and delivery being R44. In North-West India it is used for furniture, carvings, and other purposes. In Bengal and Assam it is the chief wood for making tea-boxes, but is getting scarce on account of the heavy demand. The Bhutias use it for shingles and for wood-carving; they also hollow it out for rice-pounders. It is—or rather used to be, for very large trees are now rather scarce—hollowed out for dug-out canoes in Bengal and Assam. In Bengal, Assam, and Burma it grows to a very large size, trees 20 feet in girth with a height of 80 to 100 feet of clear stem being not uncommon in forests which have been only little worked, like those in Dumsong and in some parts of the Chittagong Hill Tracts. At page 91 of the *Indian Forester* (Vol. I.), the cubic contents of four trees in the Reyang Valley, Darjeeling, are given as 211, 375, 720, and 400 cubic feet respectively; the third of these had a mean girth of 12 feet and a length of 80 feet, while the second had a girth of 20 feet. It is easily propagated from seed, but the seeds being very small and light, the seed-beds must be sheltered till the seedlings have well come on. It also coppices freely.

Dr. Dymock says that the wood resembles mahogany, and is used in Bombay for making medicine chests and surgical instrument cases. It is also employed for the hoops for sieves, for bridges, and for many other purposes.

CEDRUS, *Loud.; Gen. Pl., III., 93.*

Cedrus Deodara, *Loudon; DC. Prodr., XVI., 2, 409.* 846

DEODAR; HIMÁLAYAN CEDAR.

Syn.—PINUS DEODARA, *Roxb., Fl. Ind., Ed. C.B.C., 677.*

Vern.—*Kilan-ká-pér, kílan, deodár,* HIND.; *Dewdár, geyár, kelí, kelu, keorí, kelaí, kalain, kálon, kenwal, keoli-kelmang, kaiwal, kelmang, palurr, dadá,* PB.; *Nakhtar, lmanza,* AFG.; *Diár, deodár, dewdar, deowar, dadár,* HAZARA, KASHMIR, GARHWAL, KUMAON; *Palúdar,* HAZARA; *Kelu, keoli, kilar, kilei,* HIMALAYAN NAMES; *Deodár, dar,* KASHMIR; *Kelmang,* KUNAWAR; *Giam,* TIBET; *Devadáru, devdár, debdáru,* BENG.; *Dévdár, vánseo-deodár,* GUJ.; *Dévadárúcha-jháda, dewadar,* MAR.; *Dévdáru,* DUK.; *Dévadári-chedi,* TAM.; *Dévadári-chettu,* TEL.; *Dévatáram,* MALA.; *Dévadári-mará,* KAN.; *Devadáru,* SANS.; *Shajratud-dévdár, sanóbarul-hind,* ARAB.; *Darakhte-dévdár, sanóbarc-hindí, nashtar,* PERS.

References.—*Brandis, For. Fl., 516; Gamble, Man. Timb., 400; Stewart, Pb. Pl., 220; Voigt, Hort. Sub. Cal., 337; Pharm. Ind., 225; Moodeen Sheriff, Supp. Pharm. Ind., 199; U. C. Dutt, Mat. Med. Hind., 247, 296; Dymock, Mat. Med. W. Ind., 2nd Ed., 757; O'Shaughnessy, Beng. Dispens., 612; S. Arjun, Bomb. Drugs, 132; Baden Powell, Pb. Pr., 410, 424, 573; Cooke, Gums and Gum-resins, 128; Athinson, Him. Dist., 830; Atkinson, Gums and Gum-resins, N.-W. P., 42; Balfour, Cyclop.;. Royle, Ill. Him. Bot., 350; Browne, Forester, 374; Irvine, 28.*

Habitat.—A very large and tall tree, found in the North-West Himálaya, between 4,000 and 10,000 feet, extending east to the Dauli river (a tributary of the Alaknanda below the Niti Pass), in the mountains of Afghánistán and in North Belúchistán.

GUM. 847

Gum.—It yields a true oleo-resin, called *Kelon-ka-tel.* The preparation of this oleo-resin is thus described by **Mr. Baden Powell**:—

"First, an earthen *ghara,* or vessel with a wide mouth, and capable of containing about 4 seers, is sunk into the ground. Next, a large *ghara* of about 12 seers capacity is taken, and three small holes are drilled in its under-side; it is then filled with scraps of the pine wood, and over its mouth another smaller jar is placed, and kept there by a luting of clay very carefully applied; and then both the jars are smeared over with a coating of clay. These two jars thus stuck together are next set on the mouth of the receiver or *ghara* sunk into the ground, and the joint or seat is made tight by a luting of stiff clay. Light firewood is now heaped around the apparatus and ignited, and kept burning from four to eight hours; the *rationale* of the process being that the heat causes the tar contained in the chips inclosed in the large *ghara* to exude, and it falls through the three holes drilled in the bottom, and into the receiver sunk into the ground. When the fire is out, the ashes are raked away, the jars very carefully separated, so that pieces of dirt may not fall into the receiver, and the latter is then exhumed and the contents poured out. It is only necessary to replace the receiver, with the jars over it as before, duly charged with chips, and lute the joints up carefully, and the process can be carried on as before. With care the same jars may be made to do over and over again without cracking. One seer of wood yields about 2·6 *chitaks* of tar and 4·3 *chitaks* of charcoal. To procure a seer of tar requires 6 seers 4 *chitaks* of wood-chips to charge the pot, and 2 maunds 6 seers and *chitaks* of chips for fuel." (*Pb. Prod.*, 410.)

Oil.—"An oil is obtained from the wood by destructive distillation; it is dark coloured, thick, and resembles crude turpentine. It is used for anointing the inflated skins which are used for crossing rivers; and as a

The Deodar or Himalayan Cedar.

remedy for ulcers and eruptions, for mange in horses and sore feet in cattle." (*Gamble, 406.*)

MEDICINE. 848 **Medicine.**—The aromatic wood is employed medicinally as a carminative, diaphoretic, diuretic, and useful in fever, flatulence, inflammation, dropsy, urinary diseases, &c. It is chiefly used in combination with other medicines (*U. C. Dutt's Mat. Med. Hind., 247*). The tree yields a coarse, very fluid kind of turpentine, held in much esteem by the natives as an application to ulcers and skin diseases. The oil also enters into nostrums used by the natives in the treatment of leprosy. **Dr. Gibson** recommends the use of oil in large doses, as highly effectual in this disease. **Dr. J. Johnston** is said to have cured a severe case of *lepra mercurialis* by treating externally and internally with deodar oil It has been remarked that a drachm of the oil was as large a dose as the patient's stomach could bear. Its use may be extended to other skin diseases with advantage. **Dr. Royle** states that the leaves and small twigs of the *Deodara* are also brought down to the plains, as they are supposed to possess mild terebinthenate properties. (*Pharm. Ind.*) In Kangra the wood is pounded with water on a stone, and the paste applied to the temples to relieve headache. **Assistant Surgeon Sakharam Arjun** describes the wood as a bitter stomachic, useful in fever, costiveness, piles, and pulmonary complaints.

FOOD. 849 **Food.**—The young shoots and plants are eagerly browsed by goats, &c.

TIMBER. 850 **Structure of the Wood.**—Heartwood light-yellowish brown, scented, moderately hard. In each annual ring the outer belt of firmer and darker coloured tissue is generally narrow, and the inner belt is not very soft, but in exceptional cases, and under certain conditions which have not yet been studied, the inner belt is soft and spongy. This peculiarity has nothing to do with the rate of growth or with the altitude, as fast-grown trees possess hard tissue in the spring wood. Medullary rays fine and very fine, unequal in width. No vertical resinous ducts, as in *Pinus*, but the resin exudes from cells which are not visible to the naked eye. On the edge of certain annual rings are frequently found concentric strings of dark-coloured pores or intercellular ducts, which are prominent on a vertical section as dark lines, and in the vicinity of which the wood is sometimes more resinous.

In common with most species of the Order, the *Deodar* has well-marked annual rings which, there is little, if any, reason to doubt, each represents the growth of a year. More information has, perhaps, been collected on the subject of the rate of growth of *Deodar* than of any other species of Indian tree, though we have as yet no such complete series of trees of known age to deal with as were available at Nilambur for the question of the rate of growth of teak. The geographical range of *Deodar*, specially in altitude, is very wide, and this circumstance, considering that some specimens may be obtained from sheltered places in comparatively warm valleys, while others come from exposed and high situations, makes it doubtful whether much value can be attached to general deductions from data collected from many quarters, and whether it should not usually be the practice to take only for use in any forests, the experiments made on trees in that or neighbouring localities. But the experience we have hitherto gained is very valuable, and it will be best to put together the items of information available. In **Brandis'** *Forest Flora of North-West and Central India*, pp. 520 to 524, a large amount of information is collected, to which reference can be made. It is there stated that the *Deodar* forests may be classified in three great divisions, *viz.* :—

1st—Those in a dry climate in the vicinity of the arid zone of the

TIMBER.

inner Himálaya, having usually the age of trees 6 feet in girth above 140 years;

2nd—Those in the intermediate ranges and valleys, having 6 feet in girth for an age of between 110 and 140 years;

3rd—Those in the outer ranges under the full influence of the monsoon, and having the age of trees 6 feet in girth usually below 110 years.

Deodar wood is extremely durable, being by far the most durable of the woods of the Himálayan conifers. It is the chief timber of North-West India, and is used for all purposes of construction,—for railway sleepers, bridges, and even for furniture and shingles. (*Gamble.*)

CELASTRUS, *Linn.; Gen. Pl, I., 364.* 851

The *Flora of British India* raised Wight and Arnott's sub-genera (1) EUCELASTRUS and (2) GYMNOSPORIA to the rank of genera. This was at first followed by the authors of the *Genera Plantarum*, but subsequently (Vol. I., page 997) was corrected back to the original position. The former embraces some four species of unarmed climbers, and the latter fifteen armed shrubs or small trees. Accepting the above restoration we shall, in this instance, depart from the nomenclature established in the *Flora of British India;* but this need occasion little or no inconvenience, since the principal synonymy of the economic species will be found below.

Celastrus emarginata, *Willd.;* CELASTRINEÆ. 852

Syn.—GYMNOSPORIA EMARGINATA, *Roth., in Fl. Br. Ind., I., 621;* CELASTRUS EMARGINATA, *W. and A., Prod., 160; Roxb., Fl. Ind., Ed. C.B.C., 208;* CATHA EMARGINATA, *G. Don.*

C. oxyphylla, *Wall.* 853

Syn.—GYMNOSPORIA ACUMINATA, *Hook. f.; Fl. Br. Ind., I., 619.*

C. paniculata, *Willd.; Fl. Br. Ind., I., 617; Wight, Ic., t. 158.* 854

BLACK OIL; THE OLEUM NIGRUM PLANT.

Syn.—CELASTRUS ALNIFOLIA, *Don.;* C. DEPENDENS, *Wall.;* C. MULTIFLORA and NUTANS, *Roxb.*

Vern.—*Mál-kangni, mál kungí,* HIND.; *Sankhú, sankhii* (leaves, *kotaj, kuter*), PB.; *Málkákni,* OUDH, KUMAON; *Mál kangni,* BENG.; *Kujari, kujri,* SANTAL; *Kujúri,* KOL.; *Chiron,* MAL. (S. P.); *Kákundan rangul, wahrangur,* C. P.; *Kanguni, mál kangní,* BOM.; *Malkangana,* GUJ.; *Málkángóni, mál kánganitela, kangani, pigavi,* MAR.; *Ruglim,* LEPCHA; *Atiparich-cham, válulvai,* TAM.; *Málkanguni-vittulu, gundumeda, bavungie, maneru, moierikata, maiyala erikut,* TEL.; *Váluzhuva,* MALA.; *Kariganne,* KAN.; *Jiotishmati,* SANS.; *Myin khoung-na-young,* BURM. The vern. names of Oleum Nigrum: *Málkangní-kajantar,* DUK.; *Váluluvai-tailam,* TAM.; *Málkanginitailamu,* TEL.

References.—*Roxb., Fl. Ind., Ed. C.B.C., 209; Brandis, For. Fl., 82; Kurz, For. Fl. Burm., I., 252; Gamble, Man. Timb., 86, xiii.; Thwaites, En. Ceylon Pl., 72; Dalz. & Gibs., Bomb. Fl., 47; Stewart, Pb. Pl., 40; Aitchison, Cat. Pb. Pl., 31; Voigt, Hort. Sub. Cal., 166; Pharm. Ind., 56; Moodeen Sheriff, Supp. Pharm. Ind., 97; Dymock, Mat. Med. W. Ind., 144; O'Shaughnessy, Beng. Dispens., 271; Murray, Drugs and Pl., Sind, 147; Bidie, Cat. Raw Prod., Paris Exh., 24, 61, 104; S. Arjun, Bomb. Drugs, 30; Baden Powell, Pb. Pr., 336; Cooke, Oils and Oil-seeds, 35; Birdwood, Bomb. Prod., 18; Lisboa, U. Pl. Bomb., 216; Balfour, Cyclop.; Treasury of Bot.; Mysore Cat., Cal. Exh., 42; Irvine, Mat. Med. Patna, 66; Kew Cat., 31.*

The Oleum Nigrum.

Habitat.—A scandent shrub of the outer Himálaya, from the Jhelum to Assam, ascending to 4,000 feet; Eastern Bengal, Behar, South India, and Burma; in Ceylon it is common up to an elevation of 2,000 feet.

OIL. 855 **Oil.**—The SEEDS yield by expression a deep scarlet or yellow oil, used medicinally. The oil deposits a quantity of fat after it has been kept a short time. Its odour is pungent and acrid, and treated with sulphuric acid it turns of a dark bistre colour. It is much admired as an external application along with a poultice of the crushed seeds. It is also burnt in lamps, and employed in certain religious ceremonies. The seeds submitted to destructive distillation yield the "Oleum Nigrum," an empyreumatic black oily fluid employed medicinally in the treatment of *beri-beri* (*Cooke*). According to **Dr. Dymock**, the seeds are distilled along with benzoin, cloves, nutmegs, and mace. This oil is manufactured in the Northern Circars, the best in Vizagapatam and Ellore, where it is sold in small blue or black bottles, each containing about ¼ oz., at prices from 12 annas to one rupee a bottle.

MEDICINE. Oil. 856 **Medicine.**—The red seeds are used medicinally, principally for cattle. They are given in rheumatism and paralysis. An empyreumatic oil is obtained from the seeds by a rude form of distillation, which is applied externally. This oil, under the name of "Oleum Nigrum," was brought forward by the late **Dr. Herklots** as a sovereign remedy in *beri-beri.* When administered in doses of from ten to fifteen drops twice daily, its action as a powerful stimulant is generally followed in a few hours by free diaphorisis not attended by exhaustion. It is specially efficacious in recent cases, and where the nervous and paralytic symptoms predominate. (*Baden Powell; Pharm. Ind.*) It will be seen from the note below that **Dr. Moodeen Sheriff** holds a strong opinion regarding the merits of this oil. The SEEDS are considered by the Muhammadans to be "hot and dry, aphrodisiacal and stimulant, useful both as an external and internal remedy in rheumatism, gout, paralysis, leprosy, and other disorders which are supposed to be caused by cold humours. They may be administered in such cases commencing with a dose of one seed, to be gradually increased to fifty by daily increments of one; at the same time the oil may be applied externally, or the crushed seeds combined with aromatics. The latter application is said to be very efficient in removing local pains of a rheumatic or malarious nature" (*Dymock, Mat. Med. W. Ind., 144*). **Stewart** says the LEAVES also are used medicinally. "The oil of the seeds is a diuretic and has been used successfully in healing sinuses and fistulæ" (*Bomb. Gaz., VI., 15*). The **Rev. A. Campbell** remarks that the Santals use the oil in disorders of the stomach.

Seeds. 857

Leaves. 858

Special Opinions.—§ "The compound and empyreumatic oil obtained by a destructive distillation of the seeds of **Celastrus paniculata,** which is commonly known as Oleum Nigrum—Black Oil—is quite different from the oil of the same seeds extracted by compression. The former is black and thick, with a strong and peculiar aromatic smell; and the latter, yellow and of the consistence of oil. The black oil manufactured at Vizagapatam and Masulipatam is the best. It is a good diuretic, diaphoretic, and nervine stimulant. It is certainly the best remedy for *beri-beri.* I have seen many cases which did not benefit for weeks or months under the use of other medicines, but began to improve at once when this oil was employed. The first good effect of this medicine is generally the increase in the quantity of urine, and with this the dropsical effusion begins to disappear. A relief in paralytic symptoms accompanying *beri-beri* is also noticed in some cases about this time, but generally much later than this period. During the use of this medicine the native practitioners invariably enjoin a very low and strict diet, giving nothing to the patient except milk and bread—a restriction which is as injurious as unnecessary in my opinion. The patient labouring under *beri-beri* requires a very liberal and nourishing diet. I have also used Oleum Nigrum in some cases of simple and uncomplicated dropsy, and with good and encouraging results. *Dose:* From 10 to 30 minims as a diuretic, from 5 to 15 minims as a diaphoretic and nervine stimulant" (*Honorary Surgeon Moodeen Sheriff, Khan Bahadur, Triplicane, Madras*). "The seeds boiled in milk are used by natives in nervous affections. They are also used as food for quails" (*Assistant Surgeon Bhagwan Dass, Rawal Pindi, Panjáb*). "Said by some natives to be useful as a nervine tonic, specially in impotency, but the fact seems to be doubtful" (*Surgeon-Major C. J. McKenna, Cawnpore*). "Called in Telegu *Mál-kangni;* an oil is obtained by distillation from the red seeds. The oil is given in ten-drop doses two or three times a day on betel leaf as a vehicle. During the time the patient is under this treatment he should eat meat roasted. I have seen two or three cases of *beri-beri* cured by this treatment, and have also given it, with a fair amount of success, in dropsy from anæmia" (*Surgeon-Major Lionel Beech, Cocanada*). "The juice of the leaves mixed with that of the leaves of **Hydrocotyle asiatica**, and powdered spikenard, is considered a cooling application in inflammatory brain affections" (*Assistant Surgeon Sakharam Arjun, Bombay*). "The 'black oil' obtained by destructive distillation, in combination with several aromatic substances, is a powerful diuretic, and used successfully in *beri-beri* and dropsy. The diet, while using it, should consist exclusively of wheaten cakes and flesh of sheep" (*Honorary Surgeon P. Kinsley, Ganjam, Madras Presidency*). "An oil extracted by heat is a specific in the treatment of *beri-beri* with marked success. Also used in dropsical affections with invaluable results. Is a stimulant and diuretic; under the administration of this drug, *strict* diet should be observed, chiefly of *wheat, chappatties,* with *fried meat,* and *milk,* and nothing else should be taken. Is an invaluable remedy among the people of the Northern Circars, especially of those of the malarious tracts" (*Surgeon-Major E. W. Levinge, Rajamundry, Godavery District*). "Said to be useful as an aphrodisiac" (*Surgeon-Major D. R. Thompson, Madras*).

MEDICINE.

Food for Quails.

Structure of the Wood.—Pinkish yellow, soft.

TIMBER. 859

Celastrus senegalensis, *Lam.*

860

Syn.—GYMNOSPORIA (Celastrus) MONTANA, *Roxb., as in Fl. Br. Ind., I., 621;* C. MONTANA, *W. & A., Prod., 159;* CATHA MONTANA, *Don.*

Vern.—*Sherawane,* TRANS-INDUS; *Talkar, dajkar, mareila, kingaro, kharái,* PB.; *Baikal, gajachinni,* C. P.; *Mál kangoni,* BOM.; *Danta, babur,* GONDI; *Dhatti,* BHIL; *Bharatti, yekal,* MAR.; *Danti, dantúsi, pedda chintú,* TEL.

References.—*Roxb., Fl. Ind., Ed. C.B.C., 208; Brandis, For. Fl., 81; Kurz, Fl. Burm., I., 252; Beddome, Fl. Sylvat., LXVI; Dalz. & Gibs., Bomb. Fl., 48; Gamble, Man. Timb., 87.*

Habitat.—A profusely-armed tall shrub, common in the northern dry and intermediate zones of Central, South-Western, and North-Western India; distributed to Afghánistán, Central Asia, and Australia. The *Flora of British India* distinguishes several forms: **C. montana,** *Roxb.,* comprises those forms which have the branches less profusely armed, and the leaves larger and broader; **C. senegalensis,** *Lam.,* those in which the stems are more robust, and profusely armed, and the leaves smaller and narrower.

Medicine.—The BARK, ground to a paste and applied to the head, with mustard oil, is said to destroy *pediculi.*

MEDICINE. Bark. 861

862 **Celastrus spinosus,** *Royle.*

Syn.—GYMNOSPORIA ROYLEANA, *Wall., as in Fl. Br. Ind., I., 620.*

Vern.—*Jaliddhar,* HIND.; *Dzaral,* TRANS-INDUS; *Kandu,. kamla, kandiári, kander, láp, patáki, lei, lí, phúpári, badlo, kadewar,* PB.; *Kúra, bagriwala darim, gwála darim,* N.-W. P.

References.—*Boiss, Fl. Orient., II., 11; Brandis, For. Fl., 80; Gamble, Man. Timb., 86; Baden Powell, Pb. Prod., 582; Stewart, Pb. Pl., 41.*

Habitat.—A thorny, distorted bush, abundant on the outer North-Western Himálaya (Kumaon and Garwhal, altitude 1,000 to 4,500 feet) and distributed to the Concan and thence to Afghánistán; common on the Salt Range at about 5,000 feet in altitude.

MEDICINE. Seed. 863 **Medicine.**—In the Salt Range the smoke from the SEEDS is said to be good for toothache.

TIMBER. 864 **Structure of the Wood.**—Lemon-coloured, hard and close-grained; weight 49 ℔ a cubic foot. **Gamble** says the wood deserves attention as a possible substitute for boxwood, for carving and engraving. **Baden Powell** remarks that it is used in the Panjáb for walking-sticks.

865 **Celery.** See **Apium graveolens,** *Linn.*; UMBELLIFERÆ.

CELESTITE; *Mallet, Mineralogy, 141.*

Bombay. 866 **Celestite** or **Celestine** is a natural mineral, found in rhombic or tabular crystals or in masses. It is a form of Strontium sulphate, which is used in the arts in the preparation of Strontium nitrate—a Salt employed in fireworks to give a red light. There are two localities in India where Celestite has been found—in Bombay and Sind, scattered over the surface of the Kirthar limestones; and in the Panjáb, on the tertiary red clays of the Salt Range.

Punjab. 867

CELOSIA, *Linn.; Gen. Pl., III., 24.*

For botanical characters of the genus see under **Amarantaceæ (A. 914).**

The name is derived from *kelos,* burnt, in reference to the colour of the flowers in the common garden species.

868 **Celosia argentea,** *Linn.; Fl. Br. Ind., IV., 714;* AMARANTACEÆ.

Vern.—*Debkoti, sufaid múrgha, sarwari,* HIND.; *Sirgit arak,* SANTAL; *Sarwáli, siráli, ghogiya,* N.-W. P.; *Sarwálí, salgára, chilchil, sil, sarpankha,* PB.; *Swet-múrgá,* BENG.; *Surwalí, ucha-kukur,* SIND; *Lápadi,* GUJ.; *Kudhu, kurdu,* BOMB.; *Kúrdú, kurada,* MAR.; *Gurugu, panche chettu,* TEL.; *Kirri-handa,* SING. Several of these vernacular names imply *white-cock's-comb.*

References.—*Roxb., Fl. Ind., Ed. C.B.C., 228; Thwaites, En. Ceylon Pl. 247; Dalz. & Gibs., Bomb. Fl., 215; Stewart, Pb. Pl., 181; Aitchison, Cat. Pb. Pl., 130; Murray, Drugs and Pl. Sind, 100; Baden Powell, Pb. Pr., 373; Lisboa, U. Pl. Bomb., 170; Balfour, Cyclop.*

Habitat.—An abundant weed of the fields in Central and Northern India (from Chutia Nagpur to the Panjáb), occasionally ascending to altitude 5,000 feet in the Himálaya; it is also met with in the warmer parts of Ceylon. It appears very commonly in the monsoon season.

MEDICINE Seeds. 869 Oil. 870 **Medicine.**—The SEEDS are officinal, being an efficacious remedy in diarrhœa. The **Rev. A. Campbell** says the Santals extract a medicinal oil from them.

FOOD. 871 FODDER. 872 **Food.**—The plant is used as a pot-herb in times of scarcity, and is eaten by cattle, especially buffaloes.

Celosia cristata, *Linn.; Fl. Br. Ind., IV., 715; Wight, Ic., t. 730.* 873

Vern.—*Kokan, pila-murghka, lál-murghka,* HIND.; *Mawal, taji khoros, bostán afraz, kanjú, dhúrá-drú,* PB.; *Máwal,* KASHMIR; *Lál múrgá* (the red form), *húldi-múrga* (the yellow), BENG.; *Erra-kodi- utta-totakuru, kodi-juttu-tota-kura,* TEL.; *Mayur asikha,* SANS.; *Kyet-monk,* BURM.

References.—*Roxb., Fl. Ind., Ed. C.B.C., 228; Dalz. & Gibs., Bomb. Fl., 215; Stewart, Pb. Pl., 182; Murray, Drugs and Pl., Sind, 101; Baden Powell, Pb. Pr., 373; Balfour, Cyclop.; Treasury of Botany; Spons, Encyclop., 938.*

Habitat.—Cultivated as an ornamental plant in the plains, and on the Himálaya, Kashmír (5,000 feet). In *Spons' Encyclopædia* occurs the remark that this plant is "Common all over Bengal and Northern India generally."

Fibre.—"It yields a strong flexible fibre, so highly esteemed that rope made of it sells at five times the price of jute rope." Confirmation of this fact is much required, and also samples of the plant from which the fibre has been extracted. It is known in Bengali as *Lál-múrga*, but **Roxburgh** makes no mention of the fibre; indeed, with the exception of the notice in *Spons' Encyclopædia* quoted above, no author, as far as the writer can discover, alludes to the fibre. FIBRE. 874

Medicine.—The FLOWERS are officinal, being considered astringent; they are used in cases of diarrhœa and in excessive menstrual discharges. The SEEDS are viewed as demulcent. MEDICINE. Flowers. 875 Seeds. 876

Special Opinion.—§ "Seeds demulcent and useful in painful micturition, cough, and dysentery" *(Dr. U. C. Dutt, Serampore).*

Food.—Cultivated in gardens—both the red and the yellow forms—on account of the stem, which is eaten as a pot-herb. **Professor Church** (in *Food-Grains of India)* is apparently in error when he speaks of the food-properties of the seeds of this plant. The writer can find no mention of the plant being cultivated on account of its seeds, nor indeed of these being eaten. Besides, three of the vernacular names given by the Professor are not names for this plant. *Síl* (and names derived from that word) are more correctly applied to **Amarantus paniculatus,** the seed of which is eaten, so that it seems probable **Professor Church's** account of **Celosia cristata** should be transferred to **Amarantus paniculatus.** FOOD. 877

CELSIA, *Linn.; Gen. Pl., II., 929.*

Celsia coromandeliana, *Vahl.; Fl. Br. Ind., IV., 251; Wight, Ic., t. 1406;* SCROPHULARINEÆ. 878

Vern.—*Kúkshima, koksimá,* BENG.; *Kutki,* MAR.; *Kuláhala,* SANS.

References.—*Roxb., Fl. Ind., Ed. C.B.C., 491; Thwaites, En. Ceylon Pl., 217; Dalz. & Gibs., Bomb. Fl., 176; Aitchison, Cat. Pb. Pl., 105; Voigt, Hort. Sub. Cal., 497; Pharm. Ind., 161; Moodeen Sheriff, Supp. Pharm. Ind., 97; U. C. Dutt, Mat. Med. Hind., 306; Dymock, Mat. Med. W. Ind., 482; S. Arjun, Bomb. Drugs, 191; Drury, U. Pl., 128; Balfour, Cyclop.*

Habitat.—An herb found throughout India, from the Panjáb to Pegu and Ceylon, ascending to 5,000 feet in altitude. It generally appears during the dry season as a weed, on garden or cultivated lands.

Medicine.—The inspissated JUICE of the leaves has been prescribed in cases of acute and chronic dysentery. It acts as a sedative and astringent. (*Pharm. of Ind.*) MEDICINE. Juice. 879

Special Opinions.—§ "Juice of the whole plant, including the root, leaves, and stem, squeezed out by pounding it, is used in half *chittack* doses, morning and evening, in cases of syphilitic eruptions. The juice of

MEDICINE.

the leaves mixed with mustard oil, in equal proportions, is applied as an external application for relieving the burning sensations of the hands and feet" (*Surgeon Anund Chunder Mukerji, Noakhally*). "If a little of

Root. 880 the root is chewed in fever, or when there is urgent thirst, a cooling sensation will occur and thirst be appeased" (*Surgeon William Wilson, Bogra*). "Expressed juice of the leaves, mixed with sugar and water, used as a drink in bleeding piles" (*Brigade Surgeon S. M. Shircore, Moorshedabad*). "I have been told that the roots, leaves, and flowers are used as an astringent in diarrhœa, acute and chronic dysentery. The action of the drug is said to be immediate and direct. It is also reputed to be an emetic and expectorant, being employed in capillary bronchitis of children. An Assistant Surgeon tells me he has used it with good results in diabetes, both insepidus and melletus" (*Surgeon J. French Mullen, Saidpur*). "The root is used in dysentery and as a cholagogue" (*Brigade Surgeon J. H. Thornton, Monghir*).

CELTIS, *Tourn.; DC. Prodr., XVII., 168.*

881 **Celtis australis,** *Linn.; DC. Prodr., xvii., 169, 170, 179;* URTICACEÆ.

THE EUROPEAN NETTLE-TREE; THE HONEY-BERRY TREE.

Syn.—It is probable **Brandis** is correct in viewing all the following species as but varieties of **C. australis,** at least all those with glabrous fruits.

Vern.—*Kharak, kharika, khirk,* N.-W. P.; *Kharak,* SIMLA, KUMAON; *Brimlu, khirk, khalk, khark, khirg, kú, roku, choku, bramji, batkar, kái, bigni, biúgli,* PB.; *Kar,* KUNAWAR; *Tagha,* SIND; *Tagho, takhúm,* AFG.

References.—*Brandis, For. Fl., 428; Gamble, Man. Timb., 343; Aitchison, Cat. Pb. Pl., 139; Treasury of Botany.*

Habitat.—A moderate-sized, deciduous tree, found in the Suliman and Salt Ranges, and throughout the Himálaya from the Indus to Bhután, ascending to 8,500 feet; also in the Khásia Hills. Extensively cultivated in South Europe.

FOOD. Fruit. 882 FODDER. 883

Food and Fodder.—The tree is largely planted for fodder; cows fed on the leaves are supposed to give better milk. The FRUIT is also eaten. "It is remarkably sweet, and is supposed to have been the Lotus of the ancients, the food of the Lotophagi, which **Herodotus, Dioscorides,** and **Theophrastus** describe as sweet, pleasant, and wholesome, and which **Homer** says was so delicious as to make those who ate it forget their native country. The berries are still eaten in Spain, and **Dr. Walsh** remarks that the modern Greeks are very fond of them" (*Treasury of Botany*). It is nowhere grown as a fruit tree in India, although, as **Atkinson** adds, it is eaten by all classes and is esteemed.

A dark-purple form of the fruit is called *roku* and a smaller yellow form *choku.*

TIMBER. 884 **Structure of the Wood.**—Grey or yellowish grey, with irregular streaks of darker colour. Weight 47lb per cubic foot. It is tough and strong, and is used for oars, whip-handles, and for other purposes requiring toughness and elasticity (*Gamble*).

DOMESTIC. 885 **Domestic Uses.**—"The branches are extensively employed in making hay-forks, coach-whips, ramrods, and walking-sticks" (*Treasury of Botany*).

886 **C. caucasica,** *Willd.; DC. Prodr., xvii., 170.*

Vern.—*Batkar, brúmij, brimdú, brimla, bigní, biúgu, kharg, khark, khirk, karik, kharak, khalk, kú, takhum, tágho, wattamman, kanrak, kirki, kar, kargam, taghum, takpun, kúrg, kanghol mirch* (the fruit), PB.; *Túghar,* PUSHTU.

References.—*Brandis, For. Fl., 428, 429; Gamble, Man. Timb., 344; Stewart, Pb. Pl., 209; Aitchison, Cat. Pb. Pl., 139; Baden Powell, Pb. Pr., 574; Balfour, Cyclop.*

Habitat.—A moderate-sized tree of Afghánistán, Beluchistán, the Salt Range, Hazára, and Kashmír. **Aitchison** says it is cultivated near shrines and graveyards in Afghánistán.

Fibre.—The bark is made into cordage (*Baden Powell*). FIBRE. 887

Medicine.—The FRUIT is officinal, being given as a remedy in amenorrhœa and colic. MEDICINE. Fruit. 888

Food.—The FRUIT, a small drupe, is eaten by the natives, who regard it as sweetish, but it has almost no flesh. (*Roxb.*) FOOD. Fruit. 889

Structure of the Wood.—Light yellow, hard to very hard. Structure resembling that of **C. australis.** **Mr. J. H. Lace** writes that he has "been told Biluchís use the wood for gun-stocks." It is, however, subject to the attack of insects, and is accordingly chiefly used for charcoal and fuel. TIMBER. Fuel. 890

Domestic Uses.—**Dr. Bellew** mentions that in the Peshawar valley the wood is often made into charms to keep off the evil eye from man and beast, and **Dr. Cleghorn** states that its bark is used for sandals (*Stewart, Pb. Pl., 209*). DOMESTIC. Charms. 891 Sandals. 892

Celtis cinnamomea, *Lindl.; Kurz, For. Fl. Burm., II., 472.* 893

Syn.—C. DYSODOXYLON, *Thw.*

Vern.—*Gúrenda*, SING.

References.—*Gamble, Man. Timb., 343; Thw., En. Ceylon Pl., 267; Trimen, Cat. Ceylon Pl., 83; Dymock, Mat. Med. W. Ind., 748.*

Habitat.—An evergreen tree, frequent in the forests of the Eastern Peninsula, from Assam and Chittagong to Pegu and Martaban; also common in Ceylon and the Malayan islands.

Medicine.—A light-brown wood, sold in India under the name *Narakya-úd* (or Hell's Incense), is used as a charm against evil spirits. This was described by **Dr. W. Dymock** in the 1st edition of his *Materia Medica of Western India* under its vernacular name. The writer's attention having been drawn to this, a correspondence was instituted. **Dr. Dymock** stated that the Bombay supply came from Ceylon. A reference was then made to **Dr. H. Trimen**, to which the following reply was received: "I send you a dried specimen of **Celtis cinnamomea** and a piece of the wood. This is the *Gúrenda* of the Sinhalese, a name which implies the odour of the wood. I should be glad if you would satisfy yourself that it is really the wood sold in Indian bazaars. I cannot get any trace of its being exported from Ceylon; indeed, such export is denied here by persons who would be likely to know." The wood furnished was of a pale grey colour and had nothing like the degree of odour of the wood sold in India. MEDICINE. Wood. 894

In his 2nd edition of the *Materia Medica of Western India*, **Dr. Dymock** adds additional information regarding the wood and under the name of **Celtis dysodoxylon.** It appears to be known to the Tamil-speaking people as *pudacarpan.* **Thumberg** says of it: "The tree was called by the Dutch *strunthout*, and by the Singalese *urenne*, on account of its disgusting odour, which resides specially in the thick stem and the larger branches. The smell of it so perfectly resembles that of human ordure, that one cannot perceive the smallest difference between them. When the tree is rasped and the raspings are sprinkled with water, the stench is quite intolerable. It is nevertheless taken internally by the Singalese as an efficacious remedy. When scraped fine and mixed with lemon juice it is taken internally as a purifier of the blood in itch and other cutaneous eruptions, the body being at the same time anointed with it externally."

MEDICINE. Price. 895 Dr. Dymock states: "The peculiar odour is probably due to the presence of *napthylamine.* The price of the wood in Bombay is R30 per candy of $7\frac{1}{2}$ cwts. The Portuguese call it *Pao de merda* and *Pao Sujo.*"

It has thus still to be proved that the *Narakya-úd* is derived from **Celtis cinnamomea**, but should this be found correct, it is probable India may get its supplies from Assam or Burma, or perhaps from the Malayan Peninsula, instead of from Ceylon. The various opinions given above have been here recorded as a basis of further investigation, since the Indian trade in the wood is of some importance.

896 **Celtis eriocarpa,** *Dcne.; DC. Prodr., XVII., 179.*

Vern.—*Akata, katáia,* HIND.; *Batkar, bat tamanku,* PB.; *Tagha,* AFG.

References.—*Brandis, For. Fl., 429; Gamble, Man. Timb., 343; Baden Powell, Pb. Pr., 574; Balfour, Cyclop.*

Habitat.—A moderate-sized, deciduous tree, found in the Suliman and Salt Ranges, from 2,000 to 3,000 feet, and distributed along the Himálaya from the Indus to Nepal, ascending to 4,500 feet.

DOMESTIC. **Domestic Uses.**—The bark is used for making shoes (*Baden Powell*).

897 **C. orientalis,** *Linn.* See **Sponia orientalis,** *Planch.*

898 **C. Roxburghii,** *Planch.; Brandis, For. Fl., 429.*

Syn.—C. TRINERVIA, *Roxb., Fl. Ind., Ed. C.B.C., 262.*

Vern.—*Kharak, batkar, brúmaj, brúndu,* PB.; *Cheri chara, kathúniár,* C. P.; *Bowmaj,* BOMB.

References.—*Bedd., Fl. Sylv., CCCXII.; Gamble, Man. Timb., 343; Dalz. & Gibs., Bomb. Fl., 273; Lisboa, U. Pl. Bomb., 131.*

Habitat.—A glabrous, evergreen tree, common in the forests of South India, the Central Provinces, the Panjáb, Bengal, and Burma; also in the Kumaon Himálaya. Uses similar to those of the preceding species.

TIMBER. 899 **Structure of the Wood.**—Grey, hard, and close-grained. The Pathans use the wood for churn-sticks.

900 **C. tetranda,** *Roxb.; DC. Prodr., XVII., 179.*

EUROPEAN MYRTLE TREE.

Vern.—*Adona* (?), HIND.; *Kúmsúm, sungsúm,* LEPCHA; *Haktapatia,* ASS.; *Tagho,* SIND.; *Takkum,* AFG.

References.—*Roxb., Fl. Ind., Ed. C.B.C., 262; Brandis, For. Fl., 429; Kurz, For. Fl. Burm., II., 472; Gamble, Man. Timb., 344; Murray, Pl. and Drugs, Sind, 145.*

Habitat.—A tall tree of the outer Himálaya, from Kumaon eastward, to the Ava Hills in Burma; also on the Western Ghâts.

TIMBER. 901 **Structure of the Wood.**—Greyish white, moderately hard. Used in Assam for planking and canoes.

C. trinervia, *Roxb.* See **C. Roxburghii,** *Planch.*

902 **C. Wightii,** *Planch.; DC. Prodr., XVII., 184; Wight, Ic., t. 1969.*

Syn.—SOLENOSTIGMA WIGHTII, *Bl.; Kurz, For. Fl. Burm., II., 471.*

Vern.—*Vella-thorasay,* TAM.; *Tella-káká-mushti,* TEL.

References.—*Gamble, Man. Timb., 343; Thwaites, En. Ceylon Pl., 267; Balfour, Cyclop.*

Habitat.—A small evergreen tree of the mountains of South India and the Andaman Islands; is also met with in the hot dry parts of Ceylon.

TIMBER. 903 **Structure of the Wood.**—Greyish white, very hard, close-grained. Weight 53 ℔ per cubic foot. Annual rings indistinctly marked by a narrow belt without pores (*Gamble*).

CEMENTS. 904

CIMENTS, *Fr.;* CAMENTE, KITTE, *Ger.*

The term "Cement" is applied to a class of substances used for uniting two bodies, and which ultimately harden and bind them together. The following classification of these substances from *Spons' Encyclopædia* may be here given: (*a*) Calcareous cements, (*b*) Gelatinous cements, (*c*) Glutinous cements, (*d*) Resinous cementing compounds, and (*e*) Non-resinous cementing compounds. Interesting information regarding the Cements of India will also be found in *Balfour's Cyclopædia of India.* See also *Baden Powell's Panjáb Products.*

Calcareous. 905
(*a*) CALCAREOUS CEMENTS.—These are of mineral origin, and are limited in number. The mixture of lime and sand is an important cement of this class which is commonly known as mortar. (See **Carbonate of Lime.**) There are also a few called *hydraulic* cements, such as *Portland* cement, which have the property of setting or becoming hard under water. "Common lime does not possess this property; but limestones containing from 10 to 25 per cent. of alumina, magnesia, and silica, yield a lime, on burning, which does not slake when moistened with water, but forms a mortar with it, which hardens in a few days when covered by water." (*Page.*) "Portland cement is now made in Calcutta from argillaceous *kankar,* to which a fat limestone is added in the proper relation with the argillaceous constituents. Hitherto this fat limestone has been obtained in Calcutta, at a cheap rate, as it is brought out as ballast." (*Ball, Econ. Geology;* see also *Professional Papers on Indian Engineering,* published at Rurki; *Indian Economist, Vol. II., p. 276, &c.*) Cocoa-nut water and also jaggery is in India often mixed with lime in making polishing cements. (See **Cocoa-nut Juice under Cocos nucifera.**)

Gelatinous. 906
(*b*) GELATINOUS CEMENTS.—These have their origin in the substance known as "gelatine" obtained by boiling animal tissues in water. It is separated from water by simple evaporation, when it is converted into a dry hard substance called by different names, such as "glue," "size," "isinglass," &c., according to the sources from which they are derived. Of these, "glue" and "size" are employed as cements, and in India a strong and useful glue, made from cartilage obtained from fish, is used by every jeweller and gold-leaf beater.

Glutinous. 907
(*c*) GLUTINOUS CEMENTS.—The base of this class of cements is a substance containing a large proportion of gluten, such as the flour of rice, wheat, &c., which is commonly known as "paste." Special rices are grown in Burma for this purpose and largely exported. (*See* **Rice.**) A strong cement in common use in India is made from the gluten of rice mixed with a small quantity of pure lime.

Resinous. 908
(*d*) RESINOUS CEMENTING COMPOUNDS.—The cementing properties of this class of substances are due to the presence of resin, gum-resin, or gum, such as common rosin, india-rubber, gutta-percha, gum-arabic, &c. The following are a few of the Indian plants which are known to afford substances used as cements:—

Adenanthera pavonina (seeds).
Ægle Marmelos (glutinous and tenacious matter).
Artocarpus hirsuta (juice).
A. integrifolia (juice).
Balsamodendron Roxburghii (gum-resin).
Bauhinia retusa (gum).
Borassus flabelliformis (juice).
Cratæva religiosa (fruit).
Dichopsis elliptica (gum).
Euphorbia Cattimandoo (milky juice).
E. Royleana (juice).
Feronia Elephantum (gum).
Tamarindus indica (seeds).
Typha angustifolia (down of the ripe fruit).

Resinous. The resin from the *Sál*, **Shorea robusta,** is employed by the Santals to repair metal cooking-pots.

See also the list of plants under **India-rubber** and **Gutta-percha.**

Non-resinous 909 (*e*) NON-RESINOUS CEMENTING COMPOUNDS.—The cements under this class are too numerous to be mentioned here. The reader is referred to the list given in *Spons' Encyclopædia*, pp. 626-627.

CENCHRUS, *Linn.; Gen. Pl., III., 1105.*

Cenchrus catharticus, *Del.; Duthie, Fodder Grasses, 15;* GRAMINEÆ.

Syn.—C. ECHINATUS, *Rich.*

Vern.—*Bhurt*, HIND.; *Dhaman, argana*, N.-W. P.; *Basla, leá, lapta, bhort*, PB.; *Bharbhunt*, JEYPORE; *Bharout*, AJMIR; *Kukar*, BANDA.

References.—*Stewart, Pb. Pl., 252; Aitchison, Cat. Pb. Pl., 163; Murray, Pl. and Drugs, Sind, 10, 13; Duthie, List of Grasses, N.-W. P., 9.*

Habitat.—This grass is met with in arid ground in the plains of the North-West Provinces and of the Panjáb.

FODDER. 910 **Fodder.**—Eaten when young by cattle in the hot weather; nutritious shoots are given out during the hottest season (*Crooke quoted by Duthie*). By some it is considered excellent fodder, by others only middling. The seeds are eaten in times of scarcity (*Stewart*).

911 **C. montanus,** *Nees.*

This fodder grass is known as the *anjan* and *dháman* in the Panjáb, and is considered by some one of the most nutritious of grasses and makes good hay.

912 CENTAUREA, *Linn.; Gen. Pl., II., 477.*

Centaurea Behen, *Linn.;* COMPOSITÆ.

THE WHITE BEHEN OR WHITE RHAPONTIC.

Vern.—*Bahman safaid, suffaid bahman*, HIND., BOMB.; *Behen (or bahman) abiad*, ARAB.; *Bahman-i-suffaid*, PERS.

References.—*Dymock, Mat. Med. W. Ind., 379; S. Arjun, Bomb. Drugs, 77; Baden Powell, Pb. Pr., 355; Birdwood, Bomb. Prod., 49; Balfour, Cyclop.; Hanbury, Sc. Papers, 290.*

Habitat.—A native of the Euphrates Valley. The root is largely imported into India, reaching Bombay from the Persian Gulf. It is always to be found in native druggists' shops.

CENTIPEDA, *Lour.; Gen. Pl., II., 430.*

913 Centipeda orbicularis, *Lour.; Fl. Br. Ind., III., 317; Wight, Ic., [t. 1610;* COMPOSITÆ.

Syn.—ARTEMISIA STERNUTATORIA, *Roxb., Fl. Ind., Ed. C.B.C., 600.*

Vern.—*Nakk-chikní, nagdowana, pachittie*, HIND., BENG., and BOMB.; *Mechitta*, BENG.; *Nákacinkaní*, MAR.; *Afkar*, ARAB., SIND.

References.—*Dymock, Mat. Med. W. Ind., 362; Murray, Pl. and Drugs, Sind, 184; S. Arjun, Bomb. Drugs, 75; Birdwood, Bomb. Prod., 47.*

Habitat.—A common plant throughout the plains of India and Ceylon in moist places, appearing in fields during the latter part of the cold season. It forms procumbent, densely-branched tufts.

MEDICINE. Seeds. 914 **Medicine.**—"The minute SEEDS are used as a sternutatory by the Hindús, also the powdered herb. The plant does not grow in this part of India, but the dry herb, both entire and in powder, is always to be obtained in the druggists' shops." (*Dymock, Mat. Med. W. Ind.*) "The pow-

Leaves. 915 dered LEAVES are used in affections of the head, such as colds, &c., as sternutatory. Boiled to a paste and applied to the cheeks, it is employed in the cure of tooth-ache" (*Murray*). MEDICINE.

Special Opinions.—§ "*Nak-chikni*, sulphur, vinegar, and the leaves called *chitta*, mixed together, are used for pityriasis versicolor" (*Surgeon-Major C. W. Calthrop, Morar*). "It is used for hemicrania" (*Surgeon-Major J. Robb, Ahmedabad*).

CEPHAELIS, *Swartz.; Gen. Pl., II., 127.*

Cephaelis Ipecacuanha, *Rich.; Fl. Br. Ind., III., 178; Bot. Mag., [t. 4063;* RUBIACEÆ. 916

IPECACUANHA ROOT, *Eng.;* RACINE D'IPÉCACUANHA ANNELÉE, *Fr.;* BRECHWURZEL, *Germ.*

Syn.—C. EMETICA, *Pers.;* CALLICOCCA IPECACUANHA, *Brot.;* IPECACUANHA OFFICINALIS, *Arruda.*

References.—*Kurz, For. Fl. Burm., II., 5; Gamble, Man. Timb., 219; Pharm. Ind., 115; Flück. & Hanb., Pharmacog., 370; Ainslie, Mat. Ind., 543; O'Shaughnessy, Beng. Dispens., 379; Year-Book of Pharm., 1873, 233; Atkinson, Him. Dist., 884; Balfour, Cyclop.; Hanbury, Sc. Papers, 343, 423; Kew Cat., 82; Kew Reports for 1877, 1882; Jour. Ag. Hort. Soc., Vol. V., p. 47.*

Habitat.—A native of Brazil, introduced into India and Burma, being cultivated at the Government Cinchona plantations with scanty success. There are two wild members of the genus, however, met with in Malacca.

CULTIVATION OF IPECACUANHA IN INDIA.—"To the late Dr. Anderson, Superintendent of the Royal Botanic Gardens near Calcutta, must be attributed the honour of having conceived the possibility of the cultivation in India of this most valuable drug. An interesting sketch of the early efforts in this direction is given in the following passages. The importance in India of ipecacuanha as a remedy for dysentery, and the increasing costliness of the drug, have occasioned active measures to be taken for attempting its cultivation in that country. Though known for several years as a denizen of botanical gardens, the ipecacuanha plant has always been rare, owing to its slow growth and the difficulty attending its propagation. CULTIVATION.

"With regard to the acclimatisation of the plant in India, much difficulty has been encountered, and successful results are still problematical. The first plant was taken to Calcutta by Dr. King in 1866, and by 1868 had been increased to 9; but in 1870-71 it was reported that, notwithstanding every care, the plants could not be made to thrive. Three plants, which had been sent to the Rungbi plantation in 1868, grew rather better; and by adopting the method of root propagation, they were increased by August 1871 to 300. Three consignments of plants, numbering in all 370, were received from Scotland in 1871-72, besides a smaller number from the Royal Gardens, Kew. From these various collections, the propagation has been so extensive, that on 31st March 1873, there were 6,719 young plants in Sikkim, in addition to about 500 in Calcutta, and much more in 1874.

"The ipecacuanha plant in India has been tried under a variety of conditions as regards sun and shade, but thus far with only a moderate amount of success. The best results are those that have been obtained at Rungbi, 3,000 feet above the sea, where the plants, placed in glazed frames, were reported, in May 1873, as in the most healthy condition" (*Pharmacographia, 372*).

Dr. King reported to the Director of the Royal Botanic Gardens, Kew, in 1877, that he had distributed plants from the Calcutta Botanic Garden to Ceylon, Singapore, Burma, and the Andaman Islands, and also stated

CULTIVATION.

that "the peculiarly slow growth of this plant tends to prevent the cultivation of it from being taken up with spirit by European planters. The insignificant struggling appearance of the plant is, besides, little calculated to excite enthusiasm, or even interest, among the planting community." **Mr. Cantley** reported from Singapore, in 1882, that the ipecacuanha plants grown in partial shade under some trees were transplanted into pots, and the change was found to be highly beneficial to their vigorous growth (*Kew Reports for 1877, 1882*).

In communication with **Messrs. P. Lawson and Son** of Edinburgh, **Dr. Anderson** arranged for the propagation of seedlings, and in 1870-71 had a few experimental plants sent to India. Some of these were cultivated in the Calcutta gardens and the others sent to Madras. Of the latter **Colonel Beddome** early reported that the higher regions of the Nilgiri hills were not found to be suitable. About this stage the Bombay Government became anxious that a consignment of plants should be furnished to that Presidency for cultivation at the Cinchona plantations at Mahábaleshwar. The first definite consignment of **Messrs. Lawson's** seedlings was entrusted to **Mr. W. Walton** of the Cotton Department, Bombay. The Wardian case, under the care of that gentleman, contained 12 seedlings, all of which **Dr. King**, in 1871, reported as having arrived in Calcutta in a healthy condition. These were sent to Darjeeling, one plant having died on the journey. Shortly after, several other Wardian cases, containing seedlings, were received at Calcutta, both from **Messrs. Lawson** and from the late **Professor Balfour**, Superintendent of the Edinburgh Botanic Gardens.

From the extensive official correspondence and reports which the writer has been permitted to peruse, it would appear that the process of acclimatisation has been attended with a certain amount of success. As early as 1874, it was reported there were at the Rungbi plantation near Darjeeling 63,292 plants. These were mostly, however, small root-cuttings, and **Dr. King** (*Journal, Agri-Horti. Soc., 1874, Vol., V. p. 47*) wrote of them: "The recent success in propagating has been entirely due to the discovery that this plant, unlike most others, can be propagated freely by root-cuttings, while from the slowness of the plant's growth, materials for stem-cuttings are yielded very sparingly. Propagation has all along been carried on in glass-covered frames and at an elevation of about 3,000 feet above the sea. Our efforts have naturally been confined hitherto to increasing the number of plants, so as to get a sufficiently large stock for experiment, with the view of determining the conditions under which Ipecacuanha can be grown as a crop. The work has been carried on by the Cinchona establishment, and very little, if any, special expenditure has been incurred on its account.

"When this experiment in acclimatization was first begun, very little was known regarding the plant and the conditions required for its growth. We have now learnt from experience, that it is a humble creeping undershrub, of peculiarly slow growth, that it apparently requires a thoroughly tropical climate, by which I mean a pretty equal day and night temperature, the absence of a decided cold season and an atmosphere pretty steadily and thoroughly saturated with moisture. We have proved that it cannot stand exposure to a hot sun, and that it is apparently impatient of stagnant moisture at its roots. We do not as yet know what sort of soil best promotes the development of the root (the medicinal part), but experiments are now going on with the view of settling this point.

"As already stated, what remains to be done is to find out how to grow Ipecacuanha profitably as a crop. As a first step towards this, patches of plants have been put out at different elevations and under different

CULTIVATION.

conditions as to soil, moisture, and shade. We have not even now a sufficiency of large enough plants to do this on a large scale, for it must be remembered that the great majority of the plants above returned are still tiny things, under two inches high, and which, with their slow rate of growth, will not be much more than double that height a year hence.

"In conclusion I would remark that no part of Sikkim has a tropical climate, that of the bottom of the lowest valleys being no more than sub-tropical. It may, therefore, be found necessary to afford the plants, during the cold season at any rate, some cheap and rough kind of shelter."

In an official communication (dated February 1888), **Dr. King** says: "It does not appear to me that the cultivation of Ipecacuanha in India is a matter of very great importance. Fears were freely expressed, some twenty years ago, that the supply of the drug from South America would fail, and that the price would rise in consequence. These fears have, however, fortunately not been realized, and the drug is now obtainable at pretty much the same price as twenty years ago."

In *South India* cultivation seems more hopeful than in Sikkim. The late **Mr. McIvor**, in May 1870, planted a few Ipecacuanha plants in the Botanic Gardens at Barliyár. These succeeded fairly well, but in 1881-82, **Mr. Lawson**, the present Superintendent of the Botanic Gardens, reported that he did not think the plant could be there grown as an article of commerce. Later on, he seems to have attained more confidence in the possibility of its successful cultivation, and his reports are accordingly more hopeful. In 1885-86, he reported having 200 plants in a healthy condition, and states he had seen a bed of Ipecacuanha "growing very vigorously in the teak forest at Nilambur, the climate of which seemed to suit it much better than that of Barliyár." The last account gives the plants in the Government plantations of South India as having increased to 700.

In the official communication from **Dr. King**, to which reference has been made above, that gentleman says of the South Indian experiments: "During the year 1878, I had an opportunity of seeing some plants that were being cultivated at the Government teak plantations at Nillambore in the Madras Presidency. These were very healthy indeed, and I pressed on the overseer in charge of them the advisability of growing Ipecacuanha there on a large scale. The matter has, I believe, been lately taken up by **Dr. George Bidie, C.I.E.**, Surgeon General of Madras. Some good results from the impetus given to the cultivation in Madras by that officer may, therefore, I think, shortly be forthcoming." **Dr. King** (in the letter already quoted) says: "The growth is so very slow, and the protection required in the cold season is so considerable, that I found I could not produce the drug in any quantity at the usual market rate (from 4 to 5 shillings per pound), at which it can be bought in London."

In an official communication dated May 1887 **Dr. Bidie** writes hopefully. "It appears to me," he says, "that the time has arrived for India to cultivate her own supplies of the precious root as in the case of cinchona, so as to be no longer dependent on Brazil, from which unforeseen circumstances, such as war or epidemic, may at any time interrupt the supply."

PROPAGATION.

Propagation.—The discovery of the ease with which the plant may be raised by root-cuttings has already been alluded to. That Ipecacuanha can be grown in India has been shown, but with the exception of the locality in South India mentioned above, so far no other district has been shown to afford the hope that it can become an important commercial product. There are doubtless, however, many other similar regions where it might be grown. The plant grows slowly, and has little in it to attract the attention of the cultivator, so that it may be doubted *when* private enterprise may be prepared to relieve the Government of its present

PROPAGATION.

efforts. **Dr. King**, in his paper read before the Agri-Horticultural Society, indicates clearly the peculiarities and necessities of the plant, and in his more recent communication (the official papers referred to above) he reiterates more strongly the same opinions. "There can be no doubt that the occurrence of a distinctly marked cold season is disadvantageous to the growth of Ipecacuanha. I sent plants of it for trial to the Andaman Islands and Singapore, both being localities where there is no cold season. But at neither place has the cultivation been much of a success. I had an opportunity of seeing, in the Singapore Garden, during the year 1879, the Ipecacuanha plants which I had sent from Calcutta, a year or two previously. And contrary to my expectations, I found them growing very indifferently. The plants sent to the Andamans I have never seen, but I understand that they did not come to much."

Large numbers of plants have been freely distributed to private cultivators, but it may be concluded that it still remains to be demonstrated whether or not the medicinal properties are preserved in the Indian cultivated stock. These may improve as in the case of some of the Cinchonas, but on the other hand, they may decline, so that it must be concluded Ipecacuanha in India is even now but in its most early experimental stage.

MEDICINE. Root. 917

Medicine.—In large medicinal doses Ipecacuanha is emetic; in small doses expectorant, diaphoretic, and alterative; and in intermediate doses nauseant. The powdered Ipecacuanha, in the form of ointment, acts as a counter-irritant. In dysentery it is by modern use regarded as a specific. "The treatment of this disease by large doses of Ipecacuanha (grs. xxx to grs. lx), of late years re-introduced, has been found most effectual. In diarrhœa, and in some forms of dyspepsia, especially when connected with functional derangement or torpidity of the liver, it acts beneficially. As an expectorant it is in common use in catarrhs, chronic bronchitis, asthma, phthisis, the early stages of hooping-cough, &c. In hœmorrhages, especially in uterine hœmorrhages and in menorrhagia, it has proved an effectual remedy. For removing crude and indigestible matter from the stomach, Ipecacuanha acts with certainty and safety as an emetic, without inducing nearly the same amount of subsequent depression that follows tartar emetic; it is especially adapted for childhood and for persons of a delicate constitution. As a counter-irritant (2 drs. of powdered Ipecacuanha incorporated with 2 drs. of olive oil and 4 drs. of lard, rubbed into the skin for a few minutes, once or twice daily), it has been advantageously used in hydrocephalus, chronic chest affections, &c. The powder, moistened and made into a paste, applied locally to the stings of venomous insects, is said often to allay the pain and irritation in a remarkable manner. As an emetic, from 15 to 30 grains, its action being promoted by warm diluents. As a nauseant and expectorant, from 1 to 3 grains. As an alterative, from a quarter to half a grain" (*Pharm. Ind.*, *115*).

"The re-introduction of Ipecacuanha in large doses in the treatment of acute dysentery is due to **Mr. E. J. Docker** (*Lancet, July and August, 1858*). Much valuable information on this mode of treatment, drawn mainly from facts supplied by the medical officers of the Madras army, is furnished by **Dr. W. R. Cornish** (*Madras Quart. Med. Journ., Jan. 1861, p. 41*). A full *résumé* of the subject has been published by **Dr. J. Ewart** (*Indian Ann. of Med. Sc., 1863, Vol. VIII., p. 396*). The results, on the whole, have been of the most satisfactory description" (*Pharm. Ind.*, *454*).

CHEMISTRY. 918

Chemical Composition.—"The peculiar principles of Ipecacuanha are *Emetine* and *Ipecacuanhic Acid*, together with a minute proportion of a fœtid volatile oil. The activity of the drug appears to be due solely to the alkaloid, which, taken internally, is a potent emetic.

CHEMISTRY.

"Emetine, discovered in 1817 by **Pelletier** and **Magendie**, is a bitter substance with distinct alkaline reaction, amorphous in the free state as well as in most of its salts; we have succeeded in preparing a crystallized hydrochlorate.

"The root yields of the alkaloid less than 1 per cent.; the numerous higher estimates that have been given relate to impure emetine, or have been arrived at by some defective methods of analysis.

"The formula assigned to emetine by **Reich** (1863) was $C^{20} H^{30} N^2 O^5$, that given by **Glénard** (1875) $C^{15} H^{22} NO^2$, and lastly that found in 1877 by **Lefort** and **F. Würtz**, $C^{28} H^{40} N^2 O^5$.

"The alkaloid may be obtained by drying the powdered bark of the root with a little milk of lime, and exhausting the mixture with boiling chloroform, petroleum-benzin, or ether. It is a white powder, turning brown on exposure to light, and softening at 70°C. Emetine assumes an intense and permanent yellow colour with solution of chlorinated lime and a little acetic acid, as shown by **Power** (1877). A solution containing but $\frac{1}{1000}$ of emetine still displays that reaction. We found that alkaloid to be destitute of rotatory power, at least in the chloroform solution.

"The above reactions may be easily shown thus: Take 10 grains of powdered Ipecacuanha, and mix them with 3 grains of quicklime and a few drops of water. Dry the mixture in the water bath and transfer it to a vial containing 2 fluid drachms of chloroform: agitate frequently, then filter into a capsule, containing a minute quantity of acetic acid, and allow the chloroform to evaporate. Two drops of water now added will afford a nearly colourless solution of emetine, which, placed in a watch-glass, will readily give amorphous precipitates upon addition of a saturated solution of nitrate of potassium, or of tannic acid, or of a solution of mercuric iodide in iodide of potassium. To the nitrate **Power's** test may be further applied.

"If the *wood*, separated as exactly as possible from the bark, is used, and the experiment performed in the same way, the solution will reveal only traces of emetine. By addition of nitrate of potassium, no precipitate is then produced, but tannic acid or the potassico-mercuric iodate affords a slight turbidity. This experiment confirms the observation that the bark is the seat of the alkaloid, as might, indeed, be inferred from the fact that the wood is nearly tasteless.

Ipecacuanhic acid, regarded by **Pelletier** as gallic acid, but recognised in 1850 as a peculair substance by **Willigk**, is reddish brown, amorphous, bitter, and very hygroscopic. It is related to caffetannic and kinic acids; **Reich** has shown it to be a glucoside.

"Ipecacuanha contains also, according to **Reich**, small proportions of resin, fat, albumin, and fermentable and crystallizable sugar; also gum and a large quantity of pectin. The bark yielded about 30 per cent., and the wood more than 7 per cent., of starch" (*Pharmacographia, p. 374*).

Special Opinions.—§ "Applied locally to bites of venomous insects and scorpions" (*Surgeon-Major C. W. Calthrop, Morar*). "With out-door patients suffering from dysentery, Ipecacuanha in large doses was found unsuited and inconvenient. The following formula in such cases was used with much benefit: Ipecacuanha 5 grains, opium ¼ or ½ grain in one pill, and given every third or fourth hour; when for a malarious origin, quinine one grain to each pill was added" (*Honorary Surgeon Peter Anderson, Madras Presidency*). "In 3-gr. doses it is a most efficient calmative and sedative in delirium tremens" (*Surgeon-Major W. Farquhar, Ootacamund*).

CEPHALANDRA, *Schrad.; Gen. Pl., I., 827.*

919 **Cephalandra indica,** *Naud.; Fl. Br. Ind., II., 621; Wight, Ill., t. 105;* CUCURBITACEÆ.

Syn.—COCCINIA INDICA, *W. & A.;* MOMORDICA MONADELPHA, *Roxb.*

Vern.—*Bhimb, kanduri-ki-bél,* or *kandurí,* HIND.; *Kandúrí, ghol, kúndrú,* PB.; *Telá-kúchá, bimbu,* BENG.; *Golarú, kandúrí,* SIND; *Ghobe, gluru, galédu,* GUJ.; *Tendli, rántondla, tenduli, bhimb,* BOMB.; *Zidadi, tendli, tondali, bimbi,* MAR.; *Kóvai, kwe, kwai,* TAM.; *Donda, bimbiká, kâki-donda, kai-donda,* TEL.; *Kwel, gwel, kóva,* MALA.; *Tonde-balli,* KAN.; *Bimba* (or *vimba*), *bimbika,* SANS.; *Kabare-hindí,* ARAB. & PERS.; *Ken-bung, tsa-tha-khwa,* BURM.; *Kóvaká,* SING.

References.—*Roxb., Fl. Ind., Ed. C.B.C., 696; Thwaites, En. Ceylon Pl., 128; Dalz. & Gibs., Bomb. Fl., 103; Stewart, Pb. Pl., 96; Aitchison, Cat. Pb. Pl., 64; U. C. Dutt, Mat. Med. Hind., 171; Dymock, Mat. Med. W. Ind., 2nd Ed., 351; Moodeen Sheriff, Supp. Pharm. Ind., 110; Murray, Pl. and Drugs, Sind, 41; Atkinson, Him. Dist., 701; Lisboa, U. Pl. Bomb., 159; Balfour, Cyclop.; Treasury of Botany, 304.*

Habitat.—Common throughout India, often cultivated.

MEDICINE. Juice. 920 **Medicine.**—"The expressed JUICE of the thick tap-root of this plant is used by the leading native *Kavirájas* as an adjunct to the metallic preparations prescribed by them in diabetes." "The expressed juice is directed to be taken in doses of one *tola* along with a pill, every morning." (*U. C. Dutt, Mat. Med. Hind.*) The ROOT, according to **Moodeen Sheriff,** is sold as a substitute for *kabar* (**Capparis spinosa** root) in the bazars of Southern India. The leaves are of a deep green colour, and are useful as a colouring agent in preparing Savine ointment from

Root. 921 the essential oil. "The ROOT when cut exudes a somewhat sticky juice, which hardens into a reddish gum on drying, and is very astringent, but not bitter like the fruit" (*Dymock*). "The bark of the root, dried and reduced to powder, is said to act as a good cathartic, in a dose of 30

Leaves. 922 grains" (*Medical Topography of Dacca, 58*). "The LEAVES, mixed with *ghí,* are applied as a liniment to sores. The whole plant, bruised and mixed with the oil of **Euphorbia neriifolia** and powdered cummin seeds, is administered by natives in special diseases" (*Atkinson*). "The leaves are applied externally in eruptions of the skin, and the plant internally in gonorrhœa" (*Balfour*). "In the Concan the green fruit is chewed to cure sores on the tongue" (*Dymock*).

FOOD. Fruit. 923 **Food.**—"The oblong FRUIT, about 2 to 2½ inches long, green when young, scarlet-red when ripe, fleshy, smooth, is eaten both raw and cooked. The ripe fruit is sweet" (*Lisboa*). The fruit is one of the commonest of native vegetables (*Dymock*). It is eaten fresh when ripe and cooked in curries when green (*Roxb.*)

924 **Cephalocroton indicum,** *Beddome, 261;* EUPHORBIACEÆ.

A common tree in the moist forests of South India (altitude 1,500 to 4,000 feet); yields a timber useful for building purposes.

CEPHALOSTACHYUM, *Munro; Gen. Pl., III., 1213.*

(See Vol. I., B 69, No. 9.)

925 **Cephalostachyum capitatum,** *Munro;* GRAMINEÆ.

Vern.—*Gobia, gopi,* NEPAL; *Payong,* LEPCHA; *Silli, sullea,* KHASIA.

Reference.—*Gamble, Man. Timb., 429.*

Habitat.—Found in Sikkim and the Khásia Hills.

Food.—This semi-scandent and often gregarious bamboo, on flowering, produces a rice-like GRAIN, which is eaten by the natives in times of scarcity. The leaves are good for fodder. FOOD. Grain. 926

Structure of the Wood.—The stems are 12 to 30 feet long, strong, with internodes about 2½ feet, thin, yellow, used for bows and arrows by the Lepchas. It flowered in Sikkim in 1874 (*Gamble*). TIMBER. 927

Cephalostachyum latifolium, *Munro.* 928

Reference.—*Gamble, Man. Timb., 429.*

Habitat.—A species with large leaves, found in Bhután.

C. pallidum, *Munro; Kurz, For. Fl. Burm., II., 563.* 929

Vern.—*Beti,* Ass.

Reference.—*Gamble, Man. Timb., 429.*

Habitat.—A bamboo with shrubby stems. It grows in the Mishmi Hills and in Ava.

C. pergracile, *Munro; Brandis, For. Fl., 567.* 930

Vern.—*Tin-wa, kengwa,* BURM.

References.—*Kurz, For. Fl. Burm., II., 564; Gamble, Man. Timb., 429.*

Habitat.—A bamboo common in upper mixed forests of Burma; often gregarious. It has stems often 40 to 50 feet long.

CERA.

Cera alba and flava. 931

WAX (which see for further information; as also **Honey**).

Vern.—*Móm,* PERS., HIND., DEC., BENG.; *Sinth,* KASHMIR; *Mín,* GUJ.; *Ména,* MAR., KAN.; *Shama,* ARAB.; *Mozhukku, méllugú,* TAM.; *Mainam, minum,* TEL.; *Mezhuka, lelin,* MALA.; *Madhujam, siktha,* SANS.; *Itti, miettie,* SING.; *Phayouii,* BURM.

References.—*Pharm. Ind., 278; Moodeen Sheriff, Supp. Pharm. Ind., 97; Ainslie, Mat. Ind., I., 470; Bidie, Cat. Raw Prod., Paris Exb., 64; Baden Powell, Pb. Prod., 160; Fleming, Med. Pl. and Drugs, Asiatic Res., XI., 195.*

Description.—"The prepared Honeycomb. Occurs in masses; firm, breaking with a granular fracture; yellow, having an agreeable honey-like odour. Not unctuous to the touch; does not melt under 140°; yields nothing to cold rectified spirit, but is entirely soluble in oil of turpentine. Boiling water in which it has been agitated, when cooled, is not rendered blue by iodine. Yellow Wax, bleached by exposure to moisture, air, and light. Occurs in circular cakes, hard, nearly white, translucent. It is not unctuous to the touch, and does not melt under 150° F." (*Pharm. Ind.*)

Medicine.—"Honey is emollient and slightly laxative, and is often employed as a flavouring agent in cough mixtures and gargles. Wax is also emollient and demulcent, and has occasionally been prescribed in dysentery, diarrhœa, and catarrh; but its chief use is as an ingredient in ointments, plasters, and suppositories. Dose: of White Wax from ten to twenty grains suspended in a mixture by aid of mucilage." (*Pharm. Ind.*) For further information see **Bees,** also **Wax.** MEDICINE. 932

Special Opinions.—§ "The oil is used as a liniment, and is of great value in muscular and chronic rheumatism" (*Surgeon-Major A. S. G. Jayakar, Muskat, Arabia*).

Ceramic Manufactures, *see* **Earthen-ware.**

Cerasus cornuta, *Wall.*, see **Prunus Padus,** *Linn.*

CERATONIA, *Linn.; Gen. Pl., I., 574.*

933 **Ceratonia Siliqua,** *Linn.; DC. Prodr., II., 486;* LEGUMINOSÆ.

THE LOCUST-TREE; THE CAROB TREE; ST. JOHN'S BEAN, OR BREAD OR LOCUST BEAN; ALGAROBA *of Spain;* CARRUBIO, *It.;* CARUBA, *Ger.*

Vern.—*Kharnúb, kharnúb núbti* (the pods), PB.; *Kharnúb shámi* or *khirnúb nubti,* ARAB.

References.—*Roxb., Fl. Ind., Ed. C.B.C., 361; Brandis, For. Fl., 166; Gamble, Man. Timb., 133, 145; Dalz. & Gibs., Bomb. Fl. Suppl., 28; Stewart, Pb. Pl., 62; Aitchison, Cat. Pb. Pl., 52; DC., Origin of Cult. Pl., 334; Voigt, Hort. Sub. Cal., 246; Ainslie, Mat. Ind., I., 364; S. Arjun, Bomb. Drugs, 47; Baden Powell, Pb. Pr., 342; Atkinson, Him. Dist., 885; Birdwood, Bomb. Prod., 28; Lisboa, U. Pl. Bomb., 154; Balfour, Cyclop.; Morton, Cyclop. Agri.; Kew Cat., 50; Treasury of Botany; Smith's Dict. Econ. Pl., 93; Church, Food-Grains of India; Duthie in Agri.-Hort. Soc. Ind. Journal, VI. Vol., New Series, 98.*

Habitat.—A slow-growing, evergreen tree, indigenous in Spain and Algeria, the eastern part of the Mediterranean region, and in Syria; now almost naturalised in the Salt Range and other parts of the Panjáb.

CULTIVATION. **Cultivation.**—"The carob grew wild in the Levant, probably on the southern coast of Anatolia and in Syria, perhaps also in Cyrenaica. Its cultivation began within historic times. The Greeks diffused it in Greece and Italy; but it was afterwards more highly esteemed by the Arabs, who propagated it as far as Morocco and Spain. In all these countries the tree has become naturalised here and there in a less productive form, which it is needful to graft to obtain good fruit. The carob has not been found in the tufa and quaternary deposits of Southern Europe. It is the only one of its kind in the genus CERATONIA, which is somewhat exceptional among the LEGUMINOSÆ especially in Europe. Nothing shows that it existed in the ancient tertiary or quaternary flora of the south-west of Europe" (*DeCandolle's Orig. of Cult. Pl., 337*).

Mr. J. E. O'Conor published in 1876 a very exhaustive paper on the subject of the Carob tree. This gives an abstract of all that is known on the subject, while at the same time it deals fully with the efforts which have in India been made to introduce the plant.

The experimental cultivation has been carried on in most provinces, but chiefly in the North-West Provinces, the Panjáb, and Madras.

934 *In the North-West Provinces* it was first introduced by **Dr. Royle** in 1840, and again "was introduced by **Dr. Jameson** from Malta in 1861, and by 1863 it was extensively propagated and distributed in the Dún. The trees, though they flourish well, do not seem to give pods in such quantities as they yield in Malta and Italy. In 1866 the same report was received, and in 1880 it was decided to try to improve the quality of the pods by grafting, which, in Italy, not only produces better fruit, but gives a yield in a much shorter space of time. The trees appear to be unaffected by any extremes of temperature or excessive moisture" (*Atkinson, Him. Dist., 885*). **Mr. G. Ricketts** of Allahabad made experiments at Benares and Cawnpore, and found that the tree grew extremely well in the latter district. **Mr. Duthie** is doubtful of the extent to which the Carob is likely to be able to stand "the soaking condition of the ground during the rainy season."

CULTIVATION. 935

In the Panjáb, considerable quantities of seed have been sown from as early as 1844, in the districts of Panipat, Gurgaon, Rohtak, and Delhi, with little or no success. In 1862 some of the seed imported by **Mr. George Ricketts** were tried at Lahore and Ferozepore. The tree was found to thrive, "though it does not grow rapidly, and does not yet ripen its seed, or indeed produce pods, except in rare instances. One or two female trees existed in one of the Lahore gardens, and were cut down by the owner, Vandal-like, probably because he did not care to be bothered by questions from the Agri.-Horticultural Society as to their progress" (*Stewart, Pb. Pl.,* 63). **Mr. Ricketts** was of opinion that the seeds should be well soaked before planting, and the trees when planted out should not be too far from each other to ensure their fruiting.

936 *In Madras,* the experiments were made in various localities, but the general result was anything but satisfactory. The seeds did not germinate in some cases, and in others, the seedlings soon died off.

937 *In Bombay and Sind.*—"During the last two years, District Forest Officers in the Bombay Presidency have been engaged in carrying out experiments with carob seed, but the results do not appear to have been very promising. In Sind the Conservator states that all the plants were protected by mats from the frost during the cold season, and adds that when once these plants have established themselves in the soil, they should be able to exist without artificial irrigation or protection; at present they are too small, and it would be premature to express an opinion as to their flourishing in Sind or not. The Superintendent of the Economic Garden at Haidarabad, Sind, also states that, though the plant will grow, the slowness of growth will prevent its being of much use except as an ornamental shrub in gardens. The reports from other stations in all parts of the Presidency are of a similar nature. At some stations the plants have died, and at others the growth is very slow. In the Government gardens at Poona there are two trees about fifteen years old. The peculiarity of this tree is that it has the sexes on separate individuals. From the female tree in the Poona gardens about 71lb of fairly good fruit were obtained in May last year, and the crop would have been heavier if protected from parrots" (*Indian Daily News, 1883*).

938 *In Oudh,* the tree did remarkably well at Lucknow. **Dr. Bonavia** reported that some of the trees attained a height from 18 to 20 feet and were in a very healthy condition. **Mr. Duthie** recommends the tree should be planted on well-drained soils.

The Lower Provinces of Bengal are, according to **Dr. King**, unsuitable for the cultivation of Carob, although experiments in Hazaribagh were reported on favourably. It was there discovered that the germination was facilitated by carefully peeling off a portion of the seed-coat.

The North-Western Provinces, Panjáb, and Oudh are recommended as the best localities for the purpose; but it must be admitted that on the whole the efforts to introduce the tree into India have not been successful.

MEDICINE. Pods. 939

Medicine.—Mr. Baden Powell says that the pods are used by the natives in coughs attended with much expectoration. They are said by **Ainslie** to be viewed by the Arabians as cold, dry, and astringent. The husk of the pods has been considered as antacid, purgative, pectoral, and astringent. The author of the *Makhzan-ul-Adwiya* alludes to them as medicinal.

FOOD. Pods. 940

Food.—The pods, full of sweet, nutritious pulp, are a common article of food in the Mediterranean for man, horses, pigs, and cattle, and are imported into the Panjáb under the name of *Kharnub-nubti* (*Brandis*). They form an important constituent in the patent cattle-foods. They are supposed to be the "husks" of the Prodigal son, and the "Locusts" of John the Baptist.

In the *Treasury of Botany* occurs the following account of Carob pods as a food-stuff: "These pods contain a large quantity of agreeably-flavoured, mucilaginous, and saccharine matter, and are commonly employed in the south of Europe for feeding horses, mules, pigs, &c., and occasionally, in times of scarcity, for human food. During the last few years, considerable quantities of them have been imported into England and used for feeding cattle; but although they form an agreeable article of food, they do not possess much real nutritive property, the saccharine matter belonging to the class of foods termed carbonaceous or heat-givers, the seeds alone possessing nitrogenous or flesh-forming materials, and these are so small and hard that they are apt to escape mastication. They form one of the ingredients in the much-vaunted cattle-foods at present so extensively advertised, the green tint of these foods arising from this admixture. Some years ago they were sold by chemists at a high price, and were used by singers, who imagined that they softened and cleared the voice. By fermentation and distillation, they yield a spirit which retains the agreeable flavour of the pod." **Professor Church** in *Food-Grains of India* (p. 170) states that "The nutrient ratio is here about 1 : 8·5, and the nutrient value 68. As sugar, pectose, gum, &c., occupy the place of starch in these pods, the starch-equivalent cannot be calculated in the ordinary way, for the sugar, &c., are of less nutrient worth than starch, containing for a given weight less carbon." "The tree flourishes in a dry and poor soil." In Cyprus, where the tree grows luxuriantly several varieties are distinguished; the pods of the best kinds are less astringent than those of the wild sort.

TIMBER. 941 **Structure of the Wood.**—Hard, heavy, excellent as fuel, and valued for cabinet-work. (*Brandis.*)

DOMESTIC. Seeds. 942 **Domestic Uses.**—"The SEEDS are said to have been the original carat weight of jewellers" (*Notes, Dept. Rev., Agri. and Commr., 1871-1879*). This seems, however, an error, as there is little to lead to the conclusion that the Asiatic plant **Abrus precatorius** afforded the *ratti* seeds.

CERBERA, *Linn.; Gen. Pl., II., 699.*

Cerbera Manghas, *Linn.*, see **Tabernæmontana dichotom,** *Roxb.;* [APOCYNACEÆ.

943 **C. Odollam,** *Gærtn.; Fl. Br. Ind., III., 638; Wight, Ic., t. 441.*

Syn.—C. LACTARIA, *Ham.;* TANGHINIA ODOLLAM, LACTARIA, and LAURIFOLIA, *Don.*

Vern.—*Dabúr, dhakur,* BENG.; *Kada má, kat-arali, kadaralai, kadu,* TAM.; *Odallam,* MALA.; *Gon-kaduru,* SING.; *Ka-lwah,* BURM.

References.—*Roxb., Fl. Ind., Ed. C.B.C., 232; Brandis, For. Fl., 322; Kurz, For. Fl. Burm., II., 171; Gamble, Man. Timb., 262; Thwaites, En. Ceylon Pl., 192; Dalz. & Gibs., Bomb. Fl. Supp., 53; Voigt, Hort. Sub. Cal., 531; Pharm. Ind., 139; Dymock, Mat. Med. W. Ind., 2nd Ed., 508; Cooke, Oils and Oil-seeds, 36; Lisboa, U. Pl. Bomb., 99; Balfour, Cyclop.; Treasury of Botany; Kew Cat., 96; Rheede, Hort. Mal., t. 39.*

Habitat.—A small tree of the salt swamps, or of the coasts of India, Ceylon, and Burma; common in the South Konkan.

FIBRE. Bark. 944 **Fibre.**—A fibre prepared from the BARK is said to have been sent by the Forest Department of Madras to the Amsterdam Exhibition of 1883. (See *T. N. Mukharji's Amsterdam Exhib. Descriptive List.*)

OIL. Seeds. 945 **Oil.**—The SEEDS yield an oil which is used for burning, and by the Burmese to anoint the head. (*Kurz, Gamble, &c.*)

MEDICINE. Sap. 946 Leaves. 947 **Medicine.**—"Emetic and purgative properties are assigned to the MILKY SAP and to the LEAVES; but their use is to be condemned, as the number of safe and efficient medicines of both classes is quite large enough, and there is reason for believing that this tree, even in moderate quantities, is possessed of poisonous properties." (*Pharm. Ind.*) DeVry has separated from the seeds a crystalline poisonous glucoside similar to that obtained from **Thevetia**, which see. The NUT is narcotic and poisonous. The GREEN FRUIT is employed to kill dogs (*Balfour*). The BARK is purgative. MEDICINE. Nut. 948 Fruit. 949 Bark. 950

Special Opinions.—§ "The kernel of the fruit is an irritant poison, producing, when taken internally, vomiting and purging, soon followed by collapse and death" (*Surgeon-Major J. M. Houston, Travancore: John Gomes, Esq., Medical Storekeeper, Trevandrum*).

Structure of the Wood.—Grey, very soft, spongy. Annual rings marked by a sharp line; weight, 21℔ per cubic foot. It is only occasionally used for firewood. TIMBER. 951

Domestic Uses.—The poisonous JUICE of the fruits was formerly used in Madagascar as an ordeal in cases of suspected crime or apostacy (*Kew Cat., 96*). DOMESTIC. Ordeal Nut. 952

Cerbera Thevetia, *Linn.*, see **Thevetia neriifolia,** *Juss.*

CEREALS. 953

The term "Cereal" is applied to all edible grains obtained from the family of grasses (GRAMINEÆ). The following are the principal cereals cultivated in India:—RICE, WHEAT, BARLEY, OATS, INDIAN-CORN, and the various species of MILLET. As these grains are treated separately, the reader is referred to their respective places in this work for detailed information. The term "Pulses" is applied to edible leguminous seeds, such as the pea, and "Other Grains" to all edible seeds that do not fall into Cereals or Pulses, such as buckwheat, amarantus, &c.

CEREVISIÆ FERMENTUM.

Cerevisiæ Fermentum. 954

YEAST PLANT OF TORULA CEREVISIÆ.

Reference.—*Pharm. Ind., 262.*

The history of yeast is replete with interest, even although many of the details of the action of the plant in the process of fermentation are unexplainable even at the present day. There is little doubt but that the discovery of the peculiar effect of yeast upon sugary liquids, in converting these into alcoholic beverages, has been known from antiquity, and that too by the most remote and diverse members of the human family. While the knowledge of the practical utility of this agent has therefore existed for ages as an established principle with the inhabitants of the world in the preparation of beverages, it is only within recent years that its action has been shown to be due to the growth within the malted liquid of a microscopic and fungoid plant, belonging to the genus **Torula**. It is 955
highly probable, indeed, that we shall discover that a good many other fungi besides **Torula** have this action, and in fact there seems abundant evidence that many other vegetable substances, not in the least related to the fungi, also exercise a like influence over sugary liquids. The action seems to be almost mechanical—simple contact of the fermentation agent with the sugary liquid. This must be viewed as a closely allied phenomenon to the effect of sulphuric acid on starch, contact converting the latter into sugar, while the acid itself remains unchanged in quantity or

CEREVISIÆ Fermentum.

The Yeast Plant

chemical nature. In the process of beer-brewing two manifestations of the same kind are met with. The grain from which the beverage is to be prepared is first moistened either with hot water or by being placed in a warm confined atmosphere. As the result, it sprouts or germinates. The chemistry of this action consists in the fact that in a warm moist atmosphere the simple contact of a substance known as *diastase* with the starch of the grain converts the latter into sugar. Diastase may be defined as a transformed condition of gluten produced within the seed during the first stage of germination, and no sooner is the diastase formed than it immediately commences to act upon the insoluble starch. This is a wise provision of nature. The embryo plant is imbedded in a mass of starch. The base of the embryo contains gluten, but both starch and gluten are insoluble, and cannot be transformed into the structure of the germ until rendered soluble. On the seed falling into the ground and on being subjected to moisture, it germinates or sprouts. A portion of the gluten degenerates into diastase, and the simple contact of this substance when thus formed causes the insoluble starch to pass through one or two transformations, until it is ultimately reduced to grape sugar. This new substance is rapidly absorbed, and for the first period of its existence the infant plant feeds upon the food stored up for it within the seed. It produces first a root and then a stem, and by the time the nourishment contained within the seed has been exhausted, the root has commenced to absorb food from the soil. In fermentation this curious property is taken advantage of. The grain is first germinated, and when by simple contact the resulting production of diastase has converted the starch of the grain (or malt as it is now called) into sugar, the germination is stopped by the malt being dried. After breaking the grain, the soluble and insoluble starch products are washed out of the husk with warm water, when the diastase completes its action on the still insoluble starch. It has been found that for every 100 parts of starch, in good malt, 1℔ of diastase is produced, but that quantity will suffice to convert the starch of 1,000℔ of grain into sugar. Hence has come into existence the practice of mixing, in the second stage of brewing, malted with unmalted grain. (Compare with the account of Vinegar given under **Acetum, Vol. I., A. 356.**) When the diastase has completely converted the starch into sugar, the liquid is heated to the boiling point, to kill, as it is called, the diastase, and the sweet liquid is now ready for the yeast. Before supplying this, the brewer filters the wort, for the boiling has not only killed the diastase, but has coagulated it, as also all the other albuminous matter, and by filtration the turbidity is removed.

956 The yeast is now applied and the liquid kept for five or six days at a fixed temperature. The fungus rapidly grows and multiplies. What nourishment these minute plants take has never been clearly established, but through their simple presence or contact with the sugar they cause that substance to break up into two new compounds—alcohol which remains in the liquid, and carbonic acid gas which escapes into the atmosphere. A curious fact has been recorded, that yeast propagated time after time on the same kind of beer loses its power; hence the practice prevails of fermenting one brew with yeast reared on another. The modern system of *Pasteurising* beer by heating it in carbonic acid gas is practised with beers fermented at low temperatures. These beers, containing no yeast, are clear, and are at the same time found to stand the climate of India in some respects better than the beers that used formerly to come to this country in such large quantities. The yeast is killed by the process of heating to 60°. In the brewing of beer only about a quarter of the fermentable substance is converted into alcohol, the remainder giving the

sweet flavour to the beverage. The yeast lives and increases in the fermenting liquid, but appears to abstract nothing from it; and just as contact of diastase has changed starch into sugar, so contact of yeast with sugar produces alcohol.

It has already been said that there would appear to be other substances which similarly produce fermentation. Through the kindness of **Mr. C. B. Clarke** the writer received from the Khásia Hills a small cake prepared from a fungus found growing on the flowering heads of what appears to be a cyperaceous plant (probably **Rhyncospora aurea**), and is used like yeast to produce fermentation. This is an exceedingly curious and important discovery, for there is perhaps no commodity more difficult to procure than bread, in the camp life which many Indian officers have to endure for months together. If these convenient cakes are found suitable for baking, the Khásia yeast may be put to a much more extended use than hitherto. While travelling in Mánipur in 1880, the writer discovered at a small village called Seugmoi a large native distillery. After some little trouble he was allowed to inspect this, and had every detail of the process of making a sort of rice-whiskey explained to him. The malting and preparing of the liquid while crude was similar to the European method, but he was greatly astonished on being shown flat white cakes used for fermentation. These were made from rice-flour and a powder prepared from the wood of an extensive climber, the ingredients being baked with a little water and sun-dried. Apparently the particles of the wood act the part of a ferment, or probably give birth in the malt to a species of fungus which does so. The writer was afterwards shown the climber which appeared to be a new species of **Cnestis**, and he provisionally named it **Cnestis potatorum.** The climber was not in flower, however, and he was unable to name it for certain, so that it may even prove a species of **Millettia** and not a **Cnestis** at all. 957

Throughout the hill tracts of India, beer and spirits of various kinds are prepared, and nearly always through the aid of some astringent plant which may either play the part of hops or be connected in some way with causing or facilitating fermentation. The substances chiefly used in this way are **Anamirta Cocculus** (fruits), **Acacia arabica, A. ferruginea,** and **A. leucophlœa** (the bark), the fruits of **Phyllanthus Emblica,** leaves and pods (*bhang*) of **Cannabis sativa,** and **Datura fastuosa** (the seeds burned on a charcoal fire, over which empty vessels are placed to get impregnated with the poisonous smoke before being filled with date-palm juice). In the last two instances the additions are made to make the beverage more intoxicating and poisonous, and there are doubtless many other substances used for this purpose. It seems probable, for example, that in Upper India the Darnel **(Lolium temulentum)** is used to render liquor intoxicating. The bark of **Ligustrum robustum** has the reputation of accelerating fermentation. In the Santal country *mahua* flowers are regularly fermented and distilled. The flowers are placed in earthen vessels and mixed up with a powder produced from the barks of the following trees: **Terminalia belerica, T. tomentosa, Phyllanthus Emblica, Anogeissus latifolia, Shorea robusta,** and the roots of common rice. After a time the *mahua* ferments and is distilled, but the distiller carefully preserves the earthen vessels for future use, having discovered that if not washed out these vessels will cause the *mahua* flowers to ferment without the aid of the astringent barks. **Rev. A. Campbell** informs the writer that the Santals use **Ruellia suffruticosa,** *Roxb.* (the *chaulia*), when they wish to prepare a pleasant beverage from rice, but add to this **Clerodendron serratum,** *Spreng.* (the *Saram lutur*), to make the beverage intoxicating. According to some authors, an alcoholic beverage is prepared from the juice of **Calotropis** 958

gigantea, but by others this plant is only used as an adjunct with alcoholic liquors. However used, this fact is interesting, since the plant is not far
959 removed botanically from **Sarcostemma,** the genus supposed by some to yield the *Soma* of the Sanskrit writers. The *Homa* of the Parsís is the dried twigs of **Ephedra vulgaris,** which to this day are used in the preparation of a sacred beverage. It seems highly probable that the *Soma* of the *Vedas* was an auxiliary used in the preparation of an alcoholic beverage, and may possibly have been the *Homa* of the Parsís, the jointed character of **Ephedra** stems answering considerably to the descriptions of the *Soma.*

Our knowledge of the subject of yeast and yeast substitutes is too imperfect to justify more than suggestive observations, and it may be shown that the substances indicated are after all only flavouring ingredients or at most auxiliaries to fermentation; but in that case the true yeast plant must spontaneously attack the sweet liquids, since, with the exception of the Khásia fungus alluded to, no other instance is known of fungi being directly used.

960 The Angami Nagá prepares his *zú* beer from rice. A log of wood 20 feet in length and 3 feet in thickness is hewn out into a large trough. This is placed in the centre of the village, constituting the communal brewery from which every man can take at will his allowance of *zú.* A large quantity of rice is placed in the trough, and over this hot water is poured. After a day or so, a further quantity of hot water is added, when on the third day the beer is ready. At this stage it is a milky liquid, having a bitter sweet and refreshing flavour. On the fourth day it becomes intoxicating. In most parts of India a spirit is distilled from rice (known as *Surá*), but the writer is not aware of any record having hitherto been published of a rice beer. He has on several occasions drawn attention to a remarkable similarity that exists in the names of plants and modes of using them between Japan and the Nagá Hills. The *zú* rice beer almost bears out this idea with the *sake* rice beer of Japan.

961 The visitor to Upper Sikkim has presented to him a pleasant drink of hot *marwá* beer as a token of friendship. This is sucked through a straw from a bamboo jug. The Angami drinks his *zú* from a similar natural tankard, but he also sups with a bamboo spoon the white sediment of rice-flour. Men with bamboo jugs drinking and supping the *zú* in the public square constitute a striking feature of the Angami village. An alcoholic beverage is in India prepared from most species of millet (*Paisht*) and from *mahua* flowers, from the fruits of **Eugenia Jambolana** and from the flower-heads of **Anthocephalus Cadamba,** in Sind from dates and in Afghanistan from raisins. But apparently wheat and barley are but rarely used for this purpose, the liquor from the former being called *Madulika* and from the latter *Kohala.*

962 In India the favourite beverages are prepared from the juices of trees, chiefly palms (*Váruni*), or from sugar-cane (*Sidhu*). For this purpose the juice is extracted from the cocoanut, the date, the palmyra, **Caryota urens,** and the *ním* tree. Fermentation is generally set up in these beverages by means of fermentation seed. This consists of rice saturated in a former fermentation, the grains of rice retaining apparently the germs of the yeast plant. Yeast from the *tari* beverage is largely used in India for baking bread.

For further information see "**Spirits,**" also "**Wines.**"

MEDICINE. 963 **Medicine.**—The ferment obtained in brewing beer is successfully used as a stimulant in the adynamic forms of fever and dysentery. It is chiefly used as a poultice. In India, where yeast is rarely procurable, the toddy (*tari*) poultice, in a great measure, answers the purpose. (*Pharm. Ind.; see also the fermentation seed of* **Borassus, B. 689.**)

CERIOPS, *Arn.; Gen. Pl., I., 679.*

Ceriops Candolleana, *Arnott; Fl. Br. Ind., II., 436; Wight, Ic., [t. 240;* RHIZOPHOREÆ. 964

THE MANGROVE.

Vern.—*Kirrari, kiri, chauri,* SIND; *Gorán,* BENG.; *Madá,* AND.

References.—*Brandis, For. Fl., 218; Kurz, For. Fl. Burm., I., 448; Beddome, Fl. Sylv. Anal., Pl., XIII., Fig. 5; Gamble, Man. Timb., 176; Thwaites, En. Ceylon Pl., 120; Aitchison, Cat. Pb. Pl., 59; Murray, Pl. and Drugs, Sind, 190.*

Habitat.—A small, evergreen tree, met with on the muddy shores and tidal creeks of India and the Andaman Islands. Common in Sind.

Dye.—The BARK is used for tanning. This and the next species are economically not distinguished, both being used under the name of *garán* or *gorán*. They are exceedingly valuable tans, imparting a good red colour to leather; they seem to deserve to be brought prominently to the notice of European tanners. They, no doubt, to a small extent, reach England under the name of Mangrove Bark. This, according to **Murray**, is said to be superior to oak, completing in six weeks an operation which, with the latter, would occupy at least six months. Sole leather so tanned is also reported to be more durable than any other. DYE. 965 TAN. 966

Medicine.—The whole PLANT abounds in an astringent principle. A decoction of the BARK is used to stop hæmorrhage, and is applied to malignant ulcers. On the African coast, a decoction of the SHOOTS is used as a substitute for quinine. MEDICINE. Plant. 967 Bark. 968 Shoots. 969

Structure of the Wood.—Red, hard; weight, 63℔ per cubic foot. Used in Sind for the knees of boats and other similar purposes; in Lower Bengal for houseposts and for firewood. TIMBER. 970

Domestic Uses.—The bark is used as a litter for cattle. Litter for cattle. 971

C. Roxburghiana, *Arnott; Fl. Br. Ind., II., 436.* 972

Vern.—*Garán* or *Ghorán,* BENG.; *Kabaing, kyabaing, ka-pyaing,* BURM.

References.—*Kurz, For. Fl. Burm., I., 448; Gamble, Man. Timb., 176; McCann, Dyes and Tans, Beng., 133, 158, 458.*

Habitat.—A large shrub of the coast of Chittagong, down to Tenasserim (*Kurz*).

Tan: Dye.—The BARK is used in tanning leather. This and the preceding species might be supplied to any extent and very cheaply; there seems a good future for *Garán* barks in tanning. They also yield a good colouring matter. In Balasore the *Garán* grows abundantly on the seashore; a good dye is prepared from the bark in that district, and is used to give a brown colour. It is supposed to strengthen ropes and boatmen's cloths (*McCann*). TAN. Bark. 973 DYE. Bark. 974

Structure of the Wood—Weight of the wood, 46℔ per cubic foot. TIMBER. 975

CERIUM.

This metal is used medicinally in India. Minerals supposed to contain it have been collected in the Karnal district, in Madras, and in Nepal (*See Ball's Econ. Geology*). 976

CEROPEGIA, *Linn.; Gen. Pl., II., 779.*

Ceropegia Arnottiana, *Wight; Fl. Br. Ind., IV., 74;* ASCLEPIADEÆ. 977

Vern.—*Uta-long,* BURM.

Reference.—*Balfour, Cyclop.*

Habitat.—Grows in Khásia Mountains, Burma, and Tenasserim.

978 **Ceropegia bulbosa,** *Roxb.*, var. **esculenta**; *Fl. Br. Ind., IV., 67; [Wight, Ic., t. 845.*

Vern.—*Khappar kadu,* HIND.; *Patalatum bari,* BOMB.

References.—*Roxb., Fl. Ind., Ed. C.B.C., 250; Dalz. & Gibs., Bomb. Fl., 153; Voigt, Hort. Sub. Cal., 534; Dymock, Mat. Med. W. Ind., 2nd Ed., 526; Lisboa, U. Pl. of Bomb., 165; Balfour, Cyclop.*

Habitat.—Met with in the Panjáb and in the Bombay Presidency.

FOOD. Tubers. 979 Leaves. 980 Roots. 981

Food.—TUBERS and LEAVES are used as pot-herbs in Multan and Sind. Shepherds are fond of eating the tubers, which they consider to be tonic and digestive. "Every part of this plant is eaten by the natives, either raw or stewed in their curries. The fresh ROOTS taste like a raw turnip" (*Roxburgh*).

982 **C. tuberosa,** *Roxb.; Fl. Br. Ind., IV., 70.*

Syn.—C. ACUMINATA, *Dalz. & Gibs., l.c., not of Roxb.*

Vern.—*Khapper-kadu,* BOMB.; *Pátál tumbdi,* MAR.; *Commú-madu,* TEL.

References.—*Roxb., Fl. Ind., Ed. C.B.C., 251; Dalz. & Gibs., Bomb. Fl., 153; Dymock, Mat. Med. W. Ind., 436; Murray, Pl. and Drugs, Sind, 162; S. Arjun, Bomb. Drugs, 85.*

Habitat.—Met with in the Deccan Peninsula from the Konkan southwards.

MEDICINE. Tubers. 983

Medicine.—"The starchy, somewhat bitter TUBERS, are used as a nutritive tonic in the bowel complaints of children." (*Dymock, Mat. Med. W. Ind.*) They are also eaten. It is probable the economic information given under **C. bulbosa** and this species has been confused or is equally applicable to both plants and perhaps to one or two other species such as **C. juncea** and **C. acuminata.**

Cetaceum, see **Physeter macrocephalus,** *Linn.*; MAMMALIÆ.

984 **Cervidæ,** the family of the deer, of interest economically for their antlers and their skins. See "**Horns**" and also "**Skins.**"

CETRARIA.

985 **Cetraria islandica,** *Achar.*; LICHENES.

ICELAND MOSS.

References.—*Pharm. Ind., 258; Flück. & Hanb., Pharmacog., 737; O'Shaughnessy, Beng. Dispens., 672.*

MEDICINE. 986

Medicine.—Imported into India and sold in chemists' shops.

Cevadilla or **Sabadilla,** see **Asagræa officinalis,** *Lindl.*; LILIACEÆ.

Ceylon Moss, see **Gracillaria (Plocaria) lichenoides,** *Greville*; ALGÆ.

CHÆTOCARPUS, *Thw.; Gen. Pl., III., 323.*

987 **Chætocarpus castaneæcarpus,** *Thw.; DC. Prodr., XV., 2, 1127;* [EUPHORBIACEÆ.

Vern.—*Búlkokra,* BENG.; *Palakuna, sadavaku,* TAM.; *Hedóka, hédawaka,* SING.

References.—*Kurz, For. Fl. Burm., II., 409; Gamble, Man. Timb., 366; Thwaites, En. Ceylon Pl., 275; Trimen, System. Cat., Ceylon Pl., 82.*

Habitat.—A moderate-sized tree, found in the Khásia Hills, Eastern Bengal, Burma, the Andaman Islands, and Ceylon.

Structure of the Wood.—Light-red, moderately hard, close-grained; weight 58lb per cubic foot; used in Ceylon for building. TIMBER. 988

CHAILLETIA, *DC.; Gen. Pl., I., 341.*

Chailletia gelonioides, *Hook.; Fl. Br. Ind., I., 570;* CHAILLETIACEÆ. 989

Syn.—MOACURRA GELONIOIDES, *Roxb., Fl. Ind., Ed. C.B.C., 264.*

Vern.—*Moakurra,* SILHET, BENG.; *Balu-nakuta,* SING.

References.—*Kurz, For. Fl. Burm., I., 230; Gamble, Man. Timb., 80; Bedd., Fl. Sylv., 59; Thwaites, En. Ceylon Pl., 79; Trimen, System. Cat. Ceylon Pl., 17; Dalz. & Gibs., Bomb. Fl., 52; Lisboa, U. Pl. Bomb., 47.*

Habitat.—A small subdiœcious tree, commonly met with in the hilly eastern parts of Bengal and Silhet, in the forests of Madras, and in the Western Peninsula on the Ghâts from the Konkan southwards; it is also met with in the moister parts of Ceylon up to an elevation of 3,000 feet.

Structure of the Wood.—This is one of the timber trees specially mentioned by **Dr. Lisboa** in his *Useful Plants of the Bombay Presidency,* but very little of a definite character can be learned regarding the value of the wood. TIMBER. 990

Chalcedony, see **Carnelian.**

Chalk, see **Carbonate of Lime.**

CHAMÆROPS. 991

Chamærops Ritchieana, *Griff.; Gen. Pl., III., 924;* see **Nannorhops Ritchieana;** PALMÆ.

Chamois Leather, see **Leather & Skins.**

Chamomile or **Camomile,** see **Matricaria Chamomilla,** *Linn.;* COMPOSITÆ.

Chánáy Kéléngu, see **Tacca pinnatifida** (?)

Chank shells, see **Shells** and also **Pearl Fisheries.**

CHARA.

Chara involucrata, *Roxb.; Fl. Ind., Ed. C.B.C., 648.* 992

Vern.—*Jangli pátá,* HIND.; *Jhanj,* BENG. (These vernacular names are applicable to all **Charas,** indeed to most submerged plants.)

Habitat.—There are a large number of species both of **Chara** and **Nitella** found in tanks and pools of water near Calcutta during the cold and hot season.

Domestic Uses.—"Used to purify water and clarify sugar." (*E. F. T. Atkinson.*) In Bengal **Hydrilla verticillata,** *L. C.* (**Vallisneria verticillata,** *Roxb.*), is used for this purpose, and the writer has never seen **Chara** so employed. **Roxburgh** says of **Hydrilla**: "The Barhampur sugar refiners use this herb, while moist, to cover the surface of their sugars, as clay is used in the West Indian islands, and in two or three days the operation is finished exceedingly well." DOMESTIC. Clarify sugar. 993

Charcoal, see **Carbon.**

994 **CHARCOAL, Timbers used for—**

Abies Smithiana.
Acacia arabica.
A. Catechu.
A. modesta.
Adhatoda Vasica (gunpowder).
Albizzia procera.
A. stipulata.
Anacardium occidentale.
Anogeissus latifolia.
Betula cylindrostachys.
Boswellia serrata.
Butea frondosa (gunpowder).
Cajanus indicus (gunpowder).
Callicarpa arborea.
Calotropis gigantea.
Casearia glomerata.
Cassia Fistula.
Castanopsis tribuloides.
Colebrookia oppositifolia (gunpowder).
Corchorus capsularis (gunpowder).
Cornus macrophylla (gunpowder).
Cynometra polyandra.
Daphne mucronata (gunpowder).
Dillenia indica.
D. pentagyna.
Echinocarpus dasycarpus.
Ehretia Wallichiana.
Elæocarpus lanceæfolius.
Eucalyptus Globulus.
Eugenia tetragona.
Euphorbia antiquorum.
Excæcaria Agallocha.
Ficus cordifolia.
F. infectoria.
F. religiosa.
Hippophæ rhamnoides.
Juniperus excelsa.
Lagerstrœmia parviflora.
Mangifera indica.
Mimosa rubicaulis (gunpowder).
Phyllanthus Emblica.
Pieris ovalifolia.
Pinus excelsa.
P. longifolia.
Premna latifolia.
Prosopis glandulosa.
P. spicigera.
Quercus Ilex.
Q. incana.
Q. semecarpifolia.
Q. spicata.
Rhododendron arboreum.
Salix tetrasperma (gunpowder).
Semecarpus Anacardium.
Sesbania ægyptiaca (gunpowder).
Sponia orientalis (gunpowder).
S. politoria (gunpowder).
Stereospermum suaveolens.
Tamarix articulata.
Terminalia myriocarpa.
T. tomentosa.
Xylosma longifolium.

995 **Dr. Schlich**, in a note (dated January 1883), regarding the supply of fuel for the Barwai iron-works (in Holkar's territory) near Nimar, estimated that the Punassa and Chandgarh reserves contain 44,384,000 maunds of fuel, chiefly of **Boswellia thurifera, Hardwickia binata, Anogeissus latifolia, Odina Wodier,** and **Pterocarpus marsupium** trees, which, with the exception of **Anogeissus** and **Boswellia**, are not specially mentioned by writers on the subject as being good for fuel. These trees may, however, be added to the above list. **Dr. Schlich**, in his note, estimated that to produce 15 tons of pig iron a day, 372,604 maunds of charcoal would be annually required, or say 1,800,000 maunds of firewood.

Chaulmugra, see **Gynocardia odorata,** *R. Br.*; BIXINEÆ.

Chavannesia esculenta, *A. DC.*, see **Urceola esculenta,** *Benth.*

Chavica Betle, *Miq.*, see **Piper Betle,** *Linn.*; PIPERACEÆ.

C. officinarum, *Miq.*, see **Piper officinarum,** *C. DC.*

C. Roxburghii, *Miq.*, see **Piper longum,** *Linn.*

Chay root, see **Oldenlandia umbellata,** *Linn.*; RUBIACEÆ.

Cheep, see Shells.

Cheeronjee (chíronjí or chíraulí) oil, see Buchanania latifolia, *Roxb.;* [ANACARDIACEÆ.

Cheese, see Ghí.

Cheilanthes tenuifolia, *Sw.;* FILICES. 996

Vern.—*Nanha, dodhari,* SANTAL.

The Reverend A. Campbell writes that the Santals prescribe a preparation from the roots of this fern for sickness attributed to witchcraft or the evil eye.

CHEIRANTHUS, *Linn.; Gen. Pl., I., 68.*

Cheiranthus Cheiri, *Linn.; Fl. Br. Ind., I., 132;* CRUCIFERÆ. 997

THE WALL-FLOWER.

Vern.—*Todri surkh,* HIND.; *Khueri,* BENG.; *Todri surukh, todri nafarmáni* (seeds), *todri siyáh,* PB. The vernacular name *todri* (or *towdri*) has been given to this plant by Stewart and Murray, but Dr. Dymock affirms that the parcels of *todri* seed imported from Persia contain corymbs of small pods, and cannot, therefore, be obtained from this plant. They are probably derived, as he suggests, from a species of Iberis, which see.

References.—*Stewart, Pb. Pl., 13; O'Shaughnessy, Beng. Dispens., 186; Irvine, Mat. Med., Patna, 114; Murray, Drugs and Pl., Sind, 49; S. Arjun, Bomb. Drugs, 11; Dymock, Mat Med. West. Ind., 2nd Ed., 56; Year-Book of Pharm., 1874, p. 622; Baden Powell, Pb. Pr., 327; Balfour, Cyclop.; Treasury of Botany.*

Habitat.—Cultivated in gardens in North India, but is not indigenous; known as "Viole gialle," or yellow violets.

Oil.—The FLOWERS are employed to make a medicated oil. For this purpose they are boiled in olive oil; this prepared oil is much used for enemata (*Year-Book of Pharmacy, 1874, 622*). OIL. Flowers. 998

Medicine.—FLOWERS said to be cardiac, emmenagogue, and aphrodisiac, employed in paralysis and impotence. "The dried PETALS are much used in Upper India as an aromatic stimulant" (*O'Shaughnessy*). "The SEED is also used as an aphrodisiac" (*Irvine*). MEDICINE. Flowers. 999 Petals. 1000 Seeds. 1001

Special Opinion.—§ "Used as an aphrodisiac" (*Surgeon J. Anderson, M.B., Bijnor*).

CHENOPODIUM, *Linn.; Gen. Pl., III., 51.* 1002

A genus of annual or perennial herbs, belonging to the Natural Order CHENOPODICEÆ (χήν, a goose, and πούς, a foot).

Erect or prostrate herbs. *Stem* angled. *Leaves* alternate, entire lobed or toothed. *Flowers* minute, 1-5 merous. *Ovary* free, depressed or compressed. *Styles* 2-3. *Seed* horizontal or vertical, *testa* crustaceous, *albumen* floury.

There are about 50 species of the genus, met with in the world. These are distributed in all climates. India possesses seven species, with perhaps numerous varieties and cultivated forms of most of these.

Chenopodium album, *Linn.; Fl. Br. Ind., V., 3;* CHENOPODIACEÆ. 1003

THE WHITE GOOSE-FOOT.

Syn.—C. VIRIDE, *Linn.; Roxb. Fl. Ind., II., 58.*

The White Goose-foot.

Vern.—*Bathú-sag* or *bethuá sák, chandan betú,* BENG. and HIND.; *Bathúa, báthú, jauság, lúnak,* PB. PLAINS; *Irr* (CHENAB VALLEY), and *Em* (LADAK), PB.; *Bethuwa, charái, jau-ság, bhutwa,* N.-W. P.; *Bhatua arak',* SANTAL, and *Khartua sag,* HIND. in SANTAL PERGANAS; *Chakwit,* BOMB.; *Jhil,* SIND; *Khuljeh ke baji,* DUK.; *Parupu kire,* TAM.; *Pappu kura,* TEL.; *Vastuk,* SANS.; *Kulf,* ARAB.

References.—*Roxb., Fl. Ind., Ed. C.B.C., 260; Stewart, Pb. Pl., 178; Aitchison, Cat. Pb. Pl., 126; Voigt, Hort. Sub. Cal., 321; U. C. Dutt, Mat. Med. Hind., 123; O'Shaughnessy, Beng. Dispens., 523; Murray, Drugs and Pl., Sind, 103; Baden Powell, Pb. Pr., 372; Lisboa, U. Pl. Bomb., 169; Atkinson, Him. Dist., 696, 708, 731; Buck, Dyes and Tans, N.-W. P., 9; Balfour, Cyclop.*

Habitat.—Common throughout the tropic and temperate Himálaya from Kashmír to Sikkim, ascending to 12,000 feet above the sea, and in Tibet to 14,000 feet. General in the plains of India from the Panjáb to Bengal, Western and Southern India. Wild and also cultivated.

There are various cultivated and wild forms of this plant. **Voigt** describes three of these: (α) **album** proper, *chandan betú* of Bengal; (β) **viride**, the *bettú shak*, entirely green: and (γ) **purpureum**, the *lál bethí*, a form with "the angles of the stem and branches of a fine purple colour: leaves and the mealy panicles somewhat reddish."

Slewart describes what appears to be a form of this plant as a **Chenopodium** which he was unable to identify. He gives the following vernacular names for it, and expresses the opinion that it is quite equal to **C. Quinoa**:—

Vern.—*Mustakh,* KASHMIR; *Gaddí siúngar, bajarí banj, ratta,* RAV.; *Siriári,* BIAS; *Bithú, báthú, tákú,* SUTLEJ; *Gniú,* LADAK, PB.

The leaves of this plant "are eaten as a pot-herb on the Sutlej, but the plant is chiefly cultivated for its grain, which is considered better than buck-wheat."

DYE. Plant. 1004

Dye.—A decoction of the PLANT is added to the indigo solution, to aid the fermentation process to which the dye is subjected before it is applied to cloth. This practice prevails at a certain town named Chibraman in the Farakhabad district (*Buck, Dyes and Tans, N.-W. P., 9*). Compare with the use of **Cassia Tora, C. 798.**

MEDICINE. 1005

Medicine.—Said to be used "in special diseases and as a laxative in spleen and bilious disorders" (*Atkinson*). It is also given "in bile and worms" (*Baden Powell*). **Dymock** (in *Mat. Med. W. Ind., 879*) remarks that the drug known as "*bazr-el-katif* (Arab.), *tukm-i-sarmak* (Pers.)," may be the seeds of "a kind of spinach: some say that it is the *bathúa* of Hindustan, which is **Chenopodium album.**" It is "deobstruent and diuretic."

Special Opinion.—§ "Considered laxative and recommended for use by Sanskrit writers in the form of pot-herb in piles" (*U. C. Dutt, Civil Medical Officer, Serampore*).

FOOD. Plant. 1006

Seeds. 1007

Food.—Cultivated by the Hill tribes on the higher western Himálaya, and occasionally in other parts of India. The wild plant is also regularly collected and eaten as a pot-herb and green vegetable. The SEED of the cultivated plant is the principal product, but the leaves and twigs are also eaten as a spinach. **Atkinson** (*Him. Districts, p. 697*) says "it is entirely a rain crop and attains a height of six feet. The seeds ripen in October." The plant is often injuriously present in the cold-weather crops of the plains.

Professor Church (*Food-Grains of India*) says the leaves of **C. album** "are rich in mineral matters, particularly in potash salts. They likewise contain a considerable amount of albumenoids and of other compounds of nitrogen." The seeds are said to be superior to buck-wheat.

Domestic Uses.—**Baden Powell** says that this plant is used in the Panjáb "to clean copper vessels preparatory for tinning them." DOMESTIC. 1008

Chenopodium ambrosioides, *Linn; Fl. Br. Ind., V., 4.* 1009

THE SWEET-PIGWEED; MEXICAN TEA.

Syn.—C. VALPINUM, *Wall;* AMBRINA AMBROSIOIDES.

Vern.—*Herba Santa Maria* in Brazil. In Chili this is known as *Culen.*

References.—*Dalz. and Gibs., Bomb. Fl. Suppl., 73; Bent. and Trim., Med. Pl., 216.*

Habitat.—An old world, widely-spread species, now introduced into America, common in many parts of India, such as Bengal (**Voigt** says it is completely domesticated about Serampore), Silhet, the Deccan, and Coimbatore, &c. It has a weaker and less offensive smell than **C. anthelminticum**, from which it may be distinguished by having its flowers in leafy racemes.

Medicine.—This is said to afford an essential oil to which the tonic and antispasmodic properties of the plant are attributed. It is commonly reported that this plant is used as a substitute for the officinal **C. anthelminticum**, having in a milder degree the anthelmintic properties of that plant. It is employed in pectoral complaints and enjoys the European reputation as a useful remedy in nervous affections, particularly chorea. Officinal preparation an infusion. MEDICINE. Oil. 1010

It is somewhat remarkable that the properties of this plant should be practically unknown to the people of India, although it is probable that the writers on the drugs of India have dealt collectively or generically with the species, and that they may be so used by the people of India, the various species not being distinguished.

Food.—This plant affords the Mexican tea. FOOD. 1011

C. Blitum, *Hook. f.; Fl. Br. Ind., V., 5.* 1012

Syn.—BLITUM VIRGATUM, *Linn.*

Vern.—*Súndar* (J.), *kúpald* (C.), PB.

References.—*Stewart, Pb. Pl., 177; Von Mueller, Extra-Tropical Plants.*

Habitat.—North-Western India: Kashmír, altitude 8,500 feet and Western Tibet at 12,000 to 14,000 feet. **Stewart** found the plant wild in the Jhelam, Chenáb, and Rávi basins and in the Trans-Indus at altitudes from 7,000 to 10,000 feet.

Dye.—According to **Von Mueller**, "the fruits furnish a red dye." DYE. 1013

Food.—**Stewart** remarks that "the extremely insipid FRUIT is sometimes mistaken by Europeans for a kind of strawberry, and which it much resembles. In Ladák the LEAVES are eaten as a pot-herb." FOOD. Fruit. 1014 Leaves. 1015

C. Botrys, *Linn.; Fl. Br. Ind., V., 4.* 1016

THE JERUSALEM OAK.

Syn.—C. ILICIFOLIUM, *Griff. Notul., IV., 337.*

References.—*Dalz. & Gibs., Bomb. Fl. Suppl., 73.*

Habitat.—Temperate Himálayas from Kashmír to Sikkim, at altitudes from 4,000 to 10,000 feet: Tibet 11,000 to 14,000 feet. **Stewart** says it occurs at Peshawar, and **Dalzell** that it was originally introduced into Bombay but has now gone wild. A weed of fields.

Medicine.—Reported to be used as a substitute for **C. anthelmenticum** and to possess the same properties as **C. ambrosioides.** According to *U. S. Dispensatory* it has been used in France with advantage in catarrh and humoral asthma. The officinal preparation is an oil. MEDICINE. Oil. 1017

1018 **Chenopodium murale,** *Linn.; Fl. Br. Ind., V., 4.*

Vern.—*Bátú, kúrúnd, kharatua,* PB.

References.—*Stewart, Pb. Pl., 178.*

Habitat.—General in many parts of India from the Panjáb to the Gangetic Valley, the Deccan, and South India.

FOOD. 1019 **Food.**—Used as a pot-herb in the Panjáb.

1020 **C. Quinoa,** an American species, has once or twice been tried in India, but apparently with little success (*See Church, Food Grains of India, p. 110*).

Cherry, see **Prunus Cerasus,** *Linn.*; ROSACEÆ.

Chestnut, Horse, see **Æsculus indica,** *Colebr.* **(A. 567),** and **Æ. Hippocastanum,** *Linn.* **(A. 573)**; SAPINDACEÆ.

Chestnut, Sweet, see **Castanea vulgaris,** *Lam.*; CUPULIFERÆ.

Chestnut, Water, see **Trapa bispinosa,** *Roxb.*, and **T. nutans,** *Linn.*; ONAGRACEÆ.

CHICKRASSIA, *A. Juss.; Gen. Pl., I., 339.*

1021 **Chickrassia tabularis,** *Adr. Juss.; Fl. Br. Ind., I., 568; Beddome, Fl. Sylvat., t. 9*; MELIACEÆ.

THE CHITTAGONG WOOD.

Syn.—SWIETENIA CHICKRASSIA, *Roxb., Fl. Ind., Ed. C.B.C., 370*; C. NIMMONII, *Grah.; Dalz. & Gibs., Bomb. Fl., 38.*

Vern.—*Chikrassi, pabba, dalmara,* BENG.; *Boga poma,* ASS.; *Pabha pubha,* BOMB.; *Pabba, palara, núl,* MAR.; *Aglay, agal, agle-marum eleutharay,* TAM.; *Madagari vembu, chittagong chettu, chittagong karru, cheta kum karra,* TEL.; *Dovedah,* MALA.; *Ganti malle,* SALEM; *Dalmara, lal, devdari,* KAN.; *Maín,* HYDERABAD; *Saiphra, sey barasi,* MAGH.; *Chegarasi,* CHAKMA; *Yimmah, yeng-ma, yimma, nga-bai, taw-yeng-ma, taw yimma, zimma,* BURM.; *Arrodah,* AND.; *Hulan-ghuk-gass, hulan-hick-gala, hulodi,* SING.

References.—*Brandis, For. Fl., 66; Kurz, For. Fl. Burm., I., 227; Gamble, Man. Timb., 76; Thwaites, En. Ceylon Pl., 61; Dalz. & Gibs., Bomb. Fl., 38; Voigt, Hort. Sub. Cal., 137; O'Shaughnessy, Beng. Dispens., 250; Drury, U. Pl., 131; Cooke, Gums and Gum-resins, 13; Atkinson, Him. Dist., 814; Birdwood, Bomb. Prod., 325; Lisboa, U. Pl. Bomb., 45; Balfour, Cyclop.; Treasury of Botany; Kew Cat., 29.*

Habitat.—A large tree, native of the hills of Eastern Bengal, South India, and Burma, and also found in the warmer parts of Ceylon.

GUM. 1022 **Gum.**—It yields a transparent, amber-coloured GUM, said to have been sent from Madura to the Indian Museum in 1873 (*Spons' Encycl.*) The gum "consists of irregular tears, amber-coloured or light brown, somewhat transparent, but not brittle" (*Cooke, Gums and Gum-resins, 13*).

DYE. Flowers. 1023 **Dye.**—The FLOWERS yield a red and a yellow dye.

MEDICINE. Bark. 1024 **Medicine.**—The BARK is powerfully astringent, though not bitter.

TIMBER. 1025 **Structure of the Wood.**—Heartwood hard, varying from yellowish brown to reddish brown, with a beautiful satin lustre; seasons and works well; sapwood of a lighter colour (*Gamble*). Weight from 40 to 52lb per cubic foot. The wood is used for furniture and for carving. "It is the Chittagong wood of commerce, and from its fresh cedar-like smell is called *lal* or *devdari* in Kanara. The wood is dark-coloured and close in the grain. It is used for every purpose, and is much valued" (*Bomb. Gaz., XV., 66*).

"The wood is well known in Madras and easily procured, and is extensively used in cabinet-making, coming under the denomination of

'Chittagong wood,' being imported from that district, though it is abundant in the mountainous parts of the peninsula. It is close-grained, light-coloured, and delicately veined, makes beautiful and light furniture, but is apt to warp during the season of hot land-winds. According to Dr. Gibson, it is a fine straight-growing tree, rather common in the southern jungles of the Bombay Presidency, but much less so in the northern. Its wood could easily be creosoted. It is valuable for cabinet and house purposes, and is used in the Madras Gun-Carriage Manufactory to make plane tables and for furniture work. It furnishes one of the deodars of Malabar. It is found also in Kanara and Sunda, in the tall jungles near and on the Ghâts, particularly at Gunesh Wood. Wood there whiter, but tough and close-grained; and, from its general situation, it is hardly known to the carpenter. It grows in the warmer parts of Ceylon." (*Balfour, Cyclop.*)

Chicory, see **Cichorium Intybus,** *Linn.*; COMPOSITÆ.

China Root, see **Smilax china,** *L.*; LILIACEÆ.

Chionanthus albidiflora, *Thw.*, see **Linociera albidiflora,** *Thw.*

C. zeylanica, *Linn.*, see **Linociera purpurea,** *Vahl.*; OLEACEÆ.

Chiréta, see **Swertia Chirata,** *Ham.*; GENTIANACEÆ.

Chloride of Ammonium, see **Ammonium chloride.**

Chloride of sodium, see **Sodium chloride.**

CHLORIS, *Sw.*; *Gen. Pl.*, *III.*, *1165*.

Chloris barbata, *Swartz*; *Duthie, Fodder Grasses, 53*; GRAMINEÆ. 1026

Syn.—ANDROPOGON BARBATUS, *Linn.*

Vern.—*Gandi, gavung, paluah, jargi, konda-pulla,* N.-W. P.; *Ganni, jharna,* PB.; *Phundi,* AJMIR; *Prenji,* MERWARA; *Chhinkri,* JEYPUR; *Bűrdiya, phulkia,* C. P.; *Botya jharo,* BERAR; *Konda-pulla,* SOUTH INDIA; *Mayil-kondai-pullu,* TAM.

References.—*Roxb., Fl. Ind., Ed. C.B.C., 111; Thwaites, En. Ceylon Pl., 371; Dalz. & Gibs., Bomb. Fl., 296; Aitchison, Cat. Pb. Pl., 167; Murray, Pl. & Drugs, Sind, 12; Bidie, Cat. Raw Prod., Paris Exh., 76.*

Habitat.—Very common in Northern India, Sind, and Ceylon; grows in large tufts on pasture ground, especially on sandy soils.

FODDER. 1027

Fodder.—Cattle eat it up to the time of flowering, after which they do not seem to touch it (*Roxburgh*; *Duthie*).

C. Roxburghiana, *Edgew.*; *Bâmna,* AJMIR; *Mathaniya,* LALITPUR; *Hika gadi, sala-kodam gadi,* C. P.; a grass, not uncommon in Northern India, is in Ajmir considered good fodder. 1028

C. tenella, *Roxb.*; *Kagya,* AJMIR, *Morbhaga,* UDAIPUR; a grass common in Rajputana, Bundelkhand, and Central Provinces, is also considered good fodder.

CHLOROPHYTUM, *Ker.*; *Gen. Pl.*, *III.*, *788*.

Chlorophytum breviscapum, *Dalz. in Kew Journ., II., 142,* [LILIACEÆ. 1029

Vern.—*Bimpól,* SING.

References.—*Dalz. & Gibs., Bomb. Fl., 252; Thwaites, En. Ceylon Pl., 339; Baker, Linn. Soc., XV., 321; Treasury of Botany, II., 1280.*

Habitat.—Frequent in the Malwan District, Bombay, in rocky situations. **C. Heynei,** *Baker*, a nearly allied species, met with in the southern and central parts of Ceylon, at no great elevation.

MEDICINE. Bulb. 1030 **Medicine.**—Used medicinally by the Singhalese (*Thwaites, En. Ceylon Pl., 339*). There are several other species of this genus met with in India, and it seems probable their medicinal properties have been overlooked. **C. tuberosum** is general throughout India, from Bombay to Prome, ascending the Himálaya to 3,000 feet in altitude. **C. nepalensis** occurs in the eastern sub-tropical Himálayas, while **C. arundinaceum** occurs on the sub-tropical Himálaya and on Parisnath in Behar, altitude 4,000 feet.

CHLOROXYLON, *DC.; Gen. Pl., I., 340.*

1031 **Chloroxylon Swietenia,** *DC.; Fl. Br. Ind., I., 569; Bedd., Fl. Sylvat., t. 11; Wight, Ic., t. 56;* MELIACEÆ.

THE INDIAN SATIN-WOOD.

Syn.—SWIETENIA CHLOROXYLON, *Roxb., Fl. Ind., Ed. C.B.C., 370.*

Vern.—*Dhoura, bhırra, girya,* HIND.; *Behru, biluga, bhayrú, bheyri,* URIYA; *Behra, girya, behru, bihri, bhirra, bihra,* C. P.; *Sengel sali,* KOL.; *Bharhúl,* KARWAR; *Bhira,* GOND; *Bhirwa* BAIGAS; *Hulda, billú, hardi, bheria,* BOMB.; *Halda, bheria,* MAR.; *Múdúdad, burús, purúsh-múdudad-marum, purus-burus, vummray, múdúda, vummaai-porasham, kodawah-porash, kodawah-porasham, vummay-maram, kodawa, purrh,* TAM.; *Billu, billuda, bilgu, biluga, billu-chettu, billa kora, billu kura, bhallú-chettu,* TEL.; *Mashudla,* KAN.; *Huragalu,* MYSORE; *Bilu,* KURNUL; *Búrútch gala, búrúta, burute* (*mutirai,* TAMIL, in Ceylon), *ma, burute,* SING.

References.—*Brandis, For. Fl., 74; Gamble, Man. Timb., 77; Thwaites, En. Ceylon Pl., 61; Dalz. & Gibs., Bomb. Fl., 39; Voigt, Hort. Sub. Cal., 137; Dymock, Mat. Med. W. Ind., 2nd Ed., 177; Drury, U. Pl., 131; Cooke, Gums and Gum-resins, 25, 115; Atkinson, Gums and Gum-resins, 34; Atkinson, Him. Dist., 814; Lisboa, U. Pl. Bomb., 46; Balfour, Cyclop.; Treasury of Botany; Kew Cat., 29.*

Habitat.—A moderate-sized, deciduous tree, found in Central and South India, and Ceylon. Common in the forests of the Konkan, Deccan, and Coromandel, flower in March.

GUM. 1032 **Gum.**—"Satin-wood gum was contributed by **Dr. Cleghorn** to the Madras Exhibition of 1855. The specimen in the collection from Salem (1873) referred to this source is in irregular, fractured, and agglutinated tears, very variable in size, brittle, with a shining resinous fracture, translucent, brown, somewhat resembling *bábúl* gum, but more brittle, soluble in water, tasteless or slightly sweet. Mucilage rather turbid, dark mahogany colour, with an odour as of fusil oil. It was a peculiar and remarkable phenomenon which the mucilage of this sample exhibited, in that its surface was in an hour or two covered by a thick pellicle of gum, the upper surface of which became quite dry, as if, by rapid evaporation of the water in which it was dissolved, it was returned to the solid state. Although this pellicle was broken up, it continued daily to re-form on the surface of the solution.

"Another sample in the reference collection is from Ceylon, paler in colour, and in definite, rounded, shining, amber-coloured tears" (*Cooke, Gums and Gum-resins, 25*).

DYE. 1033 **Dye.**—"Yields a yellow dye" (*C. P. Gaz., 103*).

OIL. 1034 **Oil.**—The tree yields a wood-oil (*Beddome*).

MEDICINE. Bark. 1035 **Medicine.**—"The astringent BARK is prescribed sometimes by Hindú physicians, but it is not in common use, nor is it to be met with in the shops" (*Dymock, Mat. Med. W. Ind., 144*). The LEAVES are applied to Leaves. 1036 wounds (*Beddome*). *Behra* is also used in rheumatism (*C. P. Gaz., 118*).

TIMBER. 1037 **Structure of the Wood.**—Very hard, yellowish brown, and close-grained; the inner part of a darker colour, with a beautiful satiny lustre, but there is no distinct heartwood; seasons well. Annual rings distinct. Weight, 56lb per cubic foot.

SATIN-WOOD.

It is durable and excellent for turning; used for agricultural implements, cart-building, furniture, and picture-frames. It is, however, very liable to warp and split if not well seasoned in the shade. In Madras it is prized for ploughs and oil-mills, and also used for naves of gun-carriage wheels; it is found to stand well under water. It has been tried as a substitute for boxwood in engraving, but has not been found suitable. It is imported into England for cabinet-work and the backs of brushes, **Dr. Dymock** says that the wood is oily and turns well, making nice stethoscopes, &c. **Balfour** says that the Peradenia bridge in Ceylon—a single arch of 205 feet—is made of this wood.

"Small quantities are taken to Europe, but the consumption there is very small, and the value of the wood for furniture is not known as much as it merits. The market is at present glutted with an over-supply, and the brokers, who were selling wood twelve to fifteen months ago at £20 a ton, cannot now get £6. In Ceylon, satin-wood is used for building, furniture, &c. Old satin trees are frequently hollow-hearted to a height of 8 to 10 feet from the ground, and these hollow logs fetch high prices as kotties or well pipes—*i.e.*, as a casing for surface wells. In the Batticaloa district and in other parts of the island no other wells are used. The *kotties* part of the satin-wood cut is exported to Madras, where it is used for furniture and general building purposes" (*Indian Forester, X., i. 38*).

Chocolate nut and **bean,** see **Theobroma Cacao,** *Linn.*; STERCULIACEÆ.

CHONEMORPHA, *Don*; *Gen. Pl., II., 720.*

Chonemorpha macrophylla, *G. Don*; *Fl. Br. Ind., III., 661*; [*Wight, Ic., t. 432*; APOCYNACEÆ. 1038

Syn.—ECHITES MACROPHYLLA, *Roxb., Fl. Ind., Ed. C.B.C., 246.*

Vern.—*Gar badero,* HIND.; *Yokchounrik,* LEPCHA; *Harki,* SYLHET.

References.—*Brandis, For. Fl., 328; Kurz, For. Fl. Burm., II, 187; Gamble, Man. Timb., 261; Dalz. & Gibs., Bomb. Fl., 146; Voigt, Hort. Sub. Cal., 523; Balfour, Cyclop.*

Habitat.—A large climber with milky sap, met with in North and East Bengal and Burma.

Gum.—Yields a kind of Caoutchouc, which see. GUM 1039

Medicine.—**Balfour** alludes to a plant (**C. malabarica**) "the leaves of which, rubbed up in rice water, are applied to carbuncles; and the roots used in fever with dried ginger and coriander seed." The *Flora of British India* alludes to that plant as a doubtful species. MEDICINE. 1040

Chowlí, or **Chaulí,** see **Vigna Catiang,** *Endl.*; LEGUMINOSÆ.

CHROMIUM AND CHROMITE. 1041

The metal Chromium occurs to a limited extent in India in the form of chrome ochre (chromite) in Salem in Madras and Spiti and Kashmír in the Panjáb Himálaya. It is the colouring principle of many minerals, such as the emerald, serpentine, olivine, &c. It is employed in the arts in the manufacture of pigments, the commonest and best known of which is the yellow chromate and the red bicarbonate of potash. For further information see *Ball's Econ. Geology, 332; Mallet, Mineralogy, 53; Balfour's Cycl., 717.*

CHRYSANTHEMUM, *Linn.*; *Gen. Pl., II., 424.* 1042

There are three wild species belonging to this genus met with in Western Thibet and one in upper Sikkim—all alpine in their character, never occurring below 9,000 feet. The Chrysanthemums of Indian pharmacy are the two garden species.

1043 **Chrysanthemum coronarium,** *Linn.; Fl. Br. Ind., III., 314; Bot. Mag., t. 1521;* COMPOSITÆ.
CHRYSANTHEMUM.

Syn.—C. ROXBURGHII, *Desf.;* PYRETHRUM INDICUM, *Roxb., Fl. Ind., Ed., C.B.C., 604;* MATRICARIA OLERACEA, *Ham. in Wall., Cat., 3229.*

Vern.—*Gúl-chíní,* HIND., DEC.; *Akur kurra, gúl dáudí,* HIND.; *Gúl-daudí,* BENG.; *Pitho garkah,* ASS.; *Zænil, bagaur,* PB.; *Kalzang,* LADAK; *Seoti,* BOMB.; *Tursiphal, gule-sewati,* MAR.; *Gúl dáudí,* GUZ.; *Shámantip-pú,* TAM.; *Chámanti,* TEL.; *Hale,* KAN.; *Shévantiká, chandra-mallika, seunti, swenti,* SANS.; *Gule-daudi,* PERS.; *Lawúlú-gas,* SING. *Gúl-chíní* is also applied to **Plumiera acutifolia,** *Poiret,* APOCYNACEÆ.

References.—*Dalz. & Gibs., Bomb. Fl. Supp., 48; Aitchison, Cat. Pb. Pl., 77; Pharm. Ind., 127; Moodeen Sheriff, Supp. Pharm. Ind., 99; Dymock, Mat. Med. W. Ind., 371; Murray, Pl. and Drugs, Sind, 183; S. Arjun, Bomb. Drugs, 79; Drury, U. Pl., 133; Balfour, Cyclop.*

Habitat.—A native of the Mediterranean region, only known in India under cultivation as an ornamental garden plant. There are several very distinct varieties, some large, others small flowered, and white, yellow, or orange coloured. The foliage also varies considerably, some forms having large and coarse, others small leaves. Two of the coarser forms seem almost naturalised in India, and to such an extent that **Roxburgh** viewed them as "natives of Bengal."

MEDICINE. Flowers. 1044 Root. 1045

Medicine.—"The FLOWERS are stated by **Dalzell** and **Gibson** to form a tolerable substitute for Chamomile for medicinal purposes. The ROOT, chewed, communicates the same tingling sensation to the tongue as pellitory, and might doubtless be used as a substitute for it. The people of the Deccan administer the plant, in conjunction with black pepper, in gonorrhœa (*Dr. Walker, Bombay Med. Phys. Trans., 1840, p. 71*)." (*Pharm. Ind.*)

"*Akur kurra* is a drug commonly used for toothache, and assigned by **Jameson** to **Spilanthes oleracea.**" (In *Flora of British India,* **S. Acmella,** *Linn.,* var. **oleracea,** *Clarke; Roxb., Fl. Ind., III., 410.*) "It is probably derived from different plants in different places. It is prescribed largely in infusion, in conjunction with the lesser galangal and ginger, by native practitioners; and by itself in European practice, for colic, hysterical affections, pain in the head, and lethargic complaints; also in typhus fever. In paralysis of the tongue it has been used as a local application with advantage; also in apoplexy, chronic ophthalmia, and rheumatic affections of the face. By the Persians it is considered discutient and attenuant, and according to **Celsus** it was an ingredient in the famous cataplasm which, in his time, was employed as a resolvent and for maturing pus; also as an agent for opening the mouths of wounds" (*Murray, Plants and Drugs of Sind*).

Garlands. 1046

Sacred Uses.—"The beautiful yellow fragrant flowers of this plant are made into garlands and offered at the shrines of *Vishnu* and *Siva*" (*Balfour*).

1047 **C. indicum,** *Linn.; Fl. Br. Ind., III., 314; Bot. Mag., t. 327, 2042, 2556.*
THE COMMON GARDEN CHRYSANTHEMUM OF INDIA.

Syn.—PYRETHRUM INDICUM, *DC. Prodr., VI., 62;* CHRYSANTHEMUM INDICUM, *Willd. in Roxb., Fl. Ind., Ed., C. B. C., 604.*

Vern.—*Gúl dáudí,* HIND., a name applied, according to **Roxburgh,** to all the varieties; *Gendi, bágáur* (*genda* is the Hindustani for **Tagetes erecta**), PB.; *Kalzang,* LADAK; *Chevati, akurkura,* BOMB.; *Shevati,* MAR.; *Akkara carum,* TAM.; *Chámunti,* TEL.

References.—*Roxb., Fl. Ind., Ed. C.B.C., 604; Clarke, Compositæ Ind., 146; Dalz. & Gibs., Bomb. Fl. Supp., 48; Stewart, Pb. Pl., 124; S. Arjun, Bomb. Drugs, 192; Baden Powell, Pb. Pr., 358; Birdwood, Bomb. Prod., 50.*

Habitat.—Commonly cultivated in Indian gardens, and is in fact only known in a garden state.

Medicine.—It would appear that this and the preceding plant are not distinguished from each other by the natives of India, and the vernacular names apply to both. In medicinal properties they, more or less, resemble each other. **Baden Powell** says of this species that it is considered by the natives heating and aperient, and useful in affections of the brain and calculus, and also to remove depression of spirits. **Drury** says the "natives of the Deccan administer the plant, in conjunction with black pepper, in gonorrhœa." MEDICINE. Flowers. 1048

Sacred Uses.—The flower-heads are sacred to *Vishnu* and *Siva.* Garlands. 1049

CHRYSOPHYLLUM, *Linn.; Gen. Pl., II., 653.*

Chrysophyllum Roxburghii, *G. Don; Fl. Br. Ind., III., 535; Bedd., Fl. Sylv., t. 236;* MELIACEÆ. 1050

THE STAR APPLE.

Syn.—C. ACUMINATUM, *Roxb., Fl. Ind., Ed. C.B.C., 201.*

Vern.—*Petakara,* BENG.; *Pithogarkh,* ASS.; *Hali, hali-maru,* KAN.; *Tarsi, tarsiphala,* BOMB.; *Tarsi,* MAR.; *Lawúlú,* SING; *Thankya, than-kya-pen, thagya,* BURM.

References.—*Kurz, For. Fl. Burm., II., 118; Gamble, Man. Timb., 242; Thwaites, En. Ceylon Pl., 174; Dalz. & Gibs., Bomb. Fl., 138; Voigt, Hort. Sub. Cal., 340; Lisboa, U. Pl. Bomb., 88; Balfour, Cyclop.*

Habitat.—An evergreen tree of Bengal, Burma, the Western Ghâts, and Ceylon.

Food.—FRUIT edible. **Roxburgh** says: "The fruit ripens in October, and is greedily eaten by the natives, though to me the taste is by no means agreeable: the pulp being almost insipid, and, though tolerably firm, uncommonly clammy, adhering to the lips or knife with great tenacity." FOOD. Fruit. 1051

Structure of the Wood.—White, close-grained, moderately hard; pores small, in short radial lines between the numerous, very fine, medullary rays. "The wood is used for building, but is not by any means in general use." (*Bomb. Gaz., XV., pt. i., 66.*) TIMBER. 1052

CHRYSOPOGON, *Trin.; Gen. Pl., III., 1135.*

Chrysopogon aciculatus, *Trin.; Duthie, Fodder Grass, 39;* GRAMINEÆ. 1053

Syn.—ANDROPOGON ACICULATUS, *Linn.* (*? Retz.*); *Roxb., Fl. Ind., Ed. C.B.C., 88;* A. ACICULARIS, *Kunth.*

Vern.—*Súrwala, lampa,* HIND.; *Chor-kántá,* BENG.; *Kate chettu, katle gaddi,* TEL.; *Kudira-gullu,* MALA.; *Shunkhini, chorapushpi, keshini,* SANS.; *Tuttiri,* SING.; *Gnung-myit,* BURM.

References.—*Thwaites, En. Ceylon Pl., 366; Trimen, System. Cat., 108; Dalz. & Gibs., Bomb. Fl., 303; U. C. Dutt, Mat. Med. Hind., 295; Balfour, Cyclop.; Watt, in Report, Calcutta Intern. Exhbn.*

Habitat.—A small, coarse grass, growing on barren, moist pasture ground throughout Bengal, also in the North-West Provinces, Central Provinces, and in the warmer parts of Ceylon. Along with **Cyperus rotundus** and **Imperata arundinacea** this constitutes the characteristic turf of Bengal.

Fodder.—Cattle do not seem to like it. Its thin, straight culms, 1 to 2 feet high, flower, and the small spikelets of awned, barbed, fruits which follow, are troublesome to those who walk through the grass, as they stick FODDER. 1054 to the stockings and produce until removed a pricking and itching sensation. As soon as the spikelets appear cattle refuse to eat the grass.

1055 **Chrysopogon cœruleus,** *Nees; Duthie, Fodder Grasses, p. 39.*

Syn.—RHAPHIS CŒRULEA, *Nees.*

Vern.—*Dhaulian,* PB.; *Khar,* SALT RANGE; *Dhaula,* SIWALIK RANGE; *Ghweia,* KUMAON; *Tigri,* BUNDELKHAND; *Pálla paggar gadi,* CHANDA; *Jhingra-ka-jhara, khidi,* BERAR.

Habitat.—A common grass on the hilly tracts of Northern India, usually on stony or sandy soils.

FODDER. 1056 **Fodder.**—On the Siwalik range it is extensively used as fodder.

1057 **C. gryllus,** *Trin.; Duthie, Fodder Grasses, 40.*

Syn.—C. ROYLEANUM, *Nees;* ANDROPOGON GRYLLUS, *Linn.*

Reference.—*Aitchison, Cat. Pb. Pl., 176.*

Habitat.—The plains and hills of the Panjáb and N.-W. Provinces.

FODDER. 1058 **Fodder.**—Mueller says it is a useful fodder grass in Australia.

1059 **C. montanus,** *Trin.; Duthie, Fodder Grasses, p. 40.*

Syn.—C. PARVIFLORUS, *Benth.;* ANDROPOGON MONTANUS, *Roxb.*

Vern.—*Ballak,* RAJ.

Habitat.—The hilly parts of Northern India (Mount Abu).

FODDER. 1060 **Fodder.**—In Rajputana it is said to be viewed as excellent fodder, and the grain is also sometimes collected and eaten by the natives.

Cicca disticha, *Linn.*, see **Phyllanthus distichus,** EUPHORBIACEÆ.

Cicendia hyssopifolia, *W. & A.*, see **Enicostema littorale,** *Blume;* [GENTIANACEÆ.

1061 **CICER,** *Linn.; Gen. Pl., I., 524.*

Cicer arietinum, *Linn.; Fl. Br. Ind., II., 176; Wight, Ic., t. 20.* [LEGUMINOSÆ.

THE COMMON GRAM OR CHICK PEA; CECE, *It.;* GARBANZOS, *Sp.*

Vern.—*Cholá, bút, but kalái,* BENG.; *Chana, chunna,* HIND.; *But,* SANTALI; *Channa, cholá,* PB.; *Cholá, chaná,* RAJPUTANA; *Chana, harbara,* BOMB.; *Chenna,* DUK.; *Kadli,* KARNATICK; *Chahna, chano,* SIND; *Chania, chana,* GUJ.; *Harbara,* MAR.; *Kadalai,* TAM.; *Sanna-galu, harimandhakam,* TEL.; *Kudoly, kempu kadale, kari kadale,* KAN.; *Humez,* ARAB.; *Nakhud,* PERS.; *Chanaka, chennuka,* SANS.; *Kalapai,* BURM.; *Homos,* EGYPT.

The vernacular names of the acid liquid are: *Chané-ká-sirkah, búnt-ká-sirkah,* HIND.; *Harbare-ká-sirká,* DUK.; *Chana-amba, amba,* BOMB.; *Búnt-nu-zirko, khári, ambu,* GUJ.; *Kadalai-pulippu, kadalai-kádi,* TAM.; *Shan-aga-pulusu, shanagakádi,* TEL.; *Kadale-kádi,* MALA., KAN.; *Chana-kámla,* SANS.; *Khallul-himmas,* ARAB.; *Sirkahé-nakhúd,* PERS.

References.—*Roxb., Fl. Ind., Ed. C.B.C., 567; Stewart, Pb. Pl., 63; Aitchison, Cat. Pb. Pl., 45; DC., Origin of Cult. Pl., 323; Voigt, Hort. Sub. Cal., 226; Pharm. Ind., 80; Moodeen Sheriff, Supp. Pharm. Ind., 99; U. C. Dutt, Mat. Med. Hind., 149; Dymock, Mat. Med. W. Ind., 2nd Ed., 256; Ainslie, Mat. Ind., 56; Murray, Drugs and Pl., Sind, 120; Bidie, Cat. Raw Prod., Paris Exhbn., 72; Baden Powell, Pb. Prod., 241, 342; Atkinson, Him. Dist., 693, 709, 732; Birdwood, Bomb. Prod., 293; Lisboa, U. Pl. of Bom., 152, 277; Duthie & Fuller, Field aud Garden Crops, pt. I., 33; Balfour, Cyclop., 236; Treasury of Botany; Kew Official Guide to Museums, 42; Church, Food-Grains of India, 128.*

Habitat.—Extensively cultivated, as a rabi crop, throughout India, especially in the northern provinces.

This is the **Cicer** of the Romans, and the parched seed, as an article of food with the poor, is alluded to by Horace (*Cicer frictum*). It is also the ερ-εβινθος of Dioscorides. The botanical specific name owes its origin to a not altogether fanciful resemblance of the seed, when first forming in the pod, to a ram's head (the *krios* of the Greeks). The English name "gram" is applied to a totally different product in the Madras Presidency, where it denotes the seed of the plant known in the other provinces as *kurti* (**Dolichos biflorus**)" (*Duthie and Fuller, Field and Garden Crops, I., 33*). In Madras **D. biflorus** is more correctly horse-gram, two forms of **Phaseolus Mungo** being known as "black and green gram," and **Cicer** as "Bengal gram." These terms are, however, unknown in other provinces, where the word "gram" is exclusively given to the pea of **Cicer**.

HISTORY.

History.—The chick-pea was thus known to the Greeks in Homer's time under the name *Erebinthos*, and to the Romans as *Cicer;* and the existence of other widely different names shows that it was early known and perhaps indigenous to the south-east of Europe. It is supposed that the chick-pea has been cultivated in Egypt from the very earliest times of the Christian era, and was perhaps considered common or unclean, like the bean and the lentil. But it is most likely that the pea was introduced into Egypt as well as amongst the Jews from Greece or Italy. Its introduction into India is of more early date, for there is a Sanskrit name and several other names in modern Indian languages. "The Western Aryans (Pelasgians, Hellenes) perhaps introduced the plant into Southern Europe, where, however, there is some probability that it was also indigenous. The Western Aryans carried it into India. Its area may have extended from Persia to Greece, and the species now exists only in cultivated ground, where we do not know whether it springs from a stock originally wild or from cultivated plants." (*DC., Orig. Cult. Pl.*)

CULTIVATION.

CULTIVATION.

N.-W. P. Large. 1062 Small. 1063 Cabuli. 1064

N.-W. Provinces.—The varieties grown in the North-Western Provinces are classed as large-grained and small-grained, the former of a reddish and the latter of a light-brown colour. There are also a black-grained variety and a white-grained, known as "Cabuli." Gram is grown either alone or mixed with other crops, namely,—wheat and barley. The area under cultivation in the temporarily-settled districts is estimated at about 42⅘ lakhs of acres. It is sown from the middle of September to the middle of October at the rate of 80 to 100℔ to the acre, generally in a soil which lay fallow during the preceding kharif; the crop is gathered in March, April, and May. The soil for gram varies from the heaviest clay to the lightest loam, but it is found to prefer the former. It does not require so fine tillage as wheat and barley do, nor much irrigation, and a deep rather than well-pulverised seed-bed is all that is necessary. The tops of the shoots are picked off with a view to make the plants bushy and strong, and increase the outturn of grain.

The cost of cultivation, according to **Messrs. Duthie and Fuller**, is as follows:—

	R	a.	p.
Ploughing (four times)	3	0	0
Seed (80℔)	2	0	0
Sowing	0	14	0
Reaping	1	9	0
Threshing	2	0	0
Cleaning	0	6	0
TOTAL	9	13	0
Rent	3	0	0
GRAND TOTAL	12	13	0

CICER arietinum.

The Common Gram

CULTIVATION.

The approximate average outturn for unirrigated land in the several divisions varies from 5 to 8 maunds per acre in the case of gram, and from 6 to 9 maunds in the case of gram-barley and gram-wheat. For irrigated land the outturn is estimated at 12 maunds for gram alone, 14 for gram-barley, and 13 for gram-wheat.

C. P. 1065 *The Central Provinces.*—In these provinces gram is described as one of the principal rabi (winter) crops. It is sown in October and November and harvested in March and April. To ascertain the yield 33 experimental harvestings were made in 1886 in eleven districts. The highest return was in Narsinghpur, where 873℔ to the acre were obtained, and the lowest, 237℔, in Chanda. Taking the mean of all the returns in the eleven districts the yield may be expressed at 557℔. In the Chanda Settlement Report, it is stated that two kinds of gram are grown—the grey and the white. It is remarked that gram is not a popular crop in the Wardah District.

BOMBAY. *Bombay.*—There are 692,295 acres under this pulse, and in Sind 34,166 acres. The crop experiments made in the Bombay Presidency reveal the following results: In Kaira District a large form of gram gave 738℔ to the acre, the total value of the crop having been R14-15-6, the assessment being 31·38 per cent. on the return. In this experiment 54℔ of seed were given to the acre, and the remark is made that it was a dry crop following rice, poorly cultivated, on a black soil, and without manure. In another experiment made in the Panch Mehals, the yield was 750℔ with 1,200℔ straw from 40℔ seed, the assessment being 13·73 on the value of the crop. This crop is said to have followed as a second crop on a field that had been manured for maize. As much as 985℔ are also recorded as the yield, but on the other hand returns as low as 139℔ are mentioned.

Large. 1066

Small. 1067

The following extracts from the Bombay Gazetteers will be found interesting, especially the very general opinion which prevails in Bombay, that the gram crop kills weeds, and at the same time enriches the soil. This opinion, held to a large extent throughout India, may have something to say to the very general association of gram and wheat or gram and barley grown on the same field. The idea that it does improve the soil is one well worthy of careful scientific investigation. Such examples as the associated cultivation of tomatos in protecting cabbage and cauliflower from the attacks of caterpillars are well known to the gardener, and it is possible the association of this pulse with cereal crops is based upon established experimental results of a more sound character than that hitherto advanced,—a safeguard against failure, one crop succeeding should the other fail. With this as a possibility it would seem unwise to discourage the cultivator from the practice of such mixed crops, until the point here raised has been disposed of. The gram crop ripens before the wheat, and the admixture of the pea with wheat so loudly complained of by European merchants is the consequence of either of two things—*1st*, the wilful purchase of such admixture; for the natives of India regularly eat the two grains mixed, and to meet this demand the Indian corn merchant always has in stock a supply of mixed gram and wheat or mixed gram and barley. The Indian Agents of the European firms make the mistake of buying this Indian marketable mixture, and having done so the firms they represent are loud in their condemnation of gram as an adulteration of wheat. But, *2nd*, taking advantage of this fact, and of the fact also that gram and wheat are grown together on the same field, there seems every reason to suppose that a certain amount of wilful—one might almost say criminal—admixture of gram takes place in wheat sold as pure wheat. Such admixture is mainly, if not entirely, effected by the dealer not by the cultivator.

Kills weeds. Improves soil.

Justification of mixed crops.

Wheat and Gram.

Of Poona it is stated that the *chana* or *harbhara* (gram) is the most CULTIVATION.
largely grown of all the pulses, but chiefly in the east of the district. It
requires good black soil and is sown in November without either water
or manure and is harvested in February. The leaves are said to be used
as vegetables. The grain is eaten green, is boiled as a vegetable, and is
parched, when it is known as *hola*. When ripe it is split into *dál* and eaten Hola.
boiled in a variety of ways and in making a sweet-meat called *puran-poli*. 1068
It is slightly soaked, parched in hot sand, and called *phutanás*, which are Dal.
sometimes flavoured with turmeric, salt, and chilies. The grain is largely 1069
given to horses, and the leaves and stalks are dried as fodder (*Gaz., XVIII.*, Puran-poli.
p. 42). In Nasik it is sown in October and November and reaped in 1070
March. It is stated to be admirably suited for cultivation on new lands Phutanas.
as the oxalic acid of the leaves is supposed to kill the weeds. In Sátára 1071
there are said to be several varieties grown either on dry land or on manured and irrigated lands. In Belgaum gram is known as *kadli* in the Karnátak language, but the Marhatta name *harbhara* is that by which gram is best known in Bombay. In Kolhápur it is said to be sown in September and the beginning of October, taking five months to ripen. In black soil it is sown as a first crop, and in rice and garden lands it is raised as a second crop following rice. Gram is considered the best *bevad* or preparatory crop for **Sorghum vulgare** and cotton. "It certainly checks weeds. But it as certainly benefits the land in other ways also, which are not yet satisfactorily known. The average acre outturn is 650lb" (*Bomb. Gaz., XXIV., p. 169*). "As it takes very little out of the soil and checks weeds, gram is grown more to clear the ground than for profit, the returns seldom more than covering the cost of tillage" (*Bomb. Gaz., XII., p. 151*).

In the Panjab, as, indeed, in all wheat-producing provinces, gram is PANJAB.
grown. In the Bannu District it is stated to be cultivated on light sandy 1072
soils. Previous to sowing, the land requires fewer ploughings than for wheat. The sowings are generally begun and fairly completed in October. The amount of seed to the acre various from 32 to 48lb. Rain in March and April, so beneficial for wheat, and indeed abundant rain or prolonged cloudy weather at any time after germination, is injurious for gram, as it causes the plant to sprout too exuberantly, and to flower prematurely. The crop ripens about 15 days before wheat, and is generally plucked by the hand before wheat-reaping commences. "In Marwat it is rotated with wheat, the people alleging, as a reason for the practice, that gram leaves catch the dust of the spring dust-storms, so common in that tract, while the high even surface of wheat hardly intercepts it" (*Pb. Gaz., Bannu Dist., p. 142*). In Jhang gram is grown upon almost any soil, but the inundated tracts are considered the best. The habit prevails of turning on the cattle to graze down the plants. By this treatment they branch more freely and are supposed to stand the injurious effects of rain better. The seed is sometimes sown (45lb to the acre) with a drill plough but more frequently it is scattered broadcast after one imperfect ploughing of the soil. Rain in March to April causes the pods to be attacked by caterpillars. In Montgomery 30 to 40lb of seed to the acre are given, but, as
in all other parts of the province, the seed is sown broadcast after a Red.
careless ploughing has been given to the soil. As a fodder the stalks and 1073
leaves are considered injurious to milch-cattle, and little better than poison Black.
to horses. Three kinds of gram are said to be grown—red, black, and 1074
white; the last is, however, rare. It is known as *Cábulí chhola*. It is White.
softer, parches better, and yields a better *dál* than the others. Con- 1075
fectioners use it as it does not require to be peeled before use. Gram is Cabuli.
injured by lightning and rain. Of the Karnal District it is stated that the 1076

CULTIVATION. gram grows best on the stiff soils but is exceedingly sensitive to frost. A green worm called *sundí* attacks the seed, especially if the Christmas rains are late. In Hoshiarpur it is believed a line of linseed around the gram field is supposed to protect the crop from the injurious effects of lightning. In Gurgaon the people also believe lightning is injurious to the gram crop when in flower; in Gujranwala hares are very destructive to the gram crop. Of Dera Ismail Khan it is said gram fails altogether one year out of every three. In Muzaffurgarh the young leaves are eaten as a vegetable, being
Phalli. known as *phallí*. The pods are roasted and eaten under the name of
1077 *ámín* and *dhadhrí*. *Amín*, plural *ámíán*, is used in the north, *dhadrí* in
Amin. the south. The word *amín* is said to take its origin from an expression
1078 in allusion to gram ripening first of the *rabí* crops. The effect of gram
Improves soil. in improving the soil is known in Multan. "The crop is not only profitable, but it is also said to act as a manure and improve the land for the next *kharif* crop."

RAJPUTANA. *In Rajputana and Central India,* gram is also grown, and especially
1079 along with wheat. There is nothing, however, of a special nature to
CENTRAL INDIA. record.

1080 *Bengal.*—Gram, except in the wheat-producing districts, is not very
BENGAL. extensively cultivated. The Director of Agriculture reports that "There
Straw-coloured. are two varieties grown, *viz.*, the straw-coloured and the white, or
1081 *Kábulí*. Gram requires a heavy soil, does best in the clay or wheat
Kabuli. soil, can be grown in loam, but not in a sandy soil, comes after the *kele*
1082 *paddy*, a connecting link between the *aus* paddy and the *amun*. Five or six ploughings suffice to prepare the land, fine pulverisation of the soil not being required."

"Gram may be sown alone or mixed with wheat; in the first case seven seers and in the other five seers, to the *bigha*." The sowing time extends from the second week in October to the first week in November. "No after-cultivation is required." Harvest time is, February to March. "Threshing is effected by beating with a stick or treading under bullocks' feet. At the first beating or treading only the pods come out, the second and the third beating or treading gives the seed. The outturn is from $2\frac{1}{2}$."

BURMA. *In Burma.*—Mason says gram is grown extensively by the Burmese.
1083 GRAM AS A ROTATION WITH WHEAT.—In a recent lecture, on Indian agriculture, delivered before the agricultural students of the Edinburgh University, Professor Wallace, while stating his opinion that wheat cultivation could not be greatly extended in this country, alluded to the beneficial effects of leguminous crops cultivated in rotation with wheat. It has already been shown in the remarks under gram cultivation in the Bombay Presidency, that this fact is fully recognised by the Indian cultivator. The Professor anticipates a ruinous reduction of pulse cultivation in India, but admits that, although the scientific principle of a rotation of crops is not thoroughly understood by the Indian cultivator, the habit of cultivating pulses, and particularly gram, as a mixed crop with wheat, or in rotation with wheat, in a measure meets this necessity. It should be borne in mind, however, that seasonal peculiarities force on the Indian farmer a rotation. He has at least two if not three crops every year—the *rabí* and *kharíf*, the former reaped in spring and the latter in autumn. The majority of the pulses belong to the latter crop and are thus cultivated in the season when wheat cannot be grown, and are on that account not likely to be seriously displaced by an extended wheat cutivation. Gram is in fact the only leguminous crop that might suffer in this direction, and hence it seems desirable that as little as possible should be urged against the practice of growing that pulse as a mixed crop with wheat or barley. From

what has been said, it may be inferred that adulteration of gram with wheat grain is more an accident than a necessity of the habit of mixed cultivation.

CULTIVATION.

Gram recommended as an article for European Cattle.

GRAM AS AN ARTICLE OF CATTLE DIET.—In an address delivered before the Society of Arts the writer took occasion to recommend the extended importation of gram into England as an article of diet for horses. Throughout India it may safely be said gram is the staple article of horse food. In Madras another pulse takes the place of gram, but horse diet in this country has always a much larger percentage of pulses in it than in Europe. The animals thrive admirably on such a diet, and the opinion may be advanced that where muscular strength is required a diet that contains a distinct and rational proportion of nitrogenous matter is a more wholesome one than the over-starch diet given in Europe. The writer stated in the paper alluded to: "Chemically, a horse diet which consists exclusively of cereals, cannot possibly be so good for the animal, nor so likely to produce muscular strength, as a diet with a liberal admixture of some kind of peas. Husked gram contains of albuminoids 21·7 per cent., and of starch 59·0 per cent. Indian corn contains only 9·5 per cent. to 70·7 per cent. of starch. When it is recollected that the albuminoids are the muscle-forming constituents of diet, it becomes apparent that a diet which would contain oats and gram, or Indian corn and gram, would be more nutritious and strength-giving than the modern English food for horses of oats and Indian corn. To obtain the indispensibly necessary amount of albuminoids from an English diet, the animal has to eat a greatly excessive and injurious amount of starch." This opinion was supported by some of the gentlemen present at the meeting of the Society of Arts, **Sir Joseph Fayrer** remarking that gram "was a much better food than that used in England, being more nutritious. It would be a great advantage if the English would feed their horses on gram rather than oats and crushed Indian corn." As in a measure opposed to these views—views that would lead to the supposition of a possible future large export trade in Indian gram—**Professor Wallace**, in a paper read before the Farmer's Club of London, while admitting that a large trade seemed possible with India and England in gram and other pulses, remarked: "These foods, however, are not to be adopted without caution. No less than two diseases in horses are attributed to gram. The one may be called Principal Williams' Manchester Horse disease, a blood poison which I believe to be identical with the *anthrax* of India. The other is a nerve disease, described by **Principal McCall** of Glasgow, in which the tongue becomes paralysed." Without attempting to dispute these high opinions, it may be said that our horses do not appear to suffer more than those of Europe, and that it hardly seems possible gram-feeding has in England been tried to the extent to justify either a favourable or unfavourable opinion. The writer has, besides, consulted the fairly complete library that now exists on the subject of cattle and cattle diseases in India, and in no instance is there the slightest allusion to gram as the cause of any disease. Indeed, anthrax would appear to occur far more frequently among cattle not fed on gram than among those that get a regular amount of that pulse in their diet. In the small Native State of Manipur, where gram is not grown, as food for cattle, *anthrax* or a closely allied disease, is a very common cause of death among the rice-fed ponies. The disease alluded to is in India attributed to a sudden and large supply of fresh grass after periods of scarcity—an annual occurrence due to the periodicity of the rains following a hot season when all grass is burned up. May it not be that the pulse viewed as "gram" by the above mentioned authorities was not gram at all but the injurious seed of **Lathyrus sativus**, the properties of which, in causing paralysis, are well known?

These remarks regarding anthrax have, however, been made in this place mainly to prevent undue alarm, until Professor Wallace's suggestions regarding a possible connection between it and gram-feeding have been proved correct.

CHEMISTRY. 1084

CHEMICAL PROPERTIES OF GRAM.

Professor Church, in his *Food-Grains of India*, gives an interesting account of this pulse, but is in error in too prominently restricting the name *gram* to the forms of **Phaseolus Mungo.** This is the case only in the Madras Presidency; throughout the rest of India the terms black and green gram are practically unknown, the word gram signifying the pulse **Cicer arietinum**, although the term horse-gram is sometimes applied to the pea of **Dolichos biflorus.** In Madras it might fairly well bear that name, since it takes the place of **Cicer arietinum** as a food for horses. The Professor gives a valuable table as the result "of nine analyses of the unhusked peas and of four analyses of the peas from which the husk has been removed."

"COMPOSITION OF THE CHICK-PEA.

IN 100 PARTS.

	Husked.	With Husk.	In 1 lb. Husked.
			Oz. Grs.
Water	11·5	11·2	1 367
Albuminoids	21·7	19·5	3 207
Starch	59·0	53·8	9 192
Oil	4·2	4·6	0 294
Fibre	1·0	7·8	0 70
Ash	2·6*	3·1†	0 182

* 1·1 of Phosphoric Acid.
† 0·8 of Phosphoric Acid.

"The nutrient ratio in the unhusked peas is 1 : 3·3; the nutrient value is 84."

The unhusked peas are therefore more nutritious than the husked, and it may be concluded that the process of steeping them in water before being mixed with the oats or other cereal both softens the pea and removes entirely the dust and mud associated with the pulse. This is an important point, for a large amount of mud mixed with the food must of necessity prove injurious to cattle. It is a common practice in India for clubs and messes to specially feed up their sheep on gram, *gram-fed mutton* having a high reputation.

TRADE. 1085

TRADE AND PRICES.

Very little can be learned regarding the internal trade in gram. It is extensively eaten by the natives in every part of the country, and there must therefore exist a very considerable internal trade in the pulse. The grain could be most conveniently obtained from Bombay, Karachi, or Calcutta, the supply being drawn from Cawnpore, Patna, and Lahore, among many other centres of agricultural produce. Of the Bombay Presidency gram is largely grown in Sátára, Ahmadnagar, and Nasik. In Madras gram mostly means other pulses than that presently under consideration.

TRADE.

The foreign trade is at present not very extensive. The following were the exports during the past five years :—

	Cwt.	R
1882-83	312,953	8,28,647
1883-84	392,694	11,99,796
1884-85	314,965	9,28,848
1885-86	338,129	10,74,771
1886-87	306,979	9,84,046

The exports in 1870 were only 23,171 cwt., valued at R94,900; but it seems probable these returns include with the true gram the peas of **Dolichos biflorus,** and perhaps also those of **Phaseolus Mungo.** During the late Colonial and Indian Exhibition the writer took the opportunity to ask several grain merchants the commercial terms used in Europe for the various Indian pulses. The majority of these gentlemen agreed in calling the peas of **Phaseolus Mungo** (the black and white grams) *múttar,* a word which in India would at once be accepted as an English corruption of the native name for grey-peas—**Pisum** (which see). It therefore seems desirable, in the view of a future extended trade, that greater care should be taken in distinguishing the various pulses of India; and the vernacular name for the true gram, *chaná,* would perhaps be a safer one to use than any other.

PRICES. 1086

Prices.—In a recent number of the publication issued by the Department of Finance and Commerce under the title of *Prices and Wages in India,*" Mr. O'Conor has published tables which afford perhaps the most trustworthy data for arriving at a knowledge of the price of gram; his figures represent seers (2lb) to the rupee. **Mr. O'Conor's** results of average prices may be thus summarised :—

	I 1873 to '76.	II 1877 to '80.	III 1881 to '84.	IV 1873 to '80.
Madras	23·63	17·77	32·05	20·7
Bombay and Sind	17·06	11·47	18·45	14·27
Bengal	20·58	15·31	21·77	17·94
North-Western Provinces and Oudh	26·61	18·36	24·53	22·48
Panjab	30·04	18·29	26·7	24·16
Central Provinces	31·02	18·1	27·25	24·56

It would, perhaps, be unsafe to carry these figures further; but the mean of Column IV. might give the reader an average approximation of the retail price of gram in India. But it must not be lost sight of that "gram" as presently exported means more than the pea of **Cicer arietinum,** and includes (as perhaps do the above figures) pulses that have a lower value than the true gram.

It may, however, be said that gram could be landed at a price considerably below that at present paid in England for horses' food. Referring to the recent provincial publications (*Crop Experiments, and the Agricultural Department Reports*), some useful facts of prices may here be placed before the reader, with the remark that these will be found to differ in some respects from those given above by **Mr. O'Conor.**

C. P. 1087

In the *Central Provinces* during 1886 the price of gram varied considerably, according to the district and season of the year. It would be unsafe to attempt to strike an average for all these prices, but the following exhibits the two highest and two lowest quotations, the prices being

CICER arietinum. The Common Gram

PRICES. seers to the rupee, in which of course a larger quantity for the sum mentioned would mean cheapness and a less quantity dearness:—

Districts.	August 15th.	November 15th.	February 15th.	May 15th.
Mandla	45·	42·	40·	40·
Damoh	39·	27·	29·8	40·
Sambalpur	15·	19·8	19·8	...
Wardha	20·	22·	21·	24·

The difference between the prices at which the cultivators sell the produce of their fields to the dealers, at harvest time and at other periods throughout the year, is not as a rule very great, still the prices are a little more favourable after harvest. Gram being a *rabi* crop it is harvested from February and March to April, and a mean of the quotations for the Central Provinces gives the average price in May as 26·8 seers to the rupee or 53·8℔ for, say, 1*s*. 5*d*. at present rate of exchange.

BENGAL. 1088 *Bengal* is not a large gram-growing province, and it is accordingly dearer there than in most other parts of India. The Director of Agriculture, in his report for 1886, gives the price of gram at 24 seers to the rupee after harvest and 20 seers at other seasons. Taking a high exchange, these quantities would represent 48 to 40℔ for 1*s*. 5*d*.

BOMBAY. 1089 *Bombay*.—The quotation has been given in one of the *Crop Experiments* of 60 seers to the rupee, or, at the rate of exchange adopted in the preceding estimates, 120℔ for 1*s*. 5*d*. It is probable, however, that this figure is much too low, and that the average price in the Western Presidency bears a closer approximation to that given for the Central Provinces and Bengal.

PANJAB. 1090 *Panjáb*—In the Lahore district, according to the *Gazetteer*, gram is stated to be sold at 100℔ to the rupee (= 1*s*. 5*d*.). In the Mooltan district, the average price for the past 20 years is given as 60℔ and in the Jhelam district for the past 44 years as from 68 to 110℔ according to the various parts of the district.

N. W. P. 1091 *In the North-West Provinces* gram is variously quoted in the *Gazetteers;* thus, in Bulandshahr 26 seers; in Meerut since 1850 to the present date it has ranged from 55 seers to 20, and in 1869 fell to 9½ seers; in Muzaffarnagar since 1821 the price has varied from 70 seers the highest to 14 the lowest; in Budaun it is given at 30·8 seers; in Bijnor about the same; in Bareilly it is much more expensive, and in Gorakhpur gram is considerably dearer than wheat.

DYE. 1092 **Dye.**—The leaves are said to give indigo. This curious fact is known to the Chinese. The dye is allied to the Assam so-called green, obtained from **Vigna Catiang,** which see.

MEDICINE. Seeds. 1093 **Medicine.**—In medicine the SEEDS are considered antibilious. The chief interest medicinally is, however, in the ACID LIQUID obtained by collecting the dew-drops from the leaves. The fact that the drops of dew are thus chemically changed through contact with a living plant is a point of great botanical interest not at present fully understood. The liquid

Gram Vinegar. 1094 is found chemically to contain oxalic, acetic, and malic acids. This vinegar is mentioned by the old Sanskrit writers as a useful astringent, which might with advantage be given in dyspepsia, indigestion, and costiveness.

One of the earliest European writers who describes "Cicer Vinegar" was the Polish explorer **Dr. Hove,** who spent the greater part of two years in the Bombay Presidency in 1787-88. His report was some 70 years

MEDICINE.

afterwards published in the Records of the Bombay Government (XVI. 1855): at page 57 he says:—"On the road to Dowlat" (a village about 7 miles from Dholka), "we met with numerous women who gathered the dew of the grain called by the inhabitants *chana* or gram, by spreading white calico cloths over the off-spring, which was about 2 feet high, and so drained it out into small hand-jars. They told me that in a short period it becomes an acid, which they use instead of vinegar, and that it makes a pleasant beverage in the hot season, when mixed with water; as likewise they used it as an antidote for the venom of pernicious snakes, of which there is a great number in the wet season. I tasted the dew but found it of no particular taste, except rather softer than common water, as it is peculiar to the dew." Further on at p. 63, he observes that the natives "attributed many different qualities to this dew, but especially that against the venom of snakes. Although it was fresh off the plant it had a particular mineral acid peculiar to itself. I compared that which I had gathered a few days ago, which had likewise already acquired a mineral acid, but not quite so powerful."

Sir George Birdwood gives the following account of the acid liquid in his Catalogue of the Bombay Products: "When at Sholapore, some years ago, my *munshi* asked me to lend him some towels to gather 'a spirit' which he stated fell at night on fields of growing gram, and which with water formed an agreeable drink in the hot season. The cloths were laid over the tender gram, and by the morning were saturated with dew, having an intensely acid taste. This was wrung off and bottled. Though unfamiliar to Europeans, this substance is well known to natives, and is mentioned by **Royle** and others. The acid is said to be oxalic."

Dr. Moodeen Sheriff gives an interesting account of the collection of this liquid. "A piece of clean cloth is tied to the end of a stick and the pulse crop is brushed with this in the early morning, so as to absorb the dew. This is then wrung out and preserved." "The genuine drug can only be obtained from persons who own fields of gram; what is sold by native druggists is dilute sulphuric acid slightly tinged with some colouring matter." It is useful in diarrhœa and dysentery, and is given as a drink with water in sunstroke. The boiled leaves are applied as a poultice to sprains and dislocated limbs. The fresh juice of the leaves mixed with crude carbonate of potash is administered with success in dyspepsia (*S. Arjun, Bomb. Drugs, 9, 193*). The acid liquid is employed as a refrigerant in fever. It is much used in the Deccan in the treatment of dysmenorrhœa; the fresh plant is put into hot water and the patient sits over the steam. **Dr. Walker** observes that this is another way of steaming with vinegar. (*Pharm. Ind.*) "The free use of the vegetable, owing to the abundance of oxalic acid, is apt to do harm to persons liable to calculus, as it leads to the formation of oxalate of lime in the bladder." (*Drury, U. Pl.*) It is said "to increase the secretion of the bile; also, when roasted like coffee, is considered aphrodisiac; also used in cases of flatulency, and in retention of urine and cutamenia. It serves as a substitute for coffee." (*Baden Powell, Pb. Prod.*)

Special Opinions.—§ "The liquid obtained from macerating the seeds in water is used as a tonic among the natives" (*Assistant Surgeon Nil Rutton Banerji, Etawah*). "Is used to allay vomiting" (*Surgeon-Major D. R. Thomson, Madras*). "Cold infusion of *chhola* is also considered to be antibilious" (*Surgeon Anund Chunder Mukerji, Noakhally*). "The vinegar (*Chana-amba*, Bom.)—that sold in the bazars—is generally dilute sulphuric acid coloured with sugar" (*Surgeon-Major W. Dymock, Bombay*). "It is used with the tender leaves of *ním* in cases of leprosy. The water in which it has been macerated is used as a remedy for bilious- Chana-amba.

MEDICINE. Chana-khar. ness" (*Brigade Surgeon J. H. Thornton, B.A., M.B., Monghir*). "The vinegar, which is known here as *chána khar*, is used for enlarged spleen." (*Surgeon-Major J. Robb, Ahmedabad*). "In bronchial catarrh, the seeds, eaten in a parched condition at night, followed by a cup of warm milk, give great relief" (*Surgeon-Major A. S. G. Jayakar, Muskat, Arabia*). "It is believed that the plant exhales acid vapour which is absorbed by the dew. It is also collected by spreading muslin cloth on the plants overnight, and wringing out the moisture from it early in the morning. The acid solution thus obtained is useful in vomiting and dyspepsia" (*Native Surgeon T. Ruthnam Moodelliar, Chingleput, Madras Presidency*). "The dew-drops are used to check nausea and vomiting successfully: also in cholera" (*Surgeon-Major J. J. L. Ratton, Salem*). "It is believed to have alterative properties" (*Aligarh*).

CHEMISTRY. 1095 **Chemical Composition.**—The seeds contain, according to **Balfour**, moisture 10·80 per cent., fatty matter 4·56 per cent., nitrogenous matter 19·32 per cent., mineral constituent (ash) 3·12 per cent., and starchy matter 62·20 per cent. **Dr. Warden**, however, gives the following composition: "One hundred parts without husk contain water 11·39, nitrogenous matters 22·7, fat 3·76, starch 63·18, and mineral matter 2·60 (*Parkes*)." (*Conf. with Church's Analysis of Pulse on a previous page.*)

FOOD. 1096 **Food.**—Gram forms the chief food for horses. Amongst the poorer classes of natives parched gram (*chabena*) is much eaten. **Masson** informs

Parched Gram. 1097 us that in the Panjáb it is made into bread, which was a favourite article of food with the Sikh sirdárs. The natives also eat it boiled in the form of

Ragout. 1098 *ragout*, seasoned with a little pepper or capsicum. The YOUNG PLANTS are sold and eaten throughout the Deccan, Madras, and Gujarat, either raw

Young plants. 1099 or roasted in hot ashes. The STALK and LEAVES, after the seed is threshed out, constitute one of the most valued kinds of *bhúsa* for fodder (*Stewart*;

FODDER. 1100 *Murray*). In Oudh the young leaves are used like spinach. **Dr. Christie** remarks, in *Madras Journal of Science*, No. 13, that the acid exudation from all parts of the plant is collected by the ryots and used in their curries instead of vinegar.

The following account of gram given in the *Treasury of Botany* may be quoted here: "In India the seeds form one of the pulses known under the name of 'Gram,' and are greatly used as an article of food by the natives, being ground into meal, and either eaten in puddings or made into cakes. They are also toasted or parched, and in this state are commonly carried for food on long journeys. Rolled in sugar-candy, these toasted peas form a rough sort of comfits, and gram-flour made up with sesamum oil and sugar-candy is an Indian sweetmeat."

Cicer Lens, *Willd.*, see Ervum Lens, *Linn.*

1101 C. soongaricum, *Steph.; Fl. Br. Ind., II., 176.*

Vern.—*Tizhú, jawáne, banyarts, sárri, serri*, PB.

References.—*Stewart, Pb. Pl., 63; Murray, Drugs and Pl. Sind, 120; Church, Food-grains of India, p. 131.*

Habitat.—Met with in the Western Himálayas, temperate and alpine region, altitude 9,000 to 15,000 feet; Piti, Lahoul, Kumaon, Yarkand, Tibet.

FOOD. Seeds. 1102 **Food.**—"Said to fatten cattle quickly. Its SEEDS were sent to the Agri.-Horticultural Society many years ago (having been first found in the Himálaya by **Captain Munro** about 1844-45), with information that the

Shoots. 1103 grain is eaten by the people. The YOUNG SHOOTS are prepared as a pickle by the Chinese, and a vinegar is made from the leaves. The latter are often covered by a viscid exudation, with a strong aromatic odour.

Aitchison states that in Lahaul shoots are used as a pot-herb, and that the peas are eaten there, as they are, both raw and cooked, in parts of Ladak" (*Stewart, Pb. Pl., 63 ; Hinderson, Mission to Yarkand*).

CICHORIUM, *Linn.; Gen. Pl., II., 506.*

Cichorium Endivia, *Linn.; Fl. Br. Ind., III., 391 ;* COMPOSITÆ. 1104

THE GARDEN ENDIVE.

Vern.—*Kasini,* HIND., BOMB., BENG.; *Kashini-virai,* TAM.

References.—*Kurz, For. Fl. Burm., 78 ; Aitchison, Pb. Pl., 81 ; DC., Origin of Cult. Pl., 97 ; Dymock, Mat. Med. W. Ind., 2nd Ed.; Lisboa, U. Pl. Bombay, 163; Irvine, Top. Ajmere, 142; Treasury of Botany.*

Habitat.—"The Endive is generally considered to be a native of Persia, Northern India, China, and perhaps Egypt. Be this as it may, there is no doubt of its having been used as an esculent food from a very early period by the Egyptians, through whom the Greeks and Romans probably became acquainted with it (*Treasury of Botany*). The Arabs call it *Hindyba,* evidently a corruption of *Intyba,* and the Persians *Kasni,* a name which is current all over India, proving the introduction of the plant from the West. Muhammadan writers mention several varieties of Endive which are probably the same as those known to us." (*Dymock, Mat. Med. W. Ind.*)

Medicine.—"Endive is much valued by the *hakims* as a resolvent and cooling medicine, and is prescribed in bilious complaints much as taraxacum is with us. The SEEDS are one of the four lesser cold seeds of old writers, and as such are still in use in the East" (*Dymock*). The ROOT is "considered warm, stimulating, and febrifuge; given in '*munjus,*' the diluent taken preparatory to purging; the seed is used in sherbets" (*Med. Top. Ajmere, 142*).

MEDICINE. Seeds. 1105 Root. 1106

Food.—"Endive, radishes, and succory are mentioned by **Ovid** as forming part of a garden salad; and **Pliny** states that endive in his time was eaten both as a salad and pot-herb. As such it has been used in Great Britain for three centuries, and it is a singular fact that the manner in which it was prepared for winter use, as described by **Gerarde** in 1597, differs but little from the mode that is often practised at the present day" "It is cultivated solely for the stocky head of leaves, which, after being blanched to diminish their bitterness, are used in salads and stews during winter and spring" (*Treasury of Botany*).

FOOD. 1107

C. Intybus, *Linn.; Fl. Br. Ind., III., 391 ;* COMPOSITÆ. 1108

THE WILD OF INDIAN ENDIVE, CHICORY, OR SUCCORY.

Vern.—*Kasni,* HIND., PERS.; *Hindyba,* ARAB.; *Kashini-virai,* TAM.; *Kasini-vittulu,* TEL.; *Hand, gúl, suchal, kásni,* PB.; *Kásani,* GUJ.

References.—*Brandis, For. Fl., 77 ; Kurz, For. Fl. Burm., 77 ; Stewart, Pb. Pl., 124; Aitchison, Pb. Pl., 81 ; DC., Origin of Cult. Pl., 96 ; O'Shaughnessy, Beng. Dispens., 408 ; Murray, Drugs and Pl. Sind., 186; S. Arjun, Bomb. Drugs, 77 ; Year-Book of Pharmacy, 1874, 626; Baden Powell, Pb. Prod., 355 ; Birdwood, Bomb. Prod., 49 ; Balfour, Cyclop., 236 ; Morton, Cyclop. of Agri., 457 ; Ure, Dict. of Arts, &c., 770 ; Kew Official Guide to Museum, 87 ; Johnstone, Common Life (Ed., Church), 159 ; Smith's Dict., Economic Plants, 112 ; Bell, Chemistry, Foods, 60.*

Habitat.—North-West India, Kumaon, distributed westward to the Atlantic.

§ "In the plains of the Panjáb it is cultivated by natives as a pot-herb (*ság*), and may be an escape; truly wild at 4,000 to 11,000 feet" (*Surgeon-Major J. E. T. Aitchison, Simla*).

The Wild or Indian Endive.

HISTORY.

History.—"The wild perennial chicory, which is cultivated as a salad, as a vegetable, as fodder, and for its roots, which are used to mix with coffee, grows throughout Europe, except in Lapland, in Morocco and Algeria, from Eastern Europe to Afghánistán and Beluchistán, in the Panjáb and Kashmír, and from Russia to Lake Baikal in Siberia. The plant is certainly wild in most of these countries; but as it often grows by the side of roads and fields, it is probable that it has been transported by man from its original home. This must be the case in India, for there is no known Sanskrit name" (*DC. Origin of Cult. Pl., 96*).

CULTIVATION. 1109

Cultivation of Chicory.—Chicory flourishes on any kind of soil, and has been found to be abundantly profitable upon poor sandy lands as well as on richer and more productive soils. The cultivation of chicory as a fodder plant is simple enough. The seed is sown broadcast upon land that has been dug or deeply ploughed, from seven to twelve pounds per acre. This is the way the plant is grown in the best meadows in the south of France and in Lombardy. "The best mode of culture, however, for a fodder or herbage crop, is as follows: Prepare the soil, by thorough cleansing and pulverization, as early in the spring as the season will admit; apply a good coat of partially decayed fold-yard dung, and drill in the seed during March, 4℔ per acre, at about nine-inch intervals between the rows. When the plants are about five inches in height, carefully hoe them and single out, leaving them about six inches apart, after the usual method in turnip culture,—that is, by boys following the hoers. Some recommend that the seed be sown in a bed, and when the plants are fit for transplanting—which will be when about five inches high—they are to be set out in rows nine inches apart, and at six-inch intervals from plant to plant in the rows. In either case, the land must be kept clean, and well hoed, particularly in the first season; ordinary attention will afterwards suffice, and the crop will continue luxuriant and profitable for five years at least, and frequently from eight to ten. When the plants begin to exhibit symptoms of failure, the ground should be cleared of the roots, and the course of cropping pursued for a few years, and it may then be again sown or planted with chicory.

"In preparing the land for a *root crop*, deep ploughing is recommended; but, unless the soil is very deep, it is probable that subsoil ploughing will answer better. The surface must be well worked; indeed, it cannot be reduced to too fine a mould. As the plants are a long time in coming up, generally five or six weeks from the time of sowing the seed, it is necessary that the land should be very clean, or the weeds (particularly chickenweed) are liable to overtop and smother the young plants. The time of sowing varies in different districts; in the midland and eastern counties, the second or third week in May is considered best, fot if sown earlier (when cultivating for the root), many of the plants will run to seed; in which case they are called 'runners,' or 'trumpeters,' and must be carefully dug out and destroyed, when the time for taking up has arrived; because, if allowed to become mixed with the bulk, they will spoil the sample. The best crops have been obtained when the seed has been sown broadcast; but the preference is usually given to drilling, the crop being easily hoed and cleaned. The rows are generally from nine to twelve inches apart, and about 3 or 4℔ of seed per acre is the quantity used. Most of the cultivators of chicory single out the plants so as to leave spaces between them in the rows, each about six or eight inches long; but there are many who do not do this, fancying that four or five small plants produce more weight of root than one large plant. The expediency of this, however, is very questionable, as it does not allow of the land being nearly so well cleaned as when the practice of singling is adopted" (*Morton, Cyclop of Agri., I., 457*).

Chicory and Coffee. **CICHORIUM Intybus.**

CULTIVATION.

In India.—Very little of a trustworthy character can be learned regarding the cultivation of chicory in India. It is alluded to in the Kangra Gazetteer as cultivated on account of its seeds, which are used medicinally as an alterative. It seems probable that the plant is also grown as a fodder in some parts of the Panjáb plains, but although a large trade might easily be done in the root, this fact seems quite unknown to the Indian cultivator. It would seem desirable that some effort be made to introduce the industry of chicory root cultivation into India, a crop that would give a good return, while affording a useful and much needed fodder at the same time. **Baden Powell** alludes to the root as grown in the Panjáb as a drug and selling at 2 annas a seer. He mentions specimens of root and of seed as sent to the Lahore Exhibition from nearly every district.

Great Britain imports annually close upon 200,000 cwts. of the root. It is extensively grown in England, but the best roots are imported from Belgium and Holland.

MEDICINE. 1110

Medicine.—"Has tonic, demulcent, and cooling properties. The seeds are considered carminative and cordial. A decoction of the seed is used in obstructed menstruation" (*S. Arjun*). The root is bitter and used medicinally in the Panjáb. It contains nitrate and sulphate of potash, mucilage, and some bitter extractive principle. An infusion of chicory mixed with syrup causes a thickening of the liquid (*Balfour*).

Special Opinions.—§"Used as a substitute for taraxacum" (*Assistant Surgeon Nehal Sing, Saharunpore*). "Found to act usefully on the liver in cases of congestion" (*Surgeon F. Perry, Jullunder*). "A strong infusion of powdered seeds proves highly useful in checking bilious vomiting" (*Surgeon E. W. Levinge, Rajabundry, Godavery District*). "Much used by natives as a cooling medicine in fevers" (*Assistant Surgeon Bhugwan Dass, Rawal Pindi*).

FOOD. 1111

Food.—"The young plant is in some places employed as a vegetable." (*Roxb.*) "It is used as a salad by the French under the name of *Barbe du capucine*, the young leaves being blanched like endive. Its roots are roasted, ground, and mixed with coffee to flavour it. They are sometimes used as a substitute for coffee, as was the case in France during the suspension of the foreign trade of that country" (*O'Shaughnessy*). Chicory is used largely by the Egyptians, and it is well known that both the leaves and roots once constituted half the food of the poorer classes, as they probably do at the present day. "Within the last few years, grocers mixing chicory with coffee are bound to affix a label on the outside of the package announcing the admixture, so that purchasers can now have pure coffee, or coffee mixed with chicory, as they prefer, for there are some who like the mixture. It need hardly be said that chicory is entirely destitute of those properties which render coffee an agreeable and nutritive beverage, while, on the other hand, it possesses medicinal properties closely like those of dandelion, and which therefore render it unwholesome for constant use" (*Treasury of Botany*). The working of the Act regulating the admixture of chicory with coffee in England has by many persons been viewed as inferior to the French system, where the grocer is ordered to sell the chicory by itself. All that the English grocer requires to do is to sell pure "coffee" when he advertises such. His special "coffee mixture" may be anything he pleases to make it. The sale of chicory separate from coffee has been strongly recommended by **Sir James Elphinstone.** "The root tastes at first sweetish and mucilaginous, and then very bitter; the bitterness is greater in summer than in spring. For over a hundred years chicory has been used as a substitute for, and admixture with, coffee. In preparing chicory, the roots are washed, cut into small pieces and kiln-dried, and then roasted and ground. Roasted chicory

Chicory in Coffee.

FOOD. contains a volatile empyreumatic oil, to which its aroma is due, and a bitter principle. It contains no caffeine. Infused in boiling water it yields a drink allied in flavour and colour to coffee. It is largely used in Belgium. In some parts of Germany, the women are said to be regular chicory topers (*Parry*)." (*Surgeon C. J. H. Warden, Prof. of Chemistry, Medical College, Calcutta*).

The following extract, relating to the fact of the chicory roots being a new source of alcohol, was published in the *Tropical Agriculturist* of 1st December 1882, *page 495; also p. 57*:—

"According to **Erfindungen** und **Erfahrungen**, the celebrated coffee substitute, chicory, seems likely to become of importance as a source of alcohol. The root contains an average of 24 per cent. of substances easily convertible into sugar, and the alcohol obtained by its saccharification, fermentation, and distillation, is characterised by a pleasant aromatic taste and great purity" (*Chemist and Druggist*).

ADULTERATIONS. 1112 **Adulterations.**—"Roasted chicory is extensively adulterated. To colour it, Venetian red and, perhaps, reddle are used. The former is sometimes mixed with the lard before this is introduced into the roasting machine; at other times it is added to the chicory during the process of grinding. Roasted pulse (peas, beans, and lupines), corn (rye and damaged wheat), roots (parsnips, carrots, and mangold wurzel), bark (oak-bark tan), wood-dust (logwood and mahogany dust), seeds (acorns and horse-chestnuts), the marc of coffee, coffee husks (called coffee-flights), burnt sugar, baked bread, dog-biscuit and baked livers of horses and bullocks (!), are substances which are said to have been used for adulterating chicory. A mixture of roasted pulse (peas usually) and Venetian red has been used under the name of *Hambro' powder* for the same purpose" (*Ure's Dict., Art and Manuf.*) A recent examination of certain "coffee mixtures" revealed the fact that roasted cockroaches and iron rust were employed as adulterants. (*See* **Coffea arabica**, *para. Adulterants.*)

CIMICIFUGA, *Linn.; Gen. Pl., I., 9.*

1113 **Cimicifuga foetida,** *Linn.; Fl. Br. Ind., I., 30;* RANUNCULACEÆ.

Vern.—*Jiunti*, PB.

References.—*Stewart, Pb. Pl., 2; Treasury of Botany; Kew Official Guide to the Museum, 8.*

Habitat.—Found in the temperate Himálaya, from Bhután to Kashmír; altitude 7,000 to 12,000 feet.

MEDICINE. Root. 1114 **Medicine.**—The ROOT is said to be poisonous. In Siberia it is used to drive away bugs and fleas. Under the name of a nearly allied plant (**Actæa spicata**), the writer has already referred to this plant, and chiefly with the view of attracting attention to these useful but apparently neglected plants.

Garrod, in his *Materia Medica*, calls **Cimicifuga racemosa,** *Linn.*, the Black Snake Root, and remarks that it is a remedy much used in America. He gives the dose of the tincture as 30 to 40 minims. He remarks: "Its use is said to have been attended with much success in rheumatic fever, in chorea, in lumbago, and in some forms of puerperal hypochondriasis." The *Pharmacographia* gives the history of **C. racemosa.** It was first made known to Europe in 1696, and was scientifically identified and named by **Linnæus** in his *Materia Medica* in 1749. In 1823 it was introduced into medical practice in America, and to England in 1860.

There seems every reason to expect that the Indian species, which differs from **C. racemosa** only very slightly, will be found to possess all its medi-

MEDICINE.

cinal virtues. **C. racemosa** is chiefly prescribed in the form of tincture, and employed in rheumatic affections, dropsy, the early stages of phthisis, and chronic bronchial diseases. Externally, a strong tincture has recently been used to reduce inflammations (*See Year-Book of Pharmacy, 1872*). A section of the root exhibits a central pith with broad radiating plates subdividing the wood into 3-5 wedge-shaped sections, with a thick brittle bark surrounding the wood. It contains a resinous active principle which has been termed Cimicifugin or Macrotin. In its action this drug resembles hellibore on the one hand and colchicum on the other. It is most useful in acute rheumatism, and a powder of the root is perhaps the best mode in which to give the drug, in doses of 20 to 30 grains (*Royle's Mat. Med. ed. by Harley*).

Special Opinion.—§ "A poultice prepared of the fresh leaves is used here, and said to be very useful in rheumatic affection of joints" (*Surgeon C. J. W. Meadows, Burrisal*).

CINCHONA, *Linn.; Gen. Pl., II., 32.* 1115

Cinchona, *Linn.;* RUBIACEÆ.

CINCHONA BARK, PERUVIAN BARK, JESUIT'S BARK, COUNTESS'S BARK; ECORCE DE QUINQUINA, *Fr.;* CHINARINDE, *Germ.*

References.—*Howard's Illustrations of the Neuva Quinologia of Pavon; Howard's Quinology of the East India Plantations, 1869; Gamble, Man. Timb., 223; Vrij, Kinologische studiën, 1868; Pharm. Ind., 114, 448; Ainslie, Mat. Ind., I., 72; O'Shaughnessy, Beng. Dispens., 383; Moodeen Sheriff, Supp. Pharm. Ind., 101; Flück. & Hanb., Pharmacog., 338-70; U. S. Dispens., 15th Ed., 425; Bent. & Trim., Med. Pl., 140, 143; Bidie, Cat. Raw Pr., Paris Exh., 10; Martindale and Westcott, Extra Pharm., 130; Pharm. Jour., Sept. 6th, 1873, 181; Howard's Analysis and Observations, Pharm. Jour., XIV. (1855), 61, 63; Papers Showing Results of Medical Trials of Cinchona Febrifuge, 1878; Broughton in Pharm. Jour., 1873, 521; Report on Quinological work in Madras by D. Hooper; Drury, U. Pl., 134; Royle, Ill. Him. Bot., 238-40; Müller, Extra-Trop. Pl., 78; Propagation and Cultivation of the Medicinal Cinchonas, by W. G. McIver, 1867; The Introduction of Cinchona Cultivation into India, also Notes on the Culture of Cinchonas, by C. R. Markham, 1859; Manual of Cinchona Cultivation in India, by G. King, M.B., 1880; Notes on the Quinquinas by Weddell, 1871; Cultivation of the Cinchonas in Java by K. W. Van Gorkom, 1870; Cinchona Committee's Report on the Nilgiri Plantations, by Walker, 1878; Report on Collections and Seeds of the Chinchonas of Pitayo, by Cross, 1871; Quinquina Plantations of Java by Jungbuhn and Vrij, 1861; Agri. Hort. Soc. Ind., Vols. VI., Pt. II., 1; IX., 140-146; XI., Ap. p. LXXVII. to LXXXVII.; XII., 265, 271, and Ap. XXXV.; XIV., 209 and 129; Indian Forester, X., 177; Year-Book Pharm., 1871, 85; 1874, 19-21, 150; 1873, 443-447; 1875, 12, 153; Elborne in Pharm. Soc. Jour.; Gazetteers — Burma, I., 124; Bengal, Darjeeling District; Official Correspondence and Reports from 1852 to 1887; Indian Agriculturist, VIII., 64, 105, 143, 224, 225; Tropical Agriculturist, 1883, 704; Administration Reports, Bengal, 1882-83, 25, 280; Madras, 99; "Cinchona Cultivation in India," in Calcutta Review, 1866 (No. 84, Art. VI., 384); Balfour, Cyclop.; Smith, Dic., 116; Treasury of Bot., 284; Ure, Dic., Indust., Arts, and Man., 732, 401; Kew Reports, 1877, 15; 1879, 13; 1880, 11-13; 1881, 10; 1882, 18-19; Kew Offl. Guide to the Mus. of Ec. Bot., 33; Kew Offl. Guide to Bot. Gardens and Arboretum, 74, 75; Simmonds, Trop. Agri., 38, 78.*

Dr. King of Calcutta, and **Mr. Lawson** of Madras, each contributed a historical account of the Cinchona cultivation of India, in connection with the samples shown by them at the Colonial and Indian Exhibition held in London in 1886. The writer has availed himself of these notes in compiling the present article, but has at the same time verified the historic and other facts by consulting the works enumerated above.

Habitat.—Dr. King says: "The trees producing the medicinal barks are all natives of tropical South America, where they are found in the dense forests of the mountainous regions of the western parts of that continent at a height of from 2,500 to 9,000 feet above the level of the sea, and in an equable but comparatively cool climate. The Cinchona-producing region forms a crescentic zone which follows the contour of the coast line, but nowhere actually touches it, beginning at 10° N. and extending to 20° S. latitude. The crescentic belt is nowhere much above a hundred miles in width, but its length (following its curve) is more than two thousand. During its course, it passes through the territories of Venezuela, New Granada, Ecuador, Peru, and Bolivia."

"It must not be supposed that each of the medicinal species is to be found growing throughout the whole length of the zone just described; on the contrary, the distribution of the various species is very local, not only as regards latitude, but as regards elevation above the sea. The species found in the region between 10° N. and the equator (the barks of New Granada) were described by **Mutis** in the last century, and more recently by **Karsten** in his *Flora Colombiæ.* **Mutis'** notes remained in manuscript until 1867, when **Mr. Clements Markham** succeeded in unearthing and printing them; and both his notes and drawings have still more recently been published at Paris by **M. Triana** in his *Nouvelles Etudes sur les Quinquinas.* The Cinchonas of the region between the line and 14° S. (the barks of Ecuador and Northern Peru) were first examined by **Ruiz** and **Pavon**, and a magnificient work founded on **Pavon's** specimens was published by **Mr. J. E. Howard** in 1862; while those indigenous in the region from the fourteenth parallel of south latitude to the extremity of the zone in 20° S. were described by **M. Weddell** in his splendid monograph published at Paris in 1849."

HISTORY.

HISTORY OF THE INTRODUCTION OF THE DRUG INTO EUROPE.

"The introduction of the medicinal Cinchona bark to Europe was effected by the Countess of Chinchon, wife of a Spanish Viceroy of Peru. This lady having been cured by its use of an attack of fever contracted while in that country, brought a quantity of the bark to Europe on her return from South America, about the year 1639. Jesuit missionaries appear also to have taken an active part in its introduction. Hence the early names given to the medicine were *Peruvian* or *Jesuit's bark,* and *Countess's powder.* Nothing, however, was known to science of the tree producing this bark until 1739, when **La Condamine** and **Jussieu,** members of a French exploring expedition then in South America, obtained plants with the intention of sending them to the *Jardin des Plantes* at Paris; but the whole collection unfortunately perished in a storm at sea near the mouth of the River Amazon. The first living Cinchonas ever seen in Europe were some *Calisaya* plants raised at the *Jardin des Plantes* from seeds collected by **Dr. Weddell** during his first journey to Bolivia in 1846. In 1742 **Linnæus** established the botanical genus CINCHONA, a term which continues to be employed by the majority of botanists, although some writers (more particularly **Mr. C. R. Markham, C.B.**) prefer the name Chinchona, as more accurately perpetuating that of the noble lady who introduced this invaluable remedy to Europe" (*King*).

The ALKALOIDS.

1116 HISTORY OF THE ALKALOIDS.—"The most important and at the same time peculiar constituents of Cinchona barks are the alkaloids

HISTORY OF THE ALKALOIDS.

enumerated in the following table :—

Alkaloid.	Chemical composition.
Cinchonine	$C_{20} H_{24} N_2 O$
Cinchonidine (quinidine of many writers) . . .	Same formula.
Quinine	$C_{20} H_{24} N_2 O_2$
Quinidine (conquinine of Hesse)	Same formula.
Quinamine	$C_{20} H_{26} N_2 O_2$

"There are other cystallizable alkaloids, but they have no medicinal value so far as is yet known, and there is a non-crystallizable alkaloid which has febrifugal power. These alkaloids exist in the bark in combination with certain organic acids called *kinic, cincho-tannic,* and *quinovic.* Of the alkaloids above mentioned the most valued is undoubtedly *quinine.* Although Cinchona barks have been employed in Europe as febrifuges for the past two centuries, it was not until the year 1820 that any of the several active principles to which they owe their efficacy was obtained in a separate form. The first to be so separated were quinine and cinchonine. Quinidine was discovered in 1833, and cinchonidine not until 1847. Quinamine was discovered so recently as 1872 by Hesse in bark of **C. succirubra** grown in Sikkim.

"Soon after the discovery of quinine, the sulphate of that alkaloid began to be used by the faculty as a medicine in cases where some preparation of 'bark' was required, and gradually the new salt drove out of fashion to a very large extent the powder, tinctures, and decoctions of bark which formerly enjoyed such reputation in medical practice. Until the discovery of quinidine and cinchonidine, commercial sulphate of quinine consisted really of a mixture of the sulphates of all the Cinchona alkaloids, the outward appearance of these being alike. With the separation of the new alkaloids, chemical tests for their recognition began to be inserted in the various *Pharmacopœias*, and pure quinine began to be insisted on in medical practice. The other alkaloids fell therefore into unmerited neglect, and they were not, until quite recently, included in the British *Pharmacopœia.* Their excellence as febrifuges, as will be subsequently related, has now been thoroughly established by the trials given to them by officers of the medical services of the three Indian Presidencies. Cinchona bark still continues to be rated by the European quinine-makers in proportion to the percentage of quinine it contains, the other alkaloids being counted for little or nothing as marketable products. These unsaleable alkaloids have accordingly been accumulating in the hands of makers in Europe, and are purchaseable at a comparatively low price. Regarding the proportion of these alkaloids in Cinchona bark, the learned authors of the *Pharmacographia* write as follows :—

"This is liable to very great variation. We know from the experiments of Hesse (1871), that the bark of **C. pubescens,** *Vahl.*, is sometimes devoid of alkaloid. Similar observations made near Bogota upon **C. pitayensis,** *Wedd.*, **C. corymbosa,** *Karst.*, and **C. lancifolia,** *Mutis*, are due to Karsten. He ascertained that barks of one district were sometimes devoid of quinine, while those of the same species from a neighbouring locality yielded 3½ to 4½ per cent. of sulphate of quinine.

"Another striking example is furnished by De Vry in his examination of quills of **C. officinalis** grown at Ootacamund, which he found to vary in percentage of alkaloids from 11·96 (of which 9·1 per cent. was quinine) down to less than 1 per cent. . . . . . .

"Among the innumerable published analyses of Cinchona bark, there are a great number showing but a very small percentage of the useful principles, of which quinine, the most valuable of all, is not seldom altogether wanting. The highest yield, on the other hand, hitherto

HISTORY OF THE ALKALOIDS.

observed, was obtained by **Broughton** from a bark grown at Ootacamund. This bark afforded not less than 13½ per cent. of alkaloids, among which quinine was predominant.

"The few facts just mentioned show that it is impossible to state even approximately any constant percentage of alkaloids in any given bark. We may, however, say that good *Flat Calisaya Bark*, as offered in the drug trade for pharmaceutical preparations, contains at least 5 to 6 per cent. of quinine.

"As to *Crown* or *Loxa* bark, the *Cortex Cinchonæ palidæ* of pharmacy, its merits are, to say the least, very uncertain. On its first introduction in the seventeenth century, when it was taken from the trunks and large branches of full-grown trees, it was doubtless an excellent medicinal bark; but the same cannot be said of much of that now found in commerce, which is to a large extent collected from very young wood. Some of the Crown Bark produced in India is, however, of extraordinary excellency, as shown by the recent experiments of **DeVry**.

"As to red bark, the thick flat sort contains only three to four per cent. of alkaloids, but a large amount of colouring matter. The quill Red Bark of the Indian plantations is a much better drug, some of it yielding 5 to 10 per cent. of alkaloids, less than a third of which is quinine and a fourth cinchonidine, the remainder being cinchonine and sometimes also traces of quinidine (conquinine).

"The variation in the amount of alkaloids relates not merely to their total percentage, but also to the proportion which one bears to another. Quinine and cinchonine are of the most frequent occurrence; cinchonidine is less usual, while quinidine is still less frequently met with, and never in large amount. The experiments performed in India have already shown that external influences contribute in an important manner to the formation of this or that alkaloid; and it may even be hoped that the cultivators of Cinchona will discover methods of promoting the formation of quinine, and of reducing, if not of excluding, that of the less valuable alkaloids" (*Flück. & Hanb., Pharmacog., 361*).

History of the Introduction of Cinchona into India.

INTRODUCTION INTO INDIA.

Dr. King writes: "The practice of the bark collectors in the wild regions in which Cinchonas naturally grow involved the destruction of each tree felled for its bark, yet no measures were ever taken by the owners of either public or private forests to secure supplies for the future by conservancy or re-planting. Meanwhile the consumption of bark in Europe steadily increased, and, as a natural result, prices rose, and fears began to be entertained that the supply would ultimately fail. The British and Dutch Governments being, by reason of their tropical possessions, the largest consumers of Cinchona barks and of the alkaloids prepared from them, their attention began to be seriously attracted to the increasing price and scarcity of the drug."

Mr. Lawson remarks: "**Dr. Royle**, Botanist to the Indian Government, was the first to suggest that the several species of the genus Cinchona, which yield the celebrated fever-curing barks, should be introduced into India for cultivation. In June 1852 in a report on the subject he writes: "Among the vast variety of medicinal drugs produced in various parts of the world, there is not one, with probably the single exception of opium, which is more valuable to man than the quinine-yielding Cinchonas. The great value of Peruvian bark as a medicinal agent was universally acknowledged very shortly after it became first known in Europe. Its utility and employment have been greatly increased ever since its active principle has been separated in the form of quinine. So greatly indeed has the consumption increased, and so little care has been bestowed upon the preservation of the natural forests, that great fears have been entertained that the supply might altogether cease, or be obtainable only at a price which would place it beyond the reach of the mass of the community."

And again in the same report: "The probability of entire success in the cultivation of the Cinchona trees in India seems to admit of hardly any doubt if ordinary care is adopted in the selection of suitable localities. I myself recommended this measure many years ago, when treating of the family of plants to which the cinchonas belong. I inferred from a comparison of soil and climate with the geographical distribution of cinchonaceous plants, that the quinine-yielding Cinchonas might be cultivated on the slopes of the Nilgiris and of the Southern Himálayas in the same way that I had inferred that Chinese tea plants might be cultivated in the Northern Himálayas."

HISTORY OF THE INTRODUCTION INTO INDIA.

"**Dr. Royle's** recommendations, although approved of, were not at the time acted upon, but were allowed to remain in abeyance until 1859, when the increasing demand for the Cinchona drugs, combined with their constantly increasing dearness, forced the subject again upon the attention of Government. Indeed, things had come to such a pass that it seemed almost certain that, in the course of a very few years, the wholesale destruction of the trees which was going on in America, would reduce the supply of bark to almost nothing." **Dr. King** carries the history to more recent dates by furnishing a brief sketch of **Mr. Markham's** mission to the Andes. "In 1850 **Dr. Grant**, the Honourable East India Company's Apothecary-General in Calcutta, urged this measure; and in 1852 **Dr. Falconer**, then Superintendent of the Calcutta Botanical Garden, recommended that an intelligent and qualified gardening collector should be deputed for a couple of years to the mountains of South America for the purpose of exploring the Cinchona forests, and of procuring an ample stock of young plants and seeds of all the finest species.

"**Dr. Falconer's** proposals were, however, not approved of, and instead, an unsuccessful attempt was made to procure seeds through the agency of Her Majesty's Consuls on the western coast of South America. Three years after **Dr. Falconer's** suggestion had been made and disapproved, **Dr. T. Thomson** (his successor at the Calcutta Garden) again pressed the matter, as also did the late **Dr. T. Anderson.** The Medical Board supported the proposals of these officers in an elaborate minute. It was not, however, until 1858 that the despatch of a special agent to South America was sanctioned by the Secretary of State for India.

"The agent selected for this work was **Mr. C. R. Markham, C.B.**, who organised a triple expedition to the forests of the Andes for the purpose of collecting seeds and plants. **Mr. Markham** himself undertook to collect seeds of the *Calisaya* or yellow bark tree (the most valuable of all the Cinchonas) in the forests of Bolivia and Southern Peru, where alone it is to be found. He arranged that **Mr. Pritchett** should explore the grey bark forests of Huanaco and Humalies in Central Peru, and that **Messrs. Spruce and Cross** should collect the seeds of the red bark tree on the eastern slopes of Chimborazo, in the territory of Ecuador. **Mr. Markham** has narrated his adventures in an interesting volume in which he has, besides, collected much valuable information concerning the inhabitants and flora of regions he traversed. Landing at Islay in March 1860, **Mr. Markham**, accompanied by **Mr. Weir** (a practical gardener), proceeded inland in a north-easterly direction, crossed the two chains into which the Andes are there divided, and, after considerable hardship, arrived in one of the series of long valleys which stretch along the western slopes of the snowy range of Caravaya, and descended to the

HISTORY OF THE INTRODUCTION INTO INDIA.

great plain of western Brazil. **Mr. Markham** penetrated this valley (called Tambopata) to a point beyond that reached by the distinguished French traveller, **M. Weddell**, and by the Dutch Agent, **M. Hasskarl**; and, notwithstanding that his proceedings were prematurely cut short by a failure in his food supplies, he was successful in collecting 497 plants of **Cinchona Calisaya** and 32 of the less valuable species **ovata** and **micrantha.**

"Instead of sending these plants direct to India, **Mr. Markham** was compelled by his orders to take them to India *viâ* Panama, England, the Mediterranean and the Red Sea, and thus to expose them to transhipments and alterations of temperature which ultimately killed them all.

"About the time **Mr. Markham** was exploring the yellow bark forests of Southern Peru, **Mr. Pritchett** was collecting seeds and plants of the species producing grey bark in the forests near Huanaco, in the northern part of the same territory, and was successful in bringing to Lima in the month of August a collection of seeds and half a mule-load of young plants of the three species **C. micrantha, peruviana,** and **nitida.** The task of collecting seeds and plants of the red bark was undertaken by **Messrs. Spruce** and **Cross. Mr. Spruce** had a previous knowledge of the Andes, and he was thus enabled very speedily to form at Limon a nursery of young plants of **Cinchona succirubra,** which were ultimately conveyed safely to India by **Mr. R. Cross.** A quantity of seeds of this species was also collected and sent to India by post. **Mr. Cross** was subsequently commissioned to procure seeds of the pale barks in the forests near Loxa, and this commission he executed with great success. A third expedition to New Granada was made by the same collector with the object of securing seeds of the Carthagena bark, **Cinchona lancifolia** and **pitayensis.** The seeds obtained by **Mr. Cross** were sent to Kew, where they germinated well, and the resulting plants were sent out to India."

"While these arrangements for collecting seeds and plants were being carried out in South America, it had been settled that the experiment of cultivating Cinchona in British India should be begun in the Nilgiri Hills, and a patch of forest-land, fifty acres in extent, situated behind the Government garden at Ootacamund, was accordingly taken up and prepared for the first Cinchona experiment. **Mr. Markham's** consignment of **Calisaya** plants, having reached England in a promising state, continued in that condition until they reached Alexandria; but on their arrival in the Nilgiris in October, they were all in a dying state. Some cuttings were, nevertheless, made from them, but not one of these struck root. **Mr. Pritchett's** plants of grey bark were quite as unfortunate, for they reached India either dead or dying. **Mr. Cross's** plants of **succirubra,** raised from cuttings at Limon, together with six **Calisayas** which had been raised at Kew in 1862, were the only living Cinchona plants collected by **Mr. Markham's** triple expedition that reached India in good condition. The supplies of seeds procured by the three expeditions were more fortunate than the plants. These were sent, in the first instance, to the Royal Garden at Kew, where some were retained and sown. A few of the plants brought from South America were also retained at Kew, so that a sort of reserve depôt was formed there in case of failure in India. For the successful introduction of Cinchona into India and other British possessions, Government are largely indebted for advice, as well as for more active assistance, to **Sir William** and **Sir Joseph Hooker,** the illustrious botanists, father and son, with whose names the fame of the great national institution at Kew has for half a century been identified.

"The seeds not retained at Kew were sent to India. Those of the grey barks arrived in the Nilgiris in January 1861, and those of the red barks two months later. In the month of December 1861, **Dr. Anderson** delivered over to **Mr. McIver** at Ootacamund the plants he had brought from the Cinchona plantation which the Dutch had just succeeded in establishing in Java. **Dr. Anderson** had been sent by the Government of India to visit these plantations, and by the courtesy of the Dutch authorities he was allowed to take away with him 50 plants of **Cinchona Calisaya,** four plants of **lancifolia,** and 284 plants of **Pahudiana.** On the 4th March 1862, **Mr. Cross'** collection of pale or crown bark seeds from Loxa arrived, and the introduction of Cinchona to India became thus an accomplished fact" (*King*).

HISTORY OF THE INTRODUCTION INTO INDIA.

South India.

Introduction into South India.—"The success of **Cinchona succirubra** and **officinalis** on the Nilgiris has been remarkable. Not only do the trees grow luxuriantly, but their bark is richer in alkaloids than much of the Cinchona bark imported from South America. The Government plantations there, according to the returns for 1884-85, contain 1,618,744 trees of sorts. The Nilgiri plantations were under the superintendence of **Mr. McIver** until his death, since which they have been under **Mr. M. A. Lawson.**

"Encouraged by its success on the Nilgiris, Cinchona cultivation was warmly taken up by European residents in the other high lands and hill ranges of the Madras Presidency. The coffee planters of Wynaad put out a good many red bark trees on their estates, and these are found to grow well. In South Canara a small plantation was formed in 1869, at a place called Nagooli, above the Koloor Ghát, and at an elevation of 2,500 feet above the sea; but the experiment there was pronounced by the Madras Government as unlikely to be productive of useful results, and was abandoned. On the Mahendra Mountain, in the Ganjam district, the opening of a small plantation was sanctioned by the Madras Government early in 1871; but this also was given up. Under the Forest Department, an attempt was made to introduce Cinchona on the Nulla Mully Hills, but the first hot weather killed all the plants (red barks), and, a similar fate overtaking a second supply planted in 1867, the experiment was abandoned. In Coorg, Travancore, the Pulney, Tinnivelly, and Shevaroy ranges in the Madras Presidency, the planting of Cinchona was taken up to a greater or less extent, both by private planters and the Government" (*King*).

Mr. Lawson enlarges on the history of the Nilgiri cultivation. He writes of **Mr. Markham's** mission and its results :—

"After a series of adventures, often attended with much danger and more discomfort, the first consignment of plants, consisting of **C. succirubra,** was despatched from Guayaquil on the 2nd January 1861, under the superintendence of **Mr. Cross,** to England, and from thence they were transported through the Red Sea to India. Here 463 arrived in good condition. These were taken to the Nilgiri Hills, the district previously selected by **Dr. Royle** as that in which the different varieties would most probably thrive best. For the hardier kinds **Mr. Markham** selected a site near the top of Dodabetta, the highest rounded knoll of which is about 8,700 feet above the level of the sea, while for the more tender sorts he selected a tract of country about Naduvatam, a small Toda village which lies on the edge of the hills facing the west, and which ranges between 5,500 and 6,000 feet. The plants, on their arrival, were handed over to **Mr. W. G. McIver,** who, for some time previously, had held the appointment of Superintendent of the Government Gardens at Ootacamund, and it is to his care and sagacity that the rapid, enormous increase of the plants is due. Easy as it is now found to propagate and rear the different kinds of Cinchona, it ought never to be forgotten that

HISTORY OF THE INTRODUCTION INTO INDIA.

this ease is the result of the patience and intelligence which Mr. McIver brought to bear upon their cultivation, at a time when nothing about it was known, and everything had to be discovered by experimentation." Mr. Lawson concludes his account by furnishing a list of the species and commercial forms that have been introduced into the Madras plantations. Of these the following are the more important:—

(1) **C. officinalis.**
(2) **C. succirubra.**
(3) **C. Calisaya.**
(4) **C. Ledgeriana.**
(5) **C. javanica.**
(6) **C. Santa Fe (com. form).**
(7) **C. morada (com. form).**
(8) **C. verde (com. form).**
(9) **C. zamba morada (com. form).**
(10) **C. carthagena (com. form).**
(11) **C. Pahudiana.**
(12) **C. Humboldtiana.**
(13) **C. Pitayensis.**
(14) **C. micrantha.**

He adds: "Of these, the only kinds which are largely grown in the Government plantations are **C. officinalis** and **C. succirubra,** and innumerable varieties which are believed to be hybrids between these two species. The other kinds are kept only as botanical curiosities, for they are either worthless as quinine-yielders, or are such as the elevated climate of Dodabetta and Naduvatam does not suit."

Bengal.

Introduction into Bengal.—Dr. King gives the following interesting sketch of the Sikkim plantations:—

"Cinchona cultivation in the Bengal Presidency was begun under the direction of Dr. Thomas Anderson, Superintendent of the Royal Botanical Garden, Calcutta. Subsequently to the death of Dr. Anderson, the Sikkim plantation has been under the charge of Dr. Anderson's successors, *viz.*, Mr. C. B. Clarke, during 1870 and 1871, and Dr. George King, since the latter date. Since 1866, the Sikkim plantations have been under the executive charge of Mr. J. Gammie, the Resident Manager. The first Cinchona seeds received by Dr. Anderson were some sent by Sir W. J. Hooker to the Botanical Gardens, Calcutta, in 1861. In December of that year, these had produced thirty-one plants. During the same year, the Government of Bengal and the Supreme Government of India had taken up the matter in earnest, and accordingly, in the month of September, Dr. Anderson was sent to Java with the double object of familiarising himself with the Dutch mode of cultivation, and of conveying to India the plants which the Governor of that colony had generously offered to the Government of India. Dr. Anderson returned from Java in November, bringing with him 412 living Cinchona plants and a quantity of seeds of **Cinchona Pahudiana.** Shortly after his return from Java, Dr. Anderson proceeded to Ootacamund, and there made over to Mr. McIver 50 of the **Calisaya,** 284 of the **Pahudiana,** and 4 of the **lancifolia** plants which he had brought from Java. In return, he took to Calcutta from Ootacamund 193 plants of **succirubra** and of the species yielding grey bark. Some of the Java plants died in Calcutta, and on the 19th January 1862 the total stock in the Botanical Gardens there from all sources consisted of 289 plants.

"Dr. Anderson recommended that these should be sent to Sikkim, that being the part of the Himaláya which offered, in his opinion, the greatest hope of success. The Sikkim plantations of Government have been largely increased, and at 31st March 1885 their contents were as follows:—

HISTORY OF THE INTRODUCTION INTO INDIA.

	Red (Cinchona succirubra).	Yellow (Cinchona Calisaya and Ledgeriana).	Yellow (Cinchona Calisaya, verde and morada).	Hybrid (unnamed variety).	Other kinds.	Total of all sorts.
Mungpoo Division	2,132,000	801,118	134,300	345,100	25,593	3,438,111
Sittong " .	1,100,000	70,000	15,000	40,000	...	1,225,000
Rungjung "	...	2,15,000	34,000	...	...	249,000
GRAND TOTAL OF ALL KINDS .	3,232,000	1,086,118	183,300	385,100	25,593	4,912,111

"A Cinchona plantation was begun by a private company in Sikkim almost simultaneously with that belonging to Government, and more recently a second such plantation has been opened out in Bhotan. Patches of Cinchona were also planted in several tea gardens in the district, but the cultivation has not commended itself to private enterprise to the same extent in Sikkim as in the South of India and Ceylon."

Khasia hills.

"*Attempts to introduce Cinchona into the Khasia hills.*—In 1867, a Cinchona plantation was opened at Nunklow on the north-western slopes of the Khâsia Hills. The trees throve very well, but on account of the scarcity and dearness of labour, the plantation was abandoned.

N.-W. Provinces.

"*Into North-Western Provinces.*—The cultivation also received a very patient trial for several years in the North-Western Provinces of India, and plantations were begun at various altitudes from 2,000 to 6,500 feet above the sea; but the plants all ultimately perished from frost. A similar result followed the spirited attempt of **Colonel Nassau Lees** to grow Cinchona in the Kangra valley.

Bombay.

"*Into Bombay.*—The cultivation was attempted also on the Mahableshwar Hills in the Bombay Presidency; but there it failed from the excessive moisture of the climate.

Burma.

"*Into Burma.*—In Burma several attempts were made to grow Cinchonas. The most successful of these dates from 1870, and experience has shown that the red barks thrive moderately well; but the difficulty in Burma as in the Khâsia Hills is scarcity of labour" (*King*).

In the Forest Administration Report of Burma for 1881-82, there occurs the following passage regarding Cinchona: "Much money has been spent on the Cinchona plantation at Thandoung, some miles to the north of Toungoo, and about 54,000 plants are now alive. But the plantation does not thrive so well as could be wished, and it is desirable that the advice of an expert should be obtained as to the best course to be taken. It was hoped that **Dr. King** would have visited Burma, but as yet he has been unable to do so. If the Government of Bengal can spare him, perhaps he will be able to come in May 1883. At Pyoonchoung the cultivation of Cinchona has done so poorly that orders have been given to abandon further outlay on the experiment there. About 300℔ of Cinchona bark were recently received from Thandoung, and arrangements have been made for testing its medicinal value. A portion of the Thandoung plantation has been sacrificed to make room for a sanitarium for the troops in garrison at Toungoo."

Ceylon.

Introduction into Ceylon.—"In Ceylon the cultivation of Cinchona was begun in 1861 under the late distinguished botanist, **Dr. Thwaites**, Director of the Botanic Garden at Peradeniya. It was subsequently

HISTORY OF THE INTRODUCTION INTO INDIA.. taken up with great vigour by the very spirited planting community of that then most flourishing colony, and to such an extent was the cultivation carried, that in the year 1881 no less than three millions of pounds of dry Cinchona bark were exported from that island to England, and in subsequent years the exports have materially increased" (*King*). During the years 1885-86-87, **Dr. King** informs the writer the annual exports from Ceylon touched 15 million pounds.

THE SPECIES OF CINCHONA.

There are between 30 and 40 species of Cinchona plants with numerous hybrids and varieties. Indeed, so readily do these species hybridize that some doubt may be entertained as to the desirability of retaining as species many forms that have been described as such. The commercial barks are obtained from about a dozen species, **Calisaya** and **Ledgeriana** being the chief quinine-yielding forms. In a further paragraph reference will be made to the commercial Indian hybrids, but in this place it will be necessary only to allude to the better known species and varieties which are cultivated in India.

1117 **Cinchona Calisaya,** *Weddell;* RUBIACEÆ.

THE CALISAYA BARK OF YELLOW BARK OF COMMERCE, a term also applied to the bark of C. LEDGERIANA.

Vern.—*Búrak,* DEC.; *Shurappattai,* TAM.; *Jradap-patta,* TEL.

References.—*Kew Reports, 1877, pp. 14, 28; 1879, pp. 12, 13; 1880, pp. 11, 25, 32; 1881, 25; 1882, pp. 18, 19, 38; Trop. Agriculturist, 1883, 706.*

Habitat.—A very variable tree, with a trunk twice as thick as a man's body when well grown. Cultivated in Sikkim at moderate elevations. **Dr. King**, in a report dated 1872, says: "This plant yields the yellow bark of commerce, and is a sort second to none in value; it promises to do well in Sikkim. From the difficulty of propagating this species artificially, the progress made hitherto has been slow." Since the above was written the cultivation of this species has been so successfully extended that it is at most only second to **C. succirubra** in point of importance in the Sikkim plantations. In a Resolution of the Bengal Government dated March 1888, it is stated that **Mr. Wood** was of opinion that good quinine barks could be grown in Sikkim. "**Dr. King**, the Superintendent of the plantations, was very strongly of this opinion, and in 1875 he recommended that all further planting of red bark trees should cease. This recommendation was not acted upon for some time. Full effect has, however, been given to it of recent years, and *succirubra* has been supplanted by *Calisaya* to the extent of about a million trees." On the other hand, the attempt to cultivate this species in the Nilgiri hills has been practically abandoned. *Calisaya* was discovered by **M. Weddell** in 1847; it is a native of Bolivia and South Peru. The supply of bark from natural sources is uncertain.

MEDICINE. Bark. 1118 Powder. 1119 Leaves. 1120 **Medicine.**—This yields one of the most valuable of the Cinchona barks, rich in alkaloids, among which quinine forms $\frac{1}{2}$ to $\frac{4}{5}$. The BARK and POWDER form the officinal parts, being powerfully antiperiodic, tonic, and astringent: the two former properties are due to the presence of quinine. The LEAVES also possess tonic and antiperiodic properties.

§ "Useful in conjunctivitis" (*Surgeon Major G. Hunter, Karachi*).

Mr. W. Elborne, in the *Pharmaceutical Journal*, remarks of this bark: "Two varieties of CALISAYA bark are distinguished in commerce,—*flat* and *quilled*. *Flat Calisaya* bark is flat or nearly so. It is generally

MEDICINE.

uncoated, consisting almost entirely of liber; is $\frac{1}{8}$ to $\frac{1}{2}$ inch thick. Its texture is compact and uniform; the transverse fracture is finely fibrous. Externally the colour is slightly brownish, tawny-yellow, frequently interspersed with darker patches; the surface is marked by shallow longitudinal depressions which are caused naturally on separating the periderm from the liber and not by the instruments used in detaching it, as formerly supposed. Internally the surface has a wavy fibrous appearance. The taste is very bitter, the bitterness being gradually developed on chewing. The bark of the root is readily known by occurring in short, more or less curved or twisted pieces. *Quill Calisaya* occurs in tubes $\frac{3}{8}$ to $1\frac{1}{2}$ inches thick, often rolled up at both edges. They are coated with a thick, rugged, corky layer, marked with deep longitudinal and transverse cracks, the edges of which are elevated. Small specimens are difficult to distinguish from Loxa bark (*Pharmacog.*, *353*). *Calisaya* bark is the best of all the Cinchona barks—good qualities yielding at least 5 or 6 per cent. of quinine. *Var.* **Ledgeriana**, which yields a bark of extraordinary richness in alkaloids, is more especially cultivated in Java and Sikkim. Supplies of **Calisaya** bark are chiefly derived from India and Java, South America now supplying but very little. This species yields the *Cortex Cinchonæ Flavæ* of the *Pharmacopœia*.

TIMBER. 1121

Structure of the Wood.—Reddish-grey, moderately hard, even-grained. Pores small, in short radial lines. Medullary rays fine, closely packed.

VARIETIES OF C. CALISAYA.

Josephiana. 1122

Zamba. 1123

Morada. 1124

Verde. 1125

Blanca. 1126

Numerous varieties and hybrids have been distinguished of this species, especially by **Weddell**. The best known are *var.* **Josephiana** (named after its discoverer, **Joseph de Jussieu**), and *var.* **Ledgeriana**; but **C. zamba, morada, verde,** and **blanca** are recognised and are being experimentally grown at the Naduvatum plantation, Nilgiri hills. **Dr. Van Gorkum,** the Director of the Cinchona culture in Java, wrote in 1873: "Our plantation consists mostly of **C. Calisaya,** in which quinine is the chief alkaloid." "The Java Cinchona barks are celebrated in Europe for their superior outward appearance and have been able to command a high price. I do not know how far that superior outward appearance may be dependent on the manner of harvesting, drying, and packing, but certain it is that their treatment is highly spoken of." "There are numerous varieties of **C. Calisaya,** but we possess one with which we have become acquainted, especially from the numerous analyses of **Mr. Moens,** and which produces a superior manufacturing bark."

The variety known under cultivation as **C. Ledgeriana** may now be separately alluded to.

Cinchona Ledgeriana (a cultivated form). 1127

This is by many writers viewed as but a form of the preceding species. In cultivation it has many peculiarities that entitle it to a separate position, at least commercially. **Mr. Lawson** says, of the South Indian plantations, after having discussed **C. officinalis** and **C. succirubra**:—

"The only other kind of Cinchona which is being grown at all extensively is the **C. Ledgeriana.** This kind flourishes well at the lower elevation of the Wainád (3,000 feet), where it yields a very high percentage of quinine. On this account it is deservedly a great favourite with many planters. It is, however, a small tree when compared with other kinds of Cinchonas, and consequently the amount of bark harvested in a given number of years is much smaller than that taken from other kinds. The bark also, when it is renewed, is less rich in quinine than the natural bark, so that the trees, instead of having their bark improved by the process of stripping, as is the case in the other kinds of Cinchona, decrease in value. These two circumstances make it doubtful if plantations of **C. Ledgeriana** will, in the long run, be much more profitable to the planter than those formed of the more robust kinds, although the bark of the latter may have a lower percentage of quinine."

During the Colonial and Indian Exhibition several Cinchona experts spoke in the highest terms of this plant. It was urged that its cultivation was certain to prove more remunerative than that of any other species. It could be propagated at lower altitudes than the others (scarcely growing above 4,000 feet), and was, from this point alone, a more economical plant. In Sikkim **C. Ledgeriana** grows well as low as 2,500 feet.

The learned authors of the *Pharmacographia* say of this plant: "Towards the middle of the year 1865, **Charles Ledger**, an English traveller, obtained seeds of a superior Cinchona, which had been collected near Pelechuco, eastward of the lake Titicaca, about 68° west longitude and 15° south latitude, in the Bolivian province of Canpolican. In the same year the seeds arrived in England, but were subsequently sold to the Dutch Government, and raised with admirable success in Java, and a little later in private gardens in British India. The bark of **C. Ledgeriana** has since proved by far the most productive in quinine of all Cinchona barks. The tree is a mere form of **C. Calisaya**. **Mr. Hooper**, Quinologist to the Madras Government, in a recent report, remarks: "In the **Ledger** bark it will be noticed that there is a steady rise of quinine up to the age of between five and six years, after which there is no apparent increase."

1128 **Cinchona carthagena.** (Commercial name.)

This has been successfully introduced into the Nilgiri hills within the past few years, and **Mr. Lawson** alludes to it in his reports. In 1881-82 he says that up to date "the propagation of this valuable Cinchona was carried on with most satisfactory results." Again, in 1882-83, the plants "continue to make a very satisfactory growth."

1129 **C. officinalis,** *Hook.*

LOXA OR CROWN BARK; the Pale Bark of Commerce.

Syn.—C. CONDAMINEA, *Humb.*

References.—*Year Book of Pharm.*, *1873, 447; 1875, 161; 1878, 444.*

Habitat.—A native of Ecuador and Peru. Cultivated at high elevations on the Nilgiris, in Ceylon, and in Sikkim, but not extensively. **Mr. Lawson** says of the Nilgiri plantations: "the **C. officinalis,** in its most luxuriant condition, forms a weak straggling tree, which rises at the very most to something less than 20 feet. It has dark glossy leathery leaves, which give to a plantation consisting of this species a black and sombre appearance." The cultivation of this species' has been practically abandoned in Sikkim, as the climate is found to be too moist.

MEDICINE. Loxa Bark. 1130 **Medicine.**—The chief alkaloid afforded by this species is quinine. This affords the *Cortex Cinchonæ Pallidæ* of the Pharmacopœia.

TIMBER. 1131 **Structure of the Wood.**—Yellowish-grey, similar in structure to that of **C. Calisaya.**

Mr. W. Elborne describes the bark of this species:—

"The bark breaks easily with a fracture which exhibits very short fibres on the inner side. The Loxa bark of commerce is chiefly produced by this species, though occasionally other species of Cinchona contribute to furnish it. At the present day it is scarcely possible to obtain genuine Loxa or Crown bark from South America; India, Ceylon, and Jamaica being the chief sources of the bark in commerce.

"The South American Crown bark yields on an average ·5 to 1 per cent. of alkaloids, while the Indian bark yields as much as 4·30 to 5 per cent., consisting principally of quinine, and next in order cinchonidine and cinchonine."

Cinchona succirubra, *Pavon.* 1132

RED BARK.

References.—*Year-Book of Pharm., 1873, 70—73, 447; 1874, 19—20, 150—154; 1875, 12, 159; Kew Report, 1877, 28.*

Habitat.—Cultivated on the Nilgiris and other hills of South India; at the plantations of Rangbí and Poomong in Sikkim; on the hills east of Toungoo, in Burma; and in parts of the Satpura Range in Central India.

Mr. Lawson writes of South India, while speaking of **C. officinalis**: "The **C. succirubra**, on the other hand, has a bold sturdy stem, which, in rich soil and sheltered situations, grows to the height of 50 feet or more. The leaves are a bright apple-green in colour, and a plantation made up of this species looks as light and bright, as that of the **C. officinalis** looks dark and gloomy."

This and **C. Calisaya** and the Sikkim hybrid are the principal kinds grown in Bengal, and **C. officinalis,** while practically a failure in Sikkim, is the chief species grown on the Nilgiri hills, and after that **C. succirubra,** and third in importance **C. Ledgeriana.**

MEDICINE. Red Bark. 1133

Medicine.—This species thrives at a lower elevation than the others, but is comparatively poor in quinine, though rich in cinchonine and cinchonidine. It yields its best bark when eight years old. From it is chiefly derived the "Cinchona Febrifuge," which is now largely manufactured at the Government Plantation of Rangbí. **Mr. W. Elborne** remarks (*Pharm. Soc. Jour.*): "The experiments of **Mr. J. E. Howard** and others have proved that the bark of the root contains a larger proportion of alkaloids than that of the stem, and that the proportion of alkaloid diminishes upwards to the branches." **Mr. David Howard** has also shewn that the nature of the alkaloid varies according to the part of the tree from which the bark has been taken.

In the opinion of pharmacists the bark most suitable for medicinal use is the **Cinchona succirubra.** The cause of this preference, as pointed out by **Mr. Holmes**, are the following: (1) the red bark supply will probably be always equal to the demand on account of its growing on a much lower elevation and consequent distribution over a much wider area; (2) if of a good quality it contains all the alkaloids with the exception of *aricine*; (3) it is less liable to be mixed with hybrids on account of its characteristic appearance. **Professor Fluckiger** suggests the use of **Cinchona succirubra** in the *Pharmacographia*; and both **Mr. Umney** and **Mr. R. W. Giles** have repeatedly pointed out the unsatisfactory results of using the yellow barks for pharmaceutical preparations.

Red Cinchona bark is generally coated, and consists of liber, the cellular and tuberous coats, and usually more or less of the epidermis; its outer surface is rough, furrowed, and frequently warty; the colour of the epidermis varies from reddish brown to chestnut brown, cryptogamic plants are not so frequent as on some other kinds of bark. The cellular coat of the flat pieces is very thick and spongy. The inner surface of the quills is finely fibrous, giving a comparatively smooth fracture, while the fracture of the flat pieces is both fibrous and splintery. (*Elborne.*)

As to the proportion of alkaloids in red bark, the thick flat sort contains from 3 to 4 per cent. of alkaloids, but a large amount of red colour-

MEDICINE. ing matter. The brick-red colouring matter is not found in the growing plant but in the dried bark, and **Mr. J. E. Howard** considers that it is really an excretory product of vegetation, a part used up and brought by contact with the air into a state in which it can no longer be serviceable to the living plants, and from which it still degenerates by a still further degeneration into *humus.* It is by a process of *eremacausis* that the red bark acquires its colour, the cinchotannic acid in which it abounds having become oxidised and changed into cinchona red, and under these conditions the alkaloids also appear to undergo some corresponding alterations. They are now implicated with resin which appears to have also become oxidised so as to act the part of an acid, and is with difficulty separated. But the most remarkable feature is the altered condition of the alkaloids themselves. Quinine, which formed a considerable portion of the whole, is now diminished, while cinchonine and cinchonidine remain much the same. The quill red bark of Indian plantations is a much better drug, some of it yielding 5 to 10 per cent. of alkaloids, less than a third of which is quinine and a fourth cinchonidine, the remainder being cinchonine and sometimes traces of quinidine. (*Elborne.*)

TIMBER. 1134 **Structure of the Wood.**—Yellow, moderately hard. Pores small in radial lines; medullary rays, closely packed, fine and very fine.

HYBRIDS. 1135

Hybrids of Cinchona.

Kuntze, after examining the living Cinchonas in the Indian plantations and working through the collections of dried specimens in the Herbaria and the literature of the species Cinchona, proposed to reduce all to four forms. It has been admitted by most writers that considerable reductions will have to be effected, since in few genera do the species manifest a greater tendency to variation and hybridization than do the plants referred to the genus Cinchona. **Mr. J. Broughton,** in a report submitted to Government in July 1871, furnishes interesting information as to the tendency to hybridism among the species of Cinchona. On this same subject he also communicated a paper in 1870 to the Linnæan Society of London. Speaking of a hybrid between **C. succirubra** and **micrantha** he remarks: "This plant was picked up a seedling under a tree of the latter. I analysed its bark, and found its yield was poor, but represented a mean between the qualities of the two species. Examination among seedling trees led to the discovery of many other examples of hybridism, especially cross-breeds between **C. succirubra and officinalis.**" "I cannot but think that this ready hybridism between the species of Cinchona affords an explanation of the occurrence of the numerous varieties which have been recognised by botanists." In the course of this examination of hybrids **Mr. Broughton** picked up an interesting form which he called "**lanceolata**"—a name afterwards corrected into "**angustifolia.**" The analysis of

Angustifolia. 1136 the bark of this form revealed the fact that it was extremely rich in alkaloids. It has been established that this form is a hybrid from **officinalis**

Bonplandiana 1137 allied to the form **Bonplandiana.** From the fact that it is reproduced by seed, **Mr. Howard** suspects that it may be a species not a hybrid. Be that as it may, it yields a very high percentage of pure quinine. **Mr. Broughton** on analysis found the bark of this form to contain "a total of 8 per cent. of alkaloids (of which 7·15 proved to be quinine)." "I do not think I am exaggerating," observes the same chemist, "by adding that the quinine product of this bark is the finest yet obtained, possessing as it does all the qualities that constitute excellence." Suffice it to say that this form is now extensively propagated on the Nilgiri hills.

About the same period a valuable hybrid appeared in Sikkim among plants reared from Ceylon seed. This is known as "the hybrid" to distinguish it from the numerous self-sown hybrids that are constantly appearing in the plantations. Of this form **Mr. C. B. Clarke** wrote in 1871, that the gardener took it for **C. pitayensis.** **Mr. McIver** thought it was **C. uritasinga,** an opinion which **Mr. Broughton** could not concur in. It has continued to be alluded to ever since as the Sikkim hybrid and is now extensively cultivated. **Dr. King**, in his report for 1874, says: "The analysis of the bark" of this hybrid or species "shows it to contain much quinine. Since the discovery of this fact, every effort has been made to propagate this variety. Experience, moreover, proves that it grows well in Sikkim and at a higher level than **Calisaya.**" Then, again, in a Resolution of the Government of Bengal, in the following year, it is remarked: "The hybrid plants promise great success; they yield a bark rich in alkaloids and are of a vigorous growth."

Although repeated reference is made by writers to the Nilgiri and Sikkim hybrids down to the present date, the subject seems to have lost the interest it created in the early stages of Cinchona cultivation. More is now said of the cultivation of the recognised specific forms described in the preceding paragraphs than of the hybrids, while it has been suggested by many writers that the true future of the industry lies in the improvement, by hybridization or otherwise, so as to produce a plant that will give the maximum of quinine or other alkaloid desired to be obtained.

Chemical Peculiarities of the Cinchona Plants.

CHEMICAL PECULIARITIES. **1138**

We may conclude this account of the forms of Cinchona grown in India by displaying their chemical peculiarities in the following table of comparative analysis taken from **Mr. Lawson's** report:—

The Analyses of the different kinds of barks grown on the Government estates given below, have been made during the past year by Mr. Hooper, the Government Quinologist.

		Quinine.	Cinchonidine.	Quinidine.	Cinchonine.	Amorphous alkaloids.	Total.	Sulph. quinine.
1	C. officinalis, natural . .	2·77	1·57	·16	·39	·50	5·39	3·72
2	,, mossed . .	3·40	1·50	·20	·45	·62	6·17	4·57
3	,, renewed . .	4·21	·85	·22	·65	·70	6·63	5·66
4	C. angustifolia, natural . .	3·97	1·32	·12	·12	·87	6·40	5·34
5	,, mossed . .	5·60	1·41	·33	·04	·97	8·35	7·53
6	,, renewed . .	4·91	·89	·38	·19	1·14	7·51	6·60
7	C. succirubra, natural . .	1·91	2·11	...	1·14	·88	6·04	2·57
8	,, mossed . .	1·69	2·03	...	1·68	·98	6·38	2·27
9	,, renewed . .	1·84	1·48	...	1·25	·71	5·28	2 47
10	,, branch . .	1·38	2·28	...	1·59	1·16	6·41	1·85
11	,, root . . .	1·24	·77	·41	1·43	1·27	5·12	1·66
12	,, renewed shavings.	2·30	1·16	...	2·06	1·45	6·97	3·09
13	C. robusta, natural . . .	1·43	2·08	...	1·58	·31	5·40	1·92
14	,, mossed . . .	1·92	3·16	...	·77	·35	6·20	2·58
15	,, renewed . . .	4·40	2·54	...	·51	1·65	9·10	5·92
16	,, branch . . .	1·64	2·71	...	1·17	·50	6·02	2·20
17	C. micrantha, natural . .	...	...	...	1·92	·40	2·32	...
18	,, renewed . .	tr.	2·45	...	1·12	1·02	4·59	...
19	,, branch . .	...	...	...	1·60	·45	2·05	...
20	C. Calisaya, natural . . .	1·21	2·32	...	2·13	·29	5·95	1·62
21	,, branch . . .	·59	·73	...	1·93	·48	3·73	·79

CINCHONA. **Chemical Peculiarities of the Cinchonas.**

CHEMISTRY.

Analyses of different kind of barks grown on Government estates, &c.—contd.

		Quinine.	Cinchonidine.	Quinidine.	Cinchonine.	Amorphous alkaloids.	Total.	Sulph.quinine.
22	C. Calisaya *var*. Anglica, natural	·81	·88	·29	1·49	·44	3·91	1·09
23	,, ,, branch	tr.	tr.	·25	2·04	·36	2·65	...
24	C. Ledgeriana, natural . .	5·49	1·33	...	·82	·88	8·52	7·38
25	,, branch . .	2·21	·49	...	1·07	·50	4·27	2·97
26	C. javanica, natural . . .	...	...	1·32	2·64	·48	4·44	...
27	,, branch . . .	...	...	1·43	1·49	·45	3·37	...
28	C. Humboldtiana, natural . .	2·24	1·55	tr.	·49	·90	5·18	3·01
29	,, renewed . .	1·28	·64	...	·43	1·07	3·43	1·72
30	C. pitayensis, natural . .	2·34	·56	1·10	1·93	·39	6·32	3·14
31	,, mossed . .	3·81	·95	·63	1·91	·37	7·67	5·12
32	,, renewed . .	2·50	·52	·78	2·33	·55	6·68	3·36
33	C. nitida	1·42	2·45	...	1·45	·67	5·99	1·91
34	Pahudiana, natural . . .	·04	·10	...	·39	·43	·96	·05
35	,, renewed . . .	·51	1·19	...	·28	·87	2·85	·68

Dr. King furnishes the following analysis of the yellow and hybrid barks of Bengal:—

"The Sikkim plantations produce red and yellow barks. Of the yellow barks the following four analyses may be taken as characteristic:—

Yellow Bark—(Sikkim).

Crystallized Sulphate of Quinine . .	3·93	4·83	6·04	3·49
Ditto of Cinchonidine .	0·36	0·51	0·97	0·32
Ditto of Quinidine . .	traces.	0·06	0·04	0·85
Cinchonine	0·17	0·21	...	...

"But besides red and yellow bark the Sikkim plantations now produce a large quantity of hybrid bark, the composition of which may be seen from the following analysis of four samples:—

Hybrid Barks—(Sikkim).

Crystallized Sulphate of Quinine . .	6·12	3·99	3·12	3·24
Ditto of Cinchonidine .	2·46	3·33	1·21	2·46
Ditto of Quinidine . .	traces.	traces.	0·30	...
Cinchonine (alkaloids)	0·55	0·57	0·71	0·52"

CULTIVATION.

Climate, Situation, and Soil suitable for Cinchona Cultivation.

Dr. King's account of these is as follows:—

In Bengal.
1139

"With regard to the climate suitable for Cinchonas, it may be laid down as a universal rule that none of the medicinal species will stand frost. They prefer rather a cool climate, in which the contrast between summer and winter and between day and night temperatures is not very great. These conditions are in some measure obtained in the Nilgiris and in Sikkim. At Ootacamund, about 7,500 feet above the sea, the minimum lowest temperature in the shade, calculated on an average of the three years, is about 49° and the maximum 69° Fahrenheit; and at Neddiwattum, situated about 2,000 feet lower, the minimum, calculated also over three years, is found to be about 54° Fahrenheit, and the maximum 66° Fahrenheit. Observations taken in 1866 and 1867 at an elevation of 3,332 feet

CULTIVATION.

in the Rangbí valley, in Sikkim, show a minimum temperature of 40° and 41° Fahrenheit, and a maximum of 88° Fahrenheit; the mean minima for the two years being 59·20° and 57·53°; the mean maxima 71·7° and 72·28° Fahrenheit; and the mean temperatures 65·6° and 64·89°, respectively. The latter figures give an idea of a climate fairly suitable for **succirubra**, but rather cold for **Calisaya**. A more congenial climate for both species is indicated by the figures obtained at a lower station (elevation above the sea 2,556 feet) which, for the years 1866 and 1867, are as follow :—

Minimum temperature	40° and	41°	Fahr.
Maximum ,,	92·3° ,,	94°	,,
Mean minimum temperature	59·3° ,,	60·94°	,,
,, maximum ,,	80·6° ,,	81·59°	,,
,, temperature	70·1° ,,	71·26°	,,

"In various parts of Ceylon a favourable climate for Cinchona is obtained, as will be seen from the following extract from a most reliable local publication :—

"In the Dimbula district, for example, there is a mean temperature of 65·8° Fahrenheit, with nothing colder in the shade in winter than 44·5° (12° above freezing point) and nothing hotter in the shade in summer than 89°, both extremes being exceptional; and the latter helping to produce a maximum temperature favourable to coffee cultivation, equally so to tea and Cinchona without being injurious to human health. Dismissing the rarely occurring extremes, we get a mean maximum in the shade of 73·2° Fahrenheit, against a mean minimum of 58·4° Fahrenheit, resulting, as we have already noticed, in a mean shade temperature of 65·8° Fahrenheit.

"In the matter of moisture, the peculiarities of the Cinchonas were at first rather misunderstood, their preference for incessant rain and mist having been exaggerated. It is found, especially on the Nilgiris, that all the species (and particularly the red barks) withstand long droughts. All the species assume a yellowish tint during the rains (indicating an excessive supply of moisture), and (in the Nilgiris) all make their most vigorous growth during the seasons in which sunshine and showers alternate. After a continuance of dull steamy days all the species seem to become tender, and a sudden change to bright sunny weather affects the plants in a most marked way, causing their leaves to flag. In Sikkim, **succirubra** makes its most vigorous growth during the latter half of the rains, but both on the Nilgiris and Himálayas the trees continue to grow for two months after the rains cease.

"Observations which have been made show that (calculated on the returns of five years) there are at Ootacamund no fewer than 218 dry days in the year and at Neddiwattum about 240 dry days. The rainfall of the former locality (on an average of three years) is about 44 inches per annum, and that of Neddiwattum 105 inches. The rainfall in Sikkim is much heavier than on the Nilgiris, but is much affected by locality. At Rangbí (altitude 5,000 feet) during 1872, 165·55 inches of rain were registered, while at Rishap (3,000 feet lower and 4 miles distant) only 120·6 inches fell.

"As regards elevation above the sea, it is found that in the Nilgiris **succirubra** succeeds best at altitudes of from 4,500 to 6,000 feet. An elevation of 7,000 feet is found to be too high, the growth being too slow to be profitable. Pale or crown barks thrive in a zone above this, and seem to succeed well even up to 8,000 feet. **Calisaya** on the Nilgiris has not been a success at any elevation; but it does rather better, as also do the grey barks, within the **succirubra** zone than at higher elevations. In Sikkim, 16° further north, experience has shown that **succirubra** and the grey

CULTIVATION.

braks thrive well from 1,500 to 3,500 feet, and can be got to grow both as low as 800 feet and as high as 5,000 feet; **Calisaya** thrives between 1,500 and 3,000 feet; **officinalis** does not thrive at any elevation.

"All the species are most impatient of stagnant moisture at their roots, and therefore require an open subsoil, a sloping exposure, and the other conditions of perfect drainage. They cannot be got to grow on flat land. Like most other plants, they prefer a rich soil, and for this reason they do better in newly-cleared forest than in grass lands of the sort so extensive in the Nilgiris. The brown or pale barks, however, are more tolerant than the others of a soil poor in vegetable humus, and grow fairly well on grass land. The freer and more friable the surface soil the better; but an open well-drained subsoil is above all things indispensable to their successful growth. As soon as the roots of a Cinchona tree get down into subsoil in which there is any tendency for moisture to collect, the plant most certainly begins to sicken and die. The basis of the soil of the Nilgiris is decomposed gneiss; in Sikkim it is composed both of gneiss and of decaying mica schist."

PROPAGATION. Bengal. 1140

METHODS OF PROPAGATION IN INDIA.—**Dr. King** writes:—

"*In Bengal.*—The Cinchonas may be propagated by cuttings or seeds. But as seeds, even of the most valuable sorts, are now obtainable, it is unnecessary to describe the modes of artificial propagation. The seeds germinate best at a temperature varying from 65° to 70° Fahrenheit; but they will germinate at a temperature as low as 55° Fahrenheit. The most efficient mode of sowing them is in open beds which are sheltered by thatched roofs. The seeds must be sown in fine, rich, thoroughly-decayed vegetable mould, either pure or mixed with an equal volume of clean sharp sand which does not feel clayey or sticky when a little of it is taken up and compressed between the fingers. Mould of this sort can usually be easily collected in the forest, and is specially abundant at the base of old clumps of bamboo. After being sifted, the soil so collected should then be spread in layers about two or three inches in depth and five feet wide on beds of ground which have been previously well cleared and which should slope to one side, so that no water whatever may lodge in them at any season. These beds should be protected from rain and sunshine and from all drip by a single sloping thatch. The surface of the seed-bed should from the first be smooth and even, but not hard and compressed. The seed should then be scattered pretty thickly on the surface, and afterwards a very little fine earth or sand may be sprinkled above it. It is not desired to cover the seeds, but merely to steady them by a little earth above them here and there, so as to get them into proper contact with soil. Water should be given by means of a very finely drilled syringe. The seeds will germinate in from two to six weeks. When the seedlings have got two or three pairs of leaves, they should be transplanted into nursery beds formed in every respect like the seed-beds, but with a thicker layer of vegetable soil. They should be pricked out in lines at distances of a little less than 1½ inches, with a space between the lines of about 2 inches. After having been pricked out, the plants should remain untouched until they are about 4 inches high, when a second transplantation will be necessary. On this second occasion they should be planted out in the same manner as before, only at distances of about 4 to 4 inches each way. When from 9 to 12 inches in height, the seedlings are ready for transplanting into the situations they are permanently to occupy.

"In the early Nilgiri planting the trees are put out at distances of twelve feet apart; subsequently at distances of eight feet; and latterly at six by six feet. In Sikkim, the earlier planting stands six by six feet, but for the past twelve years a distance of four by four feet has been

PROPAGATION.

adopted. The red bark, even in South America, is never a large tree; **Cinchona officinalis** is but a big shrub, and it is doubtful whether in India **Calisaya** will ever attain any very great size. Wide planting is therefore obviously an error. All the Cinchonas, moreover, have the habit of throwing out a quantity of superficial rootlets, and young Cinchona plantations do not thrive until the soil between the trees is sufficiently protected from the sun to allow these superficial rootlets to perform their functions freely. The growth of weeds is also checked by shade. By close planting, therefore, two desirable objects are speedily obtained, and moreover, the trees are encouraged to produce straight clean stems. As the trees begin to press on each other, they can be thinned out, and a quantity of bark may thus be got at a comparatively early period, with positive advantage to the plants that are allowed to remain on the ground."

In Madras. 1141

In Madras.—**Mr. Lawson** gives the following account of the method pursued in South India: "All the Cinchonas may be propagated very readily by seed or cuttings. The former mode is usually adopted for the sake of cheapness, while the latter is only resorted to when it is desired to obtain a stock of some well-known variety very rich in particular alkaloids. The seed is sown broadcast in beds specially prepared and made of rich leaf-mould. They are protected from the sun by light *pandals*, that is, by a thatch of ferns or mats raised 3 feet above the beds, or by branches of ferns simply stuck in the ground, sufficiently thick to completely shade the soil. When about an inch in height, these seedlings are pricked out into other beds at a distance of 3 inches apart. When they have grown 9 inches or a foot in height they are ready to be planted out in the estate. This is always done in wet and cloudy weather, and each plant is immediately protected with a little dome of fern. If this is not done, and the sun scorches the plants before they are well-rooted, their destruction is certain. For each plant a pit 2 feet cube is dug some months beforehand, so that the soil, when it is returned to the pit, is well ærated and pulverized. As all Cinchonas are lovers of rich food, their well-being in the early stages of their growth is greatly enhanced if the pits are liberally manured. After the first winter is past and the spring showers have set in, the plants, if healthy, are no longer in any danger of destruction from the ordinary climatic changes, and, at the expiration of four or seven years, according to the species, they will yield their first harvest of bark."

Modes of collecting the Bark.

COLLECTION.

Bengal. 1142

In Bengal.—"Various methods of harvesting the bark crop have been adopted. On the Sikkim plantations, the most profitable has been found to be the complete uprootal of the trees, and the collection of the whole of the bark from root, trunk, and branches. A modification of this, which has also been practised there as well as on some of the plantations in South India and Ceylon, is coppicing. It does not, however, by any means invariably happen that the stools yield coppice; for they not unfrequently die, in which case the whole of the root-bark is lost: for the bark of any dead part of a Cinchona tree is always destitute of alkaloids.

"So long ago as 1863, the late **Mr. McIver** discovered that, if a portion of the bark of a living Cinchona be carefully removed so as not to injure the young wood of the tree, the removed bark will, provided certain precautions are taken, gradually be renewed. At the same time he discovered that bark so renewed is richer in alkaloids than natural bark, and he further discovered that the exclusion of light from growing bark materially increases its richness in alkaloid. These discoveries gave rise to the practice of harvesting Cinchona bark by stripping off a certain proportion of their bark (from a third to a half) from living trees, and of covering the

COLLECTION.

stems that had been operated upon with a coating of moss or straw in order to exclude light. The results of this process were very satisfactory both in the Nilgiris and Ceylon. It was also discovered that, provided natural shade be afforded, it is not necessary to coat the partially decorticated trees with moss or straw. **Mr. Moens**, Director of the Dutch plantations in Java, suggested a modification of this process which consists in shaving off the superficial layers of bark from the whole surface of the stem, care being taken that at no point shall the young wood be laid bare. **Mr. Moens** found that the bark of trees thus treated gradually acquires its former thickness, and that the renewed bark is richer in alkaloids than the original bark. This process has been successful in Java, and also to some extent in the Nilgiris and Ceylon" (*King*).

"The practice of stripping bark from living trees was not resorted to during the past year, as it was found that the renewal of the bark under moss was rendered impossible in Sikkim by the attacks of ants" (*Resolution of the Government of Bengal, dated October 1876*).

In Madras. 1143

In Madras.—The following account deals in greater detail (than in the above references) with the practice pursued in Madras :—

"The ordinary way of taking the bark from the trees on the Government plantations is that known under the name of stripping. The barker, with the sharpened point of an ordinary pruning knife, makes several cuts running down the stem parallel to each other, about an inch apart, and then with the blunt back of his knife, he raises every alternate narrow strip and removes it from the tree, being very careful all the while not to crush through the layer of delicate cambium cells which lies between the strip of bark he has to take off and the wood of the tree. If the operation is performed carefully and the cambium cells are not injured, a new layer of bark will be formed in the place of that which has been taken away. If, on the other hand, the layer of cambium cells is crushed or scratched off by clumsy workmanship, no new bark will be formed. In order to facilitate this new formation of bark the stem is covered with moss, grass, leaves of the New Zealand flax (**Phormium tenax**) or some other similar substance, and this is kept in its place by being bound round by rope made from coir yarn. At the end of a few weeks, if recuperation has gone on satisfactorily, there will be seen on the wounded surface small corky pustules; these gradually increase in size and eventually coalesce, so as to form a new bark. The tree should then be manured, if possible, and allowed to remain for three years, after which those intervening strips of bark which were left on the tree are removed. And this process is repeated every third year, until the tree dies or it becomes too old to renew its bark sufficiently quickly to allow of its being profitably grown. In the latter case it is either up-rooted and a young plant put in its place, or it is cut down and one or more shoots are allowed to spring up from its stool.

"Instead of taking the bark from the tree in narrow strips, it is the practice with some planters, and especially with those in Java, to shave the trees. This is done by an ordinary pruning knife or by an American spokeshave. The experiments made in the Madras Government plantations, with the view of discovering which of the two processes is the best, have been either very faulty or show most decidedly that, while the shaving process may be suitable in the case of young trees, the older trees suffer much less severely when the bark is taken from them in strips" (*Lawson*).

TREATMENT.

Treatment of the Removed Bark.

Bengal. 1144

In Bengal.—"After removal from the trees, Cinchona bark has to be carefully dried, and on the best modes of doing this careful experiments have been made. From these it has been found that exposure to a high

TREATMENT OF BARK.

temperature, especially in a moist atmosphere, causes bark to become almost worthless. Even the sun's rays are hurtful, if bark is long exposed to them. To secure it in the best possible condition, bark should be taken off the trees in large pieces, and these should be arranged on drying stages, under shelter from the light and heat of the sun's rays, but freely open to the access of air. The pieces should be frequently turned. Bark should be taken off in dry weather only. If allowed to become mouldy and to ferment, as is apt to happen if it be taken off during wet weather, deterioration more or less serious surely occurs. Dry bark, on the other hand, will keep unchanged for many months. **Mr. Broughton** calculates that trunk bark loses from 70 to 74·8 per cent. of weight in drying, and branch bark from 75 to 76 per cent. The Sikkim experience goes to show that trunk red bark loses 73 per cent., and twig bark 75 per cent" (*King*).

In Madras. 1145

In Madras.—" After the bark is removed from the trees it is dried by the sun or by artificial heat. It is then packed in gunny bags, forming bales containing 100℔ of the bark. It is then despatched for sale, and sold either locally in Madras or in London" (*Lawson*). **Mr. Broughton** urged in his reports that the yield of alkaloids is injured by exposing the bark to sunlight. He remarks: " The experimental evidence of this, already adduced, appears to me to be quite conclusive of the fact, so that further proof is scarcely needed. Further proof appears, however, in the circumstance, of which I have been for some time aware, that the bark of opposite sides of the same tree differs in yield of alkaloids. This is, of course, only fully apparent in trees that are equally exposed to sunlight on each side, which from the site of the plantations does not generally occur." He then gives the analysis of bark in support of this, and shows that the bark from the north exposed side of a tree—a side which for four months was more exposed to the sun than the south—afforded 68 per cent. less alkaloid.

DISEASES OF THE CINCHONA TREES.

DISEASES. 1146

" Cinchona trees are liable to a kind of canker, which often destroys the terminal and lateral branches, and not unfrequently kills the plants outright. This canker is most abundant in situations where the subsoil is badly drained. In Java and Ceylon an insect pest, a species of *Heleopeltis*, causes considerable injury to the expanding leaf-buds " (*King*). In the official correspondence which the writer has perused, the question of the diseases to which the Indian cultivated Cinchonas are subject is hotly contested. This was made one of the subjects of special enquiry by the Cinchona Commission which sat in 1871. The late **Mr. McIver** reported to the Madras Government that he anticipated the cultivation of Cinchona in Sikkim would soon come to an end, as disease to an alarming extent had made its appearance. Several of the persons examined by the Commission stated that the supply of plants received from the Nilgiri hills manifested the disease referred to on their arrival. This **Mr. McIver** disputed, and advanced the argument that the disease might have been acquired during the journey to Bengal. About this time, however, **Mr. Davison**, an officer of the Madras Cinchona staff, admitted that the disease in question was also known on the Nilgiri hills, but that the trees on manifesting it were cut down. The improvements in cultivation pursued in Bengal soon proved that, however obtained, the alarmist statements made by **Mr. McIver** were unfounded, and that the Cinchona industry of Bengal had if anything less to fear from disease than almost any other branch of agricultural enterprise. **Dr. King**, in the correspondence alluded to, established two diseases: " one, a constitutional malady affecting the whole plant and usual-

DISEASES.

ly fatal; the other local and by no means fatal. The former disease is confined entirely to trees which have been originally planted in damp situations, or in situations which have become damp subsequently by the oozing of drainage water in the way already explained. Disease first attacks the roots of such trees. Its existence becomes apparent by the discolorization of their leaves, which ultimately fall off. Gradual shrivelling of the cortical and woody tissues then takes place from the root upwards, and before this process has gone far the death of the plant has begun. This disease is in fact apparently nearly identical with that known to gardeners in England as 'Canker.' It is not in any way infectious or contagious, as some appear to think. It depends entirely on a local cause, namely, excess of moisture in the soil; and where that does not exist, it cannot occur." "The second form of disease does not affect the entire constitution of the plant, but manifests itself in patches on the stem and branches. The appearance of one of these patches is as if some escharotic had been dropped on the bark, which is of a dark unnatural colour, shrivelled, dry, and brittle; occasionally these appearances extend to the wood, but as a rule they do not. In size the patches vary; many are about the size of a shilling, others are much larger. They are not numerous on one tree and are often confined to a single branch. When small, no apparent affection of the general health of the plant occurs, and growth goes on unchecked. When, however, a large patch occurs on a small tree, involving the bark pretty nearly all round the stem, death results. Death from this disease is, however, as far as my observations go, not common, and it is a well-established fact that a tree which has been extensively affected will, when cut down, throw up from its stump perfectly healthy shoots, while in hundreds of trees at Rangbí, I have seen illustrations of recovery, the little patches of diseased bark being thrown off and replaced by perfectly healthy tissue, and the plant apparently as robust as if it never had been attacked." **Dr. King** adds that the disease is most prevalent during the rains, and that he is not prepared with any theory as to its cause. "This disease is not confined like the last to certain spots, but is found on plants in all parts of the plantation."

A careful examination of all that has been written and of the evidence recorded before the Cinchona Commission, leads to the conclusion that the two diseases distinguished by **Dr. King** were by the earlier observers viewed as one and the same. If anything, **Mr. McIver** and most other writers allude to the second disease,—the professional gardeners and Cinchona planters assigning as a cause the damp soil to which **Dr. King** attributes the first disease. The late **Mr. Scott**, in his evidence before the Commission, attributed, as a probable cause of the disease of the bark, the excessive humidity of the atmosphere checking the transpiration and retarding thereby the circulation of the sap—an effect which he thought might cause extravasation of sap into the tissue, and thus produce the isolated patches of disease. This explanation would be in keeping with **Dr. King**'s observation, that it is more prevalent during the rains, and would at the same time point to the conclusion that in point of humidity Sikkim possesses about the maximum that the Cinchona plant can be successfully propagated under. This idea receives further support from the fact that, while **Cinchona succirubra** and **C. Calisaya** can be readily propagated in Sikkim, **C. officinalis** cannot, but that species has been most successfully grown on the less humid slopes of the Nilgiri hills.

It may be concluded that, with care in the selection of sites and the more perfect system of cultivation now pursued, all danger from disease has been practically removed.

ANNUAL YIELD OF BARK.

YIELD.

Bengal. 1147

In Bengal.—The outturn of bark from the Government plantation was, in 1885-86, 339,201lb, bringing the total yield of bark up to 3,256,927lb. Almost the whole of this large amount has been used up in the manufacture of the Government Cinchona Febrifuge—a medicine of which, during the past eleven years, 68,473lb has been used up in India (for the effect of these on the imports of Quinine see the two concluding paragraphs of this article). The yield of bark shown above for Bengal, and that to follow for Madras, do not include the bark from private plantations, the particulars regarding which are not available.

Madras. 1148

In Madras.—"The average amount of bark taken annually from the several Government estates is about 100,000lb or 1,000 bales, and the price realized per bale about R100; but in the course of a few years, when the estates have been restored to their former prosperous condition, the amount of bark annually taken will be greatly increased" (*Lawson*).

RESPECTIVE VALUE OF THE ALKALOIDS.

VALUE OF ALKALOIDS.

"As has been already explained, the medicinal cystallizable alkaloids contained in the bark are quinine, cinchonidine, quinidine, and cinchonine, together with an amorphous alkaloid. A fifth called aricine is occasionally found, but has never been used in medicine. **M. Hesse** has also recently announced the existence of another alkaloid occurring only in the **succirubra** bark grown in Sikkim. This base has received the name of quinamine. As everybody knows, it is the first named of these which has hitherto formed the specific for malarious fever. Bark for the manufacture of this alkaloid consequently brings a price in direct proportion to the amount of quinine contained in it. The barks of **Calisaya officinalis** and **Calisaya** contain the largest proportions of quinine, and are consequently the most valuable to a quinine-maker, who, in buying a bark, takes account only of the quinine in it, and allows little or nothing for the other alkaloids. Cinchona barks, however, have long had a recognised value otherwise than as sources of quinine. They have long been regarded as most valuable tonics, and as such are used for making various pharmaceutical preparations, decoctions, tinctures, &c. For the manufacture of such preparations red bark, as being the richest of all in its yield of *total alkaloids*, has always been much esteemed, and of late years (since it began to get scarce) has brought a price as high or even higher than that got for the barks richer in quinine" (*King*).

GOVERNMENT CINCHONA FEBRIFUGE AND QUININE.

FEBRIFUGE. 1149

"It had for many years been suspected that the other alkaloids in which red bark is so rich are nearly, if not quite, as efficacious febrifuges as quinine. The settlement of this point naturally demanded attention at an early stage of the Cinchona experiment. In order to settle it by actual trial, Commissions of medical officers of Government were appointed, and the result of an extended series of trials instituted by them may be given in the following extracts from their reports:—

"In regard to the relative effects of the three new alkaloids, and with them chemically pure sulphate of quinine, the evidence derived from their use shows that with the exception of sulphate of cinchonine, as already stated, they, in a remarkable degree, so closely resemble each other in therapeutical and physiological action as to render distinctive description of little or no practical utility. In a large proportion of cases in which they were tried chemically, pure sulphate of quinine and sulphates of quinidine and cinchonidine appeared to indicate nearly equal febrifuge

FEBRIFUGE.

power, and in equal circumstances their use produced almost the same physiological results.

"The result confirms the general opinion expressed by the Commission last year, and likewise conclusively established beyond doubt, that ordinary sulphate of quinidine, chemically pure sulphate of quinine, and sulphate of quinidine possess equal febrifuge power, that sulphate of cinchonidine is only slightly less efficacious, and that sulphate of cinchonine, though considerably inferior to the other alkaloids, is, notwithstanding, a valuable remedial agent in fever.

"There is no longer room to doubt that the alkaloids are capable of being generally used with the best effects in India. They have been compared with quinine, a drug which possesses, more than any other that can be named, the confidence of medical practitioners here; and have been found by more than one observer, to supplement this sovereign remedy in some of its points of deficiency. The risk attending their use is clearly not greater than in the case of quinine, nor such as to be in any way deterrent; while the diversities of opinion on their relative usefulness and potency are no more than will be found between opinions concerning any three drugs of the Pharmacopœia examined by separate observers.

"Red bark is rich in total alkaloids, but not very rich in quinine; and the quinine in it is difficult of extraction. Crown bark is, on the other hand, rich in crystallizable quinine, and is nearly as highly valued by the quinine-maker as good American yellow. The establishment of the therapeutic excellence of these alkaloids largely increased the value of the red bark plantations in India, and made much easier of solution the problem of supplying its fever-stricken population with a cheap and effectual febrifuge. And for the solution of this problem the Government very speedily took active steps, by appointing **Mr. J. Broughton**, a skilled chemist educated in England, as Quinologist to the Nilgiri plantations. **Mr. Broughton**, after making some valuable observations on the chemistry of living Cinchonas and initiating a process for extracting the whole of the alkaloids from **succirubra** bark, retired from the service of Government about 1877. The manufacture of **Mr. Broughton's** amorphous quinine was, however, discontinued on the departure of **Mr. Broughton**, and since then the whole of the bark produced on the Nilgiri plantations has been disposed of by sale. In 1873, **Mr. C. H. Wood** was appointed Quinologist to the Government plantation in Sikkim, and by him a process of manufacture was indicated by which the mixed alkaloids of red bark are extracted in the form of an amorphous white powder. This powder is called CINCHONA FEBRIFUGE, and up to the 31st March 1885, 70,491℔ of this drug had been manufactured at the Sikkim plantation. This drug is disposed of only in India, and is issued to Government medical institutions and to private purchasers for charitable distribution at the rate of R16-8 per pound. Although not such an elegant preparation as sulphate of quinine it has been proved (during the very extended trials to which it has been submitted) to be an equally efficient febrifuge. The methods in use for the extraction of the alkaloids from Cinchona bark depend, first, on the displacement of them from their natural combination with quino-tannic acid by prolonged maceration with lime or an alkali; secondly, on their removal in a free state by dissolving them in spirit or mineral oil; thirdly, by their separation from the spirit or oil in combination with hydrochloric or other suitable acid; and finally, by their crystallizations as sulphates or tartrates. The details of the processes by which these results are effected vary in different factories" (*King*).

Mr. Wood, in his report for 1875-76, described his process of preparing the febrifuge as follows:—"The dry bark is crushed into small pieces

C. 1149

FEBRIFUGE.

(but not powdered) and is put into wooden casks, where it is macerated in the cold with very dilute hydrochloric acid. The liquor is then run off into wooden vessels and mixed with an excess of a strong solution of caustic soda; a precipitate forms, which is collected on calico filters, and well washed with water. The precipitate is then dried at a gentle heat and powdered. It constitutes the crude febrifuge which is next submitted to a process of purification. In the latter process a certain weight of the crude product is dissolved in dilute sulphuric acid, and a small quantity of a solution of sulphur in caustic soda is added to the liquor. After the lapse of 24 hours the liquor is carefully filtered. The filtrate is mixed with caustic soda, and the resulting precipitate collected on calico and washed with a small quantity of water; dried and powdered it is then ready for issue, and is sent out under the name of CINCHONA FEBRIFUGE."

QUININE. 1150

QUININE.—By a Resolution of the Government of Bengal (1888) publicity has been given to an important discovery of a simple means of isolating pure quinine and the other alkaloids. The history of this discovery may be here given:—"During a visit which he paid to Holland in 1884, **Dr. King** acquired some hints as to a process of extraction by means of oil. And now, benefiting by the advice of some chemical friends, **Mr. Gammie** has been able to perfect this process, with the result that the whole of the quinine in yellow bark can be extracted in a form undistinguishable, either chemically or physically, from the best brands of European manufacture. This can be done so cheaply that, as long as the supply of bark is kept up, quinine need never cost Government much above twenty-five rupees per pound. It is true that, at the present moment, quinine is obtainable in the open market at rates not very different from this; but that is due to entirely exceptional causes. For some time back the Ceylon planters have been up-rooting their Cinchona trees, both to save them from disease, and to make way for tea-planting which appears now to be becoming the principal industry of that Colony; and Cinchona bark has actually been sold in London below the cost of its production in Ceylon. Indeed, so far has the fall in price gone, that South American bark has been practically driven out of the market. This is a state of matters which cannot continue very long, and which is not likely to recur. In the ordinary course, therefore, quinine might be expected soon to rise to what may be considered its normal price. The object of making public the process now discovered is to check this rise in the price of a drug of such general utility."

Mr. Gammie describes his process as follows:—

Method of extraction of the alkaloids from Cinchona bark by cold oil as used at the Government Cinchona Factory in Sikkim.

"In order that the oil may speedily and effectually act on the Cinchona bark, the latter is reduced to a very fine powder by means of Carter's Disintegrator; and to get the powder of a uniform fineness, it is passed through a scalper, which is a machine commonly used for sifting flour. The scalper is in the form of a box enclosing a sloping, six-sided, revolveing chamber, covered with silk of 120 threads to the lineal inch. It is driven at the speed of about thirty revolutions to the minute. Any particles of the powder which may be too coarse to pass through the silk meshes drop out at the lower end of the revolving chamber and are again passed through the disintegrator.

"2. A hundred parts of the finely-powdered bark are then set aside to be mixed with 8 parts of commercial caustic soda, 500 parts of water, and 600 parts of mixture composed of 1 part of fusel oil to 4 parts kerosine oil. If the caustic soda be of inferior quality, a little slaked lime

QUININE.

(about 5 parts) may be used in addition to the 8 parts of caustic soda; or caustic soda may be altogether omitted, and 15 parts of slaked lime may be used instead of it. The caustic soda is dissolved in the water and mixed with the bark. Then the oil is added, and the whole is kept thoroughly intermixed in an agitating vessel. Should lime be used, it is mixed in fine powder with the dry bark before adding the water and oil.

"3. The agitating vessels in use at Mungpoo are barrels with winged stirrers revolving in them vertically, and with taps on the sides for drawing off the fluids. The first stirring is carried on for four hours, and then the whole is allowed to rest quietly in order that the oil may separate out to the top of the watery fluid. When the oil, which has now taken up the greater part of the alkaloids, has cleared out, it is drawn off by a tap placed just above the junction of the two fluids. The oil is then transferred to another agitator, and is there thoroughly intermixed with acidulated water for five or ten minutes, the mixture being again allowed to rest for the separation of the oil. It will now be found (if sufficient acid has been used and the stirring has been thorough) that the alkaloids have been removed from the oil to the acidulated liquor. The oil is again transferred to the bark mixture, and is kept intermixed with it for two or three hours; the oil is again drawn off in the same way, washed as before in the same acidulated liquor; and this process is repeated a third or a fourth time or until it is found, by testing a small quantity of the oil, that the bark has been thoroughly exhausted of its alkaloids. Each stirring subsequent to the second, need not be continued for more than an hour. The quantity of acid required to take up the alkaloids from the oil will entirely depend on the quality of the bark operated on. If the bark contains 4 per cent. of alkaloids, about 2℔ of either sulphuric or muriatic acid mixed in twenty gallons of water should be sufficient, and so on in proportion.

"4. The after-treatment of the acidulated water containing the alkaloids depends on the product desired, and on the kind of acid that has been used. Should sulphate of quinine be desired and sulphuric acid have been used, the liquor is filtered (if necessary), heated, and made neutral by adding a very weak solution of either caustic soda or liquor ammonia. It is then allowed to cool, and as it cools the crystals form out. These crystals are afterwards separated from the mother liquor by draining through a cloth filter. After they have been thus obtained, the crystals are dried. They are next dissolved in about fifty times their weight of boiling water. The resulting liquor is filtered hot through a little animal charcoal. On cooling after filtration the crystals again form out, and they are separated as before from the mother liquor by filtration through a cloth. The crystalline mass obtained by filtration is then placed in small lumps on sheets of white blotting paper stretched on slabs of plaster of Paris. By this means they are practically dried. They are afterwards thoroughly dried by being laid on blotting paper in a room heated to about 10 degrees above the temperature of the open air.

"5. If Cinchona febrifuge is wanted, the alkaloids are exhausted from the oil by muriatic acid, the solution being neutralized and filtered in the same way. On an excess of caustic soda solution being added, the alkaloids are precipitated. After standing some hours the whole bulk of liquor and precipitate is passed through cloth filters; and when the alkaline liquor has drained off, the precipitate is washed with a little plain water, dried, and powdered. The powder is Cinchona Febrifuge ready for use."

TRADE. 1151

TRADE IN CINCHONA.

PRESENT CONDITION OF THE BARK TRADE.—Dr. King has kindly furnished the following paragraph on this subject:—"The present condition

TRADE.

of the Cinchona bark trade is one of depression. This is by no means due to any diminution of the demand for the Cinchona alkaloids, but in a great measure to the fact that an entirely new source of quinine has of late been discovered in the northern parts of South America. This new source lies in a species of *Remija*, the bark of which is found to contain quinine in comparatively small proportions, but in an easily-extracted form. *Remija* trees are at present found growing in situations quite close to the ports of shipment. There is therefore little expenditure in collecting this bark, and as the trees are plentiful, it has within the past few years been poured into the London market in enormous quantities under the designation of *Cuprea* bark. The depression is also greatly due to the enormous exports from Ceylon, where cinchona is everywhere being up-rooted to make way for Tea. The effect of these flushings has been temporarily to swamp the market, the *Cuprea* crushing out the more costly Cinchona barks. The Cinchona planter, however, has only (if he can afford it) to play a waiting game; for, if the importation of *Cuprea* bark goes on much longer at the present rate, *Remija* trees will soon become scarce in all easily accessible spots; and the exports from Ceylon must soon diminish. With the extension of civilization, and with the increase of wealth in tropical countries, the consumption of quinine must steadily increase; at any rate, as long as malarious fevers continue to exist in these countries."

CUPREA OR REMIJA BARK.
1152

Remija plants have only recently been introduced into India. Plants are being grown in the Sikkim plantations, and **Mr. Lawson** alludes to those in the Nilgiri plantations as too young to advance any opinions regarding the success of this new undertaking. It seems probable, however, that it may be found possible to cultivate the Cuprea-bark plant in regions where labour may be less expensive than is the case with the Cinchona plantations. **Remija purdieana** and **R. pedunculata** yield the Cuprea bark of commerce.

In the official correspondence regarding Cinchona, various opinions have been given as to future prospects. **Mr. J. E. Howard**, in a letter addressed to the Secretary of State in 1872, remarked: "It remains that the planters should not over-supply the demand of the world; this, indeed, is a *possibility*, but one so remote that it may be dismissed from all thought for at least the present generation, and the range of altitude above the sea level and the climate under which the Cinchona can be profitably grown are at best extremely limited, as **Mr. Broughton's** reports abundantly shew, and it will be found eventually that the really productive plantations are not too numerous for profit." It must, however, be admitted that the spirit with which the Ceylon planters took up and are now abandoning Cinchona cultivation, and that the discovery of a new American source of supply, have precipitated **Mr. Howard's** *possibility* in little more than fifteen years. An experienced Ceylon planter stated at a meeting of the Royal Pharmaceutical Society that the price now paid for bark had fallen so low that profit had become problematic.

Indian Foreign Trade in Cinchona and Quinine.

The earliest notice of Indian-grown Cinchona bark in the London market occurs in 1867, but it was some years later before the bark assumed a commercial position. Ten years later, in the Review of Trade for 1875-76, **Mr. J. E. O'Conor** remarks: "The total value of the imports of quinine in 1875-76 was R1,91,619; but it would seem that the removal of the import duty in August 1875 has stimulated the imports which, in the nine months of the current year, are valued at R2,28,978. It is manifest that as yet, even with the aid given by Government in the shape of imported quinine and the alkaloids of Cinchona produced in India at the cost of the State, this valuable febrifuge can reach only a fraction of the population."

TRADE IN CINCHONA. From the value shown in 1875-76 the imports steadily increased, till in 1883-84 they became 9,936℔ valued at R7,25,227, since which date they appear to have steadily declined in value, being last year 12,088℔ valued at R3,62,466. These facts indicate the depreciation in the value of quinine which has taken place, for in quantity it will be seen the imports are greater than at any previous year. The exports of Indian Cinchona bark have steadily increased for years past. In 1882-83 they amounted to 641,608℔ valued at R7,90,861, and last year 1,286,900℔ valued at R14,56,381. Thus, both in quantity and value the exports are double what they were five years ago. These facts would seem to almost point to an opposite conclusion to that alluded to in the concluding sentence of the last paragraph, for, while the imports of quinine have nearly doubled, the declared value of these is only half that of the imports of 1882-83. The exports of Cinchona bark from India are mainly from Madras, the bulk of that produced in Sikkim being used up in the manufacture of the Government Cinchona febrifuge. Commenting on the present Cinchona trade, **Mr. J. E. O'Conor,** in his last *Review of the Trade of British India,* says: "This trade has not developed to the extent which was at one time hoped, and indeed it has been somewhat a disappointment to those who invested in the business with expectations of large fortunes in the no distant future. The fall in prices and the competition of other countries have restricted the trade; but though its dimensions are still relatively small, the trade has been increasing."

In addition to the imports of quinine as a commercial article, reference must be made to the quantities included among Government stores. It is satisfactory to observe, that, as a direct result of the manufacture of the febrifuge, the imports of quinine have fallen off from an amount valued in 1876-77 at R2,69,734 (£26,000) to R683 (£68). If alongside of this fact be placed a statement of the financial position of the Government plantations, the immense benefit conferred on the people of India by the Government effort to provide the only trustworthy specific against the malaria which carries off annually its thousands of the population. In a note written for the Colonial and Indian Exhibition Catalogue **Dr. King** says: "The total cost of the Sikkim plantation has been under eleven lakhs of rupees" (£110,000). "The actual profit on the sale of Cinchona products from the beginning of the plantation to the present time (1884-85) amounts to R4,95,513 (£45,513), while the saving to Government by substitution in its Dispensaries, and Hospitals of cinchona febrifuge for quinine amounts to over twenty-five lakhs of rupees" (£250,000).

"The Government plantations on the Nilgiris contained on 31st March 1885, 1,618,744 cinchona trees of various sorts. During the official year of 1884-85 these plantations yielded a crop of 118,017℔. The financial results of the Nilgiri plantations since their commencement shows a net surplus of profit of R5,51,743 (£55,174)."

CINNABAR.

1153 Cinnabar is a sulphide of mercury, known in the vernacular as *Shingarf.* It is used in dyeing, but more for domestic use than by the professional dyer. It is said to be found in Central India and to be also produced artificially; it sells for R140 a cwt.

See **Mercury.**

CINNAMOMUM, *Blume; Gen. Pl., III., 155.*

Cinnamomum Camphora, *Nees; Fl. Br. Ind., V., 134; Wight, Ic., t. 1818;* LAURINEÆ. 1154

JAPAN CAMPHOR of Commerce is obtained from this tree.

Syn.—CAMPHORA OFFICINARUM, *Nees;* LAURUS CAMPHORIFERA, *Kamp.; Roxb., Fl. Ind., Ed. C.B.C., 340.*

Habitat.—A tall tree, with smooth, shining leaves, a native of China, Japan, and Malay Islands; introduced into the Botanic Garden at Calcutta in 1802. This is one of the sources of camphor. For further information see **Camphor.**

C. glanduliferum, *Meissn.; Fl. Br. Ind., V., 135.* 1155

THE NEPAL CAMPHOR WOOD; THE NEPAL SASSAFRAS.

Syn.—LAURUS GLANDULIFERA, *Wall, in Act. Ser., Med. and Phys., Cal., I., 45.*

Vern.—*Malligiri, marisgiri,* NEPAL; *Rohu,* LEPCHA; *Gunserai,* MECHI, ASS.; *Gundroi,* CACHAR.

References.—*Brandis, For. Fl., 376; Gamble, Man. Timb., 306; Voigt, Hort. Sub. Cal., 308; Pharm. Ind., 196.*

Habitat.—A large tree of South Himálaya from Kumáon eastwards to Assam, the Khásia Hills, and Sylhet.

MEDICINE. 1156

Medicine.—In the *Indian Pharmacopœia* this plant has been recommended as worthy of more attention than has been hitherto paid to it. The wood may be used as a substitute for sassafras. (*Home Dept. Cor.*)

TIMBER. 1157

Structure of the Wood.—Rough, pale brown, highly scented, with a strong smell of camphor when fresh cut; has a certain lustre. Weight 38·5℔ per cubic foot. It distantly resembles that of an **Albizzia** on a vertical section, but is rougher; it is soft to moderately hard, even-grained. Durable, easily worked, and is not touched by insects. Used in Assam for canoes and boat-building; in Sikkim for boxes, almirahs, and other articles; also for planking. It is being tried for sleepers (*Gamble*).

C. iners, *Reinw.; Fl. Br. Ind., V., 130; Wight, Ic., t. 130, 122, 135.* 1158

Syn.—LAURUS NITIDA, *Roxb., Fl. Ind., Ed. C.B.C., 338.*

Vern.—*Jangli-dárchíni, dar chini,* HIND.; *Jangli dálchíni,* DEC.; *Ránácha-dál-chini, ran-dal-chini,* MAR.; *Tikhi,* BOMB.; *Káttu-karuváp pattai, sembela puli pilla,* TAM.; *Adavi-lavanga-patta, pachaku,* TEL.; *Kattu-karuvátoli, kot-karva, kát carva,* MALA.; *Adavi-lavanga-patte, dalchini, yellagada, dálchini yanne, lavanga yale, lavangada yale, cuddú-lavanga,* KAN.; *Sikiyabo, tikyobo, lúleng-kyaw, thit-kyam-bo, len-kyan, nalingyaw,* BURM.

References.—*Brandis, For. Fl., 375; Kurz, For. Fl. Burm., II., 287; Gamble, Man. Timb., 305; Voigt, Hort. Sub. Cal., 307; Pharm. Ind., 195, 460; Moodeen Sheriff, Supp. Pharm. Ind., 102; Flück. & Hanb., Pharmacog., 528, 533; Waring, Bazar Med., 44; Drury, U. Pl., 137; Birdwood, Bomb. Prod., 71; Lisboa, U. Pl., Bomb., 111; Balfour, Cyclop.; Treasury of Bot.; Simmonds, Trop. Agricult., 490; Kew Cat., 110.*

Habitat.—A tree of Eastern Bengal, South India, and Burma. It is "a native of Sumatra; from thence Dr. Charles Campbell sent plants in 1802 to the Botanic Garden at Calcutta, under the Malay name *Kúlit manew.* After seven years the young trees blossomed in February and ripened their seeds in May." (*Roxb.*)

MEDICINE. Bark. 1159

Medicine.—"The INNER BARK possesses in the fresh state a powerful cinnamonic odour and taste, and by careful drying and preparation appears capable of affording *cassia lignea* of good quality. Dr. Æ. Ross

MEDICINE. states that this tree is very abundant in the Balaghat jungles of North Kanara, and that it was from this locality that the *cassia bark*, once so largely exported from that district, was obtained. The smaller BRANCHES when carefully prepared, he pronounces to be nearly equal to that of **C. zeylanicum.** At his recommendation, **Dr. Ross** states, the Bombay Government now farms out these trees, and by this means a very considerable addition has been made to the revenue. It may be used as a substitute for cinnamon, to which, adds **Dr. Ross,** it can hardly be reckoned inferior." (*Pharm. Ind.*) "The SEEDS, bruised and mixed with honey or sugar, are given to children in dysentery and coughs, and combined with other ingredients in fevers." (*Drury, U. Pl.*)

Branches. 1160

Seeds. 1161

FOOD. Bark. 1162 Leaves. 1163

Food.—The BARK and the LEAVES are used for curries. (*Lisboa.*) See **Cassia buds** under **C. Tamala.**

TIMBER. 1164

Structure of the Wood.—Billets of this tree are often sold, together with other kinds of firewood, by the wood-cutters.

1165

Cinnamomum obtusifolium, *Nees; Fl. Br. Ind., V., 128; Wight, Ic., t. 139.*

Syn.—LAURUS OBTUSIFOLIA, *Roxb., Fl. Ind., Ed. C.B.C., 339*; L. CASSIA, *in Herb. Ham.*

Vern.—*Tezpat, ramtezpat, kinton,* BENG.; *Bara singoli,* NEPAL; *Nupsor,* LEPCHA; *Patichanda,* ASS.; *Dupatti,* MECHI; *Krowai,* MAGH.; *Lú-leng-kyaw,* BURM.

References.—*Brandis, For. Fl., 375; Kurz, For. Fl. Burm., II., 287; Gamble, Man. Timb., 305; Voigt, Hort. Sub. Cal., 307; Flück. & Hanb., Pharmacog., 528; Balfour, Cyclop.; Simmonds, Trop. Agri., 490; Kew Cat., 110.*

Habitat.—An evergreen tree, with grey aromatic bark, quarter inch thick, native of the outer North-East Himálaya, ascending to 7,000 feet, and of Eastern Bengal, Burma, and the Andaman Islands.

FIBRE. 1166

Fibre.—The "Muga" silkworm (*Antheræa assamæ*) sometimes feeds on its leaves.

MEDICINE. Bark. 1167

Medicine.—**Dr. Kurz** says the aroma of the BARK is variable, and the bark of the root of the Martaban plant is as aromatic as the best Ceylon cinnamon. **Dr. Gimlette** says the bark is "collected in Dunabaisia, a valley adjacent to that of Nepal proper; it is used in dyspepsia and liver diseases."

FOOD. Leaves. 1168

Food.—LEAVES are aromatic; used in curry. In Assam the dried leaves are used as a spice.

TIMBER. 1169

Structure of the Wood.—Reddish grey, moderately hard, shining, mottled on a vertical section by the medullary rays, the pores containing a gummy substance which exudes copiously on the wood being wetted. Weight, 41lb per cubic foot. **Balfour** says that the wood is useful for various purposes.

1170

C. Parthenoxylon, *Meissn.; Fl. Br. Ind., V., 135; Wight, Ic., t. 1832.*

THE MARTABAN CAMPHOR WOOD.

Syn.—LAURUS PORRECTA, *Roxb.*; PARTHENOXYLON PORRECTUM, *Blume*; LAURUS PARTHENOXYLON, *Jack.*; SASSAFRAS PARTHENOXYLON, *Nees.*

Vern.—*Kayo-gadis,* MAL.

References.—*Kurz, For. Fl. Burm., II., 289; Gamble, Man. Timb., 305; Pharm. Ind., 196; Moodeen Sheriff, Supp. Pharm. Ind., 103; Balfour, Cyclop.*

MEDICINE. Fruit. 1171 Oil. 1172

Habitat.—A native of South Tenasserim, to Penang and Sumatra, Java and China.

Medicine.—The FRUIT yields an OIL used in rheumatic affections. An infusion of the root is also employed as a substitute for sassafras.

Cinnamomum pauciflorum, *Nees; Fl. Br. Ind., V., 129.* 1173

Syn.—LAURUS RECURVATA, *Roxb., Fl. Ind., Ed. C.B.C., 338.*

Vern.—*Dinglatterdop,* KHASIA.

References.—*Gamble, Man. Timb., 305; Flück. & Hanb., Pharmacog., 528; Simmonds, Trop. Agri., 490.*

Habitat.—Met with in the Assam Valley, Khásia Hills, and Sylhet. Bark. 1174

Roxburgh says the leaves and bark, while possessing the generic peculiarities as to aromatic flavour, are inferior to the other species. Leaves. 1175

Structure of the Wood.—Light red, very aromatic, beautifully mottled on a radial section by the medullary rays: rough, hard. Weight, 39℔ per cubic foot. TIMBER. 1176

C. sp. 1177

Vern.—*Hmanthin,* BURM.

Reference.—*Gamble, Man. Timb., 307.*

Habitat.—Met with in South Tenasserim.

Structure of the Wood.—White, with a pink tinge, shining, moderately hard, highly scented. Weight 36 to 43℔ per cubic foot. It is plentiful at Tavoy and Mergui, where it is used for building. TIMBER. 1178

C. sp., perhaps **C. Parthenoxylon,** *Meissn. (Kurz, II., 289),* or **Aperula Neesiana,** *Bl. (Brandis, 383).* 1179

Vern.—*Ka away,* BURM.

Reference.—*Gamble, Man. Timb., 307.*

Habitat.—Met with in South Tenasserim.

Structure of the Wood.—Orange-brown, scented, moderately hard, oily to the touch. It resembles the wood of **C. glanduliferum** in structure. Weight 43 to 46℔ per cubic foot; durable, used for house-building and shingles. TIMBER. 1180

C. sp. (This is probably **C. iners,** *Reinw.,* which see.) 1181

Vern.—*Sinkozi,* BURM.

Reference.—*Gamble, Man. Timb., 307.*

Habitat.—Met with in South Tenasserim; found by the late Mr. Lee in Mergui, but rather scarce.

Structure of the Wood.—Red, soft, strongly scented. TIMBER. 1182

C. Tamala, *Fr. Nees; Fl. Br. Ind., V., 128.* 1183

THE CASSIA LIGNEA OR CASSIA CINNAMON.

Syn.—LAURUS CASSIA, *Roxb., Fl. Ind., Ed. C.B.C., 337, 339* (C. CASSIA, *Don, not of Linn.*); C. ALBIFLORUM, *Nees. in Wall, Pl. As. Rar., II., 75.*

Vern.—*Dálchini, kirkiria, kikra, talisputar, talíspatri, barahmi, silkanti, sinkami (taj kalmi, taj kalam,* bark, and *tezpat, tajpat,* leaves), HIND.; *Tejpát,* BENG.; *Chota sinkoli,* NEPAL; *Nupsor,* LEPCHA; *Dopatti,* ASS.; *Tejpát, barmi,* DEC.; *Darchini, tamálá,* BOMB.; *Taj, tamálpatra,* GUJ.; *Dalchini tiki,* MAR.; *Tálisha-pattiri,* TAM.; *Tállisha-patri,* TEL.; *Lavanga patte, lavangada patte, dalchini,* KAN.; *Thit-kya-bo,* BURM.; *Tamál, tespatra,* SANS.; *Zarnab,* ARAB.

References.—*Brandis, For. Fl., 374; Kurz, For. Fl. Burm., II., 288; Gamble, Man. Timb., 306; Stewart, Pb. Pl., 187; Voigt, Hort. Sub. Cal., 308; Pharm. Ind., 196; Moodeen Sheriff, Supp. Pharm. Ind., 102; Flück. & Hanb., Pharmacog., 528; U. C. Dutt, Mat. Med. Hind., 224; Dymock, Mat. Med. W. Ind., 2nd Ed., 670; O'Shaughnessy, Beng. Dispens., 543; S. Arjun, Bomb. Drugs, 116; Baden Powell, Pb. Pr., 373; Atkinson, Him. Dist., 705; McCann, Dyes and Tans of Beng., 143; Birdwood, Bomb. Prod., 72; Piesse, Perfumery, 106, 112; Balfour, Cyclop.; Smith, Dict., 119; Treasury of Botany; Trop. Agri., 1882-83, 453, 533, 827, 876; Simmonds, Trop. Agri., 490; Fleming, Med. Pl. and Drugs, As. Res., Vol. XI., 170; Kew Cat., 111.*

CINNAMOMUM Tamala.

The Cassia Lignea.

Habitat.—A moderate-sized evergreen tree on the Himálaya, sparingly from the Indus to the Sutlej; common thence eastward, between 3,000 and 7,000 feet, to Eastern Bengal, the Khásia Hills, and Burma.

DYE. Leaves. 1184 **Dye.**—The LEAVES are commonly used as a condiment, but they are also employed in calico-printing in combination with myrabolans. **Dr. McCann** says that in Lohardaga, Chutia Nagpur, the bark (*taj*) is used as an auxiliary with **Mallotus philippinensis.** About 33 tons of the leaves and 24 tons of the bark are annually exported from the tract between the Ramganga and the Sarda. **C. Tamala** is most likely to yield the *Taj* (*Atkinson*) and *Tejpát* of the North-West Provinces and Panjáb, but in Bengal the leaves and bark of **C. obtusifolium,** *Nees,* more commonly bear these names. In fact, the leaves of any species of the genus would be at once called *Tejpat* by a native, but for economic purposes **C. Tamala** is superior to any of the other Indian species. The bark of this plant is the *Cassia Lignea* of Indian commerce. The *Cassia Cinnamon* of Europe is obtained from China, the source of which is still obscure. It is chiefly, however, attributed to **C. Cassia,** *Bl.*, which it seems may be proved but a form of **C. Tamala,** *Nees* (**Gamble** reduces it to be a synonym). The true cinnamon is, however, **C. zeylanicum,** *Breyn.* The roots of **C. zeylanicum,** as also, sparingly, of **C. Tamala** and **C. obtusifolium,** yield camphor, but the true camphor plant of commerce is **C. Camphora,** *Nees,* a native of Japan.

OIL. 1185 **Oil.**—The outer bark of the plant yields on distillation an essential oil. From one cwt. of bark, ¾ of a pound of oil is obtained. The oil has a pale yellow colour and the smell of cinnamon, although very inferior to it in quality. It is chiefly used in the manufacture of soap, especially what is called "military soap." (*Piesse, Perfumery, 10.*)

Sources of Cassia Lignea.—The cinnamon known as Cassia Cinnamon or Cassia Lignea of Indian commerce is obtained from this plant. It is coarser and sold in larger pieces than the true cinnamon or bark of **C. zeylanicum,** for which it is often used as an adulterant. **Kurz** says the bark of the root is quite as good as the true cinnamon bark. In Manipur the writer found the natives on the eastern frontier regularly in the habit of collecting the root-bark instead of the stem-bark.

By some authors the **C. Cassia,** *Bl.* of China is kept distinct from **C. Tamala,** *Nees.* The *Flora of British India* retains doubtfully a plant collected by **Kurz** in Burma under that name. In medical works it is generally stated that the Cassia cinnamon is derived from **C. Cassia,** but on this point **Flückiger** and **Hanbury,** in their *Pharmacographia,* say: "Although it is customary to refer it (Cassia bark) without hesitation to a tree named **Cinnamomum Cassia,** we find no warrant for such reference; no competent observer has visited and described the Cassia-yielding districts of China proper and brought therefrom the specimens requisite for ascertaining the botanical origin of the bark." China is the chief source of the *Cassia Lignea* of commerce, it being exported from Canton. All doubt as to the source of the bark, buds, &c., has recently been removed by the collections made by **Mr. C. Ford. Mr. Dyer** read a paper on this subject before the Linnæan Society, in which he was able to establish from **Mr. Ford's** specimen that the China plant was, as long supposed, **C. Cassia,** *Bl.* No doubt whatever exists as to the source of the Indian *Cassia Lignea.* It is obtained from **C. Tamala,** *Nees,* and from **C. obtusifolium,** *Nees.* **Gamble** remarks that the bark of **C. Tamala** is largely collected and sold under the name of *Taj.*

CASSIA BUDS. 1186 **Cassia Buds or Flores Cassiæ.**—These are the immature fruits of trees yielding *Cassia Lignea.* They are something like cloves, and consist of the unexpended flower-heads. They possess properties similar to those of the bark. It appears from a very old writing that the cassia buds were employed in preparing the spiced wine called *Hippocras* (*Pharmacog.; Treasury of Botany*). **Dr. Dymock** alludes to "Kálá nágkesar (known in Europe as Cassia buds)" as the immature fruits of **Cinnamomum Cassia**, *Blume*, and **C. iners**, *Reinw.*, "imported into Bombay from China and Southern India." He further adds: "Two kinds are found in the Bombay market, Chinese and Malabar; they are used as a spice by the Muhammadans." CASSIA BUDS.

Description of Cassia Bark.—"Chinese *Cassia Lignea*, otherwise called Chinese cinnamon, which, of all their varieties, is that most esteemed, and approaching most nearly to Ceylon cinnamon, arrives in small bundles about a foot in length and a pound in weight, the pieces of bark being held together with bands of bamboo. CASSIA BARK. 1187

"The bark has a general resemblance to cinnamon, but is in simple quills, not inserted one within the other. The quills, moreover, are less straight, even and regular, and are of a darker brown; and though some of the bark is extremely thin, other pieces are much stouter than fine cinnamon,—in fact, it is much **less uniform**. The outer coat has been removed with less care than that of Ceylon cinnamon, and pieces can easily be found with the corky layer untouched by the knife.

"Cassia bark breaks with a short fracture. The thicker bark cut transversely shows a faint white line in the centre running parallel with the surface. Good cassia in taste resembles cinnamon, than which it is not less sweet and aromatic, though it is often described as less fine and delicate in flavour.

"An unusual kind of cassia lignea is imported since 1870 from China and offered in the London market as China cinnamon, though it is not the bark that bears this name in Continental trade. The new drug is in unscraped quills, which are mostly of about the thickness of ordinary Chinese cassia lignea; it has a very saccharine taste and pungent cinnamon flavour.

"The less esteemed kinds of cassia bark, which of late years has been poured into the market in vast quantity, are known in commerce as Cassia Lignea, Cassia vera, or Wild Cassia, and are further distinguished by the names of the localities whence shipped, as Calcutta, Java, Timor, &c.

"The barks thus met with vary exceedingly in colour, thickness, and aroma, so that it is vain to attempt any general classification. Some have a pale cinnamon hue, but most are of a deep rich brown. They present all variations in thickness, from that of cardboard to more than a quarter of an inch. The flavour is more or less that of cinnamon, often with some unpleasant addition suggestive of insects of the genus Cimex. Many, besides being aromatic, are highly mucilaginous, the mucilage being freely imparted to cold water. Finally, we have met with some thick cassia bark of good appearance that was distinguished by astringency and the almost entire absence of aroma" (*Pharmacographia*, 530).

Medicine.—The BARK is given for gonorrhœa, and the LEAVES are used in rheumatism as a stimulant (*Stewart*). The latter "are supposed to have furnished the **Folia malabathri** (see *Tamolpathri, or Indian Leaf of Old Pharmacologists*). They were held in considerable repute by the ancients for their stomachic and sudorific properties (*Dr. Adam's Translation of Paulus Ægineta*, Vol. III., p. 238), and they still constitute an article of the Indian Materia Medica under the name of *Tij* or *Tidj-pat*, the *Tudje* of the *Taleef Shereef* (p. 55, No. 291). They partake of the aroma and pungency, and probably also of the carminative properties, of cinnamon." (*Pharm. Ind.*) They are used in flatulant colic, diarrhœa, and other diseases arising from the disordered state of the bowels. They resemble cloves MEDICINE. Bark. 1188 Leaves. 1189

MEDICINE. closely in medicinal properties, for which they may be substituted. **Baden Powell** says that the leaves are considered by the natives hot and cardiac, and that they are useful in colic, indigestion, and nausea. The bark is prescribed by the hakims in debility of the stomach, enlargement of the spleen, affections of the nerves or heart, pains in the womb, also in retention of urine and catamenia, and bites of serpents and poisoning by opium. "An aromatic oil extracted from the fruit and leaves is used as a medicine" (*Bomb. Gaz., XV., 66*).

Special Opinions.—§ "*Dalchíní*, used in dispensary in place of true cinnamon; equally efficacious" (*Assistant Surgeon Nehal Sing, Saharunpore*). "The leaves in Kashmír, *Barg-i-Taj*, are employed as a substitute for **Chavica Betle**, *Retz*" (*Surgeon-Major J. E. T. Aitchison, Simla*). "Used with long-peper and honey in coughs and colds, also in bronchitis and hay asthma" (*Brigade Surgeon J. H. Thornton, Monghyr*). "Given in decoction or powder in suppression of lochia after child-birth, with much benefit" (*Surgeon-Major J. J. L. Ratton, Salem*). "Is used in coughs, flatulence, and fevers" (*Surgeon-Major D. R. Thomson, Madras*).

CHEMISTRY. 1190 **Chemical Composition.**—"Cassia bark owes its aromatic properties to an essential oil, which, in a chemical point of view, agrees with that of Ceylon cinnamon. The flavour of cassia oil is somewhat less agreeable, and, as it exists in the less valuable sorts of cassia, decidedly different in aroma from that of cinnamon. We find the specific gravity of a Chinese cassia oil to be 1·066, and its rotatory power in a column of 50 mm. long, only 01.° to the right, differing consequently in this respect from that of cinnamon oil.

"Oil of cassia sometimes deposits a stearoptene, which when purified is a colourless, inodorous substance, crystallizing in shining, brittle prisms. We have never met with it.

"If thin sections of cassia bark are moistened with a dilute solution of perchloride of iron, the contents of the parenchymatous part of the whole tissue assume a dingy brown colour; in the outer layers the starch granules even are coloured. Tannic matter is consequently one of the chief constituents of the bark; the very cell-walls are also imbued with it. A decoction of the bark is turned blackish green by a persalt of iron.

"If cassia bark (or Ceylon cinnamon) is exhausted by *cold water*, the clear liquid becomes turbid on addition of iodine; the same occurs if a concentrated solution of iodide of potassium is added. An abundant precipitate is produced by addition of iodine dissolved in the potassium salt. The colour of iodine then disappears. There is consequently a substance present which unites with iodine, and, in fact, if to a decoction of cassia or cinnamon the said solution of iodine is added, it strikes a bright blue coloration, due to starch. But the colour quickly disappears, and becomes permanent only after much of the test has been added. We have not ascertained the nature of the substance that thus modifies the action of iodine; it can hardly be tannic matter, as we have found the reaction to be the same when we used bark that had been previously repeatedly treated with spirit of wine and then several times with boiling ether.

"The mucilage contained in the gum-cells of the thinner quills of cassia is easily dissolved by cold water, and may be precipitated together with tannin by neutral acetate of lead, but not by alcohol. In the thicker barks it appears less soluble, merely swelling into a slimy jelly" (*Pharmacographia, 531*).

The leaves are known as *Tejpat*, and the bark as *Taj*.

FOOD. Bark. 1191 Leaves. 1192 **Food.**—The BARK and the dried LEAVES are used to flavour dishes. It is much employed to adulterate true cinnamon.

Structure of the Wood.—Reddish grey, splits and warps, moderately hard, close-grained, slightly scented; not used. Weight 39 lb per cubic foot. TIMBER. 1193

Introduction of Chinese Plant.—Dr. King, in his report of the Botanic Gardens of Calcutta, 1883-84, alludes to plants received from Hong-Kong. "The plant seems a slow grower, and some time must elapse before anything can be gathered as to the likelihood of its being grown to a profit in Bengal as a source of Cassia bark." 1194

FOREIGN TRADE OF CASSIA LIGNEA. TRADE. 1195

Year.	IMPORTS.		EXPORTS AND RE-EXPORTS.	
	Quantity.	Value.	Quantity.	Value.
	cwt.	R	cwt.	R
1880-81	19,660	4,68,576	4,487	1,18,248
1881-82	9,705	1,90,891	3,865	94,408
1882-83	13,240	2,61,543	2,211	45,921
1883-84	19,917	3,84,491	5,365	1,05,310
1884-85	14,769	2,48,344	4,692	81,394

Imports for 1884-85.

Presidency to which imported.	Quantity.	Value.	Country from which imported.	Quantity.	Value.
	cwt.	R		cwt.	R
Bombay . .	12,308	2,01,944	Aden	...	3
Bengal . . .	2,226	41,460	China—Hong-Kong	13,557	2,24,805
Madras . .	235	4,940	Straits . . .	1,212	23,536
TOTAL . .	14,769	2,48,344	TOTAL .	14,769	2,48,344

Re-exports for 1884-85.

Presidency from which exported.	Quantity.	Value.	Country to which exported.	Quantity.	Value.
	cwt.	R		cwt.	R
Bombay . .	4,675	81,114	Persia . . .	2,785	48,826
Bengal . .	13	225	Arabia . . .	980	17,051
Sindh . . .	4	55	Turkey in Asia .	715	11,956
			Other Countries .	212	3,561
TOTAL . .	4,692	81,394	TOTAL .	4,692	81,394

Dr. Dymock (*Mat. Med., W. Ind., 2nd Ed., 667*) alludes to Cassia Lignea under the name **C. Cassia,** *Bl.* He states that a considerable trade is done in Bombay in the Chinese imported article, and that *Kalfah* or Malabar Cassia is also largely used as a substitute for the Chinese. The former, he says, is sold at $3\frac{1}{2}$ to 4 annas per lb, the latter at about R5 for $37\frac{1}{2}$lb. In a farther page he alludes to **C. Tamala,** so that, apparently, the Malabar Cassia is, according to **Dr. Dymock,** different from **C. Tamala.** Definite information regarding the Indian trade in **C. Tamala** cannot be obtained, but it seems probable very little if any of the truly Indian bark is exported.

1196 **Cinnamomum zeylanicum,** *Breyn.; Fl. Br. Ind., V., 131; Wight, Ic., t. 123.*
TRUE CINNAMON.

Syn.—LAURUS CINNAMOMUM, *Willd.; Roxb., Fl. Ind., Ed. C.B.C., 336.*

Vern.—*Dalchini, qalami-dár-chíni,* HIND., BENG., DUK.; *Dárchíní, kirfa,* PB.; *Taj, dalchini, tikhi,* BOMB.; *Taj, dálchíní,* GUJ.; *Dálchíní, dara-chini,* MAR.; *Karruwa, lavangap-pattai, karuváp-pattai,* TAM.; *Sana-linga, lavanga-patta, sanna-lavangapatta,* TEL.; *Cheriya-ela-vanna-toli, lavanga-patta,* MALA.; *Lavanga-patte, dála-chini,* KAN.; *Rassu kúrúndu,* SINGH.; *Lulingyaw, thitkyabo, simbo-sikiyabo, timbo-likyobo,* BURM.; *Gudatvak,* SANS.; *Qirfahe-sailániyah, dársini,* ARAB.; *Tali khahe, sailániyah, dár-chini,* PERS.

References.—*Brandis, For. Fl., 375; Kurz, For. Fl. Burm., II., 287; Beddome, Fl. Sylv., 262; Gamble, Man. Timb., 305; Thwaites, En. Ceylon Pl., 252; DC., Origin of Cult. Pl., 146; Voigt, Hort. Sub. Cal., 307; Pharm. Ind., 194; Moodeen Sheriff, Supp. Pharm. Ind., 103; Bentl. and Trim., Med. Pl., 224; Flück. & Hanb., Pharmacog., 519; U. C. Dutt, Mat. Med. Hind., 224; O'Shaughnessy, Beng. Dispens., 539; Murray, Drugs and Pl. Sind, 110; Bidie, Cat. Raw Prod., Paris Exb., 15; Waring, Bazar Med., 43; S. Arjun, Bomb. Drugs, 116; Baden Powell, Pb. Pr., 373; Buck, Dyes and Tans, N.-W. P., 37; Birdwood, Bomb. Prod., 71; Lisboa, U. Pl. Bomb., 110, 170, 224; Piesse, Perfumery, 106, 112; Balfour, Cyclop.; Smith, Dict., 118; Treasury of Botany; Trop. Agri., 423; Simmonds, Trop. Agri., 490, 492; Irvine, 26, 28; Med. Top. Ajmere, 133; Fleming, Med. Pl. and Drugs, As. Res., Vol. XI., 170; Kew Cat., 111; Kew Official Guide to the Botanical Garden and Arboretum, 37; U. S. Agril. Dept. Report, 1881-82, p. 224.*

Habitat.—It is a native of the Ceylon forests, but now cultivated on the western coast of that island; it is also said to be met with in the forests of Tenasserim, and is experimentally being cultivated in South India.

CAMPHOR. 1197 **Camphor.**—The root yields camphor (*Kurz*). **Lisboa** says that camphor is prepared from the root-bark.

DYE. 1198 **Dye.**—"*Tejpát* is imported largely from Nepal and from the North-West Provinces forests, and is used, together with myrobalan, chiefly in calico-printing, apparently as a clarifier" (*Buck, Dyes and Tans of N.-W. P., 38*). It seems probable this passage should be referred to **C. Tamala** and not to "**Laurus Cinnamomum.**"

OIL. 1199 **Oil.**—The liber of this plant yields the essential oil of cinnamon, an oil of considerable importance. Three oils are obtained from this plant, one from the bark to the extent of ½ to 1 per cent. Distillation is carried on extensively in Ceylon and occasionally in England. The oil is of a golden-yellow colour, with the powerful odour of cinnamon, sweet and aromatic, but with a burning flavour. It is largely used in perfumery. The leaves yield a brown, viscid, essential oil, of clove-like odour, sometimes exported from Ceylon and sold under the name of "clove oil." The third oil is obtained from the root, of yellow colour, specifically lighter than water, with an odour of camphor and cinnamon, and a strong camphoraceous taste. A fatty oil expressed from the fruit is also noticed by early writers, but it is at present unknown.

Description of the Drug.—"Ceylon cinnamon of the finest description is imported in the form of stick, about 40 inches in length and ⅜ of an inch in thickness, formed of tubular pieces of bark about a foot long, dexterously arranged one within the other, so as to form an even rod of considerable firmness and solidity. The quills of bark are not rolled up as simple tubes, but each side curls inwards so as to form a channel with incurving sides—a circumstance that gives to the entire stick a somewhat flattened cylindrical form. The bark composing the stick is extremely thin, measuring often no more than $\frac{1}{1000}$ of an inch in thickness. It has a light brown, dull surface, faintly marked with shining wavy lines, bearing here and there scars or holes at the points of insertion of leaves or twigs. The inner surface of the bark is of a darker hue. The bark is brittle and splintery, with a fragrant odour peculiar to itself and the allied barks of the same genus. Its taste is saccharine, pungent, and aromatic" (*Pharmacographia, p.* 525).

MEDICINE. Bark. 1200

Medicine.—Cinnamon is aromatic, stimulant, and carminative. It is used in flatulence, flalulent colic, spasmodic affection of the bowels, atonic diarrhœa, and gastric irritation; and has been supposed to act as a stimulant of the uterine muscular fibre, and hence employed in menorrhagia and in tedious labour depending upon insufficiency of uterine contractions. It is also given as a stimulant with other medicines in advanced stages of fever. The volatile oil obtained from the leaves by distillation, known as "clove oil," has been examined by **Dr. Stenhouse**, who found it to be, like the oil of cloves, essentially a mixture of engenic acid and a neutral hydrocarbon having the formula $C_{20} H_{16}$. It also contains a small quantity of benzoic acid. In medicinal properties and uses it resembles closely the oil of cloves. (*Pharm. Ind.*) "Cinnamon is largely used in compound prescriptions. A combination of cinnamon, cardamoms, and *tejapatra* leaves, passes by the name of *trijataka,* these three aromatics being often used together" (*U. C. Dutt*). As a powerful stimulant it is given in cramps of the stomach, toothache, and paralysis of the tongue (*Murray*). **Baden Powell** notices the use of cinnamon in low fever and vomiting, and also as an addition to purgatives to prevent griping. Cordial and astringent properties are also ascribed to it.

Oil. 1201

Special Opinions.—§ "Powdered cinnamon in 20-grain doses is a reputed medicine in dysentery" (*Assistant Surgeon T. N. Ghose, Meerut*). "Appears to be useful in certain forms of amenorrhœa when chewed or as Ol Cinnamoni" (*Surgeon-Major G. Y. Hunter, Karachi*). "The bark ground up with water into a paste is applied to the temples in neuralgia and severe headache" (*K. N. A., Dacca*). "Warm stomach cordial, carminative and astringent, useful in flatulence and diarrhœa. Cinnamon oil applied locally in very small quantity gives great relief in neuralgic headache" (*Surgeon C. M. Russell, M.D., Sarun*).

CHEMISTRY. 1202

Chemical Composition.—"The most interesting and noteworthy constituent of cinnamon is the essential oil which the bark yields to the extent of ½ to 1 per cent., and which is distilled in Ceylon, very seldom in England. It was prepared by **Valerius Cordus**, who stated, somewhat before 1544, that the oils of *cinnamon* and *cloves* belong to the small number of essential oils which are heavier than water, '*fundum petunt.*' About 1571 the essential oils of *cinnamon,* mace, *cloves, pepper,* nutmegs, and several others, were also distilled by **Guintherus** of Andernach, and again, about the year 1589, by **Porta.**

"In the latter part of the last century it used to be brought to Europe by the Dutch. During the five years from 1775 to 1779 inclusive, the average quantity *annually* disposed of at the sales of the Dutch East India Company was 176 ounces. The wholesale price in London between 1776 and 1782 was 21*s.* per ounce, but from 1785 to 1789 the oil fetched 63*s.* to 68*s.*, the increase in value being doubtless occasioned by the war with Holland commenced in 1782. The oil is now largely produced in Ceylon, from which island the quantity exported in 1871 was 14,796 ounces; and in 1872, 39,100 ounces. The oil is shipped chiefly to England.

"Oil of cinnamon is a golden-yellow liquid having a specific gravity of 1·035, a powerful cinnamon odour, and a sweet and aromatic but burning taste. It deviates a ray of polarized light a very little to the left. The oil consists chiefly of *Cinnamic Aldehyde,* $C_6 H_5 (C H)_2 C O H$, together with a variable proportion of hydrocarbons. At a low temperature it

CINNAMOMUM zeylanicum.

True Cinnamon.

CHEMISTRY. becomes turbid by the deposit of a camphor which we have not examined. The oil easily absorbs oxygen, becoming thereby contaminated with resin and cinnamic acid, $C_6 H_5 (C H)_2 CO OH$.

"Cinnamon contains sugar, mannite, starch, mucilage, and tannic acid. The *cinnamomin* of **Martin** (1868) has been shown by **Wittstein** to be very probably mere mannite. The effect of iodine on a decoction of cinnamon will be noticed under the head of 'Cassia Lignea.' Cinnamon afforded to **Schätzlar** (1862) 5 per cent. of ash consisting chiefly of the carbonates of calcium and potassium" (*Pharmacog.*, 526).

Adulteration.—The authors of *Pharmacographia* remark that "Cassia lignea being much cheaper than cinnamon, is very commonly substituted for it. So long as the bark is entire, there is no difficulty in its recognition, but if it should have been reduced to powder, the case is widely different. We have found the following tests of some service when the spice to be examined is in powder: Make a decoction of powdered cinnamon of known genuineness, and one of similar strength of the suspected powder. When cool and strained, test a fluid ounce of each with one or two drops of tincture of iodine. A decoction of cinnamon is but little affected, but in that of cassia a deep blue-black tint is immedintely produced. The cheap kinds of cassia, known as *Cassia vera,* may be distinguished from the more valuable *Chinese Cassia,* as well as from cinnamon, by their richness in mucilage. This can be extracted by cold water as a thick glairy liquid, giving dense ropy precipitates with corrosive sublimate or neutral acetate of lead, but not with alcohol."

FOOD. Bark. 1203

Food.—It is chiefly used as a condiment and for flavouring confectionery; also in curry, and enters into the preparation known as *pán.*

TRADE. 1204

FOREIGN TRADE OF CINNAMON.

Year.	IMPORTS.		EXPORTS AND RE-EXPORTS.	
	Quantity.	Value.	Quantity.	Value.
	℔	R	℔	R
1879-80	1,785	484	202	24
1880-81	7,707	3,511	19,432	4,833
1881-82	2,244	512	67,466	14,436
1882-83	18,731	3,641	27,768	11,068
1883-84	13,687	2,640	35,181	9,330

Detail of Imports, 1883-84.

Province into which imported.	Quantity.	Value.	Country whence imported.	Quantity.	Value.
	℔	R		℔	R
Bengal	916	437	Straits Settlements	11,924	2,034
Madras	12,547	2,143	Other Countries	1,763	606
British Burma	224	60			
TOTAL	13,687	2,640	TOTAL	13,687	2,640

False Pareira Brava. CISSAMPELOS Pareira.

TRADE.

Detail of Exports, 1883-84.

Province from which exported.	Quantity.	Value.	Country to which exported.	Quantity.	Value.
	lb	R		lb	R
Bengal . .	4,032	860	United Kingdom .	30,334	8,328
Bombay . .	715	122	Mauritius . .	3,472	690
Madras . .	30,434	8,348	Other Countries .	1,375	312
TOTAL .	35,181	9,330	TOTAL .	35,181	9,330

CISSAMPELOS, *Linn.; Gen. Pl., I., 37, 962.*

Cissampelos Pareira, *Linn., Fl. Br. Ind., I., 103;* MENISPERMACEÆ. 1205
FALSE PAREIRA BRAVA.

Syn.—C. HERNANDIFOLIA, *Wall., Cat., 49, 79, partly; Roxb., Fl. Ind., Ed. C.B.C., 742.*

Vern.—*Akanádi, dakh nirbisi, pári, harjeuri,* HIND.; *Akanádi, nemuká,* BENG.; *Tejo malla,* SANTAL; *Batúlpoti,* NEPAL; *Katori, tikri, parbik, patáki, bat bel, zakhmi halyát, zucum-yeat, batindú púlh* (leaves), *pilíjari, pilajur,* and *katori* (root), PB.; *Katori* (root), *belpath* (leaves), SIND; *Nirbisi* (root), DUK.; *Venivel,* BOMB.; *Parayel,* GOA; *Pomúshtie, pún múshtie, váta-tirupie,* TAM.; *Pata,* TEL.; *Ambashthái páthá,* SANS.; *Deyamitta, weni-wæla,* SING.

References.—*Brandis, For. Fl., 10, 571; Gamble, Man. Timb., 11; Thwaites, En. Ceylon Pl., 13, 399; Dalz. & Gibs., Bomb. Fl., 5; Stewart, Pb. Pl., 6; Aitchison, Cat. Pb. and Sind Pl., 3; Pharm. Ind., 7; Flück. & Hanb., Pharmacog., 28; U. C. Dutt, Mat. Med. Hind., 103; Dymock, Mat. Med. W. Ind., 22; Ainslie, Mat. Ind., II., 315; O'Shaughnessy, Beng. Dispens., 200; Murray, Pl. and Drugs, Sind, 38; Year Book Pharm., 1873, pp. 23, 450, and 1874, p. 85; S. Arjun, Bomb. Drugs, 193; Drury, U. Pl., 138; Atkinson, Him. Dist., 732; Birdwood, Bomb. Prod., 4; Balfour, Cyclop.; Smith, Dict., 313; Treas. of Bot.; Hanbury, Science Papers, 11, 382; Kew Offl. Guide to the Museum, 10.*

Habitat.—A lofty climber, common both to the Old and New Worlds. In India it is met with in the tropical and subtropical provinces from Sind and the Panjáb to Ceylon and Singapore, ascending in the hotter valleys of the Himálaya to about 5,000 feet. Common below Simla at that altitude. It furnishes the *Radix Pareiræ,* or False Pareira Brava of druggists. The true drug is, however, derived from **Chondodendron tomentosum,** *Ruiz et Pav.,* growing in Peru and Brazil. **Cissampelos Pareira** was, for a long time, believed to have been the source of the true drug.

MEDICINE. Root. 1206

Description of the Drug.—"The dried ROOT occurs in the form of cylindrical, oval, or compressed pieces, entire or split longitudinally, half an inch to four inches in diameter, and from four inches to four feet in length. Bark greyish brown, longitudinally wrinkled, crossed transversely by annular elevation; interior woody, yellowish grey, porous, with well-marked, often incomplete, concentric rings and medullary rays. Taste at first sweetish and aromatic, afterwards intensely bitter." (*Pharm. Ind.*)

In distinguishing the true from the false drug, the following facts have to be borne in mind: "In the root of **Chondodendron** there is a large well-marked central column composed of wedges diverging from a common axis, round which are arranged a few concentric rings intersected by

MEDICINE. wedge-shaped rays, which are often irregular, scattered, and indistinct. The axis is not often eccentric. In **Cissampelos Pareira** the root and stem are nearly alike in structure, and in transverse section there are concentric rings." "(*Year-Book of Pharm.*, *1873, 30.*)

Root. 1207 Bark. 1208 Leaves. 1209 **Medicine.**—The dried ROOT and BARK are used as mild tonics and diuretics in advanced stages of acute and chronic cystitis and catarrhal affections of the bladder; also exercises apparently an astringent and sedative action on the mucous membranes of the genito-urinary organs. They are generally administered in the form of decoction and extract. The leaves are applied to abscess. **Ainslie** writes: "The leaves of this plant are considered by the *vytians* as of a peculiarly cooling quality, but the root is the part the most esteemed; it has an agreeable, bitterish taste, and is considered as a valuable stomachic. It is frequently prescribed in the latter stages of bowel complaints, in conjunction with aromatics. **Cissampelos Pareira** has been very highly extolled by several writers for its medical virtues, particularly by **Sloane, Marcgraaf, Barham,** and **Wright.** The first speaks of the efficacy of the leaves as a vulnerary for a green wound; the second recommends the root given in decoction, in the stone. **Lunan** notices its powers as an antidote against poisons. **Barham,** as quoted by the gentleman last mentioned, has this remarkable sentence respecting it: 'I knew a physician who had performed great cures on consumptive persons, who informed me that his remedy was simply a syrup made of the leaves and root of this plant, for which he had a *pistole* a bottle."

According to **Dr. W. Wright,** the roots "are black, stringy, and as thick as sarsaparilla, agreeably aromatic and bitter, and have been ordered in nephritic disorders, in ulcers of the kidneys and bladder, in humural asthma, and in some species of jaundice. A decoction of them is used for pains and weakness of the stomach." (Quoted by *Ainslie*).

Special Opinions.—§ "Used locally in cases of unhealthy sores and sinuses. Natives bind a twig of this plant as a charm a little above any wound-sore" (*Civil Surgeon J. H. Thornton, B.A., M.B., Monghir*). "Root given for pains in the stomach and for dyspepsia, diarrhœa, dropsy, and cough; also for prolapsus uteri, and applied externally in snake-bite and scorpion sting" (*Rev. A. Campbell, Santal Mission*).

CHEMISTRY. 1210 **Chemical Composition.**—" **Fenueilli** states that the root contains resin, a yellow bitter principle, a brown colouring matter, starch, an azotised substance, and various salts of ammonia and lime" (*O'Shaughnessy*). **Wiggers** discovered in this root the substance *pelosina,* which exists to the extent of about ½ per cent.

§ "Contains a bitter principle, *Buxine,* which, according to **Fluckiger,** is probably identical with *Berberine*" (*Surgeon C. J. H. Warden, Professor of Chemistry, Calcutta*).

Cissus carnosa, *Lam.,* see **Vitis carnosa,** *Wall,* AMPELIDEÆ.

C. discolor, *Blume,* see **V. discolor,** *Dalz.*

C. edulis, *Dalz.,* see **V. quadrangularis,** *Wall.*

C. pedata, *Lamk.,* see **V. pedata,** *Vahl.*

CITRULLUS, *Schrad.; Gen. Pl., I., 826.*

1211 **Citrullus Colocynthis,** *Schrad.; Fl. Br. Ind., II., 620; Wight, Ic., t. 498;* CUCURBITACEÆ.

COLOCYNTH, *Eng.*

Syn.—CUCUMIS COLOCYNTHIS, *Linn., Roxb. Fl. Ind., Ed. C.B.C., 700.*

Vern.—*Indráyan, mákál,* HIND.; *Makhal, indráyan,* BENG.; *Indraun-maraghúne, khártuma, ghúrúmba, kúrtamma, túmbi, ghorúmba, vish-lúmba* (*hanzal* and *indráyan,* fruit; *tukkhm túmma,* seeds), PB.; *Tru-jo-gosht, tru-jo-par,* SIND; *Trúná deda, tras, indrávaná, indrávena, indrak,* GUJ.; *Indrayan, kaddu-kankri,* BOMB.; *Henzil, indrawan,* DUK.; *Thorli indráyan, indráyán, indravana, kadu vrindavana,* MAR.; *Paycúmuti, pey-ko-mattitumatti, peyt-tumatti, verit-tumatti, veriecúm-uttie,* TAM.; *Putsa kayachoythú-putsa, eti-puch-cha, papara búdama, veri-púch-cha pútsa-kaia, chitti-pápara,* TEL.; *Tumti kayi, páva-mekke, kayi,* KAN.; *Peycom muttie,* MALA.; *Indra-váruni, vishala, indral-varuni, mákhál,* SANS.; *Habzal, aulqum, hamzal,* ARAB.; *Hindavanahe-talkh, khar-buzahe rúbáh, khar-buzahe-talkh, kabiste-talkh,* PERS.; *Kiyási, khiá-si, khiáti,* BURM.; *Sheti-putsa, yak-komadú, tittacommodú,* SING.

References.—*Thwaites, En. Ceylon Pl., 126; Dalz. & Gibs., Bomb. Fl., 101; Stewart, Pb. Pl., 96; Aitchison, Cat. Pb. and Sind Pl., 64; Pharm. Ind., 94; Moodeen Sheriff, Supp. Pharm. Ind., 103; Flück. & Hanb., Pharmacog., 295; Bentley & Trim., Med. Pl., 114; U. C. Dutt, Mat. Med. Hind., 172; Dymock, Mat. Med. W. Ind., 279; Ainslie, Mat. Ind., I., 84; O'Shaughnessy, Beng. Dispens., 344; Murray, Pl. and Drugs, Sind, 39; Bidie, Cat. Raw Products, Paris Exb., 9, 62; Year-Book Pharm., 1873, p. 57; S. Arjun, Bomb. Drugs, 57; Drury, U. Pl., 139; Baden Powell, Pb. Pr., 388; Cooke, Oils and Oil-seeds, 37; Atkinson, Him. Dist., 701, 732; Birdwood, Bomb. Prod., 37; Lisboa, U. Pl. Bomb., 254; Stocks on Sind; Duthie & Fuller, Field and Garden Crops, N.-W. P., II., 57; Balfour, Cyclop.; Smith, Dictionary, 128; Treasury of Botany; Kew Offl. Guide to the Museum, 71; Fleming, Med. Pl. and Drugs in As. Res., Vol. XI., p. 164.*

Habitat.—An annual found wild in waste tracts of North-West, Central, and South India. It is the wild gourd of the Book of Kings.

The plant cannot be said to be systematically cultivated anywhere in India; the fruits are collected from plants which grow wild on certain desert tracts of North-West India (*Duthie and Fuller*).

Oil.—Yields, according to **Ainslie**, a clear, limpid oil, used in many of the southern provinces for burning in lamps. (See below.) It is said to be used to dye the hair. **OIL. 1212**

Medicine.—The *Pharmacopiœa of India* describes Colocynth as a hydragogue cathartic, useful in constipation, hepatic and visceral congestions, dropsical affections, and other cases requiring purgatives. Sanskrit writers describe the FRUIT as "bitter, acrid, cathartic, and useful in jaundice, ascites, enlargements of the abdominal viscera, urinary diseases, rheumatism, &c. An OIL prepared from the seeds of Indian Colocynth is used for blackening grey hairs. A poultice of the ROOT is said to be useful in inflammation of the breasts." (*U. C. Dutt, Mat. Med. Hind*). According to the Muhammadan writers, Colocynth is a drastic purgative, removing phlegm from all parts of the system. They recommend the fruit, leaves, and root to be used in costiveness, dropsy, jaundice, colic, worms, elephantiasis, &c. It acts as an irritant on the uterus, and its fumigation brings on the menstrual flow. The author of the *Makhzan* describes a curious mode of administration. "A small hole is made at one end of the fruit and pepper-corns are introduced; the hole is then closed, the fruit enveloped in a coating of clay and buried in the hot ashes near the fireplace for some days; the pepper is then removed and used as a carminative aperient. A similar preparation is made with rhubarb root instead of pepper" (*Dymock, Mat. Med. W. Ind.*). Murray, in his *Apparatus Medicaminum,* recommends the use of the tincture of Colocynth in cases of gout, rheumatism, violent headaches, and palsy, in doses of fifteen drops, morning and evening. **Dr. Kirkpatrick** states that the rind with rhubarb is used by the native practitioners in suppression or repression of urine.

MEDICINE.

Fruit. 1213
Oil. 1214
Root. 1215

CITRULLUS Colocynthis.

Colocynth.

Colocynth is rarely employed alone; it is generally given in combination with other purgatives and carminatives. It commonly causes griping when used alone; in excessive doses it produces inflammation of the intestines and even death. The principal efficient forms for the use of this drug are the compound extract of Colocynth, compound Colocynth pill, and Colocynth and henbane pill. (*Bentley and Trim., Med. Pl., 114.*) From the pulp a watery extract is prepared, which is much employed as a purgative in the form of pills.

According to **Dalzell** and **Gibson**, a compound extract of Colocynth is prepared in large quantities at Hewra, for the supply of the medical stores. In Panjáb the fruit is extensively employed as a purgative for horses. The pulp of the fresh fruit mixed with warm water, or the dried pulp with *ajwain*, is reckoned a special remedy in cholera. The dried root reduced to powder is given as a purgative. (*Bellew.*) **Stocks** says the root and the juice are both used medicinally in Sind. In a report of the drugs shown at the late Colonial and Indian Exhibition from Baroda, the properties of the fruit and root are given in very nearly the same terms as above, so that the knowledge of this drug seems very extensively diffused over India.

Special Opinions.—§ "Used in dropsy and amenorrhœa" (*Native Surgeon T. Ruthnam Moodelliar, Chingleput, Madras*). "First-rate medicine for asthma" (*V. Ummegudien, Mettapollian, Madras*).

CHEMISTRY. 1216 **Chemical Composition.**—"The bitter principle has been isolated in 1847 by **Hübschmann.** He observed that alcohol removes from the fruit a large amount of resin. By submitting this solution to distillation, the bitter principle remains partly in the aqueous liquid, partly in the resin, from which the *Colocynthin* is to be extracted by boiling water. The whole solution was then concentrated and mixed with carbonate of potassium, when a thickish viscid liquid separated. **Hübschmann** dried it and re-dissolved it in a mixture of 1 part of strong alcohol and 8 parts of ether. After treatment with charcoal, the solvents were distilled and the remaining bitter principle removed by means of water. This on evaporating afforded 2 per cent. of the pulp of a yellow, extremely bitter powder, readily soluble in water or alcohol, not in pure ether. Colocynthin is precipitated from its aqueous solution by carbonate of potassium. Colocynthin was further extracted by **Lebourdais** (1848) by evaporating the aqueous infusion of the fruit with charcoal, and exhausting the dried powder with boiling alcohol.

"Again, another method was followed by **Walz** (1858). He treated alcoholic extract of colocynth with water, and mixed the solution, firstly, with neutral acetate of lead, and subsequently with basic acetate of lead. From the filtered liquid the lead was separated by means of sulphuretted hydrogen, and then tannic acid added to it. The latter caused the colocynthin to be precipitated; the precipitate washed and dried was composed by oxide of lead, and finally the colocynthin was dissolved out by ether.

"**Walz** thus obtained about ¼ per cent. of a yellowish mass or tufts which he considered as possessing crystalline structure, and to which he gave the name *Colocynthin.* He assigns to it the formula $C_{56}H_{84}O_{23}$, which in our opinion requires further investigation. Colocynthin is a violent purgative; it is decomposed, according to **Walz**, by boiling dilute hydrochloric acid, and then yields colocynthein, $C_{44}H_{64}O_{13}$, and grape sugar. The same chemist termed colocynthitin that part of the alcoholic extract of colocynth which is soluble in ether but not in water. Purified with boiling alcohol, colocynthitin forms a tasteless crystalline powder.

"The pulp, perfectly freed from seeds and dried at 100° C., afforded us 11 per cent. of ash; the seeds alone yield only 2·7 per cent. They have, even when crushed, but a faint, bitter taste, and contain 17 per cent. of fat oil.

"The fresh leaves of the plant, if rubbed, emit a very unpleasant smell" (*Pharmacog., p. 296*).

Food.—The *Year-Book of Pharmacy* (1873) gives the following account of the fruit as a food substance:— FOOD. Fruit. 1217

"The FRUIT, which is about as large as an orange, contains an extremely bitter and drastic pulp, from which colocynth is obtained. This pulp is said to be eaten by buffaloes and ostriches, but is quite unfit for human food. The seed-kernels, however, which contain but a very small quantity of bitter principle, are used as food by some of the natives of the African desert. For this purpose the seeds are first freed from pulp by roasting and boiling, and subsequent treading in sacks, and then deprived of their coatings, which are also decidedly bitter, by grinding and winnowing. A single kernel, thus separated, has only a mild oily taste, but several, if tasted together, exhibit a distinct bitterness. The kernels contain about 48 per cent. of a fatty oil and 18 per cent. of albuminous substances, besides a small quantity of sugar, and may therefore be regarded as a sufficiently nutritive esculent."

"The KERNELS are heated to boiling, then washed with cold water, dried and powdered, and eaten with dried dates, or used in other ways as food" (*Bentley and Trimen*). "The fruits are often used as food for horses in Sind, cut in pieces, boiled, and exposed to the cold winter nights. They are made into preserves with sugar, having previously been pierced all over with knives, and then boiled in six or seven waters, until all the bitterness disappears." (*Drury, U. Pl.*) Kernels. 1218

Domestic Uses.—The fresh ROOT is used as a tooth-brush. The people of the Berber upon the Nile obtain a TAR from the fruit which they use for smearing leather water-bags. The bad smell of the tar prevents the camels from cutting open the water-bags (*Flück. & Hanb., Pharmacog., 297*). DOMESTIC. Tooth-brushes. 1219 Tar. 1220

Citrullus vulgaris, *Schrad.; Fl. Br. Ind., II., 621.* 1221

THE WATER-MELON.

Syn.—CUCURBITA CITRULLUS, *Linn.; Roxb., Fl. Ind., Ed. C.B.C., 700.*

Vern.—*Tarbuza, tarbuz, turmúz, karbuj, halinda, hindwana, samanka,* HIND.; *Tarbuza, tarmuj,* BENG.; *Tarbúz, mathira, hindwána,* PB.; *Karigo, chauho, meho,* SIND; *Tarbuch, turbuch, karinga,* GUJ.; *Turbuj, kalingad, kalinga, pharai,* BOMB.; *Tarbuj, kalingada,* MAR.; *Pitcha, pullum,* TAM.; *Chayapula, tarambuja,* SANS.; *Dilpasand, kachrehn,* PERS.; *payé, Phá-yaithi,* BURM.

References.—*Dalz. & Gibs., Bomb. Fl., 102; Stewart, Pb. Pl., 95; Aitchison, Cat. Pb. and Sind Pl., 64; DC., Origin Cult. Pl., 262; U. C. Dutt, Mat. Med. Hind., 320; Dymock, Mat. Med. W. Ind., 289; Ainslie, Mat. Ind., I., 217; S. Arjun, Bomb. Drugs, 58; Baden Powell, Pb. Pr., 264, 265, 347; Cooke, Oils and Oil-seeds, 40; Atkinson, Him. Dist., 701; Birdwood, Bomb. Prod., 155; Lisboa, U. Pl. Bomb., 159; Duthie & Fuller, Field and Garden Crops, I., 56; Balfour, Cyclop.; Smith, Dic., 435; Kew Offl. Guide to the Museum, 71; Med. Top. Ajmere, 142.*

Habitat.—Cultivated very generally for its cool, refreshing fruit, especially in Upper and Northern India, and appreciated by natives as well as Europeans. It is supposed to be the Melon of Egypt, the loss of which the Israelites regretted so much. It is sown in January or February, the fruit ripening in the beginning of the hot season. The crop is often destroyed by untimely rain or hail-storms. A peculiar form (Atkinson informs us), known in the North-Western Provinces as *kálinda,* is sown in sugar-cane fields in June and ripens in October. In Sind the water-melon is said to be grown in the *kharif* season.

HISTORY. **History.—Linnæus** believed it to be a native of Southern Italy, while **Seringe** supposed it to be indigenous to India and Africa. It was afterwards discovered that it grew wild in tropical Africa. "**Livingstone** saw districts literally covered with it, and the savages and several kinds of wild animals eagerly devoured the wild fruit." It was cultivated by the ancient Egyptians, as appears from their paintings. The Chinese only received the plant in the tenth century of the Christian era (*DC., Orig. Cult. Pl., 263*).

OIL. 1222 **Oil.**—The seeds yield a clear, bland, pale-coloured, limpid oil, used for burning in lamps, and probably also as an edible oil (*Cooke*).

MEDICINE. Seeds. 1223 **Medicine.**—The SEEDS are used as a cooling medicine. **Dr. Dymock** says that they are in great demand and kept decorticated and ready for use. In Bombay they are considered cooling, diuretic, and strengthening, and are sold in the bazars along with other cucurbitaceous seeds. **Ainslie** remarks that the Vytians prescribe the JUICE of the fruit to quench thirst, and also as an antiseptic in typhus fever, in which cases he himself administered it with good results.

Juice. 1224

Special Opinion.—§ "Cooling as well as a diuretic" (*Assistant Surgeon Anund Chunder Mukerji, Noakhally*).

FOOD. Fruit. 1225 **Food.**—The FRUIT is large, ovoid, green, and smooth; the flesh is whitish yellow or red. The SEEDS are compressed and variable in shape and colour; they are sometimes dried and the kernels eaten. **Stewart** says they are eaten parched with other grain. In the North-West Provinces and Oudh the plant is largely cultivated, but statistics of the areas are wanting; the only districts for which figures are available are Bulandshahr, Jalaun, and Meerut, and these show respectively 56, 48, and 26 acres annually. "In the sands of Bikanir, water-melons occur spontaneously in such numbers as to form for some months in the year no small part of the food of the scanty population. The seeds of these and of other cucurbitaceous plants cultivated in gardens are ground during times of scarcity into a kind of flour" (*Raj. Gaz., 31*). The water-melons of the North-Western Provinces are famed all over India and are used as refrigerants, and as a sherbet ingredient.

Seeds. 1226

1227 **Var. fistulosus,** *Stocks; Duthie & Fuller, Field and Garden Crops, N.-W. P., II., 46, Plate XLVII.*

In the *Flora of British India* **C. fistulosus** has been given as a synonym to **C. vulgaris,** *Schrad.*, but it seems desirable to retain it as a variety.

Vern.—*Tandús, tendu, tind albinda, tensi,* N.-W. P.; *Tinda, albinda, dilpasand,* PB.; *Meho, trindus, dilpasand, tinda, alvinda,* SIND.

References.—*Stewart, Pb. Pl., 96; Balfour, Cyclop.*

Habitat.—In the North-West Provinces this fruit is sown some little time before the rains, the fruit ripening during the rains. "Cultivated in Sind from April to September, generally in the same plot of ground with common melons, gourds, and cucumbers. In the North-West Provinces and Oudh it is cultivated in the western districts before the rains in well-manured land, either as a sole crop or with other vegetables, and is eaten during the rainy season."

MEDICINE. 1228 **Medicine.**—**Royle** remarks that the seeds are used medicinally.

FOOD. 1229 **Food.**—"The fruit is picked when about two-thirds grown, the size and shape of a common field turnip. . . . It is pared, cut in quarters, the seeds extracted, well boiled in water, and finally boiled in a little milk, with salt, black pepper, and nutmeg. Musalmans generally cut it into dice, and cook it together with meat in stew or curries. Hindus fry it

Pickle. 1230

in *ghi* with split gram-peas (**Cicer arietinum**) and a curry-powder of black pepper, cinnamon, cloves, cardamoms, dried cocoanut, turmeric, salt, and asafœtida. It is sometimes made into a preserve in the usual manner. It is often picked when small, cooked without scraping out the seeds, and regarded a greater delicacy than when more advanced." (*Dr. Stocks, in Hooker's Journal of Botany, quoted by Duthie and Fuller.*)

CITRUS, *Linn.; Gen. Pl., I., 305.* 1231

A genus of shrubs or trees, usually spinous, belonging to the Natural Order RUTACEÆ. *Leaves* alternate, 1-foliolate, more or less serrate, and gland dotted, petiole often winged. *Flowers* axillary, solitary, fascicled or in small cymes, white or pinkish, sweet-scented. *Calyx* cupular or urceolate, 3-5-fid. *Petals* 4-8, linear-oblong, thick, imbricate. *Stamens* 20-60, irregularly united or connate into bundles and inserted hypogynously round a large cupular or annular disk. *Ovary* many-celled; style stout, deciduous; stigma capitate. *Fruit* a superior berry, with removable rind and separable sections, packed full of pulpy tissue which is composed of large cells and developed from the endocarp, projecting toward the centre, thus embracing and surrounding the seeds. *Seeds* in each cell variable in number, horizontal or pendulous; testa coriaceous or membranous; embryo sometimes 1 or more in one seed; cotyledons plano-convex, often unequal; radicle small, superior.

This genus comprises 5 tropical Asiatic species and 2 Australian.

The different varieties of the Orange, the Lemon, the Lime, and the Citron have been critically examined by a large number of patient and careful observers, but, it must be admitted, with but indifferent results. **Brandis**, after presenting a concise and pregnant account of the Indian species, concludes by explaining,—"My object in bringing these questions forward prominently in this place is to induce others, with more leisure and more opportunities of observation, to study a subject of great historical interest, which may eventually serve to bring out important results regarding the spread and changes of arborescent species under cultivation." Since these words were penned, it is feared we have not advanced very far towards a solution of the problems which hinge upon the nativity of the orange and the lemon. Shortly after the appearance of **Dr. Brandis'** *Forest Flora*, **Dr. Rice** of New York published in *New Remedies* a most interesting account of the genus **Citrus**, but without giving botanical descriptions of the species. Long anterior, however, the cultivated species were the subjects of two invaluable works, *viz.*, **Gallesio's** *Traité du Citrus, Paris, 1811*, and **Risso** and **Poiteau's** *Histoire Naturelle des Orangers, 1818*, in folio, with 109 plates. Through the efforts of **Sir J. D. Hooker** and other Indian botanists, a number of wild species have been collected and described. It has thus been possible to reduce many of **Risso's** cultivated plants to their probable wild species, but the completion, as far as may be possible, of the task which **Dr. Brandis** urged, must still rest with persons in India and China who have the opportunity and time to study the living plants. An important step has recently been taken by **DeCandolle** in his *Origin of Cultivated Plants.* Bringing to bear the results of an extensive study of early literature and philological data with botanical research, he has been able to establish the lines upon which all future study should be directed, and in doing so has fixed the limits or the regions to which the various species may be viewed as indigenous. The modern authors, to whom reference has been made, view the important forms as referable to three species, *viz.*, **C. Medica**, *Linn.;* **C. Aurantium**, *Linn.;* and **C. decumana**, *Linn.* Under the first species, **Sir J. D. Hooker**, in his *Flora of British India*, establishes four varieties, and under the second three, and gives in addition a fourth and little-known species, **C. Hystrix**, *DC.* **Kurz**, in his *Flora of British Burma*, breaks up one of the varieties of **C. Medica** into two forms, placing **C. Limetta,**

Risso, as a synonym under **C. nobilis,** *Lour.* (the Mandarin)—a species which he regards as quite distinct from **C. Medica,** *Linn.*

The specific distinctions in **Citrus** are based chiefly upon the degree to which the petiole is winged, on the colour of the flower (pinkish-white in the lemons and pure white in the oranges), and on the shape of the fruit, pear-shaped and more or less mamillate in the lemons and globular and non-mamillate in the oranges. Species characterised by the degree of development of a certain feature must naturally under cultivation become hopelessly intermixed, hybridisation rendering it almost impossible to distinguish the forms. This is true in its fullest extent with the members of the genus **Citrus,** and it is by no means an easy task to say in what respects an orange differs from a lemon. The extreme forms are readily enough recognised, but these break down when a large collection is examined side by side. The writer, however, is disposed to agree with **Kurz** that there is no advantage gained by combining the Sweet Lime (**C. Limetta,** *Risso*) with the Sweet Lemon (**C. Medica,** var. **Lumia,** *Risso*). It would seem desirable to accept **Roxburgh's** position, and to place the majority of the forms described by him under **C. acida,** *Roxb.,* along with **C. Limetta,** *Risso,* but apart altogether from **C. Medica.** The writer would even go further and view the lemons as having by no means so distinct a claim as the limes to be regarded of Indian origin. The limes appear intermediate in character between **C. Medica** and **C. Aurantium,** having the rounded fruit, white flowers and winged petioles of **C. Aurantium,** with the flavour, chemical properties, and peculiar character of the rind of **C. Medica.** Whether **Kurz** be correct in viewing the sweet lime of India as but a form of **C. nobilis,**—the Mandarin of China,—may be doubted; but these are certainly allied plants, and to this group should be added **C. Hystrix,** the three species being separated from **C. Medica** and **C. Aurantium** by their very much smaller flowers. It is usual to regard the small round, dark orange-red fruits sold at hill stations as Mandarins, and **DeCandolle** states that **Mr. C. B. Clarke** is of opinion that the cultivation of the Mandarin is extending on the Khásia hills. **Dr. Bonavia** appears to doubt the existence of the Mandarin in the Khásia hills but recommends its introduction. That author speaks of good Mandarins as occurring in Ceylon, but is unaware of any in India. The true Mandarin, in the opinion of most writers, does occur in India, but it would be interesting to have the question of its relation to the sweet lime more clearly established. According to **Kurz,** these two cultivated plants are one and the same species, **C. nobilis,** being much cultivated all over Burma. This conclusion may not, however, be regarded as satisfactory, from the fact that the Mandarin is chiefly characterised by the extreme thinness of the rind and deliciously flavoured pulp, whereas in the sweet lime the rind is coarse or even thick, and the pulp much inferior to that of the Mandarin. **Dr. Rice** regards the Mandarin or Maltese orange as a variety of **C. Aurantium; C. Hystrix** is the characteristic wild species of Burma.

Having now indicated very briefly the present position of this subject, and the probable changes which may be effected in the grouping of the known forms, it will not be necessary, for the purposes of the present publication, to depart materially from the attitude taken by the authors of the *Flora of British India.* The following analysis drawn from that work, with one or two additions from **Kurz's** *Forest Flora of Burma* (published subsequently), may be found useful :—

* *Young shoots and leaves perfectly glabrous; transverse vesicles of the pulp concrete.*

† A shrub; young shoots purple; petiole more or less naked; petals generally tinged with red; flowers

often unisexual; stamens 20-40; style long, thick; fruit globose, ovoid or oblong, often mamillate; rind very thick and rough . . . . . 1. **C. Medica.**

†† A tree 15 to 25 feet in height; petiole short winged; flowers small, white, usually solitary; style long, thick; fruit globose or somewhat oblong, not mamillate; rind very thin, nearly smooth, shining, yellow or orange coloured . . . . 2. **C. nobilis** (and ? **C. Limetta**).

NOTE.—If **C. Limetta** be added as a synonym of **C. nobilis** the definition of the rind would have to be modified.

††† A small shrub; leaflet smaller than the broadly winged petiole; flowers as in **C. nobilis**, only pedicillate and clustered in the axils of the leaves; style very short; fruit globose or ovoid, a little larger than the size of a walnut; rind thick, yellow . . . . **C. Hystrix.**

†††† A tree; young shoots whitish; petals more than twice the length of those in the two preceding species; flowers bisexual; stamens 20-30; style long, thick; fruit globose or flattened; pulp sweet, acid or bitter . . . . **C. Aurantium.**

** *Young shoots and under-surface of the leaves pubescent; transverse vesicles of the pulp distinct* . . . **C. decumans.**

The writer would wish it to be observed that he does not advance any positive personal opinions as to the Indian members of the genus **Citrus**. It has been found impossible for him to institute original investigations, and therefore, in writing the present article, he has restricted himself entirely to compilation from, and critical analysis of, the opinions held by others. There seems every probability that this difficult subject, when approached in a scientific manner, will be radically changed; and the account here given may help to direct attention to the points which deserve earliest investigation, while giving popular and commercial information of some practical value.

Citrus Aurantium, *Linn. (in part); Fl. Br. Ind., I., 515;* RUTACEÆ. 1232

The name *Aurantium* is not derived from the Latin *Aurum* "gold," but comes to us from the Arabic *nárandj.* This became *nárendj* (*nárang*) in the Persian, and its equivalent in Sanskrit is *nágaranga,* and in Hindustani *nárangi.* Names beginning with *nar* are generally associated with fragrance. The name for the orange first reached Europe through the Moors, and became *náranga* in Spanish, *laranga* in Portuguese, *Arancio* in Italian, and in mediæval Latin *arangia, arangius,* and afterwards *aurantium.* There seems little doubt, however, that in Sanskrit, as also in the European languages, these names were first applied to the bitter orange and were only appropriated in later times for the sweet orange. The English word orange is derived from the same root. (*Rice, DeCandolle, Yule-Burnell, &c.*)

Var. 1. **Aurantium proper (var. β. dulcis,** *Linn.*) (*For var. 2, see p. 345.*) **Var. 1st Aurantium. 1233**

Botanical Diagnosis.—Petiole naked or winged; pulp sweet, yellow, very rarely red; rind loose or adhering.

THE SWEET ORANGE, CHINA ORANGE, PORTUGAL ORANGE, *Eng.;* ORANGER, *Fr.;* ARANCIO DOLCE, PORTOGALLO, MELARANCIO, *It.;* NARANJO, *Sp.;* LARANJEIRA DE FRUCTO DOLCE, *Port.;* APFELSINE, SÜSSER POMERANZENBAUM, ORANGENBAUM, *Germ.;* PORTOGALLO, *Gr.;* LARANJAS, *Rus.*

CITRUS Aurantium.

The Sweet Orange.

Vern.—*Nárangí, sangtara, nárenj, náringi, nárúnge, sunthura, amrit-phal, kumla nebu,* HIND.; *Kamlá nembu, nárungi, nárengá,* BENG.; *Suntala,* NEPAL; *Santara, nárangi, náringi, náranj,* PB.; *Nárungi, nárangi,* GUJ.; *Náranghi cantra, naringsála, nárangi, náringi,* BOMB.; *Náringa, saku-limba, nárungasála,* MAR.; *Náringhie, orangen,* DUK.; *Kichili, chechu, kitchli, kitchili, kich-chilip-pazham, kozhunjip-pazham, kichlie-pullum, collúngie-pullum, kolinji-marum,* TAM.; *Ganjanimma, kittali, kitchlie, kich-chili-pandu, nárinja-pandu, kittali-pandu, náranga-pandu, kichidi, kichidie-pandu,* TEL.; *Kittaboippe, kittalesippe, kittale-pannu,* KAN.; *Máhura-nárenná, kólánji-narakam, jéroc-nanis,* MALA.; *Jerúc, simao-manis,* MYSORE; *Nágaranga,* SANS.; *Náranj,* ARAB.; *Nárang, nárendj,* PERS.; *Thau-ba-ya, sh-on-si, lieng-man, sunguen, shoungpang,* BURM.; *Dodang, nárang-ká, dodan, panneh-dodang,* SING.

References.—*Roxb., Fl. Ind., Ed. C.B.C., 590; Brandis, For. Fl., 53, 572; Kurz, For. Fl. Burm., I., 197; Gamble, Man. Timb., 59; Dalz. & Gibs., Bomb. Fl. Supp., 12; Stewart, Pb. Pl., 29; Aitchison, Cat. Pb. and Sind Pl., 29; Rice, Commercial Prod. Citrus, in New Remedies, 1878; DC., Origin Cult. Pl., 181; Voigt, Hort. Sub. Cal., 142; Pharm. Ind., 42; Moodeen Sheriff, Supp. Pharm. Ind., 104; Flück. & Hanb., Pharmacog., 124; U. S. Dispens., 15th Ed., 268; Bentl. & Trim., Med. Pl., 51; U. C. Dutt, Mat. Med. Hind., 127; Dymock, Mat. Med. W. Ind., 107; Ainslie, Mat. Ind., I., 281; O'Shaughnessy, Beng. Dispens., 231; Bidie, Cat. Raw Prod., Paris Exh., 4, 90; Year-Book Pharm., 1874, 623; S. Arjun, Bomb. Drugs, 21; Drury, U. Pl., 139; Baden Powell, Pb. Pr., 575; Cooke, Gums and Gum-resins, 13; Atkinson, Him. Dist., 710, 732; Atkinson, Gums and Gum-resins, N.-W. P., 16; Birdwood, Bomb. Prod., 12, 141, 191, 218; Lisboa, U. Pl. Bomb., 148; Spons, Encycl., 1025-7; Piesse, Perfumery, 159; Balfour, Cyclop.; Smith, Dic., 300; Treasury of Botany; Kew Off. Guide to Museums, 26; Kew Off. Guide to Bot. Gardens and Arboretum, 64; Journal, Agri.-Horti. Soc., old series, XIV., 199, New Series, Vol. I., p. IV., 372; The Orange Groves of Shalla; New Series, Vol. VIII., p. I., 15; Trop. Agricutst., 117, 188, 272-505, 874, 892, 962; Simmonds, Trop. Agri., 438; Fleming, Med. Pl. and Drugs, in As. Res., Vol. XI., p. 164; Mysore Cat. Cal. Exb., 21.*

Habitat.—Cultivated in most parts of India, but specially so in the valleys on the southern face of the Khásia Hills, in Nagpúr in the Central Provinces, and to a small extent in Nepál, Sikkim, and one or two other Himálayan stations. In Burma, **Kurz** says, the orange is met with, though not to any great extent. "The sweet orange does not come to perfection in all parts of India. In Lower Bengal it does not fruit at all, or does not bear freely, and the fruit is dry and austere. Calcutta is supplied from the valleys of the Khásia Hills north of Sylhet. Delhi, Nagpúr, Aurangabád, Santgur near Villore, and the Northern Circars, are famous for their oranges, but there are large tracts where none or inferior kinds only are produced. In India the fruit generally ripens between December and March, according to the climate of the locality. A variety which flowers twice a year (February and July), and yields two crops—the first from November to January, and the second from March to April—is grown at Nagpúr (*Firminger's Gardening, 2nd Ed., p. 223*)" (*Brandis*).

HISTORY.

History.—**DeCandolle** has shown that there is strong presumptive evidence that the Asiatic names now given to this form more correctly belong to the bitter orange, and that the sweet orange is most probably not originally Indian. "We come back," he says, "by all sorts of ways to the idea that the sweet variety of the orange came from China and Cochin-China, and that it spread into India, perhaps towards the beginning of the Christian era." It was, according to some authors, taken to Europe by the Portuguese about 1548, the first tree having stood for some time at Lisbon. From this point, the cultivation of the sweet orange spread to Rome and along the Mediterranean. **DeCandolle**, however, is of opinion that the sweet orange may have reached Europe before the

HISTORY.

date just given, but of inferior quality, so as not to attract the attention it deserved until later. Had the sweet orange existed in India for more than 2,000 years, it could scarcely have escaped becoming associated with Sanskrit literature, whereas the so-called Sanskrit names now given to it denote the property of the bitter orange or the sweet lime. The Greeks were also sure to have known of it through Alexander's expedition, and the Hebrews would have received it through Mesopotamia. Both the sweet and the bitter orange were unknown to the Romans. Whether or not the Portuguese deserve the credit of introducing the orange to Europe, they found it abundant in India; the Florentine, VASCO DE GAMA, having noted this fact in his account of the Mission to India. The names Portogalls, ITALIAN, Protokale, ALBANIAN, and Portoghal, KURDISH, indicate the intimate relation which Portuguese bore to the diffusion of the plant. The Dutch SINNASAPPEL or APPELSINNA and the German APPELSINE point to its Chinese origin. **Dr. Bretschneider** (*On the study and value of Chinese Botanical Works, page 55*) shows very conclusively that the orange is a native of China; the names given to the various forms are represented by a particular character which occurs in the most ancient Chinese writings, whereas the names given to the pumelo and the lime are of a much more modern character.

Dr. Bonavia has given the subject of the Indian Oranges, Limes, and Lemons more careful consideration than any other Indian authority. In an instructive paper read before the Agri.-Horticultural Society of India (*Journal, new series, 1887, Vol. VIII., pages 15 to 39*) he has brought together many facts of interest. He seems to be fully of opinion that even the sweet orange is indigenous to India. He thus holds a different opinion from that entertained by **M. DeCandolle**, and in support of his view quotes several authorities who have furnished him with facts regarding the wild or supposed wild sweet oranges of India. He gives the name '*Súntara*' as the Sanskrit for the loose-skinned sweet oranges. At Delhi an orange of this type is grown as the *Sintra*. With regard to this **Dr. Bonavia** observes: "I need hardly mention that the 'Sintra' orange has *nothing to do* with the orange of Cintra (a town of Portugal). It is merely a corruption of Suntara, which word, as I have elsewhere explained, appears to be of Sanskrit origin. The writer can discover nothing to justify the opinion that the word Súntara or more correctly Sangtara or Santara, is derived from the Sanskrit. It is, according to the best authors, a Persian corruption, and it can hardly be doubted that Santara is derived from Cintra—a town famous for its fruits. **Yule-Burnell** say: "As early as the beginning of the fourteenth century we find **Abulfeda** extolling the fruit of Cintra. His words, as rendered by **M. Reinaud**, run: 'Au nombre des dependances de Lisbonne est la ville de Schintara; à Schintara on recueille des pommes admirables pour la grosseur et la gout." That these *pommes* were the famous Cintra oranges can hardly be doubted. **Baber** (*Memoirs of Zehir-ed din Muhammed Baber, Emperor of Hindustan*) describes an orange under the name of *Sangtarah*, which is, indeed, a recognised Persian and Hindi word for a species of the fruit. And this early propagation of the sweet orange in Portugal would account not only for such wide diffusion of the name *Cintra*, but for the persistence with which the alternative name of *Portugals* has adhered to the fruit in question." Numerous passages might be quoted in support of this: "The *Sengtereh* . . . is another fruit . . . in colour and appearance it is like the citron (*Táranj*), but the skin of the fruit is smooth" (*Baber (1526), Memoirs, page 328*). **Kirkpatrick**, in his *Nepaul (1811), page 129*, speaks of the Nepaul *Santola* orange as superior to that of Silhet, a name which, he says, "I take to be a corrup-

The Sweet Orange.

HISTORY.

tion of *Sengterrah*, the name by which a similar species of orange is known in the Upper Provinces."

The sweet and the bitter cultivated oranges are, by some writers (among whom are the learned authors of the *Pharmacographia*), stated to be derived from the same stock. The authors mentioned say: "Northern India is the native country of the orange tree. In Garhwal, Sikkim, and the Khásia Hills, there occurs a wild orange, which is the supposed parent of the cultivated orange, whether Sweet or Bitter." This belief (held very generally some ten years ago) would support the opinions published in **Dr. Bonavia's** paper alluded to above. Referring to the small sweet orange of Nepal found at Bútwal and known over the North-West Provinces by that name, **Major Buller** says: "The orange is called *Suntolah*, and is not known in Nepal under the name of Bútwal orange. The orange trees grow wild, not in the valley but on the hills." **Mr. J. H. Fisher**, Collector of Etawah, refers to an incident where the Rajah of Kulabandi (a Feudatory State near Sambulpur in the Central Provinces) brought him oranges, "the produce he told me of wild orange trees, which grew certain places in the forests on his estate." **Mr. Fisher** adds, however, that as he was unable to visit the locality he "never had an opportunity of seeing these wild trees." Both the last mentioned writers appear to allude to sweet oranges, but it would be unsafe to infer, even from the existence of plants in forests far away from cultivation, that they were truly indigenous. It is significant that in Nepal these supposed wild oranges bear the same name as the oranges of many widely removed parts of India—a name, too, which there seems little cause for doubting, is derived from Cintra, a Portuguese town famous for its oranges about the time of the Saracenic conquests in Europe. It would, indeed, be difficult to account for the commonality of this name, except on the supposition, as in the case of the pine-apple, that it came with the plant. The difficulty confronts a hasty inference that a wild fruit of such merit should have been unknown to the Sanskrit writers. Authors on Sanskrit literature admit that limes and bitter oranges but not sweet oranges are alluded to by the earlier writers.

CULTIVATION.

Cultivation of Oranges in India.—There are two great centres of sweet orange cultivation in India—the Khásia Hills and Silhet on the eastern side and Nagpur in the central tracts of the country. The former meets the Calcutta market and the latter the Bombay. It would be difficult, however, to pitch on a district where some form of orange, lime or lemon was not cultivated. Nepal and the lower central Himálaya are famous for a small exceptionally sweet orange, which is largely consumed over the North-West Provinces. Delhi competes in a small way with Nagpur in the Bombay market. The Delhi supply, in addition to local orchards, is drawn from Ulwar, Gurgaon, &c. The opening up of the new Nagpur-Bengal Railway will place the orchards of the Central Provinces in touch with Calcutta and thereby probably greatly extend orange cultivation. The regions mentioned are those from which the sweet oranges of India are mainly obtained, but in many other localities these oranges are grown, the supply, for example, offered at railway stations being mostly local produce. At most hill stations oranges are to be had from the lower altitudes; in Darjeeling an orange is sold that much resembles the so-called green oranges of Ceylon.

Dr. Bonavia refers the sweet oranges to four cultivated races, two of which should most probably be referred to **C. nobilis**, namely, the Mandarin and the blood-red Maltese-like orange found at Gujranwala. The Maltese orange proper has recently been introduced into India, and is being cultivated at Jounpore and other localities. From an industrial or economic point of view, it is of little consequence whether, a sweet orange be referable to **C. Aurantium** or **C. nobilis**; we may therefore follow **Dr. Bonavia**, since that authority has very strikingly exemplified the manner in which continental India might have a continuous supply of oranges with extended facilities of communication.

CULTIVATION.

THE RACES OF SWEET ORANGES.

Race 1st, Santara. 1234

Race 1st, Santara.—These are defined as rind smooth, of a yellowish-orange colour, the skin or jacket being loose. The centres of Santara orange cultivation are Nagpur, Delhi, Ulwar, Gurgaon, Lahore, and Mooltan; turning west, Poona, the Shevaroy hills, Madras, Coorg, and Ceylon; and east, Silhet, Bhutan, and Nepal. It seems probable that both this form and that to follow are grown in the Assam plantations, but the former most abundantly. The Santara orange is plucked in November, December, and January.

Vern.—The following are the special vernacular names mentioned for this form. *Kamalá moglái* (Sweet forms), *náringhi, káki, khátajamir* (bitter), BENG.; *Uso santra, uso mongar* (sweet), *uso sim, uso yanpriang, uso komphor* (bitter), KHASIA; *Rabáb* (*tenga*), *jora tenga*, ASS.; *Naga tenga*, NAGA HILLS; *Suntolah, butwal* (the former in Nepal and the latter in N.-W. P. for the same orange); *Suntara*, C.P. (near Wardha two crops are obtained; one ripens in spring known as the *miragbahar*, and the other ripens in winter, the *ambiabahar*); the former is merely an after-crop known in most other parts of India as *dumrez*); *Sintra* or *rungtra* (the latter being the name of the village where mostly grown), DELHI; *Karna* or *Sungturu* (a pear-shaped form), LAHORE; *Suntara*, POONA; *Kithlí*, MAD.; *Konda nárun*, SING.

Mr. Morris (in his Godavery District, Madras Presidency) says: "a very luscious orange with a loose skin is grown among the hills; but it is so different from the thick-skinned species, that it has received an entirely different name in Telegu—*Kamalá pandu.*" This is a significant fact, the word *Kamalá* being thus claimed as both a Bengali and a Telegu word. **Dr. Bonavia** says that in Assam the word *Kamalá* is believed to be derived from the river of that name, and in a foot-note the Editor of the Agri.-Horticultural Journal suggests that it may come from Kumilla, the capital of Tippera. Presuming that one of these derivations is correct, the inference would then have to be drawn that this loose-skinned orange of the central tracts of India came from Assam, and carried its name with it. The plant could scarcely have been indigenous to both localities and received accidentally the same name in two languages.

It may, however, be doubted how far it is correct to throw all the above oranges together. The Delhi orange, for example, has a thick rind and is very spongy, more so than either the Khásia or Nagpúr orange. The orange with a thick rind, met with in the Godavery District, **Mr. Morris** informs us, was introduced by the Dutch, and to this day bears the name *Batái náringa pandu*, a name suggestive of Batavia.

Race 2nd, Keonla. 1235

Race 2nd, Náringhi or Keonla.—The natives make a distinction between these; but for practical purposes they may be viewed as identical and distinguishable from the preceding by the rind being rougher, of a darker colour, thinner, and adhesive (*e.g.* jacket not loose). This is the orange that comes latest into the Calcutta market. It is plucked about January and February. The Keonla orange is, perhaps, more extensively diffused over India than the Santara. It can stand a greater amount of heat and is therefore the orange of the isolated and private orchards over the greater part of the country. It is never so sweet as the Santara orange, but its bitter sweet flavour is perhaps all the more grateful at the season of the year at which it is available.

Vernacular names in the various provinces of India for this peculiar form are not available.

The Sweet Orange.

RACES OF SWEET ORANGES.

Before proceeding to discuss the third class of sweet oranges referred to by **Dr. Bonavia** it may be as well to refer to another author. **Mr. Atkinson** says of Kumaon: "The sweet orange is the form most usually cultivated, and there are several local varieties, some named after the localities in which they are produced, and others according to specific local distinctions in size and flavour. The three more common varieties cultivated in the plains are the *Santara, Nárangi*, and *Kaunla* or *Kumla*. The last is the smallest and most esteemed." The writer feels strongly inclined to suspect that *Kaunla, Kumla, Kaonla*, and even *Kamalá* are names derived from a common source, and that the oranges they represent should be isolated from those designated *Santara* or some derivative from that word. If this supposition prove correct, grave doubts may be entertained of either form being indigenous to India. It is a coincidence not met with (so far as the writer is aware of) in connection with any other unmistakably indigenous cultivated plant, that names so much alike as those given above, should occur in the most remote parts of India and be used by peoples as distinct anthropologically as they well could be. It may be further suggested that the thick skinned oranges may be found to correspond to **Mr. Atkinson's** second class *Nárangi*. That writer concludes his account of the Himálayan oranges as follows: "The orange has been found wild or apparently wild with unwinged petioles at Bágeswar in Kumaon (by **Strachey** and **Winterbottom**), and with globose fruits, naked or margined petioles, and oblong-lanceolate, acuminate leaves in Garhwál (by **T. Thomson**)." It is almost impossible to avoid the conviction that too strong opinions have, by all writers, been advanced as to the Himálayan home of the sweet and bitter oranges. The two discoveries referred to by **Mr. Atkinson** are those upon which the Himálayan habitat has been mainly founded. The fact that these wild (or only supposed wild plants) have not winged petioles or only margined petioles is to say least suggestive of **C. Medica** more than of **C. Aurantium**. This idea, taken in conjunction with the peculiarities mentioned regarding the vernacular names as given to the various forms of the Indian sweet oranges of cultivation (and even to the supposed wild oranges of Nepal), is sufficient to justify the conclusion that the whole subject is still involved in the utmost obscurity. A scientific exploration of the reputed regions of the wild oranges of India and a careful scrutiny of the cultivated forms and the names given to them would seem likely to upset nearly everything that has appeared on this subject. This is an unfortunate admission, since to India and Indian exploration must be laid the blame of the apparently mistaken notions which prevail regarding the sweet and bitter oranges.

Race 3rd, Malta.
1236 *Race 3rd, Malta or Blood-red oranges* (Conf. with **C. nobilis**).—There are several forms of this orange, some modern, others early introductions. It seems probable, however, that the large coarse green oranges (*see p. 347*) of Darjeeling and other Himálayan stations may be but degenerations from this stock, or perhaps Indian cultivated forms from the same wild species. The oranges of Burma, for example, may have been derived from the indigenous plant, a species perhaps identical with, or at least closely allied to, that from which the Maltese orange appears to have been developed. Be that as it may, a blood-red orange and also a green orange, with a coarse thick adhesive rind, is commonly met with in many parts of India, in both cases possessing a peculiar and distinctive odour which at once isolates these forms from the oranges already described. They come into season after the *Keonla* orange, and, as pointed out by **Dr. Bonavia**, were an effort to be made to extend, where possible, the cultivation of the blood-red forms, India might obtain a supply of oranges in

RACES OF SWEET ORANGES.

the hot season, the time when these fruits would be most acceptable. Speaking of the Gujranwala oranges **Dr. Bonavia** says **Colonel Clarke** introduced these from Malta in 1852—56. **Dr. Bonavia** himself introduced the same orange into Lucknow in 1863, and **Mr. C. Nickels** established the Jounpore stock in 1872. Prior to the Mutiny blood oranges were grown in Lucknow, so that there must have been earlier introductions than those mentioned above. From these centres, however, the cultivation of the red oranges has been greatly extended, so that they are now met with in most districts in Upper India. At Poona a blood orange is grown under the name of the *Mussembí*, a name given 1237
to a similar small red orange imported into Bombay from Zanzibar. At Tanjore a red orange is grown which bears a vernacular name, but from the absence of such names in other parts of India, **Dr. Bonavia** very naturally arrives at the conclusion that the better qualities of red oranges must be modern introductions.

Speaking of the blood oranges of Gujranwála, **Dr. Bonavia** says: "the specimens of blood oranges sent to me by **Mr. Steel**, Deputy Commissioner of Gujranwála, in my opinion, are the *best oranges* that I have tasted in India. The pulp is of the orange claret colour. Many of the specimens were full-blooded, and smeared externally with a blood tinge. The juice was simply nectar-like. In short, their flavour was, in my opinion, simply perfect. I thought them equal to that of the blood oranges of Malta." "**Mr. Steel** states that the soil on which they grow is a stiff clay with plenty of *kankar* in it. But the real secret, he thinks, is to bud them on sweet lime-stock."

"It appears that in Gujranwála there is a suitable soil and climate and the best orange in India. There is also skill to turn these materials to account. Here there is a chance of creating an extensive trade in blood oranges, as a speciality of Gujranwála. They are not only exquisite oranges, which if, properly packed, would bear long journeys, but they are *late* oranges, and therefore would not compete with the *Santara* oranges, which flood the Calcutta and Bombay markets from Silhet and Nagpur." **Mr. Steel** reports that in February they are "barely ripe, and would remain on the trees till the middle of March. Last year, some by careful packing were kept in good condition till July."

Race 4th, Mandarin. 1238

Race 4th, Mandarin Oranges (Conf. with **C. nobilis**).—Several writers are of opinion that the small highly-scented flat oranges occasionally met with throughout India, especially at hill stations, are Mandarins. The true Mandarin has a peculiar smell common to both fruit and leaves, which closely resembles that of the blood oranges. Indeed, by most writers the Mandarin is a special Chinese development from the same stock as the Maltese orange. In a further page particulars will be found regarding this orange; suffice it in this place to add 'that in **Dr. Bonavia's** opinion the true Mandarin, while found in Ceylon, does not exist in India. **Mr. C. B. Clarke**, on the other hand, says the cultivation of this form is rapidly extending in the Khásia hills. **Dr. Bonavia** recommends its introduction in "the highlands of Bengal," "where it would be out of the influence of the hot winds," which have killed or rendered useless all the plants grown in Upper India.

Having now briefly indicated the chief forms of sweet oranges met with in India, the present article may be completed by giving some idea of the orange industry at the two great commercial centres—Silhet and Nagpur.

Silhet. 1239

I.—Oranges of Silhet and the Khasia Hills.—A most instructive paper appeared on this subject in the Journal of the Agri.-Horticultural Society of India, from the pen of **Mr. C. Brownlow** (*Vol. I., Part IV., New*

CITRUS Aurantium.

The Sweet Oranges of Silhet.

ORANGE PRODUCTION IN INDIA.

Series, 1869, p. 372). **Mr. Brownlow** gives the fullest particulars regarding the "Orange groves of Shalla," his paper being a model after which all such reports might, with great advantage, be framed. The scenery, natural vegetation, nature of the soil and climate of Shalla are dwelt on to the degree essential to affording a conception as to the adaptability or otherwise of new localities proposed for orange cultivation. The methods of propagation, collection, and transport are next fully disposed of. Indeed, so admirably has **Mr. Brownlow** fulfilled his task that any abridgment of his paper must mar its usefulness. The limited space at the writer's disposal precludes the reproduction of the entire paper, and the reader who may be specially interested in this subject is therefore referred to the original; the following abstracts, however, may be found useful:—

Soil.

Soil.—Mr. Brownlow shows that the prevalence of water below the roots is a feature evidently favourable to orange cultivation, although this water must not be stagnant. The peculiar underlying pebbly stratum is such that the water percolates from the river below the orange groves and back again, and that the floods inundate the land in spring tide, thus manuring the soil and preserving its fertility. "The land is flat, having a slight slope away from the river; there are a few points that rise above the general level, the uniformity of which is only broken by channels and cross-channels of natural surface drainage. These depressions are wet and clayey, and may be traced by the *ekur* grass and *tehra* (a wild cardamum) that grow in them. It is scarcely necessary to say that the orange will not thrive in them, and wherever in the groves they occur they are in consequence left uncultivated. Here, in one large connected piece of perhaps 1,000 acres, is the garden that supplies a great part of eastern as well as western Bengal with oranges; I say perhaps 1,000 acres, because the area under cultivation is not known to the Khásia proprietors themselves." "One may walk for a good hour or two, always under the shade of orange trees without reaching the limits of cultivation; and when, as in December and January, every tree is laden, no sight can be more enjoyable. I have been through the Sorrento gardens, but this beats Sorrento, and the Neapolitan orange-growers would find some difficulty in selecting, out of their entire *piáno*, a piece at all approaching the size of this." Speaking of the merits of the Silhet orange, **Mr. Brownlow** says:—"Moreover, even an Italian sun is incapable of imparting that pure lusciousness and of combining the sweet, the tart, and the bitter in the same just proportions as we find here." The climate and soil, in **Mr. Brownlow's** opinion, is that eminently suited to orange cultivation, and we may therefore reproduce **Dr. Waldie's** analysis of the soil, collected for that purpose by **Mr. Brownlow**, from the Shalla plantations.

"Of the sample received 100 parts dried at 212°F. = 97·27 or 102·8 as received equal to dry 100.

Soil dried at 212°F.

Alumina	6·09
Peroxide of iron	4·93
Lime	·19
Magnesia	·13
Alkalies (by difference and loss)	·80
Silica solution	·15
These dissolved by H. Cl.	12·29
Alumina, chiefly with a little oxide of iron and a little lime	3·49
Dissolved by heating with $H_2 SO_4$	...
Organic matter and combined water lost by burning	5·66
Silica and quartz	78·56
	100

C. 1239

ORANGE PRODUCTION IN INDIA.

"It will be observed that this is a very siliceous soil, proceeding from the decomposition of siliceous rocks alone. It contains no carbonate of limes and is a very open and porous soil."

Cultivation. 1240

CULTIVATION.—The seed is sown in January and February, thickly in troughs or boxes in about 6 inches of soil. These seed-boxes are raised above the height pigs could reach them, and are often protected by nets from rats and squirrels. The seedlings are pricked out during the ensuing rains; but in doing so the boxes are broken up and the earth shaken away from the roots, so that there is absolutely no injury done to the tap-root. They are transplanted into a nursery in the grove; here they remain until retransplanted to their destined places in the grove. The system seems defective and the nursery is only once a year weeded, *viz.*, in October. Grafting is quite unknown, and no care seems to be spent on the selection of the seed.

Collection and Pruning. 1241

COLLECTION AND PRUNING.—Each collector has a ladder, about 20 feet long, made of light bamboo. A coarse net bag, held open at the mouth by a cane ring, depends on his back by a strap passed over the right shoulder and chest. Into this he throws the oranges and before descending he removes the withered leaves and dead branches, or cuts out boughs injured by the loranthus parasite that does such damage to the plants. "The orange trees receive no other handling than the above; they are never systematically pruned or thinned, and are allowed to retain just what fruit they set, and yet the crop turns out wanting neither in size, flavour, nor abundance. Contrast with this the elaborate summer and winter pruning of the French gardens and the systematic cultivation and manuring of the Genoese; and yet with all their labour they produce a fruit inferior in quality and beyond all measure dearer in price than that produced by the comparatively thriftless and indolent Khásia." Boys are employed with pellet bows to keep off the crows, squirrels, monkeys, hornbills, and other animals destructive to the crop. All the fruit which falls to the ground by wind or otherwise is gathered "every morning, peeled and given to pigs and dogs, and it is not a little remarkable to see how the dogs have come by habit to relish" this food.

Transport 1242

TRANSPORT TO THE PLAINS.—The oranges so collected are taken down the river in long canoes or dug-outs and sold at Chuttuck. They are counted in fours; 750 fours making the *son* (or 3,000); but the delicate finer qualities with thin skins are consumed locally, as they are not found to endure the rough handling of transport. Mr. Brownlow mentions that at Phalli Bazar, a little above Shalla, oranges of a slightly inferior quality are sold by bartering for rice, fish, &c., to the Muhammadan boatmen at R6 a *son*, being R4 less than the oranges at the Shalla groves, and yet this includes the cost of cultivation, labour of plucking, and carriage to the river.

TRADE. 1243

TRADE IN SILHET ORANGES.

Mr. G. Stevenson, Deputy Commissioner, Silhet, has furnished the following tabular statement:—

	Boat Traffic.	
	Quantity in maunds.	Value in Rs.
1880-81	1,20,398	2,40,796
1881-82	1,46,592	not known
1882-83	1,02,631	1,28,288
1883-84	1,14,969	2,27,062
1884-85	1,20,884	2,47,352

The Sweet Oranges of Nagpur.

TRADE. Dr. Bonavia, commenting on these figures, says :—"This averaged about 1,21,095 maunds of oranges per annum, worth from 1¼ to 1½ lakhs of rupees, in favourable years; 1,14,969 maunds of Silhet oranges are said to be equal to about 8,05,360 oranges." In a foot-note, however, Dr. Bonavia further adds: "This is the number given, but it appears too small. Taking 8,05,360 to represent the *sons* referred to by Mr. Brownlow, the figures would be 2,41,60,800, or about 210 oranges to the maund."

Nagpur. 1244 II.—Oranges of Nagpur in the Central Provinces.—We have already given several passages that refer to the so-called wild oranges both of Nepal and the Central Provinces. It will only be necessary further to give here a brief account from the pen of Mr. J. B. Fuller, as published by Dr. Bonavia, in order to place before the reader a comparative sketch of these groves to complete what has been said of the Khásia hills. These two localities represent the bulk of the orange production of India. Mr. Fuller says :—"Within the last twelve years many new orchards have been planted in Nagpur, Kamptee, and other parts of the district, and orange cultivation is now spreading rapidly in other districts of the Province. There is a great demand for the Nagpur oranges in Bombay, and considerable quantities of the fruit are annually exported to this and other places. In the year 1885, 22,609 maunds of orange fruit were exported from Nagpur station, out of which 21,400 maunds were exported to Bombay alone."

It is perhaps only necessary here to repeat that the North-West Provinces receive their supplies from Nepal, Delhi, and to some extent also from Nagpur. Panjáb, Madras, and Burma are practically dependent on local production from isolated orchards, Madras drawing largely from the Shevroys.

Properties and Uses—

GUM. 1245 **Gum.**—The orange tree is said to yield a gum of no importance. A sample was sent from Masulipatam to be shown at the Madras Exhibition in 1855.

MEDICINE. Rind. 1246 **Medicine.**—The *Pharmacopœia of India* treats the sweet and bitter varieties together, remarking that the dried outer portion of the RIND of the fruit possesses stomachic and tonic properties. It is useful in atonic dyspepsia and general debility, but it is rarely employed alone. The orange peel is generally used in the form of infusion, tincture, or syrup. The water distilled from orange flowers is employed, in one or two fluid ounces, as an antispasmodic and sedative in nervous and hysterical cases. The syrup of orange flower is chiefly used as a flavouring agent and to perfume external applications.

"The Muhammadan writers describe the best kind of oranges as large, thin-skinned, and smooth; they say that the rind and flowers are hot and dry, the pulp cold and dry, and recommend the fruit in colds and coughs, when febrile symptoms are present; it is best administered baked with sugar. The juice is valuable in bilious affections, and stops bilious diarrhœa. The orange is the safest of the acid fruits; the peel is useful for checking vomiting, and the prevention of intestinal worms. Orange poultice is recommended in some skin affections, such as psoriasis, &c. Oranges are considered to be alexipharmic and disinfectant; orange-water stimulating and refreshing. The essence is extracted by oil from the rind and flowers, and is used as a stimulating liniment." (*Dr. Dymock, Mat. Med. W. Ind.*)

Ainslie makes the following remarks: "Oranges are in great repute amongst the Hindú physicians, who suppose that they purify the blood, allay thirst in fevers, cure catarrh, and improve the appetite. A sherbet made with the juice of the ripe fruit is a favourite beverage with Europeans in India in hot weather, and is certainly much safer than that made with lemon juice, which is extremely apt to bring on cholera morbus. The rind of oranges is well known to be a useful carminative, and is a valuable addition to bitter infusions in cases of dyspepsia and flatulence." The rind pulverised and added to magnesia and rhubarb affords a grateful tonic to the stomach in gout and dyspepsia. The roasted pulp is an excellent application to fœtid ulcers. MEDICINE.

Special Opinions.—§ "The fresh rind of the fruit is rubbed on the face by people suffering from acne" (*Surgeon-Major R. Gray, Lahore*). "If the rind be mixed with a little water, and then rubbed on a part affected with eczema, much relief will be derived" (*Surgeon W. Wilson, Bogra*).

Food.—The fruit is largely imported from the Khásia Hills and distributed over Bengal, the largest quantity finding its way into the Calcutta market. The fruit has a thin rind, and is sweet and juicy. The orange grown in and about Delhi is on the average larger, but more spongy, with a much thicker rind than the preceding. The Nagpur orange is compact, and of an excellent quality. "Nagpur oranges find a market in Bombay and the Central Provinces, and pass as far even as to Allahabad. They are excellent, and may be anticipated to compete with the Khásia orange if Nagpur be connected with Calcutta by railway" (*Mr. L. Liotard*). FOOD. 1247

Orange trees are, by some authors, said to attain great age, some being stated to live for upwards of 600 years; in the orangery of Versailles a tree, known as the "Great Constable," is stated to be 450 years old. An orange tree, at the convent of St. Sabina at Rome, dates from the year 1200. The produce of one tree ranges from 500 to 6,000 fruits a year, and the tree sometimes grows to a height of 50 feet, with a trunk 12 feet in circumference.

Structure of the Wood.—Yellowish white, moderately hard, close and even-grained. TIMBER. 1248

Var. 2. Bigaradia, *Fl. Br. Ind., I., 515.* (*For var. 1st, see p. 335 and for 3rd, p. 347.*)

Botanical Diagnosis.—Petiole short-winged; flowers large, strongly scented; rind very aromatic, pulp bitter. Var. 2. Bigaradia. 1249

THE BITTER OR SEVILLE ORANGE; BIGARADIER, *Fr.*; ARANCIO FORTE, *It.*; POMERANZE, *Ger.*

Syn.—C. VULGARIS, *Risso*; C. BUXIFOLIA, *Poir.*

Habitat.—The bitter orange is very extensively grown in the warmer parts of the Mediterranean, especially in Spain and Malta. In India it does not seem to be cultivated except in gardens, but it is believed by most authors to be originally a native of the outer Himálaya from Garhwal and Sikkim to the Khásia Hills; and it is probable that its area extends also to Cochin-China. The Indian market is met almost entirely from wild sources.

Marmalade is chiefly made from the rind of this species, but it is doubtful whether Indian-made marmalade is manufactured from the true Seville orange, or simply from one of the numerous bitter indigenous oranges. The form known as *khatta* is extensively employed for grafting as a stock for the better qualities of imported bitter oranges. Definite information cannot be obtained as to the extent the Seville orange is being cultivated in India. Marmalade 1250

OIL OF NEROLI.

Oil and Perfumery.—Essential oils are obtained from most of the species of the Citrus family. **Sir W. O'Shaughnessy,** speaking of the sweet OIL. 1251

PERFUMERY. oranges, says that "the leaves are rather bitter and contain essential oil. A still more fragrant oil, called oil of *neroli* by the perfumers, is afforded by the flowers." **Piesse**, in his work on Perfumery, describes neroli oil, and says that the best quality is obtained by distillation, with water, from the flowers of **Citrus Aurantium** (the sweet variety). According to the same author, an inferior quality of neroli is derived from the blossoms of **Citrus Bigaradia** (the bitter variety). This oil is called *Essence de Néroli Bigarade*, and the oil from the flowers of the sweet variety bears the name
1252 of *Essence de Néroli Pétale* or *Néroli Louce*. This statement is opposed, however, to the opinion given by almost every other writer, the neroli otto from the sweet orange being used only as an adulterant to that from the bitter. The fresh flowers of the Bigaradia orange yield on distillation *Essence de Néroli Bigarade*, and if the sepals are carefully removed from the flowers, the essence is known as *Essence de Néroli Pétale*. The latter is finer and much more expensive than the former. From the seeds *Essence de Petit Grain* used to be manufactured, but this is now entirely distilled from the leaves and twigs: it is therefore a misnomer to call it *Essence de Petit Grain*. Similar essences are distilled from the leaves of most species of Citrus, and these are all used together with essential oil of orange leaf to adulterate neroli otto. The water which passes over with the oil during distillation constitutes, when separated from the oil, *Orange-flower Water* (*see below*).

1253 The extraction of Neroli oil is chiefly carried on at Grasse, Cannes, and Nice, in South France, also in Algeria. In France, about 20,000 cwt. of the flowers are annually distilled. The sweet variety yields but half the amount of oil which may be obtained from the bitter, as much as 0·6 per cent. being often obtained. The oil of neroli is commonly adulterated with *bergamot* and *petit grain*. According to **Flückiger**, the neroli commonly sold contains $\frac{3}{8}$ths *Essence de petit grain*, $\frac{1}{8}$th essence of bergamot, and $\frac{4}{8}$ths of true neroli.

Fine neroli oil is brownish, of most fragrant odour, and bitterish aromatic flavour; specific gravity, at 11° C., being 0·889. It is neutral to test paper. When mixed with alcohol it displays a bright violet fluorescence quite distinct from the blue fluorescence of a solution of quinine.

Neroli Camphor.
1254 **Neroli Camphor.**—The authors of the *Pharmacographia* obtained by distillation from the oil a very small amount of camphor called *Neroli Camphor*, and they state that they were unable to obtain any similar substance from the oils of bergamot, petit grain, or orange peel.

Eau de Cologne.
1255 **Uses of Neroli Oil**—Oil of neroli is employed almost exclusively in perfumery. The "petale" and the "bigarade" neroli are used to an enormous extent in the manufacture of Hungary water and Eau de Cologne and other handkerchief perfumes. The "*petit grain*" is mainly consumed for scenting soap.

1256 OTHER PERFUMES.—The flowers by infusion in a fatty body make an admirable pomatum, the strength and quality varying according to the number of infusions of the flowers made in the same grease. By digesting orange-flower pomatum in rectified spirits in the proportion of from six pounds to eight pounds of pomade to a gallon of spirit for about a month, the *extrait de fleur d'orange* is obtained, a handkerchief perfume surpassed by no other scent. In this state its odour resembles that of the fresh flowers so much that with closed eyes the best judge could not distinguish the scent of the extract from that of the fresh flowers. (*Piesse.*)

1257 ORANGE-FLOWER WATER.—This is an important article of manufacture, among the distillers of essential oils. It is largely used in pharmacy. "There are three sorts of orange-flower waters found in commerce. The first is distilled from the flowers; the second is made with distilled water and neroli; and the third is distilled from the leaves, the stems, and the young unripe fruit of the orange tree." (*Piesse.*) "As met with in commerce, orange-water is colourless or of a faintly greenish-yellow tinge, almost perfectly transparent, with a delicious odour and a bitter taste." (*Pharmacog.*) PERFUMERY.

Essential Oil of Orange peel.—"Largely made at Messina, and also the south of France. It is extracted by the sponge, or by the *écuelle* process, partly from the Bigarade and partly from the sweet or Portugal Orange, the scarcely ripe fruit being in either case employed. The oil made from the former is much more valuable than that obtained from the latter, and the two are distinguished in price-currents as *Essence de Bigarade* and *Essence de Portugal.*

"These essences are but little consumed in England, in liqueur-making and in perfumery." (*Pharmacog.*)

Var. 3. Bergamia, *Fl. Br. Ind., I., 515.* Var. 3. Bergamia 1258

THE BERGAMOT ORANGE.

Syn.—C. Aurantium, *var.* Bergamia, *W. & A. Prodr., 98;* C. Limmetta, *var. DC. Prodr., i., 539.*

Vern.—*Limún, nibú, limú,* Hind., Duk.; *Nébu,* Beng.; *Limbu,* Mar.; *Limbu, nimbu,* Guj.; *Limú,* Sind (according to Stocks); *Elumich-champasham,* Tam.; *Nimma-pandu,* Tel.; *Cheru-nárannâ, jonakam-nárannâ,* Mala.; *Nimbe-hanna,* Kan.; *Jambira-phalam,* Sans.; *Limúe-hámiz, limú,* Arab.; *Limúe-tursh, limú, limeh,* Pers.; *Dehi,* Sing.; *Lámyá-si,* or *tám buyú-si,* Burm.

References.—*Brandis, For. Fl., 54; Dalz. & Gibs., Bomb. Fl., Supp., 13; Voigt, Hort. Sub. Cal., 142; Pharm. Ind., 45; Mooden Sheriff, Supp. Pharm. Ind., 104; Flück. & Hanb., Pharmacog., 121; U. S. Dispens., 15th Ed., 1002; Bentl. & Trim., Med. Pl., 52; O'Shaughnessy, Beng. Dispens., 231; Waring, Bazar Med., 89; Drury, U. Pl., 141; Piesse, Perfumery, 100, 102; Spons, Encycl., 1417; Balfour, Cyclop.; Smith, Dict., 49; Treasury of Botany; Ure, Dict. of Arts and Manufactures.*

Habitat.—The Bergamot Orange is cultivated near Reggio in South Calabria, in Sicily, and in the south of France, but it is only rarely met with in India. It may be doubted how far the above vernacular names given to it are correct. The fruit, when full grown, is still unripe and green; they are sometimes known as green oranges. Some of the green oranges met with in India (*and already alluded to, p. 340*) may belong to this variety.

BERGAMOT OIL.

Oil.—The rind of the fruit yields on expression the oil known under the name Bergamot. For this purpose the fruits are used, and one hundred of them are said to produce about three ounces of the otto. Formerly the oil was extracted by distillation or by expressing the rasped rind, but these processes have been superseded by the *écuelle*, a special instrument described in *Spons' Encyclopædia*, page 1457. OIL. 1259

General Characters of the Oil.—The oil, as produced by the machine referred to above, is of a greener tint than that obtained by the old process. "It is a clear, limpid liquid, with a peculiar and very fragrant odour, and a bitterish, somewhat warm, aromatic taste. Its specific gravity varies from 0·86 to 0·88, and its boiling point from about 361° to 383°. It has a slightly acid reaction; is mixible with rectified spirit, oil of turpentine, and glacial acetic acid; and is dextrogyre." (*Pharmacog.*)

Chemical Composition.—The authors of the *Pharmacographia* say: "If essential oil of bergamot is submitted to rectification, the portions CHEMISTRY. 1260

CHEMISTRY. that successively distil over do not accord in rotatory power or in boiling point—a fact which proves it to be a mixture of several oils, as is further confirmed by analysis. It appears to consist of hydrocarbons, $C_{10}H_{16}$, and their hydrates, neither of which have as yet been satisfactorily isolated. Oil of bergamot, like that of turpentine, yields crystals of the composition $C_{10}H_{16}+3H_2O$, if 8 parts are allowed to stand some weeks with 1 part of spirit of wine, 2 of nitric acid (sp. gr. 1·2), and 10 of water, the mixture being frequently shaken.

"The greasy matter that is deposited from oil of bergamot soon after its extraction, and in small quantity is often noticeable in the oil of commerce, is called *Bergaptene* or *Bergamot Camphor*. We have obtained it in fine, white acicular crystals, neutral and inodorous by repeated solutions in spirit of wine. This composition, according to the analysis of **Mulder** (1837) and of **Ohme** (1839), answers to the formula $C_9H_6O_3$, which, in our opinion, requires further investigation. Crystallized bergaptene is abundantly soluble in chloroform, ether, or bisulphate of carbon; the alcoholic solution is not altered by ferric salts."

Properties and Uses.—The oil of bergamot is much employed in perfumery. It has stimulant properties, but is rarely used in medicine. It is sometimes employed to give an agreeable odour to ointments and other external applications.

Essential Oil. 1261 **Essential Oil of Leaves and Flowers.—Dr. Charles Rice** of New York, in his *Commercial Products of the Citrus Family*, says that an essential oil is distilled from the leaves which is used for adulterating oil of bergamot. The flowers also, he states, are said to yield an oil of which no information can be obtained.

MEDICINE. Juice. 1262 **Medicine.**—The JUICE of the fruit possesses properties similar to those of lemon juice (see under **Citrus Medica,** *var.* **Limonum**). It is often preferred to lemon juice, as the fresh juice can be readily obtained in nearly all parts of the tropics, and as the preserved lemon juice is less effectual. It is useful as a refrigerant drink in small-pox, measles, scarlatina, and other forms of fever. It may also be taken with advantage in cases of hæmorrhage from the lungs, stomach, bowels, uterus, kidneys, and other internal organs. (*Waring, Bazar Medicines.*)

1263 Citrus decumana, *Linn.; Fl. Br. Ind., I., 516.*

THE SHADDOCK, PUMELO, or POMPELMOS, THE FORBIDDEN FRUIT, PARADISE-APPLE, *Eng.*; POMPELMOUSE, *Fr.*; POMPELMOES, *Sp.*

The word Pumelo is a contraction of "pomum melo," the melon apple.

Vern —*Mahá níbu, chakotra, bátávi nebú, sadáphal,* HIND.; *Bátávi nebu, mahá-nimbu, chakotra, bator-nebú,* BENG.; *Chakotra,* PB.; *Bijoro,* SIND. (according to **Stocks**); *Oba kotru,* GUJ.; *Papanass, papnass,* BOM.; *Papnasa, panis, pappa nasa,* MAR.; *Bombalinas,* TAM.; *Edapandú,* TEL.; *Sakotra hannu,* KAN. (according to **Cameron**); *Bombeli-marunga,* MALA.; *Púmplemús,* MYSORE; *Parvata,* ? SANS.; *Shouh-ton-oh, shanktones,* BURM.; *Maha-maram, jambúla, maha-naram,* SING.; *Púmplemús,* MALA.

References.—*Roxb., Fl. Ind., Ed. C.B.C., 590; Brandis, For. Fl., 58, 572; Kurz, For. Fl. Burm., I., 196; Gamble, Man. Timb., 59; Rice, Citrus Family in New Remedies, 1878; Dalz. & Gibs., Bomb. Fl., Supp., 12; Stewart, Pb. Pl., 29; DC. Origin Cult. Pl., 177; Voigt, Hort. Sub. Cal., 141; Flück. & Hanb., Pharmacog, 117; U. C. Dutt, Mat. Med. Hind., 127; O'Shaughnessy, Beng. Dispens., 232; Baden Powell, Pb. Pr., 334; Cook, Gums and Gum-resins, 13; Atkinson, Him. Dist., 710; Atkinson, Gums and Gum-resins, 16; Birdwood, Bomb. Prod., 141; Lis-*

The Pumelo; The Citron.

boa, U. Pl. Bomb., 148; Smith, Dict., 375; Treas. of Bot.; Ure, Dict. Arts and Manuf., III., 765; Kew Offl. Guide to the Bot. Gardens and Arboretum, 64, 65; Trop. Agri., 117; Simmonds, Trop. Agri., 441.

Habitat.—A native of the islands of the Malay Archipelago, more particularly abundant in the Friendly Isles and Fiji. Introduced into India from Java and into the West Indies by **Captain Shaddock**; hence the name *Shaddock*. It is cultivated in most tropical countries. In India and Burma it is a common fruit tree. It is, however, more frequent in Bengal and Southern India than in the North-West Provinces. The vernacular name *Batavi nebu* suggests its having been originally brought from Batavia. "The fruit is very large, weighing sometimes ten to twenty pounds, roundish, with a smooth pale-yellow skin, and white or reddish sub-acid pulp. When the fruits attain their largest size, they are called pompoleons, or sompilmousses; those of the smallest size form the 'Forbidden fruit' of all the English markets." (*Treasury of Botany.*)

Gum.—Said to yield scantily an unimportant gum. In 1855, **Lieutenant Hawkes** sent to the Madras Exhibition a sample of this gum (*Cooho*). GUM. 1264

Medicine.—**Mr. Baden Powell** says that the FRUIT is nutritive and refrigerant. It contains sugar and citric acid, with much essential oil in the peel. The leaves are said to be useful in epilepsy, chorea, and convulsive cough. MEDICINE. Fruit. 1265 Peel. 1266 Leaves. 1267

Food.—This tree is a favourite with the natives of India, as it gives fruit all the year round; flower, unripe and ripe fruits may be seen on the same tree at once. There are two varieties: one with whitish, and the other with reddish, pulp. Besides, the individual fruits differ from one another in size, some reaching 2 feet in circumference, and also in quality according to the soil, climate, and situation. **Dr. Bonavia** (in the paper to which repeated reference has been made) says: "The best pummelows I have seen are the thin-skinned red pummelows of the Bombay market. They come in about Christmas. They are juicy, and of the colour of raw beef internally and of a globose shape. There is no reason why this fine thing should not be extensively grown. All the other varieties of pummelows I have seen in India and Ceylon are not to be compared with this, and are hardly worth propagating to any extent." FOOD. 1268

Citrus Medica, *Linn.; Fl. Br. Ind., I., 514.* 1269

The name MEDICA given to this species is derived from the fact that in one of the first unmistakeable references to it, the fruit is spoken of as the apple of Media. The acid forms at least have well-marked Sanskrit names, and there would appear to be no doubt that the cultivation of the plant, spread from India to Mesopotamia and Media, from whence it became known to Europe. The presumption is therefore that it is a native of India, and in confirmation of this opinion one or two forms of a wild plant, supposed to be the source of the citron and the lime, are common on the outer Himálaya. The citron, lemon, and lime are, however, so different from each other that it is probable, if derived from one and the same species, they are the developments of indigenous varieties which may have originated in widely different regions. The cultivated forms of the citron approach the pumelo, having a thick, coarse, and rough rind, while the lime comes nearer to some of the forms of the true orange, being spherical and smooth-skinned. It seems probable that the home of the former may be found to be the mountain tracts of Eastern Bengal, more particularly of the Khásia and Garo hills, while the latter is of a more northern character, extending along the foot of the Himálaya to the Panjáb.

The sweet lime (**C. Limetta**) appears to be the southern manifestation of the species, and the writer would be disposed to look for the lemon in the far east, if not in China, even although the Chinese names for it do not occur in the ancient writings. As a cultivated plant, it may have spread from China to India before it had attracted much attention in China itself. Although not wild, the plant is more frequent in Assam than in Bengal, and it is possible it may have entered India across the Chino-Assam frontier.

This species includes as varieties the Citron, the Lemon, the Sweet and the Sour Lime.

Var. 1 Medica.

Var. 1. Medica proper.

1270 THE CITRON, CEDRAT-TREE, ADAM'S-APPLE, *Eng.;* CEDRATIER, CITRONIER, *Fr.;* CIDRATO, CEDRO, *It.;* CIDRO, *Sp.;* CIDREIR, *Port.;* CEDRATEN, CITRONENBAUM, *Germ.*

Considerable difference of opinion prevails as to the origin of the word Citron. It is presumed that the Median apple was synonymous with the Persian apple, and by some authors this is even supposed to have been the golden apples guarded by the mythical daughters of Hesperus. By later authors the apple of Media is translated *Citrium malum*, but the word Citrus was applied in Latin to both the Citron and the Cedar. If the word Citron be of Indian origin, **Dr. Rice** suggests that it may have come from *Chitra*, 'excellent,' but he adds that it might also be supposed to come from the Chinese. **Bretschneider** gives the Chinese as *kü* (or *kiuh*), and the very common Indian ncun affix-*tra* would have produced *kutra*, which might easily have become *Chitra* and Citron.

Syn.—C. AURANTIUM, *var.* MEDICA, *W. & A. Prodr.;* C. MEDICA, *var. A., Linn.;* CITRUS MEDICA, *Risso.*

Vern.—*Bijaura, limbu, kutla, bara nimbu, turanj, nimbu, limu,* HIND.; *Beg-púra, lebu, nebu, bijaura, bara nimbu, turanj, honsá nebu,* BENG.; *Bajauri nimbú,* PB.; *Bijorú, turanj, bálank,* GUJ.; *Bijapúra, mahálunga, bijori, limu,* BOMB.; *Mavalung, limbu,* MAR.; *Turanj,* DUK.; *Elumich-cham-pazham, nárttam-pazham,* TAM.; *Nimma pandu, nára dabba, dabba, mádhipala-pandu, bija púra, pulla-dabba, lungamú, bijapuramu,* TEL.; *Nimbe hanu, limbu, máda-lada-hammu,* KAN.; *Ganapati-náranna,* MALA.; *Matulunga, phalapúrá, begapúra; víjapúra* (according to **Brandis**), SANS.; *Utraj, utrej, utroj, úturinji,* ARAB.; *Turanj,* PERS.; *Thanba-ya, shauk ta kera, shouk-ta-kwoh, sh-ousa khavá, sh-on-takhavá,* BURM.; *Sedaran,* SING.

References.—*Roxb., Fl. Ind., Ed. C.B.C., 590; Brandis, For. Fl., 51, 572; Kurz, For. Fl. Burm., I., 197; Gamble, Man. Timb., 59; Dalz. & Gibs., Bomb. Fl., Supp., 13; Stewart, Pb. Pl., 29; Rice, Citrus Family in New Remedies, 1878; DC. Origin Cult. Pl., 178; Voigt, Hort, Sub. Cal., 142; Moodeen Sheriff, Supp. Pharm. Ind., 106; Flück. & Hanb., Pharmacog., 128; U. C. Dutt, Mat. Med. Hind., 127; O'Shaughnessy, Beng. Disp., 230; Year-Book Pharm., 1874, 623; Drury, U. Pl. 142; Cooke, Gums & Gum-resins, 14; Atkinson, Him. Dist., 710; Atkinson, Gums and Gum-resins, 16; Birdwood, Bomb. Prod., 12, 142; Lisboa, U. Pl. Bomb., 149; Piesse, Perfumery, 114; Smith, Dict., 119; Treasury of Botany; Ure, Dict. Arts and Manuf., I., 807; Kew Offl. Guide to the Bot. Gardens and Arboretum, 84; Simmonds, Trop. Agri., 439, 448.*

Habitat.—Said to be wild in Chittagong, "Sitakund Hill," the Khásia and Garo hills. **Atkinson** remarks it is wild in the Bhabar and along the Sarju, under Gangoli in Kumaon. Cultivated throughout the warm moist region of India and in Sicily and Corsica; also in other parts of Italy, and in Spain, Portugal, the West Indies, and Brazil. In addition to citrons, similar to those met with in Europe, a very common Indian form much resembles a small pumelo.

The Citron; The Lemon.

CITRUS Medica.

History.—The citron is supposed to have been introduced into Greece
and Italy from Persia and the warmer regions of Asia. It is described by
Theophrastus as abundant in Media three centuries before Christ, and HISTORY OF
may have been known to the Hebrews at the time of the Babylonish THE CITRON.
Captivity. According to **Gallesio** it was introduced into Italy about the
third or fourth century. The Jews cultivated citron when under the Roman
rule, and used the fruit, as at the present day, in the Feast of Taber-
nacles; each person bringing a citron in his hand. **Dr. Royle** found the
species growing wild in the forests of Northern India, and, as already
stated, it may therefore fairly be conjectured that the original home of
the citron was in India. It has now spread over the whole of the civilised
world, and even in cold regions it is cultivated under artificial heat.

Gum.—Said to yield scantily an unimportant gum. Sent from Ma- GUM.
sulipatam to the Madras Exhibition in 1855. 1271

Oil.—The flowers yield on distillation a very fragrant oil resembling OIL.
neroli, which is chiefly used for the manufacture of Hungary water. An- 1272
other perfume known as Cedrat is obtained from the rind of the fruit, both
by distillation and expression. The extract of cedrat is only the essential
oil of citron dissolved in spirits, to which bergamot is sometimes added.
(*Piesse.*)

Medicine.—Citron RIND is hot and dry and tonic; PULP cold and dry; MEDICINE.
SEEDS, LEAVES, and FLOWERS hot and dry; JUICE refrigerant and astrin- Rind.
gent. According to **Theophrastus** the fruit is an expellent of poisons. 1273
To one who has taken a poison injurious to life, it may be given, producing Pulp.
a strong effect on the bowels, and the poison is drawn out. It also corrects 1274
fœtid breath. (*Drury.*) The distilled water of the fruit is used as a Seeds. 1275
sedative (*Year-Book, Pharm., 1874, 623*). Leaves.

Special Opinions.—§ "The rind is made into a marmalade and is 1276
an antiscorbutic" (*Surgeon-Major A. S. G. Jayakar, Muskat*). "It Juice.
is made into preserve and is used for dysentery" (*Surgeon-Major* 1277
J. Robb, Ahmedabad). Marmalade.

Food.—The FRUIT is described in the *Flora of British India* as large, 1278
oblong or obovoid; rind usually warted, thick, tender, aromatic; pulp FOOD.
scanty, sub-acid. The rind makes good comfit; the pulp is also pre- Fruit.
served in sugar. Both fruit and preserve are somewhat bitter to the 1279
taste. The rind of the fruit candied is well known as a delicate Comfit.
sweatmeat. **Atkinson** says the wild fruit is used for pickling (*khatái*). 1280
Dr. Bonavia remarks that citrons are very little used in India, except for Candied Rind.
medicinal purposes. "On the Western coast of India, they have many 1281
large varieties, and at Mangalore they eat the thick sweet skin after Pickles.
peeling off the bitter rind. In Lucknow, and in Rampur, Rohilcund, and 1282
other places they make a preserve of the thick skin of the citron, which Preserve made of skin.
they call 'Turunj.' All the citrons, both sweet and sour, have a dry pulp." 1283

Structure of the Wood.—White, moderately hard. TIMBER.

Domestic Use.—The fruit put amongst clothes keeps away moths. 1284
DOMESTIC

Var. 2. Limonum, *sp. Risso.* 1285

The word lemon is from the Arabic *limûn*, and this, through the Var. 2.
Persian, is the Hindi *limu, limbu,* or *nimbu*; probably adopted by the Limonum.
Sanskrit people. Much stress is by authors laid upon the fact that the 1286
name *nimbu* is to this day in actual use in Kashmír. It is difficult to see
why this should be viewed with such favour; *nimbu* is the generic name
for any lime or lemon throughout India, but there is a strong probability
that **C. acida,** rather than **C. Limonum,** is the plant most generally referred
to under that name, and the lemon is more frequently spoken of as the
bará nimbú or large *nimbu.*

The Lemon.

THE LEMON, *Eng.;* CITRONNIER, LIMONIER, *Fr.;* LIMONE *It.;* CITRONE, *Germ.*

Syn.—C. AURANTIUM, *var.* LIMONUM, *W. & A. Prodr., 98;* C. LIMONUM, *Wall. Cat., 6389;* C. MEDICA, *Willd.* (according to *Roxb.*), *Fl. Ind., Ed. C.B.C., 590.*

Vern.—*Jámbíra, bará nimbú, pahári-nimbu, pahári-kaghzi,* HIND., DUK.; *Karna nebu, gora nebu, bara-nebu,* BENG.; *Kimti, gulgul, khutta,* PB.; *Metá limbu, mótu-limbu, mótu-nimbu,* GUJ.; *Thóra-limbu,* MAR.; *Periya-elumich-cham-pazham,* TAM.; *Pedda-nimma-pandu,* TEL.; *Valiy, acherunáranna,* MALA.; *Doddá-nimbe-hannu,* KAN.; *Mahá-jambíra-karuná?* SANS.; *Qalambak,* ARAB.; *Kalinbak,* PERS.; *Kígisamyá-si,* BURM.; *Lokka-dehi,* SING.

Referenccs.—*Brandis, For. Fl., 52; Dalz. & Gibs., Bomb. Fl., Supp., 13; DC. Origin Cult. Pl., 179; Voigt, Hort. Sub. Cal., 142; Pharm. Ind., 43; Moodeen Sheriff, Supp., Pharm. Ind., 105; Flück. & Hanb., Pharmacog., 14, 118; O'Shaughnessy, Beng. Dispens., 230; Murray, Pl. and Drugs, Sind, 80; Year-Book Pharm., 1874, p. 623; S. Arjun, Bomb. Drugs, 22; Drury, U. Pl., 141; Baden Powell, Pb. Pr., 334; Atkinson, Him. Dist., 710; Piesse, Perfumery, 144; Smith, Dictionary, 242; Treasury of Botany; Ure, Dict. Arts and Manufactures, III., 105; Kew Off. Guide to the Museum, 25; Kew Off. Guide to the Bot. Gardens and Arboretum, 64; Journal, Agri.-Horti. Soc., old series, Vol. XIV., 99; Trop. Agri., 117, 874; Simmonds, Trop. Agri., 440; Fleming, Med. Pl. and Drugs, in As. Res., Vol. XI., p. 164.*

Habitat.—Cultivated abundantly in the south of Europe and in India. Seems highly probable that the wild form of this plant has not as yet been discovered. All the Himálayan and Khásia plants are, as far as the writer's experience goes, wild forms of the lime or citron, but not of the lemon. It is highly probable the lemon is of much more recent origin than the citron and the lime.

The question has been recently raised as to the highest altitude oranges and lemons could be grown in India. A writer in the Agri.-Horticultural Society's Journal said they could not be grown above 5,000 feet. **Madden** refers to the lemons grown at Almora, the fruit being collected in summer and ripened in straw. The altitude given above is perhaps correct for the Indian species generally.

HISTORY OF THE LEMON.

History.—**Dr. Royle** is said to have found the tree growing wild in the north of India, and **Atkinson** reports that **Madden** spoke of the *jamíra* or wild variety found in the Kota Dún of Kumáon. **Royle's** wild plants were known as *behári-nimbu* or *pahári-kaghazi* in the Dun. **De Candolle** states that the lemon was unknown to the ancient Greeks and Romans, and that its culture only extended into the West with the conquests of the Arabs. On their spreading over the vast regions of Asia and Africa, they carried with them everywhere the orange and the lemon. The latter was brought by them in the tenth century from the gardens of Oman into Palestine and Egypt. **Jacques de Vitry,** writing in the thirteenth century, very well describes the lemon which he had seen in Palestine; and doubtless it was by the Crusaders first brought into Italy, but at a date which cannot be exactly ascertained. From the north of India it appears to have passed eastward into Cochin-China and China, and westward into Europe, and it has naturalised itself in the West Indies and various parts of America" (*Treasury of Botany*).

LEMON OIL.

OIL. 1287

Oil.—The rind of the lemon, when rasped and subjected to expression, or when distilled, affords an essential oil known as "essence of lemon" or "citron-zeste," according to the method adopted. The oil is largely manufactured in Sicily, at Reggio in Calabria, and at Mentone and Nice in France. A brief account of the methods of extraction, as given in the *Pharmacographia* (*p. 119*), may be reproduced here :— Method of extraction. 1288

Sponge process.—" The workman first cuts off the peel in *three* thick longitudinal slices, leaving the central pulp of a three-cornered shape, with a little peel at either end. This central pulp he cuts transversely in the middle, throwing it on one side and the pieces of peel on the other. The latter are allowed to remain till the next day and are then treated thus: the workman seated holds in the palm of his left hand a flattish piece of sponge, wrapping it round his fore-finger. With the other he places on the sponge one of the slices of peel, the outer surface downwards, then presses the zest side (which is uppermost), so as to give it for the moment a convex instead of a concave form. The vesicles are thus ruptured, and the oil which issues from them is received in the sponge with which they are in contact. Four or five squeezes are all the workman gives to each slice of peel, which done he throws it aside. Though each bit of peel has attached to it a small portion of pulp, the workman contrives to avoid pressing the latter. As the sponge gets saturated the workman wrings it forcibly, receiving its contents in a coarse earthen bowl provided with a spout; in this rude vessel, which is capable of holding at least three pints, the oil separates from the watery liquid which accompanies it and is then decanted."

E'cuelle process.—" A stout saucer or shallow basin of pewter, about 8½ inches in diameter, with a lip on one side for convenience of pouring. Fixed in the bottom of this saucer are a number of stout, sharp, brass pins, standing up about half an inch; the centre of the bottom is deepened into a tube about an inch in diameter and five inches in length, closed at its lower end. This vessel, which is called an *écuelle à piquer*, has, therefore, some resemblance to a shallow, dish-shaped funnel, the tube of which is closed below. The workman takes a lemon in the hand, and rubs it over the sharp pins, turning it round so that the oil-vessels of the entire surface may be punctured. The essential oil which is thus liberated is received in the saucer, whence it flows down into the tube; and as this latter becomes filled, it is poured into another vessel that it may separate from the turbid, aqueous liquid that accompanies it. It is finally filtered, and is then known as *Essence de Citron au Zeste*. A small additional produce is sometimes obtained by immersing the scarified lemons in warm water and separating the oil which floats off."

" A second kind of essence, termed *Essence de Citron distillée*, is obtained by rubbing the surface of fresh lemons, or of those which have been submitted to the process just described, on a coarse grater of tinned iron, by which the portion of peel richest in essential oil is removed. This grated peel is subjected to distillation with water, and yields a colourless essence of very inferior fragrance, which is sold at a low price."

Description and Chemical Composition.—The lemon oil is of a faint yellow colour, of exquisite fragrance and bitterish aromatic taste. **Piesse** says that what is procured by expression has a much finer odour and a more intense lemony smell than the distilled product. The oil dissolves sparingly in rectified spirit, but readily in anhydrous alcohol. It mixes freely with bisulphide of carbon. The chief constituent of the essential oil "is the terpene, $C_{10} H_{16}$, which, like oil of turpentine, easily yields crystals of terpin, $C_{10} H_{16} ZO H_2$. There is further present, according to **Tilden** 1879), another hydrocarbon, $C_{10} H_{16}$, which already boils at 160° C., whereas the foregoing boils at 176°C. Lastly, a small quantity of cymene and of a compound acetic ether, $C_2 H_3 O (C_{16} H_{17} O)$, would appear to occur also in oil of lemons. The crude oil of lemons already yields the crystalline compound $C_{10} H_{16} + Z H Cl$, when saturated with anhydrous CHEMISTRY. 1290

The Lemon.

hydrochloric gas, whereas by the same treatment oil of turpentine affords the solid compound C_{10} H_{16} and H Cl." (*Pharmacog, p. 120.*)

PERFUMERY. 1291 **Properties and Uses.**—Essence of lemon is used in perfumery and as a flavouring agent. **Piesse** says: "Lemon otto may be freely used in combination with rosemary, cloves, and caraway, for perfuming powders for the nursery. From its rapid oxidation it should not be used for perfuming grease, as it assists rather than otherwise all fats to turn rancid; hence pomatums so perfumed will not keep well. In the manufacture of other compound perfumes, it should be dissolved in spirit, in the proportion of six to eight ounces of oil to one gallon of spirit. There is a large consumption of otto of lemons in the manufacture of eau de Cologne." In medicine, oil of lemon is a stimulant and carminative when given internally, and stimulant and rubefacient when applied externally. It has been used as a local application with doubtful results.

MEDICINE. 1292 **Medicine.**—There are three officinal parts of the fruit mentioned in the *Pharmacopœia of India*: (1) the outer part of the rind; (2) the essential oil of the rind noticed under the head "Oil;" (3) the juice of the ripe fruit. The rind is said to be stomachic and carminative. Lemon juice is highly valued as an antiscorbutic and refrigerant—primarily antalkaline; secondarily, antacid. It forms the best remedy for scurvy, and an excellent drink in fever and inflammatory affections. It has met with success in acute rheumatism, dysentery, and diarrhœa. It also forms an antidote to acro-narcotic poisons. (*Pharm. Ind.*) **Mr. Baden Powell** says that it is considered by natives also an antidote to animal poisons.

In bilious and intermittent fevers it is specially useful, combined with port wine and cinchona bark (*Drury*). In the *Hindú Materia Medica* fresh lemon juice is recommended to be taken in the evening, for the relief of dyspepsia, with vomiting of meals. In rheumatic affections such as pleurodynia, sciatica, lumbago, pain in the hip joints, &c., **Sarangadhara** recommends the use of lemon juice with ***yavakshara*** and honey (*U. C. Dutt*).

The best substitute for lemon juice is a solution of about eight drachms of citric acid in sixteen ounces of water, with the addition of a few drops of lemon oil. Lemon juice may also be used in preparing effervescing diaphoretic and diuretic draughts. The relative proportions of lemon juice and citric acid with the alkaline carbonates, for the formation of effervescing draughts, are as follow:—

Lemon juice—	or Citric acid—	to 20 grains of
Fl. drs. iijss . .	grs. xiv . .	Bicarbonate of Potash,
Fl. drs. vi . .	grs. xxiv . .	Carbonate of Ammonia,
Fl. drs. iv . .	grs. xvii . .	Bicarbonate of Soda.

The lemon juice, being liable to spontaneous decomposition, speedily becomes unfit for medical use. "One of the best methods of preserving the juice is to allow it to stand for a short time after expression, till a coagulable matter separates, then to filter, and introduce it into glass bottles, with a stratum of almond oil or other sweet oil on its surface. It will keep still better if the bottles containing the filtered juice be suffered, before being closed, to stand for fifteen minutes in a vessel of boiling water. Another mode is to add one-tenth of alcohol and to filter. The juice may also be preserved by concentrating it either by evaporation with a gentle heat, or by exposure to a freezing temperature, which congeals the watery portion, and leaves the juice much stronger than before." (*U. S. Dispens., 15th Ed., 849.*)

Dr. Charles Rice of New York states that the bark of the root has been used in the West Indies as a febrifuge and the seeds as a vermifuge.

§ "Lemons, as well as other fruits of the same order, contain a principle—*hesperidene.* By some chemists this substance is described as bitter and crystalline, and by others as tasteless. **Gladstone** obtained from oil of orange peel about 95 per cent. of a terpene, which he called hesperidene. A glucoside amantini has been isolated from the flowers of **Citrus decumana.**" (*Surgeon C. J. H. Warden, Professor of Chemistry, Calcutta.*) MEDICINE.

Citric Acid.—A crystalline acid is prepared from lemon or lime juice. It occurs in colourless crystals, is very soluble in water, less soluble in rectified spirit, and insoluble in pure ether. The chief use of citric acid in medicine is in the preparation of effervescing draughts and refrigerant drinks, dose being from ten to thirty grains. Citric acid. 1293

§ "The amount of free citric acid contained in Indian limes appears to be somewhat less than that found in the varieties cultivated in Europe, and varies from 25 to 30 grains of uncombined citric acid per fluid ounce." (*Surgeon C. J. H. Warden, Professor of Chemistry, Calcutta.*)

Lemon Syrup.—In the *Pharmacopœia of India* the following directions are given for the preparation of this substance: "Take of fresh lemon peel two ounces; lemon juice, strained, one pint; refined sugar, two pounds and a quarter. Heat the lemon juice to the boiling point, and having put it into a covered vessel with the lemon peel, let them stand until they are cold, then filter and dissolve the sugar in the filtered liquid with a gentle heat. The product should weigh three pounds and a half and should have the specific gravity 1·34." Syrup. 1294

Special Opinions.—§ "*Lime juice.*—Most useful in dysentery with sloughing of the mucus membranes. I have given 12 ounces a day in apparently hopeless cases with success" (*From a Contributor*). "Lemon oil mixed with glycerine is applied on the eruption of acne" (*Surgeon R. Gray, Lahore*). "Lemon juice and gunpowder used topically for scabies" (*Surgeon-Major E. C. Bensley, Rajshahye*). "The fruit in the form of pickle is useful in hypertrophy of the spleen" (*Surgeon J. C. Penny, Amritsar*).

Food.—The lemon juice is used largely in sherbets and cooling drinks. The fruit is also pickled. FOOD. 1295

Var. 3. acida. 1296

THE SOUR LIME OF INDIA.

Syn.—C. ACIDA, *Roxb., Fl. Ind., Ed. C.B.C., 589.* (**Roxburgh** appears to include under this not merely the Sour Lime but all Lemons.) The **C. Limetta,** *Risso,* described by many authors (*e.g.,* **Dr. Rice** in *New Remedies*) as having a "very acid, even acrid," juice, must refer to this plant and not to the South Indian sweet lime, the juice of which is sweet and pleasant. If this proves correct the synonyms may require to be rearranged.

Vern.—*Lebú, nebú, limbu, nimbú, limún, nibú, limú,* HIND.; *Lebu, nebu, limbu, nimbú, páti-nebu, kagúji-nebu, kaghzi nimbú, camral-nebú, taba, nebú,* BENG.; *Nimbú, kuttah-nimbú,* PB.; *Khata limbu, lebu, limbu-nimbu,* GUJ.; *Limbu,* MAR.; *Limún, nibú, nimbú, límú,* DUK.; *Elumich-cham-pasham, elemitchum, elimichumpullum,* TAM.; *Nimma-pandu, némmapúndú,* TEL.; *Nimbe-hannu,* KAN.; *Cheru-naranna, jonakam-naranna, jérúk, nipis, limowe, erúmitchi-narrdcum,* MALA.; *Jambíra, limpáka, nimbuka, vijapura* (according to **Dutt**), SANS.; *Límun, límúe-hámiz, nimu, límú,* ARAB.; *Límúe-tursh, límú,* PERS.; *Thanbaya, sámyá-sí, tambiyá-si,* BURM.; *Dehi,* SINGH.

References.—*Brandis, For. Fl., 52; Stewart, Pb. Pl., 29; DC. Origin, Cult. Pl., 179; U. C. Dutt, Mat. Med. Hind., 226; Ainslie, Mat. Ind., I., 193; Atkinson, Him. Dist., 710; McCann, Dyes and Tans, Bengal, 159; Kew Off. Guide to the Museum, 25; Kew Off. Guide to the Bot. Gardens and Arboretum, 64.*

CITRUS Medica.

The Sour Lime.

Habitat.—Wild in the warm valleys of the outer Himálaya, from Garhwal and Sikkim to the Khásia and Garo hills, Chittagong, and probably also the mountain tracts of the Central Provinces and of the Western Peninsula and the Satpura mountains of Central India. It ascends to about 4,000 feet, and occurs as a small, much-branched thorny bush. Is cultivated all over India and Burma, and is by far the commonest species of Citrus. In the various languages of India all the others are treated as forms of this, some being distinguished as the large lime, others as the horse lime, the sweet lime, &c.; but this is the lime itself according to native gardeners. There are many minor cultivated forms, differing chiefly in size. The fruits of all are more or less round, smooth, with a shining rind, green, or only tinged with yellow when ripe.

DYE. 1297 **Dye.**—The leaves of this plant are used in tanning in Mánbhum. This seems to be doubtful; at most, the leaves can be used only as an adjunct to the tans, imparting an odour to the leather.

MEDICINE. 1298 **Medicine.**—"Lime-juice is much used in medicine by the native practitioners; they consider it to have virtues in checking bilious vomiting, and believe that it is powerfully refrigerant and antiseptic. In the Tamool Medical Sastrum entitled *Aghastier Vytia Anyourou* there is quite an eulogy on the lime: 'It is a fit and proper thing to be presented by an inferior to a superior; it is beautiful to behold; cooling and fragrant to the smell; the juice of it rubbed upon the head will soothe the ravings of frenzy; and the rind of it dried in the sun has the power, when laid under the pillow, of conciliating affection.' **Dr. Thomson**, in his *London Dispensatory*, tells us that lime-juice, taken in the quantity of half an ounce, allays hysterical palpitations of the heart." (*Ainslie, Mat. Ind., 193.*)

Special Opinions.—§ "The *Kaghuze nimbo*, the juice of which is so universally used as a cooling and grateful drink in the form of sherbet, should be mentioned here" (*Brigade Surgeon G. A. Watson, Allahabad*). "Fresh lime-juice often proves effectual in relieving the irritation and swelling caused by musquito bites" (*Brigade Surgeon J. H. Thornton, Monghyr*).

FOOD. 1299 **Food.**—The Sour Lime of India has "flowers small, fruit usually small, globose or ovoid, with a thick or thin rind, pulp pale, sharply acid." **U. C. Dutt** says: "The fruits are cut vertically into two pieces, and the fresh juice, squeezed out with the fingers, is sprinkled on soup, *dál*, curry, &c., to which it imparts a pleasant acid taste and agreeable flavour. A

Pickle. 1300 pickle of *pati-nebu* in its own juice and salt is a popular and effectual medicine for indigestion brought on by excess in eating, or by indigestible articles of diet. The fruits are first rubbed over a stone, or their rind scraped a little so as to thin it. They are then steeped in juice obtained from other fruits of the sort, and exposed to the sun for a few days with the addition of common salt. When crisp and of a brown colour, they are preserved in porcelain vessels or glass jars. This preparation is called *jarák-nebu* (that is, digestive lemon) in the vernacular." **Atkinson** says "the *Kághazi nibu*, or thin-skinned variety of Jaunpur and Azamghar, is celebrated all over Upper India. Next in value is the *páti-níbu* or small round variety, and there is also a large variety, the *kámaráli-nibu* of Bengal, or *khatta-níbu* of the North-West Provinces. The small sour limes are used for sherbets and making lime-juice, and the larger ones for sherbets and for preserves, especially chips." **Dr. Bonavia** remarks "The true limes are the most used by the natives. They are to be found everywhere, and even where no other Citrus occurs, some kind of lime is sure to be seen. Nevertheless, it is astonishing that so common a thing, so useful a fruit, and a tree so easily raised from seed, is not to be found in the *villages* of the North-West Provinces. There is probably

FOOD.

not a *village* in the whole of India where the *kághzi-nimbú* would not readily grow." "Although they are called limes, I believe them to be an *acidless* variety of one of the *lemons* of India." "They call them *sherbetee* or *mitha-nimbú*, and also *amrutphal*. The best-flavoured sorts are grown in dry climates, such as Mooltan and Muscat." The imports into Bombay of Muscat limes were valued in 1884-85 at R1,555 and in 1885-86 at R1,695. Dr. **Bonavia** divides the sour lemons into "*Bajouras*—a sort of Citron lemon;" "lemons proper," and a group of sour **Citrus** known by the name of *gungolee* and *behari* lemons.

Var. 4. Limetta, *W. & A.; Fl. Br. Ind., I., p. 515.* 1301

THE SWEET LIME OF INDIA.

Syn.—C. NOBILIS, *Lour.*, as in *Kurz, For. Fl. Burm., I., 197; Wight, Ic., t. 958;* C. LIMETTA, *Risso.* It might be asked, has the C. LIMETTA, *Risso*, sweet or bitter fruits? if the latter, it might be viewed as a synonym of var. **acida.**

Vern.—*Mitha nebu, nembú, mitha amrit-phal,* HIND.; *Mitha nebu,* BENG.; *Mila-nimbú,* PB.; *Mitha limbu,* GUJ., BOMB.; *Elemitchum,* TAM.; *Nemma-pandu, gajanimma,* TEL.; *Erúmitchi narracum,* MALA.; *Madhukarkatiká,* SANS.; *Thanbaya,* BURM.; *Dehi,* SING.

References.—*Brandis, For. Fl., 52; Dalz. & Gibs., Bomb. Fl. Supp., 13; Stewart, Pb. Pl., 29; Rice, New Remedies, 1878; DC. Origin Cult. Pl., 179; Atkinson, Him. Dist., 710; Lisboa, U. Pl. Bomb., 140; Piesse, Perfumery, 150; Smith, Dict., 245; Treasury of Bot.; Ure, Dict., Arts and Manuf., III., 117; Trop. Agri., 1882-83, 597; Simmonds, Trop. Agri., 441.*

Habitat.—Commonly cultivated in most parts of India and Burma. Most probably a native of Southern India; **Wight** says it is indigenous at Kolagberry in the Nilgiri hills.

Botanic Diagnosis.—Leaves with winged petioles; flowers small, white; fruit globose or ovoid, shortly mamillate; rind with concave vesicles.

The limes approach much nearer to the true oranges than do any of the other forms of **C. Medica.** Indeed, it is difficult to say how far the published accounts of **C. Limetta** have become mixed up with **C. Bigaradia,** and the vernacular names given to both these forms, as well as to **C. Lumia** (the Sweet Lemon), are hopelessly confused. Many of the so-called green oranges met with in India (and referred to at pages 340 and 347) may be but sweet limes. According to some authors **C. nobilis** (the Mandarin Orange) is placed under the sweet lime, and by others under **C. Bigaradia.**

Medicine.—§ "Extensively used as refrigerant in fever and jaundice" (*Surgeon J. C. Penny, Amritsar*). MEDICINE 1302

Food.—The fruit is both eaten fresh and after being preserved or cooked in various ways, but the juice is not so much valued as that of the preceding variety. FOOD. 1303

Var. 5. Lumia, *W. & A.; Fl. Br. Ind., I., 515.* 1304

THE SWEET LEMON, *Eng.;* LUMIE, *Fr. & Germ.*

Vern.—See C. LIMETTA.

Habitat.—This form is very little known in India, and occurs only occasionally in gardens. It is probable that, with the lemon, this is not an Indian form. **Atkinson** and many Indian writers use the terms "sweet lime" and "sweet lemon" as synonymous.

Botanical Diagnosis.—Leaf petioles simply margined; flowers tinged with red; fruit bright yellow, ovoid-oblong, with a long curved mamilla; rind with convex vesicles; pulp sweet.

OIL. 1305 **Essential Oil.**—**Dr. Rice** says that this oil is prepared at Squillace in Calabria by mechanical means.

1306 **Citrus nobilis,** *Lour.*

THE MANDARIN ORANGE, sometimes also called the MALTESE ORANGE.

Syn.—CITRUS CHINENSIS and C. MYRTIFOLIUS.

Vern.—Probably the same as for C. LIMETTA; it is the *kán* of China.

Habitat.—Cultivated in China and Cochin-China, where it appears to be indigenous, distributed, apparently in modern times, to the mountain tracts of India. A small sweet orange is said to be wild on the hill tracts beyond Sudiya in Assam, which very much answers to the character of the
1307 Mandarin. **Mr. Clarke** reports that its cultivation on the Khásia hills has been greatly extended. **Dr. Bonavia** speaks in the highest terms of the blood oranges of Gujranwala and of Jaunpore. New to European gardens at the beginning of the present century, but now cultivated plentifully in Sicily and Malta, known as *tangerines* in St. Michael's.

Botanical Diagnosis.—A moderate-sized tree; fruit uneven in surface, spherical but flattened on the top; rind very thin, dark reddish-yellow; pulp almost blood red with a peculiar flavour; both leaves and fruit have the same odour.

ENCOURAGEMENT OF CULTIVATION IN INDIA.

Having briefly indicated the principal forms of Citrus met with in India, the present article may be concluded by a useful and practical suggestion offered by **Dr. Bonavia**, namely, to encourage the growth, at certain selected centres of India, of the oranges, limes, lemons, and citrons for which they are more peculiarly famous. He urges that the blood oranges of Gujranwala and Jaunpore should be fostered and developed, as these are not only the finest oranges met with in India, but would come into market in the hot season when no others are available; that the true Mandarin might be grown in the mountain tracts of Bengal and Burma;
1308 that Poona might extend its cultivation of *Keonla* and *Mussembi* oranges; that the Deccan might pay special attention to its warty *keonla*, "the best of its kind;" that Lucknow might make a speciality of "its fine large *Behari nimbu*" (a name given by **Royle** to the wild lemon) and its *Kághzi kálan*; that Lahore should give attention to its pear-shaped *karua* and the large, sour, and juicy lemon known in the Panjáb as *gulgul*; and that Bombay should prepare to meet the Indian demand for its excellent pomelos. In this way, with extended railway communication, free interchange might be made with the various provinces and a more constant and uniform supply of improved fruits kept up throughout the year. "By restricting the varieties, to be grown on a large scale, to those localities to which they are most suited, a race of growers would be trained, who would thoroughly understand the wants of that particular variety, and would grow up conversant with the best modes of dealing with it, not only with regard to the cultivation and propagation, but also with the best modes of packing and preserving the fruit for a long time."

CLAUSENA, *Linn.; Gen. Pl., I., 304.*

1309 **Clausena indica,** *Oliv.; Fl. Br. Ind., I., 505; Beddome;* RUTACEÆ.

Syn.—PIPTOSTYLIS INDICA, *Dalz.; Dalz. & Gibs., Bomb. Fl., 29;* BERGERA NITIDA, *Thw., Enum. Ceylon Pl., 46.*

Vern.—*Migong-karapichi-gass,* SING.

Reference.—*Lisboa, U. Pl. of Bomb., 33.*

Habitat.—A shrub or small tree, met with in the Western Peninsula from the Bombay Ghâts to the Anamally Hills, and also in Ceylon.

Structure of the Wood.—Close-grained and hard; adapted for the lathe. TIMBER. 1310

Clausena pentaphylla, *DC.; Fl. Br. Ind., I., 503.* 1311

Syn.—AMYRIS PENTAPHYLLA, *Roxb.; Fl. Ind., Ed. C.B.C., 321.*

Vern.—*Rattanjote, surjmukha, teyrúr,* HIND.

References.—*Brandis, For. Fl., 49; Gamble, Man. Timb., 59.*

Habitat.—A deciduous shrub, native of the Sub-Himálayan tracts, from Kumáon to Nepál, especially the *sál* forests of the Dúns and of Oudh.

Medicine.—The bruised leaves are highly aromatic, and are believed to possess medicinal properties. MEDICINE. Leaves. 1312

CLAVICEPS.

Claviceps purpurea, *Tulsane;* FUNGI. 1313

THE ERGOT, ERGOT OF RYE, HORNED OR SPIKED RYE (*Secale Cornutum*), BUNT.

Syn.—SCLEROTIUM CLAVUS, *DC.;* ERGOTŒTIA ABORTIFACIENS, *Queh;* OIDEUM ABORTIFACIENS, *Berk. & Br.*

References.—*Pharm. Ind., 251; O'Shaughnessy, Beng. Disp., 631, 673, 76; Balfour, Agri. Pests of India, 61, 115; Flück. & Hanb., Pharmacog., 740; Bentl. & Trim., Med. Pl., IV., 303; U. S. Dispens., 15th Ed., 556-7.*

Dr. R. Tytler (in the *Cal. Med. Phys. Trans., 1831, vol. V., p. 441*) reports that barley in the Upper Provinces of India is often affected with a disease very similar to, if not identical with, ergot of rye. The diseased grain is spoken of as being very poisonous. This same, or apparently the same, disease has been observed on oats, rice, and particularly **Pennisetum typhoideum.** Wheat ergot is common in Europe, and was once considered superior medicinally to ergot of rye. Wheat ergot has not, it would seem, been observed in India, but an allied disease known as *kandwa* or 'smut' is common. This same name is, however, given to the ergot 'bunt' found on the millets; but from the absence of berberry bushes in the wheat districts of India, it is almost safe to conclude that the *kandwa,* on being carefully examined, will be found not to be ergot as known in Europe.

Medicine.—The sclerotium (the compact spawn) of **Claviceps purpurea,** produced within the paleæ of the common rye, **Secale cereale,** forms the officinal part. "In medicinal doses ergot acts principally upon the muscular fibres of the uterus, causing them to contract strongly and continuously, more especially during labour and after delivery, hence it is largely used to promote contraction of the uterus in cases of tedious parturition, or to prevent flooding after delivery. The administration of ergot is also most beneficial in menorrhagia and leucorrhœa. Moreover, as ergot causes contraction of the small arteries generally by its action on their muscular walls, it is a powerful agent in checking hœmorrhage, whether from the lungs or bowels; and also to diminish congestion in affections of the cerebro-spinal membranes, and in other cases. It has likewise been employed to cause the expulsion of coagula of blood, polypi and hydatids, from the uterus. MEDICINE. 1314

"In overdoses ergot produces nausea, vomiting, colicky pains, headache, and sometimes delirium, stupor, and even death. Taken for a length of time, as in bread made with diseased rye, it acts as a poison, producing two conditions of the constitution, termed, respectively, gangrenous ergotism and convulsive ergotism, both accompanied with formication" (*Bentley & Trimen*).

1315 No effort appears to have been made to test medically the properties of the Indian ergot, and up to date the supply is drawn entirely from Europe. A good deal has, however, been written on the poisonous property of barley, wheat, and especially of millets which, to outward appearance, seem of good quality but which contain a fungus, most probably an ergot. It seems probable that Indian wheat rust may be due to a species of **Æcidium** reared on a **Euphorbia**.

1316 Some writers have attributed to an ergot the poisonous qualities which *kesari* (**Lathyrus sativus**) is said to possess. An indulgent use of this pea induces a paralysis of the lower limbs which is generally incurable.

See under **Fungoid Pests.**

CLAY.

1317 **Clay** is a hydrated silicate of alumina, which is expressed in mineralogy by the formula $H_2 Si_2 O_8 + H_2 O$ which may be said to be $Si O_2$ 46·40, $Al_2 O$ 39·68, Water 13·92.

Vernacular and other names.—*Argile*, Fr.; *Thon*, Germ.; *Gil, chikni matí, sangi-i-dalam*, Hind.; *Káli munnu* (Potter's clay), Tam.; *Banka munnú*, Tel.; *Tannab* (white clay), Mal.; *Krishna mirtika*, Sans.

References.—*Manual of the Geology of India, Part III.* (*Ball*), *Part IV.* (*Mallet*); *Ure's Dict., Arts, Manufactures, &c., I., 734; Spons' Encycl., 635; Balfour's Cycl., Vol. I., 739; Encycl. Brit.; Official correspondence; Other publications referred to in the text.*

Properties and Classification.—The pure clay, defined above, when it occurs, is generally known as "Kaolin" or "Porcelain-clay." There are, however, numerous other inferior qualities, such as fire-clay, pipe-clay, shale, clunch, loam, mud or silt, mudstone, &c., &c. Some of these would, however, be more correctly defined as soils containing more or less clay. Usually they are soft and plastic, and emit, when breathed on, the peculiar odour known as "argillaceous." They chiefly occur as superficial deposits in river-basins, estuaries, or dried-up lakes. Pure clay is derived from a decomposition of felspar, from which the silicates of potash, soda, &c., have been washed out. The purer forms of clay are derived from granite, the quartz and mica having been washed away as sand, and alumina silicate thrown down in the low-lying tracts of country alluded to. Common clay usually contains a distinct proportion of sand, but, in addition, it contains (on account of the various sources from which it may be derived) other ingredients such as oxide of iron, lime, magnesia, &c. The oxide of iron and the organic constituents impart to it its colour; the former makes red clays, and the latter dark or even almost black clays. The proportion of sand and of iron present in a clay may be said to be the chief governing principles that determine the economic value or utility of a clay. Iron may qualify a clay for one purpose but altogether disqualify it for another. Any finely-divided mineral substance, which contains from 10 to 30 per cent. of alumina in the form of silicate, and which becomes plastic on being moistened and retains the form imparted to it by a mould, even when dried or burned, is popularly termed "clay."

These facts naturally lead to an industrial classification of the clays, and in dealing with those met with in India we shall, as far as possible, take them up in the alphabetical order of their better known names in preference to attempting a scientific assortment.

I.—BRICK CLAYS.

1318 In the early part of the present century, it was thought necessary to import bricks into India from England. It was soon discovered, however, that in almost every district clays suitable for this purpose existed in abundance, for bricks were employed in many buildings in India long anterior to the arrival of the English. Some of an enormous size are found in the ancient monuments, and in more recent times others much smaller than the European type.

Ball says: "As a rule Indian-made bricks do not bear a very high reputation for strength or durability; but it has been demonstrated that good bricks can be made, and it seems probable that, in many cases where the bricks are bad, the system of manufacture, rather than the material, is to blame. Of course there are some clays so impregnated with lime *kankar* nodules that without grinding they are inapplicable to the manufacture of good bricks. In the neighbourhood of most of the large rivers in India, clays are, however, to be found tolerably free from these impurities. The largest brick factory in India is situated at Akra near Calcutta, where from 20 to 30 million of bricks are turned out annually." (For technical details regarding brick-making in India see the Rurki Professional Papers on Indian Engineering.)

II.—EDIBLE AND MEDICINAL CLAYS AND FULLER'S EARTH. 1319

In most bazars in India a fine unctuous or oily clay is sold as a drug or as an article of food eaten by *enceinte* women, or used by ladies as a cosmetic. Allied to this is the clay used to effect caste markings on the forehead. **Balfour** says such a clay "is excavated from a pit near Koluth in large quantities, and exported as an article of commerce, giving a royalty of R1,500 yearly. It is used chiefly to free the skin and hair from impurities; and the Cutchi ladies are said to eat it to improve their complexions." Throughout India the soft mud found on the banks of the rivers is used as a detergent, the hair being well rubbed with the clay before being washed in the stream. The writer found certain persons excavating a pale yellow mud from a hillock near the capital of Manipur, which he was informed was regularly eaten by the women throughout the State. **Irvine** (*Mat. Med. of Patna, 66*) says *multani* Multani. 1320
mittie—a kind of light yellow ochre—is eaten in dyspepsia in doses from 5 to 30 grs. **Sakharam Arjun** (*Bombay Drugs, p. 167*) remarks that *mulatání-mattí* "is eaten by pregnant females to relieve acidity of the stomach and is given mixed with sugar in cases of leucorrhœa." He further comments on an imported earth known as *Sang-i-Basrí* (a Persian name). "This is generally imported from Bassorah and the Persian Gulf, as its name implies. It is used in tonic preparations and in irregular menses and with benefit from the iron it contains." He states that the earth in question is a silicate of alumina with lime and iron. **U. C. Dutt** (*Sans. Mat. Med.*) after dealing with red and yellow ochre (which see) or the *geru mátí* in Beng., and *gairika* in Sans., adds: "besides *gairika* several other varieties of earth are described and occasionally used in medicine. A sweet-scented earth brought from Surat and called *Saurástra mrittiká* is regarded as astringent and useful in hœmorrhages. It enters into the composition of several medicines for relieving bleeding from internal organs." If this earth be a natural product of Surat it is nowhere (so far as the writer can discover) described in the records of that district, but in the same way, the so-called *multání mátí*, sold under that name throughout India, is not referred to in the *Gazetteer of Mooltan*. **Mr. Baden Powell** remarks of this earth (*Pb. Prod., 24*): "The name '*multání*' applied to the earth does not indicate its origin, except, perhaps, as regards the trade in it." These instances of a name being supposed to indicate the origin of a product are of considerable importance in checking a too liberal application of the principle that

the source of a product may be inferred from its name. Under his account of Rawalpindi, **Mr. Baden Powell** says of *mitti gáchni:* "This is a soft and saponine drab-coloured earth, something like fuller's earth, sold in small pieces; it is used for cleaning the hair, also in medicine; it is to be had in every bazar, where it is called '*mitti Múltání*' or '*gil-i-Múltání.*'" **Capt. F. R. Pollock**, on Dera Ghází Khán reports, "it is stated that this *Múltání mitti* is imported to Dera Ghází Khán from the interior of the western range (Sulaimaní) to the extent of 10,000 maunds." The Assistant Commissioner of Múltan writes: "although it would appear Múltan is famous for its *mitti* or earth, yet there are no mines or pits here which produce the substance. It is imported from the sandy and rocky tracts of country lying to the south and south-west of Múltán. It is of the following descriptions:—

1321 "*1st*—White *mitti*, which is termed *khajrú*, or eatable.

1322 "*2nd*—Yellow *mitti*, which is termed *bhakrí*, and is used by the poorer classes for dyeing cloths, &c.

1323 "*3rd*—Light green, or *sabz mitti*, which is chiefly used by the natives for washing and cleaning the hair."

The writer was shown an earth in the Simla bazar, said to be edible, and which bore the name of *gagní* or *garí*; the shop-keeper could, however, give no information as to its source.

The above facts and references to authors may be accepted as indicating the extent to which certain earths are used for medicinal or quasi-medicinal purposes. Suffice it to add that the earths alluded to are most probably forms of or allied to fuller's earth. Of the last-mentioned earth **O'Shaughnessy** says (*Beng. Disp., 34*) "a superior kind of 'Fuller's earth' (or *sabún matí*) occurs at Colgong." **Ball** (*Econ. Geol., 570*) gives a much more detailed account of this earth, accepting also the supposition that the edible earths of India are most probably forms of fuller's earth. His account is of so much interest that we may reproduce here the main facts from it: "Being of detrital origin fuller's earth does not possess a definite chemical composition, but in general terms it may be described as a soft unctuous silicate of alumina. It derives its name from having been used in the 'fulling' of woollens; though it is still largely employed for this purpose, other detergents are now more generally used. In India fuller's earth is employed in washing of cloths which are used in the manufacture of lac, indigo, &c., and doubtless for many other purposes. It is believed that earths of this nature afford the principal part of those which are used as comestible. The practice of eating earth is widespread over the world, and though there is not much information available on the subject, in reference to India, the fact is known that these edible clays from different localities are to be had in most Indian bazars, and it seems possible that the practice of eating them is not limited merely to pregnant women, as is sometimes stated."

"The probability is, that once acquired, the habit is not easily given up. Saucer-shaped chips, about 2 inches in diameter, of partially baked clay for eating, are sold in the Calcutta bazar; they are said to be made by potters a few miles to the north of Calcutta." **Mr. Ball** gives the following as the best known Indian sources of this earth:

Sabun Miti.

1324 *Bengal.*—The *sabún mití* or soap-earth of Colgong in the Bhagalpur Division. The earth sold in Calcutta as Rajmahal *mittí*, a comestible earth, the precise source of which is not known.

1325 *Rajputana.*—Fuller's earth used to be obtained near Ajmir.

1326 *Bikanir State.*—The *Gazetteer* of this State mentions that fuller's earth is excavated at the village of Meth near Kolath. Over 2,000 camel-loads are taken every year to Sirsa.

Fire Clay. CLAY.

1327 *Bombay and Sind.*—A pale greenish clay is found in Western Sind, which is used for washing, and is also eaten by pregnant women.

1328 *Panjáb.*—Dera Ghazi Khan and Multan already alluded to; in the Salt range at Nilawan, Mr. Wynne says a lavender-coloured clay is found which is used as a fuller's earth.

III.—FIRE CLAYS. 1329

These derive their name from their refractory nature—that is to say, from their capacity to resist very high temperatures without fusing, fissuring, or altering their shape. The essential character of such clays is that they should be as near as possible free from lime, iron, or alkaline earths which promote the fusion of silica as in glass-making. In Europe the best clays for this purpose are those from the floors or underclays frequently found below coal seams; they exhibit impressions and carbonised remains of plants. Although there can be little doubt that the coal-fields of India are of a different age from those of England, still the underlying clays are found to afford a fairly good fire-brick material. Such bricks are largely made at Raniganj and are used in the blast furnaces of the Bengal Iron Works, at Karharbari, and at Barwai. Promising-looking fire-brick clays are found in the Chanda or Wardha coal-fields, in 1330
the Umaria coal-field, and at Jubalpore; while in Maulmain (Burma) a fire-brick clay of good quality is believed to exist. **Balfour** states: "Fire-clays are procurable at Streepermatoor, Tripasoor, Chingleput, Mettapoliam, and Cuddapah; indeed, are very common in many parts of India, and bricks can be made that resist the action of great heat. A clay found at Beypore, 20 to 30 feet below the surface, is used for fire-bricks and for lining furnaces." **Ball** makes no mention of these South Indian sources of fire-clays, but he remarks that "it is probable that, with proper manipulation, some of the pottery clays" "would afford perfectly refractory materials." He accordingly alludes to these so-called fire-clays under pottery clays. (Conf. with **No. 1333.**) A similar result was found to be the case with the Colgong clay experimented with by **Sir W. O'Shaughnessy**: "the bricks and crucibles which were manufactured from it were considered to be equal to the articles imported from Europe." **O'Shaughnessy** describes his preparation as three parts of the white clay called *khari* with one part of fuller's earth called *Sabún mátí*, mixed with water and baked at a red heat. The crucibles so made, he says, are perfectly infusible and impermeable to melted metals or saline matters, and bear sudden heating and cooling without fracture.

1331 On the subject of Fire-clays the following valuable communication has been received from **Messrs. Burn & Co.** of Raniganj:—" Our experience only goes as far as fire-clay obtained in the coal measures of the Raniganj District, and this we consider as good as the best English fire-clay. For your information we beg to quote some extracts from the official report of trials made at Her Majesty's Mint, see pages 18, 19, and 20, part I, volume VIII. of Records of the Geological Survey of India, 1875, which are as follow:—

"(1) First experiment in September 1874 by **Theodore W. H. Hughes, Esq., F.G.S., A.R.S.M.,** Officiating Deputy Superintendent, Geological Survey, India.

"'The fire-bricks tested by me were furnished by the firm of **Messrs. Burn and Company.** The materials from which they are made are very refractory and capable of resisting high temperature, without sensibly fusing. That, compared with Stourbridge fire-bricks, they are somewhat superior.

"'The specimens were subjected to a temperature of over 3,000° Fht., the melting point of cast-iron being 2,786° Fht.'

"Second experiment in January 1875 by **H. B. Medlicott, Esq., M.A., F.G.S.**, Officiating Superintendent, Geological Survey, India.

"'Several of them stood the test perfectly, shewing no sign of cracking or of vitrification. These latter trials were made in the presence of **Mr. Whitelaw**, Manager of the Bengal Iron Company's proposed works and others, who agreed in the favourable estimate formed of the quality of these bricks.'

"In addition to the foregoing we beg to quote you the opinions of **D. W. Campbell, Esq.**, Locomotive Superintendent, East Indian Railway, and **J. Blackburn, Esq.**, Engineer and Manager of the Oriental Gas Company. The former, in a letter to us, dated 23rd February 1875, writes:—

"'(2) I have had the fire-bricks and fire-clay tried here, they are both very good; I will send you a requisition as soon as present stock is exhausted.'

"And **Mr. Blackburn**, in his letter of 2nd March 1875, states as follows:—

"'(3) The Gas retorts made for the Company by your firm two years ago have since been kept in constant use at a temperature of about 2,000° Fht., and they have been found fully as durable and effective as those of the best English manufacture.'

"We trust that the above extracts will be found to contain the information required by **Dr. Watt** for the Dictionary of Economic Products, but in case he wishes to analyse the clay himself, we have pleasure in sending herewith a few sample pieces obtained from the coal measures of the Raniganj District."

1332

IV.—PIPE CLAYS.

This is known as *Namam* in Tamil and *Kharra* in Dukhni; its English name is taken from the fact of its being used to manufacture tobacco-pipes. It much resembles China-clay, only that it possesses more silica. **Balfour** says: "This is found in abundance in several parts of India; the Hindus employ it for making the distinguishing marks on their foreheads, and (moistened with water) it is often applied round the eye in certain cases of ophthalmia, as well as to parts of the body that are bruised." It seems probable that the clay here referred to should be given under the heading already dealt with, *viz.*, edible and medicinal or fuller's earths. **Ball** makes no mention of pipe-clays occurring in India. **Blanford** states that a thick bed of true pipe-clay exists between Terany and Kauray in Trichinopoli.

1333

V.—POTTERY CLAYS.

These might be popularly referred to three sections or degrees of purity: (*a*) porcelain or kaolin clays; (*b*) ordinary white or glazed pottery clays; and (*c*) red or tile and flower pot clays. In every province, indeed in almost every district of India, one or other of these clays occur. There is practically no demand, however, for white or glazed pottery. The Muhammadans and Christians use a certain amount of glazed earthen-ware, but the caste system of the Hindus precludes them from using most of their earthen vessels twice. This fact has given birth to the practice of making the coarsest and cheapest pottery. Even in districts where clays exist which approach in quality to that known as kaolin, the native potter is content to evolve from his crude wheel, vessels often elegant in form, but constructed of a material little better than that used by the brick-maker. One European pottery, that of **Messrs. Burn & Co.**, of **Raniganj**, in

Pottery Clay.

Bengal, is attempting to compete with European imported articles. Under the care of the School of Art, an effort is being made to utilise the white clays or kaolins of Madras, and **Mr. George Terry** of Bombay has succeeded, in connection with the Art School of that presidency, in establishing an industry in art pottery, the ornamental designs being faithful copies of the Ajanta cave paintings. With the exception of these foreign or new industries there remains only the glazed pottery of Sind, of the Panjáb, and of the North-West Provinces as rising above the all but universal type of coarse red pottery. **Ball** very properly says:—"There is probably no part of the world, not excluding remote oceanic islands, where the use of glazed pottery is less known than is the case in many parts of India." **Mr. Kipling** (*Journal of Indian Art*) says: "No substance resembling the fine clays of Dorsetshire, Devonshire, and Cornwall, is known to the Indian potter, whose only resource, with one or two unimportant exceptions, is the brick earth of the plains and rivers. Fuel, which is of equal importance with potting minerals, is scarce, and coal has never been used by the native artizans." In Bengal, coal is used by the native brick-maker and the common potter, but **Mr. Kipling's** statement is nevertheless true of India as a whole. He proceeds to comment on the potter's craft:—"As to social status, no craft, excepting, perhaps, that of the leather-dresser, is held in lower esteem than the potter's trade in Hindustan, the Deccan, and South India." **Mr. Kipling** next distinguishes the two classes of workers in earth, *viz.*, *Kumhars* and *Kashigars*. The former are the common village potters who "produce wares which, though of little technical value as pottery and of small commercial importance, are often good in colour and form, and perfectly fitted for the purposes they are intended to serve." The latter, the *Kashigars*, are "makers of glazed earthenware who are only to be found in the Panjáb and in Sind, and within the last few years in the town of Bombay and at Khurja in the North-Western Provinces. The name of the trade is Persian, derived probably from Kashan, the earliest seat of the manufacture, and the *Kashigar* is usually a Mussulman of good caste. In India the art has been, until recently, almost entirely architectural in its character and devoted to the covering of the wall surfaces of mosques and tombs with enamelled plaques and tiles. Persia may originally have borrowed the fashion from Tartar or Chinese sources, but there seems little doubt, notwithstanding some vague traditions as to its importation direct from China, that it was introduced into India by the Mussulman invasion, and not by means of the friendly intercourse which there seems reason to believe subsisted at various times with Tibet and the further East." **Sir George Birdwood** (*Indian Arts*) has recorded a high testimony as to the merit of the artistic forms of the common red pottery—forms which are seen portrayed on some of the earliest monuments of India. He has also spoken, with the highest admiration, of the elegant adaptations of the decorative designs with the forms and uses of the vessels which are turned out by the workers in glazed pottery. It is not within the scope of the present work to enter upon these subjects. Sufficient has been said to convey a general impression of the magnitude and character of the Indian ceramic art, and we may therefore conclude the present article with a brief abstract of the published facts regarding the clays met with in the provinces of India which are suitable for pottery, omitting all reference to the third class of clays, *viz.*, the ferruginous clays suitable for the coarser articles, since such clays practically occur in every district of India. The following abstract (1st to 7th) from **Ball** gives the chief sources of the superior pottery clays of India:—

1st, South India.—The cretaceous rocks of Trichinopoli afford, according to **Mr. H. Blandford**, some fine clays, well adapted for pottery. "Fels- 1334 par and kaolin are obtainable in different parts of the district." "In the South Arcot district a fine plastic clay occurs in the Cuddalore beds near the south bank of the Guddalum," but it contains small quantities of lime and iron, the latter giving it a pinkish tint. In North Arcot the granite rocks of the district are decomposed to a certain extent, and, according to **Mr. Foote**, would yield a certain but not very considerable supply of kaolin. White goblets are made in Arcot which enjoy some reputation, but the source of the clay is not known. Fine pottery clays exist in great abundance in the district of Chingleput, more especially at Sripermatur. From the beds exposed at Coopum a supply has been taken for the Madras School of Art.

1335 *2nd, Mysore.*—For many years it has been known that kaolin earth existed in great abundance in this State, the beds extending from Bangalore to Nandydrug. When mixed with quartz these clays have been found to afford a valuable fire-clay. Specimens of a white clay sent from Mysore were favourably reported on by **Minton.**

1336 *3rd, Mangalore.*—As early as 1811 **Dr. Christie** discovered, in association with the laterite, an extensive deposit of what he conceived to be pure porcelain clay.

1337 *4th, Bengal.*—In Orissa white clays occur in the Mahanadi valley of Rajmahal age. These clays are used by the natives for ornamenting their houses and in tanning leather. The Colgong clay has already been alluded to; it is of the same age as that used at Patharghata in the manufacture of pipes. In several parts of the Rajmahal hills there are beds of white silicious clays belonging to the Barakar coal measures which are suitable for the manufacture of many articles of hard pottery, and which, with proper treatment, would afford suitable material for fire-bricks. But the best known clays of this series are the refractory and other clays now being worked by **Messrs. Burn and Co.** of Raniganj. The clay used at the pottery works is chiefly obtained from the coal-beds and consists of more or less decomposed shale, but a white lithomarge is obtained under laterite at a point about 12 miles north-east of Bankura. A certain amount of kaolin, **Mr. Ball** states, might be obtained from this area. White clays have also been reported from the Darjeeling District.

1338 *5th, N.-W. Provinces.*—In the year 1838, a **Mr. J. Jeffreys** established pottery works at Fatehgarh and produced articles with a very considerable degree of success. Black pottery is made at Azimgarh, which owes its colour to the organic matter present in the clay.

1339 *6th, Panjáb.*—According to **Mr. Baden Powell** two classes of clays occur in this province—a grey clay which burns red, and clays which burn to a yellowish white or cream colour. Reference has alredy been made to some of these, but for pottery purposes the clays of Dera Ghazi Khan, Dera Ismail Khan, and Kohat deserve special mention. There are kaolin mines at Kassumpur in the Delhi District, and also on the hills near the Kutub Minar. By washing, the quartz and mica are removed from these, and the kaolin pressed into the cakes which are sold for white-washing purposes, and may possibly also be used in pottery. Good kaolin is also reported to be found at Buchara near the Lota river in the Alwar hills.

1340 *7th, Assam and Burma.*—Rich deposits of porcelain clays have been reported to occur in Upper Assam near the Bhramakhund, known locally as *rukmanipitha*, and a fine clay for pottery purposes is also said to be found near the base of the cretaceous rocks at the western end of the Garo hills. In Burma the ordinary alluvial clay, mixed with sand, affords the material for common pottery, but a dark-coloured seam in the Irawadi valley is much sought after by the potters. Some of the upper beds in the nummulitic group are said to consist of China clay and would answer well for pottery, owing to their freedom from iron. Kaolin is also reported to exist in Tenasserim. Of the clays experimented with by **Sir William O'Shaughnessy** that from Singapore was said to be the best.

VI.—MATERIALS USED FOR GLAZING OR PAINTING POTTERY IN INDIA. 1341

The indigenous art of glazing pottery, as practised in India is crude and unsatisfactory. **Ball** says: "The varnish or imperfect glaze used for the sugar-boilers' pans, known in Bengal as *kolas*, is thus described by **Mr. Piddington**: There are two kinds of earth used, one of which is called *belutti*; it is a silicious and ochreous earth, the best being found 16 or 18 miles from Kulna. By levigation it is prepared for use, the process lasting, it is said, 15 days. The other earth is called *Úporomí*, and is a tenacious loam. The best was obtained at Monad, 20 miles west of Chinsurah, and at Panchchowki, 16 miles south-west of Kulna. Its preparation is said to take three months, and only 10 seers are obtained from one maund of the earth; two varieties of the *úporomí* are *gad* and *majarí*. Successive layers of mixtures of *gad*, *belutti*, and *úporomí* are smeared on the sun-dried common red ware which is then burnt and glazed at one firing. This varnish is capable of resisting great heat, and a very penetrating solution like that of sugar." In the experiments performed by **O'Shaughnessy** the most suitable glaze proved to be borate of lime. The black colour of pottery is often obtained from the smoke 1342
of oil-cake thrown into the kiln when the baking is complete. At other times an organic varnish is used for this purpose, except when, as mentioned in connection with Azimgarh, the clay itself contains the necessary organic matter to cause it to burn black. Artificially blackened pottery is produced at Monghir, Patna, Sarun, Chunar, and Surat. In the younger rocks of the Rajmahal series certain clays occur called *khari*. These are used as pigments. According to **Buchanan** the potters of Rajmahal use this *khari* for giving a white surface to pottery made of ordinary clays. Cheap pottery is often painted after having been baked, such as that seen at Kota, Lucknow, Benares, &c.; at other times it is powdered with mica, or by other mechanical means has a colour imparted to it. Black pottery is, for example, often etched, and a preparation of tin and mercury rubbed into the patterns in imitation of metal bidri-ware. With the exception of these miserable attempts the *kumhar* potter is innocent of the art of glazing his wares. A much more advanced knowledge is possessed by the *Kashigar*; indeed, the possession of this knowledge is the recognised characteristic of his trade. "The shades of blue which constitute the chief feature of the Sind and Panjáb pottery are produced by oxide of cobalt. The supply of this substance is limited to certain mines in Rajputana" (see **Cobalt**). Glazed pottery is made in 1343
Sind, chiefly at Hala, Hyderabad, Tattu, and Jerruck, and in the Panjáb at Lahore, Multan, Jhang, Delhi, &c. The chief places for the manufacture of encaustic tiles are at Bulri and Saidpur in Sind. **Sir George Birdwood** (*p. 307*) says, in the glazing and colouring, two preparations are of essential importance, namely, *kanch*, literally glass, and *sikka*, oxides of lead. In the Panjáb the two kinds of *kanch* used are distinguished as *angrezí kanchi*, "English glaze" and *desi kanchi*, "country glaze." "The former is said to be made of "*sang-i-safed*, a white quartzose rock, 25 parts; *sajji* or pure soda 6 parts; *sohaga telia* or pure borax, 3 parts; and *nausadar*, or sal ammoniac, 1 part. Each ingredient is finely powdered and sifted, mixed with a little water, and made into white balls of the size of an orange. These are red-heated, and after cooling again, ground down and sifted. Then the material is put into a furnace until it melts, when clean-picked *shora*

kalmi or saltpetre is stirred in. A foam appears in the surface, which is skimmed off and set aside for use." The latter is similarly made of
1344 quartzose rock and borax or siliceous sand and soda. "A point is made of firing the furnace in which the *kanch* is melted with *kikar*" (**Acacia arabica**), "*karir*" (**A. Catechu**), or "Capparis wood." "Four *sikka*, or oxides of lead, are known, namely, *sikko safed*, white oxide, the basis of the blues, greens, and greys used; *sikka zard*, the basis of the yellows; *sikka sharbati*, litharge, and *sikka lal*, red oxide." "*Sikka safed* is made by reducing the lead with half its weight of tin; *sikka zard* by reducing the lead with a quarter of its weight of tin; *sikka sharbati* by reducing with zinc instead of tin; *sikka lal* in the same way, oxidising the lead until red." "All the blues are prepared by mixing either copper or manganese, or cobalt, in various proportions with the above white glaze. The glaze and colouring matter are ground together to an impalpable powder ready for application to the vessel." "The *rita* or *zaffre* is the black oxide of cobalt found all over Central and Southern India. which has been roasted and powdered, mixed with a little powdered flint." **Sir George** further describes another process of preparing the *nila* or indigo blue glaze for use by itself, which consists in taking powdered flint 4 parts, borax 24, red oxide of lead 12, white quartzose rock 7, soda 5, zinc 5, and *zaffre* 5, burning the mixture in the *kanch* furnace as before.

1345 "The yellow glaze used as the basis of the greens is made of *sikka zard*, white oxide 1 seer, and *sang safed*, a white quartzose rock or mill-stone, or burnt and powdered flint, 4 chittaks, to which, when fused, 1 chittak of borax is added."

1346 "The green colours produced are: (1) *Zamrudi*, deep green (1 seer of glaze and 3 chittaks of *chhil tamba* or calcined copper); (2) *Sabz*, full green ...... (3) *Pistaki*, bright green ...... (4) *Dhani*, pale green" by smaller proportions of the copper. "Another green is produced by burning 1 seer of copper filings with *nimak shor*, or sulphate of soda." **Sir George Birdwood**, in his most interesting account of Indian pottery, after having described the glazes and colours used, proceeds: "The colours, after being reduced to powder, are painted on with gum or gluten. The vessel to receive them is first carefully smoothed over and cleaned, and, as the pottery clay is red when burnt, it is next painted all over with a soapy, whitish engobe, prepared with white clay and borax and **Acacia** or **Anogeissus** gums called *kharya mutti*. The powdered colours are ground up with a mixture of *nishasta*, or gluten and water called *mawa*, until the proper consistence is obtained, when they are painted on with a brush. The vessels are then carefully dried and baked in a furnace heated with *ber* (**Zizyphus**), or, in some cases, Capparis wood."

1347 VII.—CLAYS OR EARTHS EMPLOYED AS PIGMENTS OR DYES.

See "**Pigments**" for further information as to colouring of pottery.

Clearing Nut, see **Strychnos potatorum**, *Linn.*; LOGANIACEÆ.

CLEIDION, *Blume; Gen. Pl., III., 320.*

1348 **Cleidion javanicum**, *Bl.; Fl. Br. Ind., V., 444;* EUPHORBIACEÆ.

Syn.—ROTTLERA URANDA, *Dalz. & Gibs., Bomb. Fl., 230.*

Vern.—*Okúrúgass, okuru*, SING.

References.—*Kurz, For. Fl., Burm., II., 390; Beddome, Fl. Sylv., t. cclxxii; Gamble, Man. Timb., 348; Thwaites, En. Ceylon Pl., 272; Lisboa, U. Pl. Bomb., 123.*

Habitat.—An evergreen tree met with in the tropical forests of Northern and Eastern Bengal, South India, Burma, and Ceylon.

Structure of the Wood.—Uniformly white or yellowish, rather heavy, fibrous but close-grained, soft; takes good polish, but is not durable. In Madras it is used for building purposes. TIMBER. 1349

[EUPHORBIACEÆ.

CLEISTANTHUS, *Hook. f.; Gen. Pl., III., 268;*

Cleistanthus malabaricus, *Müll.-Arg.; Fl. Br. Ind., V., 276.* 1350

References.—*Gamble, Man. Timb., 357; Lisboa, U. Pl. Bomb., 120.*

Habitat.—A small tree found in the Konkan and Malabar districts of South India.

Structure of the Wood.—**Lisboa** mentions this plant amongst his useful timbers. TIMBER. 1351

C. myrianthus, *Kurz; For. Fl. Burm., II., 370; Fl. Br. Ind., V., 275.* 1352

Vern.—*Mo-man-tha,* BURM.

Reference.—*Gamble, Man. Timb., 357.*

Habitat.—A moderate-sized evergreen tree of the tropical forests of Burma and the Andaman Islands.

Structure of the Wood.—Moderately hard, reddish grey. Weight 41℔ per cubic foot. TIMBER. 1353

CLEMATIS, *Linn.; Gen. Pl., I., 3.*

Clematis barbellata, *Edgew.; Fl. Br. Ind., I., 3;* RANUNCULACEÆ. 1354

Reference.—*Gamble, Man. Timb., I.*

Habitat.—A woody climber of the western temperate Himálaya, Garhwál, and Kumaon.

C. Buchananiana, *DC.; Fl. Br. Ind., I., 6.* 1355

References.—*Kurz, For. Fl. Burm., I., 17; Gamble, Man. Timb., I.; Royle, Ill. Him. Bot., I., 51.*

Habitat.—A large woody climber, occurs throughout the temperate Himálaya at 6,000 feet.

C. Gouriana, *Roxb.; Fl. Br. Ind., I., 4; Wight, Ic., t. 933-4.* 1356

References.—*Roxb., Fl. Ind., Ed. C.B.C., 457; Kurz, For. Fl. Burm., I., 16; Gamble, Man. Timb., I.; Thwaites, En. Ceylon Pl., I.; Dalz. & Gibs., Bomb. Fl., I.; Aitchison, Cat. Pb. Pl., I.; Voigt, Hort. Sub. Cal., 2; O'Shaughnessy, Beng. Dispens., 160; Royle, Ill. Him. Bot., I., 44, 51; Balfour, Cyclop.*

Habitat.—An extensive climber found in the hilly districts from the Western Himálaya, rising up to 3,000 feet, to Ceylon and the Western Peninsula.

Medicine.—This plant and some of the other species abound in an acrid poisonous principle. The LEAVES and fresh STEMS, if bruised and applied to the skin, cause vesication. In France the **C. vitalba,** *Linn.,* is used by mendicants to cause artificial sores for the furtherance of their impostures. MEDICINE. Leaves. 1357 Stems. 1358

C. grata, *Wall.; Fl. Br. Ind., I., 3.* 1359

Vern.—*Ghantiáli, biliri,* HIND.

References.—*Gamble, Man. Timb., I.; Voigt, Hort. Sub. Cal., 2; Royle, Ill. Him. Bot., I., 44, 45, 51; Balfour, Cyclop.*

Habitat.—A climber of the sub-tropical and temperate Himálaya at 2,000 to 3,000 feet.

1360 **Clematis montana,** *Ham.; Fl. Br. Ind., I., 2.*

Vern.—*Ghantiáli,* HIND.

References.—*Gamble, Man. Timb., I.; Royle, Ill. Him. Bot., I., 45, 51.*

Habitat.—A woody climber of the temperate Himálaya, from the Indus to the Bramaputra, ascending to 12,000 feet, always above 8,500 in Sikkim, and in the Khásia Hills, Manipur, above 4,000 feet.

1361 **C. napaulensis,** *DC.; Fl. Br. Ind., I., 2.*

Vern.—*Pawanne, birri, wandak,* PB.

References.—*Stewart, Pb. Pl., 3; Royle, Ill. Him. Bot., 23.*

Habitat.—Found in the temperate Himálaya from Garhwál to Bhutan.

MEDICINE. Leaves. 1362 **Medicine.**—In Kanáwar the LEAVES are said to act deleteriously on the skin.

1363 **C. triloba,** *Heyne; Fl. Br. Ind., I., 3.*

Vern.—*Moravela, morvel, moriel, ranjáe, ránjai,* BOMB.; *Moravela,* MAR.

References.—*Dalz. & Gibs., Bomb. Fl., I.; Dymock, Mat. Med. W. Ind., 2nd. Ed., 21; S. Arjun, Bomb. Drugs, 2.*

Habitat.—An extensive climber met with in the mountains of the Malwa district of the Deccan, and West Konkan.

MEDICINE Plant. 1364 **Medicine.**—The PLANT is used as a remedy in leprosy, blood diseases, and fever (*Dymock*). The Greeks used **C. vitalba** for the same purposes as the natives of India employ this and other species.

FIBRE. 1365 Distillate. 1366 The above species of **Clematis** yield fibres which are regularly used for agricultural purposes, and although authors allude to the medicinal properties of only one or two species, they are all more or less used by the natives of hill districts. **Bracounot** has pointed out that the acrid active principle may be distilled with water and is soluble in fixed oils.

CLEOME, *Linn.; Gen. Pl., I., 105, 968.*

Cleome pentaphylla, see **Gynandropsis pentaphylla,** *DC.;* CAPPARIDEÆ.

1367 **C. viscosa,** *Linn.; Fl. Br. Ind., I., 170; Wight, Ic., t. 2.*

Sometimes called WILD MUSTARD.

Syn.—C. ICOSANDRA, *Linn.;* POLANISIA VISCOSA, *DC.;* P. ICOSANDRA, *W. & A.*

Vern.—*Kanphuti, húrhúr* (or *húlhúl*), *jangli harhar, khanphutia,* HIND.; *Húr-húria,* BENG.; *Húl húl, bugra,* PB.; *Kattori,* SIND; *Tinmani, tilwan,* GUJ.; *Húrhúriya, kánphúti, pivala-tilá-vana,* BOMB.; *Kanfodi-kánphodi, harhuria,* MAR.; *Chorie-ajowan, chúrai-ajwani, jangli hul-vul,* DUK.; *Nahi-kuddaghú, naykadughu, nayavaylie,* TAM.; *Kuka-omintaw, kukha-avalu,* TEL.; *Kat kuddaghú, aria-vila,* MALA.; *Nayi-bela,* KAN.; *Aditya bhakta, kúka váivinta, kúka-vumitie, svana-burbárá, shunaca-barbara,* SANS.; *Walaba,* SING.

References.—*Roxb., Fl. Ind., Ed. C.B.C., 501; U. C. Dutt, Mat. Med. Hind., 289; Dymock, Mat. Med. W. Ind., 2nd Ed., 61; Ainslie, Mat. Ind., II., 223; O'Shaughnessy, Beng. Dispens., 206; Murray, Pl. and Drugs, Sind, 52; Drury, U. Pl., 351; Baden Powell, Pb. Prod., 330; Cooke, Oils and Oilseeds, 37; Atkinson, Him Dist., 732; Birdwood,*

CLEOME viscosa.

or Hurhur.

Bomb. Pr., 276; Lisboa, U. Pl. Bomb., 145; Spons' Encylop., 1415; Balfour, Cyclop.

Habitat.—A common weed throughout the greater part of India, appearing in the rainy season; very common in Bengal and South India.

Oil.—The seeds yield a light olive-green-coloured limpid oil when subject to a great pressure. It seems likely that this oil would prove serviceable where a very liquid oil is required. The oil could be prepared to any extent. OIL. 1368

Medicine.—The JUICE of the leaves is poured into the ear to relieve earache. According to **Rheede**, it is useful in deafness. **Dr. Dymock** writes that the juice mixed with oil is a popular remedy in Bombay for purulent discharges from the ear, whence the Bombay name of the plant *Kánphuti*. "The LEAVES boiled in *ghi* are applied to recent wounds, and the juice to ulcers" (*Drury*). In Cochin China the whole plant, bruised, is used for counter-irritation and blistering (*O'Shaughnessy*). The bruised leaves are applied to the skin as a counter-irritant. **Ainslie** says; "The small, compressed, netted-surfaced, hottish-tasted SEEDS of this low-growing plant are considered by the Vytians as anthelmintic and carminative; they are administered in the quantity of about a tea-spoonful twice daily." They are also occasionally given in fevers and diarrhœa. MEDICINE. Juice. 1369 Leaves. 1370 Seeds. 1371

Special Opinions.—§"The juice of the fresh leaves if squeezed into the ear relieves pain" (*Assistant Surgeon Nil Ratton Banarji, Etawah*). "The seeds of this plant are occasionally an ingredient in chutney to promote digestion" (*Surgeon-Major John North, Bangalore*). "Used to relieve ear-ache and as an astringent in cases of atorrhœa: the ear should be syringed well before its application" (*Brigade Surgeon J. H. Thornton, Monghyr*). "Alterative, useful in secondary syphilis and enlargement of the liver and spleen" (*Surgeon-Major J. McD. Houston, Travancore; and John Gomes, Esq., Medical Storekeeper, Trevandrum*). "The seed made into chutney has strong digestive power" (*Native Doctor Ummegudien, Mettapolliam, Madras*).

"The seeds of **Cleome viscosa** are anthelmintic, rubefacient, and vesicant; and the leaves rubefacient, vesicant, and a useful remedy for a few diseases of the ear. The seeds are valuable in expelling round-worms, and also as a rubefacient and vesicant in all the complaints in which mustard is used. The leaves are also useful in the same way as a local stimulant, and, in addition to this, the juice possesses a curative influence over some cases of otalgia and otorrhœa, but the smarting it produces in the ear, especially in the last-named disease, is an objection to its use. The seeds are used internally in powder with sugar, and externally in the form of a poultice or paste by bruising with vinegar, lime-juice, or hot water. The leaves are also applied to the skin in the form of a poultice or paste by bruising with vinegar, lime-juice, or hot water, and their juice for the use of the ear is pressed out by bruising them without water. The dose of the powder is from thirty grains to a drachm, twice a day for two days, and followed on the third morning by a dose of castor oil or some other purgative. For children the dose is from five to twenty grains, according to their age. As a drug the leaves of **Cleome viscosa** are much superior to those of **Gynandropsis pentaphylla.** It is the former which possess a distinct fœtid smell and efficient rubefacient and vesicant properties, and not the latter. The above plants are frequently found growing together and are often confused partly from a general botanical similarity between them, and partly on account of their native synonyms being almost the same. The close similarity of their seeds adds greatly to this confusion. There will be, however, no difficulty in

MEDICINE. distinguishing the two plants if due attention is paid to the following botanical characters :—

"*Cleome viscosa.*—Siliqua flat, striated, pubescent, and sessile or short stalked; flowers yellow; stem and branches quite covered with viscid glandular hairs; smell fœtid and strong.

"*Gynandropsis pentaphylla.*—Siliqua round; glabrous, stalks long, sometimes even as long as the siliqua itself; flowers always white; stem and branches slightly covered with glandular hairs; smell fœtid, but not strongly.

"As the seeds of both of these plants are very similar, I need not describe them separately. They are as follows: small, flat, and slightly acrid or bitterish in taste. They yield a small quantity of fixed oil on expression.

"As a rubefacient and vesicant, the seeds under examination are much superior to the mustard seed in this country, and quite equal to the mustard imported from Europe. If they can be reduced to as fine a powder as Europe mustard, I think they will be found to excel the latter also in remedial value" (*Honorary Surgeon Moodeen Sheriff, Khan Bahadur, Triplicane, Madras*).

FOOD. Seeds. 1372 Plant. 1373 **Food.**—"The SEEDS of **Cleome viscosa** are much used by the natives, chiefly the Brahmins, in their curries; they are sold in all the bazars at a trifling price." (*Roxb.*) **Lisboa** says that the PLANT is eaten boiled with chillies and salt as salad.

CLERODENDRON, *Linn.; Gen. Pl., II., 1155.*

This name alludes to the variable properties of the species *kleros*, lot, and *dendron*, a tree.

[VERBENACEÆ.

1374 **Clerodendron Colebrookianum,** *Walp.; Fl. Br. Ind., IV., 594;*

Vern.—*Kadungbi*, LEPCHA.

Reference.—*Gamble, Man. Timb., 299.*

Habitat.—An evergreen shrub, with silvery-grey bark, met with in Sikkim and the Khásia Hills, 2,000 to 6,000 feet; also in Burma.

FOOD. 1375 **Food.**—The young LEAVES are eaten by the Lepchas.

TIMBER. 1376 **Structure of the Wood.**—Grey, soft.

1377 **C. inerme,** *Gærtn.; Fl. Br. Ind., IV., 586.*

Syn.—VOLKAMERIA INERMIS, *Linn.*

Vern.—*Sang-kuppi, sáng-kúpi, lán-jai,* HIND.; *Bun-jumat, bun-join, bon-joi, bán-jai, ban-juen, batraj,* BENG.; *Vana-jai,* BOMB.; *Vana-jai,* MAR.; *Isamdhárí, sang-kupi,* DUK.; *Shengan-kuppi, piná-shengam-kuppi, shangam-kupi, pinari,* TAM.; *Pishinika, úti chettu, pisangi, pisingha, tak-kólapu-chettu, nalla-kupi, eru-pichecha, eti-pisinika, penni ka, eru puchcha,* TEL.; *Shangam-kuppi, nir-notsjil, kundalí,* SANS.; *Pirolai kyout,* BURM.; *Wal-gúranda,* SING.

References.—*Roxb., Fl. Ind., Ed. C.B.C., 477; Brandis, For. Fl., 363; Kurz, For. Fl. Burm., II., 266; Gamble, Man. Timb., 299; Thwaites, En. Ceylon Pl., 243; Dalz. & Gibs., Bomb. Fl., 200; Voigt, Hort. Sub. Cal., 465; Moodeen Sheriff, Supp. Pharm. Ind., 108; Dymock, Mat. Med. W. Ind., 2nd Ed., 598; S. Arjun, Bomb. Drugs, 194; Piesse, Perfumery, 225; Balfour, Cyclop.; Bomb. Gaz., XV., i., 67; Rheede, Hort. Mal., V., t. 49.*

Habitat.—A large, ramous, often scandent evergreen shrub, common in tidal forests in Bengal, Burma, and the Andamans.

PERFUMERY. 1378 **Perfumery.**—An exquisite perfume is said to be derived from the flowers of this plant (*Piesse*).

MEDICINE. Plant. 1379 **Medicine.**—**Dr. Dymock** says that the PLANT has a reputation as a febrifuge in remittent and intermittent fevers. This fact is supported by **Dr. Sakharam Arjun**, who, upon the authority of **Dr. Hojel**, states that "the thick succulent leaves are very bitter, and on expression yield a large quantity of thickish somewhat mucilaginous juice, with a slightly saline but intensely bitter taste. Although not generally known, it has of late been used as a febrifuge and antiperiodic with marked benefit."

Clerodendron infortunatum, *Gærtn.; Fl. Br. Ind., IV., 594; [Wight, Ic., t. 1471.* 1380

Syn.—VOLKAMERIA INFORTUNATA, *Roxb., Fl. Ind., Ed. C.B.C., 478;* G. VISCOSUM, *Vent.*

Vern.—*Bhánt, bhat,* HIND.; *Bhánt, ghentú,* BENG.; *Kharbari, barni* or *varni,* SANTAL; *Kulamarsal,* KOL.; *Chitu,* NEPAL; *Kdung,* LEPCHA; *Lukunah,* MECHI; *Káli basúti,* PB.; *Kari,* BOMB.; *Bhandira, kari,* MAR.; *Bockada,* TEL.; *Peragú,* MALA.; *Bhándira, bhanti, bhantaka,* SANS.; *Ka-aunggyl, bujiphyú, khaoung gyi,* BURM.; *Gas-pinna,* SING.

References.—*Brandis, For Fl., 363; Kurz, For. Fl. Burm., II., 267; Bedd., For. Man., 173; Gamble, Man. Timb., 299; Thwaites, En. Ceylon Pl., 243; Dalz. & Gibs., Bomb. Fl., 200; Stewart, Pb. Pl., 165; Voigt, Hort. Sub. Cal., 465; Pharm. Ind., 164; Moodeen Sheriff, Supp. Pharm. Ind., 108; U. C. Dutt, Mat. Med. Hind., 293; Dymock, Mat. Med. W. Ind., 2nd Ed., 597; Murray, Pl. and Drugs, Sind, 174; S. Arjun, Bomb. Drugs, 194; Drury, U. Pl., 144; Baden Powell, Pb. Pr., 364; Atkinson, Him. Dist., 732; Balfour, Cyclop.; Rheede, Hort. Mal., II., t. 25.*

Habitat.—A pinkish-white-flowered shrub, common in waste places throughout the greater part of India and Burma, and in the damp forests of Ceylon up to an elevation of 5,000 feet. Grows gregariously, forming a dense under-vegetation, specially associated with the Bamboo. On passing into fruit the calyx becomes scarlet, and the plant is then even more attractive than when covered with its fœtidly-scented flowers.

Medicine.—"**Dr. Bholanath Bose** calls attention to the LEAVES of this plant as a cheap and efficient substitute for chiretta as a tonic and antiperiodic." (*Pharm. Ind.*) According to **Dr. Kanny Lal De, C.I.E.,** the fresh JUICE of the leaves is employed by the natives as a vermifuge, and also as a bitter tonic and febrifuge in malarious fevers, especially in those of children. **Dr. Dymock** states that he has not seen the leaves used medicinally in Bombay, but they are bitter. **Dr. Honigberger** mentions the use of the BARK in medicine by the Arabian and the Indian physicians.

MEDICINE. Leaves. 1381

Juice. 1382

Bark. 1383

Special Opinions.—§ "The expressed juice is an excellent laxative, cholagogue, and anthelmintic. It is used as an injection into the rectum in cases of ascarides. It is also a valuable bitter tonic, and the natives believe that its presence cures scabies in the locality" (*Brigade Surgeon J. H. Thornton, B.A., M.B., Monghir*). "Is said to be a very useful antiperiodic" (*Surgeon-Major E. Sanders, Chittagong*). "The juice of the fresh leaves is used as a febrifuge for infants and children" (*P. W. B., Dacca*). "The juice of leaves found to be an efficient anthelmintic" (*Surgeon C. J. W. Meadows, Burrisal*). "The decoction of the leaves is a powerful antiperiodic, and is a valuable adjunct to arsenic in the treatment of malarious fevers" (*Civil Medical Officer U. C. Dutt, Serampore*). "Decoction of the leaves is used as an antiperiodic" (*Surgeon Anund Chunder Mukarji, Noakhally*). "The fresh juice of the leaves used as a febrifuge; used also in torpidity of the liver" (*Civil Medical Officer W. Forsyth, Dinagepore*). "Used also in decoction as a febrifuge." (*Surgeon-Major E. C. Bensley, Rajshahye*).

Decoction. 1384

Domestic Uses.—**Edgeworth** mentions that this plant is used in the Ambala district to give fire by friction.

DOMESTIC. 1385

[*Ic., t. 1473.*

1386 **Clerodendron phlomoides,** *Linn.; Fl. Br. Ind., IV., 590; Wight,*

Vern.—*Urni, pírun,* HIND.; *Panjot,* SANTALI; *Gharayt,* SIND; *Irun, arni,* GUJ.; *Airan,* BOMB.; *Airanamúla,* MAR.; *Talúdalel, taludala, wada madichi,* TAM.; *Telaki, nellie, tekkali, teleki, tilaka,* TEL.; *Váta-ghní,* SANS.

References.—*Roxb., Fl. Ind., Ed. C.B.C., 477; Brandis, For. Fl., 363; Gamble, Man. Timb., 298; Thwaites, En. Ceylon Pl., 243; Dalz. & Gibs., Bomb. Fl., 200; Aitchison, Cat. Pb. Pl., 120; Voigt, Hort. Sub. Cal., 465; Dymock, Mat. Med. W. Ind., 498; Ainslie, Mat. Ind., II.; 408; Murray, Pl. and Drugs, Sind, 174; S. Arjun, Bomb. Drugs, 104; Royle, Ill. Him. Botany, 299; Balfour, Cyclop.*

Habitat.—A tall pubescent shrub, common in many parts of India, principally in the drier regions of the Panjáb, Sind, Mairwara, the Dekkan, Behar, Bengal, Oudh, Central Provinces, and also in Ceylon.

MEDICINE. Root. 1387 **Medicine.**—**Dr. Dymock** says that the natives of Western India suppose the ROOT of the plant has alterative properties, but he has never seen it used as such. "It is used as a bitter tonic, and is given in the convalescence of measles" (*S. Arjun*). The juice of the leaves is, according to **Ainslie**, considered by the Indian practitioners as alterative. It is prescribed by them in neglected syphilitic complaints in doses of half an ounce or more twice daily.

The **Rev. A. Campbell** says the Santals give this plant to their cattle to cure them of diarrhœa and worms, or when the stomach swells. **Mr. Campbell** also says the Santals rub the plant over their bodies in dropsy.

1388 **C. serratum,** *Spreng.; Fl. Br. Ind., IV., 592; Wight, Ic., t. 1472.*

Vern.—*Barangi, gant-bahárangí* (root), HIND.; *Saram lutur,* SANTAL; *Chúa,* NEPAL; *Yi,* LEPCHA; *Bharungi,* GUJ.; *Bharang, bharangi,* BOMB.; *Bháranga-mula* (root), MAR.; *Chiru dekku, vátham addakki, shimtek* (root), TAM.; *Bhárangi, brah-mari mari, gandu-bhárangí* or *gantu-bhárangi* and *gunti paringaie* (root), TEL.; *Cheru tékka, nápálu, jeru-tika, tsjeru-teka, kanta-bháranní* (root), MALA.; *Barbará* (root), SANS.; *Ken-henda,* SING.; *Bebya, baikyo,* BURM.

References.—*Brandis, For. Fl., 364; Kurz, For. Fl. Burm., II., 267; Gamble, Man. Timb., 299; Dalz. & Gibs., Bomb. Fl., 200; Aitchison, Cat. Pb. Pl., 121; Voigt, Hort. Sub. Cal., 466; Pharm. Ind., 164; Moodeen Sheriff, Supp. Pharm. Ind., 108; Dymock, Mat. Med. W. Ind., 2nd Ed., 598; Murray, Pl. and Drugs, Sind, 174; Bidie, Cat. Raw Prod., Paris Exh., 37; S. Arjun, Bomb. Drugs, 105; Drury, U. Pl., 144; Atkinson, Him. Dist., 732; Lisboa, U. Pl. Bomb., 168; Balfour, Cyclop.*

Habitat.—A blue-flowered shrub, common in the Sub-Himálayan tract, from the Sutlej eastward to the Khásia Hills, South India, and Burma.

MEDICINE. Root. 1389 **Medicine.**—"The ROOTS constitute the *Gantu bharangí*, TEL. (*Gunti Paringhie of Ainslie, Vol. II., p. 112*), which, according to **Sir Walter Elliot** (*Flor. Andhrica, p. 57*), is exported largely from Vizagapatam for medicinal purposes. It is used by the natives in febrile and catarrhal affections. It occurs in the form of a brittle, wiry, woody root, in short pieces of a thickness varying from that of a large packthread to that of a small quill, frequently swollen into hard, woody, globular excrescences. It has scarcely any odour or taste. As a medicinal agent it is probably of little value." (*Pharm. Ind.*) **Dr. Dymock** affirms that this species is used for the

Leaves. 1390 same purposes and is known by the same name as the next. "The LEAVES are boiled in oil for applications in ophthalmia, the roots boiled in water with ginger and coriander are given in nausea, and the seeds are slightly

Seeds. 1391 aperient" (*Atkinson, Him. Dist., 732*). The SEEDS boiled in butter-milk are occasionally used in cases of dropsy.

Special Opinions.—§ "Slightly aperient" (*Surgeon H. W. Hill, Manbhoom*). "Used in infusion (ʒi to xx) in bronchial affections, and as a febrifuge" (*Brigade Surgeon W. H. Morgan, Cochin*). "Used medicinally by the Santals especially in fevers" (*Rev. A. Campbell, Manbhoom*).

Food.—The tender LEAVES are eaten as vegetable (*Lisboa*). The FOOD.
flowers are also eaten as greens (*Balfour*). The ROOT is used by the Leaves.
Santals to cause the fermentation of rice-beer (*Rev. A. Campbell*). 1392
Root.
[*Wight, Ill., t. 173.* 1393

Clerodendron Siphonanthus, *R. Br.; Fl. Br. Ind., IV., 595*; 1394

Syn.—SIPHONANTHUS INDICA, *Linn.; Roxb., Fl. Ind., Ed. C.B.C., 481.*

Vern.—*Barangi, bhárangi,* HIND.; *Bamúnhatti, brahman-patta, bámanháti,* BENG.; *Arní, daw-ái-mubarik, arnah,* PB.; *Bharangi,* BOMB.; *Sarum eutur,* DUK.; *Brahmuni, brahmunu yushtika, brahma yashtiká, bhárgi,* SANS.; *Naijamphá ti,* BURM.

References.—*Brandis, For. Fl., 364; Gamble, Man. Timb., 299; Thwaites, En. Ceylon Pl., 243; Dals. & Gibs., Bomb. Fl., Suppl., 69; Stewart, Pb. Pl., 165; Aitchison, Cat. Pb. Pl., 121; Voigt, Hort. Sub. Cal., 465; U. C. Dutt, Mat. Med. Hind., 219, 294; Dymock, Mat. Med. W. Ind., 2nd Ed., 598; Baden Powell, Pb. Pr., 364; Royle, Ill. Him. Bot., 299; Atkinson, Him. Dist., 733; Balfour, Cyclop.; Treasury of Botany.*

Habitat.—A large shrub, with red calyx, white flowers, and blue berries, found in Kumaon, Bengal, and South India; it is also common in gardens in Ceylon, where it is not indigenous.

Gum.—Yields a resin (*Baden Powell, Pb. Prod., 364*). GUM.
Medicine.—Mr. Home says the WOOD is tied round the neck by the 1395
Bengalis and used as a charm against various ailments (*Gamble*). MEDICINE.
"The ROOT is considered useful in asthma, cough, and scrofulous affec- Wood.
tions. The root beaten to pulp is given with ginger and warm water in 1396
asthma. It enters into the composition of several compound decoctions Root.
for diseases of the lungs. A CONFECTION called *Bhárgíguda* is prepared 1397
with a decoction of this root and the ten drugs called *dasamula,* chebulic Confection.
myrobolan, treacle, and the usual aromatic substances. It is used in 1398
asthma. An OIL, prepared with a decoction and paste of the root in the Oil.
usual proportions, is recommended for external application in the maras- 1399
mus of children" (*U. C. Dutt, Mat. Med. Hind., 219*). Mr. Baden Plant.
Powell writes that the PLANT is slightly bitter and astringent, and that 1400
the resin is employed in syphilitic rheumatism.
Special Opinion.—§ "The expressed JUICE of the leaves and tender Juice.
branches is used with *ghí* as an application in herpetic eruptions and pem- 1401
phigus. The BRANCHES, cut into small pieces and threaded like beads, Beads.
are put on the necks of children suffering from these diseases as a charm, 1402
and it is believed by the natives that the smell of this plant is sufficient to cure these diseases" (*Brigade Surgeon J. H. Thornton, B.A., M.B., Monghir*).

CLITORIA, *Linn.; Gen. Pl., I., 528.*

[LEGUMINOSÆ.

Clitoria Ternatea, *Linn.; Fl. Br. Ind., II., 208; Bot. Mag., t. 1542*; 1403

Vern.—*Aparájitá, aprájit, shobánján, khágin, kowa, kowa-theti, kálizer, visnukranti, kavá-thénthi,* HIND.; *Aparájitá, uparájitá, nil-aparájitá, swet-aparájitá,* BENG.; *Dhanattar, aprájit, vishnú-kanté, níla ghiria, níl-kanth, kálzar, kawá-túnti, nil-isband, shámí-ka-bij, dhanttar,* PUSHT.; *Garani,* GUJ.; *Gokarna-mul, kájalí, gokaran,* BOMB.; *Gokarni, sholongá-kuspi, gokurna-mula* (root), *gokurna-bija* (seeds), MAR.; *Phiki-ki-jar-ká-jhár, ghutti-ki-jar-ká-jhár, kalizer ké kejurr, ghutti-ki-jar-ke-binj* (seeds), DUK.; *Karka-kartum, kara-kartan, kákkanan-kodi,*

CLITORIA Ternatea.

A Powerful Cathartic.

TAM.; *Dintana, tella, mella, tella-dintana, nila-dintana, nella-ghentana vayrú* (seeds), *nella-ghentána*, TEL.; *Shanga-pushpam, kákkanamkoti, káka-valli, káka-villa* (seeds), *sholongo-kuspi, shunkú-pushpa, shunkur-puspa, sholongo-kusbi, shlonga-kuspi* (in **Birdwood**), MALA.; *Vishnu-kantisoppu, kirgunna, gokarna mul*, KAN.; *Vishnu krántá, asphota, aparájitá, gokarna múl, nílaghirie kurní, níla-ghiria, khurne*, SANS.; *Mázariyune-hindí* (Indian Mezereon), *bazrulmázari yúne-hindí* (seeds), ARAB.; *Darakhte-bikhe-hayát, tukhme-bikhehayát* (seeds), PERS.; *Búkyu, pai noung ni, oung mai phyú*, BURM.; *Kattarodú, nil-kattarodú, nil-katarolu*, SING.

References.—*Roxb., Fl. Ind., Ed. C.B.C., 566; Thwaites, En. Ceylon Pl., 88; Dalz. & Gibs., Bomb. Fl., 68; Stewart, Pb. Pl., 64; Aitchison, Cat. Pb. Pl., 47; Voigt, Hort. Sub. Cal., 213; Pharm. Ind., 80; Moodeen Sheriff, Supp. Pharm. Ind., 108, 348; U. C. Dutt, Mat. Med. Ind., 147, 291; Dymock, Mat. Med. W. Ind., 2nd Ed., 235; Ainslie, Mat. Ind., II., 139; O'Shaughnessy, Beng. Dispens., 315; Bidie, Cat. Raw Prod., Paris Exh., 28, 111; S. Arjun, Bomb. Drugs, 39; Drury, U. Pl., 145; Baden Powell, Pb. Pr., 339; Birdwood, Bomb. Pr., 28; Lisboa, U. Pl. Bomb., 254; Balfour, Cyclop.; Treasury of Botany; Irvine, Nat. Mat. Med., Patna, 76; Med. Top., Ajmere, 204; Mysore Cat. Cal. Exb., 209; Home Dept. Cor., Nov. 1880; Mason's Burma, p. 412.*

Habitat.—A common garden flower, also occurs in every hedge-row all over India. The seeds were first taken to England from the Island of Ternate, one of the Moluccas; hence the specific (and former generic) name of the plant.

DYE. Seeds. 1404 **Dye.**—**Bidie** remarks that the SEEDS are said to be used by dyers. "The corollas of the blue variety are said to afford a blue dye in Cochin China, but it is not permanent; and **Rumphius** says that they are used for colouring boiled rice in Amboyna" (*Treasury of Botany*).

MEDICINE Root. 1405 **Medicine.**—The ROOT is a powerful cathartic like jalap, and has been recommended to be used along with other laxatives and diuretics in ascites and enlargements of the abdominal viscera (*Dymock*). **Ainslie** recommends it in croup as an emetic, but **O'Shaughnessy**, in *Bengal Dispensatory*, says: "We have used the root extensively in order to test its alleged emetic effects, but have never observed their occurrence. An alcoholic extract acts, however, as a brisk purgative in from 5 to 10 grain doses. But griping and tenesmus are often produced, and during the operation of the medicine the patient is feverish and uneasy. We cannot recommend the use of this medicine." **Moodeen Sheriff** is, however, much in favour of the root in the treatment of irritation of the bladder. It acts at the same time as a diuretic, and in some cases as a laxative. The SEEDS are, however,
Seeds. 1406 more useful, and have gained a certain reputation in Europe as a safe medicine, especially for children. The powdered seeds are purgative and aperient. Combined with acid tartrate of potash and ginger, they are
Leaves. 1407 administered in the same doses as jalap. The infusion of the LEAVES is used for eruptions. Great care is necessary in preventing the name *Kalizirki*, sometimes applied to the seeds of this plant, from being confused with the seeds of **Ipomæa hederacea**, for which they may be substituted.

Mr. Baden Powell, in his *Panjáb Products*, gives the following account of the medicinal properties of this plant: "The seeds are said to be cooling and to act as an antidote to poisons. The roots are used as emetics and in rheumatism; the seeds in large doses are purgative and anthelmintic, and used for weakness of sight, sore-throat, and mucous disorders; also in tumours and the affections of the skin, and in dropsy."
Juice. 1408 "The JUICE of the leaves, mixed with that of green ginger, is administered in cases of colliquative sweating in hectic fever" (*Taylor, Med. Top. Dacca, 52, 53*).

Special Opinions.—§ "There are two varieties of **Clitoria Ternatea** distinguished by the colour of their flowers, as blue and white, and the blue

MEDICINE.

again has a sub-variety, in which the flowers are double. There is no distinct difference between the action of the seeds of these varieties, or if any at all, it is in favour of the white one. The plants are in flower all the year. The seeds are not generally sold in the bazar, but when they are, they are almost always of very inferior quality, in consequence of their being collected before their maturity. They should not be removed until they are quite matured and dried on the plant. The seeds gathered with this precaution are nearly round or slightly compressed along the edges, oblong, dull green, greenish brown, or brown in colour, and minutely mottled. The ends of some seeds are round, and of others flat, as though cut off clean by a knife; taste disagreeable and acrid, and no smell. The thicker and rounder the seeds are, the more active they prove. The immature seeds are flat and dark brown in colour; the matured thick and round seeds are an efficient purgative, and produce five or six motions in one drachm or one drachm and a half doses. Their action is increased in proportion to the increase of their quantity up to two drachms, when the number of motions is generally eight or nine. The seeds are one of those drugs which act very satisfactorily when used alone, but they may also be administered in combination with cream of tartar, in equal proportion, and with a few grains of ginger in each drachm of the compound powder. The dose of the compound powder is from a drachm and a half to two drachms. The fresh root, or rather root-bark, of **Clitoria Ternatea** is a domestic medicine in this country, and is very frequently administered to children in croup and pulmonary affections by old women and midwives. It acts in children as an emetic, nauseant, expectorant, and laxative, and thus relieves the above affections in many cases. When employed in large doses in adults, it acts as a demulcent and diuretic, and relieves some of the symptoms of gonorrhœa and irritation of the bladder, as strangury, scalding of urine and frequency of micturition, and in some cases the gonorrhœal discharge itself is much abated under its use. One small root is generally a dose for children under two years, and one large root or two small ones for those between three and six years. For adults the dose is four or six roots if small, and three to five if large" (*Honorary Surgeon Moodeen Sheriff, Triplicane*). "There are two varieties of this plant: one has white and another bluish-coloured flowers; for medicinal purposes the latter variety is preferable. Juice of the leaves mixed with common salt is applied warm all around the ear in ear-aches, especially when accompanied with swelling of the neighbouring glands" (*Surgeon Anund Chunder Mukarji, Noakhally*). "Seeds purgative, root demulcent; dose, seeds powdered, 30 to 60 grains; root, one to two drachms of dry bark in powder" (*Apothecary Thomas Ward, Madanapalle, Cuddapah*). "Is used as a drastic purgative and diuretic in dropsy, also in cases of cystitis. The roots of the blue species are used as an antidote in cases of snake-bite" (*Brigade Surgeon J. H. Thornton, B.A., M.B., Monghir*). "The seeds are used as a mild purgative for children" (*Surgeon-Major J. Robb, Ahmedabad*). "There are two kinds: one has white flowers and the other blue. The root of the white is considered best" (*Native Doctor V. Ummegudien, Mettapollian, Madras*). "The root is a drastic purgative and found useful in general dropsy" (*Assistant Surgeon Shib Chunder Bhuttacharji, Chanda, Central Provinces*). "The powdered root of this plant is used occasionally in Mysore as a cathartic in dropsy" (*Surgeon Major John North, Bangalore*).

1409

1410

Sacred Uses.—The flower is held sacred to the goddess *Durga*. SACRED USES. 1411

Clover, see **Trifolium pratense,** *Linn.;* LEGUMINOSÆ.

Cloves, see **Caryophyllus aromaticus,** *Linn.;* MYRTACEÆ.

CNICUS, *Linn.; Gen. Pl., II., 468.*

1412 **Cnicus arvensis,** *Hoffm.; Fl. Br. Ind., III., 362;* COMPOSITÆ.

Syn.—CARDUUS LANATUS, *Roxb.; Fl. Ind., Ed. C.B.C., 595.*
Vern.—*Bhur-bhur,* N.-W.-P.
Reference.—*Smith, Dictionary, 410.*

Habitat.—Found throughout India, especially in cultivated fields in the Gangetic plains; the common thistle of India.

OIL. Seeds. 1413 **Oil.**—Produces small black SEEDS, which yield a large quantity of oil. The seeds are gathered by the poorer classes, and the oil expressed by them for their own use. It burns with smoke; is otherwise of good quality.

Cnidium diffusum, see **Seseli indicum,** *W. & A.;* UMBELLIFERÆ.

COAL.

1414 **Coal.**

CHARBON DE TÈRRE, *Fr.;* STEINKOHLEN, *Germ.;* CARBONI FOSSILI, *It.;* CARVVES DE PEDRA, *Port.;* CARBONES DE PIEDRA, *Sp.*

Vern.—*Kóyelah* or *kuela,* HIND.; *Kóyalá,* BENG.; *Kólsá,* DUK.; *Kari* or *Simai karri,* TAM.; *Boggu* or *Sima boggu,* TEL.; *Kari,* MAL.; *Iddallu,* KAN.; *Kóelo, kólso,* GUJ.; *Anguru,* CING.; *Fahm,* ARAB.; *Zughál,* PERS.; *Angáraha,* SANS.; *Misu-e, midu-ye,* BURM.

References.—So much has been written regarding Indian Coal that an enumeration of the publications would occupy many pages. The reader is referred to *Ball's Economic Geology,* pp. 599-604, to the *Memoirs, Records of the Geological Survey,* and to the *Journals of the Asiatic Society of Bengal.* The following works may, however, be specially mentioned:—

Final Report of the Coal Committee; Dr. T. Oldham's Report on the Coal Resources of India; Sel. Rec. Govt. Ind., LXIV.

Ball's Coal-fields and Coal productions of India; Annual Administration Reports on Railways in India.

REGIONS OF INDIAN COAL.

The following account of the coal-fields of India has been furnished by **Mr. H. B. Medlicott** for this publication:—

1415 ABSTRACT OF THE FEATURES OF INDIAN COAL.

"India possesses extensive stores of coal, though none of it belongs to the so-styled carboniferous period, and in India itself the coal-measure rocks are not all of one formation. All the coal of peninsular India occurs in the rocks known as the Gondwana system, the fossil flora of which has a mesozoic facies; and all the coal of extra-peninsular India occurs in rocks of cretaceous or tertiary age. In both cases the distribution is partial. The Gondwana coal-measures have only been found in the central and north-eastern provinces, *i.e.,* in western Bengal, the Central Provinces, and the Nizam's Territories, only skirting the south border of the North-Western Provinces, with remnants in the extreme north-east of the Madras presidency. The tertiary coal has been traced all along the outer margin of the Indo-Gangetic plains from Sind to Pegu, but it is only in Assam and Upper Burma that valuable measures have been found where a cretaceous coal occurs in workable quantity.

Coal-fields of India. (*H. B. Medlicott*). **COAL.**

"In both regions the quality of the coal varies much, as in all coal-measures; but the best in both, reaches a very high standard, almost if not quite, up to that of high class English coals. In the Gondwana (Bengal) coal the general defect is an excess of ash, and also in some an excess of moisture; while in the tertiary (Assam) coal the percentage of ash is low, but that of the volatile combustible matter is high, producing a lighter fuel. The following tabular statement exhibits these facts:—

	BENGAL.		ASSAM.	
	Average of 31.	Best.	Average of 23.	Best.
Fixed carbon	53·20	66·52	56·5	66·1
Volatile, exclusive of moisture	25·83	28·12	34·6	33·5
Moisture	4·80	·96	5·0	...
Ash	16·17	4·40	3·9	·4
	100·	100·	100·	100·

"In Bengal only the Raniganj and Karharbari fields have as yet been largely worked, and to a small extent the Daltonganj field. Several other large coal-fields are still quite untouched, owing to difficulty of communication.

"In the Central Provinces the Mohpani mines in the Narbada valley, and the Warora mines in the Wardha valley, have been for some time in work, and the* Umaria and Sohagpur fields in the Rewah State are being opened up.

"In the* Singareni and Sasti fields of the Nizam's Territories some preliminary mining has been carried out pending the establishment of railway communication.

"Vigorous mining enterprise has recently been started in the Makum coal-field in Upper Assam."

MORE DETAILED STATEMENT OF THE COAL-YIELDING DISTRICTS.

"The mineral is more particularly developed in the central eastern portion of the Peninsula. **SOUTH INDIA. 1416**

In the Madras Presidency it is found at—

"*Beddadanol.*—Lat. 17°14′; Long. 81°17′30″. The field, about 38 miles from Rajahmundry, is about 5¾ square miles in extent, and contains four seams of very poor coal, worthless as fuel. This is the most southern occurrence of coal in the peninsula.

"*Damercherla* (or *Madaveram*).—Lat. 17°36′; Long. 81°7′. Has its most important portion on the Nizam's side of the River Godavari; on the British side there are probably 25,000 tons of coal, of which only about one-half would be available.

"*Lingalla.*—Lat. 18°; Long. 80°54′. Two seams, neither of which exceed 2 feet in thickness, occur in the bank of the Godavari; and another, 5 feet thick, in its bed.

"*Singareni.*—The best field as yet known for Madras, but still in the Nizam's Dominions, is that near Singareni, lat. 17°30′30″; long. 80°20′. There are five seams: the thickness of one was not ascertained, those of the

* Since opened out.

others are respectively 6, 3, 3, and 34 feet. This coal answers well for smithy purposes and stationary engines, and was found to be a serviceable fuel when tried on the Madras Railway. Railway communication is now being rapidly pushed forward; and a colliery being started, coal reported of high quality.

"*Kamaram.*—Lat. 18°5′; Long. 80°14′. Two seams of fair coal, 9 and 6 feet in thickness respectively. The available coal is estimated at 1,132,560 tons; its position is, however, unfavourable to its development.

"*Tandur.*—Lat. 19°9′; Long. 79°30′. This village is situated about the centre of a strip of Barakar rocks, extending from Kairgura to Aksapali, and contains a 15-foot seam of fair coal.

"*Antergaon.*—Lat. 19°32′30″; Long. 79°33′. South of this place a 6-foot seam occurs, 9 inches of which are shale.

"*Sasti* and *Paoni.*– In the Nizam's Dominions, included in the Wardha area, a 50-foot seam occurs here, a considerable portion of which is of good quality: 30,000,000 tons of coal are estimated to be available from this source.

ORISSA. 1417 "*Talchir,* in Orissa.—The field is situated in the valley of the Brahmini, and it is about 700 square miles in extent. The coal is of an inferior quality. The field has not been practically explored.

BENGAL. 1418 "*Rajmahal Hills.*—Over about 70 square miles on the western margin of the Rajmahal Hills, coal measure rocks are exposed; and these doubtless extend over a vastly greater area under the younger formations. Separated by these overlying rocks, there are five distinct fields, namely, Hura, Chaparbhita, Pachwara, Mohowgurhi, and Brahmini. There is no continuity of the seams in each of these, while the data about them are very vague and incomplete. If the coal measures extend below the trap to the east, they would be close to the water carriage of the Ganges and hence transport would be cheap; but on the other hand the coal of this region is for the most part stony and bad.

"*Deogarh.*—In the Jainti, Sahajori, and Kandit Karaiah fields, coal of different qualities occurs. Some in the Jainti field is excellent, but that known from the Sahajori area is inferior.

"*Karharbari* or *Kurhurbali*, in the district of Hazaribagh.—This small field, having an area of 8 square miles, is of great importance on account of its position (about 200 miles from Calcutta by rail) and the good quality of its coal. The coal occurs in three principal seams, with an average total thickness of 16 feet; the estimated amount of coal is about 136,000,000 tons, while the available portion is estimated at about 80,000,000 tons; for steam work it is on the average superior to that of Ránigani. The chief companies possessing mines in this field are, the East Indian Railway, the Bengal Coal Company, and the Raniganj Coal Association. Should the output rise to 500,000 tons per annum, as is likely, the life of the coal-field will be 162 years.

"*Raniganj* or *Raneegunge.*—This field is situated on the rocky frontier of Western Bengal at a distance of 120 miles from Calcutta. The available coal, exclusive of waste, is estimated in round numbers at 14,000,000,000 tons. The total area exposed is about 500 square miles; but the real area is possibly even double that, as the beds dip to the east under the alluvium. This is the largest and most important coal-field in which coal is worked in India; its proximity to the main line of railway, and to the port of Calcutta, tending to give it pre-eminence over other less favourably situated localities. The principal Companies engaged here in the extraction of coal are:—the Bengal, Barakar, Equitable, New Birbhoom, and Raniganj Association, besides many minor firms and native associations. Many of the seams are of considerable thickness, one containing from 70 to 80 feet of coal. As a rule, however, the best coal is not found in the very thick seams.

"*Jharia* or *Jeriah.*—This field is situated in the valley of the Damuda river, 16 miles west of the Ràniganj field, and is nearly all included in the district of Mánbhúm. The thickness and quality of the seams vary a good deal, but there is much valuable fuel; the estimated available coal is 465 million tons, the area being about 200 square miles. The immediate future prospects of this field depend much on the new line of communication with the Central Provinces. Twenty miles of tramway would bring it into communication with the proposed railway.

"*Bokaro.*—This field is situated in the Damuda valley and commences at a point 2 miles west of the termination of the Jharia field; its area is about 220 square miles. The quality of the coal is fairly good. Some of the seams are of a large size, one being 83 feet in thickness; there is here a large store of valuable fuel available (about 1,500 million tons). Nothing has been done to develope the resources of this field.

"*Ramgarh.*—This field, situated to the south of the Bokaro field, has an area of about 40 square miles. The coal is for the most part of poor quality and limited in extent. There are, however, a good many seams. There are probably 5 million tons of available fuel. The western extremity of the field is close to the road between Hazaribagh and Ranchi, and it is believed that some of the outcrop coal is occasionally worked by the natives and carried to Ranchi for sale.

"*North Karanpura.*—Situated at the head of the Damuda valley, has an area of about 472 square miles, and the estimated amount of coal is 8,750 million tons.

"*South Karanpura.*—Situated to the south-east of the northern field, has an area of 72 square miles, and the estimated amount of coal is 75 million tons. The assays of some of the coal indicate a high calorific power.

"*Chope*—Is a small field of less than a square mile in extent. Situated on the Hazaribagh plateau.

"*Itkuri*, 25 miles north-west of Hazaribagh. A few seams of inferior coal are exposed.

"*Aurunga.*—In the district of Lohardaga, in the valley of the Koel, a tributary of the Son. The area is 97 square miles, and the estimated amount of coal is 20 million tons; but the quality of the coal as taken from the outcrop is poor.

"*Hutar*, to the west of the Arunga field, has an area of 78·6 square miles. The assays of the coal gave favourable results.

"*Daltonganj*, also in the valley of the Koel, area 200 square miles. The seams are not numerous. One, which has a thickness of 6 feet, contains excellent fuel. The estimated total available amount of coal is 11,600,000 tons.

"*Tatapani, Iria,* and *Morne.*—Situated in the valley of the Son and tributaries. These fields are portions of a large tract stretching far to the westward. Several coal seams of workable thickness and many thin ones exist. NORTH-WEST PROVINCES. 1419

"*Singrowli.*—In this area there are several outcrops; the Kota colliery, which is now abandoned, yielded coal of fair quality.

"*Sohagpur.*—The area is 1,600 square miles. There are several seams of coal, and owing to the horizontality of the strata they possess the advantage of accessibility. The proposed railway from Katni to Bilaspur passes along the western margin of the field.

"*Johilla.*—Also on the proposed Katni-Bilaspur railway. The area is 11 square miles, and a strong seam of excellent coal has been proved.

COAL. **Coal-fields of India.**

"*Umaria.*—This field is more conveniently situated as regards railway communication, and is that where successful workings have lately been established, and good coal obtained that gave excellent results. This field, with a proved area of about 3 square miles, and an estimated amount of 28 million tons of coal, is of great importance on account of its commanding geographical position (34 miles from the Katni station on the East Indian Railway), and its being the nearest source for the supply of the North-Western Provinces and the Panjáb.

"*Korar.*—Three miles north of Umaria. The area is 9 square miles, and a thick seam of good coal has been proved.

"*Jhilmili*—Is another area of about 41 square miles, in which seams of some promise have been observed.

"*Bisrampur*—Has an area of about 400 square miles occupying the central basin of Sarguja; it contains some good coal suitable for locomotives.

"*Lakhanpur*—South of the Bisrampur area, holds some seams of good coal; the area is 50 square miles.

CENTRAL INDIA. 1420

"*Raigarh, Hingir, Udaipur* and *Korba* fields in the Mahanadi valley.—With the other associated rocks, these occupy an area of at least 1,000 square miles; some of the seams are very thick, two being respectively 90 and 168 feet; but though including good coal they often contain a large proportion of shale, and the horizontal extension of the seams is sometimes irregular and uncertain. These fields will probably assume importance in connection with the line to connect Calcutta with the Central Provinces. The recent boring experiments show that the Korba area has proved most worthy of consideration; particularly at Ghordewa, 9 miles to west-north-west of Korba, where there is a 5-foot seam of good coal.

CENTRAL PROVINCES. 1421

"*Satpura Basin, south of the Narbada Valley.*—The *Mohpani* field is of importance in consequence of its position with reference to the Great Indian Peninsula Railway (95 miles by rail, west-south-west from Jabalpur). The coal is worked by the Narbada Coal Company and supplied to the railway, but the supply falls short of its requirements.

"*Shahpur* (or *Betul*) on the south of the Tawa valley.—This field contains seams of irregular thickness and inferior quality.

"*Pench Valley.*—There are many coal seams, some of which are of considerable thickness, and the coal often of fair quality.

"*Wardha-Godavari Valleys.*—The *Bandar* field—near the village of Chimur, 30 miles north-east of Warora in the Chanda District, contains three seams of coal, with a maximum total thickness of 38 feet.

HYDERABAD. 1422

"*Wardha* (or Chanda), *&c.*—Includes, with several other areas, Sasti and Paoni in Hyderabad, in which coal has been proved to exist. There are about 1,714 million tons of coal available, *viz.*:—

Warora basin	14	million tons.
Ghugus	45	
Wun	1,500	
Between Wun and Papur	50	
Between Janara and Chicholi	75	
Sasti and Paoni (Nizam's territory)	30	

The only pits worked in this wide area are at Warora, whence a special branch line conveys the coal to the Nagpur branch of the Great Indian Peninsula Railway.

BOMBAY. 1423

"*Cutch.*—There are a few thin shaly seams at Trambal (Tromba or Trombow), about 5 miles north-east of Buj, in a stream north of Sis-agad, and in a stream west of Guneri near Lakhpat. Besides these jurassic seams, there are some tertiary carbonaceous layers of no promise.

"*Sind.*—Several layers, a few inches thick of tertiary coal, occur. SIND. 1424

"*Baluchistan* and *Afghanistan.*—At Mach (or Much) in the Bolan Pass several beds of tertiary coal occur, mostly less than a foot in thickness; one has, however, a thickness of 2 feet 6 inches at one point. TRANS-INDUS. 1425

"*Shahrig.*—On the Harnai route, there are outcrops of several thin seams of tertiary coal, none being 2 feet thick, while the greater number are under 6 inches. Some of the coal is of fair quality and would be useful for local purposes. The latest reports give a 6-foot seam of coal near Kosht; but the dip is said to be as high as 45° which will militate greatly against its profitable extraction.

"*Chamarlang,* in the Luni Pathan country, about 75 miles from Dera Ghazi Khan.—There are several seams of tertiary coal, of which the principal one has a thickness of 9 inches.

"*Kanigaram,* in the Waziri country.—A narrow seam of tertiary coal exists near this place; eight small beds of coal are also said to occur in the Ghilzai country at Dobani, at His-Saruk (? Hissarlik), and at Syghan in the Hazara country. The Syghan coal ignites with difficulty.

"*Salt Range, Trans-Indus continuation of.*—At Mullakheyl and Chushmon, there are irregular strings and nests of nummulitic coal in the alum shales. At Kotki, beds, both of nummulitic and jurassic age, occur, containing coal.

"At Kalabagh nummulitic coal exists in very small quantities in the alum shales; the so-called Kalabagh coal consists of carbonized wood in a bed of jurassic shale, of which it forms $\frac{1}{20}$th to $\frac{1}{25}$th part or less.

"*Salt Range proper.*—Nummulitic coal is found at Amb (or Umb), PANJAB. 1426
Sunglewar, Chamil, Kutta, Sowa Khan, Deiwal, Nurpur (Nilawan), and Karuli, but only in small quantities, presenting no prospect of being profitably worked. At Dandot, in the neighbouhood of which coal is seen at three localities, and where thickest is 2 feet 6 inches. The later development of the field, owing to the proximity of railway communication, has resulted in the opening of the Dandot colliery, which is rapidly approaching completion, and promises, notwithstanding the thinness of the seam, and the friable and pyritous condition of the coal, to be a fairly remunerative concern. Up to the end of 1887, 10,000 tons of coal were despatched and utilized, and it is expected that within the present half year 20,000 tons may be extracted, of which it is computed that 16,000 tons of screened coal can be delivered. At Pid there is a seam of good bright fuel 3 feet thick in places. As the locality is near a good road a fair amount of fuel might be obtained, for the coal contains less pyrites than elsewhere. At Bhaganwalla, the outcrop of the seam is 3 feet 6 inches and extends for 2 miles; the coal is much cracked and jointed and contains much pyrites. By means of suitable workings good masses of bright coal might be obtained, and though the locality is difficult of access, it might be improved in this respect. The available coal is estimated at 16,20,000 maunds (60,000 tons).

"*North-West Himálayas.*—At Dandli, near Kotli, on the Punch, and HIMALAYAN. 1427
at the north-west shoulder of the Sangar Marg Mountain, there are beds of nummulitic coal, the position of which, however, seems to preclude the possibility of successful exploitation. The latter field has been recently examined, and seems to hold out a fair prospect of success.

"Coaly matter and lignite occurs sparsely in the Sivalik sandstones of the sub-Himálayas, and has frequently given rise to false hopes of the discovery of workable coal in these regions. There is, however, no probability of such being met with.

"*Sikkim.*—There is a coal-field in the Darjiling District which occupies a narrow zone stretching along the foot of the Himálayas from Pankabari to Dalingkote; the coal is of Gondwana age and is much crushed; some of it is in the form of a powder, and has assumed the character of graphite.

ASSAM. 1428 "*Dufla Hills.*—A seam of Gondwana coal, 5 to 6 feet in thickness, is known to exist, but it will probably never possess any economic value.

"*Khási and Jaintia Hills.*—Both cretaceous and nummulitic coal occurs, but on account of the high elevations at which the basins are situated there are great difficulties in the way of the transport of the coal to market. The most important of the localities are: Mao-beh-lyrka, and Langrin where the coal is of cretaceous age, and Cherra Punji, where it is nummulitic. In the Mao-beh-lyrka field the abundance of pyrites is a drawback to its use as fuel, though it has been worked to supply the station of Shillong (18 miles distant). The available quantity of coal is estimated at from 387,000 to 470,000 tons. In the Jaintia Hills, carbonaceous deposits are reported to exist at five localities, *viz.*, Am-ur, La-ka-dong, Narpur, Sha-tyng-gah, and Shermang. At La-ka-dong the coal is of nummulitic age, and is irregularly developed, but its amount is estimated at 1,500,000 tons.

"*Garo Hills.*—The Daranggiri coal-field (cretaceous) contains a 7-foot seam of coal, favourably situated for working, but at present useless for want of access.

"*Upper Assam.*—There is an important field at *Makum* which is being worked by the Assam Trading Company; it contains several seams of coal, one of which is over 100 feet thick, 75 feet being good coal. The beds are disturbed and the coal seams lie at an average angle of about 40°, so that some difficulty may be met with in working them. An approximate estimate gives 18,000,000 tons as available, supposing the workings to be nowhere carried more than 200 yards from the face or 400 feet to the deep.

"*Jaipur*, in Upper Assam.—The coal in this field is for the most part in thin seams and of poor quality, though there are a few workable seams (one 5 feet 3 inches thick) of good quality. The coal-field is estimated to have 10,000,000 tons of coal available; this is exclusive of what may be proved by borings, but is mostly of poor quality.

"*Nazira*, in Upper Assam.—Some of the seams in this field are of considerable thickness, 30 feet and over; the estimated quantity available is 10,000,000 tons.

"*Janji* and *Disai.*—Two small and unimportant fields in Upper Assam.

BURMA. 1429 "*Arakan.*—In the Arakan Division, at the Baronga Islands, on the western coast of Angara-Khyong, about 2 or 3 miles from its southern extremity, coal is said to exist at three localities below highwater mark. On the central island, Peni-kyong, at the south end, coal has been reported in a seam 1 foot thick; on Ramri Island, less than one mile W. 10° N. of Tsetama, two seams occur, one of which has a thickness of 6 feet, and the other of 2 feet 5 inches. A 2-feet 6-inch seam of similar coal occurs on the Cheduba Land.

"*Pegu.*—Coal was discovered in 1855, and a mine opened at Thayetmyo, but after a few cwts. had been extracted, the work was abandoned on account of the seam dying out: further explorations have been recently carried out. At Dalhousie, near the mouth of the Bassein river, and in other places, traces of lignite, which have at times given rise to fallacious hopes of a source of fuel, have been met with. In the Myanoung division of the Henzada district several seams of coal occur.

"*Tenasserim.*—Coal has been found in the tertiary rocks at a number of localities: those at which the coal may possess a possible value are: Thoo-hte-khyoung (or Thatay-Kyoung) on the Great Tenasserim river, where a mine was formerly worked by Government but subsequently abandoned. This seam was 11 feet 8½ inches thick, of which 6 feet 8 inches were true coal. At Hienlap (or Hienlat), about 6 miles from the last locality, there is a seam from 17 to 18 feet in thickness, and the coal is of pretty uniform character with conchoidal fracture. Three quarters of a mile north of Hienlap, at Kan-ma-pyeng, there is coal, the main seam being 8 feet thick; but pyrites is abundant. On the Lenya river, at A-Tong-wo, about 8 miles above Lenya village, coal is exposed in a small tributary (the Phlia), and forms only an irregular bed varying from 1 foot to 2 feet 6 inches in thickness. The quality is such that if found in abundance it would be a useful fuel for many purposes.

"*Upper* (or Native) *Burma.*—Near Thingadaw (on the western banks of the Irawadi) there are three coal localities. The most southerly is 10 miles west of the village of Tembiung, in a stream bed; the seam is 4 feet thick, but the coal is in a flaky and cracked condition, and rapidly disintegrates. The second locality is 5 miles further north on the upper waters of the Kibiung stream, and 5 miles west of Thingadaw. The coal, with the included shale, is 5 feet 6 inches thick; an amber-like resin occurs with it. The third locality is 8 miles north-west of Thingadaw; the coal is hard, compact, and jetty, and includes small lumps of amber-like resin. The thickness is 3 feet 9 inches to 4 feet; both floor and roof are good. On the Chindwin river near Kalewa is a 10-foot seam of cretaceous coal; it is well situated for transit purposes. On the Paulwing river there are numerous irregular thick seams of tertiary coal. ANDAMAN.

1430 "In the Andaman and Nicobar Islands coal is known to exist; but so far as they have been examined there are no grounds for belief that a valuable deposit of coal occurs. (*See Manual of the Geology of India, Part III.*)

INDIAN MINES. 1431

Dr. Walter Saise, Manager, E. I. R. Company's Collieries, has obligingly furnished the following note on Coal and Coal-mining in India which, it may here be remarked, is based on the results of 1883-84, but on returns some of which are not accessible to Government. This explanation accounts for the apparent discrepancies between the returns of production and consumption published by Government for that year and the figures here given by Dr. Saise. On a further page will be found more recent figures abstracted from Government returns which bring this brief note on Indian coal up to present date:—

INDIAN CONSUMPTION OF COAL.—"The coal and coke used in India are either imported or raised and made in the country. The foreign sources of coal and coke supply are Europe, Australia, and Africa. Taking coal first, the proportion of coal raised in the country and that imported is as under— 1432

	Tons.
Imported (1883-84)	678,000
Raised in India (1884) about	1,556,400
	2,216,000

"The value of the former is stated to be R1,09,96,047. The value of the latter at the pit's mouth may be taken at R38,45,000. The imported coal is chiefly large or steam coal. The marketable* coal raised may be taken at 1,200,000 tons yearly, the balance being either used as coke or allowed to go to waste. Of the marketable coal the largest proportion is steam and rubble which are used on railways to a large and in steamers

* See page 388.

COAL.

Coal and Coal-mining in India.

to a smaller extent. The small kinds of rubble or smithy are used in stationary engines for smithy purposes, brick-burning, and lime-burning.

"The quantity of Indian coal used on railways in 1884 was 436,804 tons, the quantity of imported coal being 197,342 tons. The imported coal is used on railways unfavourably situated as regards Indian coal-fields.

1433 QUALITY OF INDIAN COAL.—"The quality of Indian coal varies much.

"Below is a table of ultimate analyses of specimens from Karharbari and Raniganj coal-fields with analysis of English and Welsh coals for comparison :—

COAL-FIELD.	Carbon.	Hydrogen.	Oxygen and Nitrogen.	Sulphur.	Ash.	
Karharbari	78·20	4·34	7·89	0·42	9·15	Main Seam.
E. I. Railway	70·93	4·10	12·49	0·52	11·96	Upper Seam.
Raniganj (N. B. Coal Co.)	74·31	5·12	9·67	0·47	10·43	
England . Newcastle	82·83	5·32	7·13	1·17	3·55	
England . South Wales	88·47	4·59	3·02	1·25	3·09	

"It will be noticed that in several particulars Indian coal is inferior to English, *1st*, in containing more ash, and *2nd*, less carbon and hydrogen.

"In the table below the commercial analyses of many Indian coals b l the writer and **Mr. T. H. Ward, F.G.S.**, are given, as also commerciay analyses of Newcastle and Welsh coals, for comparison :—

COAL-FIFLD.	Spec. gravity.	Ash.	Fixed carbon.	Volatile matter.	Sulphur.	Heating power by Thomson's calorimeter.	REMARKS.
BENGAL. Karharbari . Lower seam	1·35	9·15	66·84	24·00	0·42	13·20	
BENGAL. Karharbari . Upper seam	1·33	11·96	60·46	27·59	0·52	12·50	
BENGAL. Raniganj. Alipore (average) .	1·389	14·63	60·86	20·81	1·36	12·89	
BENGAL. Raniganj. Barakar . .	1·327	7·27	64·26	27·63	1·56	13·89	
BENGAL. Raniganj. Dhadka . .	...	7·64	49·61	42·75	...	...	
BENGAL. Raniganj. Borrea . .	...	10·03	60·70	29·27	...	12·35	
BENGAL. Raniganj. Belrooi . .	...	9·59	53·70	36·75	...	12·40	
Tindaria	...	27·68	65·22	7·10	...	...	Not worked.
CENTRAL PROVINCES. Umaria coal-field (1885).	1·439	16·03	71·77	12·20	Trace	...	
CENTRAL PROVINCES. Johilla (1882) .	...	13·55	57·95	28·50	...	...	Not worked.
CENTRAL PROVINCES. Warora (average)	...	8·99	42·85	48·16	0·849	...	
Burmah coal (Murray Coal Co.)	1·390	13·36	50·00	36·44	7·86	13·06	
Assam	...	1·60	53·42	44·98	2·52	13·99	
Welsh	1·312	3·68	82·66	13·66	1·59	...	
Newcastle	...	3·49	63·25	33·26	1·07	...	

"The above table shows that there is great diversity in the chemistry of the coals of India, and the variations in physical features are just as marked. With the exception of Tindaria and Assam coal, all Indian coals are remarkably laminated in structure, the laminæ consisting of a dark highly

Coal and Coal-mining in India. (*W. Saise.*) COAL.

carbonaceous shale, a bright pitch-looking matter, and a mineral charcoal —a very dull charcoal-looking substance. When these laminæ are very fine, the coal appears homogenous and is of good quality. But when the laminæ are thick the coal is extremely dirty and not so good in quality.

"Tindaria coal is anthracitic in look and flaky in fracture. Assam coal has a very peculiar fracture and breaks into small pieces. Other Indian coals are cuboidal or conchoidal in fracture. Warora coal breaks like shale into long irregular splinting pieces.

"The Indian coals now in the market from Karharbari and Raniganj may be considered as very fair steam coals, especially the coal of Karharbari, suitable in all respects for locomotives and steamers; although behind the Welsh coals in heating quality, they are not far behind the Newcastle coals, and are much of the same character, possessing a large percentage of volatile matter.

COMPARISON OF INDIAN WITH IMPORTED COAL FOR RAILWAY PURPOSES.—"The Indian and imported coals have been tried on Indian Railways with the following results:— 1434

EAST INDIAN RAILWAY.

COAL.	Gross weight of trains.		℔ per mile of coal consumed.	℔ per ton mile.
	Tons	cwts.		
Karharbari	207	19	30·12	·145
Raniganj Sanctoria	212	17	32·21	·151
,, Equitable	208	1	33·68	·161
,, Ordinary	204	14	36·98	·181
North Wales	215	9	31·90	·148
South Wales, Cardiff	203	11	32·64	·160
New South Wales	207	14	31·42	·151

D. W. CAMPBELL,
Locomotive Supdt., East Indian Railway.

COAL.	Gross weight of trains.		℔ per mile of coal consumed.	℔ per ton mile.
	Tons	cwts.		
Karharbari	166	12	25·76	·155
Raniganj	181	7	33·33	·184
Barakar	170	3	30·04	·177
Fothergills (S. W.)	183	12	30·45	·165
North Wales	174	9	27·12	·156
Australia Duckenfield Merthyr	180	4	27·43	·133
Godavari	171	12	33·48	·196

F. H. TREVETHICK,
Locomotive Supdt., Madras Railway.

C. 1434

COAL.

Coal and Coal-mining in India.

"It will be seen from these results that Karharbari coal is a good steam coal, little inferior to imported coals, and that the other Indian coals (except Godavari) are of fair quality. Umaria coal, tried on the Great Indian Peninsula, gave 42·63lb per train mile with a gross load of 410 tons. This is nearly but not quite as good as Karharbari coal.

INDIAN PRODUCTION.—"The sources of Indian coal supply and the estimated yearly output are as under :—

CENTRAL PROVINCES	Warora	100,000
	Narbada	28,000
	Umaria	7,290
BENGAL	Karharbari	520,000
	Raniganj	890,000
ASSAM		50,000
		1,595,200*

As the newer fields develop this estimate will have to be increased.

DISTRIBUTION OF INDIAN SUPPLY.—"The Warora coal-field is connected with the Nagpur branch of the Great Indian Peninsula by the Wardha Coal State Railway; the Mohpani (Narbada) coal-field by a branch from Gadawara with the Great Indian Peninsular. The Umaria coal-field has been tapped by the new line from Kutni through the East Indian Railway, Jubbulpur line. The Assam coal-field is connected with the Brahmaputra river by a line from Dibrugarh.

"The coal from the collieries of the Central Provinces is used on the following railways: Great Indian Peninsular, Rajputana-Malwa, Wardha Coal State Railway, and the Nagpur-Chattisgarh, the smaller coal going to mills.

"The Bengal coal finds its way to the Panjáb railways and the railways of Bengal, as also into the manufactories of Calcutta and the large cities along the line of railway. Some is used in the steam-ship lines. Small coal is largely employed for brick-making. Comparatively little is utilized for domestic purposes. The Colliery Companies should endeavour to create a want by teaching the people how to use small coal in large towns, such as Allahabad, instead of wood and cowdung. Agencies like those in English cities could probably do this in a few years, and the large waste of small coal that goes on at present would thus be obviated.

1435

MINING IN INDIA.

"Has made considerable progress during the past few years, machinery and well-appointed heapsteads and pit frames are coming generally into use.

"In most cases the railway is brought close to the mines, and where this is difficult, tramways of various gauges, worked by locomotives, carry the coal from the mine to the railway wharves.

"The seam is generally shallow, and engine-inclines or shallow pits give outlets for the coal. The two deepest mines in India are 23D shaft of the East Indian Railway Karharbari collieries, Bengal, 429 feet deep, and the Helen Pit of the Narbada Coal and Iron Company, Central Provinces, which is 402 feet deep.

"The system of working varies very much. At Warora, Central Provinces, where 100,000 tons per annum is wound by direct acting engines out of two shafts 200 feet deep, the system most nearly approaches the

* It may be noted that it is the marketable coal that appears in the Government returns, not the actual amounts raised. In 1883-84 these were 1,200,957 tons. Conf. with p. 385.—*Ed.*

English. No women work underground, and work is constant from Monday morning to Saturday night. The work time is divided into three shifts of 8 hours each. The seams, which vary from 8 to 12 feet, are worked thus. Galleries or bords and headways are driven 12 feet wide, 6 feet in height, leaving the roof coal, and pillars 40 feet square. The coal is so hard, it has to be nicked and undercut and then blasted down. The pillars are worked by splitting each from one headway to another and then taking the far end off in slices. The roof coal comes with it.

"At the Mohpani collieries a similar system is worked. The difficulties met with in these mines, owing to the faulted and disturbed nature of strata, are probably unequalled in India.

Karharbari coal-field.—" Is the smallest field in Bengal. It is mainly worked by three Companies—the Ranigunge Coal Association, the Bengal Coal Company, and the East Indian Railway. The mines are connected with the main line wharves by metre gauge or 2-feet gauge lines worked by locomotives. During the busy season the coal-field presents a scene of great activity. As much as 50,000 tons of coal and coke have been raised and despatched in one month. The coal-field is connected with the East Indian Railway Chord line by a branch from Madhupur to Giridi, the terminus or colliery station. In mechanical arrangements for raising coal, this coal-field is well advanced. The old-fashioned gin is almost obsolete and bullock-carts have little to do. 1436

"The system here is similar to that obtaining all over Bengal. The working hours are from 6 A.M. to 6 P.M., and perhaps later when extra work is required. Only four days a week real work is done, and the consequence is that collieries must have a far greater number of working places than the same output in England would warrant. All the miner's family work with him, carrying or training his coal. Picks of English pattern and make are now universal, the crowbar and single pick having been ousted. The workings are on the bord and pillar system. Pillars vary from 12 feet to 40 feet square and 40 feet × 60 feet. In the shallow mines and thin seams (7 to 8 feet) the former size obtains, in the thick seams (from 12 to 20 feet thick) the latter. Pillars are worked in the 8-feet seam in the following manner. A 4-feet chock is placed between each pillar in the row of pillars (generally six in number) that are to come out. A chock is also placed in front of each pillar. The pillar is then attacked from the front side. When pillars are taken out the chocks are withdrawn and the roof falls.

"The remarks on the Raniganj coal-field given below apply in some measure to this field. On sinking, coal-cutting, the miners' love of holiday, lighting of mines, &c. the description in one case is a description in the other. Payments in this coal-field are weekly on Sunday mornings, the miners resorting from the pay offices to the East Indian Railway bazaar, which was established to attract local labour, and which has done so. The labourers consist of low caste Musulmans and Hindus, as also aborigines—Santhals and Kols. There are some Bauris, brought from Bengal to teach the local men how to cut coal. The local men, however, cut coal better as they have discarded the Bauri "Sabel." Local labour is more tractable, and the Bauris are not in such requisition as formerly.

"Drainage is effectively carried out by Tangye's special and lifting and forcing pumps, worked by bob-levers from horizontal engines. The machinery is of good type, and winding and hauling are done by good engines.

"Ventilation is attended to in the deep mines, mainly by furnaces or steam-jets.

Coal and coal-mining in India.

"The miners live in small villages, aggregations of huts of mud walls of bricks set in mud with thatched or tiled roof. The huts consist of one room, sometimes two, of from 6′×6′ to 10′×10′ in size. Those better off have cowsheds and granaries; these two latter with the dwelling forming three sides of a quadrangle. The larger proportion of the labourers cultivate during the rainy season and work at the collieries only in the cold and hot season, say from October to June. Some of the labourers have settled down to coal-cutting as a calling, and these work constantly, always excepting Monday, which is invariably a holiday.

"Coal-cutting is paid for by contract, at so much a tram or bucket; these are of various sizes. The price generally amounts to from 7 to 8 annas per ton for large, and 1½ to 1¾ annas per ton for small coal. All other work, as stone-cutting, sinking, rail-laying, &c., is paid for by daily wages.

"The coal is hand-picked into four kinds. Steam is larger than 2 in cube, rubble larger than ¾ in cube, smithy down to ¼ in cube, and all smaller than that is called slack or dust. This picking or screening is done by contract, and for rubble and smithy the coolies get about 4 annas per ton. Slack is not paid for. Loading is done by hand into the railway wagons. At the mines tipplers are used for discharging the coal from underground trams into the wagons that run in the narrow gauge tramways.

"A large amount of hard coke is made in open ovens in this coal-field, the amount reaching 25,000 tons per annum, of which the East Indian Railway makes and consumes the larger portion. Soft coke or half-burnt smithy for smiths' forges, &c., is also made to a large extent, about 7,000 tons per annum being the outturn.

"The following notes on the Raniganj coal-field are by **Mr. T. H. Ward**:—

1437 "'The Chord line, East Indian Railway, passes across this coal-field, and the collieries are clustered on either side and along the Barakar branch. Sidings and branches, up to 6 miles in length, built by private enterprise, connect most of the collieries to the main line. Here winding engines, wire rope guides and tipplers, and the regular paraphernalia of an English colliery are rapidly supplanting the primitive "gin" and bucket of a few years ago. These gins were (and some are still) turned by women, 25 to 30 being employed on each gin. They kept time to a monotonous chant, which they sang as they tramp round and round.

"'The sinking in the district is easy, through sound sandstones, no brickwork being required to protect the sides. Heavy water is sometimes met with.

"'The coal in the east of the field is very strong and non-caking. The sandstone roof is also very strong and comes right down into the coal. Practically no timber is required in working the coal in the manner described below. In the west of the field at Sanktoria, for instance, the coal is not so strong, though the roof is everywhere the same. From Belroie, near Sitarampore, westwards, the seams worked are all coking coals.

"'The seams worked are seldom less than 10 feet and sometimes reach 18 feet in thickness. In the Barakar Coal Company's Komerdohi colliery and the Bengal Coal Company's Liakdi colliery on the west of the Barakar, the enormous thickness, of upwards of 80 feet, has been found. This seam has, up to the present, only been quarried at its outcrop. It dips at 1 in 4 or 5 to the south.

"'The mine is laid out underground on the same plan throughout the district. This plan has been stereotyped all over the field, and is adopted without reference to its suitability to the different conditions obtaining in the various seams worked. Indeed, it has been adopted apparently more with reference to the prejudices of the native miner than from economical considerations. Galleries are excavated to the full height of the seam 12 feet to 16 feet in width, leaving square pillars of varying sizes to support the roof, many acres being thus often left on pillars. The native coolie insists (and he has his own way very much in this coal-field) on commencing operations at the roof and working downwards until the full height of the seam has been excavated. His chief and dearly-prized weapon is a 'sabal' or crowbar with a sharp point at one end. With this he smashes the coal, standing always when at work. He never grooves beyond the first 'cleat;' gangs of 4 or 5 men occupy each gallery; they are paid by the bucket or tram of steam coal or small delivered at the pit bottom. If any timber has to be set in a working place, a man of the carpenter caste (*Chútar*) who is paid a daily wage must be sent for the purpose.

"'Women and children work underground, and are principally employed in carrying the small coal and dust. They are also paid by the tram or bucket. The women often take their babes, 2 and 3 months old, down the mine, taking with them also a small cot on which the child sleeps or plays while its parents are at work.

"'Access to the mines is very generally by inclines opening to the surface.

"'In the eastern part of the district the seams are for the most part flat, in the central and western parts the strata are often steep (the general dip being southerly), and intrusions (dykes) of trap rock become more frequent. The deepest shafts are about 250 feet, the largest part of the coal get 'won' being from much less depths. Some fire damp has been met with in the western part of the district. Chanch colliery (west of the Barakar) belonging to the Bengal Coal Company was abandoned some years ago after an explosion in which several men were burnt, some of whom died. At Sanktoria, also belonging to the Bengal Coal Company, some men were burnt in 1883.

"'The quarries at Komerdobie and Liakdi have already been mentioned. Thousands of tons of coal have been won from the outcrops merely, of these magnificent seams, and thousands of tons remain still to be worked without indenting on their resources at any greater depth.

"'The 'Bauri' is the principal caste which supplies coal-cutters for the district. In some respects the Bauri collier's characteristics are amusing like those of his western prototype. He is very fond of getting drunk, especially at week ends, and very much disinclined to go to work on Mondays. For the rest he is good tempered and improvident. It is a difficult matter to persuade him, although he is always paid a 'ticca' (contract) rate for his work and could easily increase his earnings, to do more than will, with his wife's contribution, keep the household 'in rice' and himself in drink for the day. The nearly universal and very bad custom in this district is to pay each evening for the work done during the day. The collier or cooly has often to wait about until 8 or 9 P.M. for his money. He then goes cheerfully home and remains up half the night drinking and singing with his companions (he is very social in his habits) incomprehensibly happy with his tuneless 'tom-tom.' In the morning he trudges back very often 7 or 8 miles (a distance travelled of course twice a day) to work and is down the pit at 9 or 10 A.M. All day in the intervals of work he sucks the comforting 'hubble bubble.'

"'The light which the collier carries with him is exceedingly primitive. He gets an allowance of oil in proportion to the number of trams of coal he cuts. Every morning he draws at the godown sufficient for his requirements during the day, and an allowance of cotton thread or old rags to serve for wick. This oil he burns in a 'chirag' or small piece of stone hollowed out into the shape of a boat (a piece of tile from the roof of his house is often substituted). In this he places a small quantity of oil and

C. 1437

a portion of wick. Any oil he can save from his 'allowance' is his perquisite and he can carry it home. Mohawa and castor are the chief oils used. Some of the mines are lighted by kerosine, burnt in small tin lamps, holding about 2 ounces with small circular wicks. The native does not like this plan so well, as he cannot use it to rub on his body nor to season his food,—a purpose for which mohawa oil is used.

"The ventilation of the underground workings receives very little attention, and in most collieries none at all. The great freedom from fire damp and the lofty seams exploited have kept this question in the background. The ignorant native has not yet recognised that his health and longevity is in question, and he has besides helped much to prevent ventilation becoming a necessity by the wonderful power of endurance he has shown. This power of endurance enables him to work for hours at the bottom of a sinking shaft with water pouring over his naked body or to work all day long and day after day in driving a 'rise' gallery, perhaps hundreds of feet from any air current in an atmosphere which is fœtid and laden with steam. This want of ventilation is a blot on the mining of the district and ought to be speedily remedied."

1438 GENERAL CONCLUDING REMARKS BY DR. SAISE.—The coal industry in India employs about 30,000 persons, the quantity of coal raised per annum per person employed, surface and underground, being 51 tons.

"In Europe the numbers are different, varying with the thickness of seams and nature of difficulties met with :

England (average) . 348 tons per person employed underground and surface per annum.
Belgium . . . 134 Ditto Ditto.
Saarbruckin . . 187 Ditto Ditto.

There is no Government regulation of the coal industry; any person can manage a mine on any system he likes, whether or not he has experience or training. Interest has a great deal with the appointment of the managing staff, and it is to be feared that the best is not made of the splendid coal deposits, the favourable roof, and the moderate depths and inclinations of the seams."

1439

TRADE IN COAL.

The following brief note, prepared by the Revenue and Agricultural Department, gives the most recent information regarding the internal and foreign trade in Coal :—

The present consumption of coal and other mineral fuel in India may be estimated at two million tons, of which three-fourths of a million tons are imported from the United Kingdom and one and one-fourth million tons produced in India.

The following table exhibits the amounts of fuel consumed by the Indian Railways during the years 1885 to 1887 as given in the last Railway Administration Report :—

YEAR.	COAL.		Coke.	Patent Fuel.	Wood.
	English.	Country.			
	Tons.	Tons.	Tons.	Tons.	Tons.
1887	212,529	479,210	9,564	30,029	292,808
1886	240,063	460,948	9,132	26,212	259,513
1885	225,721	476,277	10,439	23,117	255,178

Trade in Coal.

COAL.

In 1886 there were 99 collieries in Bengal (of which, however, 37 were closed), 2 in the Central Provinces, 3 in Assam, and 1 in Umeria in Rewa (Central India), or 105 in all, of which 68 were actually worked. The output was returned as follows:—

	Tons.
Bengal	1,187,000
Central Provinces	117,300
Assam	70,800
Central India	13,500
TOTAL .	1,388,600

Assam has since increased its output, the figures for 1886-87 being returned at 72,000 tons. It is stated in the Railway Administration Report for 1886-87 that—

"Coal continues to enjoy the confidence of the public. Its sale to the river steamers and tea factories is increasing. It has been contracted for by the Dacca State Railway, the Kaunia-Dharlla State Railway, and the Eastern Bengal State Railway—by the latter for smithy purposes. It is being largely enquired for by the Ocean Steamer lines trading with Calcutta; also by the Eastern Bengal State Railway system. It has been found suitable to the engines of the Darjeeling-Himálayan Railway and the Northern Bengal State Railway; but the difficulty of access to these two railways from the river Brahmaputra prevents its extensive use by their administrations. The coal continues dusty, though it is being mined deep in the hill-sides. But its nature is beginning to be understood, and its friability is not found to be a drawback to its use as a steam fuel.

"The coke is found to be saleable to the tea factories of Lakhimpur to an extent of about 3,000 tons per annum. The Company is preparing by means of an increased labour force to enlarge the output of coal to 100,000 tons yearly."

Collieries have recently been opened out at Dandot (Panjáb) and Singareni (Nizam's Territory). The coal in these mines has been pronounced of good quality, and in Upper Burma coal has been found (in the Kali Valley on the Chindwin River), but arrangements have not as yet been made to work this new source of supply.

Mr. O'Conor, in his review on the Sea-borne Trade Returns for 1878-79, gives the following historic sketch of the Indian coal industry,—

"Coal mining in India is rapidly attaining considerable importance. The commencement of this industry appears to date back to 1820, when a mine was opened in the Raniganj district in Bengal. For twenty years no new mine seems to have been opened, and then only three mines were opened down to 1854. In that year the commencement of the East Indian Railway line, which was laid to run through the coal-bearing regions of the Damuda basin, gave an impetus to the mining industry and new pits were opened in larger numbers—2 in 1854, 3 in 1857, 3 in 1859, 3 in 1860, 2 in 1861, 1 in 1865, 2 in 1868, 1 in 1869, 1 in 1870, 2 in 1871, 1 in 1872, 3 in 1873, 7 in 1874, 5 in 1875, 3 in 1876, and 5 in 1877. All these were in Bengal in the Raniganj and neighbouring districts, which contain now altogether 56 mines at work. In the Central Provinces also the coal-fields of Narsingpur and Chanda have been utilized for the purposes of the Great Indian Peninsula Railway."

In the paragraph above the number of mines in 1886-87 is stated to have increased to 105.

FOREIGN TRADE. 1440

FOREIGN TRADE.—The total imports into India of coal (including coke and patent fuel, of which a small quantity is received) have more than

COAL.

Trade in Coal.

doubled themselves since 1866-67, having risen from 341,000 tons, valued at R55 lakhs, in that year to 765,000 tons, valued at R130 lakhs, in 1886-87. The United Kingdom supplies nearly all the imported coal, though Australia, which ranks next to it as a source of supply, is now sharing more largely in the imports, the value of its consignments in 1886-87 being R4·75 lakhs against R1·19 lakhs in 1866-67. Most of the imported coal is for steamers on their return journey from India and for the cotton mills in Bombay, which are too remote from the Indian coal-fields to take advantage of them. The percentage taken by each province in these imports is noted on the margin.

Bombay	71·2
Lower Burma	10·4
Bengal	9·5
Madras	4·7
Sind	4·2

INTERNAL TRADE. 1441

INTERNAL TRADE.—Statistics may now be given regarding the internal movements of coal by rail during 1886-87 between the different blocks (*i.e.*, provinces, chief towns, and Native States). The total trade amounted in quantity to 1,097,800 tons and in value to R158·83 lakhs. The position of each block as a *net* exporting or importing centre may be thus indicated:—

Exports.	Tons.	Imports.	Tons.
Bengal	743,000	Calcutta	504,000
Bombay Town	162,000	Bombay Presidency	162,000
Central Provinces	44,000	North-Western Provinces and Oudh	161,000
Karachi	7,000	Rajputana and Central India	66,000
Assam	4,000	Punjab	35,000
Madras Town	2,000	Berar	23,000
Madras	1,000	Sind	5,000
		Mysore	4,000
		Nizam's Territory	3,000

As might be expected, Bengal, where the most extensive mines in India are situated, takes the lead among the exporting centres. Of its exports, Calcutta took last year 68 per cent., the North-Western Provinces and Oudh 22 per cent., Rajputana and Central India 6 per cent., and the Punjab 4 per cent. The consignments from Bombay Town, which consist mostly of English coal, are conveyed principally to the presidency mills, the balance of the foreign imports being used by the shipping and the town mills. The exports from the Central Provinces go to Berar and the Bombay Presidency. Calcutta, the North-Western Provinces and Oudh, and the Panjáb virtually receive their entire supplies from the Bengal mines. Rajputana and Central India draw their largest supplies from Bengal. Berar imports its coal mostly from the Central Provinces, Sind from Karachi, Mysore from Madras, and the Nizam's Territory from Bombay Town.

The development of the coal industry in India is indicated by the fact that the *gross* exports from Bengal to other provinces and Calcutta have increased from 641,807 tons in 1882-83 to 755,831 tons in 1886-87, and those from the Central Provinces from 26,451 tons to 56,125 tons during the same period. Assam for the first time shows a net export (4,000 tons), in referring to which the Director of Land Records and Agriculture writes:—"This is entirely due to the increased output of the Makum coal-mines near Dibrugarh, which now supply nearly all the coal used in the Assam Valley, besides furnishing large quantities for export."

1442 **Coke.** (A note contributed by Dr. W. Saise.)

"Coke is imported and also made in India. In 1883-84 the imports amounted to 16,700 tons, valued at R4,10,738. Coke, however, is now made to a very large extent in Bengal. It is a most important industry in its relation to coal raisings, as the manufacture of coke means the utilization of small and otherwise useless coal. The industry is of recent and very rapid growth, having increased fourfold since 1875. There are two kinds of coke, called respectively hard and soft. The former is dense and is used for foundry and locomotive purposes. It is made chiefly in ovens consisting of two walls 6 to 8 feet high, 8 to 9 feet apart, and 40 feet long. In the walls are vertical flues connecting with horizontal flues running through the coal which is pressed between the walls. This primitive system is universal in Bengal, and a very fair coke, at great expenditure of coal, however, is made. One or two collieries have, or are about te have, the rectangular closed oven to produce a uniform quality of coke at less expenditure of coal. Soft coke is incompletely burnt coal, made for the purpose of supplying a more or less smokeless fuel. It supplants charcoal for cooking purposes and small coal for smithy purposes.

"The traffic returns on the East Indian Railway, which taps, with the exception of the Assam field, the whole of the coke-making districts of India, shew that in a year about 55,000 tons of coke, exclusive of foreign coke, are led over the line; add to this their own consumption, the respectable total of 77,000 tons on them per annum is arrived at. The proportion of hard coke is difficult to state, but at one large colliery which manufactures one-third of this quantity, the proportion of hard to soft coke is $3\frac{3}{4}$ to 1. Taking this all over, the quantities may be stated as under :— 1443

Hard coke for foundry blast furnaces, locomotive, &c. . .	65,800
Soft coke	11,200

per annum. The former figures mean a large consumption of small coal, a point of great importance in India. The introduction of better coking plant, of washing machinery, will ultimately drive dear English coke out of the market. The value of this coke may be taken at R5,39,000 at the collieries."

COBALT. 1444

Cobalt; *Ball, Econ. Geol., 324 & 616; also Mallet, Mineralogy, 27.*

Cobalt metal is never met with in the native form, except in small proportions as a constituent of meteoric iron. The ores of the metal occur chiefly in primitive rocks and are usually very complicated. They contain nickel, iron, and often bismuth and copper, mineralized either by sulphur or by arsenic, or by both together (*Miller*). The chief ores are Speiss Cobalt or tin, white Cobalt $Co\,As_2$, Cobalt Glauce CoSAs., and Linnæite or Cobalt Pyrites $Co\,S+Co_2\,S_3$.

Vern.—The mineral containing Cobalt, which is used by the Indian jewellers and potters, is known as *sehta* or *sáita*, HIND. In the Panjáb this substance is known by the name of *zaffre*, and **Sir George Birdwood** gives it also that of *rita*, a corruption, most probably, from the Hindustani name.

SOURCE.—A complex mineral (*sehta*) is found in various mines in Rajputána, especially in those of Babui and Bagor near Khetri. Mr. Mallet says of this substance, that it has the specific gravity of 6·00. On analysis it yielded the following composition :— 1445

Sulphur	19·46
Arsenic	43·87
Antimony	a trace.
Nickel	a trace.
Cobalt	28·30
Iron	7·83
Gangue	·80
	100·26

COBALT.

Source of Cobalt.

This substance is generally known as Cobaltite. In the *Rajputána Gazetteer*, and in the *Jury Reports* of the Exhibition of 1862, occur accounts of the Jeypur enamels, but in a recent publication, **Dr. T. H. Hendley** (*Journal of Indian Art*), gives more precise details. **Sir George Birdwood** (in his *Industrial Arts of India*) under Enamels (*pages 165—168*) and also under Pottery (*pages 301—324*), gives most instructive particulars regarding the Indian uses of Cobalt. He states: "The *rita* or *zaffre* is the black oxide of Cobalt found all over Central and Southern India, which has been roasted and powdered, mixed with a little powdered flint" (*p. 308*). **Mr. Ball** says, while speaking of the Jeypore blues in enamelling, "The production of the colours was a secret only known to certain families, except as regards the different shades of blue, which are stated" "to be produced by an oxide of Cobalt. This oxide is doubtless prepared by roasting the Cobaltite." The various authors who have described Cobaltite, in the Records of the Geological Department, seem to be unanimous in their opinion that Cobalt is only rarely met with in India, and that, too, in the mines of Rajputána alone (as far as peninsular India is concerned), and that the oxide is artificially prepared; in other words, that it does not occur naturally in Central and Southern India. The art of producing a rose colour enamel on gold with cobalt seems still to be a secret with the *minakaris* or enamellers of Jeypore. Cobalt minerals are also said to occur in two other localities—Nepal and Burma.

1446 **Economic Uses.**—Under the head of "Clays used for Pottery" (C. 1333) will be found some account of the uses of cobalt in the ceramic industry, while in the above remarks reference has been made to the nobler art of enamelling. In a work specially dealing with economic products, it is perhaps unnecessary to enter at greater detail into a substance the uses of which are so intimately associated with the higher branches of Industrial Art. **Dr. Hendley** says that the colours used by the Jeypore enamellers "are obtained in opaque vitreous masses from Lahore, where they are prepared by Muhammadan *manihars* or bracelet-makers. The Jeypore workmen state that they cannot make the colours themselves. The base of each colour is vitreous and the colouring matter is the oxide of a metal such as cobalt or iron. Large quantities of cobalt are obtained from Bhagore near Khetri, the chief town of a tributary State of Jeypore, and are used in producing the beautiful blue enamel." In these passages **Dr. Hendley** does not make it quite clear whether the Jeypore enamellers prepare their own material for the blue colour, though unable to prepare the other colours, or whether the entire mass of the crude material is conveyed to Lahore and other centres to be prepared and returned in its manufactured condition to the Jeypore workers in enamel. He, however, proceeds to say: "All the colours known can be applied to gold. Black, green, blue, dark yellow, orange, pink, and a peculiar salmon colour, can be used with silver. Copper only admits of the employment of white, black, and pink, and even of these the last is made to adhere with difficulty (this applies to Jeypore copper enamels). In the order of hardness and of application to the metals, the colours are as follows—white, blue, green, black, red. The pure ruby red is the most fugitive, and it is only the most experienced workmen who can bring out its beauty." **Hoey** (*Trade and Manufactures in N. Ind.*) gives some details regarding the *Manihars'* industry, which he divides into two sections, *viz.*, the makers of glass bangles and the makers of lac bangles. **Baden Powell** (*Panjáb Manufactures*) discusses the Múltán enamel industry and furnishes particulars regarding the *Míná* blue vitreous enamel. In the *Múltán Gazetteer* (*p. 107*) this subject is enlarged upon, and reference is also made to the Baháwalpur enamels, where, in addition to opaques, a semi-translucent sea-green and also a dark blue are produced.

In Europe Cobalt is largely used as a pigment and to colour ordinary glass.

Coccinia indica, *W.&A.*, see **Cephalandra indica,** *Nand.*; CUCURBITACEÆ.

COCCULUS, *DC.*; *Gen. Pl., I., 36, 961.* 1447

Cocculus cordifolius, *DC.*, see **Tinospora cordifolia,** *Miers*; MENISPERMACEÆ.

C. indicus (see *Flück. and Hanb., Pharm., p. 31*), a commercial synonym for **Anamirta Cocculus,** *W. & A.*, see **Vol. I., A. 1037.**

C. Leæba, *DC.*; *Fl. Br. Ind., I., 102.* 1448

Vern.—*Vallúr, illar billar, parwatti, vehri,* PB.; *Ullar-billar,* SIND.

References.—*Gamble, Man. Timb., 11; Brandis, For. Fl., 9; Stewart, Pb. Pl., 6; Aitchison, Cat. Pb. and Sind Pl., 3; Murray, Pl. and Drugs, Sind, 38.*

Habitat.—A large climber of the dry and arid zones, especially of Western India; the Panjáb, Sind, and the Carnatic.

Medicine.—**Stewart** says the stems often become as much as 3 or 4 feet in girth. It is used in Sind and Afghánistan in the treatment of intermittent fevers and as a substitute for **Cocculus indicus** (*Murray; Dymock*). MEDICINE. 1449

Food and Fodder.—In the Trans-Indus, **Stewart** says, it is browsed by goats but by no other animals. Said to be used as a partial substitute for hops in the manufacture of Indian beer (*Murray*). FOOD and FODDER. 1450 Hop Substitute. 1451

C. palmatus, *DC.*, see **Jateorhiza palmata,** *Miers.*

C. villosus, *DC.*; *Fl. Br. Ind., I., 101.* 1452

Vern.—*Jamti-ki-bel, hier, dier,* HIND.; *Kursan, zamir,* SIND; *Vasanavela,* MAR.; *Wassanwel, parwel,* BOMB.; *Káttuk-kodi,* TAM.; *Dúsari-tige, chipuru-tige, katle-tige,* TEL. In the Concan the Vaids give this plant the Sans. name of *Vanatiktika.*

This plant sometimes bears the name *Farid-búti* (a name which, more correctly, should be applied to **Pedalium Murex,** so called in remembrance of the fact that **Shaik Farid Shakar-gunj** is supposed to have lived on water rendered mucilaginous by the leaves of that plant having been shaken in it). This same property is possessed by the leaves of **Cocculus villosus.**

References.—*Gamble, Man. Timb., 11; Roxb., Fl. Ind., Ed. C.B.C., 732;* (under **Menispermum hirsutum,** *Willd.*); *Drury, U. Pl., 145; Dymock, Mat. Med. W. Ind., 2nd Ed., 32.*

Habitat.—A large climber of the dry and arid zones, Sind, Panjáb, Deccan, extending into Madras and Bengal.

Medicine.—"The JUICE of the LEAVES, mixed with water, has the property of coagulating into a green jelly-like substance, which is applied externally by the country-people under various circumstances on account of its cooling nature; and is also taken internally, sweetened with sugar, as a cure for gonorrhœa." **Roxburgh** says: "A decoction of the fresh ROOTS, with a few heads of pepper, in goats' milk, is administered for rheumatic and old venereal pains; half a pint every morning is the dose. It is reckoned heating, laxative, and sudorific." By more recent writers the root is said to be alterative and to be a good substitute for sarsaparilla. **Dymock** remarks that in the Concan the roots rubbed with Bonduc nuts in water are administered as a cure for belly-ache in child- MEDICINE. Leaves. 1453 Roots. 1454 ren; and in bilious dyspepsia, they are giver in 6-*massa* doses with ginger and sugar; they are also an ingredient, with a number of bitters and aromatics, in a compound pill which is prescribed in fever. The *Pharmacopœia of India* states that this possesses the bitterness and probably the tonic properties of *gulancha* (**Tinospora cordifolia**). **Stocks** alludes to this as a Sind drug under its bazar name of *zamır*, and remarks that it is employed in pains of the head.

FOOD. 1455 **Food.**—The leaves are made into curry and eaten by patients under treatment, with the roots or the jelly from the leaves. If suffered to stand for a few minutes, the jelly clears, "the gelatinous or mucilaginous parts separate, contract and float in the centre, leaving the water clear like Madeira wine, and almost tasteless." (*Roxb.*) With regard to this property the remark under the vernacular name *Farid-búti* should be read. In Eastern Bengal the writer repeatedly observed the milkmen carrying milk to market with a few leaves of this plant and the spine-like leaflets of the date-palm placed in the vessel. On enquiry he was told these prevented the milk fom getting bad through the heat and the shaking to which it was subjected. He has never been able to investigate this point further, but it is probable the leaves of the **Cocculus** are added more with the object of thickening the water-adulterated milk. A large amount of the milk brought into Calcutta is regularly preserved or adulterated in this manner. **Dr. Dymock** alludes to the fact that this plant was eaten during the famine of 1877-78 in the Khandesh district, and that it is always more or less eaten in Kaladgi.

FODDER. 1456 **Fodder.**—**Roxburgh** says that goats, cows, and buffaloes eat the plant.

DOMESTIC. 1457 **Domestic Uses.**—"The juice of the ripe berries makes a good, durable, bluish purple ink" (*Roxb.*)

COCCUS; *Packard, Guide to the Study of Insects, 526.*

A genus of Insects belonging to the **Coccidæ** of the Order **Hemiptera.** Several species are, by Entomologists, referred to this genus, but two only are of commercial importance,—the one a native of Southern Asia and the other of the American Continent, namely, **Coccus lacca** and **C. cacti.** The former affords the resin and the dye known as LAC (or *lakh*) and the latter the dye, COCHINEAL. In the present article lac-dye and cochineal will be more particularly dealt with, but to complete the account of the useful products derived from the above insects, the reader should consult the article LAC in a further volume, where details of the manufacture, trade, and economic uses of the Lac Resin will be found.

As having an important bearing on the industry it may be desirable to discuss here very briefly the scientific peculiarities of the insects which belong to this genus. These differ from the other members of the tribe to which they are referred, in that the females are wingless and become fixed to the plant on which they feed. When this takes place they lose all power of locomotion, and in the case of cochineal, the legs practically disappear through corpulent growth. The same loss occurs in lac by the deposit of a resinous substance around the body. The males possess 10-jointed antennæ, and have two anal bristles, while the females have 9-jointed antennæ and are covered by a flattened hemispherical scale.

1458 **Coccus cacti,** *Linn.*

THE COCHINEAL INSECT; COCHENILLE, *Fr.;* KOCHENILLE SCHARLACHWURM, *Germ.;* COCCINIGLIA, *It.;* COCHINILLA, *Sp.*

Vern.—*Kirmdana,* BENG.; *Kirmaz,* BOMB.; *Kiranda,* N.-W. P.; *Kirm,* PB.

References.—*Royle, Prod. Res. of Ind., 57; Encyclop. Britannica, VI., 97; Balfour, Cycl. of India; Liotard, Dyes and Tans of India; Wardle, Report on the Dyes of India; Buck, Dyes and Tans of N.-W. P.;*

The Cochineal Insect.

Official Papers on Pigments used in India; Crookes, Dyeing and Calico Printing, 350; Hummel, the Dyeing of Textile Fabrics, 348; British Manufacturing Industries (Dyeing and Bleaching) by T. Sims, 155; Churchill, Technological Hand-book (Bleaching, Dyeing, and Calico-Printing), 139; Baden Powell, Panjáb Prod., Vol. I., 190-195; Ure, Dict. of Arts, Manuf., &c., Vol. I., 837; U.S. Dispens., 466.

Habitat.—The Cochineal insect was first discovered by the Spaniards in Mexico in the year 1518, but it was not made known to Europe until 1523. At first it was supposed to be a seed, but in 1703 **Leeuwenhock** showed it to be an insect. In Mexico it is particularly abundant in the provinces of Oaxaca and Guerrero. It occurs in many localities in Central America, and for long has been one of the most important articles of export from Guatemala, but it is met with also in South America, and recently it has been found (or perhaps only an allied insect) in the West Indies and in the southern portions of the United States.

HISTORY. 1459

HISTORY AND INTRODUCTION.—The immense importance of the trade, early established in this insect, led to efforts for its propagation in other countries, and for many years this has been profitably prosecuted in Teneriffe, the Canary Islands, Java, Algeria, and to some extent even in Spain. According to some writers the best quality now comes from Honduras. The attention of the Court of Directors of the East India Company was directed to this subject by **Dr. James Anderson** of Madras in 1786. He forwarded to **Sir Joseph Banks** samples of a dye-yielding insect which was proved to be a species of Coccus, but not Cochineal. **Sir Joseph Banks** concluded from the existence of a species of Coccus in India, that the true Cochineal insect might be acclimatised in that country. Directions were accordingly issued for the introduction of both the insect and the plant on which it feeds. A Cochineal insect was ultimately successfully conveyed to Bengal by "**Captain Neilson** of H. M. 74th Regt., on his return to India in June 1795. When the fleet in which he sailed repaired for refreshments to the coast of Brazil, **Captain Neilson,** in his walks at Rio Janeiro, saw a plantation of **Opuntias,** and obtained several plants with the insect on them. Many of them died during the passage to Bengal; and a few only remained alive on the last plant, of which several of the leaves had withered. **Captain Neilson,** on his arrival at Calcutta, sent the survivors to the Botanic Garden, where they were placed on the several species of **Cactus** or **Opuntia.** On the China and Manilla species of the **Nopal,** and even on that from Kew, the survivors began to die fast. It fortunately occurred to make trial of the indigenous **Opuntia,** (*sic*) on which they were luckily found to thrive amazingly, and so rapidly, that **Captain Neilson** himself writes, on the 3rd August 1795, that he had the day before seen at the Company's garden near Calcutta about one thousand fine plants covered with the insects; enough to stock all India." (*Royle, Productive Resources of India, p. 60, published 1840.*)

The above passage has been reproduced here as being the earliest and at the same time most complete account of the introduction of the Cochineal insect into India. Without learning the details we are next informed of its having been successfully introduced into South India, but whether from the Bengal stock or through some fresh effort, cannot be discovered. Passing over a gap of 60 or 80 years, numerous writers refer to "the indigenous insect" in such a pronounced manner as to suggest the doubt whether or not **Captain Neilson's** stock had, during that period, overrun the whole of India and become so completely acclimatised as to be mistaken for indigenous. Even **Royle,** in the above passage, alludes to the "indigenous **Opuntia,**" whereas no member of the family to which that plant belongs (except the Ceylon **Rhipsalis**) was known in the world prior

The Cochineal Insect.

HISTORY.

to the discovery of America, and therefore no **Cactus** can be called indigenous to India. This is more than a quibble as to the correct usage of a scientific term. If the **Coccus** sent to **Sir Joseph Banks,** one hundred years ago, was found feeding on a **Cactus,** it must be regarded as but an earlier introduction than the Cochineal brought to India by **Captain Neilson.** It therefore seems probable that the Portuguese (or whoever introduced the **Opuntia**) may have intentionally or unintentionally brought the Cactus-feeding Coccus also. In 1848 **Dr Dempster** addressed a letter to the Governor General of India which afterwards appeared in the Journal of the Agri.-Horticultural Society. He there extols the superior quality of the dye obtained from "the native" or "indigenous" insect as compared with the imported. "The quality," he says, "of native Cochineal which I found capable of dyeing a certain weight of woollen cloth proves that the indigenous insects contain an amount of colouring matter not inferior to the fine Mexican cochineal." In the same year **Dr. A. Fleming** published an account of the discovery of the Cochineal insect on the **Cactus** hedges near Gindiala in the Panjáb. He writes: "I got satisfactory proof that the Indian cochineal is an *article of commerce* in the country." In his *Panjáb Products* **Mr. Baden Powell** refers to an occasion when the **Cactus** had increased so rapidly in the Jullunder Doab "as to become a nuisance; and rewards were offered for its extermination, which, however, were rendered unnecessary shortly after, as a large number of insects of some kind of **Coccus** appeared and soon effected the destruction of the plant, which is now only occasionally to be met with."

Mr. Liotard (*Memorandum on Dyes and Tans of India*) enters into considerable detail regarding what he calls "the indigenous insect," and **Mr. McClelland** says, "the insects seem to thrive on our own indigenous species of **Opuntia**; but as we have abundance of the South American plant, **O. cochinillifera,** that species may also be tried along with the several sorts of our own."

In all these instances the **Coccus** alluded to is a cactus-feeding insect; but the lac insect, as stated above, belongs to the same genus, and *it* feeds upon many widely different trees (*see a further paragraph*), but has never been recorded as feeding on the **Cactus.** From the travels of **Lieutenant Burnes** and **Dr. Gerard** (*see Journal, Asiatic Society, Bengal, II.*) we learn that a species of what they are pleased to call Cochineal was seen to flourish on the roots of a plant growing in a marsh near Herat, but that the natives, instead of using that dye, are stated to import their cochineal from Bokhara and Yarkand. Without speculating too far as to what the Herat cochineal may prove, when thoroughly investigated, it may be here remarked that the Polish cochineal (**Coccus polonicus**) feeds on the roots of a **Scleranthus** found in sandy places throughout Europe. **Mr. Baden Powell** alludes to the Bokhara cochineal as imported into the Panjáb. In numerous official and other publications, trans-Himálayan cochineal is referred to. If this should prove distinct from the cactus-feeding species, it may be found allied to the **Coccus ilicis** of Greece, an insect which has long been used as a dye under the name of *kermes, chermes,* or *alkermes.* That insect is reported to feed upon a species of oak. The Herat Coccus may, on the other hand, be allied to the **Coccus maniparus,** *Ehrenbergh,* which is found in Sanai feeding on Tamarix, and is supposed to be the cause of the gum-like exudation known as Manna.

1460 THE INTRODUCTION OF THE OPUNTIA OR PRICKLY-PEAR.—The above remarks may be accepted as disposing of the question of "the indigenous cochineal insect which feeds on the common prickly-pear." If not indigenous then, as an acclimatised insect, has it deteriorated after the lapse of 100 to 150 years? Perhaps the further question may also be suggested—was the insect derived from the best stock? If unfavourable answers have to be given to these enquiries, then it would remain to be ascertained by actual experiment whether an improved and fresh stock could be acclimatised. We shall return to this point later on, but it may perhaps be desirable to reiterate the position here advocated, that the Cactus-feeding cochineal, now met with all over India, is an introduced insect. The natives of different parts of India are known, however, to collect in a small way other insects, besides the cochineal and lac, and are stated to hold that these afford tinctorial principles. They are, in all probability, truly indigenous, and may even be well worthy the attention of commercial experts. A scarlet dye is often alluded to in the ancient writings of India, and some authors have even translated these passages so as to make them appear to allude to cochineal. If not lac-dye they may refer to the insects mentioned above. The whole subject of Indian cochineal has, in the writer's opinion, been obscured through the mistaken idea regarding "indigenous cochineal insects," and is not likely to be satisfactorily cleared up until careful investigations have been made in Madras the home of the first so-called Indian cochineal insects which were sent to Europe, and at the same time the head-quarters of the acclimatised **Opuntias**. The sudden appearance and disappearance of a Coccus in the Panjáb, mentioned by **Mr. Baden Powell**, would justify the conclusion that **Captain Neilson's** insect need not have taken more than a few years to get dispersed all over India. The prickly-pear is not very abundant in Bengal, but doubtless the heavy rains of that province account for the non-appearance of the insect. The writer, though located 12 years in Bengal, does not recollect having seen Cochineal on the Cactus hedgerows of the Lower Provinces. It is significant that **Dr. Hove** (a botanist who explored a large portion of the Bombay Presidency 100 years ago), makes no mention of having seen the Cactus in the Western Presidency. **Dalzell** and **Gibson**, in their *Bombay Flora* (published 1861), say:—"The prickly-pear, native of Brazil, now too common about most Deccan villages, where it forms a nidus for snakes, filth, and malaria of every description. It were most advisable that Government should, as a measure of sanitary police, take for its eradication more energetic and continuous steps than those hitherto adopted." "The native tradition is, that a few seeds of the plant were brought by a Sirdar in his palankin from Delhi, and verily his gift has been as noxious to the Deccan as was that of the poisonous shirt of Hercules. It is hardly found in Gujarat; there we have only noticed it in Sidhpore, between Ahmedabad and Deesa." **Dr. W. Gray** says, in his sketch of the Botany of Bombay (1886), "the prickly-pear has spread in some districts of this Presidency to such an extent as to have become a noxious weed." It was so abundant at the beginning of the present century, in the Madras Presidency, that **Ainslie** wrote of it:—"This species of **Opuntia** is indigenous in India."

Madras Cochineal Plant. 1461

Panjab Cochineal Plant. 1462

Bombay Cochineal Plant. 1463

MODERN EFFORTS TO REINTRODUCE THE COCHINEAL INSECT. 1464

There are commercially two chief kinds of this insect, but whether distinct species, or the one only the wild form of the other does not appear to have been clearly made out. The former (the so-called wild insect) is known in trade as the **Grana sylvestris**, and the latter as the **Grana fina.** From the early correspondence regarding the introduction of the cochineal insect into India, it would appear that **Captain Neilson's** stock was the **Grana sylvestris.** A voluminous correspondence has ensued since 1795 as to the desirability of introducing the superior quality, which fetches (from its greater amount of the tinctorial principle) three times the price paid for the wild insect. As late as 1882, the Madras Government had this subject brought to its attention, and instructions were given that **Dr. George Bidie, C.I.E.**, should supervise the experiment. The Agri.-Horticultural Society of Madras agreed to place at the disposal of Government a small plot of ground for the purpose of this experiment, although that Society does not appear to entertain any high hopes of ultimate success. **Dr. Bidie** addressed two letters to the Government refuting the position taken up by the Society, and his opinions and recommendations were accepted by the Government.

Forms of Cochineal.

It seems probable that the insect alluded to by **Dr. James Anderson** as found in India prior to the arrival of the Rio Janeiro supply, was also the **Grana sylvestris**, hence possibly a certain amount of the confusion that has crept into the literature of this subject—that insect from its American name of "the wild insect" having come to be viewed as wild or indigenous in India. There is no authentic information as to whether the **Grana fina** exists in this country, but it seems probable that the different qualities of the insects found may be due to the existence of breeds or races derived from both these stocks. The want of technical knowledge has prevented Indian writers, on this subject, from expressing a more definite opinion than that a superior or an inferior cochineal was found in certain districts. This would seem to point to the desirability of having a representative series of the insects met with in India collected and scientifically and tinctorially examined as the first step towards the establishment in India of a commercial industry. We read of numerous futile attempts to bring about this desired object but of no combined and systematic investigation. As often happens with economic questions, the desirability of establishing a cochineal industry in India has been periodically brought to the attention of the Government, but allowed to lapse into inactivity from many causes, chiefly the transfer to scenes of greater usefulness of the officers who interested themselves in the subject. **Dalzell** and **Gibson**, under the heading **Opuntia Toonah**, *Mil.*, say: "This is a species on which, according to **Humboldt** and **Bonpland**, the cochineal **Grana fina** is fed; others say that the false cochineal insect only feeds on trees. We have had numerous experiments regarding the introduction of this product. In the new-production-fever years, ranging from 1833 to 1845, sundry attempts were made by the late **M. Sundt** and others, but after considerable expense incurred, and a heavy amount of correspondence, as usual in such cases, the whole ended in smoke." (*Fl. Bomb. Supp.*, 40).

Grana Fina. 1465 **Grana sylvestris. 1466**

Grana fina and Grana sylvestris.—**Humboldt** was, perhaps, the earliest observer to distinguish "the *fine* from the *silvester* or wild sort of cochineal." The former insect, he says, is mealy, or covered with a white powder, while the latter is enveloped in a thick cottony substance which prevents the rings of the insect being seen. The **Grana fina** is reported to be a native of Mexico, and the **Grana sylvestris** of South America. **Dr. Balfour** remarks: "It has been mentioned that at Vizagapatam there is a great deal of the red flowering prickly pear on which the cochineal insect feeds; that the insect under propagation at Oossoor (Bangalore) has been ascertained to be the true cochineal insect, and to be procurable in several districts in South India; but it only destroys the plants with red flowers and few prickles; and that it will not propagate on the yellow flowering prickly pear or **Opuntia**. I have seen it tried at Bellary and fail." Commenting on this, **Mr. Liotard** remarks (and he has been followed by several more recent writers): "Regarding the future in India, it may be well to lay stress on the statement made by **Dr. Balfour** that

Red-flowered Opuntia. 1467

Yellow-flowered Opuntia. 1468

the true cochineal insect only destroys the prickly pear plant with red flowers and few prickles, and will not propagate on the yellow-flowered plant or **Opuntia.**" Again, "as regards the Peninsular, we learn from **Dr. Balfour** that not only the *variety* (*sic*) of plant required but the superior *species* (*sic*) of the insect also exists in parts of the Madras Presidency." Although **Dr. Balfour's** remark as to the existence of the true cochineal insect in Madras has been thus reiterated by other writers, the Madras Government in 1882 decided to make an effort to introduce the true insect from Algiers. In refutation of **Dr. Balfour's** opinion regarding the plant on which it feeds **Dr. Bidie** quotes the passages in **Roxborough, Ainslie,** and **Royle,** which all emphatically prove that the imported insect prefers the common acclimatised yellow-flowered **Opuntia Dillenii** to the more scarce and recently imported red-flowered form. There can be no doubt as to this point, but at the same time all writers seem to be agreed that **Captain Neilson's** insect, which was found to thrive best on the common **Opuntia,** was the **Grana sylvestris** and not the **Grana fina.** If **Balfour** be correct in the statement that the latter insect does actually exist in Madras, he may also be correct in his further opinion that it thrives best on the red-flowered or sub-spineless plant. It might, therefore, only lead to further confusion to advocate the claims of the yellow-flowered plant until it has been demonstrated whether the Madras insect that feeds on the red-flowered cactus is or is not a race derived from the true cochineal insect, perhaps more ancient than **Captain Neilson's** stock. The position assumed by **Mr. Liotard** of urging the extended cultivation of the red-flowered plant might, therefore, prove as subversive of success as **Dr. Bidie's** recommendation to import the Algiers insect in the hope that it would thrive on the yellow-flowered plant. **Dr. Bidie** seems to have lost sight of the fact, that, as far as can be learned, the so-called wild insect and not the semi-domesticated, has as yet been introduced into India, and that all the opinions he has quoted refer to the plant on which the former and not the latter is able to subsist. It would thus appear that the first and most natural step towards the introduction into India of a commercial industry in cochineal should be the thorough investigation of the races of coccus already existing in the country and the plants on which they feed. Such an enquiry, as already suggested, might lead to the discovery of a race derived from the true cochineal insect, but so degenerated as to fully justify the importation of a new stock. The plant on which the acclimatised insect is found to feed would naturally be that which should be fostered in anticipation of the arrival of a fresh importation. Degeneration, if established, might be accounted for by an originally semi-domesticated creature having been allowed to run wild for a century or more, or from having been forced to feed on the wrong plant. Mistakes may thus be made, but the course indicated would most probably prove the most direct, and it may happen that we possess a long-acclimatised stock which, under careful treatment, would prove more hopeful than any insect that might now be introduced.

Steps to be taken in the Development of an Indian Cochineal Industry. 1469

PECULIARITIES OF THE COCHINEAL INSECT.—This account of cochineal may therefore be concluded by referring to some of the more striking peculiarities of the insect which have a direct bearing on the question of its propagation. **Balfour** says: "There are three periods of life of the cochineal insect. It is viviparous, and at its birth is a mere speck, and at that time no difference can be detected by a microscope between the sexes; they are all equally active, seeming to profit eagerly of the short period during which motion is allowed them. After a few days they attach themselves to the cactus plant, and from that moment the female never quits her hold. A cottony coat grows over her, which falls off in 13 to 15 days." 1470

COCCUS cacti.

Propagation of the Cochineal Insect.

Male. 1471 "The male also adheres to the plant, and in about 12 days becomes enveloped in a cottony cylindrical purse, open at the bottom; the insects huddle together one upon another to appearance, so that at a little distance nothing is seen but a white patch of cotton of uneven surface; they continually increase in bulk. After remaining in this state for a month or thereabouts, the sexes become distinctly recognisable. The male becomes a scarlet fly, with two transparent wings about three times the length of his body." "He is now again become active (particularly an hour after sunrise), but rarely takes to the wing, being easily carried away by the wind; he jumps and flutters about, and, having impregnated the female, dies in a few days."

Female. 1472 "The females go on increasing in roundness. They appear generally so enormously overgrown, that their eyes and mouth are quite sunk in their rugæ or wrinkles; their antennæ and legs are almost covered by them, and are so impeded in their motions from the swellings about the insertions of their legs, that they can scarce move them, much less move themselves, and the insect to the casual observer looks more like a berry than an animal. When they are about three months old they begin to yield their young. In this state the insect is in a torpid condition, and may be detached from the plant. She had previously formed on her extremity an amber-coloured liquid globule, varying in size according to the abundance of juice in the cactus, and this is supposed to indicate the maturity of her pregnancy."

"It is remarkable that from the moment of her fixing upon the plant, she loses her eyes and the form of her head; instead of a mouth she has an extremely fine proboscis, which it is supposed she introduces into the imperceptible pores of the leaf she feeds on; and such is her excessive torpor, that once removed she will not attach herself again. After shedding the whole of her young, the mother dies and becomes a mere shell, turning black. It is therefore at the time that the female commences to shed her young that measures are taken to remove the young to other cactus leaves. A nest is formed, in the shape of a sausage or purse, of cotton gauze or other tissue pierced with small holes, in which 8 or 10 of

Cochineal nesting. 1473 the females are put, and the purse is fastened at the bottom of a leaf of cactus by means of a thorn. The young escape and spread themselves over the surface of the leaf. The mid-day is found to be the best time for this operation, to enable the newly-born insects to get rid of the glutinous matter which they bring from the parent. On this account nesting is not recommended in damp or cloudy days." "The common belief is that the cochineal insect lays eggs; this is not the case. The young insects, while contained within the mother, appear to be all connected one after the other by an umbilical cord to a common placenta, and in this order they are in due time brought forth as living animals, after breaking the membrane in which they were at first probably contained as eggs. Being thus brought forth, they remain in a cluster under the mother's belly for two or three days, until disengaged from the umbilical cord. Every cochineal mother produces above a hundred young ones; but the mortality is great, and three or four mothers are required to cover one side of a cactus leaf with sufficient young for cultivation."

1474

PROPAGATION.

In an interesting pamphlet written by I. S. C. D. and published by the Government, much useful information has been brought together regarding the various systems pursued in America and other countries in the propagation both of the insect and the plant. We cannot afford space to deal with this subject, and must accept the above abstract of the

The Cochineal Dye.

life-history of the insect as indicating the great governing factors with regard to the insect, and refer the reader to **Opuntia Dillenii** in another volume for the more important facts regarding the plant. The following abstract from the above pamphlet may, however, be found useful: "The proper manner of gathering varies according to the object to which the plants are devoted; but, as a general rule, the leaves on which the bags are placed are sharply cut off with a knife, close to the branches, and the cochineal is swept off them into broad baskets closely woven to prevent loss." Collection. 1475

"After the leaves are all cut off and swept, they are dropped into the ridges, *where they are left*; another set of gatherers carefully scrape off the insects which have passed into the branches or trunk of the plant, since leaving only one or two of these insects on the branches is fatal to the health of the plant." "The cactus cannot bear much water when not strengthened with manure." "When a plantation is reserved for the production of a winter crop, the leaves should be covered with cochineal in the month of October or November; by planting the young cochineal at this season it ripens, and is ready for gathering at the latter end of February or of March. Another part of the plantation is reserved for receiving the seed at this season; but as the plants cannot be forced to bud during the winter, the seed must be planted in March upon last year's leaves, which have the disadvantage of being tough for the insect, and this renders a winter crop more precarious than one obtained in summer." Wind and rain are very destructive: hence a region with a pronounced rainy season would either be unsuitable or the seed-stock at least would have to be reared and preserved throughout the rains under cover. This is, perhaps, the greatest disadvantage of cochineal as a crop for more than certain restricted parts of India,—one crop only being obtained. Propagation. 1476 Suitable Climate. 1477

After collecting the insects they are killed by immersion in boiling water or by exposure to a strong sun or the heat of a furnace. They are then dried. The different modes of killing and drying account for the characteristic peculiarities of the cochineal of certain regions. Treatment of Crop. 1478

Cochineal Dye.

DYE. 1479

Mr. Wardle, in his recent *Report on the Dyes of India*, mentions experiments performed by him with several samples. Of a Hyderabad sample he says, it "appears to be very good." "The Government report, in which reference is made to it, is by **Major W. Tweedie**." "It would be interesting to ascertain whether the cochineal is produced in the Hyderabad Residency, or is imported from South America." Of Shikarpur cochineal he writes: "This is a moderately good sample." Of a consignment from Kaladgi, Bombay, **Mr. Wardle** states: "This sample of cochineal was very small and poor, and the shades are consequently not the best obtainable by cochineal." Again, "the cochineal in lump consists of insects matted together by some dark-coloured substance. Both samples small and poor." Reference has already been made to **Dr. Dempster's** report on cochineal from the lower North-Western Himálayas. He says: "It is beyond all doubt a true **Coccus cacti**; and although it will probably turn out to be a distinct and separate species, it agrees very closely with the description given of the woodland or wild cochineal of Mexico." It may be observed that the word "true," used in the first clause of the sentence, somewhat contradicts the concluding words, and further, that the "wild cochineal" is not the Mexican insect. **Dr. Dempster** continues: "In the month of December the young brood were extremely numerous, very lively, and ready to leave the mother and spread themselves over the plant. Sulphate of alumina, added to an alkaline solution of the colouring matter of the native (*sic*) cochineal,

COCCUS cacti.

The Cochineal Dye.

threw down a copious deposit, which, when collected and dried, turned out a lake equal in beauty to the purple lakes found in Ackerman's colour-boxes." "My experiments in dyeing woollen cloth with Indian cochineal have been eminently successful, and have far exceeded my expectations. Using the formulæ employed in Europe for dyeing scarlet with Mexican cochineal, I substituted the indigenous colouring matter, and produced tints which, I think, will be pronounced equal in brilliancy to the best Europe-dyed scarlet broadcloth." "I find here an imported cochineal brought from Bombay, the price of which is quoted in a recent Bombay
1480 price current at R4½ a pound." **Dr. Dempster** goes on to say this imported cochineal is used by the Ludianah Kashmir-shawl dyers, but that the article obtained locally was superior. He adds: "The quantity of native cochineal which I found capable of dyeing a certain weight of woollen cloth proves that the indigenous insects contain an amount of colouring matter not inferior to the fine Mexican cochineal." This statement is so completely at variance with the opinions of all other European writers, that the inference is unavoidable that the imported dye with which **Dr. Dempster** compared the local article was not Mexican cochineal. The Doctor admits that the insect obtained in the Panjáb was the so-called "wild cochineal" or **Grana sylvestris**, an insect which affords only one third the amount of dye to be obtained from the Mexican or **Grana fina**. The imported cochineal experimented with by **Dr. Dempster** may even have been the Bokhara cochineal, large quantities of which find their way into the Panjáb and to Bombay. **Dr. A. Fleming's** account of Panjáb cochineal has also been incidentally alluded to, but a further passage may be here extracted from that observer's report: "All the roadsides and fields in this village are lined with magnificent specimens of the cactus, far superior to any I have seen since I left Ludianah, and their leaves are covered with the cochineal insect, which, it strikes me, attains here, probably from good feeding, a larger size than I have ever seen it do before. As I passed these hedges of the prickly pear, numerous Kashmiris were scraping the cochineal with a blunt iron instrument from the surface of the leaves into a basket such as the natives use for winnowing corn. On asking them what they were collecting this for, they told me it was to sell to the Amritsar dyers, who give them one rupee for the *angrazi* (English) *ser* (2℔) of the substance *when dry*. In order to dry it, they rub the cottony matter and the insect into balls of a soft consistence, and then dry this in the sun on a *sirky* mat. By this process the insects are squeezed, and their colouring matter absorbed by their cottony envelope." From this description there is little room for doubt that the cochineal insect seen by **Dr. Fleming** was the **Grana sylvestris.**

1481 PRINCIPLE OF THE DYE.—The colouring matter of cochineal, as of lac, is derived solely from the female insect, and is produced only at the period approaching parturition. If not collected at once, it is largely deteriorated. According to **Balfour**, "20,000 insects, dead and dried, make up one pound of cochineal, the ordinary value of which is R1 and 12 annas." **MacCulloch** (*Com. Dict.*) states that 70,000 are required for that weight. These two figures are almost alternately given by different writers—a fact which may be accounted for by the larger or smaller size of the different breeds of insects.

1482 APPLICATION OF THE DYE.—**Professor Hummel** says: "It is little used in cotton-dyeing, except by the calico-printer. Formerly much employed in silk-dyeing, it has now been almost entirely replaced by the use of various aniline reds; while in wool-dyeing, since the introduction of the azo-reds, its use has become more and more limited." "Two different shades of red are obtained from cochineal, namely, a bluish red, called *crimson*, and a yellowish or fiery red, called *scarlet.*" *Wool* mordanted with 2 per cent. of bichromate of potash and dyed in a separate bath receives a good purple, the colour being darkened by the addition of sulphuric acid to the mordant. **Mr. Hummel** gives particulars of the dyeing for crimson or scarlet. Wool to be dyed the former colour is mordanted with aluminum sulphate and tartar, the dyeing being effected in a separate bath. There are other methods, but the above is perhaps the best. Lime-salts are not beneficial. The latter shade is produced by the acid of stannous salt and cream of tartar or oxalic acid. The mordanting may be performed separately or along with the cochineal.

Wool dyeing. 1483

For *silk* the mordant is alum, to be worked into the fabric for half an hour and steeped overnight. The fabric is then washed and dried and dyed in a separate bath. This gives the crimson. For the scarlet, after boiling and washing, the silk is first grounded with a light yellow produced with soap and arnatto and thereafter washed. For darker shades soap should not be used. In both cases the fabric should be mordanted by the same process as described or the crimson, only using nitro-muriate of tin in place of alum. By the aid of iron mordants fine shades of lilac may be obtained.

Silk dyeing. 1484

In a recent report on the pigments used in the North-West Provinces the following particulars are given regarding cochineal. One part of cream of tartar to $\frac{1}{80}$ of alum and four parts of cochineal are used. These ingredients are each ground separately, after which the cochineal is macerated in water overnight, and the tartar and alum added in the morning. The mixture is then evaporated down to one fourth its original bulk, and poured on to a plate, the remaining water being allowed to dry up in the sun. The paint is then primed with 7 parts chalk, $\frac{3}{80}$ hirmizi, 1 part amber, and 2 parts linseed oil.

Pigments. 1485

(For Ammoniacal Cochineal see under paragraph of Chemistry.)

Cochineal as a Medicine.

Medicine.—Cochineal is used mainly as an agent for colouring drugs, but it is supposed by some to possess anti-spasmodic and anodyne properties.

MEDICINE. 1486

Chemical Composition.—As far as has been determined, cochineal and lac owe their tinctorial properties to an acid apparently identical in character. This is formed within the body of the female insect. The chemical examination of this substance has revealed somewhat conflicting results—a fact which has led certain writers to presume that its composition varies. **Pelletier** and **Caventon** isolated the acid from cochineal and called it carmine, a nitrogenous compound which they expressed by the formula $C_8H_{13}NO_5$. Subsequent observers (**Arppe, Warren de la Rue, Hugo Müller,** &c.) showed it to be an acid, and found that, in a perfectly pure state, it does not contain nitrogen, though accompanied by nitrogenous matter which it is difficult to separate from it. **John** named the colouring principle cochinilin. The acid of the authors named has been expressed as $C_{14}H_{14}O_8$, but the crystalline carminic acid isolated by **Dr. Schützenberger** is given as $C_9H_8O_5$, the same substance being expressed by **Dr. Schaller** as $C_9H_8O_6$. Most recent writers give its formula as $C_{17}H_{18}O_{10}$ (*Crookes*). It may be separated from cochineal by precipitating its aqueous extract with plumbic acetate and decomposing the washed precipitate with sulphuric acid. The solution thus obtained is alternately precipitated, and the precipitate decomposed, a second and a third time in a similar manner, employing, however, hydric sulphide to effect the final decomposition. The filtered solution is evaporated to dryness, the residue dissolved in alcohol, and the crystalline nodules of carminic acid

CHEMISTRY. 1487

COCCUS cacti.

Trade in Cochineal.

obtained on allowing this solution to evaporate treated with water (*Miller, Elements of Chemistry, P. III., 690*). This same substance has been found in the flowers of **Monarda didyma** and probably in other plants. Pure carminic acid is a purplish-red substance, which, when reduced to a very fine powder, is bright red. Its crystals taste decidedly acid, it is very soluble in water and alcohol, but only slightly so in ether. It may be heated to 136° without undergoing decomposition. The acid may for ordinary purposes be obtained by macerating cochineal in ether and treating the residue with successive portions of boiling alcohol, which on cooling deposits a part of the carminic acid and yields the remainder by spontaneous evaporation. The watery infusion of cochineal is of a violet-crimson colour, which is brightened by acids and deepened by alkalies. The colouring matter is readily precipitated. The salts of zinc, bismuth, and nickel produce a lilac precipitate, and those of iron a dark purple approaching to black. The salts of tin, especially the nitrate and chloride, precipitate the colouring matter of a brilliant scarlet; with alumina it forms the pigment known as *lake*. The pigment known as *carmine* is the colouring matter of cochineal thrown down by acids, salts of tin, &c., or by animal gelatine. Alum does not cause a precipitate in the aqueous solution unless some ammonia be next added, when carmine lake is thrown down. Neutral alkaline salts turn carminic acid to violet, while the acid salts of alkalies (bitartrate of potash, for example) render the shade more of an orange.

The chemical history of the carminates is, however, incomplete. The alkaline carminates are soluble; the others, as far as has been ascertained, are amorphous substances. The different results obtained with cochineal under the influence of chemical reagents is due to the presence of nitrogen, but as indicated, as a general rule, acids turn the colour to yellowish red, oxalic acid producing the best result, while the alkalies turn it to violets. When acted on by dilute sulphuric acid, carminic acid splits up into carminic red (carmine) and a glucose which is non-fermentable.

1488 AMMONIACAL COCHINEAL.—When a solution of carminic acid in ammonia is for some time left to itself, a modification takes place, the acid combines with the elements of ammonia, thereby forming an amide acid. This fact has been long known and taken advantage of to form beautiful violets, amaranthes, and mauves—colours which are not reduced to yellow-reds even by the action of oxalic acid. Chloride of tin, for example, gives with this double compound a deep violet. Ammoniacal Cochineal occurs in commerce in two forms, "cakes" and "paste."

For further particulars see **Carmine**.

1489 TRADE IN COCHINEAL.

The Madras Government exported, in September 1797, 21,744℔. From the reports of the sales of Indian Cochineal during the years 1797, 1798, and 1799, it appears that 55,196℔ were sold at an average of 8*s.* 8¼*d.* per pound, which was a little more than the prime cost. The Board of Trade in Madras reported in 1807 that during the past seven years 73,366⅛℔ had been sent to England, but that from the London price-current cochineal was not an article of profit to the Company. The Board, therefore, suggested the propriety of discontinuing the purchase or reducing the price to be paid to the producers. The home authorities, with the view of still further fostering the industry, directed the continuance of the purchase, even although the Company could only do so at a loss. Gradually, however, the industry declined, the exports were ultimately discontinued, and when we next read of cochineal it is as an article of import, not of export trade. Prior to 1875 the imports of dyeing materials were not separately returned, but in that year India imported 3,541lb valued at R5,18,410; 10 years later 2,138lb valued at R2,25,863, and last year only 1,994lb. The imports have fluctuated remarkably, but a decline may be fairly stated to have been established, a decline too which bears a remarkable inverse correspondence with the increased imports of aniline dyes. The imports of aniline dyes were in 1875-76 valued at R3,87,850; in 1885-86, R9,17,841. With the growth of this aniline trade lac-dye has been entirely destroyed. 1490 During the Colonial and Indian Exhibition, the writer was informed by a merchant that so completely had the lac-dye trade been destroyed by aniline that a large quantity of lac-dye was recently thrown into the Thames as worthless and unsaleable. (For the trade in lac-dye see a further page.)

Coccus lacca, *Kerr.* 1491

THE LAC INSECT, *Eng.*; LAQUE, *Fr.*; LACK, *Germ.*; LACCA, *It.*

Vern.—*Lakh*, HIND.; *Gálá*, BENG.; *Lákshá*, SANS.

Habitat.—This insect is indigenous to the forests of India, and occurs in aggregated masses around the twigs of certain trees, especially **Butea frondosa, Ficus religiosa,** and **Schleichera trijuga.** For a complete list of the plants on which it feeds, see below.

DESCRIPTION AND MODE OF GROWTH.—Lac is the resinous incrustration formed on the bark of the twigs, through the action of the lac insect. 1492 When the larvæ or grubs of the **Coccus lacca** escape from their eggs they crawl about in search of fresh sappy twigs. When satisfied, they become fixed and form a sort of cocoon by excreting a resinous substance. The male cocoon is ovoid in shape, the female circular. For about 2½ months the insects remain within their cocoons in the lethargic state, but structural changes have been accomplished by which they have reached the mature or imago condition. The male escapes from the cocoon by backing out at the ventral opening. The female has also become mature; but since it is destined to remain in its present position, it renews activity and commences to throw up around itself a more perfect coating of resin, until its body becomes completely encrusted. It is supposed that there are about five thousand females for one male. Upon the circular body of the female there are three openings, which become developed, as the incrnstation proceeds, into three filamentous tubes. One serves the purpose of an anal opening, and through it impregnation is also accomplished; the others are breathing stomata. When the male escapes from the cocoon, it at once commences to crawl over the females. The impregnated female, after depositing her eggs below her body, commences to construct cells round each with as much precision as the bee forms its comb.

The irritation caused by parasitic insects on vegetable tissues results in the formation of many curious and extraordinary structures, some of which are economically of great use to man, such as gall-nuts, lac, &c. In the case of the lac insect, the plants chosen are those naturally possessed of resinous principles, but still the insect exercises a peculiar influence over the resinous sap, changing its properties entirely. The **Coccus lacca** penetrates the bark of the twig by its proboscis or penetrator until it reaches the sap-wood; from there it sucks its nourishment and transforms the sap into the resinous excretion—lac—which it encrusts around itself. As time advances, further changes are visible; the body of the female enlarges considerably and becomes brilliantly coloured. The red colour is due to the formation of a substance intended as food for the offspring. The eggs germinate below, and the larvæ, eating their way through the body of the mother, make their escape to repeat this strange history.

COCCUS lacca.

Trees on which the Lac Insect feeds.

1493 TREES ON WHICH THE LAC INSECT IS REPORTED TO FEED.

1. **Acacia arabica,** *Willd.* (LEGUMINOSÆ). The *Babúl* or *Kikar* (*Gamble, 151*). "In Sind and Guzerat yields large quantities of lac."
2. **Acacia Catechu,** *Willd.* (LEGUMINOSÆ).
3. **Albizzia lucida,** *Benth.* (LEGUMINOSÆ). *Silkori,* BENG.
4. **Aleurites moluccana,** *Willd.* (EUPHORBIACEÆ). The *Akrot* of the plains, introduced from Malay, now almost wild, especially in South India.
5. **Anona squamosa,** *Linn.* (ANONANCEÆ). The *Ata,* a tree introduced from the West Indies.
6. **Butea frondosa,** *Roxb.* (LEGUMINOSÆ). The *Dhak* or *Palas.*
7. **Butea superba,** *Roxb.* (LEGUMINOSÆ). A climber, scarcely distinguishable from the tree **B. frondosa,** except by its habit.
8. **Carissa Carandas,** *Linn.* (APOCYNACEÆ). Var. **spinarum,** sp., *A. DC.*
9. **Celtis Roxburghii,** *Bedd.* (URTICACEÆ). Eastern Bengal, Central and South India.
10. **Ceratonia Siliqua,** *Linn.* (LEGUMINOSÆ). The Carob Tree; now almost naturalised in the Panjáb and South India.
11. **Croton Draco,** *Schlech.* (EUPHORBIACEÆ).
12. **Dalbergia latifolia,** *Roxb.* (LEGUMINOSÆ).
13. **Dalbergia paniculata,** *Roxb.* (LEGUMINOSÆ).
14. **Dichrostachys cinerea,** *W. & A.* (LEGUMINOSÆ). The *Virtuli,* a shrub of Central and South India.
15. **Dolichandrone Rheedii,** *Seem.* (BIGNONIACEÆ). A small tree of Burma and the Andaman Islands.
16. **Eriolæna Hookeriana,** *W. & A.* (STERCULIACEÆ).
17. **Erythrina indica,** *Linn.* (LEGUMINOSÆ).
18. **Feronia Elephantum,** *Correa* (RUTACEÆ).
19. **Ficus bengalensis,** *Linn.* (URTICACEÆ).
20. **Ficus comosa,** *Roxb.,* in Assam.
21. **Ficus cordifolia,** *Roxb.* (*Gamble, 335*); Assam Lac.
22. **Ficus elastica,** *Bl.* The India-rubber Tree (the *Bar*).
23. **Ficus glomerata,** *Roxb.*
24. **Ficus infectoria,** *Willd.* The *Pakar* or *Keol.*
25. **Ficus laccifera,** *Roxb.* (URTICACEÆ). A native of Sylhet, the *Ruthal But.*
26. **Ficus religiosa,** *Linn.* The *Aswat* or *Pipal.*
27. **Garuga pinnata,** *Roxb.* (BURSERACEÆ). The *Garuga* or *Kaikar.*
28. **Kydia calycina,** *Roxb.* (MALVACEÆ). A small tree, the *Pola.*
29. **Lagerstrœmia parviflora,** *Hook. f.* (LYTHRACEÆ). The *Bakli* or *Sida.*
30. **Mangifera indica,** *Linn.* (ANACARDIACEÆ). The Mango, in its wild state, often yields lac.
31. **Nephelium Litchii,** *Camb.* (SAPINDACEÆ). The Lichi.
32. **Ougeinia dalbergioides,** *Benth.* (LEGUMINOSÆ). The *Sandan.*
33. **Prosopis spicigera,** *Linn.* (LEGUMINOSÆ). The *Jhand* of the arid zones of the Panjáb and Guzerat.
34. **Pterocarpus Marsupium,** *Roxb.* (LEGUMINOSÆ). The *Bija* or *Kino* tree, a native of Central and South India.
35. **Pithecolobium dulce,** *Benth.* (LEGUMINOSÆ). The *Dakhini babúl,* a tree introduced from Mexico.
36. **Schima crenata,** *Korth.* (TERNSTRŒMIACEÆ). An evergreen tree of Burma.

Uses of Lac.

37. **Schleichera trijuga,** *Willd.* (SAPINDACEÆ). The *Kusum* or *Kusumb.* This is the most important of all the lac trees. It is a native of the sub-Himálaya, Central and South India, and Burma.
38. **Shorea robusta,** *Gærtn.* (DIPTEROCARPEÆ). The *Sál* Tree. The ease with which this plant coppices, and its power of endurance and rapid growth, make it one of the best trees for lac cultivation.
39. **Shorea Talura,** *Roxb.* A native of Mysore; sometimes called **S. laccifera** or **Vatica laccifera.**
40. **Tectona grandis,** *Linn.* (VERBENACEÆ). The *Teak-wood,* a native of Central and South India and Burma.
41. **Terminalia tomentosa,** *W. & A.* (COMBRETACEÆ). The *Saj, piasal, asan.*
42. **Zizyphus Jujuba,** *Lam.* (RHAMNEÆ). The *Ber* or *Kul.* Although the lac yielded by this tree is inferior in quality, the ease with which it may be propagated makes it a good lac-yielding tree, suited especially to the Panjáb.
43. **Zizyphus zylopyra,** *Willd.* (RHAMNEÆ). The *Kat-ber.*

PROPERTIES AND USES OF LAC.

After the larvæ escape, the old encrusted twigs are removed and cut
up into pieces 4 to 6 inches long. These form *stick-lac.* They are spread **Stick lac.**
upon a flat floor and a roller passed over them by which the resinous **1494**
crust is broken from off the twigs. The wood is carefully removed, and
the resin thrown into tubs of water, where it is either beaten with a wooden
pestle or trodden under foot. The liquid becomes red coloured, and one **Lac-dye.**
washing after another is performed. The washings are carefully preserved **1495**
and afterwards evaporated, when a red substance is obtained which is
made into small cakes and dried like indigo. This is *lake-lac* or the *lac-dye*
of commerce. By the washings the resin has been freed from its impurities, and now exists in a fine pulverized or granular state. This is the **Seed-lac.**
seed-lac of commerce. After drying, the seed-lac is placed in bags 10 **1496**
feet long and 3 or 4 inches in diameter. These are stretched across
charcoal fires until the lac begins to melt. The operators then commence
to twist the bags in opposite directions, holding them every now and again
over specially-prepared glazed porcelain troughs, or simply the stems of **Shell-lac.**
the plantain slit down the middle, and thus formed into smooth, glazed, **1497**
natural troughs. Through the pores of the cloth the melting lac is forced,
and dropping upon the troughs it spreads out in thin sheets. These are **Sheet-lac.**
removed and allowed to dry, any impurities being broken out of the thin **1498**
flakes. When packed in bags these become the thin pieces known in
commerce as *shell-lac* or *shellac.* Sometimes, and especially with the
coarser qualities used for home consumption, the melted lac is let drop into
rounded pieces about 1 to 1½ inches in diameter. These constitute *button-* **Button-lac.**
lac, and if formed into larger masses, *sheet* or *piece lac.* **1499**

QUALITIES AND PRICES OF LAC.—The quality of lac varies chiefly ac- **D. C.**
cording to the tree upon which the insect feeds. The best lac is *kusum* lac, **1500**
or lac from the *kusum* tree (**Schleichera trijuga**), next *dhak* or *palas* (the **Liver.** **1501**
Butea frondosa), then *pipal* (**Ficus religiosa**). The *kusum* is said to last for **Native**
ten years, while all other qualities are only good for two or three years. The **Orange.**
kusum lac twigs are of a light golden colour, from which orange shell-lac is **1502**
manufactured. This sells in London for about £10-12 per cwt. of best **Garnet.**
quality, *i.e.,* "fine lac orange D.C." The other qualities are known as **1503**
"liver" or "native orange," maximum price £8-9 per cwt.; "garnet" **Native-leaf.**
£7-8 maximum; "native leaf" and "button" being generally about **1504**
£3-6 or £3-8 per cwt. The best lac comes from Siam.

ADULTERATION OF LAC.—Lac is frequently adulterated with orpiment, **Adulterated**
or still more frequently with common resin, which may be detected by its **Lac.** **1505**

smell on crushing the lac. The writer was once informed by a merchant that his firm in the usual course of business imported very largely resin which he believed was used up by the native dealers in adulterating the lac which they and other merchants exported. The gentleman in question condemned strongly the process of adulteration, but justly remarked that resin was an ordinary article of trade used for other purposes which if *they* discontinued to import would only be more largely imported by *other* firms.

Varnish. 1506 Batti. 1507 Sealing-wax. 1508 Cement. 1509

USES OF LAC.—In India lac is dissolved in native spirits and coloured; in this form it is used as a varnish for carpentry and furniture; mixed with sulphur and some colouring agent, it is formed into the sticks, *batti*, like sealing wax, which are used by the toy-makers to coat their wooden wares. In Europe it is largely made into *sealing wax* and dissolved in spirits, it forms *spirit varnish*. It is made into *cement* and into lithographer's ink, and is used to stiffen hats and other articles constructed of felt.

Dye. 1510

LAC DYE.

Having now indicated the main features of the lac industry collectively, the present article may be concluded by dealing in greater detail with the subject of the dye extracted from **Coccus lacca.** The reader is referred for further particulars regarding the Europeon industry and trade in the Resin to the article **LAC.**

The natives of India from remote times have used lac-dye not only for textile purposes but as a pigment. It is by them largely used for colouring leather and in wool and silk dyeing, although aniline has affected the demand very seriously. The colour is not so bright as that derived from cochineal, but it is more intense, and has the reputation of being less easily affected by perspiration. **Mr. J. E. O'Conor** has written a special pamphlet on Lac, which contains very nearly all that is known up to date. **Dr. McCann** (*Dyes and Tans of Bengal, 49-66*) gives details regarding the dye in Bengal, and **Sir E. C. Buck** (*Dyes and Tans, N.-W. P., 24*) furnishes information regarding its use in the North-West Provinces. Owing to the existence of the resinous matter mechanically mixed with the dye, lac is not so easily worked as cochineal. All the reactions and processes we have already discussed under **Cochineal** are, however, applicable with slight modifications to this colouring agent. Taking advantage of the properties of both dyes, lac is often combined with cochineal.

1511 The physical tests for lac-dye are that the cakes should be tolerably easily broken by the fingers, the fracture uniformly exhibiting a deep colour; that they should not have a shining resinous appearance, and that they should evolve a pronounced and peculiar odour. The harder the cakes the larger presumably is the amount of shellac in them. The removal of the shellac is essentially necessary with woollen goods that require to be hot-pressed, since if shellac be present the paper will adhere.

Dr. McCann treats of the decline of the lac-dye industry. Practically speaking, it will not now pay to boil down the coloured washings obtained as a by-product in the shellac industry. Although still used to some extent in India, the article is scarcely, if at all, exported.

COCHLOSPERMUM, *Kunth.; Gen. Pl., I., 124, 971.*

1512 **Cochlospermum Gossypium,** *DC.; Fl. Br. Ind., I., 189;* BIXINEÆ.

SOMETIMES CALLED WHITE SILK-COTTON TREE.]

Syn.—BOMBAX GOSSYPIUM, *Linn.; Roxb., Fl. Ind., Ed., C. B. C., 515.*

Vern.—*Kúmbi, gabdi, ganiár, galgal, gangal,* HIND.; *Hopo,* SANTALI; *Gulgal,* KOL.; *Gangàm,* GOND; *Kontopalás,* URIYA; *Kúmbi,* PB.; *Gajra, kúmbi,* N.-W. P.; *Gúngú, kong, gondu-gogu,* TEL.; *Tanaku, kongillam,* TAM.; *Betta tovare, arisina burga,* KAN.; *Chima-púnji,* MAL.; *Ganeri,* BHIL; *Ganeri, gunglay, kathalyá gonda,* MAR.; *Katíra-i-Hindí,* PERS. and HIND. **Irvine** gives the plant the name of *narfaris.*

For the Gum.—**Moodeen Sheriff** gives the following: *Nát-ká-katérá, nát-ká-katérá-gónd,* DEC.; *Hindi-katérá,* HIND.; *Tanaku-pishin,* TAM.; *Konda-gógu-banka, konda-gogu-pisunu,* TEL.; *Shima-pangi-pasha,* MAL.

For the Cotton.—*Pili-kapás-ki-rúi, katéré-ké-jhár-ki-rúi,* DEC.; *Tanaku-parutti,* TAM.; *Konda-gógu-patti,* TEL.; *Shima-pangi-parutti,* MAL.

References.—*Brandis, For. Fl. 17; Gamble, Man. Timb., 17; Dymock, Mat. Med. West Ind., 2nd Ed., 73; Moodeen Sheriff, Mat. Med. S. Ind. 35; S. Arjun, Bomb. Drugs, 14; Pharm. Ind., 27; O'Shaughnessy, Beng. Disp., 225; Lisboa, U. P. Bomb., 6, 250; Stewart, Pb. Pl., 18, Baden-Powell, Pb. Prod., 329, 397; Atkinson, Econ. Prod., N.-W. P; Part I., 18; also Him. Dist., 733, 783; Cooke, Gums and Gum-resins, &c., 29, Drury, U. P., 146; Murray, Pl. and Drugs, Sind, 47; Forest Ad. Rep., Chutiá Nágpur, 1885, p. 28.*

Habitat.—A small deciduous tree, with short, thick, spreading branches; grows in forests at the base of the North-Western Himálaya, from the Sutlej eastward to Central India, Bundelkund, Behar, Orissa, and the Deccan; also in the Prome district of Burma. Commonly planted near temples. When the tree is devoid of leaves (in March to April) it bursts into its handsome large yellow flowers, its pendulous, pear-shaped fruits ripening before the new leaves appear.

GUM. 1513

Gum.—This is often sold in the bazaars of India as *katíra* or *kathíra* (the Persian and Arabic for Tragacanth), that name having been given to the gum of this tree by the early Muhammadan settlers in India. There are two gums usually sold as substitutes for the true *katíra,* namely, the gum of **Cochlospermum** and the gum of **Sterculia urens.** Both these gums belong, like tragacanth, to the same series, namely, the pseudo-gums. They are insoluble in water, but swell and form a pasty mass. The writer has already dealt with these gums at some detail in VOL. I, B. No. 283, under their generic name Bassora or Hog-gums. In that article it will be found that the hope has been held out that the gum from **Cochlospermum Gossypium** may prove to be the gum found, by the bookbinders of America, so useful in marbling paper and colouring the edges of books. The gum is obtained freely on the trees being tapped. It occurs in striated and twisted pieces, of a pale semi-transparent colour, transversely fissured with a tendency to split up into flat scales. While not soluble it readily becomes diffused in minute particles through a large quantity of water; hence its applicability to the art of marbling paper. The writer has found a mixture of $\frac{1}{3}$ of this gum with ordinary gum-arabic to afford a thick paste-like gum that sets readily, is not liable to crack, and is very adhesive, the spurious tragacanth apparently acting as a medium for the other gum. It is neutral and yields with alkalies a thick mucilage of a pinkish colour which, according to **Mitchell,** is not precipitated by acids. The gum is largely used by the Indian shoemakers, and, like that from **Sterculia,** could doubtless be employed to impart a polish to tasar silk.

Stewart remarks: "The *kátíra,* of which 10 maunds are stated by *Davies' Trade Report* to be imported annually *viá* Peshawar, must be entered by mistake, or be the product of a different plant." (Doubtless the true *katíra* or Tragacanth.—*Ed.*) "And, oddly enough, the same authority gives 50 maunds of this substance as exported from Ludhiána

GUM. to Afghánistan by the Bolán." (Perhaps the Indian *katíra* is exported to be used as an adulterant with tragacanth.—*Ed.*) **Moodeen Sheriff** remarks there are three forms of this gum—*white*, *red* or *brown*, and *black*. The writer has never seen the tree yield any but a pinkish white or often almost straw-coloured gum, and suspects the black to be an adulterant or substitute. **Dr. Moodeen Sheriff**, however, gives the prices of these three gums: "Best or white variety of the gum, wholesale, R5 per maund; retail or bazaar, 4 annas per pound: of the second best or red variety, wholesale, R4 per maund; retail or bazaar, 3 annas per pound: of the worst or black variety, wholesale, R3 per maund; retail or bazaar, 2 annas per pound."

FIBRE. Floss. 1514 **Fibre.**—The seeds possess a short but very soft and elastic floss, from which fact the plant has received its specific name. This floss is much too short to be of any service as a textile fibre, but, with the flosses of **Bombax malabaricum, Eriodendron anfractuosum,** and **Calotropis gigantea,** it has been classed as a "silk cotton." By some writers these have recently been designated "kapok fibres," but there is every reason to believe that the true kapok of the Dutch upholsterers is the floss of **Eriodendron anfractuosum** (*see Vol. I.*, **B. 641**). In some parts of India the floss of this tree is collected and used for stuffing pillows, for which purpose it would seem better suited than the floss from **Bombax malabaricum,** as it is not so liable to get matted. It might be found serviceable as a gun-cotton. (Conf. with **C. 175** and **Kapok** in a further volume.)

Bark. 1515 The **Rev. A. Campbell** states that the Santals prepare a good, useful cordage fibre from the bark of the tree. In the report of the Conference held on Indian fibres, at the late Colonial and Indian Exhibition, it is stated that **Mr. Campbell's** fibres from this tree were much admired, the floss being viewed as possessing the merit of elasticity—a merit which might allow of its competing favourably with the true kapok.

OIL. 1516 **Oil.**—The **Rev. A. Campbell**, Santal Mission, Chutiá Nágpur, describes a bright red oil which by hot expression he extracted in abundance from the seeds. He adds, although this property of the seeds is well known to the Santals, they never extract the oil. **Cooke** in his *Oil and Oil-seeds* alludes to this circumstance, but remarks that beyond the fact of the seeds affording an oil, nothing further is known. Samples of this oil were shown at the late Colonial and Indian Exhibition and these are now deposited in the Kew Museum. Were a use to be found for the oil it could be produced in large quantities and afford employment to the inhabitants of the dry hilly tracts where this tree abounds, at a season during which they have little to do. It could therefore be produced cheap.

MEDICINE. Gum. 1517 **Medicine.**—The gum has the properties in a mild degree of Tragacanth, for which it is proposed by **Moodeen Sheriff** and others as a substitute. It is also used as a mild demulcent in coughs.

Floss. 1518 The floss has been recommended as admirably suited for padding bandages, splints, &c., being soft and cool. On this account it has been suggested as suitable for pillows and cushions used in hospitals, &c. **Irvine** (*Mat. Med., Patna, p. 78*) says the dried leaves and flowers are used as stimulants.

TIMBER. 1519 **Structure of the Wood.**—Extremely soft, grey, but has no heart-wood, and is not apparently put to any useful purpose; weight 17lb per cubic foot.

Cockles, see **Molluscs** (edible).

Coco or **Cocoa,** see **Cocos nucifera**; **Coca,** see **Erythroxylon** and **Cocoa Nibs,** see **Theobroma.**

COCOS, *Linn.; Gen. Pl., III., 945.*

Cocos nucifera, *Linn.; Brandis, For. Fl., 556;* PALMÆ. 1520

THE COCOA-NUT PALM; THE COIR OF COCOA-NUT FIBRE; PORCUPINE WOOD; COCOSER, *Fr.;* COCOSNUSS, KAIR, *Germ.*

Vern.—*Nárel, náriyal, náriel, náríyel, náriyal-ka-pér,* HIND.; *Nárikel, náriyal, dáb, nárakel,* BENG.; *Nariel, naríyéla, nárıera, náliyer, nár-yal, jháda, náryal,* GUJ.; *Maar, naril, mahad, narel, naral-cha-jhádá, már, naural,* BOMB.; *Narela, nárula, náralmád, mád, máda, mahad, várala, narel, nárali-cha-jháda, naral, mar, tenginmar* (the juice-yielding form in Kanara), MAR.; *Nárél-ká-jhár, nárél,* DUK.; *Tenna, tenga, tennan-chedi, tenna-maram, téngáy, taynga,* TAM.; *Nari kadam, ten-káia, kobbari, goburrí-koya, ten kaya, kobri chullú, kobbari chottu, ten-káya-chottu, erra-bondula, gujju-narekadam,* TEL.; *Thenpinna, kin-ghenna, tengina, tenginá-gidá, tenginá-káyi, tengina chippu, tenginav amne, tenginararu,* KAN.; *Tenga, tenn-marum, tennd, nur, kalapa, nyor, kalambir,* MALA.; *Nur,* MYSORE; *Nari-kela, nári-kera, nári-keli, langalin,* SANS.; *Jadhirdah, shajratun nárjíl, shajratul-jouze-hindi, nárjíl, jouze-hindi,* ARAB.; *Darakhte-nárgíl, darakhte-bándinj, nárgil, badinj* (*narjible* in Ainslie), PERS.; *Pol, pol-gass, pol-gahá, pol-nawasi, tambili,* SING.; *Ong, ung, ung-bin, ón, onsí, ontí, ondi,* BURM.; *Kalapa,* JAVA.

DRY KERNEL, COPRA (KOPRA) OF COPPERAH—

Khóprá, HIND.; *Khópru,* GUJ.; *Khóprá, khópré-ki-batti,* DUK.; *Kobbarait-téngáy,* TAM.; *Kobbera, kobbera-ténkáya,* TEL.; *Kóppara,* MALA.; *Ko-bari, kobbari,* KAN.

OIL, COCOA-NUT OIL—

Khópáre-ká-tél, naríyal-ká-tél, naril-ká-tél, HIND., DUK.; *Nárikél-tail, náriyal-tél,* BENG.; *Náryal-nu-tél,* GUJ.; *Náralicha-téla, naral-tela, kóbra cha-téla,* MAR.; *Téngá-yenney, taynga-nunay, tengái-yenne,* TAM.; *Tenkáya-núne, tenkáia núnay,* TEL.; *Tenna-enna, minak, kalapu, minak-nur, nur-minak, kalambir, kalapa minak,* MALA.; *Tenginá-yanne, cobri,* KAN.; *Nárikela-tailam,* SANS.; *Dhonun-narjíl, dhonul-jouze-hindí* (*jowz-hind* in Ainslie), ARAB.; *Róghane-nárgíl, róghane-bándinj,* PERS.; *Pol-tel,* SING.; *Ón-sí,* BURM.; *Cay-dua,* COCHIN-CHINESE.

WATER—

Yelnir-ka-pani, DUK.; *Yella nir,* TAM.; *Yella-niru,* TEL.

TODDY—

Nárélí, HIND.; *Nárél-ki-séndí, narillie,* DUK.; *Téngá-kallu, tennan-kallu, tennang-kallú,* TAM.; *Tenkáya-kallu, tenkala,* TEL.; *Nargilie, nargilli,* ARAB.; *Táríye-nárgíl,* PERS.

FIBRE—

Coir ? (See first paragraph of chapter on Coir), HIND.; *Tennam nar,* TAM.; *Tenkaia nar,* TEL.

COCOA-NUT CABBAGE—

Tennam kurtu, TAM.; *Tenkaia gurtu,* TEL.; *Naril-ka-krute,* ARAB.

COTTON OF TOMENTUM—

Tenna maruttú pungie, TAM.; *Tenkeia-chettú-puthie,* TEL.; *Tennam-púppa,* MAL.

References.—*Roxb., Fl. Ind., Ed. C. B. C., 664; Kurz, For. Fl. Burm., II., 540; Gamble, Man. Timb., 422; Thwaites, Enum. Ceylon Pl., 330; Dalz. & Gibs., Bomb. Fl., 279; D.C., Origin of Cult. Pl., 429; Voigt, Hort. Sub. Cal., 643; Pharm. Ind., 247; Moodeen Sheriff, Supp. Pharm. Ind., 112; Flück. & Hanb., Pharmacog., 721; U. S. Dispens., 15th Ed., 1616; U. C. Dutt, Mat. Med. Hind., 247, 311; Dymock, Mat. Med. W. Ind., 652;* also *2nd Ed., 800; Ainslie, Mat. Ind., I., 77, 451; II., 415, 418, 419; O'Shaughnessy, Beng. Dispens., 642; Bidie, Cat. Raw Prod. Paris Exb., 46, 63, 75, 79, 93, 95, 106, 124; S. Arjun, Bomb. Drugs, 195; Drury, U. Pl., 146, & 469; Baden Powell, Pb. Prod., 511; Royle, Fibrous Pl., 102; Royle, Ill. Him. Bot., I., 395, 398; Liotard, Memo. on*

COCOS nucifera.

The Cocoa-nut Palm.

Paper-making, 5, 14-16, 30; Treloar on the Prince of Palms; Cooke, Gums & Gum-resins, 14; Cooke, Oils and Oil-seeds, 11; Crookes, Dyeing, 510; Birdwood, Bomb. Pr., 183, 214, 249, 290, 319, 338; Lisboa, U. Pl. Bomb., 136, 180, 212, 237; Spons, Encyclop., 9, 39, 1353, 1383, 1621, 1639; Balfour, Cyclop.; Smith, Dic., 123; Treasury of Botany; Ure, Dic. Ind. Arts and Manu., 841; Kew Official Guide to the Museum of Economic Botany, 23; Med. Trop. Ajmere, 141; Shortt, Monograph on the Cocoa-nut, 1885; Jackson, Treatment of Cocoa-nut in the Planters' Gazette; Trop. Agri., 1882-83, pp. 308, 568, 842, 904, 933; Simmonds, Trop. Agri., 220-244; Mysore, Cat Cal. Exb., 42; Journ. Agri.-Hort. Soc. Ind., 1843, p. 253; Bombay Gazetteers, Vols. VIII., 95; X., 34-35; XI., 27-30; XIII., I., 295; XV., I., 58-60; XVIII., 49; XXII., 303; N.-W. P. Gaz., VI., 147, 429, 703, 763, 769; Mysore Gaz., I., 131-134; II., 294; Burma Gaz., I., 131; Imperial Gaz., Vols. I., 282; VII., 372, 380; VIII., 394; IX., 230; X., 296; Administration Reports of the Andaman and Nicobar Islands; Morris, Godavery District, Madras, p. 70; Selections, Records, Madras Gov., XXIII., 13, 131; Cleghorn, Edinb. New Phil. Journal, 1861; Marshall, Natural and Economic History of the Cocoa-nut, 1832; Robinson, Report of the Laccadive Islands; Bennett's Wanderings, II., 295; Rumphius in his *Herb. Amboin., I., pp. 1-25,* gives a long account of the use of the cocoa-nut, under the name *Palma indica major; Roxb., Corom., I., 52, t. LXXIII.; Avicen. Relat., III., 2; also Hist. Rei Herbariæ, pp. 268, 269; Loureiro, Flora Cochin-chinensis, II., 566.*

Habitat.—A pinnate-leaved palm, with a straight or often gracefully curved stem, marked by annular scars; cultivated throughout tropical India and Burma, especially near the sea-coast. On the eastern and western coasts it is particularly abundant, more so towards the south. There are several cultivated varieties but all flower in the hot season, the nuts ripen ing from September to November. **Dr. Shortt** states that in South India the palm thrives at altitudes up to 3,000 feet above the sea, and he even mentions one on the Shevaroy Hills at 4,500 feet. Cocoa-nuts are abundant in Bangalore up to 3,000 feet.

Indian Region. 1521 Starting from the Bay of Bengal, the cocoa-nut palm follows the Gangetic basin inland for about 150 to 200 miles; from the western coast its cultivated distribution inland is much more limited, and in Kolaba, for example, is little more than half a mile from the beach. In very exceptional circumstances, or under the most careful garden cultivation, it may be seen further inland in Bengal than stated, and it even occurs in some parts of Assam. It is, however, essentially a plant of the coast, and luxuriates on the islands of the Indian Ocean. The Indian region of the cocoa-nut may thus be said to be the lower basins of the Ganges and the Brahmaputra, and the Malabar and Coromandel coasts. In the Brahmaputra valley it ascends to a greater distance from the sea than in the Gangetic; but in both it is an introduced tree, as it nowhere occurs in forests far away from human dwellings. On the Malabar coast, and on the islands off the coast of India, it may be different; but even in these localities it rarely exists as a forest tree, although it is self-sown. It is abundant on the Laccadive Islands, and on the Nicobar group in the Bay of Bengal, but excepting the recent efforts at cultivation, it was formerly rarely met with on the Andaman Islands, which are only 72 miles to the north. It re-appears again, however, abundantly on the Cocos Islands, a small group lying some 30 to 40 miles still further north (where it is in no way cultivated). **M. DeCandolle** states briefly the arguments in favour of an American as well as those of an Asiatic origin for this tree, and concludes by expressing the opinion that it most probably belongs to the "Indian Archipelago." Its introduction into Ceylon, India, and China, he states, does not date further back than three thousand years, "but the transport by sea to the coasts of America and Africa took place perhaps in a more remote epoch, although posterior to those epochs when the geographical and physical conditions were different from those of our day."

CULTIVATION OF THE COCOA-NUT.

CULTIVATION.

It is commonly reported that there are in India 480,000 acres under the cocoa-nut. A number of passages from Indian authors will be found scattered through the present account of the palm, which every now and again recur to the question of its cultivation. It may, however, be desirable to give here a brief abstract of the opinions published by the better known European writers, since from these may be gathered the results of scientific experiments.

Sowing. 1522

SOWING.—Ripe nuts, carefully collected, should alone be employed as seed, and for this purpose they are usually gathered from February to May. Seed from very young or very old trees should be avoided. After having been kept for a month to six weeks they should be planted. This may take place in January to April, or again in August, provided the rains are not heavy. The seed-beds should be dug 2 feet deep and the nuts planted 1 foot apart. The nuts should be laid on their sides, leaving 2 inches of their surface exposed. Ashes, or ashes and salt, should be freely placed in the trenches; these act both as a manure and as a preventative against insects. The seed-bed thus prepared should be kept moist, but not soaked. The germinated seeds may be transplanted when they are in their second to their sixth or even twelfth month. In the Godavari district they are placed in their permanent positions when three to four years old. In damp localities the transplanting may be done in the hot season, otherwise during the rains.

Transplanting. 1523

TRANSPLANTING.—The seedlings should now be put out in the plantation, pits 12 yards apart having been prepared for them. In rich soils the pits may be small, but in poor soils 1 to 2 yards wide and 2 or 3 feet deep. In cold clay soils these pits should be filled with sand. In marshy land, walls should be constructed around them. Ashes are often recommended to be freely mixed with the prepared soil to be put into the pits, as this is supposed to prevent the attacks of the beetles that prove so destructive to the trees. Cultivation of turmeric, arrowroot, &c., in the pits, along with the cocoa-nuts is believed to be beneficial. The soil round the seedlings is also often kept damp by a bed of leaves, particularly such as will not encourage, but rather check, the approach of ants into the prepared soil. If the soil be naturally poor, salt, ashes, paddy-straw, fish manure, goats' dung, and dry manures may be added during the first year.

Treatment of Plantation. 1524

TREATMENT OF PLANTATION.—By the end of the first year the normal leaves will begin to form, and at this stage the soil around the plants should be dressed and ashes added. Every succeeding year the ground should be opened out and manured about the commencement of the rains, the soil being replaced and levelled about the close of the rains. By the fourth year the stem begins to appear and has about 12 leaves; it is distinctly visible by the fifth year, when the tree has about 24 leaves. The spathes commence to be formed by the sixth year, and the stem is then 1 to 2 feet above the ground, but in exceptionally favourable climates and soils it may be three or four times that height. The first few spathes do not form fruits, but bye-and-bye they begin to do so, and in three or four more years the tree is in full bearing. **Dr. Shortt** says that in good soils and if watered the cocoa-nut begins to yield in the fifth year, but in poor soils and if not watered they only commence to yield in the seventh or not till the tenth year. About six months after flowering the fruits set, and by the end of the year they are fully ripe.

Cocoa-nut palms may be easily transplanted, and indeed often with advantage. Some of the fibrous roots should be cut away, and manure,

The Cocoa-nut Palm.

CULTIVATION.

together with a little salt, placed in the pit in which it is intended to plant the tree.

Yield. 1525

YIELD.—As a rule a cocoa-nut throws out a spathe and a leaf every month: each flowering spike yields from 10 to 25 nuts. The produce of a tree in full health and properly tended may be from 50 to 120 and even 200 nuts a year, the yield depending greatly, of course, on the suitability of the climate and soil for cocoa-nut cultivation; a safe average would be 100 nuts a year to each tree in full bearing. The cocoa-nut will continue to bear for 70 to 80 years.

1526

CULTIVATED FORMS.

There are five recognisable varieties of the cocoa-nut met with in Ceylon. These have been described as, *1st*, the *Tembili*, a plant with an oval-shaped nut of a bright orange colour; *2nd*, a more spherical form; *3rd*, a heart-shaped fruit of a pale yellow colour, with an edible inner rind, which turns red when the outer skin is removed; *4th*, the ordinary form; *5th* a small nut about the size of a turkey's egg. This last form is rare but much admired. **Spon** (*Encycl.*, 1353) says "there are some 30 varieties of cocoa-nut distinguished by the natives of the districts producing them, but many of these distinctions are obviously groundless." Repeated reference will be found throughout this article to the different forms which occur in India, but of these, with perhaps the exception of that met with in the Laccadives, scarcely any deserve special mention. The Laccadive small-fruited form, with a soft, fine, but strong coir, seems well worthy of special consideration where the object of cultivation is the production of fibre. **Dr. Shortt** says there are 30 different forms in Travancore. He adds: "The largest variety of cocoa-nut that I have seen and examined comes from Ceylon. I have occasionally seen specimens nearly as large from the Coromandel coast. There is a small dwarf variety which fruits while

Dwarf Cocoanut. 1527

it is about 2 feet high; the plant continues to grow and with age attains to a height of from 10 to 15 feet." A small form is met with in East Africa that does not possess the fibrous pericarp—(see concluding sentence of chapter on medicinal properties, page 448). In Indian newspapers announcements of branched cocoa-nuts occasionally appear, as also of branched date-palms. These are viewed with superstitious horror by the ignorant. They are most probably the result of two plants growing together, or of two or more embryos in one nut.

SOIL.—The cocoa-nut "thrives best in low, sandy situations, within

Soil. 1528

the influence of the sea breeze, and never attains the same perfection when grown inland." (*Spons' Encycl.*) **Simmonds** writes: "Soils suitable for a cocoa-nut plantation are variously described as below, particularly observing that stony grounds, or those overlying rocky foundations, are to be avoided:—

"1. Soils mixed with sand, either dark-coloured or river-washed.
"2. Where sand is mixed with clay, ferruginous earth, or black mould.
"3. Clayey soils where the under-strata consist of sand.
"4. Sand and clay, even when mixed with gravel and pebbles.
"5. The sea-shore banks of backwaters, rivers, tanks, and paddy-fields.
"6. Alluvium of rivers and backwaters, provided a yard and a half of land is to be generally seen above water-level.
"7. Marshy land even in brackish soils (but not where salt is formed in crystals by evaporation).
"8. All level lands exposed to the sea breeze where the soil is good, as the valleys between hills, tanks, and ditches which have been filled up.
"9. Lastly, even the floors of ruined houses well worked up, and any places much frequented by cattle and human beings on account

CULTIVATION.

of the ashes and salts of ammonia from the urine, &c., deposited day by day in the soil."

Simmonds further says: "The nuts for seed should not, on being gathered, be allowed to fall to the earth, but be lowered in a basket or fastened to a rope. If let fall, the polished cover to the fibres will be injured and collect damp about the nut, or the shell inside may be cracked and the water disturbed. These are fatal injuries, or even if the plants still grow, they will, on being transplanted, not make fresh shoots, but produce weak trees having their fronds constantly drying up nuts rarely matured, and often are even without kernel in those which appear perfect. If the nuts are allowed to dry on the tree before gathering, the plants are liable to be lost, not having water inside to cherish the growth of the sprout (before the actual roots shoot into the soil)."

"Nurseries should be somewhat exposed to the influence of the sun, though not too much heat: plants thus grown will even, though deficient in stature, be strong, and when transplanted will not fail, nor suffer from heat. The planting of the nuts should take place in January to April, and also in August, provided the rains are not heavy, and then the planter may expect fruitful trees to be produced when grown; but nurseries formed during the heavy monsoon will generally fail, or produce trees which will yield small nuts. Too much moisture of every kind is injurious to the plants." Speaking of soils **Dr. Shortt** says: "The cocoa-nut requires alluvial and loamy soil for its successful growth, but any soil with a free mixture of sand and clay answers fairly well. Sea-sand where procurable is recommended to be thrown into the pits when the earth is being returned around the plants. Half sand half earth is considered the best material to fill up the pits with."

Peculiarities of Indian Cultivation.

The following passages from the Gazetteers will be found instructive and of value to intending cultivators as having a special bearing on India.

I. *In Bombay* (*Kolaba District*).—Of the liquor-yielding trees of this district the cocoa-palm is the most important. "The moist climate, sandy soil, brackish water, and abundance of fish manure, make its growth s vigorous that the yield of juice is much in excess of the wants of the district. The trees are grown within walled-in or hedged enclosures, sometimes entirely given to cocoa-nut palms, in other cases partly planted with mangoes, jack, betel-nut, and other fruit trees. Every garden has one or two wells, from which the trees are watered by a Persian wheel. In starting a cocoa-nut garden, a bed is prepared, and in it, at the beginning of the rainy season, from twenty to forty large, ripe, unhusked nuts are planted 2 feet deep. The bed is kept soaked with water, and after from three to six months the nut begins to sprout. The seedlings are left undisturbed for two years. They are then, at the beginning of the rains, planted in sandy soil in rows about 18 feet apart, and with a distance of about 15 feet between the plants. For about a foot and a half round each plant the ground is hollowed 3 or 4 inches deep, and during the dry months the plants are watered daily or once in two days, and, once or twice in the year, enriched with fish manure or with a mixture of salt and *náchni*. When nine years old the trees begin to yield nuts twice a year and sometimes thrice, 120 nuts being the yearly average yield from each tree. The trees are then ready to be tapped. Each cocoa-palm, when ready for tapping, is estimated to represent an average outlay of about 18*s*. (R9).

I. Bombay. 1529

"The cocoa-nut gardens are generally owned by high-caste Hindus, who let the trees to some rich Bhandari who has agreed to supply the owner

The Cocoa-nut Palm.

CULTIVATION.

of the liquor-shops with fermented or distilled juice. The Bhandári pays the owner of the garden R1 (2 shillings) a month for every three trees" (*Kolaba Dist., Bomb. Gaz., XI., 28*). Of the Thana District it is stated—"The seed-nuts are prepared in different ways. The best and oldest tree in the garden is set apart for growing seed-nuts. The nuts take from seven to twelve months to dry on the tree. When dry they are taken down, generally in April or May, or left to drop. When taken down they are either kept in the house for two to three months to let half of the water in the nut dry, or, if the fibrous outer shell is not dry, they are laid on the house-roof or tied to a tree to dry. After the nuts are dry, they are sometimes thrown into a well and left there for three months, when they sprout. If the nuts are left to drop from the tree, which is the usual practice in Bassein, they are either kept in the house for some time and then left to sprout in a well, or they are buried immediately after they have fallen. When the nuts are ready for planting they are buried either entirely or from one half to two thirds in sweet land, generally from 1 to 2 feet apart, and sometimes as close as 9 inches. A little grass, rice-straw, or dry plantain leaves are spread over the nuts to shade them. If white-ants get at the nuts the grass is taken away, and some salt or saltish mud mixed with wood ashes and a second layer of earth is laid over the nuts. Nuts are sometimes planted as late as August (*Shrávan*), but the regular season is from March to May (*Chaitra* and *Vaishákh*), when, unless the ground is damp and their inner moisture is enough for their nourishment, the nuts want watering every second or third day until rain falls. The nuts begin to sprout from four to six months after they are planted, and when the seedlings are a year or eighteen months, or, what is better, two years old, they are fit for planting. At Bassein the price of seedlings varies from 5*d*. (3 annas 4 pie) for a one or one and a half year old seedling, to 6*d*. (4 annas) for a two-year-old plant. In planting them out the seedlings are set about six yards (12 *háts*) apart in the 2-feet-deep holes, in which about 1¼ pounds (2 *tipris*) of wood-ashes have been laid to keep off white-ants, and the garden must be very carefully fenced to keep off cattle. The plants are then watered every second day, if not every day, for the first year; every third day, if not every second day, for the second and third year; and every third day, if possible, for the fourth and fifth year. Watering is then generally stopped, though some Bassein gardeners go on watering grown trees every seventh or eighth day. For two years after they are planted out the young trees are shaded by palm leaves or by growing *mutheli* plantains. During the rains, from its fifth to its tenth year, a ditch is dug round the palm and its roots cut, and little sandbanks are raised round the tree to keep the rain-water from running off. In the ditch round the tree, 22 pounds (4 *páylis*) of powdered dry fish manure (*kuta*) is sprinkled and covered with earth, and watered if there is no rain at the time. Besides fish manure the palms get salt-mud (*khára*
1530 *chikhal*) covered with the leaves of the croton-oil plant, *jepál erand* (Croton Tiglium), and after five or six days with a layer of earth; or they get a mixture of cow-dung and wood-ashes covered with earth; or night-soil, which on the whole is the best manure. Palms suffer from an insect named *bhonga* which gnaws the roots of the tree, and from the large black carpenter-bee which bores the spikes of its half-opened leaves. When a palm is suffering from the attacks of the *bhonga*, a dark red juice oozes from the trunk. When this is noticed, a hole 3 inches square is cut in the trunk from 4 to 6 feet above where the juice is coming out, and is filled with salt, which drives away or kills the insect. To get rid of the boring bee, it is either drawn out by the hand, or it is killed by pouring into the spike assafœtida water or salt-water.

The Cocoa-nut Palm.

COCOS nucifera.

CULTIVATION.

"A well-watered and manured tree, in good soil, begins to yield when it is five years old, and in bad soil when it is eight or ten years old. A palm varies in height from 50 to 100 feet, and is in greatest vigour between the ages of twenty and forty. It continues to yield till it is eighty, and lives to be a hundred.

"When the tree begins to yield, a sprout comes out called *poi* or *pogi*, at the bottom of which is a strong web-like substance called *pisundri*. After about a fortnight the tree flowers, though few blossoms come to perfection. Many of the young nuts also fall off, and only a few reach maturity. A young nut is called *bonda*, a nut with a newly-formed kernel is called *shále*, and a fully-formed nut *nárel*. A good tree yields three or four times a year, the average number of nuts being about seventy-five" (*Gaz., XIII., I., 295*).

In the report of the Káthiawár District (*Bomb. Gaz., VIII., p. 95*), there occurs a short but interesting account of the cocoa-nut: "At Mahuva, in 1875, 1,500 acres were planted with 170,000 palms. At Khandera there is a garden with 7,000 palms, and there are about 2,000 at Bhávnagar. The advantage of the cocoa-nut over the mango is the uniformity with which it bears." "A singular fact about the cocoa-palm is that it grows freely in solid limestone, provided a hole about $3\frac{1}{2}$ feet deep by 3 feet in diameter is cut in the rock and filled with mould. All the trees at Gopnáth are planted in solid rock."

In concluding this account of the cultivation of the cocoa-nut it may be stated that, according to the various district Gazetteers, there are from 30,000 to 40,000 acres under the palm, with about 100 trees to the acre. Kánara, Ratnágiri, and Káthiawár appear to be the districts where the largest number of trees occur. Of Ratnágiri it is stated that if grown for the fruits only, each tree gives a net yearly profit of 2*s*. $4\frac{1}{2}$*d*. (R1-3).

II. Madras. 1531

II. *In Madras.*—The cocoa-nut palm flourishes in this Presidency, frequenting the banks of estuaries and backwaters, and abounding on the sandy tracts near the sea, especially along the Malabar and Coromandel coasts. Although met with in the Bombay Presidency, it is only plentiful in the southern division, and barely leaves the immediate coast. On nearing the Madras Presidency from Bombay it becomes more and more plentiful. Of its abundance on the Malabar coast an opinion may be formed from the description of the town of Cannanore, the clumps of the cocoa-nuts being said "to be seen between the officers' houses, surrounding the cantonments in every direction, and extending in the distance as far as the eye can reach; the cantonment may be said to be embedded in a forest of these trees" (*Royle*). Of South Kánara, it has been estimated that there are 80,000 acres under the cocoa-nut. Indeed, the Malabar coast and the Laccadive and Maldive Islands are pre-eminently the seats of the Indian cocoa-nut industry. The enquirer after Indian cocoa-nuts, coir, or cocoa-nut oil, need practically concern himself with no other part of the country unless he add to these the Nicobar Islands. The last-mentioned islands furnish a very large number of cocoa-nuts, but apparently the islanders are ignorant of, or too indifferent to earn, the art of making coir or expressing the oil. So far this remark seems to be almost applicable to the Maldives also, a group of large islands under a Sultan, who is subordinate to the Governor of Ceylon, and not to the Viceroy of India. This fact is of some importance, since the casual examination of the trade returns might convey the idea that the Laccadives export no coir, if the still further error might not even be committed of supposing the Laccadives to contain no cocoa-nuts at all. The Laccadives are mainly under the administration of the Collector of Malabar, and the imports from these islands are treated as if they were produce of the main-

COCOS nucifera.

The Cocoa-nut Palm.

CULTIVATION.

land, while the imports from the Maldives are returned as from foreign territory. Last year the Maldives sent 7,897,453 cocoa-nuts to India, and the Nicobar Islands 4,510,000. Of the inhabitants of these groups of islands it is not reported that they manufacture coir, and apparently they prepare only a small amount of copra, although they sell their nuts at a price far below that which prevails on the mainland of India.

1532 Writers in Europe, who have described the commercial article Coir, are in the habit of placing the coir from Cochin in the first rank. Some doubt seems to be associated, however, with the commercial term "Cochin Coir." The small Native State probably alluded is described in the *Imperial Gazetteer* as "possessing no important trade by sea or land." It seems impossible to believe that all the coir returned under the name of "Cochin Coir" could therefore come from Cochin. Indeed, the suspicion exists that the better class of Malabar and Laccadive coir, consigned to Europe, may be so designated, if not also some of the exports of coir from Cochin-China and the Straits. In the returns of the coasting trade for British India it is shown that last year the total exports of coir from Cochin by sea amounted to only 689 cwt., valued at R4,134, and manufactured coir 2,777 cwt., valued at R25,339: these were all sent to Bengal or Bombay; how much may have gone by land to Madras cannot be discovered. It is significant that **Dr. Shortt** in his Monograph on the cocoa-nut palm, which has just appeared, makes no mention of Cochin coir.

Repeated reference will have to be made, in subsequent pages, to the Laccadive and Malabar coir and the other cocoa-nut products from these regions, so that we shall here content ourselves with this brief notice of

1533 Madras, concluding only by giving the description of the cultivation given in *Morris's Descriptive and Historical Account of the Godavery District*: "Young plants of a year's growth are planted out, and watered for six years, after which they do not require much water. The trees generally bear fruit about the ninth year after transplantation. The expenses of cultivation are stated to be R668 for a *putti* of land,—namely, R140, being the price of 600 young plants, R48 being the value of the labour required for planting them, and R480 being the wages of labourers employed to water and tend the trees until they come into bearing. When the trees begin to bear fruit, the value of the produce of a tree, exclusive of the fibre, is estimated at about 12 annas a year, making the total value of the produce in a *putti* of land R300" (*p. 70*).

III. Mysore. 1534

III. *In Mysore* "there are four varieties of the cocoa-nut: *1st*, red; *2nd*, red mixed with green; *3rd*, light green; and *4th*, dark green. These varieties are permanent, but although the red is reckoned somewhat better than the others, they are commonly sold promiscuously. Their produce is nearly the same.

"The soil does not answer in the Bangalore district unless water can be had on digging into it to the depth of 3 or 4 cubits, and in such situations a light sandy soil is the best. The black clay, called *ere*, is the next best soil. The worst is the red clay, called *kebbe*; but with proper cultivation all the three soils answer tolerably well.

"The manner of forming a new cocoa-nut garden is as follows: The nuts intended for seed must be allowed to ripen until they fall from the tree, and must then be dried in the open air for a month without having the husk removed. A plot for a nursery is then dug to the depth of 2 feet, and the soil is allowed to dry three days. On the *Ugádi* feast (in March) remove 1 foot of earth from the nursery and cover the surface of the plot with 8 inches of sand. On this, place the nuts close to each other, with the end containing the eye uppermost. Cover them with 3 inches of sand and 2 of earth. If the supply of water be from

CULTIVATION.

a well, the plot must once a day be watered; but if a more copious supply can be had from a reservoir, one watering in the three days is sufficient. In three months the seedlings are fit for being transplanted. By this time the garden must have been enclosed, and hoed to the depth of 2 feet. Holes are then dug for the reception of the seedlings at 20 feet distance from each other in all directions, for when planted nearer they do not thrive. The holes are 2 feet deep and a cubit wide. At the bottom is put sand 7 inches deep, and on this is placed the nut with the young tree adhering to it. Sand is now put in until it rises 2 inches above the nut, and then the hole is filled with earth and a little dung. Every day for three years, except when it rains, the young tree must have water.

"The cocoa-nut palm begins to produce when seven or eight years old, and lives so long that its period of duration cannot readily be ascertained. Young trees, however, produce more fruit which comes forward at all seasons of the year. A good tree gives annually a hundred nuts. A few are cut green on account of the juice, which is used as drink, but by far the greater part are allowed to arrive at some degree of maturity, although not to full ripeness, for then the kernel would become useless.

"Cocoa-nut palms are planted in Chiknayakanhalli in rows round the areca-nut gardens, and also separately in spots that would not answer for the cultivation of this article. The situation for these gardens must be rather low, but it is not necessary that it should be under a reservoir; any place will answer in which water can be had by digging to the depth of two men's stature. The soil which is here reckoned most favourable for the cocoa-nut is a red clay mixed with sand. It must be free of lime and saline substances. Other soils, however, are employed, but black mould is reckoned very bad. The cocoa-nuts intended for seed are cut in the second month after the winter solstice. A square pit is then dug, which is sufficiently large to hold them, and is about a cubit in depth. In this, fifteen days after being cut, are placed the seed-nuts, with the eyes uppermost and contiguous to each other, and then earth is thrown in so as just to cover them, upon which is spread a little dung. In this bed, every second day for six months, the seed must be watered with a pot, and then the young palms are fit for being transplanted. Whenever, during the two months following the vernal equinox, an occasional shower gives an opportunity by softening the soil, the garden must be ploughed five times. All the next month it is allowed to rest. In the month following the summer solstice, the ground must again be ploughed twice; and next month, at the distance of 48 cubits in every direction, there must be dug pits a cubit wide and as much deep. In the bottom of each a little dung is put; and the young plants, having been previously well watered to loosen the soil, are taken up and one is placed in each pit. The shell still adheres to the young palm, and the pit must be filled with earth so far as to cover the nut. Over this is put a little dung. For three months the young plants must be watered every other day; afterwards every fourth day, until they are four years old, except when there is rain. Afterwards they require no water.

"Every year the garden is cultivated for *ragi, uddu, hesaru,* or whatever other grain the soil is fitted for, and is well dunged, and at the same time four ox-loads of red mud are laid on the garden for every tree that it contains, while a little fresh earth is gathered up towards the roots of the palms. The crop of grain is but poor, and injures the palms; it is always taken, however, as, in order to keep down the weeds, the ground must at any rate be ploughed, as the manure must be given, and as no rent is paid for the grain. On this kind of ground the cocoa-nut palm begins

C. 1534

The Cocoa-nut Palm.

CULTIVATION.

to bear in twelve or thirteen years, and continues in perfection about sixty years. It dies altogether after bearing for about a hundred years. They are always allowed to die, and when they begin to decay a young one is planted near the old one to supply its place.

"In this country, wine is never extracted from this palm, for that operation destroys the fruit, and these when ripe are considered as the valuable part of the produce. A few green nuts are cut in the hot season, on account of the refreshing juice which they then contain, and to make coir rope; but this also is thought to injure the crop. The coir made from the ripe nuts is very bad, and their husks are commonly burned for fuel.

"The crop begins in the second month after the summer solstice, and continues four months. A bunch is known to be ripe when a nut falls down, and it is then cut. Each palm produces from three to six bunches, which ripen successively. A middling palm produces from 60 to 70 nuts. As the nuts are gathered they are collected in small huts, raised from the ground on posts. When a merchant offers, the rind is removed at his expense, by a man who fixes an iron rod in the ground, and forces its upper end, which is sharp, through the fibres, by which means the whole husk is speedily removed. He then, by a single blow with a crooked knife, breaks the shell without hurting the kernel, which is then fit for sale and is called *koppari*. A man can daily clean 1,300 nuts. From 20 to 30 per cent. of them are found rotten" (*Mysore Gaz., I., 131-134*).

IV. Nicobar Islands. 1535

IV. *On the Nicobar Islands* the cocoa-nut palm is very abundant, although, as already stated, it exists only under recent cultivation on the Andaman Islands, but reappears still further to the north on the group of the Cocos Islands. **Sir W. W. Hunter** gives an interesting account of the Nicobar trade in cocoa-nuts which may be here quoted: "At present the principal product of these islands is the cocoa-nut palm, and its ripe nuts form the chief export." "The northern islands are said to yield annually 10 million cocoa-nuts, of which about half are exported. The estimated number exported in 1881-82 was 4,570,000. As this important product is six times cheaper here than on the coast of Bengal or in the Straits of Malacca, the number of English and Malay vessels that come to the Nicobars is every year increasing." "The trade in cocoa-nuts is carried on chiefly by native craft from Burma, the Straits Settlements, Ceylon, &c. Forty vessels of an aggregate tonnage of 6,270 tons visited the islands for cocoa-nuts in 1881-82." The Administration Report for 1885-86 gives the exports as 4,510,000 nuts and 5,730 bags of copra. In that year 49 vessels, with an aggregate tonnage of 8,218 tons, obtained permission to trade with the Nicobar Islands for cocoa-nuts, &c. The same report states that there are now 112,000 cocoa-nut palms under cultivation at Port Blair.

V. Burma. 1536

V. *Of Burma* it is reported that the cocoa-nut is "largely cultivated, and might be much more so in many places along the Arakan coast as it is in Ceylon, and as doubtless it would be but for the sparseness of population, the difficulties of approaching the coast except at a few spots, and the absence of the means of land communication between the ports and the sites fitted for the production of the trees." In the Bassein district of Pegu it has been stated that there are 10,000 acres under cocoa-nuts.

VI. Bengal. 1537

VI. *In Bengal,* while the palm is plentiful throughout the lower Gangetic basin, it exists only in garden cultivation, and the produce is not much in excess of the local demand. There are no large plantations such as have been described in Madras, Mysore, and Bombay, because in Bengal the date-palm is used as the source of palm-juice or toddy and not the cocoa-nut. It is, however, fairly abundant in Noakhally, Backerganj, Jessore, and the 24-Parganas.

CULTIVATION.

VII. Upper India.

VII. *In Upper India* the cocoa-nut is alluded to in many works, but only as an article of import and export; it is not cultivated. **Dr. Hartwig** (*Tropical World*) says: "This noble palm requires an atmosphere damp with the spray and moisture of the sea to acquire its full stateliness
and growth; and, while along the bleak shores of the Northern Ocean 1538
the trees are generally bent landward by the rough sea-breeze, and send forth no branches to face its violence, the cocoa, on the contrary, loves to bend over the rolling surface, and to drop its fruits into the tidal wave. Wafted by the winds and currents over the sea, the nuts float along without losing their germinating power, like other seeds which migrate through the air; and thus, during the lapse of centuries, the Cocoa-palm has spread its wide dominion from coast to coast, through the whole extent of the tropical zone."

VIII. Ceylon.

VIII. *Ceylon.*—Speaking of Ceylon cultivation **Mr. Treloar** says: "The ripe nuts are first planted in a nursery, where they are covered an inch
deep with sand and sea-weed or soft mud from the beach, and watered 1539
daily till they germinate. In two or three months a white shoot containing the foliaceous rudiments springs from one of the three holes in the end of the nut; the radicals emerging from the other two orifices opposite to the shoot, and penetrate the ground." This is not quite a correct description of the germination. The leaf-stalk of the cotyledon elongates and pushes the embryo bodily out of the seed. The blade of the cotyledon remains within the nut forming a sort of arm of attachment. The lower point of the projected embryo elongates and forms the roots, and from a slit in the cotyledonar sheath the plumule or stem makes its appearance. The "three holes" on the nut are all close together, not "opposite" as in the above description and are only spots not holes. But **Mr. Treloar** proceeds: "The nuts set in April, grow large enough in about four months to be planted out before the annual rains, but for the next two or three years or more the young plants require constant care. They must be watered and shaded from the glare of the sun by screens of plaited leaves from the cocoa-nut tree or the fan-shaped fronds of the palmyra."

Enemies to the Cocoa-nut.

It is commonly stated that if the soil be too rich a large grub with a reddish-brown head soon finds its way to the roots and into the stem. This eats its way through the tissues until the leaves turn yellow, the terminal bud withers, and the tree is killed. This appears to be the
beetle known as **Butocera rubus.** "In the Straits of Malacca, the chief 1540
natural enemy of the tree is a species of elephant-beetle, which begins by nibbling the leaves into the shape of a fan; it then perforates the central pithy fibre, so that the leaf snaps off; and lastly, it descends into the folds of the upper shoot, where it bores itself a nest, and, if not speedily extracted or killed, soon destroys the tree. A similar kind of
beetle is known on the Coromandel coast, and is extracted by means of 1541
a long iron needle or probe, having a barb like that of a fish-hook. By using this and by pouring salt or brine on the top of the tree, so as to descend amongst the folds of the upper shoots, the evil may be prevented or got rid of." This destructive beetle is known to entomologists as **Calandra palmarum**; but still another beetle bores round holes into the stem
itself and lives there. Rats, flying-foxes, and squirrels injure the tree and 1542
sometimes kill it by eating the tender terminal bud or cabbage. It is equally necessary to protect the trees from wild hogs, elephants, cows, porcupines, all of which graze on the young plants. But of the dangers to which the cocoa-nut is subject none are so great as the attacks of beetles, two of which are alluded to above. **Mr. Treloar** says of Ceylon: "Still

COCOS nucifera.

The Cocoa-nut Palm: Coir Fibre.

CULTIVATION.
1543 more formidable is the *cooroominyo* beetle (**Butocera rubus**), which waits to pierce the tender trunk near the ground, and to deposit its eggs in the cavity whence the young grubs, directly they are hatched, begin to eat their way up through the centre of the tree to the young leaf-ends at the
1544 top." The West Indian plantations are said to have been devastated by the attacks of a small beetle (**Passalus tridens**), and a similar calamity is reported to have occurred in Zanzibar through a species of **Oryctes.**
1545 The Burmans are great adepts at detecting the beetles in date and cocoa-nut palms and extract them as prized articles of food.

GUM.
1546

GUM.

The stem of this well-known tree is in Taheiti said to yield gum. It forms large stalactitic masses, red-brown, translucent or transparent. (*Spons' Encycl.*) **Cooke**, in his report on Gum and Gum-resins, says that this gum was sent to the Madras Exhibition of 1855 from Travancore. No other author appears to allude to this gum however, and it therefore, seems probable that if produced it is met with only in certain localities. The writer cannot recollect ever having seen a gum adhering to the stems of the palm.

DYE.
1547

DYE.

"In a patent obtained by **Mr. J. H. Baker** (No. 5139, March 29th 1825) the whole or every part of this tree is claimed as a dye-ware, especially the husk enclosing the fruit, and the foot-stalks of the leaves. The dye was to be extracted by water, cold or boiling, or by solutions of lime, potash, ammonia, &c., and was to serve for dyeing nankeens, blue-blacks, &c. The infusion was likewise to serve as a substitute for nut-galls in Turkey-red dyeing. The material does not appear ever to have come into practical use." (*Crookes.*)

Mr. Liotard says of this dye property: "Produces a dirty-brown (*khaki*) colour, and is a good deal used from its abundance. Lime and *chaula* are added as mordants." **Drury** remarks that "the shell when burnt yields a black paint which in fine powder and mixed with *chunam* is used
1548 for colouring walls of houses." Cocoa-nut oil is frequently employed in certain processes of dyeing. Thus in Mysore powdered myrobalams and sulphate of iron mixed with cocoa-nut oil are used for imparting a black colour to silk. It is not known whether any other oil might serve the purpose of the cocoa-nut,—in other words, whether or not that oil possesses special properties that assist the tinctorial actions.

1549 The natives of India generally do not seem to be aware of the dye properties. The milk is, however, said to be used by plasterers both in India and Ceylon, from an idea that when mixed with lime or colour-washes it increases the adhesive property and gives a polish. For this purpose
1550 vegetable matter of some kind is as a rule added to cements (see **No. 1626** and also The Article **Cement, C. No. 905**).

COIR FIBRE.
1551

COIR FIBRE.

The thick pericarp or outer wall of the fruit yields the valuable COIR FIBRE of commerce. The SHEATHS of the leaves are used to wrap up articles, and as paper to write upon. At the Colonial and Indian Exhibi-
Leaf-Stalks. tion the writer proposed that they should be tried as a means of strength-
1552 ening harness, the softer layers being even used for corsets and as surgical
Tomentum. splints. The FIBRE OF THE LEAF-STALKS is also prepared, and might prove
1553 useful in the manufacture of paper. A delicate TOMENTUM or COTTON is
Coir. often seen at the base of the leaf; this is used as a styptic. But the most
1554 important fibre yielded by the cocoa-nut palm is of course COIR. The name

The Cocoa-nut Palm: Coir Fibre.

COCOS nucifera.

COIR FIBRE.

of this fibre is said to come from the Malayalam *káyar* (from the verb *káyaru,* to twist) through the Portuguese corruption *coiro.* "The word appears in early Arabic writers in the forms *kánbar* and *kanbár,* arising probably from some misreading of the diacritical points (from *káiyar* and *kaiyár*)" (*Yule & Burnell*). *Kayer* is said to be also the Tamil for a rope.

Both the fibre and the rope were first exported to Europe about the middle of the sixteenth century, but it was not until the great International Exhibition of 1851 that coir ropes and coir matting attained a commercial importance in England. The merit of this modern trade is largely due to the efforts of the following firms, *viz.*, **Messrs. Chubb, Round & Co.**, and **Messrs. Treloar & Sons** of London; the Oriental Fibre Mat and Matting Co., Highworth, Wilts; and **Messrs. W. I. Sly and T. Wilson** of Lancaster, who were the patentees of improved machinery for making figured fabrics of coir.

Although a considerable amount both of coir fibre and yarn is exported from India, the article, taking India as a whole, is obtained chiefly as a by-product. It is accordingly inferior in quality and colour to the special coir obtained from Cochin, the Laccadives, Madras, Malabar, Ceylon, Singapore, &c. Locality seems to exercise a considerable influence over the quality of the fibre,—soil climate, and proximity to the sea being important influences. But there are other considerations. Certain varieties or cultivated forms of the cocoa-nut are better suited than others for the production of coir. If cultivated specially for the supply of juice or to afford fruit, the fibre would appear to be in the one case imperfectly formed and in the other overgrown. A great deal depends upon the collection of the fruit at the exact time the fibre is mature, and this followed by an accurate system of steeping, beating, and cleaning the fibre, completes the manipulation calculated to produce the superior qualities of coir. (*Conf. with Mr. Jackson's report in next para.*) "The fibre appears in the market in various degrees of fineness, depending on the age at which the cocoa-nut was cut and husked, and the care bestowed in steeping and cleaning." **Mr. Treloar** says: "The usual indications are that the commoner and coarser fibre comes from the old nuts, and the finer, lighter quality from the new; but there are, of course, essential differences in the qualities brought from each locality, and the Cochin are usually the best." "Here let it be parenthetically but emphatically remarked that *any attempt to give to cocoa-nut fibre a fairer hue by the process of bleaching is to destroy its quality if it be good, and if it be of common quality to make it almost worthless.*"

PROPERTIES OF COIR. 1555

Properties of the Fibre and Season when Mature.—"The Cochin has the purest hue and fetches the best price." On this account it has been customary to imitate this by bleaching. "Cocoa-nut fibre is tough, elastic, springy, easily manipulated within certain limits, and eminently suitable for manufactures where lightness, cleanliness, and great indestructibility are required. It will stand water, is almost impervious to wind and wave, or to damp and rain, and, as we have seen, flourishes in the saline breath of the sea; but it will not stand bleaching. It gives up when confronted with sulphuric acid, chloride of tin, or any other chemicals which are designed to convert it into a sham product. *For this reason we use none but unbleached fibre in any of our manufactures*; and although two or three varied shades of colour are frequently to be seen in one of our ornamental mats, and sometimes form a simple pleasing pattern, they are obtained by combining or incorporating in the same mat different descriptions of *natural* unbleached fibre." *Spons' Encyclopædia* has it that the fibre "is much impaired by waiting for the nuts to arrive at maturity, consequently, for fibrous purposes, the latter are usually cut at about the tenth month.

The Cocoa-nut Palm: Coir Fibre.

PROPERTIES OF COIR.

If cut earlier than this, the fibre is weak; if later, it becomes coarse and hard, requires a longer soaking, and is more difficult to manufacture." **Dr. Buchanan Hamilton** in his journey across Mysore states (*I., 156*) the green cocoa-nuts are sold for their husks, from which fibre is extracted, but the husks of the ripe cocoa-nuts are commonly burnt for fuel (*II., 50*). At the same time immense quantities of apparently ripe cocoa-nuts, in husk, are sent to Europe, the coir from the husk being there separated, cleaned, and manufactured. **Mr. Jackson** of Kew, in the *Planters' Gazette*, describing a visit to **Messrs. Chubb, Round & Co.'s** factory, gives an interesting account of the process of husking there pursued. He says: "The enormous heap of husks—which, indeed, is known in the locality as the 'mountain'—comes upon view immediately upon entering the premises, and one can scarcely, at first sight, realise the fact that the enormous pile is composed entirely of these apparently useless portions of the fruit. At the time of my visit this reserve stock of husks was estimated at considerably over a million and a half." Cocoa-nuts, or, as they are generally termed in the trade, "Cocker-nuts," to distinguish them from the **Theobroma Cocao,** which furnishes cocoa and chocolate, are shipped principally from Trinidad, Jamaica, Demerara, Tobago, several of the other Leeward Islands in the British West Indies, Ceylon, Belize (British Honduras), all round the coast of America, and the Fiji Islands. Nearly all the nuts are imported in the husk or outer covering, from which, on arrival, they are stripped by men using two fine-pointed steel chisels, and who, by constant practice, become so skilful in the art that many are able to open 1,000 to 1,200 nuts per day. The nuts themselves after being removed from the husks are generally sold to wholesale fruit dealers, who, in turn, supply the retailers, costermongers and others, &c." In the above passage **Mr. Jackson** has furnished the Indian people with new ideas. India is not enumerated by him as one of the countries that furnish cocoa-nuts to England; the fibre of what appear to be mature cocoa-nuts is actually used; the consumption of cocoa-nut kernel has in England attained a vast proportion, and the fibre can be cleaned after apparently having been kept for years on the nut. These facts open up a new field of trade of which with a little assistance the Nicobar and Laccadive Islands might profitably and without fear of any rival hope to enjoy a large share.

SEPARATION OF COIR.
1556

Separation of Coir in India.—"The removal of the fibre from the shell is effected by forcing the nut upon a pointed implement stuck into the ground; in this way a man can clean 1,000 nuts a day. The fibrous husks are next submitted to a soaking, which is variously conducted. In some places they are placed in pits of salt or brackish water, for 6 to 18 months; in other places, fresh water is used, but it becomes foul and injures the colour of the fibre. The chief point to be considered is the duration of the soaking; if it be continued too long, the fibre will be weakened; if it be curtailed, the subsequent extraction and cleansing of the fibre will be rendered more difficult. The most approved plan of conducting the soaking is in tanks of stone, brick, iron, or wood; steam is admitted to warm the water. By this means the operation is rendered very much shorter, and the fibre is softened and improved. The further separation of the fibre from the husk is largely effected by hand. After thorough soaking, the husks are beaten with heavy wooden mallets "and then rubbed between the hands." (*Spons' Encyclop.*; **Royle** and **Marshall** give the same facts.)

Robinson describes the separating and cleaning of the fibre as practised in the Laccadive Islands as follows: "When soaked sufficiently long, it is taken out of the pit and beaten with a heavy mallet. Subsequently it is said to be rubbed with the hands until all the interstitial

SEPARATION OF COIR.

cellular substance is separated from the fibrous portion. When quite clean it is arranged into a loose roving preparatory to being twisted, which is done between the palms of the hands in a very ingenious way, so as to produce a yarn of two strands at once."

"As the husk gets hard and woody if the fruit is allowed to become quite ripe, the proper time for cutting it is about the tenth month. If cut before this, the coir is weak; if later, it becomes coarse and hard, and more difficult to twist, and requires to be longer in the soaking pit, and thus becomes darker in colour. When cut, the husk is severed from the nut and thrown into soaking pits. These, in some of the islands, are merely holes in the sand, just within the influence of the salt water. Here they lie buried for a year, and are kept down by heaps of stones thrown over them to protect them from the ripple. In others, the soaking pits are fresh-water tanks behind the crest of coral. In these, the water, not being changed, becomes foul and dark coloured, which affects the colour of the coir. When thoroughly soaked, the fibrous parts are easily separated from the woody by beating. If taken out of the pits too early, it is difficult to free the coir from impurities; if left in too long, the fibre is weakened, as is said to be the case also with that soaked in fresh water." (*Robinson's Report on the Laccadives.*) In the Maldives (neighbouring islands under the suzerainty of the Governor of Ceylon) cocoanuts are very plentiful, and enormous quantities of both the nut and the fibre are exported to India and Ceylon. (*See the further paragraph on trade in nuts.*)

From what has been said in an early paragraph regarding the cultivation of the cocoa-nut palm in Mysore, it will be seen that the opinion prevails that if tapped for the juice the fruit-bearing is materially injured. If this be a fact ascertained and borne out in other parts of the country, it might be inferred that regions like the Malabar coast, where tapping is largely practised, would not be suited for the production of good fibre. On the other hand, **Royle** says: "But the fruit-bearing power of the trees may be considerably improved by extracting toddy from the blossom-shoots for the manufacture of jaggery during the first two years of its production, after which it may be discontinued." In the Konkan the opinion is held that "if tapped the trees become unproductive much sooner."

The Bombay process of extracting the fibre is briefly described in the *Bombay Gazetteer* of the Thana district: "The fibrous part of the outer coating is made into coir by the Bassein gardeners. For this purpose the fibres are stripped from the nuts, left under water for two months, and then beaten by a wooden mallet." The writer cannot discover any detailed description of the process adopted in India generally (except that of the Laccadives) for the separation, steeping, and cleaning of the fibre, but to the best of his knowledge it agrees with what has already been given; although in the Laccadives, the Malabar Coast, Ceylon, and other important coir-producing countries, the art is carried to greater perfection, the fibre being correspondingly superior to that prepared on the mainland of India. Large portions of the coast of India (see above in a passage from the *Burma Gazetteer*) are so inaccessible that the methods pursued in Ceylon and elsewhere are scarcely applicable, and, indeed, where applicable, are not pursued, from the greater importance attached to the palm as a supplier of toddy or juice. **Dr. Stocks** urges that the cocoa-nut might with advantage be cultivated on the brackish soils in the neighbourhood of Karachi, and it seems possible that a coir industry might there be developed. It has been reported that in Madras cocoa-nut cultivation has been successfully prosecuted in the reclamation of salt-impregnated lands where

The Cocoa-nut Palm : Coir Fibre.

nothing else would thrive. (*Gen. Admin. Report, p. 95.*) A curious fact in regard to cocoa-nuts grown on salt marshes is conveyed by the following passage :—

"The cocoa-nuts growing in mangrove soils, on the side of creeks, and more or less saturated with salt, have their milk brackish, and the sap is saline also. These trees do not suffer from the attacks of the rhinoceros-beetle, and are found to bear much sooner than those planted in a sandy soil" (*p. 182-83*).

TRADE IN COIR 1557

Interesting Facts connected with the Trade in Indian Coir (*Conf. with p. 435*).

Although, as suggested, the better class fibre is most likely not produced where tapping for the juice is practised, still it should not be forgotten that the Malabar ports are the chief seats of the export of coir from India. In most works (written in Europe) on the subject of coir, the statement is made that the best quality comes from "Cochin." As already stated, it is not quite clear, however, whether the Native State of Cochin or the whole of the Malabar coast is meant, or whether Cochin coir is a mere commercial term for all good coir wherever obtained. In the Indian regions alluded to above, cocoa-nut cultivation is prosecuted to a considerable extent. Of Cochin (Madras), it may be said, coir is perhaps the most important article of export from that Native State, but **Dr. Shortt** (*in his Monograph on the Cocoa-nut Palm*) does not apparently mention Cochin coir. He states that the best Madras coir comes from the Laccadives, Amindivi, Kadamat, Kiltan, and Chetlat. As indicated by the passage quoted above from **Mr. Jackson's** paper **Messrs. Chubb, Round & Co.** do not, it would seem, use any Cochin fibre but prefer a husk which they separate from a mature or at least edible nut.

In a recent report on the trade of Madras, the progress of the coir industry of that presidency for the past twenty-five years is shown. The average exports to foreign and Indian ports for the five years ending 1860-61 were 148,220 cwt., valued at R3,74,804; and for the five years ending 1880-81, they were 271,934 cwt., valued at R21,79,767, while for the year 1881-82 they were R23,54,202. Of the last mentioned valuation, the exports from the Malabar coast alone amounted to R22,43,000. From these figures a definite idea may be obtained of the immense importance of Malabar and the Laccadives as the chief seats of the Indian coir industry, since the Madras Presidency heads the list of Indian exports. This idea is borne out by the statement made by **Royle** that "the Laccadive Islands are famed for the good quality of the coir which is made there and exported to the Malabar coast." Again, speaking of the peculiar form of the palm grown in the Island of Kiltan, **Royle** observes: "It requires no attention and comes into bearing early. The tree is not so large and strong as that of the coast, and the nut about two thirds of the size only, and round in shape. The husk is smaller and less woody, and the fibre finer and more delicate but stronger than that of the coast nut. The nut is also said to be more compact and oily, and to keep better than the coast nut, although, for the sake of the coir, the nut is cut before being quite ripe." How far the exports of coir from the Malabar coast correspond to Indian-grown coir cannot be discovered. The Northern Laccadives are administered by the Collector of Malabar and the Southern by Alí Rájá of Cannanore. **Sir W. W. Hunter** in the *Imperial Gazetteer* (*VIII., 394*) says: "The article (coir) is paid for to the producers at fixed prices, and is sold on the coast at the market rates; the difference constitutes the revenue or profits of trade of the Government and Alí Rájá respectively. The latter pays a fixed tribute of R10,000 (£1,000) to the

TRADE IN COIR.

Government on account of the islands which he manages. No change has been made for many years in the price which is given by Government for the coir produced in the islands attached to Kánara." The returns of the coasting trade of India do not specify the amounts of coir sent from the Laccadives to Malabar, so that the somewhat interesting subject of how far the juice-extracting industry of the coast is combined with the preparation of fibre cannot be definitely learned. The following facts are, however, instructive.

Imports. 1558

IMPORTS of coir (manufactured and unmanufactured) into Madras from other Indian ports—

	Cwt.	R
1884-85	14,745	95,884
1886-87	13,750	81,386

Exports. 1559

EXPORTS to other Indian ports—

	Cwt.	R
1884-85	186,860	12,66,356
1886-87	128,228	7,08,255

Turning to the tables that give the details of these figures, it is shown that of raw or unmanufactured coir Madras receives none from British or foreign Indian ports, so that unless the Laccadives, which (as stated above, are treated as part of Malabar), furnish a very large amount, a considerable local production must exist on the Malabar coast to allow of the extensive exports given above, and the still larger exports shown in the returns of foreign trade. In the *Imperial Gazetteer* it is stated of the Malabar district alone that "the value of exported cocoa-nut products is estimated at nearly a million sterling annually."

In a previous page some indication of the extent of the Nicobar trade in cocoa-nuts has been given. There does not, however, appear to be any trade in coir, although it seems possible that one of the inducements that bring the native and other crafts from Ceylon and the Straits, for the cheap Nicobar cocoa-nuts, may be to use these in supplementing their home supplies—supplies which are in much demand in the coir market. It may be worth while suggesting that an effort might with advantage be made to instruct the natives of the Nicobars in the art of preparing the coir fibre—an art so profitably practised by their neighbours, the islanders of the Laccadives. This is indeed one of the most hopeful aspects of a possible enhanced Indian trade in coir, until such time as the cultivation of the palm can be more vigorously prosecuted along the Coromandel coast to Burma. It seems remarkable that the cheap cocoa-nuts sold in the Nicobar Islands should attract traders from Ceylon and the Straits, while India appears to make little or no effort to participate in the advantages of that trade. This is perhaps due to the administrative arrangement which has associated the Maldive Islands with Ceylon instead of India, the Ceylon traders calling at the Nicobar as well as the Maldive Islands

YIELD PER NUT OF FIBRE AND PRICE.

YIELD OF FIBRE. 1560

Mr. Robinson, in his *Report on the Laccadives*, states that the difference in the quantity of coir manufactured from a coast nut and from an island nut is very considerable. We may premise that 40 cocoa-nuts are said to yield 6℔ of coir in Ceylon. Mr. Robinson remarks: "Three large coast nuts will yield 1℔ of coir measuring 22 fathoms; whereas, ten small, fine island nuts go to about 1℔ of coir, but this will measure 35 fathoms: 2℔ of such yarn, measuring from 70 to 75 fathoms, are made up into *sooties*, of which there are 14 to a bundle, averaging about a maund of 28℔. A

COCOS nucifera.

The Cocoa-nut Palm: Coir Fibre.

PRICE. 1561 Mangalore candy of 560lb will thus be the produce of 5,600 nuts, and should contain about 20,000 fathoms of yarn. The actual price of coir received by the islanders is about R13 per candy. The value of the coir produce of a tree is calculated to be from 2 to 2½ annas; and that of the produce of 100 trees from R13 to 15." "The average value of the total raw produce of a tree bearing fruit would then be seven annas to half a rupee; and that of a plot of 100 trees, R45." For the nuts which they export to the Malabar coast they get from R7 to 10 per thousand, or rather 1,100, as 10 per cent. is always allowed for luck in these sales. The islanders export from 300,000 to 400,000 nuts annually. The natives bring their coir to the coast in March and April, which is then received into the Government godowns. Until the year 1820 all coir was paid for at the rate of R21-14-0 per Mangalore candy, or R25 per Calicut candy of 640lb. After that year the coir was divided into three classes. Since then the average price paid for a Mangalore candy of Ameendevy and Kadamat coir has been R20-2-0 (or R23 per Calicut candy of 640 lb). But for the Kiltan and Chetlat coirs, which are the best, an average of R20-12-7 or R23-12-0 per Calicut candy is paid. Up to A.D. 1825-26, the Bombay and Bengal Governments took almost the whole of the coir brought from these islands, and credited the Mangalore Collectorate with R25 per candy. The price has since fallen very much during the last twenty years. It has been frequently below the price paid to the islanders, and at best has never yielded above 12 to 20 per cent. profit. The average imports of coir have been from 500 to 600 candies. **Mr. Morris**, in his account of the Godavery district, Madras, gives the following brief statement regarding the production and yield of coir:—

"The cocoa-nut tree yields an excellent fibre. The quantity of fibre in the above extent of land (a *putti*) is estimated at 150 maunds, yielding R93-12-0, at 10 annas a maund. The fibre is prepared by the outer covering of the cocoa-nut being moistened and beaten with wooden mallets, after the fibre has thus been loosened. The coir thus obtained is twisted into ropes. The fruit is exported, but very little of the fibre" (*Morris's God. Dist.*, 70).

Spons' Encyclopædia gives the London prices of coir as "Cochin—good to fine, £19 to £25 a ton; coarse, £16-10*s*. to £19-15*s*. Yarn—good to fine, £26-10*s*. to £46 a ton; medium, £21-5*s*. to £28-10*s*.; common, £14 to £22-10*s*.; roping, £18 to £24."

USES OF COIR. 1562

Uses of Coir.

"The fibrous husk of the cocoa-nut is not its least valuable product, and gives rise to a very large trade, both in the East and in Europe. At first it was only used in this country (England) for stuffing mattresses and cushions, but its applications have been enlarged and its value greatly increased by mechanical processes, and in a small pamphlet issued by **Mr. Treloar**, more than twenty years ago, he stated that its natural capabilities having been brought out, coir has been found suited for the production of a variety of articles of great utility and elegance of workmanship—table-mats, fancy baskets, and bonnets, &c. Instead of being formed into rough cordage only, and mats made by
1563 hand, by means of ingeniously-constructed machinery the fibre is rendered sufficiently fine for the loom, and matting of different textures and coloured figures is produced, while a combination of wool in pleasing designs gives the richness and effect of hearth-rugs and carpeting. Brushes and brooms for household and stable purposes, matting for sheep-folds, pheasantries, and poultry yards, church cushions and hassocks, hammocks, clothes lines, cordage of all sizes, and string for nurserymen

USES OF COIR.

and others, for tying up trees and other garden purposes; nosebags for horses, mats and bags for seed-crushers, oil-pressers and candle-manufacturers, are only a few of the varied purposes to which the fibrous coating of the cocoa-nut is now applied." (*Simmonds, Trop. Agri., 234.*) The uses of coir are of course so varied and extensive that it is scarcely necessary to enter upon them in greater detail than indicated in the above passage. To the natives of India it is invaluable as lasting in a damp climate. It is accordingly universally employed in tying the bamboos used in the construction of their huts. 1564

Fibrous Sheaths.

FIBROUS SHEATHS OF THE LEAVES AND COCOA-NUT COTTON.—A brief reference has been made to these in an early part of this article. The finer ones are used as filters and sieves, but the coarser are apparently put to no purpose, although they have been proposed as suitable for paper-making. They might be used to strengthen saddlery, and even for ladies' corsets and splints. **Knox** says of Ceylon that "the filaments at the bottom of the stem of the cocoa-nut may be manufactured into a coarse cloth called *gunny*, which is used for bags and similar purposes." 1565

On the young sheaths and petioles a brown-coloured cotton or tomentum will be seen similar to that already described under **Borassus flabelliformis** (B. 680). This is sometimes collected and used by the natives to stop bleeding from wounds. A good sample of it was shown at the Colonial and Indian Exhibition.

Cadjans.

CADJANS.—"The leaves are plaited into mats and screens and also made into baskets, and combs are said to be made of the midrib of the leaflets in the Friendly Islands. In the Laccadive Islands mats are made of the cocoa-nut leaf. These mats are of fine quality and much esteemed when exported. In the islands they are employed for the sails of the smaller boats." "The Singalese split the FRONDS in halves, and plait the leaflets neatly, so as to make excellent baskets, and under the name of *cadjans* they form the usual covering of their huts, as well as of the bungalows of the Europeans." "The dried fronds are sometimes used as torches or for fuel; their midribs, tied together, are sometimes used as brooms for the decks of ships, as the fibres of the stalk are woody, brittle, and difficult to clean." (*Royle.*) 1566

Fronds. 1567

COLLECTIVE TRADE IN COCOA-NUT PRODUCTS.

TRADE IN COCOA-NUT PRODUCTS.

This trade, as with every other article of Indian produce or manufacture, is referable to three great sections: (*a*) internal trade or local consumption; (*b*) inter-provincial trade adjusting the balance of local demand; and (*c*) foreign trade (*e.g.*, imports and exports) to and from India and other countries. Where the cocoa-nut grows it is of such importance and enters so largely into the daily life of the people, that little or nothing can be ascertained of the actual consumption. The returns of road, river, and rail traffic throw some light on this, and the coasting trade affords another means of arriving at an approximate estimate of a certain proportion; but even these returns fall far short of establishing a tangible conception of the total local consumption. Wherever the palm grows, each villager, as a rule, has some trees, the produce of which is used up by himself or sold to his less fortunate neighbours, without having to go many yards from the spot where produced. At the same time, a considerable amount of the inter-provincial exchange must necessarily figure again under foreign exports, or at most re-exports, so that while the returns of foreign trade indicate but a very small proportion of the production, it would be unsafe to reckon these up with the available returns of coasting and inter-provincial trade. To give some idea of the present position and 1568

COCOS nucifera.

Trade in Cocoa-nut Palm Products.

TRADE.

growth of the trade in the cocoa-nut palm it will not be necessary to go further back than the year 1850. Royle, in his *Fibrous Plants of India*, gives the imports and exports for that year compiled from the records of the Statistical Department of the India Office. Accepting these as correct (remarking only that a certain amount of the coasting trade is included by that author with the foreign trade) we may summarise his results in the following statement:—

All published Imports and Exports for 1850.

	Imports.	Exports.
	R	R
Nuts	5,24,889	10,140
Kernels	8,66,120	4,31,008
Coir and rope	2,31,934	2,84,514
Oil	76,648	1,51,843
Shells	5,970	*Nil*
Cadjans	2,990	*Nil*
TOTAL	17,08,551	8,77,505

This gives a grand total of R25,86,056; that is to say, less than the foreign imports of last year. To compare with the above statement of TOTAL TRADE, the following table of the FOREIGN TRADE for 1886-87 (exclusive of all internal and inter-provincial or coasting traffic) may be given:—

Foreign Imports and Exports for 1886-87.

	Imports.	Exports.
	R	R
Nuts	5,98,203	8,462
Copra (or kernels)	11,76,799	79,836
Coir (unmanufactured)	6,839	77,391
,, (manufactured but exclusive of ropes)	1,50,701	19,14,448
Oil	7,54,515	13,24,589
TOTAL	26,87,057	34,04,726

If to the above table of foreign trade we were to add the returns (included by Royle) of coasting trade from Malabar, the Laccadives, Coromandel, Konkan, &c., the comparison would be brought out still more forcibly, but, as the tables stand, the foreign imports and exports to and from India of cocoa-nut products were last year valued at over 60 lakhs of rupees, whereas in 1850 (removing approximately the items of coasting trade) they were considerably less than one third that amount. The most striking feature of this development has been the growth of the trade in manufactured coir and in oil. As far as the returns of foreign trade indicate, it would appear that the imports and exports of cocoa-nuts and of *copra* or *khoprá* (a commercial name for the kernels) have remained stationary during the past forty years. How far the returns of foreign trade can be accepted as an indication of total trade may be learned from the following statement

C. 1568

The Cocoa-nut Palm.

COCOS nucifera.

TRADE.

of the values of the coasting trade in cocoa-nut products during the year 1886-87:—

Coasting Trade in	Imports.	Exports.
	R	R
Nuts	24,21,941	16,88,773
Kernels (copra)	35,31,115	23,00,958
Coir	12,20,749	9,27,302
Oil	20,60,067	20,74,455
TOTAL .	92,33,872	69,91,488

The table furnished by Royle for the trade in 1850 practically corresponds to the two last-given tables conjointly so that instead of 25 lakhs the present sea-borne trade has expanded until it may now be returned as valued at 223 lakhs of rupees. It has been remarked that a certain amount of the inter-provincial coasting traffic may reappear as exports to foreign countries or figure in the road, river, and rail traffic to interior parts of the country. While, therefore, the estimate of 223 lakhs must include duplex if not multiplex returns (*e.g.*, Bengal imports from Malabar figuring again as exports to foreign countries or re-exports to other Indian provinces), Bengal at the same time produces cocoa-nuts and the rail and river traffic from Bengal to Assam, the North-West Provinces, the Central Provinces, Central India, and even the Panjáb, may be viewed as very considerably in excess of the overlappings of sea-borne traffic. Take by way of illustration one item of this internal trade: Bengal sent to Assam in 1883-84 cocoa-nuts to the number of close upon two millions, valued at R69,000. In a like manner Bombay imports cocoa-nut products from Madras, Ceylon, Zanzibar, &c., and distributes doubtless a large proportion of these inland, competing with Bengal for the central tracts of India where the palm does not grow, but at the same time Bombay produces a large quantity of cocoa-nuts locally, and these are added to the foreign imports in the competition of the Indian internal trade. The importance of the internal trade may further be demonstrated by the fact that while Madras exports far more to foreign countries than all the other provinces put together (and indeed it takes the largest slice in the coasting export trade), it receives by sea practically no cocoa-nut products. Its entire supplies of Malabar and Laccadive fibre and copra seem to be conveyed to the port of shipment by internal means of transport. India is itself perhaps the largest consuming country in the world for cocoa-nut products, so that, recollecting this fact, a conception of the total trade may be had by adding to the sea-borne traffic an allowance for local production. Even when this has been done, a very imperfect idea will have been obtained of the value of the tree to the people of India. The mere returns of trade cannot give a just conception of the importance of a product which, like the cocoa-nut, to a large population, may be said to be their source of wealth as well as their food, drink, and occupation.

TRADE IN COIR, MANUFACTURED AND UNMANUFACTURED.

In all the returns of this subject care is taken to explain that these do not include ropes—coir ropes and cords being placed under a general heading with all vegetable cords.

I. The exports of RAW COIR are, however, so insignificant that a false impression is likely to be conveyed. The so-called manufactured coir, which figures extensively in the returns, appears to be largely crude

COCOS nucifera.

The Cocoa-nut Palm.

TRADE.

coir yarn which is dressed and employed by the European manufacturers, but of course a considerable trade is also done in mats, rugs, carpets, and other such manufactures. Glancing at the figures of the foreign trade in Coir (unmanufactured), the trade would seem to have practically remained stationary for many years past, and to be too small to justify the conclusion that India participates anything like to the extent it might in meeting the home market. The exports have averaged from 10,000 to 15,000 cwt. for the past twenty years: they were last year 12,347 cwt., valued at R77,391; but in 1883-84, they reached to 20,098 cwt., valued at R1,59,683. The foreign imports of coir are from Natal and Ceylon, and the bulk of these go to Bengal. The coasting trade last year conveyed from one Indian port to another the following quantities of unmanufactured coir: Imports 18,052 cwt. and exports 17,733 cwt. Of this trade, Madras exported 15,586 cwt. and imported only 309 cwt., Bombay exported 2,146 cwt. and imported 8,836 cwt., while Bengal exported only 1 cwt. but imported 8,335 cwt. The bulk of the Bombay and Bengal supplies came from Madras (*viz.*, 5,756 cwt. and 7,645 cwt. respectively). Of the exports to foreign countries the United Kingdom received 10,215 cwt. of last year's production, and of that amount 8,940 cwt. were consigned from Madras.

II. Of Manufactured Coir (excluding ropes) India imported last year (18,709 cwt.) valued at R1,50,701 and exported 208,622, cwt., worth R19,14,448. Of the imports, Ceylon sent 17,657 cwt., of which Bengal received 11,956, valued at R1,22,552. Of the exports, Madras sent to foreign countries 168,678 cwt., valued at R15,69,774, Bombay and Bengal each sending about 20,000 cwt. Of these exports the United Kingdom received 186,395 cwt., valued at R17,32,815, and next in importance followed France, 9,836 cwt., the United States, 2,621 cwt., Australia, 2,485 cwt., and Arabia, 2,545 cwt., &c.

Of the coasting trade in manufactured coir the imports and exports from one province to another were—imports 150,396 cwt., valued at R11,16,957, and the exports 134,665 cwt., valued at R8,36,427. Of these, Bengal received 60,500 cwt., Bombay 74,561 cwt., Sind 1,776 cwt., Madras 13,441 cwt. The Bengal and Bombay imports came mainly from Madras and Travancore, Cochin ranking next. The importance of Travancore as a seat of the coir-manufacturing industry may be demonstrated by its imports into Bengal and Bombay; Madras sent 30,185 cwt., valued at R2,61,199, and Travancore 27,613 cwt., valued at R2,86,277; to Bombay, Madras sent 50,264 cwt., valued at R2,72,567, and Travancore 17,327 cwt., valued at R1,40,260. At the same time Madras last year sent a large amount to Travancore, *viz.*, 14,283 cwt., valued at R1,36,810. Of the total exports in the coasting trade (*viz.*, 134,665 cwt.) Madras sent to other ports 112,642 cwt, and Bombay, next in importance, exported only 21,647 cwt. Of the total coasting trade in imports (*viz.* 150,396 cwt.) Bombay generally heads the list; it received last year 74,561 cwt., while Bengal took 60,500 cwt., being followed by Madras with 13,441 cwt. Sind and Burma are unimportant; the former received only 1,776 cwt. *Thus it will be seen that both in foreign and internal trade the coir industry is mainly concentrated in the Madras Presidency.*

COIR ROPES. 1569

Coir Ropes.

Nothing can be learned as to the extent of the foreign and internal trade in coir ropes and cords, since the trade returns for these are published jointly with those of all other ropes. It has been said, however, that coir string is universally employed by the natives of India in the construction of their bamboo huts. For this purpose alone the consump-

COIR ROPES.

tion must be enormous. The merits of coir as a rope fibre are now fully appreciated throughout the world, the elasticity and lightness of the fibre making it eminently suited for this purpose. But to these properties has to be added its great power of withstanding moisture even under continued actual submersion. On these grounds it is in great demand for maritime purposes as hawsers, although its roughness renders it unserviceable for standing riggings, its elasticity being for such purposes a disadvantage. It is, however, better suited for running riggings, its lightness being taken advantage of. In the *British Manufacturing Industries* (on Fibres and Cordage) it is stated, "Coir is one of the best materials for cables on account of its lightness, elasticity, and strength. It is durable and little affected when wetted with salt water. Numerous instances have been related of ships furnished with this light, buoyant, and elastic material riding out a storm in security, while the stronger-made, though less elastic, ropes of other vessels have snapped in two. Of coir and coir-made rope, about 9 or 10 million pounds are annually shipped from India. Much of it is prepared in Ceylon, but Cochin is noted as the port of shipment for the best quality of yarn, and many thousand cwt. are annually exported from there." This statement of many thousand cwt. being shipped from Cochin should be compared with the trade returns quoted and the remarks made in an early paragraph. In *Spons' Encyclopædia* the following opinion is given of coir fibre for the purpose of cordage: "In **Dr. Wight's** experiments, coir cordage broke at 224lb. Though not superlatively strong, the elasticity of the fibre, and the capacity it exhibits of withstanding the action of sea-water, render it valuable for cordage purposes, to which it is widely applied locally, besides being less extensively imported into this country for a similar use." **Messrs. Harton & Co.**, of Calcutta, placed in the Colonial and Indian Exhibition a trophy of ropes of which a striking feature was the arches of hawsers, 12 inches in diameter, thrown across the path; some of these were made of coir.

OIL.

OIL. 1570

The sliced kernel, dried at ordinary temperatures, either in the sun or artificially, contains from 30 to 50 per cent. of oil. The method of extracting this oil in India, especially when it is required to be colourless, is as follows: The kernel is boiled with water for a time, then grated and squeezed in a press. The emulsion thus obtained is next boiled until the oil is found to rise to the surface. The ordinary commercial oil is expressed by rude oil-mills worked by oxen.

The oil is white and nearly as fluid and limpid as water in tropical climates. It has a sweet and, according to some tastes, an agreeable odour when fresh, but is liable to become rancid in a short time.

In Europe the oil is chiefly used in the manufacture of candles and soap. In India it is employed in cooking, and as medicine when fresh, and for burning, painting, soap-making, and anointing the body when rancid.

Regions where Oil is Produced.—While in the above sentences a brief abstract has been given of cocoa-nut oil, it is necessary to deal with this subject in greater detail. Enquiries are frequently addressed to the Government of India by merchants interested in the trade in this substance, so that it has become necessary to put on record as complete an account as can be collected from the scattered publications that exist, even should that prove but a statement of the littleness of our knowledge. One of the earliest and to this day the most satisfactory descriptions of the Indian cocoa-nut oil industry is that written by **Lieutenant H. P. Hawkes** and published in 1857. Gazetteer writers have contented themselves with

COCOS nucifera.

The Cocoa-nut Palm: Its Oil.

OIL.

treating the subject as too well known to call for any detailed description, and at most only the meagrest accounts have been given. To the merchant desirous of starting a new or extending an existing trade, the question of primary importance to which he calls for a reply is the province or district with which he should open up dealings. The chief products of the cocoa-nut are coir-fibre, oil, and toddy, or the juice from which sugar and spirits may be prepared. We know that in Bombay the juice is largely extracted from the tree, that in Mysore the fibre is the chief preparation, and that in Madras and Travancore enormous quantities of both fibre and oil are exported; while Bengal, on the other hand, imports immense numbers of cocoa-nuts and a large quantity of copra, but exports very little of the products of the palm. It can nowhere, however, be discovered whether any two of these primary products or all of them, can be derived from the same trees or even prepared by the same cultivators—certain plants or portions of the plantation being periodically set apart for these several industries. Under coir fibre it has been said that the green or unripe cocoa-nut is alone used for that purpose, while most writers seem to agree that the ripe kernel is necessary for the oil. It would be most instructive to know if cultivation had resulted in the production of certain races of cocoa-nuts famous for their oil-yielding properties, just as the inhabitants of the Laccadive Islands appear to have developed a small-fruited one with a specially good fibre. In connection with commercial reports on cocoa-nut oil it is generally stated that the finest qualities are obtained from "Cochin." (**Spon** places Cochin after Ceylon.) It will be recollected that this same statement occurs regarding the fibre derived (or supposed to be derived) from that Native State. The writer has failed to discover any account of the Cochin oil industry, and is almost forced to the opinion that by "Cochin cocoa-nut oil," as with "Cochin coir," may be meant the superior qualities of the oil derived from the Madras Presidency. If ripe cocoanuts are essentially necessary for the preparation of the oil, then the Maldive and Nicobar Islands might be looked to as the great seats of the oil industry. But while these islands export perhaps little short of from 15 to 20 million ripe cocoa-nuts a year, they do not appear to manufacture cocoa-nut oil, *and the ripe husks are of no use for fibre*. So, in a like manner, the Laccadives would not be looked to as a source of oil: these islands are famous for their coir, the inhabitants growing a peculiar cocoa-nut that would seem to be inferior to the Malabar either as an oil-yielding or an edible nut. The imports from the Maldives and Nicobar Islands into Madras are very unimportant as compared with those recorded against Bengal, yet Madras, and not Bengal, appears to control the cocoa-nut oil market. This fact would lead to the inference that the locally grown nuts of Madras were largely employed for the expression of oil—the very considerable imports from the Laccadives affecting mainly the coir industry. But if this inference be correct, there remains the difficult position that the ripe nuts, serviceable for oil-making, yield no fibre. The presumption would therefore appear to be that a very much larger amount of the Madras coir comes from the Laccadives than we have any definite knowledge of at present, or that a large proportion of the coast cocoa-nuts or those of certain localities only are always or periodically set apart for oil-yielding. It may, of course, be the case that the trees are, so to speak, pruned by the removal for coir of so many green nuts from each tree, the remainder being allowed to ripen for oil purposes or as articles of diet.

This brief review, from want of definite information, may be accepted as indicating the direction that future reports might assume; but it may safely be concluded that, as with coir, so with cocoa-nut oil, Madras is the

Oil.

chief seat of the trade. Certain writers familiar only with Bengal (with the waving feathery clumps of cocoa-nuts dispersed through its suburban jungles or surrounding its mango topes) have advocated the claims of the Lower Provinces as a future region of oil production. This would appear to be a pure hallucination which the enormous imports of ripe nuts should have prevented. It is extremely doubtful if Bengal is ever likely to do more than meet the local and internal demand for ripe nuts and oil. The European oil merchant, if he finds the suggestion impracticable which has been offered in an early paragraph,—*viz.*, to call in the aid of the Maldive and Nicobar Islands,—will do well to concentrate his attention on the Madras Presidency.

Mode of Preparation of the Oil.—The ripe kernel is cut out of the shell in various ways, and either dried by exposure to the sun or by artificial means. It is then known as copra (*khoprá*). **Royle** says: "The Malabar method of making the oil is by dividing the kernels into two equal parts, which are ranged on shelves made of laths of the betel-nut palm or split bamboo, spaces of half an inch wide being left between each lath. Under 1571
these a charcoal fire is lit, and kept up for two or three days, in order to dry them, after which they are exposed to the sun on mats, and when thoroughly dried are subjected to pressure in an oil-press." **Balfour** remarks: "The purest oil is obtained by gathering the kernel and depositing it in some hollow vessel, to expose it to the heat of the sun during the day, and the oil drains away through the hollow spaces left for the purpose." **Hawkes** 1572
states that "the oil is generally prepared from the dried kernel of the nut, by expression in the ordinary native mills." The *Gazetteer of Thána* 1573
mentions three processes of making the oil. The first, giving origin to the oil there known as *khobrel*, is, as in the above quotations, a process of cold, Khobrel.
dry expression. The method of drying is described thus: "To make 1574
khobrel the kernel is taken from the shell by cutting the nut in half, called *váti*. After drying in the sun for a week the kernel is cut in thick pieces, which are crushed in the oil-mill."

But a hot wet process is also adopted by which an oil is obtained which seems to possess different properties from that prepared by cold expression. The *Thána Gazetteer* describes two such oils: "To make *ável* the Avel.
fresh kernel is scraped on an iron blade set in a wooden footstool. The 1575
scrapings are then put in a copper vessel over a slow fire, and after boiling are squeezed; sometimes instead of boiling them the scrapings are rubbed on a stone with a stone roller, and from time to time a little water is thrown over them. The scrapings are then squeezed and the juice boiled in a copper vessel, when the oil rises to the surface and is skimmed off. To make *muthel* dried kernels are cut into thick pieces and boiled Muthel.
in water. The pieces are then crushed in water and the whole is again 1576
boiled over a slow fire, when the oil rises to the surface and is skimmed off." It is worthy of careful observation that practically the difference between *ável* and *muthel* oil is, that the former is made from fresh kernel instead of from copra. **Dr. Shortt** says: "Boiled oil is obtained by bruising the kopra or the fresh cocoa-nut, mixing it with an equal quantity of water, and then boiling the mixture. As the water evaporates the oil rises to the surface. It is poured off, and the *débris* of the kernel is compressed by handfuls, so that any oil that remains may be extracted. Two 1577
quarts of oil are produced, on an average, from 15 to 20 nuts." In Borneo an oil expressed from the fresh cocoa-nut is used as a hair-oil, and is supposed, for that purpose, to be superior to oil obtained from copra. **Hawkes** says of the hot expression oil: "When required 1578
for edible purposes, the kernel of the fresh nut is taken, rasped and mixed with a little boiling water. This yields by pressure a milky fluid 1579

OIL.

which, on being boiled until all the water has evaporated, produces a clear edible oil. Only just sufficient water to moisten the pulp should be added, as a larger proportion prolongs the operation and deteriorates the product. When fresh prepared, this oil is comparatively free from smell, but speedily acquires an unpleasant odour; many attempts have been made to divest the oil of this smell, which renders it inapplicable for the perfumer's use, but only with partial success." Nearly every writer describes a different mode of preparing the oil obtained by the hot moist process. The reader is referred to a further page where this subject will be found to be dealt with under the head of *The Oil as a Medicine.*

As far as the writer's personal experience goes, the bulk of the oil met with in commerce is obtained by cold dry expression from copra or sun-dried kernel. There are no large mills for extracting the oil, but the crude oil-press used for oil-seeds generally, is employed, and every village in India has a few of these. They are accordingly to be found scattered not only throughout the cocoa-nut area, but even far into the interior where locally-produced or imported copra is expressed to meet the local demand for the oil. The production and consumption can therefore be judged of, and only approximately, by the foreign exports and internal traffic in the oil; of the local production and consumption nothing can be learned, except that the oil is universally used by the bulk of the people of India for some purpose or other.

In the Jury Reports of the Madras Exhibition interesting information regarding the extraction and yield of cocoa-nut oil has been recorded. "Half a hundredweight of the dried kernel is a charge for a full-sized *checko*" (or country mortar-like oil-mill), "and a pair of stout well-fed bullocks will get through four such charges in a day; so that twenty mills are required to get through two tons in the twenty-four hours. The man who drives has usually a boy to assist him in taking the oil, which is got out of the mortar by dipping a piece of rag into the fluid and squeezing it into an earthen vessel." So much oil do the cocoa-nuts contain that they
1580 are sometimes fixed on the end of poles and ignited for illuminations. For this purpose they are well suited, since both the fibrous husk and oil of the kernel burn brilliantly.

Properties of the Oil.—Cocoa-nut oil is a fixed fatty or greasy or non-drying oil, the specific gravity of which, according to **Royle**, is 0·892. When pure and freshly made it is of a pale-yellow colour, liquid in India, but having the consistence of lard in a temperate climate, and being then of a fine white colour. It becomes solid between 40° and 50° F. [according to *Spons' Encycl.* it solidifies at 16° to 18°C. (*e.g.*, 61° to 64½° F.), and according to **Shortt** at "65° to 70° F."] and liquid at about 80° F. It has a bland taste, and a peculiar, not disagreeable, odour. It is readily dissolved in alcohol. It consists mainly of a fatty principle designated *cocinin* with small amounts of *olein*. The *cocinin* when saponified with alkalies yields its glycerin and a cocinic acid (cocostearic acid, $C_{13} H_{26} O_2$). The oil has also been found to contain several solid and volatile acids, as *caproic, caprylic, capric,* and *pichuric* acid. (*U. S. Dispensatory.*) **Dr. Dymock** says that **Oudemans** regards the fatty principle as "a mixture of several glycerides. He has, moreover, shown that the fatty acid of the oil consists principally of *lauric acid* melting at 43° C., mixed *palmic acid* melting at 62° C., and myristic acid melting at 53·8° C. The oil also yields by saponification *caproic* and *caprylic* acids. It contains no *oleic* acid."

In *Spons' Encyclopædia* it is stated that "Its principal fatty acid is lauro-
1581 stearic, together with oleic, palmic, myristic, and some others of less importance all combined with glycerine." One of the most remarkable fea-

tures of this oil is that it will take up a larger amount of water than any other commercial oil. This makes it eminently suitable for soap-making, and but for the smell which such soap leaves on the skin the oil would be even more extensively employed by the soap-maker than it is. OIL.

Industrial and Domestic Uses of the Oil.—This oil has now for many years been largely used in the candle trade. CANDLES. 1582 **Messrs. Price & Co.** introduced in 1840, on the occasion of Her Majesty's marriage (when for illumination a cheap self-snuffing candle was required), a new composite candle, which was a mixture of stearic acid and cocoa-nut stearine. This was subsequently greatly improved until at one time cocoa-nut oil was the chief feature of **Price's** patent candles. The immense improvements which have taken place in the European candle industry have to some extent lessened the demand for the oil, but it is still largely employed. "It is an excellent illuminator, in both candles and lamps, as it emits no smoke."

Of no less importance is cocoa-nut oil to the soap-maker. "It forms a hard and very white soap, more soluble in salt-water than any other SOAP. 1583 kind made on a commercial scale. During the last fifteen years its consumption for soap-making in England has been greatly reduced by the competition of palm-kernel oil extracted here." (*Spons'.*) **Watt** (*The Art of Soap-making*, 27) says of this oil: "It is extensively used in soap-making, especially for the inferior kinds of soap, and will bear a large admixture of water, in combination with silicate of soda and other substances, and yet form a hard soap. All soaps made with even a small percentage of cocoa-nut oil impart an offensive smell to the skin after washing with it. This oil is very extensively used in the manufacture of artificial mottled soaps, but more especially in the north of England, where enormous quantities of it are consumed annually." "One of the most important additions to the list of fatty matters, suitable for soap-making, was the vegetable substance called *cocoa-nut oil* or *cocoa-butter*, which, from its extreme whiteness and capability of forming a hard soap, soon became an acceptable substitute in some degree for the more costly tallow. Soap made from this oil, or vegetable butter, is capable of taking up a larger percentage of water—and still forming a hard soap—than any other known fatty matter. The soap made from it, moreover, is more soluble in saline or "hard waters," even sea-water, and from this reason it has long been made into soap called *marine soap* for use on board ship." The odour which it imparts to the skin or garment washed with it will last for several hours. The odour resembles that of infants' vomit. On this account it should never be added to the ingredients used in the manufacture of a toilet soap. It does not readily saponify with caustic soda leys by itself, but does so readily when mixed with tallow or palm oil.

A large amount of the native-made soap used in India is prepared 1584 from this oil. This soap is made by boiling the oil with *dhobie's* earth, salt, saltpetre, quicklime, and water. The oil is also extensively employed by the richer people as lamp oil. It enters into the dietary of the mass of the inhabitants of India, and by them is largely employed to anoint the body or as a hair-oil; and variously-prepared forms of the oil are extensively used medicinally.

Prices and Yield of the Oil.—Speaking of the year 1854 **Hawkes** 1585 says: "The prices of this oil vary most considerably in different parts of the country. For the quarter ending 31st October, the maximum and minimum prices at nineteen large stations in all parts of the Madras Presidency were, R8-5-4 at Jubbulpore, and R1-14-0 at Bangalore, per maund. The average of 21 large stations in the presidency gives R4-9-5 per maund, or about £41 2*s.* per ton. The market value of "Cochin oil" in London in January 1855 was £46 10*s.* per ton, the average being from £46 to £48."

COCOS nucifera.

The Cocoa-nut Palm: Its Oil.

OIL.

Further on he adds that the "Cochin is usually 20*s*. per ton more than the Ceylon or Coromandel coast article." **Dr. Dymock**, in his last edition of the *Vegetable Materia Medica of India*, says: "The value of cocoa-nut oil in Bombay ranges from R16 to 20 per cwt."

The *Gazetteer of Kolhapur District* states that 48℔ of dried copra yield 26℔ of oil and 18℔ of oil-cake. This would be about 54·1 per cent. of oil. Another writer puts the yield down at 36 per cent. There are so many different modes of preparing the oil that, apart from the possibility of there being superior and inferior oil-yielding forms of the plant, it must necessarily be difficult to fix definitely what may be regarded as the yield. It may, however, be accepted as somewhere between 30 to 50 per cent. **Hawkes** states that each tree is calculated to yield at least 2½ gallons of oil per annum, and the coir obtained from the nuts is estimated to yield one fourth of the value of the oil, whilst the oil-cake is very valuable for cattle as a manure." It will be observed the idea seems to be conveyed in the above passage that the coir from the ripe or copra-yielding nut is of value. No other writer appears to support this opinion.

Royle says that 2 quarts of oil may be expressed from 14 to 15 cocoa-nuts. *Spons' Encyclopædia* states that in the ordinary country oil-mill 180℔ of copra will yield 40 quarts of oil, and that about 40 nuts are required to yield a gallon of oil. The trees grown on salt marshes are stated to yield much less oil than those grown on mixed sandy and loamy soils.

1586

Trade in Cocoa-nut Oil.

Royle remarks that the imports into Great Britain of cocoa-nut oil were in 1850, 98,039 cwt., of which India furnished 85,096 cwt. **Hawkes** states: "The average annual quantity exported from the Madras Presidency from 1850-51 to 1854-55 is about 1,410,963 gallons. Of this by far the largest proportion is sent to the United Kingdom and France, the remainder finding its way to Arabia, Mauritius, Bombay, and the French (Indian) ports." In 1850, as in the present day, the cocoa-nut oil trade almost entirely centred in Madras, so that the above passages may be taken as approximately indicating the extent of the foreign demand for the oil forty years ago. In 1880-81 the foreign exports amounted to 1,888,122 gallons valued at R20,90,797, Madras alone having shipped to foreign countries 1,690,520 gallons, and sent in addition by coasting trade to other Indian ports 1,493,756 gallons. In 1886-87 the exports were 1,099,864 gallons valued at R13,24,589, and the imports 556,562 gallons valued at R7,54,515. The bulk of the exports (*viz.*, 689,087 gallons) went to the United Kingdom, Madras alone shipping 1,090,480 gallons of the total exports. The imports were mainly from Ceylon (438,144 gallons), Bengal taking by far the largest proportion of these imports (*viz.*, 350,437 gallons). If to these facts an abstract of the coasting traffic be added, some idea of the present position of the cocoa-nut oil trade may be had. The imports coastwise were last year 1,567,486 gallons valued at R20,60,067; the exports were 1,942,809 valued at R20,74,455. These were the amounts of the oil that went to and from the various ports of India; but the full meaning of these figures will be brought out by giving some of the particulars of this exchange. Of the imports, Bombay received 794,577, Burma 338,056, Bengal 131,463 gallons, and these quantities were almost entirely obtained from Madras. Cochin sent to Bombay 15,789 gallons and to Madras 13,188 gallons. The other items to make up the total coastwise imports were unimportant. Local production added to these imports would constitute the supply from which the exports could be made, and in the case of Madras it is noteworthy that that presidency imported

OIL.

practically no cocoa-nut oil, so that her exports to foreign countries and to other Indian ports were drawn exclusively from local supplies. With the exception of the small amounts obtained from Cochin, Bombay, &c., and some 6,000 gallons from Ceylon and other foreign countries, Madras imported no cocoa-nut oil. But she exported 1,754,701 gallons, of which 1,008,621 went to Bombay, 273,347 to Burma, 191,413 to Travancore, and 155,202 gallons to Bengal. But Bengal exported coastwise 8,648 gallons and Bombay 3,454. The Bengal exports went to Burma and the Bombay to Sind, Madras, Goa, Kattywar, &c. Adding the foreign exports to the coastwise exports and deducting the total of the imports, we learn that Madras exported last year 3,425,221 gallons of cocoa-nut oil—an amount which may be viewed as the surplus over local consumption. Turning to Bengal and Bombay, a very different state of affairs is found to prevail, the imports exceed the exports, in the former by 313,009 gallons and in the latter by 1,125,572 gallons. By these amounts the local production did not equal the consumption plus the internal trade from these presidencies. Cocoa-nut oil is thus a speciality of Madras trade.

COPRA. 1587

Copra or Dried Kernel.

A very imperfect idea of the supply and demand for this oil would, however, be conveyed were we to omit to examine in this place the trade in copra or dried kernel, the substance from which the oil is expressed. This is largely exported to foreign countries and sent from one province of India to another to be locally made into oil.

	1884-85.		1885-86.		1886-87.	
	Cwt.	R	Cwt.	R	Cwt.	R
Imports . .	39,653	3,95,685	105,296	10,20,841	125,222	11,76,799
Exports . .	64,323	5,34,291	21,755	1,86,800	9,337	79,836

The imports come chiefly from Ceylon and the Straits Settlements, and are almost exclusively delivered in Bengal and Bombay, only very small amounts being received by Madras. The exports, on the other hand, go mainly from Madras (8,135 cwt. of last year's exports) Bombay being next in importance. The greater part of these exports (7,149 cwt.) go to Portugal, Persia, Ceylon, Russia, and Arabia, each receiving from 300 to 500 cwt. So far for the foreign traffic. The imports and exports coastwise have now to be considered. The total imports by coasting traffic were 347,255 cwt., valued at R35,31,115, and the exports 236,250 cwt., valued at R23,00,958. Of the imports, Bombay received 219,204 cwt., Bengal 62,971 cwt., Sind 34,658, Madras 27,025 cwt. Of the exports, Madras sent to other Indian ports 182,509 cwt. Bombay 53,295 cwt., Bengal exporting only 15 cwt. The balance between imports and exports will reveal, as with the traffic in oil, the fact that Madras is the great producing region, Bombay and Bengal receiving very much larger quantities than they export.

We shall revert to this subject again under Food since it seems possible a certain number of the ripe cocoa-nuts imported by various provinces may be ultimately used up in the preparation of oil.

OIL CAKE. 1588

Oil-cake or Punac.—Before passing from the consideration of cocoa-nut oil it is necessary to say something about the oil-cake. This is viewed as an exceedingly valuable manure, especially to cocoa-nut palms grown inland. It is also largely used to fatten fowls, pigs, cows, and other

COCOS nucifera.

The Cocoa-nut Palm as a Medicine.

animals. It is sometimes exported to Europe. In Madras it sells for 3 to 4 maunds (of 25℔) per rupee.

MEDICINE.

MEDICINE.

Fruit. 1589 The GREEN FRUIT is given as a refrigerant, the FLOWERS as an astringent, and the OIL employed as a substitute for cod-liver oil. The milk of the nut, the juice from the FLOWERING SPIKE, and the tomentum from the LEAVES are all used medicinally.

Flowers. 1590 Oil. 1591 Spike. 1592 Leaves. 1593 Water. 1594 WATER OR MILK FROM THE GREEN NUT.—"The WATER (or milk) of the unripe fruit is described as a fine-flavoured, cooling, refrigerant drink, usefui in thirst, fever, and urinary disorders." (*U. C. Dutt.*) "It may be drunk to almost any quantity without injury, and is considered by the native doctors as a purifier of the blood." (*Ainslie.*) It is commonly believed in Bengal, however, that too much cocoa-nut milk induces a hydrocele swelling of the scrotum.

Edible Pulp. 1595 THE EDIBLE PULP AND THE MILK PREPARED THEREFROM.—The PULP of the young fruit is nourishing, cooling, and diuretic. The pulp of the ripe fruit is hard and indigestible but is used for medicinal purposes. **Ainslie** says: "By scraping down the ripe kernel of the cocoa-nut and adding a litte water to it, a white fluid is obtained by pressure, which very much resembles the milk in taste and may be used as a substitute for it." "**Dr. Shortt** reports having successfully employed the fresh milk—*i.e.*, the EXPRESSED JUICE of the grated kernel—in debility, incipient phthisis, and cachetic affections, in doses of from 4 to 8 ounces twice or thrice daily. It has a pleasant taste, and may be used as an excellent substitute for cow's milk in coffee; it may thus be advantageously administered even to children. In large doses it proves aperient, and in some cases actively purgative; hence it is suggested by **Mr. Wood** as a substitute for castor oil and other nauseous purgatives." (*Pharm. Ind.*, 247.)

The following is a prescription known in Hindu medicine as *Narikela-khanda*: "Take of the pounded pulp of cocoa-nut half a seer, fry it in 8 *tolás* of clarified butter, and afterwards boil in 4 seers of cocoa-nut water till reduced to a syrupy consistence. Now add coriander, long pepper, bamboo manna, cumin seeds, nigella seeds, cardamoms, cinnamon, *teja-patra*, the tubers of **Cyperus rotundus** (*mustaka*) and the flowers of **Mesua ferrea** (*nága kesara*) 1 *tola*, each in fine powder, and prepare a confection, Dose 2 to 4 *tolás* in dyspepsia and consumption." (*U. C. Dutt, Hind. Mat. Med.*, 248.)

Shell. 1596 THE SHELL.—"The cleared SHELL of the nut or portions of it are burnt in a fire, and while red hot, covered by a stone cup. The fluid which is deposited in the interior of the cup is rubefacient, and is an effectual domestic remedy for ringworm." (*U. C. Dutt, p. 248.*) The *Bombay Gazetteer of the Thána District* alludes to this in the following words: "The shell when burnt yields an oil which is used as a cure for ringworm." "In the Antilles, the cocoa-nut is the popular remedy for tapeworm, and its efficacy has been conclusively demonstrated by medical men in Senegal. A cocoa-nut is opened and the almond extracted and scraped. Three hours after its

1597 administration a dose of castor oil is given. The worm is expelled in two hours afterwards. In nine cases in which this remedy was tried by a surgeon in Senegal the result was complete.—*Natal Mercury.*" (*Trop. Agri.*, 1882-83.)

OIL. 1598 THE OIL.—A reference to the account given of the ordinary oil in another page will reveal the fact that there are three or four oils obtained from the cocoa-nut, or rather three or four methods of preparing oil from it which seem to give to the substance different properties. In the Thána district, for example, three oils are prepared from the edible portion or ker-

MEDICINE.

Shell-Oil.

nel of the nut. These are known as ***khobrel, avel,*** and ***muthel.*** A fourth
oil is, however, repeatedly alluded to, namely, an oil prepared from the
shell of the nut (*see above*). This last-mentioned oil is perfectly distinct from
the oil of the kernel, and is used only in the treatment of ringworm. Its 1599
chemical properties have never apparently been determined, nor does it seem
to have before this been pointedly made known to European medical authorities as a substance actually prepared and employed by the Indian
doctors. It is remarkable that the same properties should be assigned
to the shell by the inhabitants of other parts of the world besides India,
although they do not apparently distil the oil from it. But of the kernel
oils used medicinally, the most conflicting statements have been published both as to their action and mode of preparation. Thus: "A very
cheap, hard, white soap is prepared from the oil, suitable for pharmaceutical purposes, such as plaster-making and the preparation of soap
liniment" (*Dymock*). The *Pharmacopœia*, on the other hand, says
this oil is inferior to ground-nut oil and sesamum oil as a vehicle for
liniments. **Sakharam Arjun** remarks: "The fresh oil is prepared for
medicinal purposes by boiling the milk of the ripe cocoa-nut. It is used
as an application for burns and in baldness." **Ainslie** observes it is ob- 1600
tained by boiling the bruised kernels in water, or "on other occasions it is
obtained by expression." **Drury** says: "The oil used internally for medicinal purposes is not the common commercial oil in its crude state, but the
oleine obtained by pressure refined by being treated with alkalies, and
then repeatedly washed and distilled with water." The therapeutic properties of the oil are discussed in the *United States Dispensatory*. "In
Germany it has been used in pharmacy, to a considerable extent, as a
substitute for lard, to which, according to **Petten Kofer**, it is preferable
on account of its less tendency to rancidity, its more ready absorption
when rubbed on the surface of the body, and its less liability to produce
chemical changes in the substance with which it is associated. Thus the
ointment of iodine of potassium, when made with lard, becomes yellow in
a few days, while if made with cocoa-nut oil it remains unchanged for
two months or more. Vegetable substances also keep better in ointment
prepared with this oil than with lard. Besides, it takes up one third more
water, which is a useful quality when it is desirable to apply saline solutions 1601
externally." "A preparation has been shown to us, said to be the liquid
part of cocoa-nut oil, prepared in London, and, under the name of *cocoolein*, used, instead of the oil itself, as a substitute for cod-liver oil. The
dose of this, as well as of the oil, is half a fluid ounce three times a day."

The various processes adopted in India for preparing oil from the cocoanut result in the formation of substances that are reputed to possess widely different properties. This fact might almost be supposed to be in consequence of chemically different oils being isolated. **Dr. Dymock** says
of the so-called ***muthel*** oil: "In the Konkan the oil which separates from
the freshly-rasped kernel, alone or mixed with tamarind-seed oil, is used 1602
under the name of ***mutel*** as an application to burns and rheumatic swellings; sometimes black pepper is added to it." In the *Thána Gazetteer* a
somewhat different process of preparing *mutel* (? *muthel*) oil is given. "To
make ***muthel,*** dried kernels are cut into thick pieces and boiled in water.
The pieces are then crushed in water and the whole boiled again over a
slow fire, when the oil rises to the surface and is skimmed off."

Cocoa-nut oil is said to promote the growth of hair; "hence it is much used as a local application in alopecia, and in loss of hair after fevers and debilitating diseases." "The oil is given in plethora and as a vermifuge in Jamaica. It is given while fasting, warmed and with a little sugar, in flux. An emulsia of the oil and kernel is prescribed in coughs

The Cocoa-nut Palm as a Medicine.

MEDICINE.

and pulmonary diseases generally. Pound the kernel with water, place it to settle, and skim off the cream. This is preferable to the expressed oil."

1603 " Cocoa-nut oil was proposed by the late **Dr. Theophilus Thompson** (*Proceed. of Royal Society, 1854, Pt. III., p. 41*) as a substitute for cod-liver oil; and in this character it has been favourably noticed by **Dr. J. H. Warren** (*Boston Med. and Surg. Journ., Vol. III., p. 377*) and others. The substance used in these cases was not the ordinary commercial oil, but the oleine obtained by pressure from the crude oil (in the solid state it is met with in England), refined by being treated with alkalies, and then repeatedly washed with distilled water. In his Lethsomian Lectures **Dr. Thompson** gives the result of his treatment with this agent in 53 cases of phthisis. Of the first 30, 19 were much benefited, in 5 the disease remained stationary, and in the remaining 6 the disease continued to advance. Of the second 23, 15 were materially benefited, 3 remaining stationary, and 5 became worse. **Dr. Garrod** (*Brit. and For. Med. Chir. Rev., Jan. 1856*) has shown that it exercises a marked influence, almost equal to cod-liver oil, in increasing the weight of the body. The great advantage of its employment experienced by **Dr. Thompson, Dr. Garrod**, and also by the Editor, who instituted some trials with it, is, that under its prolonged use it is apt to induce disturbance of the digestive organs and diarrhœa. Its use is favourably noticed in the Report of **Drs. Van Someren and Oswald, and Mr. J. Wood.**" (*Pharmacopœia of India.*)

Dr. Dymock says cocoa-nut oil has been tried in Europe as a substitute for cod-liver oil, "but its indigestibility is a great drawback to its general use." **Drury** observes: " its prolonged use, however, is attended with disadvantage, inasmuch as it is apt to disturb the digestive organs and induce diarrhœa." May it not be that the unfavourable opinions formed by some writers regarding this medicinal oil proceed from the fact that nearly every author describes a different mode of preparing it and consequently that it is possible many different substances or a substance in many stages of purity or impurity may have been experimented with? In the Maldives cocoa-nut oil is esteemed a powerful antidote against the bite of poisonous reptiles.

Juice. 1604 THE JUICE.—The freshly-drawn JUICE is considered refrigerant and diuretic, and is valuable as a preparation known as toddy poultice (see also under **Borassus, B. 677**). The fermented juice constitutes one of the spirituous liquors described by the ancient writers. "A tumblerful of the fresh juice is sometimes taken early in the morning on account of its refrigerant and slightly aperient properties." (*Dymock.*)

Husk. 1605 SCRAPINGS OF THE HUSK.—"The outside SCRAPINGS OF THE HUSK and branches applied to ulcers will cleanse and heal them rapidly if soaked in proof rum; the efficacy of this application was proved by the case of two bad ulcers occasioned by the bite of a negro's teeth. The young roots boiled with ginger and salt are efficacious in fevers, the same as the bamboo." (*Royle.*)

Tomentum. 1606 THE COTTON OR TOMENTUM.—"This is a soft, downy, light-brown-coloured substance, found on the outside of the lower part of the branches of the cocoa-nut tree, where they spring from the stem, and are partially covered with what is called *panaday*, or coarse vegetable matting of the tree. The COCOA-NUT COTTON is used by the Indians for stopping blood, in cases of wounds, bruises, leech-bites, &c., for which purposes it is admirably fitted by its peculiar texture." (*Ainslie, Mat. Ind.*) (Compare with tomentum of **Caryota urens** and of **Borassus, B. 680.** See also under **Tinder.**)

The Cocoa-nut Palm as a Medicine.

MEDICINE.

Flowers. 1607

THE FLOWERS.—Are sometimes used medicinally, being said to be astringent.

Nuts. 1608

IMMATURE NUTS.—These, like the flowers, are often employed medicinally, especially as an astringent in the sore-throats of children.

Roots. 1609

THE ROOT.—"The ROOT is used as a diuretic, as also in uterine diseases." (*U. C. Dutt, 248.*) It is also employed as an astringent gargle in sore-throat.

Ashes. 1601

THE ASHES.—"The ASHES of the leaves contain an amount of potash; they are used medicinally."

Bud. 1611

THE BUD.—The tender buds of this palm, as also of **Borassus** and **Phœnix,** are esteemed as a nourishing, strengthening, and agreeable vegetable.

Special Opinions.—§ "The husk of the fruit of the **Cocos nucifera** is used in the treatment of tapeworm. It is often as efficacious as the oil of male fern when taken on an empty stomach" (*Surgeon-Major W. Nolan, M.D., Bombay*). "The juice of an immature fruit is used in acidity and gastric irritation. The volatile oil from the nut-shell is employed as a local application in ringworm" (*Civil Surgeon J. H. Thornton, B.A., M.B., Monghyr*). "Scraped cocoa-nut is used as an application in eczema of the scrotum. The milk of a young cocoa-nut drunk every morning is a popular domestic diuretic and diluent for very old men" (*Surgeon-Major John North, I.M.S., 1st Madras Cavalry, Bangalore*). "A black oil is extracted from the shell, and is used in itch and other parasitic affections" (*Surgeon-Major R. Thomson, M.D., C.I.E., Madras*). "The cocoa-nut milk of the green fruit is a cooling, refrigerant drink, containing albumen and salines. It is a good drink in cholera cases. It succeeds in checking vomiting when other means fail. Cocoa-nut oil, prepared from fresh pulp, is a good substitute for cod-liver oil. The dose I give is from 20 to 30 minims in the beginning, rising to a drachm thrice daily. An ash is prepared from cocoa-nut pulp by the *Koberajes* which is a valuable ant-acid and digestive. It is called '*Narkel khond.*' A sweet extract is also prepared, which is used for similar purposes" (*Civil Surgeon R. L. Dutt, M.D., Pubna*). "The sweet toddy obtained from this palm is very refreshing and possesses laxative properties. Its continued use (twice or thrice weekly) during pregnancy has a marked effect on the colour of the infant, which is born of a fair complexion,—*i.e.,* if of dark parents, comparatively fair; if of lighter-coloured parents, the offspring generally assumes a European complexion" (*Honorary Surgeon P. Kinsley, Chicacole, Ganjam, Madras Presidency*). "The milk of the green fruit is cooling and is given to allay vomiting in bilious fevers. Cocoa-nut oil inunction has been found to be useful in tuberculous affection of the skin. It improves the general health like cod-liver oil" (*Assistant Surgeon Shib Chunder Bhattacharji, in Civil Medical charge; Chanda, Central Provinces*). "The water contained in the green fruit is anti-emetic and soothing" (*Civil Surgeon D. Basu, Faridpur*). "Preserved cocoa-nut (*Narikel khondo*) is used by the *Kobirajes* as an alterative in cases of chronic heartburn and phthisis pulmonalis" (*Assistant Surgeon Anund Chunder Mukerji, Noakhalli*). "The oil is extensively used to fatten and is given for phthisis" (*Surgeon-Major Lionel Beech, Cocanada*). "An excellent substitute for cod-liver oil; very much used here" (*Civil Surgeon C. J. W. Meadows, Burrisal*). "The oil promotes the growth of the hair" (*Civil Surgeon J. Anderson, M.B., Bijnor*). "The oil is considered to increase the growth of hair and render it black (*A Civil Surgeon*). "If the flowers are mixed with sugar, the root of *khus-khus,* and white *chandan,* with a little water, the combination will be found good in bilious fever, will check vomiting, and produce a cooling

COCOS nucifera. **The Cocoa-nut Palm: Its Edible Products.**

MEDICINE. sensation" (*Civil Surgeon William Wilson, Bogra*). Useful "in dysentery, diarrhœa, menorrhœa, and stomatitis" (*Native Surgeon T. Ruthnam Moodelliar, Chingleput, Madras Presidency*). "**C. mamillaris**, dwarf
1612 cocoa-nut tree, Pemba, East Africa: fruit large, smooth, distinctly three-cornered: pinkish yellow when ripe: without the fibrous pericarp of the common cocoa-nut. Yields very little oil, but supplies a refreshing drink in fevers and in hot weather, and is said to produce free diuresis: used when the nut is full grown, but before it begins to ripen. *Vern.* of East Africa, *Muazi*. **C. nucifera**; *Muazi-ya* Pemba, **C. mamillaris** (*Surgeon-Major John Robb, M.D., Surat, Bombay Presidency*).

FOOD.

FOOD PRODUCTS.

Under the head of food products obtained from this palm we may note the following:—

Cocoa-nut Cabbage. 1613 *Cocoa-nut Cabbage.*—This is the terminal bud at the summit of the tree. It is used as a vegetable and also makes an excellent pickle, but to obtain it the palm is killed.

Young Cocoa-nut. 1614 *Young Cocoa-nut* (VERN. *dáb*).—This is the tender fruit, plucked off the tree for the cooling, sweetish, clear water, and the soft, cream-like pulp it contains. The water is drunk and the pulp eaten by natives of all classes.

Mature Cocoa-nut. 1615 *Mature Cocoa-nut* (VERN. *jhúna narkel*).—This is the fruit in its mature state, with its outer, thick, fibrous covering completely dried. It contains less water, but has a thicker and harder albuminous layer than the tender fruit; when dried this albuminous substance is known as copra. It is eaten with parched rice, or rasped and put into curries or made into sweetmeats. Copra is either allowed to ripen and dry within the shell, when it separates naturally and is removed entire, or the shell is broken and the copra cut out and dried either in the sun or over fires. The former exists in large pear-shaped pieces smaller than, but of the same shape as, the interior of the nut, and is known as "natural copra." The latter occurs as the irregularly-cut pieces known as "artificial copra." An oil is extracted from copra which is employed for various culinary purposes, and is also exported to a certain extent. (For further particulars regarding copra see under the chapter OIL.)

Juice. 1616 *Juice.*—The cocoa-nut also yields a juice from the flowering spike which may be drunk in its fresh state—*toddy*; or fermented and distilled—*arak*; or boiled down to sugar and eaten as—*jaggery*.

Root. 1617 *The Root.*—This is said to be chewed like areca-nut with betel-leaf in *pán*.

NUTS. 1618

THE NUTS.

The above is a brief abstract of the food products of this palm. The extent to which the unripe fruit is cut, the water and unripe kernel being consumed and the husk made into coir, may be partly inferred from what has been already said regarding the fibre. To a large population in India the cocoa-nut is almost a staple article of diet.

TRADE In nuts. 1619 TRADE IN COCOA-NUTS.—The trade in the ripe cocoa-nut is very extensive. We have repeatedly alluded to this in the foregoing remarks, and may rest satisfied by briefly reviewing here the main facts of the foreign and coasting trade in these nuts, as recorded in **Mr. J. E. O'Conor's** Statement of the Trade and Navigation of British India.

Foreign.—In the year 1886-87 India imported from foreign countries ripe cocoa-nuts to the number of 15,596,918, valued at R5,98,203: she exported to foreign countries 275,230 nuts, valued at R8,462. Of the imports to India the Maldives sent 7,897,453, the Straits Settlements 5,542,758, Ceylon

The Cocoa-nut Palm: Its Edible Products.

COCOS nucifera.

TRADE IN NUTS.

1,434,821, and East Africa 627,346. Of these imports Bengal took 8,430,229, valued at R1,75,552, Burma 5,618,949, valued at R3,72,702, Bombay and Madras each received 700,000, and Sind 86,800. Bengal exported no cocoa-nuts to foreign countries, but Bombay and Madras each sent about 150,000 to Egypt, Arabia, and Turkey in Asia. The foreign trade in ripe cocoa-nuts is therefore very unimportant, and but for the Maldives being viewed as foreign territory (while the Laccadives and Nicobar Islands are not), it would be scarcely worthy of notice. It is noteworthy that India at present takes practically no part in meeting the English market. The passage already quoted from **Mr. Jackson's** paper in the *Planters' Gazette* conveys some idea of the magnitude of the British demand. Speaking, however, of the extent to which the nuts are eaten or made into confectionery, he continues: "Cocoa-nuts are largely used in the north and west of England, and they are also in great demand at holiday times, at fairs, on race-courses, and such like gatherings in all parts of the kingdom." He also remarks that the Trinidad nuts are considered the sweetest in flavour, and are mostly preferred by the confectioners and bakers, though the Ceylon run them close.

Indian.—The coastwise trade or interprovincial exchange is, however, very important. The total imports from one port to another were last year 91,756,349 nuts, valued at R24,21,941, and the exports 68,938,032, valued at R16,88,773. Of the imports Bengal received 2,998,617, Bombay 77,984,733, Sind 1,172,819, Madras 5,541,718, and Burma 4,058,462. The imports into Bengal and Burma were largely from the Nicobar Islands and into Bombay from Madras (47,268,590), from Goa (18,825,703), and from ports within the Bombay Presidency, while the imports into Madras were from other ports within the presidency (5,162,354), presumably a large proportion of which were from the Laccadives.

Of the coastwise exports in 1886-87 Bengal sent to Burma, according to one official table of coastwise trade, 1,676,773, but according to another only 6,890. Bombay, in addition to its exchanges between its own ports, sent considerable numbers to Sind, Kattywar, Cutch, Cambay, &c. Madras gave more than three fourths of its coastwise exported nuts to Bombay, with a little over a million nuts to Bengal, the same number to Burma, and 2,591,475 to Cutch. Burma exports no cocoa-nuts, but it seems probable that some of its imports, which appear as from Bengal, may be from the Nicobar Islands. These islands being associated in trade returns with Bengal, direct exports may occasionally not appear as exports from Bengal; hence, in all probability, the disparity in the figures of imports into Burma alluded to above.

Juice from the Cocoa-nut.

JUICE. Madras. 1620

Dr. Hugh Cleghorn has described as follows the process of tapping the palm for its juice in Madras—a process which is essentially that followed in Bombay and other parts of the country: this palm is not tapped in Bengal. When the spathe is a month old, the flower-bud is considered sufficiently juicy to yield a fair return to the (*Sánár*) toddy-drawer, who ascends the tree with surprising ease and apparent security, furnished with the apparatus of his vocation. A year's practice is requisite before the Sánár becomes an expert climber. The spathe when ready for tapping is 2 feet long and 3 inches thick. It is tightly bound with strips of young leaves to prevent expansion, and the point is cut off transversely to the extent of one inch. He gently hammers the cut end of the spathe to crush the flowers thereby exposed and to determine the sap to the wounded part, that the juice may flow freely. The stump is then bound up with a broad strip of fibre. This process

COCOS nucifera.

The Cocoa-nut Palm: Toddy.

JUICE.

is repeated morning and evening for a number of days, a thin layer being shaved off on each occasion, and the spathe at the same time trained to bend downwards. The time required for this initiatory process varies from five to fifteen days in different places. The time when the spathe is ready to yield toddy is correctly ascertained by the chattering of birds, the crowding of insects, the dropping of juice, and other signs unmistakeable to the Sánár. The end of the spathe is then fixed into an earthen vessel called *kudave,* and a slip of leaf is pricked into the flower to catch the oozing liquor and convey the drops clear into the vessel. When the juice begins to flow the hammering is discontinued. A single spathe will continue to yield toddy for about a month, during which time the Sánár mounts the tree twice a day and empties the juice into his *eropetty* (a vessel made of closely-plaited palmyra fibre), and repeats the process mentioned above of binding and cutting the spathe an inch lower down, and inserting its extremity into the *kudave.* The flow is less during the heat of the day than at night. One man will thus attend to 30 or 40 trees. Forty trees will yield about 12 Madras measures (1½ to 2 gallons) of juice—7 measures in the morning and 5 in the evening. This is at the rate of about a quarter of a measure per tree. The length of time a tree will continue to yield varies from six months to a year in very favourable soil. But it is not considered prudent to draw all the juice one can from a tree, as it will then become barren all the sooner. **Dr. Shortt** says the quantity of sap a tree will yield varies according to locality and the age of the spathe; 3 to 4 quarts is the average quantity obtained in 24 hours for a fortnight or three weeks. "Sometimes this fluid is converted into what is termed *nira* by lime-washing the vessels that collect the fluid in order to neutralise the acidity. It is then sold as a sweet and refreshing drink in the bazaars." "Toddy," he proceeds to say, "is also boiled down into a coarse kind of sugar called *jaggery,* which is converted into molasses for the manufacture of spirits, or refined into white or brown sugar before fermentation sets in."

Bombay. 1621

In Bombay the cocoa-nut palm is tapped for its juice in Ratnágiri (*Gaz., X., 34*), in Kolába (*XI., 28*), in Khándesh (*XII., 321*), in Thána (*XIII., Part I., 295*), and in Kánara (*XV., Part I., 58; Part II., 205*). According to the returns the writer has had access to, there are some 3½ million trees in Bombay, of which about 30,000 to 40,000 are tapped for their juice.

The following abstract from the *Kolába and Ratnágiri Gazetteers* may be accepted as fairly representing the process of tapping pursued in Bombay, the yield, rent paid, return and profit being there shown. The cocoa-nut gardens are generally owned by Hindus, who let the trees to rich Bhandáris, who agree to supply the owner of the liquor shops with fermented or distilled juice. From the very earliest times cocoa-nut trees have been taxed, a distinction being made between trees kept for fruit and those set apart to be tapped. In the Ratnágiri district, it is stated, toddy trees let at from 2*s.* to 6*s.* (R1 to R3) a year. In addition to rent, a Government tax on trees tapped has to be paid. The maximum leviable rate was in Malabar and Deogad 2½*d.* (1 anna 8 pie) a month or 2*s.* 6*d.* (R1¼) a year on each tree tapped. Under the new system a special license is granted to tap trees, at a fixed rate for each tree, and under certain conditions as to the number of trees included in the license. The licensees are allowed to sell toddy by retail at the foot of the trees, but not to distil, the latter privilege being vested exclusively in the licensed shop-keepers for the sale of country spirit. In Kolába, it is said, the crude juice of fifteen trees costs the Bhandári about £1-2*s.* (R11) a month or 1*s.* 6*d.* (12 annas) per each tree. Besides the wages of the distiller and cost of fuel the Bhandári has to make good to the liquor-shop keeper part of the tap-

COCOS nucifera.

The Cocoa-nut Palm: Toddy.

JUICE.

ping tax he had paid to Government. Government levies from the liquor-shop keepers £60 (R600) a year for every hundred trees tapped. Three fourths of this the liquor-shop keeper pays; the remaining fourth he recovers from the Bhandári who supplies the liquor. The Bhandâri's share of the tax amounts to £15 (R150) on one hundred trees for one year,—that is, a monthly charge of £1-5s. (R12½) on the one hundred trees, or on each tree a monthly tax of 3*d.* (2 annas).

In Ratnágiri the yield is said to vary from 35 to 64 imperial gallons from each tree. In Kolába a tree is said to yield on an average 4½ pints (1¾ seers) of juice a day, or 10½ imperial gallons a month. The juice is seldom sold raw: most of it is distilled by the Bhandári and sold by him to the liquor-shop keeper. With the wages of an assistant the monthly charge for distilling the produce of one tree is about 2*d.* (1¼ annas). The cost of fuel is about 6*d.* (4 annas) more, or about 8*d.* in all. Distilling lowers the quantity of liquor by about one half,—that is, it reduces the average monthly outturn of each tree from 10½ to 5½ gallons of spirit. Taking everything into consideration, tapping, distilling, &c., the Bhandári pays about 2*s.* 5*d.* (R1-3-3) for the produce of each palm. Allowing for loss by estimating, instead of 5½, only 5 gallons, and he obtains 3*s.* (R1-8) for the spirit prepared from each palm. This leaves him a net profit of 7*d.* (4½ annas) on each tree, and if he possesses a plantation of 300 trees he makes a fairly good income.

Spirit.

Of Ratnágiri, it is said, there are ordinarily three kinds of palm spirit,
known respectively as *rási*, *phul* or *dharti*, and *phéni*; *rási* being the (Rasi.)
weakest and *phéni* the strongest. In some places a still stronger spirit 1622
called *duvási* is manufactured. The average wholesale rates at which (Phul.)
the farmers buy stock from the manufacturers are for the imperial gallon, 1623
tádi 2¾*d.* (1 anna 10 pie), *rási* 8¾*d.* (5 annas 7 pie), *phul* 1*s.* 1⅛*d.* (8 (Pheni.) 1624
annas 9 pie), *phéni* 2*s.* 6¾*d.* (R1-4-6), and *duvási* 4*s.* 9½*d.* (R2-6-4). The spirits are distilled in private stills, licensed to be kept at certain Bhandáris' houses under fixed conditions as required, in proportion to the number of trees licensed to be tapped in the vicinity. One still is usually allowed for every 100 trees, and the still-pot is limited to a capacity of 20 gallons.

TARI. 1625

Fermented and Unfermented Beverage.

This is one of the forms of the so-called palm-wine so much extolled by the early European visitors to India. From what has been said in the preceding pages regarding the juice it may have been inferred that, if left for a short time after removal from the tree, it rapidly ferments and becomes intoxicating. This is the *tari* or toddy (or in the case of the cocoa-nut more specifically known as the *níra*), a beverage very extensively consumed in India. Fermentation is said to be prevented by the addition of a little lime to the fluid. The earthen vessels into which it drains are generally powdered with lime when the fluid is to be drunk in its fresh unfermented state, or is intended to be boiled down to sugar or *jaggery.* It is also drawn early in the morning instead of being left on the tree overday. Robinson says of the Laccadive islanders that "they are still so strict in the abstinence from all fermented liquors, that the manufacture of toddy would not be tolerated in the islands." Self-fermented toddy is extensively used by the bakers in India in place of yeast. When fermented the juice may be distilled into spirits or made into vinegar. One hundred gallons of *tari* yields on an average twenty-five of *arak* by distillation.

COCOS nucifera.

The Cocoa-nut Palm: Sugar.

PALM SUGAR.

PALM SUGAR.

Instead of being fermented, the liquor may be evaporated down and its sugar thus extracted. "Eight gallons of sweet toddy, boiled over a slow fire, yield 2 gallons of a lusciously-sweet liquid, which is called *jaggery* or sugar-water, which quantity being again boiled, the coarse brown sugar called *jaggery* is produced. The lumps of this are separately tied up in dried banana leaves" (*Royle*). **Dr. Shortt** says: "The sap is poured into large pots over an oven, beneath which a strong wood-fire is kept burning, the dead fronds and other refuse of the plants being used as fuel. The sap soon assumes a dark-brown semi-viscid mass, well known as *jaggery* or *gúr*, which whilst warm is poured into earthen pots or pans for preservation. Ten to twelve seers of the sap yield one of *jaggery;* the value of a maund of this *jaggery* is about 2 rupees. In this state it is sold to *abkari* contractors, sugar refiners, or merchants. The sugar refined comprises several sorts, known in the market as moist, raw, coarse, and fine sugar. The *jaggery* is placed in baskets and allowed to drain; the watery portion or molasses dropping into a pan placed below. This is repeated, so that the *jaggery* or sugar becomes comparatively white and free from molasses. This sugar—for so it may now be called—is put out to dry, and the lumps broken up; when dry it is termed raw sugar, and weighs about 25 per cent. of the whole mass, the rest of it

Refined. 1626 being collected in the form of molasses." Thus cocoa-nut sugar is chiefly met with in the form of *jaggery*. It is well known, however, that it is capable of being refined, according to European principles, and a certain amount of cocoa-nut sugar is regularly prepared. "The success of **Dr. J. N. Fonseca** (author of the *History of Goa*), in converting toddy of the cocoa-nut tree into crystallized sugar, has been hailed with satisfaction by the press at Goa, and flattering calculations are made of the advantages that will accrue to the country from the development of this new industry" (*Bombay Gazette*). A similar sugar is prepared from the date-palm, from the palmyra-palm, and from the Indian sago-palm (**Caryota urens**). The date-palm is very largely used for this purpose in Bengal, and the cocoa-nut and palmyra palms in Madras, while in Bombay, apparently, sugar is only very occasionally made from the juices of these trees; but when extracted it is most generally prepared from the palmyra or Caryota palms. Some years ago the Government of Bombay, getting alarmed at the growth of the habit of toddy-drinking, brought Jessore sugar manufacturers to try the experiment of preparing sugar from the date-palms of the western presidency. According to the returns of the Surat district there are in that district alone 1,195,901 date-palm trees, of which 489,395 were tapped in 1867-68. But it was found that the returns from sugar manufacture were so poor, as compared to the profits from the sale of *tari*, that the experiment practically failed. It is not known whether or not sugar to any appreciable extent is actually prepared from the Bombay palms, nor even whether a license is necessary to tap trees for sap intended to be so used. Of the Thána district it is said: "Coarse sugar or *gûl* is also made by boiling the juice in an earthen pot over a slow fire." It is worth recording that, according to the Gazetteers, there are 3,500,000 cocoa-nut trees in Bombay, of which 50,000 are regularly licensed. Of palmyra palms there are said to be 47,810 trees in Surat alone, of which 16,739 are regularly tapped. Of Caryota palms there are 70,000 trees, of which about 20,000 are tapped; 48,900 of these occur in Kánara, 21,672 in Kolába, and the remainder in Ratnágiri.

In a recent report on the trade in Indian sugar issued by the Revenue and Agricultural Department, no mention is made of palm sugar being

The Cocoa-nut Palm: Sugar.

PALM SUGAR.

prepared in Bombay, so that it may be inferred the trees licensed to be tapped are employed entirely in the supply of toddy. It is noteworthy, in passing, that it should pay the Bengal and Madras people to make sugar from the palm juice, while it is viewed in Bombay as unprofitable to do so. Turning to Madras it will be found that the report on sugar alluded to above gives the area under the sugar-yielding palms in 1881-82 as—

	Acres.
Palmyra	24,900
Cocoa-nut	5,700
Date	1,600
	32,200

The writer of that report adds: "In 1884-85 and 1885-86 the area under cocoa-nut, date palms, and palmyras was 31,000 acres and 28,000 acres respectively, and the outturn 22·60 lakhs maunds and 19·98 lakhs maunds. The total quantity of jaggery produced from cocoa-nuts, &c., is apparently more than that obtained from sugar-cane." In a special report on the cocoa-nut issued by the Revenue and Agricultural Department in 1886 it was estimated that there were 7,7765 acres under that palm. Taking the customary estimate of 100 trees to the acre, we arrive at the conclusion that out of a total of 7,776,500 trees, 570,000 were tapped, or perhaps only tapped for sugar, others being tapped for toddy. There exists in all the works and reports the writer has been able to consult the greatest possible confusion as to whether or not the trees may be tapped for sugar without paying the license levied on the tappings made with the view to the preparation of the beverage. It would be instructive to know if the 5,700 acres of cocoa-nuts in the above statement of Madras are exclusively set apart for sugar, and are independent of the trees spoken of in excise reports as licensed for the preparation of toddy. If every tree tapped has to pay the heavy tax imposed on the preparation of the toddy, it might fairly be inferred that the failure to develope a palm-sugar industry proceeded to some extent from that fact. But there are many other difficulties to the creation of a large trade in palm sugar. In this respect the following passage will be found instructive:—

"From time immemorial (*sic*) the natives of Ceylon have known* how to produce crystallized sugar from the inspissated juice of the cocoa-nut spathe. About thirty years ago, in consequence of a letter from the late **Mr. J. Glanville Taylor**, of Batticaloa, asking for information as to the probable success of attempting to utilize cocoa-nut palms for sugar-making, we went fully into the matter, receiving considerable assistance from **Mr. D. C. Amesekere**, a proctor who, when we last heard of him, was practising at Kurunegala. On that occasion he sent us a quantity of crystallized cocoa-nut tree sugar, which, however, was somewhat discoloured by smoke. The result of our enquiries was that, although the juice when collected was rich in saccharine matter, yet the cost of collection would render the enterprise unprofitable. What pays natives on a small scale will not pay Europeans when the matter is entered into on commercial principles. An experiment might be tried, however, labour being economised by the use of ladders, perhaps, and a larger use than the natives make in toddy-drawing, of safe passages from tree to tree." (*Tropical Agriculturist, 1882-83, 568.*)

* **De Candolle**, quoting from **Seeman**, says, upon a rock near Point de Galle may be seen "the figure of a native prince, Kotah Roya, to whom is attributed the discovery of the uses of the cocoa-nut, unknown before him; and the earliest chronicle of Ceylon, the *Marawansa*, does not mention this tree, although it carefully reports the fruits imported by different princes."

COCOS nucifera.

The Cocoa-nut Palm: Spirit.

CEMENT. 1627

CEMENT MADE OF LIME AND COCOA-NUT JAGGERY.

So often is this subject alluded to that it deserves special notice. The practice seems to prevail through Southern and Western India, but is apparently not followed, or more likely unknown, to the natives of Bengal. In the *Thána Gazetteer* it is stated: "Mixed with lime, this palm sugar makes excellent cement." **Drury** remarks: "This jaggery is mixed with *chunam* for making a strong cement, enabling it to resist great heat and to take a fine polish." "The water of the nut is used by the bricklayers in preparing a fine whitewash, also in making the best and purest castor oil, a certain proportion of it being mixed with the water in which the seeds are boiled."

In *Spons' Encyclopædia* there occurs the following regarding Ceylon jaggery: "Amongst a variety of purposes to which it is put is that of being mixed with the white of eggs and with lime from burnt coral or shells. The result is a tenaceous cement, capable of receiving so beautiful a polish that it can only with difficulty be distinguished from the finest white marble."

This subject appears to be well worthy of chemical investigation, for there seems every reason to presume that the property of this ingredient in combination with lime might, with great advantage, be employed to replace the whitewashes commonly used, to the injury of the garments of whoever may lean against walls so coloured. (*Conf. with opening sentences under Domestic Uses, and the account given under* **Dye, C. 1547.**)

SPIRIT. 1628

PALM SPIRIT OR ARAK.

Instead of being consumed as a fermented beverage the palm wine may be distilled, thus forming palm spirit or arak. As no separate record is kept of the cocoa-nut from the other palm spirits, we shall rest satisfied with what has been said regarding the trees licensed to be tapped. Under SPIRIT will be found further particulars regarding the method of taxation and process of distillation generally pursued. The present notice of cocoa-nut spirit may therefore be concluded by the following note kindly furnished for this work:—

"**Dr. Lyon**, of Bombay, has recorded some interesting details regarding the alcoholic strength of toddy from the cocoa-nut, date and brab. In the following table is shown the average alcoholic strength of six night-collected samples of each of the three kinds of toddy at respectively three and eight hours after collection, and the average maximum alcoholic strength attained by the samples; as well as the strength of samples collected during the twelve day-hours, when examined the morning after collection:—

	PROOF SPIRIT PER CENT.		
	Cocoa-nut.	Date-palm.	Brab (Borassus).
Night samples.			
3 hours after collection	7·15	5·8	3·9
8 " " "	10·0	8·0	4·7
Maximum strength	11·9	11·0	7·9
Day samples.			
15 hours after collection	10·8	11·7	6·5

SPIRIT.

"Dr. Lyon finds that in toddy collected in pots which have previously been used, fermentation commences before the pots are removed from the tree. The toddy appears to attain its maximum strength within 24 hours after removal from the tree. The volume of toddy yielded is greater during the twelve night than twelve day hours. Comparing trees of the same class, a yield of toddy above the average is, as a rule, accompanied by attainment of alcoholic strength above the average. The rapidity of fermentation and yield of alcohol varies for different samples. (*Chemical Examiner's Annual Report, Bombay.*)"—*Prof. C. Warden, Calcutta.*

The reader is referred for further particulars to **Borassus flabelliformis (B. 683), Caryota urens (C. 714),** to **Narcotics,** to **Phœnix sylvestris,** and to **Spirits.**

VINEGAR. 1629

Vinegar from Palm Wine.—Nearly every writer who has dealt with the subject of the useful products of the cocoa-nut alludes to the vinegar prepared from the juice. "One hundred gallons of toddy produce by distillation, it is said, twenty-five of *arak.* Or it may be allowed to undergo the acetous fermentation and produce very good vinegar. Or instead of being allowed to ferment, the toddy may be made to yield *jaggery* or sugar. For this purpose a supply of sweet toddy is procured mornings and evenings, particular care being taken that the vessels employed have been well cleaned and dried." (*Royle, Fib. Pl.*)

The vinegars prepared from the juice of the various palms that yield such juices do not appear to have been carefully examined. The natives of India attribute peculiar properties to each.

STRUCTURE OF THE WOOD.

TIMBER. 1630

Outer wood close-grained, hard, and heavy. Vascular bundles black or dark purple, closely packed in the outer part of the stem or horizontal section, circular or uniform, enclosing vessels and cells.

The wood is commercially known as "Porcupine wood;" it is used for rafters and ridge-poles, house-posts, and other building purposes; for spear-handles, walking-sticks, and fancy work. When freshly cut "it possesses great elasticity, and is for this reason particularly well adapted for temporary stockades which are exposed to cannon-shot." (*Drury.*)

DOMESTIC SACRED USES.

DOMESTIC. 1631

So many of these have already been alluded to that it is scarcely necessary to attempt to enumerate the thousand and one uses to which the palm is put by the people of India. Under sugar or *jaggery* on the opposite page will be found a brief notice of the very interesting uses of the juice and milk in the preparation of polishing cements. This art is much practised in Madras, the smooth shining wall-plasters being much admired. By Hindús, the dried shell is almost universally used as the water-bowl of their smoking-pipes or *húkah.* In Madras these shells are made into elegantly-carved ornamental vases, lamps, spoons, sugar-pots, tea-pots, &c. Cocoat-nuts are largely employed as offerings to the gods by the Hindús, and cocoa-nut day (the full moon in August) is celebrated throughout the country. Entire shells are obtained by filling them with salt water and burying them in the sand for a time. By this process the kernel is destroyed and may be washed out. If thoroughly ripened before being treated in this manner, the shell will keep for many years; if not, it will soon rot. The dried kernel is sometimes cut into ornaments such as flowers or garlands; these are worn by Hindú women. The following extract gives a graphic account of the manner in which the cocoa-nut enters into the every-day life of the people of the tropics:—

Hukah Bowls. 1632

Ornamental Objects. 1633

Spoons. 1634

Sugar-pots. 1635

Tea-pots. 1636

Dickens in *Household Words* says: "To a native of Ceylon the

COCOS nucifera.

The Cocoa-nut Palm: Domestic Appliances.

DOMESTIC.

cocoa-nut palm calls up a wide range of ideas; it associates itself with nearly every want and convenience of his life. It might tempt him to assert that if he were placed upon the earth with nothing else whatever to minister to his necessities than the cocoa-nut tree, he could pass his existence in happiness and content. When the Cingalese villager has felled one of these trees after it has ceased bearing (say in its seventieth year), with its trunk he builds his hut and his bullock-stall, which he thatches with its leaves. His bolts and bars are slips of the bark, by which he also suspends the small shelf which holds the stock of home-made utensils and vessels. He fences his little plot of chillies, tobacco, and fine grain with the leaf-stalks. The infant is swung to sleep in a rude net of coir string made from the husk of the fruit; its meal of rice and scraped cocoa-nut is boiled over a fire of cocoa-nut shells and husks, and is eaten off a dish formed of the plaited green leaves of the tree with a spoon cut out of the nut-shell. When he goes a fishing by torch-light, his net is of cocoa-nut fibre; the torch, or *chule*, is a bundle of dried cocoa-nut leaves and flower-stalks; the little canoe is a trunk of the cocoa-palm tree, hollowed by his own hands. He carries home his net and his string of fish on a yoke, or pings, formed of a cocoa-nut stalk. When he is thirsty he drinks of the fresh juice of the young nut; when he is hungry he eats its soft kernel. If he has a mind to be merry, he sips a glass of arrack, distilled from the fermented juice of the palm, and dances to music of rude cocoa-nut castanets; if he be weary he quaffs 'toddy,' or the unfermented juice, and he flavours his curry with vinegar made from this toddy. Should he be sick, his body will be rubbed with cocoa-nut oil; he sweetens his coffee with *jaggery* or cocoa-nut sugar, and softens it with cocoa-nut milk; it is sipped by the light of a lamp constructed from a cocoa-nut shell and fed by cocoa-nut oil. His doors, his windows, his shelves, his chairs, the water-gutter under the eaves, are all made from the wood of the tree. His spoons, his forks, his basins, his mugs, his salt-cellars, his jars, his child's money-box, are all constructed from the shell of the nut. Over his couch when born and over his grave when buried, a branch of cocoa-nut blossoms is hung to charm away evil spirits." This is, of course, a European picture, some of the illustrations being scarcely in accordance with fact. It is, however, a true picture of the all-importance of the "Prince of Palms" to the inhabitants of the tropical regions.

In order to convey some idea of the numerous uses of the cocoa-nut palm, the following extract from the Colonial and Indian Exhibition Catalogue may be here reproduced. It is a list of certain articles prepared from the palm, exhibited by **Mr. M. C. Pereira**, Head Assistant to the Government Medical Storekeeper, Bombay:—

(1) **Coir** (*Kábál, Káthá*) —The fibre made of cocoa-nut husk; in this state it is used for stuffing cushions, pillows, beds, making rope-mats, &c.
(2) **Spoon** (*Ulkí*).—Used in the cook-rooms of Europeans, and by the natives for drinking gruel (rice *conjí*); has the advantage over the metallic one of not being corroded.
(3) **Drainer** (*Zárá*).—Used for draining food fried in *ghí* (clarified butter) or oil.
(4) **Ladle** (*Doho*).—Used for water.
(5) **Ladle, small** (*Buddi*).—Used by natives for taking out oil for daily use from an earthen vessel containing the yearly or quarterly stock. It is not corroded by the oil.
(6) **Hubble-bubble** (*Gudgudi*).—This is the *húkah* of the poorer classes.
(7) **Beads** (*Mani*).
(8) **Vinegar** (*Sirká Amti*).—Made of the juice (*toddy*) of the cocoa-nut palm.

DOMESTIC.

(9) **Pickle** (*Lonche, Achár*).—Made of the pith of the top of the fresh tree with vinegar of the juice (*toddy*) of the same palm.
(10) (*Pogí*).—The spathe of the blossom.
(11) **Rib** (*Kadí Hirkúte*).—The rib of the leaf.
(12) **Broom, Goa** (*Kersuní, Butará, Zadú*).—Made of leaf-ribs; it is much used for sweeping purposes.
(13) **Strainer** (*Mandorá*).—The sheaths by which the leaves are held firm to the tree. Used for straining cocoa-nut juice (*toddy*) and cocoa-nut milk, and for general straining in the cook-room.
(14) **Woolly floss** (*Burá*).—Much used as a styptic for cuts by the *toddy* drawers and cultivators.
(15) **Blossom** (*Kontí*).—The blossom in the state when it is tapped for drawing juice (*toddy*).
(16) **Chain** (*Sanklí Kargotá*).—Used round the waist to retain the loin cloth. The size is for a child. Set in metal may be used as a watch-guard.
(17) **Drum** (*Dholkí*).—Made of a piece of the trunk of the cocoa-nut tree.
(18) **Wood piece of rafter** (*Duroú Wánsa*).—Made of the lower part of the tree 10, 20, and 25 feet in length.
(19) **Oil** (*Khobrel*).—Oil expressed in the native mills for commerce.
(20) **Oil** (*Muthel*).—Oil extracted from fresh cocoa-nuts by rasping fine, drying, and pressing between coir and twisting with hands or by extracting the milk and separating the oil by heat. Used internally in lieu of cod-liver oil and externally for ulcers with good results.
(21) **Hair oil.**—Cocoa-nut hair oil.
(22) **Liquor** (*Daru, Rashi Urákh*).—Spirituous liquor 60° U.P., distilled from cocoa-nut juice (*toddy*) and drunk hot.
(23) **Punch** (*Queimado, Portuguese name*).—The punch is made of the liquor of the cocoa-nut palm with spices and sugar from the receipt of the Portuguese. There is no native name for it, and it is only known to the Native Christians of Bombay. Drunk hot for a cold, one or two cupfuls.
(24) **Liquor** (*Fhenidárû Port Dobrado*) (*double*).—Liquor made of cocoa-nut (*toddy*) juice by redistillation 20° U.P.; formerly much used for making medicinal tinctures and country brandy.
(25) **Cocoa-nut** (*Nárel*).—This fruit takes a year to ripen.
(26) **Sweetmeat** (*Nárlipák*).—Prepared from the kernel of the nut.
(27) **Sweetmeat.**—Prepared from the kernel with saffron.
(28) **Splints** (*Kambí*).—Made of (*pogúy*) the spathe of the blossom used for this purpose by the *toddy* drawers and natives of Goa, &c.
(29) **Door mats.**—Made of the fibre of many shapes and sizes by natives and in the jails.
(30) **Buggy mats.**—Made of the fibre of many shapes and sizes by natives and in the jails.
(31) **Carriage mats.**—Made of the fibre of many shapes and sizes by natives and in the jails.
(32) **Floor mats.**—Made in Malabar and in the Bombay jails of different sorts and colours.
(33) **Cage** (*Pinjará, Khurí*).—Made of the rib of the leaf.
(34) **Horn** (*Pipání Tontora*).—Made of the leaf of the palm; gives a loud sound when fresh.
(35) **Horn, small size** (*Dhaktí Pipání*).—Made of the leaf of the palm; gives a loud sound when fresh.
(36) **Toy parrot** (*Popat*).—Made by children of the leaf of the palm; when new it looks better.

COCOS nucifera.

The Cocoa-nut Palm : Domestic Appliances.

DOMESTIC.

(37) **Toy parrot in cage** (*Pinjaryát Popat*).—Made by children from the leaf of the palm; when new it looks better.
(38) **Leaf woven, Cudjan** (*Zavlí*) —The leaf of the tree, used for thatching houses; has the advantage over tiles of keeping the house cool.
(39) **Root** (*Mûl*).—Used medicinally, astringent, and as a gargle for sore mouth.
(40) **Rope** (*Káthá, Sumbha*).—This is extensively used.
(41) **Oil-bottle** (*Dowlá*).—Hung beneath the labour-cart with castor oil and brush in it for lubricating axles.
(42) **Nut, immature** (*Khakota*).—Used medicinally as an astringent; children are fond of it.
(43) **Trough** (*Panshira*).—Trough made of cocoa-nut tree, used for catching water drawn from a well with a Persian wheel for irrigation purposes (model).
(44) **Conduit** (*Panhál*).—A conduit put under the hole of the trough for conveying water for irrigation purposes.
(45) **Adapter** (*Nalá*).— Piece of the adapter used for connecting the native still to the condenser.
(46) (*Tuntuna*).—Native musical instrument, used by the poorer classes.
(47) **Beam** (*Báhál*).—Piece of beam of the shape used for houses. It is also used for fishing-stakes in the sea; generally two cocoa-nut trees make a stake 60 to 70 feet long.
(48) **Rosary box.**—Made of immature cocoa-nuts.
(49) **Charcoal Powder** (*Kolsá*).—Burnt shell used for preparing black and lead-coloured washes for houses.
(50) **Broom** (*Zádú*).—Made of the ribs of the leaf; used by the Bombay and other municipalities for sweeping roads, streets, yards, &c.
(51) **Broom** (*Záddû*).—Made of the stems of the blossom and nuts; used by the cultivators for collecting dry leaves for (*rab*) burning on the fields.
(52) **Crab trap** (*Kathimbrá Dhoderá*).—Made of the stem of the leaf.
(53) **Fish trap** (*Malai*).—Made of the ribs of the leaf.
(54) **String of pots** (*Mál*).—This is made of fibre of 60 or 70 feet in length, and about 50 or 60 earthen pots fixed to it and put on the Persian wheel (*rahat*), which in rotating brings up the pots filled with water and takes down the empty ones.
(55) **Violin** (*Sárangí*).—Used by the lower classes of natives, particularly the gosavies (a class of professional beggars).
(56) **Sling** (*Shinká*).—Used for keeping sundry articles of food out of the reach of cats, rats, and ants by hanging it on a hook to the ceiling. Tied to the ends of a bamboo, serves for carrying water-pots, baskets, &c. The small one is used by milkmen for carrying milk for sale.
(57) **Flesh glove** (*Hátáli*).—Used for washing and rubbing cattle and horses.
(58) **Tar with acetic acid** (*Kartel*).—Made by burning the shells in a pot with a small hole in the bottom, placed on another, heated by fire on all sides. Used by the natives for ringworm and skin diseases.
(59) **Rope** (*Dore*).—Made of various sorts and sizes.
(60) **Brush** (*Chavár*).—Made of the husk of the nut for cleaning sieves, washing baskets and rice-drainers (*Shibum*).
(61) **Sugar, molasses** (*Gúl*).—Made of the juice (*toddy*) in Goa.
(62) (*Band*).—Peeled from the outer part of the stem of the leaf. Is used as a cord by the *toddy* drawers.
(63) **Cocoa-nut gilded** (*Karyácha Nárel*).—Offered by the higher classes of Hindús to appease the sea on the cocoa-nut fair day. At weddings the bridegroom and bride carry it in their hands.

DOMESTIC.

(64) **Husk** (*Sál, Chavád, Sodan*).—Used as fuel. Especially for backing purposes also affords coir fibre.
(65) **Scoops.**—Made of the shell. The round and deep ones are used as drinking cups.
(66) **Neck belts** (*Pattá*).—Used for yoking bullocks and buffaloes to carts, ploughs, oil-mills, &c.
(67) **Sack** (*Thoili Jalí*).—Used for sending out articles; a somewhat similar one is attached to the cart for carrying straw or grass.
(68) **Tooth-brushes** (*Dáton*).—The pedicels of the blossom are used as tooth-brushes.
(69) **Brushes** (*Kunchá, Kuchrá*).—The peduncles of the blossom are used for whitewashing houses, &c.
(70) **Blind** (*Dol-Dhápan*).—Used for blinding bullocks and buffaloes while yoked to the Persian wheel, oil-mill, &c.
(71) **Nest** (*Gharta, Gharbá*).—Made by birds out of the fibre of the leaf.
(72) **Soap** (*Sabu*).—Made of cocoa-nut oil; has larger percentage of water than any other soap.
(73) **Puzzles and toys.**—Rings, whips, neckties, rattles, crosses, &c.
(74) **Bats for cricket.**—Made of the wood (cocoa-nut).
(75) **Oil-cakes** (*Pend*).—Oil-cake from the native mill.
(76) **Patimar (ship)** (*Fatemárí*).—Toy made by the boys of the fishermen class.
(77) **Boat, fishing** (*Hodke*).—Toy made by the boys of the fishermen class.
(78) **Kernel** (*Khobre*).—Dry kernel.
(79) **Stem** (*Jhintár*).—Used as broom.
(80) **Charpai, Cot** (*Khát, Báj*).—Used by the natives (model).
(81) **Potash (crude)** (*Khár*).—The ash of the stem of the leaves; they produce 20 per cent. of ash.
(82) **Cocoa-nut, abortive** (*Vánzá Nárel, Váhil*).—Used as floats for beginners in swimming.
(83) **Spadix.**—The spadix prepared for drawing juice (*toddy*). A thin slice is cut from the palm stem three times a day. The juice flows from this and drips down into an earthen pot suspended on purpose. A small piece of the leaf is fixed above to prevent the bottom of the pot from touching the point, the sheath of the leaf covering the mouth of the pot to keep out flies.

Codilla.—A commercial term for the refuse separated on cleaning hemp or flax fibres. 1637

CODONOPSIS, *Wall.; Gen. Pl., II., 557.*

[*t. 60, fig. 3;* CAMPANULACEÆ.

Codonopsis ovata, *Benth.; Fl. Br. Ind., III., 433; Royle, Ill., 253,* 1638

Vern.—*Lúdút.*

Habitat.—A herbaceous plant common in the N. W. Himálaya from Kashmír to Gurhwál at altitudes from 8,000 to 12,000 feet, distributed into Afghánistán.

MEDICINE. 1639

Medicine.—Aitchison (*Kuram Valley Flora, in Linn. Soc. Jour., XIX., 147*) says:—"The roots and leaves of **Codonopsis** are made into poultices and employed in the treatment of bruises, ulcers, and wounds."

FOOD. 1640

Food.—"The large tap-root is ground into flour and eaten in Lahoul" (*Stewart; Aitchison*). In Kuram it is said to be eaten raw or cooked.

COFFEA, *Linn.; Gen. Pl., II., 114.*

[RUBIACEÆ.

1641 **Coffea arabica,** *Linn.; Fl. Br. Ind., III., 153; Wight, Ic., t. 53;*

COFFEE, *Eng.;* CAFÉ, *Fr.;* KAFFEE, *Germ.*

Vern.—*Bun* (the berry), *Kahwa* (the same roasted and ground). *Kawa, bun, bún, coffee, coffi,* HIND.; *Kápi, kava,* BENG.; *Bund, cappi,* GUJ.; *Kawa, bun, kahwa, búnd, caphi, caffi,* BOMB.; *Kaphi, kan, bund, bún,* MAR.; *Búnd, tochém-keweh, cahwa,* DUK.; *Kapi-kottai, cápie-cottay, capi,* TAM.; *Kapi-vittulu, capi,* TEL.; *Kappi-kura, kawa, kopi,* MALA.; *Kaphi, Bonda-bíja, kápi-bija,* KAN.; *Bun, qahvá, kahwa, kuehwa,* ARAB.; *Bun, qahvá, kahwa, Tochém-keweh, cahwa,* PERS.; *Ka-pwot, káphi-si,* BURM.; *Kópi-atta, copi cottá,* SING.

The Arabic term *Kahwa,* according to some writers, was originally applied to wine, but by others it is a corruption from Kaffa, the name of a district in Abyssinia where the plant is said to be indigenous. If the latter supposition be correct, then cavé, café, and coffee are remarkably near the original name. But *kahwa* and coffee are terms applied to the beverage. The name *Bun* is generally given to the plant. It is the vernacular name in Shoa, and in Yemen as applied to the berry.

References.—*Roxb., Fl. Ind., Ed. C.B.C., 181; Brandis, For. Fl., 276; Kurz, For. Fl. Burm., II., 27; Gamble, Man. Timb., 231; Dals. & Gibs., Bomb. Fl., 44; DC. Origin Cult. Pl., 415-418; Ainslie, Mat. Ind., 81; O'Shaughnessy, Beng. Dispens., 402; Moodeen Sheriff, Supp. Pharm. Ind., 113; U. S. Dispens., 15th Ed., 306; Bent. & Trim., Med. Pl., 144; Bidie, Cat. Raw Pr., Paris Exh., 81; Year-Book Pharm., 1874, 301; Drury, U. Pl., 151; Lisboa, U. Pl. Bomb., 162; Birdwood, Bomb. Pr., 207; Royle, Prod. Res., 184; Royle, Ill. Him. Bot., I., 4, 162, 240; Christy, New Com. Pl., I., 3; Crookes, Dyeing, 513; Spons' Encyclop., 422, 691-722, 1407, 1420; Balfour, Cyclop., 767 777; Smith, Dic., 125; Treasury of Bot.; Morton, Cyclop. Agri., 490; Ure, Dic. Indust. Arts and Manu., I., 842; Kew Reports, 1877, 17 & 28; 1879, 20 & 30; 1880, 18 & 34; 1881, 16 & 29; 1882, 24 & 39; Kew Off. Guide to the Mus. of Ec. Bot., 34; Kew Off. Guide to Bot. Gardens and Arboretum, 47; Trop. Agri., 1882-83; Simmonds, Trop. Agri., 27-79; Agricultural Gazette of India; Indian Newspapers; Rev. J. Berkley in Colombo Observer; Branson in Jour. Soc. of Arts, XXII., 1874, p. 456; Proceedings of Agri.-Hort. Soc. of Madras and Journals and Proceedings, Agri.-Hort. Soc. of India (numerous articles and correspondence); Official correspondence in the Proceedings of the Revenue and Agricultural Department; Hiern, in Linn. Soc. Jour., April 1876; Shortt, Handbook to Coffee-Planting in South India, 1864; Bidie, Coffee-Planting & Report on the Ravages of the Coffee Borer, 1869; Coffee Leaf Disease by Harman, 1880; Cooke, Report for the India Office on the Diseased Leaves of Coffee and other plants, 1876; Correspondence respecting the Coffee-leaf disease in Ceylon, Colonial Office, 1875; Correspondence between Madras Govt. and the Wynaad Planters' Association concerning Labour-Law and Coffee-Stealing Prevention Act, 1879; D. Morris, Handbook of Coffee-leaf Disease; H. Pasteur, Reports of Colonial and Indian Exhibition, 167; Bell, Chemistry of Food, 40; Church's Ed. of Johnston's Chemistry of Common Life, 148; Welter-Essai sur l'Histoire du Café, 1868; Beverages we Infuse, Blackwood's Magazine, No. 459; Madras Exhibition Jury Reports; Hassal, on Food; Hull, on Coffee-Planting in South India and Ceylon; Nietner, on the Enemies of the Coffee Plant; Colonel Onslow, on Mysore Coffee-Planting; Tennent's Ceylon; G. Anderson, Coffee Culture in Mysore (Bangalore, 1879); R. H. Elliott, Planter in Mysore (London, 1871); W. H. Middleton, Manual of Coffee-Planting (Natal, 1866); W. Sabonadière, Coffee Planter of Ceylon (London, 1870); A. R. W. Lascelles, Nature and Culture of Coffee (London, 1865); Van Delden Laërne, Coffee Culture in Brazil and Java (London, 1885); R. B. Tytler, Prospects of Coffee Production (Aberdeen, 1878); W. G. McIvor, Laborie's Coffee Planter of St. Domingo (Madras, 1863).*

Coffee Cultivation.

CULTIVATION.

Habitat.—Most authors seem to agree that the coffee plant is indigenous to Abyssinia, the Soudan, and the coasts of Guinea and Mozambique. "Perhaps in these latter localities, so far removed from the centre, it may be naturalised from cultivation. No one has yet found it in Arabia, but this may be explained by the difficulty of penetrating into the interior of the country. If it is discovered there it will be hard to prove it wild, for the seeds, which soon lose their faculty of germinating, often spring up round the plantations and naturalise the species. This has occurred in Brazil and the West India Islands, where it is certain the coffee plant was never indigenous" (*De Candolle*).

It is a small, much-branched tree or bush 15 to 20 feet in height, with whitish bark and white orange-like flowers. The fruit, which is red on ripening, is about the size of a small cherry, and contains two seeds, closely united. These, on being separated, constitute the coffee berries of commerce; and on being roasted and ground, the coffee of the shops.

1. In India **Coffea arabica**—the coffee plant—is largely cultivated, but other species are also met with.

2. **C. bengalensis**, *Roxb.*, occurs from Kumáon to Mishmi, also in Bengal, Assam, Sylhet, Chittagong, and Tenasserim. Fruit ovoid-oblong. (*Harína* in Chittagong, *see Agri.-Hort. Soc. Ind. Proceedings, Oct. 1865.*)

3. **C. fragrans**, *Korth.*, found in Sylhet and Tenasserim. Fruit much like the two last.

4. **C. Jenkinsii**, *Hook. f.;* Khási Mountains. Fruit and seeds different from the last, being ellipsoid.

5. **C. khasiana**, *Hook. f.;* Khási and Jaintia hills. Fruit ¼ inch in diameter, smooth; seeds ventrally concave.

6. **C. travancorensis**, *W. & A.;* occurs in Tranvancore. Fruit broader than long.

7. **C. Wightiana**, *W. & A.;* the Western Peninsula, in arid places from Coorg to Travancore. Fruit much broader than long, with a deep furrow.

With the exception of the first these species are not of any special economic importance; and very little coffee is grown in the tracts in which they are reported to be found. The coffee-cultivating region in this country is Southern India, and the enterprise has there gained much importance. It at present not only supplies most of the coffee consumed in India, but exports large quantities to other countries.

(For Liberian Coffee see the concluding paragraph of this article.)

HISTORY OF COFFEE CULTIVATION AND OF THE HABIT OF COFFEE-DRINKING.

COFFEE CULTIVATION. 1642

The regions best suited for coffee cultivation lie between 15° N. and 15° S. latitudes, but it is grown as far as the 36° N. to the 30° S. in regions where the temperature does not fall beneath 55° F. (13° C.). The area of its cultivation is in fact very nearly the same as that of cotton. Within the tropical region it may be cultivated at the level of the sea or even much further to the north and south of the equator than has been indicated. The plant manifests, in other words, a remarkable power of endurance, but it does not follow that where it may be grown as an ornamental garden bush it may there afford the commercial product. Within the tropics it will yield profitable returns only if cultivated on the hills at altitudes between 1,000 and 5,000 feet. It requires a climate with a temperature ranging from 60° to 80° F. (15° to 27° C.); as to humidity, in a suitable region there should be no month devoid of rain, but the annual fall should not exceed 150 inches. A temperate

COFFEA arabica.

Habit of Coffee-drinking.

HISTORY.

climate within the tropics is that required. An atmosphere resembling that of an English hot-house produces the finest crops, but it is inimical to the planter and favourable to weeds. The most suitable climate is therefore that which Europeans prefer to live in, and not the unhealthy climate essential for tea cultivation. Heavy clouds are very objectionable, and strong winds blow away the flowers and make 50 per cent. difference in crop. If too hot and dry, the plants require shade, and if strong winds prevail during the flowering season, belts of forest have to be left to protect the plantation. This is regarded an important consideration in clearing land for a coffee plantation. **Dr. Shortt** says: "In low countries there is not sufficient moisture in the soil, and when shaded and irrigated, it produces a coarse and uneven bean devoid of the peculiar aroma essential to good coffee." While the coffee plant does not seem to luxuriate on the immediate coast and under the direct influence of the sea breezes, still it is a noteworthy fact that in India the best gardens (such as those of the Nilghiris, the Wynaad, Mysore, Coorg, Mungerabad, and Shevaroys) bear a certain relation to the coast; indeed few good plantations occur beyond the limits of marine influence. On this account the recommendations of the early advisers of the Government of India to prosecute experimental coffee cultivation on the lower Himálaya from Darjíling to Kumáon have been abandoned. The occurrence of certain wild species on the mountains of Northern and Eastern India has been shown to afford no criterion of the possible regions where the African plant might be successfully grown. Coffee-planting has in fact been practically concentrated on the lower mountain slopes of South India, a region which like Ceylon has many features in common with the Abyssinian and other African regions where the wild coffee abounds. Some parts of the Nilghiri hills are, however, found to be too high, the plants growing well, but not maturing their seeds.

It has been stated that the coffee plant of commerce is truly wild in Abyssinia, and that it is there called *bun* or *boun*. This name appears to have followed it into Egypt and Syria. **Bellus** and **Alpin** both write of it under that name, and state that the Egyptians extract the drink called *cavé* from the seeds. A reference to the vernacular names in a preceding paragraph will show that *both* these names are used in India and occur also in the Arabic and Persian languages. **Yule** and **Burnell** remark: "There is very fair evidence in Arabic literature that the use of coffee was introduced into Aden by a certain **Sheikh Shihabuddin Dhabhani**, who had made acquaintance with it on the African coast, and who died in the year H. 875, *i.e.*, A.D. 1470, so that the introduction may be put about the middle of the fifteenth century—a time consistent with the other negative and positive data. From Yemen it spread to Mecca (where there arose after some few years, in 1511, a crusade against its use as unlawful), to Cairo, to Damascus and Aleppo, and to Constantinople, where the first coffee-house was established in 1554. The first European mention of coffee seems to be by **Ranwolff**, who knew it at Aleppo in 1573." (Conf. with remarks in a further page regarding introduction into India.)

The habit of coffee-drinking spread but slowly from Arabia Felix, but in the Mahomedan countries through which it became gradually diffused, it soon met with the opposition of the priests, owing to the coffee-houses having become more popular than the mosques. To check this, the article was heavily taxed. The first mention of a coffee-shop in Great Britain occurs in 1652. (Tea was publicly sold in London in 1657.) **Mr. D. Edwards**, a Turkey merchant, acquired the habit of drinking coffee and imported a Greek servant, **Pasqua Rossie**, for the purpose of preparing his favoured beverage. His friends grew so fond of it that to prevent their

Consumption of Coffee.

HISTORY.

too frequent visits to his house he recommended **Rossie** to start a public coffee-shop. This was opened in St. Michael's Alley, Cornhill. Coffee-shops rapidly multiplied, but the beverage (although from a very different reason) soon met with as much official opposition in London as it had sustained in Constantinople. Charles II. (in 1675) viewed these shops as the meeting-places for disaffected persons, and a royal proclamation was issued for their suppression. Coffee is spoken of as being in use in France in 1640, and the first public café was opened in Paris in 1669. Shortly after, it became general throughout Europe. It may be here added that of the three great dietary beverages Cocoa was the first to make its appearance in Europe, coming from South America through the Spaniards; coffee followed, coming from Arabia by way of Constantinople; and tea, the latest of the series, came from China through the Portuguese and the Dutch (*Encycl. Brit., VI., 110*). Coffee-drinking rapidly attained great proportions in Britain, and by 1847 it had reached its maximum. In 1833 the monopoly, granted by the British Parliament and held for 180 years, prior to that date, by the Honourable the East India Company, to be the sole importers of tea into England, was rescinded. With free trade in tea, the price of that article fell considerably, and the duty was steadily lowered, the result being the growth of an immense tea trade which by 1847 checked the further development of the demand for coffee. There are doubtless many causes that may have contributed to bring this about; chief amongst them may be placed the facility with which coffee can be adulterated, the greater consumption of cocoa, and the ease with which tea may be prepared. Indeed certain admixtures with coffee have been legalised, chicory, the most important of these, being made to bear a duty. To the present date this system has prevailed in England, while in France and some other European countries the dealer is prohibited from preparing or selling mixtures, but he may sell chicory separately, for the consumer to mix with his coffee, just as he may sell sugar. These legislative measures appear to have had much to say to the growth of a greater coffee consumption in continental countries than in England, or rather to the decline of coffee consumption manifested in Great Britain with the growth of the tea demand.

BRITAIN. Decline in Consumption. 1643

DECLINE OF CONSUMPTION IN BRITAIN.—The consumption of coffee in Great Britain was, in 1847, 37,441,373℔; in 1857, 34,518,555℔; in 1867, 31,567,760℔; but in 1874 it had declined to 31,859,408℔, and slightly improved in 1880, being in that year 32,480,000℔. These figures must not be confused with the imports of coffee. Great Britain does an immense trade in importing and re-exporting the beans or in exporting special preparations of coffee. The imports into Great Britain average from 130 to 170 million pounds. The total European annual consumption has been estimated at 360 million pounds, and the world's production 1,000,000,000℔. The consumption to head of population has been variously stated by different writers owing to the years they have selected. In Great Britain, for example, from 1857 to 1859, it was 1¼℔, from 1865 to 1867 it was 1℔, and from 1875 to 1877 it had fallen to ¾℔. Even where the consumption is in the ascendant (in non-coffee-producing countries) the increased consumption is not proportioned to the increase of population, so that in Europe at least the demand for coffee is not materially progressing. The German Empire consumes the greatest amount. Holland takes 21℔ per head, Denmark 14℔, Belgium 13½℔, Norway 9¾℔, Switzerland 7℔, Sweden 6℔, France 2¾℔, Austro-Hungary 2℔, Greece 1½℔, Italy 1℔, the United Kingdom ¾℔, and European Russia ¼℔. The United States of America are supposed to use on an average 8℔ per head of population per annum. **Mr. H. Pasteur**, in his report on the coffee shown at the

COFFEA arabica.

Coffee Cultivation Extended.

HISTORY.

Colonial and Indian Exhibition in London, 1886, wrote: "The total production of coffee in the world is roughly estimated at about 600,000 to 650,000 tons, of which Brazil alone produces between 340,000 and 380,000 tons, and Java 60,000 to 90,000 tons; the proportion of British-grown coffee being only about 35,000 tons, of which India contributes 15,000 to 18,000 tons, Ceylon 10,000 to 12,000 tons, and Jamaica 4,000 to 5,000 tons. Although numerically very small, the productions of our Colonies and of India occupy the front rank, owing to their excellence. Nowhere is finer coffee grown than in India and Jamaica, and its value, as well as that of Ceylon, is firmly established above that of all other kinds, even of Mocha, which at one time stood above all others."

EXTENDED CULTIVATION. 1644

EXTENDED CULTIVATION.—The cultivation of the coffee plant began to extend towards the end of the seventeenth century, being carried on in various countries possessing a sub-tropical climate, such as India, Java, Ceylon, Jamaica, and Brazil. Down to 1690, the only source of coffee-supply was Arabia, but in that year the Governor General (**Van Hoorne**) of the Dutch East Indies received a few seeds from the traders who plied between the Arabian Gulf and Java. These he sowed in a garden at Batavia where they grew and flourished so abundantly that the culture of the plant on an extended scale was immediately commenced in Java. One of the first plants grown in that island was sent to Holland as a present to the Governor of the Dutch East India Company. It was planted in the Botanic Gardens at Amsterdam, and seedlings raised from this plant were sent to Surinam in 1718. (**Crawford** (*Indian Archipelago, I., 486*) says, however, that coffee was not introduced into Java until 1723.) Ten years later coffee was introduced into the West Indies through a plant presented to Louis XIV. from the Amsterdam stock, and from that date it spread throughout the New World until now the progeny of the single plant sent from Java produces more coffee than all the other plants in the world. In Brazil coffee is completely acclimatised, and there are said to be 530 million plants under careful cultivation. Coffee is also extensively grown in Costa Rica, Guatemala, Venezuela, Guiana, Peru, and Bolivia with Jamaica, Cuba, Porto Rico, and the West Indian Islands generally. Its cultivation has long been pursued in Queensland, and in various other parts of Australia it has been found possible of cultivation. In Sumatra, Borneo, Siam, the Straits Settlements, and Fiji its cultivation has also been prosecuted with some degree of success; but after Brazil and Java, Ceylon and India are the countries where its introduction has assumed an important commercial character.

CEYLON Introduction. 1645

INTRODUCTION INTO CEYLON.—According to **Dr. Shortt** and others it was introduced into Ceylon by the Arabs, prior to the invasion of that island by the Portuguese. Its systematic cultivation about 1690 was undertaken by the Dutch, but on the cession of their territory its cultivation was continued by the natives of Ceylon. In 1825 the impetus to fresh effort was given by **Sir Edward Barnes** in the establishment of an upland European plantation. In 1877 it was estimated that the capital invested in Ceylon coffee was nearly 14,000,000*l*. The scourge of leaf-disease, a fungus (HEMILEIA VASTATRIX), however, made its appearance in 1869 and spread over the island, "weakening the trees, undermining the crop-bearing capabilities, and leading to the gradual extinction of the plantations over many of the best districts." "One cannot avoid a feeling of sadness and regret at the thought." that the samples of coffee shown at the Colonial and Indian Exhibition "represent only the fast vanishing remains of what was but nine years ago the most extensive and flourishing of the coffee crops raised on British soil by British enterprise and capital. The production, which in 1873 amounted to nearly 1,000,000 cwt., declined to 665,000

Introduction of Coffee Cultivation into India.

HISTORY. INDIAN. Introduction. 1646

cwt. in 1876, to 312,000 cwt. in 1884, and to 230,000 cwt. in 1885" (*Pasteur*).

INTRODUCTION INTO INDIA. The history of the introduction of coffee into India is very obscure. Most writers agree that it was brought to Mysore some two centuries ago by a Muhammadan pilgrim named **Baba Budan**, who, on his return from Mecca, brought seven seeds with him. This tradition is so universally believed in by the inhabitants of the greater part of South India, that there seems every chance that there may be some foundation for it. **Jan Huygen van Linschoten**, a native of Holland, who, under the protection and in the service of the Portuguese, visited India in 1576 to 1590 (and wrote a most instructive account of his travels), while describing all the important products of the Malabar Coast from Sind to Cape Comorin, makes no mention of coffee. In a chapter devoted to Japan he alludes to the remarkable practice the Japanese had of drinking hot water in which they "seethe the powder of a certaine herb called chaa." His contemporary **Dr. Paludanus** adds a note on this subject to the effect that in the same way the Turks prepare a beverage from "the fruit, which is like unto the Bakelaœ" (laurel berry), "and by the Egyptians called Bon or Ban." This note proves that **Linschoten** did not hear of coffee being used in South India during his visit. But the coffee plant now grows very readily on the mountains which, to this day, bear the pilgrim's name. **Royle**, in his *Productive Resources of India*, says: "The plant has long been introduced into India, and coffee of a fine quality is cultivated on the coast of Malabar; also to a considerable extent in Coimbatore, and the cultivation might, no doubt, be easily extended elsewhere. It was tried in the Calcutta Botanic Gardens, where it succeeded remarkably well under the shade of the teak plantations, and nothing could be more healthy looking or in better bearing than these coffee plants when seen by the author in 1823. **Dr. Roxburgh** had long previous to this ascertained that two middling plants, at the age of six or seven years, produced a crop in the Botanic Gardens, in one year, of 7lb of the dry berries, which gave three pounds of clear coffee equal to the Jamaica produce." **Dr. Wallich**, in his evidence before a Select Committee of the House of Commons, stated: "I will say for myself I never used to drink good coffee except that produced in the Company's garden at Calcutta." Subsequent writers have, however, shown that while the plant can be grown in the plains of India, the care necessary and the expense entailed render the attempt unprofitable on a commercial scale. The earlier volumes of the Journals and Proceedings of the Agri-Horticultural Society of India contain many letters and reports on coffee cultivation in Assam, Bombay, Bengal (Chutia Nagpur, the Rajmahal Hills, and Chittagong), but while many of the results there published seem to have justified a continued effort, the experiments have nevertheless been practically abandoned. There are at present some 10 acres under coffee in Lohardugga (Chutia Nagpur) and about an acre in Chittagong, with perhaps close on 100 acres scattered over Assam and the same amount in Bombay. Coffee-planting in India at the present day is concentrated in the Madras Presidency, and as a European industry it may be said to date from **Mr. Cannon's** plantation at Chikmúglúr, in Mysore. This was established in 1830, but as a curiosity **Major Bevan** grew coffee in the Wynaad in 1822. It was cultivated by **Mr. Cockburn** on the Shevaroys in 1830; **Mr. Glasson** formed a plantation at Manantoddy in 1840; the plant was taken to the Nilghiris in 1846; to Belgaum a little earlier; to Darjiling, by **Captain Smaller**, in 1856, and to Chittagong subsequently. It has been reported to yield 9 maunds an acre in Chittagong, and that there are

HISTORY.

thousands of acres of good suitable land for coffee near navigable rivers where manure and labour are cheap.

Coffee has also been introduced into Burma. For some time the effort to open out plantations seemed to be doubtful; and **Mr. Petley**, speaking of the garden on the Karen Hills, north-east of Toungoo, reported recently that much damage had been done by a mole cricket. Since then, however, the construction of a railway from Rangoon through a hopeful coffee region has given birth to new expectations. The Agri-Horticultural Society of Burma, in their annual report for 1887, say, the demand was great for seedlings, both of Arabian and Liberian coffee. Large numbers are reported to have been sent to Upper Burma. It is added that "it is noteworthy that the Arabian variety does best on the Toungoo Hills, while at Tavoy the Liberian variety is alone thought worthy of cultivation." "Local demands, too, are increasing, as land is being taken up along the lines of railway between Rangoon, Prome, and Toungoo, and gardens have been formed whereon small grantees are now cultivating fruit and other useful trees as well as coffee."

METHODS.
1647

METHODS OF CULTIVATION.

Space cannot be afforded to deal with every feature of this subject: the reader is referred to the numerous special publications quoted under the paragraph of references; only the more salient features will be touched upon, and especially those which have a bearing on the future expansion of the industry.

LOCALITIES, CLIMATES, AND SOILS SUITABLE FOR COFFEE CULTIVATION AS AN AGRICULTURAL PRODUCT.—Under the heading "History of Coffee," the subject of the region of coffee cultivation and the climate necessary have been discussed. **Dr. Shortt** says of SOIL: "This should be rich, abounding in moisture and containing much humus or vegetable mould; consequently we find that the plant thrives best on either red or black clay, containing combinations or preparations of iron, and covered over with humus formed by the decay of vegetable matter produced by dense forests. When these points are overlooked, the results are soon seen in the rising plantation. The planter, perhaps, instead of choosing forest land, has taken up a poor grassy or stony situation, and however much water he may have access to, his plants are stunted and soon become yellow, unless he resorts to heavy manuring at a very early stage, which materially increases the expense of the concern. In hard rocky soils the pits require to be deeply excavated to permit of the tap roots of the plant striking perpendicularly down, and even when every precaution is taken, it will be found that estates opened out on poor soils will always prove more expensive than those on forest land, and are not so lasting. The berry produced on rich ferruginous clay is found to contain more aroma and the bean is heavier when compared with those of other localities. This fact is so well known to coffee-brokers generally that, in London, a new importation is frequently weighed after being roasted." Some difference of opinion prevails as to the degree of moisture the soil should contain. In *Spons' Encyclopædia* there occurs the following: "The points which determine the value of a plot for coffee culture are—1, elevation; 2, aspect; 3, shelter from winds; 4, shelter from wash; 5, temperature; 6, rainfall; 7, proximity to a river; 8, character and richness of soil. Most of these are necessarily subject to variation according to locality. Shelter from wind is perhaps of paramount importance and should not be sacrificed for richer soil, as the latter can be artificially obtained much quicker than the former. In wooded country the estate may be laid out in blocks of 50 acres, encircled by

METHODS.

natural belts of forest. Flat land must be avoided; and wet soil is fatal to coffee, and flat lands would entail great expenditure for drainage. Steep slopes, on the other hand, are objectionable, on account of the wash occasioned by rains carrying away soil and manure and exposing the roots of the shrubs. The surface soil must be fairly good, the subsoil may be poor but must never be stiff clay; the shrub is essentially a lateral feeder. As a general rule virgin forest land has been found most suitable to break up for coffee estates; it has become naturally enriched by decayed vegetable matters, and the burning to which it is subjected frees it from insects and from weeds." Not only therefore do the opinions expressed in these two passages differ as to the degree of moisture which the soil should contain,—**Dr. Shortt** saying it should "abound" and the writer in **Spons'** holding that moisture is "fatal"—but **Dr. Shortt** remarks, the planter "must be in the enjoyment of robust health, to be able to withstand the deadly effects of a damp atmosphere, for, in all probability, he will have to spend his time surrounded by the direst malaria, &c." **Spons'**, on the other hand, says:—"The most suitable climate is precisely that which Europeans prefer. Frost, even though it be only at night and for a short period, is fatal." It seems probable that opinions have greatly changed since **Dr. Shortt** wrote his *Hand-Book of Coffee-Planting in South India.* The writer's limited practical experience of coffee-planting and knowledge of coffee plantations would incline him to the opinion that the remarks just quoted from **Dr. Shortt's** work are much more applicable to Tea than to Coffee.

Nursery and Seed.—Having selected the site for a plantation, cleared and burned the trees (taking care, where necessary, to have protecting belts against prevalent winds), laid out the roads and carried the water-supply to the coffee-house, it next becomes necessary to select and prepare the spot for a nursery. The soil should have a gentle slope, be well drained but retentive of moisture, rich and within access of artificial or natural irrigation. The land should be thoroughly ploughed up or trenched to a depth of 18 to 24 inches and the weeds entirely exterminated. Manure at the rate of from 3 to 5 tons an acre should be worked into the surface soil. The seed-beds may be shaded, but not to the exclusion of the sun, nor to such an extent as to allow dripping from the protecting trees. Each bed should be raised to allow drainage, and separated from the others by narrow paths. If on sloping ground, a deep trench should be run round the top portion of the nursery so as to divert the surface water. Nursery. 1648

The seeds should be sown in rows 6—9 inches apart and about 2 inches in depth, the seeds being carefully deposited along these lines about 1 inch apart from each other. They should then be lightly covered with mould and mats or by branches thrown over the beds. Watering should be done in the morning or after sunset. Seeds. 1649

The selection of seed is of great importance. The stock should be taken from carefully cultivated, healthy, and vigorous plants from 7 to 10 years old and the seed should not be gathered until fully ripe. "A bushel of seed should give 20,000 to 30,000 plants, the best is *parchment* coffee, picked when fully ripe, pulped by hand, unfermented, unwashed, and dried in the shade" (*Spons*).

"A bushel will rear 10,000 plants covering 10 acres." (*Balfour, Cyclop. Ind.*) "They should be fully ripe when plucked off the branches, and sown when fresh, at a depth of 1 inch, and dibbled in the soil in drills 10 to 12 inches apart from each other, so as to give the plantings plenty of room to grow, and subsequently enable the planter to remove them with facility from the nursery to the plantation; or the seeds may be sown in drills,

COFFEA arabica.

Coffee Cultivation—Planting.

METHODS.

and as the seedlings begin to grow the drills should be thinned out to the same distance. The seeds may be even scattered broadcast in the beds, and as they sprout should be thinned out to the regulated distance; care should be taken to let the plantings grow free of each other, which will make them vigorous." (*Shortt.*)

"When the plants have two to four leaves they should be carefully transplanted, in damp, cloudy weather, from the seed-beds to the nurseries, and placed 9 to 12 inches apart. Care must be taken not to double up the tap-root, and not to leave a space for water to accumulate and rot the roots. If the tap-root is very long, it is best shortened by an oblique cut, when it soon shoots again. When transplanting from the seed-beds to nurseries is not practised, the plants are left in the seed-bed until they have grown larger; but **Stainbank** and others strongly recommend the former plan, as, by checking the growth, the young wood becomes hardened, and better able, when finally planted out, to resist insects and unfavourable weather. A practical suggestion for preventing young seedlings being eaten off at the surface of the ground by grubs, is to lightly wrap round a piece of paper about 3 inches broad, where the stem joins the root, on planting." (*Spons.*)

Planting out. 1650

LINING AND PLANTING OUT.—Soon after being cleared the estate is "lined out" for the reception of the plants. The two following methods are in vogue: (1) "A base line is laid down, as nearly as possible, straight up and down the slope; a cross line is set off exactly at right angles: on this line, stakes are driven into the ground at the distances determined upon for the position of the plants: to each stake a rope is fixed, and stretched parallel with the base line and as straight as possible: small stakes are provided along these lines: a rope is finally held across them at succeeding stages of equal width, as guided by measuring poles, and the small stakes are put in where the moveable rope crosses the fixed ones, each stake indicating the site for a plant. (2) A rope is furnished with bits of scarlet rag at the distance fixed upon between the plants; it is stretched across the plot and stakes are inserted at each rag; the rope is then moved forward a stage at a time, gauged by measuring-rods. The first plan is the better, especially in broken ground, but is more laborious; the second is available on even grass-land, but the stretch of the rope must be estimated and allowed for."

When about a year old the seedlings are planted out in their permanent places in the plantation; but if dull weather be selected for transplantation, many coffee planters prefer to have two-year old seedlings. Much difference of opinion prevails as to the distance apart, and the question hinges mainly on the character of the soil, the degree of shade, and nature of the climate. In cold climates, when the plants are not likely to attain any great size, close planting is indicated, the reverse being the case under influences that would cause vigorous growth. In India the distance adopted varies between 4 and 8 feet each way—7 feet being very common, or 6 feet between the plants and 7 feet between the rows. This would give 1,037 plants to the acre. Before the plants are removed from the nurseries, the holes for their reception are made at the points fixed by the pegs during lining. A ball of earth is taken up with each seedling, and a number of these, on a tray, is carried to the plantation. If exposed to the air for any very long period in the process of transplanting, each ball should be wrapped round with damp moss. The earth should be firmly packed around the seedlings so as to prevent water lying and soaking into the roots.

Cultural operations. 1651

CULTURAL OPERATIONS.—The further treatment may be briefly reviewed. *Weeding*, or the removal of all wild plants from the plantation so

METHODS.

as to prevent the young seedlings from being choked. *Staking*, or supporting the plants by canes against winds: this may not be enough, and then it will become necessary to *shelter* the seedlings. The degree to which this may be necessary is a matter which depends upon the nature of the climate and the extent to which the removal of trees has deprived the plantation of the natural protection which belts of trees would have afforded. According to many planters, however, all trees should be removed and shade procured through the cultivation of the charcoal tree **(Sponia Wightii).** In two years this forms an ample shade, but as it grows older the leaves are shed, so that it requires to be renewed. This is easily done, the timber coming in useful. **Marshall Ward**, in his report on the coffee-leaf disease, urges the advantage of belts of trees in helping to check the diffusion of the spores of the fungus. "It is a matter for regret," he adds, "that such immense unbroken areas of coffee exist without break of any kind, and one can trace the swaying backwards and forwards of the spore-laden winds in consequence." *Draining.*—Nothing is more important than a complete system of drains and roads. If the operations in this direction have not been completed up to date, the energies of the planter during the first two years may very appropriately be turned to these considerations. Drift surface-water not only removes the soil, but may altogether wash away the plants. A proper system of drainage becomes essential, not only to remove the water from damp and cold water-logged soils, but to provide against the dangers of sudden torrents of rain. A system of *trenching* and *manuring* has also to be organised. The former is strongly recommended as saving the necessity of *terracing* to protect the soil from the floods of rain. These trenches answer the purpose of refuse pits for the accumulation of manure. Most planters are strongly in favour of weeds being burned or exposed on the roads to the full force of the sun. It weeds, &c., be placed in the trenches, it is generally recommended that these should be cleaned before the setting in of the rains, so as to afford every means for catching the fine soil washed into them by the surface water. The trenches should be fed by catch-drains, and have outlets by which the water can be run off into the drainage channels. Coffee being an exhaustive crop, *manuring* becomes essential, but the exhaustion is more the result of the peculiar method of cultivation than due to the crop. The precautions against the removal of surface soil, if fully carried out, are generally found more efficient than even the most expensive system of artificial manuring. If the soil does not contain lime it becomes necessary to supply it. The best nitrate manure is well-rotted dung, but, if necessary, bones, oil-cake, and night-soil may be resorted to. The manures and the methods of applying them to one plantation are not always applicable to another, so that no general rule can be laid down, and the indications afforded by the soil itself must be followed. Most planters urge the necessity of *forking* the soil at least once a year. This consists in softening the hard-trodden soil by digging it up by means of an iron fork to a depth of 12 to 18 inches.

Pruning. 1652

PRUNING AND TREATMENT OF THE PLANTS.—Having attended to the soil, the planter's attention will now have to be turned to the treatment of the plants. Where seedlings have failed or look sickly or have been injured, healthier ones should be substituted. The kind of pruning first indicated is generally that known as "topping." The age and height at which this should be performed are much-debated points. They are generally solved by local circumstances. The operation of topping consists in nipping off the central bud so as to check a too great upward growth and thus cause the plant to branch profusely. In Ceylon this is generally performed when the plants are 12 or 18 months old. **Sabonadiere** "prefers to

COFFEA arabica.

Coffee Cultivation—Pruning.

METHODS.

postpone the operation till the shrubs have borne their maiden crop, even though extra staking be required to withstand the wind. His plan is to remove the two primaries at the required height, by a sloping outward cut close to the stem, and then to remove the top by an oblique cut, so that the stumps resemble a cross, and a firm natural knot remains to guard against the stem splitting down. **Hall** (Ceylon) contends that the plants should be topped as soon as they have reached the required height, when the soft wood is easily severed by a pinch between the finger and the thumb. In Natal the shrubs are topped either at their full height—4½ to 5 feet—or at 3 feet, allowing a sucker to grow up on the weather side to complete the height. The latter plan is preferred. There is much advantage gained in limiting the height to 5 feet; not only is the crop gathered more easily and without damage to the tree, but it is actually heavier, and the shrubs are more readily made to cover the ground." (*Spons' Encyclop., 696.*) **Dr. Shortt** says: "Pruning consists of various operations connected with either arresting the height of the plants to cause them to spread out laterally, or in removing the additional growth of wood, to encourage the plants to push out new fruit-bearing shoots. These various operations come under the different heads of topping, pruning, and handling." With regard to topping he adds: "It is undoubtedly called for on all plantations that lie exposed and are likely to suffer from gales, &c., but in sheltered localities it does not matter so much, though it will save a good deal of trouble, for ladders will be required to pick the fruits off untopped trees." Again, there are two principal objects to be borne in mind with reference to topping,—*viz.*, to give security to the plant in the soil against wind and storm; and, secondly, to give facility for the collection of the crop. Of the two, the latter is the more important, for it applies without reference to locality, whereas in the former, locality is the chief point on which the question turns." The first result of topping is to induce the growth of masses of shoots; these are removed by what is technically called *handling*. "The first to appear are vertical suckers or 'gormandisers' from under the primary boughs: these are immediately rubbed off without injuring the bark. From the primaries spring secondary branches, in pairs, and at very short intervals. All such appearing within six inches of the main stem are removed at once, so that a passage of at least a foot is left in the centre of the tree for the admission of air and sun. The object of pruning is to divert the energies of the plant from forming wood and to concentrate them upon forming fruit. The fruit of the coffee tree is borne by young wood; and as the secondaries are reproduced when removed, they are cut off as soon as they have borne, and a constant succession of young wood is thus secured." (*Spons.*) This removal of secondary twigs from the primary boughs is what the planters call "pruning." The practical effect of the treatment briefly indicated above is to cause a plant about 5 feet in height to develope horizontally primary branches or boughs at intervals of about 6 inches throughout the height of the stem, and to form along these boughs a constant supply of secondary fruit-bearing twigs. All ascending or cross-wise branches or twigs are at once removed, so as to force the plant into the arbitrary and unnatural type of horizontal spreading branches which have the advantage of exposing to the sun and light a large surface from which the crop can with ease be removed. When practicable, the bushes should be handled twice before the crop, and all secondary fruiting twigs pruned off after removal of the crop. The pruning should be finished before the ensuing flowers begin to form; but where this has been neglected, and it is apparent that a flush of so heavy a character as to weaken the plant has set in, it will be necessary to sacrifice

Coffee Cultivation—Season.

METHODS.

this by pruning the plant down to the extent it may be expected to fruit without injury. The lateral or primary boughs should not be allowed to grow more than 2½ feet, otherwise they will droop and exclude the light from those below. In pruning, it is often recommended to leave the opposite lateral to that removed, so as to allow of its fruiting next year. By thus cutting the secondaries every other year a continuous crop is secured. All tertiaries should be systematically nipped off; broken, diseased, or dead branches should be cut off.

Catch-crops. 1653

CATCH-CROPS.—Much has been written for and against the growing of other crops along with coffee. In Darjeeling it was tried to grow tea and coffee together, but with little or no success, in spite of the fact that the out-door labour and manufacture of these crops so fit into each other that economy might be effected. In Natal and other countries, plantains, yams, and cocoa have been grown, and in some cases the coffee has even been treated as the secondary crop. But the only supplementary crop of which favourable reports exist is Indian corn. This grows rapidly and affords shade at the very time that it is most needed.

Seasons. 1654

SEASONS FOR COFFEE-PLANTING AND MANUFACTURING OPERATIONS.—The industry being chiefly in South India, the seasons for operations very closely correspond with those of Ceylon. The season for commencing agricultural operations is about October, and the buildings require to be finished by January. The best time for firing the felled trees is the beginning of February, the trees having been allowed to dry for about two months. About the same time the land should be lined and pitted, this work continuing till the rains set in. The nursery is usually made in May or June, so that, by the time the land is pitted and the rains have set in, the one-year old seedlings are ready for transplantation. The blossoms appear in March of the second and third year, and continue every year after. About October every preparation should be complete for the collection of the crop and the manufacture of the berries. The fruits commence to ripen in October or early in November and continue till January. Thus from flowering to harvest occupies about eight months. None but fully ripe berries (technically known as "cherries") should, according to **Dr. Shortt**, be collected, the women and children going over the plantation periodically to remove all the bright or blood-red ones, while carefully leaving the others to mature; once ripe, the sooner collected the better. **Mr. Pasteur** says: "The usual course, however, is to pick the cherry before complete maturity, when it is of a deep red or cherry colour, the berry inside being then found to be of a fine dark-green or bluish green, which it is the endeavour of the planter to preserve as carefully as possible, the value of his coffee depending chiefly on the depth and brightness of the colour." The more gradually the bloom fades the better. Rain at this season is unfavourable, but after the fruit has set a shower is beneficial. Immediately the crop is gathered in, the boughs of the bushes should be tied up to allow of the droppings of birds, &c., that contain berries to be picked up, also the berries that have fallen to the ground. This forms what is generally known as "jackal coffee." Before the boughs are opened out again, the ground around each plant is *manured* and *forked*.

The preparing or manufacturing of the "cherry" into the "berry" will be found dealt with in a further page.

INDIAN. Area and outturn. 1655

INDIAN AREA UNDER, AND OUTTURN OF, COFFEE.

The cultivation of coffee is practically confined to Southern India. During the three years 1883, 1884, and 1885 the average area under mature

COFFEA arabica.

Area of Coffee Cultivation in India.

AREA AND OUTTURN.

plants was returned at 186,500 acres, and the average yield at 31¼ million pounds, which were thus distributed :—

	Acres.	lb
Mysore	82,100	7,110,000
Madras	55,100	13,160,000
Coorg	42,300	9,330,000
Travancore	4,800	820,000
Cochin	2,200	830,000
TOTAL .	186,500	31,250,000

These statistics, which are in all probability defective, have been taken from the Statistical Tables of British India published by the Department of Finance and Commerce up to 1887. These tables include the Native States of Cochin, Travancore, and Mysore, and hence the area given is greater than that returned (119,142) in the Agricultural Statistics of British India published by the Department of Revenue and Agriculture. The total area taken up for coffee cultivation is 354,331 acres, of which 39,618 acres are under plants which have not as yet commenced to give a return.

Considerable areas suitable for coffee cultivation are still available in British India: of the Nilghiris it has been said that there exists 200,000 acres of reserve suitable for coffee. The port of shipment for Nilghiri coffee is Calicut, to which the crops are conveyed for a considerable distance by water. The Shevaroy Hills are more inland, and cultivation does not seem to have progressed so much on these hills, the distance from sea being probably unfavourable. Land is, however, still available on these hills, and the ease with which the produce can be removed by the Madras and Raipore Railways should tell much in their favour. In Mysore, coffee cultivation is not likely to extend very much, as all the available coffee land has been taken up. Of the total area, *viz.*, 81,754 acres under mature plants, 81,543 are in the Kadur district: the port of shipment being Mangalore. Mysore for the Wynaad and Coorg. The rainfall of Coorg seems to be too great for coffee-planting progressing much further than at present, except on the sheltered tracts.

"A northern aspect is best, being moist during the dry season, and possessing the most uniform temperature; but it will be modified either eastwards or westwards according to the locality, so as to suit the prevailing winds. On the western slopes of the coast-ranges, the south-west monsoon bursts with such force that coffee cannot withstand it; in that situation, therefore, an easterly tendency of aspect is imperative. Further inland, the drier and hotter climate will compel a westerly deviation, so as to catch as much as possible of the monsoon rains. In the western or wetter districts, shade is inadmissible; in the eastern or drier districts, it becomes a necessity." (*Spons' Account of the Coffee District of Mysore.*)

The following passages regarding the seats of Indian coffee cultivation may be found useful:—

MYSORE. 1656

In Mysore the cultivation is limited almost exclusively to the Kadur District. In Vol. II., page 410 of the *Mysore Gazetteer* published in 1876, it is stated that "the coffee cultivation of Southern India may be said to have had its origin in this district; for the plant was first introduced about two centuries ago by a Mahomedan pilgrim named **Baba Budan**, who, on his return from Mecca, brought a few berries in his wallet, and, taking up his abode on the hills that now bear his name, planted them near his hut. It was not, however, till about sixty or seventy years ago, that the cultivation extended beyond his garden, and not above forty years since European enterprise was first attracted to it. One of the earliest European

Area of Coffee Cultivation in India.

AREA AND OUTTURN.

planters was Mr. Cannon, who formed an estate on the high range immediately to the south of the Baba Budangiri, where the original coffee-plants are still in existence flourishing under the shade of the primeval forest.

"The success of Mr. Cannon's experiment led to the occupation of ground near Aigur in South Manjarabad by Mr. Green in 1843. During the last fifteen years, estates have sprung up between these points with such rapidity that European planters are settled in almost a continuous chain of estates from the northern slopes of the Baba Budans to the southern limits of Manjarabad, not to mention Coorg and Wynaad beyond."

The above account of the introduction of coffee into Mysore was first published by Colonel Onslow, from whom all subsequent writers have borrowed their information without materially adding to or correcting any one feature of the original statement.

MADRAS. 1657

Madras Presidency.—The following extract taken from pages 290 and 291, Vol. I. of the Madras Manual published in 1885, gives interesting particulars regarding the cultivation of coffee in the Madras Presidency: "The principal coffee tract of Southern India is along the western coast, and coffee estates extend in nearly an unbroken line along the summits and slopes of the Western Ghauts, from the northern limits of Mysore down to Cape Comorin. The only portions of the area within the limits of the Madras Government are the Wynaad tract and the Nilgiri Hills, the rest being in Mysore, Coorg, and Travancore."

South Wynaad 1658

Of the early plantations the Madras Manual adds: "Nearly all the land taken up at this period was what is known as grass or bamboo land, and in consequence most of the estates proved unprofitable. Of many of them not a trace, except the ruins of bungalows, remains at the present day. After the first attempts, coffee cultivation was transferred to South Wynaad. For ten or fifteen years it made little progress. In 1855 and 1856 a number of new estates were opened out, some too hastily, and consequently with little success. In 1862 the return showed 9,932 acres under cultivation. In 1865 there were 200 estates covering 14,613 acres. An official enquiry was made on the subject of Wynaad coffee in the year 1868, and, according to the returns then made, the acreage was 29,909·08, of which 21,479·54 acres were held by Europeans and 8,429·54 acres were held by natives. At the present moment there are in Malabar about 24,000 acres of matured plants, 2,900 acres of immature plants, and 26,000 acres of land taken up for planting but not yet planted. The average yield per acre in the Wynaad is said to be about 156℔, and the cost of cultivation about R250 per acre, which at present prices would not give a good return. The table below, showing the quantities of Wynaad coffee shipped on the Malabar coast during a period of twelve years, indicates nearly all the crops, as very little passes out by Mysore or Coimbatore:—

	Cwt.
1856-57	32,658
1857-58	16,204
1858-59	36,934
1859-60	49,680
1860-61	48,742
1861-62	91,080
1862-63	43,907
1863-64	91,947
1864-65	110,548
1865-66	125,891
1866-67	66,552
1867-68	128,011

COFFEA arabica.

Area of Coffee Cultivation in India.

AREA AND OUTTURN. Nilghiris. 1659

"Coffee cultivation on the Nilghiris was reported on in 1872. A large area of land on the Nilghiris has proved to be admirably suited for the cultivation of the coffee shrub. Not less than 22,897 acres are now under coffee plantations, besides 12,231 acres taken up for planting. Twenty-five years ago the area under coffee did not much exceed 500 acres. This great increase is entirely the result of private enterprise, and has added much to the prosperity of the Nilghiris, while at the same time benefiting the districts immediately adjoining. In the establishment of these coffee estates a property has been created worth about 5 millions of rupees. Of the total expenditure, about one third is for the payment of wages to coolies, and most of this is carried into the low country, either in payment for food-grains consumed by plantation coolies, or as cash carried by the coolies themselves when they return to their homes. Estimating that the sum sent into the low country in this way represents annually R6,00,000, this will support about 14,000 families of labouring people. Moreover, in carrying coffee to the coast, and sorting, packing, &c., a large amount of other labour is employed. Until a few years previous to 1850 the coffee plantations on the Nilghiris were found only on the eastern slopes, but they have now been extended to the southern, northern, and north-western slopes; there are also some extensive plantations in the Ouchterlony Valley and in the neighbourhood of Coonoor. Coffee cultivation is also carried on on the Shevaroy Hills in the Salem District, where nearly 6,000 acres are under the crop, and an area of 4,680 acres has been taken up for planting; on the Pulney and Shiroomullay Hills in Madura, where nearly 4,400 acres have been planted and a considerable area has been taken up for planting; and in the Tinnevelly and Coimbatore Districts, in the former of which there are about 2,000 acres under coffee and in the latter about 800 acres."

Coorg. 1660

In Coorg coffee is also extensively produced, for there are but few Europeans and natives there who are not interested in its cultivation.

The following information is gathered from the Administration Report of Coorg for the year 1882:—

"There are in the province 212 coffee plantations owned by Europeans, and 4,594 by natives, comprising an area of 77,474 acres, or a little more than one thirteenth of the area of the whole district.

"The area of land held by the former is 41,507 acres on an assessment of R76,129, and by the latter 35,967 acres paying an assessment of R66,440: besides which, much coffee is grown by the latter on their *banes* (plots of forest land attached to rice-fields free of assessment). The average size of each estate held by Europeans is 196 acres and by natives 8 acres. Of the whole area 40,350 are bearing, producing 6,125 tons of coffee, or on an average 3 cwt. the acre; but the average yield in most European estates, which are much better cultivated than native estates, reaches 7 cwt. the acre. Taking the average cost of cultivation at R120 per acre on European estates, and R40 on native, each cwt. of coffee costs R27. The cost of cultivation at the rate per acre assumed above comes to nearly 32 lakhs of rupees. Of this not less than 60 per cent. on an average may be estimated as having been paid to labourers in wages. Calculating that 26,893 labourers, which is about the average number employed throughout the year, received R6 each per mensem, upwards of 19 lakhs of rupees were expended for labour during the year. The value of the coffee produced, taking the selling price to be, on the average, R30 per cwt. on the spot, was about 36 lakhs of rupees." (*Madras Weekly Mail.*)

Travancore. 1661

Travancore.—The area under coffee in the former State in 1885 was 4,013 acres, and in the latter 2,407 acres. The area under coffee in Travancore seems to have declined considerably within the past few

AREA AND OUTTURN.

years, or the returns are more nearly correct than they used to be. In 1883 there were said to be 6,268 acres under coffee, with 4,353 acres taken up but not yet planted. That area comprised 56 estates distributed over the State at altitudes from 753 to 4,800 feet above the sea. The cost of cultivation per acre was returned on the average as R45 and the yield 87℔, the minimum yield being 35℔ and the maximum 237℔.

Speaking of the future prospects of this State, the *Madras Manual* says of Travancore: "To the north, the mountains rise to an elevation of 8,000 feet with plateaux over 7,000 feet. The more important of these is part of the group known as the Anamallays." "The plateaux, by reason of their good clime, rich soil, abundant timber and water-supply, are likely to become better known as the demand for coffee-land increases. One plateau alone (Eroovimullay, or Hamilton's Valley) is 6 miles long by 3 wide, and contains about 10,000 acres of excellent tea and coffee land."

COCHIN. 1662

In Cochin there were, in 1883, 17 gardens, and these gave the return of 342℔ to the acre at a cost of R24.

Technical Terms. 1663

TECHNICAL TERMS USED BY THE COFFEE PLANTERS.—The ripe coffee fruit is termed the "cherry." The succulent outer coat of the fruit is the "pulp," the inner adhesive layer the "parchment." The seed-coat within the parchment, which adheres closely to the seed, is called "the silver skin." The pulp is usually removed at the plantation, but it is a common practice for planters to send the "berry" or seed enclosed in its parchment to the coast town or even to Europe, in order that by special and expensive appliances it may be deprived of its parchment. This has been strongly recommended within recent years, as the extra cost of transport has been found to be more than compensated for by the better quality of the produce and the great facilities afforded in Europe for working the complicated machinery necessary for this purpose.

MANUFACTURE.

PREPARATION OR MANUFACTURE.

The preparation of the "berry" from the "cherry" may be said to be accomplished in the following stages: (1) *Pulping*; (2) *Fermenting*; (3) *Drying*; (4) *Peeling, Milling,* or *Hulling*; and (5) *Sizing* and *Winnowing*.

A volume might be written on the various systems and mechanical appliances that have been or are now employed during the various stages of coffee preparation. The primitive native system is to sun-dry the cherry, then to pound it in the common rice-pounder and winnow away the fragments of the dry pulp and parchment separated from the berry. Besides being tedious, ineffective, and expensive, this process does not secure the uniformity of colour and freedom from injury to the beans which secures a high return. This fact has led to the periodic improvement and perfecting of the methods and machinery now in use.

Pulping. 1664

PULPING.—The operation known by this name consists in the removal of the pulp which surrounds the beans. This is most easily and effectively accomplished if the collections of ripe cherries made each day are passed through the machinery at once. If unavoidably delayed, it may be necessary to ferment the cherries before they can be pulped. The most simple machine in use is that known as the "disc pulper." This consists of rotating discs the surfaces of which are covered with sheet copper roughened by having projections punched forward. A "single pulper" of this description will pulp 20 to 25 bushels an hour and may be worked by three coolies. A "double pulper" of this type has two such discs and is furnished with a feeding roller. It will pulp 40 bushels an hour, and may be worked by from four to six coolies, and double that amount if worked by

COFFEA arabica.

Coffee Manufacture.

MANUFACTURE.

steam. The discs work against smooth iron beds so adjusted that the complete cherry cannot pass between. They are torn upwards against the beds, and the projections on the discs tear off the pulp, allowing the beans to drop into one receiver and the fragmentary pulp to be carried into another. The disc pulper is in fact somewhat like the cotton gin which drags the fibre forward and drops the seed behind. The "cylinder pulper" is an older invention in its conception, but has been improved and perfected to a much greater extent than the disc; the latter, being light and cheap, is more generally used in new than in well-established plantations. In the construction of a pulping-house it is generally recommended to secure a hill-side against which an excavation can be made for the house. This should consist of three storeys—a loft in which the cherries are spread out—the pulping floor or platform, and the cisterns. By constructing this building against an embankment or steep cliff, the cherries may be carried direct into the top loft without requiring to be raised. A good supply of water has also to be conveyed to the loft so as to descend with the cherries into the pulping machine in a continuous stream.

Space cannot be afforded for a discussion of all the inventions and contrivances now in use for pulping the cherry; suffice it to say that the cylinder pulper consists of a drum with sharp teeth working against a *breast* fixed to within a certain distance of the cylinder, or by having two sets of *chops* similarly adjusted. If *chops* be used, the bottom one should have a sharp edge, so as to sever the pulp entangled by the teeth from the bean. The teeth fix in the pulp and thus drag it forward while the beans are carried into the cisterns. By means of sieves the cleaned beans are separated from the partially-pulped cherries, the latter being made to pass once more through the pulper. The stream of water and cherries is carried from the loft of a tube which dips to the bottom of a basin known as the *hopper*. Stones subside in the hopper, while the continuous stream from above causes the hopper to discharge a uniform supply of cherries and water to feed the pulper.

Fermenting 1665

FERMENTING. -The parchment coffee, which may or may not have been assorted by contrivances in the pulper and sieves, has now to be fermented to remove from it the saccharine matter. If this be not accomplished it is difficult to dry the beans. By taking advantage of the descending flow of water, the beans are carried into tanks, and these tanks must in their turn be higher than the drying platforms on to which the fermented beans have finally to be dispersed. There are generally four fermenting tanks—two in which the fermentation actually takes place, and two in which the beans are washed. One of each is used for the produce of one day's pulping. All the coffee pulped in one day is allowed to remain in the front or receiving cistern until fermentation has set in. The period necessary for this will depend greatly on the temperature of the atmosphere, but from 12 to 18 hours will generally suffice. The contents of the fermenting vat are then run into the washing cistern, and the receiving vat rendered available for another day's produce. By having two sets of these tanks the pulping operation may be carried out continuously, each day's collection being disposed of so as to have the pulper ready for the next day's work. When properly fermented the beans are easily deprived of their saccharine matter by being driven from the fermenting vat by a goodly supply of water and thoroughly washed in the washing tanks. The size of the fermenting and washing cisterns will of course depend on the size of the plantation. When possible they should be constructed of wood, the planks being not less than 2 inches thick. Wooden tanks are not so cold as stone or brick tanks,

Coffee Manufacture.

MANUFACTURE.

and are accordingly preferred. The tanks should slope towards the discharge openings.

Drying. 1666

Drying.—The washed beans should now, through the flow of water, be carried to the drying floors or platforms and be exposed to the influences of the sun and atmosphere. The drying floors are usually made of concrete, but sometimes asphalt is employed. A simpler process is to harden the ground and cover it with a coir matting. This has the advantage of admitting of the surplus matting being thrown over the beans in the event of an occasional shower, but shed accommodation into which the beans may be rapidly conveyed is essential. During the drying, the beans have to be turned over repeatedly either by rakes or by the coolies' feet. The difficulties against which the planter has to guard at this stage of the manufacture are too rapid drying cracking the beans, or a disproportioned drying through reckless turning or racking. To secure a better and more steady slow drying, various artificial contrivances have been invented which are now employed by many planters, but the result is the same,—namely, the drying of the beans. **Mr. Pasteur** says: "On gardens and plantations cultivated by Europeans the cherry is removed as quickly as possible after being picked, by being put through pulpers and undergoing a very careful and delicate process of mashing and washing, until the berries are left with their parchment envelope perfectly clean. In many cases, however, there are neither appliances, time nor labour, to put the fresh-gathered fruit through this process, and under a tropical sun the cherry dries quickly, and has then to be pounded to the great detriment of the colour as well as the quality of the bean; hence the difference between *unwashed* or ordinary pale and *washed* or coloured or plantation coffee,—the taste of the *washed* coffee being, as a rule, much more delicate, and free from the earthiness and common rough flavour of the *unwashed*.

Peeling. 1667

Peeling or Milling.—This consists of the removal of the parchment and silver from the beans. As already stated, this operation is now chiefly effected by the dealers, at the port of shipment, and not by the planters. Indeed, much has been written in favour of the beans being sent to Europe in parchment, and milling machinery is now in use in London for this purpose. The following passage from **Mr. Pasteur's** report will be read with considerable interest, and may be viewed as indicating a possible new direction of coffee enterprise:—

"Among the samples of Wynaad coffee, those from the Eva Estate deserve special attention, one half of that crop having been despatched in parchment to be peeled and sized in London. The experiment has proved quite successful, the coffee represented by the sizes, 1st, 2nd, and peaberry, being fully equal in colour and appearance to the corresponding sizes prepared in India. The whole was sold at the same public auction—the London-cured realising a rather better price than the other half. Similar and more recent experiments made with some shipments from Costa Rica, Guatemala, and New Granada have shown startling results, the portion prepared in London having realised from 10*s.* to 14*s.* per cwt. more than that cured in Central America. These experiments would tend to show that the parchment preserves in a remarkable degree the colour and the quality of the berry against the incidents or accidents of a land and sea transport. In the case of the Costa Rica and New Granada shipments cured in London, the berries seemed fuller and of better shape and weight than the others, as if (which is by no means improbable) the parchment left for two or three months longer than usual around the berries had acted as a kind of natural preserver, inside of which the berry had time, as it were, to mature more completely than when deprived of its outer and inner

COFFEA arabica.

Coffee Manufacture.

MANUFACTURE. coating almost immediately after being picked. The curing requires machinery, motive power, drying grounds, delicate manipulation, and constant supervision; where any of those requisites fail, the coffee suffers in appearance, and consequently in value. Suitable machinery for treating parchment has been erected at two of the London wharves, and there is every reason to hope that this is only the beginning of a new and profitable home industry. Growers will not be slow to perceive that the small increase of freight which they have to pay on parchment is more than compensated for by the enhanced price which the improvement in the quality of their coffee will enable them to obtain." In the *Kew Bulletin* for May 1888 the advantages of milling coffee in Europe, instead of at the plantation, are strongly urged. The cost of doing so is stated to be only 2*s*. and 6*d*. per cwt. (*Report on the Col. and Ind. Exhibition*, page 169.)

The greatest danger in peeling consists in the fact that before being passed through the mill the beans require to be again heated. On the plantation this is generally done by exposure to the sun. The extent to which this is necessary depends greatly on the nature of the beans, and long experience is required to determine this point. As a practical hint it is generally laid down that they should be dried till they resist the pressure of the thumb-nail, but no two samples are alike, and overdrying will result in serious loss of weight and underdrying will deteriorate the colour.

Sizing. 1668 SIZING AND WINNOWING.—The peeled coffee as it comes from the mill is subjected to a fan which (as with the chaff from corn) effectually drives off the parchment and skin, leaving the clean coffee behind. After this it is separated into various sizes for the market. This has the effect of not only meeting the special demands of the consumers, but in furnishing a bean of uniform size that will admit of uniform roasting. Formerly this used to be done by the hand, but mechanical contrivances are now universally employed.

Packing. 1669 PACKING.—Having followed all the precautions and adopted all the most approved methods and appliances, the coffee producer, to secure the success of his labours, has now only to attend to packing. The beans must be saved from exposure to the air, or from being packed in cases that would impart a false aroma. This is usually done by packing the produce in casks, care being taken to select timber that will not taint the coffee. Bags are sometimes employed, but are inferior to casks, and the shipments of coffee should not be made along with cargoes of merchandise likely to injure the coffee.

ADULTERANTS. 1670

ADULTERANTS AND SUBSTITUTES FOR COFFEE.

Adulteration is never effected by the planter: indeed, it is practically impossible. Until the beans have been ground, mechanical impurities such as mud and stones are the only admixtures that may exist in the coffee as it leaves the plantation. While this is so there is perhaps no other dietary article that is so much and so persistently adulterated as coffee. This in a large measure appears to be due to the legislative system which has permitted a mixture to be sold so long as it is declared to be such. Criminality consists alone in selling as *pure coffee* an article that contains anything but coffee. Legally "chicory" may be the roasted chicory root itself or the root of an allied plant or other vegetable substance applicable for the same purpose as chicory. No questions are therefore raised as to the ingredients of a mixture; and indeed, if further protection to the manufacturer be necessary, such mixtures may even be registered as patent medicines. This fact, together with the long-established custom of mixing chicory with coffee, has given origin

Adulteration of Coffee.

ADULTERANTS.

to a gigantic system of adulteration. The substances which are most generally employed are—

"*1st*—Roots such, as chicory, dandelion, mangold-wurzel, turnips, parsnips and carrots, &c.

"*2nd*—Seeds such as beans, peas, date-stones, malt, rye, &c.

"*3rd*—Burnt sugar, biscuits, locust-beans, figs, &c." (*Bell, Chemistry of Foods.*)

During the proceedings of a Coffee Protection Association formed in London in 1886 the writer had the opportunity of examining certain well-known mixtures and of seeing some of the practices of adulteration. One of the most curious which was brought to his attention was the use of artificially-prepared beans in so close imitation of the real article that the mixture of the spurious with the true coffee beans might be fearlessly ground in the purchasers' presence and sold as *pure coffee.* This subject has already been alluded to under **Chicory** (*see* **Cichorium Intybus, C. Nos. 1107 & 1108**), and need not be elaborately dealt with in this place. A largely consumed adulterant of coffee is a substitute for chicory known as *mochara.* This consists of ripe figs dried, roasted, and pulverised. Burnt sugar is sometimes added to coffee in small quantities to give colour to the mixture, and from an idea that it preserves the aroma. Three or four pounds to the hundredweight might be admissible without being viewed as an adulterant. When, however, roasted sugar or a sugar-yielding root (known as caramel) is added to a large extent, it becomes a serious adulterant, and perhaps one of the most extensively used of all adulterants. It is to the roasted sugar contained naturally in chicory (caramel) that that ingredient owes its bitter flavour and aroma—properties which recommend an admixture of chicory to some consumers as a desirable addition to the beverage. This fact allows of extensive adulteration, since the sugar contained in any other root will yield, when roasted, caramel bitter. Were saccharine roots the only adulterants employed in coffee, there might be less ground for urging the adoption of the French system which permits the grocer to sell separately chicory or any other substance which the consumer desires to mix with his coffee, but prohibits the vendor from manufacturing special preparations or mixtures. Roasted flour coloured with ferruginous earth is to some extent used as a coffee adulterant, and even roasted liver and other objectionable animal substances are said to have been found in coffee mixtures. A simple mode of detecting the presence of chicory or other caramel admixtures in ground coffee is to throw a little on the surface of a glass of clear water. The readily solvent nature of the particles of caramel will at once impart coloured streaks to the water, while only after some minutes will pure coffee give its colour to the water.

Caramel.

Date-seeds were at one time supposed to be likely to come into use as a coffee substitute, and a company was actually formed to carry out this idea, without sufficiently reflecting on the means of procuring and collecting the seeds, supposing even that when roasted and ground they were found to possess in a sufficient degree the flavour and aroma of coffee. The seeds of several species of **Cassia** have for centuries and are even now used by the inhabitants of tropical countries in place of coffee. These do, as a matter of fact, afford, when roasted and ground, a decoction which closely resembles coffee. The reader is referred to the account given under **Cassia occidentalis (C. No. 784)** for particulars of a coffee substitute which would seem to deserve more careful consideration. India could produce, at a nominal price as compared to coffee, immense quantities of the so-called "Negro Coffee," if that article should be found to commend itself as a wholesome and cheap substitute for true coffee.

Negro Coffee.

COFFEA arabica.

Trade in Coffee.

ADULTERANTS.

The decline in the use of coffee which seems to have set in throughout the world seems to be largely due to the difficulty of procuring the pure article. If this be so, it would seem high time for the coffee planters and others interested in coffee to take the necessary steps to remove this injurious reputation, and to place in the hands of the consumer a cheap and pure coffee.

COMMERCIAL TERMS. 1671

COMMERCIAL TERMS AND QUALITIES.—The coffee bean usually consists of two oval plano-convex seeds, though sometimes there is but one, which from its shape is known as the "peaberry." The commercial value of coffee depends upon many circumstances,—form, size, colour, smell, flavour, age, and uniformity within the sample. Form to some extent, though not always, depends upon the source: there are three commercial types as to form—*Mocha*, small round peaberry; *Bourbon*, pointed and medium-sized; and *Martinique*, large and flattened. Colour depends entirely on the degree of ripeness when plucked and the care taken in the preparation, but, as a general rule, the Old World coffees are inclined to yellow and the New World to green. Weight decreases with age or is lessened by over-drying in the preparation. Odour is, however, the most important test, but it can be judged of only by professional dealers.

PRICES. 1672

PRICES OF INDIAN COFFEE.

Mr. Pasteur, writing of the crop of 1885, says: "Taking 90*s.* per cwt. as the average value of the *bulk* from the estates of true Mysore type, the *Coorg* Mysore estates would be worth 80*s.* for *bulk*, the Nilghiri 83*s.*, the Coorg 82*s.*, the Wynaad 78*s.*, and the Travancore 70*s.* per cwt.; whilst native Mysore of average quality would be worth 63*s.*, and native Coorg or Wynaad 60*s.* per cwt. The finest qualities of Mysore range in value from 100*s.* to 135*s.* per cwt." The same writer gives Ceylon as *large size* ranging from 90*s.* to 105*s.*, and *bulk* 75*s.* to 85*s.*, the average being 80*s.* None of the other British-grown coffees, with the exception of Jamaica, were valued as high even as those of Ceylon; and, as stated in another paragraph, Mr. Pasteur, one of the highest commercial authorities, gives the first place to the Indian coffees in point of value or merit. "Nowhere," he says, "is finer coffee grown than in India and Jamaica, and its value, as well as that of Ceylon, is firmly established above that of all other kinds, even of Mocha, which at one time stood above all others."

The rise in the prices paid for coffee during the first twenty years of its cultivation by Europeans led to its rapidly extended cultivation. Dr. Bidie says that "a hundredweight of coffee at the present day (1869) is worth double what it was some years ago. In 1848, in Ceylon, a hundredweight of native coffee was sold for the same price as a bushel of rice, *viz.*, R4½ and, about the same time, estate coffee from the Wynaad was selling on the coast for R17 the hundredweight. At the present day (1869), estate coffee is worth from R30 to 34 the hundredweight, and although the working expenses of estates have greatly increased of late, still the rise is insignificant as compared with the marvellous improvement in prices."

TRADE. 1673

TRADE IN INDIAN COFFEE.

"India now stands first and foremost among British possessions, both for the quality and quantity of its production" Disease has, however, "in many places affected the vitality and shaken the strength of the trees, so that they have been less able to resist periods of drought or of heavy monsoon weather, and small and irregular crops have been the consequence. It would seem, however, as if plantations were gradually recovering their former strength, and with good cultivation and manuring

Indian Trade in Coffee.

TRADE.

and fair seasons India may hope to maintain its position as our largest and best field for the production of fine coffee. A hopeful sign for the future may be gathered from the superior average quality of the crop of 1885-86 to that of the two previous ones." (*Reports, Col. and Ind. Exh.*)

There were in 1885-86 in British India, including the Native States of Mysore, Cochin, and Travancore, 44,985 coffee plantations. Since 1875-76 the number of plantations given in official returns has fluctuated from 47,000 to 38,000. This has been accounted for by the fires which destroyed certain gardens, the imperfect returns, and the amalgamation of small gardens. The bulk of the coffee exported from India is washed coffee prepared under European supervision, many of the small native planters selling their produce to neighbouring European planters or to the special firms that do a considerable trade in pulping and peeling coffee. At the same time, there is by no means an inconsiderable trade in unwashed or native coffee,—that is, coffee prepared by the crude native process to which reference has been made. **Mr. Pasteur**, in his report of the coffees shown at the Colonial and Indian Exhibition, deplored the paucity of the samples shown of native Malabar coffee, and this subject would seem to commend itself to the attention of Government, since paying industries, adaptable to the cultivator with small means, seem to be much wanted. The European market receives an immense amount of native or *unwashed* coffee. **Mr. Pasteur** says: "A large portion of the crops from Brazil, Java, St. Domingo, and to a less extent Central America and Guatemala, —in fact, fully three fourths of the world's production,—are prepared as *unwashed* or pale coffee; whilst nearly the whole of the Ceylon crop, three fourths of the Indian, and one fourth to one third of the Java, are prepared as *washed* or green coffee." At another place **Mr. Pasteur**, speaking of the native Malabar coffee, says: "It is a matter of regret that shippers from the Malabar coast have not sent any specimens of those kinds to the Exhibition; they are quite suitable for our home consumption, and form an important item of the Indian production." The returns for the coffee districts of India show Madras to have nearly a third of its coffee area owned by natives, Coorg about one half, and Mysore fully four fifths. These facts give some idea of the extent of the probable production of native or unwashed berry in India.

It is necessary to point out, before proceeding to discuss the returns of the foreign trade in British Indian coffee, that the town of Cochin itself is treated as British India, in the official trade returns, but the territory surrounding it as Native State. It therefore becomes necessary to add to the returns of foreign exports those from Cochin State and from Travancore State in order to obtain a correct idea of the total trade. The exports from these States during the past five years have averaged 20,376 cwt.

No statistics are available regarding the Indian inland trade in coffee. As regards the external trade, the average *imports* during the five years ending 1886-87 amounted in quantity to 25,300 cwt. and in value to R6,87,000. The average *exports* during the same period came to 352,600 cwt. valued at R1,38,54,000. To these foreign exports from India should, however, be added those shown above as sent from the Native States of Cochin and Travancore, namely, 20,376 cwt., making a grand total of coffee sent from India for each of the past five years of 372,976 cwt. Of the imports, Netherlands India through the Straits is the main source of supply; and next to it come Ceylon and Aden. Bombay receives most of this coffee, a little going also to Madras and Burma. Over 90 per cent. of the total exports proceed from Madras; the remaining 10 per cent. is shipped chiefly from Bombay, but this apparently consists of the re-exports of foreign and Madras coffee. The United Kingdom and France are the

C. 1673

COFFEA arabica.

Trade in Coffee.

TRADE.

two largest consumers of Indian coffee. During the past five years the coasting trade, which consists chiefly of despatches from Madras to places within the presidency and to Bombay, has averaged in quantity 70,000 cwt. and in value R22 lakhs.

Towards the close of the account given, on preceding page, of the History of Coffee, **Mr. Pasteur's** statement regarding the decline of the Ceylon trade has been quoted. With the discontinuance of a large portion of the Ceylon cultivation the greatest hopes were entertained of a bright future for the Indian coffee industry. Prices revived from 1885 to 1887, and during that period the exports to foreign countries maintained a higher level than during any previous consecutive period. During the past fourteen years the exports have averaged 344,785 cwt., but for 1885-86 they were 371,000 cwt., and fell slightly in 1886-87. Whether this high level will be continued, lost, or augmented, the future must be left to reveal; but upon this result will depend the further question whether India is to take advantage of the decline of the Ceylon industry. The Indian foreign trade in coffee has chronically fluctuated. It attained its highest recorded point in 1875-76, the exports in that year amounting to 371,900 cwt.; it fell to 302,500 cwt. in 1876-77, and to 297,300 cwt. in 1877-78. The bulk of the exports go from Madras (*viz.*, 90 per cent.), so that the growth of the trade since 1867-68 down to the present date may be seen by a comparison with the Madras exports (given at page 473) from 1856-57 to 1867.

COST OF CULTIVATION AND YIELD.

COST. 1674

So much has been written on this subject that it scarcely falls within the scope of the present article to deal with the various conflicting opinions that have been advanced. According to some writers the profits on coffee cultivation in India are problematic; according to others, the industry, after passing through an incubation of risk and danger, in which severe losses have been sustained, is now firmly established. **Dr. Shortt** in his useful Hand-book estimates the cost of opening 200 acres of forest land for the cultivation of coffee, including purchase of land, tools, felling, clearing, lining, holing, planting, road-making, building planter's house and coolie lines, and keeping the same in order for three years, as follows:—

	R
1st year	7,160
2nd year	3,300
3rd year	4,460
Instruments	700
Buildings and roads	1,830
TOTAL	17,450

This estimate, he states, is applicable to Coorg and Wynaad, more especially the former, but he only allows R125 a month for European supervision. He proceeds to state that "the third year is supposed to make a return. The average produce of an acre is estimated at 7 cwt., but we could not do better than keep on the safe side and take the produce of an acre at 5 cwt. The 200 acres will yield 1,000 cwt. of coffee beans, and if we take the value of a cwt. at R28 (that is giving R7 to the maund of 25fb), the return will be R18,000, giving a profit of cent. per cent. After the third year the average expense will not exceed R5,000 on a well-managed plantation, and the profit subsequently will be something fabulous." No allowance is made for the purchase of pulping machinery,

COST.

the erection of a pulping-house, and other accessaries to the preparation of the bean, but Dr. Shortt adds with reference to this that "these will at best form but a small item." But he has omitted apparently to estimate for the purchase of grass and forest land, and to take into consideration the cost of the labour of preparing the beans.

The author of the valuable article on coffee-planting in *Spons' Encyclopædia* gives several estimates both for India and for Ceylon. He states: "The following estimate (in rupees) for coffee cultivation in South India is based on the purchase of 300 acres of forest land at R50 and 200 acres grass land at R25, bringing 200 acres of the former into full bearing; labour, 4 annas a day, exclusive of maistries' wages." Then follows a balance sheet, the main facts of which may be expressed as follows:—

The 200 acres by the seventh year are brought under full bearing, and have not only cleared off the expense of the purchase and cultivation of the estate up to date, but the plantation has given its owner over and above R15,971. To continue to work it an expenditure of R23,645 would be entailed, but the return from the crop would be about R54,000 a year, so that with a portion of this the estate might now be extended to its full limits, 300 acres. This estimate has not only been framed to cover the charge of building all the necessary houses, but to furnish those with pulping and other machinery, and to stock the yard with 100 head of cattle and provide a horse for the superintendent. The capital necessary to organise such an estate (without having to obtain loans on crops) would thus be about R75,000, or say £5,000, and during the fifth, sixth, and seventh years that sum would be recovered. Interest on capital is the only feature that seems to have been omitted, but this can easily be provided against, and if the undertaking were started on borrowed capital, a loan of R36,000 would be required up to the end of the second year, of R15,000 during the third, and R24,000 during the fourth: the loans and interest being refunded by the eighth year. The above calculation seems to be a liberal one, every necessary for successful cultivation being provided for. Dr. Shortt's statement may be viewed as an indication of what an owner with smaller capital might do by working his own estate. The writer is, however, unable to verify these estimates; but since they have been framed by high authorities, they may be viewed as approximately indicating the possibilities of the Indian coffee industry when, with average seasons and fair prices, the speculation is entrusted to careful and skilful supervision. The hopeful prospect thus presented might, however, prove visionary through causes which not even a just and fair estimate could have taken into consideration. The highest hopes were once entertained of Indian coffee-planting, and yet large sums of money have been lost. It is therefore desirable to place alongside of these estimates, opinions of a very different character. Dr. Bidie says: "From ten to twelve years ago (1857-1859), the high price of land, and the flourishing state of coffee culture in Ceylon, induced planters from that island to come over to India, and their presence and efforts gave a great impetus to coffee culture. The demand for land rapidly grew in every part of the coffee districts, and as capitalists had full confidence in the success of planting, estates multiplied and extended their acreage very rapidly. Throughout 1860 and 1861, there was a perfect mania for planting—extravagant ideas of the profits to be reaped from it having taken possession of the public mind, while such contingencies as bad seasons appear to have been entirely forgotten. Men of moderate means joined in buying land, and invested their hard-won savings without a doubt as to success, while others of greater ambition established joint stock companies, spent lakhs in the purchase of

C. 1674

PROFITS.

ready-made estates, and pleased their own minds and those of the other shareholders with visions of 50 or 60 per cent. of profit. As might have been foreseen, such extravagant hopes have never been realised, the anticipated fortunes having retreated far away into the future, and the 50 or 60 per cent. dwindled down to 5 or 6. In many cases, indeed, these adventures have, from various causes, proved complete failures, the balance always being on the wrong side; and, taking them as a whole, the results have been such as to render the public distrustful of coffee culture as a safe or profitable investment, and to lower greatly the value of estates." (*Report on the Ravages of the Borer on Coffee Estates.*)

DISEASES.

DISEASES OF THE COFFEE PLANT.

1675 The number of diseases to which the coffee plant is liable are numerous, many depending on climate, soil, and method of cultivation. These are all more or less local, and, as a rule, readily curable. To this class belong the *Canker* which does such havoc in Natal. This is generally believed to be due to want of depth of soil, but climate and bad cultivation may have also to do with it. *Rot* or the withering of the young leaves is due to wet and cold.

There are, however, certain specific diseases some of which have practically baffled both the planter and the scientists, and have proved so disastrous as to have ruined the plantations in large tracts of country. This has been the case with Ceylon, the leaf blight having there proved so far incurable as to have caused the planters to substitute tea for coffee on their estates. Numerous reports have been published, such as those by **Marshall Ward, Nietner, Bidie, Harman, Forbes Watson, Morris, Cooke, Balfour, &c.** To review even briefly all that has been written on the diseases of the coffee plant would take up far more space than can be afforded in the present outline of the coffee industry. It may be said that the specific diseases are referable to two sections—*Fungoid* and *Insectiform*.

I. The chief FUNGOID diseases are:—(*a*) *Leaf-blight.*—This is a fungoid disease which is supposed to have first made its appearance in Ceylon in 1869, and to have appeared in South India two years later. It has since appeared in Java and Sumatra, but does not seem to occur beyond the limits of the Indian Ocean. It is caused by the fungus **Hemileia vastatrix**, an organism allied to mould. It attacks the underside of the leaves, in the form of spots or blotches, at first yellow, but which ultimately turn black. These spots are covered with a pale-yellow powder. They eventually extend over the whole surface of the leaf, which then drops from the plant, thus depriving it of the power of growth, or of developing or maturing its fruit. This blight is present throughout the year, but in its early stage it generally occurs from December to February in the form of invisible filamentous threads which cover every part of the bark and leaves.

Every effort has been made to find a cheap and effectual cure, but with little success. If powdered sulphur, alone or mixed with caustic lime, be blown over the plants and scattered on the ground below the boughs, the disease is prevented and the coffee plants seem at the same time to be benefited. This is, however, expensive and is more a preventative than a cure. When once the disease has taken hold of the leaves, nothing has yet been discovered that will destroy it without at the same time killing the leaves.

(*b*) *Leaf-rot* or *Candelillo* is a disease attributed by **Dr. Cooke** to the fungus **Pellicularia Koleroga**, *Cooke*. It is prevalent in Mysore plantations in July, the leaves, flowers, and berries becoming covered with a shiny

DISEASES.

gelatinous substance which turns black about the time that the affected parts fall from the plant (*Kew Reports, 1879, 30 ; and 1880, 35*).

II. Of the INSECTIFORM diseases met with in India the following are those which give most trouble :—

(*c*) *Borer.*—This pest used to be known as the "worm" and "coffee fly." It is most troublesome in South India, especially in Coorg and the Wynaad, where in 1865-66 it destroyed whole estates. It has been determined as the beetle **Xylotrechus quadrupes.** It is red or yellow, with black in transverse lines. It damages the trees by boring holes into the stem usually a few inches above the ground. These passages are at first transverse, but soon ascend spirally to the growing tip where the larvæ are matured. The plant early shows signs of death, and ultimately withers down to the point where the beetle entered. This pest is most prevalent in hot exposed gardens, and may be kept in check by free irrigation.

(*d*) *Bugs.*—Various insects are by the planters called bugs. They belong to the same family as the lac and cochineal, *viz.*, **Coccidæ.** There are three pests of this nature, known as the "brown," "black," and "white" bugs. The brown bug has been determined as **Lecanium coffeæ.** This establishes itself on the young shoots and buds, which it covers with a scaly incrustation in which the larvæ are developed. This causes the destruction of the parts to which it adheres, the flowers and young fruits falling freely. The pest does not do much harm, however, until it has been two or three years on an estate. It prefers cold damp plantations at about 3,000 feet in altitude. This bug may be first recognised as brownish wart-like bodies. These are the females each of which produces some 700 eggs. Fortunately this pest is freely attacked with parasites which greatly help the planter.

The black bug is known as **Lecanium nigrum.** Like the preceding this attaches itself to the tenderest shoots; it also prefers gardens at high altitudes in damp situations. The female somewhat resembles a scollop-shell. When the eggs are incubated the twigs become covered with an injurious black soot-like powder and the insects destroy the young berries.

The white bug or **Pseudococcus adonium** is somewhat like a wood-louse, though much smaller, it being only $\frac{1}{18}$ inch in size. It is flat, oval, and covered with white down in parallel ridges running across the back. It seems to prefer hot dry plantations and disappears with the rains, only to return in time to destroy the setting of the fruits. It is found on the roots about a foot below the surface of the soil, in the axils of the leaves, and among the clusters of flowers and young fruits. It may be easily recognised by the white excretion formed around the larvæ.

All these and the other less known coffee-bugs have a strong dislike to tobacco juice. They may be prevented from developing to an injurious extent by brushing the twigs with tobacco. Some planters recommend saltpetre and quicklime in equal proportions dusted on to the affected parts, or a washing with a preparation of soft soap, tar, tobacco, and spirits of turpentine. **Mr. Neitner** says that a bug of some kind exists on all estates: "Am I wrong," he asks, "in saying that if there was no bug in Ceylon, it would, at a rough guess, produce 50,000 cwt. of coffee more than it actually does?" **Balfour** explains the action of the bug as stopping up the pores through which respiration and transpiration take place, thus preparing the way for "the fungus which never fails to attend on the bug." "**Mr. Neitner** tells us that several means of checking the extension of the bug have been proposed and tried. Amongst these the introduction of the red ant; but their bites are so fierce and painful that the coolies refused to go amongst the trees while the ants were there. Rubbing off the bug

COFFEA arabica.

Diseases of the Coffee Plant.

DISEASES.

by hand has been tried, but it can only be attempted upon young trees without crop; and **Mr. Nietner**, although allowing that an immense quantity of bug is thus destroyed, is nevertheless of opinion that the effect is but trifling. He thinks that the application of tar to the roots is a good suggestion, although he is obliged to admit that hitherto no important results have been achieved by it. He adds that high cultivation would appear to have the effect of throwing it off. But as the bug seems to depend on locality, **Mr. Nietner** does not look for any beneficial result so long as the physical aspect is unchanged. He thinks that if the open, warm, airy patteras were cultivated, which the experiments on a large scale tried at Passelawa show that they can be, the brown bug, which is the great destroyer, would not find the conditions favourable to its existence; or perhaps if estates, as a rule, were made smaller than they generally are, and if the reduction in acreage were counterbalanced by a higher system of cultivation universally carried out, the bug would not be so numerous as it now is." (*Balf. Cyclop.*)

(*e*) *Grub.*—The larvæ of the moth **Agrostis segetum** are very destructive: this disease is known to the planter as "Black Grub." It appears about August to October. It lives in the ground, but during night comes out to feed and does much harm when very plentiful. It is, however, local, preferring certain parts of the estate, but does not confine its ravages to the coffee plant only, as it eats any cultivated plant—vegetable or fruit tree—but despises weeds. It is very destructive to young plants. **Mr. Nietner** states that he lost as much as 25 per cent. of his seedlings through this pest. The "White Grub:" this includes the larvæ of several species of **Melolonthidæ** or Cockchafers. These do much damage by eating the roots of the trees. **Mr. Gordon** considers them as one of the greatest enemies to coffee-planting.

(*f*) *Other Pests.*—The *Locust* does of course much injury when present to any great extent, but this is more an accidental and occasional than a regular pest. *Weevils* often do considerable damage, but fortunately they are not very prevalent. Rats, squirrels, monkeys, and jackals eat the pulp and drop the berries. These are collected and, as already stated, such berries form the so-called Jackal Coffee.

COFFEE-LEAF TEA. 1676

Coffee-leaf Tea.

It has long been known that coffee leaves, if cured by a process similar to that adopted with tea leaves, afford a beverage which contains sufficient caffeine to entitle it to a position as a cheap substitute for tea or coffee. Indeed, according to some writers, the leaves contain more caffeine than the berries. A decoction from the leaves is said to be regularly used by the inhabitants of Sumatra, especially at Padang. A **Mr. John Gardener** of London even patented a process for manufacturing and partially roasting the leaves, from the belief that they were likely to come into use in Europe. Unfortunately, however, the leaves have an unpleasant senna-like flavour which greatly militates against their chances of European popularity. But perhaps the chief objection to coffee-leaf tea rests on the fact that the plants will not afford both a crop of leaves and fruits, and the latter is therefore never likely to be subordinated to the former as a commercial article. But for this fact coffee-leaf might be sold at 2*d*. a pound as compared with tea at 10*d*.

The following note has been furnished for this work by **Prof. Warden** of the Calcutta Medical College:—

"Coffee contains about ½ to 2 per cent. of a white crystalline principle caffeine, which is similar in composition to the alkaloids, theine, contained in tea. A small quantity of volatile oil is contained in coffee, but during

The Uses of Coffee.

the roasting of the berries a larger amount is developed, to which the aroma is due. Caffeine appears to act as a stimulant to the nervous system. Coffee leaves have been used as a substitute for the berries: they contain caffeine. **Mr. N. M. Ward** of Padang writes regarding the use of the coffee leaves as follows: "I was induced, several years ago, from an occasional use of the coffee leaf, to adopt it as a daily beverage, and my constant practice has been to take a couple of cups of strong infusion with milk in the evening as a restorative after the business of the day.........As a beverage the natives universally prefer the leaf to the berry, giving, as a reason, that it contains more of the bitter principle, and is more nutritious." The best mode of roasting is by holding the leaves over a fire made of dry bamboo or other wood which gives little smoke. When sufficiently roasted the leaves have a buff colour; they are ground to a powder and used in the same way as coffee. (*Hanbury.*)

COFFEE PULP. 1677

COFFEE PULP.

It has long been known that the ripe pulp of the coffee cherry contains an amount of sugar which might with advantage be converted into alcohol. At present the washings from the pulping machine are run off and no advantage taken of the sugar they contain. Several writers have urged the planters to utilise this by-product, but as yet no definite steps have been taken in that direction. It is indeed even questionable whether or not it would pay the planter to divert his attention to a perfectly distinct enterprise. The tendency of the present day is to enable the manufacturer in every branch of industry to compete to the last degree by affording him the means of deriving additional revenue from the waste or by-products of his industry. In this light it seems possible that coffee pulp may come to be put to some useful purpose. It contains much mucilage, with gum and sugar. It is said that in Arabia the pulp is actually employed in the preparation of a pleasant beverage. The pulp is allowed to dry on the fruit and then husked. This husk is employed in the preparation of the infusion known as *kahwe* or *kischer*. **Dr. Shortt** states that according to his experiment 8 oz. of dried husk, when steeped in water until fermentation sets in, yielded on distillation 1 oz. of spirits. If not employed in this manner, might not the dried husk find a demand as an auxiliary to cattle food?

OIL. 1678

OIL.

The term "Coffee-oil" is in the trade given to palm-oil in which the kernels have been more or less burnt during the process of extraction. The oil thus obtained possesses the odour of coffee: hence the name. At the same time the roasted beans of coffee possess an essential oil to which indeed they owe their aroma. During the process of roasting a large proportion of this essential oil is given off, and it has often been proposed that the drums employed in coffee-roasting should be connected with an exhauster so as to condense their oil in a receiver. By this means the aroma might be restored to the coffee or employed to flavour liqueurs. This empyreumatic oil is formed during the roasting, and probably at the expense of caffeine and other constituents of the coffee (see under **Chemistry**).

MEDICINE 1679

MEDICINE.

Coffee, while not officinal in the British Pharmacopæia, is so in that of the United States of America. Many medical men, however, recommend its use in England for mild affections. Its dietary property, as a

COFFEA arabica.

The Uses of Coffee.

MEDICINE.

stimulant to the nervous and vascular system, is that upon which its claims to medicinal recognition depend. It produces a feeling of buoyancy and exhilaration resembling the first effects of alcohol, but it is not followed by depression and collapse. It increases the frequences of the pulse, and stimulates the system to throw off feelings of fatigue, or to sustain prolonged and severe muscular exertion. It has even been contended that caffeine has the power of checking the waste of the tissues: **Lehmann** found that the distilled oil had this effect in quite as strong a degree as tea. The well-established property of coffee in preserving wakefulness depends upon its stimulating property on the nervous system. When swallowed it produces a warming cordial impression on the stomach, quickly followed by a diffused agreeable nervous excitement which extends itself to the cerebral functus giving rise to increased vigour of imagination and intellect without any subsequent stupor such as follows on the use of most other stimulants. **Moleschott** found that it influences the imagination less than the reasoning powers. Its extensive use in diet has doubtless prevented coffee taking a more pronounced position as a drug. There seems no doubt but that it might with advantage be prescribed in nervous affections, especially where symptoms of deficient energy of the brain are manifested without congestion or inflammation. In light nervous headaches, not proceeding from derangements of the stomach, it often proves immediately effectual. It has acquired much reputation as a palliative in the paroxysms of spasmodic asthma, and has been recommended in hooping-cough and in hysterical affections. "**Hayne** informs us that in a case of violent spasmodic disease, attended with short breath, palpitation of the heart, and a pulse so much increased in frequency that it could scarcely be counted, immediate relief was obtained from a cup of coffee, after the most powerful antispasmodics had been used in vain for several hours. By the late **Dr. Dewces** it was highly recommended in cholera infantum, and it has even been used with asserted advantage in cholera. It is said also to have been used successfully in obstinate chronic diarrhœa" (*United States Dispensatory*).

Coffee is much less astringent than tea, and hence it does not cause constipation so readily.

Wood states that "upon those who use it habitually, its characteristic influence is not fully evinced, as it has either lost its power in a great measure by repetition, or the secondary are so mingled with the primary effects, that the latter are not readily distinguished." "Tea differs in its effects from coffee mainly in degree. It is less stimulant to the nervous system, less apt to oppress the stomach, probably quite as efficient as a tonic to the digestive organs, and more astringent in consequence of the amount of tannic acid it contains. Certain it is that tea, especially black tea, may be taken habitually with impunity by persons who cannot use coffee without suffering, and that it sits more lightly on the stomach. In febrile diseases, a cup of tea is often not only tolerated, but agreeable to the patient, and refreshing in its effects; while coffee, however much it may be relished in health, is usually repulsive to the patient in a fever, and not well accepted by the stomach or the system" (*Therapeutics and Pharmacology, I., 625*).

Use in Typhoid Fever.—"**Dr. Guillasse**, of the French Navy, reports that in the early stages of typhoid fever coffee is almost a specific. Two or three tablespoonfuls of strong black coffee every two hours, alternating with one or two teaspoonfuls of claret or Burgundy wine, produce a most beneficial effect. Citrate of magnesia daily, and after a while quinine, is the treatment followed by **Dr. Guillasse**." (*Madras Mail, 1883.*) In **Johnston's** *Chemistry of Common Life* it is stated: "The great use of

Chemical Composition of Coffee.

MEDICINE.

coffee in France is supposed to have abated the prevalence of gravel in that country. In the French colonies, where coffee is more used than in the English, as well as in Turkey, where it is the principal beverage, not only gravel, but gout, is scarcely known."

Unroasted coffee has been employed in intermittent fever, but it is much inferior to quinine. Roasted coffee is said to have the effect of destroying offensive and noxious effluvia from decomposing animal and vegetable substances, and therefore to be capable of beneficial application as a disinfecting and deodorising agent.

Special Opinions—§ "The powder of the roasted coffee, burnt in the wards of a hospital early in the morning, is a deodoriser, and a very fragrant one" (*P. Kinsley, Honorary Surgeon, Chicacole, Ganjam, Madras Presidency*). "Is also an antisoporific; when consumed in large quantities, is supposed by the Arabs to have an anaphrodisiacal effect" (*A. S. G. Jayakar, Surgeon-Major, I. M. D., Muskat, Arabia*). "Dried coffee roasted in an open vessel is a useful deodorant" (*Henry David Cook, Surgeon-Major, Calicut, Malabar*). "Is an antidote in opium-poisoning" (*G. A. Watson, Allahabad*).

CHEMISTRY. 1680

CHEMISTRY.

The roasting or torrefying of the coffee beans, combined with the pulverising they are afterwards subjected to, induces certain changes to which in a large measure the flavour and aroma of the coffee are due. The woody tissue becomes friable, and at the same time certain chemical changes take place. The chief organic constituents of raw coffee are caffeine, fat, caffeic acid, gum, saccharine matter, legumin, and cellulose. Payen gives the following analysis :—

Cellular tissue	34·000
Hygroscopic moisture	12·000
Fat	13·000
Starch, sugar, dextrin, and vegetable acids	15·500
Legumin	10·000
Chlorogenate of potash and caffeine	3·5 to 5·000
Nitrogenous matter	3·000
Free caffeine	0·800
Thick insoluble ethereal oil	0·001
Aromatic oil	0·002
Mineral constituents	6·697

Bell (in his *Chemistry of Foods*) gives the following table of the analysis of two samples, raw and roasted, of both Mocha and East Indian coffees. We reproduce the table, both because of its allowing of comparison between these two coffees and of indicating some of the chemical changes effected by roasting :—

Constituents.	Mocha.		East Indian.	
	Raw.	Roasted.	Raw.	Roasted.
Caffeine	1·08	·82	1·11	1·05
Saccharine matter	9·55	·43	8·90	·41
Caffeic acids	8·46	4·74	9·58	4·52
Alcohol extract, containing nitrogenous and colouring matter	6·90	14·14	4·31	12·67
Fat and oil	12·60	13·59	11·81	13·41
Legumin or albumin	9·87	11·23	11·23	13·13
Dextrin	·87	1·24	·84	1·38
Cellulose and Insoluble colouring matter	37·95	48·62	38·60	47·42
Ash	3·74	4·56	3·98	4·88
Moisture	8·98	0·63	9·64	1·00
	100·00	100·00	100·00	100·00

COFFEA arabica.

Chemistry of Coffee.

CHEMISTRY.

Should the whole of the testa of the seed (the silver skin of the planters) not have been removed, it separates during the process of roasting. This is known as the roaster's "flights:" or the "fibre:" it should be removed from the beans before submitting these to the grinding mill. On being roasted the beans swell up and lose from 15 to 20 per cent. of their weight. There is perhaps no operation of so much importance as that of roasting. It should be performed in a covered vessel, over a moderate fire, and the seeds should be kept in constant motion. If mixed sizes are roasted together, the coffee will be much inferior to that obtained by roasting carefully picked and assorted beans. The degree of roasting required for one class of coffee is not the same as that for another. The heat should not be greater than is sufficient to impart a light brown colour to the bean. When roasting is carried too far, a disagreeable smell and a bitter and acrid taste gradually mingle with the essential aroma, and thus lessen the merit and value of the coffee. By reducing the beans to charcoal the aroma and flavour are entirely destroyed. When the roasting has been effected to the right extent, the volatile oil is produced at the expense of some of the other constituents. A glance at the table above will show that nearly the whole of the saccharine matter has disappeared. This is not the case with the sugar in chicory or other roots, a large proportion remaining as sugar, and hence the rapid colouration imparted to water by a coffee powder containing chicory or other cane-sugar-yielding roots, as compared with pure coffee. There is something altogether peculiar in the behaviour of the sugar of coffee under the influences of torrefication. How the volatile oil is formed seems to be a puzzle. This oil has been termed Caffeone, and it is the aromatic principle of coffee. It is wholly the product of torrefication, the materials of which it is formed being obtained by the destructive influence of heat on the other constituents of coffee. Though present only in minute quantities, this empyreumatic oil exercises a powerful influence upon the animal economy. "This activity of the volatile oil of coffee justifying us in concluding, as I have already said, that the similar oil produced in tea by the roasting, takes a similar share in the effects which the infusion of tea as a beverage produces" (*Church*). Caffeine ($C_8 H_{10} N_4 O_2$) is, however, the principle upon which the dietetic property of coffee depends, and it does not appear to be altered by torrefication. It is identical with the alkaloid found in tea. Weight for weight, tea yields about twice as much theine as the roasted coffee-beans yield caffeine. On this account a greater amount of coffee is necessary to make a cup of beverage than of tea. The "grounds" generally thrown away contain about 13 per cent of nutritious gluten. Amongst Eastern nations the habit prevails of drinking (as is done with cocoa) the grounds as well as the decoction, thus the full nutritive property of the bean is secured. Several writers have strongly advocated the adoption of this practice, but it seems doubtful whether this is ever likely to be followed more than that the tea leaves should be eaten which contain the glutinous matter of tea. Several prosecutions have been made against persons who have collected the rejections of coffee, and with flour and other ingredients made these up in the form of imitation beans.

The English system of boiling a small quantity of coffee is far inferior to the more liberal French method: The wholesale system of roasting in stock pursued in England, packets of the ground coffee being sold to the consumer which may be years old, is far inferior to the continental system of the consumer roasting and grinding his own coffee in small quantities as required.

TIMBER. 1681

Structure of the Wood.—Wood white, moderately hard, close-grained. Pores very fine and extremely fine; medullary rays very fine, numerous.

LIBERIAN COFFEE.

LIBERIAN COFFEE. 1682

This is the **Coffea liberica,** *Hiern.*, a native of Liberia, Angola, Golungo, and Alto, and probably also of several other parts of West Tropical Africa. It is a taller and stronger plant than **C. arabica**, yielding also a larger leaf and berry. It was first made known to Europe about the time the coffee-leaf disease made its appearance in Ceylon. Its hardier growth led to the opinion that it might be able to withstand the action of the fungus, and on this account demands poured in to the Royal Botanic Gardens of Kew for plants or seeds to be experimentally tried. Fortunately the Director of the Gardens was fully able to meet these demands until the question of seed-supply was taken up by certain recognised merchants. The Kew Reports are full of the most interesting details regarding the success which attended the experiments made in almost every part of the tropical regions of the globe. Whether the high hopes, entertained by some writers, of its supplanting **Coffea arabica** are likely or not to have even an approximate realisation remains still to be seen. **Mr. Thomas Christy,** for example, says: "Now, however, that the vigorous growth, excellent yield, large berry, and freedom from disease of the Liberian coffee are becoming known, it is being tried in all coffee-growing countries, and the success which has hitherto attended its cultivation promises soon to cause it to supplant the **Coffea arabica.**" The coffee planters of Ceylon have chosen to supplant their coffee by tea, and while the reports issued by the Superintendent of the Nilghiri Gardens continue favourable, the enthusiasm with which Liberian coffee was first received seems to have toned down considerably, leaving the matter still in an experimental position.

COIX, *Linn.; Gen. Pl., III., 112.*

A group of grasses belonging to the tribe MAYDEÆ, and popularly known as "Job's Tears." Under that designation is included not merely the species of COIX but of CHIONACHNE, and probably also of POLYTOCA. The latter are not of such importance as to justify their separation in a work treating purely of economic products, and therefore the popular or rather practical view of these plants will be adopted in the following brief account of the species of "Job's Tears."

Coix gigantea, *Koen.; Duthie, Fodder Grasses, N. Ind., 18;* GRAMINEÆ. 1683

Vern.—*Kesai,* BERAR; *Danga gurgur,* BENG.

Reference.—*Roxb., Fl. Ind., Ed. C. B. C., 650.*

Habitat.—A tall, erect, aquatic grass, with large broad leaves, found throughout the plains of India. It is distinguished from the next species by the florets of the male spikes being in threes, the central one stalked. **Roxburgh** found it in the valleys of the Circar mountains, but from his description of this and of the form named by him **C. aquatica,** it seems probable that both are referable to one species, if, indeed, they should not be treated as varieties of **C. lachryma.**

It seems probable also that **C. gigantea** and **C. aquatica** are the wild states of the cultivated plant, **C. lachryma.** At all events, no one seems to have observed them under cultivation, and thus, while the grains are not apparently eaten, the other properties of **Coix lachryma** are applicable to the above.

C. Kœnigii, *Spreng.; Duthie, Fodder Grasses, 19.* 1684

Syn. for CHIONACHNE BARBATA, *R. Br.* (the COIX BARBATA, *Roxb.*)

Kurz in his report on Pegu refers to this plant under the Burmese name of *Kyaip*. It is also known in India, where it bears the following vernacular names: *Gurgur*, BENG.; *Bhus, kirma-gilâram gadi*, CHANDA; *Kadpi*, BALAGHAT, C. P.; *Varival*, MAR.; *Ghella gadi*, TEL.

FODDER. 1685 **Fodder.**—Duthie says that in Balaghát in the Central Provinces, it is said to be used as fodder when in the young state. **Roxburgh**, however, remarks that, owing to its coarse nature, cattle do not eat the grass.

1686 **Coix lachryma**, *Linn.; Duthie, Fodder Grasses, 18.*

JOB'S TEARS.

Syn.—C. ARUNDINACEA, *Lamk.*; LITHAGROSTIS, LACHRYMA JOBI, *Gærtn.*

Vern.—A recent correspondence between the Government of India and the various provincial Governments has brought to light new and interesting information regarding this plant. It has been shown that Coix is much more extensively cultivated than was formerly supposed, and that there exists a very extensive series of wild and cultivated forms of Job's Tears, which the writer has placed under the above species. Should this be proved incorrect, a certain redistribution of the vernacular names, here attributed to the various species of Coix, would become necessary. One of the most remarkable of the forms of **Coix lachryma** has been figured in the last part of *Hooker's Icones Plantarum, Pl., 1764*, as **C. lachryma**, *var.* **stenocarpa**, and an effort, in the following remarks, will therefore be made to indicate, as occasion occurs, the vernacular names that more properly belong to that form. *Sankrú*, HIND.; *Gurgur* or *kunch*, BENG.; *Jargadi*, SANTAL; *Kassai-bija*, BOMB.; *Ránjondhala, rán-makkai*, MAR.; *Jondhali*, POONA; *Sánklu* (Sabathu Hills), PB.; *Dabhir* (Mount Abu), RAJ.; *Sankru*, (according to **Royle**) *baru* (at Sáharanpur), N.-W. P.; *Ganddula, garun* (Bundelkhand), *kasei, galbi, gadi* (Chanda), *galu* (Seoni), *gurlu* (Balaghat), C.P.; *Sohriu* (Khásia and Jaintia Hills), ASSAM; *Jargadi*, SANS.; *Kyeit, kalithi* or *cheik* (the white form), *kyeikphun, sakyeik* (the edible), BURM.; *Bé* (large form), *be-ma* (the small), KAREN; *Kudhia thia* or *kudhati* (the black form), *so-tsa* (the white), and *ke-sí* or *kasí* (collective or generic name), NAGA HILLS; *Mung*, MANIPUR; *Kirindi-mana*, SING.; *Ee-jin, ee-yin*, a name used in China and Malacca.

The Latin Coix and probably the Greek Κοιξ were applied apparently by early writers to an Ethiopian palm, but according to **Theophrastus** to a reed-like plant which may possibly have been the modern Coix. It is worthy of note that the name *kasi* by the Nagas given generically to the cultivated forms should reappear all over India: thus *kasei* in Chanda, *kassai* in Bombay, *kesai* in Berar (for **C. gigantea**); and that many of the names in use in Burma should have so strong a resemblance to that word, *viz.*, *kyeit, kyeithishe, kulésé, kalinsee, &c., &c.* It seems probable that the expression Job's Tears is wrongly applied to Coix. The Arabs have a tradition that Job used the flowers of **Inula dysenterica** to heal the sores on his body, hence according to them **Inula** and not **Coix** would be the true Job's Tears. That expression, as applied to Coix, rests most probably, however, on the shape of the grains, and **Gerarde** says "every graine resembleth the drop or teare that falleth from the eye." The very application, however, of the name Job's Tears would suggest an ancient cultivation; but the fact that in the wild state the plant yields a much larger grain than either rice or wheat may be taken as suggestive of Coix having at an early date been cultivated by the inhabitants of the countries where it was found; and it would seem as if it had been displaced from popular favour when the superior properties of other food-grains became known.

References.—*Roxb., Fl. Ind., Ed. C.B.C., 649; Thwaites, En. Zey. Pl., 357; Dymock, Mat. Med. W. Ind., 2nd Ed., 853; Balfour, Cycl. Ind.; Hooker's Him. Jour., II., 289.*

Habitat.—Met with on the plains of India, and on the warm slopes of the hills from the Panjáb to Burma. It seems probable, as stated under the preceding species, that the forms referable to this type are mostly cultivated; they are less aquatic in character than those referred to **C. gigan-**

tea, and appear to occur at higher altitudes. They are also more stunted in growth, and the involucre (or shell around the grain) is looser, softer, and apparently always furrowed—at least this is so with all the cultivated forms.

FORMS OF. 1687

THE FORMS OF JOB'S TEARS.—There are three or four well-marked forms of Job's Tears met with in India, which differ from each other in shape, colour, and degree of hardness, and in the presence or absence of grooves or furrows along the length of the hardened involucre. As to shape there are three types—a long *cylindrical* or tubular (*var.* **steno-carpa**), the normal *pear-shaped* condition, and a *flattened spherical* form about the size, and not unlike the shape, of the fruits of **Malva rotundifolia**, only smooth and polished.

The writer has had the pleasure to examine a large collection of samples made in Burma and Assam, and would offer the following remarks regarding these.

1st—The cylindrical form is returned as frequently cultivated, and also wild in the Pegu Divisions of Burma (in the following districts—Prome, Pegu, Hanthawaddy and Tharawaddy); also in the Tenasserim Division (in the following districts—Amherst, Taung-ngu, Tavoy, Mergui, and Salween). It is said to be wild but not cultivated in the Naga Hills of Assam. From no other part of Assam, however, have samples of this form been received, but in the part of *Hooker's Icones Plantarum* (to which reference has been made above) it is stated that "Mr. R. Bruce of Balipara" forwarded samples to the British museum, with a note to the effect that the "involucres are known to 'the Assamese and the Míris, and called by them the *cowrmonee* or crow-bead, from the fondness of these birds for this berry.'" It would appear, therefore, that the cylindrical grain may occur in the Míri country, but up to date (in connection with the present enquiry) no information corroborative of this fact has been received from Assam, and the plant does not appear to occur in any other part of India, so that it may safely be viewed as a native of Burma, and possibly distributed into the mountain tracts of Upper Assam and Cachar. The cylindrical grain is always of a white colour, smooth, polished, not furrowed, but constricted towards both extremities, and whether wild or cultivated, is collected for ornamental purposes only, and not as an article of food.

2nd—Of the pear-shaped form there are numerous sorts, varying in size and colour—some pale and bluish white; others grey, yellow, or brown-black. They are often constricted at the base into a disk-like annulus, and in all the samples said to be collected from cultivated stock, the grains are more or less deeply furrowed, and in the slate-coloured samples the bottoms of the furrows are of a brown shade. The cultivated forms are also loose-shelled and flattened on one side, somewhat obliquely, like the smaller cardamom. The wild forms are smooth-shelled, the shell being often so thick and hard that it can scarcely be broken. The cultivated forms are frequently grey, brown, or even black, although in point of abundance the straw-white form is that most extensively met with. They rarely ever have the shining, polished appearance of the grains returned as collected from wild plants. From Assam has been received the most extensive and varied series of cultivated Coix.

3rd—Of the flattened spheroidal kind all are smooth, hard-shelled, and look like artificial beads. They are often yellow or even pink, and do not possess the disk or swelling at the base. They appear never to be cultivated, and have been sent in the greatest abundance from Burma. It seems probable these belong to a different plant from the forms described above.

Job's Tears.

FORMS OF.

It is somewhat remarkable that in all the cultivated forms the shell is loose and furrowed. These forms vary considerably as to size from that of a grain of rice or small grey pea to a haricot bean. A sample from Akyab is fully half an inch long, and is pronouncedly flattened obliquely with a large swelling at the base. It seems probable that these loose-shelled forms belong to **Coix lachryma** proper, while some of the smooth, hard-shelled kinds (and especially the spheroidal forms without the basal annulus) may be from **C. gigantea** or some undescribed species. Of the flattened spheroid forms sent from Hanthawaddy some are extremely pretty, are pale pink, smooth, and shining, with a natural central perforation, making them look like artificial beads. A brown sample from Akyab is so hard and shining as to closely resemble small marine shells. The Deputy Commissioner says the globular form grows wild in low marshy land and is not eaten. This may be the plant Roxburgh described as **C. aquatica.**

The following brief abstract of the reports received from Burma will convey some idea of the forms of Coix there met with, while it will afford the means of recording the vernacular names that are in use with reference to the various wild and cultivated plants.

PEGU DIVISION.

BURMA. Pegu. 1688

In the Pegu District five forms exist: a large pear-shaped kind known as *cheik* or *kyeikthi* which grows wild plentifully, but is not collected either for food or for ornamental purposes. There are two forms of this, one white, the other brown grey, both are smooth, polished, and very hard. A brown edible form is cultivated—a polished grain with the characteristic furrows and basal annulus. Lastly, there are two forms of *var.* **stenocarpa,** both of which are cultivated—the one called 'male *cheik*,' long, thin, and quite cylindrical, and the other 'female *cheik*,' shorter and slightly swollen in the middle. Separate names are not given to distinguish the cylindrical from the pear-shaped forms. The best quality is said to come from the upper valley of the Pegu river.

In Hanthawaddy District some seven or eight forms exist in a wild state or are cultivated. One only is grown as an article of food, namely, a slaty brown irregular grain, of a dull colour, furrowed, and with an annulus. This is found only on the plains, is called *Kyeikthi,* and is sold for 8 annas a basket. All the others are wild or cultivated, but collected purely for ornamental purposes. One is a medium-sized steel grey seed, smooth, shining, and pear-shaped. Three are pinkish-brown, small, of the flattened spheroidal form, and the most perfect beads in the whole collection of Coix seeds before the writer. These have been lettered B. D. and E., but do not appear to bear separate names, although they are said to be "grown" by the Karens. The sample D. would most probably command a large sale in Europe, as it is of a rich colour, smooth, and polished, with an almost artificially regular perforation, and so hard that it is impossible to break the grains without the aid of a hammer. Two samples of the cylindrical seeds complete the list of the Hanthawaddy consignment of Coix. These correspond exactly to the two forms described above under Pegu, the sample marked G. agreeing with the so-called "male," and C. with the "female" form.

In the Prome District both spherical and cylindrical forms are said to occur, wild and cultivated. Of the samples forwarded along with the report, one is a small steel gray, furrowed and loose-shelled form, which, judging from all the other samples of Coix examined by the writer, must be cultivated, and most probably not wild. The other two samples furnished are the forms of the cylindrical described above, only that the longer form is nearly as much swollen in the middle as the shorter. The Deputy

FORMS OF.

Commissioner deals in his report with a much more extensive series than he has furnished samples of. He says the forms of Coix are known collectively by the name *Kyeikthi*. The cylindrical being *Kyeikshe* (literally, long *Kyeik*; of the globular form there are names to distinguish certain recognised types thus :— *Kyeikphun*, white *Kyeik*; *Sakyeik*, edible *Kyeik*; *Pyaung*, or maize-like *Kyeik*; and *Kyeikni*, or red *Kyeik*.

In the Tharrawaddy District the Deputy Commissioner says that all the forms are known by the Burmese name *Kyeikthi*, but that a large round edible form is known to the Karens as *Bè*, and is cultivated, while another smaller round kind is known as the *Bè-ma* (or female *Bè*) and is collected for ornamental purposes. He further forwards a sample of the cylindrical grain, and says it is known as the *Bè-kwa*.

ARAKAN DIVISION.

Arakan. 1689

In the Akyab District the pear-shaped form is both wild and cultivated. From the town of Akyab, the Deputy Commissioner has furnished three samples of the wild plant, the seeds being smooth, polished, and very hard, especially a brown form. He states that these forms grow in the low marshy lands and are not eaten. He, however, furnishes a sample of a cultivated form obtained from Myohaung—the largest Coix grain yet examined—which fully supports all that has been stated above. It is steel grey, deeply grooved, with a loose shell and pronounced basal swelling. The Deputy Commissioner describes this as "the cylindrical form," but while it is certainly longer than the Akyab grain, it is not the cylindrical form (*var.* **stenocarpa**) described above, but is a monster form of the ordinary cultivated pear-shaped grain.

In the Kyauk-pya District three forms of Coix occur—two wild and one cultivated. The writer has not seen any specimens of these, yet has no reason to doubt but that they would answer very much to the types described under Akyab. One of the wild forms is larger than the other and is known as *jaisee* or *kalinsee*, while the smaller form is the *chitsee*. The edible form is also known as *chitsee*, and is both eaten and made into beer.

TENASSERIM DIVISION.

Tenasserim. 1690

In the Amherst District both the round and cylindrical forms are grown, the former being eaten, and the latter used for ornamenting ladies' dresses. A wild round form is said also to exist. Samples have not been communicated, but the Deputy Commissioner reports that both are known as *kyeit*.

In the Shwe-gyin District no form of Coix is known.

In the Taung-ngu District it is stated that the cylindrical form grows wild, while the globular is cultivated: both are known as *kyeit*: the former is used for ornamental purposes, and the latter is grown as an article of food and for making beer.

In the Tavoy District the round and the cylindrical forms both exist, wild and cultivated; they are known as *kyeit* and *kalithi*.

In the Mergui District, and chiefly in the Palaw township, cylindrical and round Coix are both cultivated and wild. The Deputy Commissioner forwards seven samples, of which Nos. 1, 2, and 4 are cultivated, the others wild, while Nos. 3, 5, 6, and 7 are used for ornamental purposes, and No. 4 is extensively eaten. It is worthy of note that of these samples only those cultivated, *viz*, Nos. 1, 2, and 4 have the shell or involucre furrowed—the others are smooth and shining.

(1) *Kaleik* is a dark brown or bluish black polished grain of the pear-shaped series.

COIX lachryma.

Job's Tears.

FORMS OF.

(2) *Kaleik Kauk-nyin*, the same as the last so far as the appearance of the grain goes.

(3) *Kaleik Sinzwè* is the narrow cylindrical grain already alluded to as "male." The so-called "female" is No. 6 of the present list.

(4) *Kaleik Pauk-pauk.*—This is the smallest pear-shaped and furrowed grain in the Burmese series. It is almost round, with an apical elongation; is pale, straw-coloured, and pronouncedly furrowed. The Deputy Commissioner says it is "used extensively as a food grain."

(5) *Kaleik Yingwè.*—This is a very small form of the flattened spheroidal, grain of a dirty milky white colour, a little smaller than the Hanthawaddy sample marked D, but of the same shape. The seeds are less than a ¼ inch in diameter and not much more than half that size in thickness through the central perforation.

(6) *Kaleik Yaing*, the form of **stenocarpa** that has been described as "female," a short cylindrical grain with a central swelling.

(7) *Kaleik Kyauk* is a large white or straw-coloured pear-shaped grain devoid of surface furrows. This is the largest straw-coloured grain in the Burmese collection, as No. 4 above is the smallest. Many of the steel grey whites are quite as large as No. 7, but few of the straw-coloured ones approach it in size.

In the Salween District both the globular and the cylindrical form is cultivated, but the former exists also in a wild state. They are known in Burmese as *kyeit*, the cylindrical being *kyeithishe*, and the globular *kyeitthilon*. In the Shan language they are *Malweleitayaung*, the cylindrical and *Malweleitamun*, the globular. In Karen, *Baw-kwa* the cylindrical, and *Bowma*, the globular; also in Karenni *Kulèsè* the cylindrical and *Tabusè* the globular. Both forms are extensively grown in the Shan States, where the cylindrical is sold for R1 a bushel and the globular from 4 to 6 annas.

The following abstract of available information regarding Coix cultivation in Assam may be here given to complete this brief review of the subject :—

ASSAM. 1691

Sir J. D. Hooker remarks : "A great deal of Coix is cultivated in the Khásia hills; the shell of the cultivated sort is soft, and the kerrel is sweet, whereas the wild Coix is so hard that it cannot be broken by the teeth; each plant branches two or three times from the base, and from seven to nine plants grow in each square yard of soil; the produce is small, not above 30 to 40 fold." **Mr. McCabe**, the Deputy Commissioner of the Naga Hills, reports : "The Nagas of this district cultivate six varieties of Job's Tears. The generic name is *Ka-si*, and the varieties are as follows :—

"'*Sibu.*'—The seed is of a bluish grey colour and pear-shaped in form. This is a medium-sized form, with soft shell deeply furrowed with brown lines : oblique and with a basal swelling. This is smaller but somewhat like the large grain obtained from Akyab.

"'*Kerengiza-si.*'—Of the same colour as *Sibu*, but more cylindrical in shape. Hardly to be distinguished, in fact, from *Sibu*, except in being more oblique and in having a pronounced apical and basal constriction.

"'*Sipia.*'—A small grain about half the size of *Sibu*, of a russet brown colour. This might also be said to be a more regularly-formed grain, with a harder shell than either of the preceding, but not sufficiently hard to admit of its being used for ornamental purposes.

"'*Sámáprè.*'—Pear-shaped in form resembling *Sipia*, but smaller in size. This dark brown regular grain looks at first sight remarkably like some of the forms of black rice. It is about the same size and is pointed at both extremities. It is considerably like an elongated caraway.

Job's Tears. COIX lachryma.

"'*Kadáthá.*'—Almost globular in form, of a mottled brown and grey colour. The most marked peculiarity of this grain is that it is dark brown like the *Sipia* form in the lower half and yellow or straw-coloured in the upper. FORMS OF.

"'*Kasi.*'—Globular in form of a light grey or yellow colour. This is the most common variety."

The Naga hill samples, examined by the writer, fully support the opinion formed on examining those from Burma,—namely, that the cultivated races have all a loose easily breakable shell, which is also deeply furrowed. None have smooth polished hard shells like the wild forms which are collected in Burma and other parts of India to be used for ornamental purposes. It may also be added that the average elevation of the Naga and Khasia hills may be put down at from 3,000 to 5,000 feet, whereas the smooth-shelled forms are met with chiefly in the marshes of the plains of India and Burma. The white forms of the Khásia hills are harder, more polished, and less furrowed than the cultivated white forms from any other part of India, but they still preserve the characters assigned collectively to the cultivated forms. From the Khásia and Jaintia hills two samples of Coix have been received both of the milky white kind. A large and a small grain from the latter resembles very much the small white grain obtained from Mergui (No. 4 above), only that it is a little larger. In the report which accompanies these samples it is stated that four kinds of Coix are grown in these hills, but that "none of the four are wild, all are cultivated exclusively as an article of food. The cylindrical form" (*var.* **stenocarpa**) "is unknown to the Khásias." The dark coloured forms are said to boil softer than the white and the smaller of the two white forms "is slightly better flavoured than the larger." Naga Hills. Khasia Hills.

Food.—This curious grain might almost be said to be unknown to the natives of India generally, except as a weed of cultivation. To the hill tribes on the eastern frontier, however, it is an important article of food; with the Tankhul Nagas of Manipur it might, indeed, be almost described as the staple article of diet. In several districts of Burma it is also regularly grown as an article of food. **Mason** says the esculent Coix cultivated by the Red Karens is parched like Indian corn. Of the Bassein district **Mr. W. T. Hall** *(Director of Land Records and Agriculture)* reports that it is sown in gardens, the crop ripening in November. The produce sells for R2 to R3 a bushel. That officer has also forwarded to the writer numerous reports received from the Commissioners of the various Divisions, from which the following account of the method of cultivation may be here reproduced:—"The mode of cultivation is as follows:—*1st*, before the seeds are put in the ground they are tied in a piece of cloth and watered every day for about 7 to 8 days, when whitish roots appear. They are then placed in the ground. In some cases the roots do not appear till 10 or 15 days. *2nd*, at the place where the plants are to be grown furrows are formed and the seeds are laid on the earth which is first mixed with cow's dung, afterwards the seeds are covered up with a little earth. Another method is to dig a hole where dung and decayed leaves are burnt and plant the seeds in these places. This method is considered the most successful. When the plants bear fruit and the latter becomes mature or grows white, the branches should be broken off. This will cause the plants to yield another crop and thus to last much longer." Speaking of the cultivation pursued in Akyab the Deputy Commissioner writes (of the Myohaung township) with reference to the form which he calls "the cylindrical," but which, according to the samples discussed above, is a large loos-shelled grain of the pear-shaped series:— FOOD. 1692

COIX. lachryma.

Job's Tears.

FORMS OF.

"The cylindrical is sown by the wild hill tribes on *Kaing* land or on the slopes of hills. They do not till the land for this purpose; the seeds are thrown broad-cast, and no care is taken of them. In times of scarcity of food the cylindrical are eaten, but now they are only used as ornaments for their dresses." The Deputy Commissioner of Kyaukpyu writes regarding a beautiful hard round form which is collected from the wild plant and used for ornamental purposes. Of the cultivated forms he says this is known as *Chitsee*. "It grows in June and July and dies in November and December. The plant is 4 or 5 feet high and like a reed." But a smaller, more delicate, variety is also cultivated, which he remarks is eaten and also used in the manufacture of the small beer known as *Khanag*." He adds "The seed has to be cleaned and has the taste of maize." Of the two kinds grown he says: "The plants, however, differ widely in other respects, and I am unable to say if they belong to the same variety or not."

CHARACTER OF THE EDIBLE GRAIN.—On breaking the outer shell, a cowry-shaped grain is obtained which, **Professor Church** says, bears on being cleaned the proportion of 1 to 4 to the total weight of the unhusked article. The Professor gives the following analysis—

Composition of Job's Tears (Husked).

	In 100 parts.	In 1 lb.
Water	13·2	2 oz. 49 grs.
Albuminoids	18·7	2 „ 434 „
Starch	58·3	9 „ 143 „
Oil	5·2	0 „ 364 „
Fibre	1·5	0 „ 105 „
Ash	2·1	0 „ 147 „

"The nutrient-ratio is here 1 : 3·8, the nutrient-value 89." From these facts it may be inferred that the grain is not likely to prove of greater economic value in the future than it is at present to the poor hill tribes who are under the necessity of growing this cereal, since, in consequence of their imperfect agricultural system and poor soil, nothing else will grow even so successfully as Coix. **Dr. Smith** says "it is larger and coarser than pearl barley, but is equally good for making gruel. As it is sold for five pence per Chinese pound, it makes an excellent diet-drink for hospital patients in China." It is worthy of note, however, that from the extensive series of cultivated forms which exist, and the occurrence of a long list of names for the plant and grain in nearly every vernacular language of India and Burma, an indication is given of an ancient cultivation which may have taken its birth in China and spread through the Malayan regions into Burma and thence to Assam and to India generally. If this, on further investigation, be found to be a correct supposition, it is possible that Coix may have preceded rice and been in the plains of India abandoned in favour of the more wholesome grain. Even the wild plant has so large a grain as to favour the idea of its having been early adopted as a plant to be cultivated. This idea of distribution into India is partly supported by the coincidence of the vernacular names, and may also be accepted as receiving favour from the fact that in the Indo-Burman region the plant is met with largely in a wild state, and at the same time continues to be cultivated and exhibits a greater number of forms than occur anywhere in India proper. Indeed, Coix can hardly be said to be cultivated anywhere in India at the present day. In the Khásia and Naga Hills some five or six forms of the loose-shelled and furrowed kind are grown, but the plant is said to be rarely, if ever, met with in the wild state, while the cylindrical is reported as wild in the Naga Hills but never

C. 1692

Job's Tears.

COIX lachryma.

FORMS OF.

cultivated. Unfortunately, samples of this wild plant have not been communicated, but **Mr. McCabe**, who writes regarding it, must know it well from the extensive use to which it is put by the Angami Nagas amongst whom he is living. He describes it as the cylindrical grain. His words are:—"The cylindrical form is only found in the wild state and is called *sikrá*. This plant is never cultivated but is found growing on the edges of terraced cultivation, and in the small gardens in the villages. The leaves resemble closely those of the cultivated species, but the plant is smaller and the stem much tougher. The seed is used, in place of cowries, in decorating the kilts and ornaments worn by young men. The yield is not so great as that of *kasi*, and the young plants suffer from the depredations of rats." He gives an interesting saying regarding the *kasi* which would show that the tribes on the Naga Hills apparently think that the rotund form was produced from the cylindrical by cultivation. "In the beginning of the world rats brought paddy and *sikra* from Japvo Mountain. Man, on seeing these products, took the paddy for himself and left the *sikra* for the rats." Japvo is the highest peak of the Naga system, where neither wild rice nor wild coix occur. The writer does not recollect having ever seen the cylindrical form in the Naga Hills, although he collected numerous samples of the globular; but all under such conditions as to lead him to the opinion that they were cultivated forms or at most only escapes from cultivation.

Medicine.—In some parts of India medicinal properties are assigned to the grain, as, for example, by the Santáls, who affirm that "the root is given in strangury, and the menstrual complaint known as *Silka*" (*Rev. A. Campbell*). **Dr. Dymock** says the *Kassai-bija* is used as a diuretic. MEDICINE. 1693

Domestic Uses.—In many localities the wild, hard, dry, spherical grain is extensively used by the aboriginal races for ornamental purposes. Necklaces of these seeds are frequently worn, and baskets and other ornamental articles are occasionally decorated with them, especially those made in the Nepal Tarái. The Karens cover their dresses with the narrow cylindrical form in embroider-like designs, and the Angami Nagas construct elegant earrings in which a rosette of these seeds surrounds a greenish beetle wing. The various grains which we have in the present article treated of popularly as forms of Coix or Job's tears, seem to stand a good chance of coming into use in Europe in the construction of artificial flowers, laces, bugle-trimmings, and other such purposes for which glass beads are now used, and possibly also in Catholic countries for the manufacture of Rosary beads. If found capable of being dyed a deep black colour, there might be an extensive demand for them, since they would be much more durable than glass. During the late Colonial and Indian Exhibition, several merchants, especially from France, enquired after seeds suitable for the above purposes. The writer was not able at the time to furnish these gentlemen with samples of the cylindrical seed to which repeated reference has been made above, but he gave them samples of the ordinary edible pear-shaped form. They seemed to think there might be some prospect of even that form coming into use. On being shown the Karen ornamented dresses they professed a firm conviction that the cylindrical grain would find a ready sale. This led the writer to show these garments to **Mr. W. T. Thiselton Dyer**, Director of the Royal Botanic Gardens, and in consequence a requisition was in due course forwarded to the Government of India asking that a thorough investigation should be instituted. The greater part of the information contained in the present article is the result of the enquiry now in course of being carried out. Specimens of the cylindrical grain and the plant yielding these were early furnished by the Government of Burma, and were at first

DOMESTIC. Necklaces. 1694

Earrings. 1695

Artificial flowers. 1696

Laces. 1697

Bugle-trimmings. 1698

Rosary beads. 1699

DOMESTIC. identified as **Polytoca Wallichiana,** but have since been determined as **C. lachryma** *var.* **stenocarpa.** Subsequently, numerous samples of Job's tears, from every district in Burma, were obtained, and it has transpired that of the spherical form there are several small hard grains which seem quite as likely to find a market in Europe as the cylindrical. At the Exhibition only the coarser loose-shelled edible form was shown, but there would seem every prospect that the wild forms specially collected by the hill tribes of Burma for decorative purposes are those which should be offered as most suitable for the European market. Along with these the cylindrical form would afford the manufacturer of laces, &c., a choice of two forms which might be elegantly combined.

PRICE. 1700 PRICE OF COIX GRAIN.—This has been variously estimated at from 8 annas to R4 a basket, but it seems probable that were a regular demand to arise, a fixed rate would soon be established, which would probably rule considerably below that of rice. It would have, however, to be discovered whether the hard forms could be cultivated without losing their characters which recommend them as decorative articles. The writer has offered the suggestion that some of these may be the produce of a distinct species from that of the true Job's tears **(Coix lachryma),** and if so it might be found possible (as with the cylindrical) to cultivate them without softening the pericarpium or involucre. The cultivation of the wild walnut or of the hazelnut distinctly softens the shell, and were this to happen with the forms of Job's tears recommended above, their merit would to a large extent be destroyed. On the other hand, the price would be greatly lowered were the plants found capable of being cultivated without losing their hard glossy outer covering. At present the plants that yield these beads are abundant in Burma, and perhaps also in lower Nepal, to such an extent that no fears need be entertained of the demand, for some time to come, exceeding the supply.

Coke, see **Coal.**

COLA, *Schott.; Gen. Pl., I., 218.*

1701 **Cola acuminata,** *R. Br.;* STERCULIACEÆ.

Syn.—STERCULIA ACUMINATA, *Beauv.*

References.—*Kew Reports, 1880, p. 14; 1881, p. 10; Christy, New Commercial Plants, No. 8, p. 5; Treasury of Botany, p. 311; Smith, Dict. Econ. Pl., p. 127; Balfour, Cycl. of India; U. S. Disp., 15th Ed., p. 1754; Pharmaceutical Society Journals.*

This large West Tropical African tree has been experimentally introduced into India, but it is not known with what degree of success. From it is obtained the COLA NUT, which continues (and apparently) deservedly to attract attention as a substitute for cocoa and chocolate (**Theobroma Cacao**). It has been said the beverage made with **Cola** paste is ten times more nutritious than chocolate made with cocoa. The reputation of this substance in sustaining the system against fatigue is such that it is meeting with consideration from the military authorities of the world as an article to be given to soldiers during active service.

The bean has been analysed by **Messrs. Heckel and Schlagdenhauffen,** by **Dr. Attfield,** and others.

There are many tracts of country in India that seem likely to prove suitable to **Cola** cultivation, and doubtless this subject will in the future receive a greater degree of attention than it has as yet obtained from the Indian planters.

COLCHICUM, *Linn.; Gen. Pl., III., 821.*

Colchicum autumnale, *Linn.;* LILIACEÆ.

OFFICINAL COLCHICUM; MEADOW SAFFRON or AUTUMN CROCUS. 1702

References.—*Pharm. Ind., 243; Flück & Hanb., Pharmacog., 699; U. S. Dispens., 15th Ed., 469, 470; Bentley & Trim., Med. Pl., 288; Dymock, Mat. Med. W. Ind., 835; Ainslie, Mat. Ind. Preface, xxi.; O'Shaughnessy, Beng. Dispens., 658; Year Book of Pharmacy, 1874, p. 630; Royle, Ill. Him. Bot., I., 385; Spons, Encyclop., 808; Balfour, Cyclop.; Smith, Dic., 128; Treasury of Botany; Morton, Cyclop. Agri., 490.*

Habitat.—The plant grows in the meadows throughout Europe. Attempts have been frequently made to introduce several species into India, but with very little success. **Mr. Baden Powell** says that in the Panjáb a species of **Colchicum** is known as *Harantutiya.*

The fresh corms and the seeds of **Colchicum** are officinal.

C. sp. 1703

Vern.—*Súringán, talkh, shirín,* PB.; *Loabale-barbari, súringán,* HIND., BOMB., BENG., TAM., and ARAB.; *Aaknak,* PERS.

Mr. Baden Powell gives this the name of **C. illyricum,** The HERMODACTYL or "FINGER OF HERMES." **Dr. Moodeen Sheriff** says there are two kinds of the drug—*Súrinjáne-shirín* (sweet *Súrinján*) and *Súrinjáne-talkh* (bitter *Súrinján*). **Dymock** speaks of these as the tasteless variety and the bitter, but adds a third form or rather substitute which he says is the sliced bulbs of **Narcissus tazetta,** which are imported from Persia and sold as a bitter *Súrinján.* The learned authors of the *Pharmacographia* (and also **Dr. Cooke**) are of opinion that the bitter HERMODACTYL is not the produce of a **Colchicum** at all; while **Professor Planchon,** and following him several other authors, attribute the drug to **Colchicum variegatum,** *Linn.*, a native of the Levant and not known to be found in Kashmir or Persia. **Planchon** in his account of *Súrinján* gives a figure of **C. variegatum,** *Linn.*, in the *Bot. Mag., t. 1028.*

References.—*Royle, Ill. Him. Bot., 385; Baden Powell, Pb. Pr., 381; Irvine, 95; Dymock, Mat. Med. West. Ind., 2nd. Ed., 835; Pharmacopæia of India, 246; Moodeen Sheriff, Supp. Pharm. Ind., 153; O'Shaughnessy, Beng. Dispens., 661; Makhzan, article No. 1053; Pereira, Mat. Med., Vol. II., Pt. I., 167; U. S. Dispens., 15th Ed., 1663; Planchon, Ann. Des. Sciences Nat. Bot., IV. (1855), 132; Cooke in Pharm. Journal, April 1871.*

Habitat.—The plant from which this medicinal product is obtained is said to be found in Kashmir, but the Indian supply, according to **Dymock,** is imported into Bombay from the Red Sea ports and from Persia.

History.—**Dr. Dymock** gives the following account of this drug:—"The Hermodactyl, or 'Finger of Hermes,' was unknown to the early Greeks; it appears to have been first used medicinally by the Arabs or later Greek physicians; it was first mentioned by **Alexander** of Tralles, who flourished A.D. 560 (*Libr. XI.*). It is deserving of special notice that under the name of Surugen or Hermodactyl, **Serapion** comprehends the κολχικον and εφημερον of **Dioscorides** and the ερμοδακτυλος of **Paulus Ægineta** (*Pereira, Vol. II., Pt. I., p. 166*). **Masih** and other early Arabian writers describe three kinds of Hermodactyl—the white, the yellow, and the black; in this they are followed by most of the more recent Muhammadan writers. According to **Ibn Sina,** the flower of the *Súrinján* is the first flower which appears in spring in the moist valleys beneath the mountains; the leaves, he says, lie flat upon the ground, the flowers are yellow and white. HISTORY. 1704

HISTORY. Mir Muhammad Husain tells us in his *Makhzan* that the white is the best, and that it is not bitter, next the yellow; both may be used internally; the black, he says, is poisonous and only to be used externally. He describes the Hermodactyl plant as having leaves like a leek and a yellow flower; it is called in Persia *Shambalíd*; the black variety, he says, has red flowers. Muhammadan physicians consider the drug to be deobstruent, alterative, and aperient, especially useful in gout, rheumatism, liver, and spleen. In gout they combine it with aloes; with ginger and pepper it is lauded as an aphrodisiac; a paste made of the bitter kind with saffron and eggs is applied to rheumatic and other swellings; the powdered root is sprinkled on wounds to promote cicatrization. Two kinds of *Súrinján* are met with in Indian shops, bitter and sweet. European physicians in India who have tried the drug consider the sweet Hermodactyl to be inert or nearly so, and the bitter to have properties similar to **Colchicum.**

MEDICINE. 1705 **Medicine.**—§ "Purgative, diuretic, sedative, chologogue, doses 2 to 8 grains, used in acute gout and rheumatism, irites, periostitis, kidney and heart diseases attended with gout, uric acid diathesis, chronic bronchitis, constipation, gonorrhœa, jaundice, synovetis, dysmenorrhœa." (*Chuna Lall, 1st class Hospital Assistant, in charge of City Branch Dispensary, Jubbulpore*). "Two varieties are found in the bazar—sweet and bitter; the latter is officinal and useful in rheumatic affections" (*T. N. Ghose, Assistant-Surgeon, Meerut*).

Colchicum luteum, *Baker*, according to **Aitchison,** in a note furnished to the writer, "occurs in early Spring in the Panjáb from Campbellpore, across to Abbottabad, the Gullies, at Murree, and in Kashmir extending to Zoja pass.

Probably it is the root of this that is *Harán-tutiya*. But the root of **Merendera Persica,** *Bois.* (*Syn.* **Aitchisonii,** *Hooker*) may be mixed with it.

SUBSTITUTES. 1706 SUBSTITUTE OF SÚRINJÁN.—Dr. **Dymock** says that the sliced bulbs of the true **Narcissus (N. tazetta)** which are imported into India from Persia as a substitute for *Súrinján* are easily recognisable. He remarks this drug "may be at once detected by its larger size and tunicated structure. The taste is bitter and acrid, the substance amylaceous and very similar to that of the Hermodactyl. It is used as an external application, and, according to the author of the *Makhzan*, has properties very similar to those of *súrinján-i-talkh*. Value, annas 3 per ℔.

COLDENIA, *Linn.; Gen. Pl., II., 841.*

1707 **Coldenia procumbens,** *Linn.; Fl. Br. Ind., IV., 144;* BORAGINEÆ.

TRAILING COLDENIA.

Vern.—*Tripungkhi, tripunkhi, tripungki,* HIND.; *Bursha,* SIND; *Tripakshi,* BOMB.; *Seru-padi, siru-padi,* TAM.; *Hamsa-padu, hama-padi,* TEL.; *Tripakshi,* SANS.; *Serappadi,* TAM. in CEYLON.

References.—*Roxb., Fl. Ind., Ed. C.B.C., 150; Voigt, Hort. Sub. Cal., 445; Thwaites, En. Ceylon Pl., 215; Dalz. & Gibs., Bomb. Fl., 171; Aitchison, Cat. Pb. Pl., 93; Ainslie, Mat. Ind., II., 435; Dymock, Mat. Med. W. Ind., 2nd Ed., 576; S. Arjun, Bomb. Drugs, 96; Murray, Pl. & Drugs, Sind, 170; Drury, U. Pl., 153; Balfour, Cyclop.; Treasury of Botany.*

Habitat.—A small annual weed, usually quite flat, common throughout tropical India; it generally grows on dry rice-fields during the cold season, disappearing about the beginning of the periodical rains. It is common in the hot dry parts of Ceylon. Distributed to Asia, Africa, Australia, and America.

Medicine.—As a medicine, equal parts of the dry PLANT and fenugreek SEEDS rubbed to a fine powder, and applied warm to boils, quickly brings them to suppuration (*Ainslie*). The fresh leaves, ground up, are applied to rheumatic swellings (*Murray*). MEDICINE. Plants. 1708

COLEBROOKIA, *Sm.; Gen. Pl., II., 1180.*

A Himálayan genus, comprising only one species, and that one of the commonest and most abundant plants in the Lower Himálaya and mountains of India, ascending to 4,000 feet in altitude. Leaves. 1709 1710

Colebrookia oppositifolia, *Sm.; Fl. Br. Ind., IV., 642;* LABIATÆ. 1711

Vern.—*Pansra*, HIND.; *Shakardána, phísbekkar, dúss, samprú, súáli, briáli, barmera, shakardána* (TRANS-INDUS), PB.; *Dulshat*, KUMAON; *Dosúl*, NEPAL; *Bhainsa, barsa pakor*, SANTAL.

In some parts of the Panjáb called *basúti*, a name which is more correctly applied to **Adhatoda Vasica.**

References.—*Roxb., Fl. Ind., Ed. C.B.C., 467; Voigt, Hort. Sub. Cal., 452; Kurz, For. Fl. Burm., II., 277; Gamble, Man. Timb., 300; Darjeeling List, 63; Dalz. & Gibs., Bomb. Fl., 209; Stewart, Pb. Pl., 167; Aitchison, Cat. Pb. & Sind Pl., 115; Baden Powell, Pb. Pr., 575; Balfour, Cyclop.; Treasury of Botany.*

Habitat.—A shrub with grey bark, common on the outer Himálaya, from the Indus to Bhután, and Ava, also on the lower hills of India, flowering in February and March.

The form described by Dr. Roxburgh under the name of **C. ternifolia** seems to be peculiar to Western India, extending along the Ghâts to Mysore. It is now viewed as not even worthy of separate recognition as a variety.

Medicine.—The leaves are applied to wounds and bruises (*Stewart*). "The down is used by the Paharias to extract worms from bad sores on the legs (*Gamble*). A preparation from the root is used by the Santáls in epilepsy (*Campbell*). MEDICINE. 1712

Fodder.—The leaves are used as fodder for cattle (*Balfour*). FODDER. 1713

Structure of the Wood.—Greyish-white, moderately hard, close-grained. Weight 46lb per cubic foot. It is used for gunpowder charcoal. TIMBER 1714

COLESEED or COLLARD, see **Brassica campestris,** *Linn., var.* **Napus, B. No. 810.**

COLEUS, *Lour.; Gen. Pl., II., 1176.*

Coleus aromaticus, *Benth.; Fl. Br. Ind., IV., 625;* LABIATÆ. 1715

COUNTRY BORAGE.

Syn.—C. AMBOINICUS, *Lour.; Voigt, Hort. Sub. Cal., 450;* PLECTRANTHUS AROMATICUS, *Roxb.; Fl. Ind., Ed. C.B.C., 466.*

Vern.—*Páthor chur*, HIND.; *Pátér chúr*, BENG.; *Páthor chur, pathúr chúr, owa*, BOMB.; *Pathúr chúr*, MAR.; *Páshána bhedi*, SANS. In *Flora Andhrica, karpúra-valli* is applied to this plant, but **Dr. Moodeen Sheriff** is of opinion, that the name is more in use for **Anisochilus carnosus,** than any other name.

References.—*Dalz. & Gibs., Bomb. Fl. Supp., 66; Pharm. Ind., 168; Moodeen Sheriff, Supp. Pharm. Ind., 114, 51; U. C. Dutt, Mat. Med. Hind., 313; Dymock, Mat. Med. W. Ind., 505; Drury, U. Pl., 153; Lisboa, U. Pl. Bomb., 168; Royle, Ill. Him. Bot., I., 303; Balfour, Cyclop.*

Habitat.—A native of the Moluccas, cultivated in gardens throughout India; has a pleasant aromatic odour and pungent taste.

MEDICINE. Plant. 1716 **Medicine.**—The PLANT "is employed in Cochin China, according to **Loureiro** (*Flor. Cochin, p. 452*), in asthma, chronic coughs, epilepsy, and other convulsive affections. **Dr. Wight** (*Illust., Vol. II.*) speaks of it as a powerful aromatic carminative, given in cases of colic in children, in the treatment of which the expressed juice is prescribed mixed with sugar or other suitable vehicle. In his own practice he observed it produce so decidedly an intoxicating effect that the patient, a European lady, who had taken it on native advice for dyspepsia, had to discontinue it, though otherwise benefiting under its use." "The **Rev. J. Long** (*Jour. of Agri.-Hort. Soc. of India, 1858, Vol. X., p. 23*), also notices its intoxicating properties, and states that the people of Bengal employ it in colic and dyspepsia." (*Pharm. Ind.*) **Dr. Dymock**, however, remarks that he has never heard of this plant producing the intoxicating effects noticed in the *Phormacopœia of India*, and that, if it does, it must be when taken in a much larger quantity than is usual in Bombay.

Juice. 1717 **Special Opinions.**—§ "Expressed JUICE of the LEAVES is considered as an anodyne and astringent, and applied over and around the eyelids, in cases of conjunctivitis" (*Anund Chunder Mookerjee, Assistant Surgeon, Noakhally*). "Said by Sanskrit writers to have a specific action on the bladder and to be useful in urinary diseases, vaginal discharges, &c." (*U. C. Dutt, Civil Medical Officer, Serampore*). "Useful in chronic dyspepsia" (*S. M. Shircore, Civil Surgeon, Moorshedabad*).

FOOD. Plant. 1718 **Food**—According to **Dalzell** and **Gibson**, the PLANT forms an agreeable addition to the cooling drinks used in the hot season. **Roxburgh** says that "the leaves, and indeed all parts of the plant, are delightfully fragrant; they are frequently eaten with bread and butter, also bruised and put into country beer, cool tankards, &c., being an excellent substitute for Borage."

1719 **Coleus barbatus,** *Benth.; Fl Br. Ind., IV., 625; Wight., Ic., t. 1432.*

Vern.—*Garmal*, BOMB.

References.—*Voigt, Hort. Sub. Cal., 449; Thwaites, En. Ceylon Pl., 238; Dalz. & Gibs., Bomb. Fl., 205; O'Shaughnessy, Beng. Dispens., 491; Drury, U. Pl., 154; Lisboa, U. Pl. Bomb., 168; Royle, Ill. Him. Bot., I., 101, 103; Balfour, Cyclop.*

Habitat.—A native of the Peninsula, Gujrát, Behar, and of the subtropical Himálaya, from Kumáon and Nepal; ascending to 8,000 feet. Common on rocky places at an elevation of 2,000 to 5,000 feet; it is also met with on the dry barren hills about Bangalore, whence it was introduced into the Botanic Gardens at Calcutta, where it grows luxuriantly and blossoms during the cold season.

FOOD. 1721 **Food.**—This plant "is commonly cultivated in gardens of the natives at Bombay for the roots, which are pickled (*J. Graham*)." (*Drury*). **Lisboa** says that the pickled root is much used by the Gujarátis.

1721

COLLOCALIA.

It would appear that there are two or three species of Swiftlet which form edible nests. **Dr. Jerdon** is of opinion that the best nests are obtained from **Collocalia linchi**, which builds in the Nicobar Islands, and along the coast of the Bay of Bengal, to Arakan and southwards to Java. Several other species occur on the Malabar Coast, and even in the Eastern Archipelago, as far as New Guinea, one even occurring in Mauritius. The writer is unable to discover the synonym of these species, and has, therefore, thrown the economic facts procurable under the names below, which are commonly given to the "Edible Bird's Nests." FOOD.

Collocalia nidifica, *Gray ;* CYPSELIDÆ. 1722

C. linchi, *Horsfield.*

THE EDIBLE BIRD'S NEST, SALANGANE, *Eng.;* NIDS DE TUNQUIN, *Fr.;* INDIANISCHE-VOGEL-NESTER, *Germ.;* NÍDI-DI-TUNCHINO, *It.;* NIDOS DE LA CHINA, *Sp.*

Sometimes called Edible Swallows' Nests; the bird is more properly a Swift than a Swallow.

Vern.—*Ababil-ka-ghoslah,* HIND.; *Sarong-burong,* MAL.; *Hikai,* NICOBAR; *Gnathiet,* BURM.; *Yen-wo,* CHINESE; *Susyh,* JAVA; *Larvet,* JAPANESE; *Salangana,* MALAY ARCHIP.

References.—*Forbes Watson, Ind. Survey of Ind., 344; Balfour's Cyclopædia of India, 365; Bomb. Gaz., X., 62; Report of the Administration of the Andaman and Nicobar Islands, 1885-86, pp. 4 & 33-35; Official Correspondence, Proc. Agri. Rev. & Com., July 1871, p. 2; Mason, Burma, 201.*

Habitat.—The birds which construct the edible nests are found chiefly on the Pigeon Islands (North Kánara), Vingorla Rock, and Sacrific Rocks on the coast of Malwan (Ratnagiri District), at Tavoy and Mergui, and in the Andaman and Nicobar Islands. **C. nidifica** is known also to occasionally visit Darjeeling, Assam, and the Nilgiri Hills, and to breed on most of the islands on the coast of the Malabar and the Concan round the Bay of Bengal to the Burmese Coast, and the Malay Peninsula. **C. linchi** is the principal species of Java, known locally as *tintyc.* In most of the regions where these birds are found, there are caves which afford shelter and protection. These caves occur chiefly in limestone formations, and are often several miles from the coast; at other times they have to be entered by boats.

ANDAMAN ISLANDS.—**Mr. Portman,** in his report of the Andaman Island Edible Birds' Nests, says:—"I have observed two kinds of swallows, both of which build in the caves. The larger bird has more white in his plumage, and builds a nest of twigs and grass, &c., glued together, and attached to the rock by a peculiar mucilaginous matter. The smaller bird builds a nest of white mucilaginous matter entirely, and it is this nest which is so much sought after. The nest is built in the form of a small bracket attached to the side or roof of the cave, of a semi-circular form, with a radius of about 1½ inches, and regarding the matter of which it is composed opinions differ." "The caves at present known are Passage Island, where a small quantity of the best nests are procurable, but which is only approachable in the calmest weather. North Cinque Island, to which the same remarks apply. Chirya Papu, one cave. North Coast of Rutland Island, opposite Yaratan, one cave. Jolly-Boy Island, north side, one cave. Montgomery Island in Port Campbell, one cave on west side. Neill Island, one cave on north-east coast, very difficult of approach. John Lawrence Island, east coast, opposite East Island. The cave is hidden by a mangrove swamp. Strait Island, South Point, one cave. South Button Island, several caves, yielding the best quality of nests. About three miles inland, at the north end of Stewart's Sound, large caves are to be found in a hill, from which the greatest quantity of our nests are obtained." "In Borneo, from which country China obtains the majority of her birds' nests, the better qualities of nests are found in caves in the interior in crystalline limestone rock, only an inferior quality of nests being found on the seashore. These remarks apply equally to the Andamans; and I have no doubt that when the interior of the islands ANDAMAN ISLANDS. 1723 is explored, many more nest-yielding caves will be found. All our present knowledge is derived from the Malays, who, through fear of the Andamanese, did not dare to search the interior. The explorations should be confined to hilly country, where the crystalline limestone formation predominates."

NICOBAR ISLANDS. 1724

NICOBAR ISLANDS.—**Mr. deRœpstorff**, in his official report of the Nicobar Edible Birds' Nests, remarks: "The best nests I found at Katchall. They were entirely snow-white, and of the best quality. The next best quality I have got were from the Island of Bomboka. This island I have not personally visited;" but he adds, the nests from it "are quite free from foreign matter, and have not the same snow-white beautiful colour as the ones from Katchall. The nests from Katchall are round and egg-formed, while those from Bomboka are long, like the section of an orange."

"The third quality I have is from Sambelong. This is white enough, but intermixed with little weeds or granual stalks. These nests are of good quality, but need cleaning to separate the stalks. The fourth quality I got from the Car Nicobar from a cave in 'Dryad's Bay' in de Roepstorff's bluff in the north end of this island. These nests were entirely worthless for purposes of trade, consisting of the little weeds which are mentioned in the nests from Sambelong. These nests are, however, fastened together by exactly the same glutinous matter which forms the nests first mentioned."

"The Island of Katchall is mostly formed of coral, limestone, and sandstone in all different stages, old, flinty, and yet forming. The island has gone through a series of volcanic revolutions and convulsions, and presents a very pretty landscape, many rents and tearings, ravines and caves extending far under the earth. In these caves dwell the bats and the little swallows. The light of the sun never shines there. The ground is soft to tread on. If you lift it up and inspect it under the torch-light it is seen to contain the wings of the insects, that have fallen a prey to the bats, glimmering like a thousand little rubies; the soil is moist, spread it a little, and you see the little long-shaped excrements of the swallows together with the feathers fallen from the roosting birds. This is the guano. The swallows' nests are not easily seen, but if you lift the torch up to the arched roof by the side of the alabaster-like transparent stalactites white like these, the black head of the little mother appears out of her white little nest."

BURMA. 1725

IN BURMA.—**Mason** says of **C. fuciphaga (C. linchi)**: "This particular species occurs abundantly on parts of the coast of the Malayan Peninsula, in the Nicobar Islands, and the Mergui Archipelago, and so high as on certain rocky islets off the southern portion of the coast of Aracan, where the nests are annually gathered, and exported to China. From all this range of coast we have seen no other species than **fuciphaga**, nor does it appear that any other has been observed; and I have examined a multitude both of the adults and of the young taken from the nests, collected in the Nicobars and preserved in spirit, all of which were of the same species. Still, what appears to be **C. nidifica** inhabits the mountains far in the interior of India, though hitherto unobserved upon the coasts; and it is worthy of notice that **C. fuciphaga** does not appear to have been hitherto remarked inland in this country" (*Staunton quoted by Mason*).

"It may be here added that **C. fuciphaga** is constantly seen inland in these provinces. The Karens in the valley of the Tenasserim in the latitude of Tavoy are well acquainted with the bird, and they say it crosses the mountains to and from the interior every year. That it is the same species there can be no doubt, for the Karen name of the bird is 'the white swallow,' from its white belly."

In the Burma Gazetteer a list of the birds found in the province is given, and among these are included three species of **Collocalia**, *viz.*, **C. innominata,** *Hume*, **C. spodiopygia,** *Peale*, and **C. linchi,** *Horsf.*

MALABAR COAST.—Very little of a definite nature can be learned regarding the edible swallows' nests collected on the western coast. They are said to be found in Ratnágiri, North Kánara, and even in Mysore. According to the Gazetteer of the Ratnágiri District the species found on the Vingorla Rock is **C. unicolor,** *Jerdon, No. 103.* "The rock on which the nests are found is about four miles long." **MALABAR COAST. 1726**

PECULIARITIES OF THE NESTS AND THE MODE OF COLLECTING THEM.—The greatest difference of opinion prevails regarding the nature of the material of which the nests are formed. Early writers used to contend that they were made of a sea-weed which the bird collected for the purpose and chemically changed in some mysterious way. **Ure** (*Arts, Manufactures, and Mines*) says: "The nests are made of a particular species of sea-weed which the bird macerates and bruises before it employs the material in layers so as to form the whitish gelatinous cup-shaped nests so much prized as restoratives and delicacies by the Chinese." On the other hand, many recent writers discredit this theory and believe that the gelatinous material is either the natural saliva of the bird or a substance brought up from the stomach for the purpose and derived from the natural food of the swift, *viz.*, insects. In support of this opinion they point out that the better qualities of the nests are found in caves far removed from the sea. Some of the nesting caves of Borneo are 140 miles from the sea. **Mr. deRoepstorff** points out that there are no edible nests in the Nicobar settlement, but a few miles off in a richer tract of country where insect life abounds they are plentiful. "It is thus," he says, "in places where the food of the swallow is plentiful, that they exist under the most favourable circumstances, and where the nests are best." In the Ratnágiri District Gazetteer it is stated "the swiftlets breed in March and April, in caverns of the rocks, the nests being made of inspissated saliva, in the form of white gelatine, pure white when fresh, but when old, brownish and mixed with extraneous substances." **Mr. Portman** remarks: "The swallow is supposed by some to make this matter, which resembles isinglass, from a species of sea-weed (*fucus*) resembling Carrageen, an Iceland moss. I have often seen this sea-weed, but have never seen the birds on the sea-shore gathering it. Another theory is that the bird excretes this matter from his own throat during the breeding season." "I am unable to give any decided opinion in the matter, but the natives have a theory that the birds bring it down from the sun." **COLLECTION. 1727**

Mr. Portman publishes an interesting account of collecting the nests as pursued in the Andaman Islands. "Before the arrival of the swallows, and as soon as the weather is sufficiently settled, say, about the first week in November, all the caves in the islands should be visited and thoroughly cleaned, the portions of old nests and debris being removed. After the arrival of the birds, and as soon as it is ascertained that they have built their nests, all the caves should be visited and the nests collected and brought in. The date of this visit, and, indeed, the number of collections during the season, are fixed by the time at which the north-east monsoon rain ceases. Being unusually late this year (1885-86), we did not commence nest-collecting till the end of February, but with a dry December the collection might commence on the 15th January. As the collection of nests from the present known caves takes about a month and the swallows rebuild their nest in six weeks or so, the collectors should wait about 10 days in Port Blair, and then go out again, taking care to observe exactly the same order in their rounds. The nests may be col-

COLLOCALIA nidifica.

Edible Birds' Nests.

COLLECTION. lected until the commencement of the rains, when the collection should cease, and the birds be left to breed. Although the great demand is for the white nests, still it may be remarked that the *fucus* attachments of the grass nests, and the old nests gathered in the November cleaning, may be sold locally at R5 per seer, and should, therefore, be collected. Each collection averages about 52℔ of nests." He then proceeds to state the number of men employed by Government to collect the nests, adding: "The six collectors are supplied with torches, rough ladders, axes, and *dahs*, also with a large clean bag lined with linen slung to the side, and an iron implement, about a foot long, with three prongs at one end, and the other end being shaped like a cold chisel. These men detach, with this implement, nests from the sides and roofs of the caves, placing them carefully in their bag, from which, at the end of the work, they are transferred to a box provided with a lock.

"The greatest care is necessary in detaching the nests from the caves, that they should not be broken or soiled. After being brought into the settlement, they are cleaned and packed in circular bundles about a foot in diameter, and four inches thick, ready for export. The refuse from the cleanings should be saved and sold."

Cooking Nests. 1728 COOKING THE NESTS.—"They are first soaked in cold water for two hours, when they swell up and become soft. They are then easily picked to pieces and cleaned. After this they are boiled in clear chicken-broth until dissolved, a process occupying about two hours longer. The usual allowance is one nest (value R1) to a teacupful of soup. Any clear soup may be used. The nest is absolutely tasteless and flavourless, and I have not found that it is particularly strengthening or useful in any way."

TRADE. 1729 TRADE IN EDIBLE NESTS.—Particulars are not available regarding the full extent of the trade in Indian nests. The merchants are Chinamen who reside in Rangoon. They recognise three classes:—

"No. 1, large, pure, white nests, averaging from R110—115 per viss= $3\frac{1}{2}$℔;

No. 2, clean, but slightly coloured nests averaging from R100—140 a viss;

No. 3, more discoloured and dirtier nests averaging

The refuse sells at from R5—15 a seer."

Balfour states that $8\frac{1}{2}$ million nests are annually imported into Canton, and that nests of the first quality fetch £5—£6 the pound; those of the second 9*s*. $4\frac{1}{2}$*d*.; those of the third 3*s*. 1*d*. **McCulloch** says the second quality fetches £4-14*s*., and the third £2-15*s*. The bulk of the more expensive nests are sent to Pekin for the use of the Court. The Japanese do not use the nests but they prepare from a sea-weed an artificial nest called *Dschin-schan*, which they export to China. Of the Ratnágiri district it is stated the right to collect nests is farmed out to Goanese, and fetches about R$28\frac{1}{2}$ a year. The Andaman contractor used to pay R3,000, but last year, owing to the contractor having thrown up his contract, the Government worked the nesting and realized R4,900.

GUANO. 1730

GUANO IN THE SWALLOW CAVES.

An inquiry was instituted into this subject, and **Mr. deRoepstorff** reported: "The guano is so plentiful that ship-loads could be supplied and the landing on the sandy beach would be easy." "I am certain that I could produce bird nests and guano to the value of at least one lakh of rupees per annum." This opinion was expressed regarding the Nicobar islands only, so that if to this be added the possible supply from the Andaman Islands, there would appear to be no reason why India might not at least meet all its own demands for guano manure if not open up an export trade in the article.

Collodion, see under **Gossypium**.

1731 **COLOCASIA**, *Schott.; Gen. Pl., III., 974.*

DeCandolle states that the **Colocasia** of the ancient Greeks was most probably the sacred lotus, but that the name became transferred to the AROID, to which it is now applied by modern writers. He also suggests the possibility of *colcus* or *kulkas* having come from *kachu*, and been introduced into Egypt through the Arabs, and from Egypt made known to the Greeks. It is thus doubtful which derivation of the word may be accepted as carrying with it the greatest degree of probability. There seems little doubt but that the Egyptian-cultivated **Colocasias** came from India, although it is probable that the cultivation of these plants was commenced in more centres than one, and that, too, independently of each other, such as in India, the Malay Peninsula, Japan, and the Fiji Islands. "The Malay names *kelady*, *tallus*, *tallus*, *taloo*, or *tulves*, perhaps gave origin to the well known name of the Otahitans and New Zealanders—*tallo* or *tarro*, *dalo* of the Fiji Islanders" (*DeCandolle*). If this be accepted then may not the Bombay name *terem* be admissible as coming from the same root?

[*Wight, Ic., t. 786*; AROIDEÆ.

1732 **Colocasia antiquorum**, *Schott.; DC., Mono. Phanerog., II., 491;*
TARO, EDDOES, SCRATCH-COCO, EGYPTIAN ARUM, COCO, KOPEH.
Sometimes but incorrectly called YAM.

Syn.—ARUM COLOCASIA, *Willd.; Roxb., Fl. Ind., Ed. C.B.C., 624.*

Vern.—*Kachú, gori-kachú, ashú-kachú, arvi, ghoya, ghuya, avois, ghuiya, auri, ghwiya, arwi*, HIND.; *Kachú, kuchú, ashú-kuchú, kalo-kuchú, char-kuchú, bun-kuchú, gúri*, BENG.; *Dzu* (cultivated) and *kirth* (wild), ANGAMI NAGA; *Ráb, álú, kasauri, gágli, givian, kachálú, ghuyan*, PB. (in Kangra there are said to be three forms known as *kachálú, gandiali*, and *arbi*); *Kachu álu, terem*, BOMB.; *Alú*, MAR.; *Arvi, chamkúré-ka-gaddah*, DUK.; *Saru*, URIYA; *Shámak-kizhangu, shema-kalenga*, TAM.; *Shámá-thúmpa, chama-kura, chama-kuru, chāma-gadda, chama-dumpa, chema*, TEL.; *Chémpa-kizhanna, kaladi*, MALA.; *Sháme-gadde, keshavaná-gadde*, KAN.; *Kachchi, katchú, kachwi, kachwæ*, SANS.; *Kalkas, qulqás, kur*, ARAB.; *Má hu ya pein*, BURM.; *Gahala, tadala, habarala*, the young cultivated tubers being known as *kandalla*, or *tadala*, SING.; *Imo*, JAPANESE.

References.—*Voigt., Hort. Sub. Cal., 686; Thwaites, En. Ceylon Pl., 335; Stewart, Pb. Pl., 247; Aitchison, Cat. Pb. and Sind Pl., 143; DC., Origin of Cult. Pl., 73; Pharm. Ind., 250; Moodeen Sheriff, Supp. Pharm. Ind., 114; U. C. Dutt, Mat. Med. Hind., 301; S. Arjun, Bomb. Drugs, 195; Atkinson, Him. Dist., 704, 733; Drury, U. Pl., 154; Lisboa, U. Pl. Bomb., 182; Birdwood, Bomb. Pr., 186; Royle, Ill. Him. Bot., I., 406, 407; Balfour, Cyclop.; Smith, Dic., 403; Treasury of Botany; Morton, Cyclop. Agri., I., 491; Irvine, Med. Top. Ajmere, 207.*

Habitat.—Wild over the greater part of tropical India, and also cultivated throughout India on account of its corms, which are used as an important article of diet when boiled. **Stewart** says: "It is grown at places in the hills to a considerable elevation; I have seen it at nearly 7,600 feet in Chumba and Kúllú." **DeCandolle**, in his *Origin of Cultivated Plants*, writes: "Since the different forms of the species have been properly classed, and since we have possessed more certain information about the floras of the South of Asia, we cannot doubt that this plant is wild in India, as **Roxburgh** formerly, and **Wight** and others have more recently, asserted, likewise in Ceylon, Sumatra, and several islands of the Malay Archipelago."

Engler (in *DC., Mono. Phanerogm., vol. II.*) describes some seven varieties of this plant, three of which are apparently met with in India:—

α. **typica**; *Wight, Ic., t. 786;* **Arum colocasia,** *Roxb. Fl. Ind., Ed. C.B.C., 624.* *Chámakúra* or *Chéma-kúra, kúra* or *dumpa,* TEL.

Of this form **Roxburgh** describes two cultivated conditions, the *gúri-kachú*, the corms of which ripen in Bengal during February and March, and the *ashú kachú*, which does not ripen until the end of the year.

ε. **esculenta** (*Schott., Syn., 41*). This is the **Arisarum esculentum,** *L.*, as in *Rumph., V., t. 110, f. 1,* but not the **Calla calytrata,** *Roxb., Fl. Ind., Ed. C.B.C., 631,* which he says is **Arisarum esculentum,** *Rumph., Amb., V., t. 111, f. 1,* cultivated form.

ζ. **nymphæifolia** (**Arum nymphæifolium,** *Roxb., Fl. Ind., Ed. C.B.C., 624; Wight, Ic., t. 786; Rheede, Mal., XI., t. 22*). This is the *Sar-kachú* of Bengal and the *Sepa-kilangu* or *Senei-kilangu* of Coimbatore District, Madras; *Wel-ala yakutala,* SING. **Roxburgh** says that this is merely a more aquatic state; that it is rarely cultivated, but is found wild in abundance on the borders of lakes and tanks. "The root, or rather the subterraneous stem, often grows to the length and thickness of a man's arm. The petioles, scape, and leaves, are of a reddish colour, and the plants considerably larger than any of the varieties of **Colocasia**" (*var.* **typica** above), "yet the leaves are narrow in proportion to their breadth." The only good character by which to know this form "is the shortness of the club of the spadix." "Every part of this plant is eaten by the Hindus."

A good deal has been written regarding the cultivated species of **Colocasia,** but it has been found impossible to discover what species, still less which varieties, are alluded to. On this account it has been deemed desirable to compile the economic information here given from such authors as could be depended on for the accuracy of their general information, and to thus leave for future research a more detailed description than will be found here.

The following facts seem to refer to *var.* **typica.**

MEDICINE. 1733 **Medicine.**—The pressed juice of the petioles is styptic, and may be used to arrest arterial hœmorrhage. **Dr. Bholanath Bose** reports very highly in favour of this property, and states that the wound heals by first intention after its application. (*Pharm. Ind.*) It is sometimes used in emache and otorrhœa, and also as an external stimulant and rubefacient by the natives.

Special Opinions.—§"The juice expressed from the leaf stalks of the black species is used with salt as an absorbent in cases of inflamed glands and buboes. The juice of the corm of this species is used in cases of alopecia. Internally, it acts as a laxative, and is used in cases of piles and congestion of the portal system, also as an antidote to the stings of wasps and other insects" (*Surgeon J. H. Thornton, Monghyr*). "I have seen remarkable instances of its styptic properties (juice); if applied to fresh and clear wounds, it enables the tissues to unite by first intention within a few hours" (*Surgeon D. Basu, Furridpore*).

FOOD. 1734 **Food.**—The plant has large heart-shaped leaves, borne on long footstalks, rising from a short farinaceous corm. This corm forms an important article of food to the natives throughout India, being largely cultivated, but rarely if ever eaten from the wild state of the plant, which occurs everywhere as a weed of damp places. The wild condition of the plant is by the Angami Nagas called *Kirth.* "The young leaves may be eaten like

FOOD.

spinach; but, like the root, they require to be well cooked in order to destroy the acridity peculiar to Aroids. A considerable number of varieties are known, some better adapted for puddings, some for bread, or simply for boiling or baking. The outer marks of distinction chiefly rest upon the different tinges observable in the corm, leaf-stalks, and ribs of the leaves,—white, yellowish, purple" (*Seeman, Flora Vitiensis*). **Atkinson** says: "The tuber of the cultivated variety is long, white, carrot-shaped, often weighing several pounds, and forms an important article of food among the lower classes, where quantity and not quality is a desideratum. It is usually served fried in *ghí* or boiled and pounded into a paste, and also in curries. There are varieties that are very small, hardly weighing more than a quarter of a pound." In the Manual of Coimbatore it is stated that the corms (apparently of *var.* **nymphæifolia**) often weigh as much as 70 to 80lb each, and that an acre will yield 250 maunds (of 25lb), worth 12 annas a maund. The tubers are used by the natives of Bombay in curries, &c. They form the common food of the inhabitants of Travancore. The Malays hold it in high estimation (*Balfour*).

§ "Is considered very nutritious by the natives, who use it in their curries" (*Honorary Surgeon P. Kinsley, Chicacole, Madras*).

Colocasia cucullata, *Schott.* 1735

Syn. for ALOCASIA CUCULLATA, *Schott.*

C. indica, *Engl.; DC., Mono. Phanerog., II., 494.* 1736

Syn. for ALOCASIA INDICA, *Schott.*, which see, **A. 809.**

This plant is said to be specially cultivated in Brazil for its esculent stems and small pendulous tubers. It is known as *Man saru* in Orissa, and is there used in the treatment of piles.

C. macrorrhiza, *Schott.* 1737

Syn. for ALOCASIA MACRORRHIZA, *Schott.*

A species met with in Eastern Bengal and Sylhet, also in Ceylon (the *habarella*). Often cultivated, and the leaves of the very young plant also eaten (*Thwaites, En. Ceyl. Pl., 336*). It has been found impossible to obtain definite information as to the extent this plant is cultivated in India, and also as to whether or not it can be viewed as indigenous. **DeCandolle**, in his *Origin of Cultivated Plants*, refers to it as wild in Otahiti and in Ceylon. It is known in the former as *apé* and in the Friendly Islands as *kappé*. **Ainslie** (*Mat. Ind., II., 463*) gives its Chinese name as *dea-vew* and the *verrughung kalung* in Tamil, and the *Hastid carnid* (?) in Sanskrit. He remarks: "This root, in its raw state, like most of the arums, possesses a degree of acrimony; in conjunction with gingelly oil, the native practitioners prepare a kind of liniment with it, which, they allege, when rubbed on the head, sometimes cures intermittent fevers after every other remedy has failed." The active principle is very volatile, so much so that by the application of heat or by simple drying, the roots become innocuous.

C. virosa, *Kunth.; DC. Mono. Phanerog., II., 495; Roxb., Fl. Ind., Ed. C.B.C., 632 (under* calla). 1738

Vern.—*Bish Kachú.*

This plant, which is a native of the Lower Provinces, is the only member of the genus which the natives of India regard as poisonous. It is sometimes used medicinally, but is never eaten.

COLOCASIA virosa.

Poisonous Properties of Aroids.

CHEMISTRY. 1739

Chemistry.—Through the kindness of **Messrs. Pedler** and **Warden** (*Professors of Chemistry in the Calcutta University*), the writer has had the pleasure to receive an advance copy of their paper* on the chemical properties and medicinal uses of the species which, by the early botanists, were all treated as belonging to ARUM, but which, by modern authors, have been thrown into some half a dozen genera. Their object in writing the paper was to investigate the *Toxic Principle* possessed by these plants, and the enquiry was suggested on receiving from the Civil Surgeon of Dibrugarh "some portions of raw *Bish Kachu* tubers and leaves with the following statement: 'A cooly woman administered some of the fried *kachu* to another sick cooly on the same garden, but the man, experiencing a burning sensation in his mouth, instantly spat it out. A pig ate what was so thrown away and died in an hour. A second pig was experimented on with some of the same stuff, and fatal results also supervened.' During the course of the same year a second case of poisoning by *kachu* was referred to the Chemical Examiner's Department; in this case slices of *kachu* tubers were introduced into a jar containing 'goor.' The symptoms induced were sufficiently urgent to necessitate admission of the person into the Medical College Hospital: the stomach-pump was used, as the symptoms were those of irritant poisoning."

A sample of the corms and leaves of the *bish kachu* sent from Dibrugarh were forwarded to **Dr. King** for identification; but as a flower had not been furnished he was unable to name the plant further than that it was a species of **Alocasia** or **Colocasia**. **Roxburgh** and all subsequent writers on economic botany say that the *bish kachu* is **Colocasia virosa**, and accepting this to have been, in all probability, the plant **Pedler** and **Warden** experimented with, their results may be here briefly summarised:—In peeling the tubers "considerable irritation was experienced about the hands, but there was a complete absence of any irritative action on the olfactory organs or conjunctivæ. This fact appeared to us to point towards the non-volatile nature of the active principle." An alcoholic extract was prepared and found to have no poisonous effect. The same result followed on the administration of a distillate which was found to have no acrid taste, and, as with many other vegetable substances distilled with water, it was found to contain a trace of hydrocyanic acid. "It is possible, however, that certain varieties of ARUM may contain a larger amount of prussic acid, as, for example, the **A. seguinum** of the West Indies, which is stated to furnish a juice, two drachms of which has proved fatal in a few hours. The tubers left in the retort after distillation with water were still physiologically active, indicating that the active principle was not dissipated by mere boiling with water. Natives, in using ARUM for culinary purposes, frequently add an acid vegetable or fruit such as tamarind. We tried the action of certain acids on the fresh tubers, and ascertained that boiling with water acidulated with hydrochloric acid for a very short period, rendered the tubers quite inert when a fragment was applied to the tongue. Dilute nitric acid also acted in a similar manner. The action of acetic acid, on the other hand, was very much feebler, and the acid had to be stronger in order to produce any decided diminution in activity." "A rough analysis of the ash indicated the presence of a large amount of potassium and magnesium; calcium was also present, but we failed to obtain indications of sodium. The acids consisted of carbonic, phosphoric, hydrochloric, with traces of sulphuric, acid. We also obtained from the dried tubers very marked quantities of potassic nitrate, so that when they had been incinerated they behaved very like

*See *Jour. Asiatic Soc. Beng., LVII., Pt. II., No. 1 for 1888.*

CHEMISTRY.

tinder, containing saltpetre. The examination of the ash thus failed to afford us any clue to the physiological action of the fresh tubers."

"It now occurred to us that possibly the painful effects produced by ARUM when in contact with the tongue, &c., might be due to mechanical causes. A microscopic examination of a section of a tuber revealed the presence of very numerous bundles of needle-shaped crystals, and we also found similar crystals in the leaves and stems. These crystals were seen under the microscope to be insoluble in cold acetic acid but easily soluble in cold diluted nitric or hydrochloric acid." "There appears to us to be no reason to doubt the fact, that the whole of the physiological symptoms caused by ARUMS are due to these needle-shaped crystals of oxalate of lime, and that the symptoms are thus due to purely mechanical causes. Bearing in mind the action of re-agents on calcic oxalate, the reason why mere boiling in water failed to deprive them of their activity is explained by the insolubility of oxalate of lime in water. Again, the action of dilute acetic acid, even at temperatures of 100° C., in slightly lessening the activity of the tubers, is due to the very slight solubility of oxalate of lime in that acid. And, lastly, the complete loss of all physiological action when the tubers were treated with dilute nitric or hydrochloric acid is evidently due to the ready solubility of calcic oxalate in those mineral acids. And these assumptions, as we have already indicated, were fully demonstrated by the microscopic examination of sections of the tubers treated with the reagents we have mentioned. One point, however, remains to be explained: we observed that, on drying, the tubers lost practically the whole of their physiological activity. Clearly there could have been no loss of oxalate of lime on desiccation, and, as a matter of fact, we found as many crystals on microscopic examination of dried ARUMS as we had found in the fresh tubers. We explain this apparent anomaly in the following simple manner. In the fresh condition of the tubers, the bundles of crystals of oxalate of lime are cone-shaped, more or less, the sharp points covering a wide area, and forming the base, but, in the drying of the tubers, the needles appear to arrange themselves more or less parallel to one another, and the sharp points thus cover a smaller area. And thus, instead of each crystal acting as a separate source of irritation and penetrating the tissues, the bundles act as a whole."

The poisonous effects of certain aroid tubers are therefore the result of mechanical irritation, similar to that produced by cowage (**Mucuna pruriens**) or to chopped hairs criminally mixed with food. It would be interesting to have this line of enquiry carried to its final issue in a systematic examination of all the plants, like rhubarb, which contain raphides. It is just possible that the crystals of oxalate of lime may have more to say to the therapeutic property of that much valued domestic medicine, rhubarb, than we have hitherto conceived. The active constituent of the root has been supposed to reside in the yellowish red contents of the medullary rays. Various substances have been extracted and chemically analysed, but it may be said we have not advanced much nearer a full understanding of the chemistry of rhubarb connected with its physiological action than we were before. It is thus probable that the results of Pedler and Warden's analysis of the aroid tubers may have a more extended influence on therapeutic science than they seem to have realized.

Colocynth, see **Citrullus Colocynthis,** *Schrad.;* CUCURBITACEÆ.

Colombo (or Calumba) Root, see **Jateorhiza Calumba.**

COLUTEA, *Linn.; Gen. Pl., I., 505.*

[*103;* LEGUMINOSÆ.

1740 **Colutea arborescens,** *Linn.;* var. **nepalensis;** *Fl. Br. Ind., II.,*

THE BLADDER SENNA; NEPAL BLADDER SENNA.

Syn.—C. NEPALENSIS, *Sims.; Bot. Mag., t. 2622.*

Vern.—*Bráa,* LADAK, AFGHANISTAN.

References.—*Brandis, For. Fl., 136; Gamble, Man. Timb., 118; Stewart, Pb. Pl., 64; O'Shaughnessy, Beng. Dispens., 294; Flück. and Hanb., Pharmacog., 221; U. S. Dispens., 15th Ed., 1298, 1617; Murray, Pl. and Drugs, Sind, 131; Royle, Ill. Him. Bot., I., 195, 198; Treasury of Botany.*

Habitat.—A shrub of the temperate west Himálaya, Kunawar, Tibet, Nipal, &c., at an altitude of 8,000 to 11,000 feet.

MEDICINE. Leaves. 1741 **Medicine.**—The leaves of this plant are purgative, and are used to adulterate officinal senna, and in some parts of Europe as a substitute for senna, though comparatively feeble in their action. They are administered in infusion or decoction in the dose of about half a pint (*U. S. Dispens., 1617*).

Colza Oil, see **Brassica campestris,** *Linn.;* var. **Napus, B. No. 810.**

COMBRETUM, *Linn.; Gen. Pl. I., 688.*

[COMBRETACEÆ.

1742 **Combretum decandrum,** *Roxb.; Fl. Br. Ind., II., 452;*

Vern.—*Dhobela,* CHINDWARA; *Punk,* GONDA, OUDH; *Arikota,* TEL.; *Kali-lara,* NEPAL; *Pindik,* LEPCHA.

References.—*Roxb., Fl. Ind., Ed. C. B. C.; Brandis, For. Fl., 221; Gamble, List of Darjeeling Climbers, &c.*

Habitat.—Abundant in Bengal, at altitudes up to 3,000 feet. Very common in the North Deccan plateau, in the North-Western Provinces, Tenasserim, and the Andamans.

1743 Is said to be used medicinally, but very little is known regarding the uses of the plant. The Santáls, who call it *atená,* make baskets from its long thin stems (*Campbell*).

1744 **C. nanum,** *Ham.; Fl. Br. Ind., II., 457.*

Vern.—*Dant játhi, pharsia,* N.-W. P. and PB.

References.—*Brandis, For. Fl., 221; Baden Powell, Pb. Pr., 350; Royle, Ill. Him. Bot., I., 209.*

Habitat.—A decumbent, low shrub of the Himálayan terai, from Sikkim to the Panjáb.

MEDICINE. 1745 **Medicine.**—Mr. **Baden Powell** mentions this plant among his medicinal plants of the Panjáb.

1746 **C. ovalifolium,** *Roxb.*

Vern.—*Bandi káttu tige, yádala chettu, bandi kóta,* TEL. (the buffalo-calf tree).

A common climber throughout the Deccan Peninsula, probably eaten by buffalos.

WOODS FOR COMBS, &c. 1747

COMBS, fans, brush-backs, and other smaller articles—Woods used for:—

Adina cordifolia (combs).
Alangium Lamarckii (cattle-bells).
Albizzia stipulata (cattle-bells).
Artocarpus integrifolia (brush-backs).
Bauhinia Vahlii (umbrellas, rain-caps).
Buxus sempervirens (instruments, combs, small boxes).
Carissa diffusa (combs).
Casearia tomentosa (combs).
Chloroxylon Swietenia (picture-frames, brush-backs).
Cordia Macleodii (picture-frames).
Coriaria nepalensis (small articles).
Corypha umbraculifera (fans, umbrellas).
Cratæva religiosa (combs).
Elæodendron glaucum (combs, picture-frames).
Gardenia costata (combs).
G. latifolia (combs).
G. lucida (combs).
Gmelina arborea (picture-frames).
Olea ferruginea (combs).
Platanus orientalis (pen-cases).
Psidium Guava (instruments).
Pyrus Pashia (combs, tobacco-pipes).
Schrebera swietenioides (combs and weavers' beams).
Stephegyne parvifolia (combs).
Sterculia urens (guitars).

COMMELINA, *Linn.; Gen. Pl., III., 847.* 1747

The genus of the Spider-worts is named in honour of the Dutch botanist **Commelin.**

Commelina benghalensis, *Linn.; DC., Mono., 159; Clarke, Comm. et Cyrt., 14, Pl. IV.; Wight, Ic., t. 2065;* COMMELINACEÆ. 1748

Vern.—*Kanshura,* HIND.; *Kanchura, kanuraka, kanshira, káchrádám, kánchará,* BENG.; *Kana arak',* SANTAL; *Chura, kanna,* PB.; *Khanna,* SIND; *Kanchata,* SANS.; *Deya-mainaireya* or *diya-menériya,* SING.; *Ho-tan-tu,* CHINESE.

References.—*Roxb., Fl. Ind., Ed. C.B.C., 57; Voigt, Hort. Sub. Cal., 676; Thwaites, En. Ceylon Pl., 321; Dalz. & Gibs., Bomb. Fl., 253; Stewart, Pb. Pl., 236; Aitchison, Cat. Pb. and Sind Pl., 148; Trimen, Syst. Cat., 95; DeCandolle, Mono. Phanerogam, III., 159; Rev. A. Campbell, Descript. Cat. of the Pl. Chutia Nagpur; U. C. Dutt, Mat. Med. Hind., 303; Murray, Pl. and Drugs, Sind, 22.*

Habitat.—A native of wet places all over Bengal (*Roxb.*). It also occurs in the peninsula of India generally, and in Sind, Salt Range, and the Deccan. **Dalzell** and **Gibson** say that it is common everywhere in Bombay. Distributed to Burma, Malay, and China.

FOOD. Leaves. 1749

Food.—LEAVES eaten by the poor people as a pot-herb, especially in times of scarcity. The fleshy rhizomes of some of the species of this genus contain much starch, mixed with mucilage, and are therefore wholesome food when cooked. Starch. 1750 **Balfour** says **C. polygama** (a name which would appear to be a synonym for **C. benghalensis**) is cultivated in China as a pot-herb eaten in spring. Pigment. 1751 "The juice of the flower is used as a bluish pigment in painting upon transparencies" (*Smith*).

C. communis, *Linn.; DC., Mono. Phanerogam, III., 170.* 1752

Vern.—*Kena,* BOMB.; *Wek kyup,* BURM. **Stewart** says that this, as also **C. benghalensis**, are in the Panjáb known as *Chura, kanna.* **Balfour** gives the following names: *Kanang kirai, kunnu katti pillu,* TAM.; *Venna-devi kura, niru kassuvu, venna mudra, venna vedara,* TEL.; *Vatsa priam,* SANS.

It may be here recorded of the vernacular names given to this and, in fact, to all the species of **Commelina,** that they require to be verified and assorted under the modern scientific names for the species of this genus.

References.—*Voigt, Hort. Sub. Cal., 677; Dalz. & Gibs., Bomb. Fl., 252; Stewart, Pb. Pl., 236; Aitchison, Cat. Pb. and Sind Pl., 148; Balfour's Cyclopædia of India.*

Habitat.—A native of the hot damp regions of China and Japan. From Chittagong, plants are said to have been sent to the Botanic Gardens, Calcutta, by **Mr. W. Roxburgh.** (*Roxb.*) But it is feared a good deal of the economic information published under **C. communis** should be recorded under **C. obliqua** or **C. nudiflora,** *Linn.* The information that could not be established as referable to either of these plants has for the present been left in the present position.

FOOD. Seeds. 1753 Leaves. 1754 **Food.**—"The rugose SEEDS contained in oblong capsules were largely consumed in the Sholápur District during the famine" (*Lisboa*). **Balfour** says the "succulent LEAVES are used by the Hindus for feeding young calves when they wish to wean them from their milk." "The leaves are eaten by the natives mixed with other greens."

[*Com. and Cirt. Table I.*

1755 **Commelina nudiflora,** *Linn.; DC. Mono., III., 144; C. B. Clarke's*

Syn.—C. CÆSPITOSA, *Roxb., Fl. Ind., Ed. C. B. C., 58;* C. NUDIFLORA, *Linn.*, as described in *Roxb., Fl. Ind., Ed. C. B. C.*, is ANEILEMA NUDIFLORUM, *Linn.*, the *Kundali* of Bengal.

Habitat.—Frequent in Bengal, and distributed to Burma, Ceylon, and the Malay, also to Africa, Madagascar, Mauritius, Sandwich Islands, and Australia, &c.

Compare this with the remarks under **C. communis,** *Linn.*, and **C. obliqua,** *Ham.*

1756 **C. obliqua,** *Ham.; Clarke, p. 19, pl. IX.*

Syn.—C. COMMUNIS, *Roxb., Fl. Ind., Ed. C.B.C., 57.*

Vern.—*Kanjurá, kána,* HIND.; *Jata-kanchura, jata-kanshira,* BENG.; *Korna, kána,* BIJNOR; *Kanjura,* KUMAON.

Habitat.—This species is common over the low moist parts of India, flowering during the rainy season chiefly. It also occurs on the lower Himálaya (ascending even to 7,000 feet in altitude), and is distributed to Ceylon, Burma, and the Malay.

MEDICINE. Root. 1757 **Medicine.**—'The ROOT is useful in vertigo, fevers, and bilious affections, and as an antidote to snake-bites" (*Atkinson*).

FOOD. Root. 1758 **Food.**—**Balfour** says that the ROOT is edible; **Atkinson** that the leaves and stems are used as greens during seasons of scarcity.

1759 **C. salicifolia,** *Roxb.; Fl. Ind., Ed. C.B.C., p. 58.*

Vern.—*Jalapippali languli,* SANS.; *Paní-kánchirá,* BENG.; *Jalpipari,* HIND.; *Bir kana arak',* SANTAL.

References.—*DeCandolle, Mono. Phanerog., III., 157; U. C. Dutt, Mat. Med. Hind., 300.*

Habitat.—Common in wet places in the peninsula of India, especially in Bengal, Coromandel, and Bombay. Distributed to Burma.

FODDER. 1760 **Fodder.**—Cattle are said to be fond of this plant.

C. scapiflora, *Roxb.;* see **Aneilema scapiflorum,** *Wight.;* **A. 1122.**

1761 **C. suffruticosa,** *Bl.; DC., Mono. Phanerog., III., 188.*

Vern.—*Dare orsa,* SANTAL.

Habitat.—A native of Bengal.

MEDICINE. Root. 1762 **Medicine.**—The root is by the Santáls applied to sores (*Campbell*).

Conch Shell, a species of **Turbinella**; see **Shells**; also **Beads, B. 381.**
Condiments, see **Spices.**
Conessi Bark, see **Holarrhena antidysenterica,** *Wall.*; APOCYNACEÆ.

CONGEA, *Roxb.*; *Gen. Pl., II., 1159.*

[*t. 1479*; VERBENACEÆ.

Congea tomentosa, *Roxb.*; *Fl. Br. Ind., IV., 603*; *Wight, Ic.,* 1763

Vern.—*Tamakanwe, ka-yan,* BURM.

References.—*Kurz, For. Fl. Burm., II., 256*; *Roscoe in Roxb. Fl. Ind., Ed. C. B. C., 477.*

Habitat.—A large climber in Chittagong and Burma; distributed to Siam. Roxburgh says it is found also in Coromandel, where it flowers in the cold season, the Chittagong plant flowering in March. The *Flora of British India* describes a variety—**Azurea**—as cultivated in North India. All the species of this elegant genus are characterised by their purple bracts.

C. villosa, *Wight, Ic., t. 1479, fig. B.*; *Fl. Br. Ind., IV., 603.* 1764

A large climber of Pegu and Mergui, the leaves of which are used medicinally (*Mason, O'Shaughnessy, &c.*)

CONIUM, *Linn.*; *Gen. Pl., I., 883.*

Conium maculatum, *Linn.*; *DC., Prodr., IV., 242*; UMBELLIFERÆ. 1765

SPOTTED HEMLOCK, HEMLOCK, *Eng.*; CIGUË, *Fr.*; SCHIERLINGS, *Germ.*

Vern.—*Showkrán,* ARAB.; *Kirdamána,* BOMB.

References.—*Pharm. Ind., 104*; *Ainslie, Mat. Ind., Preface p. XII*; *O'Shaughnessy, Beng. Dispens., 369*; *Dymock, Mat. Med. W. Ind., 2nd Ed., 363*; *Flück. & Hanb., Pharmacog., 299, 301*; *U. S. Dispens., 15th Ed., 194, 484*; *Bent. & Trim., Med. Pl., 118.*

Habitat.—Met with in Europe and temperate Asia; common in England.

MEDICINE. 1766

Medicine.—Although this drug is commonly used in Indian pharmacy, and largely imported, no effort seems to have been made to cultivate the plant in the temperate regions of India. It appears to have been the κώνειον of the Greeks (the State poison of Athens), and the *Cicuta* of the Romans (*Birdwood*). Dymock says "the *Kírdamána* of the Bombay shops does not appear to have been utilized by Europeans; it comes from Persia." "The seed is sold for 8 annas per ℔."

CONNARUS, *Linn.*; *Gen. Pl., I., 432, 1001.* 1767

Very little is known regarding the Indian species of **Connarus,** and the following notes have been thrown together more by way of indicating the direction of future enquiry than as affording any definite information. The timber of the arborescent forms is valued, and the seeds of most afford a useful oil.

Connarus monocarpus, *Linn.*; *Fl. Br. Ind., II., 50*; CONNARACEÆ. 1768

Vern.—*Súnder,* BOMB.; *Ká-dat-ká-let, at-ká-let, ta-le-té,* BURM.; *Radaliya,* SING.

References.—*Beddome, Fl. Sylv. App. LXXXII.; Wight and Arnott, Prod. Fl. Pen. Ind. Or., 143; Thw., En. Cey. Pl., 80; Kurz, Pegu Report; Bomb. Gaz., XXV., 330; Dalz. and Gibs., Bomb. Fl., 53; Rheede, Mal., VI., t. 24.*

Habitat.—A small tree or shrub of the Western Peninsula, from the Concan to Travancore; common on the Southern Gháts; very abundant in Ceylon. Flowers yellow, fruit long, bright red; the tree becoming very ornamental when in fruit.

L. 1769 **Oil.**—The seeds yield an OIL.

TIMBER. 1770 **Structure of the Wood.**—The timber of this, as of most other species of the genus, is much valued for ornamental purposes.

1771 **Connarus nitidus,** *Roxb., in Hort. Beng., 49.*

References.—*Voigt, Hort. Sub. Cal,, 265; Gamble, Man. Timb., 114.*

Habitat.—Said to be found in Sylhet and British Burma.

OIL. 1772 **Oil.**—**Dr. McLelland** says that in Rangoon the seeds of this plant yield a quantity of sweet oil. The name **C. nitidus** is not referred to by the *Flora of British India*, but it may be presumed that the plant which yields the oil in question is **C. paniculatus.**

1773 **C. paniculatus,** *Roxb.; Fl. Ind., Ed. C.B.C., 505; Fl. Br. Ind., II., 52.*

References.—*Kurz, For. Fl. Burm., I., 327; Gamble, Man. Timb., 114; Wight, Ill., t. 64.*

Habitat.—**Roxburgh,** followed by **Voigt** and **Kurz,** describes this as "a large timber tree;" but **Hooker,** in the *Flora of British India,* says it is "a large climber" met with in Sylhet and the Khasia hills, to Chittagong."

1774 **C. speciosus,** *McLell.*

Vern.—*Gwedoak, kadon-kadet,* BURM.

Habitat.—Said to be a large tree of Rangoon, Pegu, and Toungho.

OIL. 1775 **Oil.**—**McLelland** says that the seeds yield an abundance of sweet oil.

The above has been extracted from **Dr. Cooke's** *Report on Oil Seeds.* The name **C. speciosus,** *McLell.,* was taken apparently from *Balfour's Cyclopædia.* It seems probable that the tree here alluded to is **C. gibbosus,** *Wall*—a large tree met with near Rangoon and in Tenasserim, Penang, and Singapore. The Burmese name *Gwe* (**Spondias mangifera**) seems very near to the above.

TIMBER. 1776 **Structure of the Wood.**—**Balfour** says of **C. speciosus:** "It has a large, heavy, and strong timber, white coloured, adapted to every purpose of house-building."

Conocarpus acuminata, *Roxb.;* see **Anogeissus acuminata,** *Wall.,* COMBRETACEÆ; **A. 1146.**

C. latifolia, *Roxb.;* see **Anogeissus latifolia,** *Wall.,* **A. 1149.**

Construction and Railway purposes—Timbers suitable for, see **Cart and Carriage Building, C. 632.**

CONVOLVULUS, *Linn.; Gen. Pl., II., 874.*

1777 **Convolvulus arvensis,** *Linn.; Fl. Br. Ind., IV., 219;* CONVOLVULACEÆ.

DEER'S FOOT BIND-WEED.

Syn.—C. MALCOLMI, *Roxb., Fl. Ind., Ed. C. B. C., 159.*

Vern.—*Veri* (?), *harin-pádi*, or by some writers *hiran paddi*, PB., HIND.; *Hirn-pug*, SIND.

References.—*Voigt, Hort. Sub. Cal., 362; Dalz. & Gibs., Bomb. Fl. 163; Stewart, Pb. Pl., 150; Aitchison, Cat. Pb. and Sind Pl., 98; O'Shaughnessy, Beng. Dispens., 502; Murray, Pl. and Drugs, Sind, 164; Year Book Pharm., 1879, 467; Medical Top. of Ajmir, 150; Baden Powell, Pb. Pr., 367.*

Habitat.—An abundant weed of cultivation all over the plains of the Panjáb and Western India, from Kashmir to the Deccan, ascending to 10,000 feet in the Himálaya. Flowers large, deep rose-coloured, sweetly scented: they appear in the cold season; very common on the black soil of Gujarat and the Deccan.

Medicine.—The officinal *hiran paddi* (or *harin pádi*) appears to be this plant. The roots possess cathartic properties. Murray says the roots are sometimes used by the Sindis as jalap. MEDICINE. Root. 1778

Fodder.—*Veri* is a dark green weed, usually found in wheat fields. It is said to be greedily eaten by goats and cattle, and is gathered by village children as a fodder. FODDER. 1779

Convolvulus Batatas, *Linn.*; see **Ipomœa Batatas,** *Lamk.*

C. parviflorus, *Vahl.; Fl. Br. Ind., IV., 220.* 1780

Vern.—*Alaranji*, TEL.

A native of Assam, the Deccan Peninsula, and Ceylon, but largely cultivated throughout India.

C. pentaphylla, *Linn.*; see **Ipomœa pentaphylla,** *Jacq.*

C. pluricaulis, *Chois; Fl. Br. Ind., IV., 218.* 1781

Vern.—*Porprang, gorakh pánw, baphalli, dodak*, PB.

References.—*Stewart, Pb. Pl., 150; Aitchison, Cat. Pb. and Sind Pl., 99.*

Habitat.—A common plant in many places throughout the plains of Panjáb, Hindustan, and Behar.

Food and Fodder.—"It is eaten by cattle and is reckoned cooling, and used as a vegetable or given in sherbet" (*Stewart*). FOOD and FODDRR. 1782

C. reptans, *Linn.*; see **Ipomœa aquatica,** *Forsk.*

C. Scammonia, *Linn.; DC. Prodr., IX., 412.* 1783

SCAMMONY.

Vern.—*Mahmúdah* (?), *sakmunia*, PB.; *Sugmonia, sák múnia*, HIND., SIND, ARAB., PERS.

References.—*Kurz, For. Fl. Burm., II., 212; DC. Origin Cult. Pharm. Ind., 153; O'Shaughnessy, Beng. Dispens., 500; Dymock, Mat. Med. W. Ind., 2nd Ed., 567; Fleming, Med. Pl. and Drugs, as in As. Res., Vol. XI., 189; Flück. & Hanb., Pharmacog., 438; U. S. Dispens., 15th Ed., 1286; Bent. & Trim., Med. Pl., 187; S. Arjun, Bomb. Drugs, 93; Murray, Pl. and Drugs, Sind, 164; Medical Topog. Ajmír, 151; Irvine, Mat. Med. Patna, 72.*

Habitat.—A climbing perennial, native of Syria, Asia Minor, and Greece. Cultivated in some parts of India.

Gum-resin.—A gum-resin imported into India. It is obtained by incision from the living root. It occurs in irregular pieces of an ash-grey colour and rough exterior. When broken, it presents a resinous surface, and of a shining black colour when dry. Thin pieces are translucent and GUM-RESIN. 1784

greenish. It has a cheesy odour and flavour. The bazar *Scammony* in Bombay, **Dr. Dymock** states, is all false, and is made at Surat.

[*DC.;* COMPOSITÆ.

Conyza alopecuroides, *Lam.;* see **Pterocaulon alopecuroideum,**

C. anthelmintica, *Linn.;* see **Vernonia anthelmintica,** *Willd.*

C. balsamifera, *Linn.;* see **Blumea balsamifera,** *DC.*

1785 **Cooawanoo Oil.**

This oil is said to be prepared from the Chelonian reptile **Caouna olivacea,** *Gray*—see **Turtles.**

Cookia punctata, *Hask.;* see **Micromelum pubescens,** *Blume,* Var. 1st; RUTACEÆ.

1786 **Copal Gum,** or **Gum Anime.**

A hard, transparent substance, resembling Amber, found as a natural exudation from certain trees. This substance is chiefly obtained from Zanzibar, the produce of **Trachylobium Hornemannianum,** a plant belonging to the LEGUMINOSÆ. It is yielded by the trees at the present day, but the commercial substance may be said to be in a half-petrified condition. This is known as Fossil Copal, and is regarded commercially as much superior to that obtained from living trees. It occurs in immense masses, found buried in the sand, far away from any living trees, and chiefly in the coast sands. There are other Copals sometimes met with. Brazilian Copal is obained from **Hymenæa Courbaril.** Madagascar Copal from **Trachylobium verrucosa.** West African Copal is furnished by **Guibourtia copalifera,** and Indian Copal from **Vateria indica,** which see. The Australian and New Zealand Copal is the produce of **Dammara australis** (CONIFERÆ). This forms large solid masses, often found in places where the trees do not now occur, and in New Zealand is known as *Kawri* and in European Commerce as DAMMAR or COWDIE PINE.

Copper, see **Cuprum.**

1787 **Coppice** or **Copse—**Plants suitable for—

The following, among many others, are plants specially mentioned as suitable for this purpose; but those given under **Hedges** and under **Pollard** may also be added :—

Acacia arabica.
Acer Campbellii.
Albizzia Lebbek.
Anogeissus pendula.
Bauhinia Vahlii.
Carissa diffusa.
Castanopsis indica.
C. tribuloides.
Casuarina equisetifolia.
Cedrela serrata.
C. Toona.
Celtis australis.
Dalbergia latifolia.
Helicteres Isora.
Heritiera littoralis.
Lagerstrœmia parviflora.
Lebdiereopsis orbicularis.
Mœsa montana.
Odina Wodier.
Pithecolobium dulce.
Populus euphratica.
Prosopis spicigera.
Quercus acuminata.
Q. semecarpifolia.
Streblus asper.
Teucrium macrostachyum.

Copra or **Khopra**—The dried kernels of the cocoa-nut, see **Cocos nucifera.**

COPTIS, *Salisb.; Gen. Pl., I., 8, 953.* 1788

The name COPTIS has been given in allusion to the much-cut leaves of the plants which have been referred to this genus.

Coptis Teeta, *Wall.; Fl. Br. Ind., I., 23;* RANUNCULACEÆ. 1789

COPTIS OR GOLD THREAD, COPTIDIS RADIX, OR MISHMI TITA.

Vern.—*Títá,* ASS.; *Mamírá,* or *Mámírán* (DYMOCK), HIND.; *Mahmira,* SIND; *Píta-karosana,* SING. **Rice** says that *títá* is a corruption of *tikta,* SANS., "bitter."

References.—*Voigt, Hort. Sub. Cal., 3; MacIsaac, Trans. Med. and Phys. Soc. Calcutta, III., 1827, III., 432; Teeta & Púcha Pat in Med. & Phys. Soc. Calcutta, 1836, VIII., 85; Ind. Ann. Med. Sci. 1856, III., 397; 1858, V., 621; Jour. Agri.-Hort. Soc. Ind., 1858, X., App. 6; Pharm. Journ., 1851, XI., 294; Simpson, Phar. Jour., 1854, XIII., 413; Pharm. Ind., 4, 435; Ainslie, Mat. Ind., II., 400; O'Shaughnessy, Beng. Dispens., 102; Moodeen Sheriff, Supp. Pharm. Ind., 114; Dymock, Mat. Med. W. Ind., 2nd Ed., 18; Flück. & Hanb., Pharmacog., 3; U. S. Dispens., 15th Ed., 1260; Bent. & Trim. Med. Pl., 3; S. Arjun, Bomb. Drugs, 3; Murray, Pl. and Drugs, Sind. 73; Davies, Report on the trade of the countries on the N.-W. boundary of India, 1862; Cooper, Mishmi Hills, p. 214; Perrins, Jour. Chem. Soc., XV., 339.*

Habitat.—A small, stemless nerb, with perennial root-stock, met with in the temperate regions of the Mishmí Hills, east of Assam. **Cooper** says that the plants grow on the ground among the moss around the stems of trees. "From each root," he remarks, "springs a single stem, about four inches high, bearing three serrated leaves, attached to the head of the stalk-like elongated trefoil."

HISTORY. 1790

Pereira (*Pharm. Jour., XI., 1852, p. 204*) was the first to suggest that *teeta* root might be the Μαμιράς or the Μαμηρά of the early European writers on medicine. He founded this opinion mainly on the fact that *mahmírá* is the name of a drug used in Sind in the treatment of eye diseases, a purpose identical with that for which the Μαμιράς was employed. **Ainslie** (*Mat. Ind.*) alludes to a drug imported into India from China under the name of *sou-line* or *chyn-len,* which, he says, possessed "stomachic virtues." Both the Sind *mahmírá* and the Chinese plant (*sou-line, chuen-lien, choulin, chouline* or *hwang-lien, &c.,*) have by modern writers been recognised as **Coptis**. **Dymock** says *mámirán* is noticed by the early Arabian writers as a kind of turmeric (*urúk*). "The plant is described by **Mir Mahammad Hussain** as having leaves like the ivy; it is said to grow near water in the hilly parts of India, China, and Khorassán. The Indian kind is described as yellow with a brown tinge; the Chinese as yellow; the Khorassán as greenish brown; and the seed is said to be like sesamum. The best kind is the Chinese, which should be small, yellow, hard, and knotty. It is said to keep good for 20 years. Whether the three kinds here described are all forms of **Coptis** it is impossible to decide. Indian writers say that *mámirán* used as a collyrium clears the sight, and as a snuff the brain, and that it relieves toothache. Internally it is given in jaundice, flatulence, and visceral obstructions" (*Mat. Med. West. Ind., 2nd Ed., 18*).

Dymock further remarks that two kinds of the drug are at the present day met with in Bombay. The best quality is only about the thickness of a crow-quill or a little thicker; it is a yellowish rhizome, hav- 1791

HISTORY.

ing spinous projections where the roots have been broken off. The whole rhizome is jointed, but the upper end is often more distinctly so, and the remains of the sheathing leaf-stalks are often attached. The second kind is considerably thicker and covered with thin wiry rootlets: it often branches at the crown into two or three heads, which terminate in tufts of lear-stalks crowded together, and not separate as in the first kind. Both of these rhizomes are contorted, and have a short fracture; the centre is spongy, and the surrounding portion bright yellow and woody:
0000 taste purely bitter. "The first kind corresponds with the description of **Coptis** root in the *Bengal Dispensatory.* The second kind with the description of that drug in the *Pharmacographia.*" While accepting this opinion it may be here stated that considerable confusion still exists in the European literature of the subject.

It is an interesting feature in the history of this drug that it continues to be imported from China, even although the Bengal supply reaches India through Assam. Indeed, it may be doubted how far the Chinese imports correspond to the roots of **Coptis Teeta.** It is customary to read that the Chinese *chuen-lien,* and probably also the *mu-lien,* are **Coptis Teeta**; but it seems open to grave doubt whether that plant is wild or even cultivated in China proper, although abundant information exists regarding its occurrence in a limited portion of the hills that separate Assam from the Chinese frontier. It may, therefore, be safely asserted that we do not know the plant which yields the Chinese drug. In Japan **Coptis anemonœfolia** affords a medicinal root, and it is, therefore, just possible that a portion of the Chinese drug may be obtained from one of the allied genera **Coptis Isopyrum** or **Helleborus,** although possibly an undescribed species. Mr. **Christy** (*New Com. Pl. and Drugs, No. 4, p. 53*) says:—"The Japanese character ('*oh-ren,*' meaning yellow *ren*), is exactly the same as the Chinese one for '*hwang-lien,*' which is the rhizome of **Coptis Teeta,** *Wall.*, and not a **Justicia** as stated by **Dr. Smith** in his Chinese Materia Medica." May it not be possible that the **Coptis Teeta** to which **Christy** alludes is the drug as described in the *Pharmacographia?* **Dr. Dymock's** account of the imported Chinese thicker form of the *mamirá* of Bombay recalls, however, some of the forms of a drug sold in Bengal under the name of *Katki* or *kurú* (*Katúka,* SANS.)—a drug now generally recognised as obtained from **Picrorhiza Kurroa. Dr.**
1792 **Dymock** thinks there is but one root sold in India under the name of *kurú*, but in connection with the Calcutta International, and again with the Colonial and Indian Exhibitions, London, the writer had three or four widely different roots consigned to him under the name of *kurú.* He is, therefore, of opinion that at least two forms of the drug must be regularly sold in the Bengal drug shops. **Gentiana Kurroa,** chips of the root of **Coscinium fenestratum, Swertia Chirata,** and other substances are frequently offered as *kurú.* May it not be possible that one of the roots known in lower India as *kurú* is in the upper and western provinces sold as *títá.* This suggestion carries with it additional strength from the well-known fact that a considerable trade is done from Kumáon and also
1793 from the Khásia Hills in the root of **Thalictrum foliolosum**—*pílíjarí*—as a substitute for **Coptis**; and along with this it seems likely that **Actœa spicata** may also be used as a substitute. Both these are abundant plants, and have dark yellowish, bitter roots. It is sufficient for the present to know that substitutes are regularly sold for *mishmí títá.* But it may be doubted if the true *títá* comes from China at all, and, indeed, as already stated, if even the plant exists in any part of the Chinese empire. The true *títá* sold in Upper and Western India may thus be *mishmí-títá* that may have found its way by re-exportation into the returns of the Chinese

HISTORY.

drugs imported into India, or may have been conveyed overland from the Indo-Chinese frontier to Chinese ports. Hence, as far as our present information admits of conclusions being drawn, there exists a strong probability that the bulk of the Chinese drug is not **Coptis Teeta** at all, but the root of some more easily procurable plant.

Sir J. D. Hooker, while in Upper Sikkim, received from salt traders, met with near the frontier, a present "of a handful of the root of one of the many bitter herbs called in Bengal *teeta*;" but he adds "the present was
that of **Picrorhiza**, a plant allied to the Speedwell, which grows at from 12,000 1794
to 15,000 feet elevation, and is a powerful bitter called *hoonling* by the Tibetans." The suggestion above that much of the *títá* sold in India might be **Picrorhiza** was made before the writer thought of consulting **Sir Joseph Hooker's** *Himálayan Journals*; and it is, therefore, almost safe to add that the Tibetan name *hoonling* may have been the original of the Chinese *honglane, hwang-lien, sou-line*, &c., and hence **Dr. Pereira** may have been mistaken in referring the Μαμιράς of the ancients to **Coptis Teeta**, since it is this imported Chinese drug that is the *mamírán* of Upper India. Further, it seems even probable that the knotty, yellow, often ramified rhizomes of **Picrorhiza**—according to modern writers the spurious *mamírán* of the Indian bazars—may have been the drug originally so called, or at least been the Indian drug which most closely resembled the Μαμιράς, and from that circumstance received from the Muhammadans the name *mamírán* which, in Upper India, it bears to this day. This view receives support from the fact that **Picrorhiza** occurs throughout the Alpine Himálaya, while **Coptis Teeta** is confined to a small area inhabited by one of the wildest of hill tribes. But there is nothing in all this to justify the inference that, in ancient times, there may have existed a much larger export in *títá* than takes place at the present day. It is much more likely that a drug found throughout the Himálaya would have been in early times carried to the drug-shops of Central, Northern, and Southern Asia rather than that the root of a plant found only within a very limited area of an inaccessible country should have come to be in extensive demand. It is possible, however, that in later times the Chinese supply may have been drawn from the Assam frontier, and ultimately consisted, to some extent, in the admittedly superior root of **Coptis Teeta**, until modern writers came to view the *mamírán* as **Coptis** and not **Picrorhiza**. **Dr. Aitchison**, in his
second paper on the Flora of the Kuram Valley, says: "I, this year, col- 1795
lected **Corydalis ramosa**, a plant employed medicinally by the natives in
the treatment of eye-diseases, simply, I believe, because it has a yellow 1796
watery juice, as every plant with a yellow juice seems to be by them considered a sovereign medicine, and all are called indiscriminately *mamírán*." He further states that the roots of **Geranium Wallichianum** were shown to him as a medicine called "*mam-i-ran*."

It has been pointed out by chemists that both **Coptis** and **Berberis** contain a large quantity of the alkaloid *berberine*; and the somewhat significant fact has to be added that the drugs obtained from these plants are used in catarrhal and rheumatic affections of the conjunctiva very much after the same fashion as the Μαμιράς of the ancients. But *berberine* is present in a great many other yellow and bitter substances, and it
may therefore have been a mere coincidence (suggested by external ap- 1797
pearances) that the root now called *mamírán* and the Μαμιράς came to be used for the same purpose. Indeed, **Picrorhiza**, on being chemically examined, may also be found to possess that alkaloid, since *berberine* is one of the most frequently met with of all the alkaloids present in vegetable substances. But even should it not possess *berberine*, that could scarcely

COPTIS Teeta.

Coptis or Mishmi Teeta.

HISTORY.

be viewed as militating against its having been adopted as a substitute for a drug for which **Coptis** would have proved more suitable. At the same time the Indian use of *mamírán* in the treatment of eye affections is but a recent application of the drug. The Hindu authors on Materia Medica who wrote in Sanskrit were alike ignorant of *títá* and *mamírán*. The word *mámírán* is of Muhammadan introduction into India, while the drug **Picrorhiza** was known to the earliest Sanskrit writers. The late **Dr. U. C. Dutt** gives a detailed account of it as used by the Sanskrit physicians, but while he, personally, must have been quite familiar with the drug now sold as *títá*, he makes no mention of it in his work on Sanskrit Materia Medica. Not only, therefore, were the words *títá* and *mamírán* unknown to the Sanskrit writers, but it seems conclusively established that even the drug **Coptis Teeta** is but of modern introduction into India. The Muhammadans were so little familiar with **Picrorhiza** that they frequently confused it with **Hellebore**, and may thus be readily believed to have given to **Picrorhiza** or to **Coptis**, when separately presented to them, the name of *mamírán*—the name of a drug which either or both may possibly have closely resembled. The Hindus are uniformly precise and accurate in their information regarding **Picrorhiza**, but say nothing of **Coptis**. The earliest writers on Indian Materia Medica who allude to **Coptis** attribute to the indigenous and imported Chinese drugs tonic properties of remedial value in the treatment of nervous diseases and in debility after fever; they rarely make any mention of its use as a collyrium in eye affections. The tonic properties of **Coptis** are possessed in a scarcely less degree by **Picrorhiza**; and it may be concluded that **Mir Muhammad Hussain's** description of the leaf of one of the forms of *mamírán* agrees admirably with **Picrorhiza**, but is altogether inapplicable to **Coptis**, and **Dr. Dymock's** description of the branching rhizome of a form of *mamírán* is explainable only on the supposition that it may be **Picrorhiza**, but quite inexplicable if viewed as **Coptis**. Besides which, the frequency in our bazars of ancient Greek names given by Muhammadan merchants to Indian drugs, suggests a caution against the founding of too strong opinions on the bare existence of a name clearly not of Indian origin which may now be applied to a very different plant from that to which it was at first given.

Collection. 1798

COLLECTION OF, AND TRADE IN, COPTIS TEETA.—**Cooper** says:—"As we neared the highest elevation, scattered trees and shrubs seemed to grow from a thick bed of dry moss, and here, for the first time, I saw the *títá* plant growing abundantly. The roots (from which, when brewed and steeped in hot water, the famous febrifuge is made) are embedded in moss. From each root springs a single stem, about four inches high, bearing three serrated leaves, attached to the head of the stalk-like elongated trefoil. The Mishmees gather the roots towards the end of the rainy season, and carry them packed in tiny wicker-work bamboo baskets to Sadiya, where they are eagerly bought by Assamese and Bengali merchants." Regarding the extent of the trade in *Mishmí títá*, the Secretary to the Assam Government reported in 1883: "I have the honour to say that the Deputy Commissioner of Kámrúp has been requested to forward to you a parcel containing ¼ seer of *Mishmí títá*." "The Deputy Commissioner of Lakhimpur informs me that the annual sale of *Mishmí títá* at Sadiya is estimated at a maund or a maund and a half. It is brought down in small open bamboo baskets, weighing about ½ a chittack each, and is sold at one pice a basket." This is, of course, the price at which the people sell it to the dealers (*e.g.*, about ½ penny an ounce), but the smallness of this trade and absurdly low price paid to the producers is out of all proportion to the extent of the trade in India and the retail price which the drug fetches. **Dr. Dymock** says of the Bombay supply: "Both kinds of the drug come from China *vid* Singapore, in bulk. The first is worth R$3\frac{1}{4}$ per lb.; the second R2." **O'Shaughnessy** says: "**Coptis Teeta** has found its way through the drug-shops of Bengal, and is even occasionally exposed for sale in the Upper Provinces."

MEDICINE.

MEDICINE. 1799

Therapeutic and Chemical Properties.—Coptis trifolia, a creeping plant, found in the northern temperate and arctic regions of America and Europe, was formerly officinal in the United States. In the foregoing remarks almost enough has been said regarding the properties of **Coptis Teeta**—a plant still officinal in India. The presence of the alkaloid *berberine* renders its use as a collyrium desirable in certain eye affections, but its pure bitter tonic properties are those which command for it the high reputation it enjoys. When chewed it tinges the saliva yellow, but it has no odour, nor is it astringent. While not a febrifuge, it is a valuable remedy in the debility that follows fever, in atonic dyspepsia, and even in mild forms of intermittent fevers. **Sir W. O'Shaughnessy** says: "It brings a high price, and is a remedy of the greatest value. It has been used in the General Hospital by the late **Mr. Twining**, who reported that its influence in restoring appetite, and increasing the digestive powers, was very remarkable, and that it might be said to possess all the properties of our best bitter tonics." "Further trials in the College Hospital were equally satisfactory. It did not seem to exercise any febrifuge virtue, but under its influence several patients, recovering from acute diseases, manifestly, and very rapidly, improved in strength. The dose was 5 to 10 grs. of the powder, or an ounce of the infusion thrice daily." **Dr. K. L. De, C.I.E.**, says: "In this indigenous article, though a costly one, we have an adequate substitute for Columba root, which it resembles not only in its medical effects, but also in its physical properties. An essence of this drug has been recently brought forward for use by **Messrs. Bathgate and Co.**, of Calcutta." 1800

Coptis affords no evidence of containing gum, or gallic, or tannic acids but it is a simple pure bitter, which bears a close resemblance to quassia in its mode of action, and is applicable to all cases in which that drug may be indicated. **E. Z. Cross** found **C. trifolia** to contain *resin, albumen, fixed oil, colouring matter, lignin, extractive*, sugar, *berberine*, and another alkaloid which seems to resemble *hydrastia* very closely, which **Cross** named *Coptina*. "The colouring matter in which the rhizome of **Coptis** abounds is quickly dissolved by water. If the yellow solution obtained by macerating it in water is duly concentrated, nitric acid will produce with it an abundant heavy precipitate of minute yellow crystals, which, if redissolved in a little boiling water, will separate again in stellate groups. Solution of iodine also precipitates a cold infusion of the root. These reactions, as well as the bitterness of the drug, are due to a large proportion of *berberine*, as proved by **I. D. Perrins**. The rhizome yielded not less than $8\frac{1}{2}$ per cent., which is more than has been met with in any other of the numerous plants containing that alkaloid" (*Pharmacographia*, 5). 1801

§ **Professor Warden** has furnished the writer with the following brief note regarding some of the properties of this plant: "The root contains a bitter, yellow alkaloid—berberin—to the extent of $8\frac{1}{2}$ per cent.

The following plants among others also contain that alkaloid— 1802

The Barberry.	**Xanthoxylon fraxoneum.**
Columba root.	**Menispermum canadense.**
Hydrastis canadensis.	**Coscinium fenestratum.**
Xanthorrhiza apiifolia.	**Coptis trifolia.**

MEDICINE. "**Thalictrum foliolosum**, *DC.*, common at Mussooree and throughout the temperate Himálaya at 5,000 to 8,000 feet, as well as on the Khásia hills, also affords a yellow root, which is exported from Kumáon under the name *Momiri*, and which it is possible may have been mistaken for **Coptis Teeta**." "In Kashmir the roots of a **Swertia** are collected and tied up in bundles and are passed off as a substitute for **Coptis**. They resemble the true root greatly." (*Surgeon-Major J. E. T. Aitchison, Simla.*) See a previous paragraph, where a Coryadalis and a Geranium are stated to have both been found to be used, in Afghánistan, as a drug called *mamírán*.

CULTIVATION. 1803 Cultivation of Títá.—In concluding this brief account of *títá* it may be remarked that little or no difficulty would be experienced in cultivating the plant in many parts of India, but that up to the present date no attempt appears to have been made to do so, although the retail price paid for the drug would apparently justify the suggestion that it would be found a remunerative crop.

1804

CORAL.

A calcareous structure formed by certain minute animals, which belong to the section or class of the Cœlenterata known as the *Actinozoa*. The type of this class may be taken to be the sea-anemone, an animal that consists of an erect tubular body with a sucker-like base and a flattened apex, the latter surrounded by a whorl of simple tubular tentacles. In the middle of the apex occurs the mouth, which opens downwards into a stomach, and this below into a body chamber. The external body consists of two walls (the *Ectoderm* and the *Endoderm*), which are thus separated from the stomach by the body cavity which surrounds the stomach, and is divided into a number of compartments or cells by a series of vertical or horizontal partitions. Upon the faces of these partitions (and thus in the interior of the animal, instead of on the exterior, as in the Hydrozoa) are borne the sexual organs in the form of bud-like ovaria. But reproduction is frequently effected by means of a process of budding which
1805 gives origin to compound organisms. The coral-forming Actinozoa are in the majority of instances not separate polypes, like the sea-anemones, but are colonies produced by budding that conjointly form for each species, a fixed and definite compound organism, in which, however, the structural peculiarities are preserved, each individual being a sea-anemone-like animal bound inseparably to a number of other individuals of the same species. The Actinozoa have no nervous system (except in the Ctenophora), nor are there any traces of a vascular system. The body-walls are, in the sea-anemone, fleshy, but in other Actinozoa, calcareous deposits take place in either the ectoderm or endoderm, giving origin to a solid skeleton which might be described as external (somewhat resembling the shell of a Crustacean) or internal, according to whichever wall becomes calcareous. It is this calcareous skeleton that bears the name of Coral, and corals may be of three kinds: the skeleton of an independent polype or a compound coral formed by external or by internal skeletons. To the two last mentioned belong all the corals known to commerce. Coral reefs are the result of budding on the fleshy surface of an internal skeleton-forming polype, the multitude of skeletons being consolidated into rock by a calcareous basis secreted for the purpose.

The Actinozoa are referred to four orders, *viz.*—The Zoantharia, the Alcyonaria, the Rugosa, and the Ctenophora. The Rugosa are entirely
1806 fossil corals, which occur only in the palæozoic or oldest rocks, while the Ctenophora (or free-swimming marine polypes) do not form a calcareous skeleton. Of the Zoantharia two tribes, the Zoantharia sclerodermata and the Z. sclerobasica, form coral; the others like the sea-anemone, are, in the majority of instances, built up entirely of fleshy materials. In dealing with Indian coral we are concerned mainly with the skeleton forming Zoantharia and with the Alcyonaria. But before proceeding to discuss these groups of coral-forming polypes, it becomes necessary, in order to understand their differences, to explain a little more fully the formation of a calcareous skeleton. The internal, or as it is technically called the *Sclerodermic*, skeleton is the result of the secretion of carbonate of lime in the inner body wall (the endoderm). It is therefore truly within the body of the animal, and in the case of a simple polype

may be described as covered externally by the outer fleshy wall and terminated
by a cup-shaped recess within which is located the apex of the animal, with its
tentacles, mouth, and stomach, &c. Budding from the surface of such an animal
would give origin to ramifications of the fleshy wall, the ectoderm of the simple
polype thus becoming the *cœnosarc* of the compound, so that the offspring would
be so far the continuation of the parent; but within this common fleshy wall
would be reproduced all the structural peculiarities of the individual, the inner
wall and partition of each new polype becoming calcareous. Such an accumulation, consolidated by a further secretion of lime in the common fleshy wall or 1806
cœnosarc, would form a coral mass or rock like that met with in coral reefs. In
the case of external skeletons the animal might be described as inverted. The
calcareous deposits take place within the outer wall or cœnosarc itself, and the
skeleton is thus truly outside the vital organism, and is common to all the individuals of the colony. In this case the coral is said to be *sclerobasic*. The
skeleton forms, in other words, an axis or basis over which the actinosoma is
spread. No further consolidation is required; the new individual is formed on
the parent, and the common cœnosarc is converted into the common skeleton
or sclerobasic coral. Such a coral can therefore alone be produced in a compound organism. In the sclerodermic coral each polype has a complete skeleton of its own, and may hence exist independently or be combined into a colony
by the subsequent formation of calcareous matter in the common fleshy wall—
the cœnosarc. Coral reefs are sclerodermic, and the ornamental corals, of
which the red coral may be accepted as a type, are sclerobasic.

To return now to the classification: a small tribe of the Zoantharia are
sclerobasic, but by far the most important tribe, the reef-forming corals, are
sclerodermic, and the Alcyonaria, or ornamental corals, are sclerobasic. Both
these groups, which we may accept as represented popularly by the terms 1807
"Reef-forming" and "Ornamental" Corals, are met with in the Indian Ocean.
The former fringe the coast of India or form atolls or reef-encircled islands such
as those of the Maldives, Laccadives, and Nicobars. The latter exist in deeper
water, more especially in the Persian Gulf and Red Sea, or are distributed into
the seas of temperate regions far beyond the limits of the coral-reef-forming
area. At the same time on the mainland of India extensive beds of ancient
coral reefs exist, such as those along the Malabar coast, Tinnevelly, &c., and
fossil coral of still more ancient character occurs in various parts of the country,
as, for example, in Sind. The ancient coral beds are valued as a source of
building stone, or as affording lime, but they might be extensively employed
as a source of manure.

Coral. 1808

CORAIL, *Fr.*; KORALLEN, *Germ.*; KORAALEN, *Dutch*; CORALLO, *It.*; CORAL, *Port. & Sp.*; KORALLU, *Rus.*; CORALLIUM, *Lat.*; Κοραλλιον, GREEK.

Vern.—*Murján, múnga*, HIND.; *Bekh-i-marján* (fragments of red-coral used medicinally), *sang-i-marján*, PB.; *Gúlli*, DEC.; *Pávalam, nuraikal* (foam-stone), TAM.; *Págádam*, TEL.; *Vidruma, prabála, pravála*, SANS.; *Béséd*, ARAB.; *Murján* or *merján*, PERS.; *Búbálo*, SING.; *Ky-a-vekhet*, BURM.; *Poálam, karang*, MALAY.

References.—*Man. Geology of India, I., 376, II., 735, III., 470, and IV., 150; Mem. Geol. Surv. Ind., XVII., 119, 125, and 181, XX, 70-73, XXI, 51, 59; C. R. Markham, 'The Tinnevelly Pearl Fishery' in Jour. Soc. Arts, XV., 747 (1867); Caldwell's Tinnevelly, 75; Forbes Watson, Indust. Surv. Ind., I., 400, 404; Dana, Corals and Coral Islands; Darwin, Voyage of a Naturalist; Darwin, Structure of Coral Reefs; Mr. J. Murray, The Structure, Origin, &c., Coral Reefs, &c., Royal Institution; Various Papers and Letters in "Nature," XXXVII., No. 956 to XXXVIII., No. 968 (1888); Maury, Physical Geog. of the Sea; Intern. Fisheries Exhib. Report, IV., 422, V., Pt. II., 177, 197, 360-361, 391, VI., 359, XIII., 39-41; Gosse, Marine Natural History; P. L. Simmonds, Com. Prod. of the Sea, 436; M. Lacaze-Duthiers, Natural History of Coral (Paris, 1864); U. C. Dutt, Mat. Med. Hind., 93; Ainslie, Mat. Ind., I., 90; Mason, Burma and its People, 394-396, 573; Bomb. Gaz., VIII. (Káthiáwár), 93; Baden Powell, Pb. Prod., 99, 154; J. E. O'Conor, Review of Maritime Trade of British India, 1881-82, p. 33; 1882-83, p. 44; 1884-85, p. 36.*

CORAL.

Coral.

Habitat.—The Coral zone extends on either side of the Equator for about 1,800 miles. **Mr. J. Murray**, of the *Challenger* Expedition, has pointed out, however, that within this area the corals abound most on the western side of the Atlantic and Pacific Oceans, a circumstance accountable for by the trade winds carrying westward the surface film of warm water and thus leaving on the eastern shores too cold a medium for the growth of the reef-forming polypes. The degree of warmth in which the coral luxuriates requires to have a surface-water temperature of 70°Fh., and to never vary from this more than a limit of 12°Fh. There are a few outlying reefs beyond the area indicated, such as the Bermuda reef, which finds a congenial medium in the Gulf Stream. Even under the Equator deep-water has a temperature much below that in which the reef-forming corals can live, and this fact may therefore be one of the governing influences that confines the corals not only within certain geographical regions but fixes each species within its area to a certain depth of water in which alone it is found to grow. Beyond the area of the reef-forming corals, the ornamental corals occur, and luxuriating, under lower temperatures, they are found in tropical seas at much greater depths than the reef-forming. The latter class of corals grow between 5 and 30 fathoms of water. They are killed by exposure to the sun, and must therefore be below low-water level. On a land subsiding they will accordingly build vertically so as to preserve their favourite depth, and on a land ascending they will extend horizontally, advancing into the requisite depth of water as the older landward and exposed portions are killed by being carried above the level of the water. This was the theory established by **Darwin**, and universally accepted for a quarter of a century, the atolls being viewed as monuments erected by the Actinozoa to a vast Pacific continent which had gradually sunk beneath the ocean. While this *may* take place, a new school has advanced the theory that it is by no means essentially necessary that to construct an atoll, the island which it encircles need be subsiding. Growth is attributed to the food materials being most abundant along the face of the reef, the approaching water being richer than that within the lagoon. It is even further explained that the chemical action of the sea-water decomposes the dead coral, thus excavating the shallow basin (or lagoon) that exists between the growing face of the reef and the land. But if this theory be admitted we have to explain the fact that once upon a time a coral laying the foundation of the present face of the reef must have existed in a depth of water under which we have no evidence of its having the power to live, or then presume the growing rim of the reef to be advancing cup-like from a peduncle situated at a depth in which the first portion of the colony found it possible to live. At the same time, however, facts exist in India that may at least be viewed as calling for careful investigation if the subsidence theory be still maintained. The whole western side of India is unquestionably rising, and reefs of various ages are now considerably above the level of the sea, whereas a few miles seaward from these dead reefs, atolls are being formed around the islands of the Indian Ocean.

REEFS. 1809

A.—Coral Reefs.

In the Manual of Geology of India, it is stated that the coral reefs of the Andaman Islands should become a source of cheap lime for Calcutta. "The idea has been suggested more than once during the past twenty years, and it is supposed that the only objection to it arises from the necessity for the presence of coasting vessels which would be involved, and the consequent risk of the convicts escaping; but with so pure a source of

lime, abundant fuel, and labour at command, there can be little doubt **CORAL REEFS.**
that Calcutta might be supplied with excellent lime at a comparatively small cost, and a useful and profitable occupation would be thus afforded for the convicts."

"In 1882 some experiments were made by the Public Works Depart- **Andamans.**
ment with lime, at Barrackpore, from coral brought up as ballast from the 1810
Andamans. The cost of the lime when burnt, exclusive of freight and collection, was from R35 to R45 per 100 maunds, as against the market price of Sylhet lime from R85 to R90 per 100 maunds."

"Opinions differ slightly as to the relative merits of the two limes, but on the whole the coral lime was considered equal to the other; whether it would answer best to burn the lime in the Andamans and bring it up slaked like the Sylhet lime, or to burn it where fuel is more expensive, can only be determined by actual trial."

In the Nicobar Islands upraised coral reefs are found on the coast of
all the islands and on the Car Nicobar, Bompoka, and several other islands **Nicobar.**
these coral banks are of great thickness, and are raised 30 or 40 feet 1811
above the sea. The atolls around the Maldive and Laccadive, as well as the reefs on the coast of the Nicobar Islands, might all afford a supply of
lime. But in Sind and in other parts of the western coast of the mainland **Sind.**
of India ancient and even fossil beds of coral occur. Mr. W. T. Blanford 1812
writes:—"Some miles south of Lohari Lang, and near Murad Khan's *'band'* (dam) across the Habb river, a thin bed composed of corals appears a few feet above the base of the Gáj group. This bed can be traced for many miles to the south. All the species of coral (five or six) are encrusting forms or small branching kinds. A. **Pachyseris**, or some closely allied form, and two or three species of **Hydnophora**, are specially common." So again near Nari he writes of coral beds: "The marly shales pass up into light yellow and brown limestone, with a coral zone abounding in
several species of coral." Mr. Fedden, in his account of the Geology of **Bombay.**
Káthiáwár, remarks: "Many masses of weathered-out coral lie scattered 1813
about the surface of the rocky ground between the villages of Nandána and Rán, indicating a coral zone not far above the local base of the group." The species of coral found in Káthiáwár have been worked out by Professor P. Martin Duncan and W. Percy Sladen (see *Palæontologia
Indica, Series XIV., Vol. I., part 4, pp. 80—91*). But Mr. Fedden conti- **Cutch.**
nues: "The whole of the sea-board facing the Gulf of Cutch from Na- 1814
wánagar westward, including the islands off the coast, is fringed with dead coral reefs, the surfaces of which are much exposed at low spring tides. In some cases the coral floor extends inland up to high tide level, as at Saláya, the *bandar* for the town of Khambhála. The coral has very fine and uniform texture, and has been worked as a substitute for stone for building, but not with very satisfactory results, owing to salt impregnation. The existence of these dead coral reefs is, of course, a proof that the country has been rising during late times." Far to the south
Mr. Foote, in his account of the Geology of Madura and Tinnevelly, states **Madura.**
that he found extensive upraised coral reefs, and upon these he lays 1815
stress as proving the rise of that portion of India. Writing of the scarp
of coral near the zemindar's bungalow on Rameswaram island, he says: **Tinnevelly.**
"Of its true coral reef origin there can be no possible doubt, as in many 1816
places the main mass of rock consists of great globular meandrinoid corals or of huge cups of a species of PORITES which, beyond being bleached by weather action, are very slightly altered, and still remain in the position in which they originally grew. The base of the reef is not exposed as far as I could ascertain, not having been sufficiently upraised along the beach, but in a well-section, a little to the south of the Gandhamána Parvattam

CORAL. Coral Reefs.

CORAL REEFS.

Chattiram, the thickness of the coral reef exposed above the surface of the water is at least 10 feet, and probably much more." Further on he remarks: "At the Pamban end of the raised reef it shows a slight northerly dip, and masses of dead coral, apparently *in situ,* protrude through the sand below high water mark. Reefs of living coral fringe the present coast, but these I was unable to examine, so cannot say whether the corals now growing there are specifically allied to those which formed the reef now upraised, but all the mollusca and crustacea I found, occurring as fossils in the latter, belong to species now living in the surrounding sea." "All the small islands occurring along the Tinnevelly and Madura coast appear to consist of sand based upon coral reefs which are largely exposed at low tide. The published large scale charts of Pamban Straits show extensive coral reefs surrounding the five most easterly islands; Moossel, Munnauli, Pullee, Pulleevansel, and Cooresuddy. The only one I was able to visit, that on which stands the Tutikorin lighthouse, shows no coral on the surface, which is sandy; but the island immediately to the north supplies large quantities of dead coral, which are used in the town as a rough building stone. Similarly, large quantities of dead coral are brought over to the mainland from several of the central group of *tivus* (Thevoo) or islands along the Madura coast." Concluding his long and interesting account of these sub-recent marine beds, **Mr. Foote** adds: "It is impossible to resist the speculation that it was this upheaval which gave rise to the formation of what is known to the Hindus as Rama's bridge, and to Mussulmans and Christians as Adam's bridge, the long narrow isthmus which once united Ceylon to India."

Mr. H. F. Blanford, in an address read before the Simla Natural History Society, thus describes one of the ancient coral reefs of South India:—

Trichinopoly. 1817

"Years ago, when camping on the undulating plain that rolls away to the north of Trichinopoly and the Coleroon, as, standing on an old coral reef of the cretaceous sea, I looked westwards on the swelling outlines of the Pachamullay and Kolamullay hills, I was conscious that there, before me, rose the land of that remote age; worn and wasted, it may be, in the sequence of the myriad centuries that have since rolled over it, but in its essential features unchanged.

"And, indeed, with no great effort of the imagination, I might have fancied myself transported back to that distant time. Not a few of those details that give reality to mental imagery were visible at hand. The coral reef on which I stood might have been one of the water-wasted reefs described by **Darwin, Jukes,** or **Dana** in the Pacific Isles; so fresh and unaltered was the creamy whiteness of the fractured rock, with half-embedded corals jutting here and there from its weathered surface. The shells, newly disembedded from a decaying limestone, were as fresh and polished as those one might gather on the sandy shore of the neighbouring coast, with even their zig-zag colour-markings still preserved; and hard by stretched a bank of coarse shingle that differed in no essential respect from a modern beach.

"But though, to an uncritical eye, the shells of that old sea might seem very like the volutes, olives, cowries, and ark-shells now thrown upon the Madras sands (and perhaps, indeed, they were their remote ancestors), it needed but to look on the great coiled ammonites scattered here and there in the broken ground, to know of a surety that around me lay the relics of a cretaceous sea. When these enclosed and protected living organisms, the coarse sandy deposits that underlie our English chalk, were slowly accumulating in a shallow ocean, where now spread the cornfields and hop gardens of Surrey and Kent, and on the sinking sea bottom of South-Eastern England, Northern France, and Belgium, the thousands of feet of white calcareous mud that, long since upheaved and hardened into chalk, greets the homeward-bound Indian in the Dover Cliffs, had yet to be slowly extracted through long ages from the sea water by minute organisms long since extinct."

B.—Ornamental Corals.

ORNAMENTAL CORALS. 1818

Very little can be learned for certain of the indigenous living ornamental corals. Indeed, it seems probable that in some of the passages already quoted, reference has been made to coral in a generic sense, some of the forms there mentioned being, strictly speaking, members of the series we have designated ornamental in contra-distinction to reef-forming. It appears desirable in every effort to develop a trade in the Indian indigenous corals that the distinction here made should be preserved, since, for ornamental purposes, it is only the sclerobasic polypes that form a calcareous substance of sufficient consistence to admit of being cut **White.** into ornamental structures. 1819 In the dealer's shops of Europe corals of various forms, shapes, and colours are met with, such as "white coral" **Brainstone.** (**Oculina virginea**), "brainstone coral" (**Meandrina cerberiformis**), the 1820 "organ-pipe coral" (**Tubipora musica**), the "sea-pens" (**Pennatula**), **Organ-pipe.** the "sea shrubs" (**Gorgonida**), the "black coral" (**G. Antipathes**), and 1821 last but by far the most valuable of all the "Red Coral" (**Corallium rubrum**). **Sea-pens.** Most of these genera are temperate, but the **Gorgonida** attain their greatest development in tropical seas. 1822 "White coral," of no market value, is **Shrubs.** common both on the Navánagar coast and along the west and south-west 1823 as far as Kodinár. **Black.** Red coral (*sic*) is sometimes found in small quantities at Mangrol" (*Káthiáwár Gazette, page 93*). 1824 Mason, of Indian writers, **Red.** gives the fullest account of the species of coral met with in the portion of the 1825 country of which his work treats, namely, the coast of Burma. **Burman.** He says of the coast of Amherst and Mergui that elegant specimens of Actinia 1826 are very rare, but he describes a species of Meandria which he calls "club-shaped Porites." He also says:—"I have noticed in the bazars, though I have never gathered it on the coast, a curious species of coral resembling the horse-tail Isis. It is branched like a tree with white striated stony joints and black horny smaller joints between, which render the whole flexible." It may be here remarked that many of the sclerobasic corals have alternating portions of a calcareous and horny axis. But **Mason** describes "a scarlet coral, composed of cylindric tubes united together." A "star coral" is very abundant on the coast in several distinct species, some studded with large embossed stars, others sculptured with regular indented stars, and still others with minute meshes giving it the appearance of lace. "Tree corals" are plentiful "on the **Tenasserim.** Tavoy Coast, and some of the specimens very beautiful, presenting superb 1827 sea-groves of various hue and form." "A handsome coral, like a tuft of long moss, also occurs, and 'black coral,' of which beads are made, is brought from the Mergui Archipelago." Of Tenasserim **Mason** further says:—"A tree coral two feet long, of a deep scarlet, is found on the coast, which the residents often call 'red coral,' but it is not the red coral of commerce; it does not grow like that, and the red colour is confined to the epidermis, the substance of the coral within being grey."

In concluding this brief review of the literature of the Indian ornamental corals, it must be admitted that we are grossly ignorant of the 1828 subject. There are no coral fisheries in India, and we do not know whether or not this is due to the absence of corals of commercial value, nor do we possess any knowledge as to the likelihood of the more valuable corals succeeding, if introduced into Indian waters. No effort has as yet been made to propagate new species or improve the existing Indian corals.

TRADE. 1829

Trade in Coral.

Some conception may be arrived at of the magnitude of the trade in Coral when it is recollected how many races of people in India regularly wear necklaces of coral. How far the prized ornaments may be derived from Indian seas, it is impossible to tell. The finest red coral is obtained from the Mediterranean; the larger pieces of a pale colour, are said to be often worth twenty times their weight in gold. The operation of preparing these as jewels consists of three stages—cutting, piercing, and rounding; but in accomplishing these operations there is generally an immense waste. The rejected pieces and inferior qualities are exported to Asiatic countries, and India receives a very large proportion of this—the inferior red coral.

In 1868-69, India received £93,126 worth of red coral, and for some years the imports steadily declined, being in 1872-73 only £40,013, but this may have been partly caused by the imposition of a duty of 7½ per cent. in April 1870. Ten years later the imports had increased to £195,936 and the year following £231,166, last year (1886-87) they were £175,018. Of

Prepared. 1830

these imports India last year re-exported £7,468 worth, and it is somewhat remarkable that of these re-exports Italy received back again £2,000 worth, and £3,820 worth were sent to the Straits Settlements. Of the re-exports Bengal issued £6,152 worth.

Italy supplied last year £161,741 worth of the total amount received by India, and of these imports, Bengal took £154,848.

Beads. 1831

In the Review of Indian Maritime Trade **Mr. J. E. O'Conor** says of 1882-83—"There has been an enormous increase in this trade, the imports having almost doubled in five years: thus, 79,643℔ in 1878-79,

Imitation. 1832

152,372℔ last year. The condition of this trade is a very accurate gauge of the condition of the agricultural classes in the North-Western Provinces, Rájputana, and Sub-Himálayan tracts. The bulk of the imports is bought by these classes to be worn as necklaces, the coral beads, when a man is prosperous, alternating with gold beads. Almost all the coral we receive is brought to Calcutta, whence it is distributed over the provinces mentioned, to be sold chiefly at the larger fairs. It is principally imported from Naples, the fishers being themselves in many, if not in most, cases the exporters, and with the caution of small traders placing a fixed price upon the goods below which the importers have no authority to sell." There is a very considerable trans-frontier trade, the Indian imported coral finding its way to Thibet. How far the re-exports consist of Indian prepared corals it is impossible to say, but there is doubtless a large community employed in preparing the special ornaments worn by the Hindus. A favourite necklace consists of gold and coral beads alternating, but in the Khásia Hills imitations of large beads of this kind form the favourite ornament. There seems to exist an extensive trade in imitation corals, since only an inconsiderable proportion of those offered for sale at fairs are real.

MEDICINE. 1833

Medicine.—In addition to being used for adornment ornamental corals have been used in Hindu medicine from a very ancient time and are mentioned by **Susruta**. "Coral," **Dr. Dutt** says, "is purified by being boiled in a decoction of the three myrobalans." Both pearls and corals are used for the same purpose, namely, in "urinary diseases, consumption, &c., and to increase the nutrition and energy of weak persons." **Ainslie** remarks that "the Tamil practitioners prescribe the red coral when calcined in cases of diabetes and bleeding piles."

CORALLOCARPUS, *Welw.; Gen. Pl., I., 831.*

Corallocarpus epigœa, *Hook. f.; Fl. Br. Ind., II., 628; Wight,* 1834
[*Ic., t. 503;* CUCURBITACEÆ.

Syn.—BRYONIA EPIGŒA, *Rottler.;* B. GLABRA, *Roxb.;* ARCHMANDRA EPIGŒA, *Arn. in Hook., Jour. Bot., III., 274.*

Vern.—*Rákas-gaddah, ákás-gaddah,* HIND.; *Karwinai,* BOMB.; *Rakkas-gaddah, garaj-phal,* DEC.; *Gollan-kóvaik-kizhangu, ákásha-garudan,* TAM.; *Nága-donda, muru-donda,* TEL.; *Kollan-kóva-kizhauna,* MAL.; *Ákásha-garuda-gadde,* KAN.; *Go-palangá,* SING. **Ainslie** says that—"In Persian the plant is called *Lufa,* and in Arabic *Asanulfil.* The Telegu names allude to its property of living on the air.

References.—*Roxb., Fl. Ind., Ed. C.B.C., 702; Ainslie, Mat. Ind., II., 158; Dalz. & Gibs., Bomb. Fl., 100; Dymock, Mat. Med. W. Ind., 2nd Ed., 353; Murray, Pl. and Drugs, Sind, 42; Moodeen Sheriff, Supp. Pharm. Ind., 78; O'Shaughnessy, Beng. Disp., 347; Pharm. Ind., 96; Walker in Bomb. Med. Phys. Trans., 1848, p. 60; Drury, U. Pl., 87; Trimen, Syst. Cat., Ceylon Pl., 38.*

Habitat.—A herbaceous climber, met with in the Panjáb, Sind, Gujrát, and south along the Deccan Peninsula to Belgaum. Distributed to Ceylon. **Ainslie, Rottler,** and **Drury** say it is also a native of Coromandel.

Medicine.—"The root is of varying thickness and length, and much resembles that of **Momordica dioica,** being in shape not that unlike a (MEDICINE. Root. 1835) badly grown turnip, but much larger; externally it is yellowish-white, and marked with raised circular rings; the taste is bitter, mucilaginous, and sub-acid. When cut it exudes a viscid juice, which soon hardens into an opalescent gum" (*Dymock*). (Juice. 1836) A drug valued by the natives of India as an alterative tonic useful in syphilitic cases. According to **Ainslie** the Vytians of South India esteem, (and in **Ainslie's** opinion justly) the merits of this drug. They prescribe it in the latter stages of dysentery and old venereal complaints. It is usually administered, he says, "in powder, which is of a very pale colour, in doses of a pagoda (about one drachm) weight in the twenty-four hours, and continued for eight or ten days together; this quantity generally produces one or two loose motions. The root, when dried, very much resembles the Columba root, to which it approaches also in medicinal qualities." **Ainslie** also states that for external use in chronic rheumatism it is made into a liniment with cummin seed, onions, 1837 and castor oil. It is considered an anthelmintic and deobstruent, and in the Deccan and in Mysore it enjoys the reputation of being a valuable remedy for snake-bite, being administered internally and applied externally to the bitten part. The authors of the *Pharmacopœia of India* concur with **Ainslie** that this drug deserves to be more carefully examined and its properties tested.

Chemistry.—A bitter yellow uncrystallizable substance has been found (CHEMISTRY.) in the root which is probably allied to *Bryonin,* the bitter principle of 1838 **Bryonia dioica.** (For the Medicinal and Chemical Properties of Bryonia, see *Dymock, Mat. Med. W. Ind., 2nd. Ed., p. 353,* and also *U. S. Dispens., 302.*) *Conf. with* **Bryonia, B. 94.**

Coral plant, see **Jatropha.**

Coral tree, see **Erythrina.**

Coral-wort, see **Dentaria bulbifera.**

JUTE.

1839

CORCHORUS, *Linn.; Gen. Pl., I., 235.*

The generic name for this group of annual plants is derived from the property of the leaves (κορη the pupil of the eye, and κορηω to purge or clear). There are about 36 species distributed throughout the tropics, of which India possesses 8. But so uniformly are these plants met with in Asia, Africa, and America, that geographical evidence does not come to the aid of the student who may desire to work out the history of the cultivated species. The following are the Indian species regarding which economic information is available, but any one of them may be used for the fibre contained within its bark, since all are rich in fibre. Writing of the edible properties of the species of CORCHORUS, **Mr. Hem Chunder Kerr** says: "Some of the plants yield leaves that are excessively bitter; others slightly so. The former, called *tikta nálitá*, according to the *Rójavallabha*, a modern medical treatise, is a specific against red eruptions, worms, and leprosy. The latter, *madhura*, cures cold, paralysis, phlegm, and wind; both of them are esteemed as good tonics. The middle and higher classes of the people take them boiled with other vegetables in the form of soups as stomachics or appetizers; the lower classes use them as articles of food." **Sir Walter Elliot** gives **C. olitorius** the Telegu name of *Périnta kúra* but makes no mention of **C. capsularis**, but *Patti* he says is the vernacular for **Gossypium herbaceum** and *Pátta* for **Cissampelos Pareira.** The confusion that exists in the literature of Corchorus and of Jute has suggested the desirability of giving the characters which the writer accepts as separating the more critical species. It seems highly probable that hybrid or intermediate forms exist, however, connecting most of the Indian types. The peculiarities of the seed would seem to afford a better means of distinguishing the species than the form of the beak or number of cells and presence or absence of septæ in the capsule.

1840

Corchorus acutangulus, *Lam.; Fl. Br. Ind., I., 398; Wight,* [TILIACEÆ.

Syn.—C. FUSCUS, *Roxb., Fl. Ind., Ed. C.B.C., 429, Ic. t. 739.*

Vern.—*Titápát*, BENG.

References.—*Dalz. and Gibs., Bomb. Fl., 25; Kurz, Contrib. Burmese Fl., 130; F. von Müeller, Sel. Extra-Trop. Pl., 88.*

Botanic Diagnosis.—Stem hairy along certain sides between the nodes (not all round), the whole petiole having spreading hairs, and being woolly along the upper surface, both surfaces of the leaf hairy, those of the upper adpressed, margin often minutely ciliate; nervules reticulate (not parallel anastomising, as in **C. olitorius**). Capsule short (1 inch long at most); winged, beak cleft into 3-4 spreading arms each, often bifid; base of capsule contracted, position of faded flower indicated by a sharp groove. Seeds small, broader than long, squarish; hilum a large thickened patch in one corner.

A very distinct, and perhaps the most abundant wild species in India, often mistaken for **C. olitorius** when not in fruit.

Habitat.—An annual herb, met with throughout the hotter parts of India and Ceylon. **Roxburgh** remarks that it flowers during the rainy and cold seasons, is never cultivated, and differs from **C. tridens**, *L.*, in having only one style; and from **C. trilocularis**, *L.*, in having only one row of seeds in each cell. **Dalzell and Gibson** say that in Bombay it is a common weed, and **Roxburgh** that it is a native of various parts of India. **Kurz** states that it is frequent in the leaf-shedding forests all over Burma up to 3,000 feet elevation. **Stewart** apparently treats **C. acutangulus** as a synonym for **C. trilocularis**, and gives *báphallí* as the vernacular for the plant, and *isband* as the bazar name for the seeds. Not only is the above association of the scientific names incorrect, but *báphallí* is the name given to **C. Antichorus**, and *isband*, the seeds, to **C. trilocularis.**

JUTE.

Mr. Hem Chunder Kerr speaks of this as "the species **C. fuscus**, or the *títá* variety of **C. capsularis.**" It would almost seem possible that from **C. acutangulus** and **C. trilocularis** the cultivated forms of jute might have been produced. At all events, **C. olitorius** so closely resembles this plant in foliage that many of the reputed instances of **C. olitorius** having been found wild may be traceable to a mistake made in viewing **C. acutangulus** as **C. olitorius.** The beak in Clarke's **C. olitorius** No. 23,613 is hooked, the tips spreading somewhat as in **C. acutangulus.** Duthie's 7,121 has the foliage, capsules, and hairs of **C. trilocularis** with the seeds of **C. olitorius.**"

Fibre.—A coarse fibre is sometimes extracted from this species and Mueller alludes to this plant as an occasional source of jute. FIBRE. 1841

Corchorus Antichorus, *Ræusch.; Fl. Br. Ind., I., 398; Wight, Ic.,* [*t. 1073.* 1842

Syn.—Corchorus humilis, *Munro;* Antichorus depressus, *Linn.*

Vern.—*Baphuli,* Hind.; *Duphálli, kúrand, bophalli, bahúphalli, babuna,* Pb.; *Múdhiri,* Sind.

References.—*Dalz. & Gibs., Bomb. Fl., 25; Murray, Pl. & Drugs, Sind, 65.*

Habitat.—A common prostrate, shrubby, plant, wild in Upper India, from the N.-W. Provinces to the Panjáb and Sind, and south-west to Káthiáwár, Gujrát, and the Deccan—a member of the Indian desert flora. Distributed to Afghánistan, Aden, Tropical Africa, &c.

Fibre.—A very indifferent fibre may be prepared from this species. FIBRE. 1843

Medicine.—Stewart says the plant is rubbed down and given as a cooling medicine; Murray that in Sind its mucilaginous property makes it a valued demulcent, which is used in the treatment of gonorrhœa. MEDICINE. 1844

Fodder.—Stocks mentions this plant as a desert fodder-plant, eaten by camels. FODDER. 1845

C. capsularis, *Linn.; Fl. Br. Ind., I., 397; Wight, Ic., t. 311.* 1846

Vern.—*Ghi-nalitá-pat* (according to Roxburgh); *Narchá* according to U. C. Dutt), Beng. The last mentioned author in the Glossary to his Mat. Med. of the Hindus gives this plant the Sanskrit name *kálasáka.*

In Bengal the words *pát* and *koshtá* are often given to both the jute-yielding species, or to the fibre obtained from them, the latter word being apparently derived from the Sanskrit *kosha,* a sheath, in allusion to the fibres surrounding or sheathing the stem. But if this be a correct derivation for the word, it is in its meaning equally applicable to hemp and sunn-hemp, in fact to any fibre obtained from the bark, or even to silk in the cocoon, and can have no specific application to jute. From *Nádika* has most probably been derived Nálitá, the spinach prepared from the leaves. (*Conf.* **C. olitorius.**) The word, *kosta* in Dacca means a handful of jute, and *kahla,* a corruption from it, is given to the fibre dried on the stems instead of being separated by retting. One writer even suggests that *koshta* may be derived from Kooshtia, the name of a district in Eastern Bengal, where a considerable amount of jute is produced. This is probably fanciful, but it is worth adding that the word *koshta* is only used in Eastern Bengal, and that the name Kooshtia is probably not a very ancient one.

Harrana is a name given in Sháhjahanpur District, N. W.-P. (according to Mr. J. F. Duthie), for this species. Mr. Hem Chunder Kerr says that in Orissa, for both the jute-yielding species, the following names are used: "*kowria* and *nálitá,* other names being *naskarkáni, kostra, kangra, kanta, títa, beral,* and *hanuman*; but whether they are mere local names of these plants or of their different varieties, or of other plants confounded with the jute-producing species, it is uncertain." Dr. Bidie failed to find Tamil or Telugu names for **C. capsularis** during an enquiry instituted in 1874 into the subject of the jute cultivation in Madras. 1847

CORCHORUS capsularis.

The Round Fruited Corchorus.

JUTE.

References.—*Roxb., Fl. Ind., Ed. C.B.C., 429; Loureiro, Fl. Cochin Ch., VI., 408; Rumph., v. t. 78, f. 1; Voigt, Hort. Sub. Cal., 127; Brandis, For. Fl., 37; Gamble, Man. Timb., 52; Kurz, Contrib., Burm. Fl., 130; Lisboa, U. Pl. Bomb., 230; U. C. Dutt, Mat. Med. Hind., 302; Thwaites, En. Cey. Pl., 31; Proceedings, Madras Govt.* (*Rev. Dept.*) *1874, No. 49; DeCandolle, Orig. Cult. Pl., 131.*

Botanic Diagnosis.—Alone distinguishable from **C. olitorius** by the short rounded capsule—a very unimportant character. **Gamble's** No. 15,912 has one capsule nearly round, while the others are distinctly those of **olitorius**, but some are 4-valved, others 5-valved. **Kurz's** No. 1231 of **C. acutangulus** has both 4- and 5-valved capsules, and **Clarke's** No. 24,899 has a 3-valved capsule. **Clarke's** No. 31,637 of **C. trilocularis**, has a 4-valved capsule, and **Hooker and Thomson's** sample of that species, from the Panjáb, has a 3-valved capsule. The capsule is thus variable.

Habitat.—A common plant "throughout the hotter parts of India." This statement, originally made by **Roxburgh**, is current in the literature of jute. While it need not necessarily be implied that a plant is wild (*e.g.*, indigenous) in the area where it is common, still that is the opinion popular writers have derived from the above carefully worded botanical description. The major portion of all we have learned regarding **Corchorus capsularis**, during the past century, leads to the opposite conclusion. There are, however, a few notices of the plant that point either to its being indigenous in India or indicate acclimatisation so successful as to have deceived modern botanists. **Mr. J. F. Duthie** has, for example, favoured the writer with a note to the effect that he found **C. capsularis** on the banks of the Gumpti near Judalpur in what appeared a wild condition. A Native of the place gave the plant the name of *Harrana*, a word which has no relation to any of the names given to the Indian species of **Corchorus** in other parts of the country. **Mr. W. A. Talbot**, in a list of the Kanara plants (*Bomb. Gaz. XV., I., 428*), states of this species that it is "found on road-sides sparingly throughout North Kanara." On the other hand **Dr. Prain** (Officiating Superintendent of the Botanic Gardens) has
1848 forwarded to the writer, for personal inspection, every sheet of **Corchorus** from the Calcutta Harbarium on which the word "cultivated" has not been inscribed. It is a remarkable fact that **Dr. Prain** should have had to say: "All our **C. capsularis** are plainly cultivated, except perhaps one by **Kurz** from the Pegu Yomah, Burma, which, however, may be an escape." **Kurz** himself says of **C. capsularis** (*Contrib., Knowledge Burmese Flora, p. 130*)—"Cultivated all over Burma and frequently seen in deserted tongyas, along the borders of forests, around villages, &c." It is indeed significant that in the Great Indian Herbarium, which represents the collections and labours of all past botanists—**Roxburgh, Wallich, Griffith, Gibson, Hooker, Kurz, Clarke, King**, &c., there should not be a specimen marked "wild," if indeed the plant be truly wild in India. **Mr. Talbot** does not say the plant is wild in Kanara, and **Dr. Gibson's** specimens of **C. capsularis** from Bombay, now in the Calcutta Herbarium, have not been sent to the writer, so that they are presumably marked as cultivated. **Mr. Duthie's** specimen alluded to above is undoubtedly the round fruited
1849 condition to which the name **capsularis** is given, and it is thus the only specimen the writer has seen that is stated to have been gathered from "what appeared a wild condition." **Roxburgh** has no hesitation in pronouncing the non-jute-yielding species as natives of India. **C. decemangularis** of **Roxburgh** has, by several botanical writers, been reduced to **C. olitorius**, but of his plant **Roxburgh** states that it is a native of Bengal. While describing **C. olitorius**, so as to distinguish it from his **C. decemangularis**, he carefully avoids committing himself to a definite opinion as to

JUTE.

its nativity. **Edgeworth** says of the Banda district, N.-W. Provinces, that this plant is found "in the fields." A special enquiry was, in 1873-74, instituted in Madras into the whole subject of wild or cultivated jute found in that Presidency, and while **C. olitorius** was reported to have been discovered both wild and cultivated, it is expressly stated in the report that **C. capsularis** does not occur in Madras. **DeCandolle**, after enumerating all the countries where the plant is cultivated (*viz.*, the Sunda Islands, Ceylon, India, Southern China, the Philippine Islands, and Southern Asia generally) says: "I am not convinced that the species exists in a truly wild state north of Calcutta, although it may perhaps have spread from cultivation and have sown itself here and there." The writer spent many years in Bengal, and botanised over the greater portion of that Presidency, but he cannot say that he ever came across either **C. capsularis** or **C. olitorius** in what he could regard as a wild or rather indigenous condition. The latter may possibly be wild in some parts of Western India, but grave doubts may be entertained as to either being natives of Bengal,—the province where they are now mainly cultivated, and where they exist frequently enough as weeds around the cultivated jute fields. The suggestion is offered, that, by experimental cultivation, it might be found possible to produce forms of **Corchorus** from some of the truly wild species which would closely approximate to **C. capsularis** and **C. olitorius.** With the imperfect knowledge we possess of this subject, the writer would be much more willing to admit the possibility of some such theory, to account for the cultivated jutes, rather than believe that manifest escapes from recent cultivation are the sole survivals of the wild forms of these plants. The scientific distinction based on the length of the fruit vessel (round in **C. capsularis** and elongated in **C. olitorius**) is, to say the least, scarcely worthy of as much consideration as the peculiarities recognised by the cultivators in distinguishing (within each of these) the various cultivated forms that yield the commercial qualities of the fibre. A similar distinction in the shape and the number of cells of the fruit was made to give origin to certain species of **Brassica,** all of which can be produced from the seeds of any one by careful cultivation. 1850

It is noteworthy that definite Sanskrit names should not exist for these most useful plants, while other plants of far less value have assigned to them names so precise as to distinguish their varieties, to separate their wild from their cultivated forms, and to indicate every possible structural peculiarity. There are neither Arabic nor Persian names for the species of **Corchorus,** known to the people of India, and the greatest uncertainty exists regarding one or two Sanskrit synonyms that have been assigned to the jute-yielding species. Indeed, it seems highly probable that these names are entirely incorrect or rather are referable to other fibre-yielding plants. **DeCandolle** says: "No Sanskrit name for the two cultivated species of **Corchorus** is known." The word *jútá*, in one of its meanings, is at most, the name for a fibre which may or may not have been the modern jute. **Mr. Ram Sunker Sen** gives, however, a very different derivation for the word jute which would remove it altogether from *jútá*. He says that at the silk filatures of Eastern Bengal refuse silk is known as "*jhuta-jhut*," evidently a corruption of the Sanskrit *uchchista*, signifying refuse. This word *jhút* being in constant use among the owners (chiefly Europeans) of silk factories in the districts of Burdwan, Nuddea, Rajshahye, Moorshedabad, Pubna, and Malda, and the fibre of the plant known as *pát* amongst the natives—bearing a strong resemblance to the same—the term *jute* in its softened form came gradually to be applied to *kosta* in dealings and shipments at Calcutta. But, as opposed to this, it may be 1851

CORCHORUS capsularis. **The Round Fruited Corchorus.**

JUTE.

urged that when **Roxburgh** was told that the plant grown in the Botanic Garden was jute, there were in all probability no such dealings in the fibre between Calcutta and Eastern Bengal. Besides, **Mr. Kerr** rejects this derivation of the word, on the ground that jute is in no way a waste, rejected, by-product or remnant, as would be implied by the word *uchchista*. At the same time **Mr. Sen's** idea would simply be that it was in appearance like the first few threads drawn from the cocoons—the waste known in Europe as "ort" and which in India is made into *chasam*—but was not itself necessarily a waste, or as **Mr. Kerr** puts it, "an offal material like ort." It must be admitted that the long golden bands of jute
1852 fibre bear a close resemblance to the ribbands of waste silk or *chasam*, and that there are many much more unlikely derivations for words in common use than **Mr. Sen's** explanation of Jute from *jhut-a-jhut*, the Hindi *jhutthá*, the more so since the fibre is mainly produced in the very districts where the people were familiar with *jhut-a-jhut*. But, on the other hand, the word *jhot* is the only name by which jute is known in Balasore, and *jhont*, *jhot*, are common names for the fibre throughout Orissa. It has been pointed out that the gardeners employed in the Royal Botanic Gardens under **Roxburgh** were most probably, as at the present day natives of Orissa, and that, therefore, the name jute given by **Roxburgh**, the first European writer who used that name, was in all probability a softened form of *jhot*, a word which may be admitted to have come from the Sanskrit *jhuta*, unless we presume **Mr. Sen's** derivation of the word to have prevailed all over Orissa prior to **Dr. Roxburgh's** discovery of the plant.

The Sanskrit word *Nádika* is said by **Dutt** to have been given to **C. olitorius** and *kálasáka* to **C. capsularis**, but while **Dr. Dutt's** work is devoted to the Materia Medica of the Hindus and is compiled from Sanskrit medical works, he only gives the above names in a Glossary at the end, and does not attribute to the plants, to which he says they refer, any properties as known to the Sanskrit writers, while the modern Hindus use the leaves of jute and the species of **Corchorus** generally, both as food
1853 and medicine. **Dr. Moodeen Sheriff**, a high authority on vernacular names, does not give Sanskrit, Arabic, or Persian names, to the species of **Corchorus**, nor, indeed, can any of the local names that are in use in the provinces of India (of a restricted or specific character) be exhibited as derived from classical synonyms. *Patta*, a Sanskrit word given to jute by some writers, reappears throughout India in some form or other, being applied first to one fibre and then to another. In its early usage it simply means a "shining fibre," and was most probably originally given to silk, although in the *Mahábháratta*, presents of garments are mentioned as *patta-jam* (*patta*-produced) and also *kita-jam* (insect-produced), thus relieving the word *patta* from being silk. *Patta* also occurs in the Institutes of Manu in such a form as to leave no doubt that it was then applied to a vegetable fibre, probably jute. *Kosha*, if it be the root from which the name used in Eastern Bengal for jute, has been derived, nowhere appears in Sanskrit literature as a name for that fibre, nor indeed
1854 for any other fibre, while *Nádiká* refers to the edible property of jute and not to the fibre. Among the early synonyms for *patta* may be mentioned that of *Rája sana*, the large or noble sana, *kakkhata patraka*, "the rough leaved," and *sanni*, the sunn-like—names which would suggest a later introduction than **Crotalaria juncea** to which *patta* is compared. This idea receives further support from the fact that while *sana* occurs in the most ancient Sanskrit works, *patta* appears in the comparatively recent. In one of the references to *patta*, it is spoken of as the *chimi* (probably a misspelling for China) *pát*, a fact which would point to the cultivated jute plant having come to India from China. **Mr. Hem**

JUTE.

Chunder Kerr reviews all the reports and early books of travel that refer to fibre or to rope-making in India, and finds that in none of these publications does there occur any mention of the word jute until 1796. In several works *pát* is, however, mentioned as a fibre viewed in India as a form of hemp, but which by the home authors was pronounced to be more nearly allied to flax. By the beginning of the present century the word *pát* was completely superseded by jute in all commercial correspondence.

With an array of facts of this kind before us we are almost justified in believing that Jute plant is either a comparatively modern development from some wild stock which was unknown to the Sanskrit writers, or that the cultivation of the plant has been introduced from some other country and most probably subsequent to the date of even the most recent Sanskrit works. If a modern development, we can scarcely admit that the stock from which it was derived could have disappeared, while numerous wild plants closely allied to **Corchorus capsularis** and **C. olitorius** are abundant and truly wild plants, all of which yield good fibres only inferior to jute to an extent easily accountable for by cultivation. The seat of the Indian cultivation of this plant is from the Hooghly district through Eastern Bengal, especially on the islands and lowlying lands of the Meghna and Brahmaputra Rivers. **C. olitorius**, on the other hand, occurs chiefly on the lowlying lands on the western side of the Hooghly river, more especially in the Burdwan district and in Western and Southern India.

Although there are numerous references to *Patta*, *Júta*, &c., in early Indian writings, enough has been said to show that the greatest caution is necessary in founding too strong convictions that these names allude to the Pát and Jute of the present day, the more so since jute cultivation in every district of Bengal is spoken of by the local authorities as of modern origin, at least in its present form. In one district its introduction is fixed at 1872, in another at 1865, in a third before the date of the British rule, and in a fourth it is put down at 400 years ago. In all districts it is spoken of, however, as a crop regarding which some period could be fixed, while no such language is used with regard to rice, cotton, sunn-hemp, or any other crop of an importance at all comparable with Jute. (*Conf. with* **C. olitorius** *in a further page.*)

FIBRE. 1855

Fibre.—See a further page, and also **Jute.**

MEDICINE. 1856

Medicine.—The leaves dried are used medicinally, being eaten at breakfast-time with rice in cases of dysentery. The cold infusion is also administered as a tonic in dysenteric complaints, fever, and dyspepsia.

OIL 1857

Oil.—"The seed when fried over the fire yields an oil chiefly used for lighting purposes" (*Ramshunker Sen, Agri. Gaz., 163*).

1858

Corchorus fascicularis, *Lam.; Fl. Br. Ind., I., 398.*

Vern.—*Hirankhori, bhauphali*, Bomb.; *Jangli* or *ban-pát, bilnalita*, Beng.

Dymock points out that the name *bhauphali* must not be confused with the Maratha *bhaphali*, a name given to an umbelliferous plant, and it may here be suggested that the N.-W. P. name *ban-phal* given to **C. olitorius** may have arisen from a misreading of *bhauphali*. The name *bhauphali* or *báphullí* is also given to **C. Antichorus.**

References.—*Roxb., Fl. Ind., Ed. C. B. C., 429; Dymock, Mat. Med. W. Ind., 2nd Ed., 115.*

Botanic Diagnosis.—Capsules small (½-¾ inch) almost cylindrical, very hairy, beak 3-4, splitting with the dehiscence of the capsule. Seeds triangular or diamond shaped, more pointed at the lower end and very similar to those of **C. olitorius** but smaller.

Habitat.—A common wild plant throughout the hotter parts of India from the Panjáb to Bengal, and westward to Bombay (common, for example, at Surat). Distributed to Ceylon.

FIBRE. 1859 **Fibre.**—The fibre extracted from this plant is employed in Sind in the manufacture of ropes.

MEDICINE. 1860 **Medicine.**—**Sakharam Arjun** mentions the fact that the whole of this plant is mucilaginous, and states that in Bombay "a watery extract mixed with sugar-candy is taken as a nutritive tonic. It is also given in seminal weakness." **Dymock** remarks that in Bombay the "whole plant is sold in the shops; it is very mucilaginous and somewhat astringent, and is valued as a restorative." The name *hirankhorí* given to it, means deer's hoof.

1861 **Corchorus olitorius,** *Linn.; Fl. Br. Ind., I., 397.*

JEW'S MALLOW.

Vern.—*Pát, koshta* (*bhunji pát,* according to **Drury,** and *bhungí,* in *O'Shaughnessy's Beng. Disp.*), *lalita pát, bhunji-pát, bhungí* or *ban-pát* (according to **K. L. De**), BENG.; *Singin janascha, koshtá* (according to **Benson**), HIND.; *Peratti-kirai, punaku cheddy,* TAM.; *Parinta, périnta-kúra;* TEL.; *Ban-pát,* SIND.; *Ban-phal,* in N.-W. P. (**Atkinson**) and PANJAB (**Stewart**); *Nádika* (according to **Dutt**), *patta* (according to **Roxb.**), and *Sing giká* (according to **Ainslie**), SANS.

The word *Nálitá* (a corruption of the Sanskrit *Nádiká*) is correctly speaking, the name for the pot-herb prepared from any Corchorus, the species so used being distinguished by a prefix, thus *ghi-nálitá* (**C. capsularis**), *tili-nálitá* (**C. acutangulus**), and *bíl-nálitá* (apparently **C. Antichorus**). *Nálitá* by itself may probably have been the spinach prepared from **C. olitorius**, the species most generally eaten. *Nálitá* becomes *nutia* in Mymensing, and *narich sag* in Chittagong, thus apparently breaking down **Dutt's** *narchá* as the Bengali for **C. capsularis,** and *Nádika* the Sanskrit for **C. olitorius. Sir Walter Elliot** alludes to this species but makes no mention of **C. capsularis,** and neither assigns *Jútá* nor *Patta* to Jute.

1862 **Ainslie** was perhaps the first European writer who assigned to this plant the Hind. name *singgin-janascha,* and while this has been reproduced by several subsequent authors, the word does not appear to be in use in India at the present day, at least not in Hindustan proper. The Sanskrit names given above have already been commented on under **C. capsularis. Mr. Hem Chunder Kerr** points out that the word *bhunjí* (given by various authors as a Bengali name for this plant) is not employed at the present day. It is derived from the Sanskrit *bhangá* (**Cannabis sativa**), and thus recalls in a remarkable way **Rumphius'** name for jute, *gunja* or *gania* (may not *gunny* or *guni* as now applied to jute cloth have come from the same source ?). In ever so many ways at all events the early literature of jute is mixed up with that of hemp and sunn-hemp, so much so as to suggest that it *may be an introduced plant, which at first was viewed as a form either of hemp or sunn-hemp.*

References.—*Buchanan-Hamilton's Dinagepore; Ainslie, Mat. Ind., II., 387; Roxb., Fl. Ind., Ed. C.B.C., 429; Griff., Not. Dist., 512; Dalz. & Gibs., Bomb. Fl., 25; Drury, U. Pl., 157; Baden Powell, Pb. Prod., 333; Atkinson's Him. Dist., 733; U. C. Dutt, Mat. Med. Hind., 311; Moodeen Sheriff, Supp. Pharm. Ind., 114; Murray, Pl. & Drugs, Sind, 64; Benson, Saidapet Exper. Farm. Man., 63; DeCandolle, Origin Cult. Pl., 132.*

Botanic Diagnosis.—Glabrous, except the upper half of the petiole, and the primary veins on the under surface, where woolly hairs occur; nervules transverse, nearly parallel, pellucid, and anastomosing. Capsule very long and glabrous, beak straight; remains of the flower forming a thick scar. Seeds somewhat triangular, pointed at both extremities, but much more so to the hilum, surface often roughened, so as to appear as if minutely hairy.

JUTE.

Habitat.—In the *Flora of British India,* it is stated of this species that it is "indigenous in many parts of India," and is generally distributed by cultivation in all tropical countries. The chief seat of its Indian cultivation is near Kulna in the Burdwan district of Bengal, but it is probably cultivated also in western and southern India; the following botanical writers allude to the plant:—**Dalzell** and **Gibson** say that it is common in Bombay, and **Talbot** (a botanical observer whose opinion must carry considerable weight) remarks: "Abundantly wild about Yellapur." **Dr. Gibson** has left a specimen of this species in the Calcutta Herbarium which was collected in Western India; from the *Herb. Ind. Or. Hooker fil., and Tomson,* there is a sheet of specimens from Mysore and the Carnatic; **Gamble's** No. 15912 is from South India; **Clarke's** No. 23613 from Kangra; and **Duthie's** No. 7121 from Dera Ismail Khan. With the exception of the last mentioned, all the above specimens (from the Calcutta Herbarium) are undoubtedly the form called **C. olitorius,** but they are not stated whether collected from wild, acclimatised, or cultivated plants. **Duthie's** specimen has the seeds of **C. olitorius,** but instead of being glabrous the capsules are hairy along the angles and have a few of the peculiar tufted hairs of **C. trilocularis,** as well as the long narrow capsules of that species. It has also the thick and somewhat linear, coarsely serrated, leaves peculiar to that plant, but the leaves are not only hairy but have a few of the tufted glandular hairs on the under surface as well as on the fruit. **Kurz** gives the habitat of **C. olitorius,** as far as Burma is concerned, as "Ava, Pegu, cultivated and wild in rubbishy places and agrarian lands." **Atkinson** says that it is found in "Dehra Dún," but in this connection it may be added that in the Saharunpur Herbarium, while there are specimens of the allied species, **C. acutangulus,** from various localities in the North-Western Provinces and the Panjáb, there are none of **C. olitorius.** One specimen of **C. acutangulus** is marked as collected at Dehra Dún, and it is probable this may be the **C. olitorius** alluded to by **Atkinson, Stewart,** and other writers on the Flora of Northern India. In the report (to which reference has been made under **C. capsularis**) on jute cultivation in Madras, it is stated that a considerable amount of **C. olitorius** is grown in Ganjám, Godavery, Kistna, and Nellore, but not for its fibre. The Collectors of Ganjám and Godavery say it is wild in their districts. The only district in the southern parts of the Madras Presidency where the plant was discovered was Salem, the Collector having found a specimen on the margin of a field, which **Dr. Bidie** identified as **C. olitorius.** A sample of **C. trilocularis** is, however, in the Saharunpur Herbarium named **C. olitorius,** and this was apparently collected by **Mr. J. S. Gamble** in the Kistna District; it bears the number 12662. The merest possibility of such a mistake existing regarding the Kistna samples reported on above may be admitted as sufficient to throw a doubt on the indigenous character of **C. olitorius** in even the northern districts of Madras. At all events, it is now admitted that the so-called "jute" of Madras commerce is Sunn-hemp and Hibiscus fibre, but not **Corchorus** at all. The Agri.-Horticultural Society of Madras submitted in 1873 a report on the jute cultivation and manufactures of that Presidency, but in the following year wrote and informed Government that they had now discovered that the plant that yielded the so-called jute of their former communication was a species of **Crotalaria** and not of **Corchorus.** **Roxburgh** points out in the *Flora Indica* that there is a wild form of the plant known in Bengal as *ban-pát* or wild *pát* which has reddish stems. In his *Hortus Bengalensis,* he speaks of two varieties of **C. olitorius,** a green form (the *pát*) and a reddish (the *ban-pát*). This opinion is accepted by **Ainslie** and by

1863

1864

JUTE.

O'Shaughnessy, both of whom call the green variety **C. olitorius** and the reddish **C. capsularis**. The term *ban* or *janglí pát* is, however, at the present day, applied in Bengal to **C. fascicularis**, a distinct species from either of the above. **Stewart** remarks that **C. olitorius** is found wild in the Panjáb; but he does not give its Panjábí names, while he says it is the *ban-pát* of Bengal, a circumstance that would seem to justify the inference that **Stewart's** wild **C. olitorius** should be corrected into **C. fascicularis**, the more so since that species is undoubtedly wild in the Panjáb, although not alluded to by **Stewart**. (*For another error committed by* **Stewart** *see the remarks under* **C. acutangulus.**) At the same time the writer, on looking over the Saharunpur Herbarium collections found one specimen, apparently correctly named **C. olitorius**, which was discovered by **Dr. Aitchison** (No. 476), and on which the note occurs, "occasional from Thul to Kuram." The Saharanpur Herbarium, as already remarked, does not, however, possess a sample of **Corchorus olitorius** as found in the Panjáb proper.

1865 If, after carefully considering these somewhat conflicting opinions, we still believe that **C. olitorius** is indigenous to India; if, indeed, we accept it as a truly wild form and not a product of cultivation (possibly from **C. acutangulus** and **C. trilocularis**) escaped and assumed a semi-wild condition, then it might almost be safe to believe that it was the parent of all the cultivated forms of jute. In the writer's opinion, however, its claim to being viewed as indigenous rests at present on doubtful evidence, but it may at least be confidently asserted that it is not wild in the districts where it is now or ever has been known to be cultivated for its fibre. Indeed, there is a strong probability that as a cultivated plant **C. capsularis** came to India from China or Cochin-China, and that **C. olitorius** may have been produced in India. Its extensive use as a pot-herb might explain its acclimatisation over so extensive an area as has been indicated. But more can certainly be said in favour of a possible Indian origin for **olitorius** than for **capsularis**. The latter would appear to have been cultivated in China before the date of its having been authentically known to the people of India. It has been grown, for example, in the neighbourhood of Canton for many centuries, and **Roxburgh** says it is there called *Oi moa*. **Mr. Hem Chunder Ker** suggests the strong resemblance of this name to the Sanskrit "au-ma" signifying "flaxen." The Malays call **C. capsularis**, *Rami-tsjima* or Chinese hemp. But in the same way **C. olitorius** has been known to the Egyptians and Syrians for a very long time, their acquaintance with it being possibly prior to the date of the evidence of a positive character, that a knowledge of the properties of the plant was possessed by the inhabitants of India. The

1866 Greek Κορχορος was applied to a pot-herb, but in all probability the plant alluded to was not the Corchorus of the present day. Accepting the derivation of the Greek word as implying a drug useful in the treatment of eye diseases, it may be pointed out that no such property is claimed for the species of Corchorus. It is perhaps only a fanciful idea, but this property of a collyrium associated with *μαμηρα* and *μολοχινα* with *títa* and *tikta* recalls the properties of **Coptis Teeta** or **Picrorhiza Kurroa** as possibly in some strange way connected with the edible and medicinal properties of Κορχορος. There is no good Hebrew name for jute; the word *malluach* (*Job, XXX., 4, 1520 B. C.*) has been translated mallows, and probably correctly, since in Egypt the mallow is extensively cultivated and used as a pot-herb. At the same time, it is well known that **C. olitorius** has for centuries been cultivated near Aleppo as a potherb, hence, says **Rauwolf**, the name *Olus judaicum* which the French have translated *Mauve de Juif*, and the English have rendered as *Jew's*

Mallow. It began apparently to be cultivated in Egypt about the beginning of the Christian era. It is there known by an Arabic name *melokych,* a word which seems in Crete to pass into *maulchia* (Conf. **DeCandolle**). It will at once be seen that these Arabic names (if indeed they be Arabic) bear no relation to the vernacular synonyms given even by the Muhammadans of India (still less the Hindus) to any form of Corchorus. This fact would point to the Muhammadans not having known it by its Arabic names prior to or during their successive invasions of India, which were continued for a thousand years from the 7th 1867
century. In consequence of this long period of Muhammadan influence, India obtained the Persian and Arabic names given to her plants and animals, but there being no names in these languages (in the forms in which they are now preserved in India) for the species of Corchorus, the inference is practically unavoidable that the properties of these plants were not known to the early Indian Muhammadans. To this line of reasoning may be added the further consideration that it was not until the middle of the 18th century that the Europeans then living in India came to hear of the plant. **Roxburgh** about the beginning of the present century was the first author to announce the name Jute as that given to a hempen fibre of great value prepared from either of two species of Corchorus. And, indeed, the paucity of vernacular names for the various forms of Corchorus is perhaps one of the most striking evidences of the knowledge of the properties of these plants being of a comparatively modern date. The reader would do well to compare the names given for **C. capsularis**—the jute plant—with those under **Cocos nucifera**—the cocoa-nut,—or for **Crotalaria juncea**—the sunn-hemp—in order to realize the degree of importance that should be attached to this suggestion and to appreciate the spirit of caution indicated as necessary before too sweeping conclusions are derived from the accidental observations of certain writers who have asserted that both forms of the jute plant are natives of Bengal, because they are plentiful weeds in cultivated situations. (*Conf. with* **C. capsularis.**)

JUTE.

FIBRE.
1868
Fibre.—See a further page and under **Jute.**

MEDICINE.
1869
Medicine.—**Ainslie** says that **Dr. Francis Hamilton** (the **Buchanan-Hamilton** of later writers) had brought to him, while in Behar, specimens of this plant as an herb used medicinally by the Hindus. "Fresh or dry after being toasted and reduced to ashes it is mixed with a little honey, and given daily in *petai* (obstructions of the abdominal viscera)." **O'Shaughnessy** remarks, "an infusion of the leaf is much employed as a fever drink among the natives of the Lower Provinces, and **Mr. Twining** mentions the practice with approbation in the second volume of his admirable work on the diseases of Bengal." **Mr. Atkinson** says: "The leaves are emollient and used in infusion as refrigerant in fevers and special diseases. The dried plant toasted and powdered is used in visceral obstructions."

1870
Dr. K. L. Dé, C.I.E., says: "The dried leaves of this plant are sold in the market. A cold infusion is used as a bitter tonic, and is devoid of any stimulating property. **Mr. Simon** of Assam informs me that it can be safely given to patients recovering from acute dysentery to restore the appetite, and improve the strength. Six grains of the powder, combined with an equal quantity of **Curcuma longa,** has been used, in several instances, with much success, in acute dysentery. It forms a cheap domestic medicine in a Hindu household." **Dr. Bidie** alludes to the dried plant being used in South India as a demulcent.

FOOD.
1871
Food.—Throughout India this plant is more or less cnltivated as a pot-herb, although chiefly so in Eastern Bengal. The Santals have a

JUTE. peculiar form which may prove an undescribed species; it is known to them as a useful pot-herb under the name of *bir-narcha* (*Rev. A. Campbell*), a name most probably derived from the Bengali *narchá* (**C. capsularis**), hence of some importance historically, since it would indicate that the knowledge of the plant was derived from the Bengalis instead of being anciently possessed by this primitive aboriginal race. **Mr. Atkinson** in his *Economic Products* gives (*Part V.*) a complete list of all the wild or cultivated vegetables, greens, &c., used by the people of the N.-W. Provinces, but makes no mention of any species of Corchorus.

DOMESTIC. 1872 **Domestic Uses.**—The stalks, after the removal of the fibre, are used for making gun-powder charcoal, and are also employed in the manufacture of baskets, &c.

1873 **Corchorus tridens,** *Linn.; Fl. Br. Ind., I., 398.*

Botanic Diagnosis.—Much more nearly related to the next species than to **C. acutangulus.** Seed larger and raphe-like cord more distinct than in **C. trilocularis,** capsule with glandular hairs in tufts.

Habitat.—The *Flora of British India* says of this species: "Generally distributed."

FIBRE. 1874 **Fibre.**—**Murray** specially mentions this species as affording a cordage fibre in Sind.

1875 **C. trilocularis,** *Linn.; Fl. Br. Ind., I., 397.*

Vern.—*Kurú chunts*, BOMB.; the seeds are in the bazars sold under the name of *Rája-jira*; *Kaunti*, SANS.; *Tandassir*, KAN. (according to **Lisboa**); the seeds are known as *Isbund* in Sind (according to **Murray**).

Reference.—*Dymock, Mat. Med. W. Ind., 2nd Ed., 115.*

1876 **Botanic Diagnosis.**—Stems, petioles, and under-surfaces of the leaves hairy (as in **C. acutangulus**), but upper surface often almost quite glabrous. Capsule long thin straight angled, beak straight; hairs on the fruit short ascending tufted, 3-6 spreading from a thickened gland which is often persistent on the old fruits. Seeds black, smooth irregularly square on section, obliquely and sharply truncate at both extremities, hilum large with a raphe-like cord thrown from it to the top of the seed crossing one of the angles. The writer would be disposed to unite **C. tridens** and **C. trilocularis,** and bring with these, into a section characterised by the seeds, the species **C. urticæfolius.** He can put no reliance on the presence or absence of a short style or of a spreading stigma, as he has found both these conditions on the same plant. The fruits of the species of Corchorus are more variable than any other part of these plants.

Habitat.—The *Flora of British India* states that this species is met with in the N.-W. Provinces, the Panjáb, Sind, and south to the Nilgiri hills. **Roxburgh,** however, says that it is a native of Bengal, and flowers about the end of the rains, and **Lisboa** that it is found in Gujarát, Sholápur, and other high dry ranges. **Dymock** remarks that it appears along with **C. olitorius,** from which it may be distinguished by its oblong, lanceolate leaves, trilocular capsules, and small seeds.

FIBRE. 1877 **Fibre.**—"From the fibres good rope is manufactured" (*Murray*).

MEDICINE. 1878 **Medicine.**—**Dymock** says: "In Bombay the seeds of **C. trilocularis,** which are bitter, are administered in doses of about 80 grains in fever and obstructions of the abdominal viscera. A bitter Corchorus was known to the Greeks. **Theophrastus** says ὀπαροιμιαζόμενος διά τὴν πικρότητα κόρχορος (H. P., 77). **Pliny** (21, 32, and 25, 13) also mentions it as a poor kind of pulse growing wild." **Murray** states that "the plant macerated in water for a few hours yields a mucilage which is prescribed as a demulcent, and the seeds as a specific in rheumatism." (*Pl. and Drugs, Sind, 65.*) JUTE.

The *Ulfaz Udwiyeh*, by **Noured-din Mahomed Abdulla Sherazi**, uses the name of *isbund* for a species of what appears to be mustard seed.

JUTE.

JUTE. 1879

In connection with the reports of the Calcutta International Exhibition the writer published the greater portion of the facts which will be found in the present account of the fibre obtained from the species of **Corchorus**. In a further volume the commercial aspects of jute will be given (see **JUTE**), while in the following pages an effort is made to present a general and historic sketch of the subject together with certain facts of economic interest connected with the species of **Corchorus**. It may here be stated that the commercial fibre Jute is obtained from either one or both of the following species of **Corchorus**, *viz.*, **C. capsularis**, *Linn.*, grown in Northern, Central, and Eastern Bengal, and **C. olitorius**, *Linn.*, raised in the vicinity of Calcutta. Little or no jute is produced in the other provinces of India, its place being generally taken by **Cannabis sativa, Crotalaria juncea**, or **Hibiscus cannabinus**. A futile effort was, however, made to establish in Bombay a trade in **Malachra capitata**. The reader is, therefore, referred to the accounts given of the above mentioned fibre-yielding plants to complete the present article on the history of, and trade in, jute, and the uses of the various species of **Corchorus**. The information contained in the writings of the early authors is often so confused that a definite knowledge of any one of these fibres can only be obtained by a careful study of all.

Comm. and Vern. Names.—Jute, or Jew's Mallow, Eng.; *Jute, mauve des juifs, corde textile*, Fr.; *Jute*, Germ.; *Pát*, Beng. **Roxburgh** says that "the Bengalis call it jute," but **Royle** enters into an explanation of the origin of the word, which he makes out to be a corruption of *choti*, the name of a coarse cloth formerly made from this fibre. In Orissa, this cloth was called *jhut, jhóto, jhútó*, from which probably **Roxburgh** derived *Jute*, seeing that the native gardeners generally found in Bengal are inhabitants of Orissa. *Phetcwoon*, Burm.; *Patta, júta*, or *jata*, Sans. **Professor Skeat**, at a meeting of the Cambridge Philosophical Society, gave three meanings to the Sanskrit word *júta*, one of which might be accepted as giving origin to the Orissa names from which **Roxburgh** constructed the form of the word now in commercial use. The plant when used as a pot-herb and dried as a medicine is in Bengal called *Nalita*. The fibre is *Pát or Koshta*, and commercially, *Jute*.

The cloth, which was once largely worn by the poorer classes, although now almost superseded by European goods, is called *Tat*. The coarser cloth made into bags and used for bedding was called *Choti*. The word *gunny* is perhaps derived from "*gun*," a sail; or from "*goni*," a South Indian name for coarse sackcloth, made originally, as it would appear, from *Sunn* not from *Jute*. (See para. 1793 and 1800, also **Crotalaria juncea.**)

References.—*Hem Chunder Kerr's Report on Jute and other Fibres in Bengal, 1877; Babu Ram Comal Sen, Trans. Agri.-Hort. Soc., Vol. II., 91; Royle, Fibrous Plants Ind., 240—252; Col. L. Conway-Gordon's Report on the Jute Traffic in Eastern Bengal, 1885; Report on Indian Fibres by Cross, Beavan, &c., page 35; Spons' Encycl., 940; Roxb., Fl. Ind., Ed. C. B. C., 429; Ainslie, Mat. Ind., ii., 387; Drury, U. Pl., &c., &c., and to the references already given under each of the species of Corchorus.*

HISTORY OF THE JUTE INDUSTRY.

HISTORY. 1880

The history of the modern Jute industry is exceedingly interesting and intimately associated with the British rule in India. There can be no doubt that jute was known to the people of India from compa-

HISTORY.

ratively remote periods, but, as indicated under **C. capsularis** and **C. olitorius**, from the confusion which existed down to the present century in the words *sunn, pat* or *patta, bhanga,* and *hemp,* &c., names applied to certain Indian fibres, it is difficult to determine for certain many of the fibre-yielding plants referred to by ancient writers. The probability is that *sunn-hemp* (the fibre of **Crotalaria juncea**) was better and earlier known to the ancient Hindus than *jute,* and that the true hemp (**Cannabis sativa**) was known to them, if not brought to India by their invading and conquering ancestors. It is almost safe to assume that in very remote times *sunni, patta,* and *bhangi* were synonymous and generic terms for fibre and coarse cloth, without much regard to the plant from which the fibre was obtained. If so, about the beginning of the present century, the word *pat* became fixed and associated with the fibre of **Corchorus olitorius** and **C. capsularis.** Prior to that date the Government returns of exports from India mention *hemp fibre*; this must have been either *sunn* or *jute,* since the true hemp fibre has not been cultivated for centuries at least, and modern experiments have shown that the plant is not capable of cultivation as a source of fibre in the plains of India.

1881 With the advance of civilization came an increased demand for cloth, at first as a luxury, and latterly as a necessity. Jute probably met this demand, and, indeed, the poorer people, little more than half a century ago, were largely clad in jute cloth of home manufacture, such as, at the present day, is used by the aboriginal tribes. The increased facilities for the importation of cheap European piece-goods checked, however, the development of this indigenous industry; but with the rapid progress in every other branch of enterprise, there opened up a foreign trade in jute which the agriculturalist found remunerative. The resources of the rich plains of India, Burma, and China, and latterly of America, Australia, and Egypt, were, by the British mercantile fleet, made available for the supply of grain. Bags were required for this trade, and thousands of rough gunnies were greedily bought up. The high price obtained was a powerful incentive to increased activity, and thus the gunny-bag trade rapidly became a recognised part of the Bengal peasant's work. By and by, however, European machinery began to compete with manual labour, and in due time it gained the day. Jute was exported to Europe for cordage, and ultimately for the manufacture of the bags required in the grain trade. The first commercial mention of the word "jute" is in the customs returns of the exports for 1828, when 364 cwt. were sent to Europe. Soon the agriculturist found that his time would be more profitably spent in preparing an extra quantity of fibre, than in manufacturing bags to compete with steam and mechanical appliances; the preparation of fibre speedily outstripped the demand for home manufacture, and a large export trade was established in raw jute to feed the Scotch mills. Thus transferred from its original home, the gunny trade took a new start in Dundee, and down to the year 1854 little or no effort was made to improve the Indian manufacture by the application of European machinery. In that year, however, the "Ishera Yarn Mills Company" was established at Ishera near Serampore by Mr. George Ackland, a large owner of coffee plantations in Ceylon, and non-official member of the Legislative Council of that Island: these mills were afterwards called the "Ishera Company, Limited," and are now known as the "Wellington Mills." Three years later (1857) the "Borneo Company, Limited," a Company originally established to exploit the Island of Borneo, founded the mills now known as the "Baranagore Jute Mills." In 1863-64 the Gouripore Jute Factory came into existence. Following these factories sprang up rapidly in every direction around Calcutta. In the Trade Returns for

HISTORY.

1869-70 the exportation of manufactured jute was 6,441,863 gunny bags manufactured by power and hand looms, and brought into competition with the Dundee bags. This trade developed steadily, and in 1879-80, ten years later, over 55,908,000 gunnies were exported from India. The relative importance of the export trade in raw jute, as compared with the exports in manufactured jute of all kinds, may be seen by a careful examination of the tables (given in another volume), but the result may be summarised by saying that in 1886-87 the exports of raw jute amounted to £4,869,814, whereas for the same year the entire exports from India of power and hand-loom jute manufactures amounted to only £1,149,296. This is of course a comparison between the total exports of raw jute and a portion of the Indian manufactures. In a further page the relative amount of Indian manufactured jute exported as such and the amount used up locally or devoted to the export trade in grain will be found. But speaking purely of India's foreign trade in jute and jute manufactures it would seem that even with 24 large European factories at work in India, and the hand-looms which still survive, scattered over the country, her raw jute interests are four times as valuable to India as her manufactures. A comparison between the exports of Indian "power-loom" as compared with "hand-loom" manufactures will still further show the extent to which the jute manufactures have passed out of the hands of the Indian peasants, who alone, little more than 40 years ago, met the demand for gunny bags. This is seen very clearly when the above figures are compared with the exports of 1850-51. At that time the value of the gunnies exported was greater than that of the raw jute,—the former being £215,978, the latter, £197,071. There were no European factories in India in 1850, so that the market was supplied by the Indian peasant's hand-loom. Steadily the exports increased, the demand for gunnies calling into existence the Dundee mills, and soon after the Indian factories. Nothing could demonstrate the development of the jute trade more than a careful examination of the exports of raw jute and manufactured jute from 1854 to 1887. During that period 24 factories, larger than the average jute factories of Europe, have come into existence, and have gradually commenced to pour their manufactures into the market, largely, if not entirely, meeting the home (Indian) consumption. While this has been taking place, the foreign exports of raw jute have continued to increase uninterruptedly, each year exceeding the preceding, and apparently quite unaffected by the powerful Indian competition with the Dundee and other foreign manufactures.

1882

CULTIVATION AND PREPARATION OF THE FIBRE.

CULTIVATION. Area. 1883

Area and Extent of Jute Cultivation.—Jute is largely cultivated in the northern and eastern districts of Bengal and to a smaller extent in the central tracts of the province. In Assam it is grown in Goalpara. The area under the crop in these two provinces during 1886-87 has been approximately estimated at $1\frac{1}{3}$ million acres and the outturn at 20 million maunds. Of this area Assam has from 15,000 to 16,000 acres, with a production of 237,000 maunds of fibre. It has been ascertained that more than half the annual yield of fibre is exported to foreign countries and mainly to Great Britain and the United States of America, the proportion respectively to these countries being 73 to 17 per cent. of the total despatches from India.

The following extract from the jute forecast issued by the Agricultural Department of Bengal for 1887 shows the chief districts where the crop is grown and the approximate areas under it, the latter being in acres:—Mymensingh 250,000, Dacca 170,000, Rungpore 162,000, Pubna 150,000,

CULTIVATION.

Tipperah 117,000, Furreedpore 85,000, Rajshahye 45,000, 24-Parganas 44,000, Dinagepore 40,000, Bogra 34,000, Nuddea 30,000, Jessore 30,000, Khoolna 30,000, Purneah 24,000, Hooghly 19,000, Goalpara 15,000.

In other provinces, jute, though occasionally cultivated, is rarely so on account of its fibre, but to a limited extent the wild, acclimatised or cultivated jutes are resorted to for the supply of the fibre used locally. In various official and public reports reference is made to Madras jute. In 1873 the Agri.-Horticultural Society of that Presidency submitted a report to Government on certain samples of jute produced in Madras, but in the year following the Society corrected this statement and informed Government that the whole of the Madras so-called jute was Sunn-hemp. It would thus appear that the jute mill now working in Madras draws its supply of fibre from Bengal, or uses up a certain amount of Sunn-hemp or Hibiscus fibre in the preparation of bags and cloth which it issues. In the Madras Manual (*Vol. I., 361*), it is stated that a portion of the jute used by **Messrs. Arbuthnot & Co.** is produced locally, "but it is hoped that before long the supply will be drawn entirely from the district." Recent experiments have, however, been made in order to discover whether the true jute plant could be profitably grown in Southern India. **Mr. Benson** (in his *Saidapet Experimental Farm Manual and Guide, page 63*), gives the result, arriving at the conclusion that, unless some parts of the Northern Division be more suitable, jute cannot be grown in Madras. So in a like manner it has been tried in Bombay and Burma, with apparently the final verdict that, in these provinces, it cannot be produced at a price to compete with Bengal. The plant can be grown most successfully in Burma, but the cost of labour has proved fatal to any idea of an extensive commercial industry. In 1872-73 **Mr. Hem Chunder Kerr** estimated that there were one million acres under jute in Bengal and Assam, distributed over 37 million acres of country, and that should the demand be doubled, the production of that amount would absorb only one-eighteenth part of the available and suitable land. The estimate given above for 1887 shows a considerable increase, but it must be added that, as the areas under the various Bengal crops have not, as yet, been surveyed, little absolute reliance can be put on these returns. Forecasts of the jute crop must, therefore, assume the form of prognostications as to reputed extension or contraction from previous years without conveying any definite idea of the normal area. Such forecasts can, however, receive confirmation from the returns of imports of jute into Calcutta and Chittagong. Allowing a percentage for local consumption, the area which was under each year's crop during past years may be arrived at by dividing the total maundage* imports by 15, that being the average yield of fibre per acre. Thus the total imports of jute into Calcutta and Chittagong were in 1884-85, 1,51,15,940 maunds, Assam having furnished of that 1,85,518 maunds. Deducting the latter and adding 25 per cent. for local consumption the Bengal production would in that year have been 1,86,63,028 maunds (or 13,330,734 cwt.) We thus arrive at the area as 1,233,913 acres. Upon the same line of reasoning the annual average for the years 1880 to 1884 would have been 1,120,160 acres, and for the period from 1876 to 1880, 861,671 acres. The year 1876 was the first in which the imports of jute into Calcutta were carefully recorded, and the above figures may therefore be accepted as indicating the expansion of the area under jute in Bengal. As confirmatory of this general conclusion, based on the pub-

Impossible in Madras. 1884

Actual area. 1885

* An effort has been made to correct returns in maunds into cwt. as being more likely to be understood by European readers; but where this has not been done, the result may be arrived at by the following simple rule: maunds × $\frac{5}{7}$=cwt.

lished figures of imports into Calcutta and Chittagong, it may be here added that **Mr. Finucane** (Director of Land Records and Agriculture in Bengal), in his report of 1886, reviews the figures furnished him by an influential jute merchant, **Mr. Field Wilson** of Naraingunge. He says: "This estimate gives the number of bales of raw jute of 400℔ each exported year by year since 1877-78, to which **Mr. Wilson** adds the quantity estimated to have been consumed by the jute mills in Bengal; but the estimate does not apparently take into account the local consumption of the jute in the interior. Making allowance for this omission, **Mr. Field Wilson's** figures, *though it will be observed they are not based on the Custom House returns, nor on the returns of import traffic regis-* 1886
tration stations, yet closely accord with the estimates above given, and afford confirmation of their substantial accuracy." The writer is responsible for the italics in the above quotation. It is desirable to draw attention to the fact that the record of the jute trade preserved by merchants bears a close approximation to that tabulated by Government from the very extensive and complicated returns of road, river, and railway traffic, the concentration in the ultimate centre thus being seen to preserve a distinct relation to the far-reaching ramifications of the stream of supply. But **Mr. Finucane** concludes his review of **Mr. Wilson's** figures as follows:—" If the annual average of the eight years ending 1884-85 be taken into consideration, the difference between the two sets of figures is not considerable, the estimate worked out in this office from the data above described being only 3·97 per cent. less than that of **Mr. Wilson.**"

Soil.—Jute seems to be capable of cultivation on almost any kind of soil. It is least successful and almost unprofitable, however, upon laterite (Soil. 1887) and open gravelly soils, and most productive upon a loamy soil, or rich clay and sand. The finest qualities are grown upon the higher lands (*suna*) in the vicinity of the homestead upon which the *aus* paddy, pulses, and tobacco generally form the rotation. The coarser and larger qualities are grown chiefly upon (*sali* lands), *i.e.*, the churs or mud banks and islands formed by the rivers; and, indeed, these qualities may also be found upon submerged lands, and may be said to luxuriate in the salt-impregnated soil of the Sunderbans.

Climate.—A hot, damp climate, in which there is not too much actual rain, especially in the early part of the season, is the most advantageous; (Climate. 1888) in exceptionally dry seasons one frequently finds crops standing through the cold season which the cultivator did not regard as worth cutting down.

Preparation of Soil.—It may be stated that, when the crop is to be raised on low lands, where there is danger of early flooding, ploughing (Preparation of Soil. 1889) commences earlier than upon the higher lands. The more clay in the soil, the more frequently it is ploughed before sowing. The preparation thus commences in November or December, or not till February or March; the soil is generally ploughed from four to six times; the clods are broken and pulverised; and at the final ploughing the weeds are collected, dried, and burned.

Seed.—No special attention is paid to the selection of good seeds, nor do the cultivators buy and sell their seeds. In the corner of the field a (Seed. 1890) few plants are left to ripen into seed, and these are, next year, sown broadcast. The sowings, according to the position and nature of the soil, commence about the middle of March and extend to the end of June.

Harvest.—The time for reaping the crop depends entirely upon the date of sowing; the season commences, with the earliest crop, about the (Harvest. 1891) end of June, and extends to the beginning of October.

CULTIVATION.

The crop is considered to be in season whenever the flowers appear, and past season, with the fruits. The fibre from plants that have not flowered is weaker than from those in fruit; the latter is coarser and wanting in gloss, though stronger. It is late reaping that is chiefly accountable for the coarse fibre found in the market.

Crop. 1892

Crop.—The average crop of fibre per acre is a little over 15 maunds, but the yield varies considerably, being as high as 30 to 36 in some districts and as low as 3, 6, or 9 in others, and it is also very dependent upon the season. In the experiments performed at the Saidapet farm, Madras, the yield was 599lb of fibre per acre if reaped close to the ground, and 703lb if pulled up by the root—less than a half of the average yield in Bengal.

Retting. 1893

Separation of Fibre by Retting.—At present, as practised by the natives, the fibre is separated from the stems by a process of retting in pools of stagnant water. In some districts the crop is stacked in bundles for two or three days, to give time for the decay of the leaves, which are said to discolour the fibre in the retting process; in others the bundles are carried off and at once thrown into the water. There is some ground for thinking that, if the drying of the leaves by stacking does not prevent the discoloration of the fibre, the fibre itself is likely to be benefited by the process, since it is found to separate more readily from the stems, and is thereby saved from the danger of rotting from over-maceration. In some districts the bundles of jute stems are submerged in rivers, but the common practice seems to be in favour of tanks or road-side stagnant pools. The period of retting depends upon the nature of the water, the kind of fibre, and condition of the atmosphere. It varies from two to twenty-five days. The operator has therefore to visit the tank daily, and ascertain, by means of his nail, if the fibre has begun to separate from the stem. This period must not be exceeded, otherwise the fibre becomes rotten and almost useless for commercial purposes. The bundles are made to sink in the water by placing on the top of them sods and mud. When the proper stage has been reached, the retting is rapidly completed. The cultivator, standing up to the waist in the fœtid water, proceeds "to remove small portions of the bark from the ends next the roots, and, grasping them together, he strips off the whole with a little management from end to end without breaking either stem or fibre. Having brought a certain quantity into this half-prepared state, he next proceeds to wash off; this is done by taking a large handful; swinging it round his head, he dashes it repeatedly against the surface of the water and draws it through towards him, so as to wash off the impurities; then, with a dexterous throw, he spreads it out on the surface of the water and carefully picks off all remaining black spots. It is now wrung out so as to remove as much water as possible, and then hung up on lines prepared on the spot, to dry in the sun" (*Mr. Henley, in Royle's Fibrous Plants, 248*).

Extraction by Machinery. 1894

Extraction of Fibre by means of Machinery.—There seems little doubt but that the retting weakens the fibre very considerably. Could a simple mechanical contrivance be invented for the purpose of extracting the dry jute fibre, and sold so cheaply that it might be procured even by the poorer cultivators, new and at present undreamt-of industries might spring into existence. It is to be feared, however, that machinery will, for some time to come, be beyond the means of the cultivator, and that the principal improvement may be looked for in the application of natural, mineral, or chemical appliances somewhat on the lines of the Ekman Patent process for the separation of the fibre. A machine deserves attention which is known as Garwood's Patent: it does no more than separate the bark from the stem, and the fresher the stem, the more easily is the bark separated.

Mr. W. Cogswell, however, who is an undoubted authority on all questions connected with jute, expressed in December 1881 his opinion that a softer fibre was obtained by the old process (*vide A. H. Society's Proceedings, December 1881*).

PROPERTIES OF JUTE FIBRE.

PROPERTIES OF JUTE. 1895

Chemical and Microscopic.—"The fibre, as found in commerce, consists of the fibre-bundles separated from the cortical parenchyma. The *bundles* contain 6 to 20 fibres. The fibres are firmly coherent in the bundle, the cohesion taking the form of fusion of contiguous walls, the line of fusion being very apparent. The *ultimate fibres* are of the normal fusiform type, 1·5—3 mm. in length. In section they are seen to be thick-walled and polygonal. *Reactions*, characteristic of the jute-allied group of fibres, are brown with iodine; deep yellow with aniline sulphate; purple with phloroglucol and hydrocloric acid; a strong affinity for the basic colouring matters. *Mercerised fibre—Microscopic features.* Concentrated solutions of the alkalies have a remarkable action on fibres of this group. They resolve the bundles more or less completely, and cause the fibre wall to swell so as to almost obliterate the cavity. The filaments, in addition to being made finer, are much softened in texture, and develop a wavy outline, giving the fibre very much the appearance of wool" (*Cross, Beavan, King, and Watt, Report on Indian Fibres, p. 36*). The chemical analysis, as given in the report just quoted, may be here briefly reviewed. Jute, in point of percentage of cellulose (perhaps the best criterion for judging of the value of a fibre), is about equal with **Urena** 77·7, **Calotropis** 76·5, **Abutilon** 75·0 and **Agave** 75·8, and follows after **Abroma** 80·0, **Rhea** 80·3, **Flax** 81·9, **Sida** 83·1, **Crotalaria** 83·0, **Marsdenia** 88·3 and **Girardinia** (Nilgiri nettle) 89·6. Jute possesses 76·0 per cent., and is thus in point of cellulose about the eighth most valuable fibre in India. It is noteworthy that of the fibres enumerated—**Abutilon, Urena, Abroma, Sida,** and **Jute** are obtained from closely allied plants and yield very similar fibres. But of these jute is the next to the last in point of chemical merit, **Sida** being the first of the series. This is a fact of the greatest importance, when it is added that the experts who examined these fibres at the Colonial and Indian Exhibition pronounced **Sida** by a long way superior to jute, being finer in point of fibre, possessing a better colour and thus more suited for many of the higher textile and new purposes to which jute is being applied. Should it be found possible to produce **Sida** at a price anything like that of jute, a formidable rival to the great Bengal fibre would be soon appear in the market.

Mercerised. 1896

Cellulos. 1897

Jute contains 10·3 per cent. of moisture and leaves 1·1 of ash; by hydrolysis or boiling for (*a*) 5 minutes, in a solution of caustic soda (1 per cent. Na_2O), it loses 13·3 and (*b*) for an hour, 18·6. As compared with these results European flax loses (*a*) 14·6 and (*b*) 22·2 and **Sida** (*a*) 6·6 and (*b*) 12·2, while other less valuable fibres lose 50 or 60 per cent. By mercerising, *e.g.*, wetting the fibre with a concentric solution of the alkali (33 per cent. Na_2O), jute loses only 11·0 per cent. As stated above, this has a remarkable effect on the jute and jute-allied fibres, causing the bundles to split up into their ultimate fibres, while at the same time obliterating the cell cavity completely, thus causing the filaments to become much finer and softer in texture. By nitration jute gains in weight, becoming 128, being in this respect inferior to any of its allied fibres, but it is found to contain 47 per cent. of carbon having the highest amount of any recorded Indian fibre; **Sida**, for example, possesses 45·2, flax 43·0, and **Bauhinia** fibre only 40·7.

Ash. 1898

PROPERTIES OF JUTE.

The results of the chemical and microscopic investigation of jute, instituted by **Messrs. Cross, Beavan, and King**, may be briefly stated to be that much more might be made of jute than has as yet been accomplished, especially in the direction of altering chemically its properties and thus adapting it for perfectly new purposes. One sample experimented with was made to resemble tasar-silk so closely that some care was necessary in distinguishing these substances; another looked remarkably like wool.

Strength.

1899 **Strength and Industrial Properties.**—**Royle** remarks: "Jute is certainly characterised by fineness, silkiness, and facility of spinning; but it is less strong than many other Indian fibres, which are possessed of similar properties with greater strength, as we hope to be able to show among the mallow and other nearly allied tribes of plants." This opinion has been fully confirmed above by the results of **Messrs. Cross and Beavan** in their chemical and microscopic examination of jute and the allied jute fibres. Accordingly, at the conferences held in connection with the Colonial and Indian Exhibition, the highest expectations were held out of a formidable rival to jute in the fibre obtained from **Sida rhombifolia.** In the recent report of experiments with Bengal fibres issued by the Agri.-Horticultural Society of India, it is stated that the Society had "arrived at the conclusion that the cultivation of **Hibiscus, Abutilon, Sansiviera,** and **Sida** has no advantage over that of jute." This result must be admitted to be somewhat disheartening, but it should not be forgotten that jute has been cultivated for centuries, that it is in consequence more amenable to the cultivator's necessities and the manufacturer's wants. The question is not, therefore, one as to whether jute or **Sida** is more easily cultivated and gives the better result in point of yield of fibre, but whether the intrinsic superiority of **Sida** fibre would justify its experimental and systematic cultivation until a stock was produced that could be grown as readily and admit of as rapid decortication as is the case with jute. The plant is wild to-day, and it is unfair to compare the yield of fibre from such a plant with results obtained from jute. After careful cultivation for 10 or 20 years it would be fair to compare the ease of cultivation and yield of fibre in **Sida** with that of jute, and during this experimental stage remunerative returns might easily be obtained since there can scarcely be two opinions as to the superiority of **Sida** over jute for the finer textile purposes. **Roxburgh** found in his comparative tests of the fibres of India that a "dry line" of **Corchorus capsularis** broke with a weight of 164℔ and a "wet line" with the same weight, whereas **Corchorus**
1900 **olitorius** gave way with 113 and 125℔ respectively, the wet line gaining 11℔ in weight. This fact of the superiority of the fibre of **capsularis** over **olitorius** is well known in modern commerce. To compare with these results it may be mentioned that, under the same test, a "dry" and a "wet" line of sunn-hemp broke with 160℔ and 209℔, respectively, the latter gaining 31℔ in weight. Testing jute in another way by macerating in water for 116 days, white, tanned, and tarred lines, **Roxburgh** found **Corchorus olitorius** white and fresh, to break with 68℔, after maceration, to give way with 40℔; **C. capsularis** 67℔ and 50℔. Very little difference was observed in the tanned ropes, but the tarred seemed to preserve their strength considerably; the line fresh and tarred broke with 61℔, and after maceration for 116 days bore a weight of 60℔.

1901 The defect of jute is the difficulty to spin the higher counts, 20 being about the finest made, commercially, and when manufactured the fabric lasts well, so long as it is not submitted to a damp influence, but rots rapidly when damp and exposed to the atmosphere.

PRICE OF CULTIVATION. 1902

PRICE OF CULTIVATION.

No trustworthy figures are available of the prime cost to the cultivators of raising and extracting a maund of jute fibre. But the following figures, which have been kindly furnished by a mercantile firm, lead to the rates paid to the growers. Jute landed in Calcutta cost as follows, per maund, in the four years ending 1883:—

Qualities.		1879-80.			1880-81.			1881-82.			1882-83.		
		R	a.	p.	R	a.	p.	R	a.	p.	R	a.	p.
Narainganj.	Fine	5	2	9	5	0	3	4	15	10	3	7	6
	Medium	4	9	6	4	6	9	4	3	4	2	15	2
	Common	4	0	9	3	13	7	3	10	4	2	7	6
Serajganj	Fine	5	4	0	5	2	0	5	1	0	3	9	0
	Medium	4	11	0	4	8	0	4	4	0	3	1	0
	Common	4	2	0	3	15	0	3	12	0	2	9	0

The average prices for the last four years were as follows:—

	Bengal.			Assam.		
	R	a.	p.	R	a.	p.
1883-84	3	12	0	4	0	0
1884-85	3	4	0	2	13	0
1885-86	3	4	0	3	1	0
1886-87	3	10	0	3	2	0

The charges per maund incurred from the time the jute is purchased from the producer to the time it is landed in Calcutta are approximately as follows:—

	Narainganj.			Serajganj.		
	R	a.	p.	R	a.	p.
Freight to Calcutta	0	8	0	0	8	0
Drumming, shipping, &c.	0	2	0	0	2	0
Aratdari	0	2	0	0	2	0
Bepari's profit	0	5	0	0	5	0
TOTAL	1	1	0	1	1	0

Deducting the charges just shown from the cost of the jute landed in Calcutta, will give the rates paid to the grower, thus:—

Qualities.		1879-80.			1880-81.			1881-82.			1882-83.		
		R.	a.	p.	R	a.	p.	R	a.	p.	R	a.	p.
Narainganj	Fine	4	1	9	3	15	3	3	14	10	2	6	6
	Medium	3	8	6	3	5	9	3	2	4	1	14	2
	Common	2	15	9	2	12	9	2	9	4	1	6	6
Serajganj	Fine	4	3	0	4	1	0	4	0	0	2	8	0
	Medium	3	10	0	3	7	0	3	3	0	2	0	0
	Common	3	1	0	2	14	0	2	11	0	1	8	0

The prime cost to the cultivators must be something lower than the figures shown in this last statement; and assuming that the data fur-

PRICE OF CULTIVATION.

nished are near the truth, if not correct, they lead to the following important inferences, *viz.*, (*a*) that the price of jute has declined considerably during the past few years, and (*b*) that while the profits of the middlemen have not varied, those of the growers have fallen proportionately with the fall of prices in Calcutta. The price of jute fluctuates very considerably; a good year induces an indiscriminate extension of the area which must of course be attended the following year by a fall in price, and from heavy losses this has the opposite effect of an undue contraction of the jute area. In 1883-84 the price averaged from R3 to R8 a maund. On the other hand, prices in Dacca fell in 1882-83 to 12 annas a maund, and, in consequence, the cultivators suffered much, although in ordinary years they are the most prosperous people in Bengal, and can often earn as much as 10 to 12 annas a day. Scarcity of rain, at the sowing season, produces a bad crop, and a caterpillar often does great damage. In the forecast for 1887, issued by the Revenue and Agricultural Department, Government of India, the following passage occurs:—"The trade statistics of the year have shown that the importation of raw jute to Calcutta from all sources was practically the same as in the previous year; while the value of the exports from Chittagong was twenty-seven lakhs more than that of the previous year. It thus appears that the crop was a larger one than that of the previous year. Owing, however, to the lowness of exchange, and to a brisker demand in Europe, prices were on an average 15·4 per cent. higher than in the preceding year.

1903 "For this reason a larger area than usual has been sown this season, save in limited tracts which had suffered from floods in the two previous years. The prospects of the crop were generally excellent to the end of May, when the young plants were seriously damaged by floods which accompanied the cyclone, especially in the districts of Rungpore, Rajshahye, Dinagepore, Bogra, Julpigoree, and parts of Hooghly. These localities, however, excepting Rungpore, are not of first-rate importance as jute-growing districts.

"On the whole, so far as can be judged at present, it may be said that the area sown this year is about 10 per cent. above that of last year; and, taking into consideration the facts that the area sown is above the normal, and that the deficient outturn caused by floods in some districts will be counterbalanced by the bumper yield in others, it may be expected that the total outturn will be a full average. Much will, however, depend on the distribution of rainfall in the latter half of July and beginning of August."

The following table, extracted from **Mr. Finucane's** Report (to which frequent reference has been made), shows the average wholesale price of jute per maund since 1876, and at the same time gives a key to the valuations returned by the Custom House:—

	Average wholesale price in 12 selected districts in Bengal.			Average declared value as per Custom House Returns.		
	R	a.	p.	R	a.	p.
1876-77	3	0	0	4	4	0
1877-78	3	0	0	4	12	0
1878-79	4	0	0	4	10	0
1879-80	4	10	6	4	13	0
1880-81	4	8	0	4	14	0
1881-82	4	8	0	4	14	0
1882-83	3	8	0	4	1	0
1883-84	3	12	0	4	12	0
1884-85	3	4	0	4	1	0

COMMERCIAL VARIETIES.

COMMERCIAL VARIETIES. 1904

There are several well-known commercial VARIETIES of jute fibre, of which the following, arranged in the order of their commercial importance, may be mentioned: *Uttariyá, Deswál, Desi, Deorá, Serajganji, Narainganji, Bákrabadi, Bhatial, Karimganji, Mirganji*, and *Jungipuri*.

For convenience of reference we shall discuss these in alphabetical order, those of importance being marked *.

1. **Bakrabadi.**—A beautiful soft fibre, one of the finest qualities from the Dacca district, being raised on the *churs* of the Megna river.
2. **Bhatial.**—A coarse strong fibre, chiefly exported to Europe for rope manufacture. It is grown on *churs* and obtained from the south of Narainganj; hence the name, from *bhati*, tidal.
3. * **Deora** (in commerce *Dowrah*).—A strong useful fibre, used chiefly in rope manufacture. It derives its name from a village near Faridpur, where there was formerly a large mart for this variety of jute. The name is given to all the jute from Backerganj and Faridpur.
4. * **Desi** (in commerce *Daissee*).—This is a useful and good fibre, largely used for gunnies; it is long, soft, and fine, but it has a bad colour and is pronounced "fuzzy." It is produced in the districts around Calcutta, such as Hugli, Burdwan, Jessore, and the 24-Parganas.
5. * **Deswal.**—A fine bright-coloured fibre, much admired on account of its strength. After the *Uttariyá* this is, commercially, the most important variety. It comes from the neighbourhood of Serajganj, and is said to consist of two kinds or sub-varieties:—
   (*a*) Bilan Deswál, or fibre from the crop grown over *bhíls* or marshes.
   (*b*) Charna Deswál, or fibre from the crop grown on *churs*.
6. **Jangipuri.**—A poor fibre, short, weak, and more suited for paper manufacture than for spinning. It comes from the Pubna district.
7. **Karimganji.**—A fairly good fibre, very long, and of good colour. It comes from the Mymensingh district, taking its name from a small village.
8. **Mirganji.**—Generally an inferior fibre; the worst kind coming from Mirganj, a village on the Teesta. The fibre generally comes from the Rungpore district.
9. * **Narainganji** (in commerce *Naraingunge*).—This is an excellent fibre for spinning, being long and soft. It comes from the Dacca district, and is exported to Calcutta from the Narainganj marts.
10. * **Serajganji** (in commerce *Serajgunge*).—Produced in the Pubna and Mymensingh districts.
11. ** **Uttariya.**—This is regarded as the finest variety; it is long, has a brilliant colour, is strong and easily spun, but it is not up to *Desi* or *Deswál* in softness. It comes into the market in November. It receives its name on account of its coming from the northern portions of Serajganj and that neighbourhood. The following are the localities from which it is obtained: Rungpore, Goalpara, Bogra, parts of Mymensingh, Kuch Behar, and Julpaiguri.

These 11 qualities, and others of minor importance, are in commerce generally grouped under four leading classes represented by the *Serajganj, Narainganj, Desi*, and *Deora*; and these, again, are classed as "Fine," "Medium," and "Common," according to the qualities of the fibres. Mr. James Duffus, in a letter addressed to the writer, says of this 1905

COMMERCIAL VARIETIES.

subject: "Every small mart in Eastern Bengal has a jute of its own, quite as worthy of mention as many of the minor forms alluded to above." This remark has an interest beyond that of commerce, for we must either infer that this extensive series of qualities of fibre indicates distinct forms of the plant or is due to methods of cultivation and cleaning of the fibre. If the former, this fact might be accepted as an indication of long cultivation or even of extensive hybridisation.

FOREIGN TRADE. 1906

FOREIGN TRADE IN JUTE AND JUTE MANUFACTURES.

For full particulars of this trade up to date see **JUTE** in another volume. The present article is intended more as a historic sketch of the jute industry in which an attempt is made to give the main facts of the cultivation of the plant, and of the Indian manufactures.

INTERNAL TRADE. 1907

INTERNAL AND COASTING TRADE.

A good deal of the details of this trade will be discussed under the headings "Home Consumption of Raw Jute" and "Home Consumption of Jute Manufactures." It may not be out of place here to indicate very briefly the relative share participated in the trade by the various existing modes of conveyance. In a special Report on this subject **Colonel L. Conway-Gordon, C.I.E.**, gives the figures from 1880 to 1885, but he refers apparently to raw jute only. During the last year dealt with by him he states that "the amounts of jute exported by sea from Calcutta and Chittagong" were 7,158,868 cwt., and 1,033,733 cwt.,* respectively, the latter port thus taking 12·6 per cent. of the total. This would give the foreign exports as 8,192,601 cwt., whereas in the statement of the Trade of British India for that year the foreign exports were put down at 8,368,686 cwt. and the coasting trade at 1,267,034 cwt., making a total of jute shipments from Indian ports of 9,635,720 cwt. **Colonel Conway-Gordon** gives the total imports into Calcutta as 9,392,813 cwt., of which 3,579,062 cwt. were conveyed by native boats; 1,969,237 cwt. by steamers; 3,482,522 cwt. by the Eastern Bengal Railway; 148 cwt. by the South Eastern State Railway; 356,496 cwt. by road; and 5,348 cwt. by sea. Thus the COUNTRY BOATS head the list, carrying to the sea-board 38·1 per cent. of the total jute supply—the EASTERN BENGAL RAILWAY carrying 37·0 per cent., and the INLAND STEAMERS only 20·9 per cent. The bulk of the BOAT TRAFFIC comes from Serajganj on the Bhramaputra, *viz.*, 894,920 cwt., Pubna sends 302,413 cwt., Nassirabad (in Mymensingh) 298,736 cwt., Madaripore 281,729 cwt., and Narainganj 223,889 cwt. The STEAMERS draw the bulk of their traffic, 1,169,656 cwt., from Narainganj
1908 and from Serajganj 602,468 cwt., while the RAILWAY traffic is mostly from Goalundo, represented by 1,300,580 cwt., from Pangsa 242,082 cwt., and from Kooshtea 219,763 cwt.

These figures give a fair conception of the local trade in jute, and from the large share still carried by native craft, an additional proof may be obtained of the value of the jute crop to the people of India. Were it possible to estimate the number of people who make it an important part of their duty to either cultivate, clean, buy, or sell the fibre, and were that number to be added to those employed in conveying it from the fields to the mills, it would be seen that jute is of importance to a far larger number of persons than to the 50,000 who find daily employment in the

* For the purpose of allowing of comparison with the returns of foreign trade, **Colonel Conway-Gordon's** figures of maunds have been converted into cwt.

European factories. But even this estimate would leave out of all consideration the indigenous hand-looms that are still able to compete with steam in the production of jute cloth, bags, and cordage. **HOME MARKET.**

RAW JUTE.

EXPORTATION AND HOME CONSUMPTION.

EXPORTS.

The following abstract of the EXPORTS OF RAW JUTE FROM CALCUTTA will be found interesting, as showing the steady and constant increase and development of the jute trade. The mean exportations for each period of five years, during the 55 years commencing with 1828, will be seen to have, in round numbers, almost doubled those of the preceding period. It should be carefully noted, however, that these figures represent but a portion of the jute industry,—namely, the exports :— 1909

Up to	Average of five years, in cwt.
1832-33	11,800
1837-38	67,483
1842-43	117,047
1847-48	234,055
1852-53	439,850
1857-58	710,826
1862-63	969,724
1867-68	2,628,110
1872-73	4,858,162
1877-78	5,362,267
1882-83	7,274,000

The foreign exports of raw jute were, in 1882-83, 10,348,909 cwt. valued at R5,84,69,259, since which they have declined considerably, being in 1886-87 only 8,306,708 cwt. valued at R4,86,98,146. The exports of 1882-83 were the highest on record.

The rapid, yet constant, increase in the jute trade, which the above 1910
figures show, from 364 cwt. in 1828 to 10,348,909 cwt. in 1882-83 representing an increase in value from R620 to R5,84,69,259 in the short period of 55 years (*e.g.*, from £62 to £5,846,925 for exported raw jute alone) speaks volumes for the noble fleet of merchant vessels trading with our Indian ports. **Mr. Hem Chunder Kerr**, in his valuable *Report on the Cultivation of, and Trade in, Jute in Bengal*, has laid much stress upon the Russian war in 1854-55 as a cause of the development of the jute trade of India. It doubtless was a cause, but perhaps only an insignificant one as compared with the demand for a cheap fibre when combined with the internal administrative reforms and the engineering enterprise which, by railway, road, and canal, brought the resources of India into the field of European commerce.

The figures of Indian trade show that the exportation of jute steadily increased from 1,092,668 cwt. in 1860-61, to 3,754,083 cwt. in 1870-71; that in 1871-72, it suddenly rose to 6,133,813 cwt., and during the past 5 years has preserved an average of about 7,274,000 cwt.

In 1882-83 Indian commercial men calculated that on an average 1911
Scotland consumed over 18,400 bales (73,600 cwt.) a week. Of these **Messrs. Cox Brothers** take 2,200; **Messrs. Gilroy & Sons**, 750; **Messrs. Malcolm, Ogilvie, & Co.**, 650; **Mr. John Sharp**, 700. In England the weekly consumption is over 1,860 bales, the largest consumers being the **Barrow Company**, 600. In Ireland the total weekly

EXPORTS.

consumption is about 730 bales, the largest firm consuming under 300 bales a week. Thus Great Britain requires over 21,000 bales or 84,000 cwt. a week, or 4,200,000 cwt. a year to keep her existing jute factories employed. These figures, when compared with the hand-loom consumption in Bengal, show how completely the gunny trade has passed out of the hands of the Indian peasant. The entire hand-loom consumption of jute in Bengal has been returned as 2,23,000 maunds a year, but allowing 50,000 maunds more to cover imperfections, this would give an annual consumption of 195,000 cwt. The Scotch power-looms alone consume 73,600 cwt. a week, or 3,710,000 cwt. a year. Although in some respects this estimate has been disturbed, it is relatively correct for the present year 1887-88.

1912 France requires 4,000 bales a week, its largest consumer, Saint Freres, requiring 700 bales; Germany requires 2,170 a week, of which the Brunswick Jute Spinning Company consume 770 bales; Belgium requires 845 bales a week; Austria, 580; Spain, 250; Holland, 400; Norway, 100. Taking annual figures for the whole of Europe it is found that Great Britain and the Continent of Europe require 1,800,000 bales a year, or 6,428,580 cwt. It may be here stated that as merchants adopt the calendar year, and Government the financial, *e.g.*, from April to March, considerable difficulty has been experienced in comparing the Government Statistical Tables of Exports with those kindly supplied by one or two well known jute firms in Calcutta.

1913 Comparing with the above figures the 22 Indian factories at work in India in 1882-83, which on an average consumed each 500 bales per week, or 600,000 bales a year, equivalent to 2,142,948 cwt., it would appear that to keep these factories working, about 8,571,428 cwt. of raw jute were required; and adding to this amount the quantity annually consumed by America, Australia, and other foreign countries, *viz.*, 600,000 bales, or 2,142,498 cwt., not included in the above calculation, the annual consumption was little short of 3,000,000 bales, or 10,714,476 cwt. These were the estimates framed for 1882-83, but in an early page it has been stated that this year's production is probably close on 20 million maunds, thus showing a very considerable expansion, although the exports of raw jute have declined somewhnt during the past five years.

Annual Capital.

1914 Looking at the exportation of raw jute, of manufactured jute, and the home (Indian) consumption known to our commercial men, the statement that the jute trade is at least represented at the present date by an annual consumption of over 15,000,000 cwt. of raw jute does not seem to be far from correct. This is roughly equivalent to an annual turn over of capital equal to about 12—14 millions of pounds sterling as compared with the exports in 1828 of £62.

MANUFACTURES.

THE MANUFACTURES OF JUTE AND THEIR EXPORTATION FROM INDIA.

1915 In the vicinity of Calcutta, since 1857, 22 jute factories have sprung up in rapid succession, with one (a small one) at Vizagapatam, and another at Cawnpore, where it commenced work in 1887. A mill was started also in Bombay in 1885, but in the following year it was removed to Calcutta. The nominal capital of the mills worked by Joint Stock Companies is stated in the returns at 285 lakhs, which, at the conventional exchange of 10 rupees to the pound sterling, would be £2,850,000. The others are private factories, but their capital may be put down at 30 to 40 lakhs of rupees. These 24 factories have 7,164 looms and 135,593 spin-

MANUFACTURES.

dles, and they give employment to 29,660 men, 11,198 women, 5,113 young persons, and 3,044 children. The Madras private jute company employs about 878 persons. Thus, up to the present date, there are in all India 24 Jute factories, which give employment to 49,015 persons and use up 2,869,088 cwt. of jute. They are almost exclusively employed in the gunny bag or cloth trade, three only doing a small business in cordage, floor cloth, or other manufactures.

In 1879 there were in England 12 factories, in Scotland 99, in Ireland 6, in all 117 factories, with 212,676 single and 7,492 double spindles, and 11,288 looms, giving employment in all to 36,354 persons. In India there are only 24 factories, but these employ 49,015 persons.

It is difficult to make a reliable comparison without the details of
every individual factory. Judging from the published statistics of jute
factories in Scotland during the year 1879, and comparing a fixed number of these with the Indian factories for the same year, we may, however,
conclude that the Indian mill workman was inferior to the Scotch work- 1916
man in the ratio of 3 to 7. That is to say, it requires 7 persons to work
one loom in an Indian factory, against 3 workmen in a Scotch factory.
This conclusion is arrived at by dividing the total number of persons
employed in a factory by the number of its looms, and obtaining the average for all Scotch factories and the average for all Indian factories. Of
course this calculation is open to the error of the Indian and English
factories not manufacturing the same class of goods; but relatively it
may be accepted as giving some sort of comparison.

Foreign Trade in Manufactures.

Foreign Trade in Manufactures. 1917

Prior to 1857 the exports of Jute manufactures from India represented hand-loom fabrics. In 1850 these were valued at £215,978, whereas the trade in raw jute was only £197,071. Fifteen years later the manufactured jute, exported to foreign countries, was valued at R18,27,983 (£182,798) and the raw jute at R75,06,690 (£750,669). In 1870-71 the exports were of manufactured jute R34,24,249 (£342,424) worth and of raw jute R2,57,75,526 (£257,755). But the revival in the exports of manufactured jute indicated by these figures, as also the partial decline of the foreign raw jute trade, was at once the death of the old hand-loom industry and the birth of the new power-loom. Ten years later (1880-81) the total exports of manufactured jute were valued at R1,13,06,716 (£1,130,671), of which the hand-looms produced R2,69,553 (£26,955), and last year they were valued at R1,15,18,577, (£1,151,857), of which the hand-looms produced R89,220 (£8,922). These figures indicate unmistakeably the growth of the Indian power-loom *foreign* trade and the decline of the hand-loom. In a further page some idea will be given of the extent of the *home* market for jute goods.

Local or Home Consumption.

Local Consumption. 1918

It should be carefully observed that the returns published by Government show only the exports, properly so called, of bales of prepared gunny-bags, gunny-cloth, or jute rope as such. They do not include the millions of gunnies, &c., which annually leave the ports of India containing grain or other produce, nor those used for home purposes or sent to other parts of India. These figures do not, therefore, show the whole outturn of gunnies annually manufactured in India. In fact, from January to December 1882, 119,042,771 gunnies were actually made by power-looms, of which only 41,523,607 were exported; so that the exports were barely one-

CORCHORUS. The Jute Fibre

MANUFACTURES. Home Consumption.

third of the number actually manufactured. The following table will show the relations of the home consumption to the exports more clearly :—

Statement of Home Consumption and Exports of GUNNIES *from 1st January to 31st December 1882.*

1919

Burma	13,312,306	
Straits	9,153,233	
Bombay and Persian Gulf	20,001,308	
Madras and Malabar	1,064,848	
Coromandel Coast	3,609,950	
Ceylon	177,777	
Up-country by rail	11,351,000	
Used in the export trade of Calcutta	11,848,742	
Total of Home Consumption	...	77,519,164
Australia	11,372,387	
New Zealand	5,060,160	
Cape of Good Hope	706,308	
Mauritius	119,078	
Egypt	691,078	
America	20,554,251	
Hongkong (not Hessians)	413,700	
Great Britain	516,417	
Europe	90,231	
Total of Foreign Exports	...	41,523,607
Grand Total of Home Consumption and Foreign Exports	...	119,042,771

The total number of gunny-bags brought to, and carried from, Calcutta during the past three years may be here given and alongside of these the foreign exports :—

	1884-85.	1885-86.	1886-87.
Imports	18,196,002	20,626,541	23,586,402
Total Exports (to other provinces of India and to foreign countries)	137,870,318	127,084,964	124,957,225
Foreign exports only	82,779,207	63,760,546	64,572,157

1920 The difference between the total exports from Calcutta and the foreign exports approximately represents the home (Indian) consumption, although there is doubtless a balance between the total of production + imports and the exports, which would represent the Calcutta local consumption. This in 1882 was estimated to be over 11 million bags, so that last year the total production of gunny-bags in Bengal was perhaps little short of 150 millions, of which 64½ millions were sent to foreign countries and 85½ millions used up in India. This may be accepted as representing the bags employed in the home, cotton, oil-seed, rice, and wheat trade, and in the export trade of India.

But in addition to gunny-bags India exported last year 12,799,225 yards of gunny-cloth, valued at R9,80,741, and this exclusively of the interportal trade which amounted to 5,728,858 yards (nearly the whole of this quantity going to Bombay), making a total of 18,480,001 yards as against 25,267,418 yards in 1885-86, and 19,923,884 yards in 1884-85. But in addition to these returns of gunny-cloth conveyed by sea, the Report of the River-borne

Traffic of Bengal for 1887 states that 605,846 pieces were sent up-country by river "direct from the jute mills without passing the Port Commissioner's wharves." A piece of power-loom gunny is equal to 80 yards, of hand-loom, to 22 yards, so that this power-loom trade alone represents close upon 50,000,000 yards of cloth, an amount that was conveyed to Durbhunga, Monghyr, Bhagulpore, Patna, Moorshedabad, Nuddea, Purneah, Chumparun, Rajshahye, the list concluding with an entry marked "for all other districts," 12,625 pieces. The districts are enumerated in the order of importance, Durbhunga having received 221,630 pieces, or fully a third of the total amount, while Monghyr took 164,556. Large though these quantities are we are left no other inference than that there must be a very considerable trade conveyed by railway, road, and river to the other provinces. Accepting the published figures as they stand, however, we thus learn that the internal trade in gunny-cloth is three or four times as great as the foreign exports. Some difficulty exists in working out the total internal trade, as in many of the returns of railway traffic gunny-bags and gunny-cloth are treated collectively, while in others they are given in maunds, or again in yards, &c. But enough has been said to convey a tolerably clear conception of the extent of the internal trade both in bags and cloth. It may be added, however, that the bulk of the hand-loom industry is conducted in Dinagepore, Purneah, Rungpore, Julpaiguri, and Tipperah; Julpaiguri turned out last year 2,336,660 and Rungpore 1,222,410 hand-loom made bags.

MANUFACTURES. Home Consumption.

CLASSIFICATION OF THE JUTE MANUFACTURES.

CLASSIFICATION OF MANUFACTURE. 1921

The manufactures from *jute* or *pát* may be referred to three primary sections :—

I.—CLOTH of different qualities ranging from substitutes for silk to shirtings, curtains, carpets, and gunnies.

II.—PAPER chiefly prepared from the "rejections" and "cuttings."

III.—CORDAGE from the coarser and stronger qualities.

These three sections may each be referred to a number of sub-divisions, which for convenience may be arranged in two leading groups, *viz.*, native and indigenous manufactures, "hand-loom," and European or "power-loom" manufactures, whether made in Europe or in India. We shall first enumerate the indigenous manufactures, since these bear on the history of the industry.

INDIGENOUS MANUFACTURES.

Indigenous Manufacture. 1922

Indigenous Cloth.—Every homestead in Bengal has suspended from a beam in the roof of the verandah a few bundles of jute fibre, which, while talking pleasantly with a neighbour, the peasant twists, with various kinds of spindles, into twine of varying thickness, intended for domestic purposes or for the yarn from which the women prepare the home-spun cloth or gunny-bags. Babu Ramcomal Sen, in the Transactions of the Agri.-Horticultural Society, describes three different modes of preparing twine or yarn in Bengal. The first is by means of a reel, called a *dhera*, the second by the *takur*, and the third by the *ghurgurra*. The first is said to be used in making yarn for gunnies, the second for fine yarns intended for cloth, and the third for twine to be afterwards made into ropes.

The natives weave three distinct kinds of jute cloth :—

1st, Thick cloth used for making gunny-bags. Of this there are three qualities, the best being known as *amrabati*. These correspond to the three qualities of hand-loom gunnies in commerce.

The Jute Fibre.

CLASSIFICATION OF MANUFACTURES.

2nd, Fine cloth.—This is generally known by the name of *mekli dhokrá*, and is chiefly used as a cloth to sleep on; it is often beautifully striped blue or red.

3rd, Coarse cloth.—This is largely used for making the sails of country boats (*gún*), and also for bags to hold large seeds or fruits.

The following are the principal districts in Bengal where indigenous jute manufactures (hand-looms) may be said to exist to any considerable extent:—Hugli, consuming about 1,20,000 maunds of jute a year; Dacca, 90,000; Rungpore, 50,000; Moorshedabad, 38,000; Malda, 25,000; Julpaiguri, Pubna, &c., smaller quantities.

European Manufactures. 1923

EUROPEAN MANUFACTURES.

Cloth made in Factories.—Jute is now largely used in the manufacture of carpets, curtains, shirtings, and is also mixed with silk or used for imitating silk fabrics. It has been applied extensively as a substitute for hemp: for this purpose the fibres are rendered soft and flexible by being sprinkled with water and oil, in the proportion of 20 tons of water and 2½ tons of train oil to 100 tons of jute. Sprinkled with this the jute is left for from 24 to 48 hours, when after being squeezed by rollers and heckled, the fibres become beautifully soft and minutely isolated, and thereby suited for a number of purposes unknown a few years ago.

The history of this trade is exceedingly interesting. In the year 1820 the fibre was first experimented with, but the result was unfavourable; and, in consequence, brokers were required to certify that sales of hemp and other fibres were not adulterated with jute. In 1832 an enterprising Dundee manufacturer experimented once more on the fibre, and the result was that he was able to show that it might be used as a substitute for hemp. From that date jute gained rapidly in public favour. It is one of those fibres that are capable of the most minute separation or subdivision, but only within the past few years has it been extensively used in the finer textile industries. For a long time the difficulty of bleaching seemed insurmountable, and the trouble experienced in dyeing the material appeared likely to nullify every effort to utilise it. All these stumbling-blocks have, however, been removed, and there cannot be a doubt that, but for the want of durability, jute would soon rank as the most valuable of all fibres. Its perishable nature, however, is fatal to its obtaining a position much higher than it has already attained, and probably admixture of jute in certain articles, such as sail-cloths, must sooner or later be viewed as a criminal offence. For information regarding the most recent developments of the European applications of jute see *Bull. Soc., Mulhouse, 1881*. The manufactures which occupy the attention of our Indian companies are almost exclusively the various forms of gunnies.

JUTE WHISKEY. 1924

JUTE WHISKEY.

In concluding this account of jute it may be mentioned as a curiosity that it has been proposed to utilize the jute ends in the preparation of a spirit which somewhat resembles the whiskey made from grain. The waste fibre is by means of sulphuric acid converted into sugar and the resulting product thereafter fermented and distilled.

CORDIA, *Linn.; Gen. Pl., II., 838.*

1925 **Cordia fragrantissima,** *Kurz; Fl. Br. Ind., IV., 139;* BORAGINEÆ.

Vern.—*Kalamet, toungkalamet,* BURM.

References.—*Kurz, For. Fl. Burm., 207; Gamble, Man. Timb., 271.*

The Sebesten Fruit.

Habitat.—A deciduous tree of Burma, chiefly in the hills of Martaban and Tenasserim.

Structure of the Wood.—Wood moderately hard, reddish-brown with darker streaks, beautifully mottled, has a fragrant scent; should be better known. It has a handsome grain, and its fresh, fragrant odour makes it very pleasant to use. Pieces sent to London for sale in 1878 realized £4-10 per ton (*Gamble*). TIMBER. 1926

Cordia latifolia, *Roxb.*; see **C. obliqua,** *Willd.*

C. Macleodii, *Hook. f. & Th.; Fl. Br. Ind., IV., 139.* 1927

Vern.—*Dhengan, dhaman, dháian, dewan, dahi, dahipalás, dihgan,* HIND.; *Reuta, porponda,* KOL.; *Bharwar, belaunan,* KARWAR; *Jugia,* SANTAL; *Dhaiwan,* SATTARA; *Dhaiwan, dhaman, daiwas, dhaim, bhoti,* MAR.; *Bot,* GOND; *Lauri kassamár,* KURKU; *Gondu,* RAJ.; *Godela,* MERWARA; *Gadru,* AJMERE.

References.—*Brandis, For. Fl., 337; Gamble, Man. Timb., 271; Duthie, Report on Bot. Tour in Merwara, 17; Griffith, Calc. Jour. Nat. Hist., III., 363; Baden Powell, Pb. Pr., 575; Lisboa, U. Pl. Bomb., 105.*

Habitat.—A middling-sized deciduous tree of Central India, the Concan, and Belgaum.

Gum.—Mr. E. A. Fraser (*Assistant Political Agent*) says that in Ráj-putána this tree affords a gum. GUM. 1928

Medicine.—The Santáls use the bark medicinally in jaundice (*Campbell*). MEDICINE. 1929

Structure of the Wood.—Heartwood light-brown, beautifully mottled with darker veins, even-grained, very hard, strong, tough, and elastic; seasons well and works easily. It is used for furniture, picture-frames, and other ornamental work; also for fishing-rods, which are said to be excellent. It deserves to be better known and more used. The Santáls value the timber for making bullock yokes. TIMBER. 1930

C. Myxa, *Linn.; Fl. Br. Ind., IV., 136; Wight, Ic., t. 169.* 1931

This fruit is known as the SEBESTEN by Anglo-Indians.

Vern.—*Lasora, lasúrá, bhokar, gondi,* HIND.; *Bohari, buhal, bahubara, boho-dari,* BENG.; *Laswara,* PB.; *Borla, bairala, baurala,* KUMAON; *Embrúm,* KOL.; *Buch,* SANTAL; *Boeri,* NEPAL; *Nimat,* LEPCHA; *Doba-kari,* MECHI; *Gondi,* URIYA; *Lasora,* RAJ.; *Lesúri, gidúri,* SIND; *Motá bhokar, bhokara, sapistán, bargund, lesúri gedúri, vargund, gedúri, sepistar, pistan, semar, goden, gondan,* BOMB.; *Gundo moto, race gundo, gundo, ráya gundo, vad-gunda, lepistan, pistan,* GUZ.; *Bhokar, montá bhokar, bhokara, bhokur, bargund, vargund, semar, goden, gondan,* MAR.; *Vidi, verasu,* TAM.; *Pedda boku, virgi, nakkera, irki, iriki, pedda-baketu,* TEL.; *Chotte, chelutimara, chella, challemara,* KAN.; *Koda,* N.-W. P.; *Selte,* GOND.; *Silu,* KURKU; *Lasséri,* BAIGAS; *Bahuváraca,* SANS. (according to Fleming); *Búkampadáruka,* SANS. (according to Ainslie); *Dábk,* ARAB.; *Súgpistan,* PERS.; *Chaine,* MAGH; *Thánat, toung thanat,* BURM.; *Lotú,* SING.

References.—*Roxb., Fl. Ind., Ed. C.B.C., 198; Brandis, For. Fl., 336* (in part); *Kurz, For. Fl. Burm., II., 208; Beddome, Fl. Sylv., 245; Gamble, Man. Timb., 270; Thwaites, En. Ceylon Pl., 214; Dalz. & Gibs., Bomb. Fl., 173; Rheede, Mal., IV., t. 37; Fleming, Asiatic Res., Vol. XI.* (1810), *164; Official correspondence in Home Department regarding Indian Pharmacopœia, p. 2937; Elliot, Flora Andhrica, 23, 123; Ainslie, Mat. Ind., II., 466; S. Arjun, Bomb. Drugs, 95; Liotard, Memo., Paper Mat., 10; Forbes Watson, 53; Campbell, Descriptive Cat. of Plants of Chutia Nagpur; Baden Powell, Pb. Pr., 368, 575; Atkinson, Him. Dist., 733, 794; Lisboa, U. Pl. Bomb., 248, 156; Birdwood, Bomb.*

CORDIA Myxa.

The Sebesten Fruit.

Pr., 169; Sind Gaz., 559; Bomb. Gaz., XV., 66; XIII., 23, VII., 42; Ind. For., VII., 82, IX., 216; Smith, Dic., 374; Kew Off. Guide to the Mus. of Ec. Bot., 98.

Habitat.—A moderate-sized deciduous tree, met with in the Salt Range sub-Himálayan tract, from the Chenab to Assam, ascending to 5,000 feet, the Khásia Hills, Bengal, Burma, Sind, Western (North Kanara), Central, and South India.

Mr. Atkinson says it is cultivated throughout the plains: is wild along the Himálayas, and flowers in March and April, the fruit ripening in May to July.

GUM. 1932 **Gum.**—Said to yield a gum in Rájputána.

DYE. 1933 **Dye.**—**Dr. McCann** states in his *Report on the Dyes of Bengal (pp. 32, 35, and 143)* that the green leaves of this tree are in Darjiling used in dyeing, along with **Morinda tinctoria.** In the N.-W. Provinces the juice of the fruit is used as a dye (*Atkinson, Econ. Prod., N.-W. P., V., 81*).

FIBRE. 1934 **Fibre.**—The bark is made into ropes, and the fibre is used for caulking boats; fuses are also made from it. **James,** in his report of Chanduka (1847), says "that from the inner bark is obtained a fibre, from which the coiled match of the native fire-arms is made."

MEDICINE. 1935 **Medicine.**—The fruit, *Sebestan,* is officinal and given for coughs. It is very mucilaginous, and the mucilage of the fruit is demulcent and used in diseases of the chest and of the urethra, and also as an astringent gargle. "The natives pickle the fruit of both **C. Myxa** and **C. obliqua.** Medicinally the dried fruit is valued on account of its mucilaginous nature and demulcent properties." "In large quantities it is given in bilious affections as a laxative." "Both kinds of fruit when dry are shrivelled, and of the colour of a dry prune." The pulp of **C. obliqua** can be separated from the nut, that of **C. Myxa** cannot; on sawing through the nut a heavy disagreeable smell is observed" (*Dymock*). The kernels are a good remedy for ringworm. **Mr. Baden Powell** says the leaves are useful as an application to ulcers and in headache. **Mr. Atkinson** remarks: The juice of the bark along with cocoa-nut oil is given in gripes. The bark and also the unripe fruit are used as a mild tonic. **Mr. Campbell** says that the Santáls use a powder of the bark as an external application in prurigo. According to **Horsfield** the bark of **C. Myxa** is used by the Javanese as a tonic, and is one of their chief remedies in fever. The leaves are also employed medicinally by the Santáls.

Special Opinion.—§ "According to Sanskrit writers the bark is useful in calculus affections, strangury, and catarrh. The ripe fruits are sweet, cooling, and demulcent" (*U. C. Dutt, Civil Medical Officer, Serampore*).

FOOD. Fruit. 1936 **Food.**—The fruit grows in clusters and consists of a drupe, the pulp of which is soft and clammy.

"The fruit when ripe is eaten by the natives and also pickled * *: the smell of the nuts when cut is heavy and disagreeable: the taste of the kernels is like that of filberts" (*Drury*).

In a report on Chanduka in Sind (1847), it is stated that the fruit, which "contains a great deal of mucilage, is eaten by the natives: it is also used in the preparation of spirituous liquors." **Mr. Atkinson** says the unripe fruit is pickled, and the ripe fruit eaten raw or stewed. **Dymock** mentions that the fruits were eaten during the famine of 1877-78 in the Nasik District.

FODDER. 1937 **Fodder.**—The leaves are given to cattle as fodder. The lac insect feeds on this plant (*Indian Forester, VIII., 82*).

TIMBER. 1938 **Structure of the Wood.**—Wood grey, moderately hard. In spite of its softness, it is fairly strong, and seasons well, but is readily attacked by insects. It is used for boat-building, well-curbs, gun-stocks, and agricultural implements; in Bengal for canoes. It might be tried for tea-boxes. It makes an excellent fuel. In a report of Chanduka in Sind (1847), it is stated that "the wood is used for sword sheaths." The Santáls regard the wood as specially useful for yokes, as it does not irritate the necks of the bullocks (*Rev. A. Campbell*).

Domestic Uses.—The leaves are apparently used in Burma as cheroot wrappers. In the Burma Forest Administration Report for 1881-82 there occurs the following entry: "R57-6 were received on 11½ million cheroot leaves (from **Cordia Myxa**) in the Tharawaddy Division." **Ainslie** says the wood is used to procure fire by friction. **Mr. Atkinson** says of the North-Western Provinces that the leaves are used as plates, and that the viscid pulp of the fruit is used as bird-lime. DOMESTIC. 1939

Cordia obliqua, *Willd.* 1940

This is the larger SEBESTEN according to **Stocks, Dymock, Birdwood,** &c., **C. Myxa** being the lesser, but the vernacular names given would imply the reverse to be the case.

Syn.—C. LATIFOLIA, *Roxb.*

Vern.—*Chhótá-lasórá, chhotá-laslasá*, HIND.; *Chhoto-bohnaári*, BENG.; *Mokhátah* or *mukhitah*, ARAB.; *Sagpistán, sapistán*, PERS.; *Chhóti-góndni*, DEC.; *Nanu-gúndi, gadgundi, vargúnd*, GUZ.; *Gidúri*, SIND; *Bargúnd, shiru-naruvili*, TAM.; *Chinna-botuku, chinna-mekkera-chettu*, TEL.; *Kottá*, MALAY.; *Tana, tanusi*, BURM.

The name Sebesten is a European adaptation of the name *Sapistán*, itself derived from *Sag-pistán*, which in Persian means Dog's nipples. **Sir Walter Elliot** gives this plant the Telegu name of *Kicha viri chettu*, and remarks that its synonym *Sléshmataka* is correctly translated "phlegm-dispeller."

References.—*Roxb., Fl. Ind., Ed. C.B.C., 198; Brandis, For. Fl., 336,* (in part) *; Thwaites, En. Ceylon Pl., 213; Dalz. & Gibs., Bomb. Fl., 173; Sind Gaz., 603; Bomb. Gaz., V., 27; Dymock, Mat. Med. W. Ind., 2nd Ed., p. 570; Atkinson, Him. Dist., 733; Birdwood, Bomb. Pr., 58, 169; Smith, Dic., 374.*

Habitat.—Found in Western India (especially Guzerát), from the Panjáb and Hindustan to Ceylon. The variety **C. Wallichii** (the *Dhaiwan* of Bombay) differs only in the degree of stellate tomentum on the leaves.

Medicine.—The fruit is used as an expectorant and astringent, and regarded as valuable in lung diseases: **Stocks** says that in Sind it is 'regarded as a demulcent.' MEDICINE. 1941

Special Opinion.—"The fruit in its raw state contains a gum used beneficially in gonorrhœa" (*Asst. Surgeon T. N. Ghose, Meerut*).

Food.—The fruit is eaten, and in the Deccan is generally known as *bhokar*. **Dr. Dymock** says the flowers and fruit were eaten in Khandesh during the famine of 1877-78. FOOD. 1942

Structure of the Wood.—Very much like that of the other species. **Stocks** remarks that in Sind it is regarded as tough, and is in considerable demand. TIMBER. 1943

C. Rothii, *Röm & Schult; Fl. Br. Ind., IV., 138.* 1944

Syn.—C. ANGUSTIFOLIA, *Roxb.* (*Fl. Ind., i., 595*).

Vern.—*Gondi, gondní, gundi*, HIND.; *Gondi*, PB.; *Gondui, gundi*, BOMB.; *Gondani*, MAR.; *Gundi*, GUZ.; *Liár, liári*, SIND.; *Narvilli*, TAM. **Elliot** gives *chinna botuka* as the TEL. for this tree.

References.—*Beddome, Fl. Sylv., 338; Gamble, Man. Timb., 271; Dalz. & Gibs., Bomb. Fl., 174; S. Arjun, Bomb. Drugs, 195; Baden Powell, Pb. Pr., 368, 575; Lisboa, U. Pl. Bomb., 102, 166, 233; Birdwood, Bomb. Pr., 58, 169; Royle, Fib. Pl., 311; Bomb. Gaz., V., 27, 285; VII., 41; Sind Gaz., 559; Central Prov. Settle. Rep.* (*Upper Godavery Dist., p. 37*).

Habitat.—A small tree of the dry zones of North-West, Central, and South India; plentiful in Rájputána. **Stocks** says that it is sometimes to be seen in Sind gardens.

GUM. 1945 **Gum.**—The bark, when wounded, yields a gum which is reported to be prepared at Coimbatore. In the Bombay Gazetteer of Baroda District, it is stated "fruit eaten by the poor and pickled, as is the gum which exudes from it."

FIBRE. 1946 **Fibre.**—The liber or inner bark yields a coarse grey, white bast fibre, which is made into rope. **Buchanan**, in his 'Journey through Mysore,' mentions having seen ropes of the bark of the *narwuli* or **Cordia angustifolia,** which he found common near Severndroog.

MEDICINE. 1947 **Medicine.**—The decoction of the bark possesses astringent properties, and is used as a gargle.

FOOD. 1948 **Food.**—The fruit is eaten by the poorer classes and is also pickled. In Sind this is regarded as an important fruit tree.

TIMBER. 1949 **Structure of the Wood.**—Wood grey, compact, hard. Used for fuel, in Sind for building, and in Cutch for agricultural implements. **Baden Powell** remarks that the wood is tough and is employed for making carriage poles. **Stocks** says the wood of the *liyar* is much used in Sind.

1950 **Cordia vestita,** *Hook. f. & Th.; Fl. Br. Ind., IV., 139.*

Syn.—GYNAION VESTITUM, *DC.*

Vern.—*Kúmbi, karúk,* PB.; *Kúm paimán, pín, indák, chinta, ajánta bairula, berula,* HIND.

References.—*Brandis, For. Fl., 338; Gamble, Man. Timb., 271; Atkinson, Econ. Prod., N.-W. P., V., 81; Baden Powell, Pb. Pr., 575.*

Habitat.—A small deciduous tree of the sub-Himálayan tract, from the Jhelum to the Sarda River and Oudh.

MEDICINE. 1951 **Medicine.**—Fruit used similarly to the other species, and when ripe is an article of food; it is considered better than that of **C. Myxa. Mr. Atkinson** states the flowers appear in spring and the fruit ripens in the rains. He remarks that the fruit is full of a gelatinous pulp which is commonly eaten and considered refreshing.

TIMBER. 1952 **Structure of the Wood.**—The wood is very similar in appearance to that of **C. Macleodii,** except that the concentric lines are occasionally interrupted; it is strong and is used for wheel and well-work.

1953

CORDAGE AND ROPES.

Many fibres are used for this purpose; in fact, the natives of India are never at a loss when in the forests to find a plant the bark of which will serve the purpose of a string or rope. The majority of such plants are more or less used locally in the preparation of ropes or cords; a considerable number are of commercial importance. Against the names in the following list have been placed one or in some cases two * to indicate the fibre-yielding plants frequently used for cordage, or the fibres which hold a position of commercial importance (* * indicating greater importance than *):—

*** Abroma augusta.**
Abutilon asiaticum.
A. Avicennœ.
**** Agave americana.**
Alnus nitida (bridge ropes).
Artocarpus Lakoocha.
Arundo Karka.
Bauhinia anguina.
B. racemosa.
*** B. Vahlii.**
Bixa Orellana.
Bœhmeria macrophylla (fishing nets).
**** B. nivea.**
Bombax malabaricum.

Borassus flabelliformis.
Broussonetia papyrifera.
Butea frondosa.
Calamus Rotang.
* Calotropis gigantea (string).
** Cannabis sativa.
Careya arborea.
Caryota urens.
Chamœrops Ritchiana.
** Cocos nucifera (coir).
* Corchorus, sp. (jute).
Cordia Myxa.
C. Rothii.
Crotalaria Burhia.
** C. juncea (Sunn-hemp).
Daphne papyracea.
Debregeasia bicolor (fishing lines).
D. leucophylla.
D. longifolia.
* Desmodium tiliæfolium.
Dombeya umbellata.
Edgeworthia Gardneril.
Eriolæna spectabilis.
Ficus bengalensis.
* Gerardinia heterophylla.
Gnetum scandens (fishing nets).
** Gossypium, sp. (cotton).
Grewia asiatica.
G. oppositifolia.
* Hardwickia binata.
Helicteres Isora.
** Hibiscus cannabinus.
H. esculentus.
H. tiliaceus.
Holostemma Rheedei.
*Ischœmum angustifolium(=Pollinia eriopoda).
Laportea crenulata.
Leptadenia Spartium.
* Linum usitatissimum (flax).
* Malachra capitata.
Maoutia Puya (fishing nets).
Marsdenia Roylei.
M. tenacissima (fishing lines).
Melochia velutina.
Memorialis pentandra.
Moringa pterygosperma.
** Musa textilis (Manilla hemp).
Ocimum Basilicum.
Odina Wodier.
Orthanthera viminea.
Pœderia fœtida.
Pandanus odoratissimus.
Parrotia Jacquemontiana (bridge ropes).
Periploca aphylla.
Phœnix paludosa.
P. sylvestris.
* Phormium tenax.
Pouzolzia viminea.
* Saccharum Munja.
S. spontaneum.
* Sansevieria zeylanica.
Sarcochlamys pulcherrima.
* Sesbania aculeata.
S. ægyptica.
Sida rhombifolia.
Silk—Tasar and Eri are sometimes used for fishing lines.
* Sterculia villosa.
Thespesia Lampas.
T. populnea.
Urena lobata.
Villebrunia appendiculata (ropes, strings—fishing lines).
Yucca gloriosa (lines).

CORIANDRUM, *Linn.; Gen. Pl., I., 926.*

The name of this genus comes from Κόρις a bug, in allusion to the peculiar smell of the plant of the common Coriander. This fact caused the plant to be viewed as poisonous during the middle ages. The ripe fruits (popularly called seeds) are quite free from this objectionable smell and were accordingly used as a spice by the Jews and the Romans and by other races as a drug, from almost the remotest times. The spice was well known in Britain prior to the Norman conquest. (*Pharmacog.*)

[*t. 516;* UMBELLIFERÆ.

Coriandrum sativum, *Linn.; Fl. Br. Ind., II., 717; Wight, Ic.,* CORIANDER. 1954

Vern.—*Dhanya* or *dhaniá,* HIND.; *Dhane,* BENG.; *Dhanya, dhana* (the seed), *Kothamira* (the plant), BOMB.; *Dhanya, khotbir, khotmir* or *kothmir,* MAR.; *Dháno,* SIND.; *Danga,* NEPAL; *Dhanyáka,* or (according to Dutt) *dhányaka,* SANS.; *Kuzbarah, kurbuzah,* ARAB.; *Kushniz,* PERS.; *Kotamalli,* TAM.; *Danyalu, kotimiri,* TEL.; *Kotambári, havija,* KAN.; *Nau-nau,* BURM.; *Ussú,* BHOTI.; *Dhánąk-chi,* TURKI.

Coriander.

References.—*Roxb., Fl. Ind., Ed. C.B.C., 272; Voigt, Hort. Sub. Cal., 23; Dalz. & Gibs., Bomb. Fl., Supp., 41; Stewart, Pb. Pl., 105; Flora Andhrica by Sir W. Elliot, 46, 99; Pharm. Ind., 101; Ainslie, Mat. Ind, I., 91, 595; O'Shaughnessy, Beng. Dispens., 371; Beng. Pharm., 39; Moodeen Sheriff, Supp. Pharm. Ind., p. 115; U. C. Dutt, Mat. Med. Hind., 173, 296; Dymock, Mat. Med. W. Ind., 375; Fleming, Med. Pl. and Drugs, as in As. Res., Vol. XI., 164; Flück & Hanb., Pharmacog., 329—331; U. S. Dispens., 15th Ed., 493—494; Bent. & Trim., Med. Pl., t. 133; S. Arjun, Bomb. Drugs, 63; Murray, Pl. and Drugs, Sind, 200; Bidie, Cat. Raw Pr., Paris Exh., 10, 86; Medical Topography of Ajmir, 132; Irvine, Mat. Med., Patna, 25; Baden Powell, Pb. Pr., 301, 352; Atkinson, Him. Dist., 705, 733; Econ. Prod., V., 29, 43; Drury, U. Pl., 39, 221; Lisboa, U. Pl. Bomb., 161; Birdwood, Bomb. Pr., 39, 161, 221; Descriptive Account of Godavery Dist., Madras, by Morris, 8; Manual of Cuddapah Dist., Madras, 199; Manual of Coimbatore Dist., Madras, by Nicholson, 225; Settlement Report, Kumaon, App., 34; Bombay Manual, Rev. Accts., 103; N. W. Gaz., VI., 589, 702, 784; Pb. Gaz., Montgomery Dist., 104; Aligarh Dist. Statistical Acct., 376; Official Report of Kumaon by Batten, 279; Spons, Encyclop., 1420, 1808; Balfour, Cyclop., 831; Treasury of Bot., 331; Morton, Cyclop. Agri., 545-547; Ure, Dic. Indus., Arts and Manuf., 907.*

Habitat.—A cultivated plant found all over India. It seems to be sown at various seasons in the different provinces and regions of India. In Bengal it is grown during the cold season: **Roxburgh** says this is the case "over India." **Voigt** remarks it is sown in the cold season, the fruits ripening in the hot season, or from March to May. Of Bombay, **Dalzell** and **Gibson** state:—"Cultivated as a rainy season crop in India, especially in the Deccan; is never irrigated." Of Madras various opinions have been given, but the account furnished in the Manual of Coimbatore is, perhaps, the most interesting and may be here quoted in full:—"Coriander is grown only on black soil in the three black soil *taluqas*; it is mixed with *uppam* cotton and sown broadcast in October and ripens in January; occasionally it is grown as a garden crop from June to September, watering once a week being sufficient. The seed is about 10 to 12℔ and the outturn is said to be only 288; as a dry crop the outturn is somewhat less, but is supposed to pay the cost of cotton cultivation."

In the Panjáb it is said to be grown in every district, and **Edgeworth** remarks that it is "frequently seen in the fields in a *quasi*-wild state." **Atkinson** and several other writers allude to it as a crop met with in the North-Western Provinces, and in Kumáon it is stated to ripen in May. Nepal grows the plant to a large extent, and the imports from that country regularly figure in the reports of the Basti District, North-Western Provinces. This trade is also a very old one. **Roxburgh** and **Ainslie** allude to it as existing in the beginning of the present century, the seed (or to be more correct the dry fruits) being known as *danga*. So, in a like manner, a by no means inconsiderable import trade is done in the seed from Afghánistán into the Panjáb (**Davies** says 25 maunds a year).

In England Coriander is often grown as a mixed crop with caraway, the yield being about 15 cwt. the acre (*Morton, Cycl. Agri.*). It is also grown in various other parts of Europe and in North Africa, but a large proportion of the world's supply seems to be now, and to have been for centuries, drawn from India. **Ainslie** states that in the beginning of the present century Egypt got her supplies of the spice from India, and that in Egypt it was then called *Kurbara shamie*. **Dymock** remarks that "Indian Coriander is much larger than that grown in Europe, and is of an ovoid form."

OIL. 1955 **Oil.**—The fruits yield from 0·7 to 1·1 per cent. of a volatile oil on distillation in water. This oil is colourless or yellowish, and has the odour and the flavour of Coriander. They also contain an essential oil which has been indicated by the formula $C_{10}H_{18}O$, and is therefore isomeric with *borneol*. By abstraction of the elements of water (by means of phosphoric anhydride) this is converted into an oil having an offensive odour $C_{10}H_{16}$ (*Pharmacog., p. 330*). But in addition to these oils **Trommsdorff** found Coriander seeds to contain 13 per cent of a fixed oil.

§ "Coriander fruit contains about ⅛ per cent of volatile oil; isomeric, with *borneol*, a fixed oil, is also present. The fruit should be bruised before being submitted to distillation" (*Professor Warden, Calcutta*).

Medicine.—The medicinal properties atributed to this plant are many,—namely, carminative, refrigerant, diuretic, tonic, and aphrodisiac. The dried fruit and the volatile oil are used as an aromatic stimulant in colic. The seeds are chewed to correct foul breath, and the roasted seeds are largely used in dyspepsia. **Dymock** writes: "A cooling drink is prepared from the fruit pounded with fennel fruit, poppy seed, *kanchan* flowers, rosebuds, cardamoms, cubebs, almonds, and a little black pepper; it is sweetened with sugar. Muhammadan writers describe the seeds as sedative, pectoral, and carminative; they prepare an eye-wash from them which is supposed to prevent small-pox from destroying the sight, and to be useful in chronic conjunctivitis. Coriander is also thought to lessen the intoxicating effects of spirituous preparations, and with barley meal to form a useful poultice for indolent swellings. It is the *Kuzbara* of the Arabs and *Kishniz* of the Persians, who identify it with the *Koriyun* of the Greeks." MEDICINE. 1956

Special Opinions.—§ "As a paste it relieves pain in cephalalgia, is used as a gargle in thrush, and as a poultice to chronic ulcers and carbuncles" (*John McConaghey, M.D., Civil Surgeon, Shahjahanpore*). "The juice of the fresh plant is used as an application to erythema caused by the application of marking nut; the bruised plant is a cooling application in cases of headache" (*Sakharam Arjun Ravat, L.M., Assistant Surgeon, Girgaum, Bombay*). "Also called *Behan* in Panjáb—*Bistduab*. It is applied by Natives in the form of a paste to relieve headache in fever with good results" (*Bhagwan Dass (2nd), Assistant Surgeon, General Hospital, Rawal Pindi, Panjáb*). "The roasted fruit is generally used" (*Dr. Bensley, Civil Surgeon, Rajshahye*). "A strong decoction of the seeds with milk and sugar to taste, is given in cases of bleeding piles" (*D. R. Thomson, M.D., C.I.E., Surgeon Major, Madras*). "Useful as aromatic, stimulant, and carminative" (*S. M. Shircore, Civil Surgeon, Moorshedabad*). "It is reputed as an antibilious remedy" (*T. N. Ghose, Assistant Surgeon, Meerut*). "Cold infusion of seeds found to be very useful in colics of children, powder of fried seeds" (*Shib Chunder Bhattacharji, Assistant Surgeon, In Civil Medical Charge Chanda, Central Provinces*).

Food.—Eaten by the natives as a vegetable. The seeds are universally used as a condiment, and form one of the ingredients in curry. FOOD. 1957
They are also employed in confectionery, and for flavouring spirits.

CORIARIA, *Linn.; Gen. Pl., I., 429.*

Coriaria nepalensis, *Wall.; Fl. Br. Ind., II., 44;* CORIAREÆ. 1958

Vern.—*Masúri, makola,* HIND.; In the Panjáb-Himálaya, **Stewart** says it bears the following names: *Gúch,* JHELAM; *Tadrelú balel,* KASHMIR; *Shálú, baulu,* CHENAB; *Kande, shalá, rau,* RAVI; *Ratsuhara, armúra, phapharchor,* BIAS; *archálwá, shere, lichakhro,* SUTLEJ; *Raselwa, archarru, pajerra,* SIMLA; *Bhojinsi,* NEPAL.

References.—*Brandis, For. Fl., 128; Kurz, For. Fl. Burm., II., 281; Gamble, Man. Timb., 113; Stewart, Pb. Pl., 39; Aitchison, Cat. Pb. and*

Coriaria.

Sind Pl., 36; O'Shaughnessy, Beng. Dispens., 270; Flück. & Hanb. Pharmacog., 221; U. S. Dispens., 15th Ed., 1622;* Baden Powell, Pb. Pr., 336, 575; Atkinson, Him. Dist., 749; Balfour, Cyclop., 813; Treasury of Bot., 331.*

Habitat.—A deciduous shrub or small tree of the outer Himálaya from the Indus to Bhutan, ascending to 8,000 feet in the North-West and to 11,000 feet in Sikkim. Distributed to Manipúr, Burma, and Yunan.

In Simla this common shrub flowers in February and March, but in Burma not till May. The abundance of this plant seems to have been the cause of the name Mussoorie being given to the North-Western Provinces Hill station; Almora, the capital of Kumaon, being in a like manner the vernacular name for **Rumex acetosa.**

TAN. 1959 **Tan.**—All parts of the plant are rich in astringent acids which might be used for tanning or for dyeing.

FOOD and FODDER. 1960 **Food and Fodder.**—"The branches are browsed by sheep. The fruit is very insipid but is eaten, although at times it is reputed to cause thirst and colic" (*Dr. Stewart*).

MEDICINE. 1961 **Medicine.**—Leaves are said to be used to adulterate senna, and to act as a powerful poison when given in large doses. The seeds are stated to sometimes produce symptoms like tetanus.

It is not known how far these opinions (which occur in the works of many Indian authors) are, strictly speaking, applicable to **C. nepalensis,** or may be the result of reading of the Mediterranean and New Zealand species, both of which are highly poisonous. The writer has seen horses eat the leaves freely without any injurious after-effects, and the natives regularly eat the fruit, maintaining only that the seeds should be rejected. The silkworm may be fed on the leaves (see **A 666**). **Stewart** mentions that in one district of the Panjáb the plant is viewed as highly poisonous. No mention is made of the leaves of the Indian species being used as a tan similar to the application of **Coriaria myrtifolia**—the Currier's sumach or *Redoul* of France. That species is frequently grown as an ornamental species in French gardens, and its leaves are often employed as a black dye, and were at one time extensively used an an adulterant in Senna. Much has been written of the poisonous properties of the New Zealand species, the Toot-poison—**Coriaria ruscifolia.** **Mr. Lander Lindsay** gives an elaborate account of the properties of that plant in the British and Foreign Medico-Chirurgical Review (1865, p. 153, and 1868 p. 465). **M. Riban** attributes the poison of the fruit to an active principle, which he has called coriamyrtin, the composition of which is represented by the formula $C_{30}H_{36}O_{10}$ a substance ranked with the glucosides.

The inhabitants of New Zealand extract an intoxicating beverage from the pulp of the fruit.

Professor Warden of Calcutta has furnished the following brief note regarding **Coriaria**:—"The **Coriaria ruscifolia** seeds contain a resinous substance and a green oil—5 minims of the oil administered to a cat, after 12 hours' fast, produced vomiting and convulsions, from which, however, the animal recovered. These symptoms agree with those exhibited by cattle when poisoned by the Toot-plant of New Zealand."

TIMBER. 1962 **Structure of the Wood.**—Grey, hard, beautifully mottled; no heartwood. It takes a good polish, and is very handsomely marked; it might be used for boxes and small articles. At present it is only used for firewood, but as such to a large extent in the Simla District.

* References to the Mediterranean or New Zealand species.

Corn—a term often specifically applied to **Avena sativa**, but generically given to all cultivated grasses which yield farinaceous grains, such as Wheat, Maize, Barley, Oats, &c. When ground, Corn is designated flour or meal. See **Avena Vol. I., 1631.** 1963

Corn-flag, see **Iris.**

Corn-Indian, see **Zea Mays.**

Corn-silk—the silky stigmata of **Zea Mays**, from which a medicinal preparation is made. See **Zea.** 1964

CORNUS, *Linn.; Gen. Pl., I., 950.* 1965

[*t. 122;* CORNACEÆ.

Cornus capitata, *Wall; Fl. Br. Ind., Vol. II., 745; Wight, Ill.,* 1966

Syn.—BENTHAMIA FRAGIFERA, *Lindl.*

Vern.—*Thummal, tharbal, tharwar, thesi, bamaur, bamora,* HIND.; *Tumbúh,* LEPCHA; *Tharwar, thesi,* PB.; *Bamaurá,* KUMAON.

References.—*Brandis, For. Fl., 253; Gamble, Man. Timb., 212; Stewart, Pb. Pl., 111; Ainslie, Mat. Ind., II., 454;* O'Shaughnessy, Beng. Dispens., 375; O'Shaughnessy, Beng. Pharm., 40; Atkinson, Econ. Prod., V., 75; Treasury of Bot., 333.*

Habitat.—A small deciduous tree of the Himálaya, from the Beas to Bhután, between 3,500 and 8,000 feet: met with also in Khásia hills, where it is glabrous or nearly so.

The Himálaya, in April and May, often becomes almost yellow from the conspicuous cream-coloured bracts which surround the flower-heads of this plant. In the North-West Himálaya, it is particularly abundant in the lower hot valleys growing along with the berberry.

Food.—Dr. **Stewart** says that the ripe fruit is sweetish, and is apparently made into a preserve and eaten by the natives. It resembles a strawberry somewhat in external appearance, and ripens in October. FOOD. 1967

Structure of the Wood.—Whitish, with reddish-brown heartwood, warps in seasoning, very hard, close-grained; used only for firewood. WOOD. 1968

C. macrophylla, *Wall; Fl. Br. Ind., Vol. II., 744.* 1969

Vern.—*Kasir, kachir, haleo, allian, haddú, harru, nang, kandara, kaksh kachúr, kochan, kágsha, rúchia,* HIND.; *Kandar,* HAZARA; *Haléo,* PB.; *Patmoro,* NEPAL; *Kagshi, ruchiya,* KUMAON.

References.—*Brandis, For. Fl., 252, t. 52; Gamble, Man. Timb., 212; Stewart, Pb. Pl., 111; O'Shaughnessy, Beng. Dispens., 375; O'Shaughnessy, Beng. Pharm., 40; Baden Powell, Pb. Pr., 575; Atkinson, Econ. Prod., V., 75.*

Habitat.—A tree, 40 to 50 feet high, frequent in the Himálaya, from the Indus to Bhután, between 3,000 and 8,000 feet; found by the writer in Manipur. It flowers in May and June.

Oil.—A species closely allied to the **C. sanguinea**, and may, like that species, be found to afford an oil from its fruits. OIL. 1970

Food and Fodder.—Goats feed on its leaves, and the natives eat the fruit. FODDER. 1971

Structure of the Wood.—Pinkish-white, hard, close-grained; warps badly, and has an unpleasant scent; yields good gunpowder charcoal. WOOD 1972

* **Cornus florida,** alluded to as having a medicinal bark, very similar in its properties to the bark of **Melia Azadirachta.**

1973 **Cornus oblonga,** *Wall; Fl. Br. Ind., II., 744.*

Vern.—*Kagshi,* SUTLEJ; *Dab,* KUNAWAR; *Kasmol, bakár, ban-bakúr, halá,* HIND.

References.—*Brandis, For. Fl., 253; Kurz, For. Fl., I., 545; Gamble, Man. Timb., 212; Stewart, Pb. Pl., 111; O'Shaughnessy, Beng. Dispens., 375; O'Shaughnessy, Beng. Pharm., 39; Baden Powell, Pb. Pr., 576.*

Habitat.—A small tree of the outer Himálaya, from the Indus to Bhután, between 3,000 and 6,000 feet; met with also in the Martaban Hills, Burma, between 4,000 and 7,000 feet (*Kurz*).

WOOD. 1974 **Structure of the Wood.**—Pinkish-white, hard, even-grained; warps and has an unpleasant scent.

1975 **C. sanguinea,** *Linn.; Fl. Br. Ind., II., 744.*

THE DOGWOOD, DOGBERRY, or HOUNDS' TREE, a name given in consequence of a decoction of the bark having been formerly used for washing mangy dogs; sometimes also called the CORNEL TREE.

References.—*Brandis, For. Fl., 253; Gamble, Man. Timb., 212; O'Shaughnessy, Beng. Dispens., 375; O'Shaughnessy, Beng. Pharm., 39; Cooke, Oils and Oilseeds, 38; Smith, Dic., 156.*

Habitat.—A shrub or small tree found in Europe, Siberia, and in Kashmír; in the last-mentioned country at 7,000 feet in altitude. The writer found the plant also growing near a village in Chumba State, but it may there have been only cultivated. The young shoots are red in spring, and the leaves turn of that colour in autumn; hence the specific name given by botanists.

OIL. 1976 **Oil.**—The pericarp of the fruit contains oil (*Brandis*). From the black fruits an oil is extracted in France which is used for burning in lamps and for soap-making. The red berries of the Cornelian cherry—**Cornus mascula,** a shrub of Europe and Northern Asia—also contain an useful oil. These facts would seem to suggest that the Indian species should be more carefully examined, as *they* also may be found to afford oils.

WOOD. 1977 **Structure of the Wood.**—Hard, much valued in Europe for the manufacture of small articles, such as tooth-picks, butchers' skewers, &c. It is valued as affording an admirable charcoal for gunpowder.

Coromandel or **Calamander-Wood,** see **Diospyros quæsita** and **D. hirsuta.**

Coroxylon Griffithii, a misprint which appears in *Balfour's Cyclopædia* and in the writings of other authors. See **Caroxylon** and also **Haloxylon.**

Corrosive sublimate, see **Mercury.**

1978 **Corundum.**

EMERY STONE, *Eng.;* L'EMERI, *Fr.;* SCHMERGEL, *Germ.;* SMERIGLIO, *Ital.*

Vern.—*Kurund,* HIND.; *Samada,* GUJ.

This, the industrial form of the mineral, is a granular alumina, with which a small amount of magnetic iron is associated. It is very freely distributed among the crystalline rocks of Southern India; but the localities where it is sufficiently abundant for industrial work are few and

far between. The finest quality of **Corundum** is perhaps that obtained between Pipra and Kadopani in the Rewah State, where the supply is considered by Mr. Mallet to be practically inexhaustible (*Records, G.S.I., V., p. 20; and Manual of Indian Geology, Part III., p. 428*). In Part IV. of the *Manual of Geology*, Mallet says there "are two distinct varieties: crystallized **Corundum**, which is abundant in the metamorphic rocks of many parts of South India: and granular massive, of which an immense deposit exists in South Rewah." Emery stone is also reported as occurring at Nongrynien, Khásia Hills. In Southern India, the localities are Travancore State, Coimbatore district, Salem district, Mysore State, Punyghee in the Bellary district, North Arcot district, Kistna and Godavari, and Hyderabad territory, and on into the Central Provinces. 1979
"The uses to which **Corundum** is put, when powdered, are well known. The consumption in India must be considerable, though possibly it was larger formerly than it is at present, as the trade of the native armourer is perhaps not so active as it used to be. A large quantity is employed by the cutters and polishers of stones, both precious and ornamental, who are to be found scattered throughout India. To what extent Indian **Corundum** is used in European countries is not very well known, but it could doubtless be applied to many of the purposes for which the emery of the Greek Islands is now used, and which, owing to a monopoly at one time, reached the high price of £30 a ton in London" (*See Manual of Geology of India, Part III., p. 422; also Part IV., 46-49; Manual of Coimbatore, p. 23*). Emery is said to be largely exported to Bombay (*Madras Manual of Administration, II., 38; Settlement Report of Upper Godavery Dist., 42; Balfour, Cyclopædia of India, 816*).

CORYDALIS, *Linn.; Gen. Pl., I., 55.*

Corydalis Govaniana, *Wall; Fl. Br. Ind., Vol. I., 124; Royle, [Ill., t. 16, f. 2;* FUMARIACEÆ. 1980

Vern.—*Bhútkis, bhutkesi,* HIND. & BENG.; *Bhutakesi,* SANS. (*Dutt, Mat. Med. Hind.*)

Some doubt seems to prevail as to the source of the *budkhes* of the drug shops. **Stewart** says that in the Ravi basin that name is given to the root of a **Ptychotis**.

References.—*Stewart, Pb. Pl., 10, 109; Pharm. Ind., 23; O'Shaughnessy, Beng. Dispens., 185; U. C. Dutt, Mat. Med. Hind., 294.*

Habitat.—A small herbaceous plant, found in the North-West Himálayas; altitude 8,000 to 12,000 feet. Common on Háttú near Simla and on the Chór; the flowers appear in April and May.

Medicine.—The root contains a principle, *Corydalia.* **Sir W. O'Shaughnessy** recommended this drug to be more fully investigated and to be experimented with as a tonic and antiperiodic. **O'Shaughnessy** describes the root "as long, fibrous, tough, and exceedingly bitter; dark brown externally, yellow within. The tincture, rendered alkaline by ammonia, and allowed to evaporate spontaneously, deposits abundant crystals of the alkali, termed *Corydalia.*" MEDICINE. Root. 1981 Corydalia. 1982

"*Corydalia* occurs in pearly crystals, is soluble in acids, with which it forms salts which do not crystallize; and which are intensely bitter to the taste. Nitric acid communicates a deep red, permuriate of iron, a rich blue colour to the alkali, and its salts. Twenty grains have been given in solution to dogs without inconvenience."

"The **Corydalis tuberosa** and **fabacea** in Europe have a bitter acrid root, usually sold as ARISTOLOCHIA root, and used chiefly as an external

application to indolent tumors. The small quantity in our possession alone prevented the *Corydalia* and its salts from being extensively tried in the treatment of ague. The chemical properties of the salts are closely analogous to those of morphia and anarcotine; an interesting fact, as it strengthens the resemblance already detected by botanists between the PAPAVERACEÆ and FUMAREÆ." It might be added also that the relation of these orders to the RANUNCULACEÆ, through **Coptis** and to BERBERIDEÆ through the berberry or *rasout* extract, is similarly borne
1983 out by their chemical and medicinal properties. (See the next species and compare with the remarks under **Coptis Teeta, C. No. 1789,** and **Berberis Lycium, B. No. 460**; also **Picrorhiza Kurroa**).

The Turkey-corn or Turkey-pea (**Corydalis formosa**) contains in its roots, according to **Mr. W. T. Werzell**, the alkaloid *corydaline*, formic acid, bitter extractive, an acrid resin with volatile oil, a tasteless resin, brown colouring matter, starch, albumen, arabin, bassorin, collulose, and various inorganic salts. He describes the crystals of *Corydaline* as slender, four-sided prisms, inodorous, tasteless, insoluble in water, soluble in alcohol, ether, and chloroform, reddened by nitric acid, and capable of forming soluble salts with acids. **H. Wicke** gives the formula C_{18},H_{19},NO_4 for the alkaloid (*Corydaline*) found in the European species—**Corydalis tuberosus.**

MEDICINE. 1984 The roots of all these plants are supposed to be tonic, diuretic, and alterative, and are prescribed in syphilitic, scrofulous, and cutaneous affections, in the dose of from 10 to 30 grains. The drug is also often used in the form of a decoction or tincture.

Corydalis ramosa, *Wall; Fl. Br. Ind., I., 125.*

Dr. Aitchison, in his *Flora of the Kuram Valley* (*Linnæan Soc. Jour., XIX., page 145*), says that in Kuram this common Himálayan scrambling annual is employed medicinally by the natives in the treatment of eye diseases, like all other plants with yellow sap. It is there called *mamírán*. It would be interesting to know if this plant is used medicinally in other parts of the Himálaya, but these properties are not attributed to it in Kulu, where the plant is abundant. (See remarks under the preceding species and compare with the account of **Coptis Teeta C. No. 1789**)

CORYLUS, *Tourn.; Gen. Pl., III., 406.*

1985 **Corylus Avellana,** *Linn.*; CUPULIFERÆ.

THE EUROPEAN HAZEL.

Vern.—*Findak, bindak*, HIND., PERS.; *Chalgoza*, PERS.

References.—*Brandis, For. Fl., 494; Gamble, Man. Timb., 390; O'Shaughnessy, Beng. Dispens., 609; U. S. Dispens., 15th Ed., 977; Baden Powell, Pb. Pr., 268, 385.*

Habitat.—Found in England, France, and eastward to the Caucasus and in Asia Minor. Alluded to by some authors as cultivated on the Himálaya. Although this might easily enough be the case, it is probable that all the hazel nuts met with in India are obtained from wild or semi-cultivated states of **Corylus Colurna.**

MEDICINE. Nuts. 1986 **Medicine.**—The nut yields an oil used for coughs, &c. It is tonic, stomachic, and aphrodisiac.

FOOD. Nuts. 1987 **Food.**—English hazel nuts are imported into India and sold in the sea-port towns. Those carried into the towns of Upper and Central India are probably all obtained from the next species.

Corylus Colurna, *Linn.* 1988

Syn.—C. LACERA, *Wall.*

Vern.—*Urni*, JHELAM; *Winri, wiri, warawi, wúriya, thangi, thankoli*, KASHMIR and CHAMBA; *Jangi*, CHENAB; *Shurli, sharoli, ban pálu, geh, ban dilla*, SUTLEJ; *Kapási, bhotia badám*, KUMAON; *Shirol*, GARHWAL; *Jhangi*, KANGRA.

References.—*Brandis, For. Fl., 494; Gamble, Man. Timb., 390; Stewart, Pb. Pl., 201; Indian Forester, IX., 197; Baden Powell, Pb. Pr., 576; Atkinson, Him. Dist., 716; Cooke, Oils and Oilseeds, 38.*

Habitat.—A moderate-sized tree of the North-West Himálaya, between 5,500 and 10,000 feet. The flowers appear in March and April, and the fruit ripens in the rains. "The trees bear every third year, and yield a crop sufficient for export to the plains" (*Atkinson*).

OIL. 1989 **Oil.**—There seems no reason to doubt but that an oil could be prepared from this species of hazel as well as from the European nut. No mention is, however, made of the natives of India extracting oil from it, although the plant is sufficiently abundant in the temperate forests, so much so as to bestrew the ground for miles with the nuts.

MEDICINE. Nuts. 1990 **Medicine.**—The nuts are not uncommon in drug-sellers' shops, being considered tonic.

FOOD. Nuts. 1991 **Food.**—The nuts are smaller than the European variety, but are nearly as good, and are largely eaten, being exported from the various hill stations in the Himálaya. The hazel nuts from Afghánistan and Kashmír are much more like the European nut, and are recognised by the natives of the plains as distinct from the Himálayan form. It is thus probable that they are either obtained from **C. Avellana** or from a cultivated superior stock of **C. Colurna**. As seen in the forests in the Simla district, the actual nuts are small and rarely mature their kernels, but they are encased in a large coarse outer coat and form large succulent heads.

WOOD. 1992 **Structure of the Wood.**—Pinkish-white, moderately hard. It is only used locally, but it is well grained and does not warp, and deserves to be better known, especially as many specimens shew a fine shining grain resembling Bird's-eye Maple.

C. ferox, *Wall; Gamble, Man. Timb., 390.*

Vern.—*Curri*, NEPAL; *Langura*, BHUTIA.

Habitat.—A small tree of Nepal and Sikkim, 8,000 to 10,000 feet.

FOOD. Nuts. 1993 **Food.**—The fruit is covered with a prickly cup; the kernel is edible.

WOOD. 1994 **Structure of the Wood.**—Pinkish-white, moderately hard, even-grained.

CORYPHA, *Linn.; Gen. Pl., III., 922.*

Corypha umbraculifera, *Linn.;* PALMÆ. 1995

THE TALIPOT PALM OF CEYLON AND THE FAN-PALM OF SOUTH INDIA.

Vern.—*Tali, bajar-battuler, tara, tallier, tarit*, BENG.; *Codda-pani, talip-panai, kottaip-panai*, TAM.; *Shritalam*, TEL.; *Kotap-pana*, MALAYAL.; *Bajar-battú, tali*, MAR.; *Biné, shritale, tále*, KAN.; *Tala*, SING.; *Pebin*, BURM.; *Tali, sritálam* (according to Sir W. Elliot), SANS.

References.—*Roxb., Fl. Ind., Ed. C.B.C., 298-299; Voigt, Hort. Sub. Cal., 640; Brandis, For. Fl., 549; Kurz, For. Fl. Burm., II., 524; Thwaites, En. Ceylon Pl., 329; Dalz. & Gibs., Bomb. Fl., Supp., 94; Rheede, Mal., III., t. 1-12; Rumph., II., 174, l. t. 8; Sir Walter*

Elliot, Flora Andhrica, 169; Madras, Man. Admin., 27; Moodeen Sheriff, Supp. Pharm. Ind., 116; Drury, U. Pl., 159; Royle, Fib. Pl., 98; Kew Offl. Guide to the Mus. of Ec. Bot., 71; Kew Offl. Guide to Bot. Gardens and Arboretum, 33.

Habitat.—A large tree of Ceylon and the Malabar Coast; cultivated in Bengal and Burma. But **Roxburgh** says it is "a native of Bengal, though scarce in the vicinity of Calcutta. Flowering time, the beginning of the hot season. The seeds ripen about nine or ten months afterwards." Reported to be very common in the moist regions of the Madras Presidency. This tall and handsome tree, **Sir E. Tennet** says, has leaves 16 feet in diameter, each covering an area of 200 superficial feet. It flowers but once and thereafter dies, and the natives firmly believe that the opening of the spadix is accompanied with a loud explosion.

Considerable confusion exists, in the writings of authors on economic subjects, between this palm—the Talipot of Ceylon—and the Palmyra or Talipot of Bengal and Madras—**Borassus flabelliformis.** Many of the uses of these palms are identical, and the vernacular names are accordingly misleading.

FIBRE. Leaves. 1996

Fibre.—The leaves are made into fans, mats, and umbrellas, and are used for writing on. They are also largely employed for thatching. **Knox,** a writer quoted by **Royle,** says: "Of this, the leaf, being dried, is very strong and limber, and most wonderfully made for man's convenience to carry along with them; for though this leaf be thus broad (enough to cover 15 or 20 men) when it is open, it will fold close like a lady's fan, and then it is no bigger than a man's arm; it is wonderfully light." **Roxburgh** remarks the leaves "are used to tie the rafters" of native houses, as they are "said to be strong and durable." It seems probable that, after removing the

Fibre-bundle. 1997

edible pulp from the interior of the stem, the long fibro-vascular cords might be used as a substitute for *kittul,* similar to the fibres extracted from the stem of **Caryota urens** (**C. 712**). These fibres are reported to be softer and more pliable than those found at the bases of the leaves. **Drury** states that "the leaves alone are converted by the Singhalese to purposes of utility. Of them they form coverings for their houses, and portable tents of a rude but effective character. But the most interesting use to which they are

Paper (olas). 1998

applied is a substitute for paper, both for books and ordinary purposes. In the preparation of *olas,* which is the term applied to them when so employed, the leaves are taken whilst tender, and after separating the central ribs, they are cut into strips and boiled in spring-water. They are dried first in the shade and afterwards in the sun, then made into rolls and kept in store, or sent to the market for sale. Before they are fit for writing on they are subjected to a second process. A smooth plank of areca palm is tied horizontally between two trees: each *ola* is then damped, and a weight being attached to one end of it, it is drawn backwards and forwards across the edge of the wood till the surface becomes perfectly smooth and polished, and during the process, as the moisture dries up, it is necessary to renew it till the effect is complete. The smoothing of a single *ola* will occupy from 15 to 20 minutes." The writer cannot discover any description of the preparation of the palm leaves as adopted in India, and in the case of the Palmyra palm (see **B. 719**), oil is employed to give the polish. The whole subject of these prepared slips of

Braids. 1999

palm leaves is worthy of more attention, since they are coming into European commerce in the manufacture of ornamental braids and in the

Hats. 2000

construction of straw or Leghorn hats.

FOOD. Sago. 2001

Food.—A kind of sago is yielded by the pith. Little information of a definite kind can be discovered as to the extent in which this starch is used in India as an article of food, nor as to the methods adopted in its preparation. **Knox** says of Ceylon that the people "beat it in mortars to flour, and bake cakes of it, which taste much like white bread; it serves them instead of corn before their harvest is ripe."

WOOD. 2002 **Structure of the Wood.**—Soft, with a hard rind composed of black vascular bundles. The vascular bundles in the centre of the stem are soft. **Roxburgh** remarks: "I do not find that the wood is put to any useful purpose."

The tree often grows to a great size before flowering; one whose measurements were given in the *Indian Agriculturist* for November 1873 as flowering at Peradeniya, Ceylon, measured: height of stem 84 feet, of flower panicle 21 feet, total 105 feet; girth at 3 feet from the ground round the persistent bases of the leaves 13 feet 9 inches, at 21 feet from the ground 8 feet 3 inches; age about 40 years. The leaves are very large, often 10 to 16 feet in diameter.

DOMESTIC. Beads 2003 **Domestic and Economic Uses.**—In addition to what has been said of the leaves being used for fans, umbrellas, &c., it may be here added that the fruit is hard like ivory, and is extensively employed in the manufacture of beads (see **B. 837, No. 21**) and are known in trade as *Bazarbuld* nuts. A considerable trade is done in these nuts from **Bombay**, the supply coming apparently from North Kanara and Ceylon. They are sold at R20 to R25 per candy of 616lb. They are also sometimes coloured red and sold as coral, or are made into small bowls and other ornaments. In Europe they are now largely employed in the manufacture of buttons. The trade in these nuts is chiefly carried on by Arabs.

Ornaments. 2004

Buttons. 2005

2006 **Corypha Taliera,** *Roxb.; Cor. Pl., t. 255.*

A closely-allied species to the preceding, which bears most of the vernacular names given above, and is put to the same industrial purposes; is a native of the north-eastern coast of Madras, especially in Coromandel. A third species may here be mentioned by name **C. elata,** *Roxb., Fl. Ind., 298,* a stately palm and native of Bengal, where it is known as *bajúr*, but **Roxburgh** views **C. umbraculifera** as the intermediate form between **Taliera** and **elata,** so that even if future botanists continue to view all three as distinct species, for industrial purposes, they may be regarded as but forms of one plant. It would, indeed, be impossible to separate under these plants the various properties assigned to them.

COSCINIUM, *Colebr.; Gen. Pl., I., 35.*

[MENISPERMACEÆ.

2007 **Coscinium fenestratum,** *Colebrooke; Fl. Br. Ind., Vol. I., 99;*

Vern.—*Jhár-ki-haldí* or *jhádi-haladi,* DEC.; *Haldi-gach,* BENG.; *Maramanjal,* TAM.; *Mánu-pasupu,* TEL.; *Marada-arishiná,* KAN.; *Darvi* (Ainslie), *dárú-haridrakam* (Moodeen Sheriff), SANS.; *Venivel,* SING.

References.—*Voigt, Hort. Sub. Cal., 332; Thwaites, En. Ceylon Pl., 12; Pharm. Ind., 10; Ainslie, Mat. Ind., II., 461; Moodeen Sheriff, Supp. Pharm. Ind., 116; Materia Medica of Madras, 11; Dymock, Mat. Med. W. Ind., 2nd Ed., 34; U. S. Dispens., 15th Ed., 321, 1540; Bidie, Cat. Raw Pr., Paris Exh., 19, 107; Correspondence in the Home Dept. regarding the Pharm. Ind., 238; Perrins in Pharm. Jour., XII., 180—500; Drury, U. Pl., 160; Christy, Com. Pl. and Drugs, pt. 9, p. 65; McCann, Dyes and Tans Beng., 91; Liotard, Dyes, App., I.; Mysore Cat. Prod. shown Cal. Exhib., 47; Kew Offl. Guide to the Mus. of Ec. Bot., 9.*

Habitat.—An extensive climber, met with in the forests of the Western Peninsula, and distributed to Ceylon and the Straits.

DYE. 2008 **Dye.**—In *Dr. U. C. Dutt's Materia Medica of the Hindus, Darvi* is given as the Sanskrit for **Berberis, sp.** Neither **Brandis** nor **Gamble** record that name, nor any apparent derivatives from it to the species of **Berberis,** nor is it so given apparently by any other author. **Ainslie,** on the other hand, gives *Darvi* as the Sanskrit for **Coscinium fenestratum.** Both **Coscinium** and **Berberis** yield a yellow dye; are valuable medicines; and the chips of the wood, but for structural peculiarities, could not be distinguished. **Ainslie** apparently was labouring under one mistake; he took the *Maramanjal*, TAM., as different from the *Vinivel-getta,* Ceylon specimens of which were sent to **Roxburgh** for identification. **General Macdowall** viewed the Ceylon specimens of this species as **Colomba** root, but **Roxburgh** corrected him. Speaking of *Mara-manjal* **Ainslie** says, "it is sometimes used as a yellow dye," but this was apparently unknown to **Roxburgh.**

Dr. Bidie remarks: "This wood contains much colouring matter, akin in properties to that of turmeric," hence the name *j r-ki-haldí* or *ghach-haldí*. **Dr. McCann,** and also **Mr. Liotard,** allude to the properties of this dye as closely resembling turmeric. The former author says of the Chittagong district that the *bark* (sic) (? wood) is imported from Koladyne in Arracan. The use of this dye-stuff he describes as follows: "The bark should be scraped so as to clean it. It is then broken up and steeped in water for nearly 2 hours, then crushed in a rice-husking machine, after which the dye is squeezed out of it. The cloth to be dyed is steeped in the dye three times, and dried in the shade after each steeping." It may also be combined with turmeric and other dye-stuffs.

MEDICINE Root. 2009 **Medicine.**—**Ainslie** says: "*Mara-manjal* is the Tamil name of a round, yellow-coloured, bitterish root, common in the bazar, about one inch in circumference, employed in preparing certain cooling liniments for the head, and is also used as a yellow dye; it is brought from the mountains, but I have endeavoured in vain to ascertain the plant." At present the root is extensively used in the hospitals of the Madras Presidency as an efficient bitter tonic. A writer quoted by **Christie** says of Ceylon that this root is viewed as "a very good substitute for *Calumba.* I have used it with good results in the form of tincture and infusion. It has also antiseptic properties to a great extent, and can be used for dressing wounds and ulcers." **Moodeen Sheriff,** in his new work on the Materia Medica of Madras (proof sheets of which he has kindly furnished the author with), says that the action of the drug is "antipyretic, antiperiodic, tonic, and stomachic," and that it is useful "in slight cases of continued and intermittent fevers, in debility, and certain forms of dyspepsia." He further states that it may be used in place of cinchona, gentian, or calumba, and that the doses are the same as with the preparations from the root of **Berberis aristata.** Chemically it has been found by **Perrins** to contain *Berberine.*

2010 The drug is sometimes sold as calumba root or for berberry, from which it may easily enough be distinguished by the peculiar structure of the wood. Bright, greenish yellow, with open porous structure, devoid of concentric rings, but having pronounced medullary rays. It is, besides, lighter and softer than berberry wood. **Dymock** remarks: "I have not met with any account of it in native works; but there is reason to believe that it has sometimes been confounded with *Darhalad,* the stem of the berberry. It is sometimes mentioned in the drug sales of Europe as False Calumba or Tree Turmeric, the latter being literally a translation of many of the vernacular names of the plant.

2011 **Special Opinions.**—"Used in diabetes. It is also stomachic" (*Surgeon-Major D. R. Thomson, M.D., C.I.E., Madras*). "Used also in cases of suppression of lochia" (*Surgeon-Major J. J. L. Ratton, M.D., M.C.,*

Salem). "This has been in use for some years in the hospital and found to be a fairly useful medicine in certain cases of dyspepsia. I think it a fairly good substitute for calumba. It has been used in the form of powder and infusion. Preparations, &c.—The same as calumba." (*Apothecary J. G. Ashworth, In Medical charge, Kumbakonum*).

Trade.—The root is sold in Madras at R$1\frac{1}{2}$ per maund, and retailed at 2 annas a pound. There are no foreign exports of the root from India but it may be had in every large bazar throughout the country, so that there must be a considerable local demand. TRADE. 2012

Cosmetic Bark, see **Murraya exotica,** *Linn.*

COSTUS, *Linn.; Gen. Pl., III., 646.*

Costus arabicus, see **Saussurea Lappa and hypoleuca**; COMPOSITÆ.

C. speciosus, *Sm.; Wight, Ic., 2014;* SCITAMINEÆ. 2013

Vern.—*Kúst, keú,* BENG., HIND.; *Orop,* SANTAL; *Gudárichákánda, kemuka,* BOMB.; *Pinnga, penva,* MAR.; *Bomma kachika,* TEL.; *Tsjana kua,* MAL.; *Keyu, keuli, kúlshirin* (root), N.-W. P.; *Kemúka,* SANS. (**Sir W. Elliot** gives the following as Sanskrit synonyms: *Pushkara mulaka* and *kásmíra.*) *Tebu,* SING.; *Pálán toung-wæ,* BURM.

This seems to be the **Tjanakua** of **Rheede,** *Mal., XI., 15, f. 8;* the **Tsana speciosa,** *Gmelin, IX.;* and the **Herba spiralis hirsuta** of **Rumph.,** *Amb., VI., 143, t. 64, f. 1.*

References.—*Roxb., Fl. Ind., Ed. C. B. C., 20; Voigt, Hort. Sub. Cal., 572; Kurz, Report on Pegu; Thwaites, En. Ceylon Pl., 320; Dalz. and Gibs., Bomb. Fl., 274; Stewart, Pb. Pl., 122, 238; Sir W. Elliot, Fl. Andhrica, 30, 88, 91, 99, 101, and 160; Ainslie, Mat. Ind., II., 167; O'Shaughnessy, Beng. Dispens., 652; U. C. Dutt, Mat. Med. Hind., 304; Dymock, Mat. Med. W. Ind., 779; Fluck. & Hanb., Pharmacog., 382; Baden Powell, Pb. Pr., 380; Atkinson, Him. Dist., 733; Drury, U. Pl., 161; Birdwood, Bomb. Pr., 86;*

Balfour, Cyclopædia of India, under **Costus** deals purely with **Saussurea Lappa,** but he alludes to **Costus speciosus** as a native of South India, Cochin-China, the Moluccas, and Sanda Islands.

Habitat.—One of the most elegant plants of this family; its spirally-twisted stem carries its glossy leaves and white flowers above the brushwood in the Indian tropical jungles. It is common everywhere throughout the country, and especially so in Bengal, where it frequents moist, shady places, and in the Concan and Coromandel it is equally abundant.

Perfumery.—Piesse says of it: "I have made some experiments with a sample of *kúsht*; it appears to be scarcely as odorous as Orris Root. The tincture has an agreeable smell, and would be useful, but no quantity has as yet been seen in our markets." An unlimited quantity might easily enough be exported from Bengal were some effort made to bring this root before the perfumers of Europe. There is a strong probability, however, that **Piesse** is referring to the root of **Saussurea Lappa** or **S. hypoleuca,** members of the COMPOSITÆ, which were formerly called **Aucklandia Costus.** It is remarkable that, while associated with the word **Costus,** both these widely different plants should have the same vernacular names, and it would be interesting to know for certain which of the two actually possesses the odour resembling the Orris, a plant nearer allied botanically to **Costus speciosus** than to **Saussurea.** There seems little doubt however, that the latter and not the former is the drug sold in Indian bazars; but it is curious how the mistake of confusing two so widely distinct plants could ever have occurred. It has been deemed PERFUMERY. 2014 desirable to leave the available information in its present form, since it is by no means established that **Costus speciosus** is not used as a substitute for **Saussurea.**

2015 § "**Piesse's** remarks must apply to **Aplotaxis** (= **Saussurea**), not to this plant, as it has no odour, and the large tuberous roots are quite insipid." (*Dr: W. Dymock in a letter to the author.*)

MEDICINE. Tubers. 2016

Medicine.—The **Costus** or *kust* root is given as a depurative and aphrodisiac. But whether or not the *kust* root should be always viewed as **Saussurea** there seems no doubt but that a certain amount of the tubers of **Costus speciosus** are regularly used by the natives of India both as food and medicine. The late **Dr. U. C. Dutt** wrote on the margin of a copy of the writer's Catalogue of the Medicinal Products of India, shown at the Calcutta International Exhibition, and opposite **Costus speciosus** (where a brief review of the conflicting opinions regarding **Costus** and **Saussurea** is given), :—"This root is said to be bitter, astringent, and digestive, and to be useful in catarrhal fevers, coughs, skin diseases, &c." **Thwaites** remarks of it :—"The Singhalese use the rhizomes as a medicine." **Atkinson** remarks of the North-Western Provinces : "From the root a strengthening tonic is made, and it is also used as an anthelmintic." The **Revd. A. Campbell** states, the root is prescribed by the Santals for "pain in the marrow." The plant referred to was identified by the writer and was **Costus**, not **Saussurea.** Referring to the Catalogue above alluded to, **Dr. Dymock** says :—"In the Calcutta Exhibition Catalogue the root is described as depurative and aphrodisiac; similar properties are attributed to it in the Concan, where it is very abundant in moist situations." This admission that it is used medicinally in the Concan confirms the general line of argument adopted in the present article. The *kust* of the drug shops is often most probable not **Costus** but **Saussurea,** but for some unexplainable reason the roots of these plants have been confused (in the literature of the subject, although they bear no resemblance to each other) perhaps for the past 200 years, but at the same time there is a certain amount of **Costus speciosus** root deliberately used, and not from any idea of adulteration with the supposed **Costus** of the ancients.

Sir Walter Elliot gives several Sanskrit synonyms for **Costus speciosus.** He may have been mistaken as to these synonyms, but he clearly recognised what the **Costus speciosus** of botanists meant, as he describes the plant. He refers to *Roxburgh's Flora Indica, Vol., I., p. 50,* and to the Coromandel plants, page 126, and states that while **Roxburgh** in these works gives *Bomma kachchika* as the Telegu for **Zingiber roseum** "in Vizagapatam, it (that name) is invariably given to **Costus speciosus,** which abounds in the forests of that province. The Sanskrit synonyms *Pushkara malaka* in **Wilson's** Sanskrit Dictionary, p. 545, and *Kásmíra* (*Wilson, p. 219*) and **Brown's** Telegu Dictionary, p. 224, are both applied to **Costus.**" He further gives *Kásmíramu* as another Sanskrit synonym for the plant, but should that word, as also *Kásmíra,* be viewed as derived from Kashmír the confusion between **Costus** and **Saussurea** might be regarded as rendered doubly perplexing. **Irvine,** in his *Materia Medica of Patna,* says of what he calls **Costus arabicus** that it "differs wholly from the real *Kút* or *Patchuk.*" He adds that it is the root of a plant found near water and is (*sic*) used in massalas, inodorous, and tasteless." Here there seems no reason to doubt we have an allusion to **Costus** and not to **Saussurea.**

FOOD. Tubers. 2017

Sweetmeats 2018

Food.—The tuber is cooked in syrup and made into preserve in some parts of India ; the natives consider it wholesome. This information regarding India was first published by **Roxburgh,** but **Ainslie** drew attention to the fact that in *Brown's Hortus Jamaic., Vol. II., p. 281,* the root stock is said to be used as a substitute for ginger. **Dr. Dymock**, commenting on this statement, remarks: "The rhizome resembles the great Galangal in growth and structure, but has no aromatic properties, the taste being mucilaginous and feebly astringent; it could only be used as substitute for ginger by being preserved with a quantity of that root sufficient to flavour it." The **Revd. A. Campbell** says the root is eaten by the Santals.

COTONEASTER, *Medik.; Gen. Pl., I., 627.*

[ROSACEÆ.

Cotoneaster acuminata, *Lindl.; Fl. Br. Ind., Vol. II., 385;* 2019

Vern.—*Riú, ráuns, riús, ruinish,* HIND.

References.—*Brandis, For. Fl., 209; Gamble, Man. Timb., 171.*

Habitat.—A deciduous shrub of the Himálaya, from the Beas to Sikkim, and occurring between 4,500 and 13,000 feet.

Structure of the Wood.—Hard, like that of **C. bacillaris**; used for walking-sticks. WOOD. 2020

C. bacillaris, *Wall; Fl. Br. Ind., Vol. II., 384.* 2021

Vern.—*Ri, riú, lin, linú, lehan, kháriz, lúni, ráu, reúsh, reús, rish, síchú, kheroa, kherbaba,* PB. HILLS; *Ruinsh,* JAUNSAR BAWUR; *Sichú, jalidar,* SALT RANGE; *Rauns,* KANGRA; *Kharwé,* PASHTU.

References.—*Brandis, For. Fl., 208; Gamble, Man. Timb., 171; Stewart, Pb. Pl., 79; Indian Forester, 1885, XI., p. 3; Kangra Gaz., 30.*

Habitat.—A small deciduous tree of the Salt Range, above 1,500 feet; of the North-West Himálaya, from the Indus to the Sarda, between 5,000 and 10,000 feet; and of Sikkim and Bhután.

Structure of the Wood.—White, turning light-red towards the centre, smooth, very hard, close and even-grained, but splits and warps much. Used for making walking-sticks; the "Alpen stocks" sold at Simla are usually made of this wood, and there is a considerable trade done in exporting it to the plains from many points along the Himálaya. This is the **Cotoneaster obtusa** alluded to in the Settlement Report of the Simla district, in which it is said the hill tribes use the sticks as goads (*chunta*). The larger pieces are made into jampan poles, axe handles, &c. **Baden Powell** suggests that it is suitable for turning. WOOD. 2022

C. microphylla, *Wall; Fl. Br. Ind., II., 385.* 2023

An ornamental plant introduced into gardens. It is known as *Khariz lúni* in Kashmír and *Garri* in Kumaon. The wood is valued for many purposes similar to the two preceding species. Used for making baskets, and mixed with **Parrotia** in the construction of twig bridges as in Kashmír; the fruit is also sweet. Fruit. 2024

Cotton and **Cotton Manufactures,** see the article **Gossypium** in Vol. III.

COTULA, *Linn.; Gen. Pl., II., 428.*

2025

Cotula anthemoides, *Linn.; Fl. Br. Ind., III., 316;* COMPOSITÆ.

Vern.—*Babúna,* PB., HIND.

Habitat.—A small herbaceous plant found in the Gangetic plain, from Rajmahal and Sikkim westwards to the Panjáb.

MEDICINE. Flowers. 2026 **Medicine.**—It furnishes part of the officinal *babúna,* which is heated with oil and applied externally in rheumatism, &c. Compare with **Anthemis nobilis,** *Linn.,* **A. 1185.**

§ "The infusion is used as an eye wash, in most diseases of the eye (*Surgeon-Major C. W. Calthrop, M.D., Morar*).

Country Borage, see **Coleus aromaticus,** *Benth.;* LABIATÆ.

Cotyledon laciniata, *Roxb.;* see **Kalanchœ laciniata,** *DC.*

COUSINIA, *Cass.; Gen. Pl., II., 467.*

2027 **Cousinia minuta,** *Boiss.; Fl. Br. Ind., 359;* COMPOSITÆ.

Syn.—C. CALCITRAPIFORMIS, *Jaub & Spach.;* C. AVALENSIS, *Bunge.*
Vern.—*Lakhtei, polí kandieri,* or *kandiári,* PB.
Reference.—*Stewart, Pb. Pl., 125.*

Habitat.—A small rigid herb, found in a wild state in some parts of the Western Panjáb plains, and distributed to Afghánistan, Baluchistan, and Persia.

FOOD. 2028 **Food.**—The young plant is used as a vegetable in the Salt range (*Stewart*).

Covellia glomerata, see **Ficus glomerata,** *Roxb.;* URTICACEÆ.

Cow-itch or **Cowhage,** see **Mucuna pruriens,** *DC.;* LEGUMINOSÆ.

Cowrie, Kawrie or **Cowdie Pine,** commercial name for **Dammara australis,** see under **Dammar, Hopea,** and also **Canarium, C. 273.**

Cowrie or **Cowry,** see **Shells,** also **Beads, B. 380.**

2029 **Cow Tree.**—Many plants, with milky sap, receive the name of Cow Tree. Perhaps the only peculiarity that more especially justifies that name is when the sap contains very little Caoutchouc and is wholesome. The Cow Tree of most writers is **Brosimum Galactodendron,** to which **Humboldt** was the first to draw special attention. It is a member of the Bread-fruit family (**Artocarpeæ**). . Several fruitless efforts have been made to introduce this plant into India, see the *Indian Forester, IX., 517.*

Crab's Eye, see **Melia Azedarach**; also **Abrus precatorius, A. 73.**

Crab Tree, see **Pyrus Malus,** *Linn.;* ROSACEÆ.

Crabs, see **Crustacea.**

CRAMBE, *Linn.; Gen. Pl., I., 98.*

2030 **Crambe cordifolia,** *Stev.; Fl. Br. Ind., I., 165;* CRUCIFERÆ.

Habitat.—A tall herbaceous annual, with leaves nearly a foot in diameter. Frequent in the North-West Himálaya, Quetta, Western Tibet, &c.; altitude 8,000 to 14,000 feet.

FOOD. 2031 **Food.**—The young leaves are, in the Sutlej Valley, eaten as a pot-herb (*Stewart*), and in Baluchistan the root is eaten (*Stocks*).

CRATÆGUS, *Linn.; Gen. Pl., I., 626.*

Cratægus Clarkei, *Hook. f.; Fl. Br. Ind., II., 384;* ROSACEÆ. 2032

A species of hawthorn met with in Kashmír, which may be viewed as intermediate in type between the two following species.

C. crenulata, *Roxb.; Fl. Br. Ind., Vol. II., 384.* 2033

THE HIMALAYAN WHITE THORN.

Syn.—C. PYRACANTHA, *Persoon;* MESPILUS CRENULATA, *Don.*

Vern.—*Gingárú, giánru,* HIND.; *Gengáru,* PB.

References.—*Roxb., Fl. Ind., Ed. C.B.C., 406; Voigt, Hort. Sub. Cal., 198; Brandis, For. Fl., 208; Gamble, Man. Timb., 170; Dalz. & Gibs., Bomb. Fl., Supp., 132; Baden Powell, Pb. Pr., 576; Drury, U. Pl., 208; Balfour, Cyclop., 836; Treasury of Bot., 344.*

Habitat.—A large spinescent shrub of the Himálaya, from the Sutlej to Bhután; found at altitudes from 5,000 to 8,000 feet, but in Kumáon at 2,500 feet.

Structure of the Wood.—White, hard, very close and even-grained; used as axe handles, staves, &c. WOOD. 2034

C. Oxyacantha, *Linn.; Fl. Br. Ind., II., 383.* 2035

THE HAWTHORN.

Vern.—*Ring, ringo, ramnia, pingyát,* or *pinyál, phindák, patákhan, ban-sanjli, sursinjli,* or *sinjli,* PB. HIMALAYAS; *Ghwansa,* or *ghwardza,* TRANS-INDUS; *Durána,* AFGH.

Habitat.—A small tree (20-30 feet), met with in the North-West Himálayas, from Quetta to the Rávi basin. Cultivated eastwards near villages, and in Afghánistan is a favourite tree planted near tombs.

Food.—Cultivated because of its FLOWERS and edible FRUIT "which is much better than that of the European hawthorn" (*Brandis*). "On the Chenáb, particularly, the fruit is large and really decent eating" (*Stewart*). FOOD. Flowers. 2036 Fruit. 2037

Structure of the Wood.—Hard and durable, used for the same purposes as the preceding. WOOD. 2038

CRATÆVA, *Linn.; Gen. Pl., I., 110.*

Cratæva religiosa, *Forst.; Fl. Br. Ind., Vol. I., 172;* CAPPARIDEÆ. 2039

Syn.—CAPPARIS TRIFOLIATA, *Roxb.;* C. ROXBURGHII, *Ham.;* C. NURVALA, *Ham.*

Vern.—*Barna, barun, bilási, bila, biliana,* HIND.; *Barún, tikto-shak,* BENG.; *Tailadu, bunboronda,* MECHI; *Purbong,* LEPCHA; *Barna, barnáhi,* PB., RAJ.; *Bela, bel,* C. P.; *Váyavarná, bhátavarná, hádavarná, kúmla, waruna, karvan,* BOMB.; *Kúmla, karwan,* MAR.; *Maralingam, marvilingá, narvala,* TAM.; *Nirválá, vitusi,* KAN., MAL.; *Uskia, usiki, usiki mánu, ulimidi, urimidi, urumitti, tella ulimidi, tella vúle,* TEL.; *Nirujani,* COORG; *Kadet, katat,* BURM.; *Varuna, asmarighna,* SANS. **Roxburgh** says that it is the *Tikta-shaka* of Sanskrit writers.

History.—**Linnæus,** and following him **Ainslie,** confused this plant with **Ægle Marmelos**; the *bel* fruit was named by the "Father of Botany" as **Cratæva Marmelos.** To this day, in many parts of India, **Cratæva** bears the same vernacular names as **Ægle,** as, for example, in the Central Provinces and in the Concan. Dr. **Moodeen Sheriff** (*Supp. to Pharm.* HISTORY. 2040

Cratæva or Bel.

HISTORY.

Ind.), under **Cratæva religiosa**, gives the following vernacular names as applied to medicinal leaves, which, on procuring samples, he found to be the leaves of **Ægle** and not of **Cratæva**:—*Bél-patrí*, HIND.; *Vilvap-pattiri, bilva-ilai*, TAM.; *Bila-patri*, KAN.; and *Bilva-patram*, SANS. In his forthcoming work (the proofs of which the writer has been obligingly furnished with) he gives the names for the **Cratæva** leaves, used in South India, as *Barmé-ké-patte*, DUK.; *Mávalingam-ilai*, TAM.; *Mavalinga-máku*, TEL. He adds that while the bark is sold in the larger bazars of India, "the leaves and root-bark" are not sold. It would thus appear that the medicinal leaves sold at the present day are those of **Ægle** not of **Cratæva.**

A brief review of the confusion which exists, in the literature of Indian Materia Medica, between **Ægle** and **Cratæva**, may prove suggestive without being supposed to commit the writer to any very pronounced personal opinion. It is significant that **Ainslie**, one of the medical officers of South India and a most painstaking observer, should have written in 1826 that he had never seen the *bel*-fruit tree—**Ægle Marmelos.** At page 86 of his work he describes it under the old Linnæan name of **Cratæva Marmelos.** He quotes **Miller's** botanical description for it, gives it the Sanskrit name of *Bilva*, the Malay name of *Tánghulo*, and the Singhalese *Beli.* "**Roxburgh**," he adds, "speaks of it under the appellation of **Ægle Marmelos**; *he* tells us that it is 'a pretty large tree, from the rind of which the Dutch in Ceylon prepare perfume, &c." At page 188 **Ainslie** next describes **Ægle Marmelos**, quotes the same botanical description, the same passage from **Roxburgh**, and gives it the same vernacular names, only calling the **Ægle Marmelos** there described by a slightly different Sanskrit name—*Bivalva.* In both articles he affirms that the plant dealt with is the **Covalum** of the *Hort. Mal.* (*Part III.*, 37). There would seem to be no doubt that he alludes in both places to the *bel*-fruit. Are we, therefore, to conclude that in the beginning of the present century the much-prized *bel* tree was not cultivated in South India. We can hardly doubt **Ainslie's** meaning when he says "The species in question I have never seen;" nor can we presume that he was labouring under the idea that **Cratæva Marmelos** was a different plant from **Ægle Marmelos**, seeing that in his two articles upon the medicinal product discussed he quotes the self-same passages. In the sixteenth century, **Garcia de Orta** (physician to the Portuguese Governor of Goa) wrote an account of the fruit under the name of *Marmelos de Benguala* (Bengal Quince), and gave it the names of *Sirifole* (= *Sriphala*) and *Beli*; the former, as he says, being the physician's name for the plant. It is worth noting that the use of the word 'Bengal' practically implies that the Madras supply was imported from that province. **Roxburgh** wrote his *Flora Indica* about the same time as **Ainslie** produced his *Materia Indica*, and the latter author frequently admits that he had seen the MS. of **Roxburgh's** work. In the *Flora Indica* it is stated of **Ægle Marmelos** that it is a native of the mountains of Coromandel, "and is also found sparingly, in the low lands." Is it thus possible that, before the *bel* fruit was cultivated to the extent it now is, **Cratæva** took its place (at least as a medicine-yielding tree) and was displaced from popular favour, the *bel*, as we now know it, receiving many of the older names? If so, the botanical name **religiosa** may rest on a stronger basis than the mere fact that the tree is grown near temples and tombs. **Lisboa** says: "So far as my enquiries go, it is not mentioned in Hindu religious books, nor used in their worship." But does this opinion rest on the existence in classical literature of descriptions that refer unmistakeably to **Ægle Marmelos**, or simply on the words *Bilva, Bel*, &c.

HISTORY.

Botanical evidence would point to **Ægle** being almost insular in its character, and it may be doubted if it is even grown to any extent in the present day beyond the limits of peninsular India; it does not succeed, for example, in Northern Panjáb. But **Cratæva** is more continental in its distribution, and is therefore more likely to have been known to the ancients.

The writer's object, however, in suggesting a doubt regarding the *bel* fruit will be gained if greater attention is paid to the two most useful plants—**Ægle Marmelos** and **Cratæva religiosa.**

References.—*Roxb., Fl. Ind. Ed. C.B.C., 426; Brandis, For. Fl., 16; Kurz, For. Fl. Burm., I., 66; Gamble, Man. Timb., 15; Dalz. & Gibs., Bomb. Fl., 8; Stewart, Pb. Pl., 17; Rheede (Nürvala), Mal. III., p. 49, t. 42; Elliot, Flora Andhrica, pp. 180, 185, 187; Pharm. Ind., 25; Ainslie, Mat. Ind., II., 86, 197, 459; O'Shaughnessy, Beng. Dispens., 9; Moodeen Sheriff, Supp. Pharm. Ind., 117; U. C. Dutt, Mat. Med. Hind., 115, 323; Dymock, Mat. Med. W. Ind., 2nd Ed., 62; S. Arjun, Bomb. Drugs, 13; Pl. and Drugs, Sind, 53; Year-book of Pharm., 1873, p. 197; Buchanan-Hamilton, Journey through Mysore and Malabar, Vol. II., p. 343; Baden Powell, Pb. Pr., 576; Lisboa, U. Pl. Bomb., 5, 290; Birdwood., Bomb. Pr., 7; Cen. Prov. Gaz., 59; Raj. Gaz., 21.*

Habitat. A moderately-sized, distorted, unarmed tree, with deciduous 3-foliate leaves. Met with here and there under cultivation, from the Rávi eastwards to Assam, Manipúr, and Burma. Also in Central and Southern India and Bengal. Probably wild in Malabar and Kánara. A favourite tree near temples and tombs.

VARIETIES.

Varieties.—The *Flora of British India* refers the forms of **Cratæva** to two varieties, which seem in a measure to correspond with the species of that genus alluded to by authors on Economic Botany.

Var. 1 t, Nurvala. 2041

Var. 1st, **Nurvala.** *Leaflets ovate-lanceolate, taper-pointed; berry ovoid-oblong.*—This appears to be the **C. Nurvala** of **Hamilton** and the **Nurvala** of **Rheede. Dalzell** and **Gibson** say, this form is the true "**Varvunna,**" and is met with in the Caranjah Hill, Warree country. **Wight** and **Arnott** (in their *Prod. Floræ Penins. Ind. Or.*) speak of it as "frequent in rich moist soil on the banks of ditches and rivers on the Malabar coast; also in Mysore, where it grows to the height of 15 or 20 feet." They also state that it is the **C. Tapia,** *Burm.* (*in part*), and also the **C. inermis,** *Linn.* (*in part*).

With the exception of the middle paragraph (which alludes to **Ægle Marmelos**) this is the form of **Cratæva** described by **Ainslie** (*Mat. Ind., I., 459*), in which he says the leaf is medicinal, being known as "*Veelvieelley* in TAM.; *Bel-ká-pát,* DUK; *Bilva-aku,* TEL.; and *Vilwa patra,* SANS. These are the names already quoted from **Moodeen Sheriff,** which that writer states are now-a-days in Madras at least, given to the leaves of **Ægle Marmelos.** There should be no difficulty in distinguishing even the most fragmentary leaf of **Ægle** from **Cratæva,** the presence of the pellucid glands in the tissue would be proof positive of the leaf not being **Cratæva. Ainslie** further states, however, of his plant that "the root, as it appears in the bazars, has a singular sub-aromatic and bitterish taste, and is supposed to possess an alterative quality." He further observes "Our article is the leaf of the **Nurvala** of **Rheede,** and the *lunu-warna* of the Cyngalese." **Trimen,** in his *Catalogue of Ceylon Plants,* affirms that the "*lunu-warana*" of Ceylon is the next variety. This is, therefore, the only serious mistake made by **Ainslie** in his attempt to distinguish the two forms of **Cratæva.**

Var. 2nd, Roxburghii. 2042

Var. 2nd, **Roxburghii.** *Leaves small, ovate-lanceolate, abruptly acuminate; berry globose.*—This is **C. Roxburghii,** *Br.,* and the **C. odora, religiosa,** and **unilocularis** of **Hamilton,** and the **Capparis trilocularis** of

CRATÆVA religiosa.

Forms of Cratæva.

VARIETIES.

Roxburgh. **Dalzell** and **Gibson** say it is common on the banks of the Nerbudda; **Roxburgh**, that it is frequent throughout India, flowering at the beginning of the hot season. **Ainslie** clearly alludes to this plant in his article (*Mat. Ind., I., 197*) "**Cratæva Tapia,** *Linn.*," which, he says, is the *Mavilinghum puttay,* TAM.; *Birmiké-chawl,* DUK.; *Maredú patta,* TEL.; *Tapia,** HIND.; *Varuna,* also *Varana,* SANS. He further calls it the "Smooth Tapia or Garlic Pear," the latter name, as he explains, being due to the fruit having "a strong smell of garlic, which it communicates to the animals which feed on it." Speaking of the medicinal properties of this species **Ainslie** remarks: "The juice of the astringent bark of this tree, though **Dr. Buchanan** says it is useless, the Vytians prescribe as a tonic in intermittent fever and in typhus; a decoction of the bark itself is also used for a similar purpose; of the latter the dose is half a tea-spoonful twice or thrice daily." **Sir Walter Elliot** alludes to this form in his *Flora Andhrica* (*pp. 180, 185, 187*), and gives it the Telegu names of *ulimidi, usiki mánu, tella-ulimidi.*

Leaves. 2043 Bark. 2044 Fruits. 2045

It may be worth pointing out that it is the leaves of variety **Nurvala** and the bark of variety **Roxburghii** that are mainly deemed medicinal. The fruits are apparently not used medicinally, but the two forms may be readily recognised by their fruits, *ovoid* in the former, *globular* in the latter. This fact is alluded to by **Ainslie**, but it is somewhat surprising that he does not tell us whether or not the natives of India were in his day aware of the rubefacient properties of the leaves of variety **Roxburghii.** He states, however, that in Jamaica, where that form also grows, "**Braham** says, the fruit is cooling, and the leaves are applied externally to take away inflammations about the anus, and also for the ear-ache." Of another Jamaica species, **C. gynandra,** he says "that the root blisters like cantharides."

These facts are of the greatest importance, in the confirmation which they afford to the opinions, expressed on a further page, by **Dr. Moodeen Sheriff,** as to the rubefacient properties of the leaves. It would be instructive to learn whether these properties were common to both forms of **C. religiosa,** or only possessed by the form which bears **Dr. Roxburgh's** name. There is also another point of some importance. **Ainslie** in his article on "**Cratæva Marmelos**" (*Mat. Ind., I., 86*), which is clearly an account of **Ægle Marmelos,** and again, in the 2nd paragraph of his article on "**Cratæva religiosa,**" refers to a resin found within the fruits, which he regards as of great value "in clearing foul ulcers." It is also used, he informs us, "in the arts as a cement." This resin and cement is well known to be produced in the fruits of **Ægle Marmelos** (around the seeds)

Cement. 2046

(see **A. 536**). In **Stewart's** *Panjáb Products,* however, it is stated that the fruits of **C. religiosa** are in Jhelum "mixed with mortar to form a strong cement." This fact, if confirmed, only shows in how many different ways the confusion between two so widely different plants as **Ægle** and **Cratæva** becomes possible.

MORDANT. 2047

Gum and Dye.—"**Aitchison** states that at Jhelum the fruit is mixed with mortar to form a strong cement, and the rind as a mordant in dyeing" (*Stewart*).

MEDICINE.

Medicine.—From what has been said it may be inferred that some doubt still exists as to whether the medicinal products of **Cratæva** can be spoken of as afforded by the one species or two species. The writer must

* A name which does not appear now to be in use in Hindustan, although mentioned by the older writers.

content himself with having indicated that, according to the literature of the subject, the medicinal BARK to be had in the bazars is obtained apparently from variety **Nurvala** and the rubefacient LEAVES from variety **Roxburghii.** If this proves correct it would become of the utmost importance, in all efforts towards more extensively utilizing these drugs, that the two forms should be carefully distinguished. It may, in fact, be due to an oversight of this nature that the drug **Nurvala** has fallen into discredit. MEDICINE. Bark. 2048 Leaves. 2049

Dr. Dymock, in his account of **Cratæva,** alludes apparently to both forms collectively. He says: "in Bombay the leaves are used as a remedy for swelling of the feet, and a burning sensation in the soles of the feet, a common complaint of a somewhat obscure nature. The leaf-juice is given in rheumatism in the Concan in doses of ½ to 3 *tolas,* mixed with cocoa-nut juice and *ghi.* In caries of the bones of the nose the leaf is smoked and the smoke exhaled through the nose. The bark and the leaf pounded and tied in a cloth are used as a fomentation in rheumatism." Juice. 2050 Fomentation. 2051

"The bark of the stem and root of this plant forms the principal medicine for calculus affections. It is said to promote the appetite, increase the secretion of the bile, act as a laxative, and remove disorders of the urinary organs" (*U. C. Dutt*). **Irvine** (*Mat. Med., Patna, p.* 128) says of the *barún,* **Cratæva Tapia**: "The fruit and bark are used in embrocations in rheumatism; not given internally." In the *Manual of Trichinopoly* (*p.* 77), it is stated of "**Cratæva** (*nurvala*) **religiosa**" the "*Marilingai,* TAM.," that "the leaves, bark, and roots are used medicinally." But the most complete account of the medicinal virtues of **Cratæva** will shortly appear in **Dr. Moodeen Sheriff's** Materia Medica of Madras. That author says: "The bark is sold in some large bazars of India, not the leaves and root-bark."

Special Opinions.—§ The following brief note has been kindly furnished for the present publication :—

"The bark of **Cratæva religiosa** is demulcent, antipyretic, sedative, and alterative tonic, and the fresh leaves and root-bark are rubefacient and vesicant. The bark is also useful in some cases of urinary complaints and fever, and in some mild forms of skin diseases in which sarsaparilla is generally resorted to. It also relieves vomiting and symptoms of gastric irritation. It is administered in decoction prepared by bruising and boiling four ounces of it with one pint and a half of water till the liquor is reduced to one pint and strained when cool. The dose of the decoction is from two to four ounces. Bruised well with a little vinegar, lime-juice, or hot water, and applied to the skin in the form of a poultice or paste, the fresh leaves of **C. religiosa** act as a rubefacient and vesicant so efficiently that I do not hesitate in saying that they are not only much superior to the mustard seeds in this country, but also quite equal if not superior to the flour of that drug imported from Europe. From 5 to 15 minutes is the time required for them to produce their full effect as a rubefacient, and if kept longer than this in contact with the skin they begin to act as a vesicant. The possession of one or two trees of **C. religiosa** by each hospital will certainly save it from the cost of the supply of Europe mustard for external use. The plant grows well with ordinary care. The fresh root-bark is also a very good rubefacient and vesicant, but it is rather too dear and not procurable in large quantities" (*Moodeen Sheriff, Khan Bahadur, Honorary Surgeon, Triplicane Dispensary, Madras*).

Food.—The FRUIT is said to be sometimes eaten (*C. P. Gaz.*, 59). FOOD. Fruit. 2052

Structure of the Wood.—Yellowish white, when old turning light-brown, moderately hard, even-grained. Used for drums, models, writing-boards, combs, and in turnery. In Trichinopoly it is also used "for making planks and as firewood." WOOD. 2053

CRATOXYLON, *Blume ; Gen. Pl., I., 166.*

[HYPERICINEÆ.

2054 **Cratoxylon formosum,** *Benth. et Hook.; Fl. Br. Ind., I., 258;*

A large tree, met with in the Andaman Islands; yields a useful timber, but the tree is rare (*Kurz, For. Fl. Burm., I., 84*).

2055 **C. neriifolium,** *Kurz ; Fl. Br. Ind., I., 257.*

Vern.—*Baibya,* BURM.

Habitat.—A moderate-sized tree, found in Chittagong and Burma.

WOOD. 2056 **Structure of the Wood.**—Dark-grey, hard, close-grained. According to **Kurz**, it is used for building purposes, for ploughs, handles of chisels, hammers, and other implements.

CRESSA, *Linn.; Gen. Pl., II., 881.*

2057 **Cressa cretica,** *Linn.; Fl. Br. Ind., IV., 225;* CONVOLVULACEÆ.

Vern.—*Gún,* SIND; *Khardi,* BOMB.; *Chavel,* NASIK (BOMB.); *Uppu sanaga,* TEL. (**Sir Walter Elliot** remarks regarding the above Telegu name that "the plant is so called from frequenting salt-lands near the sea, where it has much the look of young *Chenna* or **Cicer.**)"

References.—*Roxb., Fl. Ind., Ed. C.B.C., 265; Dalz. and Gibs., Bomb. Fl., 162; Voigt, Hort. Sub. Cal., 363; Grah., Cat. Bomb. Pl., 133; Dymock, Mat. Med. W. Ind., 2nd Ed., 569; Walter Elliot, Flora Andhrica, 186; Bomb. Gaz. (Cutch), V., 27; Stocks, Account of Sind; Aitchison, Cat. Pb. and Sind Pl., p. 98; Sakharam Arjun, Bombay Drugs, 93.*

Habitat.—A small erect shrub, common throughout the warmer parts of India, especially near the coast from Múltán, Baluchistán, and Sind, through Gujarát southwards to the Coromandel coast, and distributed to Ceylon. Appearing in the fields after the rains.

FOOD. Seeds. 2058 **Food.**—**Stocks** mentions that in Sind the seeds of this plant are ground into flour and made into cakes, pure or mixed with wheaten flour. No other author alludes to this fact, but **Dr. Dymock** mentions that in the Násik District, Bombay, the plant was eaten during the famine of 1877-78.

MEDICINE. 2059 **Medicine.**—**Dr. Sakharam Arjun** says: "It is used as a tonic and is believed to possess expectorant properties." **Dr. Dymock** remarks: "It is found in Greece, and is supposed by some to have been one of the two kinds of ἀνθυλλίς described by **Dioscorides.**"

2060 CRINUM, *Linn.; Gen. Pl., III., 726.*

A genus so named from the Greek κρίνον, a lily (*Theophrastus*). It contains about sixty species, mostly natives of the tropical regions in the old and new world. They have been worked up by **Herbert** (AMARYLLIDEÆ) and by **Kunth** (*Enum. Pl., V., 547*), but no recent work dealing with the Indian species has appeared, so that considerable doubt exists as to the synonymy of the species regarding which economic information is available. The facts here collected together as also the vernacular names given may have to be considerably rearranged when the species have been more fully worked out.

All the species are used medicinally by the people of India, but the imperfect knowledge we possess, as to the individual merits of each, has rendered it desirable to collect all known economic information mainly under the commonest and best understood species (**C. asiaticum,** *Linn.*), thus leaving the restriction and redistribution of facts to future observers.

[*Kunth, Enum., V., p. 562 ;* AMARYLLIDEÆ.

Crinum amœnum, *Roxb., Fl. Ind., Ed. C.B.C., 283 ; Herbert, 255 ;* 2061

Vern.—*Gócinda,* SYLHET.

References.—*Drury, Fl. Ind., III., 454; Voigt, Hort. Cal., 590.*

Habitat.—A native of Nepál, Sylhet, and Burma, flowering nearly all the year, but mainly in the hot and rainy seasons; the flowers are large and white.

C. asiaticum, *Linn.;* var. **toxicarium,** *Herbert, Bot. Mag., 1073.* 2062

Syn.—C. TOXICARIUM, *Roxb.,* and of *Wight, Ic., 2021 ;* C. BREVIFOLIUM, *Roxb. ;* C. DECLINATUM, *Herbert,* and of *Willd. ;* C. ASIATICUM, *Linn. ; Bot. Mag., 1073, 2908, 2239 ;* of *Herbert, p. 243,* and of *Kunth, Enum., V., p. 547* (non C. ASIATICUM, *Roxb., Fl. Ind.*).

Vern.—*Chindar, kanwal, pindar, kanmu,* HIND. ; *Nagdown* (according to **Dymock**), BOMB. ; *Nagdamani,* GUZ. ; *Nagadavana,* MAR. ; *Nagin-ka-patta,* DUK. ; *Bara-kanur, nagdaun, boda-kanod (gaer-konar-patta,* according to **Bidie, Dymock, &c.**), BENG. ; *Visha-mungil,* TAM. ; *Kesar-chettu, visha mungali, lakshminárayaná chettu,* TEL. ; *Kóyángi,* BURM. ; *Tolabo,* SING. (*Visha-mungil,* TAM. in Ceylon) ; *Man-sy-lan,* COCHIN-CHINA (see *Loureiro's Flora, Cochin-China, I., 198 ; Vishaman dala,* SANS.

References.—*Roxb., Fl. Ind., Ed. C.B.C., 283 ; Voigt, Hort. Sub. Cal., 588 ; Thwaites, En. Ceylon Pl., 324 ; Trimen, Cat. Ceylon Pl., 93 ; Dalz. & Gibs., Bomb. Fl., 275 ; Aitchison, Cat. Pb. and Sind Pl., 148 ; Elliot, Flora Andhrica, 90, 106, 193 ; Rheede, Hort. Mal., XI., t. 38 ; Rumph., Amb., VI., t. 69 ; Drury, Handbook of the Indian Flora, III., 452 ; Grah., Cat. Bomb. Pl., 216 ; Pharm. Ind., 234 ; Ainslie, Mat. Ind., II., 464 ; O'Shaughnessy, Beng. Dispens., 655 ; Moodeen Sheriff, Supp. Pharm. Ind., 35, 39, 118 ; Dymock, Mat. Med. W. Ind., 2nd Ed., 820 ; Bent. & Trim., Med. Pl., 275 ; S. Arjun, Bomb. Drugs, 196 ; K. L. De, Indigenous Drugs, Ind., 125 ; Murray, Pl. and Drugs, Sind, 19 ; Bidie, Cat. Raw Pr. Paris Exh., 18 ; Home Dept. correspondence regarding Pharmacopœia of India, 325, 437, 232, 225 ; Rev. A. Campbell's Report on Econ. Prod., Chutia Nagpur, 31 ; Report on Pegu, by Kurz ; Drury, U. Pl., 162 ; Lisboa, U. Pl. Bomb., 256, 270 ; Birdwood, Bomb. Pr., 89 ; Balfour, Cyclop., 837 ; Treasury of Bot., 348.*

Habitat.—A fairly abundant cultivated plant, its erect stems with their crown of large graceful leaves forming almost a characteristic feature of Indian gardens. It may be doubted, however, as to its being a native, although many writers have alluded to it as found in a semi-wild condition in various parts of the country. **Dr. Hove** apparently met with it in the Thana district, Bombay, in 1787. **Roxburgh** urges the importance of the erect stem in distinguishing it from **C. defixum,** and he expresses the opinion that it may be a native of Ceylon. Speaking of that region **Thwaites** remarks that "it is very abundant on the sea-coast of the island," and "frequently planted as a fence for native gardens near the sea."

Although thus not established as an indigenous plant, from the confusion in the synonymy of the Indian **Crinums,** many writers on Indian Economic Botany give the facts they publish under the name **C. asiaticum.** This idea has been followed in the present article, but it is probable future investigation may relegate to **C. defixum, C. amœnum,** or **C. pratense** much of what is here given under the popular name **C. asiaticum.**

Medicine.—**Ainslie** wrote in 1826: "The succulent bitterish leaves of this plant, which are about 2 inches broad and 3 feet long, the natives bruise and mix with a little castor-oil, so forming an application which they think useful for repelling whitlows, and other inflammations that come at the end of the toes and fingers; the juice of the leaves is employed

MEDICINE Leaves. 2063 Juice. 2064

MEDICINE. Root. 2065 for the ear-ache in Upper India. In Java, by **Horsfield's** account, this plant is reckoned one of the most satisfactory emetics the inhabitants have." "It is the root (? bulb) chewed that is the emetic, provided a little of the juice is swallowed." **Sir William O'Shaughnessy**, who wrote some 20 years later, says: "From two to four drachms of the recent root or stem, bruised into a paste, and this squeezed through cloth, affords a juice which proves emetic after a few minutes; in smaller quantities it is nauseant and diaphoretic; we have never known it to occasion any untoward symptoms. The dried sliced roots are also an efficient emetic, but require to be given
Extract. 2066 in double the dose of the recent article." The extract, whether watery or alcoholic, is very uncertain in its action. In the form of a syrup it may probably be found to retain the native principles of the recent plant. The tincture of the fresh plant does not succeed, doubtless in consequence of the large quantity of spirit counteracting the emetic effect by its stimulating energy.

These two passages express all that has since appeared, as, for example, in the *Pharmacopœia of India*; **Drury, Murray, K. L. De**, and indeed most subsequent writers, repeat in other sentences the same facts. **Dr. Dymock** adds: "I have not met with any account of this drug in native works on Materia Medica, nor does it appear to be used in Bombay, though the plant is very common. In the Concan the leaves smeared with mustard oil or *mutel* are warmed and bound round inflamed joints." Alluding to **Ainslie's** remark regarding the juice being used for ear-ache, **Dr. Dymock** gives as a footnote: "A well-known popular use of the plant; the leaves are slightly roasted, and the juice is then expressed and a few drops poured into the ear."

Bulb. 2067 The bulb of the so-called **Crinum asiaticum** is made officinal in the *Indian Pharmacopœia* as an emetic, nauseant, and diaphoretic.

Special Opinions.—§ "The juice expressed from the leaves has a soothing effect in cases of ear-ache; it should be applied locally" (*Civil Surgeon J. Anderson, M.B., Bijnor*). "Used in whitlows and other local inflammations" (*Dr. H. W. Hill, Manbhoom*).

2068 **Crinum defixum,** *Ker.* (and of *Gawl.*); *Herbert, p. 255; Bot., Mag.,* [*2208.*

Syn.—C. ASIATICUM, *Roxb.* (non *Linn.*), *Fl. Ind., Ed. C.B.C., 283;* C. ROXBURGHII, *Dalz., Fl. Bomb., 275;* BELUTTA POLA TALY, *Rheede, XI., t. 38;* RADIX TOXICARIA SECUNDA, *Rumph., VI., 156.*

Vern.—*Suk-darshan,* BENG.; *Nagdown,* BOMB.; *Kesar chettu,* TEL.; *Hintolabo,* SING. (according to **Ainslie**).

References.—*Dalz. & Gibs., Bomb. Fl., 275; Lisboa, U. Pl. Bomb., 204.*

Habitat.—A native of the Concan, of Coromandel, and of many parts of Bengal, as, for example, the Sunderbands. Flowers large, sessile, white, fragrant during night; flowering time, the close of the rainy season. **Dalzell** and **Gibson** say it is common on the banks of the Deccan rivers. It delights in swampy situations where mud abounds. **Roxburgh** lays stress upon the fact that this species is stemless in distinguishing it from **C. asiaticum,** *Linn.* (**toxicarium,** *Roxb.*). It produces long stoloniferous roots from the top of the bulb which penetrate the mud.

MEDICINE. 2069 **Medicine.**—**Lisboa** says the bulb is boiled and eaten as *shak-baji.*

2070 **C. pratense,** *Herbert; Amaryll., 256.*

Syn.—C. LONGIFOLIUM, *Roxb., Fl. Ind., Ed. C.B.C., 284;* C. LAURIFOLIUM, *Herbert & Roxb.;* C. ELEGANS, VENUSTUM, and CANALIFOLIUM, *Carey.*

Vern.—*Pa-taing,* BURM.

References.—*Voigt, Hort. Sub. Cal., 590; Bot. Mag., t. 2592 and 2121.*

Habitat.—A native of the interior of Bengal, Sylhet, Pegu, &c., flowering time the rainy season. Flowers large, white, fragrant. A variable plant, some of the names given above belonging to what may prove recognisable forms.

A closely allied plant to **C. defixum,** being more elegant but does not possess the long spindle-shaped root of that species (*Roxb.*). The form described by **Roxburgh** under the name **C. laurifolium** occurs in Pegu: it has very long weak recumbent leaves (2 inches by 5 feet). 2071

Crinum, sp. (found in Chutia Nagpur.) 2072

Mr. C. B. Clarke writes of this plant that he is unable to name it, and presumes it may be an undescribed species. In that case it should bear the discoverer's name—the **Rev. A. Campbell. Mr. Clarke** also informs the writer that he has collected another species in the tanks of Chutia Nagpur which flowers in November; he views this as distinct from the common Sunderband species, which flowers in May.

Vern.—*Sikyom baha,* SANTAL.

Habitat.—High and dry situations in Chutia Nagpur, flowering during the hot season before the leaves appear. In some respects, this resembles **C. latifolium** as described in **Roxburgh's** *Flora Indica.* 2073

Medicine.—**Mr. A. Campbell** says: "The bulb is sometimes as large as a good-sized turnip, and of the same shape. A decoction prepared from it is given internally and pounded and made into a paste; it is also applied externally by the Santals in dropsy. It is used for the diarrhœa of cattle." MEDICINE. Bulb. 2074

C. zeylanicum, *Linn.; Wight, Ic. 2019-20.* 2075

Syn.—C. ORNATUM, *Herbert;* C. ZEYLANICUM, *Roxb.;* C. LATIFOLIUM, *Roxb.;* C. MOLUCCANUM, *Roxb.;* C. HERBERTIANUM, *Herb., p. 263; also Wall., Pl. As. Rar., 2, p. 145.*

Vern.—*Súkh-darsan,* BENG.; *Gadambikanda,* BOMB.; *Goda-mánil,* SING.

References.—*Tulipa Javanica, in Rumph., Amb., V., t. 105; Sjovanna-pola-tali in Rheede, Hort. Mal., XI., t. 39; Bot. Mag., 1171, 2217, 2292, and 2466; Roxb., Fl. Ind., Ed. C.B.C., 286; Grah., Cat. Bomb. Pl., 216; Kunth, Enum., 573; Thwaites, En. Ceyl. Pl., 324; Trimen, Cat. Ceylon Pl., 93; Drury, Handbook, Fl. Ind., III., 454; Dymock, Mat. Med. West Ind., 2nd Ed., 822.*

Habitat.—A very variable plant, some of the above synonyms corresponding to well marked varieties, which, in a work on economic products, may, however, with safety, be treated collectively. It is fairly plentiful throughout the Peninsula of India—the Concan, Bengal, Coromandel, Burma, &c. It flowers in the rainy season, frequenting low, rich, uncultivated damp grounds; the flowers are fragrant, white, streaked with pink, or more or less reddish. The form which **Roxburgh** named **C. latifolium** flowers in April, is stemless, and has a spherical bulb often 2 feet in circumference.

Medicine.—**Dymock** remarks of this species: "The bulb is extremely acrid, and is used for blistering cattle, a slice being bound upon the skin. When roasted it is used as a rubefacient in rheumatism." MEDICINE. Bulb. 2076

CROCODILE (CROCODILUS, *Cuv.*).

Crocodilus palustris, *Less.* 2077

THE COMMON CROCODILE, often vulgarly called in India, the Alligator—an American Reptile.

There are apparently two other species besides the above met with in India, *viz.*, **C. porosus**, *Schneid.*, and **C. trigonops**, *Gray*. The long snouted Gavial lives on fish and turtles, and frequents the rivers of India along with the Crocodile.

Vern.—*Magr, kumhir*, HIND.; *Sisan*, SIND.

Habitat.—Found throughout India and Ceylon, affecting rivers, lakes, marshes, and even the sea coast. It may be recognised by its shorter and broader snout than that of the Gavial, and by the first and the fourth tooth of the lower jaw fitting into the upper.

Although held sacred in many parts of India (and sometimes even tamed so far as to come for food when called, as, for example, at the Mugger Pier), the Crocodile is the terror of the rural inhabitants of India along the basins of the great rivers, not even the stakes placed around bathing places proving an effectual protection. The Crocodile often attains a great size, being from 15 to 30 feet in length, and although it is reported to eat the dead bodies thrown into the rivers, it lives mostly on live animals, taking human beings when pressed for other food.

Economic Products.—OIL, SKIN, MUSK, and FLESH.

2078 **Crocodile Flesh.**—It many parts, Crocodile flesh is said to be eaten or used medicinally.

2079 **Crocodile Skin.**—See **Hides.**

2080 **Crocodile Musk.**—The natives of Africa appear to regularly extract the musk-glands of the Crocodile. **Dr. Forbes Watson**, in his *Industrial Survey of India*, page 392, alludes to a sample of this substance procured from Travancore.

2081 **Crocodile Oil.**—The oil of the Indian Crocodile contains a larger quantity of solidifiable fat than either neat's-foot or any fish-oil. It is prepared by the Sanif tribe, in the Panjáb, who eat crocodile flesh, and is also said to be procurable in abundance at Agra (*Spons' Encyclop.*, *5136*).

2082 CROCUS, *Linn.; Gen. Pl., III., 693.*

This is the *κροκος* of **Dioscorides.** It is not alluded to by the earlier Sanskrit writers, but Arabian authors speak of it as cultivated in the tenth century at Darband and Ispahan, and Chinese writers state that it was introduced into their country by the Muhammadans in the Yuen dynasty (A. D. 1280).

2083 Crocus sativus, *Linn.; Royle, Ill. Him. Bot., t. 90;* IRIDEÆ.

SAFFRON.

Vern.—*Jáfrán*, BENG.; *Késar, zafran*, HIND.; *Safran, kessar, kecara*, BOMB.; *Kecara*, MAR.; *Keshar*, GUZ.; *Kunkuma, kásmírajanmá* (Ainslie), *kumkuma* (Dutt), *saurab* (Dymock), SANS.; *Zaafarán*, ARAB., PERS.; *Kungumapu*, TAM.; *Kunkum apave*, TEL.; *Thanwai*, (**Mr. Oliver**, Forest Officer in Burma, informs the writer that this is the name for Turmeric not Saffron. The word appears in various works under Saffron and is therefore given here for the present), BURM.; *Kong*, KASHMIR; *Kurkum*, BHOTE; *Zafar*, TURKI (according to **Aitchison**).

References.—*DC., Orig. Cult. Pl., 166; Pharm. Ind., 235; Ainslie, Mat. Ind., I., 354; O'Shaughnessy, Beng. Dispens., 654; Moodeen Sheriff, Supp. Pharm. Ind., 118; U. C. Dutt, Mat. Med. Hind., 306; Dymock, Mat. Med. W. Ind., 796; Flück & Hanb., Pharmacog., 663; U. S. Dispens., 15th Ed., 501; Bent. & Trim., Med. Med. Pl., 274; S. Arjun, Bomb. Drugs, 142; K. L. De, Indig. Drugs, 42; Murray, Pl. and Drugs, Sind, 20; Year Book Pharm., 1874, p. 135, 1876, 14; Baden Powell, Pb. Pr., 381; Lisboa, U. Pl. Bomb., 177; Birdwood, Bomb. Pr., 88, 305; Liotard, Dyes, 94, 96, App., IV., VI.; Spons, Encyclop., 866; Simmonds, Trop. Agri., 379.*

SAFFRON.

Habitat.—The European supply of this plant comes from France, Spain, and Italy. It is cultivated in Kashmír at Pámpúr near the capital. In **Honigberger's** time the cultivation of Saffron appears to have been a State monopoly in Kashmír. The Indian supply is obtained from France, China, Kashmír, and a small quantity from Persia, in the form of cakes known as *Kesar-ki-rote.*

DYE. 2084

Dye.—It is chiefly used in Europe as a dye, and to colour cheese, puddings, &c., but very little as a medicine. In India it is too expensive to be used as a dye-stuff. It is, however, held in high esteem as a medicine. The product is obtained from the stigmas of the flowers, 4,000 of which are required to produce an ounce of saffron.

MEDICINE. 2085

Medicine.—As a medicine it is used in fevers, melancholia, and enlargement of the liver. It has also stimulant and stomachic properties, is highly thought of as a remedy for catarrhal affections of children, and is used in certain Indian dishes as a colouring agent. Mullahs (priests) make a kind of ink with this substance with which they write charms (*Dr. Emerson*). In over-doses it is generally reported to act as a narcotic poison. **Ainslie** gives perhaps the most complete account of the native uses of this drug, and of the opinions which prevailed among Anglo-Indian physicians at the beginning of the present century, when it was viewed as an antispasmodic and emmenagogue; it is now chiefly used as a colouring agent. Dr. **Dymock** states that the Persian and Kashmír Saffron is rarely met with in Bombay, the Chinese and European imported article being deemed much superior. **Ainslie** and **O'Shaughnessy** speak of the Indian Saffron as adulterated with safflower (**Carthamus tinctorius**).

2086

CHEMISTRY. 2087

Chemistry.—§ "The colour of saffron is due to the presence of a glucoside polychroit, which is decomposed by acids, with the formation of a new colouring principle *Crocin*" (*Prof. Warden, Calcutta*). For full particulars as to the chemistry of this drug see the *Pharmacographia, p. 666.*

TRADE. 2088

Trade in Saffron.—The imports of foreign saffron were in 1882-83, 226 cwt. valued at R4,25,124, and in 1886-87, 268 cwt. valued at R5,50,383. Of the Indian imports the bulk comes from France.

CROPS.

2089

An important feature of Indian Agriculture is the fact that, through the presence of extensive montane tracts, India possesses considerable areas that are under temperate influences, as well as vast expanses that are purely tropical. Between these two conditions almost every possible gradation exists in which the tendency to extreme humidity or extreme aridity modifies the general character. From this point of view alone India may be said to be capable of producing the crops of the arctic, the temperate, or the tropical regions, or of the deserts and swamps of the world. But superadded to geographical peculiarities, it possesses soils diversified from that of the sandy desert through every grade of rich fertile loam and humus to the recently-formed muddy swamp. Then, again, the periodicity of the climate enforces a system of agriculture quite dissimilar to that of Europe. In the greater part of India the rains occur during certain months, so that the year may be divided into three seasons—the Hot season, Rainy season, and Cold season; and as a rule each of these seasons produces its characteristic crops, so that the cultivator has usually two and sometimes three harvests a year. This is modified in certain provinces through the rains not occurring at the same period. Thus, in Bengal, Bombay, the greater part of the Central Provinces, and in Berar, the rains occur in June, July, August, and September, being preceded by the hot season, and followed by the cold. In the Panjáb, while rain falls during those months, it is not so heavy as in December, January, and February. Rain during March in the Panjáb and North-West Provinces would be most injurious. The Panjáb, North-West Provinces, and Rajputana have two seasons of rain—July, August, and September, and again December, January, and February. In Madras, while showers fall during June, July, and August, and on the western coast constitute as well marked a rainy season as in Bengal, still the greater part of the Southern Presidency does not obtain its proper rainy season until October and November. Indeed, the commencement of the rains in Madras is the truest indication of the close of the rainy season of Northern India. Western Rajputana, a large portion of Sind, and the Southern Panjáb, have no regular rains, and are collectively often spoken of as the rainless area of India. It will thus be seen that to study the crops of India, the closest attention must be paid to this shifting of what in each province is generally called its rainy season. In the regions indicated as possessing two periods of rainfall, two well-marked crops exist—the spring or *rabi* crop, and the autumn or *kharif*. The temperate mountains within these regions have accordingly a fall of snow during winter (corresponding to the winter rains at lower altitudes), and but for the existence of a mid summer to early autumn rainy season they approximate to the conditions which prevail in Europe. Taking thus into consideration the varied nature of the soil, and the climatic peculiarities, India may practically be said to be capable of producing any known vegetable product. Even in the plains (in a large portion at least of India), the winter temperature is such that temperate annual crops may be raised. The following may be given as a brief classification of the chief crops, but fuller particulars will be found regarding each in its alphabetical place in this work.

2090 *1st*, CEREALS.—This includes Wheat, Rice, Oats, Barley, Indian-corn, Millets (various kinds), and Coix (Job's tears). (Conf. with **Cereals.**)

2091 *2nd*, PULSES.—Such as Gram, Peas, Beans, Lentils, &c. (Conf. with **Pulses.**)

2092 *3rd*, OTHER GRAINS.—Buckwheat, Amarantus, Chenopodium, &c. This practically embraces all seeds which are ground into flour or eaten boiled as a staple article of diet, but which do not belong to the GRAMINEÆ (Cereals), or to the LEGUMINOSÆ (Pulses). (Conf. with **Grains.**)

2093 *4th*, SPICES AND CONDIMENTS.—Turmeric, Ginger, Cumin, Coriander, Caraway, Pepper, Betel-leaf, Capsicum, Cardamum, &c., &c. (Conf. with **Spices.**)

2094 *5th*, STARCHES AND SUGAR.—Sugar-cane, Arrow-root, Sago, &c. (Conf. with **Starches.**)

2095 *6th*, GARDEN PRODUCTS AND VEGETABLES.—Potatoes, Yams, Colocasia, Cabbage, Gourds, Melons, Cucumbers, &c., &c. (Conf. with **Vegetables.**)

The above might be grouped as edible products, but there are other crops some of them of even great importance, such as—

2096 *7th*, FIBRES.—Cotton, Silk, Jute, Sunn-hemp, and many others, the fibre from **Hibiscus cannabinus** being, after sunn-hemp, the next most important of fibre crops. (Conf. with **Fibres.**)

2097 *8th*, DYES.—Indigo, Safflower, Al (**Morinda tinctoria**), Madder, &c. (Conf. with **Dyes** and **Tans.**)

2098 *9th*, NARCOTICS.—Opium, Ganja, Tobacco, Tea, and Coffee. (Conf. with the separate accounts of each of these products and with the article **Narcotics.**)

10th, OIL-SEEDS.—Ground-nut, Rape, Mustard, Cotton-seed, Linseed, Opium-seed, Castor-oil, Gingelly or Sesame-oil, &c. (Conf. with **Oils.**) 2099

These are the principal crops of India, but the agriculturists have often other industries to occupy their attention, such as the collection of forest or jungle produce,—*e.g.*, Lac, Cutch, Myrobalams, Wild silks, Gums and Resins, and Medicinal substances, or the extraction of the Palm juice, used either in the distillation of spirits or in the preparation of sugar. Fruit trees are rarely grown as a source of revenue, although every villager has, as a rule, a few of most of the following: Plantain, Mango, Guava, Ber (the plum of the plains), Pine-apple, Custard-apple, Oranges, Lemons, &c.

CROTALARIA, *Gen. Pl., I., 479.* 2100

A genus of plants closely allied to the Broom, the generic name being derived from the Greek κροταλον (a castanet), in allusion to the rattling noise made by the loose seeds within the inflated pods. This same idea, according to **Sir Walter Elliot**, is implied by the Sanskrit name *Ghantár avamu.*

Crotalaria Burhia, *Hamilt.; Fl. Br. Ind., II., 66;* LEGUMINOSÆ. 2101

Vern.—*Sis, sissái, meini, pola, khippi, búta, khep, khip, khif, bhata, búi láthia, khársan kauriála,* PB.; *Ghagari,* MAR.; *Ghugharo,* GUZ.; *Drunnú,* SIND.

References.—*Dalz. & Gibs., Bomb. Fl., 54; Stewart, Pb. Pl., 64; Aitchison, Cat. Pb. and Sind Pl., 37; Müeller, Select Extra-tropical Plants, 86; Murray, Pl. and Drugs, Sind, 112; Baden Powell, Pb. Pr., 508; Lisboa, U. Pl. Bomb., 231; Birdwood, Bomb. Prod., 130; Royle, Fib. Pl., 272; Rajputana Gaz., 30.*

Habitat.—A low under-shrub, abundant in the sandy plains of Sind, Panjáb, Rajputana, and Cambay, ascending to 4,000 feet in altitude.

Fibre.—Is said by **Mr. Baden Powell** to yield a good fibre for cordage; used, to some extent, in the Panjáb in place of the *Sunn*-hemp (**C. juncea**) of other provinces. FIBRE. 2102

Medicine.—The branches and leaves are used as a cooling medicine. MEDICINE. Branches. 2103

Fodder.—The Rajputana Gazetteer states that the plant is much valued as a fodder. FODDER. 2104

C. juncea, *Linn.; Fl. Br. Ind., II., 79.* 2105

SUNN or SUNN-HEMP or INDIAN HEMP, FALSE HEMP, BROWN HEMP, BOMBAY or SALSETTE HEMP, WUCKOO-NAR (or TRAVANCORE FLAX), JUBBULPUR HEMP, &c., &c.

Syn.—C. TENUIFOLIA, *Roxb.*

Vern.—*San, sanai, sani* (or *sun, shon*), HIND., BENG.; *Ausá, suilá,* ASSAM; *San, phulsan, arjha san,* N.-W. P.; *San tág,* BOMB.; *Sini, tág-san,* SIND; *San, suna,* GUZ.; *San, ghagharu tág,* MAR.; *San, tág, janab,* DEC.; *Jenappa, janumú, janopa* (or *shanapam*), TAM. and TEL.; *Janapa, pulivanji, vakkavanji, chanam,* MAL.; *Wuckú,* TRAVANCORE; *Sanabu shanabiná, pundi,* KAN.; *Pan* (?), *paikhsan, paik piven,* BURM.; *Hana,* SING.; *San,* PERS.; *Sana* (*ghantáravamu,* a generic name according to **Sir Walter Elliot**), SANS.

According to some writers the name *Ambádí* or *ambárí* is, in Western India, given to this plant, but it seems probable that that name should be restricted to **Hibiscus cannabinus**. Indeed, it has been found difficult to arrive at any definite idea regarding the present area under *sunn*-hemp cultivation from the fact that the above **Hibiscus** appears to be confused with it. In Bengal, and indeed in some parts of the N.-W. Provinces,

FIBRE.

sunn or *san* is applied to the true hemp plant, a prefix being given to distinguish the fibre of **Crotalaria,** such as *Phul-sunn, Bhágá-sunn,* and *Bádál-sunn.* In Bengal this is known as *Chunpát.* But *pátsan* and *mestapát* are generally applied to **Hibiscus cannabinus.** Even the term *san* is thus often loosely used, being sometimes given to **Hibiscus cannabinus,** and at other times to jute. Indeed, *Sanai* or *Saní* is a safer name to use for this plant than the one in general use, *sunn* (*san* hemp). In the publications issued by certain provinces information occurs regarding "Bombay hemp" as distinct from "*sunn*-hemp;" in others these are returned as one and the same substance, while **Cannabis sativa** and **Hibiscus cannabinus** are separately reported. It would thus appear that the term "Bombay hemp" is often, though incorrectly, given to the Ambádí fibre, **Hibiscus cannabinus.** It is thus unfortunate that, in modern commerce, the term "hemp" should ever have come to be applied to any but the true hemp plant, as, by this usage, widely dissimilar products have been almost hopelessly confused. The *sunn* is a bush closely allied to the English broom or the Indian *dál,* while the *ambárí* is a Hibiscus or cotton-looking plant with sharply-cut leaves not unlike those of the hemp plant,—hence the specific name **cannabinus.** The true hemp has its nearest affinity, of fibre-yielding plants, in the common nettle. The hemp fibres thus afforded by these three plants have little or nothing in common.

References.—*Roxb., Fl. Ind., Ed. C.B.C., 545; Voigt, Hort. Sub. Cal., 206; Kurz, For. Fl. Burm., I., 331; Kurz, Contrib. to Burmese Fl., 266; Gamble, Man. Timb., 117; Gamble, List of Pl., Darjeeling, 25; Thwaites, En. Ceylon Pl., 81; Dalz. & Gibs., Bomb. Fl., 54; Stewart, Pb. Pl., 64; Aitchison, Cat. Pb. and Sind Pl., 37; Wight and Arnott, Prod. Fl. Peninsulæ Ind., &c., I., 183; Hem Chunder Kerr, Culivation and Trade in Jute, p. 6; Flora Andhrica, by Walter Elliot, 59, 60, 73; Moodeen Sheriff, Supp. Pharm. Ind., 119; U. C. Dutt, Mat. Med. Hind., 316; S. Arjun, Bomb. Drugs, 40; Murray, Pl. and Drugs, Sind, 112; Bidie, Cat. Raw Pr., Paris Exh., 117; Baden Powell, Pb. Pr., 342; Drury, U. Pl., 163; Duthie & Fuller, Field and Garden Crops, I., 82; Lisboa, U. Pl. Bomb., 231, 290; Birdwood, Bomb. Pr., 317; Royle, Fib. Pl., 270—293; Liotard, Paper-making Mat., 17, 27; Spons, Encyclop., 946; Balfour, Cyclop., 771, 843; Smith, Dic., 211; Treasury of Bot., 350; First Ann. Rep. Director of Agri., Bengal, lxxiv.; Manual of Trichinopoly by Moore, 72; Manual of Coimbatore Dist., Nicholson, 238-239; Mysore Gas., I., 95; Bomb. Gas., II., 63; III., 54; XI., 97; XII., 163; XIII., pt. I., 290; XVIII., 48; XXIV., 172; Bengal Ad. Rept., 1882-83, 15; Panjáb Gas. (Kangra Dist.), p. 153; N.-W. P. Gas. (Meerut), 233; (Cawnpore), 27; Agri-Hort. Soc. Ind. Jour., Vol. VII., pt. III. (New Series), pp. 224-227; Sett. Rept., Upper God. Dist., C. P., p. 36; Ramshunker Sen, in Agricultural Gazette, January 1874, 162-163.*

Habitat.—The *Flora of British India* gives the habitat of this plant as "Plains from the Himálaya to Ceylon, but often planted for its fibre." The writer is not aware of **Crotalaria juncea** having been recorded as found in a wild state anywhere in India, although it may sometimes exist as an escape from cultivation. **Kurz** says of **C. juncea** in Burma "like wild along the banks of the larger rivers, especially the Irrawaddi," and **Griffiths** that **C. juncea** is met with in Afghánistán. **Roxburgh** describes a form (by modern botanists reduced to the present plant, *viz.*, **C. tenuifolia**) which he states is a native of Coromandel. Many writers, however, familiar with the living plants, still affirm that **C. juncea** and **C. tenuifolia** are distinct. They seem at least to be cultivated recognisable states which, owing to the reputed superiority of the fibre of **C. tenuifolia,** it might, for economic purposes, be desirable to treat of separately. As a historic point of some interest it may be here added that, after **Roxburgh** reported the discovery of **C. tenuifolia** in a wild state in Coromandel, it was next found under cultivation at Jabalpur, where it seems to continue to be cultivated

C. 2105

FIBRE.

to this day, although as yet it has not been reported as found anywhere between these remotely distant regions. At the same time **C. juncea** is cultivated more or less in every province of India, competing for popular favour with **Hibiscus cannabinus**, until in some parts of the Panjáb and Sind its place is taken by the wild species **Crotalaria Burhia**, which yields a fibre of such quality as to render the cultivation of **C. juncea** superfluous.

SUNN (or SAN) HEMP FIBRE.

2106

Under the heading **Cannabis sativa** the suggestion has been offered that the Greek and the Latin *cannabis* may have been derived from the Arabic *kinnab*, a name which seems to have been adopted from the Persian *kanab*. The author of the *Makhzan-el-Adwiya*, while giving that derivation, adds that the Syrian for the plant is *kanabira*. But the word cannabis, however derived, has less of interest associated with it than the Sanskrit *Sana*. There are, for example, three fibre-yielding plants which appear in early literature to bear that name or some derivative from it. It even may be doubted if the *sana* of the earliest writers is the hemp of modern commerce. In the unmistakable references to hemp in Sanskrit, care is taken to associate with the plant qualifying and descriptive epithets that convey the idea of the well known narcotic properties of the plant. Even the Hebrew *shesh*, generally translated flax, is suggestive of intoxication, and hence the possibility of its having been used for hemp rather than flax. **DeCandolle** has established very conclusively that a form of flax was grown in Europe before the time of the Western Aryan migration, although the plant now cultivated for flax seems to have followed the path of that great civilization, having been apparently cultivated by the Aryans prior to their invasion of Europe. This conclusion is founded mainly on the fact that the root of the word "linen" did not exist in Europe prior to the period indicated, and he adds that it does not occur in the Aryan languages even of India. The mummies of Egypt are draped in garments of flax or linen, but there is nothing to justify the supposition that the Egyptians of that day knew of the hemp plant although they were familiar with flax. Thus even the history of flax is in some instances involved with that of hemp, such names as *shesh* implying an intoxicating power—a property of the hempen fibres possessed alone by **Cannabis sativa.** The *sana* fibres of the Sanskrit authors are **Crotalaria juncea** (*sunn* hemp), **Hibiscus cannabinus** (*sanpát*), and **Cannabis sativa** (true hemp of modern commerce). As already stated, there would seem to be every chance that the earliest writers allude under *Sana* to the fibre of **Crotalaria juncea,** but that, as the true hemp became known, care was taken by subsequent authors to distinguish it by the use of qualifying and descriptive words that implied its intoxicating properties. It seems possible also that the narcotic of **Cannabis sativa** was known before the discovery of the fibre yielded by the plant, although the name *bhánga* signifies to break, an idea probably suggested from the breaking or isolating of the fibrous bark from the stems, but that idea might have arisen from jute, sunn hemp, or any other bark fibre. In many instances no room for doubt is left as to the peculiar *sana* referred to. Thus in the *Ain-i-Akbari* (Gladwin's Translation) the plant is described as bearing yellow flowers—a fact that clearly fixes **Crotalaria juncea** as the *sana* there alluded to; but in further discussing the hempen fibres there occurs the following passage: "One species bears a flower like the cotton shrub, and this is called in Hindustani *sanpát*. It makes very soft rope." Here we have a distinct allusion to **Hibiscus cannabinus.**

Kshauma. 2107

Frequent reference is made in the *Rámáyana* and the *Mahábhárata* to a garment called the *kshauma*. *Kshumá* is by most writers viewed as a name for linen, but the resemblance to the Chinese *chu-má* (or *schou-má*),

FIBRE.

the name for grass-cloth, is worthy of mention. Certain synonyms occur for the *kshauma* which convey some idea of the material of which it was made. Thus *umá* or *atasí* (the *atasí-vastram*) would be the linen-made, the *patta-vastram* the Hibiscus-made, and *sana-vastram* in all probability the sunn-hemp-made garment. Later writers speak of *sana* garments as being used as sackloth and worn as a mark of punishment or mortification. A prophecy in the *Vishnu Purana* speaks with scorn of the Kaliyaga (or iron) age as one of degeneration, "when the garments of men will be like the *Saní*." It thus seems established that a gradual evolution in popular opinion took place until (as in the present day) *san* and *san-pát* were no longer used as textile fibres, but were relegated to the position they at present hold as useful fibres for cordage and sacking. The hill tribes of the North-West Himálaya weave a proportion of their clothing of hemp, but although the plant springs up wild all over the plains of India its fibre is but very rarely extracted. The people during all modern historic times have been clad chiefly in cotton, and it is somewhat remarkable that this should not have been the case from remote times, seeing that, as far as we can learn, one of the floss-yielding species of Gossypium (cotton) is truly a native of India. In the *Institutes of Manú* (*Book II., 44*) we have an early allusion to cotton. "The sacrificial thread of the Brahmin must be of cotton, and so as to go over his head in three strings; that of a Cshatriya of *sana* thread only; and that of a Vaisya of woollen thread." It is believed that the substitution of cotton for the *sana* has been carried, at the present day, to the extent of violating even this injunction. **Lisboa** (*Bombay Useful Plants, p. 290*) states: "It appears that Manú being a Brahmin, always tried to keep this distinction, and claimed superiority for his class. But, nowadays, the sacred threads of almost all the Hindus are made of cotton."

Sacred Threads. 2108

While **Cannabis sativa** is found at the present day in what appears to be a wild state over the greater part of India there is little to justify the opinion, held by popular writers, that it is indigenous, nor can it be even said to be a native of Persia, though it may possibly be of China, as it is of Russia, Siberia, and Kirghiz. On the other hand, **Crotalaria juncea,** while met with to-day almost exclusively under cultivation, would appear to be a native of India, and possibly also of Central Asia; many other species of the same genus are abundant wild plants. This fact, added to the inferences to be drawn from Sanskrit literature, point to the conclusion that it is highly probable the *sana* of the early writers was **Crotalaria** and not **Cannabis**. It is scarcely possible that the people of India as a whole should have lost all knowledge of the properties of a fibre once in general use and obtained from a plant which grows so freely, if in early times the true hemp was indigenous, or even was as abundant as it is now. There is, indeed, no evidence of hemp ever having been used in comparatively modern times as a regular textile fibre, and with the exception of the limited extraction of fibre pursued by the hill tribes, the natives of India do not utilise the wild or even the cultivated hemp plant as a source of fibre, but prefer to cultivate *sunn* hemp (**Crotalaria juncea**) or *san-pát* (**Hibiscus cannabinus**) for the cordage and sacking required for agricultural purposes. There is still a further consideration, and one of some importance,—*viz.*, that on the plains of India the hemp plant does not produce fibre of any value. Unless, therefore, we are to presume that it has degenerated, or that the climatic conditions of India have altered, the ancient people of the plains were not likely to have obtained their *sana* fibre from **Cannabis sativa.**

We may conclude this brief historic review of the hemp plants by giving the opinions that prevail regarding the origin of our word "hemp."

FIBRE.

Royle in his *Fibrous Plants of India* traces hemp from *sana*. Speaking of *sunn*-hemp he says: "Its name, Shanapam or Janapa on the Madras side, is not very unlike Canapa, Hampa, Hennip, and Hanf. From these we derive our own name 'Hemp.'" In Mysore it is known as *sanabu* and in Ceylon as *hana*. On the other hand, the root of the word *an* or *ang* may be traced from *bhánga*, a synonym for *sana*, met with in the Atharva Veda. **De Candolle** says this root occurs in the names of the true hemp in all the Indo-European and modern Semitic languages, thus: *bhánga* and *gánja* in Indian languages, *hanf* in German, *hemp* in English, *chanvre* in French, *kanas* in Keltic and modern Breton, *cannabis* in Greek and Latin, and *kannab* in Arabic.

Cultivation.

CULTIVATION. 2109

Sunn is grown by itself or at times is cultivated in strips or around the margins of fields. It is never cultivated as a mixed crop. Throughout India as a whole it is a *kharif* crop,—that is to say, it is sown about the commencement of the rains and cut at the end of September or beginning of October. It is thus off the ground to allow of being followed by a *rabi* crop in the same year. But in some parts of India there are two crops of *sunn* hemp. Thus in the Thana District of Bombay it is sown in November after the rice harvest, and the stalks are pulled up by the root in March. "It is also sown as a rainy-season crop in sandy soils" (*Gaz., XIII., I., 290*). This system has prevailed in Thana and Surat for at least the past 100 years, for **Dr. Hove**, writing in 1787, says: "The Crotalaria was sown very thick and grew to the height of 10 feet, remarkably straight and slender. I understand that it was sown in the beginning of November after the grain had been gathered in." In Khandesh it is sown in June and reaped in October. In Kolaba it is sown in November, after the rice is harvested, and the stalks are uprooted in March. In Kolhapur it is sown in August and harvested in December by being cut when the plants are full grown. In Poona it is sown in July and ripens in October. In the Central Provinces and the North-West Provinces it is a *kharif* crop, being sown with the advent of the rains; but in Bengal it is sown a little earlier, namely,—from the 15th April to 15th June; in Madras the sowings take place even still earlier. In the experiments performed at the Saidapet farm, Madras, *sunn* was sown on the 2nd of February. In the *Ain-i-Akbari* the plant is described as bearing its yellow flowers in spring—a fact which **Mr. Hem Chunder Kerr** (writing of Bengal) expresses some astonishment at, since "it now flowers in the rainy and cold seasons." **Roxburgh** says it is sown in Bengal in May and June, and flowers by August,—that is to say, towards the end of the rainy season. In the last Agricultural Report of Bengal it is stated that the crop is harvested from the 15th of August to the 15th of September. **Royle** remarks that at Commercolly there are two kinds grown—one sown in June and cut in August, the other sown in April. But this second kind, he adds, is in Dacca sown in October. Thus while the mean period of sowing is about the beginning of the rains (or in June), *sunn* hemp may be sown in almost any month and occupies the soil for $4\frac{1}{2}$ to 5 months. This is an important feature in view of the possibility of securing a continuous supply of fresh fibre throughout the whole year. It remains to be ascertained, however, what effect this varying period of cultivation has on the quality and quantity of fibre produced. Indeed, it is probable that (as is the case with rice and other crops sown at two or more seasons each year) there may be different cultivated forms of the plant produced as the result of ancient cultivation. We are ignorant of this subject, and it seems desirable that a thorough investigation should be made. Although, as stated, everything points to *sunn* hemp being a

CULTIVATION OF FIBRE.

native of India, it may be doubted if the plant has ever been found in a truly wild state. And the existence of distinct cultivated forms might not only help to confirm the opinions given of an ancient cultivation, but might also establish the superiority of certain crops over others for textile purposes. To what extent the form **C. tenuifolia** is cultivated is not known, still less do we know how far it affords the superior *sunn*-hemp referred to by writers on this subject.

Soil. 2110

Nature of the Soil recommended for Sunn-hemp.—It requires a light but not necessarily rich soil, and it cannot be grown on clay. It is therefore sown on the high sandy lands less suited for the more important crops. This is the opinion which prevails in Bengal, but **Messrs. Duthie and Fuller**, writing of the North-West Provinces, say: "Authorities differ as to whether a rich soil is necessarily required, and although there can be no doubt that fertility in the soil is necessary to promote great luxuriance in its vegetation, yet it cannot be contested that *sanai* will grow on poorer land than almost any other crop. One possible explanation of this may lie in the theory that plants of this order" (the pea family) "can assimilate nitrogen direct from the atmosphere, and are hence less dependent on the soil for nourishment; and another explanation may be deduced from the fact that its roots penetrate deeper than those of most other crops, and can hence draw supplies from a larger body of soil." At the same time the practical experiments performed at the Saidapet farm, Madras, tend to prove that the plant would not produce so much fibre on rich as on poor soil. Speaking of these experiments **Mr. Benson** says: "The seed germinated well, and the plants grew with great luxuriance, but when they had reached the time for cutting, there was no fibre whatever in their stems. The soil of this plot was a sandy loam, and probably the high cultivation and watering were unfavourable to the production of fibre." A second experiment was performed, the seed being sown on "a light and very sandy loam, recently levelled." The land was manured with "12 loads or about 4 tons per acre" of horse-manure, and the results were most favourable. In the *Mysore Gazetteer* it is stated that the best soil for *sanabu* is the red or black used for *ragí* cultivation. **Wisset** remarks that clay soils are injurious, but that on a rich soil the fibre is of a coarser quality than that grown on dry high situations. On the other hand, **Roxburgh**, while speaking of the cultivation in the Northern Circars, says it (this may be **C. tenuifolia**) is sown towards the end of the rains (October or November), and that a strong clayey soil suits it best.

Rotation. 2111

Effects of Sunn Cultivation and the Rotation of Crops Pursued.—It is all but universally believed by the Indian cultivators that *sunn*, like *gram* (see **Cicer, C. No. 1067**), improves the soil. In the *Bombay Gazetteer* (Kolhapur District, p. 172) it is stated: "As it is supposed to refresh the exhausted soil, it is considered a good *bevad* or preparatory crop, and is grown as such every second or third year in some of the fields required for sugar-cane, tobacco, and other rich crops. Sometimes it is sown as a second crop and ploughed in when young as a green manure." From Poona it is reported that the leaves are considered "excellent manure." In gardens and occasionally in dry-crop lands it is grown solely for manure, the plants being ploughed into the soil when ready to flower." The Director of Agriculture in Bengal states: "It is considered by the people of the Lower Provinces to be a renovating crop, and is sometimes used as a green manure to enrich poor paddy land and land that has been infested with weeds." He adds: "It comes after one of the pulses or mustard, and is followed by a pulse, sometimes by *shara* onions. When *sunn* is grown on good soil, it is sometimes followed by potatoes. It is not necessary to prepare the land well for *sunn*. Three or four ploughings are sufficient." . . ." Sometimes also paddy and *sunn* seeds are sown together in the same field. When the plants have properly grown, the field is lightly ploughed and the ladder (a kind of harrow) is passed over it. The paddy plants mostly recover themselves, but the tender and juicy *sunn* is buried underground and dies. A few *sunn* plants remaining are removed at the time of weeding and buried in the soil. The ryots say this green manuring does as much good to the paddy as the application of a maund of oil-cake per bigha. This practice is not, however, very widely extended, though the advantages of it are well known in Burdwan. It involves early ploughing of the paddy lands, and some extra labour which the ryot is not always able or willing to give." **Messrs. Duthie and Fuller** say of the North-West Provinces: "Ploughing in a green crop of hemp is known to add considerably to the fertility of the surface soil by increasing its stock of nitrogen, and it is extraordinary that this is not a general practice with native cultivators." In Bombay *tág* (*sunn*) is not considered a good green manure for wheat. **CULTIVATION OF FIBRE.**

Tillage, Sowing, and Harvesting.—As indicated above, the opinion prevails all over India that high cultivation is not necessary for sunn-hemp. Of Kolaba (*Bomb. Gaz., XI.*, 97) it is said: "The soil is roughly ploughed twice and the seed sown broadcast." In Bengal "the seeds are sown broadcast. It is necessary to have the plants grown thick, otherwise they become bushy and coarse and give very inferior fibres." "There is nothing more required after sowing till harvest time." In the North-Western Provinces "two ploughings at most are given, and the seed is sown broadcast at the rate of one maund to the acre and ploughed in. It germinates quicker than any other crop, the seedlings showing above ground within 24 hours after being sown. Irrigation, even when necessary, is rarely given, and no weeding is required." In the experiments made in Madras, to which reference has already been made, it was apparently sown in drills. "The land was prepared for an ordinary crop by ploughing and harrowing until it was reduced to a proper state, and the seed was then sown with the drill in rows 9 inches apart at the rate of 12lb per acre." In the Coimbatore district, where excellent *sunn* hemp is grown, the seed is sown broadcast, then ploughed in and the land formed into plots and watered twice a week from July to November (*Manual of Coimbatore*, by Nicholson, *pp.* 238-39). Of Mysore it is stated: "It is allowed no manure; and the seed is sown broadcast on the ground, without any previous cultivation, at the season when the rains become what the natives call male,—that is to say, when they become heavy. After being sown the field is ploughed twice, once lengthwise and once across; but receives no further cultivation. At other times the *sanabu* is cultivated on rice ground in the dry season, but it must then be watered from a canal or reservoir." **TREATMENT. 2112** Bombay. Bengal. N.-W. Provinces. Madras. Mysore.

Seed.—The amount of seed to the acre is variously stated. In the above passage from a report of experiments in Madras only 12lb to the acre were used, but in the North-Western Provinces about one maund (or 80lb) to the acre is general. In Bengal 20 seers (40lb) to the *bigha* (⅓ of an acre) is the customary amount of seed. **Roxburgh** states that from eighty to a hundred pounds weight to the acre were used in his time. The plants should not be more than 2½ to 3 inches apart each way, and hence thick sowing is desirable. **SEED per ACRE. 2113**

Harvesting and Yield.—It is customary to read that the crop is harvested after the flowers have appeared, but in certain localities the plants are left on the field until the fruits have begun to form, and in some instances even until they are ripe. In most cases the plants are pulled up by the roots, in others the stems are cut with a sickle close to the ground. Of the Poona district, Bombay, it is stated that the crop is "left standing for about a month after it is ripe, that the leaves, which are excellent manure, **HARVESTING. 2114** With Flowers. With Fruits.

CULTIVATION OF FIBRE.

Left standing for a month.

Steeped at once.

may fall on the land." It is not clear whether the crop is left on its roots,—that is to say, not reaped,—or whether it is cut and stacked on the fields—the latter more probably. The greatest difference of opinion prevails as to whether the cut crop should be dried before being steeped, or, like jute, be carried at once to the retting tanks. But even with jute some cultivators dry the plants sufficiently to allow of the leaves being rapidly stripped since these are supposed to injure the colour of the fibre if allowed to rot in the water of the tank. With regard to *sunn* hemp, the general rule may be almost safely laid down that in moist regions, like Bengal, rapid submersion is preferred, and in dry regions, like Madras, stacking the crop is practised. **Roxburgh** from actual experiments arrived at the opinion that "steeping immediately after the plant is pulled is the best, at least in Bengal during the rains, for then it is very difficult to dry it, and the fibre becomes weakened and the colour injured." **Royle** (*Fib. Pl.*), on the other hand, states that the strongest opinions have been expressed in favour of first drying the plants before retting, the probability being, as indicated above, that both theories are correct, but applicable to different climatic conditions. Of the Coimbatore district, Madras, for example, it is stated: "The cuttings are left on the field until thoroughly dried, then shaken to get rid of the leaves and pods, bundled, and stacked in thatched heaps till required." Here the stems are not only dried but left indefinitely, instead of having their fibres removed and cleaned at once. It is worth adding, in connection with this remarkable practice, that Coimbatore is supposed of all Madras districts to produce the finest *sunn* hemp.

Fibre not removed from bark till required.

PRODUCE. 2115

THE PRODUCE PER ACRE.—Is so variously stated that it is feared little reliance can be put on the figures. **Wisset** says that it varies from 3 cwt. to 10 cwt. per acre, or, say, a medium of 700℔. In the Kolhapur District Gazetteer, on the other hand, it is stated: "The average acre outturn of Bombay hemp is 150 pounds." In the Madras experiments made at the Saidapet farm the results were for plants in flower, cut level with the ground, on the 4th December, 300℔, pulled up by the roots on the same day 325℔; on the 15th December, when the seed-pods had partly matured, cut level with the ground 425℔, pulled up by the roots 487·5℔; and on the 24th December, when the seeds were ripe, 437·5℔. The average given by **Wisset** is thus most likely to be a high one and the Kolhapur returns incorrect. **Duthie** and **Fuller** say of the North-West Provinces: "The average outturn is about 8 maunds (or 640℔) of clean fibre to an acre, worth about R20."

640 lbs. per acre.

COST. 2116

COST OF CULTIVATION AND LOCAL PRICE OF FIBRE.—Messrs. **Duthie** and **Fuller** give the cost of cultivation in the North-West Provinces, including R5 for rent of land, at R15-6, thus leaving a profit of R4-10. In the Madras experiments "the total cost of cultivating an acre" is given as R14, but this does not appear to include rent of land. **Royle** says: "The cultivation was said to yield a tolerable profit, inasmuch as the plant requires scarcely any attention and consequently little labour or expense; and it may be off the ground in time to allow this to be prepared for any cold-season crop. But the expenses and the profits are as variously stated as the produce. The price is also given as varying from R1-8 and R1-12 to R3 per maund." **Duthie** and **Fuller** state that 640℔ are worth R20, or, say, R2-8 a maund, but that "the value of *arjhasan* has suffered great fluctuations in late years. The Settlement Officer of Allahabad writes that in 1877 its price was as high as 6 seers (12℔) per rupee, whilst a few years back it stood at 20 seers. The Calcutta price is about R5 a maund. **Dr. Buchanan Hamilton** describes two crops of *sunn*-hemp as grown in his time in Mysore. Of the one he remarks the seed is sown any time after the rains, and rather thick, the quantity used being two bushels to the

SEPARATION OF FIBRE.

acre. The produce was sold by the cultivators to the Telinga Chitties or manufacturers by the thousand handfuls of the dried stems; tall plants fetched two rupees per thousand handfuls, and short plants a rupee and a half. But another crop, he says, was sown in January. This crop had to be watered and more labour spent upon it, but the produce was more valuable. An acre, he says, required $4\frac{8}{10}$ bushels of seed, and its produce was sold for about £1 2s. $10\frac{1}{2}$d.

AREA. 2117

N. W. P. 40,000 acres.

AREA UNDER SUNN-HEMP.—As may be inferred from what has been stated regarding the ambiguity in the Indian literature of this subject, it is next to impossible to discover the extent of *sunn*-hemp cultivation. **Messrs. Duthie and Fuller**, from special returns furnished for their *Field and Garden Crops*, state that in the North-West Provinces there are about 40,000 acres under the crop. But in the *Land Administration Report for 1885-86* (page 163 A) it is stated that there were 198,728 acres under "*Sanai* or *Til* (sic)." But it is further remarked that the total area under "fibres other than cotton and jute" was in that year only 123,403 acres. This last return would include hemp (proper), *sanai* and **Hibiscus cannabinus.** The Settlement Reports of Oudh show about 800 acres under *sanai*. In *Spons' Encyclopædia* it is stated that there are 50,000 acres in the Panjáb. It is not known from what source that statement was derived, but it seems highly improbable that there is more *sunn* grown in the Panjáb than in the North-Western Provinces. The returns of the Panjáb give about 40,000 acres under "hemp," but how much of that may be the true hemp plant, how much **Hibiscus cannabinus**, and what balance remains as *sunn* hemp, it is impossible to discover. Last year there were 26,614 acres of brown hemp (**Crotalaria juncea**) grown in Bombay. Full particulars regarding Madras cannot be obtained, but of the districts for which returns are available there were last year 775 acres under "sunn" and 83 acres under "Bombay hemp." What this Bombay hemp may be cannot be learned, but in most works on the subject Bombay hemp is a synonym for *sunn*-hemp. In 1884-85 there were 380 acres of "Bombay hemp," and in 1885-86, 330 acres, so that its cultivation would appear to be declining. Of *sunn* cultivation in Coimbatore it is reported: "It can be grown anywhere and to any extent if a demand is made by agents with money in hand." In Travancore a very superior quality of fibre is produced, but it is not known to what extent the plant is cultivated. In the Central Provinces there were 24,800 acres under "False or San hemp," and in Mysore 5,076 acres. In Berar there were last year 569 acres under "hemp or **Hibiscus cannabinus**," but **Dr. Hume** in his recent report explains that there are in Berar two kinds of hemp—*ambada* and *san*. The former is in all probability **Hibiscus cannabinus** and the latter **Crotalaria juncea**. In Burma and Assam there are about 500 acres, in each province, of land entered as under "fibres other than cotton and jute." No returns are available for Bengal, but from personal observation the writer would be disposed to think there must be as much in the Lower, as in the North-West Provinces.

Panjab 50,000 acres.

Bombay 26,614 acres.

Madras.

Travancore.

C. P. 24,800 acres.

Mysore. 5,076 acres.

Berar.

Burma. 500 acres.

Bengal.

It will thus be seen that the actual area under *sunn*-hemp cannot be absolutely determined, since the fibre is not included among the agricultural products regarding which regular annual statistics are furnished. But it seems probable that there are at least 150,000 acres annually under the crop in India as a whole.

India. 150,000 acres.

SEPARATION. 2118

SEPARATION OF THE FIBRE.

The question as to whether the plant should or should not be dried before being placed in the retting tanks having been discussed above, there remains to be given here a brief account of the various modes of retting or of peeling the fibre and of cleaning and boiling it after it has been separated from the stems. In some localities the stems are recom-

SEPARATION OF FIBRE.

mended to be buried in the mud at the margin of the tanks; in others, to be submerged in the water by being weighted. In others stagnant water is condemned as destroying the colour and lustre of the fibre, running streams being urged as preferable (*Gibson's account of the Bombay fibre*). But practical and comparative experiments not having been performed in the other provinces similar to those made at the beginning of the present century by **Roxburgh**, in Bengal a definite opinion for or against the different methods pursued cannot be offered. After removal from the ground, the stems are tied in bundles (20 to 100 in each), but the leaves are generally stripped off and left on the field. When the stems are left until quite dry, the leaves either fall off naturally or are removed by the stems being beaten. It is a common practice to place the bundles of stems erect in 2 or 3 inches of water for 24 hours, so as to give the thicker and lower ends a longer submersion. But the length of time required for retting depends largely on the temperature of both the atmosphere and the water. In August and September two to three days will generally suffice. **Messrs. Duthie** and **Fuller** say of the North-West Provinces: "When stripped of the leaves they are ready for retting. The stalk is made up into bundles and placed upright for a day or two in water about a couple of feet deep, since the bark on the butts of the stalk is thicker and more tenacious than that on the upper portion, and requires, therefore, longer exposure to fermentation. The bundles are then laid down lengthways in the water and are kept submerged by being weighted with earth. The time required for retting varies from three days in hot and damp weather, to seven days if the temperature be cool and the air dry." It can generally be ascertained when the retting is complete by the bark of the lower ends of the stems separating easily, but too long fermentation, while it whitens the fibre, injures its strength. **Roxburgh**, writing of the period of retting, says: "All that seems necessary is to caution the cultivators against oversteeping the plant, which they are apt to do because it renders the separation of the bark from the stalks easier, but weakens the fibre. Small pools of clear water, well exposed to the sun's beams, seem best suited for steeping in, because heat hastens maceration, consequently preserves the strength of the fibres, while the clean water preserves their colour. Deep water, being cooler, requires more time for the operation." In the same way running water, although recommended by some writers, would seem to be objectionable owing to the longer time necessary for the degeneration of the connecting tissues. Damp mud on the margins of tanks, referred to by some, is even more objectionable, as it seems impossible to adopt this mode of retting without serious loss to the colour of the fibre.

Margin notes: Leaves stripped. — Length of submersion. — Stems placed erect in water, then horizontal. — Deep water. Running water. — Damp Mud.

Cleaning of Retted Fibre.

Having discovered that the necessary degree of retting has been attained, the cultivator, standing in the water up to his knees, takes a bundle of the stems in his hand, and threshes the water with them until the tissue gives way and the long clean fibres separate from the central canes. According to some writers, the retted stems, after being partially washed, are taken out of the water and placed in the sun to dry for some hours before being beaten out in the way described. This practice, while it is followed in some parts of the country, is condemned in others as injurious, or at least as a useless delay. In Bengal this system is only followed when the operator is afraid he may not be able to overtake the task of washing before the stems would be over-retted. This partial drying of the stems with adhering fibre would correspond to the sweating of hemp pursued in some parts of Europe; but it seems probable that if sweating be necessary, it could better be accomplished as a further process after the fibre has been separated and approximately cleaned.

In Salsette Island and other parts of Bombay, little or no retting is employed. "The plant while moist is peeled by the hand, and immediately dried in the open air or under cover, according to the state of the weather. By peeling, the fibres are better kept in their natural state of arrangement, and give support and strength to each other; whereas, by the process of the Bengalese, they get so materially entangled that a great loss is always sustained. If they are restored to their natural situation by the heckle, there is a loss of nearly one half of the original quantity, which renders the heckled *sunn* of Bengal of a high price" (*Royle*). The writer cannot discover any recent description of this Bombay process of separating the fibre without retting, but, as **Roxburgh** stated, the superior quality of Bombay over Bengal *sunn* hemp seems likely to be due to the fact that the fibre has not been subjected to strong fermentation.

SEPARATION OF FIBRE. Not Retted.

Washing the fibre is very tedious, and a man rarely works for more than three hours at a time but is relieved by turns; he will clean 15 seers a day, which represents the fibre obtained from 5 or 6 maunds of stems. Of Khandesh it is said a man earns R1 for cleaning 40℔ of fibre.

Wages for cleaning.

Reference has incidentally been made to the period when the crop should be cut, and before proceeding to discuss the further treatment of the fibre it may be as well to add here that the period of cutting will depend on the purpose for which the fibre is required. A softer and more delicate fibre will be obtained from stems cut just as the flowers appear than if allowed to pass into the fruiting stage. A few plants are always left by the cultivators to mature seed for the next year's crop, and from the stems of these they extract a strong, though coarse, fibre. On the other hand, it seems to be the habit of some cultivators (the Wunjaras of Bombay) to allow the whole crop to ripen its seeds, this coarse fibre being all they desire, together with the seeds, which are valued as a food for buffaloes. Old stems require a much longer period of retting.

Period of cutting.

Soft fibre.

Strong Coarse fibre.

Seed used as buffaloe food.

FURTHER CLEANING. 2119

Further Process of Cleaning the Fibre.—When the fibre has been separated and thoroughly washed, it is the usual custom to hang it up over bamboos to be dried and bleached in the sun. When dry it is combed if required for textile purposes or for nets and lines, but if for ordinary use—*e.g.*, ropes and twine—it is merely separated and cleaned by the fingers while hanging over the bamboo. In this primitive way the *sunn* hemp receives all the treatment it gets of the class known to European hemp growers as "breaking" and "scutching." European machinery for cleaning is never used. It is commonly admitted that it is in cleaning the fibre that the Native generally fails most. The process of washing after separation from the stems does not seem to be carefully done. **Royle** quotes a report of a sample of *sunn* hemp experimented with at Hull, of which it was stated that "by using more care in the steeping and exposure, it will be *fully equal* to the Baltic." Such opinions are current in the reports of this fibre which appeared, while the error existed of supposing it to be Indian-grown hemp or **Cannabis sativa**. It is impossible to avoid the impression that *sunn* hemp fell into disfavour when this error was exploded. An expert in 1842, for example, says: "Your hemp is very clean—a material point,—but it wants more beating and dressing; and I think the natives have not proper implements to do it with. You cannot improve in your mode of packing; it is decidedly superior to the Baltic. I do not despair of seeing the produce of the Baltic supplanted by that of India; as that defect appears to me solely to arise in the management of it: it stands too long before it is pulled or cut, or is too much steeped or exposed, to get the fibre to separate from the stalk" Unfortunately the advances of scientific exploration told all such writers that the defects they complained of were due to the fact that Bombay hemp was not hemp at all, and instead of the fibre supplanting

Breaking.

Scutching.

Said to be nearly as good as Baltic Hemp.

the Baltic hemp it is to-day in the same position commercially as it was a hundred years ago. While not hemp, it is a hemp substitute that deserves a better position than it has as yet obtained.

PROPERTY OF FIBRE. 2120

PROPERTY AND STRENGTH OF SUNN HEMP.

Unlike the golden shining jute, which occurs in long straight bundles, *sunn* hemp, as usually met with in the market, exists in tangled masses of a dull, greyish-white colour. According to **Roxburgh's** experiments, it loses one third of its mass in heckling, but the tow obtained is useful for other purposes, such as for paper-making. **Royle** states that when heckled it is "light coloured and clean, with the fibres lying parallel to one another, and showing them as well fitted for spinning. Parties who have seen it have pronounced it well worth £35 a ton." At the beginning of the present century the East India Company endeavoured to improve the quality of the fibre by growing and manufacturing it carefully, and **Royle** mentions a sample of heckled fibre sent to London by the Company that measured four to five feet in length: the ordinary length of the heckled fibre, he adds, is only about three feet long.

£35 a ton.

EARLY RECORDS. 2121

The first European mention of *sunn*-hemp occurs in **Rheede's** *Hortus Malabaricus*, V., IX., t. 26, but **Ironside** in the *Philosophical Transactions of London*, LXIV., page 99, also describes it. **Roxburgh** devoted much attention to this fibre, his first paper on the subject appearing in the Journal of the Society of Arts. Reference has been made to the fact that the East India Company, towards the close of the last century and the first few years of the present, cultivated the plant. The earliest definite mention we have of the fibre having been exported was in the year 1791-92. Although numerous favourable reports appeared shortly after this date, the whole interest in the fibre gradually died out, and the European methods of cleaning it met with a like fate; at the present day the natives nowhere practise any system of growing the plant, or cleaning the fibre that can be traced to European influence. One of the last outcomes of the desire to obtain flax and hemp from India was the issuing of instructions to the late **Dr. Roxburgh** to institute a thorough enquiry into the subject of Indian fibre-yielding plants. His report is the basis of all that has since appeared, for from that day to this, while much money has been spent in making collections of fibres and testing the properties of fibre-extracting machinery, no new and original investigations have been conducted in India with the fibres themselves. The following table gives the results of some of **Roxburgh's** experiments with sunn-hemp:—

First Exported.

No.	Names of the Plants.	Average weight each line broke with when *dry*.	Average weight each line broke with when *wet*.	Average weight gained by wetting the lines.
4	Sunn (**Crotalaria juncea**) cut before the plants were in blossom, and steeped immediately .	112	158	41
5	The same as No. 4, but dried, or rather kept some time before they were steeped	60	78	30
6	Sunn cut when in full blossom, and steeped immediately	130	185	42
7	No. 6 kept drying for some time	100	166	66
8	Sunn, winter crop cut when the seeds were ripe and steeped immediately	150	203	35
9	The same as No. 8, but dried	110	163	48
10	Sunn, winter crop cut when the seeds were ripe, and steeped immediately	160	209	31

Properties of Sunn-Hemp.

PROPERTY OF THE FIBRE.

No.	Names of the Plants.	Average weight each line broke with when *dry*.	Average weight each line broke with when *wet*.	Average weight gained by wetting the lines.
1	Hemp, the growth of the year 1800 from the Co.'s Hemp farm near Calcutta	158	190	20
2	Jeetee (**Marsdenia tenacissima**) . . .	248	343	38
29	A line made of 15 threads of sail twine (*Calloee*, **Bœhmeria nivea**)	240	278	16

Comparative value.

From these experiments there would appear to be no room for doubt as to the superiority of the rapid steeping as compared with the drying before retting. Further, the winter crop gave the best result. Indeed, casting one's eye down the long list of Indian fibres experimented with by **Roxburgh**, there are seen to be but two fibres that exceeded in strength the winter crop of *sunn, viz.* (No. 10), **Marsdenia tenacissima** and *rheu*, **Bœhmeria nivea.** The results of the experiments with these fibres, together with the experiments performed with hemp (presumably the true hemp), have been given in the above abstract from **Roxburgh's** table so as to allow of comparison. The sample of true hemp (Indian grown) gave a worse result than the winter *sunn* hemp prepared by immediate retting.

East India Company's efforts.

So far as they go, these experiments are of importance, but they leave the question of the comparative study of the *sunn* hemps of other provinces in a position of uncertainty. Of Bengal *sunn*, the rapid retting by Roxburgh's experiments seemed decidedly superior to the previous drying, but is this the case in Bombay and in Madras? and is the Bengal *sunn* inferior or superior to that of these Presidencies? This enquiry the East India Company do not appear to have taken in to consideration. Their attention was first directed to the fibre in Bengal, and without ascertaining whether or not Bengal was the best field for experimental cultivation, they prosecuted the effort to improve the Bengal hemp, and failing, allowed the whole subject to drop into the oblivion from which it is only now beginning to recover; but the new trade is from Bombay, not Bengal.

Roxburgh's experiments.

Roxburgh tried the properties of *sunn* hemp in another way in order to ascertain the power of endurance which cords made of it had under maceration in water for a considerable period. At the same time he tested the advantages or otherwise of tanning or of tarring the fibres. The following abstract from his report may be here given—

Names of the Plants.	Average weight at which each sort of line broke.					
	When Fresh.			After 110 days' maceration.		
	White.	Tanned.	Tarred.	White.	Tanned.	Tarred.
English hemp, a piece of new tiller-rope.	105	...	...	Rotten, as was also the English log line.		
Hemp from the Company's farm near Calcutta.	74	139	45	All rotten.		
Sunn hemp of the Bengalese.	68	69	60	rotten.	51	65
Jute (*Bunghi-pát*) . .	68	69	61	40	49	60

Properties of Sunn-Hemp.

PROPERTY OF THE FIBRE.

Deterioration with age.

Removal of Export Duty.

According to these experiments *sunn* hemp stood the action of the maceration better than did either of the samples of true hemp. It has further been shown that a cord 8 inches in size of best Petersburgh hemp broke with 14 tons, 8 cwt., 1 qr., while a similar rope of *sunn* only gave way with 15 tons, 7 cwt., 1 qr. Dr. **Wight** found that a rope of coir of a certain thickness broke with a weight of 224lb, of cotton with 346lb, of American aloe with 362lb, of *sunn* hemp with 407lb, of **Calatropis gigantea** with 552lb, and one of Ambárí (**Hibiscus cannabinus**) with 290lb. **Royle** has shown the slight deterioration which *sunn* hemp undergoes in the following statement: "A rope made in 1803 broke with a weight of 6 tons, 0 cwt., 3 qrs., whereas in 1806 it gave way under a tension of 5 tons, 17 cwt., 0 qr. It is of historic interest to add in this place that the trade in *sunn*-hemp lulled until the year 1867, when the export duty was removed. From that year returns of the trade of India were regularly published, and it is noteworthy that from about the middle of the present century the bulk of the exports of raw hemp (? *sunn* hemp) went from Bombay and not from Bengal, in spite of the efforts made a few years before that date to create a Bengal trade. This would seem to point to a superiority possessed by the Bombay as compared with the Bengal *sunn* hemp. It seems probable that had this fact been realised by the East India Company, their efforts to establish an Indian hemp industry would have been more successful than was the case with their attempts in Bengal.

RECENT EXPERIMENTS. 2122

Injured by Jute.

Future Prospects.

In a Report on the Indian Fibres by **Cross, Bevan, King, and Watt**, recently published by **E.** and **F. Spon**, the following passage occurs: "It is impossible to urge too strongly the claims of this much-neglected fibre—a fibre which seems to have suffered severely through the immense success of jute obscuring for a time the properties of all other fibres. At the beginning of the century *sunn* hemp occupied a much more important position than it holds to-day, and all efforts to cultivate the true hemp in India were abandoned, from the impression that the so-called indigenous hemp was, for all practical purposes, of equal merit. During the Colonial and Indian Exhibition numerous enquiries were made as to why it was that so little of the better qualities of *sunn*-hemp were procurable. **Mr. Collyer** and several other Brokers and Merchants stated that their only difficulty in pushing the trade in *sunn* hemp was their inability to procure a uniform and a large enough supply. From these and many other considerations we strongly commend this well-known fibre to the attention of the Government authorities of Western and Southern India. What encouragement they may be able to render may hopefully be anticipated to lay the foundation for a textile industry that may yet come to bear a creditable comparison with the jute trade of Bengal." The opinion expressed in the above passage that the *sunn*-hemps of Bombay and Madras are superior to those of Bengal is current in the literature of this subject, but it should be recollected that while we do possess the results of actual experiments with Bengal *sunn*-hemp, we have only general statements regarding those of Southern and Western India; but one fact remains in support of the supposed better quality of Bombay hemp,—namely, that the trade has migrated to the Western Presidency. At the same time **Roxburgh** seems to have been of opinion that Salsette *sunn* was superior to the *sunn*-hemp of Bengal, but he attributed the superiority to the mode of separating the fibre by peeling it off the stem instead of retting it. On the other hand, the *sunn*-hemp of Coimbatore (which obtained the medal at the Madras Exhibition), according to **Roxburgh's** experiments would have been improved had it been retted at once instead of being dried and stacked for an indefinite period before cleaning. If this be found by actual experiment not to be the case, then there must be something in the climate or soil of Madras and of Bombay more favourable to *sunn* hemp than exists in Bengal. FIBRE.

Chemical and Microscopic Peculiarities of Sunn. 2123

Messrs. Cross and Bevan (in the report quoted above) state that *sunn*-hemp treated "with iodine and sulphuric acid gives mixed blue and brown; with aniline sulphate, yellow streaks showing lignification. This fibre may be regarded, both chemically and structurally, as intermediate between the divisions of fibres A and B,"—*e.g.*, those suitable for higher and those for lower textile purposes. In their table giving the results of their analyses they show that when boiled for five minutes in a solution of caustic soda, it loses 8·3 per cent., and after an hour only 11·7 per cent. Among Indian fibres it occupies the third or fourth place in point of amount of cellulose. According to this classification, **Girardinia** or Nilgiri nettle heads the list with 89·6 per cent., then **Marsdenia** with 88·3, and after that **Crotalaria juncea** and **Sida rhombifolia** equal, each with 80·0 per cent. of cellulose. "The percentage yield of cellulose of the raw fibre is the most important criterion of its composition and value." It may be worth stating here by way of comparison that jute was found to possess 76·0 per cent.; **Msa**, or plaintain fibre, 64·6; **Anona** (custard apple fibre), 62·3; **Sesbania** (*dancha*), a fibre of great strength according to **Roxburgh**, 58·6; **Maoutia Puya**, a form of *rhea*, 32·7; and **Daphne**, the Nepal paper-plant, only 22·3 per cent. of cellulose. The last two have hitherto been popularly viewed as amongst the strongest fibres of India; and it would therefore appear that strength may not necessarily be a consequence of high percentage of cellulose. Without attempting to express an opinion opposed to **Messrs. Cross** and **Bevan's** results, the writer would throw out the suggestion that there apparently exists in some fibres a principle that may have been removed in the process of the analysis adopted by these distinguished chemists. While, therefore, it may be accepted as proved that such fibres as **Daphne** have ultimate fibriles that are too short to admit of their being bleached and thereafter spun in the form of the finer textiles of Europe, there still remains the practical fact that, under the crude methods adopted in India, they are valued as strong and durable fibres. It will be received with no small surprise by many that so humble a position should be assigned to the famed Poya fibre of Assam, and thus in concluding these remarks a possible explanation may be sought in the mode of hydrolysis (or washing and bleaching) employed. The Poya was found to lose 62·7 per cent. by being boiled in caustic soda, the residue being the cellulose upon which the low opinion of its properties is based. May it not be that under some other system of hydrolysis it would lose little or nothing, and even retain the property of great strength and durability for which it is justly esteemed by the fishermen of Assam for their lines and nets? The writer has for some time felt that one of the features of the exploration of unknown fibres should consist in the establishment, for each, of the peculiar mode of hydrolysis that injured the fibre least, and in chemical reactions that would check the natural degeneration it is liable to undergo. It seems scarcely fair to condemn or to praise a fibre according to its behaviour with one process of hydrolysis, and such a chemical result is likely to be often opposed to actual practical experience. It is satisfactory, however, to note that under a strong alkaline hydrolysis *sunn* hemp retains all its properties, and under nitration attains a great weight (150·5), being in this respect third in the list of the Indian fibres experimented with by **Messrs. Cross** and **Bevan**. A writer in *Spons' Encyclopædia* says of *sunn* hemp: "Samples of the fibre, exposed for two hours to steam

Percentage of cellulose.

CHEMISTRY of the FIBRE.

at 2 atmospheres, boiled in water for 3 hours, and again steamed for 4 hours, lost only 2·93 per cent. by weight, as against flax, 3·50; Manilla hemp, 6·07; hemp, 6·18 to 8·44." This hydrolysis (without the aid of an alkali) so far confirms the results given above, that in point of durability under moisture and under caustic alkali (processes of washing and bleaching), *sunn* hemp compares most favourably with other fibres, and deserves a far higher position in commercial enterprise than it has hitherto attained.

MICROSCOPIC FORM. 2124

Mr. King, who worked out the microscopic measurements of the Indian fibres in the recent report (quoted above), states that the fibre bundles consist of from 20-50 ultimate fibres which are not easily separated. He continues: "Fibres polygonal, cavity small. Fibre substance shows well marked concentric rings." Of the ultimate fibres he adds: "Length, 3-5 mm.; ends tapering abruptly; characteristic markings spiral." The writer in *Spons' Encyclopædia* gives microscopic results, however, that do not agree with the above—a fact which might suggest either that a wrongly-named plant may have been examined, or that *sunn* hemp varies to an alarming extent. He says: "The dimensions of the filaments are—length: max., 0·472 in.; min., 0·157 in.; mean, 0·30 in.;—diameter: max., 0·0020 in.; min., 0·001 in.; mean, 0·0015 in. These measurements are in round numbers double those given by **Mr. King**—a fact that would suggest the desirability of the subject being looked into once more, especially by having samples of Bengal, Bombay, and Madras *sunn* examined and compared. It would also be of great interest to have microscopic and chemical experiments made with fibres obtained from plants in flower and from those in all stages of maturity of seed, both separated by the immediate retting process and by the process of drying before retting.

Re-examination desirable.

TRADE IN SUNN HEMP.

TRADE. 2125

Little or nothing can be learned of a definite nature regarding the extent of the trade in this fibre. It is grown in every province, and nearly universally used by the people of India; but, as already stated, definite information is not procurable owing to the confusion which exists in the use of the word "hemp" (*sunn* hemp in one case, Hibiscus in another, and true hemp in a third, being the fibre alluded to). For this same reason we are unable to discover the extent of the foreign trade in *sunn* hemp. It seems probable, however, that of the exports to foreign countries shown under Indian-grown hemp the bulk is *sunn* hemp, while of the imports from other countries the bulk of the "raw hemp" may be Manilla hemp (**Musa textilis**), and of the "hemp manufactures" articles made of true hemp. We have said that shortly after the East India Company discontinued their efforts in Bengal to cultivate the plant and to extract the fibre according to European methods, a trade in *sunn* sprang up and gradually developed into a position of importance in Bombay. The exports of Indian-grown hemp* were, in 1867-68, valued at R1,04,127, but, by Act XVII. of 1867, the export duty was repealed, and in the following year they were R2,91,355, and in 1869-70, R5,07,159; of the last-mentioned exports the United Kingdom received R4,35,930 worth, France R17,274, America R5,621, and the Persian Gulf R34,029. Bombay furnished R3,12,111 worth and Bengal only R78,926, the remainder going from Madras. From these facts it will be seen how important a slice of these so-called exports of Indian hemp goes to Britain, and of how much greater importance is Bombay than Bengal in this trade. From 1869-70 down to 1884-85 the exports of raw hemp stood practically stationary, but in the following year

Exports. 2126

* Presumably *sunn* hemp or *sunn* hemp along with a certain amount of the fibre of Hibiscus cannabinus—*sanpát* or *ambádi*.

they developed to R6,88,825, and last year attained the by no means inconsiderable proportion of R16,41,384 (£164,384). Throughout the whole of that period Bombay continued to furnish from ½ to ⅔ of the total amount exported, and her share last year was valued at R8,80,494, while the Bengal share amounted to R6,72,128. It remains for the future to reveal whether this remarkable advance is due to a turn in the tide of popular favour; but it would seem desirable that no opportunity should be lost in pressing the claims of this fibre. With this object in view, it would be of the greatest practical utility to procure more carefully prepared returns both of the foreign and internal trade, care being taken to separate the quotations of *sunn* hemp from all the other fibres hitherto grouped with it under that most inappropriate and misleading name "Hemp." FIBRE, TRADE IN.

In 1866-67 a trade was first reported in exports of Indian Manufactured Hempen Goods other than cordage. This continued to expand until, in 1870-71, when it was valued at R1,64,433, of which Bengal had assigned to it R1,53,330. The bulk of these exports went to the Straits Settlements, Ceylon, and Mauritius. From 1871-72, this trade began, however, to steadily decline, and in 1874-75 was valued at R1,19,327, of which Bengal claimed R1,15,875, and Bombay would appear to have taken no share. Next year these exports fell to R5,299, of which the Bengal portion was valued at R2,500; in 1880-81 they had still further declined to R2,230, but expanded again in 1883-84 to R6,510 (Bengal taking no part in the trade), and last year they amounted to R262 only. The exports consisted chiefly of *sunn* hemp cloth and sacks, and it seems probable that this native industry may have been ruined by the remarkable success of the Bengal jute industry. A difficulty exists in tracing out the extent of the hemp trade in ropes and cordage, both the foreign imports and the Indian exports, since all returns for Ropes and Cordage, of whatever material made, are given collectively. Hempen Goods. 2127 Ropes and Cordage. 2128

IMPORTS OF HEMP AND HEMPEN GOODS.—It seems probable that the bulk of the raw fibre so reported may be the Manilla hemp used up in the Indian rope factories, and of the hempen goods, canvas and other fabrics of true hemp. This trade is not extensive; last year (1886-87) only 7,641 cwt. of hemp fibre, valued at R1,71,795, was imported; with, in addition, "hemp cloth and sacks to the value of R43,000. Under the heading of Government stores" will generally be found an item of canvas as imported (? chiefly flax); last year this was valued at R50,801. Of the imports of foreign hempen goods, a small re-export trade is done to other countries, but a very much more extensive shipping trade consists in carrying coastwise Indian hemp (*sunn* hemp)—*e.g.*, from one port of India to another. This is the coasting trade which, along with the rail, river, and road-borne traffic, completes the account of interprovincial dealings. The total value of the imports returned as received at Indian ports from other Indian ports amounted to R7,00,534—a trade which has shown a steady increase since 1882-83, when it was valued at R3,90,850. In another series of tables published by Government are shown the exports of Indian hemp from all Indian ports to other Indian ports, and these are returned as valued at R6,24,303, the trade having steadily increased since 1882-83, when it was valued at R96,087. Imports. 2129 Re-exports. 2130

USES TO WHICH SUNN HEMP IS PUT.—The chief purpose for which this fibre is utilised at the present day is the manufacture of a coarse cloth (*tát putí*) or canvas used chiefly for sacking. A large amount of the fibre is annually used up in the native cordage trade, for which it is well adapted, and large quantities of the fibre are also consumed by the European rope-makers in India. The waste tow and old materials are made into paper. It is stated that in Jaipur and in many districts in the North-West Provinces USES OF. 2131 Canvas. 2132

FIBRE. Paper. 2133 paper is regularly made of this material, and large quantities are annually
used up by the Indian paper-mills. The paper made by the natives of
Bombay is principally of *sunn* hemp; 3 parts of *sunn* to one part of cotton
is a common mixture. In *Spons' Encyclopædia* occurs the following remark
regarding *sunn* paper: "The average weight sustained by slips of sized
paper, weighing 39 grs., made from "raw" fibre, was 64℔, as compared
with Bank of England note pulp, 47℔. One batch was reported to
make a nice, clean, smooth paper, of good colour, but not taking ink
well."

Hemp & Flax Substitute. 2134 For European purposes the fibre may be used as a substitute for
hemp or for flax. Speaking of the special form of the fibre produced in
Travancore, Dr. Royle says: "The appearance of this fibre is totally different from any other which comes from India, as it is in the state
Travancore Sunn. 2135 as if prepared for spinning into thread, and must have been combed
or heckled. The fibres are brownish in colour, about 3 to 4 feet in length,
clean and shining, not so fine as flax, but still resembling some of the
coarser kinds. A very competent judge informed the author that it
might be sold for the purposes of flax, or as a kind of flax, and was
worth £35 a ton, so some specimens sent to Dundee were valued at the
same sum, and it was said could be used for the same purposes as flax,
though rather too dry." So, again, "This hemp, when prepared with
the patent liquid, became soft, white, and so fine when heckled as to bear
the closest comparison with flax at £80 per ton. It is better than any
Russian flax for fine spinning. Bombay hemp, rough and dark, and
valued at £20 per ton. This article, being similarly prepared, was
considered equal in value with the Madras hemp.

STALKS. 2136 *Sunn* stalks (after removal of the fibre) are used chiefly as firewood. But of the Kolaba district, Bombay, it is stated: "Hemp torches
Torches. 2137 are made by tying together, in four or five places, about 200 stalks with
their fibres, each torch being about three feet long and ten inches round.
Matches. 2138 Hemp matches are also made by Bohorás, who cut each stalk into about
six pieces and dip the ends into a *solution of sulphur*."

C. P. Fibre. 2139 Before concluding this account of *sunn* hemp it may be as well to
emphasize the fact that all writers agree in pronouncing the fibre obtained
Bengal. 2140 from Travancore and that from the Central Provinces (the latter being
the fibre most probably of the form known as **Crotalaria tenuifolia**) as
Bombay. 2141 superior to the ordinary *sunn* hemp. We possess so little definite knowledge regarding the cultivated forms of the *sunn* plant that it can only be
Madras. 2142 added that perhaps the first and most necessary step to be taken in any
effort to increase the *sunn* hemp industry would be to establish the
N. W. P. 2143 areas under each recognisable form of the plant, and to collect such botanical specimens in flower and fruit as would admit of their careful deter-
Immediate Process. 2144 mination. Were such specimens to be accompanied with samples of the
fibre extracted when in flower and again when in fruit, each being prepared according to both the immediate and the deferred process of ret-
Deferred Process. 2145 ting, it would then be possible to inform the commercial authorities where
the best *sunn* hemp was to be had, and it might also be found possible to
extend the cultivation of the better races of the plant and to encourage
the natives to adopt the process of preparation of the fibre which was
found the most appropriate.

It may be added that there are three or four other species of Crotalaria
which, whether wild or cultivated, are used for fibre. It is highly probable
some of these may prove to be the peculiar *sunn* hemps of certain districts. The brief notes given in the following pages regarding the more
common species, it is hoped, may help the reader to recognise these little-
known fibre-yielding plants.

Food and Fodder.—It has already been incidentally remarked that in some parts of India the seeds of this plant are collected and given to cattle. **Roxburgh** says: "This plant—and it is the only one—is also cultivated by the natives of some parts of the Northern Circars to feed their milch-cows with during the dry season. I have found that it is very nourishing, and causes them to give more milk than most other food. It only bears two or three cuttings; after that the plants perish." FODDER. Seeds. 2146

Medicine.—The seeds are sometimes used medicinally, having the reputation of being useful to purify the blood. MEDICINE. Seeds. 2147

Crotalaria laburnifolia, *Linn.; Fl. Br. Ind., II., 84.* 2148

A shrubby plant met with in the Western Peninsula, particularly in the South Concan. Properties similar to those of the next species. It is known in Hindustani as *muna*, the *pedda-galli-gísta* of Telegu. **Sir Walter Elliot** gives it the further Telegu name of *Chiri giligichcha*, and the plant is often seen in gardens on account of its flowering throughout the year.

C. Leschenaultii, *DC.; Fl. Br. Ind., II., 76.* 2149

An abundant plant on the Nilghiri Hills and higher portions of the Western Ghats. This is alluded to by **Mr. J. H. Grant** as the plant used in Satara for paper-making. It is there known as *dingoda*. **Dalzell** and **Gibson** say it is the *dingala*, and is common on the higher ghâts. Satara Paper. 2150

C. medicaginea, *Lank.; Fl. Br. Ind., II., 81.* 2151

Vern.—*Gulabi*, Pb.

A diffuse perennial abundant in the tropical regions of India from Kashmir to Burma, ascending to 6,000 feet in altitude.

Medicine.—This plant is officinal in the Panjáb being sold in the bazárs under the name of *gulábí* (*Baden Powell, Pb. Pr., 343*). MEDICINE. 2152

C. prostrata, *Roxb.; Fl. Br. Ind., II., 67.* 2153

A slender creeping weed, common on the drier plains of India ascending to 6,000 feet.

This is known to the Santals as *Nanha jhunka* or *Katic'jhunka*, and by them it is used medicinally in derangements of the stomach. It is known in Bengal as *Choto-jhunjhun* (small *jhunjhun*, *see Voigt, p. 207*). **Roxburgh** says this is known in Telegu as *Seri-gally-gista*. 2154

C. retusa, *Linn.; Fl. Br. Ind., II., 75.* 2155

A robust under-shrub, 3-4 feet in height, with stout striated branches and wedge-shaped leaves; common on sandy soils, flowering in February and March; not uncommon in the tropical regions of India from the Himálaya to Malacca and Ceylon. Also met with in China, North Australia, and tropical Africa and America.

In India this plant is often cultivated for its fibre, which is sold as a form of *sunn* hemp. To what extent it is cultivated and how far the fibre may be passed off as the true *sunn* hemp it is impossible to say. **Dr. Wright** states that in South India the fibre of this plant is regularly sold as hemp and is made into cordage and canvas. FIBRE. 2156

If the enquiry suggested in the concluding paragraph above regarding sunn-hemp, be instituted, this plant would doubtless find a place among the specimens collected as *sunn*-hemp-yielding plants, and it would be most instructive to possess definite information as to the comparative value and property of this fibre with the true sunn-hemp. In Bengal it is known as *Bíl-jhunjhun*, in Madras as *Potu-galli-gísta* (TEL.), in Bombay *Ghagrí* (MAR.), and in Ceylon as *Kaha-andana-hiriya*, SING. (*Conf. with Roxb., Fl. Ind., Ed. C.B.C., 549; Dalz. and Gibs., Bomb. Fl., 55, &c.*)

2157 **Crotalaria sericea,** *Retz.; Fl. Br. Ind., II., 75.*

A plant very much like the preceding and found over the same region. **Stewart** says it is cultivated in the Panjáb as a garden flower, and is known as *Sanní*, but apparently never cultivated as an agricultural product,

FIBRE. 2158 although its fibre is sometimes prepared. **Kurz** alludes to the plant in his report of Pegu. It is the ***pípúli-jhunjhun*** of Bengal, and **Roxburgh** remarks it bears the Sanskrit name of *Ghuntaruva*, but **Dr. Udoy Chand Dutt** corrects this into *Ghantáravá*, and gives it the Bengali name of *Jhanjhania*. It flowers in the cold season.

2159 **C. striata,** *DC.; Fl. Br. Ind., II., 84.*

A low-growing shrub, with robust, sulcate, thinly silky branches, and large yellow flowers striped with red. Fairly abundant throughout the

FIBRE. 2160 warmer parts of India.

The **Rev. A. Campbell** states that this is cultivated by the Santals in Chutia Nagpur on account mainly of its fibre. The plant is known to them as *Son jhunka*, and to the Hindustani-speaking people of that region

Charms. 2161 as *Son, San*. He adds that the root or a small portion of the stem is tied to the wrists and neck of a person suffering from dropsy. **Roxburgh** remarks this is known to the Telegu-speaking people of Madras as *Munga*.

2162 **C. tenuifolia,** *Roxb.; Fl. Ind., Ed. C.B.C., 546.*

This has been reduced by most botanists to a synonym for **C. juncea,** *Linn.*, which see.

2163 **C. tetragona,** *Roxb.; Fl. Br. Ind., II., 78.*

A stiff, very handsome shrub, often 68 feet in height, met with on the lower Himálaya (up to 3,500 feet in altitude) from Kumáon to Assam and Pegu. **Kurz** alludes to this plant and gives it the Burmese name of *Chu Yain*. The shrub flowers in October and November. **Mr. Gamble**, in his *List of the Trees and Shrubs &c., of the Darjeeling District*, says it is known by the Paharia names of *Kengeni, kotulkasub*, and to the Lepchas as *Suhutúng rúng*.

C. verrucosa, *Linn.; Fl. Br. Ind. II., 77.; Wight, Ic., t. 200.*

Vern.—*Ban-san*, HIND. and BENG.; *Vuttei-khilloo-khilloopie* (*Vutti-khillo-killupi*), TAM.; *Ghelegherinta*, TEL.; *Kiligilippe*, SING.; according to **Ainslie**). **Sir Walter Elliot** gives the following Telegu names for this species *Allogiligich-cha, gila góranta; Nil-andana-hiruja* SING. according to **Trimen.**

Habitat.—A copiously branched half-shrubby plant, 2-3 feet in height, with blue, white, or yellow flowers; found in tropical regions but ascending the Himálaya to 2,000 feet in altitude, and distributed east to Burma, the Malaya, and China. Also met with in Africa, Mauritius, and tropical America.

MEDICINE. Juice. 2164 **Medicine.**—**Ainslie** says: "I have given this a place here, on the authority of **Rheede**, who informs us that the juice of the leaves is supposed to be efficacious in diminishing salivations" (*Anslie, Materia Indica, II., p. 305*). On a further page he adds: "The slightly bitter, but not unpleasant-tasted juice of the leaves and tender stalks of this low-growing plant, is prescribed by the Tamil doctors, both internally and externally, in cases of scabies and impetigo" (*p. 478*).

CROTON, *Linn.; Gen. Pl., III., 293.* 2165

The generic name Κρότων (a tick) was given by **Linnæus** to this assemblage of plants in allusion to the shape of the seed. The chief medicinal species, **C. Tiglium**, was first made known to Europe in the sixteenth century, and for some time it was in demand; but in the seventeenth century it fell into disuse, until about 70 to 80 years ago, when its properties were again made known by **White, Marshall, Thomson, Daly, Ainslie, Perry, Frost,** and other Indian medical officers. Most of the species mentioned below are used medicinally by the natives of India, but very little of a special character can be said regarding their individual properties.

Croton argyratus, *Bl.; Fl. Br. Ind., V., 383;* EUPHORBIACEÆ. 2166

Syn.—C. BICOLOR, *Roxb.*

Vern.—*Chonoo*, BURM.; *Talib-dá*, AND.

References.—*Roxb., Fl. Ind., Ed. C.B.C., 687; Gamble, Man. Timb., 359; Kurz, For. Fl. Burm., II., 372.*

Habitat.—A moderate sized or small evergreen tree of Martaban, Tenasserim, and the Andaman Islands.

Structure of the Wood. Hard, yellow, close and even-grained, seasons well. It is worthy of notice and weighs 46 to 48lb per cubic foot. TIMBER. 2167

C. aromaticus, *Linn.; Fl. Br. Ind., V., 388.* 2168

Syn.—C. LACCIFERUS, *Linn.*; ALEURITES LACCIFERA, *Willd.*

Vern.—*Welkeppitiyá*, SING.; *Vid-puné*, TAM. (names used in Ceylon for **C. aromaticus**, the form **C. laccifera** being *Keppitiyá* in SING.).

References.—*Beddome, Forester's Man., 204; Wight, Ic., t. 19, 15; Lisboa, U. Pl. Bomb., 121; Trimen, Cat. Ceylon Pl., 81; Gamble, Man. Timb., 358; O'Shaughnessy, Beng. Disp., 553.*

Habitat.—An aromatic shrub or small tree, met with in the Dekhan from the Concan southward.

Medicine.—Said to be used medicinally. **Thwaites** remarks that the lac obtained from **C. lacciferus** "is employed by the Singalese for medicinal purposes." MEDICINE. 2169 Lac. 2170

C. caudatus, *Geisel.; Fl. Br. Ind., V., 388.* 2171

Syn.—C. DRUPACEUS, *Roxb.*

Vern.—*Nan bhantúr*, BENG.; *Takchabrik*, LEPCHA; *Wusta*, URIYA.

References.—*Roxb., Fl. Ind., Ed. C.B.C., 688; Voigt, Hort. Sub. Cal., 156; Kurz, For. Fl. Burm., II., 375; Gamble, Man. Timb., 358–359 and XXIX.*

Habitat.—A large straggling, more or less scandent, shrub of Bengal, Assam, Burma, and South India; found chiefly on the banks of streams. **Roxburgh** states that it is a native in the country about Dacca, and flowers in March, the seeds ripening in September.

Medicine.—Mr. **Home**, Conservator of Forests, writes, the leaves are applied as a poultice to sprains. MEDICINE. Leaves. 2172

Structure of the Wood.—White or yellowish-white, hard, close-grained. **Home** says it is used for fuel. TIMBER. 2173

C. Eluteria, *Bennett,* affords Cascarilla Bark,—an imported drug. 2174

C. Joufra, *Roxb.; Fl. Br. Ind., V., 387.* 2175

Vern.—According to **Roxburgh** *Joufra* is in Sylhet the name of this small tree or shrub.

References.—*Kurz, For. Fl. Burm., II., 373; Gamble, Man. Timb., 358; Medical Top. Ajmir, 140; Voigt, Hort. Sub. Cal., 156.*

Habitat.—A small shrub very similar to **C. oblongifolius**, but with smaller more accuminate leaves; met with in the Eastern Peninsula—Sylhet, Sibsagar, Pegu, Upper Burma, &c. Flowering time March and April.

MEDICINE. 2176 **Medicine.**—Like most other species, the leaves, seeds, and root of this species are occasionally spoken of as used medicinally.

2177 **Croton lacciferus,** *Linn.*, a form reduced to **C. aromaticus,** *Linn.*, by the *Flora of British India.*

2178 **C. malabaricus,** *Beddome; Fl. Br. Ind., V., 386.*

References.—*Beddome, Ic., t. 171, & Forester's Man., 204; Gamble, Man., Timb., 359; Lisboa, U. Pl. Bomb., 121.*

Habitat.—A small tree common in the western forests, ascending to 4,000 feet in altitude; Malabar, &c.

MEDICINE. 2179 **Medicine.**—Said to be used by the natives of India for medicinal purposes.

2180 **C. oblongifolius,** *Roxb.; Fl. Br. Ind., V., 386.*

Vern.—*Chucka,* PATNA (according to **Irvine**); *Bara gach,* BENG. (according to **Brandis** = large plant); *Arjunna,* OUDH; *Ach,* NEPAL; *Kurti, konya, kuli, poter,* KOL; *Putri,* LOHARDUGGA; *Gote,* SANTAL; *Kote, putol,* MAL.; *Burma parokupi,* ASS.; *Bhutan kusam,* TEL.; *Gonsur,* GOA; *Ganasur,* BOMB.; *Ganasura,* MAR.; *Thityin, the-yin,* BURM.

References.—*Roxb., Fl. Ind., Ed. C.B.C., 688; Voigt, Hort. Sub. Calc., 156; Brandis, For. Fl., 440; Kurz, For. Fl. Burm., II., 373; Beddome, Forester's Man., 204; Gamble, Man. Timb., 358-359, XXIX; Thwaites, En. Ceylon Pl., 276; Dalz. & Gibs., Bomb. Fl., 231; Pharm. Ind., 201; Dymock, Mat. Med. W. Ind., 682; Flück. & Hanb., Pharmacog., 567; S. Arjun, Bomb. Drugs, 122; Irvine, Gen. Med. Topog. of Ajmir, 128; Lisboa, U. Pl. Bomb., 120, 255; Cooke, Oils and Oilseeds, 38.*

Habitat.—A small tree found in the sub-Himálayan tract from Oudh eastward and in South India, the Deccan Peninsula, Burma, and Ceylon. **Roxburgh** remarks that it is common in the forests about Calcutta, flowering in the beginning of the hot season.

OIL. 2181 **Oil.**—The seeds afford an oil (*Gamble*).

MEDICINE. **Medicine.**—**Brandis** says that the bark, leaves, and fruit are used externally in native medicine. The seeds are purgative; **Dr. Irvine**
Seed. 2182 says the FRUITS are purgative, dose gr. ⅛ to gr. iii. **Dr. Dymock** writes "when on a visit to Goa in 1876, my attention was drawn by the native
Fruit. 2183 doctors to the root-bark of a small tree as being one of the most valuable medicines they possessed; this plant, unknown to me, at the time, proved
Root-bark. 2184 on subsequent investigation to be **C. oblongifolium.** The Goanese and inhabitants of the southern Concan administer the bark in chronic enlargements of the liver and in remittent fever. In the former disease it is both taken internally and applied externally. As an application to
Root. 2185 sprains, bruises, rheumatic swellings, &c., it is in great request. In large doses it is said to be purgative. **Flückiger** and **Hanbury** (*Pharmacog.*, 510) state that the "seeds are said to be sometimes substituted for those of **C. Tiglium.**" The **Rev. A. Campbell** remarks that the Santals use the "bark and root as a purgative and as an alterative in dysentery."

It would appear that the early writers on Hindu Materia Medica do not allude to this plant, and many of its vernacular names would point to the properties having been but recently understood. There is no good Hindi nor a Bengali name for the plant. It is not referred to by **U. C. Dutt** nor by **Ainslie**, and while **Roxburgh** describes it he makes no mention of its medicinal products. On the other hand, there is nothing to justify a contrary opinion but that the Kol and Santal names are very ancient, and that, therefore, the properties of the plant have long been known to the aboriginal tribes who inhabit the mountains of Central and Western Bengal. **Dalzell** and **Gibson** write that it is "used medicinally by the natives to reduce swellings." This is perhaps one of the earliest notices by European writers.

Structure of the Wood.—Whitish to yellow, close-grained, moderately hard and heavy; liable to crack in seasoning. **TIMBER. 2186**

Domestic Uses.—The plant is frequently employed for fences. **DOMESTIC. 2187**

Croton polyandrus, *Roxb.*, see under **Baliospermum montanum,** *Muell., Vol. I., B. 28.* **2188**

Hooker, in the *Flora of British India, V., 461*, reduces this to **B. axillare,** *Blume.* Consult also *O'Shaughnessy's Bengal Dispens., 555; U. C. Dutt's Mat. Med. of the Hindus, 229*; and *Dymock's Materia Medica, West Ind., 2nd Ed., 688*; the last work has appeared since the issue of the 1st volume of this publication.

C. reticulatus, *Heyne; Fl. Br. Ind., V., 386.* **2189**

Syn.—C. HYPOLEUCUS, *Dalz.*; C. ZEYLANICUS, *Muell.-Arg.*

Vern.—*Pándhari* or *pándharisálá*, MAR.

References.—*Dymock, Mat. Med. West. Ind., 2nd Ed., 684; S. Arjun, Bomb. Drugs, 122; Thwaites, En. Ceyl. Pl., 276; Dalz. and Gibs., Bomb. Fl., 231; Lisboa, U. Pl. Bomb., 121.*

Habitat.—A shrub with slender branches, met with in the Dekhan Peninsula from the Koncan southwards; distributed to Ceylon.

Medicine.—**Sakharam Arjun** says the bark is "used as a bitter and stomachic." **MEDICINE. Bark. 2190**

C. sebiferum, *Linn.*, and **Sapium sebiferum,** *Roxb.*, are synonyms for **Stillingia sebifera,** the Chinese Tallow Tree. This is now cultivated to some extent in India, and, according to **Roxburgh**, is known in Bengal as *Momchina.* **2191**

C. Tiglium, *Linn.; Fl. Br. Ind., V., 393.* **2192**

THE PURGING CROTON.

Syn.—C. PAVANA (or PARANA), *Hamilton.*

Vern.—*Jayapála, kanakaphála* (in **Ainslie** *dunti, bija*), SANS.; *Jaypál*, BENG.; *Jamál-gota*, HIND.; *Jamalagota, jepál, geyápal*, MAR.; *Nepál*, GUZ.; *Nepála*, KAN.; *Nerválam*, TAM.; *Nepála-vitua*, TEL.; *Nirválam*, MAL.; *Kanako*, BURM.; *Bori*, MALAY.; *Cheraken*, JAVA. *Dund* is given by **Ainslie** as PERS., and *Batú* as ARAB. *Habbussalátín, dand, dátún*, ARAB.; *Bed anjire-khatai, habbe-khatái*, PERS. (according to **Moodeen Sheriff.**)

References.—*Roxb., Fl. Ind., Ed. C.B.C., 688; Voigt, Hort. Sub. Cal., 156; Brandis, For. Fl., 440; Kurz, For. Fl. Burm., 374; Gamble, Man. Timb., 358, 359; Thwaites, En. Ceylon Pl., 277; Murray, IV., 149; Rumph., Amb., IV., t. 42; Pharm. Ind., 200; Ainslie, Mat. Ind., I., 101, 8, 596, 599; O'Shaughnessy, Beng. Dispens., 553; Moodeen Sheriff, Supp. Pharm. Ind., 120; U. C. Dutt, Mat. Med. Hind., 228; Dymock, Mat. Med. W. Ind., 2nd Ed., 684; Fleming, Med. Pl. and Drugs, as in*

The Purging Croton.

As. Res., Vol. XI., 164, 1840; Flück. & Hanb., Pharmacog., 565; U. S. Dispens., 15th Ed., 978, 1051; Bent. & Trim., Med. Pl., 239; S. Arjun, Bomb. Drugs, 123; Murray, Pl. and Drugs, Sind, 149; Waring, Bazar Med., 49; Year book of Pharm. 1874, p. 86; Irvine, 26, 42; K. L. De, Beng. Drugs, 43; Medical Top. Ajmír, 133, 139; Baden Powell, Pb. Pr., 374, 375; Drury, U. Pl., 164; Lisboa, U. Pl. Bomb., 121, 255; Birdwood, Bomb. Pr., 77; Cooke, Oils and Oil-seeds, 39; Kew Off. Guide to the Mus. of Ec. Bot., 118; Kew Off. Guide to Bot. Gardens and Arboretum, 67; Simmonds, Trop. Agri., 424.

Habitat.—A small tree (15 to 20 feet high) met with under cultivation throughout the greater part of India; probably indigenous or only naturalised in Eastern Bengal and Assam and southward to Malacca, Burma, and Ceylon.

OIL. Nuts. 2193 **Oil.**—The nuts yield an oil which is orange yellow or sherry-coloured, of the consistence of nut-oil, has a slight odour resembling that of jalap, and an acrid flavour. This is a valuable medicinal oil, which is used as a drastic purgative, especially when it is desired to act speedily and powerfully on the bowels, and when only a small volume of medicine can be administered, as in cases of obstinate constipation, in dropsy, in apoplexy, in paralysis, and in cases, when the patient cannot or will not swallow, when the oil may be dropped on the tongue. As prepared in India it is frequently so much adulterated, that it finds no sale in Europe. The nuts are exported chiefly from Bombay and Cochin (often being also Chinese re-exports), and the oil is expressed in England. **Dr. Dymock** informs the writer that the oil is expressed at the Government Medical Store Depôt at Bombay. It costs about 12 annas a ℔, whereas in 1825, the same oil was sold for about 10 shillings an ounce in England. The plant used to be grown for the purpose of its seeds at Hewra, but the supply is now imported from China *viâ* Singapore. The nuts sell for R51 per maund of 41℔.

Bombay. 2194

Cochin 2195

Chinese. 2196

European. Expressed. 2197

It is necessary to be cautious in handling the nuts or the oil, owing to their blistering the skin. The oil is frequently used for colds in the chest as an external application, causing a severe blister. It is much

2198 resorted to as a domestic cure but is not recommended by the profession.

§ "The drastic principle of the oil has not yet been isolated; it appears to exist not only in the seeds but also in the leaves and wood" (*Professor Warden, Calcutta*).

MEDICINE. Seeds. 2199 Oil 2200 **Medicine.**—The SEEDS are used as a powerful drastic purgative, and the OIL is regarded as a valuable medicine. In overdoses they act as an acro-narcotic poison. When externally applied the oil is a stimulant rubefacient and counter-irritant. Croton oil is said to possess powerful hydragogue cathartic properties. It is also useful in dropsy, obstinate constipation, and apoplexy. The ancient Hindu books make no mention of the oil, the nuts boiled in milk or roasted in a pellet of cow-dung, appear (as at the present day) to have been used. One seed is a sufficient dose, and, according to many writers, the skin of the seed, as also the contained cotyledons (or seed leaves), are poisonous. The boiled or torrefied albuminous substance, mashed up and deposited in the interior of a raisin, is the form in which natives generally prescribe the drug, but it is often combined with astringents, such as myrobalams, cutch, &c., these additions checking the acrimony of the nut and preventing griping.

2201 **Waring** says that should the administration of the nut cause griping, vomiting, or too violent purging, a good large draught of lime-juice is the best remedy; and it may safely be repeated in half an hour if the vomiting, &c., continue. **Dutt** remarks that, according to Hindu literature, the seeds are "useful in fever, constipation, intestinal worms, enlargements of the abdominal viscera, ascites, anasarca, &c."

MEDICINE. Grana Tiglia. 2202

Dr. Fleming (*in the Asiatic Researches, 1840*) writes :—
"The seeds of this plant were formerly well known in Europe, under the names of *Grana Tiglia* and *Grana Molucca.* They were employed as hydragogue purgatives; but, on account of the violence of their operation, they have been long banished from modern practice. For the same reason, they are seldom used by the *Hindu* practitioners, though not unfrequently taken, as purgatives, by the poorer classes of the natives. One seed is sufficient for a dose. It is first carefully cleared from the membranaceous parts, the rudiments of the seminal leaves, that adhere to the centre of it; by which precaution, it is found to act less roughly, and then rubbed with a little rice gruel, or taken in a bit of the plantain fruit."

2203 **Ainslie** quotes (in the first edition of his work published in 1813) the opinions of a few Indian medical officers who re-made known the properties of this drug at about the beginning of the present century or the close of the last. Practically all subsequent writers have but slightly altered the sentences used by these early observers without having added any thing of consequence to the literature of the subject. The discovery of other drugs may be viewed as having thrown into the shade croton oil and croton nuts. **Dr. Dymock** remarks of the expressed oil: "**Ainslie** (*Mat. Indica, Vol. I., p. 105*) notices the use of the EXPRESSED OIL (*Nervalum unnay*) by the Tamils as an external application in rheumatic affections, but it does not appear to have been used for internal administration until the year 1821." Completing **Ainslie's** own account of it, after stating that the oil is used for external application, he goes on to say, "as a purge it has been of late years often resorted to in England, and is thought to have still more powerful effects as a hydragogue than the torrefied seeds. **Mr. Thomson** tells us that, in some cases, merely touching the tongue with a drop has produced many loose stools; and in others, doses of one or two minims have excited the most frightful hypercantharsis; although some individuals have taken it to the extent of even ten minims without any very sensible effect. He adds from his own experience, that he would be very cautious in exhibiting the oil at first in larger doses than one or two minims, to adults; in apoplexy, convulsions, and mania the croton oil is likely to prove a medicine of great value; a very good mode of giving it is, rubbed up with the mucilage of Acacia gum, sugar, and almond emulsion, by which means its acrimony is blunted." **Ainslie** adds that **Mr. R. Daly** of Madras found the oil highly useful as an emmenagogue.

Root. 2204 "**Rumphius** informs us that the ROOT of the plant is supposed, by the inhabitants of Amboyna, to be a useful drastic purgative, in cases of dropsy, given rasped in doses of a few grains, or as much as can be held between the thumb and finger." "**Rheede**, who speaks of the plant under the name *cádél avánácu*, says, that the LEAVES rubbed and soaked in water also are purgative; and when dried and powdered are a good external application in cases of bites of serpents" (*Ainslie*).

Leaves. 2205

CHEMISTRY. 2206 **Chemistry.**—The principal constituents of croton seeds are a fatty fixed oil, tiglinic acid, crotonic or quartenylic acid, and crotonol. **Tuson** has detected the presence also of an alkaloid analogous to *Cascarillin.* (See *Flück. & Hanb., Pharmacog., and Bent. & Trim., Med. Plants.*)

2207 **Special Opinions.**—"§ Drastic purgative, used in obstinate constipation and dropsical affections. I have known instances of extreme prostration, amounting to collapse, produced by seeds, administered by native *boids* in Bengal and the North-Western Provinces" (*Assistant Surgeon Shib Chunder Bhattacharji, Chanda, Central Provinces*). "In addition to their uses as a drastic purgative the seeds are applied in the form of liniment to the penis in cases of impotence and have a high reputation in this disease amongst the natives" (*Lal Mahomed, 1st Class, Hospl. Asstt.,*

MEDICINE. 2208 *Mani Dispensary, Hoshangabad, Central Provinces*). "The seeds, half roasted over a lamp or candle flame, and the smoke inhaled through the nostrils, relieves a fit of asthma" (*Surgeon-Major R. Thomson, M.D., C.I.E., Madras*). "I have found the oil diluted with 9 or 10 parts of mustard oil or olive oil to be a very useful liniment in infantile bronchitis" (*Doyal Chunder Shome*). "Have used it as a diuretic, purgative, and rubefacient" (*D. Picachy, Civil Medical Officer, Purneah*). "The seed is frequently applied over the temples for headache and eye affections" (*Surgeon-Major Robb, Civil Surgeon, Ahmedabad*).

2209 **Croton tinctorium,** *Turnsol*, see **Crozophora (Chrozophora) tinctoria,** *A. Juss.*

Crown Bark, see **Cinchona Condaminea,** *Huml.*; RUBIACEÆ. **C. 1129.**

2210 **CROZOPHORA,** *A. Juss.; Gen. Pl., III., 305.*

By an unfortunate oversight, the old error in the spelling of the name given to this genus was not corrected when arranging the material for the present volume, and this has had the effect of placing it in the wrong alphabetical position. Being derived from χρωζω the word should of course be **Chrozophora** as corrected by **Necker.**

2211 **Crozophora (Chrozophora) plicata,** *A. Juss.; Fl. Br. Ind., V., 409;* EUPHORBIACEÆ.

Syn.—C. ROTTLERI, *A. Juss.;* C. PLICATUS, *Vahl.;* C. ROTTLERI, *Geisel.;* C. TINCTORIUS, *Wall.; Burm.;* C. PLICATUM, *Willd.* (*in Roxb., Fl. Ind.*).

Vern.—*Shadevi, súbali, sonballi*, HIND., SIND. and *Okharada*, GUZ.; *Khúdi-okra*, BENG.; *Pango nari*, SANTALI; *Suryavarta*, SANS.; *Pút kanda, nílkhantí, nil-ak-raí*, PB.; *Neal boti*, TANK; *Gurugu chettu, linga miriyam*, TEL.

References.—*Roxb., Fl. Ind. Ed. C.B.C., 687; Thwaites, En. Ceylon Pl., 443; Dalz. & Gibs., Bomb. Fl., 232; Stewart, Pb. Pl., 192; Elliot, Fl. Andhrica, 66, 107; Revd. A. Campbell, Descrip. Cat. Econ. Prod., Chutia Nagpur, 18; Ainslie, Mat. Ind., II., 398; Dymock, Mat. Med. W. Ind., 2nd Ed., 716; S. Arjun, Bomb. Drugs, 123; Murray, Pl. and Drugs, Sind, 34; Drury, U. Pl., 165; Lisboa, U. Pl. Bomb., 269; Royle, Ill. Him. Bot., I., 329.*

2212 **Sir Walter Elliot** remarks of this plant: "This is the Indian Turnsol—*Royle, Ill., I., 329.* Misled by the English name **Wilson, Brown, Piddington,** and others have imagined the plant to be the sun-flower, and still further to increase the confusion, they have turned the old Greek name **Chrozophora tinctoria,** *L.* (ἡλιοτρòπιον μικρον) into the modern **Heliotrope,** and explained the various Indian names of **Croz. plicata by Helitropium (Tiardium), indicum,** *Lindl., Veg. King., p. 281.*" This mistake has been repeated by **O'Shaughnessy,** who says that **Chrozophora tinctorium,** the Turnsol (Turnsole) is the Ηηλιιτροπον μικρον of **Dioscorides.**"

Habitat.—There are two well marked forms of this plant—(*a*) a small procumbent annual, found in sandy damp situations, such as on the banks of rivers and in the bottoms of dried-up tanks, (*b*) an erect perennial bushy form. These have apparently been reduced to one species by the *Flora of British India.* They both occur here and there throughout the warmer parts of India, from the Panjáb to Bombay, Madras, Bengal, Burma, and Ceylon. In the drier regions of Upper India the bushy condition chiefly occurs, and this is probably doubtfully distinct from **Chrozophora tinctoria.** The procumbent form is more abundant in Bengal,

Madras, and Burma, and is of no interest from an economic point of view, since the properties described below are alone applicable to the erect plant, and to **Chrozophora tinctoria.** The confusion alluded to by **Sir Walter Elliot** may be accounted for by the fact that the crumpled leaves of the procumbent plant are remarkably Boraginaceous in their general appearance; the plant is also frequently found growing along with **Heliotropium indicum** (the *hatti-súra* of Bengal).

DYE. 2213

Dye.—**Roxburgh** was, perhaps, the first author to draw attention to the fact that the fruits of this Indian plant afford a purplish-blue dye. **Ainslie,** who saw the manuscripts of **Roxburgh's** *Flora Indica,* says:—"It would appear that, cloth, moistened with the juice of the green capsules, becomes blue after exposure to the open air; they, no doubt, contain colouring matter, which might be turned to good account in the arts." **O'Shaughnessy,** who wrote 20 years later still, says—"The summits of this plant and the fruits serve for the preparation of the dye named *turnsol,* used for giving a blue and a red tint." **Voigt, Drury, Murray,** and all other modern Indian writers have simply repeated **Roxburgh's** original statement regarding the dye property of these fruits without adding any new information to the literature of the subject. It is probable that even **O'Shaughnessy's** remark was suggested from the knowledge that a species of **Chrozophora** was in some parts of Europe cultivated as a source of dye, his statements having thus a European more than an Indian significance. For further information regarding Turnsole dye see the next species.

FIBRE. 2214

Fibre.—The Santals prepare a strong and useful rope fibre from the bark, but it is difficult to separate (*Campbell*).

MEDICINE. Ashes. 2215 Leaves. 2216 Seeds. 2217 Root. 2218 Dry Plant. 2219

Medicine.—The ASHES of the root are given to children in coughs. The LEAVES are considered depurative, and are officinal under the name *nílkhanthí.* The SEEDS are used as a purgative. The **Revd. A. Campbell** states that the Santals mix the ROOT with that of **Carissa Carandas** for blistering purposes. "This is a plant which **Dr. F. Hamilton** (MSS.) had brought to him in Behar, as one of those which was supposed to have virtues in leprous affections; the dry plant is made into decoction, to which is added a little mustard" (*Ainslie*).

TIMBER. Fuel. 2220

Timber.—The stems of both this and the next species are regularly collected as fuel. **Dr. Stewart** says of **C. tinctoria**: "It is cut and carried into the city of Lahore to be used as fuel in ovens." This fact may be accepted as proving that the bush forms here alluded to are both perennial bushy plants 1-3 feet in height and not "prostrate annuals." The prostrate form would appear to be perfectly distinct, and to be most probably the **Croton plicatum** described by **Roxburgh** as met with in rice fields of Bengal, as distinct from the bushy perennial found in Chutia Nagpur and Upper India.

Crozophora tinctoria, *A. Juss.; Fl. Br. Ind., V., 408.* 2221

TURNSOLE, *Eng.*

Vern.—*Shadevi, sonballi, subali,* HIND. & SIND; *Tappal búti, nilan, kukronda,* PB.; *Kap-o-chist,* in the Hari-rud Valley, Afghánistan (**Aitchison**).

Habitat.—Common in the Panjáb, Sind, and the Deccan; distributed eastward through Afghánistan to northern Africa and the Mediterranean; cultivated in the south of France. The specimens of this plant collected in Afghánistan by **Aitchison,** in Quetta by **Lace,** and in Gilgit by **Giles,** have small, almost ovate-deltoid, leaves, on long thin petioles, and show little tendency towards the large irregularly-lobed leaves of the erect form of **C. plicata** (met with in India). The last mentioned plant has much less woolly leaves than either **C. plicata** (procumbent form) or **C. tinctoria**, but is covered with a granular mealy substance.

DYE. Blue. 2222 **Dye.**—Although it seems probable that most Indian authors who allude to having observed the fruits of **Chrozophora** yielding a purplish dye, speak of the erect perennial form of **C. plicata,** still **C. tinctoria** doubtless affords the same dye in this country as it is cultivated for in France. Apparently no advantage is taken in India of the dye principle yielded by either plant, and it may therefore be of some practical utility, in any possible future efforts to establish an industry in this dye-stuff, to give here a brief abstract regarding its European uses and methods of preparation. The researches of **Dr. Joly** (*Ann. de Chim. et de Phys., VI., 111.*) have shown that the dye principle occurs in all parts of the plant and not in the fruits only. It is also present during every stage of the growth of the plant and abounds in the cellular tissue occurring as coloured particles. As with indigo green, so with this substance, by oxidation it becomes

Yellow. 2223 blue. When the fruit "is immersed in twice its bulk of water and heated to from 50° to 60°, that liquid assumes a rather deep violet blue colouration, and deposits, on being evaporated, a beautiful azure-blue resinous

Green. 2224 substance. Acids turn the colour of the aqueous solution to a yellowish

Litmus on Rags. 2225 red which is not rendered blue again by alkalies but becomes greenish. By this reaction, therefore, the "litmus on rags" is distinguished from the litmus of commerce. The researches of **Dr. Langdale and Dr. Martius,** made with the juice of the plant just described, have proved that it dyes, without the aid of mordants, a violet-red upon wool, silk, and cotton tissues, and that this colour may be rendered fast by steaming and the simultaneous action of ammonia vapours, which, however, turn the colour

Powder. 2226 more blue" (*Crookes, Hand-book of Dyeing, &c., 383*). "This dye is called Turnesole, and is obtained by grinding the plants—little herbs seldom more than a foot high—to a pulp in a mill, when they yield about half their weight of a dark green coloured juice, which becomes purple by exposure to the air or under the influence of ammonia. It is chiefly exported to Holland, and is prepared for exportation by soaking coarse linen rags or sacking with it, the rags being previously washed clean. After soaking they are allowed to dry, and are exposed to the influence

Sacking Impregnated. 2227 of ammonia by being suspended over heaps of stable manure. They are then packed in sacks and are ready for shipping to Holland" (*Treasury of Botany*). "The red colour of the outer crusts of some kinds of Dutch cheese is due to the presence of some lactic and butyric acids in that substance. No good substitute for this 'litmus on rags' for the last named purpose has as yet ever been found. A sum of £10,000 is annually paid by Dutch farmers, chiefly to the inhabitants of Grand-Gallargues, for a commodity which, at first sight, no one would take to be any thing else but dirty rags, best suited for paper-making after having been bleached. A portion of the rags, after having been used to rub cheese with, are sent back, because it has been found that the old rags take up and develope the colourable matter more readily than new ones" (*Crookes*).

TRADE. 2228 It would thus appear that **Chrozophora** affords a colouring principle closely allied to Orchil and Litmus, but in the method of its preparation it is closely allied also to Indigo. How far this dye is capable of meeting other markets cannot at present be foretold, but there would seem every reason to suspect that a very extensive trade might be done in it. The plant is wild everywhere on the waste lands of India, luxuriating on both dry sandy tracts and river margins; it might be grown at a small cost anywhere, and the subject thus seems well worthy of attention, as there are many purposes to which it might be put in India. The writer can discover no evidence of its ever having been utilised by the natives of India, but it is a remarkable coincidence that in Bengal, at least, it bears a name (*okra*) now given to several introduced American plants. Dr. Buchanan Hamilton's remarks regarding the introduction of **Bixa Orellana** having displaced an indigenous dye-yielding plant might be even viewed as having reference to **Chrosophora.** In connection with the Calcutta International Exhibition the author published, in his Catalogue of the Dye-Stuffs there shown, some interesting facts regarding a reputed green dye found in the leaves of **Jatropha glandulifera.** (*See also Agri-Hort. Soc. Proc., 1861, XXVII.*) When it is recollected how closely allied that plant is to **Chrozophora** an additional justification for a thorough investigation of the dye properties, both of **Chrozophora** and of **Jatropha,** may be admitted as highly desirable, and it is perhaps not too much to say that the Indigo planters might find it worth their while to be the pioneers of this unexplored subject. Both these plants could at very little cost be grown as hedges around their indigo fields, thus affording a possible extra revenue, while serving a purpose for which they are eminently suited, since no herbivorous animal has as yet been observed to browse either on **Jatropha glandulifera** or **Chrozophora tinctoria.**

TURNSOLE-DYE. 2229 Of Interest to Indigo Planters. 2230

CRUSTACEA. 2231

Although the Indian waters (marine and fresh) and also marshy places abound in examples of the group of animals that may be said to be represented by the common crab, the lobster, the crayfish, the shrimp and water-flea, only one or two are of any economic interest. The small fresh-water prawn (*chingrá*) is often very plentiful in tanks, and on certain occasions may be seen to multiply in a perfectly marvellous manner—a tank sometimes suddenly appearing full of them and as suddenly empty. Although largely caught, the natives of India do not appear to fish systematically for Crustacea. Dr. D. MacDonald says of Bombay: "The Crustacea, especially prawns, are very numerous, but mostly get caught along with real fish in the nets, and, except the crab-hook" (used at low water for catching crabs in the crevices of the rocks) "no particular gear is used in their capture. There are no lobsters, although large crayfish are commonly sold by that name in the Bombay markets, and none of the numerous crabs attain the size and quality of those of northern seas. Crab and lobster pots are unknown." Ainslie gives the following vernacular names: *Ingrha,* HIND.; *Agni matsya,* SANS.; *Eerál,* TAM.; *Roiclú,* TEL. He remarks that "the prawns in India are excellent, especially on the Coromandel Coast. As food, they are considered, by the Hindus, as stimulating and aphrodisiac, and to possess virtues in diabetes, which they, and perhaps with some reason, suppose to be often produced by an insufficient quantity of animal food."

FOOD. Crabs. 2232 Prawn. 2233 Lobsters. 2234 Cray fish. 2235 Shrimps. 2236 2237 MEDICINE. 2238

CRYPTERONIA, *Bl.; Gen. Pl., I., 782.* 2239

[*Man. Timb., 199;* LYTHRACEÆ.

Crypteronia pubescens, *Blume; Fl. Br. Ind., II., 574; Gamble,* 2240

Vern.—*Ananbo,* BURM.

Habitat.—A tree 30 feet in height, met with in Burma.

Structure of the Wood.—Pale to reddish brown, fibrous, close but not straight, rather heavy, and annular rings narrow (*Kurz*). **Brandis** says it is used for cart wheels and other such purposes, but is mainly in demand for fuel.

TIMBER. 2241

CRYPTOCARYA, *R. Br.; Gen. Pl., III., 150.*

Several species afford valuable timber.

2242 **Cryptocarya amygdalina,** *Nees; Fl. Br. Ind., V., 118;* LAURINEÆ.

Vern.—*Patmaro,* NEPAL; *Kaledzo,* LEPCHA.

Habitat.—A tree with spreading branches, found from Nepal eastwards to the Khasia hills and south to the Andaman islands.

TIMBER. 2243 **Structure of the Wood.**—Strong and useful.

2244 **C. ferrea,** *Bl.; Fl. Br. Ind., V., 119.*

2245 **C. Wightiana,** *Thwaites; Fl. Br. Ind., V., 120; Wight, Ic., t. 1829; [Lisboa, U. Pl. Bomb., 113.*

Vern.—*Golu-mora,* SING.

Habitat.—A tall tree, frequent in the Dekhan peninsula from Kanara southwards to Ceylon.

TIMBER. 2246 **Structure of the Wood.**—Strong and durable, useful for building purposes.

CRYPTOLEPIS, *R. Br.; Gen. Pl., II., 740.*

2247 **Cryptolepis Buchanani,** *R. & S.; Fl. Br. Ind., IV., 5; Wight, [Ic., t. 494;* ASCLEPIADEÆ.

Syn.—NERIUM RETICULATUM, *Roxb.*

Vern.—*Karanta,* HIND.; *Utri dudhi,* SANTAL; *Guruga-pála-tige, adavi-pála-tige, madana séku,* TEL. (At Sinhachalam it is called *Málati*-like climber; **Elliot.**)

References.—*Roxb., Fl. Ind., Ed. C.B.C., 244; Brandis, For. Fl., 330; Dalz. & Gibs., 148; Gamble, Man. Timb., 265; Kurz, For. Fl. Burm., II., 199; Elliot, Fl. Andh., 11, 67, 109; Campbell, Cat. Econ. Pl., Chutia Nagpur, 49; Rheede, Hort. Mal., IX., t. 11; Grah., Cat. Bomb. Pl., 113.*

Habitat.—A climbing plant, met with throughout India from Kashmir to Assam, Burma, Coromandel, Travancore, &c., ascending the Himálayas to 4,000 feet in altitude; distributed to Ceylon.

FIBRE. 2248 **Fibre.**—**Sir Walter Elliot** says the hill people of Vizianagram make cordage and a kind of cloth from the fibre derived from this plant.

MEDICINE. 2249 **Medicine.**—The **Rev. A. Campbell** states that the Santals make a preparation from the plant which they give to children to cure them of rickets. They also combine it with **Euphorbia microphylla,** *Heyne* (the *dudhia phul*), in the formation of a medicine to be given to women "when
2250 the supply of milk is deficient or fails." Both the plants so used having a milky sap, it may be presumed the properties attributed to them by the Santals rest on the "Doctrine of Signatures."

CRYPTOMERIA, *Don; Gen. Pl., III., 428.*

2251 **Cryptomeria japonica,** *Don;* CONIFERÆ.

Habitat.—A handsome tree, native of China and Japan, but largely cultivated throughout the districts of Darjeeling, Simla, and occasionally in other hill stations.

TIMBER. 2252

Structure of the Wood.—White, soft, with a brown, often almost black, heart-wood; very uniform, with narrow bands of darker and firmer tissue at the edge of each annual ring.

CRYPTOSTEGIA, *R. Br.; Gen. Pl., II., 742.*

[ASCLEPIADACEÆ

Cryptostegia grandiflora, *R.Br.; Fl. Br. Ind., Vol. IV., 6;* 2253

Vern.—*Vilarjuti vakundi,* MAR. (according to **Dr. Sakharam Arjun** in a letter to the author); *Palay,* MAL. (according to **Sir George Birdwood**).

Habitat.—An extensive climber, cultivated in various parts of India; supposed to be a native of Africa or Madagascar.

CAOUTCHOUC. 2254

Caoutchouc.—Dalzell and Gibson (*Bomb. Fl. Sp., 55*) say "the whole plant abounds in a milky caoutchouc juice, which is like India-rubber, but hardly elastic." A considerable effort is being made to extend the cultivation of this plant both in Madras and Bombay (*See Agri.-Hort. Soc. Jour., Mad., 1883-84, and Rep. Bot. Gard. Hyderabad, Sind, 1882, p. 7; also Rep. Dir. Agri. Bomb., 1883-84, p. 16*). A sample of the Sind-prepared Caoutchouc, obtained from the plants grown in the Botanic Gardens, was reported on in August 1883, as follows, by **Mr. T. P. Bruce Warren**, Analytical Chemist to the Indian Rubber, Gutta Percha and Telegraph Company, Limited, Silvertown: "The sample herein referred to was enclosed in a cloth wrapper. The weight of rubber when received was 3℔ 5½ oz. net. Some portions of it had become very sticky and much blackened by oxidation; a very small portion only had retained the light colour of Ceára rubber. The whole had become agglomerated by the adhesiveness of the little separate masses of which the sample was composed.

"The sample was carefully torn to pieces and examined, a separate examination being made of the lighter and darker portions. The only difference found is in the much larger quantity of moisture met with in the lighter portions.

"It might have been possible to have given some assurance on this point if the time was stated how long this sample had been collected. In its present condition it is hardly equal to Ceára rubber from Brazil, although its general qualities are very encouraging.

"Digested in alcohol both the lighter and dark portions gave up a considerable quantity of soluble matter; the darker kind became partly whitened, the greater portion being unaffected in this respect.

"On washing and drying, the lighter portions lost 15·6 per cent., the darker portions lost only 2·9 per cent. The amount of ash obtained from the lighter portions was before washing 4·3 per cent., after washing 2·7 per cent. The darker portions yielded before washing 4·2 per cent., after washing 2·3 per cent. 2255

"Mixed with the suitable proportion of sulphur and heated, both portions vulcanized remarkably well. It might have been expected that the least oxidized portions would have yielded a tougher and harder product when vulcanized, as compared with the darker portions, but in this respect no difference could be perceived."

The Conservator of Forests, Northern Circle, Bombay Presidency, wrote on the 16th January 1888, that **Cryptostegia grandiflora** "is cultivated in gardens in nearly every station in India, and can be easily propagated. The cost of collecting the sap would be so great that a plantation is not 2256 likely to be commercially successful. The plant grows wild in the Western Ghâts."

Crystal Rock, see **Carnelian, C. 616.**

CTENOLEPIS, *Hook. f.; Gen. Pl., I., 832.*

2257 **Ctenolepis Garcini,** *Naud.; Fl. Br. Ind., II., 630;* CUCURBITACEÆ.

Vern.—*Gudi muralú,* TEL.

References.—*Roxb., Fl. Ind., Ed. C.B.C., 703; Dalz. & Gibs, Bomb. Fl., 99; Atkinson, Econ. Prod., V., p. 12.*

Habitat.—An annual climber, met with in Bundelkhand and the Dekhan. Grows on rubbish heaps and hedgerows.

MEDICINE. 2258 **Medicine.**—**Atkinson** says the fruit, seeds, and roots are used in medicine.

Cubeba officinalis, *Miq.*, see **Piper Cubeba,** *Linn.;* PIPERACEÆ.

Cubebs, see **Piper.**

2259

CUCUMIS, *Linn.; Gen. Pl., I., 826.*

A genus of climbing herbaceous plants embracing some 26 species, of which half are natives of Africa; a few occur in the tropical regions of Asia, Australia, and America; and several are of doubtful origin though widely cultivated. **Elliot** says the Telugu word *Budama* is applied generically to all species of CUCUMIS. The botanical generic name (which was the Latin specific name for the Cucumber) probably arose from *curvus* (Latin) in allusion to the shape of the fruit.

HISTORY. 2260 **History.**—Much confusion still exists regarding the Indian so-called wild and cultivated species and varieties. **Roxburgh** was the first author who systematically examined and described the Indian forms. In his *Flora Indica* he gives the distinctive characters of what he regards as nine species, two of which, by all subsequent botanists, have been removed to other genera, and the remaining seven reduced to three species. **De Candolle**, however (*Orig. Cult. Pl., p. 259*), seems to be of opinion that they represent but two species—**C. Melo,** *Linn.* (embracing all the wild and cultivated Indian, African, and American forms of the Melon) and **C. sativus,** *Linn.* (the Cucumber). Referring to **Roxburgh's** nine species, **Ainslie** says they are all natives of India "except the Melon, which is a native of Persia." Modern Indian authors speak of **Roxburgh's C. Melo** as the Melon, but designate the other kinds as forms of the Cucumber. To **Roxburgh's** list must be added an Arabian and African species **C. prophetarum,** *Linn.*, collected in Sind and Baluchistan by **Stocks,** and in Belgaum by **Ritchie.** It may here be pointed out that **DeCandolle** is scarcely correct when he says there is no Sanskrit name for the Melon. His words are:—"No Sanskrit name is known, but there is a Tamil name, probably less ancient, *molam,* which is like the Latin *melo.*" There are Sanskrit, Persian, Hindustani, and many other vernacular names for most of the forms of **Cucumis Melo,** both wild and cultivated; and, indeed, it seems probable that *molam* or *mulam-pandu* is but a modern corruption from the English word melon. There are, however, many ancient and pure names for the forms of the melon grown in India, as, for example, those given by **Wilson, Elliot, Dutt,** and other writers.

The experiments of **Naudin** with the various forms of **Cucurbita** and **Cucumis** go some way towards establishing a physiological classification of these plants. He concludes that where it is possible to cross fertilize

HISTORY.

with the production of fertile seeds, the plants so experimented with may be viewed as varieties or even only cultivated races derived from a common species. The opposite inference, he advances, should be drawn when it is not possible to cross fertilize. **DeCandolle**, commenting on this opinion, says: "I have already spoken of the physiological principle on which he (**Naudin**) believes it is possible to distinguish those groups of forms which he terms species, although certain exceptions have occurred which render the criterion of fertilization less absolute. In spite of these exceptional cases, it is evident that if nearly allied forms can be crossed and produce fertile individuals, as we see, for example, in the human species, they must be considered as constituting a single species." Without attempting to contest the value of **Naudin's** experiments, it may at least be stated that a too liberal acceptance of the test of the production of fertile seed by cross fertilization would materially upset many well established species. For example, it might not be difficult to show that many of the recognised and constant forms of cotton, grown in India, are hybrids between the species **Gossypium herbaceum** and **G. barbadense.** So also it is commonly stated that a fertile mule exists between the two species of Camel—**Camelus dromedarius** and **C. backtrianus**—but the progeny is more unmanageable than the mule itself, and is accordingly very little bred (see article on **Camel, C. 203**). But **Naudin's** physiological classification seems to be supported by recognisable structural peculiarities and may thus be accepted as systematically correct. At the same time, in a work devoted to Economic Products, it is not desirable to abandon well recognised forms which are grown or collected from the wild state for independent purposes. In the following account, therefore, of the Indian forms of **Cucumis**, **Roxburgh's** species have been retained (to a large extent) as the names of forms under the species established in the *Flora of British India.* 2261 2262

Cucumis Melo, *Linn.; Fl. Br. Ind., II., 620; Cogniaux, in DC., Mono. Phanerog., III., 482;* CUCURBITACEÆ. 2263

The SWEET MELON (**Stewart** and also **Baden Powell** call this the Musk Melon, but by giving it at the same time the name *Kharbuza* they remove the suspicion of **Cucurbita moschata.** The information furnished by these authors under "**C. Melo,** *L.*—musk melon" has accordingly been compiled under this species).

Vern.—*Kharbúja* or *kharbujá, khurbúj* or *kharbuza,* HIND.; *Kharmuj,* BENG.; *Tarbuj,* SANTAL; *Dungra,* C. P.; *Khurbúza,* KANGRA (in Settl. Rept., 25); *Kharabúja, kharbuj, chibúda,* BOMB.; *Chibunda,* MAR.; *Tarbucha,* GUZ.; *Gidhro* (*Sariyu chibhars*—the oil—according to **Sakharam Arjun**), SIND; *Zaghún,* LADAK; *Sarda* or *sirda paliz,* AFGHAN.; *Vellari-verai,* TAM.; *Mulampandu,* "*Karbujá dósa* (according to **Elliot**)," TEL.; *Kharbúzeh,* PERS.; *Kharvujá,* SANS.; *Re-mó,* NAGA. It seems probable that in Bombay *Tarbuja* and *kharbuja* are applied to distinct forms of the melon.

References.—*Roxb., Fl. Ind., Ed. C.B.C., 701; Voigt, Hort. Sub. Cal., 58; Thwaites, En. Ceylon Pl., 127; Dalz. & Gibs., Bomb. Fl., 103, Supp., 36; Stewart, Pb. Pl., 96; Aitchison, Cat. Pb. and Sind Pl., 63; DC., Orig. Cult. Pl., 258; Naudin, Ann. des Scien. Natur., 4th Series, Vol. XI. (1859), 34; Stocks, Account of Sind; Campbell, Econ. Prod., Chutia Nagpur, 63; Elliot, Flora Andhrica, 83; Ainslie, Mat. Ind., I., 218; O'Shaughnessy, Beng. Dispens., 343; U. C. Dutt, Mat. Med., Hind., 171, 297, 298, and 305; Flück & Hanb., Pharmacog., 297; Irvine, Med. Top. Ajmir, 142; Trans. Agri.-Hort. Soc. Ind., I., 52; Kurz, Jour. As. Soc. Beng. (1877), Part II., 102; Jhang Settlement Report, 84, 97, and 98; Baden Powell, Pb. Pr., 347; Atkinson, Him. Dist., 701; Econ.*

The Sweet Melon.

Prod., V., 9-10; Drury, U. Pl., 166; Duthie & Fuller, Field and Garden Crops, II., 51; Lisboa, U. Pl. Bomb., 158, 218; Birdwood, Bomb. Pr., 155, 283; Royle, Ill. Him. Bot., t. 47, p. 220; Firminger, Man. Gard. in India, 189; Kew Off. Guide to the Mus. of Ec. Bot., 70; Simmonds, Trop. Agri., 423.

Habitat.—Extensively cultivated on account of its fruit in the sandy basins of rivers. Said to be a native of North-West India, Baluchistan, and west tropical Africa (*DC.*). **Ainslie** wrote in 1826 that **C. Melo** "has been said to be a native of Calmuc Tartary, an opinion adopted by **Willdenow**; in India it is cultivated by seed brought from Persia (see *Tavernier's Travels in Persia, IV., Chap. II.*), where it is much prized and is called *khurbúzeh*. The Arabians term it *batíkh*. The Dukhanie and Hindustanie name is also *khurbúzah, bacacoy,* also *smagha* (MALAY); *molam pullum* (TAM.); *popone* (IT.)." It includes numerous varieties which present differences both in shape, size, and properties. (For methods of cultivation see under a further paragraph. A good plate of this plant occurs in **Duthie and Fuller's** *Field and Garden Crops.*)

OIL. 2264 **Oil.**—The flattened and elliptic seeds yield a sweet, edible oil. In fact, the seeds of most of the members of the melon, pumpkin, cucumber, and gourd family, contain oil, but the only kinds which are utilised to any considerable extent are those of the Sweet-melon (**Cucumis Melo**) and the Water-melon (**Citrullus vulgaris**). From West Africa large quantities of melon seeds are exported to France. China also does a considerable trade in them, but in India the fruit is chiefly eaten as such, and not allowed to ripen its seeds, and accordingly the supply of melon oil is not extensive.

MEDICINE. Seeds. 2265 **Medicine.**—The seeds are used as a cooling medicine. They are edible, nutritive, and diuretic, and are given in painful discharge and suppression of urine. This may be said of the seeds of all the species of CUCUMIS; and it may thus be doubted if medicinally they are distinguishable. The seeds of **C. Melo,** along with those of **C. utilissimus, Benincasa cerifera,** and **Citrullus vulgaris** are largely sold in mixture all

Mixed seeds. 2266 over India. The natives consider this combination cooling, diuretic, and strengthening. It sells for about 12 annas to one rupee a pound. (Compare with the remarks under *var.* **utilissima.**)

Special Opinions.—§ "Bruised seeds applied to the abdomen in cases of tympanites in children" (*Surgeon-Major J. J. L. Ratton, M.D., M.C., Salem*). "Not only the seeds but the pulp of the fruit is a powerful

Pulp. 2267 diuretic, very beneficial in chronic, and also in acute eczema. I can, from personal experience, recommend those subject to chronic eczema to eat a whole fruit daily when procurable. The seeds, dried in the sun, keep perfectly well in a bottle and should be used when the fresh fruit is not in season" (*Civil Surgeon S. M. Shircore, Moorshedabad*).

FOOD. 2268 **Food.**—From an agricultural point of view this is the most important species of the family. It is extensively cultivated on the sandy banks of rivers. Of the North-West Provinces it has been said—"So soon as the sand-banks are exposed by the falling of the river, operations commence by enclosing small plots with grass fences in order to protect them from the inroad of drifting sand. A plentiful stock of manure is then carried to the spot, and large holes dug at regular intervals throughout the plot, into which the manure is distributed. The melons are sown over the manure in the holes, which act therefore in the same manner as forcing beds. This is the practice in growing melons in the beds of rivers such as the Ganges and Jumna, which consist wholly of white sand. Where the river deposit is of richer quality and contains a mixture of organic matter, a much less amount of manure is required, and it is

FOOD.

reported that occasionally manure is altogether dispensed with. The melon beds commence fruiting in April and continue yielding until they are overwhelmed by the rise of the rivers in June" (*Duthie and Fuller*). The area under melons in the North-West Provinces may be estimated at 23,000 acres annually.

In the Chandwara Settlement Report it is stated that melons are generally grown in the sandy beds of rivers during the hot months. These beds are termed *dungras*, and are cultivated by men of the Dhumur caste only. **Mr. Campbell** remarks of the Santals that they cultivate **C. Melo** during both the cold and hot seasons. **Sir W. W. Hunter**, in his *Orissa*, alludes to the melon, *kharbuj*, as grown in the Puri District. It is also grown in many parts of Bombay and of Poona; for example, it is said:—"The fruit is round, green or yellowish, the skin covered with a net-work of raised brown lines. It is eaten uncooked in a variety of ways." Of the Panjáb, **Stewart** writes:—"Cultivated all over the Panjáb plains, some of the kinds being excellent, especially some of those of Múltan and Jhang. Those of the latter have been compared to the best Egyptian. Those of Kashmír are stated to be rather watery; but **Moorcroft** declares the people fatten on them 'as horses are said to do in Bokhara.' **Vigne** states that the melons of Tibet, where they are grown to 10,500 feet, are small but good. In reality those of Ladák are very similar to those of the plains, &c., but with less flavour. In Afghánistan (where *palís* appears to mean the CUCURBITACEÆ generally, and not melon *fields* as **Mason** puts it), several varieties of melon are extensively grown, and **Davies'** Trade Report states that 300 mule-loads are annually imported thence *viá* Pesháwar. The best known and most valued of these is the *sarda*, which, by express, reaches the North-West Panjáb, in good condition. It has been frequently raised in the Panjáb, but is said speedily to degenerate to the ordinary standard." A long and interesting account of melon cultivation will be found in the Jhang Settlement Report in which the writer says they are sometimes found growing wild. 2269

In Manipúr the melon is cultivated by the Nagas and is of a spherical form with ten segments. The pulp of the fruit is usually sweetish and pleasant, and is eaten by Europeans as well as natives.

CULTIVATION. 2270

Cultivation.—**Firminger** refers to two good forms of melons, one of which—the Afghan—has been alluded to above. He says "the kind which ranks as finest of all, called the *surdah*, is a native of Cabul, and has not, that I am aware, been cultivated with success in any part of India." "The seeds of this kind are at once to be distinguished from those of any other, being fully four times larger." "The next kind, second perhaps only to the *surdah*, and superior to any other with which I am acquainted, is, I believe, also from Cabul. Like the *surdah*, too, it is of the green-flesh sort. It is of a large oval form, with very smooth, pale-green exterior, traced here and there with a delicate network. This succeeded most satisfactorily at Ferozepore, and was the one which I cultivated exclusively. The seeds of this also may be known by the largeness of their size." Quoting from the Agri-Horticultural Society's Journal **Mr. Firminger** gives an account of a melon sent from Buxar by a **Mr. W. H. Bartlett**, who writes "with culture in a manured soil, the smaller of these melons may be grown to a size somewhat larger than a large goose's egg, with a bright yellow rind. The flavour is slightly sub-acid, exceedingly pleasant with the addition of a little sugar. The time for sowing is June, though I think it might be sown earlier in Bengal, say, April and May, and watered. The beds should be raised like those of a tea-nursery, and watered if the weather is dry; it fruits from July to September." A **Mr. Chew**, resident at Entally near Calcutta, advocated a 2271

CULTIVATION.

system by which the Cabul melon might be grown. It was, however, troublesome and expensive though attended with success. The chief features of this system were the selection of an open situation even by growing in gumlahs on the roof of a house; the soil $\frac{1}{8}$th sand to $\frac{7}{8}$th clay; the holes to be 2 feet deep and 2 to $2\frac{1}{2}$ feet in diameter and 4 to 6 feet apart; the compost with which the holes were filled to be half well decomposed horse or cow manure and the remainder earth; to be sown in March, a great point being the steeping of the seeds in warm water for 24 hours; afterwards retaining them in wet ashes or a wet cloth until they sprout; as soon as sprouted to be sown about a foot apart and an inch and half deep; lastly, to be deluged with water every day from sowing until the plants are two inches above ground.

2272 **Mr. Firminger** comments on the watering that it should be withheld when the plants are in blossom, given freely after they set fruit, and withheld again at the time the fruit is ripening. In Persia pigeon's dung has from time immemorial been sought after for manuring melons. Iron in the soil is fatal to melon cultivation. Many writers prefer a stiff clay to a sandy soil. French writers affirm that the fruits produced nearest the root are the best, hence a system of severe pruning is recommended, each shoot from the tap root being allowed to produce only one or two fruits. The melon, like the gourd, cucumber, and other cucurbitaceous fruits, is, in the early stage of its growth, subject to the depredations of a small red beetle. The usual preventatives adopted by the native gardeners is to dust the plants with wood ashes. This must, however, be highly injurious, and since in most cases with age the plants cease to be attacked by the beetle a better course is to cover the seedling plants with a muslin frame.

The following two forms are the cucumber-like plants which, by modern European botanists, are treated as melons, and are not even allowed the position of varieties from the type.

2273 (1) **Cucumis Melo,** *Linn.; var.* **Momordica.**

This form does not appear to be referred to in the *Flora of British India*, but it is one of the most easily recognised of the conditions of **C. Melo.** It is the **C. Momordica,** *Roxb.* (*Fl. Ind., Ed. C.B C., 700*), and which by **Gogniaux** (*in DeCandolle's Mono. Phanerog. III., 484*) has been placed as a synonym along with **C. utilissimus,** *Roxb.*, and **C. aromaticus,** *Royle*, under **C. Melo,** *Linn.*, var. **culta,** *Kurz*. The fruit is cylindrical, quite smooth, not fluted (instead of being like the melon furrowed and spherical-ovoid) but it is frequently mottled. As **Roxburgh** says, the plant is more like the cucumber than the melon, except that it is less scabrous and larger. **Atkinson** remarks: "**C. Momordica,** *Roxb.*, the *phúnt* or *túti* of the plains, and *kachra* (unripe) or *phunt* (ripe) of the hills and sub-montane tracts, appears to be also a mere variety of **C. sativus,** which it resembles in all respects, except that it is less scabrous and larger. It is cultivated in cotton or maize fields as a favourite food and good sub-
2274 stitute for common cucumber, and derives its vernacular name from the fruit bursting as it ripens. Finally, even **Kurz** (in his '*Contributions to a Knowledge of the Burmese Flora*') treats this as a synonym of **C. sativus.**" Thus according to many writers it is not only distinct from the melon but more nearly approaches the cucumber, and so is well worthy of the independent position here assigned to it.

There are several forms, but two are readily recognised—the one grown in the rains and the other in the hot season. The fruit bursts spontaneously when ripe; it is then from a foot to 2 feet long and from 3 to 6 inches in diameter, and weighs 4 to 8lb. The seeds are smaller than

Indian Forms of the Melon.

those of the common melon. A good drawing is given of the plant by **Duthie** and **Fuller** in *Field and Garden Crops.*

Vern.—*Phut* or *phunt* (ripe), *kachra* (when unripe), *tuti*, HIND.; *Phuti*, BENG.; *Kakari-kai*, TAM.; *Pedda-kai, pedda-dosrai*, TEL.; **Dr. U. C. Dutt** says this is the *Ervâru* of Sanskrit writers. **Kurz** in his Report on Pegu gives *Tha khwahumway* as the Burmese.

Habitat.—Cultivated here and there throughout India: **Roxburgh** remarks that in the Carnatic it is a cold season crop. According to **Duthie and Fuller** there are, in the North-West Provinces, about 600 acres under this vegetable. **Firminger** says that it is of the size and form of a large cocoa-nut, perfectly smooth and of a pale yellow colour when ripe. It is cultivated in the same way as the melon.

Oil.—The seeds yield an oil. OIL. 2275

Medicine.—The seeds are used as a cooling medicine. MEDICINE. 2276

Food.—**Roxburgh** writes:—"The fruit is much eaten both by Natives and Europeans; when young they are a good substitute for the common cucumber, and when ripe (after bursting spontaneously) with the addition of a little sugar they are scarcely inferior to the melon, and reckoned very wholesome." FOOD. 2277

(2) Cucumis Melo, *Linn.*; *var.* **utilissima.** 2278

Syn.—C. UTILISSIMUS, *Roxb.*

Vern.—*Kakri, kakni*, HIND.; *Kâhúr*, or *kánkur* (*Kakri*, according to **Firminger**), BENG.; *Kukri*, KANGRA (in Settl. Rep., 25), *Dosray, velliri* (in Man. of Trichinopoly), also *kakkarik*, TAM.; *Kákadi*, BOMB.; *Kákdi*, DECCAN; *Tárkákdi*, POONA; *Karkati*, SANS.; **Sir Walter Elliot** says this is the *pandili dosa* (*dosa* of the Telegús), and that *naka dosa* is a form largely grown by the ryots in their grain fields; *Bazrul-quissá, tukhme-khiyare-daraz, tukhme-khiyarzah, tukhme-khiyár* (the seeds), PERS.; *Takhva*, BURM.

References.—*Roxb., Fl. Ind., Ed. C.B.C., 701; Firminger, Man. Gard. in India, 128; Moodeen Sheriff, Supp. Pharm. Ind., 122; U. C. Dutt, Mat Med. Hind., 171; S. Arjun, Bomb. Drugs, 59; Baden Powell, Pb. Pr., 265; Birdwood, Bomb. Pr., 156.*

Description.—The various writers who have described the Indian melons, cucumbers, &c., give somewhat conflicting accounts of this fruit. **Messrs. Duthie and Fuller** (*Field and Garden Crops*) say: "this is another of the extreme forms or varieties of the melon, differing in the shape of the fruit, and the uses to which it is applied. The fruit varies from short oval or cylindrical to elongate, and is either straight or curved like some varieties of cucumber. Some specimens, grown this year in the Saharanpur Garden, measured over a yard in length. They also vary in colour from dark green to nearly white, usually changing to a bright orange colour when ripe. The seeds, like those of *phunt*, are rather smaller and more slender than true melon seeds. **Firminger** describes the fruit as a "*bright red, prickly* (sic) gourd of the size and form of an ostrich egg. When young of a cylindrical form, and in that state eaten much by Europeans in the North-West Provinces in lieu of cucumbers, being in season long before that vegetable, but not to be compared with it in flavour. The seed is sown in March and the vegetable is in use in the hot season" (Conf. with a further para. on cultivation). DESCRIPTION 2279 Seeds. 2280 Fruits. 2281

Habitat.—Cultivated in Bengal, the North-West Provinces, and the Panjáb during the hot weather and the rains. "The fruit varies from short oval or cylindrical to elongate, and is either straight or curved like some varieties of cucumber. It varies in colour from dark green to nearly white, usually changing to a bright orange colour when ripe" (*Duthie* *and Fuller*). In the Gazetteer of the Khandesh District, Bombay, it is stated of this plant that "It is perhaps the most valuable of the gourd tribe, is alike easy of culture in the field or garden during the rains, and under irrigation during the dry season. It is eaten both raw and cooked, and is considered particularly wholesome." Of Poona it is said that "this melon is usually grown in river-beds in the cold and hot weather. The seed is planted in the moist sand and the plant is manured when about three weeks old. It ripens in about two and a half months. The fruit, which is smooth and about 2 feet long, is much eaten both raw and cooked." **Stewart** remarks of the Panjáb: "This gourd attains 2 or 2½ feet, and is stated to reach the extraordinary length of 5 feet." He adds that it is cultivated throughout the Panjáb plains, but that he has seen it in the Ravi basin at an altitude of 6,000 feet.

CULTIVATION. 2282 **Cultivation.**—"This species of cucumber (*sic*) has fruits from one to two feet long. When in a young state they are covered with soft, downy hairs, and are then of a pale green colour. When fully ripe the colour changes to a brilliant orange. It is a true hot season vegetable and will not succeed, in the North-West Provinces at least, during any other season. It should be sown in the end of February and at any time during March. It prefers a dry loose open soil. After manuring, the ground should be laid out in beds, and three or four seeds sown in patches 3 feet apart. Water should be given once in 10 days" (*Indian Forester, IX., 161*).

OIL. 2283 **Oil.**—The seeds yield an oil. **Roxburgh** describes it as a mild oil which the natives "use in food and burn in their lamps."

MEDICINE. 2284 **Medicine.**—"The seeds of this useful species of CUCUMIS are described as cooling, edible, nutritive, and diuretic, and are used in painful micturition and suppression of urine. Two drachms of the seeds, rubbed into a pulp with water, are given alone or in combination with salt and *Kánjika*" (*U. C. Dutt*). **O'Shaughnessy** says "the powder of the toasted seeds is described as a powerful diuretic, and serviceable in promoting the passage of sand or gravel."

FOOD. 2285 **Food.**—*Kakri* is an important article of food with the poorer classes during the hot weather months. **Roxburgh** gives the following account of the fruit:—"This appears to me to be by far the most useful species of CUCUMIS that I know; when little more than one-half grown, they are oblong, and a little downy; in this state they are pickled; when ripe they are about as large as an ostrich's egg, smooth and yellow; when cut they have much the flavour of the melon, and will keep good for several months, if carefully gathered without being bruised, and hung up; they are also in this stage eaten raw, and much used in curries, by the natives.

"The seeds, like those of the other cucurbitaceous fruits, contain much farinaceous matter, blended with a large portion of mild oil; the natives dry and grind them into a meal, which they employ as an article of diet; they also express a mild oil from them, which they use in food and to
2286 burn in their lamps. Experience, as well as analogy, prove these seeds to be highly nourishing, and well deserving of a more extensive culture than is bestowed on them at present."

Mr. Baden Powell says of this fruit: "It is extensively eaten by natives, who eat the whole, skin and all, raw. Europeans make a salad of it with vinegar, which is very like the cucumber, but has not so much flavour."

2287 **Cucumis sativus,** *Linn.; Fl. Br. Ind., II., 620.*

THE CUCUMBER.

The larger forms of this fruit, but for the spinescent structures on the young state, often closely resembles **C. Melo,** *var.* **Momordica,** and also *var.*

utilissima, more nearly in fact than they approach the melon. Hence a certain confusion in the vernacular names.

Syn.—C. HARDWICKII, *Royle, Ill.*, 147.

Vern.—*Khíra,* HIND.; *Kaknai,* ORISSA; *Sasa, khirá,* BENG.; *Khíra, Khiyar,* PB.; *Kakri,* SIMLA; *Kakri, kankri,* BOMB.; *Kakdi,* MAR.; *Kakari,* GUZ.; *Muhevehri,* TAM.; *Dosa-kaia,* TEL.; *Sante kayi,* KAN.; *Trapusha* (according to **Dutt**), *Sukasa* (according to **Piddington**), SANS.; *Thagwa, tha-khwa-thee,* BURM.

References.—*Roxb., Fl. Ind., Ed. C.B.C., 700; DC., Origin Cult. Pl., 264; Kurz, Jour. As. Soc. Beng., 1877, II., 103; Rheede, Hort. Mal., t. 6; Med. Top. Ajmir, 142; Journ. Agri.-Hort Soc. (1875), V., 40; also IV., Part I., 120; Indian Forester, XIII., 162; O'Shaughnessy, Beng. Dispens., 32; S. Arjun, Bomb. Drugs, 58; Hunter, Orissa, II., 188; Firminger, Man. Gard. Ind., 126; Baden Powell, Pb. Pr., 347; Duthie & Fuller, Field and Garden Crops, 53; Lisboa, U. Pl. Bomb., 159; Birdwood, Bomb. Pr., 283, Plates 51, 52.*

Habitat.—There seems no doubt that the original home of the Cucumber was in North India, and its cultivation can be traced to very ancient times. **Royle's C. Hardwickii** (*Ill. Him. Bot., t. 47*) is now admitted to differ in no essential respect from the cultivated plant, and **Sir J. D. Hooker** states that this is wild from Kumáon to Sikkim. **DeCandolle** affirms that the cucumber has been cultivated in India "for at least three thousand years," but "was only introduced into China in the second century before Christ, when the ambassador **Chang-kien** returned from Bactriana. The species spread more rapidly towards the West. The ancient Greeks cultivated the cucumber under the name of *sikuos,* which remains as *sikua* in the modern language. The modern Greeks have also the name *aggouria,* from an ancient Aryan root which is sometimes applied to the water-melon, and which recurs for the cucumber in the Bohemian *agurka,* the German *gurke,* &c. The Albanians (Pelasgians?) have quite a different name, *kratsavets,* which we recognize in the Slav *krastavak.* The Latins called the cucumber *cucumis.* These different names show the antiquity of the species in Europe. There is even an Esthonian name, *Uggurits, ukkurits, urits.* It does not seem to be Finnish, but to belong to the same Aryan root as *aggouria.* If the cucumber came into Europe before the Aryans, there would perhaps be some name peculiar to the Basque language, or seeds would have been found in the lake-dwellings of Switzerland and Savoy; but this is not the case. The peoples in the neighbourhood of the Caucasus have names quite different to the Greek: in Tartar *kiar,* in Kalmuck *chaja,* in Armenian *karan.* The name *chiar* exists also in Arabic for a variety of the cucumber. This is, therefore, a Turanian name anterior to the Sanskrit, whereby its culture in Western Asia would be more than 3,000 years old." (*Orig. Cult. Pl., 266*). 2288

Oil.—The seeds yield an oil. OIL. 2289

Medicine.—The leaves, boiled and mixed with cummin seeds, roasted and powdered, are administered in throat affections. Powdered and mixed with sugar they are also powerfully diuretic, and are sold in the bazars of Upper India under the same name (*tukhmí khiyárain*) as is given to those of **C. Melo** *var.* **utilissima.** MEDICINE. 2290

"In sunstroke pieces of cucumber are put on the bed so that the patient may breathe moistened air in order to neutralize the heat of his body" (*A Surgeon*).

Food.—There are two primary forms of this species, one a creeping plant cultivated in the fields during the hot season, and the other a climber cultivated near the homesteads during the rains. The hot weather kind has small egg-shaped fruits, and is sown in February and March in any soil, preferably in a rich one, in drills. The rainy season varieties FOOD. 2291

The Cucumber.

FOOD. have much larger fruits, one of a dark green, and the other of a creamy-white colour; both when full-grown change their here to a rusty brown. The area under this variety in the North-West Provinces ranges from 15 acres in Meerut to 153 in Budaun and 183 in Allahabad (*Duthie and Fuller*).

The rainy season varieties are the most common, and are universally eaten by natives of all classes as well as by Europeans. The other varieties are also used as food, being eaten raw or cooked in curry; the small hot weather kind, and those gathered in a young state, and known as gherkins, are made into pickles. It may here be remarked that the word *gherkin* is of West Indian origin, although frequently used in India.
2292 It seems to have been derived from *agherkin*, *agurke*, Dutch, again derived from *al* and *khirgar* (Arabic) for the cucumber. By most European writers the *gherkin* is viewed as obtained from a debatable species—**Cucumis Anguria,** a plant wild or acclimatised only in the West Indies and closely allied to **C. prophetarum** and therefore not a form of **C. sativus.** It is thus probable that the small hot weather cucumber is the true *gherkin*, and if so the further suggestion might be offered that it may after all prove but a peculiar form of **Cucumis sativus.** Most if not all the forms of CUCUMIS and CUCURBITA have, in certain districts of India, hot season and cold season forms. This subject deserves to be more carefully gone into by those who may have the opportunity of doing so, and **Naudin's** experiments in cross fertilizing the two forms of cucumber alluded to above might be tried in addition to the preparation of carefully dried specimens both of the natural and hybridised plants.

CULTIVATION. 2293 **Cultivation.**—These plants are alluded to by many writers, but it is scarcely necessary to repeat all their statements. The following abstract from the *Indian Forester* (written by **Mr. Gollan,** Superintendent, Botanic Gardens, Saharanpur) gives some particulars regarding the cultivation of hot season cucumbers or gherkins:—

"This is a variety of the common cucumber, with small egg-shaped fruit, and is also a true hot season vegetable. In order to keep up the supply until the beginning of the rains, three sowings should be made, one in the end of February, one in the middle, and one in the end of March. It will succeed fairly well in any soil, but prefers a rich one. The seeds should be laid out in drills, one foot apart. The seeds should be sown along both sides of the drill, and if the soil be dry, water should be given immediately after sowing. After germination, water every ten days, but like the *kakrí* this vegetable should not be watered too often." (*Vol. IX., 162.*)

Regarding the rainy season forms **Mr. Gollan** (*Ind. For., IX., 201*) says they have much larger fruits and are more like the English cucumber; there are two forms,—"when in a young state the colour of one is a dark green, and of the other creamy-white; when full grown, both are about a foot long, and the colour changes to a rusty brown. These two, although not equal to the commonest varieties met with in England, are not to be despised. They thrive with little care and are always sure of yielding a crop."

2294 **Firminger,** in his article on Cucumber, deals fully with the two forms of the rainy season plant, but was apparently ignorant of the hot season one or did not view it as a cucumber. Speaking of the rainy season forms, he observes of the bitter sort that it "is of smaller growth and of a creamy-white colour when young, turning to a rusty colour at the ends as it ripens. This answers nearly to the description of the one called the 'White Turkey.' It is the better of the two for stewing, cooked in which

CULTIVATION.

way it affords a very delicious dish during the rains, when so few other vegetables are to be had."

Speaking of the English cucumber he states, that it cannot be grown in India as a rainy season crop like the indigenous forms; but that if sown in October it may be made to yield. This is a point of some interest, since, if derived from the Indian wild stock, cultivation in Europe has completely changed the character of the plant. A writer in 2295 the Agri-Horticultural Society's Journal (*IV., 21*) says, however, that in importing seed of cucumbers, only those grown in the open air should be got; frame cucumbers are useless for India. He recommends that they "should be sown in a box in February covered lightly with leaf mould When they have put out a strong leaf, nip off the stem above the leaf. In two or three days after this operation, they are ready for transplantation." "Disappointment may often be expected with imported seeds. They appear to be ill-suited to this country."

Domestic and Sacred Uses.—Atkinson remarks that "the juice is said to banish wood-lice and fish insects by strewing freshly-cut slices in their haunts." At page 371 of *Vratráj* it is related that *Suth* told the *Kushis*, and *Shiv* told his wife *Párwatti* to worship the plant, as by doing so females do not lose their husbands, or that these survive them. The fruit is cut into thin slices and employed in the worship of snakes on *Shravan shudh* 5th (*Nágpanchmi* day). It is likewise employed in the worship of many other gods" (*Lisboa, U. Pl. Bomb.*, 285).

DOMESTIC. 2296

C. Hardwickii, *Royle*, has been alluded to as most probably only the wild state of the cucumber. At the same time it bears separate vernacular names and is collected and sold for so very different purposes that it deserves an independent notice. It is known as the *air-álu* in Kumáon and the *páhari indráyan* on the hill tracts bordering on the plains. It is often spoken of as hill colocynth and it is used as a substitute for that drug. **Moodeen Sheriff** gives the following additional vernacular names for this plant, some of which may refer to **C. pseudo-colocynthis** :—*Malait-tu-matti*, 2297 TAM.; *Konda-puch-cha*, TEL.; *Varik-kumatti*, MAL. This is very probably the *Kirbut* of Sind, the dried fruits of which are considered emetic, and in small doses are given to children along with honey as a useful stomachic. (Conf. with account of **T. trigonus,** form **pseudo-colycinthis.**)

Cucumis trigonus, *Roxb.; Fl. Br. Ind., II., 619.* 2298

Syn.—C. PSEUDO-COLOCYNTHIS, *Royle.;* C. TURBINATUS, *Roxb.;* C. MADERASPATANUS, *Roxb.;* C. MELO, *Linn.*, var. AGRESTIS, *Naud.;* C. PUBESCENS, *Wall.;* C. ERIOCARPUS, *Boiss.;* BRYONIA CALLOSA, *Herb. Rottler.*

These are the synonyms given in the *Flora of British India*, but practically all the names given by the old authors for the Indian so called wild species of CUCUMIS, are now reduced to synonyms of **C. trigonus,** *Roxb.*

A slight modification of this view has been since advanced by **Gogniaux** (*in DeCandolle's Mono Phanerog., III., 482*), where certain of the above are referred to **C. Melo,** and the others left under **C. trigonus.** This may be indicated thus :—

C. Melo, *Linn.* 2299

Var. α agrestis, *Naud.;* SYN. C. MELO, *var.* PUBESCENS, *Kurz* (*Trans. Asiat. Soc. Beng., 1877, part 2, p. 102*); C. PUBESCENS, *Willd.; W. & A. Prod., I., p. 342; Wight, Ic., 496; Royle, Ill. Him. Bot., 220, table 47;* C. MADERASPATANUS, *Roxb.;* C. CICATRISATUS, *Stocks; in Hook. Kew Jour. of Bot., 4, p. 148;* C. ERIOCARPUS, *Boiss.;* C. TRIGONUS, *Benth., non-Roxb.*

Var. β culta, *Kurz;* SYN. C. DUDAIM, *Linn.;* C. FLEXUOSUS, *Linn.; W. & A. Prod., 342;* C. AROMATICUS, *Royle, Ill. Him. Bot., pl. 2, p. 220.*

Wild Forms of Cucumis.

C. UTILISSIMUS, *Roxb.*; *W. & A., Prod.*, 342; C. MOMORDICA, *Roxb.*; (*Conf. with syns. given under* C. MELO & C. SATIVUS).

2300 If this view be accepted a certain amount of countenance might be inferred as given to the possibility of **C. Melo**, *Linn.*, having been derived from some other plant than **C. trigonus**. The Indian wild plant, which perhaps most nearly approaches the melon, is that described by **Roxburgh** as **C. maderaspatanus**, and by **Wallich** as **C. pubescens**. But the subject is too complex for the writer to deal with it at present, further than to exhibit the opinions of the most recent authors. It may, however, be added that the natives of India recognise as distinct many of the plants indicated by the above botanical names or synonyms. Without attempting to dispute the conclusions arrived at by systematic botanists, it may therefore serve a practical or industrial purpose to refer to some of the old Roxburghian species and to give the various vernacular names that are in use for them in India, and, where possible, to indicate their economic properties. It may also be admissible in passing to suggest that the following forms may have been the sources of **C. Momordica**, **C. utilissimus**, and some of the forms of **C. sativus**, but probably had little to say to the production of **C. Melo**, provided the claims of **C. maderaspatanus**, *Roxb.*, be excluded from consideration, as the wild state of **C. Melo**, proper.

2301 1. **Cucumis trigonus**, *Roxb.*

Vern.—*Pam-budinga* (Roxburgh) and *Pulcha* (Elliot), TEL.

2302 **Botanic Diagnosis.**—This, as **Roxburgh** says, resembles most nearly **C. utilissimus**. It is never cultivated, nor is it eaten. The fruit is oval, smooth, distinctly three-sided, with the angles round and the surface streaked, with ten light and ten deep-shades of yellow.

Habitat.—The mountain tracts of Coromandel, Central Bengal, Central Provinces, and the Panjáb: flowering time the wet and cold seasons.

OIL. 2303 **Oil.**—**Dr. Ainslie** remarks that the seeds yield a fixed oil by boiling, which is used for lamps by the poorer classes. **Lieutenant Hawkes** reports that it is used for burning in lamps in some parts where the fruit abounds.

"It is extracted by boiling in water, and is procurable only in small quantities" (*Cooke*). It has been found impossible to discover to which of the plants here discussed under **C. trigonus** these notes regarding an oil obtained from the seeds of a wild CUCUMIS more especially refer.

§ "First rate oil made with the fruit for asthma" (*V. Ummegadien, Mettapollian, Madras*).

2304 2. **C. turbinatus**, *Roxb.*

Vern.—*Nulla-budinga* (Roxburgh) and *nalla-budama* (Elliot), TEL.

2305 **Botanic Diagnosis**—It is very much like **C trigonus**, but the leaves are more deeply 5-lobed and the segments bristle-toothed. It is at the same time a smaller plant, with larger flowers, and a pyriform, maculated 3-cornered, smooth fruit, which is regularly eaten.

Habitat.—According to **Roxburgh** this is a native of the same region as **C. trigonus**, and it is probably only a form of that plant and semi-cultivated.

2306 3. **C. maderaspatanus**, *Roxb.*

Syn.—C. PUBESCENS, *Wall.*

Vern.- *Ban-gumak, gomuk*, BENG.; *Takmaki*, BOMB.; *Chiber*, SIND., *Kachrí* (Stewart), *Kakrí* (Baden Powell), but *Kakri* is also **C. utilissimus** in the Panjáb. *Kodí-bu-dinga* (*Kódi-budama*, according to Elliot, who calls it also Fowl's Cucumber) TEL.; *Gong-kakirí*, SING.; *Garákshi vrikshamu* (Elliot), *Godumbá* (Dutt), SANS.

2307 **Botanic Diagnosis.**—This is almost intermediate in type between **C. Momordica** and some of the forms of **C. sativus**. The leaves are less deeply lobed than are those of **C. trigonus** or **C. turbinatus**, and in fact are almost reniform and often only 3-angled. The fruit is smooth, hairy when young, becoming hirsute with small bristles, is often mottled in colour but never 3-sided. In some respects this fruit approaches more closely to the melon proper than do the fruits of any of the other wild species met with in India.

Habitat.—Coromandel, Bengal, the North-Western Provinces, the Panjáb, Bombay, and Sind. Sometimes cultivated in Western India and Sind, and the fruits sold in the markets. **Roxburgh** says: "The fruit of this sort is used in food by the natives and much esteemed, yet they never take the trouble to cultivate the plant." **Atkinson** states of the North-West Provinces, that "**C. pubescens**, the *kachri* and *ban-gumak* of these provinces, occurs wild, and is occasionally cultivated and eaten raw or cooked. **Stewart** remarks of **C. pubescens** (*kachri*) that it occurs wild in the Panjáb plains and that the small fruit is eaten. He appears to be alluding to this plant. 2308

MEDICINE.

Medicine.—It is considered cool and astringent; it creates appetite and removes bilious disorders (*Baden Powell*). **Dr. Dymock** (*Mat. Med. W. Ind., 2nd Ed., 339*) is apparently alluding to this plant when under **C. trigonus**, *var.* **pubescens**, he says it is much less bitter than *var.* **pseudo-colocynthis**, "and is commonly used as a vegeteble after having been soaked in salt and water; the seeds of these cucumbers (*sic*) are considered cooling and are applied to Herpes, after they have been beaten into a paste with the juice of the *Darva* (**Cynodon Dactylon**)." 2309

4. **Cucumis pseudo-colocynthis**, *Royle.* 2310

Syn.—C. PUBESCENS, *Willd.*; C. ERIOCARPUS, *Boiss.*; C. CICUTRISATUS, *Stocks.*

Vern.—*Indrayan* (= colocynth), *bislúmbhi* in Northern India (**O'Shaughnessy**); *Karit*, BOMB.; **C Hardwickii** (see *ante*) is known as *páhári-indrayan*, BUNDELKHAND; *Bislombhi* of the bazars, N. W. Provinces. **Moodeen Sheriff** gives the South Indian names for what appears to be this plant:—*Hattut-tumatti*, TAM.; *Adavi-puch-cha*, TEL.

Botanic Diagnosis.—A prostrate, very scabrous plant, with gland-like hair-bearing tubercles; flowers solitary, females on long peduncles; fruit oblong and smooth, marked with eight broad stripes. 2311

From some of the descriptions of this plant the suspicion is created that it may prove a wild form of **C. sativus** (or rather of **C. Hardwickii**) or be intermediate between that and **C. prophetarum**, *Linn.*

Habitat.—A perennial met with throughout the Deccan and Sind to Baluchistan, Kashmir, and Afghanistan.

MEDICINE.

Medicine.—Pulp of the fruit is very bitter and similar in quality to colocynth for which according to **O'Shaughnessy**, it is substituted. "**Dr. Gibson**, however, expresses a doubt as to the correctness of the opinion" regarding the purgative properties of this plant and of **C. Hardwickii**. "Experiments are required to determine the point. According to the report of **Dr. J. Newton**, a decoction of the roots of these plants is used as a purgative; it is stated to be milder in its operation than the pulp of the fruit, and to cause less irritation." (*Pharm. Ind., 96.*) **Dr. Dymock** is apparently referring to this plant when he says. "*Kárit* is very common in the Bombay Presidency. The fruit is of the size and shape of a small egg, and marked with green and yellow streaks like colocynth. It is very bitter, and at the feast of the *Diwali*, or new year of the Hindus, is brought to market for sale. The Hindus of Bombay have a custom at this season of breaking the fruit under the foot and then touching the tongue and forehead with it, with the idea that having tasted bitter of their own accord, they may hope to be preserved from misfortune during the year. It is not eaten, but is used medicinally in the same way that **Citrullus amarus** is used in Sind." Could it be that the **Citrullus amarus** alluded to is **Cucumis Hardwickii**? 2312

§ "Is a purgative no doubt. The fruit is often given to horses as a purgative" (*Surgeon Major C. W. Calthrop, M.D., Morar*). 2313

"**Cucumis tuberosus**, *Heyne*," is the heading of an article in *Balfour, Cyclopædia of India.* The writer has failed to discover the plant referred to. It is said to be the *Adulay-kai, nellay piku*, TAM.; *Casara kaia*, TEL. Clearly it is not a CUCUMIS and doubtfully belongs even to CUCURBITACEÆ. The account of the plant is said to be abstracted from **Roxburgh** and from the **Rev. J. F. Kearns**, the

2314 latter in a paper which Balfour says appeared in the *Agri-Horti. Soc., Proc., 1862.*" While Mr Kearns's paper cannot be discovered in the Proceedings of the Society quoted, the information given by Balfour seems of sufficient importance to be briefly indicated here. The plant appears to be found in North Tinnevelly and to yield a tuber from which "a flour" is prepared. "A coolie load of tubers gives six large measures of fine flour, conidsered by the natives a most excellent bread stuff." "The fruit, a small capsule used in sweet-meats, is known as the *Adully*." The vernacular names above appear in *Ainslie's Materia Medica* (pub. 1813), but are not reproduced by him in his second edition of the *Materia Indica* (pub. 1826); they also occur in Forbes Watson's "Index to the names of Eastern Plants," the reference being given to *Ainslie's Mat. Ind.* Could it be possible that a Dioscorea is meant? [*Ed.*]

2315 **CUCURBITA,** *Linn.; Gen. Pl., I., 828.*

The very greatest confusion exists in the Indian publications that deal with Gourds, Pumpkins, and Vegetable Marrows; even much more so than has been indicated regarding the Melons, under CUCUMIS. All the forms met with exist in a state of cultivation only. It seems likely that in most provinces of India, **Cucurbita maxima, C. moschata,** and **C. Pepo** are grown, with **C. moschata,** in all probability, the most abundant. It has been found impossible, however, to furnish a satisfactory account of each species, and the information given below, as well as the vernacular names, will most probably have to be materially re-arranged, in which case **Lagenaria vulgaris** (the common Gourd) and **Benincasa cerifera** (the white Gourd), will have to be taken into consideration. **DeCandolle** seems to incline to the opinion that **Cucurbita maxima** may be a truly Asiatic species and the origin of "the pumpkins cultivated by the Romans, and in the Middle Ages" in Europe generally; but that **Cucurbita Pepo** is most probably a native of America, having been the source of all the American gourds and pumpkins that existed anterior to the discovery of America. **M. DeCandolle** has not ventured to assign a habitat for **C. moschata,** although he states that all writers on Asiatic and African Botany describe it as cultivated, and that "Its cultivation is recent in China, and American floras rarely mention the species. No Sanskrit name is known, and the Indian, Malay, and Chinese names are neither very numerous nor very original, although the cultivation of the plant seems to be more diffused in Southern Asia than in other parts of the tropics" (*p. 257*). The following attempt to produce a compilation of available literature regarding the Indian cultivated CUCURBITÆ fully bears out **DeCandolle's** conclusions. **C. moschata** is probably the most extensively grown, but the names for **C. maxima** seem more accurate, while those for **C. Pepo** are quite misleading, most of them probably referring to **Benincasa cerifera,** including Roxburgh's Sanskrit name *kurkarú.*

Cucurbita Citrullus, *Linn.;* see **Citrullus vulgaris,** *Schrad.;* CUCURBITACEÆ. C. 1221.

C. lagenaria, *Linn.;* see **Lagenaria vulgaris,** *Linn.*

2316 **C. maxima,** *Duchesne; R. Br., II., 622.*

MELON-PUMPKIN, SQUASH GOURD, RED GOURD.

The name GOURD is sometimes given to the fruit of this plant, but that is more correctly the name of **Lagenaria vulgaris.**

Vern.—*Mithá-kaddú, kadú,* HIND.; *Lal-bhopali, lal-dudiya,* BOMB.; *Tookm kudú,* KANGRA (Sett. Rep.); *Gaduwa,* N.-W. HIM.; *Pushini,* TAM.; *Gummadi,* TEL.; *Mattanga,* MAL.; *Kumbala,* KAN.; *Shwé-pay-on,* BURM. Sir Walter Elliot gives *Erra gummadi* as the Telegu for this fruit, and

Bagalá as Sanskrit for a species of CUCURBITA. He says that *Bagalá* does not occur in **Wilson's** Dictionary, but that *its* synonym *Karka* does; and that *Karka* is there defined as a "long gourd." **Moodeen Sheriff** gives the following names for **C. maxima**—*Mithá-kaddú*, DUK.; *Pushinik-kay*, TAM.; *Gummadi-káya*, TEL.; *Mattanga*, MAL.; *Kumbala-káyi, kumbala-hannu*, SING.; *Saphúri-komra*, BENG.

Botanic. Diagnosis.—Leaves, 5-palmate; lobes rounded, sinus, narrow; petiole, nearly as long as the blade, not prickly; fruiting peduncle, round smooth; corolla lobes, curved outwards; calyx segments, lanceolate-linear. 2317

Habitat.—Cultivated in India, and in most warm and temperate parts of the globe. This does not appear, however, to be so plentifully met with as the musk-melon. **Messrs. Duthie** and **Fuller** say that they have failed to find either **C. maxima** or **C. Pepo** in the North-Western Provinces. On the other hand a writer in the *Indian Forester* (*IX.*, *202*), and apparently, **Mr. Gollan** of Saharanpur, says—"Kudu (pumpkin) **Cucurbita maxima**" is a rainy season vegetable, "common in gardens." Much confusion exists regarding the gourds of India. Attention to the brief diagnostic characters given for the three species here dealt with, would enable district officers to speak with more certainty regarding the species grown in their respective parts of India. **Atkinson, Dutt**, and several other authors confuse the Pumpkin (**C. Pepo**) with the White Gourd (**Benincasa cerifera**). **Roxburgh** describes only two of the three species (**C. Pepo** and **C. moschata**), and **Voigt**, who wrote after **Roxburgh**, describes only **C. maxima**, to which he reduces **Roxburgh's C. Melopepo**. **Stewart** gives an account of all three plants collectively under **C. maxima**. 2318 2319

Oil.—The seed yields an oil. OIL. 2320

Medicine.—The seeds are used medicinally; the oil as a nervine tonic. The pulp of the fruit is often used as a poultice. MEDICINE. 2321

§ "Also called in Panjáb *Ghiá kaddu*. The fruit cut into small circular chips is a good application to relieve the burning of hands and feet in fevers." (*Asst. Surgeon Bhagwan Dass* (*2nd*), *Surgeon, Rawal Pindi, Panjáb*). "The pulp is used as a poultice to boils and carbuncles." (*Native Surgeon T. Ruthnam Moodelliar, Chingleput, Madras*) Hospital Assistant **Gopal Chunder Gangoolee** says that "he has used the boiled pulp of the fruit as a poultice, for unhealthy ulcers, with good effects." (*Asst. Surgeon Anund Chunder Mookerji, Noakhally*). "The part of the fruit stalk in *immediate* contact with the ripe gourd, is removed and dried, and when made into a paste by rubbing in water, is considered a specific for bites of venomous insects of all kinds, chiefly for that of the centipede." (*Honorary Surgeon P. Kinsley, Chicacole, Ganjam*). Vide "*Practitioner*," August 1878, Vol XXI., p. 128, quoting "*Medical Examiner*," June 13, 1878. "The dose recommended is an ounce and a half beaten up with sugar. I have tried pumpkin seeds such as are sold in Calcutta as a vermifuge on one patient, a European male adult. He took 4 or 5 ounces without any effect whatever except distention of the abdomen." [NOTE.—The above remark, which was furnished by a Surgeon, should most probably appear under **C. Pepo**.—*Ed.*]

Food.—This plant produces the largest known cucurbitaceous fruit, in some cases weighing as much as 240℔, and measuring nearly 8 feet in circumference. The fruit is wholesome, and when young is used as a vegetable. It is sweetish and yellow. When mature it will keep for many months if hung up in an airy place. It is largely used by natives of all classes in curry. "When very young and tender it may be employed as a pleasant vegetable for the European table, by being boiled, press- FOOD. 2322 ed down to extract the water, and served warm, with butter, salt, and pepper" (*Mr. L. Liotard*).

Mr. Gollan says of "*kudu* (pumpkin) **Cucurbita maxima**" that there are several varieties of this plant common in the gardens as a rainy season vegetable. The commonest one is a large globular gourd and of a brown colour. The young fruit resembles the vegetable marrow in flavour but the full grown fruit is also very good. The seeds should be sown from April to June. The plant requires very rich soil and the general treatment is the same as that for **Lagenaria vulgaris** (the *Al kudu.*)"

2323 **Firminger** remarks of the "Red Gourd" or *sufurí-kúmra,* also *Lál-kúmra,* "that it is a brownish-red, globular-shaped, bluntly-ribbed Gourd of enormous size, cultivated extensively by the natives for sale in the bazárs, where it is cut up and sold in slices; in my opinion the most agreeable form of any of the Indian Gourds. Dressed and cooked with boiled beef, as carrots are, it can hardly be distinguished from them either in appearance or flavour. An annual: seed sown in the rains; vegetable in use during the cold season; not often cultivated in gardens." It may be suspected that **Firminger** alludes in the above to **C. moschata** (forma **Melopepo,** *Roxb.*,) and not to **C. maxima.**

2324 The confusion between this fruit and that of the common Gourd (**Lagenaria vulgaris**) should be guarded against. Most Indian writers seem to prefer to call **C. maxima** the Gourd, and **Lagenaria vulgaris** the Bottle-Gourd. In the Settlement Report, Kumáon District, "**Cucurbita maxima** (pumpkin)" is called *Gudúa,* HIND. In another part of the same report and under the same scientific name occurs *Turbúza,* HIND., while "**Cucurbita Pepo** (Pumpkin)" is called *Bhuja,* HIND.

2325 Cucurbita moschata, *Duchesne; Fl. Br. Ind., II., 622.*

THE MUSK MELON, *Eng.;* POTIRON, *Fr.*

Syn.—C. MELOPEPO, *Roxb.*

Vern.—*Sitaphal, saphari kumhra, kumra, kaddú, mitha-kaddú,* N.-W. P.; *Kali-dudhi,* BOMB.

This is said to be the *Abobrade Guinea* of the Portuguese in India.

2326 **Botanic Diagnosis.**—Leaves as in the preceding but very often marbled with whitish blotches: petiole hairy but not prickly: fruiting peduncles angular and furrowed; calyx segments of the female flower large foliaceous.

There are two primary forms—one with the fruit smooth but mottled brown and yellow (**C. moschata** proper), and the other with the fruit torulose or fluted, with 15 to 30 ridges (**C. Melopepo,** *Roxb.*)

2327 **Habitat.**—Very extensively cultivated throughout India by the natives. As stated above there are two forms of the fruit—one smooth and somewhat oblong in shape, the other fluted and flattened spheroidal. It seems probable that the latter (the **Melopepo** of **Roxburgh**) is by many Indian writers described as **C. maxima.** The long account given by **Firminger** (*Man. Gar. for India, 128*) under the heading "**C. Melopepo,** squash" has reference to imported seed of Squash, Gourd or Vegetable-marrow, and not to the Indian cultivated fruit, **C. moschata.** He says in Bengal it should be sown in October but in the North-West Provinces not before the end of February, as the plants will not live in the cold season of these provinces. **Messrs. Duthie and Fuller** (in *Field and Garden Crops, Part II., LVII. to LX.*) give an account of **Cucurbita moschata,** but do not mention any facts regarding methed of cultivation, season, &c. They state that only the **Cucurbita** there figured appears to occur in the North-West Provinces. Their plates seem to represent the form **Roxburgh** called **C. Melopepo** and not his **C. moschata** proper, if the idea be correct that the fluted fruit is **C. Melopepo.**

Oil.—The seed yields a mild, bland, pale-coloured oil. OIL. 2328

Food.—The yellow flesh of this fruit is extensively cooked and eaten as a vegetable throughout India. There is what appears to be a form of this fruit grown in some parts of the Panjáb and North-West Provinces and known as *tendús* of Bijnor and *tendu* of the Duáb (*Atkinson*), *tindú* of the Panjáb. Regarding *tindú* **Mr. Baden Powell** says: "*tindú* (**Cucurbita tobata**?), a small round gourd when young, at which time it makes a most delicious vegetable for the table; the fruit is not bigger than a small turnip." The writer saw in the **Naga hills** a form of what appeared **C. Melopepo** which would have answered to **Mr. Powell's** description of *tindú*. FOOD. 2329 2330

Cucurbita Pepo, *DC.; Fl. Br. Ind., II., 622.* 2331

THE PUMPKIN, VEGETABLE MARROW.

Syn.—C. PEPO, *Roxb.*

Roxburgh included this plant (the pumpkin) as well as **Benincasa cerifera,** *Savi* (the white melon) under one species. **Atkinson, Drury, Dutt, Moodeen Sheriff,** and other writers have fallen into the same mistake. The two plants may be readily distinguished by observing the stamens:—in **Benincasa,** they are inserted near the mouth of the tube, anthers not united: in **Cucurbita,** the stamens are inserted below the mouth, and the anthers are more or less united. The fruits of **Benincasa** are cylindrical, 1–1½ ft. long, without ribs, at first hairy, then covered with a waxy bloom.

Vern.—*Kumra, safed kaddu, lanka, konda, kúmara, kadímah,* BENG., HIND.; *Kaula,* BOMB.; *Kohala,* MAR.; *Kumbala kagi,* KAN.; *Petha,* KANGRA; *Kurkarú,* SANS. (according to **Roxb.**); *Petha, bhunja, kumhra,* KUMAON (according to **Madden**); *Bhunga, petha* (N.-W. Himálayas), (according to **Atkinson**). **Sir Walter Elliot** says "**Cucurbita Pepo,** *Linn.*" is the *Potti gummadi,* and *búdadegummadi,* TEL.; and **Sir W. W. Hunter** that it is the *Páni-kakharu* of Orissa.

It is impossible to separate the vernacular names which belong to this plant from those applied to **Benincasa cerifera.** (*Conf. with that species,* B. 430.) **Moodeen Sheriff,** for example, gives under "**C. Pepo,** *Roxb.*," a long list of names, most of which, in all probability, refer to **Benincasa cerifera**; his Sanskrit name *Kúshpandaha* or rather *Kushmánda* certainly does.

Botanic Diagnosis.—Leaves 5-palmate, sinus, broad and segment pointed; petiole as long as the blade, the hairs of the lower surface hardened into prickles; corolla narrow towards the base, and lobes erect calyx-segments, linear-lanceolate; fruiting peduncle, woody, strongly grooved, and marked with ridges.

Habitat.—Cultivated for its fruit throughout the greater part of India. Grown in vegetable gardens, and near the huts of the natives; often allowed to spread over the roofs of their houses.

The vegetable marrow appears to be the form of this plant called by some botanists **Cucurbita ovifera.**

Oil.—The seeds yield a clear edible oil. OIL. 2332

Medicine.—The seeds are supposed to possess anthelmintic properties. The *Pharmacopœia of India* advocates trials to be made in order to establish the exact properties of these seeds. **Atkinson,** in the *Himálayan Districts,* says the leaves of this plant, as also of **C. maxima,** are used as external applications for burns. MEDICINE. 2333

MEDICINE. **Special Opinions.**—§ "The seeds are anthelmintic and used in cases for round worms though uncertain in action" (*Civil Surgeon J. H. Thornton, B.A., M.B., Monghyr*). "**Grubler** has isolated from pumpkin seeds a crystallisable variety of albumen. Hemp and castor oil seeds also contain a similar crystalline substance" (*Prof. Warden, Calcutta*).

FOOD. 2334 **Food.**—Very little more can be learned regarding the pumpkin than has been given above. It is very much to be feared that many writers on the subject have not only confused this fruit with that of **Benincasa cerifera**, but also with **Cucurbita moschata.** An Official Note on the condition of the people of Assam contains the following:—"**Cucurbita Pepo,** *Willd., Kumra*—A creeper constantly grown over the roofs of houses all over the valley. The fruit, which is large and heavy, is usually stored on the angle of the roof, and it is common to see rows of these gourds ripening along the tops of the houses. The fruit is eaten, cut up into small pieces and

Boiled. 2335 boiled with salt or in *khár* water, or fried in oil. The young tops of the tender shoots are also sometimes fried in oil or boiled in *khár* water. There are two varieties of this plant growing and used in the same way, but differing slightly, one called *boga kumra*, and the other *ranga kumra* or *chal kumra*." It is to be feared this passage refers to either **Benincasa cerifera** or **Cucurbita moschata.** The writer does not recollect ever having seen the pumpkin (**C. Pepo**) in Assam, although the two fruits named are common in Assam, Cachar, and Manipur. The system of boiling in *khár* water is, however, very interesting to whichever fruit it applies, and so

Twigs. 2336 also is the fact that the young twigs are eaten as a pot-herb. Under the names "**C. Pepo,** *DC.*, pumpkin or white Gourd—*kumhra, kúmara, kadimah, peth* (in places), *kondu*, the *lauka* and *kaddu safed* of Bijnor," **Mr. Baden Powell,** and after him **Mr. Atkinson,** record an interesting fact which most probably should be given under **Benincasa cerifera:** "A

Sherbet. 2337 sherbet is made by filling the hollow centre with sugar and exposing it to the sun until it becomes acid."

DOMESTIC. 2338 **Domestic and Sacred Uses.**—The *Vrat Kaumudi* recommends the worship of this plant, considering it a goddess. "*Dharmráj* tells *Krishna*, and *Narad* priest of the gods tells King *Chandrasen*, to observe the *Vrat* of this cucurbitaceous plant (*vide page 370 of Vratraj* in selections taken from *Padma Purán*). Its fruit is also cut with some ceremony, called *kohala muhurt*, a day or two before a marriage" (*Lisboa, U. Pl. Bomb.*, 285).

CUMINUM, *Linn.; Gen. Pl., I.,* 926.

2339 **Cuminum Cyminum,** *Linn.; Fl. Br. Ind., II.,* 718; UMBELLIFERÆ.

CUMIN, *Eng.*, the Κύμινονημερον of DIOSCORIDES; CUMINUM of HORACE and PERSIUS.

Vern.—*Zirá*, HIND.; *Jiraka, jiraka* or *ajáji* (**Ainslie**), "*Jiraka, jirana*" (**Elliot**), SANS.; *Jirá*, BENG.; *Jirú, jira-utmi*, GZ.; *Jire gire*, MAR.; *Kamún*, ARAB.; *Zirá*, PERS.; *Zéro*, SIND.; *Shiragam*, TAM.; *Jiraka*, "*jilakarra*" (**Elliot**), TEL.; *Jiringe, jirage*, KAN.; *Jirakam*, MAL.; *Duru, sudu-duru*, SING.; *Ziya*, BURM.

A considerable amount of confusion exists in the vernacular names for this plant, *Zira* or *Jira* being also applied to **Carum Carui** (See **C. 681**).

The Black Cummin of the Bible—*Melantpion* of **Hippocrates** and **Dioscorides,** and the *Gilt* of **Pliny** is **Nigella sativa.**

Habitat.—More or less cultivated in most provinces of India, except perhaps Bengal and Assam. There seems no doubt the plant is not a native of India. **Roxburgh** is silent on this point, but **Ainslie,** who wrote

about the same period says of the Calcutta Botanic Gardens (which were then under **Dr. Roxburgh**) that "the plant, however, is growing in the Botanical Gardens of Calcutta, introduced from Persia." Immediately before making this remark, the tone of which would imply that it was little known personally to either **Roxburgh** or **Ainslie** himself, he says—"Properly speaking the plant is a native of Egypt, but is cultivated now in India, though I am inclined to think that the greater part of the seed found in the bazárs is brought from the sea ports of the Red Sea." Passing over the period from 1826 to 1869, **Stewart** is the next Indian author of any importance who alludes to the plant. He states—"**Bellew** mentions this as being wild in the hills north of the Pesháwar Valley, and **Aitchison** states that it is commonly wild in Lahoul (10,000 feet), whence the seeds are largely exported towards the plains." **Davies'** Trade Report gives 500 2340
maunds as annually imported from Afghánistan through the Bolán Pass, but, *1st*, the name *Zíra* has probably caused CARUM to be mistaken for this; and *2nd*, the quantity seems enormous. The same authority also gives 25 maunds as exported by that route. **Atkinson** makes no mention of the plant in his "Kuram Valley, &c., Afghánistan Flora," but in his Catalogue of the Panjáb and Sind plants he says it occurs as an escape in Ferozepur and Jhelum. **Atkinson**, in his *Himálayan Districts of the North-West Provinces* (*also Econ. Prod. N.-W. P., Pt. V.*), says that the plant "appears wild in the hills 7—9,000 feet, and cultivated in the plains." There would thus seem some grounds for believing that in certain parts of India Cumin has become acclimatized, but we are justified in accepting the opinion expressed, both in **Hooker's** *Flora of British India*, and in **Bentham** and **Hooker's** *Genera Plantarum*, that it is a native of the Mediterranean regions, though cultivated in most warm countries on account of its aromatic seeds. The uniformity in which, through all the languages of India, the Sanskrit name *Jíraka* and the Persian *Zíra* re-appear, points at least to a common centre from which the plant spread over the east.

References.—*Roxb., Fl. Ind., Ed. C.B.C., 271; Voigt, Hort. Sub. Cal., 23; Dalz. & Gibs., Bomb. Fl., Supp., 41; Stewart, Pb. Pl., 105; Stocks, Account of Sind; Elliot, Flora Andhrica, 74; Pharm. Ind., 108; Ainslie, Mat. Ind., I., 100; O'Shaughnessy, Beng. Dispens., 1369; U. C. Dutt, Mat. Med. Hind., 172; Dymock, Mat. Med. W. Ind., 2nd Ed., 369; Fleming, Med. Pl. and Drugs, as in As. Res., Vol. XI., 165; Flück & Hanb., Pharmacog., 331; S. Arjun, Bomb. Drugs, 63; Murray, Pl. and Drugs, Sind, 199; Med. Top., Ajmir, 153; Baden Powell, Pb. Pr., 301, 351; Atkinson, Him. Dist., 705, 734; Econ. Prod., N. W. P., V., 30; Drury, U. Pl., 167; Lisboa, U. Pl. Bomb., 161; Birdwood, Bomb. Pr., 40, 221; Bomb. Gaz., III., 54; Bomb. Man. Rev. Acct., 103; Man. Caddapa Dist., 43; Settlement Report, Chanda, C. P., 83.*

Oil.—A medicinal oil is prepared from the seeds (=fruits). OIL. 2341

Medicine.—As a medicine Cumin seeds are considered aromatic, carminative, and stimulant. They are also stomachic and astringent, and useful in dyspepsia and diarrhœa. The *Pharmacopœia of India* says: "The fruit, officinal in the *London Pharm.*, are met with in bazárs throughout India, being much in use as a condiment. Their warm bitterish taste and aromatic odour reside in a volatile oil. Both fruit and oil possess carminative properties analogous to Coriander and Dill, but are comparatively rarely used in medicine by Europeans, though much valued by the natives." **Ainslie** says: "Cumin seeds are in very general use amongst the native Indians, equally as a grateful stomachic in cases of dyspepsia and as a seasoner for their curries." "It is thought to be very cooling, and on that account forms a part of most prescriptions for gonorrhœa. It is also used as an external application to allay pain and irritation. Arabian and Persian writers describe four kinds of *Kamún, viz.*, Farsi MEDICINE. 2342

MEDICINE. or Persian, Nabti or Nabathean, Kirmáni or black Cumin, which they say is the Basilikon of the Greeks and Shánú or Syrian. They consider it to have the same properties as the caraway" (*Dymock*). **Dutt** says that the Sanskrit authors recommend "a poultice made of cumin seeds with the addition of honey, salt, and clarified butter" to be applied externally for scorpion bites.

Special Opinions.—§ "Used as carminative and stomachic, half drachm doses, in combination, never alone (*Assistant Surgeon Nehal Sing, Saharunpore*). "Seeds mixed with lime-juice are used in bilious nausea in pregnant females" (*Surgeon-Major J. J. L. Ratton, M.D., M.C., Salem*). "*Safaid zira* is taken internally shortly after child-birth to increase the secretion of milk" (*Civil Surgeon R. Gray, Lahore*). "A quantity of the seeds lightly smeared with *ghi* put into a pipe and smoked relieves hiccup" (*Surgeon-Major D. R. Thomson, M.D., C.I.E., Madras*). "A reputed galactagogue." "Practitioner" (Nov. 1881, Vol. XXVII., p. 385, and p. 164 (quoting *Lancet*, 1872) however denies this action" (*G. B.*)

CHEMISTRY. 2343 **Chemistry.**—The chemistry of cumin has been dealt with fully by **Flückiger** and **Hanbury** (*Pharmacog., 332*), and their account reproduced in **Dymock's** *Materia Medica* (*2nd Ed., 369*). It is not necessary therefore to repeat the information there given, since either of the works referred to is likely to be in the hands of the student of Indian Materia Medica. **Professor Warden** has, however, contributed the following brief note for the present publication:—

"The fruit contains an essential oil, which is a mixture of Cymol and Cunimol, and other hydrocarbons. Cymol is also a product of the dry distillation of coal tar."

FOOD. 2344 **Food.**—A spice.—Cumin seeds are very like caraway, only larger and of a lighter colour, and with nine in place of five ridges on each half of the fruit. They were apparently displaced from Europe by the introduction of Caraway, but in India are still used as a spice. **U. C. Dutt** says: "Cumin seeds form an ingredient of some curry powders and pickles used by the natives.

TRADE. 2345 **Trade.**—Cumin (or Cummin) would appear to have been known to the ancients; at least there are names for it in most of the classical languages. During the middle ages it was one of the most favoured of spices. In one instance it is recorded that during 716 A.D. an annual provision was made for 150lb of Cumin for the monastery of Corbie in Normandy. Similar records might be quoted from the literature of most European countries down to comparatively modern times. It was in frequent use, for example, in England in the 13th century, and in 1453 was one of the articles of which the Grocers' Company of London had the weighing and oversight.

Foreign Trade. 2346 At the present day the European demand has greatly declined, the place of Cumin having been taken by Caraway. England receives her supplies mainly from Malta, Sicily, and Morocco, only a small amount being obtained from India. According to the returns of Sea-borne Trade issued by the Supreme Government, Cumin appears to have been first separately recorded in 1875 (instead of being lumped with "seeds of other
2347 sorts" or "spices of other sorts"), the year when a duty, formerly levied, was removed. **Flückiger** and **Hanbury**, quoting from the Reports of the Sea-borne Trade as issued by the Local Governments, state that the export of Cumin from Bombay in the year 1872-73 was 6,766 cwt.; and 20,040 cwt. from Calcutta in the year 1870-71." These are misleading quotations, since only about one-fourth of those amounts left India; the remainder represented the coasting traffic, and hence a further error, since some of the coasting imports into each of the ports named would have

reappeared again in the foreign exports therefrom. Thus of the exports from Calcutta 14,037¼ cwt. went to other Indian ports, nearly 2,000 cwt. going to Bombay, an amount which must have greatly influenced the Bombay exports of the year. These remarks have been considered necessary owing to its being customary to find India assigned a far larger share in the world's trade in Cumin than is justified by the official returns. An analysis of the figures for the year 1875-76, compared with those for 1886-87, will remove this misconception. Last year the total exports were:—Indian grown Cumin 9,051 cwt. + foreign imports re-exported 1,260 cwt., or a total of 10,311 cwt. This amount was valued at R1,41,486. In 1875-76 the total exports were 8,120 cwt., valued at R94,919. The foreign trade in Cumin has thus slightly improved, but it falls far short of what most readers would infer from the amounts quoted above as exported from two of the Indian ports. Of the foreign imports, India received in 1875-76 only 538 cwt., and last year 2,020 cwt., so that deducting the re-exports, 760 cwt. was thus added to the amount locally produced in 1886-87. But of the foreign imports 1,994 cwt. came from Persia and the remainder from Turkey in Asia. Bombay received 1,780 cwt. of these imports, the remainder going to Sind. Of the exports of Indian grown Cumin to foreign countries Bombay sent last year 6,730 cwt., Bengal 2,070 cwt., and Madras 250 cwt. Of these exports 2,827 cwt. went to the Straits and Ceylon, Arabia and East coast Africa each received a little over 1,000 cwt., France 430 cwt., and the United Kingdom only 95 cwt. TRADE. Foreign Trade. 2348

The Indian internal trade in Cumin must be at least four times as extensive as the foreign, but the ramifications of road, rail, river, and coastwise-borne traffic are so complex that it would serve no good purpose to attempt to adjust the inter-provincial exchange so as to form some sort of an idea of the actual consumption of Cumin seed in India. It may, however, be stated that, judging from the coast-wise traffic, Madras appears to consume more than any other province, and Calcutta exports far more than can possibly be produced in the Lower Provinces. These two facts would seem to point to the North-West Provinces and the Panjáb as the chief seats of Indian production, the railways carrying to Calcutta a large quantity, a portion of which is shipped to Madras to meet the South Indian market. Internal Trade. 2349

Dr. Dymock says of the Bombay traffic in Cumin that it "comes from Jubbulpore, Guzerat, Rutlam, and Muscat. Value, Rutlam, R8 to R9 per Surat maund of 37½ ℔; Muscat R6 to R6½; Guzerat, R3 to R7½; Jubbulpore, R3 to R6." 2350

Domestic and other Uses.—By the ancients smoking Cumin seeds was considered to produce pallor of the countenance. DOMESTIC. 2351

Cuprea Bark, the bark of **Ramija purdicana** or **R. pedunculata,** see Cinchona, C. 1152.

CUPRESSUS, *Linn.; Gen. Pl., III., 427.*

[*Timb., 410;* CONIFERÆ.

Cupressus funebris, *Endl.; Brandis, For. Fl., 534; Gamble, Man.* 2352

THE WEEPING CYPRESS.

Vern.—*Chandang, tchenden,* BHUTIA.

Habitat.—A handsome tree with pendulous branches, and a fibrous brown bark; often planted in Nepal, Sikkim, and Bhután, near temples and monasteries, and in China (*Gamble*).

2353 **Cupressus glauca,** *Lam.*

Habitat.—Very generally cultivated in Western India above the Ghâts (*Dalz. & Gibs., Bomb. Fl. Supp., 83*).

2354 **C. sempervirens,** *Linn.*

THE CYPRESS.

Vern.—*Sara, sarás,* N.-W. INDIA; *Farash,* SIND; *Sarúboke,* MAR.

References.—*Roxb., Fl. Ind., Ed. C.B.C., 678; Voigt, Hort. Sub. Cal., 558; Brandis, For. Fl., 533; Gamble, Man. Timb., 411; Stewart, Pb. Pl., 222; Brown's Forester, 382; O'Shaughnessy, Beng. Dispens., 621; S. Arjun, Bomb. Drugs, 134; Year Book Pharm., 1874, p. 629; Baden Powell, Pb. Pr., 378, 576; Birdwood, Bomb. Pr., 83; Balfour, Cyclop., 857; Kew Off. Guide to the Mus. of Ec. Bot., 59; Kew Off. Guide to Bot. Gardens and Arboretum, 131.*

Habitat.—A tall tree, cultivated in gardens in Afghánistan and North-West India, sometimes reaching 6 to 9 feet in girth, with 70 to 100 feet in height. **Aitchison** mentions (*Kuram Valley Fl., Pt. I., 97*) a celebrated tree near the shrine at Shálizán.

MEDICINE Wood. 2355 Fruit. 2356 **Medicine.**—WOOD and FRUIT are regarded as astringent and anthelmintic.

TIMBER. 2357 **Structure of the Wood.**—Light-brown, close-grained, moderately hard. Very fragrant, with a strong, peculiar, and pleasant scent.

It is exceedingly durable, and in the Levant and Greece is prized for trunks and boxes, the contents of which are proof against most insects. (*Brandis.*)

2358 **C. torulosa,** *Don.*

HIMÁLAYAN CYPRESS.

Vern.—*Devi-diár,* RAVI; *Deodar,* KULU, BHAJJI; *Gulla, gulrai, kallain,* SIMLA; *Leauri,* JAUNSAR; *Raisalla, sarai,* KUMAON; *Sarrú, súrah vyu,* TIBET.

References.—*Voigt, Hort. Sub. Cal., 558; Brandis, For. Fl., 533; Gamble, Man. Timb., 410; Dalz. & Gibs., Bomb. Fl., 83; Stewart, Pb. Pl., 222; Indian Forester, IX. (1883), p. 59; X. (1884), p. 2; XI. (1885), p. 5; Baden Powell, Pb. Pr., 576; Lisboa, U. Pl. Bomb., 133; Balfour, Cyclop., 857; Kew Off. Guide to Bot. Gardens and Arboretum, 146.*

Habitat.—A large tree growing on the outer ranges of North-West Himálaya, from Chamba to Nepal, scattered or in numerous isolated localities of greater or less extent, chiefly on limestone, between 5,500 and 9,000 feet. Common on the north of the Shalai, Simla, and at Naini Tal, where it often attains a height of over 100 feet with a girth of at least 6 feet, while its lowest branches all but sweep the ground (*Ind. Forester*).

RESIN. 2359 **Resin.**—The wood yields a resin, which is often burnt as incense.

TIMBER. 2360 **Structure of the Wood.**—Heartwood light-brown, with darker streaks, very fragrant, moderately hard. Has been much used at Naini Tál for building, and is sometimes used for beams on the Ravi and Sutlej. In Kulu it is made into images, and is used for the poles which carry the sacred ark. It is often burnt as incense in temples. The *Indian Forester* (*Vol. X., 63*) gives the following analysis of the ash:—

Soluble potassium and sodium compounds	0·004
Phosphates of iron, calcium, &c.	0·039
Calcium carbonate	0·044
Magnesium carbonate	0·008
Silica with sand and other impurities	0·004
TOTAL	0·099

CUPRUM or COPPER.

Cuprum; *Man. Geol. Ind., III., 239, IV., 4.* 2361

COPPER; MINERAL DE CUIVRE, *Fr.*; KUPFERERZ, KUPFER BLENDE, *Germ.*; MINERALE DI RAME, *Ital.*

Vern.—*Tanbah, tánbá, támá,* HIND., DEC.; *Tama,* BENG.; *Támra,* SANS.; *Trambú,* GUZ.; *Támbra,* KAN. & MAR.; *Nohás,* ARAB.; *Mis,* PERS.; *Shenbú (sembú),* TAM.; *Rági, támramu, shenba,* MAL., TEL.; *Kaiyení,* BURM.; "*Zangs,* BHOTE; *Miss,* TURKI; The Sulphate *Nila-túsya,* PB.; *Nila-thokar,* BHOTE; *Dina-farang,* TURKI (*Dr. Aitchison.*)"

References.—*Pharm. Ind., 378, 393-395; Ainslie, Mat. Ind., 504-508, 510, 642; Moodeen Sheriff, Supp. Pharm. Ind., 123; U. C. Dutt, Mat. Med. Hind., 62; Fleming, Med. Pl. and Drugs, as in As. Res., Vol. XI., 189; U. S. Dispens., 15th Ed., 510; Waring, Bazar Med., 46; Irvine, Mat. Med. Patna, 43, 59; Report, Intern. Trade, Punjab, 1884-85, p. 31; Jour. Agri-Hort. Soc., 1843, p. 249; Baden Powell, Pb. Pr., 9, 10, 54, 67, 68, 133; Atkinson, Him. Dist., 270, 282; Balfour, Cyclop., 805; Ure, Dic. Indus., Arts and Manu., 870, 903; Man. Cuddapah Dist., Madras, 27, 45; Bomb. Gaz., V., 123; VII., 137; VIII., 262; Monograph, Brass and Copperware, Punjab, by D. C. Johnstone, 1887.*

Consult also the numerous publications referred to by Ball (*Man. Geo. Ind., III., 611*).

DISTRIBUTION OF COPPER ORES IN INDIA.—The following brief note has been furnished for the present publication by H. B. Medlicott, Esq., F.R.S. :—"The most widely-extended copper deposits at present known to exist in Peninsular India are in the district of Singhbhúm and the State of Dhalbhúm, to work which, Companies have several times been started and given up again. At Baraganda, in the Hazáribagh district, there are copper ores and traces of old workings: and a Company has recently been started to work these ores. In Rajputana, copper ores are found in several of the independent States, and in the British district of Ajmír mining has been practised on a large scale, but is now almost extinct. In Afghánistan, copper ores have been mined to a considerable extent at various places. In the Kumaon and Garhwal districts of the North-West Provinces, copper deposits occur which have been several times unsuccessfully worked, and although smelted on a small scale by the natives would probably not repay working on a large scale. In the Darjiling district, copper ores are met with and a mine was opened some years ago at Yongri. In the Western Duars, the copper ores which occur are worked by Nepalese. Copper ores are also known to exist and to have been worked in the Karnul and Nellore districts of the Madras Presidency." DISTRIBUTION. 2362

For detailed information regarding the Indian mines and sources of copper ore the reader is referred to Ball's account in the *Manual of the Geology of India* (*Part III., pp. 239 to 280*). With a work already in the hands of the public which disposes so fully of the subject it would be superfluous to give here what at most could be but an abstract of Mr. Ball's article on copper. No new information of any importance has been brought to light since the appearance of the *Manual of Geology,* and it need, therefore, only be stated that in ancient times the natives of India appear to have worked the copper mines on a larger scale than they do at the present day. The tradition of the old diggings in Nellore, Singbhúm, and Hazáribagh is lost, but judging from the magnitude of the extinct mines they must once upon a time have given employment to a large number of people. With the appliances presently used by the native miner the access of water has always proved fatal to extended operations. Euro- 2363

CUPRUM. **Copper.**

DISTRIBUTION. pean companies have several times been started but soon dissolved, and it would appear that the hope of Indian copper mining lies in the improvement of native means and appliances. **Ball** states: "The copper ores of Peninsular India occur both in the older crystalline or metamorphic rocks and also in several of the groups of transition rocks, as, for example, in the Cuddapah, Bijawar, and Arvali groups. In extra peninsular India they are found for the most part in highly metamorphosed rocks, the precise age relations of which to those of the peninsula are not in all cases clearly made out as yet.

2364 "The ore of most common occurrence is the copper or pyrites, but towards the outcrops it is commonly altered into carbonates or oxides. The associated minerals are in general identical with those which are found under similar circumstances all the world over. Recent analyses by **Mr. Mallet** have tended to clear up much of the uncertainty which attached to two minerals which were found in Indian copper mines, and were supposed, by those who first examined and described them, to be worthy of specific distinction: these were called respectively Mysorin and Syepoorite. As a rule, to which there are probably not very many exceptions, the copper ores of India do not occur in true lodes, but are either sparsely disseminated or are locally concentrated in more or less extensive bunches and nests in the rocks which enclose them; occasionally cracks and fissures traversing these rocks have by infiltration become filled with ore which thus resembles true lodes. In not a few cases it is believed that the ores exist only as the merest traces.......At the present day, the extraction and smelting of copper ores are only carried on in the most petty manner. In the majority of cases the miners are unable to cope with the water which floods their mines, and, in spite of the fact that their earnings are small, the copper which they turn out cannot be sold at a price which would enable it to compete at the regular markets on equal terms with metal imported into India." **Mr. Mallet** writes: "Perhaps the most remarkable specimens of native copper hitherto found in India were those obtained in Kashmír, from the lower part of the Zánskar river, where it flows through tertiary rocks. In 1878, several water-worn masses of pure metal, reaching up to 22℔ in weight, were discovered in the bed of the stream, and were subsequently, when in the possession of the Governor of Ladákh, seen by **Mr. R. Lydekker**. There is a specimen in the Geological Museum (weighing about 21 oz.) cut from a lump of some 20℔. Although nearly all solid copper, it includes a little cuprite, especially on the sides of one or two cavities; 120 grains of the metal was tested for silver and found to contain a minute trace only. The source whence the nuggets came has not been traced; but recollecting how frequently native copper is connected with trappean rocks, as in the well-known Lake Superior mines, the conjecture may, perhaps, be hazarded that the vicinity of the trappean intrusions which occur between the tertiary and the carboniferous strata of the Markha valley, is one of the most likely localities for the copper to have been washed from."

FOREIGN TRADE. 2365 Foreign Trade in Copper.—The imports in 1886-87 of copper ore, old copper, unwrought and wrought copper, amounted to 615,049 cwt., valued at R1,99,40,085. For the past 20 or 30 years the imports of copper have steadily increased with the increased agricultural prosperity of the people, but within that period they have borne a marked relation to the fluctuations of agriculture. In the year 1885-86, the imports amounted to 652,973 cwt, valued at R2,09,38,405, and in 1882-83, they were 450,098, valued at R1,93,83,758. **Mr. O'Conor**, in his Review of the Sea-borne Trade of India for 1884-85, says: "The price of copper has for some time been constantly declining in England. In January

FOREIGN TRADE.

1883, Chili bars were quoted at £66-10*s*. a ton, in January 1884, at £57-5*s*., in January 1885 at £48. In March of this year it had further fallen to less than £47, and it has since gone much lower, falling below £45. The price of copper in fact is lower than it has ever been, being more than 12 per cent. below the lowest price ever known, and authorities state 30 per cent. below what the trade had previously considered a safe and moderate price. This decline is due to a greatly increased production in the United States, and it would seem to those who are in a position to estimate the conditions of future production there and elsewhere that prices must continue permanently on a low level. In Calcutta, Australian copper was quoted at R31-12 in January 1882, and it has fallen persistently since to R24-10 in January 1885." Over 50 per cent. 2366
of the imports of copper are received by Calcutta and about three-fourths of the supply is drawn from England, but Australian copper is every year being more largely imported. The sudden increase in this trade is, however, more apparent than real, as a large proportion of it is due to the fact that it comes direct to India instead of *viá* England. This direct shipment is of great value, as it means that the commercial relations of India with Australia are becoming more intimate.

Cupri Sulphas. 2367

COPPER SULPHATE OR BLUE STONE.

Vern.—*Nílá-thútha, nílá-tútá, niltá-tutiya,* HIND.; *Mór-tuttá* or *mhór-tuttah*, DEC.; *Mórtútá,* GUZ.; *Tútiyá, tutia,* BENG.; *Tuth-thanjanam, túttha,* SANS.; *Mayil-tuttam, turichu, tuttam-turichi,* TAM.; *Mayilu-tuttam,* TEL.; *Mayil-tutta, turisha,* MAL.; *Mail-tutyá,* KAN.; *Zájul-akhzar, záje-akhzar, qalqand,* ARAB.; *Záke-sabz,* PERS.; *Palmánikam,* SING.; *Douthá,* BURM.; *Túri,* MALAY.

References.—*Pharm. Ind., 378; Moodeen Sheriff's Supp. to Pharm. Ind., 123; U. C. Dutt, Mat. Med. Hind., 66; Waring's Bazar Med., 46.*

MEDICINE. Salts. 2368

Medicine.—U. C. Dutt says: "Sulphate of Copper has been known in India from a very remote period. It is prepared by roasting copper pyrites, dissolving the roasted mass in water, and evaporating the solution to obtain crystals of the sulphate. It was known as a salt of copper, for the *Bhávaprakása* says it contains some copper and therefore possesses some of the properties of that metal. It is described in this work as astringent, emetic, caustic, and useful in eye diseases, skin diseases, poisoning, &c. 2369
It is purified for internal use, by being rubbed with honey and *ghí* and exposed to heat in a crucible. It is then soaked for three days in whey and dried. Sulphate of copper thus prepared is said not to produce vomiting when taken internally. Dose, one to two grains." The *Pharmacopœia of India* says: "Hindú practitioners place much reliance on some of their rudely prepared salts of copper, which, for the most part, are obtained by the action of tamarind, lime, or other vegetable acid juice on metallic copper. None of them seem to deserve attention. The sulphate (*Níl-tutiya*) is met with in most bazars, generally of fair quality. It may 2370
be further purified, if required, by dissolving in water, filtering and evaporating to crystallization." 2371

According to European Medical practice pure sulphate of copper is tonic, astringent, emetic: in large doses an irritant poison. Locally applied in substance to a denuded or granulating surface, mildly caustic, styptic, and in solution stimulant. The article so used is imported from 2372
Europe. It is largely used in chronic dysentery, diarrhœa, epilepsy, chorea, and hysteria. Locally, it is applied in solution in gonorrhœa, leucorrhœa, purulent ophthalmia, weak ulcers, superficial hœmorrhage,

MEDICINE. and, in substance, to cancrum oris, aphthous ulcerations, exuberant granulations, and granular conjunctivitis. (*Pharm. Ind.*) **Waring** recommends an emetic of 5 grains of sulphate of copper in tepid water for Opium, Datura, Nux Vomica, Cocculus Indicus, Bish (Aconite), Arsenic, or other poisoning cases. If it does not operate in half an hour it may be repeated.

Special Opinions.—§ "Sulphate of copper can be had in all bazars. It is a ready emetic, and a powerful astringent, both internally and externally" (*Civil Surgeon G. Price, Shahabad*). "The native pure copper is calcined and reduced to the state of its oxides and is thus used as tonic, expectorant, and depressent for fevers, asthma, &c." (*Surgeon Major Robb, Ahmedabad*). "Sulphate of copper is used internally as astringent in chronic dysentery and diarrhœa in dose of $\frac{1}{4}$ to $\frac{1}{8}$ of grain, also applied externally" (*Asstt. Surgeon Nehal Sing, Saharunpore*). "Copper coins, on which there is a deposit of verdigris, are kept for an hour or two in a mixture of (ripe) tamarind and water, and then rubbed on parts of body attacked by urticaria" (*Honorary Surgeon P. Kinsley, Chicacole, Ganjam, Madras Presidency*). "Useful as an emetic in cases of poisoning" (*Civil Surgeon J. H. Thornton, B.A., M.B., Monghyr*).

Plates. 2373 "*Copper foil* (*Shabiri*, Swahili, E. Africa) cut into small pieces about an inch or more square, which are spread over the chest before and behind is the native (African) treatment of cough and all general chest troubles. Two dozen of these thin copper plates were counted in a case that came up for other treatment; their application is on the principle of a series of small blisters or counter-irritants" (Zanzibar).—*Surgeon-Major John Robb, M.D., Surat, Bombay Presidency*.

Leaf. 2374 COPPER LEAF.—A thin copper foil is sold in the Muscat bazar as an external application to unhealthy ulcers. It is applied like thin Guttapercha tissue over the surface of the ulcer and secured for days by means of a bandage.

CURCULIGO, *Gærtn.; Gen. Pl., III., 717.*

[*p. 124;* AMARYLLIDEÆ.

2375 **Curculigo orchioides,** *Gærtn.; Baker, Linn. Soc. Jour., XVII.,*

Most authors refer the native medicinal tuber known in the Panjáb as *siyáh músli* to this plant, but **Stewart** says it is obtained from **Anilema tuberosa,** *Ham.,* and **Dymock** describes it under **Hypoxis orchioides,** *Willd.,* giving **Curculigo orchioides** as a synonym. In Bengal the tuber is generally known as *Tal-lura.*

Syn.—CURCULIGO MALABARICA, *Wight, Ic., t. 2043*; HYPOXIS ORCHIOIDES *Kurz, in Ann. Mus. Lug. Bat. IV., 177*; ORCHIS AMBOINICA MAJOR RADICE RAPHANOIDEA, *Rumph., Amboin, VII., 117, t. 54, f. 1.*

Vern.—*Káli-músli, siyáh-músli, músli-kand, múshali*; HIND., BOMB.; *Tála mulí*, BENG.; *Tálmúlí*, URIYA; *Mushali, tálamúlika* (*warahi,* **Ainslie**), SANS.; *Nilap-panaik-kishangu, nelepanny kalung*, TAM.; *Nalla tady gudda, néla táti gaddalu, néla tádi*, TEL; *Nela-táti-gadde*, KAN.; *Mussulkund*, C. P.; *Hín-bin-tal*, SING.

Conf. with **Asparagus adscendens, A. 1562.**

References.—*Roxb., Fl. Ind., Ed. C.B.C., 288; Ainslie, Mat. Ind., I., 242; Roxb., Corom. Pl., I., 14, t. 13; Moodeen Sheriff, Supp. Pharm. Ind., 124; Dymock, Mat. Med. W. Ind., 2nd Ed., 818; U.C. Dutt, Mat. Med. Hind., 250; S. Arjun, Bomb. Drugs, 143, 221; Thwaites, Enum: Ceylon Pl., 324; Bomb. Gaz., VI., 14; Dalz. and Gibs., Bomb. Fl., 276; Rheede, Hort. Mal., XII., t. 59.*

Siyah Musli. **CURCULIGO orchioides.**

Habitat.—A small herbaceous plant with a rosette of radial leaves and tuberous root, native of the greater part of the hotter regions of India and Ceylon. Roxburgh says that in cultivation it flowers all the year round.

Medicine.—In most Hindu and Muhammadan works on Materia Medica frequent reference is made to two forms of a tuberous root sold as *Músli*—the one is called the *kálá* (black) *músli*, and the other the *suféd* (white) *músli*. It is now very generally admitted that these are not obtained from the same plant, and that while there are two or three adulterants or substitutes for each, the true black *músli* is **Curculigo orchioides**, and the white **Asparagus adscendens**. According to some writers the young roots of **Bombax malabaricum** constitute one of the white *múslis*, and by others the black and white forms are obtained from one and the same plant during different stages of its growth. Dr. Moodeen Sheriff remarks that in South India a false *suféd músli* is sold which is obtained from **Asparagus sarmentosus** (**A. 1577**). On the other hand Dr. U. C. Dutt says: "The roots of **Bombax malabaricum** and of **Asparagus racemosus** are sometimes sold by the native druggists of Calcutta under the name of *saffed musli*. These articles have, however, separate names and are not designated by the name of *saffed musli* in any native medical works. On the contrary, a white variety of *tála muli* or *musuli* is, as already noticed, mentioned in the *Rája Nirhantu*. The tubers of **Curculigo orchioides** become, when dry, translucent like amber. The dried roots were probably considered a separate variety, namely, the white, by the ancients." Some confusion also exists between the so-called *múslis* and the *saleps* (see **Eulophia**). It seems probable that in the different parts of India where it is met with, **Curculigo orchioides** varies considerably, forms existing which might be said to correspond with **C. brevifolia**, *Dryand*; **C. firma**, *Kotschy et Peyr*; and **C. ensifolia**, *R. Br.* Sakharam Arjun, adopting apparently this view, describes **C. brevifolia** as the *musalíkand* and **C. ensifolia** as the *kali musli*. He further states that much of the latter root sold in the Bombay Presidency is **Aneilema scapiflorum**, *Wight* (Conf. **A. 1122**). Dr. Dutt says of **C. orchioides**: "The tuberous roots of this plant are considered alterative, tonic, restorative, and useful in piles, debility, and impotence." Ainslie says that the tuberous and wrinkled root of this plant is considered "in a slight degree bitter and mucilaginous to the taste, and is supposed to possess virtues nearly similar to the last-mentioned article. It is prescribed in electuary, in the quantity of a teaspoonful twice daily; it is also considered as possessing tonic qualities, and sometimes given with milk and sugar, in doses of two drachms in the twenty-four hours, in cases requiring such medicines." Dr. Dymock says: "*Músli* prescribed for asthma, piles, jaundice, diarrhœa, colic, and gonorrhœa; it is considered to be demulcent, diuretic, tonic, and aphrodisiac, and is often combined with aromatics and bitters." Native Medical works say, the plants from which the tubers are collected should be two years old. The tubers should be washed and freed from rootlets, cut in slices by a wooden knife and dried in the shade. Dose 180 grains, beaten up with an equal quantity of sugar in a glass of milk until it forms a thick mucilage. Dymock adds that the Bombay supply "comes from Rutlam in Guzerat or from the Central Provinces. It is valued at R4 a maund of 37½℔."

MEDICINE. Black root. 2376 White root. 2377 Substitutes. 2378

2379

TRADE. 2380

Special Opinions.—§ "The tuber is regarded as a cooling medicine. Is useful in the phosphatic diathesis, and in scleroderma. It is said to possess powerful aphrodisiac properties. It is largely used in medicines by native practitioners." (*Surgeon Major J. M. Houston, Durbar Physician, Travancore, and Civil Apothecary John Gomes, Medical Store-keeper, Trevandrum*).

CURCUMA, *Linn.*; *Gen. Pl., III.*, 643.

2381 **Curcuma Amada,** *Roxb.; Fl. Ind., Ed. C.B.C., 12*; SCITAMINEÆ.

MANGO-GINGER.

Vern.—*Am-haldi*, HIND.; *Karpura-haridrá*, SANS.; *Amádá*, BENG.; *Amba-haladar*, MAR.; *Am-ki or bó-ki-adrak*, DEC.; *Mamidi-allam*, TEL.

Sir Walter Elliot (*Fl. Andh., pp. 17 & 111*) gives this plant the Telegu names of *Mamidi allam* and *Aru kanla kachóram*; but he remarks "*aru kanla*, meaning 'six eyes,' *Shad-grandhika*, 'six jointed,' are also given as synonyms of *Nalla ativasa* or **Curcuma Cæsia** and seem to be merely Sanskrit forms of the same word, both probably referring more correctly to **C. Zedoaria** or long Zedoary."

References.—*Voigt, Hort. Sub. Cal., 565; Pharm. Ind., 232; O'Shaughnessy, Beng. Dispens., 649; U. C. Dutt, Mat. Med. Hind., 257, 304; S. Arjun, Bomb. Drugs, 140; Irvine, Mat. Med. Patna, 4; Drury, U. Pl., 169; Balfour, Cyclop., 858.*

Habitat.—Found wild in Bengal and on the hills; flowering during the latter half of the rains.

MEDICINE. Tubers. 2382 **Medicine.**—The TUBERS are regarded as cooling and as useful in prurigo. They are also employed as carminative and stomachic. When fresh they possess the smell of the green mango, hence the various names above. Dr. Irvine (*Mat. Med. Patna, p. 4*) says of this root-stock that it is used as a carminative and to promote digestion; dose from ℈i to ʒii. In the *Pharmacopœia of India* it is stated that they do not possess any advantage over ginger.

External application. 2383 **Special Opinions.**—§ "Made into a paste with spirit and white of egg used as an application in chronic rheumatism, bruises, &c. Ext. Belladonna often added" (*Dr. Darasha Harmarji Baria, L.M.S., Bombay*). "Locally applied over contusions and sprains" (*Surgeon-Major Robb, Civil Surgeon, Ahmedabad*). "Roots are expectorant and astringent, useful in diarrhœa and gleet" (*Surgeon-Major J. M. Houston, Durbar Physn., Travancore, and Civil Apothecary John Gomes, Medical Store-keeper, Trevandrum*).

FOOD. 2384 **Food.**—Used as a condiment and vegetable (*U. C. Dutt*).

2385 **C. angustifolia,** *Roxb.; Fl. Ind., Ed. C.B.C., 10, 11.*

WILD OR EAST INDIAN ARROWROOT; NARROW-LEAVED TURMERIC.

Vern.—*Tikhur*, HIND.; *Ararut-ke-gadde*, DEC.; *Tavakhira*, MAR.; *Kuvegadde*, N. KANARA; *Tickar*, BOMB.; *Ararút-kishangu, kua*, TAM.; *Ararút-gaddalu*, TEL.

References.—*Voigt, Hort. Sub. Cal., 563; Dalz. & Gibs., Bomb. Fl., 274; Ainslie, Mat. Ind., I., 19; O'Shaughnessy, Beng. Dispens., 649; Dymock, Mat. Met. W. Ind., 2nd Ed., 768; Flück. & Hanb., Pharmacog., 634, 639; U. S. Dispens., 15th Ed., 1694; S. Arjun, Bomb. Drugs, 141; Jour. As. Soc. (1867), 82; Upper Godaveri District Settle. Rep., p. 40; Chanda District Settle. Rept., 109; C. P. Gazetteer, pp. 31,419 & 505; Drury, U. Pl, 168; Lisboa, U. Pl. Bomb., 175; Birdwood, Bomb. Pr., 236; Balfour, Cyclop., 858; Ure, Dic. Indus., Arts and Manuf., 972; Kew Off. Guide to the Mus. of Ec. Bot., 62.*

Habitat.—A native of the central tracts of India, from the mountains of Bengal to Bombay and Madras. Is particularly abundant in the Central Provinces, and a considerable trade is reported to be done at Raipur in the collection of the tubers. The plant is also common at Ram Ghát, Bombay. Is said to grow wild in North Canara (Bombay), but to be also cultivated (*Gaz., XV., pt. II., 20*). Mr. Atkinson remarks that, it is found wild in the North-West Himálaya. The flowers are large and yellow, longer than the bracts; they expand in the morning and wither in the evening of the same day.

Cultivation of East Indian Arrowroot.—Perhaps the most complete accounts of the cultivation of this plant are those which will be found in the Reports of the Sydapet Experimental Farm, Madras. The following passages may here be reproduced from these reports:— **CULTIVATION. 2386**

Arrowroot (**Curcuma angustifolia**). "A plot measuring 0·25 acres was planted with this crop at the end of 1879, and remained down during the year under report. It was taken up at the end of January last and yielded 986℔ of tubers, or at the rate of 3,944℔ per acre. The yield of flour obtained has generally been about 12½℔ from 100℔ of tubers, so that the above yield would represent an outturn of 493℔ of flour per acre. In another case in the College Experimental Garden, a plot, measuring 1,160 square yards, planted with this crop yielded 1,798℔, or at the rate of 7,500℔ per acre. The culture of this crop is very simple: it is only necessary to plant the sets in properly-prepared soil, and to water them occasionally during the dry season. The removal of the crop is tedious unless the tubers can be ploughed out, as potatoes are done in England, which is seldom possible owing to the dryness of the soil, so that the tubers have to be dug up. The preparation of the flour is also very simple and easy. The TUBERS have only to be reduced to pulp on a grater, after being well washed to remove soil and dirt, and then the pulp is mixed thoroughly with water so as to separate the starch completely from the fibrous matters. The whole is afterwards strained through cloth, through which the STARCH and water passes, and the fibre left behind. After this the STARCH has only to be thoroughly washed by decantation with clean water, and dried in the sun. It is then rolled on a table to break it up thoroughly into fine flour and is ready for sale. The flour can be produced at a very low price; it could be sold profitably at 4 annas per pound. And thus 400 rupees per acre could be realized. This is a remarkable return and should also be published for the information of the public." **Madras Rootstocks. 2387** **Starch. 2388** **Profits Rs. 400 an acre.**

"The following extract from a letter from the Collector of South Kanara, dated 10th March 1882, No. 517, will be found interesting: "With reference to paragraph 48 of your report on the Saidapet Farm, recorded with the Board's Proceedings dated 10th December 1881, No. 3182, I have the honour to forward specimens of ARROWROOT prepared from a plant common in the jungles of this district, and should be obliged by your informing me how it compares in quality with that grown on the Farm which you estimate to be worth about four annas a pound. I should also be glad to learn whether it is likely that cultivation would lead to an improvement in the quantity or quality, or both, and any information as to the method of preparing the soil, and the best manner of treating the plant in this district (with its annual rainfall of about 130 inches between June and November) would be thankfully received. The plant, I believe, to be the same as that experimented on by you, *viz.*, **Curcuma angustifolia**, but I send a few of the tubers for identification." **South Kanara. 2389**

"The samples were sent to **Mr. Hamilton, F.C.S.**, for examination and analysis. The following extracts are from his report thereon:—

CHEMISTRY.—"The samples of arrowroot sent are from the **Curcuma**; they exhibit under the microscope the characters of the granules peculiar to this variety of starch. This starch cannot be compared with that of the **Maranta**, which is the richest of all the feculas. The mucilage, however, yielded by sample marked '1st sort' is of a superior description and nearly as good as that of the **Maranta**. This sample is susceptible of further **CHEMISTRY, 2390** **Inferior to Maranta. 2391** improvement; it contained a number of extraneous matters, black particles, straw, &c., all of which must have been introduced during the process of drying. The other two samples were decidedly inferior. The three samples, when soaked in cold water, gave indications of the presence of slight acidity; they also exhibit to a slight extent transformation of the starch from the insoluble to the soluble form. I may add that the Farm sample

Solar heat to be avoided.

also gave the same reaction, but to a less extent. Any unnecessary exposure to the solar heat should be avoided. If the samples could be ground to a fine powder it would add to their appearance and would fit them for

Use of Caustic Soda.

immediate conversion into mucilage. I would suggest the use of a solution of caustic soda about 200 grains (half an ounce nearly) to a gallon of water for steeping the pulped roots, in lieu of plain water; this has been found useful in disintegrating and dissolving the nitrogenous matter. Thorough washing in pure spring water will remove all traces of the soda."

Cochin. 2392

The arrowroot is said to be largely manufactured at Cochin, Travancore, and Kanara. **Royle** says that "a very excellent kind called *tickar*

Travancore. 2393

is also made at Patna and Baglipore from the tubers of **Batatus (Ipomœa) edulis.**"

Substitute. 2394

Medicine.—The arrowroot is used medicinally in some parts of the country.

MEDICINE. Arrowroot. 2395

Food.—A good quality of arrowroot is prepared from the tubers especially in Travancore, where the plant grows in abundance.

FOOD. Arrowroot. 2396

Roxburgh observes that a sort of *starch* or arrowroot-like fecula is prepared,

Benares. 2397

which is sold in the markets of Benares, and is eaten by the natives. The flour, when boiled in milk, forms an excellent diet for patients or children. It is largely used for cakes, puddings, &c., though it is often complained of as producing constipation. The granules much resemble those of **Maranta** (the true arrowroot), but are flat and always stratified. **Dymock** says of this form of arrowroot that it is "a favourite

Thicken milk. 2398

article of diet among the natives, especially for children. The milkmen in Bombay use it to thicken milk which has been watered." The edible properties of the tubers of this plant are alluded to in most of the Settlement Reports of the districts comprising the Central Provinces. Of Seoni it is said they are pounded and made into gruel.

PREPARATION OF ARROWROOT. Travancore. 2399

Preparation of the Arrowroot.—**Drury** thus describes the process as practised in Travancore: "The tubers are first scraped on a rough stick, generally part of the stem of the common rattan, or any plant with rough prickles to serve the same purpose. Thus pulverised, the flour is thrown into a chatty of water, where it is kept for about two hours, all impurities being carefully removed from the surface. It is then taken out and again put into fresh water, and so on for the space of four or five days. The flour is ascertained to have lost its bitter taste when a yellowish tinge is communicated to the water, the whole being stirred up, again strained through a piece of coarse cloth, and put in the sun to dry. It is then ready for use." In the Central Provinces the root is also collected and arrowroot prepared. The process adopted in the Upper Godavari District (*Gaz.*, 505) is thus referred to: "*Tankír* or *Tíkhur* is a description of arrowroot made from the bulbs of the **Curcuma angustifolia**, which grows abundantly in the district. It is collected by the Gotés and Koís, and rubbed down on a stone, washed, and allowed to settle. It is then dried, and either sold or bartered by them to traders. The *tankír* purchased in the bazars is impure and difficult to refine, as the bulb is not pared before it is grated down. If care be taken, the flour can be made as pure as that prepared from garden arrowroot. It is strange that this root is not made so much use of as it might be, either as an article of food, or even as starch

for export." (For further particulars see the paragraph on Cultiva- **PREPARATION OF ARROWROOT.**
tion.)
Special Opinions.— §" Found in the hills of Vizagapatam and Ganjam **Vizagapatam. 2400**
Districts and used by the hill tribes as an article of diet." (*Honorary Sur-*
geon P. Kinsley, Chicacole, Ganjam District, Madras Presidency.) "Sweet- **Ganjam. 2401**
meats are prepared from *tikhur* flour, and eaten with great relish; a favourite article of diet among certain classes in these provinces" (*Assistant* **Sweetmeats. 2402**
Surgeon Shib Chunder Bhattacharji, In Civil Medical Charge, Chanda,
Central Provinces).

TRADE IN EAST INDIAN ARROWROOT.—**Drury** says the exports of
this "arrowroot from Travancore average about 250 candies annually. In **TRADE. 2403**
1870-71 were exported from Bombay 3 cwt., and from Madras, in 1869-70
3,729 cwt., valued at R14,152." **Dymock** under this species says: "Mala- **Malabar. 2404**
bar arrowroot fetches from R3 to R4 per quarter cwt. in Bombay." With
the exception of Travancore it is doubtful how far the trade returns can
be trusted as referring to this or to the true arrowroot. See **Maranta arundinacea.**

Dymock remarks of Turmeric (**Curcuma longa**) that the starch "of **Turmeric. 2405**
the young tubers at the end of the radicles, which are nearly colourless,
forms one of the East Indian arrowroots. It is to be observed that the **Starch. 2406**
tubers that yield only starch when young will yield turmeric when old; the
colouring matter and aromatic principles are deposited in the cells at a
later period of growth."

Curcuma aromatica, *Salisb; Roxb., Fl. Ind., Ed. C.B.C., 8.*

WILD TURMERIC; YELLOW ZEDOARY; COCHIN TURMERIC.

Syn.—CURCUMA ZEDOARIA, *Roxb.*

Vern.—*Jangli-haldi, ban-haldi, ban-haridra (jedwar?),* HIND.; *Ban-halud,* BENG.; *Kapur kachali,* GUZ.; *Rán hald, ambé-haldi,* BOMB.; *Kasturi-manjal,* TAM.; *Kastúri pasupa, kattu-mannal,* TEL.; *Anakúva, kattu-mannar,* MAL.; *Vanaharidrá,* SANS.; *Judwar* (according to **Roxburgh**), ARAB.; *Kastúri-arishiná,* KAN.; *Duda-kaha, wal-kaha,* SING.; *Kiyásanoin,* BURM.

References.—*Voigt, Hort. Sub. Cal., 593; Dalz. & Gibs., Bomb. Fl., 274; Ainslie, Mat. Ind., I., 490, 493; Moodeen Sheriff, Supp. Pharm. Ind., 125; U. C. Dutt, Mat. Med. Hind., 257, 322; Dymock, Mat. Med. W. Ind., 769; Year Book Pharm., 1880, 251; Home Department Corr. regarding Pharm. Ind., 240; Liotard, Dyes, App. VII.; Balfour, Cyclop., 859.*

Habitat.—**Roxburgh** says of his **Curcuma Zedoaria**: "This beautiful **Bengal. 2407**
species is a native, not only of Bengal (and common in gardens about
Calcutta), but is also a native of China, and various other parts of Asia
and the Asiatic islands. Flowering time, the hot season; the leaves
appear about the same period or rather after, for it is not uncommon to
find the beautiful, large, rosy, tufted spikes rising from the naked earth
before a single leaf is to be seen." "The plant when in flower is highly **Malabar. 2408**
ornamental, few surpassing it in beauty; at the same time it possesses a
considerable degree of delicate aromatic fragrance."

The flowering spikes are quite distinct from the leaf-bearing stems,
and the upper bracts of each are more brightly coloured than the lower,
and are sterile. **Dalzell** and **Gibson** (*Fl. Bomb.*) say that it is met with
in the Concans flowering in May when the leaves begin to appear.
Dr. Dymock remarks: "The plant which produces this drug grows wild **Concan. 2409**
in the Concan: under cultivation it produces central tubers as large as a
small turnip. I have had it under cultivation for some years, and
observe that the leaves when young have a central purple stain which

CURCUMA aromatica.

Wild Turmeric.

Mysore. 2410 Travancore. 2411 almost disappears when they attain their full size." **Drury** remarks that it is abundant in the Travancore forests. Of Mysore **Mr. D. E. Hutchins** says **C. aromatica**, the *Kad arasina*, is collected from the forests all over the province.

HISTORY. 2412 **History of Jadvar and Zedoary.**—The reader is referred to **Aconitum heterophyllum, (A 401 & 408)**, for further particulars regarding the use of the Arabic word *Jadvár*. According to certain writers (including **Roxburgh**) this is applied to a species of **Curcuma**, presumably the present species. To **Dr. Moodeen Sheriff** we are indebted for the results of much careful study on this subject, the final conclusion arrived at being that the Arabic *Jadvár* is a name which should be restricted to the roots of the non-poisonous Aconites. The confusion which exists on this subject **Moodeen Sheriff** attributes to the resemblance of the word *Jadvár* or *Zadvár* to Zedoary. *Dar-hald* and *anbé-haldi*, he adds, are in some Persian works also used as synonymous, but the former is more correctly the name for the medicinal wood obtained from a species of **Berberis.** On the other hand, **Dr. Dymock** (*2nd Ed., Mat. Med. W. Ind., 769*) writes: "This rhizome is the *Vanaharidra* of Sanskrit works, and there appears to be little doubt that it is the *Jadwár* described by **Avicenna** (Lib. II., p. 155) and figured in **Clusius'** Exotica, p. 378. At the present time the Jadwár of India is quite a different article." **Dr. Dymock** then refers the reader to his account of **Delphinium denudatum**, where he states that *Jadwár* and *Mahferfin* are Arabic names commonly given to that plant, *Zadwár* being the Persian and *Nirbisi* the Hindustani. **Dr. Royle** was the first author who affirmed that to be the case, but the writer has failed to find a native of the North-West Himálaya who gave to **D. denudatum** either of these names. Around Simla, for example, that form of Larkspur is one of the commonest of herbs, but it bears the name of *Múníla* not *Nirbisi*, and does not appear to be ever collected for medicinal or other purposes. Some short time ago the writer suggested to **Dr. Gimlette**, Residency Surgeon, Nepal, the desirability of making a collection of the Nepal Aconites and allied poisonous drugs. He had, as the result, the pleasure to receive a most instructive set of specimens.

2413 The *Kala bikh* of the Nepalese (the *Dulingi* of the Bhotias, who make a trade in collecting and selling these roots) is a very poisonous form of **Aconitum ferox**, so poisonous indeed that the Katmandu drug-sellers will not admit they possess any. *Pahlo* (yellow) *bikh* is a less poisonous form of the same plant, known to the Bhotias as *Holingi*, while *Setho* (white) *bikh* (the *Nirbisi sen* of the Bhotias) is **A. Napellus**, and *Atís* is **Aconitum heterophyllum.** The aconite adulterants or plants used for similar purposes are **Cynanthus lobatus**, the true *Nirbisi* of Nepal, the root of which is boiled in oil, thus forming a liniment which is employed
2414 in chronic rheumatism. **Delphinium denudatum**, the *Nilo* (blue) *bikh* of the Nepalese and the *Nirbisi* of the Bhotias, **Dr. Gimlette** says, is used by the *Baids* of Nepal for the same purposes as the *Setho* and *Pahlo bikh*.
2415 **Geranium collinum** (*var.* **Donianum**) is the *Ratho* (red) *bikh* of the Nepalese, and the *Nirbisi Num* of the Bhotias, and like the *Setho bikh* is given as a tonic in dyspepsia, fevers, and asthma. Lastly, a plant never before recorded as used medicinally, namely, **Caragana crassicaulis**, is known as the *Artiras* of the Nepalese and the *kúrtí* of the Bhotias; it affords a root which is employed as a febrifuge. The Nepalese name recalls *Atís* (**A. heterophyllum**) and the Bhotia, *kutkí* (= **Picrorhiza Kurroa**) —drugs also prescribed as febrifuges.

2416 Some doubt may, therefore, be admitted as resting on the assertion that **Delphinium denudatum** is the *Nirbisi* of the earlier writers. **Ainslie** urges that the "*Nirbishie*," made known by **Dr. F. Hamilton** as found in

Nepal, "must not be confounded with the word *Nirbisi*, which is the Sanskrit for **Curcuma Zedoaria.**" To the hill tribes around Simla and Kulu, at least, it is neither *Jadwár* nor *Nirbisi*, and, indeed, the roots of that plant bear but little resemblance to those of an aconite and none whatever to the rhizomes of a **Curcuma.** But at the same time **Dr. Dymock's** historic sketch of Jadwár and Zedoaria is valuable, as there seems little doubt but that many of the early authors made the mistake of viewing these names as synonymous. **Dr. Dymock** continues: "**Ainslie,** whose description of the Zedoaries is excellent, tells us that in his time the Muhammadans supposed it to be a valuable medicine in certain cases of snake-bite, administered in conjunction with golden orpiment, costus, and ajwain seeds. He also expresses his opinion that it was the Jadwár of the old Arabian writers. *Ambé-haldi* is no doubt the *Zedoaire jaune* of **Guibourt,** who tells us that the plant which produces it has been well described and figured by **Rumphius.** It is his *Tommon bezaar* or *Tommon primum*, which has been wrongly referred by most writers to **Curcuma Zedoaria** of **Roscoe.** (*Confer. Guibourt His., Nat., 6me. Ed., tom. II., p. 214.*) It would appear also that it is identical with the **Cassumunar** described by **Pareira.** (*Confer. Pareira Mat. Med., Vol. II., Pt. I., 236.*) Lastly, it would appear to be the same as the "Cochin Turmeric" noticed by **Flückiger** and **Hanbury** (*Pharmacographia. p 580*)." (*Dymock, 770.*)

Description of the Rhizomes.—"Central rhizome oblong or conical, often more than two inches in diameter, external surface dark-grey, marked with circular rings and giving off many thick rootlets; at the ends of some of them are orange-yellow tubers about the size and shape of an almond in its shell; lateral rhizomes about as thick as the finger with a few fleshy rootlets. Internally both central and lateral rhizomes are of a deep orange colour like turmeric; the odour of the flesh root is strongly camphoraceous." **Dalzell** and **Gibson** say: "The tubers of the root are palmate." **DESCRIPTION. 2417**

Dye—. It is probable that this, like the Zedoary, was formerly used in the preparation of the *Abír* powder. **Dymock** says: "Like turmeric its principal use is as a dyeing agent." **Mr. Liotard** (*Memo. Dyeing*) says it is rarely used as a dye. It gives a dirty yellow colour with the alkaline earth *chaulu*. **Ainslie** remarks: "The Native women prize it much from the circumstance that they can give with it, used externally, a particular lively tinge to their naturally dark complexions, and a delicious fragrance to their whole frame." **DYE. 2418** **Cosmetic. 2419**

Medicine.—The RHIZOMES are used medicinally, being regarded as tonic and carminative. **Thwaites** says this drug is used by the Singhalese. It holds an important place in native perfumery. **Dymock** states that "the properties of this drug are very similar to those of turmeric, but its flavour being strongly camphoraceous is not so agreeable. It is used medicinally in combination with other drugs as an external application to bruises, sprains, &c. In the Concan it is applied to promote the eruption in exanthemateous fevers; it is seldom used alone, but is combined with astringents when applied to bruises, and with bitters and aromatics to promote eruptions." **Ainslie** says the Muhammadans suppose it to be a valuable medicine in certain cases of snake-bites, administered in small doses, and in conjunction with golden-coloured orpiment, *kust* (**Costus arabicus**), and *ajúan*." **MEDICINE. Rhizomes. 2420**

Special Opinions.—§ "Used externally in scabies and the eruption of small-pox" (*Surgeon-Major Henry David Cook, Calicut, Malabar*). "Rubbed into a paste with benzoin is a common domestic application to the forehead for headache" (*Surgeon-Major John North, I. M. S., Bangalore*). "Applied to the forehead in cephalalgia, and a cosmetic."

(*T. Ruthnam Moodelliar, Native Surgeon, Chingleput, Madras Presidency.*)

TRADE. 2421 **Trade.**—"The Bombay market is supplied from the Malabar coast. Value, unpeeled R24 to R25 per candy of 5¼ cwt.; peeled R27 per candy" (*Dymock*).

2422 **Curcuma cæsia,** *Roxb.; Fl. Ind., Ed. C.B.C., 9.*

BLACK ZEDOARY.

Vern.—*Kálá haldi* or *nil-kantha*, BENG.; *Káli halada*, MAR.; *Nar-kachúra*, BOMB.; *Nar-kachúra, káli-haldi*, HIND.; *Mánu pasupu*, TEL.

Sir Walter Elliot says this is in Telegu known as *nalla ativasa*. "One of its Sanskrit synonyms is *sati*, which **Wilson** says is a kind of **Curcuma** and in Bengal **C. cæsia** is still called black *haldi* and *nila kantha*."

Habitat.—**Roxburgh** remarks: "This elegant strongly-marked species is a native of Bengal, where it blossoms in May" and just before the rains. "In the deep ferruginous purple cloud down the middle of the leaves it resembles **C. Zerumbet**, but differs widely in the colour of the roots."
Bengal. 2423 **Dymock** says it is cultivated in Bengal to supply the Indian market. He adds "through the kindness of Surgeon-Major **Peters** I have been
Dinapore. 2424 supplied with living tubers of this **Curcuma** from Dinapore; he informs me that it is common in gardens in Bengal, and is used as a domestic remedy in the fresh state much as turmeric is in this part of India."

MEDICINE. Rhizomes. 2425 **Medicine.**—**Dymock** says this is one of the two *Zerumbáds* of modern Persian writers on Materia Medica. "Strange to say, it is not noticed by most European writers on Indian drugs, though it is well known and to be found in all the shops. It is the *Tommon itam* of **Rumphius**, and the **Curcuma longa** of **Guibourt**, who classes it with the turmerics."
Cosmetic. 2426 "*Nar-kachúra* appears to have been once imported into Liverpool under the name of *Kutchú*." It is "considered to have nearly the same medicinal properties as *Kachúra*," and is chiefly used as a cosmetic.

TRADE. 2427 **Trade.**—**Dymock** says the tubers are internally very hard and horny, of a greyish black, but when cut in thin slices of a greyish-orange. The odour and taste are camphoraceous. "The drug comes overland from Bengal. Value R4 to R5 per maund of 41℔. **Guibourt** appears to have become acquainted with it from its admixture with the turmeric of commerce." **Sakharam Arjun** remarks that it is used externally as an application to bruises, for rheumatic pains, and in contusions.

2428 **C. caulina,** *Graham; Dalz. and Gibs., Bomb. Fl., 275.*

Vern.—*Chavara, chowar*, BOMB.

Habitat.—A plant common at Mahábaleshvar, Bombay, and described by the late **Mr. Graham.**

FOOD. Rhizomes. 2429 **Food.**—A form of ARROWROOT is said to be prepared from this plant. It is described by **Sir George Birdwood** and other writers, the last being
Arrowroot. 2430 **Mr. Lisboa**, who writes; "**Curcuma caulina** grows at Mahábaleshvar abundantly, and for many years the Chinese ticket-of-leave men used to manufacture arrowroot from it, and sell it to the Commissariat, and in the bazaars at Bombay. In 1878, a European prepared a few hundred pounds of it, and sent samples to be tried by **Messrs. Treacher & Co., Phillips & Co.,** and **Kemp & Co.**, but it was found wanting in nutritive properties, though no objection was made to the colour and taste. That it is inferior to West Indian Arrowroot may be gathered from its market value, 5 to 6 pounds to the rupee. During the famine of 1877, it was recommended to the suffering poor, but they never used it except in extreme scarcity."

"The preparation of Arrowroot at Mahábaleshvar is simple. The root (of which a cooly will gather 4 or 5 large basketsful a day for as many annas) is scraped, washed, and rubbed to pulp on a grater, as mortars are found to crush the globules. The pulp must then be washed no less than a dozen times at least, the sediment being stirred at each washing. The dark scum on the sediment and the muddiness of the water of the first washing slowly disappear, till when the sediment is pure-white it is allowed to harden into a cake, which is afterwards reduced to powder. A basketful of roots yields 3-4 lb of pure arrowroot."

Curcuma leucorrhiza, *Roxb., Fl. Ind., Ed. C.B.C., 10.* 2431

Vern.—*Tikor,* BENG.

Tikhur has already been given as the Central Provinces' vernacular name for **C. angustifolia. Moodeen Sheriff** says "although the name *Kúva* is correctly applicable only to **C. angustifolia** and **C. leucorrhiza,** it is often used in Malabar for **C. Amada,** because the tuberous root of that plant also yields a kind of arrowroot."

Habitat.—Roxburgh says this is a native of Behar. **Mr. J. Glass,** the Surgeon of Bhagulpore, furnished **Roxburgh** with roots of the plant, and soon after it had taken so kindly to the Botanic Gardens that **Roxburgh** wrote "it grows freely and flowers in May."

Food.—Mr. Glass wrote as follows to **Roxburgh** regarding the preparation of arrowroot from this plant, "the process for obtaining the starchy substance called *Tikor* is as follows: the root is dug up, and rubbed on a stone, or beat in a mortar, and afterwards rubbed in water with the hand, and strained through a cloth; the fecula having subsided, the water is poured off, and the *Tikor* (fecula) dried for use." **Dr. Irvine** (*Mat. Med., Patna*) alluding to this species says its "fine amylaceous farina is equal to arrowroot." FOOD. Arrowroot. 2432

C. longa, *Roxb., Fl. Ind., Ed. C.B.C., 11.* 2433

TURMERIC.

Vern.—*Haldí,* HIND.; *Halud,* BENG.; *Haldar, halja,* PB.; *Haridrá, nisá,* SANS.; *Kurkum, aurukesáfur, zarsúd,* ARAB.; *Zard-chóbah, dárzard,* PERS.; *Manjal,* TAM.; *Pasupu,* TEL.; *Mannal, marinalu,* MAL.; *Arishina,* KAN.; *Halede,* MAR.; *Halada,* GUZ.; *Kahá,* SING.; *Sanæ, tanun, hsa-nwen,* BURM. **Roxburgh** remarks this is known as *Pit-ras* in Bengal, *Pampí* in Telegu, and *Kurkum* in Hebrew. According to some writers Turmeric is the *kiang-hoang* of the Chinese.

Dymock says the best known Arabic names are *Urúk-es-subr, urúk-es-sabá-ghin* and the Persian *Zard-chubah.* This is probably the Κύπειρος of **Dioscorides. U. C. Dutt** writes that the Sanskrit *haridrádve* or the two turmerics, signifies turmeric and the wood of **Berberis asiatica, Moodeen Sheriff** says that in many books *Kurkum* is incorrectly given to saffron, and that *haridrá* is also wrongly given to yellow orpiment, that substance being in Sanskrit *Haritá lakam.*

References.—*Voigt, Hort. Sub. Cal., 565; Thwaites, En. Ceylon Pl., 316; Dalz. & Gibs., Bomb. Fl., 87; Stewart. Pb. Pl., 238; Manjella-kua, Rheede, Hort. Mal., XI., 21, t. 11; Curcuma domestica major, Rumph. Amb., V., 162, t. 37; Dar-zard of Garcias; Amomum curcuma, Gmelin and Jacq., Hort. Vind., III., t. 4; Pharm. Ind., 231; Ainslie, Mat. Ind., I., 454; O'Shaughnessy, Beng. Dispens., 649; Moodeen Sheriff, Supp. Pharm. Ind., 126; U. C. Dutt, Mat. Med. Hind., 255, 299, 311; Dymock, Mat. Med. W. Ind., 764; Fleming, Med. Pl. and Drugs, as in As. Res., Vol. XI., 165; Flück. & Hanb., Pharmacog., 638; U.S. Dispens., 15th Ed., 1629; Bent. & Trim., Med. Pl., t. 269; S. Arjun, Bomb. Drugs, 140; L. L. Dey, Beng. Drugs, 46; Murray, Pl. and*

CULTIVATION.

Drugs, Sind, 21; Waring, Bazar Med., 140; Year Book Pharm., 1873, p. 113; Medical Topog., Ajmír, 136; Mason, Burma, 513, 863; Man. Coimbatore Dist., 228, 229, and 230; Baden Powell, Pb. Pr., 299, 380; Atkinson, Him. Dist., 706, 734, 774; Drury, U. Pl., 169; Lisboa, U. Pl. Bomb., 174, 249, 288; Birdwood, Bomb. Pr., 231, 304; Liotard, Dyes, 83; Balfour, Cyclop., 859; Ure, Dic. Indus., Arts, and Manuf., Vol. III., 966; Kew Off. Guide to the Mus. of Ec. Bot., 62; Kew Off. Guide to Bot. Gardens and Arboretum, 79; Simmonds, Trop. Agri., 382, &c., &c.

Condiment Form. 2434 Dye Form. 2435

Habitat.—Turmeric is extensively cultivated all over India for its rhizomes. It is the well-known *haldí* universally used as a condiment with curry-stuffs and also as a dye, and is one of the most profitable of crops. The dye-yielding rhizome is harder and much richer in colour than the edible. These conditions are thus special adaptations which possibly point to an ancient cultivation. At the same time, though several species of **Curcuma** are undoubtedly natives of India, some of which appear to have been mistaken for the true turmeric, there is little of a positive character that would justify the supposition that **Curcuma longa** itself is a native of India. **Simmonds** (*Tropical Agriculture, p. 383*) says: "The **Curcuma longa** grows wild in the province of Mysore, and is probably indigenous to various other parts." On the other hand, **Roxburgh** and all botanical writers speak of it only as cultivated, and **Ainslie** even remarks that "The **Curcuma longa** grows wild in Cochin-China, and is there called *Kuong huynh.* **Loureiro** gives us a long list of its medicinal virtues in lepra, jaundice, and other disorders." Although there is a Sanskrit name for the plant and also names for it in most of the languages of India, the suggestion may be offered that it is most probably a Chinese or Cochin-Chinese species which may have superseded some of the indigenous **Curcumas** formerly in use and which bore the names now given to this plant, just as the true arrowroot plant is rapidly displacing the indigenous or East Indian species. **Dalzell** and **Gibson** have no hesitation in treating the plant as introduced into Bombay, and **Dr. Hove** (a botanist of high merit who explored a large portion of the Western Presidency 100 years ago) observed the plant being cultivated at Mahim, but makes no mention of its being wild. (*For History of Turmeric, see page 664.*)

CULTIVATION. 2436

Cultivation, Yield, and Soil.

Bengal. 2437

Bengal,—The earliest and to this day one of the most complete accounts of the cultivation of this plant is that given by **Roxburgh**. This may be reproduced here to admit of comparison with modern systems. "The ground must be rich, friable, and so high as not to overflow during the rainy season, such as the Bengalees about Calcutta call *danga.* It is often planted on land where sugar-cane grew the preceding year, and is deemed a meliorating crop. The soil must be well ploughed and cleared of weeds, &c. It is then raised in April and May, according as the rains begin to fall, into ridges, nine or ten inches high and eighteen or twenty broad, with intervening trenches nine or ten inches broad. The cuttings or sets, *viz.*, small portions of the fresh root, are planted on the tops of the ridges, at about 18 inches, or 2 feet asunder. One acre requires about from nine hundred such sets, and yields in December and January, about two thousand pounds weight of the fresh root." The Agricultural Department of Bengal (*Ann. Rep., 1886, p. LV.*) publishes some useful information regarding the modern system of turmeric cultivation. The Director says there are "two varieties grown—one known

Deshi. 2438

as the *deshi* or country, and the other as the Patna variety. The latter is of a richer colour and gives a better outturn. Loamy soil, even of a very inferior quality, will grow turmeric. It can be grown in shady places, but does better on open grounds. Turmeric is grown after one of the pulses. The preparation of the soil necessary for turmeric is similar to that for ginger, but lands intended for turmeric need not be worked so fine. Six or seven ploughings suffice for this latter crop. The mode of planting turmeric very much resembles that of sugarcane: the only points on which the two differ are the following:—The furrows in the sugarcane field are about 6 feet long and 22 inches apart, those in the turmeric field being twice as large and 27 inches apart. Sugar-cane cuttings are very lightly covered with earth; over 6 inches of earth is placed on the turmeric cuttings. The usual planting time is the first week of Jaistya," that is, about the 20th of May. "The plants spring up in about a fortnight. One or two weedings are necessary, and care must be taken that the fields are not inundated. In some parts of Bengal it is not considered good practice to lift the plants the first year. On the setting in of the following rains new shoots appear and the plants are tended exactly as in the first year. 'After about a year and nine months turmeric is lifted.' When it is raised the first year, as is the practice in some places, the produce is less in quantity and inferior in quality." The Director of Agriculture, Bengal, has the following estimate of the cost of cultivation:— **CULTIVATION. Patna. 2439** **Mature. 2440**

	R	a.	p.
6 Ploughings	2	4	0
3 maunds of seed at R3	9	0	0
Planting, 8 men at 4 annas a day	2	0	0
To earth up four times	4	0	0
Four weedings, 3 men at a time	3	0	0
Repairing the furrows, 4 men	1	0	0
To dig out, 6 men	1	8	0
To clean, 3 men	0	12	0
To boil, 6 men	1	8	0
To dry, 8 men	2	0	0
Earthen pots	1	0	0
Rent	4	0	0
TOTAL	32	0	0

BENGAL. 2441 It is not stated whether this is the cost of cultivating an acre or a bigha; the latter being a third of the former, the difference would be very considerable. Dr. McCann (*Dyes and Tans of Bengal*) published extracts from an extensive series of reports from all the districts of Bengal furnished in connection with the Economic Museum. In these, very contradictory statements are made. Turmeric is planted in Rájsháhí in March to April and dug up a year later; in Sáran not until June, and in Bhagulpur August, "so that, if this account be correct, the season of planting varies from February to July, and of digging up from October to April." **Season of Plants. 2442** In Hugli the estimated cost of cultivation per *bigha* is R6-8, in Rájsháhí R7-8, in Bhagalpur R15, and in Monghyr R10. **Of Harvest. 2443** The yield is variously stated at from 8 to 18 maunds a bigha, but Dr. McCann explains that the errors in the returns as to yield may be due to one district stating the yield of raw and another of the dry tubers. **Yield. 2444**

N. W. P. 2445 *N.-W. Provinces.*—Mr. Atkinson explains that turmeric is grown very generally in the North-West Provinces, and that "it is largely cultivated in Kumaon, Garhwal, and parts of Dehra-Dún, and forms one of the most important and profitable articles of export from the lower hills. It is grown in jungles where nothing else can be raised, as well as in the open Dúns and the Bhábar. The crop is planted in April-May, and the produce is gathered in November. It has been estimated that the cost of cultivating a *bísi* (40 square yards less than a British acre) to turmeric is

CURCUMA longa.

Turmeric.

CULTIVATION.
N. W. P. R36, for which one rupee goes for rent, R5 for sowing, R3 for planting out,
Cost. R20 for seed, R4½ for weeding and hoeing, and R2½ for harvesting. An
2446 acre will produce 30 maunds, worth R60, and when cured and dried about
Profit. 7½ maunds, worth R75, giving a net profit of R31 per *bísi*. **Sir E. C. Buck,**
2447 in his *Dyes and Tans of the North-West Provinces*, remarks: "According to an account of its cultivation in the Cawnpore District, it requires very plentiful irrigation, and is grown on *dúmat* (sandy loam) soil, commonly in company with *Ghuyun* (Yam, (*sic*) **Colocasia antiquorum**). In June the ground is well manured, 40 cart-loads being thrown on to one acre of land. It is then watered twice and well ploughed. When the rains set in the small roots of turmeric, to the amount of 250℔ to the acre, are then planted, one to the square foot, and so much water do they require that trenches have to be dug through the whole field only one foot apart. After the rains it has to be watered every week. The roots are ready for digging up in January."

BOMBAY. Of *Bombay* it has been stated—"In Gujarát and Kaira it is planted to-
2448 wards the end of May, and yields from 60 to 300 maunds (of 26℔ each) per
Yield. bigha." **Dalzell** and **Gibson** say it is largely grown in these parts of Bom-
2449 bay "where the garden soil is of superior quality, and water abundant." "An average crop will give a return equal to sugarcane, *viz.*, R100 per bigha." The turmeric of the Deccan is reported to be "of several kinds, the tuber in all cases being the useful part. The kind used in dyeing is the
Lokhandi. *lokhandi halad* with very hard roots." That used in medicine is "highly
2450 aromatic," and is also used in seasoning curries and *dal*.

Aromatic. *Panjáb.*—It is not apparently very extensively grown in the Panjáb; at
2451 least little of a definite nature can be learned about the Panjáb cultivation.
PANJAB. **Baden Powell** says it is grown on low moist soil and requires much care
2452 and manure. It is planted in May and is not matured till the end of November. **Stewart** writes: "This is commonly grown frequently along the edge of fields of ginger—in the Panjáb Sewálik tracts and outer hills from 2,000 to 4,000 and sometimes to 5,500 feet up to the Rávi at least, and occasionally beyond that. **Irvine** is probably mistaken in stating that it is cultivated in Peshawar and Bannu." In the Gazetteer of the Kangra District it is stated "Turmeric is reared in parts of the Hamirpur, Dehra, and Núrpúr *tahsils*. It is cultivated on low moist soils and also in the low
2453 valleys of outer Seoráj on the Sutlej, and requires much care and manure. It is planted in May like the potato, by pieces of the root, and is not matured till the end of November. The tubers are then taken up and dried, partly by the action of fire and partly by exposure to the sun. It is considered quite as remunerative a crop as sugar, and has the advantage, that it occupies the soil for six months only. A few localities supply
2454 turmeric for the consumption of the whole district." The Gazetteer further states that in the Kangra District there were, in 1880-81, 1,621 acres under this crop and in 1881-82, 1,520 acres.

MADRAS. *Madras.*—Turmeric cultivation is alluded to in various publications
2455 regarding South India, but no article has been found that deals with the Presidency collectively. Of Coimbatore it is stated that it is usually grown as a mixed crop with yams, maize, castor, brinjal, onions, &c. "The soil is thoroughly prepared by repeated ploughing and heavy manuring, municipal sweepings and ashes being a favourite manure." "In June or July, the soil having been ridged up about 2 feet apart, the rhizomes are planted, a cubit or less from one another, on the ridges and thereafter watered every three or four days until the end of December; thenceforward somewhat less often till March and April, when they are dug up. The crop is hoed and weeded several times in the first four months. The other crops are variously planted; the onions on the

sides of the ridges, the others in lines around, and through the area so as to define, shade, and in some sort protect the crop." It is explained that in some parts of the district less watering is required, and that as a rule turmeric is not grown more than once in three years and is followed by *rágí* and paddy. "The seed required is from 500 to 600 measures, and the outturn of prepared turmeric, from 3,000 to 5,000℔, value to the ryot R120 to R200. To this must be added the value of the other crops, which is very considerable; yams (*sic*) (=*Sepa kilanga* or **Caladium nymphæifolium**) will yield 250 maunds of 25℔ each, worth 12 annas per maund. Probably when these two crops are grown together the yield of each is much less. The expense of cultivation, if the labour be charged for as hired, will be something as follows :—

CULTIVATION. MADRAS.

Return. 2456

Cost. 2457

	R	a.	p.
Manure	10	0	0
Six ploughings	3	0	0
Ridging and sowing	3	0	0
Hoeing and weeding	14	0	0
Fifty waterings allowing for rain-fall (gardens only)	40	0	0
Digging out	6	0	0
Sizing and preparing	14	0	0
Seed cuttings	25	0	0
Assessment	1	8	0
TOTAL	116	8	0

"When grown on wet land the assessment is usually R6 or 8; as it is grown on the higher lands of a sandy loam character, which are seldom charged higher than stated; in this case the heavy charges for watering is eliminated. The yield on gardens is probably better than on wet lands, both in quantity and quality. That the statistics are not exaggerated is known by various leases; in a recent Karur rent suit the area of the land was 99 cents (say an acre), the crop turmeric, the rent paid by the tenant for the use of the land for this one crop was R75, and the Government assessment R6. The land-owner who pays the assessment thus cleared R69 by simply letting out his acre of land and the tenant was able to make a profit, after paying this immense rent and the whole costs of cultivation. As he probably cultivated the land himself the actual cost to him was little besides manure and seeds; but the value of the crop could not have been much under R150, and was possibly more." **2458**

PREPARATOIN OF THE RHIZOME.

PREPARATION. 2459

Various systems are apparently practised for preparing the rhizome for the market. Of Bengal it has been said:—"After the rhizomes have been dug out of the ground, they are freed from the fibrous roots and cleaned. They are then put in earthen pots, the mouths of which are to be carefully closed with earthen covers and cow-dung. These pots are then very gradually heated. The turmeric is made to boil in its own juice, a process which gets rid of the raw smell of turmeric. It is then dried in the sun, the drying taking nearly a week, during which the turmeric requires to be covered in the night to protect it from dew. In some places turmeric is boiled in water in which a little cow-dung is mixed." (*Rept. Agri. Dept., p. LV*). Of the North-West Provinces, **Sir E. C. Buck** says: "When dug up the roots are boiled and dried in the sun; in this form they are the turmeric sold in the Indian bazars. When the dye is to be used the roots are again boiled aud powdered while wet. A decoction is then made of this paste in water, in which the cloth is well steeped, being subsequently dried in the shade. In the Kumaon district

BENGAL. 2460

N. W. P. 2461

CURCUMA longa.

Turmeric.

PREPARATION. PANJAB. 2462 the roots are soaked in lime juice and borax before being powdered instead of being boiled." Of the Panjáb, **Mr. Baden Powell** says the tubers are taken up in November and dried partly by the action of fire

MADRAS. 2463 and partly by exposure to the sun. Of Coimbatore it is reported: The roots are carefully sized and separately boiled in a mixture of cow-dung and water, dried and sent to market."

AREA. 2464

Area under Turmeric.

Trustworthy particulars cannot be learned regarding the total area in India annually under this crop; but from the extensive uses of the tuber and the remunerative nature of the crop, it may be inferred to be very much more extensive than shown in the published returns. The following shows the acreage returned as under this crop:—

	Acres.
Bengal (according to **Dr. McCann**) perhaps	30,000
Madras	15,000
Bombay	6,000
Berar	2,000
Panjáb	3,500
Total	56,500

TRADE. 2465

Trade in Turmeric.

Regarding the Indian Foreign trade in this article **Mr. O'Conor,** in his Review of the Trade in 1876-77, wrote: "Turmeric was exported to the value of $10\frac{1}{2}$ lakhs of rupees, the quantity being 123,824 cwt. This article has hitherto been recorded in the returns under the heading 'Spices,' but it is more appropriately classed as a dyeing material. It is not really a spice but rather a condiment, and for this purpose it is very largely used in India; but it is also extensively employed as a dye, and almost all of that which is exported from this country is used as a dye. It is mostly exported from Bengal to England, France, and the United States." In 1879-80 **O'Conor** again wrote: "Turmeric is classed as a dye in the returns, and for that purpose it is most largely used in Europe, though it is said also to be used to adulterate mustard." In the following year he stated that the exports in Turmeric had largely fallen off, and as compared with previous years, the article was no longer of im-

Foreign. 2466 portance. In 1881-82 the exports were 70,783 cwt., valued at R3,66,047, as compared with 1877-78, when they amounted to R12,40,189. In 1885-86 the trade had so far recovered itself that the exports amounted to 156,287 cwt., valued at close on 14 lakhs of rupees. Last year they amounted to 140,994 cwt., valued at R10,32,025.

Internal. 2467 Full particulars cannot be learned as to the extent of the internal trade, but it must be very extensive, and even a trans-frontier trade exists; Kashmír receives a considerable amount. The various Indian ports last year exchanged 281,117 cwt. of turmeric valued at R24,38,260.

HISTORY 2468

History of Turmeric.

Turmeric yields a yellow dye of a fleeting character, which formerly was far more extensively employed by the natives of India than at the present day. Its chief features that recommended it for decorative purposes at marriage ceremonies, &c., were cheapness, ease of preparation, and facility of being removed. But these are conditions even more readily attained by aniline colours, while glaringly brilliant results are obtained, and, consequently, even religious injunctions have to a certain extent given place to the encroachments of the tar dyes. Writing of this subject **Dr. McCann** (*in his Dyes and Tans of Bengal, p. 85*), says: "Formerly on festive occasions an infusion of turmeric was used in dyeing garments, but magenta has now taken its place." Not only were the wedding garments formerly coloured with turmeric but even the body used to be (and still) is anointed with a paste of this substance. **Dr. U. C. Dutt** says: "The rubbing of turmeric and oil is an essential part of the Hindu marriage festival, as well as of some religious ceremonies." **Balfour** writes of the domestic uses of this dye: "The root enters into many of the religious ceremonies of the Hindus. The entire, or the corners, of every new article of dress, whether of man or woman, are stained before wearing it with a paste made of the root and water. Mixed with lime, it forms the liquid used in the Arati ceremony for warding off the evil eye. Women use it largely as a cosmetic and some smear all the body with it as a detergent. It is a mild aromatic and carminative, and is largely used as a condiment in curries; the paste is applied to foul ulcers. Clothes dyed with it are deemed a protection against fever." "With it, in conjunction with lime-juice, the Hindus of the sect of Vishnu prepare their yellow *tirúchúrnum*, with which they make the peculiar mark on their foreheads."

HISTORY.

Wedding Garments. 2469

Cosmetic. 2470

Markings on Foreheads. 2471

Sir E. C. Buck, in the *Dyes of the North-West Provinces*, says: "The dye given by turmeric is of a dull yellow colour; it is fleeting, and, except in dyeing the commoner sort of cloth, is seldom used, except in combination. The action of an alkali changes its colour to red." For the past century or more European writers have deplored the fact that no means could be discovered for making the dye fast. **Ainslie**, for example, says: "**Mr. Brande** notices the great quantity of colouring matter it yields on being digested in water or alcohol, regretting that it cannot be rendered permanent as a dye." It is somewhat remarkable that **John Huyghen Van Linschoten**, who spent several years on the Malabar coast from about the date of 1596, should describe the races of people he met with, going into every detail as to their social habits, domestic and agricultural life, marriage customs, agricultural produce, and industrial productions, but should make no mention of turmeric. He describes Cardamoms, Cumin seed, Galanga, Pepper, Cubebs, Tamarind, Ginger, Mangos, &c., &c.; but while discussing the preparation of curry and chutney makes no mention of the habit of eating turmeric or of dyeing garments with it. This might be accepted as pointing to its use having been much less general in these days (at least on the Western side of India) than at the present time. On the other hand, an ancient cultivation in India is clearly indicated by the frequent mention of the plant in the early literature of the Hindus, and by the fact that there are several well recognised or distinct cultivated forms of the plant. **Garcia de Orta**, who lived in Goa in 1563 (or shortly before **Linschoten**), describes under the name **Crocus indicus** a tuber which appears to be turmeric, and **Dioscorides** mentions an Indian plant as a kind of Cyprus (*Κύπειρος*) as resembling ginger, but having when chewed a yellow colour and bitter taste. This was most probably turmeric, but it must not be forgotten that several other species of **Curcuma** afford a yellow colour that indeed it is probable some of the so-called forms of **C. longa** may prove the tubers of different species. **Flückiger** and **Hanbury** say: "Several varieties of turmeric, distinguished by the names of the countries or districts in which they are produced, are found in the English market; but although they present differences which are sufficiently appreciable to the eye of the experienced dealer, the characters of each sort are scarcely so marked or so constant as to be recognizable by mere verbal descriptions. The principal sorts now in commerce

Dye Fleeting. 2472

 Turmeric.

HISTORY. are known as *China, Madras, Bengal, Java,* and *Cochin.* Of these the first named is the most esteemed, but it is seldom to be met with in the European market." **Linschoten**, while describing with the utmost detail the trade of Cochin, makes no mention of Turmeric, but at the same time references occur, of turmeric as employed in Europe about the time of which **Linschoten** wrote, so that it must have been exported from other parts of India or from other tropical countries. **Flückiger** and **Hanbury** say of the Cochin Turmeric of the present day that it "is the produce of some other species of **Curcuma** than **C. longa.** It consists exclusively of a bulb-shaped rhizome of large dimensions, cut transversely or longitudinally into slices or segments. The cortical part is dull brown, the inner substance is horny and of a deep orange-brown, or when in thin shavings of a brilliant yellow. **Mr. A. Forbes Scaly** of Cochin has been good enough to send us (1873) living rhizomes of this **Curcuma**, which he states is mostly grown at Alwaye, north-east of Cochin, and is never used in the country as *turmeric,* though its starchy tubers are employed for making arrowroot." (Conf. with **C. angustifolia** and other sources of East India arrowroot.)

Cochin. Doubtfully True Turmeric. 2473

TURMERIC DYE.

DYE. 2474

Dye-Yielding Rhizomes. 2475

Dye.—It has already been stated that a special form of turmeric is grown for this purpose, namely, a harder root, much richer in the dye principle than in the ordinary condiment form. This dye rhizome receives separate names in the various provinces of India, but is most generally known by the name *lok-handi haladi*; other dye forms are as *mœla-haldi, jowala-haldi,* and *amba-haldi.* Under the paragraph, above devoted to an account of the preparation of the tuber, mention has also been made of the further process which the dyer has to adopt in preparing his infusion. The employment of borax, in Kumaon, will be found to have a very considerable interest, since the system there pursued, and doubtless accidentally discovered, is dependent on an important chemical feature of the dye principle.

The colour is only deposited in the rhizome with age, and hence, in all probability, the above mentioned forms have been obtained by a process of careful selection of stock observed to produce the colour freely. It is of importance, however, that the European merchant, in purchasing for dye purposes, should see that he gets the hard dye-yielding form and not the softer aromatic condition which is used as a condiment. Although, of course, turmeric is still employed by itself as a simple and cheap dye, its more general use at the present day in India, is as an auxiliary to other dyes and in Calico printing. It is also used to some extent to impart a colour to native-made paper. Mordants are but rarely required with the dye, as it is found to attach itself readily enough to wool, silk, or cotton. Alkalies deepen the colour, making it almost red. Alum is said to purify the colour and to destroy all shades of red. The dyers of Calcutta produce a brilliant yellow, known as *basanti rang*, by mixing turmeric with *Sajimati* (Carbonate of Soda) and lemon or lime juice. **Dr. McCann** remarks of this process: "Here the acid is apparently used to correct the red tint, produced always where an alkali acts on turmeric." Myrabolams are sometimes employed with turmeric, but the chief compound colour in which turmeric plays an important part is the green shades formed along with indigo. The fabric is first dyed with indigo and then dipped in a solution of *haldí.* Turmeric is also often added to sharpen or brighten other colours, as, for example, *Singrahar* (**Nyctanthes arbortristis**), lac dye, *al* (**Morinda tinctoria**), safflower (**Carthamus tinctorius**), and *toon* (**Cedrela Toona**).

Yellow. 2476

Green. 2477

The Indian Calico-printers use turmeric by preparing a mixture of "4 gallons of water containing pomegranate rind and alum in the following proportions:—Turmeric 5lb, pomegranate rind 2lb, and alum 1¼lb. The compound is left to stand for a night, the surface water strained off, and ½lb of indigo added. It is then prepared for use by being thickened with gum, clarified butter, and flour in the usual way. The colour is greenish yellow and is fleeting." (*Buck, Dyes and Tans of N.-W. P.*, 55.) CALIC PRINTING. 2478

The rhizome is still largely used by the European dyers, though the fluctuations in the trade may be viewed as due to the development of the aniline industry. **Professor Hummel** says of it:—"Notwithstanding the very fugitive character of the colour it yields, it is still much used, especially by the wool and silk dyers for the production of compound shades—olives, browns, &c. It gives a bright yellow colour without the aid of a mordant, but when mordants are used with it, it yields other colours not unlike those obtainable from the yellow dye-woods. The colouring matter of turmeric is one of the few for which cotton has naturally a strong attraction." "The colour is not fast either to light or to alkalies; even very slight alkaline solutions,—*e.g.*, soap—change it to a reddish brown." For wool the dyeing should never be done under a higher temperature than 60° C. "Boiling should be avoided, since the bright yellow then becomes soiled by reason of impure extractive matters entering into solution. If the wool is mordanted with aluminium or tin, the colour is somewhat brighter, and in the latter case more orange. With the use of potassium dichromate and ferrous sulphate as the mordant, the colours produced are olive and brown. Silk is dyed in the same manner as wool." (*Dyeing of Textile Fabrics*, 367.) **Crookes** says:—"The colouring matter of turmeric is very sparingly soluble in water, but alcohol, ether, and fatty and essential oils dissolve it readily. The alcoholic solution of turmeric is thrown down, by the addition of tin crystals, as a red precipitate; by acetate of lead, a chestnut brown; by mercury salts, reddish-yellow; salts of iron colour the tincture brown; alkalies turn it brown; weak acids do not act upon the pigment, which is turned red by concentrated acids. The colouring matter of turmeric has received the name of *Curcumin*. EUROPEAN USES. 2479 Cotton. 2480 Wool. 2481 Silk. 2482 Curcumin. 2483

"**M. E. Schlumberger** has been the first to investigate the modifying action of boracic acid upon *curcumin*. It is well known that turmeric paper becomes brown under the joint influence of the boracic and any mineral acid, preferably the hydrochloric. Ammonia turns this colour blue. When an alcoholic solution of curcumin is boiled with boracic acid its colour turns orange, and upon the addition of water to the previously cooled solution a vermillion-coloured powder is thrown down, being a compound of the pigment with boracic acid. This combination is broken up by the sufficiently prolonged action of water; the boracic acid dissolves, and there remains a yellow resinous matter which differs from curcumin, inasmuch as it does not yield a red colouration with boracic and hydrochloric acids, and on being dissolved in alkalies gives a greenish-grey colouration. The boracic compound of curcumin dissolves with a purplish-violet colour in alkalies, but this soon turns to grey. When hydrochloric acid is added to an alcoholic solution of bromo-curcumin, and the fluid boiled, it rapidly assumes a blood-red tint, on cooling a new body is thrown down, while the boracic acid remains in solution. The substance so deposited is first washed with dilute alcohol, next with pure water, in order to eliminate all boracic acid; the residue is dried, and next dissolved in a boiling mixture of 2 parts of alcohol and 1 part of acetic acid. This fluid, being filtered while hot, deposits on cooling rosocyanin, while the pseudo-curcumin remains in solution. By pseudo- Action of Boracic Acid. Red color. 2484 Rosocyanin. 2485

EUROPEAN USES. curcumin is understood the organic resinoid substance resulting from the prolonged action of water upon boro-curcumin, just above-mentioned. The rosocyanin is first dried and next treated with ether, in order to remove the last traces of yellow colouring matter: thus purified, it is a crystalline substance, of a cantharides-like lustre, insoluble in water, ether, and benzol, but very soluble in alcohol, to which it imparts a most magnificent deep rose-red, quite comparable to fuchim solutions. This fluid becomes permanently yellow on being boiled. Ammonia turns the alco-
Blue Color. 2486 holic solution of rosocyanin a splendid blue, but this colouration changes shortly into a dirty grey; acids turn the blue ammoniacal solution red again, lime and baryta water yield blue precipitates in the alcoholic solutions, which do not contain any boracic acid at all.

Colouration of Flowers. Cyanin. 2487 "The relations existing between curcumin and rosocyanin (also called roscocyanin) and pseudo-curcumin are unknown; neither was, until July, 1870, the true composition of curcumin known. It is very probable that the phenomena of colouration as exhibited by curcumin, which turns red and blue, and then yellow again, under the action of comparatively weak re-agents, bear a relation to certain phenomena observed with flowers.

"It is not impossible that the roscocyanin of **M. Schlumberger** is identical with the red colouring matter of flowers or cyanin investigated by **M.M. Fremy** and **Cloez**, which latter substance is turned also blue by alkalies. If this suggestion proves correct, on more precise investigation turmeric could become a useful source of preparation of the red colouring matter of flowers, which it is very difficult to obtain by direct extraction.

Although turmeric is rich in colouring matter, its want of permanence is a hindrance to its application as a dye-material. Some time back the
Printing Silks. 2488 use of turmeric was almost exclusively limited to printing and dyeing silks. It is now employed to a vast extent in stuff-dyeing, forming an
Sour Browns. 2489 important constituent in certain compound colours, especially the so-called "sour browns."

MEDICINE. 2490 **Medicine.**—Used as a stimulant in native medicine; externally applied in pains and bruises, and internally administered in disorders of the blood. Its use as an external applicant in bruises, leech bites, &c., is perhaps its most frequent medicinal application. The fresh juice is said to be an anthelmintic. A decoction of the rhizomes is applied to relieve catarrh and purulent opthalmia.

A paste made of the flowers is used in ringworm and other parasitic skin diseases. **Dymock** says the Muhammadans "use turmeric medicinally in the same manner as the Hindus; they also prescribe it in affections of the liver and jaundice on account of its yellow colour." "The editor of the *Pharmacopœia of India* speaks favourably of the use of a decoction of turmeric in purulent conjunctivitis; he says it is very effectual in relieving the pain. In coryza he states that the fumes of burning turmeric directed into the nostrils cause a copious mucous discharge, and relieve the congestion." **Murray** remarks that it is "given by the native doctors in the diarrhœas which are so troublesome and difficult to subdue in atonic subjects." **Baden Powell** remarks that it is employed in "intermittent fevers and dropsy." "It contains much essential oil and starch and acts as a stimulant and aromatic tonic."

Special Opinions.—§ "The root, parched and powdered, is given in bronchitis in doses of grs. xxx to xl" (*Civil Surgeon J. Anderson, M.B., Bijnor*). "The smoke produced by sprinkling powdered *haldi* over burnt charcoal will relieve scorpion sting when the part affected is exposed to the smoke for a few minutes. A paste made of fresh rhizome is applied on the head in cases of vertigo. Fresh juice is cooling. Fumes of burning root is employed during hysteric fits" (*Assistant*

MEDICINE.

Surgeon T. N. Ghose, Meerut). "Turmeric and alum, in the proportion of one to twenty, is blown into the ear in chronic otorrhœa" (*Dr. Barasha Hormarji Baria, L.M.S., Bombay*). "For external application a paste is prepared of turmeric and caustic lime which is applied over the bruised part. It seems to avert inflammation and relieve pain. It appears also to have vermifuge properties when externally applied either as a lotion or a paste by itself" (*Civil Surgeon D. Basu, Faridpur*). "The powdered root is used as a fumigation in commencing catarrhs. The inhalation is generally taken at night and no fluid is allowed for some hours afterwards. The effect is said to be in many cases a complete cure of cold" (*Narain Misser, Kothe Bazar Dispensary, Hoshangabad, Central Provinces*). "**Curcuma longa**, the *Mungal* of Tamil, powdered and mixed with warm milk and pepper powder, will suppress an attack of catarrh with fever" (*Surgeon-Major Lionel Beech, Cocanada*). "Pounded with water to a suitable consistence and mixed with lime (the mixture has a deep-brown colour) forms an efficacious application to bruises and sprains. Pure turmeric is also useful in scabies and other skin diseases. Fumes of the burning root are resorted to by witch-drivers in hysteria. This disease is believed to depend on witch-craft, and I recollect patients treated with *mantras* while burning turmeric was held to the nostrils. A common practice with the natives is to use a rag stained with turmeric to wipe off discharge in conjunctivitis and ophthalmia" (*Assistant Surgeon Shib Chunder Bhuttacharji, Chanda, Central Provinces*).

Food.—Turmeric forms one of the indispensable ingredients in curries, and is used for colouring confections, &c. — FOOD. Condiment. 2491 Curry Powder. 2492

Chemistry of Turmeric.—Dr. Dymock gives a brief sketch of the chemical history of this subject which should be consulted. "*Curcumin*, the yellow-colouring matter of turmeric, has been examined by several chemists, whose experiments have led to the conclusion that its formula is either $C_{10}H_{10}O_3$ or $C_{16}H_{16}O_4$, that it melts at 172°, forms red-brown salts with alkalies, is converted by boric or sulphuric acid into rosocyanine, by reduction with zinc-dust into an oily body, by oxidation into oxalic or terephthalic acid, and by fusion with potash into protocatechuic acid." (*Conf. with paragraph treating of Dye property.*) — CHEMISTRY. 2493

Curcuma pseudo-montana, *Graham*. 2494

Vern.—*Sinderwani, sinderbur, sindelwan, hellounda*, BOMB.

Habitat.—Said to be a native of the Konkan, springing up at the beginning of the rains.

Food.—"The tubers, which are perfectly white inside, are boiled and eaten by the people during seasons of scarcity. Perhaps this plant, too, yields a part of East India arrowroot; that which comes from Ratnágiri is manufactured from its tubers" (*Lisboa; Dalz. and Gibs.*). — FOOD. Rhizomes. 2495 Arrowroot. 2496

C. rubescens, *Roxb.* 2497

Habitat.—"A native of Bengal, flowering time in the months of April and May, soon after which the leaves appear, and decay about the beginning of the cool season, in November. Every part has a strong but pleasant aromatic smell when bruised, particularly the root." (*Roxb.*)

Food.—Roxburgh and Voigt say the pendulous tubers of this species yield a form of arrowroot. — FOOD. Arrowroot. 2498

C. Zedoaria, *Roscoe (non-Roxb.)*; *Wight, Ic., t. 2005.* 2499

THE LONG AND THE ROUND ZEDOARY.

Syn.—C. ZERUMBET, *Roxb.*

Long and Round Zedoary.

Vern.—*Kachúra*, HIND.; *Sati, shori, kachura*, BENG.; *Sati, karchura*, SANS.; *Zurambád*, ARAB.; *Kashúr, urúk-el-káfúr*, PERS.; *Kachúra*, BOMB.; *Kich-chilick-kizhanghu, púlán-kizhanga*, TAM.; *Kich-chili-gaddala, kachóram*, TEL.; *Kach-chólam, kach-chúri-kizhanna, púlá-kizhanna*, MAL.; *Kachórá*, KAN.; *Thanu-wen*, BURM.

Fleming, Ainslie, &c., call this the *Nirbisi* of Sanscrit writers.

References.—*Roxb., Fl. Ind., Ed. C.B.C., 689; Dalz. & Gibs., Bomb. Fl., 274; Rheede, Hort. Mal., XI., t. 7; Rumph., V., t. 68; Pharm. Ind., 232; Ainslie, Mat. Ind., I., 490-492; II., 41; O'Shaughnessy, Beng. Dispens., 649; Moodeen Sheriff, Supp. Pharm. Ind., 127; U. C. Dutt, Mat. Med. Hind., 257, 317, 322; Dymock, Mat. Med. W. Ind., 2nd Ed., 771; Fleming, Med. Pl. and Drugs, as in As. Res., Vol. XI., 165; U. S. Dispens., 15th Ed., 1782; L. L. Dey, Beng. Drugs, 46; Murray, Pl. and Drugs, Sind, 21; Irvine, Mat. Med. Patna, p. 42; Kangra Gaz., I., 159; Medical Top. Ajmir, 140; Home Dept. cor. regarding Pharm. Ind., 240; Baden Powell, Pb. Pr., 300, 380; Drury, U. Pl., 170; Birdwood, Bomb. Pr., 87; Balfour, Cyclop., 859; Kew Off. Guide to the Mus. of Ec. Bot., 62.*

Habitat.—**Roxburgh** says it is a native of Chittagong, from which place the Bengal supply is derived. It is extensively cultivated in many parts of India in gardens, and, according to **Ainslie**, it is "a native of the East Indies, Cochin-China, and Otaheite." In the Kangra Gazetteer (I., p. 159), the *kachúr* is "said to be grown over the whole district, but in very small quantities, as its uses are limited."

ABIR. 2500 **Abir.**—The red powder, *Abír*, used by the Hindus at the Holi festival, is often made from the rhizome of this plant, ground to a powder and left for some time to saturate in water (*Kangra Gaz., I., 159*). The powder being purified and dried is mixed with a decoction of Sappan wood, when the red colour is obtained. *Abír* is now, however, extensively made from aniline dye. **Dr. McCann** describes the process adopted in Mymensing district, Bengal, for the preparation of the *Abír* powder; but he appears to have reversed the scientific names of the species of **Curcuma.** The *Shati* has, for the past forty years, been regarded as **C. Zedoaria,** *Roscoe,* while **Dr. McCann** gives it as **C. Zerumbet,** *Linn.,*—a name which does not exist in botanical literature. If he means **C. Zerumbet,** *Roxb.,* not *Linn.* (a synonym for **C. Zedoaria,** *Roscoe*) it is unfortunate he did not publish his economic information under the modern name, since the name **C. Zerumbet,** *Roscoe,* is applied to a perfectly distinct species.

2501 In Bengal the *Gulal* and *Abír* powders seem to be made together and sold mixed. In many parts of the country, however, this is not the case. The red powder or *Gulal* is prepared from Sappan wood and alum used to colour flour. The *Abír* or perfumed powder is not always of the same composition. In Bengal the root-stocks of **C. Zedoaria,** *Roscoe,* are used and apparently as the entire representative of the *Abír* powder of Upper and Western India. The Zedoary is also an ingredient in *Ghisi Abír* along with cloves, cardamoms, deodar, **Artemisia,** and **Cerasus.** The *Abír* most generally used, however, contains **Hedychium spicatum,** *Ham.,* instead of Zedoary combined with sandal wood. (*See* **Abir, A. 31.**)

Zedoary. 2502 **Description of the Zedoary of the shops.**—**Guibourt** says: "The round Zedoary is greyish-white, externally heavy, compact; grey and often horny internally, having a bitter and strong camphoraceous taste, like that of the long Zedoary, which it also resembles in odour. The odour of both drugs is analogous with that of ginger, but weaker unless the rhizomes be powdered, when it developes a powerful aromatic odour similar to that of cardamoms." **Dymock** adds that **Guibourt's** description "agrees exactly with the *kachúra* of India, but it is often cut into transverse slices instead of into halves and quarters."

MEDICINE. Rhizomes. 2503 **Medicine.**—The rhizomes possess aromatic, stimulant, and carminative

MEDICINE.

properties. Employed in native practice as a stomachic, and also applied to bruises and sprains. "The natives chew the root to correct a sticky taste in the mouth; it is also an ingredient in some of the strengthening conserves which are taken by women to remove weakness after child-birth. In colds it is given in decoction with long-pepper, cinnamon, and honey, and the pounded root is applied as a paste to the body" (*Dymock, Mat. Med. W. Ind., 2nd Ed., 772*). In the Kangra Gazetteer it is said of what appears to be this plant that "it is given as a carminative medicine internally and applied on the skin as a plaster to remove pains." "The root is of a pale-yellow warm and aromatic, like turmeric, but bitter."

Special Opinions.—§ "The rhizome of this plant is the *Amba-haldi* of the Bombay bazar. Bruised with alum in water, it is applied to bruised joints and other parts to remove echimoses" (*Assistant Surgeon Sakharam Arjun Ravat, L. M., Girgaum, Bombay*). "Small bits of the rhizomes are put in the mouth and chewed to allay cough" (*Assistant Surgeon Anund Chunder Mukerji, Noakhally*). "Demulcent, expectorant, and aromatic, dose about 1 drachm" (*Civil Surgeon John McConaghey, M.D., Shahjahanpore*). "The rhizome is considered to be a cooling medicine, also tonic and expectorant" (*Surgeon-Major J.M. Houston, Durbar Physn., Travancore, and Civil Apoth. John Gomes, Medical Storekeeper, Trevandrum*). "This is the *Kochora* of the bazar. It is used as an odoriferous ingredient of the cosmetics used for the cure of chronic skin diseases and internally as a mild aromatic stimulant in fever and colds" (*Assistant Surgeon Sakharam Arjun Ravat, L. M., Girgaum, Bombay*). "The roots imported into Leh as *kachúr, judwar,* and called by the Bhotes '*Bozbrga*' employed in Yarkand for washing the body, acting as a rubefacient (*Surgeon-Major J. E. T. Aitchison, Simla*). "The rhizomes are used by singers as a masticatory for clearing the throat of tenacious mucus; they are also used in cases of irritation of the fauces and upper part of the wind-pipe. The decoction is employed with sugar-candy, black pepper, and liquorice in relieving cough and bronchitis" (*Civil Surgeon J. H. Thornton, B.A., M.B., Monghyr*).

Judwar of Yarkand. 2504

Note.—The writer suspects that some of these notes allude to **Acorus Calamus**, *Linn., see* A. 430.

Perfumery.—The rhizomes of this plant constitute one of the most important articles of native perfumery. PERFUMERY. 2505

Trade.—Dymock says the Bombay supply comes from Ceylon, value R20 to R30 per candy of 7 cwt: as already stated, **Roxburgh** affirms that Bengal gets its supply from Chittagong. TRADE. 2506

Curcuma Zerumbet, *Roscoe (non-Roxb.)* 2507

The writer is unable to isolate the economic facts recorded by certain authors under this name from those given for **Curcuma Zedoaria**, and he suspects that all refer to one and the same plant, or to **Roxburgh's Zingiber Zerumbet.**

CUSCUTA, *Linn.; Gen. Pl., II., 881.*

Cuscuta reflexa, *Roxb.; Fl. Br. Ind., IV., 225;* CONVOLVULACEÆ. 2508

THE DODDER.

Syn.—C. GRANDIFLORA, *Wall.*; C. VERUCOSA, *Sweet*; C. MACRANTHA, *Don.*

Vern.—*Haldi-algusi-lutta, algusi,* BENG.; *Alagjari,* SANTAL (probably for **C. chinensis**); *Níla thári, nirádhar, ámil, sarbúti,* PB.; *Amaravela,* SANS. Bazar names for the seed.—*Akas-bel, áftímún, kasús,* HIND., PB.; *Akas-pawan, amarwel,* DEC.; *Akaswel,* GUJ.; *Nirmulí,* MAR.; *Sitama purgonalu, sitamma pógu núlu,* TEL.

CUSCUTA reflexa.

The Dodder or Cuscuta.

Some confusion exists regarding the vernacular names given to the species of **Cuscuta.** **Dymock** describes three species, two of which he has not determined botanically: he gives *Akásweli* as the local Bombay name for **C. reflexa**; *Aftímún*, a form with large fruits and seeds, the medicinal parts of which are imported from Persia; and *Kasús*, a smaller plant imported also from Persia, and generally attached to the spines and leaves of the plant on which it grows. The last mentioned may be **C. capitata,** so frequently found in the Western Himálaya, growing on the spiny plant— **Prinsepia utilis.** **Roxburgh,** who first described that species, states that it was found growing on **Crotalaria juncea.** The *Flora of British India* justly remarks that it is a puzzle to know where **Roxburgh** found it, since the species, as known to modern botanists, does not occur much below 6,000 feet. It is distributed from Simla to Kashmir, Beluchistan and Afghánistan. **Roxburgh** gives it the name of *algusi,* and calls **C. reflexa** *huldi-algusi-lata,* a name doubtless given in allusion to the yellow colour of the whole plant when mature. **Stewart** distinguishes four species, **C. macrantha,** *Don.*—the *nila thári;* **C. pedicellata,** *Led.* (the *kwiklapot, zránd* or *amlú*); **C. planiflora,** *Fenore;* and **C. reflexa,** the *ákásbel, aftímún, kasús.*

Habitat.—An extensive parasitic climber, making the trees quite hoary upon which it occurs, often growing to such an extent as to completely cover every bough and leaf. It occurs throughout the plains of India and ascends the Himálaya to about 8,000 feet.

DYE. 2509

Dye.—**Mr. Baden Powell** states that at Jhelam this plant is sometimes used as a dye. It would be a great matter if it could be utilised in this manner, as many trees are often completely covered and often killed by the plant. The dye is apparently unknown in Bengal. **Mr. Baden Powell** does not mention the colour; it is probably a yellow. **Drury** says it is occasionally used in dyeing.

MEDICINE. Plant. 2510

Medicine.—An extensive herbaceous climber, germinating in the soil, but becoming parasitic on the trees on which it is met with. It is chiefly found on **Zizyphus, Adhatoda, Ficus,** &c. The flowers are sweetly scented. The seeds are regarded as carminative, and for this purpose are boiled and placed over the stomach; they are also applied as an anodyne. A cold infusion is given as a depurative. They constitute part of the *Kasús* or purgative medicine sold in the Panjáb. (*Stewart.*) The native doctors of Sind and the Panjáb regard the seeds of this plant

Seeds. 2511

as alterative and are used along with Sarsaparilla to purify the blood. The natives having observed that the plant severs its connection from the earth, and not having discovered the existence of parasitic roots, have a proverb that he who finds the roots of this plant will become possessed with boundless wealth and of the power of invisibility. (*Murray.*) It is probable that the seeds of **Cassytha filiformis,** *Linn.*, under the vernacular name of *Akas-bel*, are sold and used indiscriminately with those of this plant.

Dymock says of the Persian dodder—*Aftímún*—that it "has a bitter

Stems. 2512

taste; in Arabic and Persian works it is described as the *Aftímun* of the Greeks, which had so great a reputation as a remedy in melancholy madness; it is still a medicine of importance with the hakíms of India, who follow in the footsteps of Jálenus (Galen)." The stems of **C. reflexa** are mentioned in the Bombay Gazetteer as specially useful in bilious disorders. **Baden Powell** says it is purgative and used externally against itch and internally in protracted fevers, retention of wind, and induration of the liver. It is also said to produce thirst.

FODDER. 2513

Fodder.—**Stewart** remarks **C. macrantha** is eaten by cattle and goats. "**Edgeworth** mentions that the mountaineers believe that crows pluck sprigs of this to drop into water, when they become snakes, and so furnish food for them."

Cus-cus (khus-khus), see **Andropogon muricatus, A. 1097.**

Cuscus seeds, see **Papaver somniferum.**

Cusparia or **Angustura bark,** see **Galipea Cusparia,** *St. Hil.*, RUBIACEÆ.

Custard Apple, see **Anona squamosa, A. 1166.**

Cutch, see **Acacia Catechu, A. 135.**

Cuttle-fish, see **Molusca.**

CYAMOPSIS, *DC.; Gen. Pl., I., 493.*

Cyamopsis psoralioides, *DC.; Fl. Br. Ind., II., 92;* LEGUMINOSÆ. 2514

Vern.—*Guár, dararhi, kuwára, kauri, syansundari, phaligawar, kachhur, khurti, khulti,* N.-W. P. and OUDH; *Guwár,* GUJ.; *Gauri, mutki, gawar,* BOMB.; *Duru raher,* SANTAL; *Pai-pázoon,* BURM. (*Kurz, Pegu Rept.*)

Habitat.—Cultivated in many parts of India from the Himálayas to the Western Peninsula. It is a robust erect annual, 2 to 3 feet high, grown as a rainy season crop.

CULTIVATION. 2515 **Cultivation.**—In the Bombay Gazetteer (Gujarat) it is said to be grown as an early kharif crop on sandy loam, sometimes by itself, sometimes mixed with *korad* (**Phaseolus aconitifolius**). "A crop of *guvár* is thought to do good to the soil." Duthie and Fuller doubt this plant being a native of India. They say a tall robust form is often grown in the North-Western Provinces as a hedge. This is known as Deoband *kawára* and is supposed to have come originally from Deoband near Saharanpur.

FOOD. Vegetable. 2516 **Horse-food. 2517** **Pulse. 2518** **FODDER. 2519**

Food and Fodder.—"*Guár* is grown in these provinces for two very different purposes,—as a vegetable for human consumption, and as a pulse for horses and cattle. For the former purpose it is invariably grown on highly manured land near villages, and assumes a much more luxuriant habit of growth than when grown for cattle. The portion eaten as a vegetable is the pod, which is plucked while green, after the fashion followed with the French beans of English gardens. As a cattle fodder it is grown for its grain and is then sown on light sandy soil, side by side and often mixed with *bájra*.

"The cultivation of guár as a vegetable is not very common, and is restricted to the market gardeners or "kachi" caste. Its cultivation as a cattle fodder is, on the other hand, of considerable importance in the districts to the west of the provinces where the agricultural cattle are of far better quality than the ordinary. Fully half of the agricultural cattle of the districts of the Meerut Division are purchased from outside, the cultivators of these districts recognizing that it is more profitable to import good animals from tracts specially fitted for breeding, than to attempt to breed them themselves on the limited grazing area at their command. The proportion of imported to home-bred cattle reaches its maximum in Meerut and steadily decreases as one goes eastward, until it becomes almost *nil* in Fatehpur and Allahabad. The value of a purchased animal is brought home more strongly to the cultivator than the value of a home-bred one, and much greater care is taken of the one than of the other. The western districts accordingly form the only tract in the provinces where crops are grown on any large scale for cattle fodder. The large cultivation of guár as a green fodder crop in the Meerut Division has been already noticed. It occupies there more than ten times as large an area as in any other Division. The cultivation of guár also reaches its maximum in the same tract, and is an indication of the care of agricultural stock which one would be glad to see extended to other parts of the provinces.

"Guár is sown at the commencement of the rains and is cut in October. Its average produce of dry pulse to the acre may be taken as 10 maunds."

Mr. Baden Powell says of the Panjáb: "Gujarat is the only district in the Panjáb proper which exhibits a sample; the pulse is stated by the Rohtak Local Committee to be made into *dál*, but to be used principally for cattle; it is boiled in a pan and then the grains are rubbed and worked about with the hand till a froth rises on the mass: a little mustard seed oil is then added; it is given to cattle to fatten them." The **Rev. A. Campbell** says the Santals eat the fruit.

CYANANTHUS, *Wall.; Gen. Pl., II., 557.*

[CAMPANULACEÆ.

2520 **Cyananthus, sp.** (? **C. linifolius**, *Wall.*); *Fl. Br. Ind., III., 434;*

Vern.—*Murra*, PB.

Habitat.—"A plant with pretty blue flowers, growing at 10,000 to 12,000 feet in Chumba."

MEDICINE. 2521 **Medicine.**—"The calyces are eaten, being mawkish-sweet, and are said to be good for asthma." (*Stewart, Pb. Pl.*)

CYANOTIS, *Don; Gen. Pl., III., 851; Wight, Ic., t. 2082 & 2089.*

2522 **Cyanotis axillaris,** *Rœm. et Schultes; DC., Mono. Phan., III., 244; Clarke's Commelinaceæ, table 35;* COMMELINACEÆ.

ONE OF THE SPIDER-WORTS.

Vern.—*Nirpulli* (**Rheede**), TAM.; *Soltraj, bagha-nulla* (**Ainslie**), HIND.; *Itsaka* (**Lisboa**); BOMB.

Habitat.—A herbaceous annual, met with in many parts of India; distributed to the Malay, China, and Australia.

MEDICINE. 2523 **Medicine.**—**Rheede** says that on the Malabar coast this is viewed as a useful remedy in timpanites, but **Dymock** writes (*1st Ed., Mat. Mad., W. India, 680,* omitted from 2nd Ed.) that although the plant is not uncommon in the western Deccan he has not known it to be used medicinally. **Ainslie** repeats **Rheede's** statement, and in a further account of the plant remarks that it was one of the plants brought to **Dr. Buchanan Hamilton**, while in Behar, as a useful medicine for external application in cases of ascites especially when mixed with a little oil. **Lisboa** says that the seeds

FAMINE FOOD. Seeds. 2524 of this, as also of **Commelina communis**, were eagerly sought for during the Bombay famine; they are wholesome and nutritious.

2525 **C. tuberosa,** *Rœm. & Schultes; DC., Monogr., Phan., III., 249.*

Syn.—TRADESCANTIA TUBEROSA, *Roxb.*; C. ADSCENDENS, *Dalz. in Hook. Jour. Bot., p. 343 (1852)*; C. SARMENTOSA, *Wight, Ic., 2087.*

Vern.—*Merom chunchi* (a name given from the resemblance of the roots to the papillæ of the goat); *Hodo jereng arak'* (the vegetable), SANTAL.

MEDICINE. Root. 2526 **Medicine.**—The **Rev. A. Campbell** says the ROOT is given in long-continued fevers and also for worms in cattle.

FOOD. Leaves. 2527 **Food.**—The LEAVES are eaten by the Santals as a pot-herb.

CYBIUM, *Cuv.; Day, Fishes of Ind., 254.*

Cybium Commersonii, *Cuv. & Val.* 2528

SEIR FISH.

Vern.—*Súrmoyi,* HIND.; *Vunjurrum* (male), *konam* (female), TEL.; *Konam, mah-wu-luachi* or *ah-ku-lah,* TAM.; *Chumbum,* MAL.

Habitat.—Seas of India, East coast of Africa, and Malay Archipelago.

Medicine.—An OIL is prepared from this fish which has been recommended as a substitute for Cod or Shark oil. **Dr. Bidie** considers (*Madras Quart. Med. Jour., Vol. V., 281*) that much of the offensive smell and taste of this and other Indian fish oils is due to the livers being allowed to putrify before the oil has been expressed. MEDICINE. Oil. 2529

CYCAS, *Linn.; Gen. Pl., III., 444.* 2530

The brief notices here given of the species of CYCAS will be found supplemented under Sago. This has been rendered necessary, from its being often difficult to discover to which plant the earlier writers refer.

Cycas circinalis, *Linn.; DC. Prod. XVI., II., 526;* CYCADACEÆ. 2531

Syn.—C. SPHÆRICA, *Roxb., Fl. Ind. Ed. C.B.C., 709;* C. CIRCINALIS, *Linn. in Thwaites En. Ceylon Pl., 294;* TODDER PANNA, *Rheede, Hort. Mal., III., 9.*

Vern.—*Orasmaro,* URIYA; *Maddú,* SING. Under **Cycas circinalis,** *Linn.,* **Ainslie** gives the following names which all appear to refer to Sago and not necessarily to **Cycas**:—*Show árisi,* TAM.; *Saouké chawal,* DUK.; *Sábudánd,* HIND.; *Zowbíum,* TEL.; *Sagu,* MAL.; *Sekuhme,* SING.; *Ságu,* JAVA; *Ságú,* BALI (*Mat. Ind. I., 361*).

Habitat.—A palm-like tree met with on the mountains of the Malabar coast and in Ceylon.

Food.—The SEEDS are ground into flour and used as food in times of scarcity. "The FLOUR obtained from the seeds of this species is made into cakes and eaten by the Cinghalese, and is reputed a remedy for some disorders." (*Enumeratio Plantarum Zeylaniæ, 294.*) FOOD. Seeds. 2532 Flour. 2533

C. pectinata, *Griff., as in Kurz, For. Fl. Burm., 503.* 2534

Vern.—*Thakal,* NEPAL.

Habitat.—An evergreen simple-stemmed, palm-like tree, found in Sikkim, Eastern Bengal, and Burma, often in *sál* or *eng* or pine forests. (*Gamble.*)

Food.—Yields a coarse sago, which, with the fruits, is eaten by the hill-people in Sikkim. (*Gamble.*) FOOD. Sago. 2535

Structure of the Wood.—Yellowish-white, in narrow wedge-shaped plates, arranged in nearly concentric rings and separated by white tissue, which, like the central pith, is full of starchy granules. TIMBER. 2536

C. revoluta, *Thunb.* 2537

Often called the SAGO-PALM OF JAPAN AND CHINA.

Habitat.—A Japanese species often cultivated in India; has a short thick stem.

C. Rumphii, *Miq.; Gamble, Man. Timb., 415.* 2538

Syn.—C. CIRCINALIS, *Roxb., Ed. C.B.C., 709.*

Vern.—*Wara-gudu,* TEL.; *Todda-maram,* MAL.; *Móndaing,* BURM.

Habitat.—A palm-like tree with a simple or branched stem; abundant in the Malabar and Cochin forests; in South Tenasserim and Andaman Islands. Often cultivated in South India.

RESIN. 2539 **Resin.**—Exudes a good sort of resin used medicinally. (*Kurz.*)

MEDICINE. 2540 **Medicine.**—Kurz says the resin is applied to malignant ulcers, and that it excites suppuration in an incredibly short time.

Scales. 2541 **Special Opinion.**—§ "The scales of the cone of the male tree, anodyne, dose 30 to 60 grains or more" (*Apothecary Thomas Ward, Madanapalle, Cuddapah*).

FOOD. Sago. 2542 Seeds. 2543 **Food.**—The interior of the stem yields a good quality of sago or starch; the nutty seeds are in Ceylon made into flour, but they are also eaten by the hill tribes of India.

2544 **Cycas siamensis,** *Miq.; Kurz, Burm. For. Fl., II., 503.*

Habitat.—An evergreen, low, stemless, palm-like tree frequent in the *eng* and dry forests of the Prome district, Burma.

RESIN. 2545 **Resin.**—Exudes a peculiar whitish gum, like tragacanth. (*Kurz.*)

CYDONIA, *Tourn.* (PYRUS, *Linn.*); *Gen. Pl., I., 626.*

2546 **Cydonia vulgaris,** *Pers.; Fl. Br. Ind., II., 368;* ROSACEÆ.

THE QUINCE.

Syn.—PYRUS CYDONIA, *Linn.*

Vern.—*Bihi* (*abi*, according to **Ainslie**), HIND.; *Bam tsúntú, bamsutu,* KASHMIR; *Shimai-madalavirai,* TAM.; *Bihi tursh, safarjal,* ARAB. **Moodeen Sheriff** gives the following names for QUINCE seeds:—*Habbus-safarjal,* ARAB.; *Bihi-dánah, beh-dánah, tukhme-ábi,* PERS.; *Beh-dánah,* HIND., DUK.; *Shimai-madalai-virai,* TAM.; *Shúne-dalimba-bíja,* SING.; *Shime-dálimba-vittulu,* TEL.

References.—*Brandis, For. Fl., 205; Gamble, Man. Timb., 161; Stewart, Pb. Pl., 80; DC., Origin Cult. Pl., 236; Home Dept. cor. regarding Pharm. Ind., 224; Ainslie, Mat. Ind., I., 332; Moodeen Sheriff, Supp. Pharm. Ind., 211; Dymock, Mat. Med. W. Ind., 2nd Ed., 302; Bent. & Trim., Med. Pl., I., 106; S. Arjun, Bomb. Drugs, 52; Irvine, Mat. Med., Patna, 10, 106; Baden Powell, Pb. Pr., 347, 597; Lisboa, U. Pl. Bomb., 119; Birdwood, Bomb. Pr., 32; Pliny, lib., XV., Cap. XI.* (*Quince used in his day was brought from Crete*).

Habitat.—Cultivated in Afghánistan and the North-West Himálaya up to 5,500 feet. **DeCandolle** says it grows wild in the woods in the north of Persia, near the Caspian Sea, in the region to the south Caucasus and in Anatolia. Naturalization may be suspected in Europe. No Sanskrit name is known for the quince, neither is there any Hebrew name, but its Persian name is *Haivah*: *aiva* is the Russian for the cultivated quince, and for the wild plant *armud.* The names in use in Europe point to an ancient knowledge of the species to the west of its original country. **DeCandolle** adds that it may have been naturalized in Europe before the epoch of the Trojan War (*Orig. Cult. Pl., 237*).

OIL. 2547 **Oil.**—**Baden Powell** mentions this as an oil-yielding plant in his *List of Panjáb Products.* **Docynia indica,** *Dcne.*, a nearly allied plant, is very plentiful in Sikkim, Bhutan, Khasia hills, and Burma. In the Naga Hills the ground at certain seasons is simply covered with the fruit left rotting under the trees. This might easily be put to some economic use, and probably as a substitute for quince.

MEDICINE. Seed. 2548 **Medicine.**—**Ainslie** says:—"The little of this article which is found in Indian bazars is chiefly in use amongst the Muhammadan practitioners,

Quince.

CYDONIA vulgaris.

MEDICINE.

who occasionally order an infusion or decoction of the SEEDS as a demulcent in gonorrhœa, and in cases of tenesmus. It is brought to India from the sea-ports of the Persian Gulf." The seeds act as a demulcent, and are used by the natives in diarrhœa, dysentery, sore throat, and fever. The dried fruit is used as a refrigerant. "The sweet and sub-acid quinces are commonly eaten as a FRUIT by the Arabs and Persians, and are considered tonic, cephalic, and cardiacal; they are also eaten baked. The LEAVES, BUDS, and BARK of the tree are domestic remedies among the Arabs on account of their astringent properties. In India they are considered cold, moist, and slightly astringent, and are one of the most popular remedies in native practice, the MUCILAGE being prescribed in coughs and bowel complaints as a demulcent; externally it is applied to scalds, burns, and blisters" (*Dymock*). **Professor Warden** sends the following note to the author:—"The seeds afford about 20 per cent. of dry mucilage, which corresponds in composition with that of linseed."

Fruit. 2549

Bark. 2550

Mucilage. 2551

The seeds coagulate 40 times their weight of water (*Pharmacographia.*)

Special Opinions.—§ "A cold infusion of the seeds forms a pleasant demulcent drink, which is much used in native practice in cases of irritation of the urinary organs" (*Assistant Surgeon Jaswant Rai, Mooltan*). "I use it as a demulcent in dysentery and catarrh as a mucilage, dose about one drachm (*Assistant Surgeon Nehal Sing, Saharunpore*). "They are known here as *muglai bedáná*, and are used specially for urinary complaints and seminal discharges" (*Surgeon-Major Robb, Civil Surgeon, Ahmedabad*). Quince seeds (*Bhidana*) if placed in water over night and then strained off the following morning make an excellent demulcent drink which if sweetened and iced is *most useful* in cases of diarrhœa for young and old (*Surgeon G. F. Poynder, A.M.D., Roorkee*).

Food.—When ripe the FRUIT is eaten; it is sweet, slightly juicy and astringent. It is also made into preserve, and, as having a powerful odour, is often used to flavour marmalade and other preserves. Wine is sometimes made from it. It is supposed by some to have been the Golden Fruit of the Hesperides. It is largely grown in Kangra (especially near Naggar), and the fruit is used in making preserves (*Gaz., p. 31*). It is also cultivated to some extent in the Peshawar Valley and at Lahore. **Stewart** says it is common in Kashmír, where the fruit is said by **Vigne** to be very fine. **Cayley** states that a small quantity is exported from Kashmír into Tibet. Abundant in Afghanistan, whence fruit and seed are largely imported into the Panjáb, and where, according to **Bellew**, the fruit is eaten fresh, candied or dried. **Irvine** states that that of Afghanistan excels all other quinces in quality (but that except them, "there is no other fruit of remarkable goodness"). **Aitchison** in his *Kuram Valley Flora* makes no mention of this plant.

FOOD. Fruit. 2552

Trade.—**Dymock** says: "Quince seeds are imported into Bombay from the Persian Gulf and Afghánistan. Value R10 to R25 per Surat maund of $37\frac{1}{2}$ ℔, according to quality." **Moodeen Sheriff** points out that *Behdánah* and *Bé-dánah* are so much alike in sound that mistakes are likely to be made. The latter is the name for a peculiar seedless raisin but is often loosely applied to all raisins.

TRADE. 2553

Cymbopogon, see **Andropogon**; GRAMINEÆ.

C. citratum, *DC.*, see **Andropogon citratus,** *DC.*; **A. 1079.**

C. laniger, *Desf.*, see **Andropogon laniger,** *Desf.*; **A. 1093.**

C. Martini, *Roxb.*; *Munro.*, see **Andropogon Schœnanthus,** *Linn.*; **A. 1117.**

Cymbopogon Nardus, *Linn.*, see **Andropogon Nardus,** *Linn.*; **A. 1107.**

2554 **CYNANCHUM,** *Linn.; Gen. Pl., 762.*

[*354;* ASCLEPIADEÆ.

Cynanchum pauciflorum, *Br.; Fl. Br. Ind., IV., 23; Wight, Ic.,*

Syn.—ASCLEPIAS TUNICATA, *Roxb., Fl. Ind., Ed. C.B.C., 253;* CYNANCHUM PAUCIFLORUM, *R. Br. in Dalz. & Gibs., Bomb. Fl., 148;* CYNOCOTONUM PAUCIFLORUM, *Decaisne, Thwaites, En. Ceylon Pl., 195.*

Vern.—*Chagul-pati,* BENG.; *Kan-kumbala,* SING.

Habitat.—A large twining shrub met with in the Deccan Peninsula, from the Concan southwards to Travancore and Ceylon. This is the region given in the Flora of British India, but according to **Roxburgh** (**Asclepias tunicata**), it is found in Bengal also.

FOOD. Leaves. 2555 **Food.**—The Cinghalese eat the young leaves of this and of many other plants of this natural family, in their curries. (*Enumeratio Plantarum Zeylaniæ, 195.*)

This does not appear to be the case in Bengal, **Roxburgh** simply remarking that its milky juice is particularly gummy.

CYNARA, *Linn.; Gen. Pl., II., 469.*

2556 **Cynara Scolymus,** *Linn.;* COMPOSITÆ.

ARTICHOKE.

Vern.—*Hati-choke,* BENG., HIND.; *Artichoke, kingin,* BOMB.

References.—*Voigt, Hort. Sub. Cal., 426; Stewart, Pb. Pl., 125; DC., Origin Cult. Pl., 92; Firminger, Man. Gard. in Ind., 160; Indian Forester, VIII., IX., and X.; Jour. Agri-Hort. Soc. Ind. (1875), V., 36; Lisboa, U. Pl. Bomb., 163; Birdwood, Bomb. Pr., 163; Smith, Dic., 25.*

Habitat.—Cultivated to a limited extent over most parts of India for the European market.

FOOD. 2557 **Food.**—The lower parts of the thick imbricated scales of the flower-heads are called artichoke bottoms, and being thick and fleshy are eaten as a vegetable. Although very generally cultivated the artichoke in India becomes larger and coarser than in Europe. **Firminger** says it is better known and more generally cultivated in India than in England. Any time from the end of July to the beginning of September is suitable for sowing the seed, which usually germinates in about 10 or 12 days. The seedlings should be transplanted when about a hand high and be placed at about 3 feet apart. They thrive best on a rich soil. The artichoke may also be propagated by suckers which should be separated from the stock in September. In the plains of India it flowers from about the beginning of May, but in the hills a little later.

CYNODON, *Pers.; Gen. Pl., III., 1164.*

[GRAMINEÆ.

2558 **Cynodon Dactylon,** *Pers.; Duthie, Fodder Grasses, N. Ind., 52;* CREEPING PANIC GRASS or DOORWA; COUCH GRASS.

Syn.—C. STELLATUS, *Willd.;* PANICUM DACTYLON, *Linn.;* PASPALUM DACTYLON, *DC.;* DIGITARIA DACTYLON, *Scop.*

Vern.—*Dúb, daurva, dubra, kabbar, khabbal, talla, tilla,* PB.; *Buráwa,* TRANS-INDUS; *Dob, nill dub,* RAJ.; *Chibbur,* SIND; *Dub, dúrbá, dúbla,*

Dub or Doorwa Grass.

BENG.; *Dhobi-ghás*, SANTAL; *Duba, káli ghas, rám ghás*, N. W. P.; *Dhupsa, hariáli*, C. P.; *Durvá*, SANS.; *Durva, karala, haryeli*, MAR.; *Arugam-pilla, hariáli*, TAM.; *Ghericha, haryali* (UPPER GODAVERY), TEL.

Mr. Baden Powell recommends that the following vernacular names for fodder plants should be carefully distinguished: *Dáb* or *dabhab* (**Andropogon muricatus**); *Dab* or *kusha* (**Eragrostis cynosuroides**); *Dub* or *khabbal*, (**Cynodon Dactylon**); and *Dib* (**Typha angustifolia**).

References.—*Roxb., Fl. Ind., Ed. C. B. C., 97; Voigt, Hort. Sub. Cal., 712; Thwaites, En. Ceylon Pl., 371; Dalz. & Gibs., Bomb. Fl., 297; U. C. Dutt, Mat. Med. Hind., 272, 297; Dymock, Mat. Med. W. Ind., 2nd Ed., 854; S. Arjun, Bomb. Drugs, 153, 347; Journal Agri. Hort. Soc. Ind. (1885), VII., Pt. III., Proc. CXI.; Report issued Agri. Dept. (1879), p. 105; Medical Top. Dacca, 60; Baden Powell, Pb. Pr., 514, 244-5; Lisboa, U. Pl. Bomb., 208, 276, 279, 283, 290; Birdwood, Bomb. Pr., 128; Royle, Ill. Him. Bot., 421; Balfour, Cyclop., 869; Smith, Dic., 157.*

Habitat.—A perennial creeping grass and flowering all the year round; grows everywhere throughout India, except perhaps in the sandy parts of Western Panjáb, where it is rare. In winter it appears scanty, at which time it may be said to be at rest. It abounds in the Sunderbuns. It is particularly abundant on road-sides, delighting apparently in the admixture of sand and gravel which it there gets along with the ordinary soil. It is readily propagated by chopping up the shoots and scattering the pieces over the prepared soil. It ascends from the plains to altitudes of 7,000 to 8,000 feet. It varies considerably both in habit and nutritive qualities, according to the nature of the soil or climate. It makes good hay keeping for several years if carefully stacked. Hay. 2559

Medicine.—In the *Athárwana Vedá* it is said: "May Dúrbá, which rose from the water of life, which has a hundred roots and a hundred stems, efface a hundred of my sins, and prolong my existence on earth for a hundred years." **U. C. Dutt** says: "This elegant and most useful vegetable has a niche in the temple of the Hindu religion. Medicinally, the fresh juice of the leaves is considered astringent, and is used as a snuff in epistaxis. The bruised grass is a popular application to bleeding wounds." It seems probable that both for sacred as well as medicinal purposes this grass is often confused with **Eragrostis cynosuroides.** The latter is the *Kash, Darbh* or *Dáb* (the Gramina of the Portuguese and the Gramen of the Romans but not the αγρωστις (**Triticum repens**) of the Greeks); it is used extensively at funeral ceremonies of the Hindus, the chief mourner wearing a ring of the grass. The latter is sacred to Ganesh. Both grasses are indiscriminately used in compound prescriptions with more powerful drugs in the cure of dysentery, menorrhœgia, &c. (*Dymock.*) **Sakharam Arjun** says:—"A white variety, which appears to be only a diseased state of the plant, is used medicinally by the native practitioners. It is acidulous and is used to check vomiting in bilious complaints." **Rev. A. Campbell** says of the Santals: "A preparation of the plant is applied in a parasitic disease, which attacks the spaces between the toes. This disease may be the same as that which is common in the West Indies, caused by **Pulex penetrans.**" MEDICINE. 2560 2561

Special Opinions.—§ "The expressed juice is astringent and is used as an application to fresh cuts and wounds. It is also diuretic and is used in cases of dropsy and anasarca, also as an astringent in cases of chronic diarrhœa and dysentery" (*Civil Surgeon J. H. Thornton, B.A., M.B., Monghyr*). "The juice of the green grass is useful in catarrhal ophthalmia, is Juice. 2562

CYNODON Dactylon.

Dub or Doorwa Grass.

MEDICINE.

astringent, used also with much benefit in hœmaturesis" (*Surgeon-Major J. M. Houston, Durbar Physn., Travancore, and Civil Apoth. John Gomes, Medical Store-keeper, Trevandrum*). "I have found the fresh juice to be a very valuable styptic in epitaxis" (*Doyal Chunder Shome*). "Antiperiodic, used as an application in scabies" (*Civil Surgeen John McConaghey, M.D., Shahjahanpore*). "The decoction of the roots is used in Mysore for secondary syphilis" (*Surgeon-Major John North, I.M.S., Bangalore*). "The decoction of the root chiefly used as diuretic" (*V. Ummegudien, Mettapollian, Madras*). "A cold infusion of *durba* grass often stops bleeding from piles. I generally give it with milk" (*Civil Surgeon R. L. Dutt, M.D., Pubna*). "Used in irritation of the urinary organs" (*Assistant Surgeon T. Ruthnam Moodelliar, Chingleput, Madras Presidency*). "Expressed juice is used by the Hakims as an injection in the nostrils for epitaxis. The bruised grass has been used by the Hindus from very ancient times as a dressing for fresh wounds, probably on account of its styptic properties" (*Assistant Surgeon Nobin Chunder Dutt, Durbhanga*). "The roots crushed and mixed with curds are used in cases of chronic gleet, dose ʒii" (*Surgeon James McCloghay, Offg. Staff Surgeon, Poona*).

FOOD. 2563 FODDER. 2564

Food and Fodder.—A cooling drink is also said to be made from the roots. It is the most common and useful grass in India, and its stems as well as its roots form a large proportion of the food of our horses and cows. **Mr. Duthie** says it varies considerably both in habit and nutritive qualities, according to the nature of the soil or climate. It makes excellent hay and will keep for years. It is by far "the most useful of all fodder grasses, especially for horses." "It is considered to be a first class fodder grass in Australia, where it is widely distributed, though in all probability introduced with cultivation. This grass is highly valued in the United States, where it is generally known under the name of Bermuda grass." (*Duthie.*) The following passage from the Madras Experimental Farm Manual will be found useful, although it would appear it is less appreciated in southern than in other parts of India:—

2565 HARIALI GRASS (**Cynodon Dactylon**).—The doob grass of Northern India—the Couch grass of Australia and America—is a valuable but overrated fodder plant, possessing great vitality and wide-spreading roots, which are capable of propagating the grass from each section of them; it is suited to our long droughts and is also capable, under high cultivation and irrigation, of producing heavy cuttings of tough wiry fodder, which, however, must possess considerable nutritive qualities; on poor soils it is liable to be crushed out by inferior types of plants, but on those of fair quality it is very persistent and difficult to eradicate; the latter point is detrimental to its use as a crop to be taken in a rotation. When highly cultivated it yields heavily under irrigation and is grown for hay near some large stations. In 1868 there was a plot of this grass on the farm measuring 3 acres; it yielded fairly without irrigation, and during the year 1870-71 gave five cuttings yielding 8 tons, 13 cwt. of hay; this hay sold for R360-13-3, whilst the cost of curing it was R105. After this the plot was kept for pasture, and in 1875 not half of the grass then growing was Hariali, the remainder being chiefly nutt-grass (**Cypersus bulbosus**) and finger-grass (**Panicum sanguinale**).

The following system is recommended for putting down this grass:—

"The land having been well cleaned should receive a dressing of fold-yard manure: when ploughing in the manure a woman should follow each plough and drop the roots in the open furrow, the succeeding plough covering them up, when its furrow is similarly planted, and the process repeated; a heavy harrowing and rolling complete the work."

FODDER. Hay.

2566 Regarding the curing of hay the following remarks with reference to this grass are of value:—

"Hariali, like most other meadow grasses, should be cut immediately the flower begins to appear; in this state the juices of the grass are more nutritious, and the hay is far superior than when made from the fully matured plant. Besides, when cut before the seed appears, the plant is more vigorous and produces another crop much sooner. Hariali hay is generally spoiled in this country by being too much exposed to the sun's rays. It is quite unnecessary to bleach the grass, in order to make it into hay. The great object should be to retain the green color of the grass by drying it as quickly as possible. Under ordinary circumstances, two days or at the most three days, should suffice for making the hay.

"Cutting should not commence until the dew is off the grass. The grass should remain on the ground for an hour or so after being cut. It should then be turned and tossed until sun-set. It cannot be tossed too much during a hot sun. To preserve the green colour and aroma of the hay it is absolutely necessary to keep it moving. At night, if the dews are heavy, it should be put up in small cocks, each containing from two to three cwt. These cocks should not be tramped, though it is advisable to beat the outside smooth with the back of a reck, in order that, should a shower of rain fall, the water may run off without penetrating the mass. A single hay rope should be passed over the cock, to prevent it from being blown to pieces by a gust of wind. Next morning after the dew is off the ground, the cocks may be opened again, and the hay spread out. It must be tossed and turned again, as on the previous day; care being taken that it is constantly kept moving. At the end of the second day, under ordinary circumstances, it will be fit to cart; though if the weather be at all damp or foggy it will be advisable to give it another day's sunning, of course putting it again into cock at night.

2567 "Hay thus rapidly made is rich in saccharine matters, and is, therefore, very liable to heat and ferment; this, to a moderate extent, does no harm; in fact it gives the hay a good flavour; however, care must be taken that it does not go too far and char the hay. If the hay is loose in a room, exposure for an hour or two in the hot sun will put it all right, or a layer or two of dry paddy or cholum straw may be put through the mass. In the stack it is equally easy to prevent too great fermentation. I have found a single line of six-inch drain pipes placed at about the middle of the stack from the centre to the outside, a capital arrangement for keeping down the temperature. A thick bamboo, or a couple of hollow pieces of the stems of palmyra or cocoanut trees, the one resting on the other so as to form a pipe, will equally effect the purpose, or, in building two or three layers of dry paddy or cholum straw placed in a stack will prevent it heating to any injurious extent."

2568 "Creeping Pannic Grass.—Of Eastern Bengal it has been said: This perennial grass is found in great abundance, and is of a superior quality to that of districts to the westward; it grows luxuriantly in the light soil along the banks of the rivers in the southern division, and affords the best pasturage in the district. The juice of the leaves is used medicinally by Hindu practitioners" (*Topography of Dacca by J. Taylor, 60*).

CYNOGLOSSUM, *Linn.; Gen. Pl., II., 848.* [Boragineæ.

2569 **Cynoglossum micranthum,** *Desf.; Fl. Br. Ind., IV., 156;*

Vern.—*Nílakrái,* Pb.; *Oudhuphulé,* Guj.; *Adhopushpi,* Sans.; *Bu-kättu, henda,* Sing.

Habitat.—Native in North India and the Himálaya, altitude 1,000 to 8,000 feet, from Kashmír to Bhutan and Pegu; common.

DYE. 2570 Several species of closely allied plants belonging to this genus are occasionally mentioned by authors as of economic value. It is doubtful how far they have been distinguished. **O'Shaughnessy** says **C. officinale** (?) yields a colouring matter of little value.

MEDICINE. 2571 **Medicine.**—The plant is officinal in the Panjáb.

CYNOMETRA, *Linn.; Gen. Pl., I., 586.*

2572 **Cynometra cauliflora,** *Linn.; Fl. Br. Ind., II., 268;* LEGUMINOSÆ.

Vern.—*Iripa,* MAL.; *Niam-niam,* MALAY.

Habitat.—A tree of the Western Peninsula, South India, Ceylon, and Malacca.

OIL. 2573 **Oil.**—It yields an oil said to be prepared in North Arcot, and used for medicinal purposes.

2574 **C. polyandra,** *Roxb.; Fl. Br. Ind., II., 268.*

Vern.—*Peng,* CACHAR, SYLHET.

Habitat.—A large evergreen tree of the Khasia Hills, Sylhet, and Cachar.

OIL. 2575 **Oil.**—*In Spons' Encyclop.* it is said that the oil which this plant yields is used medicinally.

TIMBER. 2576 **Structure of the Wood.**—Light-red, hard, close-grained. **Mann** remarks it is very useful for scantlings, and makes good charcoal.

2577 **C. ramiflora,** *Linn.; Fl. Br. Ind., II., 267.*

Syn.—C. BIJUGA, *Spanoghe.*

Vern.—*Shingr,* BENG. (as in Gamble); *Irapú,* TAM.; *Mymeng, kabeng, myeng-kabin,* BURM.; *Gal mendóra,* SING.

Habitat.—A large, evergreen tree of the Sunderbans, South India, and Burma, in tidal forests; frequent from Chittagong down to Tenasserim and the Andaman Islands (*Kurz*).

DYE. 2578 **Dye.**—Chips of the wood yield, in water, a purple dye.

OIL. 2579 **Oil.**—The seed yields an oil which is externally applied in leprosy and other cutaneous diseases.

MEDICINE. 2580 **Medicine.**—The root is purgative. A lotion is made from the leaves boiled in cows' milk which, mixed with honey, is applied externally in scabies, leprosy, and other cutaneous diseases. An oil is also prepared from the seeds, used for the same purpose. (*Rheede.*)

TIMBER. 2581 **Structure of the Wood.**—Red, hard, close-grained. **Skinner** says that it is used for house-building and carts. It is employed in the Sunderbans for posts for native huts and for fuel.

2582 **Cynosurus cristatus,** *Linn.*, is a grass which **Baron von Mueller** says is particularly valuable for withstanding drought. The roots penetrate to a considerable depth. For other species see **Eleusine.**

2583 CYPERUS, *Linn.; Gen. Pl., III., 1043.*

The roots of several species are tuberous, such, for example, as **C. corymbosus, C. esculentus, C. stoloniferus, C. rotundus, C. jeminicus, C. scariosus, &c., &c.** Several of these are edible, others afford aromatic tubers which are employed on account of that property in perfumery, in dyeing, and in medicine. **Sir Walter Elliot** says under **C. stoloniferus** that the name of *Jatámánsi* is often given to these aromatic tubers. The culms and leaves of several species are employed in sedge (popularly grass) matting, but much confusion still exists on this subject. The present account of the more important useful species is more intended as a basis on which economic facts discovered in the future may be built than as a detailed statement of their properties. The reader is referred to **Mats** and **Matting** (in a further volume) for particulars as to the species used for textile purposes, the systems adopted in preparing the fibres, and the methods of weaving the mats. The most commonly used species, it may be here mentioned, are **C. tegetum** and **C. corymosa**, but many of the others enumerated below (and regarding which special information has not been given) are also used for that purpose.

Cyperus bulbosus, *Vahl.*, see **C. jeminicus,** *Rottb.;* CYPERACEÆ. 2584

C. compressus, *Linn.; Clarke in J. Linn. Soc., XXI., 97.*

Vern.—*Chúncha,* BENG.; *Salitunga,* TEL.; *Wek-tamyet,* BURM.

References.—*Roxb., Fl. Ind., Ed. C.B.C., 65; Dalz. & Gibs., Bomb. Fl., 282; Cyperus in Griff. Itin. Notes No. 167, p. 12, and 191, p. 362; Kurz, Rept., Pegu.*

Habitat.—A common species throughout India, ascending the hills to 2,000 feet in altitude. A special form is known as *var.* **pectiniformis.** This is said to occur in Lucknow, Chutia Nagpur, and Assam. **Thwaites** says it is very common in the warmer parts of Ceylon. **Roxburgh** remarks that it "delights in a moist soil."

C. corymbosus, *Rottb.; C.B. Clarke in Jour. Linn. Soc., XXI., 158.* 2585

Syn.—C. SEMINUDUS, *Roxb., Fl. Ind., Ed. C.B.C., 63; Nees in Wight, Contrib., p. 80;* PAPYRUS PANGOREI, *Nees in Wight, Contrib., p. 88, in part.*

Vern.—*Gola methi,* BENG.; *Godú tunga kúda* (**Roxb.**) and *Goddu-tunga kodu* (**Elliot**), TEL.; *Gal ehi,* SING.

Habitat.—Found throughout the Eastern and Southern Peninsulas of India from Kumaon to Bengal, Assam, Cachar, Madras, and Ceylon. **Roxburgh** says it is a native of wet places.

Fibre.—This is the **C. Pangorie** referred to by many writers as one of the chief sources of the *Mádur* (or so called Calcutta-grass) mats. It should be observed that the name **C. Pangorie** is open to the greatest possible ambiguity. The Madras plant mentioned under that name by **Dr. Bidie, C.I.E.**, is **C. corymbosus,** *Rottb.*, var. **Pangorie,** *Rottb.*; but **C. Pangorie,** *Roxb.*, is **C. malaccenses,** *Lam.;* **C. Pangorei,** *Thw.*, is **C. tegetum,** *Roxb.*; **C. Pangorei,** *Retz.*, is **C. rotundus,** *Linn.*, var. **procerula**; and **C. Pangorei,** *Nees*, is **C. tegetum,** *Roxb.* These facts will show how impossible it is to understand what plant an author alludes to if he calls it simply **C. Pangorie.** This mistake has been made by **Drury** (*Useful Plants, p. 331*) and by **Mr. T. N. Muhkarji** in his new work (*Art-Manufactures of India, p. 310*). **Dr. Bidie** writes—"several species of sedge appear to be used for mat-making, but the one from which the finest sorts of mats are manufactured at Tinnevelly and Palghat is **Cyperus Pangorie.** Tinnevelly mats of the first quality are generally uncoloured or with one or two simple bands of red and black at each end, and they may be made so fine that a mat sufficient for a man to lie on can be rolled up and packed into the interior of a moderate-sized walking stick. The strips of the split sedge used in the Pálghát matting are not so fine as those employed in

FIBRE. 2586

MATS, 2587

Tinnevelly. 2588

Palghat. 2589

Tinnevelly, and the article is therefore heavier, coarser in texture, and not so flexible."

FODDER. 2590 **Fodder.**—"Cattle are not fond of it, and it is only eaten occasionally by buffaloes." (*Roxb.*)

2591 **Cyperus elegans,** *Linn.; C.B. Clarke, Linn. Soc. Jour., XXI., 125*

Syn.—C. MŒSTUS, *Kunth;* C. NIGROVIRIDIS, *Thw., En. Ceylon Pl., 344.*

Vern.—*Wek chan,* BURM. (*Kurz, Pegu Rept.*)

Habitat.—A native of Bengal and the Malay Pininsula; Sikkim 1,500 feet, Assam, Khasia hills 1,200 feet, Sylhet, Yunan, Chittagong, Mergui, Tenasserim, and the Andaman Islands.

2592 **C. esculentus,** *Linn.; C.B.Clarke, Jour. Linn. Soc., XXI., 178.*

Syn.—C. TUBEROSUS, *Rottb.*

Vern.—*Kaserú, díla,* PB.; *Sha-ts'au,* CHINESE.

Habitat.—There are five or six distinct forms of this plant, of which two occur in India, *viz., forma* **tuberosa** (*sp. Rottb.*) in Madras and *forma* **hindustanica** in Northern India.

MEDICINE. Root. 2593 FOOD. Root. 2594 **Medicine and Food.**—**Stewart** says: "In N.-W. Provinces the root is used as food, and is officinal as *kaserú*. The *díla* root, mentioned by Bellew as eaten in the Peshawar valley, may be the same. *Díla*, however, appears to be a generic name for the CYPERACEÆ, the roots of several of which are eaten by pigs, and their stems, &c., browsed on by cattle, as is one called *Múrg* which may be **Scirpus.**"

FIBRE. 2595 Coffee Substitute. 2596 **Fibre.**—Balfour remarks that in China the shoots of this plant are used in making huts and mats. "The toasted roots have been used as a substitute for coffee and yield a preparation resembling coffee."

2597 **C. exaltatus,** *Retz.; C.B. Clarke, Linn. Soc. Jour., XXI., 186.*

Syn.—C. UMBELLATUS, *Vahl.: according to Roxb., Fl. Ind., Ed. C. B. C., 69;* C. VENUSTUS, *R. Br.; Thwaites, En. Ceylon Pl., 432 (nec. Nees nec Kunth);* C. ALTUS, *Nees, in Wight, Contrib., 84.*

Vern.—*Pedda shaka,* TEL.

Habitat.—Commonly found in Bengal (Chutia Nagpur, Rajmahal, &c.), and in the Peninsula of India generally (Mysore, Madras, Central India, Mount Abu, Oudh, &c.), and in Ceylon. "A large species, growing in standing fresh water." (*Roxb.*)

FIBRE. 2598 Mats. 2599 **Fibre.**—This sedge is often used for matting. **Mr. C. B. Clarke** describes four forms of the plant, α the type alluded to mainly in the above notes: β *amœna* (=**C. amœnus,** *Koenig (non-Kunth)* and **C. alopecuroides,** *Roxb.*) This is met with in Calcutta and in Madras. γ *dives* a native of Egypt, but **Mr. Clarke** has also found it at Mutlah in Bengal; he remarks: "the type specimens of this at Kew agree exactly with my Calcutta example, so that if **C. dives** is a distinct species, it is an Indian one. The specimens differ from those of **Retz's exaltatus** by the much more numerous glumes to the spikelet, while they differ from those of **Koenig's amœnus** in the rigid, not tasselled, nodding spikes. **C. dives** may therefore be held a distinct species: but whatever it is considered, it must be remote from **C. alopecuroides,** *Rottb.*, which has a 2-fid style, a compressed nut, and a very thick rhachilla." δ *Oatesii* is the fourth form, and it is met with at Thyet Myo, Burma.

2600 **C. Haspan,** *Linn.; Clarke, Linn. Soc. Jour., XXI., 119.*

Syn.—C. UMBELLATUS, *Vahl., is the Pedda súka of the Telegus.*

Cyperus inundatus, *Roxb.; Clarke in Linn. Soc. Jour., XXI., 73.* 2601

Vern.—*Pati,* HIND. and BENG.

Habitat.—An aquatic species met in the jheels of some parts of Bengal, and also in China. **Roxburgh** says "it is found in great abundance on the low banks of the Ganges and rivulets near Calcutta; where the tide rises high over it, it thrives most luxuriantly and helps much to bind and protect the banks from the rapidity of the water."

Medicine.—**Irvine** (*Mat. Med. Patna, 82*) writers: "The tubers are used as a tonic and stimulating medicine." MEDICINE. 2602

C. Iria, *Linn.; C.B. Clarke, Linn. Soc. Jour., XXI., 137.* 2603

Syn.—C. PARVIFLORUS, *Nees in Wight Contrib., 87, nec. Vahl., nec.* C. UMBELLATUS, *Roxb.,* C. IRIA, *Linn. as in Roxb., Fl. Ind., Ed. C.B.C. 67.*

Vern.—*Bura chucha,* BENG.; *Wel-hiri,* SING.

Habitat.—"A native of moist cultivated lands." (*Roxb.*) Frequent in India, having been collected at Almora (1,200 feet), Mussourie, Nepal, Sikkim, Sonada (2,200 feet), Assam, Khasia Hills, Lucknow, Parisnath in Behar, Chutia Nagpur, Central India, Mount Abu, Puna, Mangalore, Ceylon, &c.

Fibre.—The culms are used in mat-making. FIBRE. Mats. 2604

C. jeminicus, *Rottb.; C.B. Clarke, Linn. Soc., Jour. XXI., 175.* 2605

Syn.—C. BULBOSUS, *Vahl.; Nees, in Wight, Contrib., 80; Dalz. and Gibs., Bomb. Fl., 284;* C. OLERACEUS, *Roxb.;* C. ROTUNDUS, *Kunth, in part.*

Vern.—*Shilandi,* TAM.; *Pura-gadi,* TEL.; *Bhadra musta,* SANS.; *Thegi,* BOMB.; *Silandi-arisi,* TAM. (in Ceylon).

Habitat.—Roxburgh says this is "a native of the dry sandy pasture ground near the sea"—he is alluding to the Coromandel Coast; **Clarke** that it occurs from Yemen, Jedda, and Beluchistan to Sind, Madras, Ceylon, Abyssinia, and Central Africa.

Food.—"The roots are used as flour in times of scarcity and eaten roasted or boiled." When roasted they have the taste of potatoes, and would be valuable for food but that they are so small. "**Dr. James Anderson,** in an excursion to the southern part of the Peninsula of India, discovered that the *shilandi arisi,* growing in sandy situations by the seaside, and requiring but little water, was the common food of the natives during a famine, and when other grains are scarce. It is nutritious, pleasant to the taste, and makes a pudding somewhat resembling that made of sago." (*Balfour.*) Some people "dry the tubers in the sun, grind them into meal and make bread of them; while others stew them in curries and other dishes." (*Drury.*) FOOD. Roots. 2606 Flour. 2607

C. longus, *Linn.; Clarke, Linn. Soc. Jour., XXI., 163.* 2608

Clarke describes five or six forms of this plant, the type of the species occurring on Mount Abu and in Cabul; β **pallescens,** *Boiss,* in Egypt, Cordofan, &c., γ **cyprica** in the island of Cyprus; δ **badia** in southern Europe, Madeira, and doubtfully in Madras; ε **elongata** in Egypt, Africa, &c.

C. malaccensis, *Lam.; Clarke, Linn. Soc. Jour., XXI., 147.* 2609

Syn.—C. MONOPHYLLUS, *Vahl.;* C. PANGOREI, *Roxb., Fl. Ind., Ed. C.B.C., 68;* C. INCURVATUS, *Roxb., p. 66;* C. TEGETIFORMIS, *Benth.;* C. GANGETICUS, *Roxb.*

Vern.—*Chumati pati,* BENG.

Habitat.—Roxburgh says of his **C. Pangorei** that it is a native "of the banks of the Ganges, and serves, with **C. inundatus**, the same useful purpose, though in an inferior degree." Of his **C. iucurvatus** remarks that it also is a native of the banks of the Ganges "flowering during the cold season." Clarke adds that it occurs at Noakhali, Calcutta, the Sunderbuns, Dacca, and is distributed to Arracan, Pegu, Singapore, Japan, and China.

2610 **Cyperus niveus,** *Retz.; C.B. Clarke, Linn. Soc. Jour., XXI., 108.*

Vern.—*Birmutha,* SANTAL.

Habitat.—Throughout India and Burma (Beluchistan, Kashmír, Panjáb, Kumaon, Simla, Kulu, Nepal, Sikkim, Assam, Bengal, Chutia Nagpur, Rajmahal, &c.), Madras, &c., &c. A native of shady moist pasture land. (*Roxb.*)

C. pertenuis, *Roxb.*, see **C. scariosus,** *R. Br.*

2611 **C. Pongarei,** *Rottb., as in Roxburgh,* see **C. malaccensis**; and for other plants named by different authors as **Cyperus Pangorei,** see **Cyperus corymbosus.**

2612 **C. rotundus,** *Linn.; C. B. Clarke, Linn. Soc. Jour., XXI., 167.*

Syn.—C. HEXASTACHYOS, *Roxb.*

Vern.—*Muthá, mothá,* BENG.; *Batha-bijir,* MUNDARI; *Utru-banda,* URAON; *Tandi sura,* SANTAL; *Mustá, gundrá, bhadra muste, mustaka,* SANS.; *Kórai,* TAM.; *Shákha tunga-véru, bhadramuste, tunga-muste, mustakamu, kaivartaka-muste, gandala,* TEL.; *Mustá, barikmoth,* BOMB.; *Kóré-ki-jhár,* DEC.; *Bimbal,* MAR.; *Motha,* GUZ.; *Kalanduru,* SING.

References.—*Roxb., Fl. Ind., Ed. C.B.C., 66; Jour. As. Soc., Pt. II. (1867), p. 82; Home Dept. Official Corres. regarding Pharm. Ind., 238; Hove's Tour in Bombay, pp. 112, 120, &c., &c.; Walter Elliot, Flora Andhrica, pp. 25, 76, 184, 120, &c.; Moodeen Sheriff, Supp. Pharm. Ind., 128; U. C. Dutt, Mat. Med. Hind., 264; Dymock, Mat. Med. W. Ind., 2nd Ed., 844; S. Arjun, Bomb. Drugs, 150; Baden Powell, Pb. Pr., 382; Atkinson, Him. Dist., 734, 808; Birdwood, Bomb. Pr., 94.*

Habitat.—A plentiful species in India occurring from Kuram Valley, Afghánistan, Gilgit, and Kashmír to Simla, Garhwal, and the Khasia hills throughout the plains (Lahore, Bengal, Madras), and ascending the mountains of the central table-land (from Mount Abu and Poona to the Nilghiri hills). **Dr. Hove,** who travelled in 1787, speaks of the plant as very abundant in Bombay.

DYE 2613 **Dye.**—Used in certain dye preparations to impart a perfume to the fabric.

OIL. 2614 **Oil.**—The rounded rhizomes are said to yield an essential oil, which the natives of Upper India use to perfume their clothes. In Bengal the tubers of this species are more largely used in perfumery than are those of **C. scariosus,** being more plentiful—in fact it is a troublesome weed. Roxburgh says that the dried and pounded root is used "as perfume at the weddings of natives."

MEDICINE. Roots. 2615 **Medicine.**—Roots are used medicinally as a diaphoretic and astringent. Stimulant and diuretic properties are also attributed to them. They are further described as vermifuge. In native practice, they are held in great esteem as a cure for disorders of the stomach and irritation of the bowels. The bulbous roots are scraped and pounded with green ginger, and in this form mixed with honey they are given in cases of dysentery in doses of about a scruple (*Med. Top. of Dacca by J. Taylor, p. 54*). "In the Concan the fresh tubers are applied to the breast in the form of *lep* (malagma) as a galactagogue. **C. rotundus** is the κύπερος of the Greeks, and is mentioned by **Dioscorides**, who says it is the *Juncus* or *Radix Junci* of the Romans, and is used as a diuretic and emmenagogue, and applied to scorpion stings, and when dried to spreading ulcers; it is also an ingredient of warm plasters. **Herodotus** (4, 71) notices it as an aromatic plant, used by the Scythians for embalming. Κύπειρον is mentioned in the Iliad (21, 351), and Odyssey (4, 603), and by **Theophrastus** in his fourth book; it appears to have been a favourite food of horses. **Pliny** (21, 18) calls it **Juncus triangularis or angulosus**; it is probably the Juncus of **Celsus** (3, 21) mentioned as an ingredient in a diuretic medicine for dropsy, although he calls it **Juncus quadratus**." (*Dymock, p. 844.*) Arabian and Persian writers describe the drug as attenuant, diuretic, emmenagogue, and diaphoretic. They state that it is prescribed in febrile and dyspeptic affections, and in large doses as an anthelmintic, and externally as applied to ulcers or used as an ingredient in warm plasters. **U. C. Dutt** says the tubers grown in moist soil are preferred. They are extensively used as an aromatic adjunct to numerous compound medicines. MEDICINE.

Special Opinions.—§ "Roots are aromatic and commonly used in indigestion of children combined with other aromatics with benefit" (*Assistant Surgeon Shib Chunder Bhattacharji, In Civil Medical Charge, Chanda, Central Provinces*). "Roots are used as an astringent in the diarrhœa and dysentery of children" (*Bolly Chand Sen, Teacher of Medicine*). "The roots are in Chutia Nagpur used in fever" (*Rev. A. Campbell*). "The fresh roots are stimulant and diaphoretic" (*Bombay Gazette, VI., p. 14*).

Fodder.—Cattle eat this so-called grass, and hogs are remarkably fond of the roots. FODDER. 2616

Cyperus scariosus, *R. Br.; C. B. C., Linn. Soc. Jour., XXI., 159.* 2617

Syn.—Cyperus pertenuis, *Roxb., Fl. Ind., Ed. C.B.C., 66.*

Vern.—*Nágar-móthá*, Hind.; *Nágar-mútha*, Beng.; *Lawála*, Mar.; *Soade-kúfi, soad*, Arab.; *Mushke-zamín*, Pers.; *Nágar-mustaka*, Sans.; *Nágar-motah*, Dec.; *Muttah-kách, kóraik-kizhangu*, Tam.; *Tunga gaddala-veru, kólatunga-muste*, Tel.; *Kóra-kizhanna*, Mal.; *Konnári-gadde*, Kan.; *Vomon niu*, Burm.

References.—*Roxb., Fl. Ind., Ed. C.B.C., 66; Med. Top. Ajmír, 147; Dymock, Mat. Med. W. Ind., 2nd Ed., 815; Irvine, Mat. Med. Patna, 75; Birdwood, Bomb. Pr., 94; Liotard, Dyes, Supp., IV.*

Habitat.—A delicate, slender grass, met with in damp places in Bengal; Oudh, and rare in the Panjáb; by no means so common a plant as **C. rotundus.**

This is apparently the *Koray kalung*, Tam., the *Nagar motha*, Duk., and *Musta*, Sans., described by **Ainslie** (*Mat., Ind. II., 162*) under the name of **Cyperus juncifolius**, *Rottler*.

Dye.—The rhizomes are used in dyeing to give a scent to the fabric, and as a perfume for the hair. **Roxburgh** describes them as "tuberous with many dark-coloured villous fibres." "Its naked delicate form, small and compound umbel, short slender leaves, and scanty involucre immediately distinguish it" from the other members of the genus. DYE. 2618

Medicine.—The root is officinal, being considered cordial, stomachic, and desiccant, and is used for washing the hair. Also regarded as diaphoretic and diuretic. "Arabian and Persian writers mention this Indian MEDICINE. Root. 2619

MEDICINE. **Cyperus**, but consider it to be inferior to **C. rotundus**." "Two kinds of *Nágarmoth* are met with in the Bombay market—Surat and Kathiawar, the first is heavier and more aromatic than the second. Value, Surat, R2 per maund of 37½lb; Kathiawar R1¼. The Surat *Nágarmoth* is probably obtained from Rájputana, where the plant is common in tanks (*Dymock*). **U. C. Dutt** says: "The root of **C. pertenuis** is somewhat tuberous with many dark-coloured villous hairs. It grows in low wet places, and is chiefly used in the preparation of medicated oils.

Special Opinions.—§ "Roots, when bruised, have a fragrant smell, and for this reason native females keep a stock of the powdered root to wash their bodies with" (*Honorary Surgeon P. Kinsley, Chicacole, Ganjam, Madras Presidency*). "Is given in conjunction with Valerian in cases of epilepsy" (*Surgeon-Major C. W. Calthrop, M.D., 4th Bengal Cavalry, Morar*). "The root is astringent, useful in diarrhœa" (*Surgeon-Major J. M. Houston, Durbar Physn., Travancore, and Civil Apoth. John Gomes, Medical Store-keeper, Trevandrum*). "A decoction is used in gonorrhœa and also in syphilitic affections," (*Surgeon J. C. H. Peacocke, I.M.D., Nasik*). "Used in fever prescriptions" (*Surgeon-Major Robb, Civil Surgeon, Ahmedabad*).

2620 **Cyperus stoloniferus,** *Retz.; C.B. Clarke, Linn. Soc. Jour., XXI., 172.*

Syn.—C. LITTORALIS, *R. Br.;* C. TUBEROSUS, *Baker.*

Vern.—*Jatámánsi,* a name given in South India to this plant.

PERFUMERY. 2621 **Perfumery.**—As with other scented species the tubers of this plant are used for various purposes of perfumery. Referring to the use of the name *Jatámánsi*, **Sir Walter Elliot** says: "In applying this botanical name, we have followed **Heyne**. But the true *Jatámánsi* is a species of *Valerian* (**Nardostachys Jatamansi,** *DC.* (*Royle, Ill., 54*), which has been successfully identified by **Sir W. Jones** with the spikenard of the ancients; *Asiatic Res., II., 405—IV., 109,* and which by Persian and Arab physicians is called *Sanbal-i-Hindi* and *Sanbal-ul-taib* and in Upper India *Jatámánsi* and *Bálch'har*. But as the true plant is only found at great elevations beyond the tropics, the term is applied in South India to the sweet-smelling tubers of various species of Cyperus, and in Upper India to the lemon-grass (**Schænanthus**) and other species of **Andropogon** which are known also under the names of *Askhar* and *Sik'húnas* (σχινος)."

2622 **C. tegetiformis,** *Roxb.; C.B. C., Linn. Soc. Jour., XXI., 157.*

Syn.—C. NUDUS, *Roxb.; Fl. Ind., Ed. C.B.C., pp. 63 and 70;* C. BENGALENSIS, *Spreng.*

Vern.—*Gúla-methi,* BENG.; *Sura,* SANTAL.

Habitat.—"A native of low wet places over Bengal; flowering during the rains." (*Roxb.*) **Clarke** mentions as localities—Calcutta, Chittagong, Noakhali, Burisal, Mymensing, Pundua, and Assam. He also states that the plant occurs in China and Japan.

FIBRE. Mats. 2623 **Fibre.**—**Roxburgh** writes: "This species is very like **C. tegetum,** and about the same size, though I am informed it is never used for mats, To know it from **C. tegetum** attend to the involucre, which in this is only about one-fourth the length of the umbel, but in that as long or longer."

2624 **C. tegetum,** *Roxb.; C.B. Clarke, Linn. Soc. Jour., XXI., 160.*

Syn.—C. CORYMBOSUS, *Koening,* in part; C. SCHIMPERIANUS, *Steud.;* C. DEHISCENS, *Steud.;* C. PANGOREI, *Thwaites* (*non-Rottb.*) *Enum. Pl. Zeyl., 344;* PAPYRUS DEHISCENS, *Nees in Wight Contrib., 89;* C. PANGOREI, *Nees* (the greater part) and C. CORYMBOSUS, *Nees.*

Mats and Matting.

CYPERUS tegetum.

Note by Mr. Clarke: "This plant, abundant in India, is the authentic C. TEGETUM, *Roxb.*; it differs decidedly from C. CORYMBOSUS in the much more distant glumes, which in the dried specimens have the margins incurved not overlapping. The spikelets are more compressed than those of C. CORYMBOSUS. The colour in India varies from pale to a high red-brown: with the more highly coloured Indian examples many African are absolutely identical; but there are other African specimens chestnut or almost black. It is far more difficult to distinguish C. TEGETUM in Africa from C. LONGUS and its various forms called C. BADIUS; the only absolute distinction appears to lie in the much longer leaves of C. LONGUS. The rhizomes of C. LONGUS differs a good deal from that of C. TEGETUM, as is evident enough when you have *the whole* of the rhizome to compare, which may be once in a hundred specimens. The narrow wing of the rhachilla is in C. SCHIMPERIANUS, as in C. LONGUS, less soluble than in the Indian C. TEGETUM, but I doubt the value of this character."

Vern.—*Mudar-ktai*, BENG.; *Wetla*, BURM.

Habitat.—A common species in India, Abyssinia, and Egypt. Mr. Clarke mentions the following localities: Almora (1,200 feet), Chumba (900 feet), Kumaon, Nepal, Sikkim (200 feet), Khasia hills, Bengal (Chutia Nagpur, Calcutta), Carnatic, Mysore, Ceylon, &c.

Fibre.—The Calcutta mats are chiefly made of this species. The culms are split into two or three, and then woven into mats upon a warp of threads previously stretched across the floor of a room. The mat-maker passes the culms with the hand alternately over and under the successive threads of the warp, and presses them home.

FIBRE. Mats. 2625

In different districts of India it is believed that two or three allied species are used for this purpose. In Madras the form C. corymbosus seems to be chiefly used. Royle repeating Roxburgh states "that the culms or stalks of the plant when green are split into three or four pieces, which, in drying, contract so much as to bring the margins in contact, in which state they are woven into mats, and thus show a nearly similar surface on both sides. Specimens of the strips of this sedge were sent to the Exhibition of 1851, as well as mats made of them." "The shining and useful mats for which the Capital of India is famous, and which are frequently imported into Europe." Since Royle's time the trade in these sedge-mats has greatly increased, and at the present day it may be said that they form a regular article of export to Europe. In the Trade Returns, however, all mats are collectively returned, so that it is impossible to give the actual figures. The exports of "mats" were last year valued at R14,416.

TRADE. 2626

Printed in the United States
By Bookmasters